Symbol	Quantity	SI Units	U.S. Customary Units	Mass-Length-Time Dimensions[a]
ε	Eddy viscosity	$Pa \cdot s (N \cdot s/m^2)$	$lb \cdot s/ft^2$	M/Lt
μ	Viscosity	$Pa \cdot s (N \cdot s/m^2)$	$lb \cdot s/ft^2$	M/Lt
ν	Kinematic viscosity	m^2/s	ft^2/s	L^2/t
ξ	Vorticity	s^{-1}	s^{-1}	$1/t$
ρ	Density	$kg/^3$	$slug/m^3$	M/L^3
σ	Surface tension	N/m	lb/ft	M/t^2
τ	Shear stress	Pa	lb/ft^2	M/Lt^2
ϕ	Velocity potential	m^2/s	ft^2/s	L^2/t
ψ	Stream function	m^2/s	ft^2/s	L^2/t
ω	Angular velocity	s^{-1}	s^{-1}	$1/t$

[a]Mass (M), length (L), time (t), and thermodynamic temperature (T).

SYMBOLS FOR DIMENSIONLESS QUANTITIES

Symbol	Quantity	Symbol	Quantity
C	Cauchy number	n	Polytropic exponent; Manning's coefficient
C_c	Coefficient of contraction	**R**	Reynolds number
C_D	Drag coefficient	S	Slope of energy line
C_f	Frictional drag coefficient	S_o	Bottom slope (open channel)
C_L	Lift coefficient	S_c	Critical slope
C_p	Pressure coefficient	**W**	Weber number
C_v	Coefficient of velocity	Y	Expansion factor
C_w	Weir coefficient	α	Kinetic energy correction factor
E	Euler number	β	Momentum correction factor
F	Froude number	η	Efficiency
f	Friction factor	κ	von Kármán's turbulence "constant"
K_L	Loss coefficient	μ_p	Poisson's ratio for pipe material
k	Adiabatic exponent	Π	Dimensionless group
M	Mach number	$\boldsymbol{\sigma}$	Cavitation number

CONVERSION FACTOR TABLE

Abbreviations used:

BTU = British Thermal Unit
cfs = cubic feet per second
ft/s = feet per second
ft = foot
gpm = gallons per minute
hp = horsepower
h = hour
Hz = hertz
in. = inch
J = joule = N·m
kg = kilogram = 10^3 gram
lb = pound force
m = metre (SI) = mile (U.S. Customary)
mb = millibar = 10^{-3} bar
mm = millimetre = 10^{-3} metre
mm^2 = square millimetre
mph = miles per hour
m/s = metres per second
N = newton
Pa = pascal = N/m^2
psi = pound per square inch
s = second
W = watt = J/s

Absolute viscosity: 1 Pa·s = 10 poises = 0.020 89 $lb \cdot s/ft^2$
Acceleration due to gravity: 9.806 65 m/s^2 = 32.174 ft/s^2
Area: 1 m^2 = 10.76 ft^2; 1 mm^2 = 0.001 55 $in.^2$
Density: 1 kg/m^3 = 0.001 94 $slug/ft^3$
Energy: 1 N·m = 1 J = 0.737 5 ft·lb = 0.000 948 BTU
Flowrate: 1 m^3/s = 35.31 ft^3/s; 1 litre/s = 10^{-3} m^3/s = 0.022 83 mgd
1 ft^3/s = 449 gpm = 0.647 mgd
Force: 1 N = 0.224 8 lb
Kinematic viscosity: 1 m^2/s = 10^4 Stokes = 10.76 ft^2/s
Length: 1 mm = 0.039 4 in.; 1 m = 3.281 ft; 1 km = 0.622 miles
Mass: 1 kg = 0.068 5 slug
Power: 1 W = 1 J/s = 0.737 5 ft·lb/s; 1 kW = 1.341 hp = 737.5 ft·lb/s
Pressure: 1 kN/m^2 = 1 kPa = 0.145 psi; 1 mm Hg = 0.039 4 in. Hg = 133.3 Pa;
1 mm H_2O = 9.807 Pa; 101.325 kPa = 760 mm Hg = 29.92 in. Hg = 14.70 psi;
1 bar = 100 kPa = 14.504 psi
Specific Heat; Engineering Gas Constant: 1 J/kg·K = 5.98 ft·lb/slug·°R
Specific Volume: 1 m^3/kg = 515.5 $ft^3/slug$
Specific Weight: 1 N/m^3 = 0.006 365 lb/ft^3
Temperature: 1°C = 1 K = 1.8°F = 1.8°R (see Section 1.4)
Velocity: 1 m/s = 3.281 ft/s = 3.60 km/h = 2.28 mph; 1 knot = 0.515 5 m/s
Volume: 1 m^3 = 10^3 litres = 35.31 ft^3; 1 U.S. gallon = 3.785 litres; 1 U.K. gallon = 4.546 litres

ELEMENTARY FLUID MECHANICS

ELEMENTARY FLUID MECHANICS

SEVENTH EDITION

Robert L. Street
Professor of Fluid Mechanics and Applied Mathematics
Stanford University

Gary Z. Watters
Professor of Civil Engineering
California State University, Chico

John K. Vennard
Late Professor of Fluid Mechanics

JOHN WILEY & SONS
New York ■ Chichester ■ Brisbane ■ Toronto ■ Singapore

ACQUISITIONS EDITOR Charity Robey
MARKETING MANAGER Susan Elbe
PRODUCTION MANAGEMENT Ken Santor and Ingrao Associates
COVER AND TEXT DESIGNER Lee Goldstein
MANUFACTURING MANAGER Susan Stetzer
ILLUSTRATION Anna Melhorn
COVER PHOTO Courtesy Office National D'études et de Recherches Aérospatiales, Châtillon, France

This book was set in Times Roman by CRWaldman Graphic Communications and printed and bound by R.R. Donnelley. The cover was printed by Phoenix Color Corporation.

Recognizing the importance of preserving what has been written, it is a policy of John Wiley & Sons, Inc. to have books of enduring value published in the United States printed on acid-free paper, and we exert our best efforts to that end.

The paper on this book was manufactured by a mill whose forest management programs include sustained yield harvesting of its timberlands. Sustained yield harvesting principles ensure that the number of trees cut each year does not exceed the amount of new growth.

Library of Congress Cataloging-in-Publication Data:

Street, Robert L.

Elementary fluid mechanics / Robert L. Street, Gary Z. Watters, John K. Vennard.—7th ed.

p. cm.

Vennard's name appears first on the 6th ed.

Includes bibliographical references and index.

ISBN 0-471-01310-2

1. Fluid mechanics. I. Watters, Gary Z. II. Vennard, John K. III. Title.

QA901.V4 1996

532—dc20 94-39840

CIP

Printed in the United States of America

10 9 8 7 6 5 4 3 2 1

PREFACE

This book had its first edition in 1940, and this, the seventh edition, retains the basic approach and style of the original that have made it appealing to many generations of students and their teachers. On the other hand, we have significantly reorganized the book and expanded its coverage in some important ways. The material originally contained in a single chapter on kinematics of fluid motion was split into two chapters [3 and 4] to more logically separate kinematics from mass conservation and to provide the bases for introduction of the Reynolds Transport Theorem, which is introduced in Chapter 4 to unify the treatment of mass conservation, work-energy, and impulse-momentum. A brief derivation of the Navier-Stokes equations is included in the chapter on real fluid flow to provide a basis for extension of the concepts of viscous flow to problems of a more complex nature. The chapter on similitude and dimensional analysis is enhanced by the addition of a section on the normalization of the governing equations of flow in terms of the significant scales of the flow. More applications were added to the chapter on pipe flow along with a new section on hydraulic transients to broaden the scope of the chapter and to give students access to modern practical pipeline problems. The chapter on lift and drag was significantly revised to bring in practical applications to baseballs, structures and road vehicles. A new chapter on fluid machinery is added to introduce the basic concepts of turbomachinery and to provide some application to both turbine and pump selection. All of the compressible flow topics were collected into one chapter and moved to later in the text. A significant effort has been made to make the Illustrative Problems stand out and be more effective in instructing the students in correct problem-solving techniques. For the instructor, the solutions manual contains problem solutions in much expanded detail. And finally, the prodigious cache of homework problems was culled, refined and enhanced.

Fluid mechanics is the study of fluids under all conditions of rest and motion. Its approach is analytical and mathematical rather than empirical; it is concerned with basic principles that provide the solution to numerous and diverse problems encountered in many fields of engineering, regardless of the physical properties of the fluids involved.

In this text, *elementary fluid mechanics* is taken to be that portion of the subject that is studied in the introductory course or two by learners who have completed differential and integral calculus and beginning engineering courses in statics and dynamics. It is not necessary for them to have studied thermodynamics or advanced mathematics. We have tried to focus the text on explication of physical concepts and principles, rather than on mathematical manipulation. It is our goal that the text will prove to be an understandable introduction to the various topics and that precious classroom time can be spent on amplification and extension of the material.

The first half of the book is focused on fundamental physical and analytical principles. The sequence leads from fundamentals of units and fluid properties through fluid statics, kinematics, systems, control volumes, conservation principles, ideal incompressible flow, impulse-momentum principles, real fluid flow, and ends with similitude, dimensional analysis and normalization of equations. The second half of the book covers applications of these principles to flow in pipes and open channels, lift and drag, fluid machinery, and compressible flow. The final chapter is an introduction to an array of fluid measurements and the instruments for making them. The appendixes include a number of useful definitions, mathematical tools, an introduction to cavitation, and some basic computer programs. In the text both differential and integral approaches are used as is typical in fluid mechanics. Furthermore, our approach is to build from the simple and specific toward the more general and complex. Nowhere is this more evident than in Chapter 4 where we introduce first the concept of systems and control volumes, and then a simple one-dimensional derivation of the key conservation of mass principle. Sequentially, this is extended to the general two-dimensional integral formulation [but it is shown how this reduces to a familiar differential form] and the very important and general Reynolds Transport Theorem.

References to written and visual materials are provided to guide the inquiring reader to more exhaustive treatment of various topics. The films listed at the ends of the chapters may be quite useful by filling in background and providing visual experience with various fluid phenomena. Over 1 300 problems are included at the ends of the chapters, and there are over 150 Illustrative Problems that show how to achieve useful engineering answers from the derived concepts and procedures presented in the quantitative articles of the text.

Recognizing the slow transition to a single international language of units, particularly in the United States, this edition again uses both the Système International d'Unités (SI) and U. S. Customary systems. The emphasis is on SI units, however, and they account for a majority of the usage.

February 1995

ROBERT L. STREET
Stanford, California

GARY Z. WATTERS
Chico, California

CONTENTS

TABLES

CHAPTER 1
FUNDAMENTALS

1.1 THE SCOPE OF FLUID MECHANICS

Perhaps it is at the shoreline, where the land meets the ocean and the sky, that we are most aware of the air and the water—the fluids in which we and the fishes, respectively, are immersed and live. At the shore we see waves—driven by the wind—rise up and assault the shore; we feel the energy in the waves as they roll ashore and thunderously break; we see the power of the water to erode the beach and transport the sand. We realize the effect of the storm winds on the shoreline. At river mouths we observe the silt and the effluent of the cities' pollution flow into the sea.

We appreciate then that there is much to learn and understand about the natural phenomena of river flow, the ocean currents, and the air motions which bring our weather. As well, there is much to be learned about human uses of fluids—about our water supply and sewerage systems, about the forces that act on the dams and breakwaters upon which we have stood, about the air-flow over and drag on vehicles, and, of course, about flying—how does a 747 stay up there?

The trend is, in fact, toward greater complexity and challenge in fluid problems. As our ancestors of pre-Christian Rome built their impressive aqueducts, they could not have imagined the scale of the water supply, irrigation, river and harbor navigation, and water power problems that are now routinely tackled. The range of new issues added in modern times is virtually infinite, including the sonic boom of supersonic planes; dispersion of man's wastes in lakes, rivers, and oceans; blood flow in veins, arteries, kidneys, hearts, and artificial heart and kidney machines; the design of super oil tankers for speed, cargo pumping, and, above all, safety; and the analysis and simulation of the world's weather and ocean currents. Thus, today's fluid mechanics has become an essential part of such diverse fields as medicine, meteorology, astronautics, solar physics, and oceanography, as well as of the traditional engineering disciplines.

This book seeks to take you on the first steps in a journey toward understanding. We will encounter the key conservation principles on which fluid mechanics is based, namely, conservation of mass, momentum, and energy. We will link up with an old friend from dynamics, Newton's second law, which equates the net force acting on a body to the product of the body's mass and its acceleration, and learn how that can be applied in fluid mechanics. Our focus is on basic flows of civil and mechanical engineering, including flow in pipes and open channels, flow around and over various structures, and flow through turbomachines—pumps, turbines, propellers, and jet engines. Armed with the key principles learned herein, readers will be more in tune with the environment surrounding them and ready, if they wish, to tackle more advanced topics, as well as broader fields such as aeronautics and astronautics or environmental fluid mechanics.

1.2 HISTORICAL PERSPECTIVE

The human desire for knowledge of fluid phenomena began with problems of water supply, irrigation, navigation, and water power. With only a rudimentary appreciation for the physics of fluid flow, we dug wells, constructed canals, operated crude water wheels and pumping devices and, as our cities increased in size, constructed ever larger aqueducts, which reached their greatest size and grandeur in the city of Rome and some of which still exist today. However, with the exception of the thoughts of Archimedes (287–212 B.C.) on the principles of buoyancy, little of the scant knowledge of the ancients appears in modern fluid mechanics. After the fall of the Roman Empire (A.D. 476) there is no record of progress in fluid mechanics until the time of Leonardo da Vinci (1425–1519). This great genius designed and built the first chambered canal lock near Milan and ushered in a new era in hydraulic engineering; he also studied the flight of birds and developed some ideas on the origin of the forces that support them. However, down through da Vinci's time, concepts of fluid motion must be considered to be more art than science.

After the time of da Vinci, the accumulation of hydraulic knowledge rapidly gained momentum, with the contributions of Galileo, Torricelli, Mariotte, Pascal, Newton, Pitot, Bernoulli, Euler, and d'Alembert to the fundamentals of the science being outstanding. Although the theories proposed by these scientists were in general confirmed by crude experiments, divergences between theory and fact led d'Alembert to observe in 1744, "The theory of fluids must necessarily be based upon experiment." He showed that there is no resistance to motion when a body moves through an ideal (nonviscous or *inviscid*) fluid; yet obviously this conclusion is not valid for bodies moving through real fluids. This discrepancy between theory and experiment, called the d'Alembert paradox, has long since been resolved. Yet it demonstrated clearly the limitations of the theory of that day in solving fluid problems. Because of the conflict between theory and experiment, two schools of thought arose in the treatment of fluid mechanics, one dealing with the theoretical and the other with the practical aspects of fluid flow. In a sense, these two schools of thought have persisted to the present day, resulting in the mathematical field of *hydrodynamics* and the practical science of *hydraulics*.

Near the middle of the last century, Navier and Stokes succeeded in modifying the general equations for ideal fluid motion to fit those of a viscous fluid and in so doing showed the possibilities of explaining the differences between hydraulics and hydrodynamics. About the same time, theoretical and experimental work on vortex motion and separated flow by Helmholtz and Kirchhoff was aiding in explaining many of the divergent results of theory and experiment.

Meanwhile, hydraulic research went on apace, and large quantities of excellent data were collected. Unfortunately, this research led frequently to empirical formulas obtained by fitting curves to experimental data or by merely presenting the results in tabular form, and in many instances the relationship between physical facts and the resulting formula was not apparent.

Toward the end of the last century, new industries arose that demanded data on the flow of fluids other than water; this fact and many significant advances in our knowledge tended to arrest the empiricism of much hydraulic research. These advances were: (1) the theoretical and experimental work of Reynolds; (2) the development of dimensional analysis by Rayleigh; (3) the use of models by Froude, Reynolds, Vernon-Harcourt, Fargue, and Engels in the solution of fluid problems; and (4) the rapid progress of theoretical and experimental aeronautics in the work of Lanchester, Lilienthal, Kutta, Joukowsky, Betz, and Prandtl. These advances supplied new tools for the solution of fluid problems and gave birth to modern fluid mechanics. The single most important contribution was made by Prandtl in 1904 when he introduced the concept of the boundary layer. In his short, descriptive paper Prandtl, at a stroke, provided an essential link between ideal and real fluid motion for fluids with a small viscosity (e.g., water and air) and provided the basis for much of modern fluid mechanics. Fluid mechanics is then a science and engineering discipline that understands the relationship between real and ideal fluids and that is based on application of fundamental laws of mechanics and thermodynamics, on the use of rational experimentation and simulation, and appropriate incorporation of fluid properties in all analyses.

In the twentieth century, fluid problems have been solved by constantly improving rational methods; these methods have produced many fruitful results and have aided in increasing knowledge of the details of fluid phenomena. The trend is certain to continue in large part because of the ever increasing power of the digital computer.

1.3 PHYSICAL CHARACTERISTICS OF THE FLUID STATE

Matter exists in two states—the solid and the fluid, the fluid state being commonly divided into the liquid and gaseous states. Actually, many would say that matter exists in four states—solid, liquid, gaseous, and plasma, the latter three being classified as fluids. The plasma state is the state of over 99% of the matter of the universe and is distinguished from the others because a significant number of its molecules are ionized. Hence, the plasma contains electrically charged particles and is susceptible to electromagnetic forces. Unfortunately, the intriguing subject of plasma dynamics is beyond the scope of this work.

Solids differ from liquids and liquids from gases in the spacing and latitude of motion of their molecules, these variables being large in a gas, smaller in a liquid, and extremely small in a solid. Thus, it follows that intermolecular cohesive forces are large in a solid, smaller in a liquid, and extremely small in a gas. These fundamental facts account for the familiar compactness and rigidity of form possessed by solids, the ability of liquid molecules to move freely within a liquid mass, and the capacity of gases to fill completely the containers in which they are placed, while a liquid has a definite volume and a well-defined surface. In spite of the mobility and spacing of its molecules, a fluid is considered (for mechanical analysis) to be a *continuum* in which there are no voids or holes; this assumption proves entirely satisfactory for most engineering problems. However, an important exception is a gas at very low pressure (rarefied gas), where intermolecular spacing is very large and the "fluid" must be treated as an aggregation of widely separated particles. It follows that the continuum assumption is valid only when the smallest physical length scale in a flow is much larger than the average spacing between molecules or the mean free path (about 1 μm or about $10^{-4} - 10^{-5}$ in.) of the molecules composing the fluid.

A more rigorous mechanical definition of the solid and fluid states can be based on their actions under various types of stress. Application of tension, compression, or shear stresses to a solid results first in elastic deformation and later, if these stresses exceed elastic limits, in permanent distortion of the material. Fluids possess elastic properties only under direct compression or tension. Application of infinitesimal shear stress to a fluid results in continual and permanent distortion. *Thus, a fluid can be defined unambiguously as a material that deforms continuously and permanently under the application of a shearing stress, no matter how small.* This definition does not address the issue of how fast the deformation occurs and as we shall see later this rate is dependent on many factors including the properties of the fluid itself. The inability of fluids to resist shearing stress gives them their characteristic ability to change their shape or to *flow*; their inability to support tension stress is an engineering assumption, but it is a well-justified assumption because such stresses, which depend on intermolecular cohesion, are usually extremely small. Although tests on some very pure liquids have shown them to be capable of supporting tension stresses up to a few thousands of pounds per square inch (perhaps 10 MPa), such liquids are seldom encountered in engineering practice.

Because fluids cannot *support* shearing stresses, it does not follow that such stresses are nonexistent in fluids. During the flow of real fluids, the shearing stresses assume an important role, and their prediction is a vital part of engineering work. Without flow, however, shearing stresses cannot exist, and compression stress or *pressure* is the only stress to be considered.

Because fluids at rest cannot contain shearing stresses, no component of stress can exist in a static fluid tangent to a solid boundary or tangent to an arbitrary section passed through the fluid. This means that pressures must be transmitted to solid boundaries or across arbitrary sections *normal* to these boundaries or sections at every point. Furthermore, if a *free body* of fluid is isolated as in Fig. 1.1, pressure must be shown as acting *inward* (p_1) and on the free body (according to the usual conventions of mechanics for compression stress). Pressures exerted by the fluid on the container (p_2) will of course act

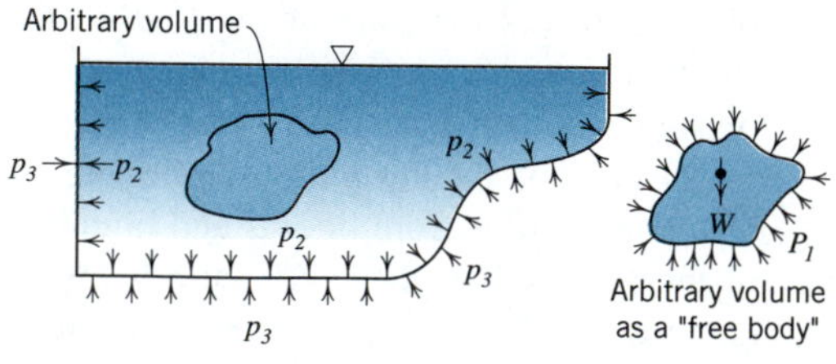

Fig. 1.1

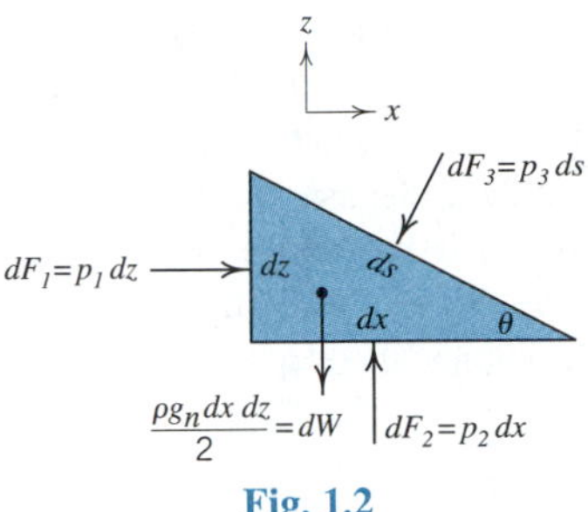

Fig. 1.2

outward, but their reactions (p_3) will act inward on the fluid as before. Another property of fluid pressure is that, at a point in a fluid at rest, it has the same magnitude in all directions; this may be proved by considering a convenient two-dimensional free body diagram of fluid (Fig. 1.2) having unit width normal to the plane of the paper. A more general proof using a three-dimensional element yields the same result as seen by solving Problem 1.2. Taking p_1, p_2, and p_3 to be the mean pressures on the respective surfaces of the element, ρ the density[1] of the fluid, g_n the acceleration due to gravity and writing the equations of static equilibrium, i.e., the summation of forces in each of the x- and z-directions:

$$\Sigma F_x = p_1\,dz - p_3\,ds\,\sin\theta = 0$$

$$\Sigma F_z = p_2\,dx - \rho g_n\,dx\,dz/2 - p_3\,ds\,\cos\theta = 0$$

From geometry, $dz = ds\sin\theta$ and $dx = ds\cos\theta$. Substituting the first of these relations into the first equation above gives $p_1 = p_3$, whatever the size of dz; substituting the second relation into the second equation yields $p_2 = p_3 + \rho g_n\,dz/2$, whatever the size of dx. From these equations it is seen that p_1 and p_3 approach p_2 as dz approaches zero. Accordingly it may be concluded that at a point ($dx = dz = 0$) in a static fluid $p_1 = p_2 = p_3$, and the pressure there is the same in all directions. Similarly, for the flow of an ideal (inviscid) fluid the same result may be proved. However, in contrast to the cases of static fluids or ideal fluids in motion, the pressure at a point is generally not the same in all directions in a viscous fluid motion owing to the action of the viscous stresses. Fortunately, this is of small consequence in most engineering flows, where the effect of the viscous normal stresses is small compared to that of the pressure (or normal stress).

With the pressure at a point the same in all directions, it follows that the pressure has no vector sense and hence is a *scalar* quantity; however, the differential forces produced by the action of pressures on differential areas are vectors because they have directions normal to the areas. The resultant forces obtained by the integration of the differential forces also are vector quantities.

Another well-known aspect of fluid pressure (which needs no formal proof) is that pressures imposed on a fluid at rest are transmitted undiminished to all other points in the fluid; this follows directly from the static equilibrium of adjacent elements and the fact that the fluid mass is a continuum. This aspect finds practical expression in the hydraulic lift used in automobile service stations.

The reader may be uneasy concerning the treatment of liquids and gases by the same principles in view of their obvious differences of compressibility. In problems where com-

[1]A summary of symbols and their dimensions is given in Appendix 1.

pressibility is of small importance (and there are many of these in engineering), liquids and gases may be treated similarly, but where the effects of compressibility are predominant (e.g., in water hammer in pipe systems and transonic or supersonic flight) the behavior of liquids and gases is quite dissimilar and is governed by very different physical laws. Usually, when compressibility is unimportant, fluid problems may be solved successfully with the principles of mechanics; when compressibility predominates, elasticity, thermodynamics and heat transfer concepts may be required as well (cf., Chapters 9 and 13).

1.4 UNITS, DENSITY, SPECIFIC WEIGHT, SPECIFIC VOLUME, AND SPECIFIC GRAVITY

The world is presently changing to the use of a single international language of units. The adopted system is the metric *Système International d'Unités* (SI). Many countries have already changed to SI units and the United States is moving toward use of the metric system in lieu of the currently used U.S. Customary system. Unfortunately, both types of units will be used for many years. Accordingly, in this edition of this book, both the metric (SI) system and the U.S. Customary (foot-slug-second) system of units are used. However, usually in illustrative examples and problems only one set of units is used within each example or problem.

For the SI and U.S. Customary unit systems, there are four basic dimensions through which fluid properties are expressed. These base dimensions and their associated units (and symbols) are:

Dimension	SI Unit	U.S. Customary Unit
Length (L)	metre (m)[2]	foot (ft)
Mass (M)	kilogram (kg)	slug (−)
Time (t)	second (s)	second (s)
Thermodynamic temperature (T)[3]	kelvin (K)	degree Rankine (°R)

In the SI and U.S. Customary systems, the above units are known as base units. There are also a number of other units that are derivable from base units and that have special names, for example,

Quantity	SI Unit	U.S. Customary Unit
Frequency (f)	hertz (Hz $= s^{-1}$)	hertz (Hz $= s^{-1}$)
Force (F)	newton (N $= kg\cdot m/s^2$)	pound (lb $= slug\cdot ft/s^2$)

[2]An early reviewer of the Fifth Edition of this book waggishly suggested that the devices which measure the flow of water, gas, and electricity into your home are *meters*, but the unit of length is the *metre*. Given that this is the *official* SI spelling, we adhere to his distinction. You'll get used to this apparent typographical error eventually!

[3]Here $T(°C) = T(K) - 273.15$ while $T(°F) = T(°R) - 459.6$. The normal freezing and boiling points of water are (0°C, 32°F) and (100°C, 212°F), respectively.

Quantity	SI Unit	U.S. Customary Unit
Energy (E), Work (W), Quantity of heat (Q)	joule (J = N·m)	British thermal unit (BTU = 778.2 ft·lb)
Power (P)	watt (W = J/s or $m^2 \cdot kg/s^3$)	horsepower (hp = 550 ft·lb/s or 550 slug·ft^2/s^3)
Pressure (p), Stress (τ)	pascal (Pa = N/m^2 or $kg/m \cdot s^2$)	pound per square inch (psi = $(lb/ft^2)/144$ or $(slug/ft \cdot s^2)/144$)
Temperature (T; practical scales)[3]	degree Celsius (°C)	degree Fahrenheit (°F)

There are several conventions used in the SI systems which should be noted. The names of multiples and submultiples of the basic and derived SI units are related to these units and are formed by means of prefixes which are the same irrespective of the units to which they are applied. For example, the prefixes used herein include mega (M) = 10^6, kilo (k) = 10^3, milli (m) = 10^{-3}, and micro (μ) = 10^{-6}. Thus, 1 megawatt = 1 MW = 10^3 kW = 10^6 W = 10^6 watts and 1 square millimetre = 1 mm^2 = 10^{-6} m^2 = 10^{-6} square metres. In addition, two other conventions are employed with SI units and are extended herein to the U.S. Customary system. First, when a compound unit is formed by multiplication of two or more units, a dot is used to avoid confusion, for example, N·m or $kg \cdot m/s^2$. Note that in general we will not use the SI notation of $m^2 \cdot kg \cdot s^{-3}$, preferring the more easily recognized $m^2 \cdot kg/s^3$. Second, while the decimal point is employed in the usual way, number spacing is accomplished with spaces in lieu of commas, for example, 14 446 or 0.000 002.

ILLUSTRATIVE PROBLEM 1.1

A hydraulic turbine that uses the power of flowing water to generate electricity has a power output of 3 MW when the flow through it is 25 m^3/s. What are the equivalent power and flowrate in U.S. Customary units, i.e., horsepower and cubic feet per second?

SOLUTION

The output power is 3 MW = 3×10^6 watts = 3×10^3 kW. Each kW is equivalent to 1.341 hp [from Appendix 1 or the equivalent tables inside the covers of the book]. Thus, the power is 3×10^3 kW $\times$ 1.341 hp/kW = 4 023 hp. ●

Each m^3/s is equivalent to 35.31 ft^3/s [from Appendix 1 or the equivalent tables inside the covers of the book]. Thus, the flowrate is 25 m^3/s $\times$ 35.31 $ft^3/s/(m^3/s)$ = 882.8 ft^3/s ●

ILLUSTRATIVE PROBLEM 1.2

In the central California valleys a typical summer temperature is 100°F. What is the equivalent temperature in °R, °C and K?

SOLUTION

There are several temperature relationships that need to be internalized. First, the absolute and practical temperatures are related by

$$T(°F) = T(°R) - 459.6$$

$$T(°C) = T(K) - 273.15$$

Accordingly, our first answer is given from

$$T(°R) = T(°F) + 459.6$$

so

$$T(°R) = 100 + 459.6 = 559.6°R \bullet$$

Second, the freezing and boiling points of water define convenient points on the Fahrenheit and Celsius scales; we have at freezing (0°C, 32°F) and at boiling (100°C, 212°F). It follows that a change of 100°C is equivalent to a change from 32°F to 212°F and so the ratio of °F to °C is

$$(212°F - 32°F)/100°C = 1.8°F/°C$$

[check Appendix 1 or the equivalent tables inside the covers of the book].

Thus, the temperature of 100°F is (100°F − 32°F) = 68°F above freezing; the equivalent Celsius temperature must be then

$$68°F/(1.8°F/°C) = 37.8°C \bullet$$

It follows that

$$T(K) = T(°C) + 273.15$$

so

$$T(K) = 37.8°C + 273.15 = 310.95 \text{ K} \bullet$$

Both the SI and the U.S. Customary systems have the advantage that they distinguish between force and mass and have no ambiguous definitions (such as pounds-mass and pounds-force).

Density is the mass, that is, the amount of matter, contained in a unit volume; specific weight is the weight, that is, gravitational attractive force, acting on the matter in that unit volume. Both these terms are fundamentally dependent on the number of molecules per unit of volume. As molecular activity and spacing increase with temperature, fewer molecules exist in a given volume of fluid as temperature rises; thus density and specific weight decrease with increasing temperature.[4] Because a larger number of molecules can be forced

[4]For example, a variation in the temperature from the freezing point to the boiling point of water will cause the specific weight of water at atmospheric pressure to decrease 4% (Appendix 2) and will cause the density of gases to decrease 37% (assuming no pressure variation).

into a given volume by application of pressure, density and specific weight increase with increasing pressure.

Density, ρ, is expressed in the mass-length-time system of dimensions and has the dimensions of mass units (M) per unit volume (L^3). Thus, $[\rho] = \text{slugs/ft}^3$ or $= \text{kg/m}^3$.

Specific weight, γ, is expressed in the force-length-time system of dimensions and has dimensions of force (F) per unit volume (L^3). Thus, $[\gamma] = \text{lb/ft}^3$ or $= \text{N/m}^3$. To the extent that this concept is used in the SI community, the term $\gamma = \rho g_n$ is called the *weight density*, not specific weight.

Because the weight (a force), W, is related to its mass, M, by Newton's second law of motion in the form

$$W = Mg_n$$

in which g_n is the acceleration due to the local force of gravity, density and specific weight (the mass and weight of a unit volume of fluid) are related by a similar equation,

$$\gamma = \rho g_n \tag{1.1}$$

Because meaningful physical equations are dimensionally homogeneous, the metre-newton-second dimensions of ρ (which are equivalent to kilograms per cubic metre) may be calculated as follows:

$$\text{Dimensions of } \rho = \frac{\text{Dimensions of } \gamma}{\text{Dimensions of } g_n} = \frac{\text{N/m}^3}{\text{m/s}^2} = \frac{\text{N}\cdot\text{s}^2}{\text{m}^4}$$

Accordingly, the equivalence of dimensions is

$$\text{kg/m}^3 = \text{N}\cdot\text{s}^2/\text{m}^4$$

or

$$1\ \text{N} = 1\ \text{kg} \times 1\ \text{m/s}^2$$

that is, a unit force produces a unit acceleration of a unit mass. Equivalently, in U.S. Customary units

$$\text{Dimensions of } \rho = \frac{\text{Dimensions of } \gamma}{\text{Dimensions of } g_n} = \frac{\text{lb/ft}^3}{\text{ft/s}^2} = \frac{\text{lb}\cdot\text{s}^2}{\text{ft}^4}$$

and

$$\text{slugs/ft}^3 = \text{lb}\cdot\text{s}^2/\text{ft}^4$$

or

$$1\ \text{lb} = 1\ \text{slug} \times 1\ \text{ft/s}^2$$

This algebraic use of the dimensions of quantities in the equation expressing physical relationship is employed extensively throughout the text and will prove to be an invaluable check on engineering calculations. A summary of quantities, their units and dimensions, and conversion factors is given in Appendix 1 and parts of that appendix are repeated inside the front covers of the book.

The specific volume α, defined as volume per unit of mass, has dimensions of length cubed per unit mass (m^3/kg or ft^3/slug). This definition identifies specific volume α as the reciprocal of density.

Specific gravity (s.g.) is the ratio of the density of a substance to the density of water at a specified temperature and pressure. Because these items vary with temperature, temperatures must be quoted when specific gravity is used in precise calculations. Specific gravities of a few common liquids are presented in Appendix 2[5], from which the specific weights of liquids can be readily calculated by

$$\gamma = (\text{s.g.}) \times \gamma_{\text{water}} \tag{1.2}$$

The specific weights of perfect gases can be obtained by a combination of Boyle's and Charles' laws known as the *equation of state*; in terms of specific weight this is

$$\gamma = g_n p/RT \tag{1.3}$$

in which p is the absolute pressure (force per unit area); T the thermodynamic temperature; and R the engineering gas constant (energy per unit mass per unit of temperature). Naturally, the equation of state can also be written in the form

$$\boxed{\rho = p/RT} \tag{1.3}$$

The gas constant is constant only if the gas is a *perfect gas*. Common gases in the ordinary engineering range of pressure and temperature may be considered to be perfect for many engineering calculations, but departure from the simple equation of state is to be expected near the point of liquefaction, or at extremely high temperatures or low pressures.

Application of Avogadro's law, "all gases at the same pressure and temperature have the same number of molecules per unit volume," allows the calculation of a universal gas constant. Consider two gases having constants, R_1 and R_2, densities ρ_1 and ρ_2, and existing at the same pressure and temperature, p and T. Dividing their equations of state,

$$\frac{p/\rho_1 T = R_1}{p/\rho_2 T = R_2}$$

results in $\rho_2/\rho_1 = R_1/R_2$. Now, the density is the mass per unit volume so, according to Avogadro's law, the number of molecules per unit volume in each gas must be the same because the temperature and pressure are the same in each gas. Thus, there are the same number of moles n in each unit volume of gas and the actual mass of each gas present is nM_1 and nM_2, where M_1 and M_2 are the molar masses, for example, in units of g/mol.[6] Therefore, $\rho_2/\rho_1 = M_2/M_1$. The molecular weight m of a substance is a dimensionless, relative number which is expressed as the mass of a molecule of the substance relative to the mass of a molecule of a standard substance (usually the mass of a molecule of oxygen, taken as 32.00). Most importantly, the molecular weight is numerically equal to the molar mass. It follows that $\rho_2/\rho_1 = m_2/m_1$. Combining this equation with the one involving ρ's and R's above gives $m_2/m_1 = R_1/R_2$, or

$$m_1 R_1 = m_2 R_2 \tag{1.4}$$

[5]The reader should refer to the International or Smithsonian Physical Tables if precise specific gravities at other temperatures are required. Students may use the values of Appendix 2 in problem solutions even though the temperatures may not be exactly the same.

[6]In the SI system, the mole (mol) is the amount of substance of a system which contains as many elementary entities, for example, atoms, as there are atoms in 0.012 kg (or 12 g) of carbon = 12. Consequently, each mole of a gas has the same number of molecules and, for example, one mole of oxygen (O_2) has a molar mass of 0.032 kg (or 32 g). The units of molar mass are, therefore, grams/mole.

In other words, the product of molecular weight and engineering gas constant is approximately the same for all gases. This product mR is called the *universal gas constant* $\mathcal{R}$ and is preferred for general use by many engineers. It has the units ft·lb/slug·°R or J/kg·K. The constancy of mR applies particularly to the monatomic and diatomic gases. Gases having more than two atoms per molecule tend to deviate from the law mR = Constant (See Appendix 2). The nominal values of $\mathcal{R}$ are 49 709 ft·lb/slug·°R and 8 313 J/kg·K.

ILLUSTRATIVE PROBLEM 1.3

Calculate the specific weight, specific volume, and density of chlorine gas at 80°F and a pressure of 100 psia (pounds per square inch absolute). Note that chlorine gas has two atoms and an atomic weight of 35.45.

SOLUTION

Given the atomic weight of 35.45, the molecular weight of chlorine gas is $2 \times 35.45 = 70.9$. From Eq. 1.4, $\mathcal{R} = mR = 49\,709$ ft·lb/slug·°R if we use the nominal value of the universal gas constant, and we deduce that

$$R = 49\,709/70.9 = 701.1 \text{ ft·lb/slug·°R}$$

In addition, T(°R) = 80°F + 459.6 = 539.6°R. Then, from a combination of Eqs. 1.1 and 1.3,

$$\gamma = \rho g_n = g_n p/RT = 32.2 \text{ ft/s}^2 \times (100 \text{ psia} \times 144 \text{ lbs/ft}^2\text{/psi})/(701.1 \text{ ft·lb/slug·°R} \times 539.6\text{°R}) \quad (1.3)$$

$$= 1.225 \text{ lb/ft}^3 \bullet$$

Accordingly,

$$\rho = 1.225 \text{ lb/ft}^3/32.2 \text{ ft/s}^2 = 0.038\,1 \text{ lb·s}^2\text{/ft}^4 = 0.038\,1 \text{ slug/ft}^3 \bullet \quad (1.1)$$

Finally, by definition, the specific volume is

$$\alpha = 1/\rho = 1/0.038\,1 = 26.3 \text{ ft}^3\text{/slug} \bullet$$

ILLUSTRATIVE PROBLEM 1.4

Calculate the density, specific volume, and specific weight of carbon dioxide gas at 100°C and atmospheric pressure (101.3 kPa = 101.3 kN/m²).

SOLUTION

The atomic weight of carbon = 12 and the atomic weight of oxygen = 16; hence, the molecular weight of CO_2 = 44. From Eq. 1.4, $\mathcal{R} = mR = 8\,313$ J/kg·K if we use the nominal value of the universal gas constant, and we deduce that

$$R = 8\,313/44 = 189.0 \text{ J/kg·K}$$

Note that this molecule has three atoms and the value of R calculated from the *universal* gas constant deviates slightly (less than 1%) from the more accurate value tabulated in Appendix 2.

Next, T(K) = 100°C + 273.15 = 373 K. Using Eq. 1.3, $\rho = p/RT$ yields

$$\rho = (101.3 \times 10^3 \text{ N/m}^2)/(189.0 \text{ J/kg·K} \times 373 \text{ K}) = 1.44 \text{ kg/m}^3 \bullet \quad (1.3)$$

By definition, the specific volume α is the inverse of the density so

$$\alpha = 1/(1.44 \text{ kg/m}^3) = 0.69 \text{ m}^3/\text{kg} \bullet$$

Finally, since Eq. 1.1 shows that $\gamma = \rho g_n$, we have

$$\gamma = 1.44 \text{ kg/m}^3 \times 9.81 \text{ m/s}^2 = 14.1 \text{ kg/m}^2\text{·}s^2 = 14.1 \text{ N/m}^3 \bullet \quad (1.1)$$

1.5 COMPRESSIBILITY, ELASTICITY

All fluids can be compressed by the application of pressure, elastic energy being stored in the process; assuming perfect energy conversions, such compressed volumes of fluids will expand to their original volumes when the applied pressure is released. Thus fluids are elastic media, and it is customary in engineering to summarize this property by defining a *modulus of elasticity*, as is done for solid elastic materials such as steel. Since fluids do not possess rigidity of form, however, the modulus of elasticity must be defined on the basis of volume and is termed a *bulk modulus*.

The mechanics of elastic compression of a fluid can be demonstrated by imagining the cylinder and piston of Fig. 1.3 to be perfectly rigid (inelastic) and to contain a volume of elastic fluid $\forall_1$. Application of a force, F, to the piston increases the pressure, p, in the fluid and causes the volume to decrease. Plotting p against $\forall/\forall_1$ produces the stress-strain diagram of Fig. 1.3 in which the modulus of elasticity of the fluid (at any point on the curve) is defined as the slope of the curve (at that point); thus

$$E = -\frac{dp}{d\forall/\forall_1} \quad (1.5)$$

The steepening of the curve with increasing pressure shows that as fluids are compressed they become increasingly difficult to compress further, a logical consequence of reducing the space between the molecules. So, the modulus of elasticity of a fluid is not constant but increases with increasing pressure.

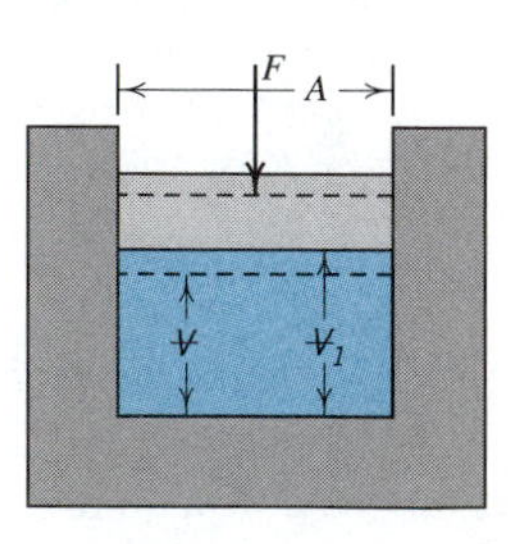

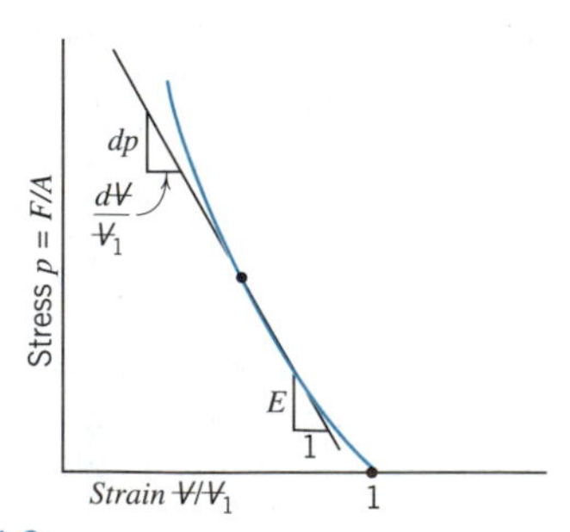

Fig. 1.3

Although the schematic curve of Fig. 1.3 applies equally well to liquids and gases, the engineer is usually concerned only with the portion of the curve near $\forall/\forall_1 = 1$ for liquids. The slope of the curve in this region is taken as the modulus of elasticity for engineering use; such values of E for common liquids are given in Appendix 2 and can be used for most engineering problems involving pressures up to a few thousand pounds per square inch or 10 MPa.

Compression and expansion of gases take place according to various laws of thermodynamics. A constant-temperature (*isothermal*) process is characterized by Boyle's law,

$$\frac{p}{\gamma} = \text{Constant} \qquad \text{or} \qquad \frac{p}{\rho} = \text{Constant} \tag{1.6}$$

whereas a frictionless process in which no heat is exchanged follows the *isentropic* relation

$$\frac{p}{\gamma^k} = \text{Constant} \qquad \text{or} \qquad \frac{p}{\rho^k} = \text{Constant} \tag{1.7}$$

in which k is the ratio of the two specific heats[7] of the gas, that at constant pressure, c_p, to that at constant volume, c_v. Values of k (frequently called the *adiabatic exponent*) for common gases are given in Appendix 2.

Expressions for the modulus of elasticity of gases may be easily derived for isothermal and isentropic processes by writing the general form of Eq. 1.5 in terms of γ or ρ. As the relative increases of γ or ρ are exactly equal to the relative decrease of volume,

$$E = \frac{dp}{d\gamma/\gamma} = \frac{dp}{d\rho/\rho} \tag{1.8}$$

When this equation is solved simultaneously with the differential forms of Eqs. 1.6 and 1.7, the results are $E = p$ for the isothermal process and $E = kp$ for the isentropic one.

In Section 6.3 it is shown that small pressure disturbances travel through fluids at a finite velocity (or *celerity*) dependent on the modulus of elasticity of the fluid. A small pressure disturbance, that travels as a wave of increased (or decreased) density and pressure, moves at a celerity, a, given by

$$a = \sqrt{\frac{dp}{d\rho}} = \sqrt{\frac{E}{\rho}} \tag{1.9}$$

The value a is frequently called the *sonic* or *acoustic* velocity because sound, a small pressure disturbance, travels at this velocity. Clearly, pressure disturbances can be transmitted instantaneously between two points in a fluid only if the fluid is inelastic, that is,

[7]Thermodynamics shows, for perfect gases, that $c_p - c_v = R$ which, in combination with $c_p/c_v = k$, yields $c_p = Rk/(k - 1)$. These specific heats also are assumed constant (with respect to temperature) for perfect gases.

if $E = \infty$. This never happens, but the assumption of an inelastic or incompressible fluid is often a convenient engineering approximation to the true state of affairs (a is about 4 860 ft/s or 1 480 m/s in water, but only about 1 100 ft/s or 335 m/s in air near sea level).

The disturbance caused by a sound wave moving through a fluid is so small and rapid that heat exchange in the compression and expansion may be neglected and the process considered isentropic. Thus, for a perfect gas, the sonic velocity may be obtained by substituting kp for E in Eq. 1.9, giving

$$a = \sqrt{kp/\rho} \tag{1.10}$$

an equation that is accurately confirmed by experiment. This equation can be put into another useful form by substituting RT for p/ρ (obtained from Eqs. 1.1 and 1.3), which results in

$$a = \sqrt{kRT} \tag{1.11}$$

and shows that the acoustic velocity in a perfect gas depends only on the temperature of the gas.

In this era of high-speed flight, you are well aware of the *Mach Number*, $\mathbf{M}$, the ratio between flow velocity and sonic velocity, and that flow velocities are defined as *subsonic* for $\mathbf{M} < 1$ and *supersonic* for $\mathbf{M} > 1$. However, $\mathbf{M}$ is also a useful criterion of relative compressibility of the fluid, which permits decisions on whether or not fluids may be considered incompressible for engineering calculations. For the flow of an incompressible (inelastic) fluid, $\mathbf{M} = 0$ because $a = \infty$. Accordingly, as $\mathbf{M} \to 0$, compressibility becomes of decreasing importance; for most engineering calculations, experience has shown that the effects of compressibility may be safely neglected if $\mathbf{M} \lesssim 0.3$.

ILLUSTRATIVE PROBLEM 1.5

Estimate the sonic velocities in the U.S. Standard Atmosphere (see Appendix 2) at the earth's surface, at the altitude of Denver, Colorado (1.6 km), and at an altitude at which a typical commercial airliner cruises (10 km).

SOLUTION

From Appendix 2, we know that the adiabatic coefficient $k = 1.40$ and the Engineering gas constant $R = 286.8$ J/kg·K ($= 286.8$ N·m/kg·K) for air. Using Eq. 1.11

$$a = \sqrt{kRT} \tag{1.11}$$

Then, from Appendix 2 we can determine the necessary absolute temperatures, as follows: At the earth's surface: T = 15°C = 288.15 K, so

$$a = \sqrt{1.4 \times (286.8\ \text{N·m/kg·K}) \times 288.15\ \text{K}} = 340.1\ \text{m/s} \bullet$$

(Note that N is equivalent to kg·m/s^2)

For the Denver altitude, we need to interpolate between the temperature of 15°C at 0 km and 2°C at 2 km; using a straight-line approximation we have 15°C − (15°C − 2°C) × 1.6/2 = 4.6°C at 1.6 km and T = 277.75 K, so

$$a = \sqrt{1.4 \times 286.6 \times 277.75} = 333.9 \text{ m/s} \bullet$$

At 10 km: T = −49.90°C = 223.25 K so

$$a = \sqrt{1.4 \times 286.8 \times 223.25} = 299.4 \text{ m/s} \bullet$$

A plane flying at 950 km/h (602 mph) moves at about 264 m/s. Near the earth's surface it is flying at a Mach number **M** of 264/340.1 = 0.78, but at 10 km the Mach number **M** = 264/299.4 = 0.88!

ILLUSTRATIVE PROBLEM 1.6

At the ocean surface and under normal atmospheric pressure, the density of sea water is 1.99 slug/ft^3. The deepest known point in the oceans is the Marianas Trench in the Pacific Ocean; there the depth is 35 640 ft and the pressure is 507.2 psia. What is the percentage change in the density of the sea water between the surface and the ocean bottom?

SOLUTION

If we rearrange Eq. 1.8, we have

$$d\rho/\rho = dp/E$$

The surface pressure is 14.7 psia so the pressure difference is 507.2 − 14.7 = 492.5 psia. The modulus of elasticity of sea water is 339 000 psi (see Appendix 2). Accordingly, the fractional change in the density is

$$d\rho/\rho = 492.5 \text{ psia}/339\,000 \text{ psi} = 0.001\,5$$

and the percentage change is 100 times that or 0.15%. •

Thus, this typical liquid is seen to be only slightly compressible!

ILLUSTRATIVE PROBLEM 1.7

Air at 15°C and 101.3 kPa is compressed isentropically so that the volume is reduced 50%. Calculate the final pressure and temperature and the sonic velocities before and after compression.

SOLUTION

From Appendix 2, the adiabatic coefficient k = 1.40 and the Engineering gas constant R = 286.8 J/kg·K (= 286.8 N·m/kg·K) for air; g_n = 9.81 m/s^2; and we may use Eqs. 1.3 and 1.7

$$\gamma = g_n p/RT \tag{1.3}$$

$$p_1/\gamma_1^k = p_2/\gamma_2^k \tag{1.7}$$

as well as equation 1.11

$$a = \sqrt{kRT} \tag{1.11}$$

From the given data, $T_1 = 15°\text{C} + 273.15 \cong 288 \text{ K}$

and $p_1 = 101.3 \text{ kPa} = 101.3 \times 10^3 \text{ Pa}$

Thus,

$$\gamma_1 = 9.81 \times (101.3 \times 10^3)(286.8 \times 288) = 12.0 \text{ N/m}^3 \tag{1.3}$$

$$\gamma_2 = 2 \times 12.0 = 24.0 \text{ N/m}^3$$

$$\frac{p_2}{(24.0)^{1.4}} = \frac{101.3 \times 10^3}{(12.0)^{1.4}} \qquad p_2 = 267.3 \text{ kPa} \bullet \tag{1.7}$$

$$\frac{9.81 \times 267.3 \times 10^3}{286.8 \times T_2} = 24.0 \qquad T_2 = 381 \text{ K } (108°\text{C}) \bullet \tag{1.3}$$

$$a_1 = \sqrt{1.4 \times 286.8 \times 288} = 340 \text{ m/s} \bullet$$

$$a_2 = \sqrt{1.4 \times 286.8 \times 381} = 391 \text{ m/s} \bullet$$

1.6 VISCOSITY

When various real fluid motions are observed carefully, two fundamentally different types of motion are seen. The first is a smooth, orderly motion in which fluid elements or particles appear to slide over each other in layers or laminae. While there is molecular agitation and diffusion in the fluid, there is virtually no large scale mixing between the layers, and this motion is called *laminar* flow. The second distinct motion which occurs is characterized by random or chaotic motion of individual fluid particles and by rapid macroscopic mixing of these particles through the flow. Eddies of a wide range of sizes are seen, and this motion is called *turbulent* flow. Both types of flow are considered in this book. Turbulent flow is, by far, the most common of the two in nature and engineering applications, but our more complete understanding of the physics of laminar flows makes them a good starting point for analysis, and a number of important laminar flow applications exist. For example, consideration of the laminar flow along a solid boundary is most instructive because it gives some early insight to key general features, such as the importance of the so-called *no-slip* condition at the solid bounding walls of continuum fluid flows and of the primary role of viscosity.

The laminar motion of a real fluid along a solid boundary is sketched in Fig. 1.4. Observations show that, while the fluid clearly has a finite velocity, v, at any finite distance from the boundary, there is no velocity at the boundary. Thus, the velocity increases with increasing distance from the boundary. These facts are summarized on the *velocity profile* which indicates relative motion between any two adjacent layers. Two such layers are shown having an infinitesimal thickness dy, the lower layer moving with velocity v, the

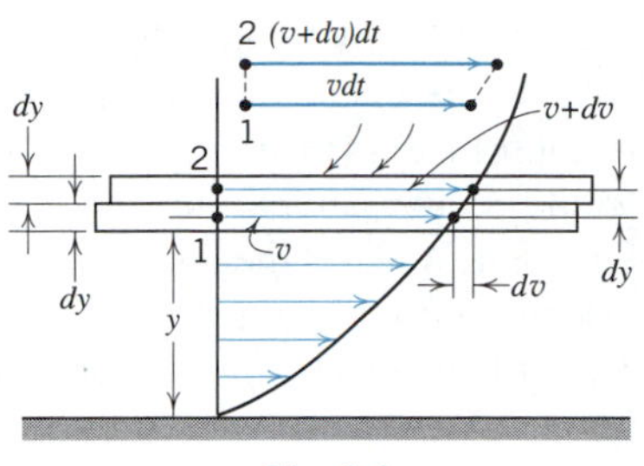

Fig. 1.4

upper with velocity $v + dv$. In Fig. 1.4, two particles 1 and 2, starting on the same vertical line, move different distances $d_1 = (v)\,dt$ and $d_2 = (v + dv)\,dt$ in an infinitesimal time dt. Thus, the fluid is distorted or sheared as the line connecting 1 and 2 acquires an increasing slope and length as t increases. In solids the stress due to shear is proportional to the strain [i.e., the relative displacement]; here

$$\text{Strain} = \frac{(d_2 - d_1)}{dy} = \frac{dv\,dt}{dy} = \frac{dv}{dy}\,dt$$

However, a fluid flows under the slightest stress, and the result of the continual application of a constant stress is an infinite strain. In fact, in fluid flow problems, the stress is related to the *rate of strain* rather than to the total strain; in this case the rate of strain is

$$\text{Rate of Strain} = \frac{(dv/dy)\,dt}{dt} = \frac{dv}{dy}$$

The frictional or shearing force that must exist between fluid layers can be expressed as a *shearing* or *frictional stress* per unit of contact area and is designated by τ. For laminar (nonturbulent) motion (in which viscosity plays a predominant role), τ is (as noted above) observed to be proportional to the *rate* of relative strain, that is, to the velocity gradient, dv/dy, with a constant of proportionality, μ, defined as the *coefficient of viscosity*.[8] Thus,

$$\tau = \mu \frac{dv}{dy} \tag{1.12}$$

All real fluids possess viscosity and therefore exhibit certain frictional phenomena when motion occurs. Viscosity results fundamentally from cohesion and molecular momentum exchange between fluid layers and, as flow occurs, these effects appear as tangential or shearing stresses between the moving layers.

Equation 1.12 is basic to all problems of fluid resistance. Therefore its implications and restrictions should be emphasized: (1) the nonappearance of pressure in the equation shows that both τ and μ are independent[9] of pressure, and that therefore fluid friction is drastically different from that between moving solids, where pressure plays a large part; (2) any shear stress τ, however small, causes flow because applied tangential forces produce a velocity gradient, that is, relative motion between adjacent fluid layers; (3) where

[8]Also termed *absolute* or *dynamic viscosity*.

[9]Viscosity usually increases very slightly with pressure, but the change is negligible in most engineering problems.

$dv/dy = 0$, $\tau = 0$, regardless of the magnitude of μ; the shearing stress in viscous fluids at rest is zero, and thus its omission in the analysis of the pressure at a point in Fig. 1.2 is confirmed; (4) the velocity profile cannot be tangent to a solid boundary because this would require an infinite velocity gradient there and an infinite shearing stress between fluid and solid; (5) the equation is limited to nonturbulent (laminar) fluid motion, in which viscous action is strong and the fluid elements are moving in straight and parallel paths. The criterion defining laminar and turbulent flow is the Reynolds number, discussed in Chapter 7. Also relevant to the use of Eq. 1.12 is the *observed* fact that the velocity at a solid boundary is zero; that is, there is no "slip" between fluid and solid for all fluids that can be treated as a continuum.

Equation 1.12 may be usefully visualized on the plot of Fig. 1.5 on which μ is the slope of a straight line passing through the origin; as seen above dv is the displacement per unit time and the velocity gradient dv/dy is the time rate of strain. Because of Newton's suggestion, which led to Eq. 1.12, fluids that follow this law are commonly known as Newtonian fluids. It is these fluids with which this book is concerned. Other fluids are classed as *non-Newtonian* fluids. The science of rheology, which broadly is the study of the deformation and flow of matter, is concerned with plastics, rubber, clay, suspensions, paints, and biological fluids (such as blood and foods), which *flow* but whose resistance is not characterized by Eq. 1.12. The relations between τ and dv/dy for two typical *plastics* are sketched on Fig. 1.5. The form of the curves is noteworthy, but there is no special significance to their relative positions. The essential mechanical difference between fluid and plastic is seen to be the shear, τ_1, manifested by the latter, which must be overcome before flow can begin. Another pair of examples shown in Fig. 1.5 are non-Newtonian, shear-affected fluids that include some suspensions and polymer solutions. The shear versus strain-rate equations corresponding to Eq. 1.12 are then, for example,

$$\tau - \tau_1 = \mu \frac{dv}{dy} \qquad \tau > \tau_1\text{: Bingham Plastic (oil paint, toothpaste)}$$

$$\tau = k\left(\frac{dv}{dy}\right)^n \qquad n > 1\text{: Shear-thickening fluid}$$

$$n < 1\text{: Shear-thinning fluid}$$

For the latter *power-law* relation, $\mu = k(dv/dy)^{n-1}$. The addition of 4% paper pulp to water reduces n from 1 to about 0.6, while for 33% lime water $n \sim 0.2$.

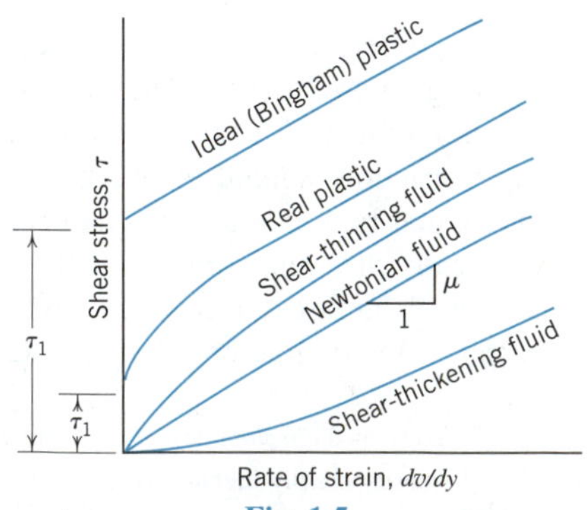

Fig. 1.5

The dimensions of the viscosity μ are determined from the dimensional homogeneity of Eq. 1.12 as follows:

$$\text{Dimensions of } \mu = \frac{\text{Dimensions of } \tau}{\text{Dimensions of } dv/dy} = \begin{cases} \dfrac{\text{lb/ft}^2}{\text{ft/s/ft}} = \text{lb}\cdot\text{s/ft}^2 \\ \dfrac{\text{Pa}}{\text{m/s/m}} = \text{Pa}\cdot\text{s} \end{cases}$$

However, from previous dimensional considerations (Section 1.4), $\text{N} = \text{kg}\cdot\text{m/s}^2$ and $\text{lb} = \text{slug}\cdot\text{ft/s}^2$, and if these are substituted above, viscosity, μ, may be quoted in kg/m·s or slug/ft·s. These dimensional combinations are equivalent to Pa·s or $\text{lb}\cdot\text{s/ft}^2$. The metric combinations times 10^{-1} are given the special name *poises* (after Poiseuille, who did some of the first work on viscosity). Viscosities quoted in *poises* or *centipoises* may be readily converted into the English system from the definitions above and by use of basic conversion factors (e.g., $1\ \text{lb}\cdot\text{s/ft}^2 = 1\ \text{slug/ft}\cdot\text{s} = 478.8$ poises).

Viscosity varies widely with temperature, but temperature variation has an opposite effect on the viscosities of liquids and gases because of their fundamentally different intermolecular characteristics. In *gases*, where intermolecular cohesion is usually negligible, the shear stress, τ, between moving layers of fluid results from an exchange of momentum between these layers brought about by molecular agitation normal to the general direction of motion. The random motion of molecules thus carries them across the direction of the flow from layers at one speed to layers at a different speed. The molecules leaving a low-speed layer collide with molecules in a high-speed layer. In the collision, the exchange of momentum tends to speed up the slower molecules and slow the faster ones. The net effect is an apparent shear force tending to reduce the speed of the higher speed layer. The reverse happens when a molecule from a fast layer collides with a molecule in a relatively slower layer. Because this molecular activity is known to increase with temperature, the shear stress, and thus the viscosity of gases, will increase with temperature (Fig. 1.6). In liquids, momentum exchange due to molecular activity is small compared to the cohesive forces between the molecules, and thus shear stress, τ, and viscosity, μ, are primarily dependent on the magnitude of these cohesive forces that tend to keep adjacent molecules in a fixed position relative to each other and to resist relative motion. Because these forces decrease rapidly with increases of temperature, liquid viscosities decrease as temperature rises (Fig. 1.6).

Owing to the appearance of the ratio μ/ρ in many of the equations of fluid flow, this term has been defined by

$$\nu = \frac{\mu}{\rho} \tag{1.13}$$

in which ν is called the *kinematic viscosity*.[10] Dimensional consideration of Eq. 1.13 shows the dimensions of ν to be square metres per second or square feet per second, a combination

[10]The wide use of the two viscosities, μ and ν, although of great convenience, may prove both surprising and troublesome to the beginner in the field. For example, water is more viscous than air in terms of μ, but air is more viscous than water in terms of ν because air is relatively much less dense than water.

of kinematic terms, which explains the name *kinematic* viscosity. In the metric system, this dimensional combination times 10^{-4} is known as *stokes* (after Sir G. G. Stokes). It is fairly common in American practice to quote kinematic viscosities in *centistokes* = 10^{-6} m^2/s = 1 mm^2/s (1 ft^2/s = 929 stokes).

Laminar flows in which the shear stress is constant (or essentially so) are the simplest ones to examine at this point; such *shear flows* were first studied by Couette and are generally known as *Couette flows*. They can be produced by slowly shearing a thin fluid

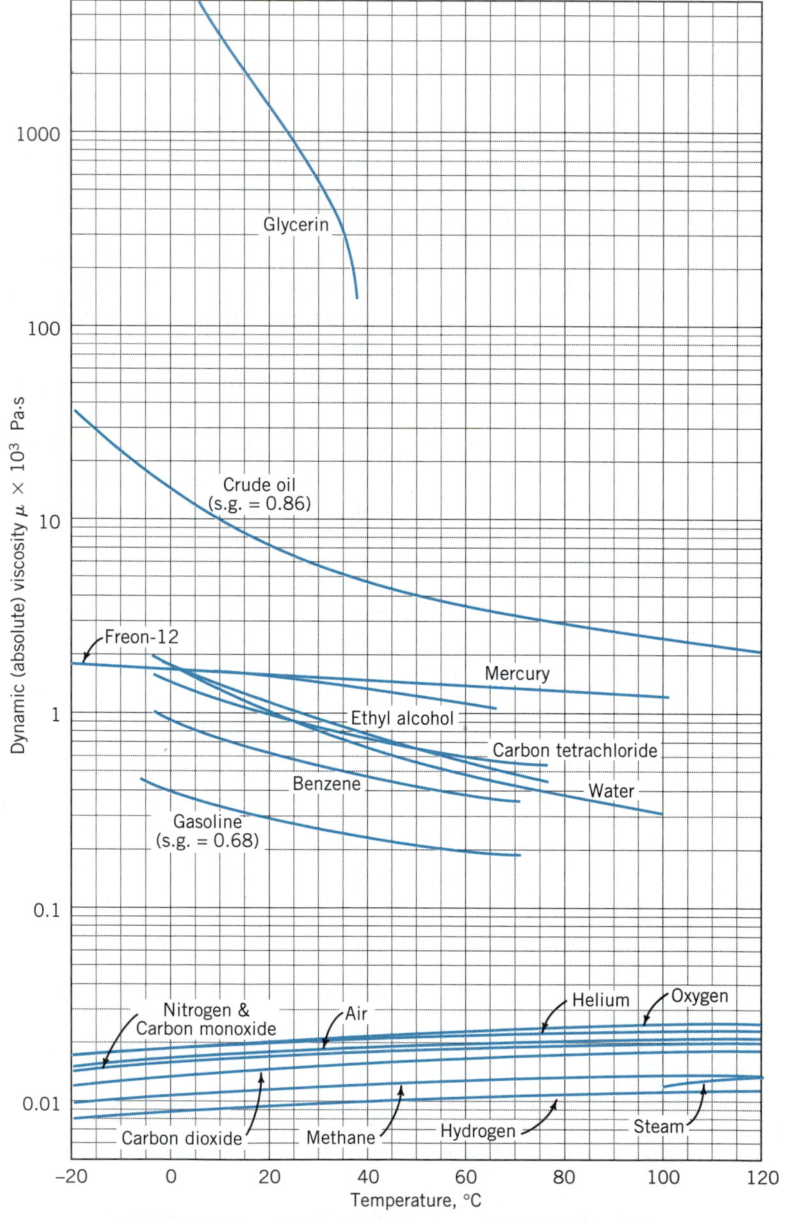

Fig. 1.6*a* Viscosities of some common fluids (SI units).

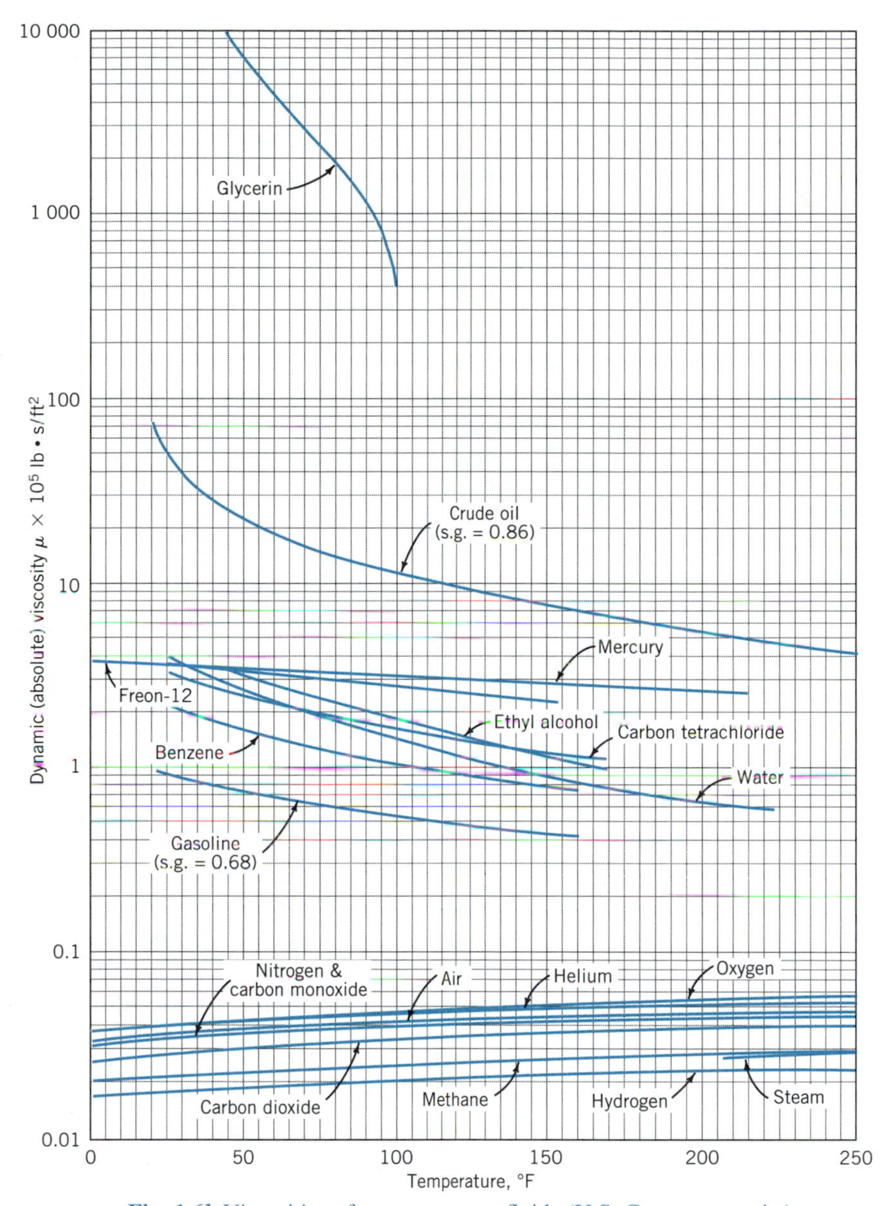

Fig. 1.6*b* Viscosities of some common fluids (U.S. Customary units).

film between two large flat plates or between the surfaces of coaxial cylinders. For constant V in Fig. 1.7, the force F is the same on both solid surfaces. If these plates have the same surface area, the shear stress on their surfaces is equal, as are the velocity gradients there. These equalities must also extend through the fluid, yielding a linear velocity profile with $dv/dy = V/h$. In this case, the rate of relative strain $(ds/dt)/dy = (dv\ dt/dt)\ dy = dv/dy = V/h$ is constant, where s is the distance a point on the upper plate moves relative to a point on the lower plate.

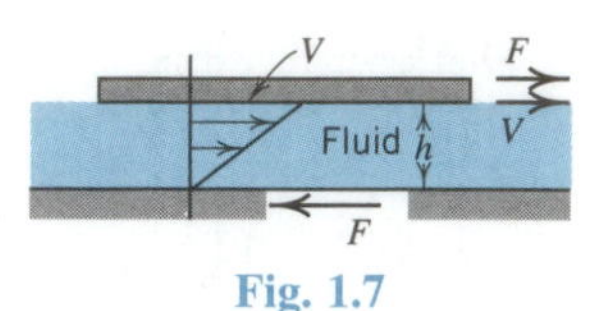

Fig. 1.7

ILLUSTRATIVE PROBLEM 1.8

Two large flat plates are separated by a thin glycerin layer that is 5 mm thick and at a temperature of 20°C. A force is applied to the top plate such that a shear stress $\tau = 26$ Pa is generated. How fast will the top plate move relative to the bottom plate after a constant velocity is attained?

SOLUTION

To a good approximation this is a Couette flow and the velocity profile between the plates is linear. The picture of Fig. 1.7 applies and from Eq. 1.12 we have

$$\tau = \mu V/h \tag{1.12}$$

Thus,

$$V = \tau h/\mu$$

Given $\tau = 26$ Pa, h $= 0.005$ m, and $\mu = 1.49$ Pa·s (from the table in Appendix 2), we have

$$\begin{aligned} V &= 26 \text{ Pa} \times 0.005 \text{ m}/1.49 \text{ Pa·s} \\ &= 0.087 \text{ m/s} \bullet \end{aligned}$$

For the coaxial cylinders of Fig. 1.8, the situation is somewhat different and more complex. Indeed, the relationship between shear stress and the velocity gradient in Eq. 1.12 is not valid here. This serves as a useful warning that the bases for an equation should be reviewed before it is used in a new situation. To understand what happens here, let us begin with both cylinders rotating at the same angular speed. After the cylinders have been rotating for a time and the start-up transients have died out, the fluid will be rotating along with the cylinders in a so-called *solid-body rotation* and without shearing. However, a velocity gradient occurs in the radial direction across the annular space even if no shear exists. In fact, the velocity profile will be linear as can be seen from the sweep of a straight line drawn from the center of rotation as the cylinders move. We have then that $\omega_1 = \omega_2 = \omega_{fluid} = \omega$ and for an angular speed ω, $v = \omega r$ and at any point $dv = \omega\, dr = (v/r)\, dr = v(dr/r)$ gives the change in velocity with small changes in r.

Suppose now that the outer cylinder rotates at a speed $\omega_2 > \omega_1$. Then, the velocity $v_2 = \omega_2 r_2$ is greater than that given by the solid body rotation. Let the extra deformation at any time be s so that the rate of deformation is ds/dt. An infinitesimal change in the velocity over an infinitesimal increment in radius dr will now be given by

$$dv = v\frac{dr}{r} + \frac{ds}{dt}$$

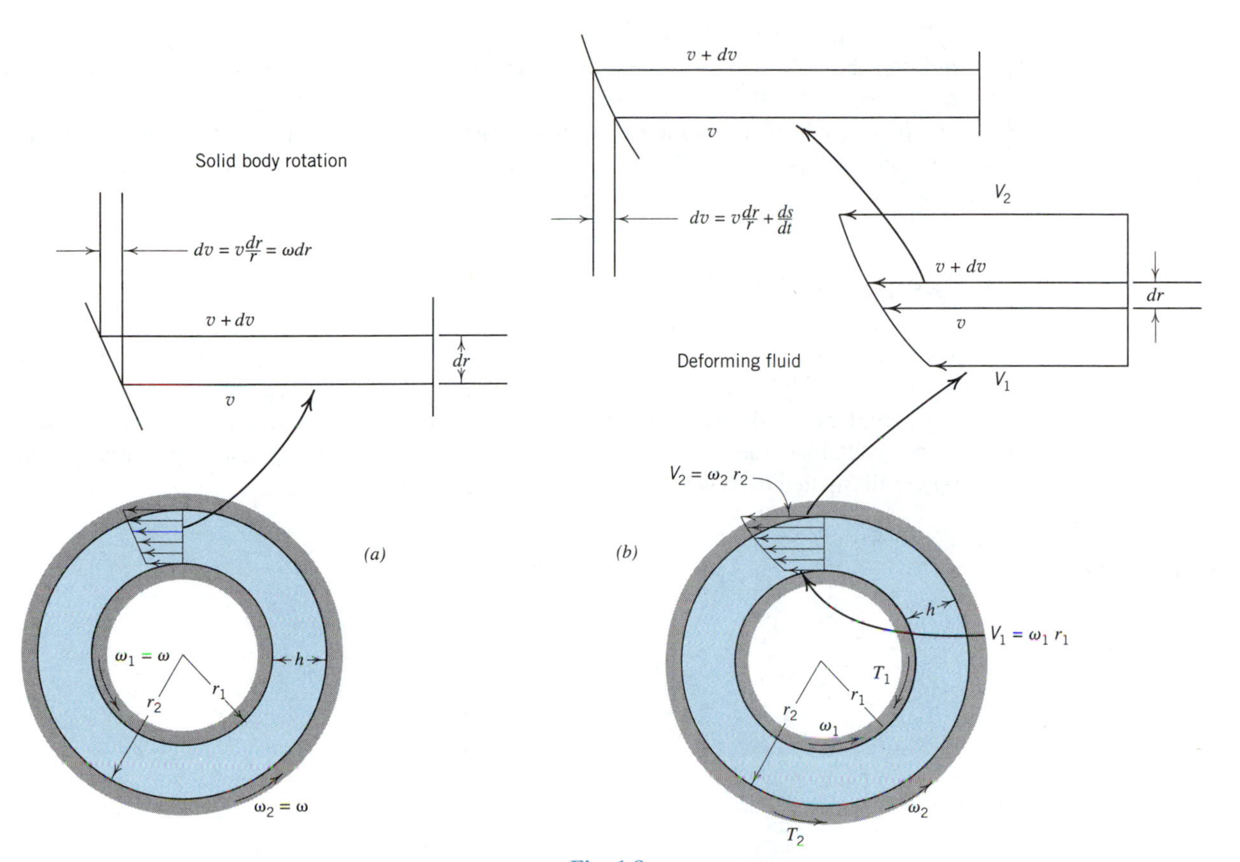

Fig. 1.8

where the first term is due to the solid body rotation and the second is due to the deformation caused by the differential rotation of the cylinders. Because strain is deformation per unit length, we divide by dr to obtain

$$\frac{dv}{dr} = \frac{v}{r} + \frac{ds/dt}{dr}$$

or

$$\frac{ds/dt}{dr} = \frac{dv}{dr} - \frac{v}{r}$$

Because shear stress τ is proportional to the rate of strain,

$$\tau = \mu \frac{ds/dt}{dr} = \mu \left(\frac{dv}{dr} - \frac{v}{r} \right) = \mu r \frac{d}{dr} \left(\frac{v}{r} \right) \tag{1.14}$$

Only if $v/r << dv/dr$, does Eq. 1.14 reduce to Eq. 1.12. Note also that the shear stress is zero for solid body rotation.

For the coaxial cylinders of Fig. 1.8*b*, the driving and resisting torques T_2 and T_1 will be equal for constant velocity, i.e., steady state. With cylinders of the same axial length, the surface area of the outer cylinder is larger than that of the inner one, and the former also has the larger radius. Accordingly, the shear and the velocity gradient at the outer cylinder will be less than the respective quantities at the inner one, and the velocity profile through the fluid will be somewhat as shown. Evidently, if $V_1 = r_1\omega_1$, $V_2 = r_2\omega_2$, and $\Delta V = V_2 - V_1$, then as $h = r_2 - r_1 \rightarrow 0$, $dv/dr \rightarrow (V_2 - V_1)/h = \Delta V/h$; equating dv/dr to $\Delta V/h$ is a common and often satisfactory engineering approximation in such problems if h is small.

ILLUSTRATIVE PROBLEM 1.9

A cylinder 86 mm in radius and 0.1 m in length rotates coaxially inside a fixed cylinder of the same length and 90 mm in radius. Glycerin (μ = 1.49 Pa·s) fills the space between the cylinders. A torque of 0.1 N·m is applied to the inner cylinder. Find the constant velocity attained, the velocity gradients at the cylinder walls, the resulting r/min, and the power dissipated by fluid resistance. Ignore end effects.

SOLUTION

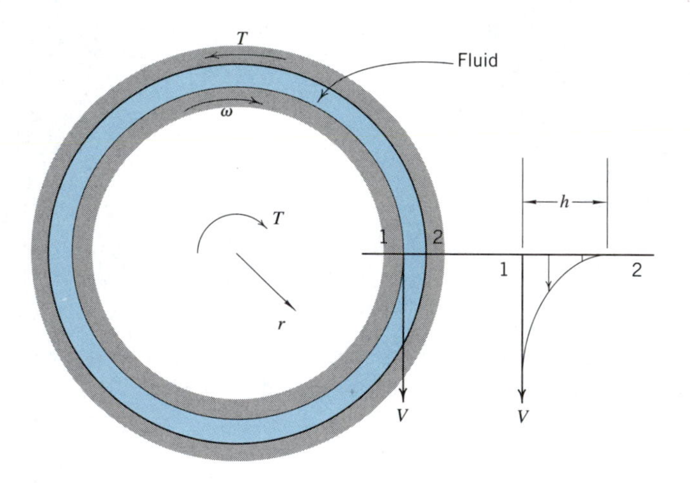

The definition sketch is shown above. This example is a good vehicle on which to develop the solutions analytically and then insert the numerical values. We have for later use:

$$r_1 = 86 \text{ mm}, \ r_2 = 90 \text{ mm}, \ \ell = 0.1 \text{ m}, \ T = 0.1 \text{ N·m, and } \mu = 1.49 \text{ Pa·s}.$$

The torque T is transmitted from the inner cylinder to the outer cylinder through the fluid layers; therefore (let r be the radial distance to any fluid layer and ℓ be the length of the cylinders)

$$T = \tau(2\pi r \times \ell)r$$

where our sign conventions yield positive torque and shear stress when the torque is applied in the counterclockwise direction to the outer cylinder or clockwise on the inner cylinder. Consequently by use of the general form of Eq. 1.14, we obtain

$$\tau = \frac{T}{2\pi r^2 \ell} = \frac{\mu r d(v/r)}{dr} \quad \text{and so} \quad \frac{d(v/r)}{dr} = \frac{T}{2\pi\mu\ell r^3} \tag{1.14}$$

Now, integrating the last relationship from the moving inner cylinder to the fixed outer cylinder yields

$$\int_{V/r_1}^{0} d\left(\frac{v}{r}\right) = \frac{T}{2\pi\mu\ell}\int_{r_1}^{r_2}\frac{dr}{r^3}; \quad -\frac{v_1}{r_1} = \frac{-T}{4\pi\mu\ell}\left[\frac{1}{r_2^2} - \frac{1}{r_1^2}\right]; \quad \frac{V}{r_1} = \frac{-T}{4\pi\mu\ell}\left[\frac{1}{r_2^2} - \frac{1}{r_1^2}\right]$$

where according to the sign conventions of Fig. 1.8, $v_1 = -V$.
From this we can calculate the numerical values:

$$V = \frac{-0.086 \times 0.1}{4\pi \times 1.49 \times 0.1}\left[\frac{1}{(0.090)^2} - \frac{1}{(0.086)^2}\right] = 0.054 \text{ m/s} = 54 \text{ mm/s} \bullet$$

The power P dissipated in fluid friction is equal to the rate of work done at the surface of the inner cylinder; we recall from mechanics that the work done for a torque is the product of the torque and the change in angle, i.e., $Td\theta$.

Thus, $P = T(d\theta/dt) = T\omega = T(V/r_1)$, where $\omega = V/r_1 = 0.63\ \text{s}^{-1}$ is the rotational speed (rad/s). It follows that $P = 0.1 \times 0.63 = 0.063\ \text{N}\cdot\text{m/s} = 0.063\ \text{W}$ •

This power will appear as heat, tending to raise the fluid temperature and decrease its viscosity; evidently a suitable heat exchanger would be needed to preserve the steady-state conditions given.

The velocity gradients can be found by use of the equation labelled 1.14 above since

$$\frac{d}{dr}\left(\frac{v}{r}\right) = \frac{1}{r}\frac{dv}{dr} - \frac{v}{r^2}$$

therefore,

$$\left(\frac{dv}{dr}\right)_1 = \frac{T}{2\pi\mu\ell r_1^2} + \frac{v_1}{r_1} = \frac{T}{2\pi\mu\ell r_1^2} + \frac{-V}{r_1} = 14.4 - 0.63 = 13.8 \text{ m/s}\cdot\text{m} \bullet$$

$$\left(\frac{dv}{dr}\right)_2 = \frac{T}{2\pi\mu\ell r_2^2} + \frac{v_2}{r_2} = 13.2 + 0.0 = 13.2 \text{ m/s}\cdot\text{m} \bullet$$

$$\text{r/min} = \left(\frac{\omega}{2\pi}\right) \times 60 = 6.0 \bullet$$

When $h = r_2 - r_1 \rightarrow 0$, then h/r_1 is small and $dv/dr \rightarrow (v_2 - v_1)/h = \Delta V/h = -v_1/h = V/h$. As a consequence, v/r is much less than dv/dr, and (compare to Illustrative Problem 1.8)

$$\tau \approx \mu\frac{dv}{dr} = \mu\frac{V}{h}$$

Assuming this linear profile case for an approximate calculation gives, for $h = 4$ mm $= 0.004$ m,

$$V = \frac{T}{2\pi\mu\ell}\left(\frac{h}{r_1^2}\right) = 0.107\left(\frac{0.004}{0.0074}\right) = 0.058 \text{ m/s} = 58 \text{ mm/s} \bullet$$

$$\text{r/min} = 6.4 \bullet$$

Because these results differ from the more accurate ones by only about 7%, the approximation may be satisfactory in this case.

1.7 SURFACE TENSION, CAPILLARITY

The apparent tension effects that occur on the surfaces of liquids, when the surfaces are in contact with another fluid or a solid, depend fundamentally on the relative sizes of intermolecular cohesive and adhesive forces. Although such forces are negligible in many engineering problems, they may be predominant in some, such as the capillary rise of liquids in narrow spaces, the mechanics of bubble formation, the breakup of liquid jets, the formation of liquid drops, and the interpretation of results obtained on small models of larger prototypes.

On a free liquid surface in contact with the atmosphere, there is little force attracting molecules away from the liquid because there are relatively few molecules in the vapor above the surface. Within the liquid bulk, the intermolecular forces of attraction and repulsion are balanced in all directions. However, for liquid molecules at the surface, the cohesive forces of the next layer below are not balanced by an identical layer above. This situation tends to pull the surface molecules tightly to the lower layer and to each other and causes the surface to behave as though it were a membrane; *hence, the name surface tension.* In fact, treating the surface as though it were a membrane capable of supporting tension is an analogy used widely in theoretical treatment of surface tension problems. When the free surface is curved, the surface tension force will support small loads. A small needle placed gently on a water surface will not sink but will be supported by the tension in the liquid surface. The surface is depressed slightly in the process which causes localized surface curvature and the consequent development of the necessary force.

The surface tension, σ, is thought of as the force in the liquid surface normal to a line of unit length drawn in the surface; thus it will have dimensions of pounds per foot or newtons per metre. Because surface tension is directly dependent on intermolecular cohesive forces, its magnitude will decrease as temperature increases (see Appendix 2). Surface tension is also dependent on the fluid in contact with the liquid surface; thus surface tensions are usually quoted in contact with air.

Consider now (Fig. 1.9) the general case of a small element $dx\ dy$ of a surface of double curvature with radii R_1 and R_2. Evidently a pressure difference $(p_i - p_o)$ must accompany the surface tension for static equilibrium of the element. A relation between

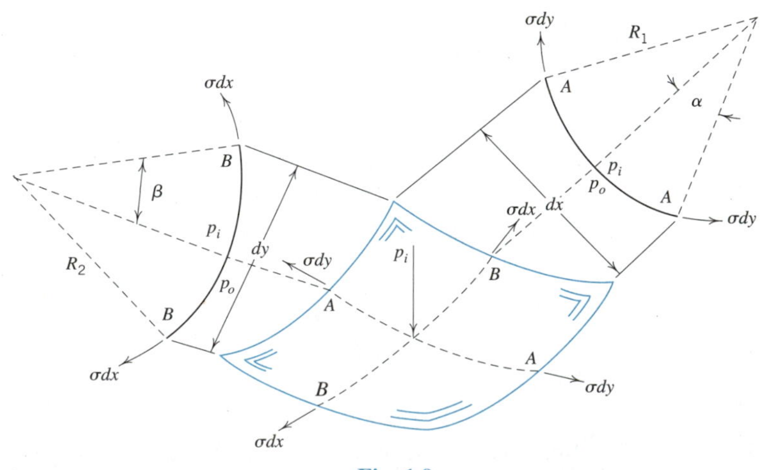

Fig. 1.9

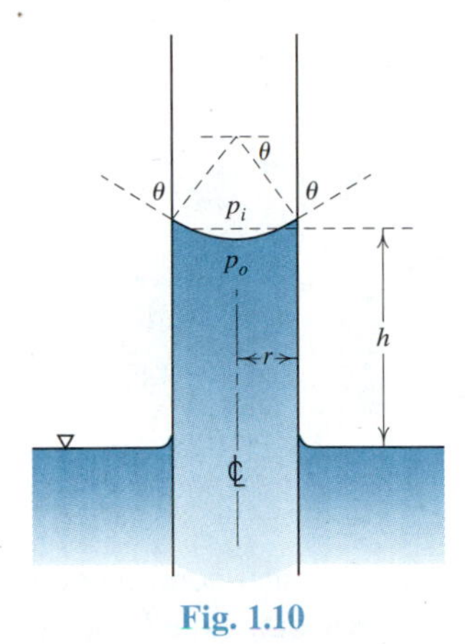

Fig. 1.10

the pressure difference and the surface tension may be derived from this equilibrium by taking $\Sigma F = 0$ for the force components normal to the element:

$$(p_i - p_o)\, dx\, dy = 2\sigma\, dy \sin\alpha + 2\sigma\, dx \sin\beta$$

in which α and β are small angles. However, from the geometry of the element, $\sin\alpha = dx/2R_1$ and $\sin\beta = dy/2R_2$. When these values are substituted above, the basic relation between surface tension and pressure difference is obtained; it is

$$p_i - p_o = \sigma\left(\frac{1}{R_1} + \frac{1}{R_2}\right) \tag{1.15}$$

From this equation the pressure (caused by surface tension) in droplets and tiny jets may be calculated and the rise of liquids in capillary spaces estimated; for a spherical droplet, $R_1 = R_2$; for a cylindrical jet, one R is infinite and the other is the radius of the jet. For the cylindrical capillary tube of Fig. 1.10 (assuming the liquid surface to be a section of a sphere), $p_o = -\gamma h$, $p_i = 0$, and $p_i - p_o = \gamma h$; also $R_1 = R_2 = R$ and $r/R = \cos\theta$. Substituting these in Eq. 1.15 yields

$$h = \frac{2\sigma\cos\theta}{\gamma r} \tag{1.16}$$

This result immediately raises several questions: the meaning of the angle θ, the limitations of the equation, and its confirmation by experiment. From the assumption of spherical liquid surface it is clear that the equation is limited to very small tubes; in large tubes the liquid surface is far from the spherical form. The angle θ is known as the *angle of contact*, and it results from surface tension phenomena of complex nature. Figure 1.11 describes the situation when mercury and water surfaces contact a vertical glass surface. Evidently the mercury molecules possess a greater affinity for each other (cohesion) than for the glass (adhesion), whereas the opposite condition obtains for the water and glass. Although the detailed character of these molecular interactions is not completely understood, the contact angles have been measured and found (for pure substances) to be as indicated.

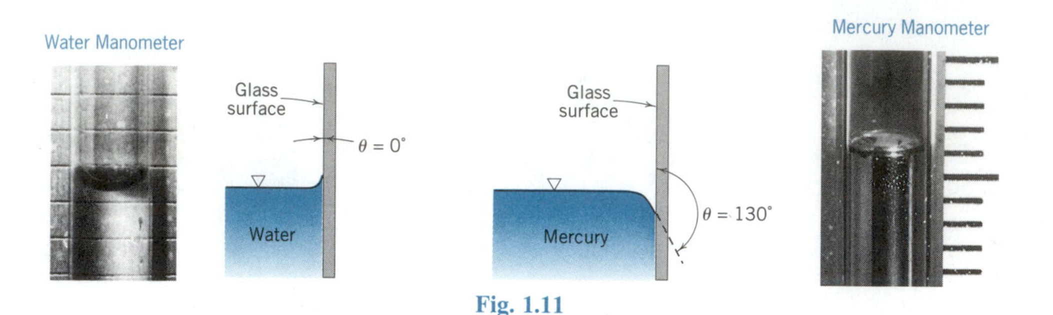

Fig. 1.11

Good experimental confirmation of Eq. 1.16 is obtained for small tubes ($r < 0.1$ in. or 2.5 mm), providing liquids and tube surfaces are extremely clean; in engineering, however, such cleanliness is virtually never encountered and h will be found to be considerably *smaller* than given by Eq. 1.16. Thus, the equation is useful for making conservative estimates of capillary errors. In the sizing of tubes for pressure measurement, the capillarity problem may be avoided entirely by providing tubes large enough to render the capillarity correction negligible for the desired accuracy.

ILLUSTRATIVE PROBLEM 1.10

Of what diameter must a droplet of water (70°F) be to have the pressure within it 0.1 psi greater than that outside?

SOLUTION

Looking at Eq. 1.15

$$p_i - p_o = \sigma\left(\frac{1}{R_1} + \frac{1}{R_2}\right) \qquad (1.15)$$

we observe that for a spherical drop, $R_1 = R_2 = R$ and $p_i - p_o = 0.1 \text{ psi} = 0.1 \times 144 \text{ lb/ft}^2$. For T = 70°F, the surface tension $\sigma = 0.004\,98$ lb/ft (from Appendix 2). Hence,

$$p_i - p_o = 0.1 \times 1.44 = 0.004\,98\left(\frac{2}{R}\right) \qquad (1.15)$$

$$R = 0.000\,69 \text{ ft} \qquad d = 0.016\,6 \text{ in.} \bullet$$

ILLUSTRATIVE PROBLEM 1.11

What is the height of capillary rise in a clean glass tube of 1 mm diameter if the water temperature is 10°C or 90°C?

SOLUTION

From Appendix 2 we find that the surface tension σ is 0.074 2 M/m and 0.060 8 N/m and the specific weight γ is 9.804 kN/m^3 and 9.466 kN/m^3 at 10°C and 90°C, respectively. The angle of contact for water on glass θ is zero degrees; hence, $\cos\theta = 1$. In Eq. 1.16 it is clear that both the surface tension and the specific weight vary with temperature; accordingly,

$$h = \frac{2\sigma\cos\theta}{\gamma r} = \frac{2 \times \sigma \times 1}{\gamma \times 0.000\,5} = 4\,000 \times \frac{\sigma}{\gamma}\text{ m} \qquad (1.16)$$

and the result is that

$$h = 4\,000 \times 0.074\,2/9.804 \times 10^3 = 0.030\text{ m} = 30\text{ mm at }10°\text{C} \bullet \qquad (1.16)$$

and

$$h = 4\,000 \times 0.0608/9.466 \times 10^3 = 0.026\text{ m} = 26\text{ mm at }90°\text{C} \bullet \qquad (1.16)$$

1.8 VAPOR PRESSURE

In the mechanics of *liquids*, the physical property of vapor pressure is frequently important in the analysis of problems. All liquids possess a tendency to vaporize, that is, to change from the liquid to the gaseous phase. Such vaporization occurs because molecules are continually projected through the free liquid surface and lost from the body of the liquid as a consequence of their natural thermal vibrations. The ejected molecules, being gaseous, then exert their own partial pressure, which is known as the *vapor pressure* (p_v) of the liquid. Because of the increase of molecular activity with temperature, vapor pressure increases with temperature; for water this variation is given in Appendix 2. *Boiling* (formation of vapor bubbles throughout the fluid mass) will occur (whatever the temperature) when the external absolute pressure imposed on the liquid is equal to or less than the vapor pressure of the liquid. This means that the boiling point of a liquid is dependent on the imposed pressure as well as on temperature.[11] In a flowing liquid, the important and potentially damaging phenomenon of *cavitation*, which is a form of boiling, may occur wherever the local pressure falls to the vapor pressure of the liquid (See Appendix 4).

A comparison of the vapor pressures of a few common liquids at the same temperature can be made by looking at the table for liquids in Appendix 2. The low vapor pressure of mercury along with its high density makes this liquid well suited for use in barometers and other pressure-measuring devices. The more *volatile* liquids, which vaporize more easily, possess the higher vapor pressures. If a liquid is placed in a sealed container with empty space above the liquid surface, the vaporization of the liquid continues until the vapor exerts the vapor pressure p_v, in the once empty space. At this stage the number of molecules escaping from the liquid exactly equals the number that is returning. However, if the space is too large, this equilibrium position is not reached and the liquid continues to vaporize or *evaporate* until the liquid is gone and only vapor remains at a pressure less than or equal to p_v.

[11]For example, water boils at 212°F (100°C) when exposed to an atmospheric pressure of 14.7 psia (101.3 kPa, absolute), but will boil at 140°F (60°C) if the imposed pressure is reduced to that at an altitude of about 39 000 ft (12 km) in the atmosphere, that is, to 2.89 psia (19.9 kPa, absolute). See Appendix 2.

ILLUSTRATIVE PROBLEM 1.12

A vertical cylinder 12 in. (~300 mm) in diameter is fitted (at the top) with a tight but frictionless piston and is completely filled with water at 160°F(~70°C). The outside of the piston is exposed to an atmospheric pressure of 14.52 psia (~100 kPa). Calculate the minimum force applied to the piston that will cause the water to boil.

SOLUTION

U.S. Customary Solution. The force must be applied slowly (to avoid acceleration) in a direction to withdraw the piston from the cylinder. Since the water cannot expand, a space filled with water vapor will be created beneath the piston, whereupon the water will boil. The pressure on the inside of the piston will then be (Appendix 2) 4.74 psia and the force on the piston $(14.52 - 4.74)\pi(12)^2/4 = 1\,106$ lb. •

SI Solution. As noted above, when the piston is slowly withdrawn, the water will boil. The pressure on the inside of the piston will then be (Appendix 2) 31.2 kPa and the force on the piston $(100 - 31.2)\pi(0.3)^2/4 = 4.86$ kN. •

REFERENCES

General and Historical

Barnes, H. A., Hutton, J. F., and Walters, K. 1989. *An Introduction to Rheology.* New York: Elsevier.

Busemann, A. 1971. Compressible flow in the thirties. *Ann. Rev. of Fluid Mechanics.* Vol. 3. Palo Alto: Annual Reviews.

Goldstein, S. 1969. Fluid mechanics in the first half of this century. *Ann. Rev. of Fluid Mechanics.* Vol. 1. Palo Alto: Annual Reviews.

McDowell, D. M., and Jackson, J. D., Eds. 1970. *Osborne Reynolds and engineering science today.* New York: Barnes & Noble.

Rouse, H. 1976. Hydraulic's latest golden age. *Ann. Rev. of Fluid Mechanics.* Vol. 8. Palo Alto: Annual Reviews.

Rouse, H., and Ince, S. 1963. *History of hydraulics.* New York: Dover.

Scott Blair, G. W. 1969. *Elementary rheology.* New York: Academic Press.

Tani, I. 1977. History of boundary-layer theory. *Ann. Rev. of Fluid Mechanics.* Vol. 9. Palo Alto: Annual Reviews.

Tanner, R. I. 1988. *Engineering Rheology.* Revised Edition; Oxford: Clarendon Press.

Taylor, G. I. 1974. The interaction between experiment and theory in fluid mechanics. *Ann. Rev. of Fluid Mechanics.* Vol. 6. Palo Alto: Annual Reviews.

Tokaty, G. A. 1971. *A history and philosophy of fluid mechanics.* London: G. T. Foulis.

Wilkinson, W. L. 1960. *Non-Newtonian fluids.* New York: Pergamon Press.

Physical Properties

Adam, N. K. 1962. *Physical chemistry.* New York: Oxford University Press, Chapters 3, 8.

Bridgman, P. W. 1943. Recent work in the field of high pressures. *Amer. Scientist,* 31, 1.

Burdon, R. S. 1949. *Surface tension and the spreading of liquids.* 2nd ed. Cambridge: Cambridge University Press.

Daniels, F., and Alberty, R. A. 1979. *Physical chemistry.* 5th ed. New York: Wiley (SI version, 1980).

Defay, R., and Prigogine, I. 1966. *Surface tension and absorption.* Trans. by D. H. Everett. New York: Wiley.

Hydraulic Institute. 1990. *Engineering data book.* 2nd ed. Cleveland: Hydraulic Institute.

Reid, R. C., Prausnitz, J. M., and Poling, B. E. 1987. *The properties of gases and liquids.* 4th ed. New York: McGraw-Hill.

Weast, R. C., Ed. 1988. *CRC Handbook of chemistry and physics.* 1st Student ed. Boca Raton, FL: CRC Press.

Units

ASME Orientation and guide for use of SI (metric) units. 8th ed. 1978. Guide SI-1. New York: The Amer. Soc. of Mech. Engin.

ASTM Standard for metric practice. 1980. No. E380-79. Philadelphia: Amer. Soc. for Testing and Mat.

FILMS

Makovitz, H. Rheological behavior of fluids. NCFMF/EDC Film No. 21613, Encyclopaedia Britannica Educ. Corp.[12]

Rouse, H. Mechanics of fluids: introduction to the study of fluid motion. Film No. U45578, Media Library, Audiovisual Center, Univ. of Iowa.

Trefethen, L. M. Surface tension in fluid mechanics. NCFMF/EDC Film No. 21610, Encyclopaedia Britannica Educ. Corp.[12]

PROBLEMS

1.1. Using a small square in two-dimensions and the same strategy employed in Section 1.3, derive a relationship for the variation of pressure in the vertical direction due to the influence of gravity in a fluid at rest and with a constant density.

1.2. Using the small wedge element shown here, prove that the pressure is the same in all directions at a point by following the same strategy employed in Section 1.3.

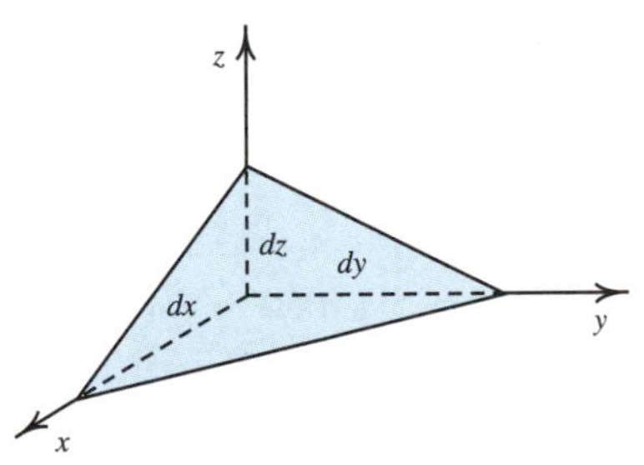

Problem 1.2

1.3. The density of a sample of sea water is 1.99 slugs/ft^3. What is the SI value?

1.4. The modulus of elasticity of mercury is 26 × 10^6 kPa. What is its value in U.S. Customary units?

1.5. A pump is rated at 50 hp; what is its kW rating?

1.6. A friend from Australia tells you that the temperature in Perth was over 35°C for 30 consecutive days in 1985. Are you impressed (i.e., how hot was it in °F)?

1.7. In the Sacramento Valley of California the summer temperature often exceeds 100°F; what is the Celsius equivalent temperature?

1.8. A fluid occupying 3.2 m^3 has a mass of 4 Mg. Calculate its density and specific volume in SI and U.S. Customary units.

1.9. If a power plant is rated at 2 000 MW output and operates (on average) at 75% of rated power, how much energy (in J and ft·lbs) does it put out in a year?

1.10. A rocket payload with a weight on earth of 2 000 lb or 8 900 N is landed on the moon where the acceleration due to the moon's gravity $g_m \approx g_n/6$. Find the mass of the payload on the earth and the moon and the payload's moon weight.

1.11. If a barrel of oil weighs 1.5 kN, calculate the specific weight, density, and specific gravity of this oil. A barrel contains 159 litres or 0.159 m^3; the barrel itself weights 110 N.

1.12. Calculate the specific weight of mercury at 60°F and at 600°F.

1.13. Calculate the specific weight, specific gravity, and specific volume of alcohol at 68°F or at 20°C.

[12]Eight millimetre silent film loops selected from NCFMF/EDC films are available as well.

1.14. Calculate the density and specific volume of liquid sodium at 538°C.

1.15. The specific gravity of a liquid is 3.0. What is its specific volume?

1.16. A cubic metre of air at 101 kPa and 15°C weighs 12.0 N. What is its specific volume?

1.17. Calculate the specific weight, specific volume, and density of air at 40°F and 50 psia.

1.18. Calculate the density, specific weight, and specific volume of carbon dioxide at 700 kPa and 90°C.

1.19. Calculate the density of carbon monoxide at 20 psia and 50°F.

1.20. Calculate the temperature of methane gas (CH_4) at 200 kPa if its density is 1.05 kg/m^3.

1.21. The molecular weight of nitrogen is 28. Use the perfect gas laws to find its density at 7.35 psia (50.6 Pa, absolute) and 212°F (100°C).

1.22. The specific volume of a certain perfect gas at 200 kPa and 40°C is 0.65 m^3/kg. Calculate its gas constant and molecular weight.

1.23. Avogadro's number is 6.023×10^{23} molecules/mol. Find the number of molecules in a unit volume of air at standard temperature and pressure (15°C; 101.3 kPa, absolute).

1.24. The derivation of the universal gas constant led to its definition as the product of molecular weight m and engineering gas constant R. Review of the derivation shows, however, that $M_1R_1 = M_2R_2$. Thus, MR = constant, where M is the molar mass. If $\mathcal{R}^* = MR$, how is this universal constant different from $\mathcal{R}$? *Hint:* Stick with SI units.

1.25. In some cases we use the term head to represent the ratio of the pressure to the specific weight so that $h = p/\gamma$, what are the dimensions of h?

1.26. Torricelli showed that the velocity of the flow of a fluid from a small hole in the wall of a tank where there is a head h of fluid above the hole is approximately $V = \sqrt{2g_nh}$. According to this formula, what are the dimensions of V?

1.27. The flux of momentum F of a flow through a pipe is $F = Q\gamma V/g_n$. Given that Q is the flowrate with dimensions of (L^3/t), what are the dimensions of F?

1.28. One cubic metre of water is placed under an absolute pressure of 7 000 kPa. Calculate the volume at this pressure.

1.29. If the volume of a liquid is reduced 0.035% by application of a pressure of 690 kPa or 100 psi, what is its modulus of elasticity?

1.30. What pressure must be applied to water to reduce its volume 1%?

1.31. Calculate the specific gravity of carbon tetrachloride at 20°C and 20 000 kPa.

1.32. When a volume of 1.021 2 ft^3 of alcohol is subjected to a pressure of 7 350 psi, it will contract to 0.978 4 ft^3. Calculate the modulus of elasticity.

1.33. One cubic metre of nitrogen at 40°C and 340 kPa is compressed isothermally to 0.2 m^3. What is the pressure when the nitrogen is reduced to this volume? What is the modulus of elasticity at the beginning and end of the compression?

1.34. If the nitrogen in the preceding problem is compressed isentropically to 0.2 m^3, calculate the final pressure and temperature and the modulus of elasticity at the beginning and end of the compression.

1.35. A gas is compressed isentropically. The measured volume and absolute pressure before and after compression are 0.30 m^3 and 50.7 kPa and 0.111 m^3 and 202.8 kPa, respectively. What are k and E for this gas?

1.36. Calculate the velocity of sound in air at 0°C (32°F) and absolute pressure of 101.3 kPa (14.7 psia).

1.37. Calculate the velocity of sound in water at 20°C (68°F).

1.38. Calculate the Mach numbers for an airplane flying at 1 130 km/h through still air at altitudes 2 and 16 km (see Appendix 2).

1.39. Calculate the kinematic viscosity of sea water at 20°C (68°F).

1.40. Calculate the kinematic viscosity of nitrogen at 40°C and 550 kPa.

1.41. What is the ratio between the viscosities of air and water at 10°C or 50°F? What is the ratio between their kinematic viscosities at this temperature and standard barometric pressure?

1.42. Using data from Appendix 2, calculate the kinematic viscosity of water at 10°C and 34 475 kPa or at 50°F and 5 000 psi.

1.43. A certain diatomic gas has a kinematic viscosity of 1.7×10^{-5} m^2/s and viscosity of 2.9×10^{-5} Pa·s at 101 kPa and 40°C. Calculate its molecular weight.

1.44. The kinematic viscosity and specific gravity of a certain liquid are 5.6×10^{-4} m^2/s and 2.00, respectively. Calculate the viscosity of this liquid.

1.45. Nitrogen at 15 psia and 100°F is compressed adiabatically to 45 psia. Calculate kinematic viscosities and acoustic velocities before and after compression.

1.46. Calculate velocity gradients for y = 0, 0.2, 0.4, and 0.6 m, if the velocity profile is a quarter-circle having its center 0.6 m from the boundary.

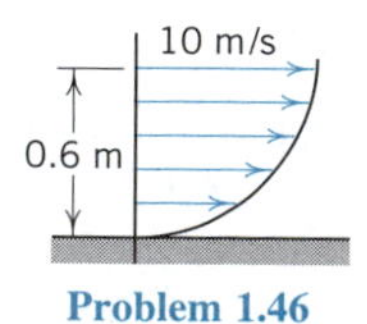

Problem 1.46

1.47. Calculate the velocity gradients of the preceding problem, assuming that the velocity profile is a parabola with the vertex 0.6 m from the boundary. Also calculate the shear stresses at these points if the fluid viscosity is 1 poise.

1.48. If the equation of a velocity profile is $v = 4y^{2/3}$(v fps, y ft), what is the velocity gradient at the boundary and at 0.25 ft and 0.5 ft from it?

1.49. The velocity profile for laminar flow in a pipe is

$$v(r) = v_c(1 - r^2/R^2)$$

where r is measured from the centerline where $v = v_c$ and R is the pipe radius. Find $\tau(r)$ and dv/dr. Plot both versus r.

1.50. The shear stress $\tau = \tau_o(1 - y)$ in a laminar flow in a pipe of radius 1 with τ_o measured at the pipe wall where $y = 0$. If the fluid is a power-law fluid so

$$\tau = k\left(\frac{dv}{dy}\right)^n$$

find the velocity profile in $0 \leq y \leq 1$. Plot and compare profiles for $n = 0.6$, 1.0, and 1.4 using $\tau_o/k = 1$. How does the centerline velocity change as a function of n?

1.51. If the viscosity of a liquid is 0.002 lb·s/ft² (0.01 Pa·s), what is its viscosity in poises? in centipoises?

1.52. If the viscosity of a liquid is 3.00 centipoises, what is its viscosity in lb·s/ft²?

1.53. If the kinematic viscosity of an oil is 1 000 centistokes, what is its kinematic viscosity in ft²/s (m²/s)? If its specific gravity is 0.92, what is its viscosity in lb·s/ft² (Pa·s)?

1.54. At what temperatures does air have a larger kinematic viscosity than crude oil?

1.55. At a point in a viscous flow the shearing stress is 5.0 psi and the velocity gradient 6 000 fps/ft. If the specific gravity of the liquid is 0.93, what is its kinematic viscosity?

1.56. A very large thin plate is centered in a gap of width 0.06 m with different oils of unknown viscosities above and below; one viscosity is twice the other. When the plate is pulled at a velocity of 0.3 m/s, the resulting force on one square metre of plate due to the viscous shear on both sides is 29 N. Assuming viscous flow and neglecting all end effects, calculate the viscosities of the oils.

1.57. Through a very narrow gap of height h a thin plate of very large extent is being pulled at constant velocity V. On one side of the plate is oil of viscosity μ and on the other side oil of viscosity $k\mu$. Calculate the position of the plate so that the drag force on it will be a minimum.

1.58. A vertical gap 25 mm wide of infinite extent contains oil of specific gravity 0.95 and viscosity 2.4 Pa·s. A metal plate 1.5 m × 1.5 m × 1.6 mm weighing 45 N is to be lifted through the gap at a constant speed of 0.06 m·s. Estimate the force required.

1.59. A cylinder 8 in. in diameter and 3 ft long is concentric with a pipe of 8.25 in. i.d. Between cylinder and pipe there is an oil film. What force is required to move the cylinder along the pipe at a constant velocity of 3 fps? The kinematic viscosity of the oil is 0.006 ft²/s; the specific gravity is 0.92.

1.60. Crude oil at 20°C fills the space between two concentric cylinders 250 mm high and with diameters of 150 mm and 156 mm. What torque is required to rotate the inner cylinder at 12 r/min, the outer cylinder remaining stationary?

1.61. A torque of 4 N·m is required to rotate the intermediate cylinder at 30 r/min. Calculate the viscosity of the oil. All cylinders are 450 mm long. Neglect end effects.

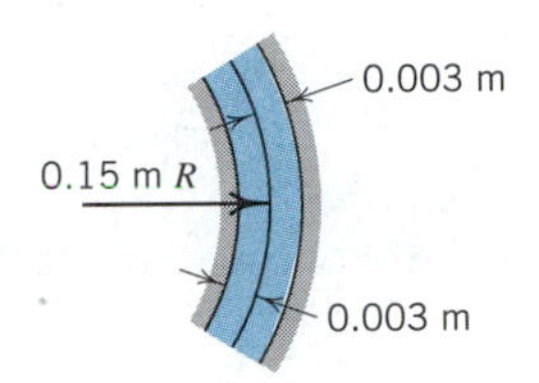

Problem 1.61

1.62. The viscosity of the oil in the preceding problem is 0.25 Pa·s. What torque is required to rotate the intermediate cylinder at a constant speed of 40 r/min?

1.63. A circular disk of diameter d is slowly rotated in a liquid of large viscosity μ at a small distance h from a fixed surface. Derive an expression for the torque T necessary to maintain an angular velocity ω. Neglect centrifugal effects.

1.64. The fluid drive shown transmits a torque T for steady-state conditions (ω_1 and ω_2 constant). Derive an expression for the slip ($\omega_1 - \omega_2$) in terms of T, μ, d, and h.

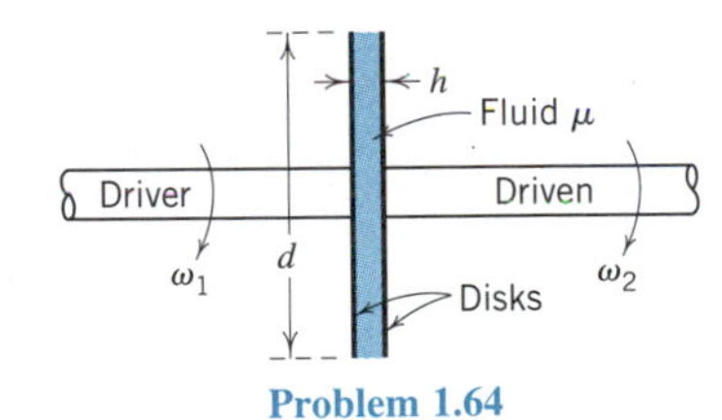

Problem 1.64

1.65. Oil of viscosity μ fills the gap h, which is very small. Calculate the torque T required to rotate the cone at constant speed ω.

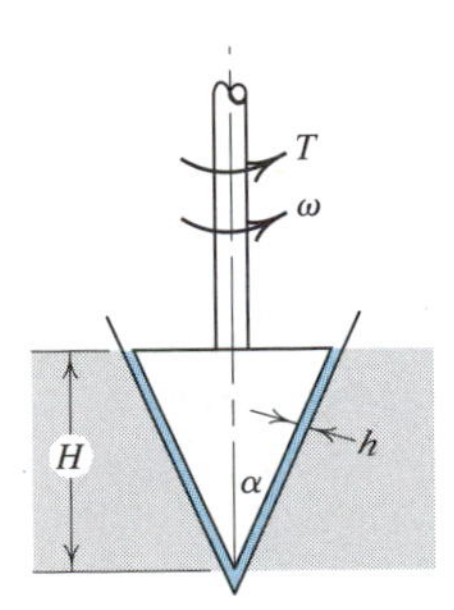

Problem 1.65

1.66. A piece of pipe 12 in. long weighing 3 lb and having i.d. of 2.05 in. is slipped over a vertical shaft 2.00 in. in diameter and allowed to fall. Calculate the approximate velocity attained by the pipe if a film of oil of viscosity 0.5 lb·s/ft^2 is maintained between pipe and shaft.

1.67. The lubricant has a kinematic viscosity of 2.8×10^{-5} m^2/s and s.g. of 0.92. If the mean velocity of the piston is 6 m/s, approximately what is the power dissipated in friction?

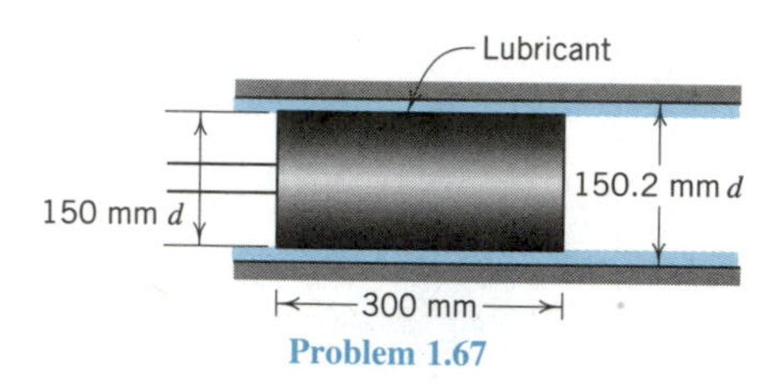

Problem 1.67

1.68. Calculate the approximate viscosity of the oil.

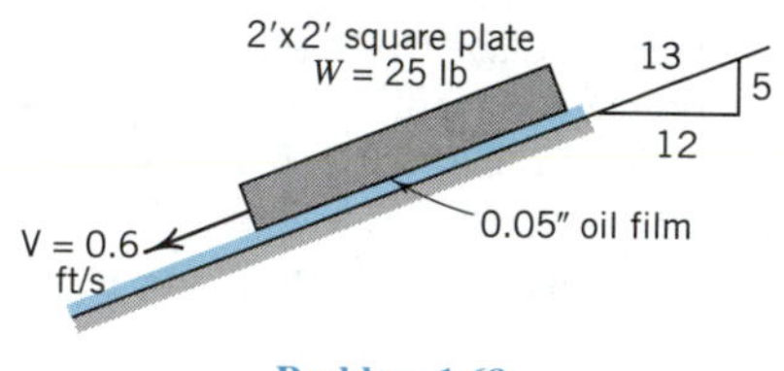

Problem 1.68

1.69. The weight falls at a constant velocity of 50 mm/s. Calculate the approximate viscosity of the oil.

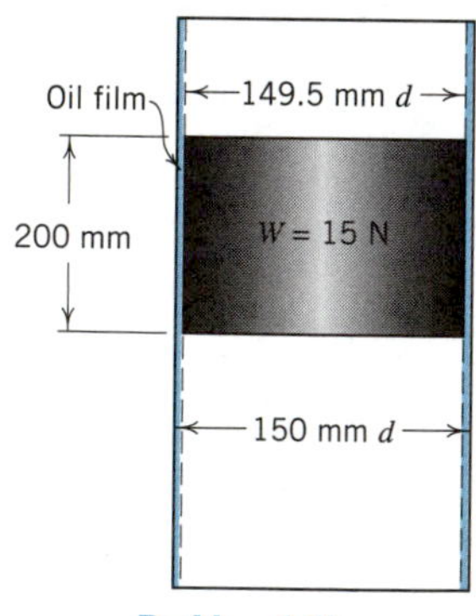

Problem 1.69

1.70. Calculate the approximate power lost in friction in this ship propeller shaft bearing.

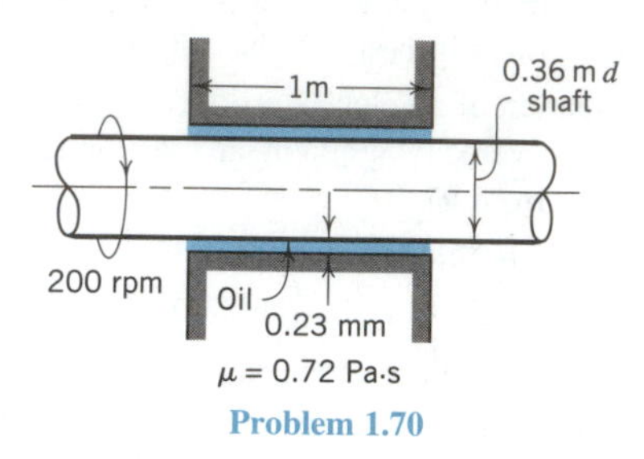

Problem 1.70

1.71. What excess pressure may be caused within a cylindrical jet of water 5 mm (0.2 in.) in diameter by surface tension?

1.72. Calculate and plot the maximum capillary rise of water (20°C or 68°F) to be expected in a vertical glass tube as a function of tube diameter (for diameters from 0.5 to 2.5 mm or 0.02 to 0.1 in.).

1.73. Calculate the maximum capillary rise of water (20°C or 68°F) to be expected between two vertical, clean glass plates spaced 1 mm (0.04 in.) apart.

1.74. Derive an equation for theoretical capillary rise between vertical parallel plates. Plot the rise as a function of separation distance.

1.75. Calculate the maximum capillary depression of mercury to be expected in a vertical glass tube 1 mm (0.04 in.) in diameter at 15.5°C (60°F).

1.76. A soap bubble 50 mm in diameter contains a pressure (in excess of atmospheric) of 20 Pa. Calculate the tension in the soap film.

1.77. What force is necessary to lift a thin wire ring 25 mm in diameter from a water surface at 20°C? Neglect weight of ring.

1.78. Using the assumptions of Section 1.7, derive an expression for capillary correction h for an interface between liquids in a vertical tube.

1.79. What is the minimum absolute pressure which may be maintained in the space above the liquid in a can of ethyl alcohol at 68°F or 20°C?

1.80. To what value must the absolute pressure over carbon tetrachloride be reduced to make it boil at 68°F or 20°C?

1.81. To what value must the absolute pressure over water be reduced to make it boil at 20°C or 68°F? At 0°C or 32°F?

1.82. At what temperature will water boil at an altitude of 20 000 ft or 6 100 m? See Appendix 2.

CHAPTER

2 FLUID STATICS

Fluid statics is the study of fluid problems in which there is no relative motion between fluid elements. With no relative motion between individual elements (and thus no velocity gradients), no shear stress can exist, whatever the viscosity of the fluid.[1] Accordingly, viscosity has no effect in static problems and exact analytical solutions to such problems are relatively easy to obtain.

It is important to point out that the skills learned in this chapter are not limited to static fluid situations. There are several instances in fluid flow problems which will be discussed later wherein pressure variations are as if the fluid is at rest.

2.1 PRESSURE VARIATION WITH ELEVATION

The fundamental equation of fluid statics relates pressure, density, and vertical position in a fluid. This relationship is derived readily by considering the static equilibrium of a differential fluid element (Fig. 2.1). The z-axis is in a direction parallel to the gravitational force field (vertical). Because equilibrium in the y-direction will yield the same results as in the x-direction, we will consider equilibrium only in the x- and z-directions. Applying

[1]See Eq. 1.12.

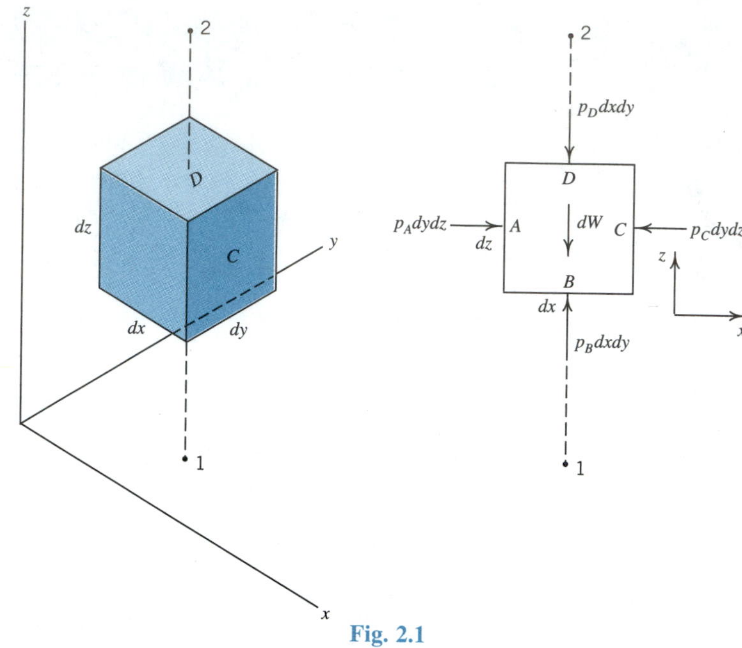

Fig. 2.1

Newton's first law to the element in the two directions ($\Sigma F_x = 0$ and $\Sigma F_z = 0$) and recalling that dx and dz are very small, we obtain

$$\Sigma F_x = p_A\, dy\, dz - p_C\, dy\, dz = 0 \tag{2.1}$$

$$\Sigma F_z = p_B\, dx\, dy - p_D\, dx\, dy - dW = 0 \tag{2.2}$$

in which p is a function of x, y, and z. In partial derivative notation,[2] the pressures on the faces of the element are, in terms of the pressure p at the center,

$$p_A = p - \frac{\partial p}{\partial x}\frac{dx}{2} \qquad p_C = p + \frac{\partial p}{\partial x}\frac{dx}{2}$$

$$p_B = p - \frac{\partial p}{\partial z}\frac{dz}{2} \qquad p_D = p + \frac{\partial p}{\partial z}\frac{dz}{2}$$

The weight of the small element is $dW = \rho g_n\, dx\, dy\, dz = \gamma\, dx\, dy\, dz$ (as dx, dy, and dz approach zero in the usual limiting process for partial differentiation, any variations in ρ and γ over the element will vanish, even though ρ and γ may vary in space). Thus, Eqs. 2.1 and 2.2 become

$$\left(p - \frac{\partial p}{\partial x}\frac{dx}{2}\right) dy\, dz - \left(p + \frac{\partial p}{\partial x}\frac{dx}{2}\right) dy\, dz = -\frac{\partial p}{\partial x}\, dx\, dy\, dz$$

and, similarly,

$$-\frac{\partial p}{\partial z}\, dx\, dy\, dz - \gamma\, dx\, dy\, dz = 0$$

[2]See Appendix 6.

Dividing by $dx\ dy\ dz$ in both cases gives

$$\frac{\partial p}{\partial x} = 0 \qquad \text{and} \qquad \frac{\partial p}{\partial z} = -\gamma = -\rho g_n \tag{2.3}$$

Because $\partial p/\partial x = 0$, there is no variation of pressure with horizontal distance; that is, *pressure is constant in a horizontal plane in a static fluid.* Therefore, pressure is a function of z only and it is permissible to replace the partial derivative in the second equation with the total derivative.

The second of Eqs. 2.3 is the basic equation of fluid statics. It can be written in the form $-dz = dp/\gamma$ and can be integrated directly to find

$$z_2 - z_1 = \int_{p_2}^{p_1} \frac{dp}{\gamma} \tag{2.4}$$

For a *fluid of constant density* (this may be safely assumed for liquids over large vertical distances and for gases over small ones), the integration yields

$$z_2 - z_1 = h = \frac{p_1 - p_2}{\gamma} \qquad \text{or} \qquad \Delta p = p_1 - p_2 = \gamma(z_2 - z_1) = \gamma h \tag{2.5}$$

permitting ready calculation of the increase in pressure with depth in a fluid of constant density.

Equation 2.5 also shows that the pressure differences $(p_1 - p_2)$ may be readily expressed as a "head" h of a fluid of specific weight γ. Thus pressures are often quoted as heads in *millimetres of mercury, feet or metres of water,* etc. The relation of pressure to head[3] is illustrated by the piezometer columns of Fig. 2.2. Liquid from the pressurized

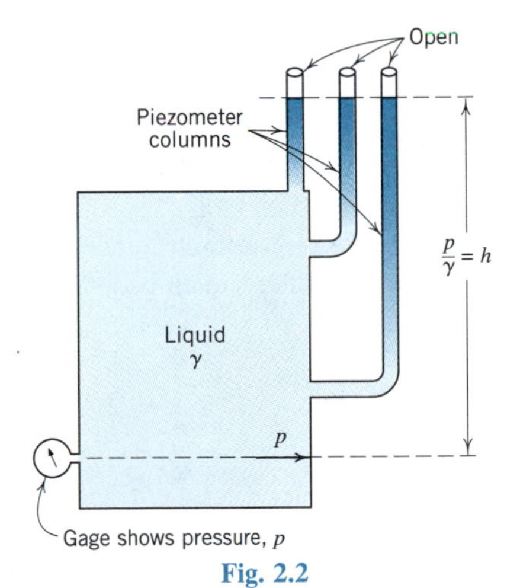

Fig. 2.2

[3]For use in problem solutions it is advisable to keep in mind certain pressure and head equivalents for common liquids. The use of "conversion factors," whose physical significance is rapidly lost, may be avoided by remembering that standard atmospheric pressure at sea level is 14.70 psia, 101.3 kPa absolute, 29.92 in. or 760 mm of mercury (32°F or 0°C), and 33.9 ft or 10.3 m of water (60°F or 15.6°C). The establishment of this standard atmospheric pressure also allows pressures to be quoted in *atmospheres*; for example, a pressure of 147 psia or 1013 kPa could be stated as 10 atmospheres.

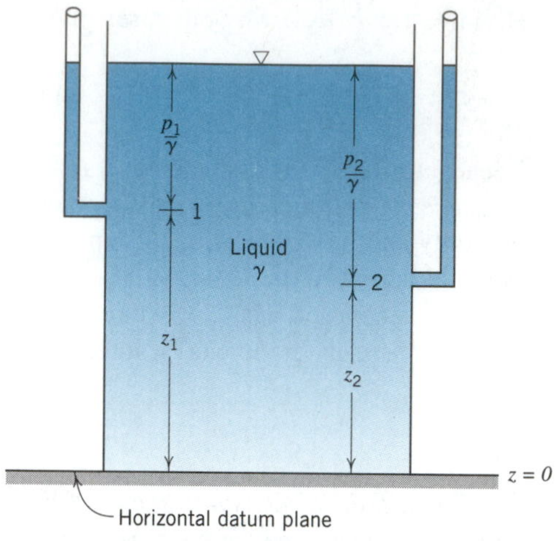

Fig. 2.3

container stands at a level in the piezometer tubes which will produce a pressure p at the gage.

Equation 2.5 can be rearranged to show

$$\frac{p_1}{\gamma} + z_1 = \frac{p_2}{\gamma} + z_2 = \text{Constant} \tag{2.6}$$

for later comparison with equations of fluid flow. Taking points 1 and 2 as typical, it is evident from Eq. 2.6 that the quantity $(z + p/\gamma)$ is the same for all points in a static liquid. This can be visualized geometrically as shown on Fig. 2.3.

ILLUSTRATIVE PROBLEM 2.1

The liquid oxygen (LOX) tank of the space shuttle booster is filled to a depth of 10 m with LOX at $-196°C$. The absolute pressure in the vapor above the liquid surface is maintained at 101.3 kPa.

Calculate the absolute pressure at the inlet valve at the bottom of the tank as the booster rests on the launch pad.

SOLUTION

The increase in pressure between the surface of the LOX and the tank bottom is given by Eq. 2.5.

$$p_1 - p_2 = \gamma(z_2 - z_1) \tag{2.5}$$

Letting point 2 be the LOX surface and point 1 the tank bottom,

$$p_1 = p_2 + \gamma(z_2 - z_1)$$

From Appendix 2, the density ρ of the LOX is 1206 kg/m^3. From Eq. 1.1,

$$\gamma = \rho g_n = 1\,206 \times 9.81 \text{ m/sec}^2 = 11\,830 \text{ N/m}^3 = 11.83 \text{ kN/m}^3$$

Now we can calculate p_1 knowing $z_2 - z_1 = 10$ m.

$$p_1 = 101.3 \text{ kPa} + 11.83 \text{ kN/m}^3 \times (10 \text{ m})$$

Recalling that kN/m^2 is equivalent to kPa,

$$p_1 = 101.3 + 118.3 = 219.6 \text{ kPa absolute} \bullet$$

For a fluid of variable density, integration of Eq. 2.4 cannot be accomplished until a relationship between p and γ is known. This problem is encountered in the fields of oceanography and meteorology. In the former, a suitable relationship can be obtained from elasticity considerations (Eq. 1.8) or an empirical relation between pressure and density, which is affected by the temperature and salinity of the sea. In the latter field, certain gas laws provide the $p - \gamma$ relationship. For gases (as in the atmosphere) the polytropic process[4]

$$\frac{p}{\gamma^n} = \text{Constant} \tag{2.7}$$

may be employed to develop relations between pressure, density, temperature, and altitude. One of the most important of these is $-dT/dz$, the rate of temperature change with altitude, termed the *temperature lapse rate*, which may be derived as follows: Inserting $\gamma RT/g_n$ for p in Eq. 2.7 and differentiating yields

$$\frac{T\,d\gamma}{\gamma\,dT} = \frac{1}{n - 1}$$

Replacing dp in Eq. 2.3 by $d(\gamma RT/g_n)$ gives

$$\frac{dz}{dT} = -\frac{R}{g_n}\left(\frac{T\,d\gamma}{\gamma\,dT} + 1\right)$$

Substituting the first of these equations into the second results in

$$-\frac{dT}{dz} = \frac{g_n(n - 1)}{nR} \tag{2.8}$$

For $n > 1$, $-dT/dz > 0$, which is the familiar situation in the lower portion of the earth's atmosphere (the *troposphere*) where temperature declines with increasing altitude. Between altitudes 11 km (36 000 ft) and 20 km (65 600 ft) in the *stratosphere*, however, the temperature has been observed to be essentially constant at -56.5°C (-69.7°F); here the atmosphere is *isothermal*, $n = 1$, and $-dT/dz = 0$. Through the troposphere the mean lapse rate has been found to be practically constant, and this has led to the definition of a *standard atmosphere* which closely approximates the yearly mean at latitude 40°. The U.S. Standard Atmosphere[5] assumes a sea level pressure of 101.3 kPa (14.70 psia) and a constant lapse rate through the troposphere of 0.006 5°C/m (0.003 56°F/ft) from 15°C (59°F) at sea level to -56.5°C (-69.7°F) at altitude 11 019 m (36 150 ft), at which the stratosphere begins. From the lapse rate above, and taking R to be 286.8 J/kg·K (1 715 ft·lb/slug·°F), n is found (from Eq. 2.8) to be 1.235, from which pressure and air density may be calculated throughout the (standard) troposphere. Although this standard (and static) atmosphere can be used for design calculations and performance predictions

[4]For the adiabatic process (Eq. 1.7), $n = k$, and, for the isothermal process (Eq. 1.6), $n = 1$.

[5]See Appendix 2.

on high altitude aircraft, the earth's atmosphere with its winds and air currents is not, of course, precisely static. However, this is usually a satisfactory approximation for the prediction of pressures and densities. For violent disturbances (e.g., tornadoes and hurricanes) the assumption of a static atmosphere is clearly untenable.

The lapse rate for an *adiabatic atmosphere* is important for purposes of comparison and may be calculated from Eq. 2.8 using $n = k = 1.40$ and $R = 286.8$ J/kg·K (1 715 ft·lb/slug·°F) providing the air is dry,[6] the result is $-dT/dz = 0.009\ 8$°C/m (0.005 35 °F/ft), which is known as the *adiabatic lapse rate* and will be shown to be a criterion of atmospheric stability. Suppose that in an adiabatic atmosphere a mass of fluid is moved from one altitude to another. If it is moved upward, it will expand almost without acceptance or rejection of heat (i.e., adiabatically) because of its poor conduction; accordingly at its new altitude it will have the same density as the surrounding air and thus possess no tendency to move from its new position. The adiabatic atmosphere is thus in a state of *neutral equilibrium* and is inherently *stable*. When this process is imagined for a lapse rate ($-dT/dz$) larger than the adiabatic, the expansion will tend to be adiabatic as before, but in its new position the density of the fluid mass will be smaller than that of its surroundings and its greater buoyancy will cause it to rise further; such an atmosphere is inherently *unstable*—and this fact leads to the expectation that a stable atmosphere can occur only for lapse rates less than the adiabatic. This expectation is confirmed when the foregoing reasoning is applied to an atmosphere with lapse rate less than adiabatic; here displacements of fluid masses produce density changes which tend to restore the air masses to their original position. Thus for diminishing lapse rates atmospheric stability steadily increases, becoming greatest in the case of an *inversion*, when the lapse rate is negative.

ILLUSTRATIVE PROBLEM 2.2

Calculate the pressure and specific weight of air in the U.S. Standard Atmosphere at an altitude of 35 000 ft above sea level using the values at sea level as a starting point.

SOLUTION

First, go to Appendix 2 and obtain the values of pressure, temperature, and specific weight at sea level which we will identify as point 1.

$$p_1 = 14.70 \text{ psia} \qquad T_1 = 59°R \qquad \gamma_1 = 0.076\ 5 \text{ lb/ft}^3$$

To construct a solution, we begin with Eq. 2.4, which must be integrated.

$$z_2 - z_1 = \int_{p_2}^{p_1} \frac{dp}{\gamma} \tag{2.4}$$

where point 2 is at 35 000 ft.

To be able to perform the integration, γ must be expressed as a function of p. To accomplish this, we use Eq. 2.7.

$$\frac{p}{\gamma^n} = \text{Constant} = \text{C} \qquad \text{or} \qquad \gamma = \left(\frac{p}{C}\right)^{1/n} \tag{2.7}$$

[6]Although atmospheric moisture is a critical factor in many meteorology problems, it is disregarded here because it does not change the sense of the development; it would, of course, change the numerical values.

Before substituting into Eq. 2.4, we need to find the constant C. To do this, we insert values of p and γ at sea level into Eq. 2.7. From the previous reading material, $n = 1.235$ for the Standard Atmosphere.

$$C = \frac{14.70 \text{ lb/in}^2 \times 144 \text{ in}^2/\text{ft}^2}{(0.076\,5 \text{ lb/ft}^3)^{1.235}} = 50.6 \times 10^3$$

Now, substituting into Eq. 2.4,

$$z_2 - z_1 = \int_{p_2}^{p_1} \frac{dp}{(p/C)^{1/n}} = C^{1/n} \int_{p_2}^{p_1} p^{-1/n}\, dp$$

Integrating,

$$z_2 - z_1 = C^{1/n} \left[\frac{p^{1-1/n}}{1 - 1/n} \right]_{p_2}^{p_1} = \frac{C^{1/n}}{1 - 1/n} [p_1^{1-1/n} - p_2^{1-1/n}]$$

With $z_2 - z_1 = 35\,000$ ft, $p_1 = 14.70 \text{ lb/in}^2 \times 144 \text{ in}^2/\text{ft}^2 = 2\,117 \text{ lb/ft}^2$ and $n = 1.235$,

$$35\,000 = \frac{(50.6 \times 10^3)^{1/1.235}}{1 - 1/1.235} (2\,117^{1-1/1.235} - p_2^{1-1/1.235})$$

$$35\,000 = \frac{6\,442}{0.190} (2\,117^{0.190} - p_2^{0.190})$$

$$1.032 = 2\,117^{0.190} - p_2^{0.190}$$

$$p_2 = (4.284 - 1.032)^{1/0.190} = 3.252^{1/0.190} = 496.4 \text{ lb/ft}^2$$

$$p_2 = 3.45 \text{ psia} \bullet$$

To find the specific weight at 35 000 ft, we use the temperature lapse rate to find T_2, then the equation of state (Eq. 1.3) to find γ_2. From the previous reading material,

$$\frac{dT}{dz} = -0.003\,56\text{°F/ft (or °R/ft)}$$

$$\int_{T_1}^{T_2} dT = -0.003\,56 \int_{z_1}^{z_2} dz$$

$$T_2 - T_1 = -0.003\,56(z_2 - z_1)$$

$$T_2 = 519\text{°R} - 0.003\,56 \times 35\,000 \text{ ft} = 394\text{°R}$$

Now, Eq. 1.3 provides an expression for γ_2.

$$\gamma_2 = g_n p_2 / R T_2 \tag{1.3}$$

The engineering gas constant R is obtained from Appendix 2.

$$R = 1\,715 \text{ ft·lb/slug·°R for air}$$

Now, substituting into Eq. 1.3,

$$\gamma_2 = (32.2 \text{ ft/sec}^2 \times 14.70 \text{ lb/in}^2 \times 144 \text{ in}^2/\text{ft}^2)/(1\,715 \times 394\text{°R})$$

$$\gamma_2 = 0.023\,7 \text{ lb/ft}^3 \bullet$$

ILLUSTRATIVE PROBLEM 2.3

In the ocean, warm patches of water that are heated by the sun induce thermals (buoyant columns of air) by heating the air above its standard atmosphere value so that the air is buoyant. If this air has a temperature of 17°C and its buoyant rise is adiabatic, at what level will the rising air be in temperature (and density) equilibrium with the standard atmosphere and stop rising buoyantly?

SOLUTION

For the Standard Atmosphere, the temperature at sea level is 15°C and the lapse rate is 0.006 5°C/m. From the previous reading material, the lapse rate for an adiabatic process is 0.009 8°C/m.

$$\text{Standard Atmosphere} \quad -\frac{dT}{dz} = 0.006\,5°\text{C/m}$$

$$\text{Adiabatic atmosphere} \quad -\frac{dT}{dz} = 0.009\,8°\text{C/m}$$

Integrating from sea level (point 1) to an altitude where both temperatures are the same (point 2),

$$\text{Standard Atmosphere} \quad \int_{15}^{T_2} dT = -0.006\,5 \int_0^{z_2} dz$$

$$T_2 - 15 = -0.006\,5z_2$$

$$T_2 = 15 - 0.006\,5z_2$$

$$\text{Adiabatic atmosphere} \quad \int_{17}^{T_2} dT = -0.009\,8 \int_0^{z_2} dz$$

$$T_2 - 17 = -0.009\,8z_2$$

$$T_2 = 17 - 0.009\,8z_2$$

Combining the two equations to eliminate T_2,

$$15 - 0.006\,5z_2 = 17 - 0.009\,8z_2$$

$$0.003\,3z_2 = 2$$

$$z_2 = 606 \text{ m} \bullet$$

2.2 ABSOLUTE AND GAGE PRESSURES

Pressures, like temperatures, are measured and quoted in two different systems, one relative (gage), and the other absolute; no confusion results if the relation between the systems and the common methods of measurement are completely understood.

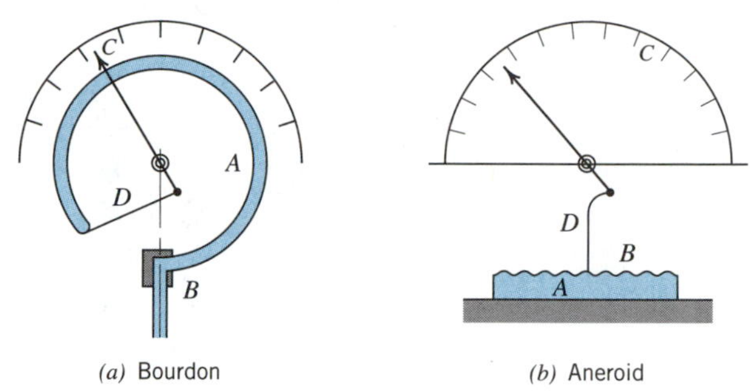

(a) Bourdon *(b)* Aneroid

Fig. 2.4 Mechanical pressure gages.

The Bourdon pressure gage and the aneroid barometer (shown schematically in Fig. 2.4) are typical mechanical devices for measuring gage and absolute pressures, respectively. In the pressure gage a bent tube (A) of elliptical cross section is held rigidly at B and its free end is connected to a pointer (C) by a link (D). When pressure is admitted to the tube, its cross-section tends to become circular, causing the tube to straighten and move the pointer to the right over the graduated scale. If the gage is in proper adjustment, the pointer rests at zero on the scale *when the gage is disconnected*; in this condition the pressures inside and outside of the tube are the same, and thus there is no tendency for the tube to deform. It is apparent that such pressure gages are actuated by the *difference* between the pressure inside and that outside the tube. For example, in the gage system of pressure measurement, if atmospheric pressure exists outside the tube, the local (not standard) atmospheric pressure becomes the zero of pressure. For pressure less than local atmospheric the tube will tend to contract, moving the pointer to the left. The reading for pressure greater than local atmospheric is positive and is called *gage pressure*, or simply *pressure*,[7] and it is usually measured in *pascals* (newtons per square metre) or psi (pounds per square inch); the reading for pressure below local atmospheric is negative, designated as *vacuum*, and usually is measured in millimetres or inches of mercury.

The aneroid gage is a device for measuring absolute pressure. The essential element is a short cylinder (A) with one end an elastic diaphragm (B). The cylinder is evacuated[8] so that the pressure therein is close to absolute zero; pressures imposed on the outside of the diaphragm cause it to deflect inward; these deflections are then a direct measure of the applied pressures, which can be transferred to a suitable scale (C) through appropriate linkages (D); here the pressures recorded are relative to absolute zero and are *absolute pressures.* Although the aneroid cylinder as conventionally used in barometers is capable of measuring only a small range of pressures, the basic idea can be applied to absolute pressure gages for more general use.

Liquid devices that measure gage and absolute pressures are shown on Fig. 2.5; these are the open U-tube and the conventional mercury barometer. With the U-tube open, atmospheric pressure acts on the upper liquid surface; if this pressure is taken to be zero, the applied gage pressure p equals γh and h is thus a direct measure of gage pressure.

[7]It has been internationally recommended that pressure units themselves should not be modified to indicate whether the pressure is absolute or gage. When the context of use leaves any doubt as to which is meant, the word ''pressure'' is to be qualified appropriately. Throughout the remainder of this book *pressure* means gage pressure unless the context or the qualified term *absolute pressure* indicates otherwise.

[8]If the cylinder could be completely evacuated, the pressure therein would be the lowest possible (absolute zero) since there would be no fluid molecules to exert pressure.

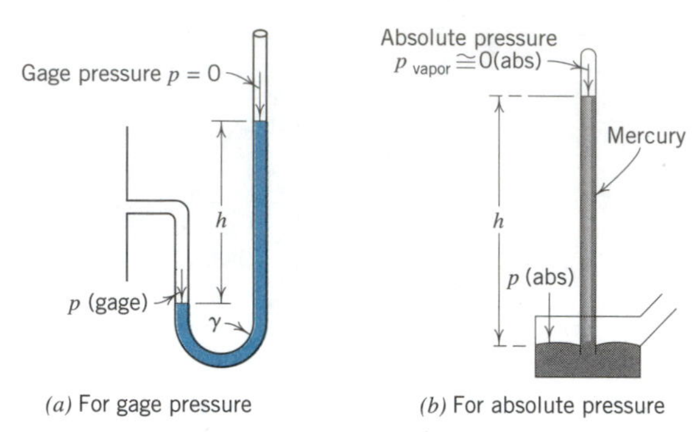

(*a*) For gage pressure (*b*) For absolute pressure

Fig. 2.5 Liquid pressure gages.

The mercury barometer (invented by Torricelli, 1643) is constructed by filling the tube with air-free mercury and inverting it with its open end beneath the mercury surface in the receptacle. Ignoring the small pressure of the mercury vapor,[9] the pressure in the space above the mercury is at absolute zero and again $p = \gamma h$; here the height h is a direct measure of the absolute pressure, p. Although conventional use of the barometer is for the measurement of local atmospheric pressure, the basic scheme is frequently used in industry for the direct measurement of any absolute pressure.

From the foregoing descriptions an equation relating gage and absolute pressures may now be written,

$$\text{Absolute pressure} = \text{Atmospheric pressure} \begin{array}{l} -\,\text{Vacuum} \\ +\,\text{Gage pressure} \end{array} \tag{2.9}$$

which allows easy conversion from one system to the other. Possibly a better picture of these relationships can be gained from a diagram such as Fig. 2.6 in which are shown two typical pressures, A and B, one above, the other below, atmospheric pressure, with all the relationships indicated graphically.

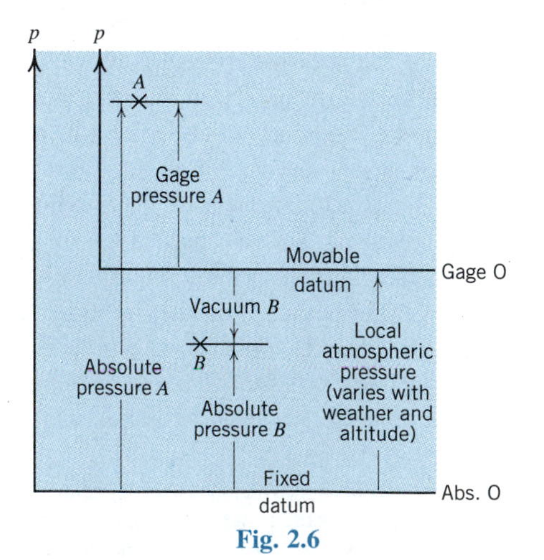

Fig. 2.6

[9]See Appendix 2.

ILLUSTRATIVE PROBLEM 2.4

A Bourdon gage registers a vacuum of 12.5 in. (310 mm) of mercury (Hg) when the atmospheric pressure is 14.50 psia (100 kPa absolute). Calculate the corresponding absolute pressure in psia and kPa absolute.

SOLUTION

From the previous reading material it is clear that absolute pressure is obtained by adding atmospheric pressure to the gage pressure or subtracting vacuum pressure from the atmospheric pressure (Eq. 2.9). In this case, since we are given a vacuum pressure, we will use the second approach.

First we have to convert the vacuum pressure head in inches and millimetres of Hg into psi and kPa. Recalling that 29.92 in. of Hg is equivalent to 14.70 psi (sea level atmospheric conditions),

$$12.5 \text{ in. Hg is equivalent to } (12.5/29.92) \times 14.70 = 6.14 \text{ psi}$$

Recalling that 760 mm Hg is equivalent to 101.3 kPa,

$$310 \text{ mm Hg is equivalent to } (310/760) \times 101.3 = 41.3 \text{ kPa}$$

Subtracting these values from the respective values of atmospheric pressure,

$$\text{Absolute pressure} = 14.50 - 6.14 = 8.36 \text{ psia} \bullet$$

$$\text{Absolute pressure} = 100 - 41.3 = 58.7 \text{ kPa absolute} \bullet$$

2.3 MANOMETRY

Bourdon and aneroid pressure gages, owing to their inevitable mechanical limitations, are not usually adequate for precise measurements of pressure; when greater precision is required, *manometers* like those of Fig. 2.7 may be effectively employed (see also Section 14.3 for a brief discussion of pressure transducers).

Consider the U-tube manometer of Fig. 2.7*a*, connected to a pipe shown in cross-section, in which all distances and densities are known, and pressure p_x is to be found. Because, *over horizontal planes within continuous columns of the same fluid*, pressures are equal, it is evident at once that $p_1 = p_2$, and from Eq. 2.5,

$$p_1 = p_x + \gamma l \qquad \text{and} \qquad p_2 = 0 + \gamma_1 h$$

Equating p_1 and p_2 gives

$$p_x = \gamma_1 h - \gamma l$$

allowing pressure, p_x, to be calculated[10]

[10]The use of derived formulas for manometer solutions is not rcommended until experience has been gained in their limitations.

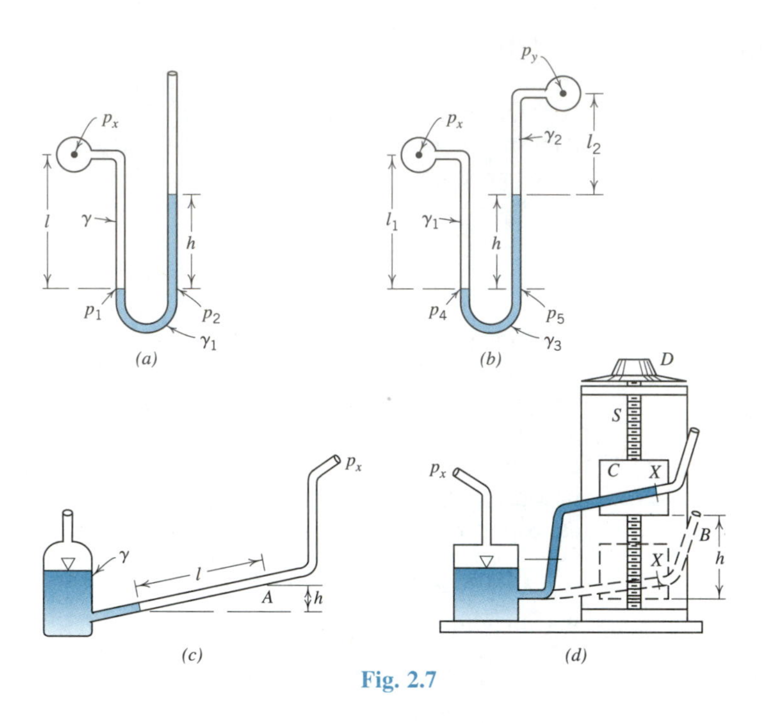

Fig. 2.7

U-tube manometers are frequently used to measure the difference between two unknown pressures, p_x and p_y, in two pipes or between two locations in a single pipe, as in Fig. 2.7*b*. Here,

$$p_x + \gamma_1 l_1 = p_4 = p_5 = p_y + \gamma_2 l_2 + \gamma_3 h$$

from which we obtain

$$p_x - p_y = \gamma_2 l_2 + \gamma_3 h - \gamma_1 l_1$$

thus allowing direct calculation of the pressure difference, $p_x - p_y$. Differential manometers of the type above are sometimes made with the U-tube inverted, with a liquid of small density existing in the top of the inverted U; the pressure difference measured by manometers of this type may be readily calculated by application of the foregoing principles. When large pressures or pressure differences are to be measured, a pressure gage or a mechanical or electrical transducer (Section 14.3) is usually used.

There are many forms of precise manometers; two of the most common are shown in Fig. 2.7. Figure 2.7*c* represents the ordinary *inclined gage* used in measuring the comparatively small pressures in low-velocity gas flows. Its equilibrium position is shown at A, and when it is submitted to a pressure, p_x, a vertical deflection, h, is obtained in which $p_x = \gamma h$. In this case, however, the liquid is forced down a gently inclined tube so that the deflection, l, is much greater than h and, therefore, more accurately read. This type of manometer, when calibrated to read directly in inches of water, is frequently called a *draft gage*.

The principle of the sloping tube is also employed in the alcohol micromanometer of Fig. 2.7*d*, used in research work. Here the gently sloping glass tube is mounted on a carriage, C, which is moved vertically by turning the dial, D, which actuates the screw, S. When p_x is zero, the carriage is adjusted so that the liquid in the tube is brought to the hairline, X, and the reading on the dial is recorded. When the unknown pressure, p_x, is admitted to the reservoir, the alcohol runs upward in the tube toward B and the carriage is then raised until the liquid surface in the tube rests again at the hairline, X. The difference between the dial reading at this point and the original reading gives the vertical travel of the carriage, h, which is the head of alcohol equivalent to the pressure p_x.

Along with these principles of manometry the following practical considerations should be appreciated: (1) manometer liquids, in changing their relative densities with temperature, will induce errors in pressure measurements if this factor is overlooked; (2) errors due to capillarity may frequently be canceled by selecting manometer tubes of uniform size; (3) although some liquids appear excellent (from density considerations) for use in manometers, their surface-tension effects may give poor menisci and thus inaccurate readings; (4) fluctuations of the manometer liquids will reduce accuracy of pressure measurement, but these fluctuations may be reduced by a throttling device in the manometer line (a short length of small tube is excellent for this purpose); and (5) when fluctuations are negligible, refined optical devices and verniers may be used for extremely precise readings of the liquid surfaces.

ILLUSTRATIVE PROBLEM 2.5

The vertical pipeline shown contains oil of specific gravity 0.90. A Bourdon pressure gage is attached to the pipe as well as an open-end manometer filled with mercury at a specific gravity of 13.57. If the oil in the pipe is currently at rest (not flowing), calculate the gage reading p_x of the Bourdon gage.

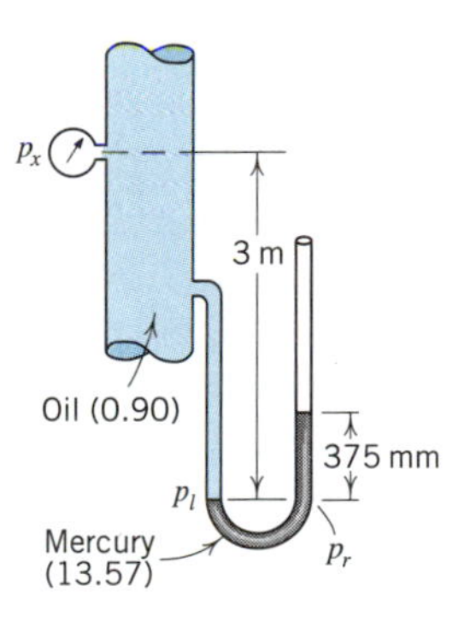

SOLUTION

Because the fluids are at rest, we know that p_r and p_l are equal because they are at horizontal locations in a continuous and homogeneous fluid. Writing expressions for each of these pressures utilizing Eq. 2.5,

$$z_2 - z_1 = h = \frac{p_1 - p_2}{\gamma} \qquad \text{or} \qquad \Delta p = p_1 - p_2 = \gamma(z_2 - z_1) = \gamma h \tag{2.5}$$

$$p_l = p_x + 9.81 \text{ m/s}^2 \times (0.90 \times 1\,000 \text{ kg/m}^3) \times 3 \text{ m}$$

$$= p_x + 26.49 \text{ kN/m}^2 \text{ or kPa}$$

$$p_r = 9.81 \text{ m/s}^2 \times (13.57 \times 1\,000 \text{ kg/m}^3) \times 0.375 \text{ m}$$

$$= 49.92 \text{ kN/m}^2 \text{ or kPa}$$

Equating these two expressions,

$$p_x + 26.49 = 49.92$$

$$p_x = 23.43 \text{ kN/m}^2 \text{ or kPa} \bullet$$

ILLUSTRATIVE PROBLEM 2.6

A differential manometer is connected to a pipe at two different locations to measure the pressure difference between the two points. Water is flowing in the pipe and the manometer fluid is carbon tetrachloride.

Find the pressure difference $p_1 - p_2$ in psi between the two points in the pipe. Identify the point with the highest pressure.

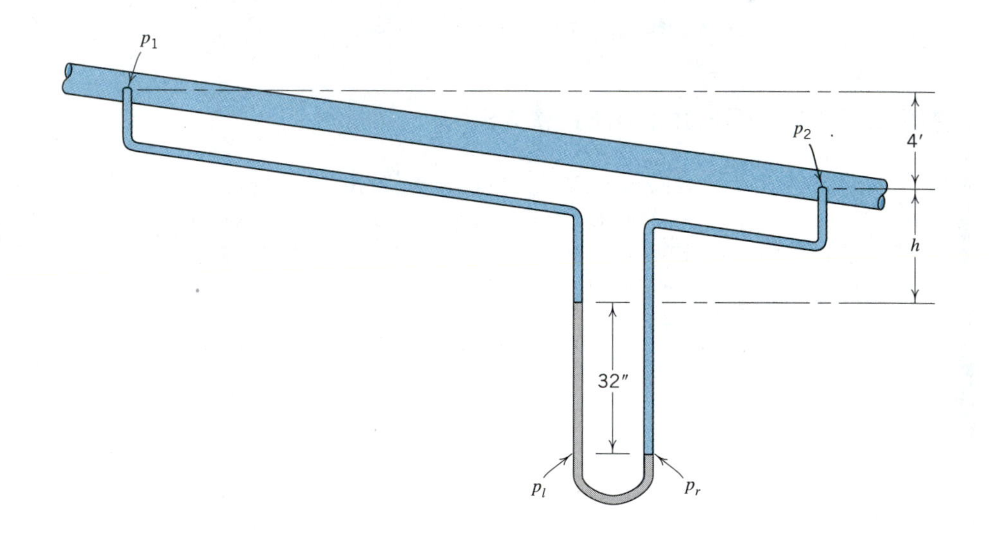

SOLUTION

The fluids in the manometer tubes are at rest so the laws of fluid statics apply. Recognizing that $p_l = p_r$, we will develop expressions for each utilizing Eq. 2.5 and equate the results.

$$z_2 - z_1 = h = \frac{p_1 - p_2}{\gamma} \quad \text{or} \quad \Delta p = p_1 - p_2 = \gamma(z_2 - z_1) = \gamma h \quad (2.5)$$

First we must go to Appendix 2 to determine the fluid properties. The specific gravity of carbon tetrachloride is found to be 1.59 and the mass density of water is 1.938 slugs/ft^3.

Water $\gamma = g_n\rho = 32.2 \text{ ft/sec}^2 \times 1.938 \text{ slugs/ft}^3 \text{ or lb-sec}^2\text{/ft}^4 = 62.4 \text{ lb/ft}^3$

Carbon tetrachloride $\gamma = \text{sp. grav.} \times \gamma \text{ of water} = 1.59 \times 62.4 = 99.2 \text{ lb/ft}^3$

Now, writing expressions for the two pressures p_l and p_r,

$$p_l = p_1 + 62.4 \times 4 \text{ ft} + 62.4h + 99.2 \times 32/12 \text{ ft} = p_1 + 62.4h + 514.1$$

$$p_r = p_2 + 62.4h + 62.4 \times 32/12 \text{ ft} = p_2 + 62.4h + 166.4$$

Equating the two expressions,

$$p_1 + 62.4h + 514.1 = p_2 + 62.4h + 166.4$$

Noting that the expressions containing the unknown values of h cancel,

$$p_1 - p_2 = 166.4 - 514.1 = -347.7 \text{ lb/ft}^2$$

or $$p_1 - p_2 = -347.7 \text{ lb/ft}^2/144 \text{ in}^2/\text{ft}^2 = -2.41 \text{ lb/in}^2 \text{ or psi} \bullet$$

It is clear from the above result that p_2 is the greater pressure.

2.4 PRESSURE FORCES ON PLANE SURFACES

The calculation of the magnitude, direction, and line of action of the pressure forces on plane surfaces is essential in the design of dams, gates, tanks, ships, and the like. For surfaces under the action of gases, the calculations are relatively simple because the pressure variation in a gas, even over a large surface, is negligible. Therefore, the pressure is uniform over the surface and the resultant force is equal to the area times the pressure and acts through the centroid of the area under pressure (known as the *center of pressure*). The same is true for liquids acting on a horizontal surface because of the fact that there is no pressure variation in the horizontal direction.

However, for the case of liquid pressure acting on a nonhorizontal surface, the situation is more complex. From Eq. 2.5 we learned that pressure in a constant-density liquid varies linearly with depth. If the liquid surface is exposed to the atmosphere as shown in Fig. 2.8, the pressure is seen to vary linearly from zero at the liquid surface to γh_2 at the bottom of the tank. This variation is depicted graphically with a *pressure distribution diagram*. Note that we do not include the atmospheric pressure because it affects both sides of the container equally and its effect cancels out. However, if the pressure acting on the free surface of the liquid is not equal to that on the exterior of the container, this pressure difference must be taken into account in the calculations (see Illustrative Problem 2.8).

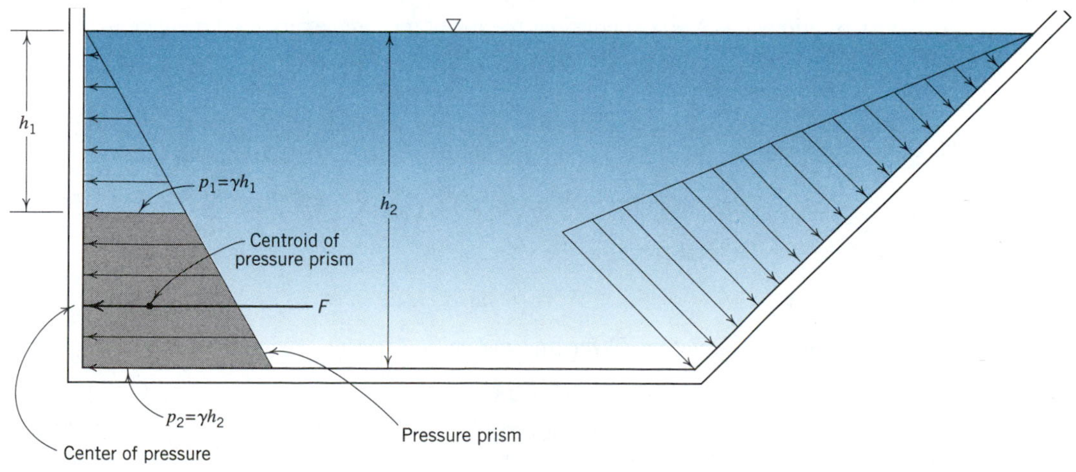

Fig. 2.8 Pressure forces on walls of a container.

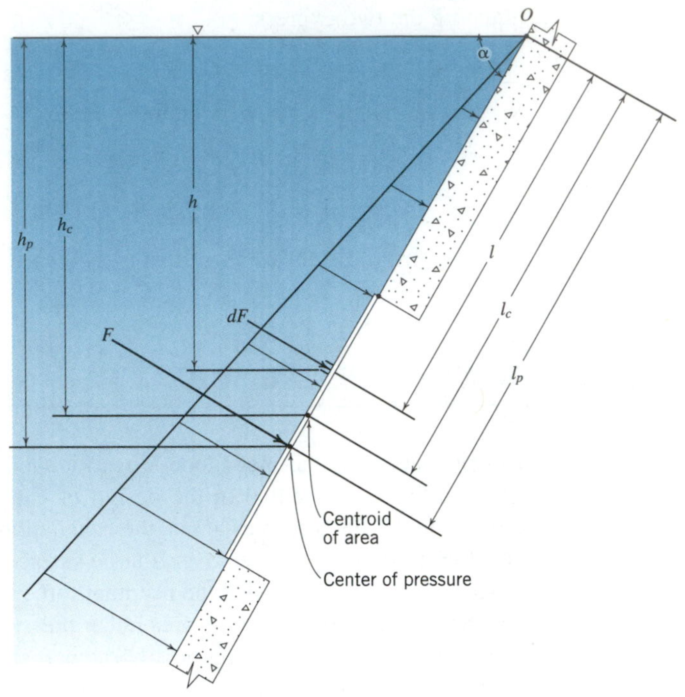

Fig. 2.9

As demonstrated in the basic mechanics course, one can visualize a volume created by the pressure acting over the plane surface. This volume is known as the *pressure prism* and it constitutes the magnitude of the resultant pressure force acting on the area and its centroid is on the line of action of the resultant force F (see Fig. 2.8 and Illustrative Problem 2.7). Unfortunately, for any surface other than rectangular, the geometry of the pressure prism is so complex and its centroid so difficult to locate that it is no longer a useful procedure.

As a consequence, we will now develop a more general approach to finding the resultant force and its line of action, which can be applied to any plane area. To develop the necessary relationships, we will consider the resultant force exerted on the gate in Fig. 2.9. The gate is inclined at an angle α from the horizontal with slant distances from the liquid surface at O measured by l.

Pressure at a depth h acts on an area dA, which is dl in height along the plane of the gate and extends horizontally the width of the gate. The differential force is

$$dF = p\,dA = \gamma h\,dA = \gamma l \sin\alpha\,dA \tag{2.10}$$

Integrating to get the total force on the gate yields

$$F = \gamma \sin\alpha \int l\,dA \tag{2.11}$$

From basic mechanics, we recall for first moments of areas,

$$\int l\,dA = l_c A$$

So,

$$F = \gamma\, l_c A \sin\alpha$$

Recognizing that $l_c \sin\alpha = h_c$,

$$F = \gamma h_c A \tag{2.12}$$

This equation shows that the resultant force on an area can be calculated if the size of the area and the location of the centroid is known. That is, the resultant force on an area equals the pressure at the centroid of the area (γh_c) times the area (A).

To complete the analysis, we must compute the location of the center of pressure where the resultant force F can be assumed to act. We again return to basic mechanics and the principle of moments, which states that the moment of the resultant about any point equals the sum of the moments of the components. Applying this principle to our situation,

$$l_p F = \int l\, dF = \int l(\gamma l \sin\alpha\, dA) = \gamma \sin\alpha \int l^2\, dA \tag{2.13}$$

If the plane area is composed of a convenient shape which can be represented by a simple functional relationship, e.g., a rectangle, the center of pressure can be found by direct integration. However, if the shape is more complex, e.g., a circular or triangular shape or some combination of rectangles, circles and triangles, a more convenient method of locating l_p may be desirable.

To accomplish this, we note that this last integral is recognized as the second moment I_o of the gate area about an axis through O. Recalling the transfer theorem for second moments,

$$\int l^2\, dA = I_o = I_c + l_c^2 A$$

Substituting the above relationship into Eq. 2.13 gives

$$l_p F = \gamma \sin\alpha\, (I_c + l_c^2 A)$$

Multiplying I_c by $l_c A / l_c A$ and factoring,

$$l_p F = \gamma\, l_c A \sin\alpha \left(\frac{I_c}{l_c A} + l_c \right) = F \left(\frac{I_c}{l_c A} + l_c \right)$$

The resulting equation for locating the vertical position of the center of pressure is

$$l_p = l_c + \frac{I_c}{l_c A} \tag{2.14}$$

This equation reveals that the center of pressure is always located below the centroid of the area although the distance between l_p and l_c diminishes with increasing depth of submergence. To locate the vertical position of the center of pressure, one only needs to know the location of the centroid of the area and the second moment of the area about its horizontal centroidal axis lying in the plane of the area. Appendix 3 gives I-values for a variety of area shapes.

The above analysis assumes that the area is symmetric about a vertical axis through its centroid or, if it is inclined from the vertical, about a similar axis in the plane of the

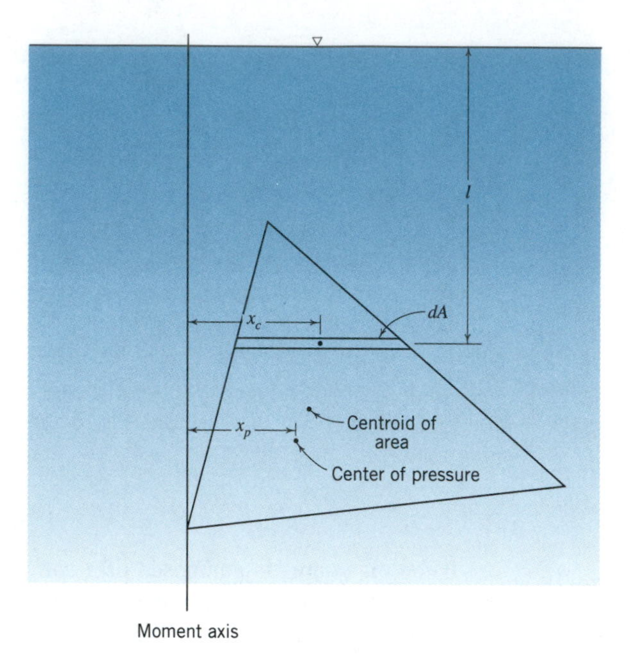

Fig. 2.10

area. If the submerged area is not symmetric as described above, additional calculations must be made to establish the lateral position of the center of pressure. Unfortunately, no general formula such as Eq. 2.14 exists for this determination. Hence, we will only present the approach to follow for this computation and leave to the reader the application to the particular case at hand.

In our analysis, we will refer to Fig. 2.10, which shows a triangular area inclined at an angle α with the vertical. Note that in profile, the diagram of Fig. 2.9 still applies and the elemental force is given by Eq. 2.10:

$$dF = \gamma l \sin \alpha \, dA$$

To begin, we select an axis in the plane of the area about which to take moments. Any axis can be chosen; however, we have arbitrarily selected one which leaves the area entirely to the right of the axis. The moment dM of the force dF about this axis is

$$dM = x_c \times dF = x_c \gamma l \sin \alpha \, dA$$

where x_c is the distance from the axis to the centroid of the elemental area dA (i.e., the midpoint of this area which, in the limit, is a rectangle).

Integrating and setting the result equal to the resultant force F times the horizontal distance x_p to the center of pressure produces

$$x_p = \frac{\gamma \sin \alpha \int x_c l \, dA}{F} \tag{2.15}$$

This integral cannot be evaluated until the specific shape of the area in question is known. Illustrative Problem 2.9 demonstrates the application of this procedure.

ILLUSTRATIVE PROBLEM 2.7

For the vertical rectangular gate shown, compute the magnitude and location of the resultant force exerted by the water acting on the gate. Find the results using direct integration, the pressure prism approach, and the method of Eqs. 2.12 and 2.14. The gate is 8 ft wide normal to the paper.

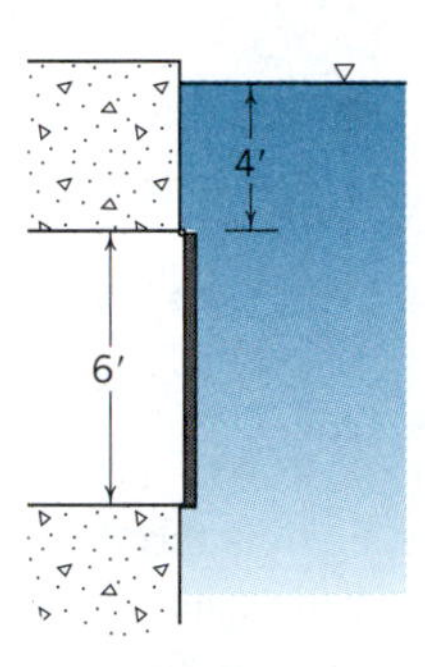

SOLUTION

Direct Integration

To compute the resultant force by direct integration, we employ Eq. 2.11.

$$F = \gamma \sin \alpha \int l \, dA \tag{2.11}$$

Because the gate is vertical, $\sin \alpha = 1$, $l = h$, and $dA = 8 \text{ ft} \times dh$.

$$F = 62.4 \text{ lb/ft}^3 \int_4^{10} h \times 8 \, dh = 62.4 \times 8 \left[\frac{h^2}{2}\right]_4^{10} = \frac{62.4 \times 8}{2} [100 - 16]$$

$$F = 20\,966 \text{ lb} \bullet$$

To locate the center of pressure we use Eq. 2.13.

$$l_p F = \gamma \sin \alpha \int l^2 \, dA \tag{2.13}$$

Again, $\sin \alpha = 1$, $l = h$, and $dA = 8 \, dh$.

$$l_p F = 62.4 \int_4^{10} h^2 \times 8 \, dh = 62.4 \times 8 \left[\frac{h^3}{3}\right]_4^{10}$$

$$= \frac{62.4 \times 8}{3} [1\,000 - 64] = 155\,750$$

$$l_p = \frac{155\,750}{F} = \frac{155\,750}{20\,966} = 7.43 \text{ ft} \bullet$$

Pressure Prism Approach

First, we break the pressure prism into two convenient parts—one resulting from a uniform pressure distribution and one resulting from a linearly varying one. The first pressure prism is a rectangular solid with a volume

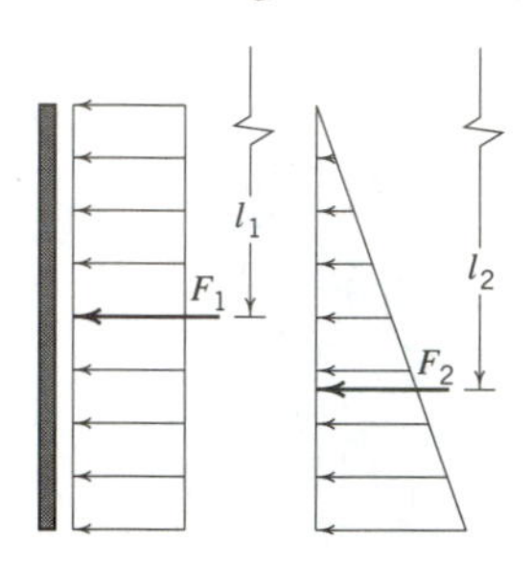

$$F_1 = 4\gamma \text{ lb/ft}^2 \times 6 \text{ ft} \times 8 \text{ ft}$$
$$= 4 \times 62.4 \times 6 \times 8$$
$$F_1 = 11\,980 \text{ lb}$$

The force F_1 acts through the centroid of the pressure prism, which is at the level of the midpoint of the gate.

The second pressure prism is a wedge-shaped volume which gives a value for F_2 of

$$F_2 = 1/2 \times 6\gamma \text{ lb/ft}^2 \times 6 \text{ ft} \times 8 \text{ ft}$$
$$= 1/2 \times 6 \times 62.4 \times 6 \times 8$$
$$F_2 = 8\,986 \text{ lb}$$

The force F_2 acts through the centroid of the pressure prism, which is two-thirds of the distance from the top of the gate to its bottom.

The total resultant force is

$$F = F_1 + F_2 = 11\,980 + 8\,986 = 20\,966 \text{ lb} \bullet$$

To find the line of action of the resultant, we employ the principle of moments and choose to take moments about the water surface.

$$Fl_p = l_1F_1 + l_2F_2$$
$$20\,966\ l_p = (3 + 4) \times 11\,980 + (4 + 4) \times 8\,986$$
$$l_p = 7.43 \text{ ft} \bullet$$

Formula Method

From Eq. 2.12, we compute the resultant force as

$$F = \gamma h_c A = 62.4 \text{ lb/ft}^3 \times (4 + 3) \text{ ft} \times (6 \times 8) \text{ ft}^2 \tag{2.12}$$
$$F = 20\,966 \text{ lb} \bullet$$

The location of the center of pressure is given by Eq. 2.14. Referring to Appendix 3 for I_c and l_c, we find

$$l_p = l_c + \frac{I_c}{l_c A} = (4 + 3) + \frac{1/12 \times 8 \times 6^3}{(4 + 3)(6 \times 8)} \tag{2.14}$$
$$= 7 + 144/336$$
$$l_p = 7.43 \text{ ft} \bullet$$

ILLUSTRATIVE PROBLEM 2.8

A circular hatch on the side of a pressurized tank is hinged at the top and held closed by a single bolt at the bottom. If the tank is filled with gasoline and pressurized to two atmospheres, find the force in the bolt necessary to counteract the fluid force on the hatch, i.e., to hold the hatch closed.

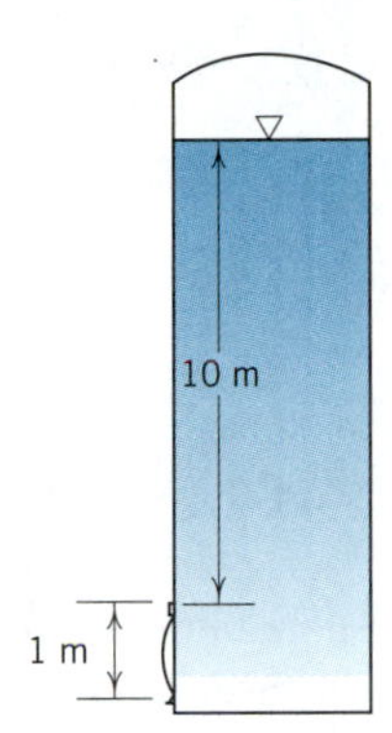

SOLUTION

First we convert the two-atmosphere pressure to kN/m^2. From Appendix 1, we see that one atmosphere is the equivalent of 101.3 kN/m^2. So the pressure exerted on the surface of the gasoline is 202.6 kN/m^2.

The approach we will take to solve the problem is to treat the effect of the liquid force F_L and the gas force F_G separately. The gas pressure force will be a uniform pressure over and above that exerted by the liquid because, as we saw in Section 1.3, the pressure imposed on a fluid at rest is transmitted undiminished to all other points.

$$F_G = pA = 202.6 \text{ kN/m}^2 \times (\pi/4)(1 \text{ m})^2 = 159.1 \text{ kN}$$

F_G acts at the center of the hatch.

Now, addressing the liquid force, refer to Eq. 2.12 and Appendix 2. From Appendix 2, the specific gravity of gasoline is found to be 0.68.

$$F_L = \gamma h_c A = (9.81 \text{ m/s}^2 \times 0.68 \times 1\,000 \text{ kg/m}^3) \times (10 + 0.5) \text{ m} \times (\pi/4)(1 \text{ m})^2$$

$$F_L = 55.0 \text{ kN}$$

To locate the line of action of F_L, refer to Eq. 2.14 and Appendix 3. From Appendix 3 we find that, for a circular area, $I_c = \pi d^4/64$. Utilizing Eq. 2.14,

$$l_p = l_c + \frac{I_c}{l_c A} = (10 + 0.50) + \frac{\pi \times 1^4/64}{(10 + 0.50)(\pi \times 1^2/4)}$$

$$l_p = 10.5 + 0.006$$

$$l_p = 10.506 \text{ m}$$

Considering the equilibrium of the hatch, we will sum the moments around the hinge H to find the force in the bolt F_B.

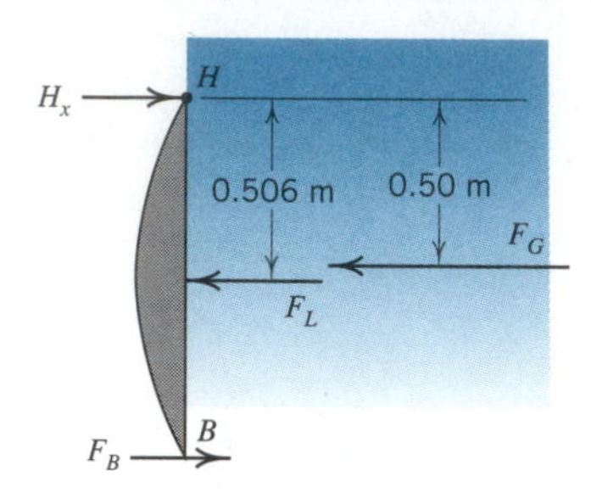

$$F_B \times 1\ \text{m} = F_G \times 0.50\ \text{m} + F_L \times 0.506\ \text{m}$$

$$F_B = 159.1 \times 0.50 + 55.0 \times 0.506$$

$$F_B = 107.4\ \text{kN}\ \bullet$$

ILLUSTRATIVE PROBLEM 2.9

A vertical gate in the shape of a quarter circle is submerged with its top edge 1 ft below the water surface. Calculate the magnitude and locate the line of action of the resultant force on the gate.

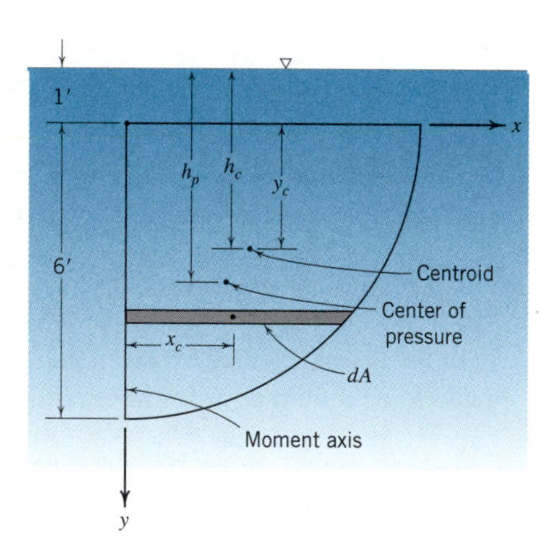

SOLUTION

We will first use Eq. 2.12 to calculate the magnitude of the resultant force. Before applying this equation, we go to Appendix 3 and obtain the location of the centroid y_c of the quarter circular area.

$$y_c = 4r/3\pi = 4 \times 6\ \text{ft}/3\pi = 2.546\ \text{ft}$$

$$h_c = y_c + 1 = 2.546 + 1 = 3.546\ \text{ft}$$

Now, from Eq. 2.12,

$$F = \gamma h_c A = 62.4\ \text{lb/ft}^3 \times 3.546\ \text{ft} \times \pi 6^2/4\ \text{ft}^2$$

$$F = 6\,257\ \text{lb}\ \bullet$$

Next, we will use Eq. 2.14 to calculate the vertical position of the center of pressure. But first, we must again refer to Appendix 3 to obtain the expression for I_c which is necessary for the use of Eq. 2.14. We note that Appendix 3 does not provide a formula for I_c but rather an I-value about an axis at the top of the gate. We will have to use the transfer theorem for second moments to arrive at a formula for I_c.

$$I_{top} = I_c + Ad^2$$

So,

$$I_c = I_{top} - Ad^2 = \frac{\pi d^4}{256} - \frac{\pi r^2}{4} y_c^2$$

$$= \pi 12^4/256 - (\pi 6^2/4) \times 2.546^2 = 254.5 - 183.3$$

$$I_c = 71.2 \text{ ft}^4$$

Now, from Eq. 2.14, recognizing that the vertical area makes $l = h$,

$$h_p = h_c + \frac{I_c}{h_c A} = 3.546 + \frac{71.2}{3.546(\pi 6^2/4)} = 3.546 + 0.710$$

$$h_p = 4.26 \text{ ft} \bullet$$

Finally, to locate the horizontal position of the center of pressure, we use Eq. 2.15. However, to integrate Eq. 2.15 we have to establish a functional relationship between x_c, l, and dA. We will use the x and y axes shown on the sketch to write the equation for the circular arc which marks the curved boundary of the gate. From analytic geometry,

$$x^2 + y^2 = 36$$

In Eq. 2.15, for our application, $\sin\alpha = 1$, $l = 1 + y$, $dA = x\,dy$, and $x_c = x/2$. With these substitutions, Eq. 2.15 becomes

$$x_p = \frac{\gamma \int (x/2)(1 + y)x\,dy}{F} = \frac{\gamma}{2F}\int x^2(1 + y)\,dy$$

Substituting the equation of the circle for x^2,

$$x_p = \frac{\gamma}{2F}\int (36 - y^2)(1 + y)\,dy = \frac{\gamma}{2F}\int_0^6 (36 + 36y - y^2 - y^3)\,dy$$

$$= \frac{\gamma}{2F}\left[36y + 18y^2 - y^3/3 - y^4/4\right]_0^6 = \frac{62.4}{2 \times 6\,257}\left[216 + 648 - 72 - 324\right]$$

$$x_p = 2.33 \text{ ft} \bullet$$

2.5 PRESSURE FORCES ON CURVED SURFACES

Resultant pressure forces on curved surfaces are more difficult to deal with because the incremental pressure forces, which are normal to the surface, vary continually in direction. There are two ways to approach the problem. One is to use direct integration by representing the curved shape functionally and integrating to find the horizontal and vertical components of the resultant force. The second method is to utilize the basic mechanics concept of a free body and the equilibrium of a fluid mass to find the two components. If the total resultant force is needed, then the two components can be combined vectorially to obtain the resultant.

Addressing first the method of integration, we recognize that the horizontal component of the resultant force is given by the equation

$$F_H = \int dF_H = \int \gamma hb\, dz$$

where b is the width of the surface and dz is the vertical projection of the surface element dL (see Fig. 2.11). The location of F_H is found by taking moments of dF about a convenient point, e.g., point C, and integrating.

$$z_p F_H = \int z\, dF_H = \int z\, \gamma hb\, dz$$

where z_p is the vertical distance from the moment center to F_H.

The vertical component is given by

$$F_V = \int dF_V = \int \gamma hb\, dx$$

where dx is the horizontal projection of the surface element. To locate F_V an approach similar to the above yields

$$x_p F_V = \int x\, dF_V = \int x\, \gamma hb\, dx$$

where x_p is the horizontal distance from the moment center to F_V. All the integrations can be performed once h as a function of x and z is found.

The technique for finding the horizontal and vertical components of the resultant force by the method of basic mechanics is demonstrated in Fig. 2.11. We choose a convenient volume of fluid on which to perform the analysis, the only requirement being that one of the fluid element boundaries coincide with the curved surface under consideration. Fol-

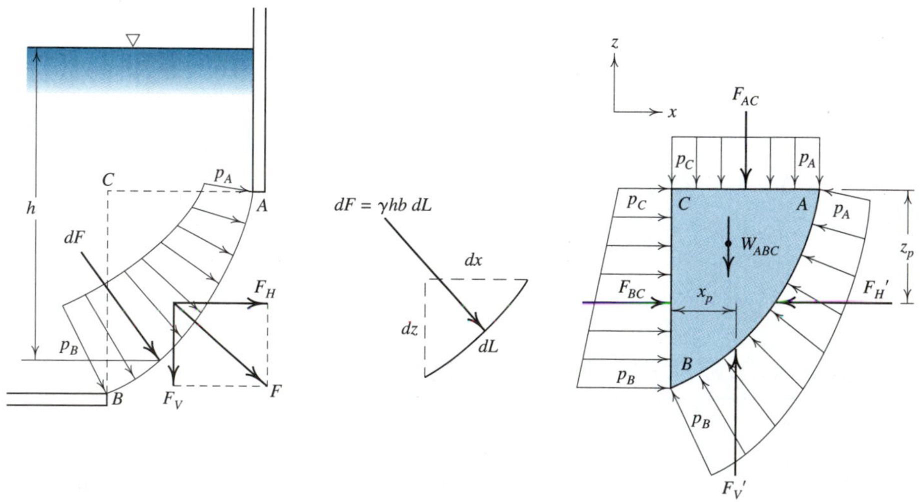

Fig. 2.11

lowing the rules for constructing a free body, we isolate the fluid mass and show all the forces acting on the mass to keep it in equilibrium. Note that the horizontal surface and the vertical surface of the free body can be treated by the methods of Section 2.4 because they are submerged plane surfaces. From the static equilibrium of the free body,

$$\Sigma F_x = F_{BC} - F'_H = 0$$

$$\Sigma F_z = F'_V - W_{ABC} - F_{AC} = 0$$

and thus $F'_H = F_{BC}$ and $F'_V = W_{ABC} + F_{AC}$. From the inability of the free body of fluid to support shear stress, it follows that F'_H must be collinear with F_{BC} and F'_V collinear with the resultant of W_{ABC} and F_{AC}. The foregoing analysis reduces the problem to one of computation of magnitude and location of F_{BC}, F_{AC}, and W_{ABC}; for F_{AC} and F_{BC} the methods of Section 2.4 may be used, whereas W_{ABC} is merely the weight of the free body of fluid and necessarily acts through its center of gravity.

When the same liquid covers both sides of a curved area but the liquid surfaces are at different levels for the two sides, the net effective pressure distribution is uniform because the effective pressure at any point on the area is dependent only on the difference in surface levels. The resultant force on such an area is obtained by application of the methods above. The horizontal component passes through the centroid of the vertical projection of the area, and the vertical component passes through the centroid of the horizontal projection.

ILLUSTRATIVE PROBLEM 2.10

Consider the cross section shown below of a typical 330 000 tonne (1 tonne = 1 000 kg) *Universe*-class oil tanker. The design of the curved portion of the hull at the lower corners of the cross section requires computation of the fluid static forces on that section. Using both direct integration and the basic mechanics approach, calculate the magnitude, direction, and line of action of the resultant force per metre of length of the vessel exerted by the seawater ($\gamma = 10\ \text{kN/m}^3$) on the surface AB, which is in the shape of a quarter cylinder.

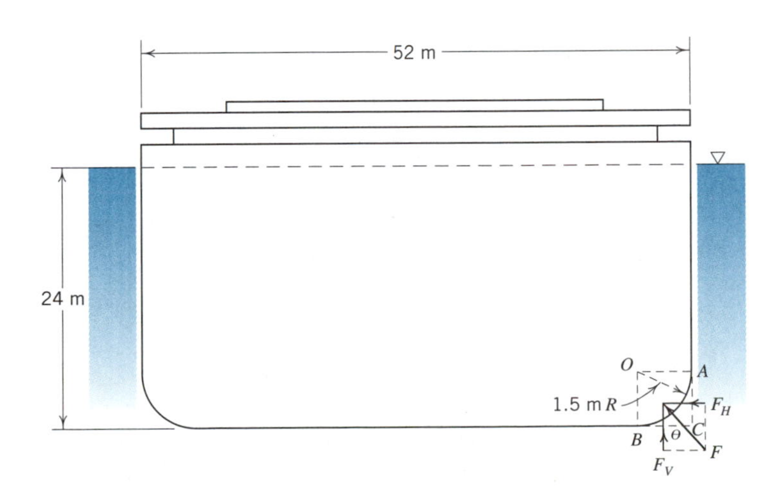

SOLUTION

Direct Integration Method

The first step in the direct integration method is to find the equation describing the curved surface. With the origin of the x-z coordinate system at O, the required equation is

$$x^2 + z^2 = 1.5^2 = 2.25$$

Recognizing that $h = 24.0 - 1.5 - z$ (z takes on negative values below O), we can write the equation for the horizontal component of the resultant force as

$$F_H = \int \gamma bh\, dz = 10\ \text{kN/m}^3 \int_{-1.5}^{0} 1\ \text{m} \times (22.5\ \text{m} - z)\, dz$$

$$= 10\left[22.5z - \frac{z^2}{2}\right]_{-1.5}^{0} = 10\left[0 - 0 - \left(22.5 \times -1.5 - \frac{(-1.5)^2}{2}\right)\right]$$

$$= 348.8\ \text{kN/m} \bullet$$

Similarly,

$$F_V = \int \gamma bh\, dx = 10\ \text{kN/m}^3 \int_{0}^{1.5} 1\ \text{m} \times (22.5\ \text{m} - z)\, dx$$

Now, from the equation of the curved surface, $z = \sqrt{2.25 - x^2}$. However, note that for the range of z and x in which we are working, x is positive and z is negative. This requires that we replace z with $-\sqrt{2.25 - x^2}$. The result is

$$F_V = 10\ \text{kN/m}^3 \int_{0}^{1.5} 1\ \text{m} \times [22.5 - (-\sqrt{2.25 - x^2})]\, dx$$

$$= 10 \int_{0}^{1.5} (22.5 + \sqrt{2.25 - x^2})\, dx$$

From a table of integrals we find

$$F_V = 10\left[22.5x + \frac{1}{2}\left(x\sqrt{2.25 - x^2} + 2.25 \sin^{-1}\frac{x}{1.5}\right)\right]_{0}^{1.5}$$

$$= 10\left[33.75 + \frac{1}{2}\left(0 + 2.25 \times \frac{\pi}{2}\right) - \left(0 + \frac{1}{2}(0 + 0)\right)\right]$$

$$= 10\left[33.75 + 2.25\frac{\pi}{4}\right]$$

$$F_V = 355.2\ \text{kN/m} \bullet$$

To locate the line of action of the horizontal component, we integrate moments of the incremental horizontal forces about O and equate the result to the moment of the resultant.

$$z_pF_H = \int z(\gamma bh)\,dz = 10\int_{-1.5}^{0}(22.5 - z)z\,dz = 10\int_{-1.5}^{0}(22.5z - z^2)\,dz$$

$$= 10\left[\frac{22.5z^2}{2} - \frac{z^3}{3}\right]_{-1.5}^{0} = 10[0 - 0 - (25.31 + 1.13)] = -264.4$$

$$z_p = \frac{-264.4}{348.8} = -0.758 \text{ m} \bullet$$

Now, to locate the line of action of the vertical component, we take moments of the vertical forces about O.

$$x_pF_V = \int x(\gamma bh)\,dx = 10\int_0^{1.5} x(22.5 - z)\,dx$$

Again replacing z with $-\sqrt{2.25 - x^2}$,

$$x_pF_V = 10\int_0^{1.5}(22.5 + \sqrt{2.25 - x^2})x\,dx = 10\left[\frac{22.5x^2}{2} - \frac{(2.25 - x^2)^{3/2}}{3}\right]_0^{1.5}$$

$$= 10[25.3 - 0 - (0 - 1.125)] = 264.4$$

$$x_p = \frac{264.4}{355.2} = 0.744 \text{ m} \bullet$$

Basic Mechanics Method

First, we construct a free body of the portion of seawater in contact with the section of the hull under investigation, showing all the forces acting and their directions.

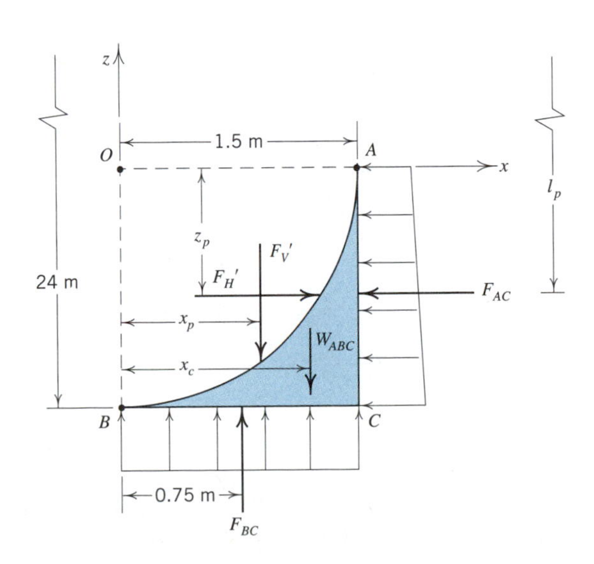

Horizontal Force Components

Recognizing that $F'_H = F_{AC}$, we proceed to calculate F_{AC} from the procedures given in Section 2.4. We note that F_{AC} is the resultant force acting on a vertical rectangular area 1.5 m high. Using Eq. 2.12,

$$F_{AC} = \gamma h_c A = 10 \text{ kN/m}^3 \times (24 - 1.5/2) \text{ m} \times (1 \times 1.5) \text{ m}^2$$

$$F'_H = F_{AC} = 348.8 \text{ kN} \bullet$$

To find the line of action of this force, we use Eq. 2.14,

$$l_p = l_c + \frac{I_c}{l_c A}$$

From Appendix 3,

$$I_c = \frac{1}{12} bh^3 = \frac{1}{12} 1 \times 1.5^3 = 0.281 \text{ m}^4$$

So, with $l_c = 24 - 1.5/2 = 23.25$,

$$l_p = 23.25 + 0.281/(23.25 \times 1 \times 1.5)$$

$$l_p = 23.258 \text{ m}$$

$$z_p = 22.5 - 23.258 = -0.758 \text{ m} \bullet$$

This value confirms the result found by integration.

Vertical Force Components

From the free body, it is clear that

$$F'_V = F_{BC} - W_{ABC}$$

Recognizing that the force F_{BC} is the resultant of a uniform pressure distribution on the bottom surface of the free body,

$$F_{BC} = (\gamma h)A = (10 \text{ kN/m}^3 \times 24 \text{ m}) \times (1 \times 1.5) \text{ m}^2$$

$$F_{BC} = 360 \text{ kN}$$

Computing the weight W_{ABC},

$$W_{ABC} = \gamma(r \times r - \pi r^2/4) \times 1 \text{ m}$$

$$= 10 \text{ kN/m}^3 \times (1.5 \times 1.5 - \pi 1.5^2/4) \text{ m}^2 \times 1 \text{ m}$$

$$W_{ABC} = 4.83 \text{ kN}$$

So, the resultant pressure force in the vertical direction is

$$F'_V = 360 - 4.83$$

$$F'_V = 355.2 \text{ kN} \bullet$$

To locate F'_V, we must locate the resultant of F_{BC} and W_{ABC} because F'_V will be equal, opposite, and collinear with that resultant. To accomplish this, we use the principle of moments from basic mechanics. However, before proceeding, we must locate the line of action of W_{ABC}. This line of action is through the center of gravity of ABC which coincides with the centroid of ABC. We find the location of the centroid from Appendix 3

$$x_c = \frac{2}{3}\frac{r}{4-\pi} = \frac{2}{3}\frac{1.5}{4-\pi}$$

$$x_c = 1.165 \text{ m}$$

Now, taking moments about B,

$$Rx_p = (1.5/2)F_{BC} - 1.165\ W_{ABC}$$

where $R = F'_V$.

$$x_p = (0.75 \times 360 - 1.165 \times 4.83)/355.17$$

$$x_p = 0.744 \text{ m} \bullet$$

So, F'_V acts 0.744 m from B, confirming the result obtained by integration.

Resultant Force

The resultant force is

$$F = \sqrt{(F'_H)^2 + (F'_V)^2} = \sqrt{348.75^2 + 355.17^2}$$

$$F = 497.8 \text{ kN} \bullet$$

The direction of F is upward to the left at an angle θ

$$\theta = \tan^{-1}\frac{F'_V}{F'_H} = \tan^{-1}\frac{355.2}{348.8}$$

$$\theta = 45.5° \bullet$$

To find the line of action of F, we take moments about O of the two components of F and set them equal to the moment of F. Let d = the distance from O to the line of action of F.

$$Fd = F'_H|z_p| - F'_V|x_p|$$

Note that in taking moments of the horizontal and vertical components, we use the absolute values of x_p and z_p because the sign of the moment depends only on the direction of rotation about the moment center.

$$Fd = 348.8 \times 0.758 - 355.2 \times 0.744$$

$$d = 0.00 \text{ m} \bullet$$

That is, the resultant force F acts through the center of curvature of the cylindrical surface.

In retrospect, it should be no surprise that the resultant force acts through the center of curvature. The pressure forces are all normal to the surface and, in the case of a circular arc, all the lines of action would pass through the center of the arc; hence, the resultant would, of necessity, also pass through the center. In the future, we can utilize this fact to more easily determine the line of action of the resultant of the pressure forces acting on a cylindrical or spherical surface.

2.6 BUOYANCY AND THE STABILITY OF FLOATING BODIES

The familiar laws of buoyancy (Archimedes' principle) and flotation are usually stated: (1) *a body immersed in a fluid is buoyed up by a force equal to the weight of fluid displaced*; and (2) *a floating body displaces its own weight of the liquid in which it floats*. These laws are corollaries of the general principles of Section 2.5 and may be readily proved by application of those principles.

A body *ABCD* suspended in a fluid of specific weight γ is illustrated in Fig. 2.12. Isolating a free body of fluid with vertical sides tangent to the body allows identification of the vertical forces exerted by the lower (*ADC*) and upper (*ABC*) surfaces of the body on the surrounding fluid. These are F_1' and F_2' with $(F_1' - F_2')$ the buoyant force on the body. For the upper portion of the free body

$$\Sigma F_z = F_2' - W_2 - p_2 A = 0$$

and for the lower portion

$$\Sigma F_z = F_1' + W_1 - p_1 A = 0$$

with which we obtain (by subtraction of the equations)

$$F_B = F_1' - F_2' = (p_1 - p_2)A - (W_1 + W_2)$$

However, $p_1 - p_2 = \gamma h$ and γhA is the weight of a cylinder of fluid extending between horizontal planes 1 and 2, and the right side of the equation for F_B is identified as the

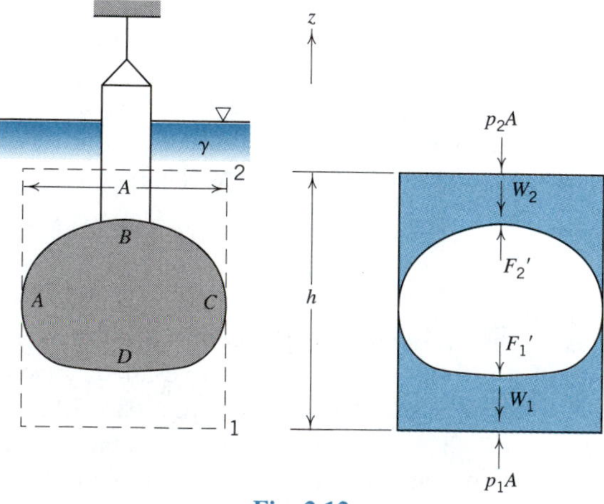

Fig. 2.12

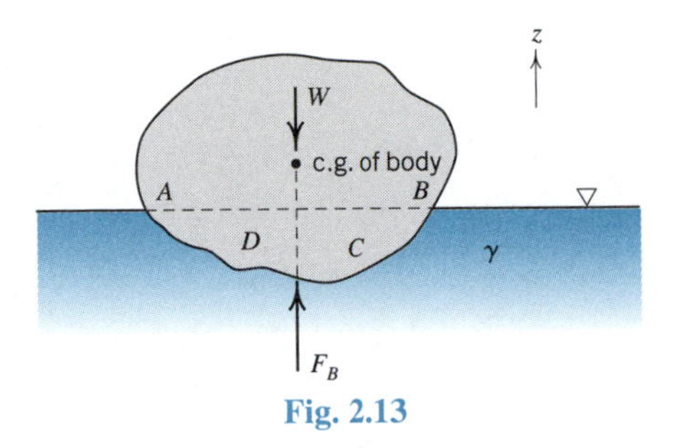

Fig. 2.13

weight of a volume of fluid exactly equal to that of the body. Accordingly

$$F_B = \gamma \times (\text{volume of submerged object}) \tag{2.16}$$

and the law of buoyancy is proved.

There are two observations which must be kept in mind when using this formula. First, this formula takes into account *all* the pressure forces acting on the submerged body so *no additional pressure force calculations are necessary*. Second, this formula presumes that a *homogeneous fluid completely surrounds the body*. If it does not, the above formula should not be used and the reader should return to the basic approach of Section 2.5 to calculate the pressure forces which provide the buoyant effect.

For the floating object of Fig. 2.13, a similar analysis shows that

$$F_B = \gamma \times (\text{volume of liquid displaced}) \tag{2.17}$$

and, from static equilibrium of the object, its weight must be equal to this buoyant force; thus, the object displaces its own weight of the liquid in which it floats.

The principles above find many applications in engineering, for example, in calculations of the draft of surface vessels, the increment in depth of flotation from the increment in weight of the ship's cargo, and the lift of airships and hot-air balloons.

The stability of submerged or floating bodies is dependent on the relative location of the buoyant force and the weight of the body. The buoyant force acts upward through the center of gravity of the displaced volume; the weight acts downward at the center of gravity of the body. Stability or instability will be determined by whether a righting or overturning moment is developed when the center of gravity and center of buoyancy move out of vertical alignment. Obviously, for the submerged bodies, such as the balloon and submarine of Fig. 2.14, stability requires the center of buoyancy to be above the center of gravity. In surface vessels, however, the center of gravity is usually above the center of buoyancy, and stability exists because of movement of the center of buoyancy to a position outboard of the center of gravity as the ship "heels over," producing a righting moment. An overturning moment, resulting in capsizing, occurs if the center of gravity moves outboard of the center of buoyancy.

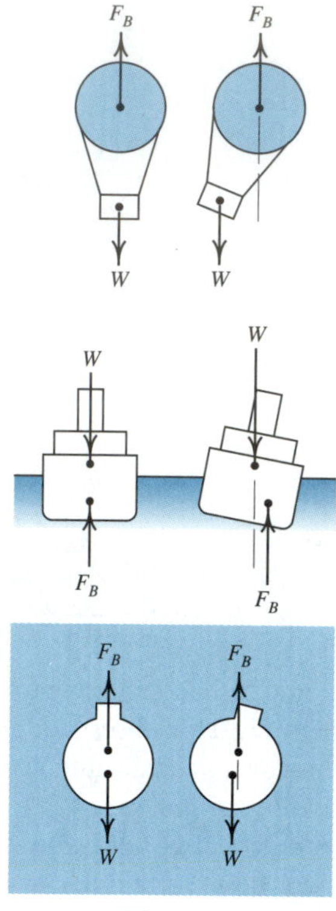

Fig. 2.14

ILLUSTRATIVE PROBLEM 2.11

A container ship has a cross-sectional area in the horizontal plane at the waterline of 32 000 ft^2 when the draft (submergence) is 29 ft. How many tons (1 ton = 2 000 lb) of containers can be added before the normal draft of 30 ft is reached?

Assume the seawater has a specific weight of 64 lb/ft^3.

SOLUTION

The ship is a floating object, hence, it displaces its own weight in liquid. Therefore, the additional weight of containers will displace their weight in liquid. If the ship's draft increases by 1.0 ft, then the volume of displaced liquid is

$$\text{Volume} = 32\,000 \text{ ft}^2 \times 1.0 \text{ ft} = 32\,000 \text{ ft}^3$$

This amounts to a weight of

$$\text{Weight} = 32\,000 \text{ ft}^3 \times 64 \text{ lb/ft}^3 = 2\,048\,000 \text{ lb}$$

In terms of tons,

$$\text{Weight of extra containers} = 2\,048\,000 \text{ lb}/2\,000 \text{ lb/ton} = 1\,024 \text{ tons} \bullet$$

ILLUSTRATIVE PROBLEM 2.12

The solid cone fits into the 1.5-ft diameter opening in the bottom of the tank to prevent water from draining from the tank. Calculate the required weight of the cone which will just prevent it from lifting away from the opening.

Volume B of the cone is a solid cylinder the same diameter as the opening. Volume A is the remaining volume of the cone above the opening.

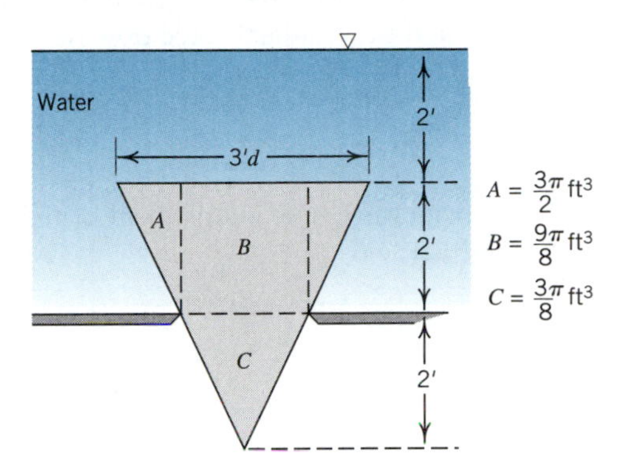

SOLUTION

This is a situation where the submerged object is not completely surrounded by a homogeneous fluid. As a consequence, we must be very careful as to how we calculate buoyancy. However, it should be noted that the buoyant force concept can be used to calculate the net upward (buoyant) force on volume A. This is possible because both the upper and lower surfaces of volume A are exposed to a homogeneous fluid.

As a consequence, we may write the net upward pressure force on that part of the cone designated as volume A as

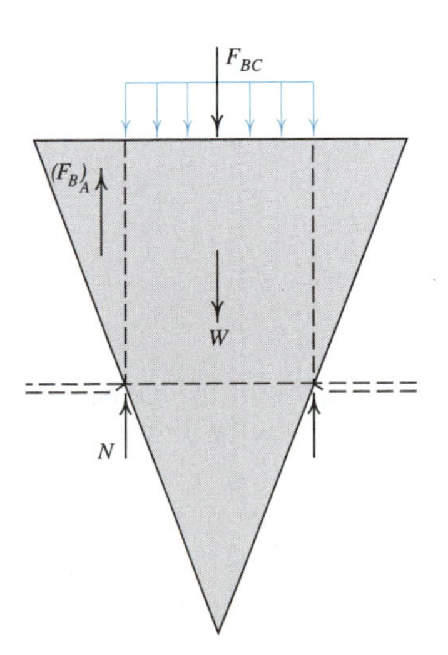

$$
\begin{aligned}
(F_B)_A &= \gamma \times (\text{displaced volume}) \\
&= 62.4\ \text{lb/ft}^3 \times 3\pi/2\ \text{ft}^3 \\
(F_B)_A &= 294.1\ \text{lb}
\end{aligned}
$$

However, the buoyant force formula cannot be used to compute the upward force on volumes B and C because the surface of the cone directly below volumes B and C is not exposed to the fluid in the tank. So, we must deal directly with the pressure force.

From the free body, we see that the pressure on the upper surface of volume B is uniform and the resultant force downward F_{BC} is

$$
\begin{aligned}
F_{BC} &= (\gamma h)A = (62.4\ \text{lb/ft}^3 \times 2\ \text{ft}) \times (\pi/4 \times 1.5^2)\ \text{ft}^2 \\
F_{BC} &= 220.5\ \text{lb}
\end{aligned}
$$

Considering static equilibrium of the cone, we note that $N = 0$ at the critical value of weight W. So,

$$
\begin{aligned}
\Sigma F_z = 0 \qquad (F_B)_A - F_{BC} - W_{crit} &= 0 \\
W_{crit} = (F_B)_A - F_{BC} &= 294.1 - 220.5 \\
W_{crit} &= 73.6\ \text{lb} \bullet
\end{aligned}
$$

2.7 FLUID MASSES SUBJECTED TO ACCELERATION

Fluid masses can be subjected to various types of acceleration without the occurrence of relative motion between fluid particles or between fluid particles and boundaries. Such fluid masses will be found to conform to the laws of fluid statics, modified to allow for the effects of acceleration, and they may often be treated by assuming a change in the magnitude and direction of g_n.

A generalized approach to this problem may be obtained by applying Newton's second law to the fluid element of Fig. 2.15 which is being accelerated in such a way that its components of acceleration are a_x and a_z. The summation of force components on such an element has been indicated in Section 2.1, and is

$$\Sigma F_x = \left(-\frac{\partial p}{\partial x}\right) dx\, dz \tag{2.18a}$$

$$\Sigma F_z = \left(-\frac{\partial p}{\partial z} - \gamma\right) dx\, dz \tag{2.18b}$$

With the mass of the element equal to $(\gamma/g_n)\, dx\, dz$, the component forms of Newton's second law may be written

$$
\begin{aligned}
\left(-\frac{\partial p}{\partial x}\right) dx\, dz &= \frac{\gamma}{g_n} a_x\, dx\, dz \\
\left(-\frac{\partial p}{\partial z} - \gamma\right) dx\, dz &= \frac{\gamma}{g_n} a_z\, dx\, dz
\end{aligned}
$$

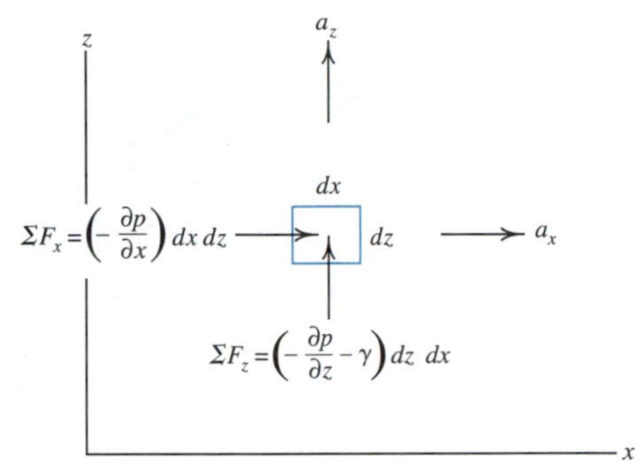

Fig. 2.15

which reduce to

$$-\frac{\partial p}{\partial x} = \frac{\gamma}{g_n}(a_x) \tag{2.19}$$

$$-\frac{\partial p}{\partial z} = \frac{\gamma}{g_n}(a_z + g_n) \tag{2.20}$$

These equations characterize the pressure variation through an accelerated mass of fluid, and with them specific applications may be studied.

One other useful generalization can be derived from the foregoing equations, namely, a property of a line of constant pressure. By using the chain rule for the total differential for dp in terms of its partial derivations,[11]

$$dp = \frac{\partial p}{\partial x} dx + \frac{\partial p}{\partial z} dz$$

and by substituting the above expressions for $\partial p/\partial x$ and $\partial p/\partial z$, we obtain

$$dp = -\frac{\gamma}{g_n}(a_x)\,dx - \frac{\gamma}{g_n}(a_z + g_n)\,dz \tag{2.21}$$

However, along a line of constant pressure $dp = 0$ and hence, for such a line,

$$\frac{dz}{dx} = -\left(\frac{a_x}{g_n + a_z}\right) \tag{2.22}$$

Thus the *slope* (dz/dx) of a line of constant pressure is defined; its *position* must be determined from external (boundary) conditions in specific problems.

From the foregoing generalizations some situations of engineering significance may now be examined.

Constant Linear Acceleration with $a_x = 0$

Here a container of liquid is accelerated vertically upward, $\partial p/\partial x = 0$, and with no change of pressure with x, Eq. 2.20 becomes

$$\frac{dp}{dz} = -\gamma\left(\frac{g_n + a_z}{g_n}\right)$$

[11]See Appendix 6.

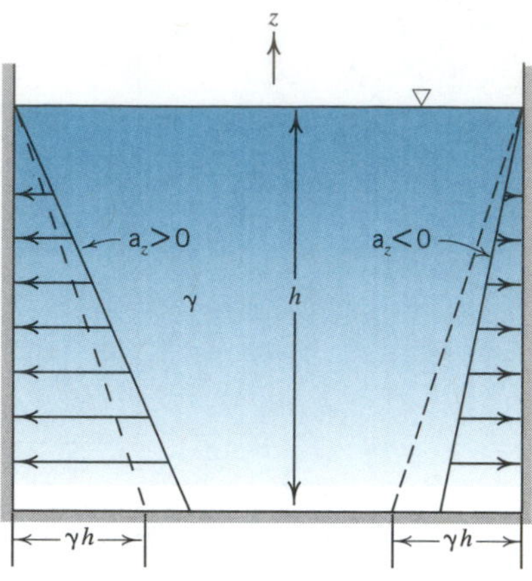

Fig. 2.16

For a_z constant, this equation shows that the characteristic linear pressure variation of fluid statics is preserved but that magnitudes of pressure will now depend on a_z. The quantitative aspects of this are shown in Fig. 2.16 for $a_z > 0$ and $a_z < 0$. The latter case is of particular interest when $a_z = -g_n$, yielding $dp/dz = 0$ and showing that the pressure is constant throughout a freely falling mass of fluid; for an *unconfined* mass of freely falling fluid the pressure is therefore equal to that surrounding it—if the surrounding pressure is zero, all pressures within the fluid mass will be zero, a fact which has many applications in subsequent problems.

ILLUSTRATIVE PROBLEM 2.13

An open tank of water is accelerated vertically upward at 4.5 m/s². Calculate the pressure at a depth of 1.5 m.

SOLUTION

This is a situation with only vertical acceleration, so we use the equation

$$\frac{dp}{dz} = -\gamma\left(\frac{g_n + a_z}{g_n}\right) = -9\,800\ \text{N/m}^3\left(\frac{9.81 + 4.5}{9.81}\right) = -14\,300\ \text{N/m}^3$$

In order to find the pressure at a depth of 1.5 m below the water surface, we need to integrate the above equation.

$$\int_0^p dp = -14\,300\int_0^{-1.5} dz$$

$$p = -14\,300(-1.5 - 0) = 21\,450\ \text{N/m}^2$$

$$p = 21.45\ \text{kPa}\ \bullet$$

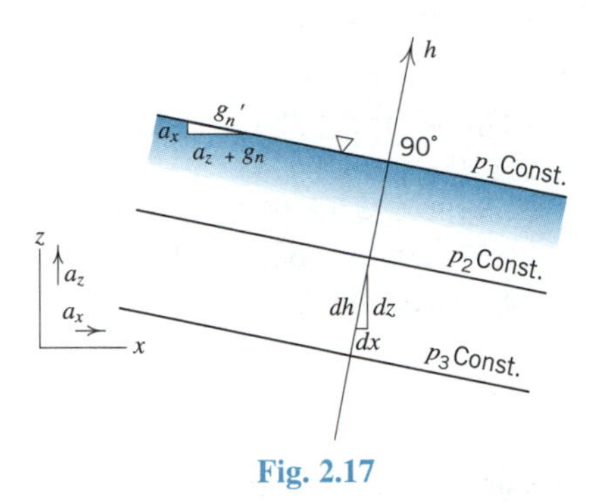

Fig. 2.17

Constant Linear Acceleration

Here the slope of a liquid surface (which is a line of constant pressure) is given (from Eq. 2.22) by $-a_x/(a_z + g_n)$ as shown in Fig. 2.17, and other lines of constant pressure will be parallel to the surface. Along x and z the (linear) pressure variations may be computed from Eqs. 2.19 and 2.20, respectively. In the direction h, normal to the lines of constant pressure, Eq. 2.21 may be employed. Dividing this equation by dh,

$$\frac{dp}{dh} = -\gamma\left(\frac{a_x}{g_n}\frac{dx}{dh} + \frac{g_n + a_z}{g_n}\frac{dz}{dh}\right)$$

but, from the similar triangles of Fig. 2.17,

$$dx/dh = a_x/g_n' \qquad \text{and} \qquad dz/dh = (a_z + g_n)/g_n'$$

Substituting these expressions above gives

$$dp/dh = -\gamma(g_n'/g_n)$$

which shows that the pressure variation along h is linear and allows computation of pressures as in statics (Eq. 2.3), $\gamma g_n'/g_n$ being used for the specific weight of the fluid.

ILLUSTRATIVE PROBLEM 2.14

An open tank, placed on an inclined plane, is filled to the top with water. The tank is then gradually accelerated up to a constant value a up the plane so that the water surface steadies at the position shown below after some spillage resulting from the movement.

Calculate the acceleration required for the water surface to remain in the position shown. Also calculate the pressure in the corner of the tank at point A both before and after acceleration.

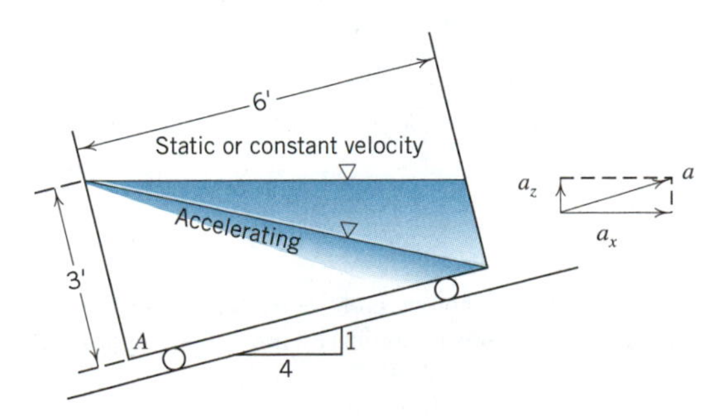

SOLUTION

We will use Eq. 2.22 to calculate the vertical component of acceleration a_z. But first, we must calculate the water surface slope dz/dx. The plane is inclined at an angle given by $\alpha = \tan^{-1} 1/4 = 14.04°$ while the water surface is tilted at an angle of $\beta = \tan^{-1} 3/6 = 26.57°$ relative to the inclined plane. The water surface slope is given by the difference in these two values.

$$\frac{dz}{dx} = -\tan(26.57° - 14.04°) = -\tan 12.53° = -0.222$$

The sign is negative because the water surface slopes downward in the positive x-direction. Now, from Eq. 2.22,

$$\frac{dz}{dx} = -\left(\frac{a_x}{g_n + a_z}\right)$$

From the slope of the incline, we note that $a_x = 4a_z$, so

$$-0.222 = -\left(\frac{4a_z}{32.2 + a_z}\right)$$

$$a_z = 1.89 \text{ ft/sec}^2$$

Since a_z is the vertical component of a, then

$$a = a_z/\sin\alpha = 1.89/\sin 14.04° = 7.80 \text{ ft/sec}^2 \bullet$$

The pressure in the corner of the tank before movement is given by the depth of water over the point. That is,

$$p_A = \gamma h = 62.4 \text{ lb/ft}^3 \times 3\cos\alpha = 62.4 \times 3\cos 14.04°$$

$$p_A = 181.6 \text{ lb/ft}^2 \text{ before movement} \bullet$$

The acceleration is always directed along the inclined plane; therefore, the acceleration normal to the plane is zero. As a consequence, the pressure variation normal to the plane is static and the pressure at A is the same as for the fluid at rest.

Radial Acceleration with Constant Angular Velocity about a Vertical Axis, and $a_z = 0$

Equations 2.19, 2.20, and 2.22 may be written[12] with radial distance, r, substituted for x and a_r for a_x to give (for $a_z = 0$): $-\partial p/\partial r = \gamma a_r/g_n$, $-\partial p/\partial z = \gamma$, and, for surfaces of constant pressure, $dz/dr = -a_r/g_n$. From kinematics, $a_r = -\omega^2 r$, in which ω is the angular velocity. Substituting this in the foregoing equations,

$$\frac{\partial p}{\partial r} = \frac{\gamma\omega^2 r}{g_n} \tag{2.23}$$

$$\frac{\partial p}{\partial z} = -\gamma \tag{2.24}$$

[12]For rigorous proof of the validity of this, a complete analysis should be made using the conventional element of polar coordinates; however (after discarding the negligible terms), the same result is obtained.

For surfaces of constant pressure,

$$\frac{dz}{dr} = \frac{\omega^2 r}{g_n} \tag{2.25}$$

The pressure gradient along r and z can be computed from the first two equations; the third can be easily integrated to

$$z = \frac{\omega^2 r^2}{2g_n} + \text{Constant} \tag{2.26}$$

showing that lines of constant pressure are parabolas (Fig. 2.18) symmetrical about the axis of rotation. The second equation shows that pressure variation in the vertical is that of fluid statics (Eq. 2.5) so that $p_3 - p_1 = \gamma h$. The second equation can be integrated (for z = constant) from the axis of rotation (where the pressure is p_c and r is zero) to any radius r where the pressure is p. The result is

$$\frac{p - p_c}{\gamma} = \frac{\omega^2 r^2}{2g_n} \tag{2.27}$$

which may also be deduced directly from Fig. 2.18.

The foregoing analysis shows the possibility of pressure being created by the rotation of a fluid mass. This principal is utilized in centrifugal pumps and blowers to produce a pressure difference in order to cause fluids to flow.

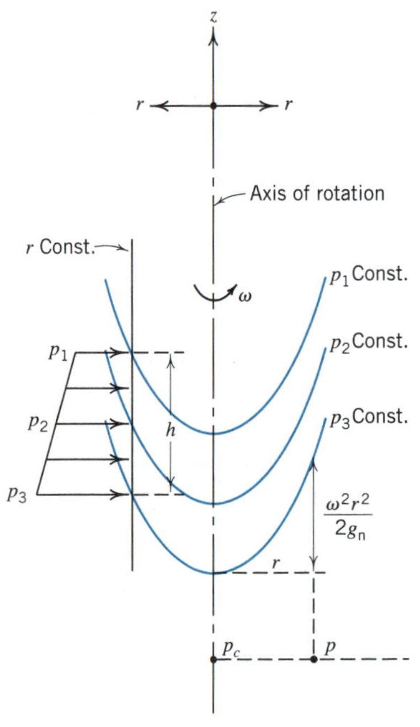

Fig. 2.18

ILLUSTRATIVE PROBLEM 2.15

The 2-ft diameter cylindrical tank shown below is fitted with three piezometer columns of the same diameter. The tank is initially filled with water to the depth 5 ft above its bottom and rotates about its central axis at a constant speed of 100 rpm.

Calculate the pressure heads at points *A* and *B*.

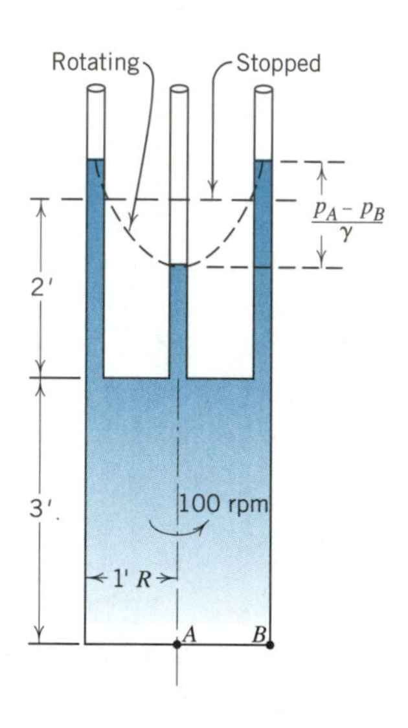

SOLUTION

We know from the previous section that surfaces of constant pressure lie on parabolic curves with their vertex at the center of rotation. Because the open piezometers constitute a constant zero-pressure surface, the water levels in the piezometers must lie on a parabolic curve. We also know that the pressure varies statically in the vertical direction so the distances from A and B to their respective liquid surfaces represent their pressure heads. The differences in pressure heads can be calculated from Eq. 2.27, knowing that the angular velocity is

$$\omega = \frac{100 \text{ rev/min} \times 2\pi \text{ rad/rev}}{60 \text{ sec/min}} = 10.47 \text{ rad/sec}$$

$$\frac{p_B - p_A}{\gamma} = \frac{\omega^2 r^2}{2g_n} = \frac{(10.47 \text{ rad/sec})^2 (1 \text{ ft})^2}{2 \times 32.2} = 1.70 \text{ ft} \tag{2.27}$$

When rotation begins, the liquid loss in the center piezometer tube will equal the liquid gain in the two outer tubes. That is, the drop in surface level in the center tube will be twice the rise in the outer tubes. Using this information,

$$\frac{p_A}{\gamma} = 5 - \frac{2}{3} \times 1.70 = 3.87 \text{ ft} \bullet$$

$$\frac{p_B}{\gamma} = 5 + \frac{1}{3} \times 1.70 = 5.57 \text{ ft} \bullet$$

ILLUSTRATIVE PROBLEM 2.16

A vertical cylinder of radius r_o contains a gas at constant temperature and rotates about its axis at a constant angular velocity ω. Derive a relationship between the pressure at the axis of rotation and that at the cylindrical surface.

SOLUTION

For a gas we can ignore the pressure variation in the vertical direction and deal only with the radial variation. This simplification allows us to write Eq. 2.23 in terms of total derivatives as

$$\frac{dp}{dr} = \frac{\gamma\omega^2 r}{g_n}$$

Substituting p/RT for γ/g_n, the above equation becomes

$$dp = \frac{p\omega^2 r}{RT}\, dr$$

Noting that T will be a constant in the container, and letting the pressures at the center of the cylinder and at the outer edge be p_c and p_o, respectively, we can integrate the above equation.

$$\int_{p_c}^{p_o} \frac{dp}{p} = \frac{\omega^2}{RT}\int_0^{r_o} r\,dr$$

$$\ln\left(\frac{p_o}{p_c}\right) = \frac{\omega^2 r_o^2}{2RT}$$

And finally,

$$p_o = p_c e^{\omega^2 r_0^2/2RT} \bullet$$

ILLUSTRATIVE PROBLEM 2.17

Many geophysical problems are concerned with flows on a rotating earth. A simple experiment of some significance involves small perturbations of a fluid in rigid body rotation. In a laboratory experiment, a circular tank of radius $r_o = 2$ m is filled with water to an average depth of $d = 75$ mm and rotated at an angular speed $\omega = 5$ rev/min about its center. Find the shape of the free surface of the water when it is in rigid body rotation.

SOLUTION

We know from Eq. 2.26 that the free surface is parabolic in shape and given by that equation as

$$z = \frac{\omega^2 r^2}{2g_n} + \text{Constant}$$

In determining the constant in the above equation, we specify that $z = z_o$ at $r = 0$. The equation becomes

$$z = \frac{\omega^2 r^2}{2g_n} + z_o$$

Of course we do not know the value of z_o, only that it is determined by the fact that the volume of water under rotation must equal the volume at rest. The equation representing this fact is

$$\pi r_o^2\, d = \int_0^{r_o} (2\pi r) z\, dr = \int_0^{r_o} (2\pi r)\left(\frac{\omega^2 r^2}{2g_n} + z_o\right) dr = 2\pi \int_0^{r_o} \left(\frac{\omega^2 r^3}{2g_n} + z_o r\right) dr$$

Integrating,

$$\pi r_o^2\, d = 2\pi \frac{\omega^2 r_o^4}{8g_n} + 2\pi z_o \frac{r_o^2}{2}$$

Dividing through by πr_o^2 gives

$$d = \frac{\omega^2 r_o^2}{4g_n} + z_o$$

Solving for z_o, recognizing that $\omega = 5 \text{ rev/min} \times 2\pi/60 = 0.524 \text{ rad/sec}$,

$$z_o = d - \frac{\omega^2 r_o^2}{4g_n} = 0.075 - \left(\frac{0.524^2 \times 2^2}{4 \times 9.81}\right)$$

$$z_o = 0.047 \text{ m} = 47 \text{ mm} \bullet$$

The water surface at the edge of the tank would be

$$z = \frac{\omega^2 r^2}{2g_n} + z_o = \frac{0.524^2 \times 2^2}{2 \times 9.81} + 0.047 = 0.103 \text{ m} = 103 \text{ mm} \bullet$$

PROBLEMS

2.1. Calculate the pressure in an open tank of crude oil at a point 8 ft or 2.4 m below the liquid surface.

2.2. If the pressure 10 ft (3 m) below the free surface of a liquid is 20 psi (140 kPa), calculate its specific weight and specific gravity.

2.3. If the pressure at a point in the ocean is 140 kPa, what is the pressure 30 m below this point? Specific weight of saltwater is 10 kN/m^3.

2.4. An open vessel contains carbon tetrachloride to a depth of 6 ft (2 m) and water on the carbon tetrachloride to a depth of 5 ft (1.5 m). What is the pressure at the bottom of the vessel?

2.5. How many inches of mercury are equivalent to a pressure of 20 psi? How many feet of water?

2.6. How many millimetres of carbon tetrachloride are equivalent to a pressure of 40 kPa? How many metres of alcohol?

2.7. One vertical metre (foot) of air at 15°C (59°F) and 101.3 kPa (14.7 psia) is equivalent to how many pascals (pounds per square inch)? Millimetres (inches) of mercury? Metres (feet) of water?

2.8. The barometric pressure at sea level is 30.00 in. (762 mm) of mercury when that on a mountain top is 29.00 in. (737 mm). If specific weight of air is assumed constant at 0.075 lb/ft^3 (11.8 N/m^3), calculate the elevation of the mountain top.

2.9. If at the surface of a liquid the specific weight is γ_o, with z and p both zero, show that, if E = constant, the specific weight and pressure are given by

$$\gamma = \frac{E}{(z + E/\gamma_o)} \quad \text{and} \quad p = -E \ln\left(1 + \frac{\gamma_o z}{E}\right)$$

Calculate specific weight and pressure at a depth of 2 miles

(2 km) assuming $\gamma_o = 64.0$ lb/ft^3 (10.0 kN/m^3) and $E = 300\ 000$ psi (2 070 MPa).

2.10. In the deep ocean the compressibility of seawater is significant in its effect on ρ and p. If $E = 2.07 \times 10^9$ Pa, find the percentage change in the density and pressure at a depth of 10 000 metres as compared to the values obtained at the same depth under the incompressible assumption. Let $\rho_o = 1\ 020$ kg/m^3 and the absolute pressure $p_o = 101.3$ kPa.

2.11. The specific weight of water in the ocean may be calculated from the empirical relation $\gamma = \gamma_o + K\sqrt{h}$ (in which h is the depth below the ocean surface). Derive an expression for the pressure at any point h and calculate specific weight and pressure at a depth of 2 miles (3.22 km) assuming $\gamma_o = 64.0$ lb/ft^3 (10 kN/m^3), h in feet (metres), and $K = 0.025$ lb/ft$^{7/2}$ (7.08 N/m$^{7/2}$).

2.12. If the specific weight of a liquid varies linearly with depth below the liquid surface ($\gamma = \gamma_o + Kh$), derive an expression for pressure as a function of depth.

2.13. If atmospheric pressure at the ground is 14.7 psia (101.3 kPa) and temperature is 59°F (15°C), calculate the pressure 25 000 ft (7.62 km) above the ground, assuming (*a*) no density variation, (*b*) isothermal variation of density with pressure, and (*c*) adiabatic variation of density with pressure.

2.14. Calculate pressures and densities of air in the U.S. Standard Atmosphere at 25 000 ft and 50 000 ft or at 8 km and 16 km. Check results with the values in Appendix 2.

2.15. Calculate the depth of an adiabatic atmosphere if temperature and pressure at the ground are, respectively, 59°F (15°C) and 14.7 psia (101.3 kPa).

2.16. Derive a relation between pressure and altitude: (*a*) for an isothermal atmosphere and (*b*) for the U.S. Standard Atmosphere to altitude 35 000 ft (10.7 km).

2.17. If the temperature in the atmosphere is assumed to vary linearly with altitude so $T = T_o - \alpha z$ where T_o is the sea level temperature and $\alpha = -dT/dz$ is the temperature lapse rate, find $p(z)$ when air is taken to be a perfect gas. Give the answer in terms of p_o, α, g_n, R, and z only.

2.18. Show that the temperature lapse rate in an adiabatic atmosphere is the reciprocal of the specific heat at constant pressure.

2.19. Assuming a linear rise of temperature in the U.S. Standard Atmosphere between altitudes of 85 000 and 100 000 ft or 26 and 30 km, calculate the value of the polytropic exponent n over this range of altitude.

2.20. Find the height of a static, perfect gas atmosphere in which the temperature decreases linearly with altitude as $T = T_o - \alpha z$. Take the height to be that altitude where $p(z) = 0$.

2.21. A mass of warm (64°F or 17.8°C), moist air is swept by the wind from sea level up the side of a coastal mountain range to the top at an altitude of 2 500 ft (762 m). Assuming the rise is adiabatic in an otherwise standard atmosphere, will the warm air continue to rise, be neutrally buoyant, or move down the leeward slope of the range? Why?

2.22. With atmospheric pressure at 14.5 psia (100 kPa, abs.), what absolute pressure corresponds to a gage pressure of 20 psi (138 kPa)?

2.23. When the barometer reads 30 in. (762 mm) of mercury, what absolute pressure corresponds to a vacuum of 12 in. (305 mm) of mercury?

2.24. If a certain absolute pressure is 85.2 kPa (12.35 psia), what is the corresponding vacuum if the atmospheric pressure is 760 mm (29.92 in.) of mercury?

2.25. A Bourdon pressure gage attached to a closed tank of air reads 20.47 psi (141.1 kPa) with the barometer at 30.50 in. (775 mm) of mercury. If barometric pressure drops to 29.18 in. (741.2 mm) of mercury, what will the gage read?

2.26. A Bourdon gage is connected to a tank in which the pressure is 40.0 psi (276 kPa) above atmospheric at the gage connection. If the pressure in the tank remains unchanged but the gage is placed in a chamber where the air pressure is reduced to a vacuum of 25 in. (635 mm) of mercury, what gage reading will be expected?

2.27. The compartments of these tanks are closed and filled with air. Gage A reads 207 kPa. Gage B registers a vacuum of 254 mm of mercury. What will gage C read if it is connected to compartment 1 but inside compartment 2? Barometric pressure is 101 kPa, absolute.

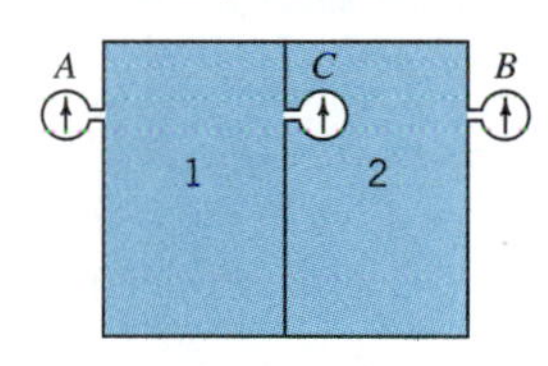

Problem 2.27

2.28. If the barometer of Fig. 2.5 is filled with a silicon oil of specific gravity 0.86, calculate h if the barometric (absolute) pressure is 101.3 kPa or 14.7 psia. Is this a practical barometer?

2.29. Calculate the height of the column of a water barometer for an atmospheric pressure of 14.70 psia when the water is at 50°F, 150°F, and 212°F.

2.30. Barometric pressure is 29.43 in. (758 mm) of mercury. Calculate h.

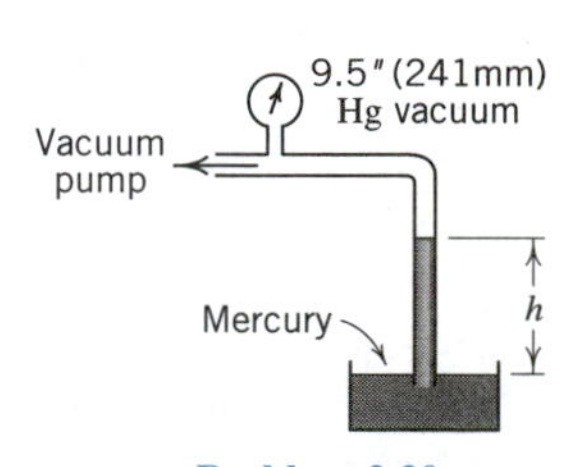

Problem 2.30

2.31. Calculate the pressure p_x in Fig. 2.7a if $l = 760$ mm, $h = 500$ mm; liquid γ is water, and γ_1 mercury.

2.32. With the manometer reading as shown, calculate p_x.

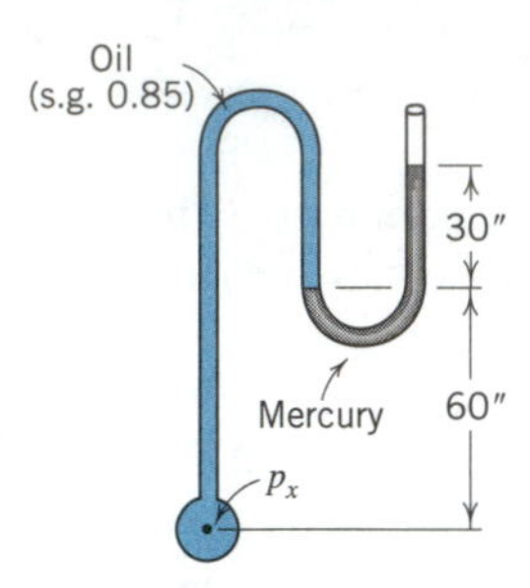

Problem 2.32

2.33. Barometric (absolute) pressure is 91 kPa. Calculate the vapor pressure of the liquid and the gage reading.

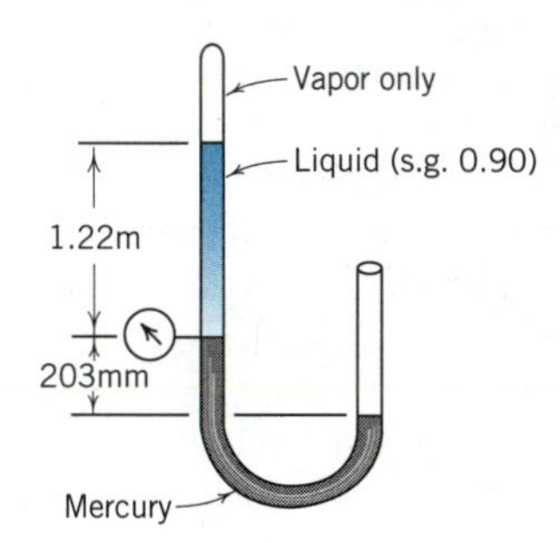

Problem 2.33

2.34. In Fig. 2.7b, $l_1 = 1.27$ m, $h = 0.51$ m, $l_2 = 0.76$ m, liquid γ_1 is water, γ_2 benzene, and γ_3 mercury. Calculate $p_x - p_y$.

2.35. Calculate $p_x - p_y$ for this inverted U-tube manometer.

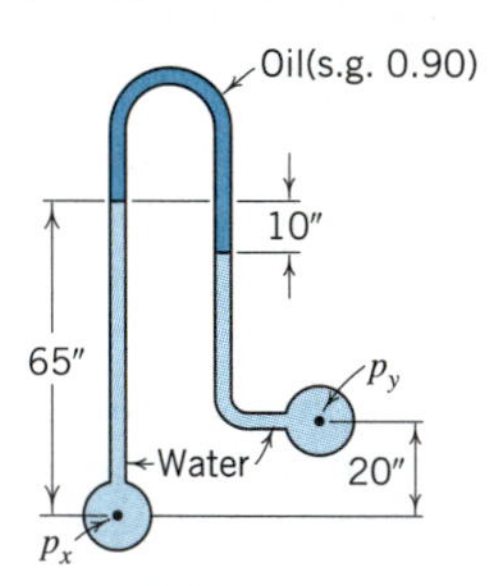

Problem 2.35

2.36. An inclined gage (Fig. 2.7c) having a tube of 3 mm bore, laid on a slope of 1: 20, and a reservoir of 25 mm diameter contains silicon oil (s.g. 0.84). What distance will the oil move along the tube when a pressure of 25 mm of water is connected to the gage?

2.37. The meniscus between the oil and water is in the position shown when $p_1 = p_2$. Calculate the pressure difference $(p_1 - p_2)$ which will cause the meniscus to rise 2 in.

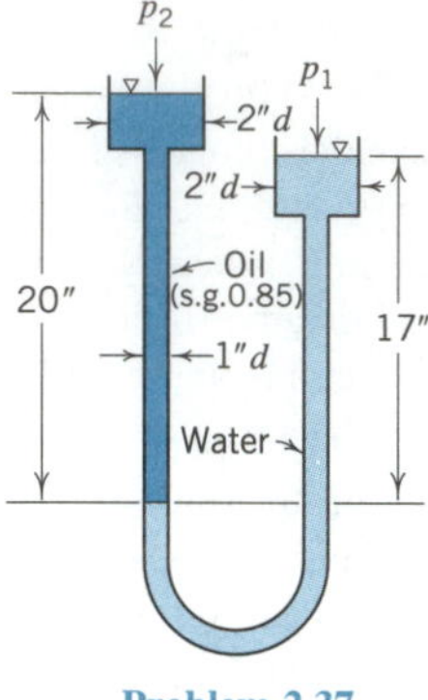

Problem 2.37

2.38. Predict the manometer reading after a 1 N weight is placed on the pan. Assume no leakage or friction between piston and cylinder.

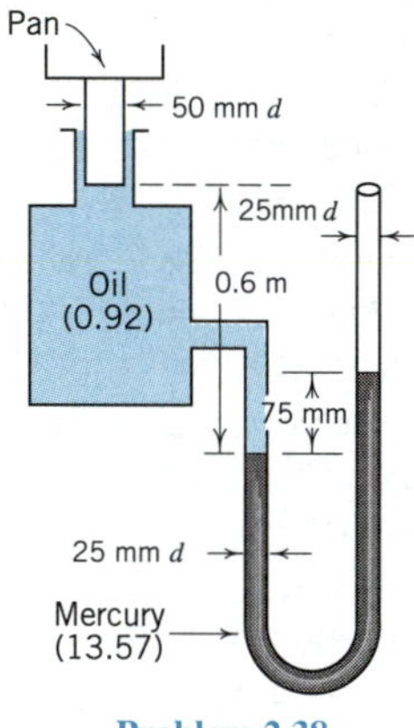

Problem 2.38

2.39. Calculate the gage reading.

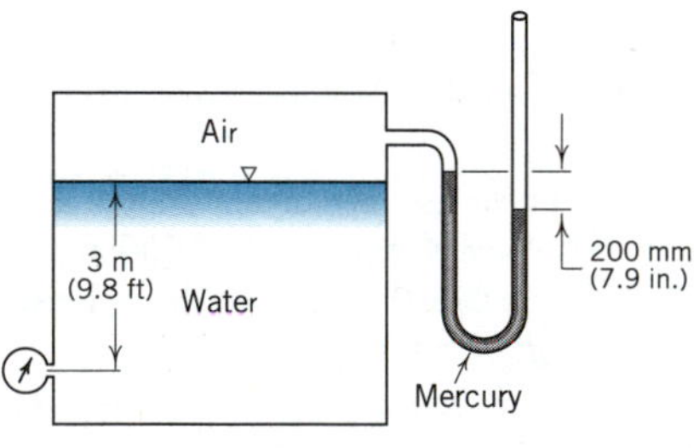

Problem 2.39

2.40. The sketch shows a sectional view through a submarine. Calculate the depth of submergence, y. Assume the specific weight of seawater is 10.0 kN/m^3.

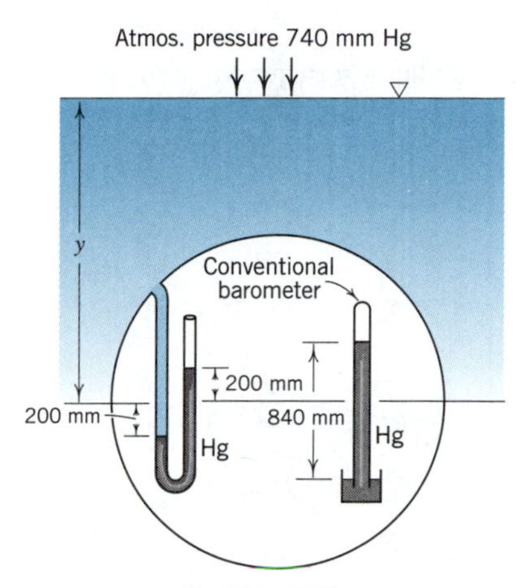

Problem 2.40

2.41. Calculate the gage reading. Specific gravity of the oil is 0.85. Barometric pressure is 755 mm of mercury.

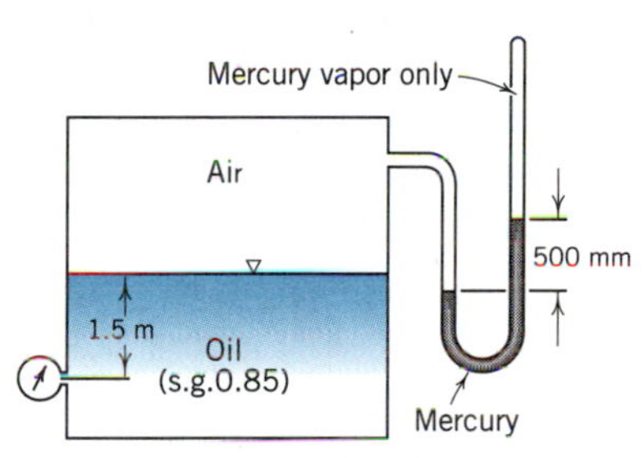

Problem 2.41

2.42. Calculate magnitude and direction of manometer reading when the cock is opened. The tanks are very large compared to the manometer tubes.

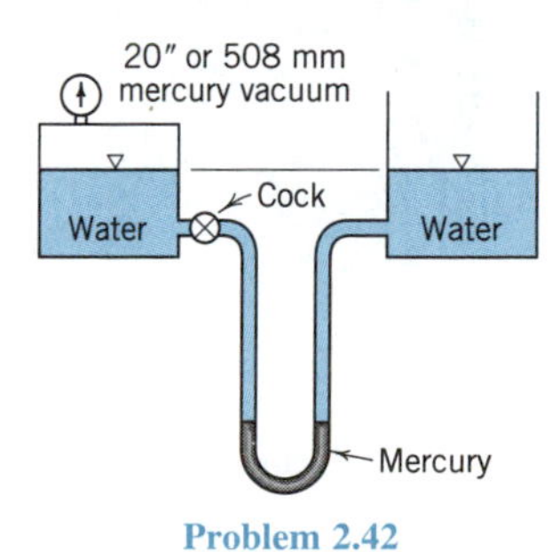

Problem 2.42

2.43. The manometer reading is 6 in. (150 mm) when the tank is empty (water surface at *A*). Calculate the manometer reading when the tank is filled with water.

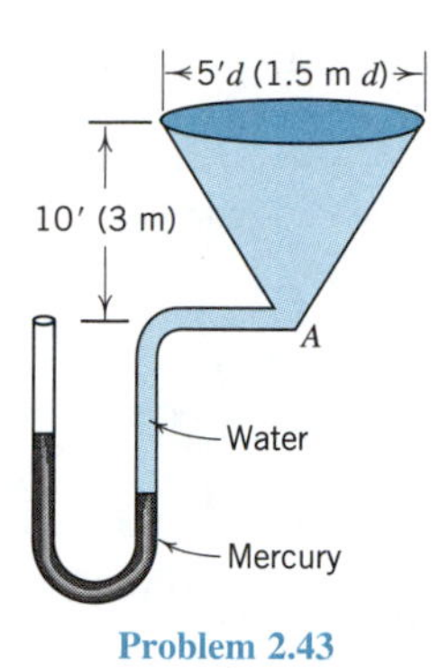

Problem 2.43

2.44. Barometric pressure is 28 in. or 711 mm of mercury. The cock is opened and the air space pumped out so that the gage reads 20 in. or 508 mm of mercury vacuum. Calculate the absolute pressure in the tank and the manometer reading. Neglect change of water surface in the tank.

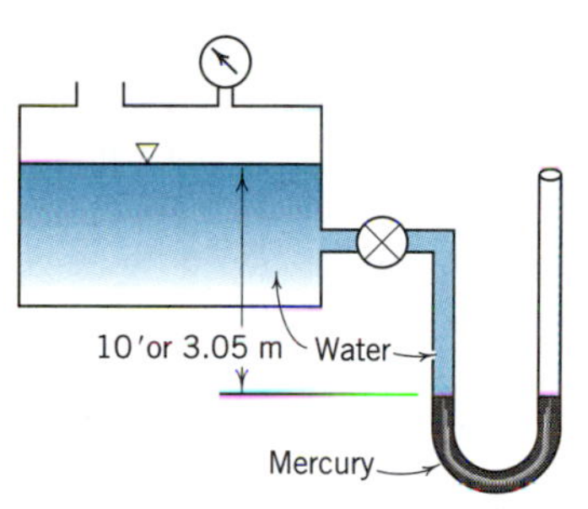

Problem 2.44

2.45. Find the pressure reading of Bourdon gage *A*.

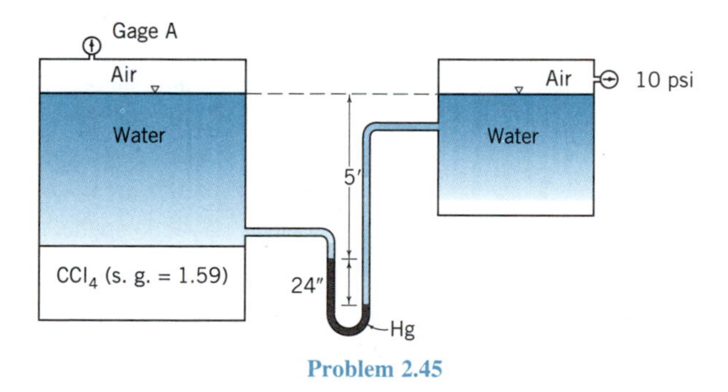

Problem 2.45

2.46. This manometer is used to measure the difference in water level between the two tanks. Calculate this difference.

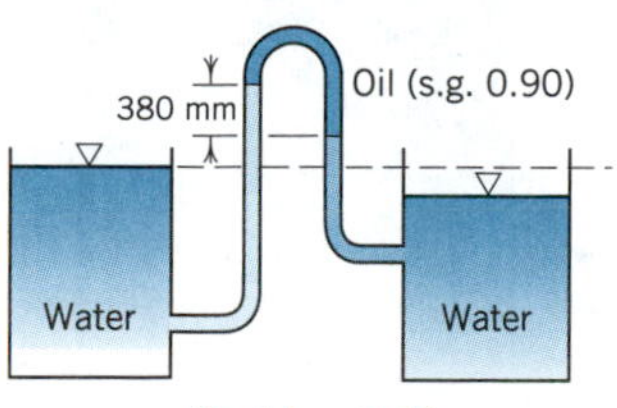

Problem 2.46

2.47. The mercury seal shown is used to support a pressure difference ($p_1 - p_2$) across the rotating disk. Calculate this pressure difference for a speed of 1 000 rpm (r/min), assuming that the mercury rotates with the disk.

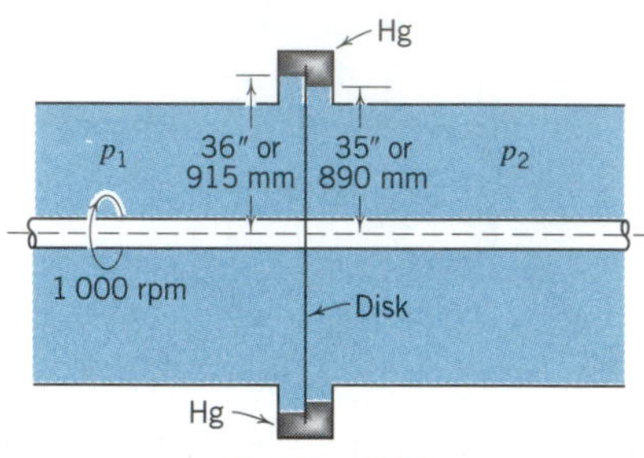

Problem 2.47

2.48. Calculate the magnitude of the force on a 24 in. (0.61 m) diameter glass viewing port in a bathyscaphe on the floor of the Pacific Ocean's Marianas Trench (depth = 35 800 feet or 10.9 km).

2.49. A rectangular gate 6 ft long and 4 ft high lies in a vertical plane with its center 7 ft below a water surface. Calculate magnitude, direction, and location of the total force on the gate.

2.50. A circular gate 3 m in diameter has its center 2.5 m below a water surface and lies in a plane sloping at 60°. Calculate magnitude, direction, and location of total force on the gate.

2.51. An isosceles triangle of 12 ft base and 15 ft altitude is located in a vertical plane. Its base is vertical, and its apex is 8 ft below the water surface. Calculate magnitude and location of the force of the water on the triangle.

2.52. A triangular area of 2 m base and 1.5 m altitude has its base horizontal and lies in a 45° plane with its apex below the base and 2.75 m below a water surface. Calculate magnitude, direction, and location of the resultant force on this area.

2.53. Compute the minimum force P necessary to keep the 3 ft high by 2 ft wide rectangular gate closed.

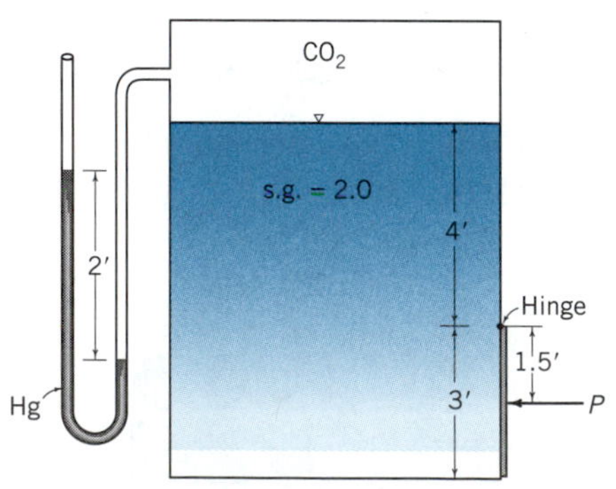

Problem 2.53

2.54. For the situation shown, find the air pressure in the tank in psi. Calculate the force exerted on the gate at the support B if the gate is 10 ft wide. Show a free body diagram of the gate with all the forces drawn in and their points of application located.

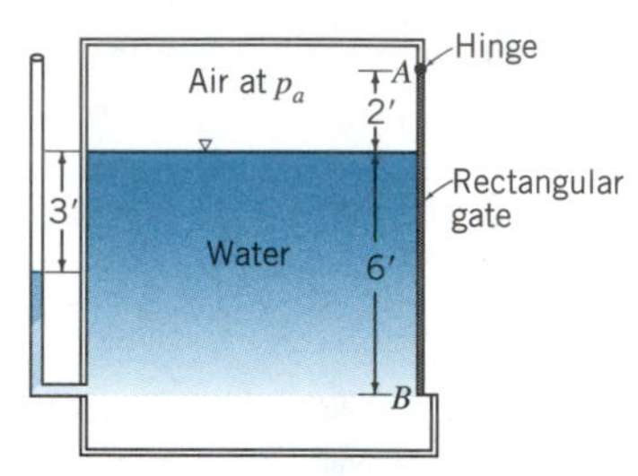

Problem 2.54

2.55. Calculate:

a) the gage pressure at the surface of the oil in psi,

b) the gage pressure at the hinge of the gate in psi,

c) the gage pressure at the bottom of the gate in psi,

d) the total force on the gate,

e) the force P necessary to keep the gate from opening.

The gate is 3 ft wide.

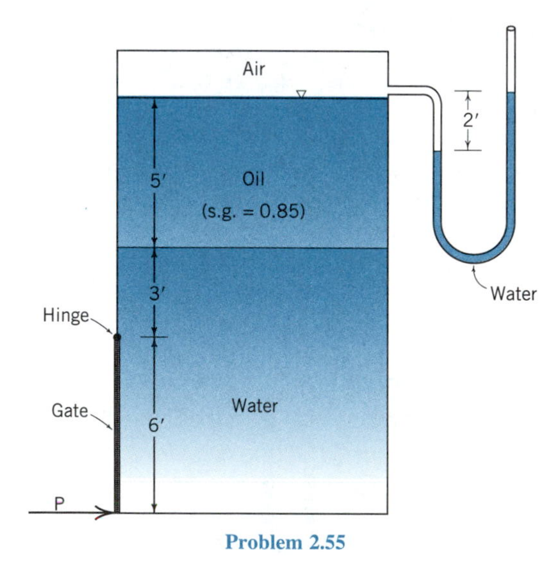

Problem 2.55

2.56. What is the pressure at A? Draw a free body diagram of the gate (10 ft wide) showing all forces and the locations of their lines of action. Calculate the minimum force P necessary to keep the gate closed.

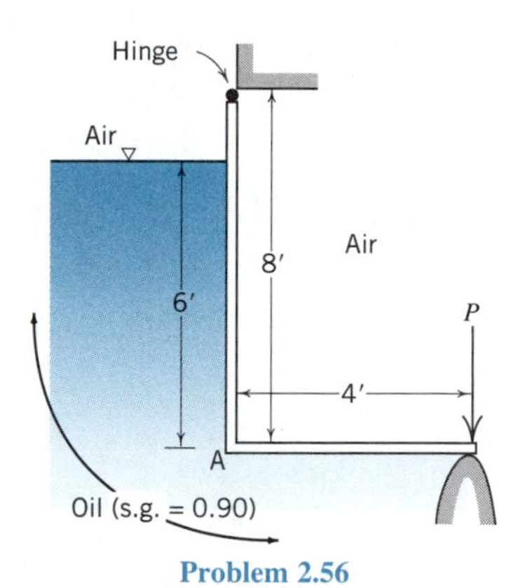

Problem 2.56

2.57. Calculate the minimum force P necessary to hold a uniform 12 ft square gate weighing 500 lb closed on a tank of water under a pressure of 10 psi. Draw a free body of the gate as part of your solution.

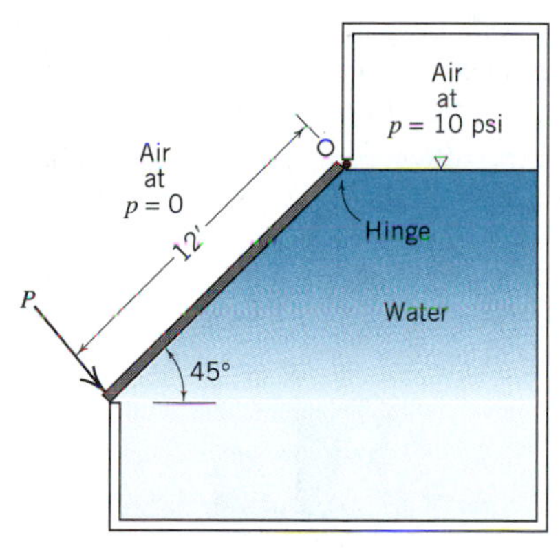

Problem 2.57

2.58. Find the magnitude, direction, and line of action of the force exerted by the liquid on the 6 ft by 10 ft gate.

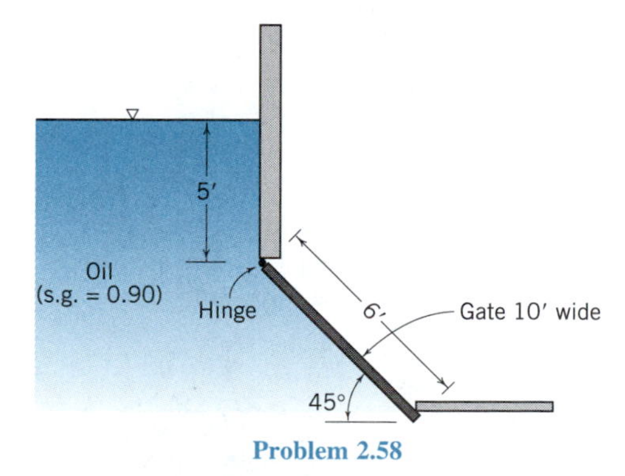

Problem 2.58

2.59. If the specific weight of a liquid varies linearly with depth h according to the equation $\gamma = \gamma_o + Kh$, derive expressions for resultant force per unit width on the rectangular gate and the moment of this force about O.

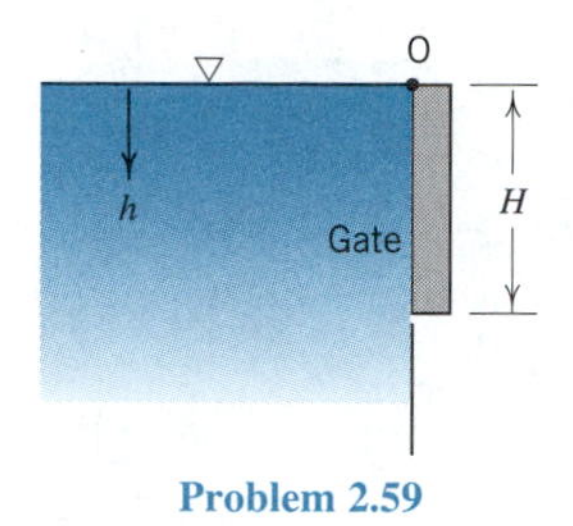

Problem 2.59

2.60. A square 9 ft by 9 ft (2.75 m × 2.75 m) lies in a vertical plane. Calculate the distance between the center of pressure and centroid, and the total force on the square, when its upper edge is (*a*) in the water surface, and (*b*) 50 ft (15 m) below the water surface.

2.61. Calculate the x- and y-coordinates of the center of pressure of this vertical right triangle.

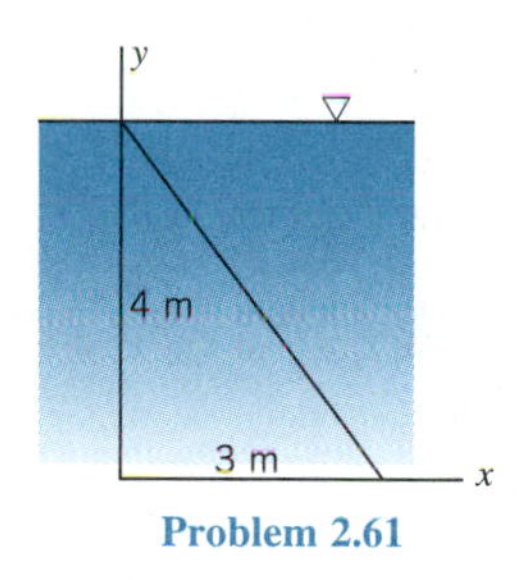

Problem 2.61

2.62. Calculate ($W \times l$) and the exact position of the pointer to hold this triangular gate in equilibrium.

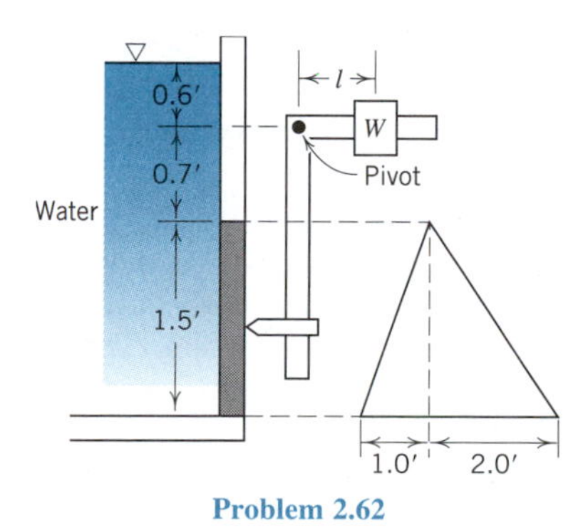

Problem 2.62

2.63. Calculate magnitude and location of the resultant force of water on this annular gate.

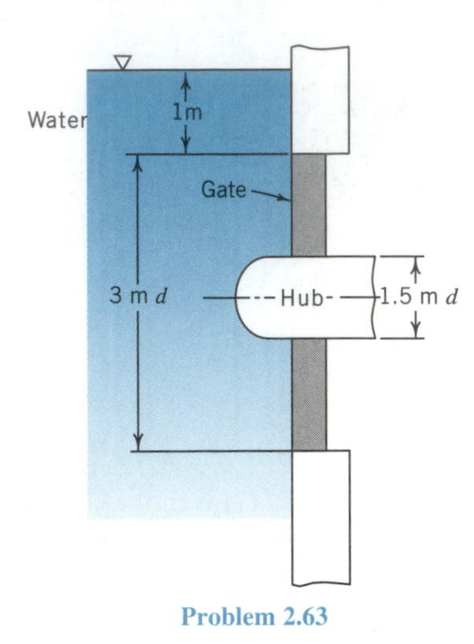

Problem 2.63

2.64. Calculate magnitude and location of the resultant force of the liquid on this tunnel plug.

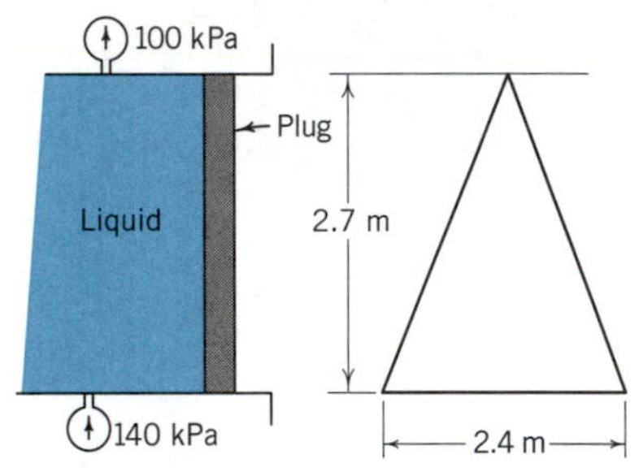

Problem 2.64

2.65. Calculate the force exerted by water on the smallest completely submerged circle (located in a vertical plane) having a distance of 1 ft or 1 m between its centroid and center of pressure.

2.66. A vertical rectangular gate 10 ft (3 m) high and 6 ft (1.8 m) wide has a depth of water on its upper edge of 15 ft (4.5 m). What is the location of a horizontal line which divides this area (*a*) so that the forces on the upper and lower portions are the same, and (*b*) so that the moments of the forces about the line are the same?

2.67. A horizontal tunnel of 10 ft (3 m) diameter is closed by a vertical gate. Calculate magnitude, direction, and location of the total force of water on the gate when the tunnel is (*a*) one-half full, (*b*) one-fourth full, and (*c*) three-fourths full.

2.68. Calculate the magnitude, direction, and location of the force of the water on one side of this area located in a vertical plane.

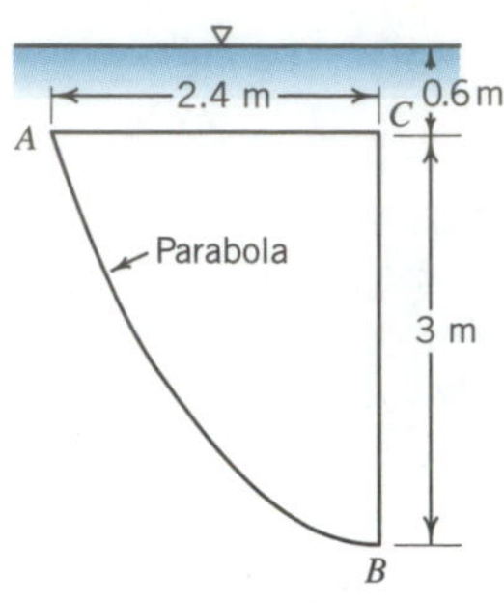

Problem 2.68

2.69. A vertical rectangular gate 2.4 m wide and 2.7 m high is subjected to water pressure on one side, the water surface being at the top of the gate. The gate is hinged at the bottom and is held by a horizontal chain at the top. What is the tension in the chain?

2.70. A sliding gate 10 ft wide and 5 ft high situated in a vertical plane has a coefficient of friction, between itself and guides, of 0.20. If the gate weighs 2 tons and if its upper edge is at a depth of 30 ft, what vertical force is required to raise it? Neglect buoyancy force on the gate.

2.71 A butterfly valve, consisting essentially of a circular area pivoted on a horizontal axis through its center (and in the plane of the valve), is 2.1 m (6.9 ft) in diameter and lies in a 60° plane with its center 3 m (9.8 ft) below a water surface. What torque must be exerted on the valve's axis to just open it?

2.72. Calculate the h at which this gate will open.

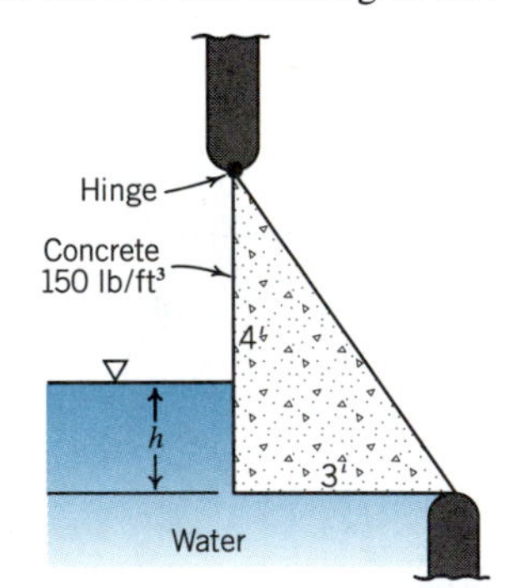

Problem 2.72

2.73. A solid homogeneous wooden block, with $\gamma = 50$ lb/ft^3, is 5 ft long normal to the paper. The block is anchored by a cable. Calculate the tensile force in the cable.

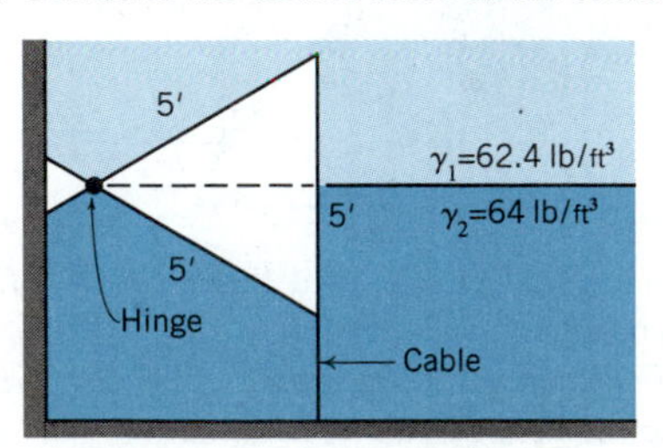

Problem 2.73

2.74. These (rectangle) miter gates at the entrance to a canal lock are 4.5 m high. Calculate the reactions at the hinges when the water surface is 0.9 m below the top of the gates.

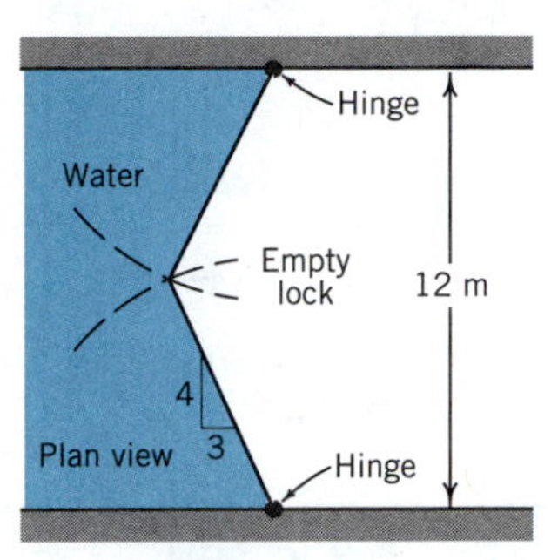

Problem 2.74

2.75. Derive an algebraic expression for the force in the wire.

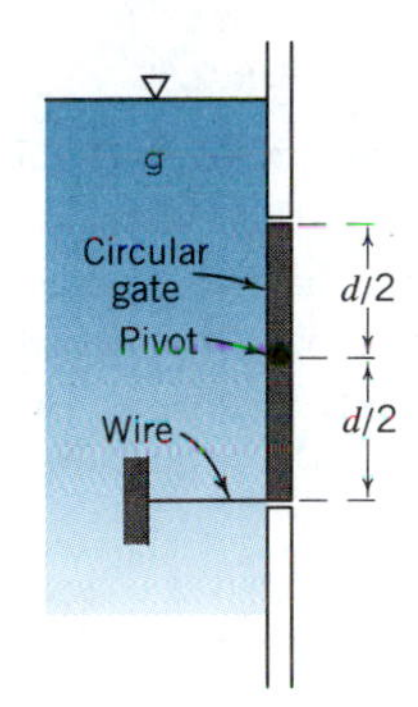

Problem 2.75

2.76. The flashboards on a spillway crest are 1.2 m high and supported on steel posts spaced 0.6 m on centers. The posts are designed to fail under a bending moment of 5 500 N · m. What depth over the flashboards will cause the posts to fail? Assume hydrostatic pressure distribution.

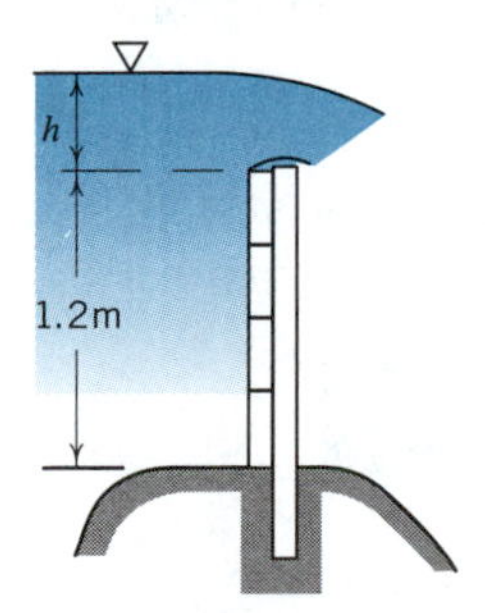

Problem 2.76

2.77. This rectangular gate will open automatically when the depth of water, d, becomes large enough. What is the minimum depth that will cause the gate to open?

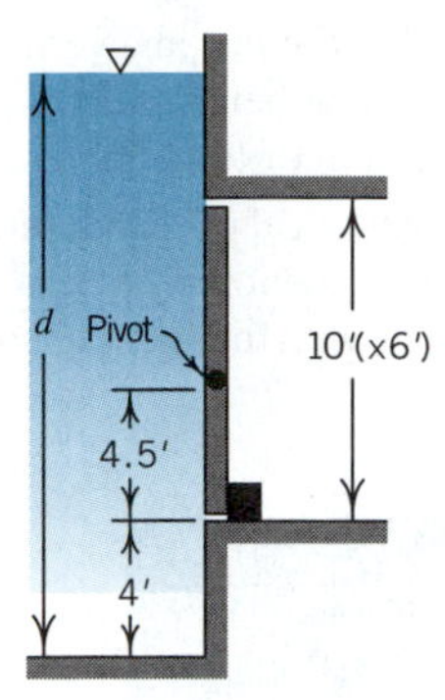

Problem 2.77

2.78. What depth of water will cause this rectangular gate to fall? Neglect the weight of the gate.

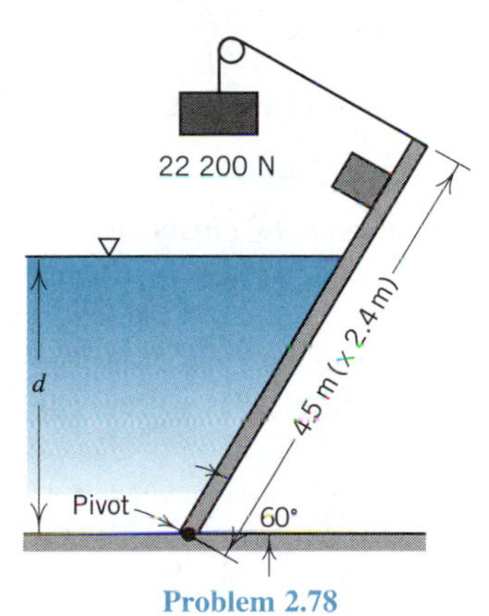

Problem 2.78

2.79. Calculate magnitude and location of the total force on one side of this vertical plane area.

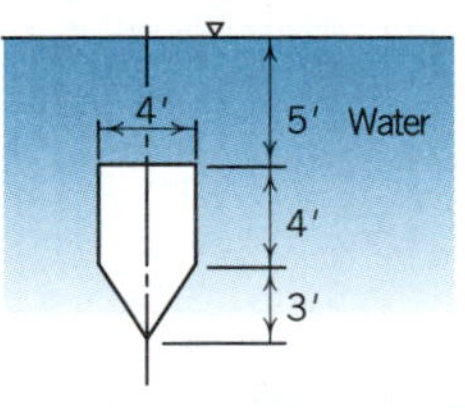

Problem 2.79

2.80. Calculate magnitude and location of the total force on one side of this vertical plane area.

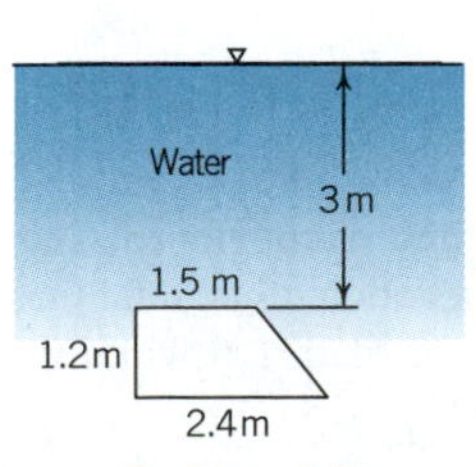

Problem 2.80

2.81. A rectangle 3 m by 4 m lies in a vertical plane with one diagonal horizontal and 7 m below the water surface. Calculate magnitude and location of total force on the rectangle.

2.82. This area lies in a vertical plane beneath a water surface. Calculate magnitude and location of the force of the water on the area. Compare results with those of the preceding problem.

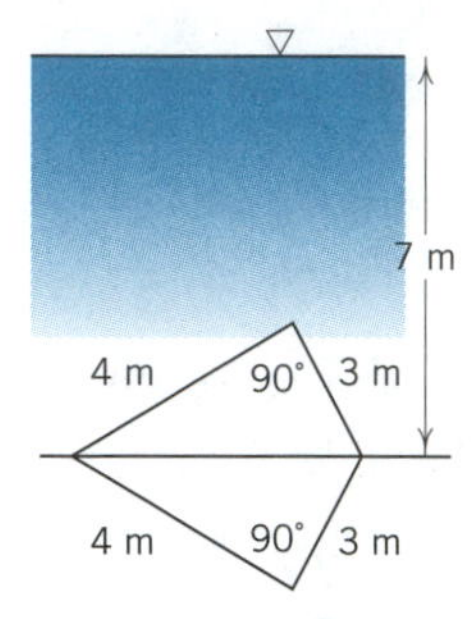

Problem 2.82

2.83. Calculate the magnitude, direction, and location of the total force on the gate of problem 2.67 when on one side the water surface is 18 ft (5.5 m) above the tunnel invert and on the other side 5 ft (1.5 m) above the tunnel invert. (The ''invert'' is the low point of the tunnel cross section.)

2.84. A vertical gate in a tunnel 6 m in diameter has water on one side and air on the other. The water surface is 10.5 m above the invert, and the air pressure is 110 kPa. Where could a single support be located to hold this gate in position?

2.85. A rectangular tank 5 ft wide, 6 ft high, and 10 ft long contains water to a depth of 3 ft and oil (s.g. = 0.85) on the water to a depth of 2 ft. Calculate magnitude and location of the force on one end of the tank.

2.86. Using force components, calculate the load in the strut *AB* if these struts have 1.5 m spacing along the small dam *AC*. Consider all joints to be pin connected.

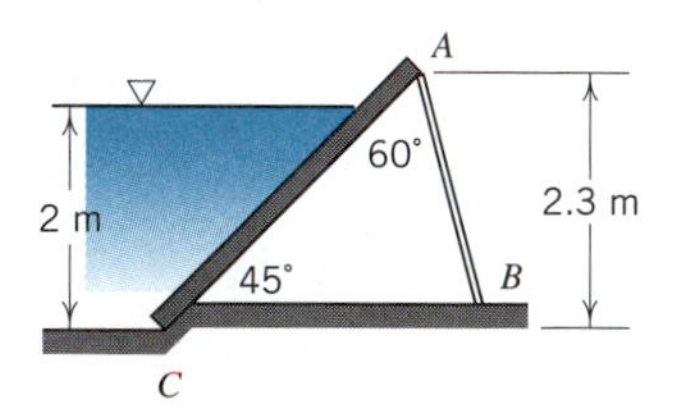

Problem 2.86

2.87. Using the method of components, calculate the magnitude, direction, and location of the total force on the upstream face of a section of this dam 1 ft wide. What is the moment of this force about *O*?

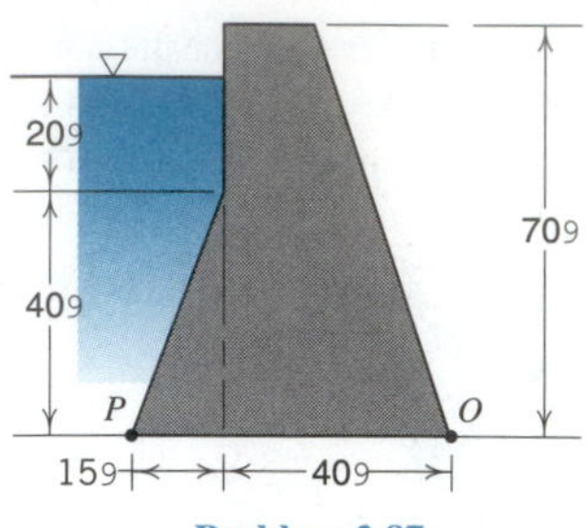

Problem 2.87

2.88. Using hydrostatic principles (not geometry or calculus), calculate the volume *BCDE* (m^3).

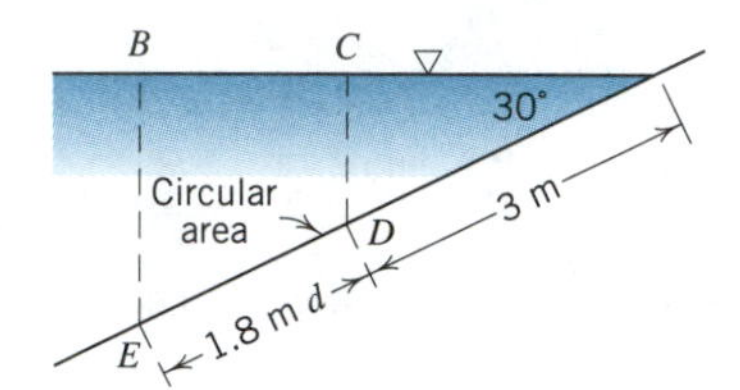

Problem 2.88

2.89. Calculate the magnitude of the total force in problem 2.52 by the method of components.

2.90. A concrete pedestal, having the shape of the frustum of a right pyramid of lower base 4 ft square, upper base 2 ft square, and height 3 ft, is to be poured. Taking γ for concrete to be 150 lb/ft^3, calculate the vertical force of uplift on the forms.

2.91. This tainter gate is pivoted at *O* and is 10 m long. Calculate the magnitudes of horizontal and vertical components of force on the gate. The pivot is at the same level as the water surface.

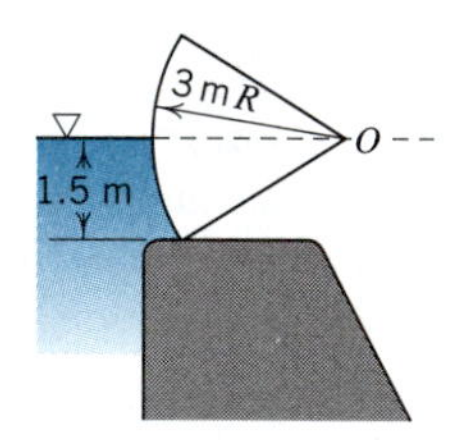

Problem 2.91

2.92. The quarter cylinder *AB* is 10 ft long. Calculate magnitude, direction, and location of the resultant force of the water on *AB*.

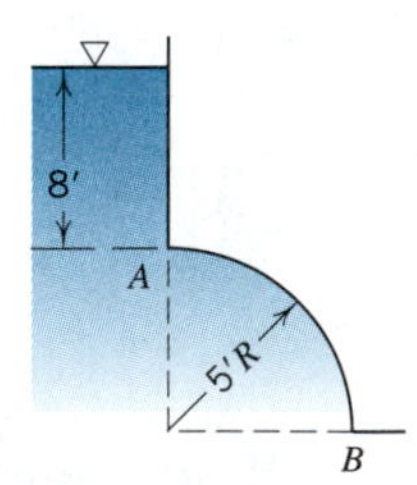

Problem 2.92

2.93. Calculate the vertical force exerted by the liquid on this semicylindrical dome *AB*, which is 1.5 m long.

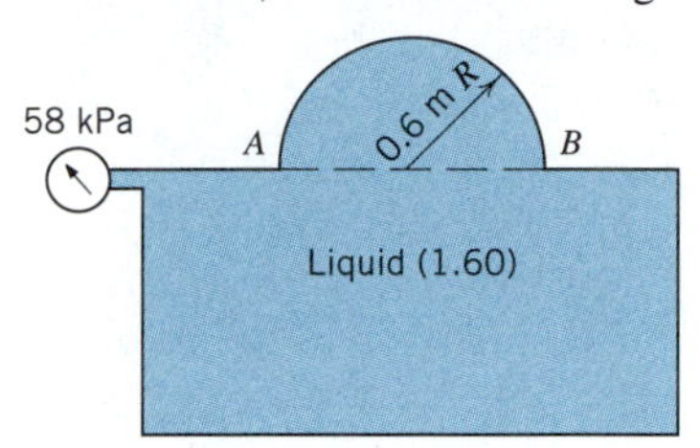

Problem 2.93

2.94. If this weightless quarter-cylindrical gate is in static equilibrium, what is the ratio between γ_1 and γ_2?

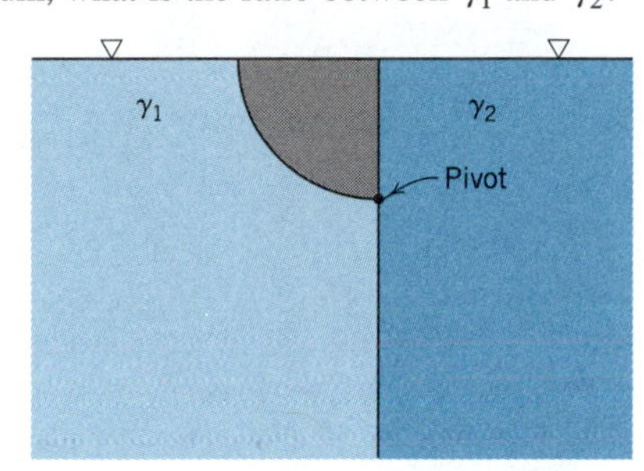

Problem 2.94

2.95. Calculate the moment about *O* of the resultant force exerted by the water on this half cylinder, which is 3 m (10 ft) long.

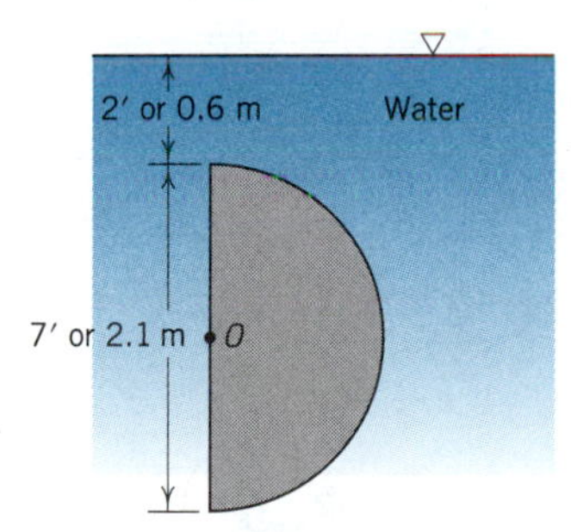

Problem 2.95

2.96. Calculate magnitude and direction of the resultant forces exerted by the water (*a*) on the end of the cylinder and (*b*) on the curved surface of the cylinder.

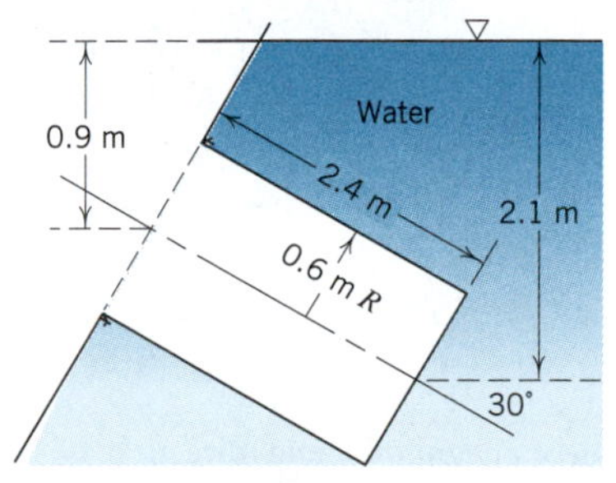

Problem 2.96

2.97. Solve problem 2.92, assuming that there is also water to the right of *AB* with a surface level 3 ft above *A*.

2.98. The cylinder is 2.4 m long and is pivoted at *O*. Calculate the moment (about *O*) required to hold it in position.

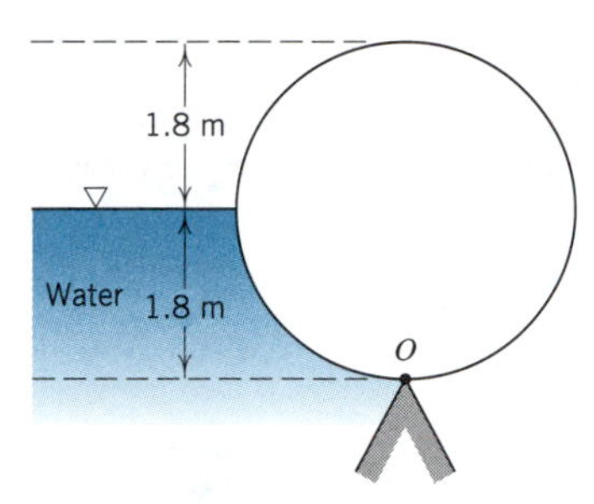

Problem 2.98

2.99. If this solid concrete (150 lb/ft^3) overhang *ABCD* is added to the dam, what *additional* force (magnitude and direction) will be exerted on the dam?

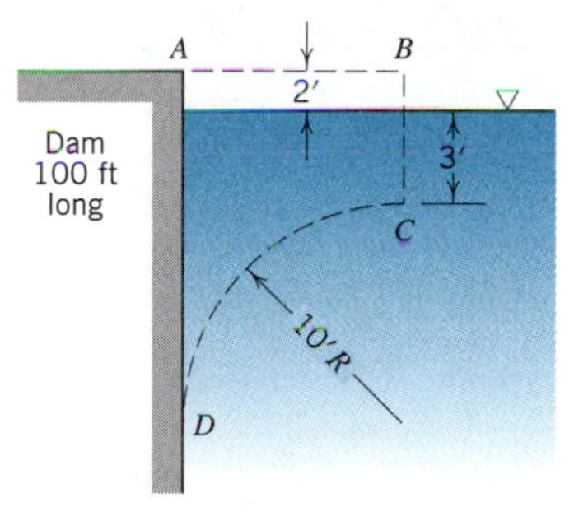

Problem 2.99

2.100. Calculate the magnitude, direction (horizontal and vertical components are acceptable), and line of action of the resultant force exerted by the water on the cylindrical gate 30 ft long.

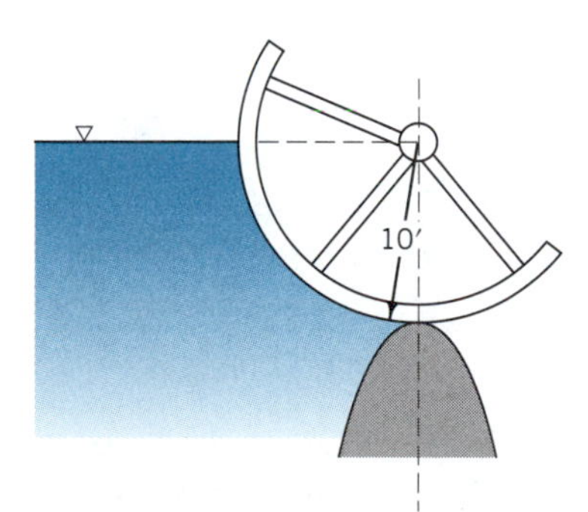

Problem 2.100

2.101. If the liquid shown above *B* in problem 2.92 consists of a 7 ft layer of water on top of a 6 ft layer of carbon tetrachloride, calculate magnitude, direction, and location of horizontal and vertical components of force on *AB*.

2.102. A hemispherical shell 1.2 m in diameter is connected to the vertical wall of a tank containing water. If the center of the shell is 1.8 m below the water surface, what are the vertical and horizontal force components on the shell? On the top half of the shell?

2.103. This half-conical buttress is used to support a half-cylindrical tower on the upstream face of a dam. Calculate the magnitude, direction, and location of the vertical and horizontal components of force exerted by the water on the buttress (*a*) when the water surface is at the base of the half cylinder, and (*b*) when it is 4 ft (1.2 m) above this point.

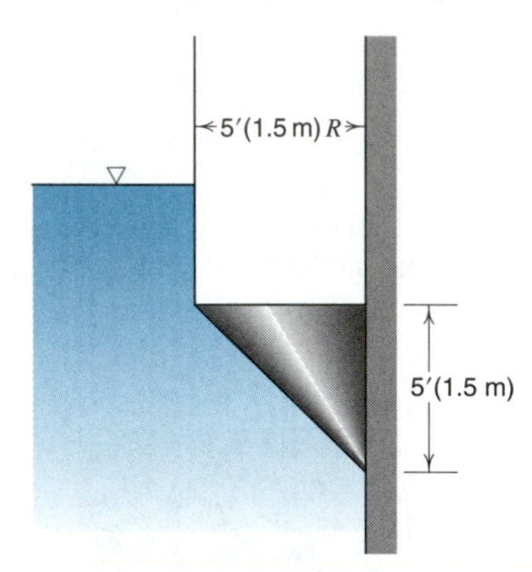

Problem 2.103

2.104. A hole 300 mm in diameter in a vertical wall between two water tanks is closed by a sphere 450 mm in diameter in the tank of higher water-surface elevation. The difference in the water-surface elevations in the two tanks is 1.5 m. Calculate the horizontal component of force exerted by the water on the sphere.

2.105. What pressure difference $p_1 - p_2$ is required to open the ball valve if the spring exerts a force of 400 N? The ball is 50 mm in diameter, while the hole in which it rests is 30 mm in diameter. Neglect the weight of the steel ball.

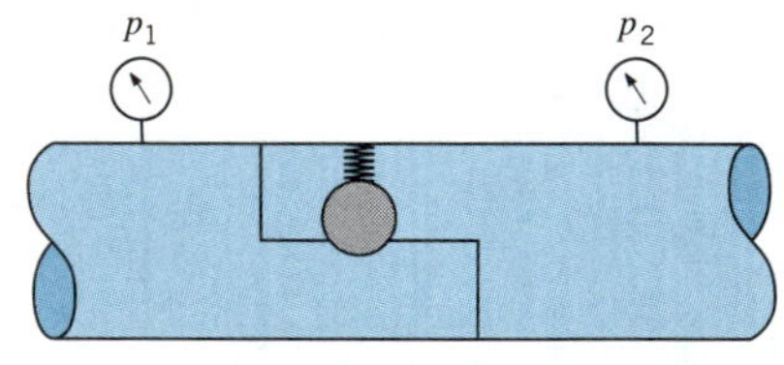

Problem 2.105

2.106. This one-eighth wooden sphere of specific weight 50 lb/ft^3 is placed in the corner of an open rectangular tank, the edges of which coincide with the *x*, *y*, and *z* axes. The joints along the lines *AB*, *BC*, and *AC* are sealed perfectly so that water cannot enter. Water is then poured into the tank to a depth of 7 ft. Calculate the magnitude, direction, and location of the resultant force (or its components) exerted by the water on the spherical surface.

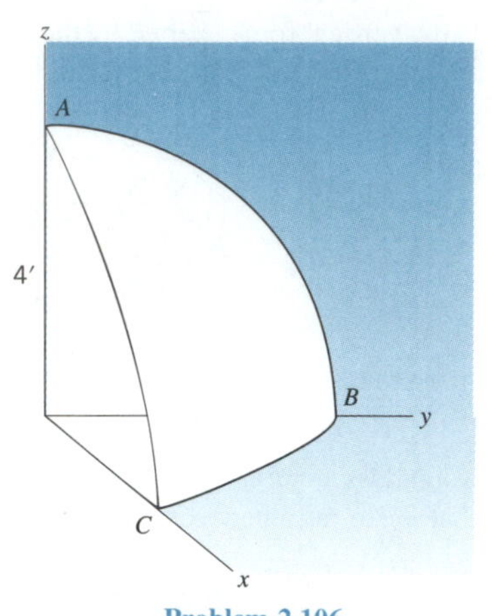

Problem 2.106

2.107. If the half cone of problem 2.103 is replaced by a quarter sphere, what are the magnitudes of the vertical and horizontal force components on the buttress?

2.108. This weightless spherical shell with attached *small* piezometer tube is suspended by a cable as shown. Calculate the total tension force in the cable. Also calculate the force by the liquids: (*a*) on the bottom half of the sphere, and (*b*) on the top half of the sphere.

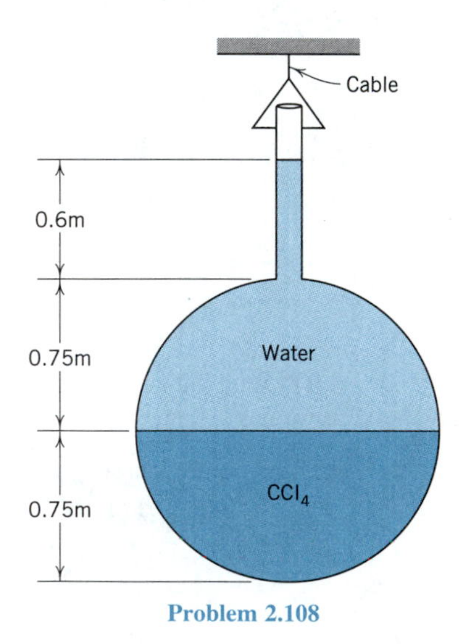

Problem 2.108

2.109. Calculate magnitude and direction of the resultant force of the water on this solid conical plug.

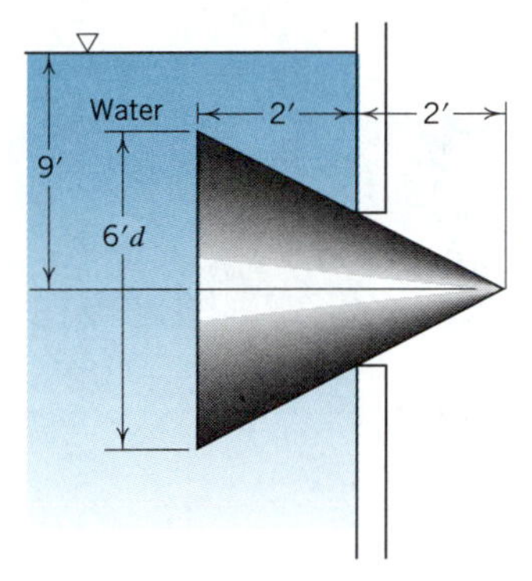

Problem 2.109

2.110. The tank is an elliptic cylinder of 2.4 m length. Calculate magnitude and direction of the vertical and horizontal components of force on AB.

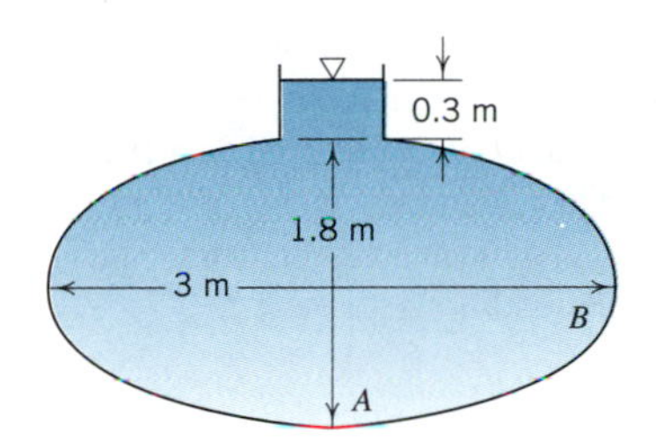

Problem 2.110

2.111. Calculate the magnitude and location of the resultant force of the liquids on the hemispherical end of this cylindrical tank.

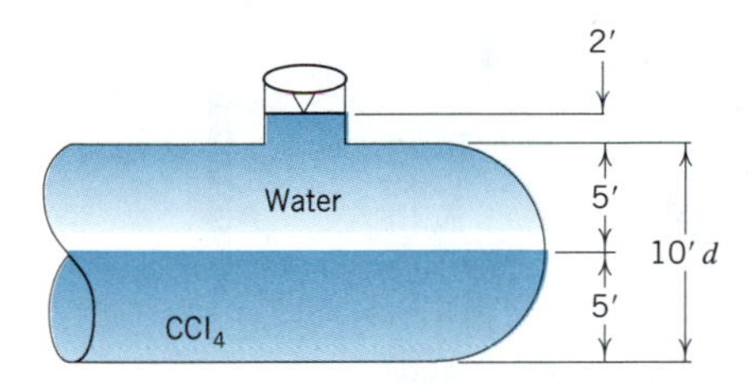

Problem 2.111

2.112. A stone weighs 60 lb or 267 N in air and 40 lb or 178 N in water. Calculate its volume and specific gravity.

2.113. Ninety-eight millilitres of lead (s.g. 11.4) are suspended from the apex of a conical can having a height of 0.3 m and a base of 0.15 m diameter, and weighing 4 N. When placed in water, to what depth will the can be immersed? The apex is below the base.

2.114. A cylindrical can 76 mm in diameter and 152 mm high, weighing 1.11 N, contains water to a depth of 76 mm. When this can is placed in water, how deep will it sink?

2.115. If the 10-ft-long box is floating on the oil-water system, calculate how much the box and its contents must weigh.

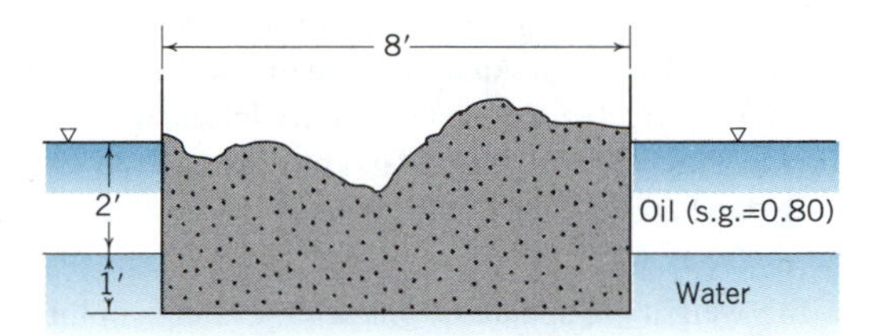

Problem 2.115

2.116. The timber weighs 40 lb/ft^3 and is held in a horizontal position by the concrete (150 lb/ft^3) anchor. Calculate the minimum total weight which the anchor may have.

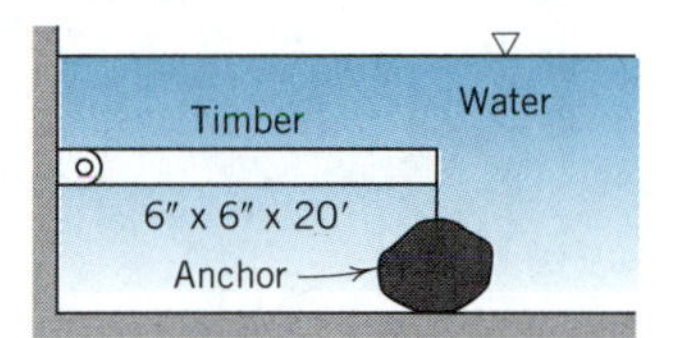

Problem 2.116

2.117. If the timber weighs 670 N, calculate its angle of inclination when the water surface is 2.1 m above the pivot. Above what depth will the timber stand vertically?

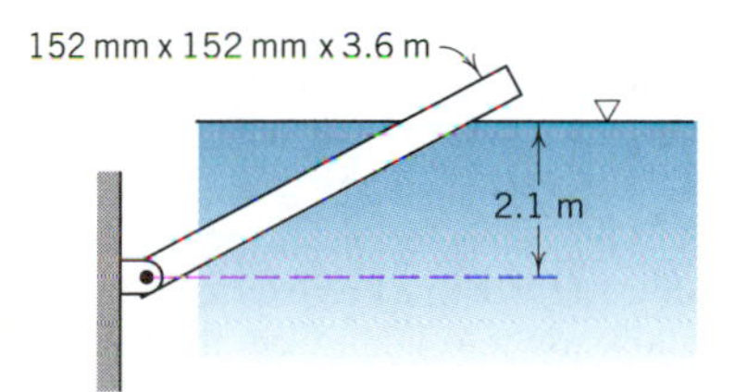

Problem 2.117

2.118. This homogeneous wooden semicylinder is 3.6 m long and floats on a water surface as shown. What moment, M, would be required to move point A to coincide with the water surface?

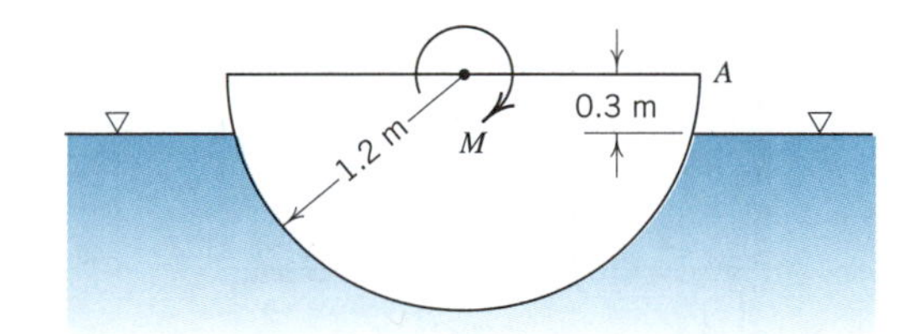

Problem 2.118

2.119. The barge shown weighs 40 tons and carries a cargo of 40 tons. Calculate its draft in freshwater.

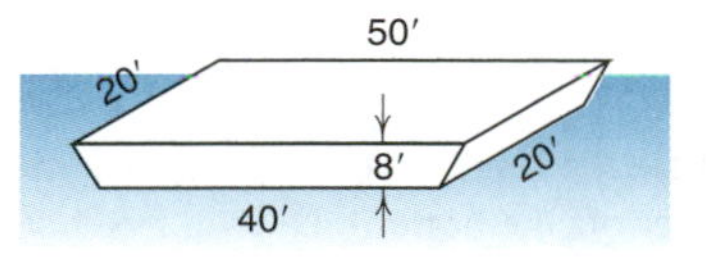

Problem 2.119

2.120. A modern, half-million tonne (1 tonne = 10^3 kg) supertanker is roughly rectangular in shape, with a length of 425 m, a width of 67 m, and a draft when fully loaded of 26 m. If the tanker carries 600 Ml (megaliters) of oil (s.g. 0.86) when fully loaded, what is its draft when empty? The specific gravity of seawater is 1.03.

2.121. The weightless sphere of diameter d is in equilibrium in the position shown. Calculate d as a function of γ_1, h_1, γ_2, and h_2.

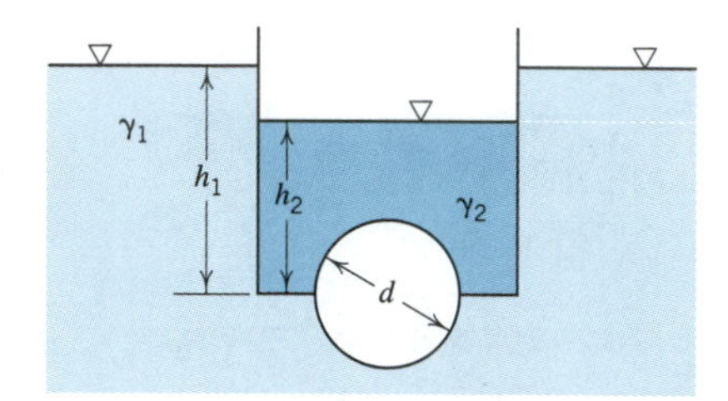

Problem 2.121

2.122. The opening in the bottom of the tank is square and slightly less than 2 ft on each side. The opening is to be plugged with a wooden cube 2 ft on a side.

a) What weight W should be attached to the wooden cube to insure successful plugging of the hole? The wood weighs 40 lb/ft^3.

b) What upward force must be exerted on the block to lift it and allow water to drain from the tank?

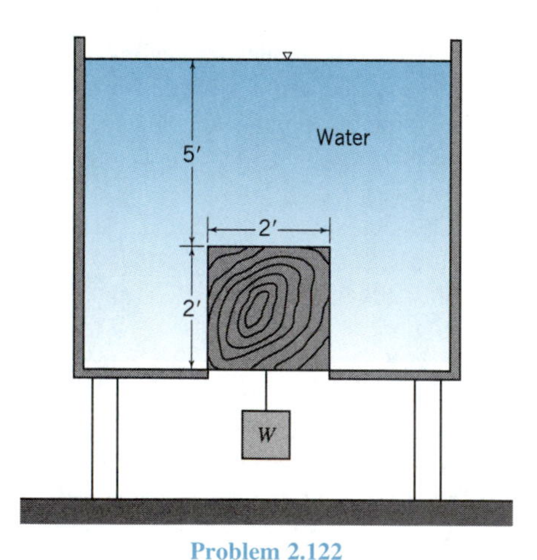

Problem 2.122

2.123. Calculate the force P necessary to lift the 4-ft diameter 400-lb container off the bottom. Draw a free body of the container to assist in the solution.

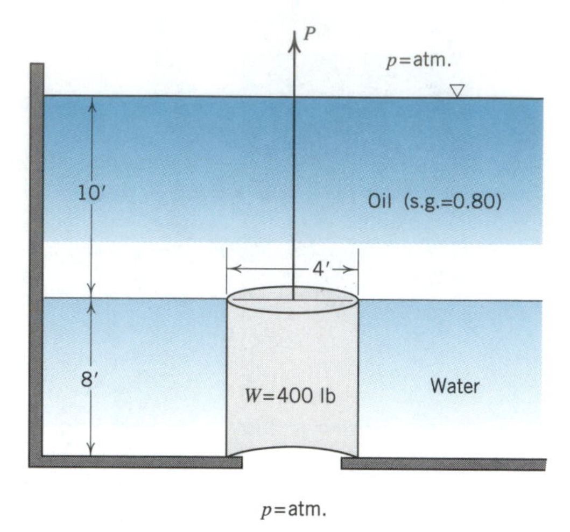

Problem 2.123

2.124. A hollow cylinder weighing 200 lb and having a metal collar around its equator is being used as a check valve in the system shown. What additional weight W_o must be added to the cylinder to just prevent leakage from reservoir 2 to reservoir 1 under the conditions shown?

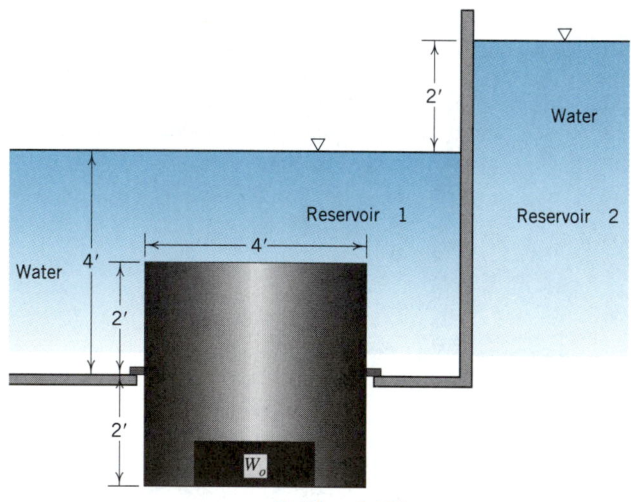

Problem 2.124

2.125. To what depth will a rigid homogeneous object (γ = 10.4 kN/m^3) sink in the ocean if the weight density therein varies with depth according to the empirical relation $\gamma(\text{kN/m}^3) = 10.0 + 0.008\,1\sqrt{\text{depth (m)}}$.

2.126. A balloon having a total solid weight of 800 lb contains 15 000 ft^3 of hydrogen. How many pounds of ballast are necessary to hold the balloon on the ground? Barometric pressure is 14.7 psia; temperature of air and hydrogen, 60°F. Assume hydrogen to be at barometric pressure.

2.127. A balloon has a weight (including crew but not gas) of 2.2 kN and a gas-bag capacity of 566 m^3. At the ground it is (partially) inflated with 445 N of helium. How high can this balloon rise in the U.S. Standard Atmosphere (Appendix 2) if the helium always assumes the pressure and temperature of the atmosphere?

2.128. An open cylindrical container holding 1.0 m^3 of water at a depth of 2 m is accelerated vertically upward at 6 m/s^2. Calculate the pressure and total force on the bottom of the container. Also calculate this total force by application of Newton's second law.

2.129. Calculate the total forces on the ends and bottom of this container while at rest and when being accelerated vertically upward at 3 m/s^2. The container is 2 m wide.

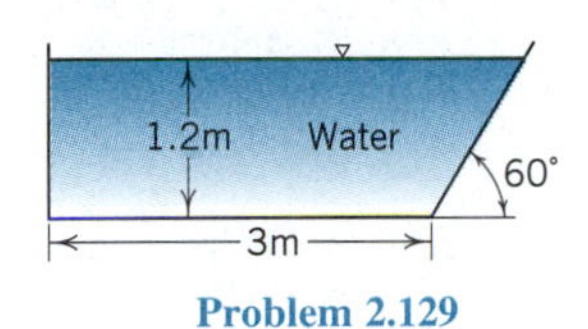

Problem 2.129

2.130. An open conical container 6 ft (1.8 m) high is filled with water and moves vertically downward with a deceleration of 10 ft/s^2 (3 m/s^2). Calculate the pressure at the bottom of the container.

2.131. A rectangular tank 1.5 m wide, 3 m long, and 1.8 m deep contains water to a depth of 1.2 m. When it is accelerated horizontally at 3 m/s^2 in the direction of its length, calculate the depth of water at each end of the tank and the total force on each end of the tank. Check the difference between these forces by calculating the inertia force of the accelerated mass. Repeat these calculations for an acceleration of 6 m/s^2.

2.132. A piece of cork weighing 4 lb and having a volume of 1.0 ft^3 is suspended from a wire attached to the bottom of the container. If the container is accelerated downward at 10 ft/s^2, what is the tension force in the wire?

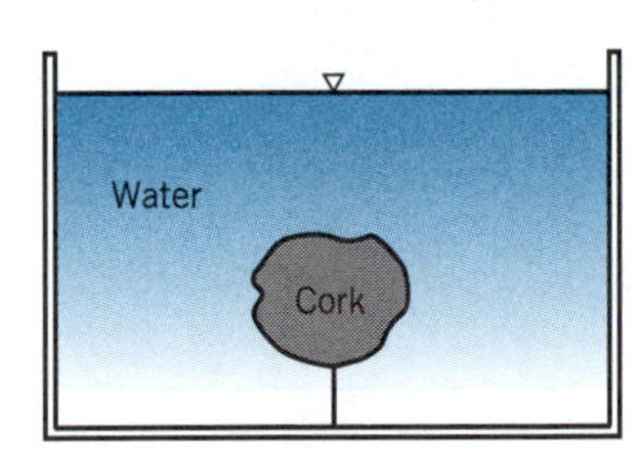

Problem 2.132

2.133. A closed rectangular tank 4 ft high, 8 ft long, and 5 ft wide is three-fourths full of gasoline and the pressure in the air space above the gasoline is 20 psi. Calculate the pressures in the corners of this tank when it is accelerated horizontally along the direction of its length at 15 ft/s^2. Using Newton's second law, calculate the forces on the ends of the tank and check their difference.

2.134. An open container of liquid accelerates down a 30° inclined plane at 16.4 ft/s^2 or 5 m/s^2. What is the slope of its free surface? What is the slope for the same acceleration up the plane?

2.135. This U-tube containing water is accelerated horizontally to the right at 3 m/s^2; what are the pressures at *A*, *B*, and *C*? Repeat the calculation for mercury. Assume that the tubes are long enough that no liquid is spilled.

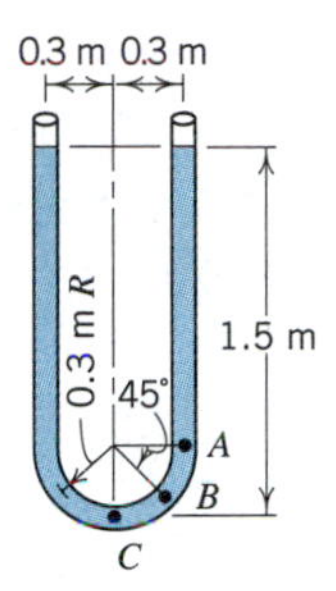

Problem 2.135

2.136. An open cylindrical tank 1 m in diameter and 1.5 m deep is filled with water and rotated about its axis at 100 r/min. How much liquid is spilled? What are the pressures at the center of the bottom of the tank and at a point on the bottom 0.3 m from the center? What is the resultant force exerted by the water on the bottom of the tank?

2.137. The tank of problem 2.136 contains water to a depth of 0.9 m. What will be the depth at the wall of the tank when the tank is rotated at 60 r/min?

2.138. The tank of problem 2.136 contains water to a depth of 0.3 m. At what speed must it be rotated to uncover a bottom area 0.3 m in diameter?

2.139. A vertical cylindrical tank 1.5 m high and 0.9 m in diameter is filled with water to a depth of 1.2 m. The tank is then closed and the pressure in the space above the water raised to 69 kPa. Calculate the pressure at the intersection of wall and tank bottom when the tank is rotated about a central vertical axis at 150 r/min. Also calculate the resultant force exerted by the water on the bottom of the tank.

2.140. The impeller of a closed filled centrifugal water pump is rotated at 1 750 r/min (or rpm). If the impeller is 1 m (3.3 ft) in diameter, what pressure is developed by rotation?

2.141. When the U-tube of problem 2.135 is rotated at 200 rpm (r/min) about its central axis, what are the pressures at points *A*, *B*, and *C*?

CHAPTER

3 KINEMATICS OF FLUID MOTION

The objectives of this chapter are to treat the kinematics of somewhat idealized fluid motion along streamlines and in flowfields. As in particle mechanics, kinematics describes motion in terms of displacements, velocities, and accelerations without regard to the forces that cause the motion. The important distinction between steady and unsteady flows is made, and the groundwork is laid for our derivation in later chapters of key dynamic equations. However, no attempt is made here to describe the kinematics of turbulence or of the motion of large-scale eddies which the reader has no doubt observed in real fluid flows. These topics are discussed later.

3.1 STEADY AND UNSTEADY FLOW, STREAMLINES, AND STREAMTUBES

There are two basic means of describing the motion of a fluid. In the *Eulerian* view, attention is focused on *particular points* in the space filled by the fluid. A description is then given of the state of fluid motion at each point as a function of time. The values and variations with time of the fluid velocity, density, pressure, acceleration and other fluid variables are determined (and perhaps recorded by instruments of an experiment) at various spatial points. We write then the velocity, acceleration, and pressure for example as

$$\mathbf{v} = \mathbf{v}(x, y, z, t);\ \mathbf{a} = \mathbf{a}(x, y, z, t);\ \text{and}$$

$$p = p(x, y, z, t),\ \text{where the velocity}\ \mathbf{v} = u\mathbf{e}_x + v\mathbf{e}_y + w\mathbf{e}_z;$$

here u, v, and w are the components of the velocity in the directions of the Cartesian coordinates (x, y, z) and $(\mathbf{e}_x, \mathbf{e}_y, \mathbf{e}_z)$ are the unit vectors along those directions [see Appendix 6].

In the *Lagrangian* view, each fluid particle is labeled (usually by its spatial coordinates at some initial time). Then, the path (i.e., the record of the coordinates of the particle at later times), density, velocity, and other characteristics of each *individual fluid particle* are traced as time passes. This view is the one that is used in the dynamic analyses of solid particles. If the position of a fluid particle is plotted as a function of time, the result is the trajectory of the particle, called a *path line*. In the Lagrangian view the particle velocity would be defined as the time derivative of its position, viz., $\mathbf{v} = \mathbf{v}(\mathbf{x}_0, t)$, where $\mathbf{x}_0$ is the initial spatial location (x_0, y_0, z_0) of the particle. The path line is clearly tangent to the instantaneous velocity at each point along the path and so the changes in the particle location over an infinitesimally small time are given by $dx = u\,dt$, $dy = v\,dt$ and $dz = w\,dt$. This means then that

$$u = \frac{dx}{dt}, \; v = \frac{dy}{dt}, \; w = \frac{dz}{dt}, \tag{3.1a}$$

and also that, along a path line,

$$\frac{dx}{u} = \frac{dy}{v} = \frac{dz}{w} = \frac{dt}{1} \tag{3.1b}$$

The Eulerian view is practical for most engineering problems (indeed, it is used in a great majority of fluid analyses) and is adopted for this introductory text. In the Eulerian view, it is easy to determine whether the fluid flow is steady or unsteady. In unsteady flow, the fluid variables will change with time at the spatial points in the flow. In a steady flow, none of the variables at any point in a flow changes with time, although the variables generally are functions of position in the space filled by the fluid. *Thus, in the Eulerian view, a steady flow still may have accelerations.*

For example, in the pipe of Fig. 3.1, leading from an infinite reservoir with a fixed surface elevation, unsteady flow exists while the valve A is being opened or closed. [Of course, we cannot really have an infinite reservoir; a large reservoir will do, but then to keep the surface elevation fixed, the same flowrate must be supplied to the reservoir as flows out of the pipe. This is not hard to do; you just supply too much and let the excess overflow somewhere! This is a key to operation of many so-called *constant head tanks*.] With the valve opening fixed, steady flow occurs and the pressure, velocities, and the like, vary only with location. Under the former conditions as the valve opens and closes, they may vary with location and time. Problems of steady flow are more elementary than those of unsteady flow and actually have wide engineering applications. In this elementary textbook, we focus primarily on steady flows, but some useful and important unsteady flows will be considered.

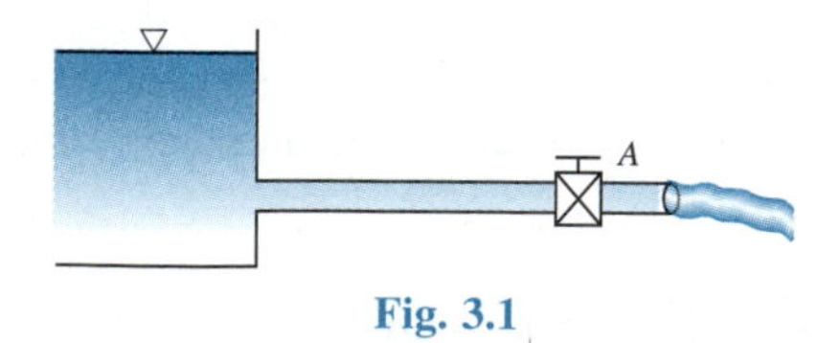

Fig. 3.1

If curves are drawn in an unsteady flow at an instant of time in such a way that the tangent at any point is in the direction of the velocity vector at that point, such curves are called *instantaneous streamlines* and they continually evolve in time. As we noted above, individual fluid particles must travel on paths whose tangent is always in the direction of the fluid velocity at any point. *In an unsteady flow, these path lines are not coincident with the instantaneous streamlines*! The injection of smoke or dye into the flow gives yet another view of the flow. If a picture is taken of all the dye or smoke particles that have passed through a particular point, i.e., the injection point, the result is a *streak line*. Such streak lines can be used to trace the travel of a pollutant downstream from a smoke stack or other discharge. In an unsteady flow the path lines, the streak lines, and the instantaneous streamlines are not coincident. However, *in a steady flow, the Lagrangian path lines are the same as the Eulerian streamlines, and both are the same as the streak lines,* because the streamlines then are fixed in space and path lines, streak lines and streamlines are tangent to the steady velocities. Thus, in a steady flow all the particles on a streamline that passes through a point in space also passed through or will pass through that point as well. In a steady flow, it follows that equations (3.1) can be integrated in space to define streamlines. Likewise at any given instant of time in an unsteady flow, equations (3.1) can be integrated in space to define a set of instantaneous streamlines. The sketching or plotting of streamlines produces a *streamline picture* or *flowfield* (Fig. 3.2*a*). Streamline pictures are of both qualitative and quantitative value to the engineer. They allow visualization of

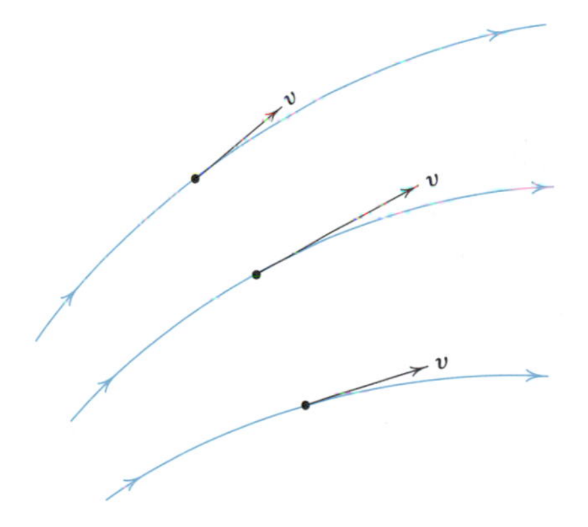

Fig. 3.2*a*

Fig. 3.2*b* Steady flow past an airfoil in a smoke tunnel (streamlines are created by introducing small jets of smoke at a number of upstream points in the flow).

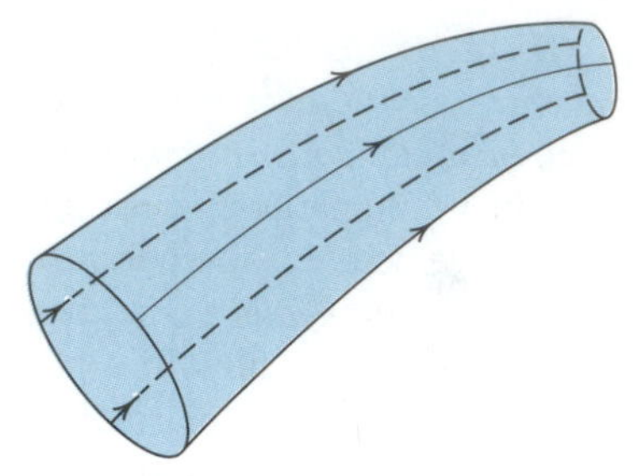

Fig. 3.3

fluid flow through mathematical and experimental determination (Fig. 3.2*b*) of the streamlines and to locate regions of high and low velocity and, from these, zones of low and high pressure, respectively. *In what follows, reference to streamlines implies a discussion of a steady flow; for unsteady flows, specific reference will be made to the unsteadiness and we will, if necessary, talk about instantaneous streamlines.*

When streamlines are drawn through a closed curve (Fig. 3.3) in a steady flow, they form a boundary across which fluid particles cannot pass because the velocity is always tangent to the boundary. Thus, the space between the streamlines becomes a tube or passage called a *streamtube*, and such a tube may be treated as if isolated from the adjacent fluid. The use of the streamtube concept broadens the application of fluid-flow principles; for example, it allows treating two apparently different problems such as flow in a passage and flow about an immersed object with the same laws. Also, because a streamtube of differential size essentially coincides with its axis (which is a streamline), it is to be expected that many of the equations developed for a small streamtube will apply equally well to a streamline.

3.2 ONE-, TWO-, AND THREE-DIMENSIONAL FLOWS—STREAMLINES AND FLOWFIELDS

In a *one-dimensional* flow, the change of fluid variables (velocity, pressure, etc.) perpendicular to (across) a streamline is negligible compared to the change along the streamline. In practice for a streamtube of finite cross-sectional area this means all fluid properties are considered uniform over any cross section. Pipe flow is usually taken to be one-dimensional and average fluid properties are used at each section. The flow in a streamtube of differential size is precisely one-dimensional because variations across the tube vanish in this limit (as the area approaches zero). Thus, flow along individual streamlines (however curved) is one-dimensional (the dimension being measured along the streamline). The concept of one-dimensional flow is an extremely powerful and practical one that produces simplicity in analysis and accurate engineering results for a wide range of problems. Later it is shown that some two- and three-dimensional flows may be treated very effectively for engineering purposes as one-dimensional in zones where the streamlines of the flow picture are all essentially straight and parallel.

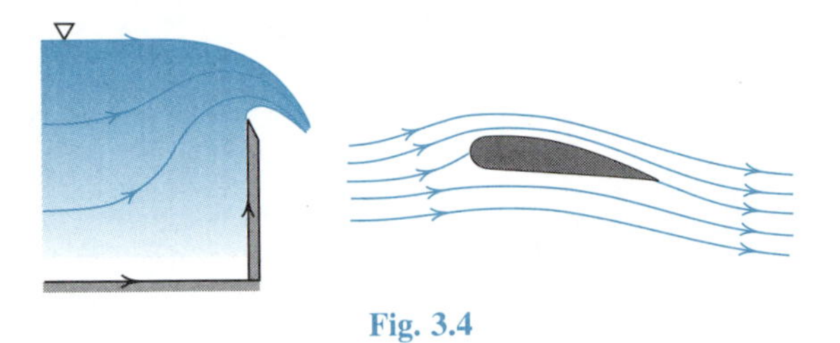

Fig. 3.4

Two- and three-dimensional flow pictures describe *flowfields*, the former when the flow is completely defined by the streamlines in a single plane, the latter in (three-dimensional) space. Examples of two-dimensional flows are shown over the weir and about the wing of Fig. 3.4. Here the velocities, pressures, and the like, vary throughout the flowfield and thus are functions of position in the field. Such two-dimensional flows are approximations to reality in that they are strictly correct only to the extent that end effects on weir and wing are negligible; this may also be visualized by assuming weir and wing to be infinitely long perpendicular to the plane of the paper. To the extent that such approximations are valid, the flow is completely described by a streamline picture drawn in a single plane. In practice, it is often possible to introduce simple corrections to the two-dimensional results to account for end effects.

Two axisymmetric three-dimensional flows are depicted in Fig. 3.5. Here the streamlines are really stream surfaces and the streamtubes are of annular cross section. On planes passed through the axis of such flows, streamline pictures may be drawn which superficially resemble two-dimensional flows; such pictures may be used fruitfully for streamline visualization, but they do not represent a reduction of a three-dimensional problem to a two-dimensional one because the mathematical descriptions of two-dimensional and axisymmetric flows are *not* the same. Nonaxisymmetric flows such as that over the fuselage and into the air inlets of a single-jet aircraft are *three-dimensional flowfields* of the most general character. These flows are more difficult to visualize and most difficult to predict. To generalize the kinematics and dynamics of the flowfield so that derived equations allow all possible flow configurations leads to mathematical complexities far beyond the scope of an elementary treatment of fluid mechanics and would obscure the physical picture so essential to a real understanding of the problem. Accordingly the treatment hereafter is restricted primarily to one-, two-, and axisymmetric three-dimensional problems.

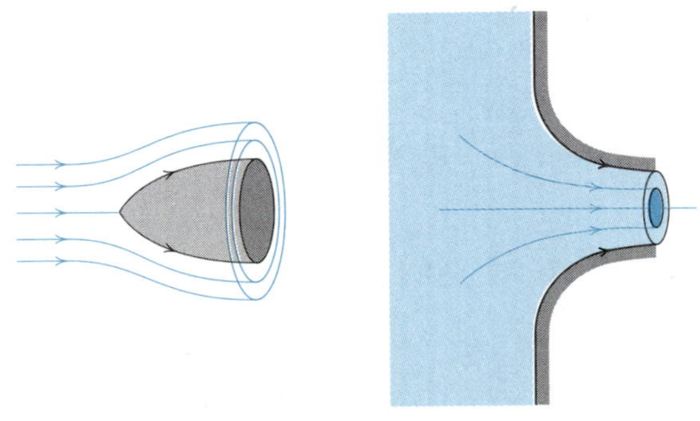

Fig. 3.5

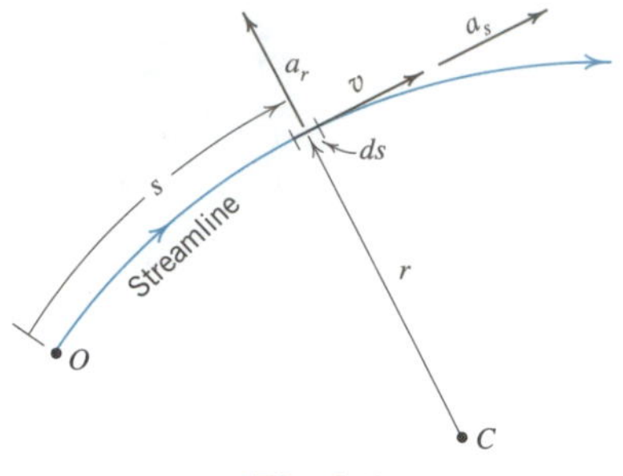

Fig. 3.6

3.3 VELOCITY AND ACCELERATION

Velocity and acceleration are vector quantities, having both magnitude and direction. However, often the direction is known or assumed and only the magnitude is to be determined. Then, the scalar components in particular directions are used. In this text most equations and analyses are in terms of scalar components, although vectors are used where they make physical principles more understandable.

For one-dimensional flow along a streamline (Fig. 3.6), velocity and acceleration may be readily defined from past experience in engineering mechanics with the motion of single particles. Select a fixed point O as a reference point and define the displacement s of a fluid particle along the streamline in the direction of motion. In time dt the particle will cover a differential distance ds along the streamline. The velocity magnitude v of this particle over the distance ds is given by $v = ds/dt$; the velocity vector is, of course, tangent to the streamline at s according to the definition of a streamline. Components of acceleration along (tangent to) and across (normal to) the streamline at s may also be written:

$$a_s = \frac{d^2s}{dt^2} = \frac{d}{dt}\left(\frac{ds}{dt}\right) = \frac{dv}{dt} = \frac{ds}{dt}\frac{dv}{ds} = v\frac{dv}{ds} \tag{3.2}$$

and, from particle mechanics,

$$a_r = -\frac{v^2}{r} \tag{3.3}$$

in which r is the radius of curvature of the streamline at s.

ILLUSTRATIVE PROBLEM 3.1

Along the straight streamline shown in the figure, the magnitude of the velocity is given by $v = 3\sqrt{x^2 + y^2}$ m/s and the coordinates x and y are given in metres. Calculate the velocity magnitude and acceleration components at the point (8, 6).

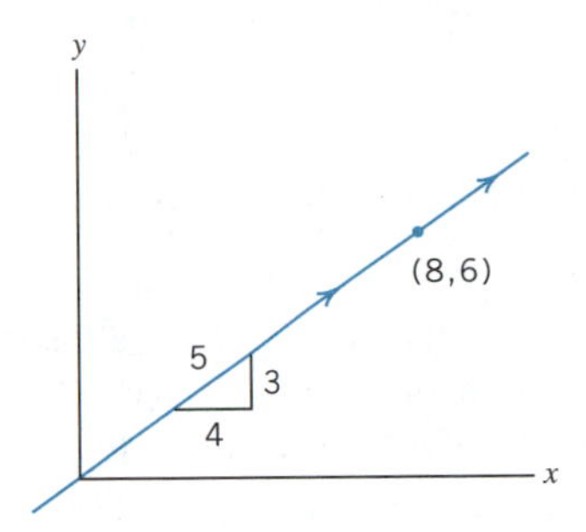

SOLUTION

Looking at Eqs. 3.2 and 3.3 and the given data, we observe that $s = \sqrt{x^2 + y^2}$ m, and so $v = 3s$ m/s and $dv/ds = 3\ \mathrm{s}^{-1}$, while the radius of curvature of the streamline is infinite. At (8, 6), $s = 10$ m; therefore,

$$v = 3 \times 10 = 30 \text{ m/s} \bullet$$

According to Eq. 3.2,

$$a_s = (3s \text{ m/s})3 \text{ s}^{-1} = 9s = 90 \text{ m/s}^2 \bullet$$

Obviously, $a_r = 0$ since in Eq. 3.3, $r = \infty$ •

ILLUSTRATIVE PROBLEM 3.2

In the geophysical experiment described in Illustrative Problem 2.17, the fluid at the wall of the tank moves along the circular streamline shown below with a constant tangential velocity component of 1.04 m/s. Calculate the tangential and radial components of acceleration at any point on the streamline.

SOLUTION

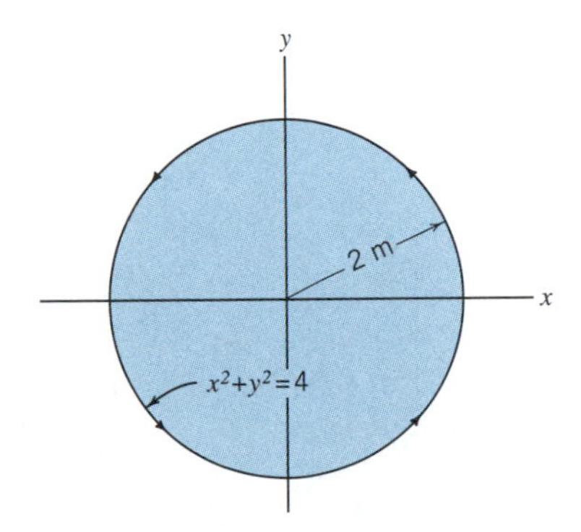

Because the velocity is constant, Eq. 3.2 tells us that $a_s = 0$. From the diagram, the radius of curvature of the streamline is 2 m and the velocity along it is 1.04 m/s; accordingly, from Eq. 3.3

$$a_r = \frac{(1.04 \text{ m/s})^2}{2 \text{ m}} = 0.541 \text{ m/s}^2 \bullet \tag{3.3}$$

directed toward the center of the tank.

In flowfields, velocity and acceleration are somewhat more difficult to define because a generalization is required which is applicable to the whole flowfield. In general, the velocities are everywhere different in magnitude and direction at different points in the flowfield and at different times. At each point, however, each velocity has components u, v, and w, which are parallel to the x-, y-, and z- axes, respectively, as we defined them in Section 3.1. Figure 3.7 gives a sketch of a three-dimensional Cartesian coordinate system and the velocity at a point. If the velocity depends on x, y, and z and on time t, its components are also functions of these variables. Written mathematically and generally in the Eulerian view,

$$u = u(x, y, z, t),\ v = v(x, y, z, t), \text{ and } w = w(x, y, z, t)$$

For steady flow, time t would not appear in these definitions, while for two-dimensional and steady flow, neither t nor z would appear for example. In terms of displacement and time (the Lagrangian view), we have, as we saw in Section 3.1, that

$$u = \frac{dx}{dt},\ v = \frac{dy}{dt}, \text{ and } w = \frac{dz}{dt} \tag{3.4}$$

where here x, y, and z are the actual coordinates of a fluid particle that is being tracked. Fortunately, the velocity at a point is the same in both the Eulerian and the Lagrangian view. The acceleration components are

$$a_x = \frac{du}{dt},\ a_y = \frac{dv}{dt}, \text{ and } a_z = \frac{dw}{dt}, \tag{3.5}$$

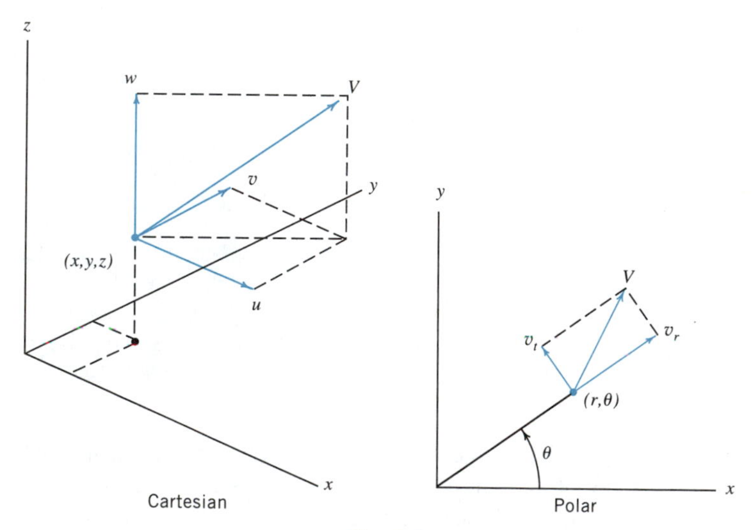

Fig. 3.7

The derivatives here are the total derivatives, also called the *substantial* or *material derivatives*, which follow a fluid particle. From Appendix 6, we can use the chain rule of differentiation for the total derivative to write

$$du = \frac{\partial u}{\partial t} dt + \frac{\partial u}{\partial x} dx + \frac{\partial u}{\partial y} dy + \frac{\partial u}{\partial z} dz$$

$$dv = \frac{\partial v}{\partial t} dt + \frac{\partial v}{\partial x} dx + \frac{\partial v}{\partial y} dy + \frac{\partial v}{\partial z} dz$$

$$dw = \frac{\partial w}{\partial t} dt + \frac{\partial w}{\partial x} dx + \frac{\partial w}{\partial y} dy + \frac{\partial w}{\partial z} dz$$

Substituting these relationships in Eqs. 3.5 and using Eqs. 3.4 yields

$$\begin{aligned} a_x &= \frac{\partial u}{\partial t} + u\frac{\partial u}{\partial x} + v\frac{\partial u}{\partial y} + w\frac{\partial u}{\partial z} \\ a_y &= \frac{\partial v}{\partial t} + u\frac{\partial v}{\partial x} + v\frac{\partial v}{\partial y} + w\frac{\partial v}{\partial z} \\ a_z &= \frac{\partial w}{\partial t} + u\frac{\partial w}{\partial x} + v\frac{\partial w}{\partial y} + w\frac{\partial w}{\partial z} \end{aligned} \tag{3.6}$$

It is immediately clear that even in steady flow, there will be accelerations of the flow if the velocity field varies in space. As before, for steady flow or for flow in two dimensions or one dimension, the appropriate terms will drop out of Eqs. 3.6.

Returning to Fig. 3.7 and looking at the sketch of the two-dimensional flow, we find that a similar analysis for a steady two dimensional flow in polar coordinates, where v_r and v_t are both functions or r and θ, yields

$$v_r = \frac{dr}{dt} \text{ and } v_t = r\frac{d\theta}{dt} \tag{3.7}$$

and, for the components of acceleration in the steady state flow,

$$a_r = v_r\frac{\partial v_r}{\partial r} + v_t\frac{\partial v_r}{r\partial\theta} - \frac{v_t^2}{r},\ a_t = v_r\frac{\partial v_t}{\partial r} + v_t\frac{\partial v_t}{r\partial\theta} + \frac{v_r v_t}{r} \tag{3.8}$$

ILLUSTRATIVE PROBLEM 3.3

For the circular streamline described in the previous Illustrative Problem 3.2 along which the velocity is 1.04 m/s and the radius of curvature is 2 m, calculate the horizontal, vertical, tangential, and normal components of the velocity and acceleration at the point P (2 m, 60°).

SOLUTION

Noting that $r = \sqrt{x^2 + y^2} = 2$ m and using similar triangles, we have

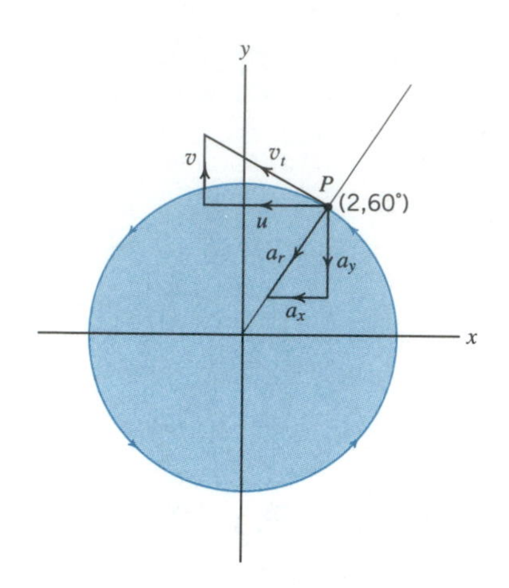

$$u = \frac{-1.04y}{\sqrt{x^2 + y^2}} \qquad \text{and} \qquad v = \frac{1.04x}{\sqrt{x^2 + y^2}}$$

At P, $x = 1$, $y = \sqrt{3}$, so

$$u = -0.90 \text{ m/s} \bullet \qquad v = 0.52 \text{ m/s} \bullet$$

Using Eqs. 3.4,

$$a_x = -\frac{1.04y}{\sqrt{x^2 + y^2}} \frac{\partial}{\partial x}\left(\frac{-1.04y}{\sqrt{x^2 + y^2}}\right) + \frac{1.04x}{\sqrt{x^2 + y^2}} \frac{\partial}{\partial y}\left(\frac{-1.04y}{\sqrt{x^2 + y^2}}\right)$$

$$= -\frac{2.16x}{8} \tag{3.6}$$

$$a_y = -\frac{1.04y}{\sqrt{x^2 + y^2}} \frac{\partial}{\partial x}\left(\frac{1.04x}{\sqrt{x^2 + y^2}}\right) + \frac{1.04x}{\sqrt{x^2 + y^2}} \frac{\partial}{\partial y}\left(\frac{1.04x}{\sqrt{x^2 + y^2}}\right)$$

$$= -\frac{2.16y}{8} \tag{3.6}$$

Substituting $x = 1$, $y = \sqrt{3}$,

$$a_x = -0.27 \text{ m/s}^2 \bullet \qquad a_y = -0.47 \text{ m/s}^2 \bullet$$

By inspection,

$$v_t = 1.04 \text{ m/s} \bullet \qquad v_r = 0 \text{ m/s} \bullet \tag{3.7}$$

Using Eqs. 3.8,

$$a_r = 0 \frac{\partial}{\partial r}(0) + 1.04 \frac{\partial}{r\partial\theta}(0) - \frac{1.04 \times 1.04}{2} = -0.54 \text{ m/s}^2 \bullet \tag{3.8}$$

$$a_t = 0 \frac{\partial}{\partial r}(1.04) + 1.04 \frac{\partial}{r\partial\theta}(1.04) + \frac{0 \times 1.04}{2} = 0 \text{ m/s}^2 \bullet \tag{3.8}$$

Note that a_x and a_y might have been obtained more easily (in this problem) by calculating them as the horizontal and vertical components of a_r. Does $a_r^2 = a_x^2 + a_y^2$?

3.4 CIRCULATION, VORTICITY, AND ROTATION

We intuitively know that tangential components of the velocity give the fluid in a flow a swirl (try to visualize the flow in Illustrative Problem 3.3 just above). A measure of this swirl has been defined and it is called the *circulation.* The circulation is designated by Γ (gamma). Γ is defined as the *line integral* of the tangential component of velocity around a closed curve fixed in the flow. For simplicity, consider a two-dimensional steady flow whose streamlines are shown in Fig. 3.8. In the two-dimensional flowfield of Fig. 3.8*a*, each streamline intersects the closed curve at an angle α and the tangential component of the velocity at the point of intersection is $V \cos \alpha$. An element of circulation $d\Gamma$ is defined as the product of the tangential velocity component and the element dl of the closed curve. Thus,

$$d\Gamma = (V \cos \alpha)\, dl$$

The sum of the elements $d\Gamma$ along the curve C marking the closed curve defines the circulation:

$$\Gamma = \oint_C d\Gamma = \oint_C (V \cos \alpha)\, dl = \oint_C \mathbf{V} \cdot d\mathbf{l} \tag{3.9}$$

in which $d\mathbf{l}$ is an elemental vector of magnitude dl and direction tangent to the closed curve at each point.

Although the calculation of circulation around an arbitrary curve in a flowfield is generally a tedious step-by-step integration, the principle can be applied easily and fruitfully to specific closed curves such as circles and squares. Calculation of the circulation around a basic differentially sized square element as shown in Fig. 3.8*b* yields a concept having great general significance because it yields the point value of the circulation in a flow. This value is needed, for example, in calculations of pressure-velocity-head relations (expressed through Bernoulli's equation) in two-dimensional ideal incompressible flows (Section 5.7).

Now, to compute the circulation, proceed from A counterclockwise around the boundary of the element, setting down the products of velocity component and distance in order; because the element is of differential size, the resulting circulation is also a differential quantity, $d\Gamma$.

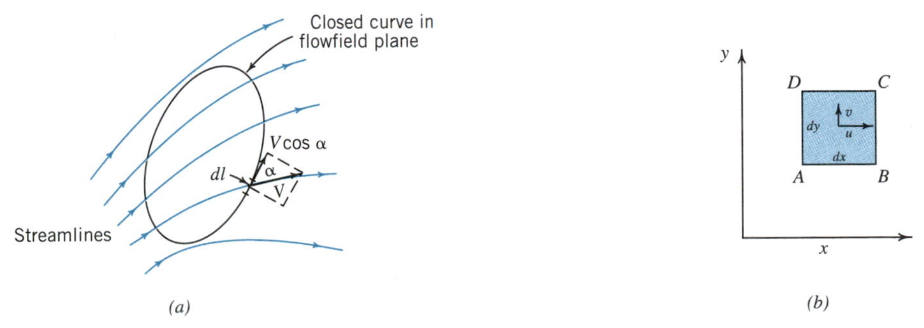

Fig. 3.8

$$d\Gamma \cong \begin{bmatrix}\text{Mean velocity}\\ \text{along } AB\end{bmatrix} dx + \begin{bmatrix}\text{Mean velocity}\\ \text{along } BC\end{bmatrix} dy$$

$$- \begin{bmatrix}\text{Mean velocity}\\ \text{along } CD\end{bmatrix} dx - \begin{bmatrix}\text{Mean velocity}\\ \text{along } DA\end{bmatrix} dy$$

$$\cong \left[u - \frac{du}{\partial y}\frac{dy}{2}\right] dx + \left[v + \frac{\partial v}{\partial x}\frac{dx}{2}\right] dy$$

$$- \left[u + \frac{\partial u}{\partial y}\frac{dy}{2}\right] dx - \left[v - \frac{\partial v}{\partial x}\frac{dx}{2}\right] dy$$

(Note $\cos \alpha = 1$ on AB and BC, but $\cos \alpha = -1$ on CD and DA.) By expanding the products and retaining only the terms of lowest order (largest magnitude), we obtain

$$d\Gamma = \left(\frac{\partial v}{\partial x} - \frac{\partial u}{\partial y}\right) dx\, dy$$

in which $dx\,dy$ is the area inside the control surface. The *vorticity*, ξ (xi), is defined as the differential circulation per unit of area enclosed, which becomes

$$\xi = \frac{d\Gamma}{dx\,dy} = \frac{\partial v}{\partial x} - \frac{\partial u}{\partial y} \tag{3.10}$$

For polar coordinates, by the same procedure,

$$\xi = \frac{\partial v_t}{\partial r} + \frac{v_t}{r} - \frac{\partial v_r}{r\partial \theta} \tag{3.11}$$

From the definition of circulation, Γ, and the methods used to calculate vorticity, ξ, the reader will sense that the latter quantity is some measure of the rotational aspects of the fluid elements as they move through the flowfield. This can be shown explicitly. Suppose two lines are drawn on the square element in a fluid flow, as shown in Fig. 3.9, so that the lines are parallel to the x and y axes, respectively. If the fluid element tends to rotate, these lines will tend to rotate also and, for the instant at which the lines are drawn on the fluid,[1] their average angular velocity can be calculated. In a small time interval dt the vertical line will rotate about the translating center of mass of the element by an amount

$$d\theta_V = -\left[\left(u + \frac{\partial u}{\partial y}\frac{dy}{2}\right) - \left(u - \frac{\partial u}{\partial y}\frac{dy}{2}\right)\right]\frac{dt}{dy}$$

[1]It is physically possible to carry out such an experiment; see, for example, J. L. Lumley, ''Deformation of Continuous Media,'' NCFMF/EDC Film No. 21608, Encyclopaedia Britannica Educational Corp.

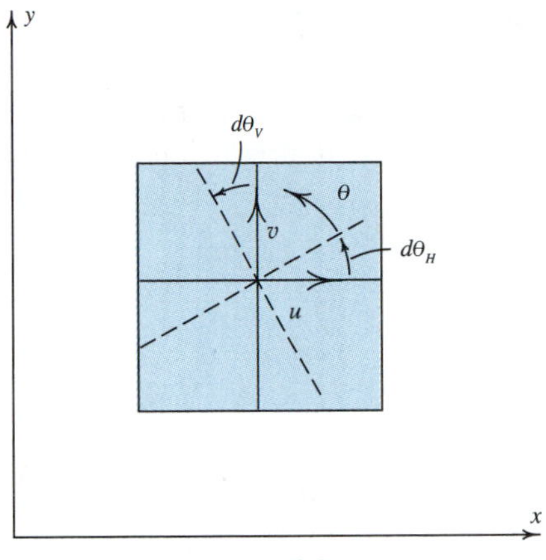

Fig. 3.9

so the angular velocity

$$\omega_V = \frac{d\theta_V}{dt} = -\frac{\partial u}{\partial y}$$

For the horizontal line

$$\omega_H = \frac{d\theta_H}{dt} = \frac{\partial v}{\partial x}$$

The *average rotation* ω of the element is then

$$\omega = \frac{1}{2}(\omega_v + \omega_H) = \frac{1}{2}\left(\frac{\partial v}{\partial x} - \frac{\partial u}{\partial y}\right) \tag{3.24}$$

and

$$\xi = \frac{d\Gamma}{dx\, dy} = 2\omega \tag{3.25}$$

The vorticity at a point is twice the rotation there, that is, twice the average angular velocity of the fluid element.

If a flow possesses vorticity (i.e., if $\xi \neq 0$), it is said to be a *rotational* flow; if a flow possesses no vorticity ($\xi = 0$), it is termed *irrotational.* In Section 5.9, it will be shown that *irrotational* flows are also characterized by a *velocity potential* and are therefore known as *potential* flows. Such definitions are adequate for present purposes but may be misleading because they imply that whole flowfields are either rotational or irrotational. Actually flowfields can possess zones of both irrotational and rotational flows. In fact, in real fluid flows (see Chapter 7) the rotational parts of the flow are often concentrated in boundary layers close to solid walls or bodies, while the flow far from the solid boundaries may be essentially irrotational.

ILLUSTRATIVE PROBLEM 3.4

Calculate the vorticity of the two-dimensional flowfield described by the equations $v_t = \omega r$ and $v_r = 0$, in which ω is a constant. A sketch of this flow is given below.

SOLUTION

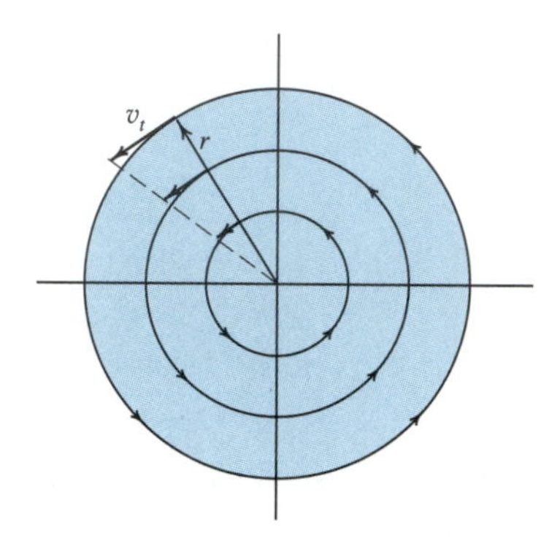

Application of Eq. 3.11 yields

$$\xi = \frac{\partial v_t}{\partial r} + \frac{v_t}{r} - \frac{\partial v_r}{r\partial \theta} = \frac{\partial}{\partial r}(\omega r) + \frac{\omega r}{r} - \frac{\partial}{r\partial \theta}(0) = \omega + \omega - 0 = 2\omega$$

Evidently this is a rotational flow possessing a constant vorticity (over the whole flowfield) of 2ω; this flowfield is well known and is called the *forced vortex*. It is clearly a *rotational* flowfield and since the radial component of velocity is everywhere zero, the streamlines of this flow are concentric circles.

ILLUSTRATIVE PROBLEM 3.5

When a viscous, incompressible fluid flows between two plates and the flow is laminar and two-dimensional, the velocity profile is parabolic and given by

$$u = U_c(1 - y^2/b^2)$$

Calculate the shear stress τ and rotation ω.

SOLUTION

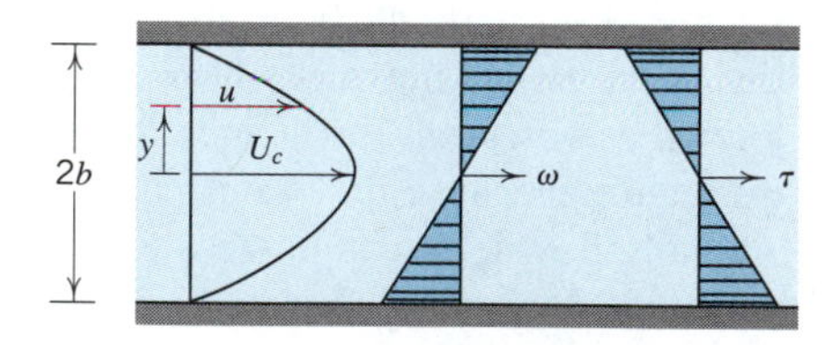

A sketch shows the velocity profile and defines the terms; then we use Eqs. 1.12 and 3.12 to calculate shear stress and rotation.

$$\omega = \frac{1}{2}\left(\frac{\partial v}{\partial x} - \frac{\partial u}{\partial y}\right) = -\frac{1}{2}[-2U_c y/b^2] = U_c y/b^2 \bullet \quad (3.12)$$

$$\tau = \mu\partial u/\partial y = -2\mu U_c y/b^2 = -2\mu\omega = -\mu\xi \bullet \quad (1.12)$$

It is interesting and significant that the rotation and vorticity are large where the shear stress is large.

FILMS

Lumley, J. L. Deformation of continuous media. NCFMF/EDC Film No. 21608, Encyclopaedia Britannica Educ. Corp.

Lumley, J. L. Eulerian and Lagrangian descriptions in fluid mechanics. NCFMF/EDC Film No. 21621, Encyclopaedia Britannica Educ. Corp.

Rouse, H. Mechanics of Fluids: Fundamental principles of flow. Film No. U45734, Media Library, Audiovisual Center, Univ. of Iowa.

Shapiro, A. H. Vorticity. NCFMF/EDC Film Nos. 21605 and 21606, Encyclopaedia Britannica Educ. Corp.

PROBLEMS

3.1. A fluid flow has the following velocity components: $u = 1$ m/s or ft/s and $v = 2x$ m/s or ft/s. Find an equation for and sketch the streamlines of this flow.

3.2. A 4 m diameter tank is filled with water and then rotated at a rate of $\omega = 2\pi(1 - e^{-t})$ rad/s. At the tank walls viscosity prevents slip of fluid particles relative to the wall. What are the speed and the tangential and normal accelerations of those fluid particles next to the tank walls as a function of time?

3.3. A fluid particle moves so that in the Lagrangian frame of reference $x = 3t$, $y = 9t^2$, and $z = 27t^3$. Find the velocity and acceleration of the particle for times from zero to five seconds.

3.4 The path of a fluid particle is given by the hyperbola $xy = 25$ while at any time t the particle position is $x = 5t^2$. What are the x- and y- components of the particle velocity and acceleration?

3.5. Calculate the accelerations in the flow of problem 3.1.

3.6. Sketch the following flowfields and derive general expressions for their components of acceleration: (*a*) $u = 4$, $v = 3$; (*b*) $u = 4$, $v = 3x$; (*c*) $u = 4y$, $v = 0$; (*d*) $u = 4y$, $v = 3$; (*e*) $u = 4y$, $v = 3x$; (*f*) $u = 4y$, $v = 4x$; (*g*) $u = 4y$, $v = -4x$; (*h*) $u = 4$, $v = 0$; (*i*) $u = 4$, $v = -4x$; (*j*) $u = 4x$, $v = 0$; (*k*) $u = 4xy$, $v = 0$; (*l*) $v_r = c/r$, $v_t = 0$; (*m*) $v_r = 0$, $v_t = c/r$.

3.7. When an incompressible, nonviscous (see Chapter 5) fluid flows against a plate in a plane (two-dimensional) flow, an exact solution for the equations of motion for this flow is $u = Ax$, $v = -Ay$, with $A > 0$ for the sketch shown. The coordinate origin is located at the *stagnation point* 0, where the flow divides and the local velocity is zero. Find the velocities and accelerations in the flow.

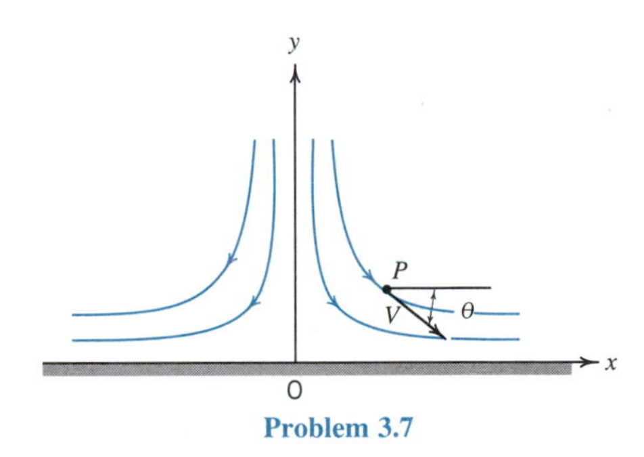

Problem 3.7

3.8. Fluid passes through this set of thin closely spaced blades. If the velocity V is 10 ft/s (or equivalently, 3 m/s), calculate the circulation for the flow.

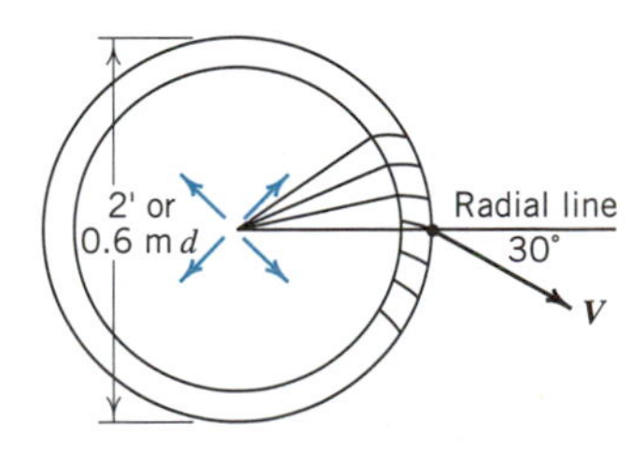

Problem 3.8

3.9. Derive the equation for vorticity in polar coordinates.

3.10 Calculate the vorticity for the flow in problem 3.1.

3.11. For the flowfields of problem 3.6 derive expressions for vorticity, and state whether the flowfield is rotational or irrotational.

3.12. For the velocity profiles shown below derive expressions for the vorticity.

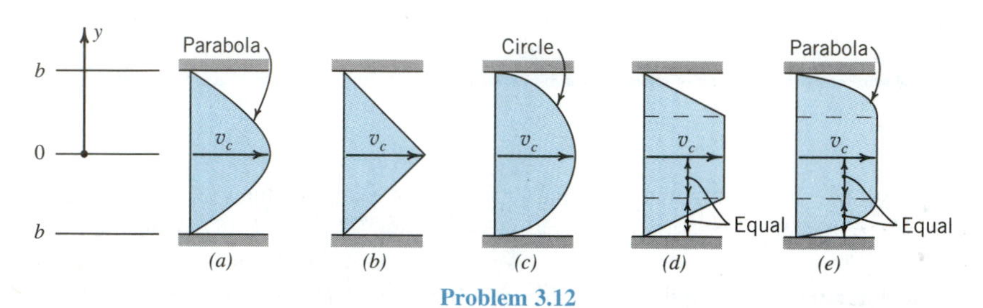

Problem 3.12

3.13. If the velocity profile in a passage of width $2R$ is given by the equation $v/v_c = (y/R)^{1/n}$, derive an expression for the vorticity.

3.14. Calculate the vorticity for the flow of problem 3.7.

3.15. For the *free vortex* flow the velocities are $v_t = 5/r$ and $v_r = 0$. Assume that lengths are in feet or metres and times are in seconds. Plot the streamlines of this flow and calculate the accelerations and vorticity. Are there any interesting points in the flow?

3.16. For the *forced vortex* flow the velocities are $v_t = \omega r$ and $v_r = 0$. Plot the streamlines of this flow and calculate the accelerations and vorticity. Are there any interesting points in the flow?

CHAPTER

4 SYSTEMS, CONTROL VOLUMES, CONSERVATION OF MASS, AND THE REYNOLDS TRANSPORT THEOREM

The objectives of this chapter are to introduce the concept of the control volume, to apply that concept to derive equations for the conservation of mass, and then to generalize these efforts to derive the Reynolds Transport Theorem, which gives a very general relationship with many applications in our later work. We begin with a discussion of the relationship between a physical system and a control volume in Section 4.1. In the following two sections, the control volume concept is used to derive the conservation of mass relationships for steady one- and two-dimensional flows. In Section 4.4, the Reynolds Transport Theorem is derived for three-dimensional and unsteady flow and the relationships from Sections 4.2 and 4.3 are recovered by simplification of the more general result. In this and subsequent chapters, we work from the simple and specific to the general, building thereby in the reader a solid foundation of understanding of the basic fluid mechanics and an ability to work from first principles.

4.1 THE SYSTEM VIS-A-VIS THE CONTROL VOLUME

A physical *system* is defined as a particular collection of matter and is identified and viewed as being separated from everything external to the system by an imagined or real closed boundary. In particle mechanics the system is a convenient physical entity. Its mass is conserved, and its energy and momentum can be defined precisely and easily analyzed. Indeed, a system-based analysis of fluid flow leads to the Lagrangian equations of motion in which particles of fluid are tracked. Unfortunately, a fluid system is both mobile and very deformable, and in the random and chaotic motion of turbulent flows even its identity may become hard to maintain over time. This suggests the need to define a more convenient object for analysis. This object is a volume in space through whose boundary matter, mass, momentum, energy, and the like can flow. This volume is called a *control volume*, and its boundary is a *control surface*.

The control volume may be of any useful size (finite or infinitesimal) and shape, provided only that the bounding control surface is a closed (completely surrounding) boundary. In general, the control volume may have a time-varying shape and/or location. *As defined here for this introductory treatment, neither the control volume nor the control surface change shape or position with time.* This approach is consistent with the Eulerian view of fluid motion, in which attention is focused on particular points in the space filled by the fluid rather than on the fluid particles. Implicit in our formulation is that the control volume is fixed in some particular coordinate reference frame. The equations that we will derive are based on the assumption that the coordinate frame is an *inertial reference frame*, i.e., one that is at rest or moving with a constant velocity. In such frames, Newton's second law applies in its usual form and the rate of change of the velocity is the total acceleration. Often, particularly in geophysical problems, it is convenient to use a reference frame that is fixed at a point on the surface of the earth; such a frame is noninertial because the earth's surface is rotating about the axis of the globe, to say nothing of the earth's orbit around the sun, etc. Thus, there is an acceleration of the reference frame and any control volume fixed in it. One consequence of this in the geophysical case is the appearance of the *Coriolis force*, caused by the rotation about the earth's axis, in the equations of motion applicable to atmospheric and oceanic motions. It has a major effect there, but is strongly dependent on the scale of the motion and so is negligible on the small scales of most engineering problems. We will not deal with noninertial reference frames here.

4.2 CONSERVATION OF MASS: THE CONTINUITY EQUATION—ONE-DIMENSIONAL STEADY FLOW

The application of the principle of conservation of mass (in the absence of mass-energy conversions) to a steady flow in a streamtube results in the *equation of continuity*, which describes the continuity of the flow from section to section of the streamtube. The derivation uses a control volume and the fluid system which just fills the volume at a particular time t. Consider the element of a finite streamtube in Fig. 4.1 through which passes a steady, one-dimensional flow of a compressible fluid. Note the location of the control volume as marked by the control surface that bounds the region between sections 1 and 2 and lies along the inner wall of the streamtube. The velocities at sections 1 and 2 are assumed to be uniform and to be consistent with the assumption of one-dimensional flow. In the tube near section 1 the cross-sectional area and fluid density are A_1 and ρ_1, respectively, and near section 2, A_2 and ρ_2. With the control surface shown coinciding with the streamtube walls and the cross sections at 1 and 2, the control volume comprises volumes I and R. Let a fluid system be defined as the fluid within the control volume $(I + R)$ at time t. The control volume is fixed in space, but in time dt the system moves downstream as shown. From the conservation of system mass

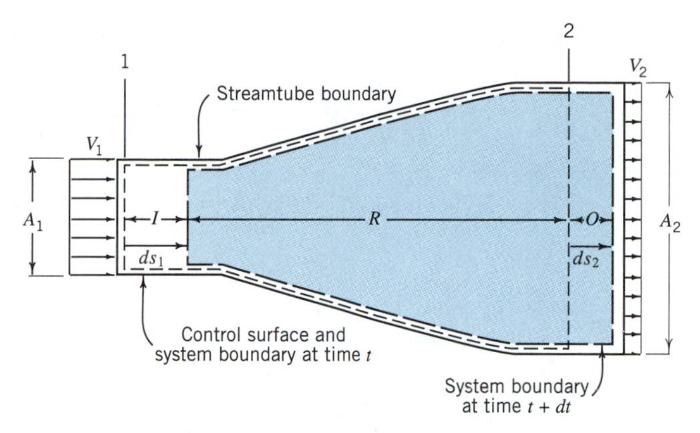

Fig. 4.1

$$(m_I + m_R)_t = (m_R + m_O)_{t+dt}$$

$$\begin{pmatrix}\text{Mass of fluid in} \\ \text{zones } I \text{ and } R \text{ at} \\ \text{time } t\end{pmatrix} = \begin{pmatrix}\text{Mass of fluid in} \\ \text{zones } O \text{ and } R \text{ at} \\ \text{time } t + dt\end{pmatrix}$$

In a steady flow, the fluid properties at points in space are not functions of time so $(m_R)_t = (m_R)_{t+dt}$ and, consequently,

$$(m_I)_t = (m_O)_{t+dt}$$

These two terms are easily expressed in terms of the mass of fluid moving across the control surface in time dt. The volume of I is $A_1\ ds_1$, and that of O is $A_2\ ds_2$; accordingly,

$$(m_I)_t = \rho_1 A_1\ ds_1 \qquad (m_O)_{t+dt} = \rho_2 A_2\ ds_2$$

and

$$\rho_1 A_1\ ds_1 = \rho_2 A_2\ ds_2$$

Dividing by dt,

$$\rho_1 A_1 \frac{ds_1}{dt} = \rho_2 A_2 \frac{ds_2}{dt}$$

However, ds_1/dt and ds_2/dt are recognized as the velocities past sections 1 and 2, respectively; therefore, if $\dot{m} = A\rho V$ is the *mass flowrate*, then

$$\boxed{\dot{m} = \rho_1 A_1 V_1 = \rho_2 A_2 V_2} \tag{4.1}$$

which is the equation of continuity. In words, it expresses the fact that in steady flow the mass flowrate passing all sections of a streamtube is constant. This equation may also be written

$$\dot{m} = A\rho V = \text{Constant} \qquad d(A\rho V) = 0 \qquad \text{or} \qquad \frac{dA}{A} + \frac{d\rho}{\rho} + \frac{dV}{V} = 0 \tag{4.2}$$

Multiplication of equation 4.1 by g_n gives the *weight flowrate*

$$G = g_n\dot{m} = A_1\gamma_1V_1 = A_2\gamma_2V_2 \tag{4.3}$$

while division of Eq. 4.2 by A gives the *mass velocity*

$$m_v = \rho V$$

which has the dimensions of slugs/ft^2·s or kg/m^2·s. The dimensions of $\dot{m}$ are slugs per second or kilograms per second. The dimensions of G are pounds per second or newtons per second; G is often a convenient and concise means of expressing flowrate of a gas in which the specific weight γ may vary along the streamtube with changes of pressure and temperature.

For many flows one is interested in the *volume flowrate*

$$Q = AV \tag{4.4}$$

which has the dimensions (L^3/s). This popular measure of flowrate is expressed in several common ways; these include: cubic metres per second (m^3/s), litres per second (l/s), cubic feet per second (ft^3/s, cusec or cfs), cubic feet per minute (cfm), gallons per minute (gpm), and millions of gallons per day (mgd).

However, from Eq. 4.2 it is clear that Q is *not* a constant along a streamtube in variable density flows. For liquids and gas flows where the variation of density is negligible the equation of continuity reduces to

$$Q = A_1V_1 = A_2V_2 \tag{4.5}$$

Thus, for fluids of constant density the product of velocity and cross-sectional area is constant along a streamtube. This quantity may, of course, be computed in compressible flow problems, but this is not generally useful because of its variation from section to section along the streamtube.

For two-dimensional flows the flowrate is usually quoted *per unit distance normal to the plane of the flow*. Taking b to be the distance between any two parallel flow planes and h the distance between streamlines, Eq. 4.3 becomes

$$\frac{G}{b} = \gamma_1h_1V_1 = \gamma_2h_2V_2 \tag{4.6}$$

in which G/b is known as the two-dimensional (weight) flowrate. The counterpart of this for the incompressible fluid (from Eq. 4.5) is

$$\frac{Q}{b} = q = h_1V_1 = h_2V_2 \tag{4.7}$$

in which Q/b (hereafter termed q) is the two-dimensional (volume) flowrate, the dimensions of which are evidently m^3/s·m or ft^3/s·ft.

Frequently in fluid flows the velocity distribution through a flow cross section may be *nonuniform*, as shown in Fig. 4.2. From considerations of conservation of mass, it is evident at once that nonuniformity of velocity distribution does not invalidate the continuity principle as presented above. Thus, for steady flow of the incompressible fluid,

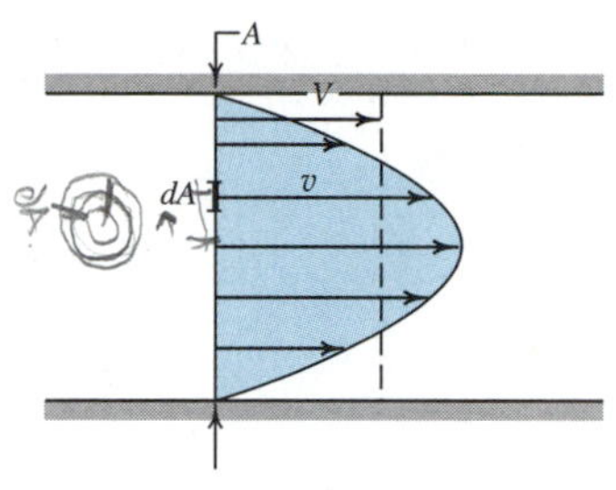

Fig. 4.2

Eq. 4.5 applies as before. Here, however, the velocity V in the equation is the *mean velocity* defined by $V = Q/A$ in which the flowrate Q is obtained from the summation of the differential flowrates, dQ, passing through the differential areas, dA. Thus, V is a fictitious uniform velocity that will transport the same amount of mass through the cross section as will the actual velocity distribution,

$$V = \frac{1}{A} \iint_A v \, dA \tag{4.8}$$

from which the mean velocity can be obtained by performing the indicated integration. With the velocity profile mathematically defined, formal integration may be employed; when the velocity profile is known but not mathematically defined (for example, when obtained from experimental measurements), graphical or numerical methods can be used to evaluate the integral.

The fact that the product AV remains constant along a streamtube (in a fluid of constant density) allows a partial physical interpretation of streamline pictures. As the cross-sectional area of a streamtube increases, the velocity must decrease; hence the conclusion: streamlines widely spaced indicate regions of low velocity, streamlines closely spaced indicate regions of high velocity.

ILLUSTRATIVE PROBLEM 4.1

Three hundred kilograms of water per second flow through this pipeline reducer. Calculate the flowrate in cubic metres per second and the mean velocities in the 300 mm and 200 mm pipes.

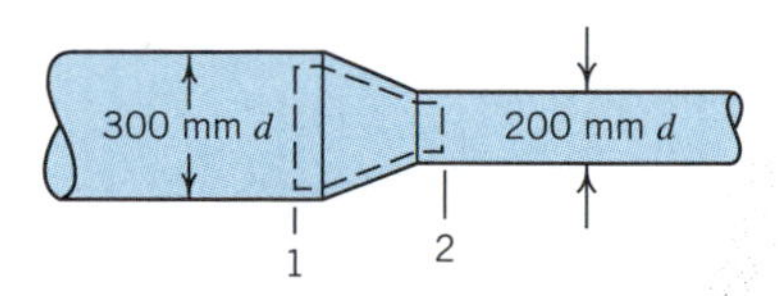

SOLUTION

On the above diagram we identify a control volume, bounded by Sections 1 and 2 in the pipes to either side of the reducer. For water, $\rho_1 = \rho_2 = 1\,000 \text{ kg/m}^3$; the mass flowrate $\dot{m} = 300 \text{ kg/s}$. Accordingly, use of Eqs. 4.1 and 4.5 yields

$$Q = \dot{m}/\rho_1 = \dot{m}/\rho_2 = (300\ \text{kg/s})/(1\,000\ \text{kg/m}^3) = 0.03\ \text{m}^3/\text{s} \bullet \qquad (4.1\ \&\ 4.5)$$

$$V_1 = Q/A_1 = \frac{0.3\ \text{m}^3/\text{s}}{\frac{\pi}{4}(0.3\ \text{m})^2} = 4.24\ \text{m/s} \bullet \qquad (4.5)$$

$$V_2 = Q/A_2 = \frac{0.3\ \text{m}^3/\text{s}}{\frac{\pi}{4}(0.2\ \text{m})^2} = 9.55\ \text{m/s} \bullet \qquad (4.5)$$

or

$$V_2 = 4.24\ \text{m/s}\left(\frac{300\ \text{mm}}{200\ \text{mm}}\right)^2 = 9.55\ \text{m/s} \bullet \qquad (4.5)$$

ILLUSTRATIVE PROBLEM 4.2

Three kilograms of air per second flow through the reducer of the preceding problem, the air in the 300 mm pipe having a specific weight of 9.8 N/m^3. In flowing through the reducer, the pressure and temperature will fall, causing the air to expand, producing a reduction of density. Assuming that the specific weight of the air in the 200 mm pipe is 7.85 N/m^3, calculate the weight and volume flowrates, the mass velocities, and the velocities in the two pipes.

SOLUTION

Again, we select a control volume, bounded at sections 1 and 2 as indicated on the diagram for the preceding problem. The specific weights, the pipe diameters, and the mass flowrate $\dot{m} = 3.0$ kg/s are known. Gravitational acceleration $g_n = 9.81$ m/s^2. Applying Eq. 4.3 produces

$$G = g_n\dot{m} = (9.81\ \text{m/s})(3.0\ \text{kg/s}) = 29.4\ \text{kg}\cdot\text{m/s}^2 = 29.4\ \text{N/s} \bullet \qquad (4.3)$$

Now, from Eqs. 1.1, 4.1, 4.3, and 4.4,

$$\rho_1 = \gamma_1/g_n = 9.8\ \text{N/m}^3/9.81\ \text{m/s}^2 = 1.0\ \text{kg/m}^3 \quad \text{and}$$
$$\rho_2 = \gamma_2/g_n = 7.85\ \text{N/m}^3/9.81\ \text{m/s}^2 = 0.8\ \text{kg/m}^3 \qquad (1.1)$$

$$V_1 = \frac{\dot{m}}{\rho_1 A_1} = \frac{3.0}{1.0 \times \frac{\pi}{4}(0.3)^2} = 42.4\ \text{m/s} \bullet \quad \text{and}$$
$$V_2 = \frac{\dot{m}}{\rho_2 A_2} = \frac{3.0}{0.8 \times \frac{\pi}{4}(0.2)^2} = 119.4\ \text{m/s} \bullet \qquad (4.1)$$

$$Q_1 = A_1V_1 = \left(\frac{\pi}{4}(0.3)^2\right) \times 42.4 = 3.0\ \text{m}^3/\text{s} \bullet$$

and

$$Q_2 = A_2V_2 = \left(\frac{\pi}{4}(0.2)^2\right) \times 119.4 = 3.75\ \text{m}^3/\text{s} \bullet \qquad (4.4)$$

or to check

$$Q_1 = G/\gamma_1 = 29.4/9.8 = 3.0 \text{ m}^3/\text{s} \qquad (4.3 \text{ \& } 4.4)$$

and

$$Q_2 = G/\gamma_2 = 29.4/7.85 = 3.75 \text{ m}^3/\text{s} \qquad (4.3 \text{ \& } 4.4)$$

Finally, the mass velocities are

$$m_{v_1} = \rho_1 V_1 = 1.0 \text{ kg/m}^3 \times 42.4 \text{ m/s} = 42.4 \text{ kg/m}^2\text{·s} \bullet$$

$$\text{and } m_{v_2} = \rho_2 V_2 = 0.8 \text{ kg/m}^3 \times 119.4 \text{ m/s} = 95.8 \text{ kg/m}^2\text{·s} \bullet$$

Compare this problem with the preceding one, noting similarities and differences.

ILLUSTRATIVE PROBLEM 4.3

Taking Fig. 4.2 to represent an axisymmetric parabolic velocity distribution in a cylindrical pipe of radius R, calculate the mean velocity in terms of the maximum velocity, v_c.

SOLUTION

We apply Eq. 4.8

$$V = \frac{1}{A}\iint_A v\, dA \qquad (4.8)$$

Take r as the radial distance to any local velocity, v, and the element of area dA, $dA = 2\pi r\, dr$, while the equation of the parabola is $v = v_c(1 - r^2/R^2)$. Then,

$$V = \frac{1}{\pi R^2}\int_0^R v_c\left(1 - \frac{r^2}{R^2}\right) 2\pi r\, dr \qquad (4.8)$$

Integration shows that for the parabolic axisymmetric profile, the mean velocity is half of the centerline velocity, that is,

$$V = \frac{v_c}{2} \bullet$$

4.3 CONSERVATION OF MASS: THE CONTINUITY EQUATION—TWO-DIMENSIONAL STEADY FLOW

For two-dimensional flow the equation of continuity may be derived by considering a general control volume and repeating the conservation of mass analysis made for one-dimensional flow. Consider the flow and control volume in Fig. 4.3. As the system, which is the fluid in the control volume at time t, moves out of the control volume into zone O, new fluid flows into the control volume filling zone I. Part of the system remains in the

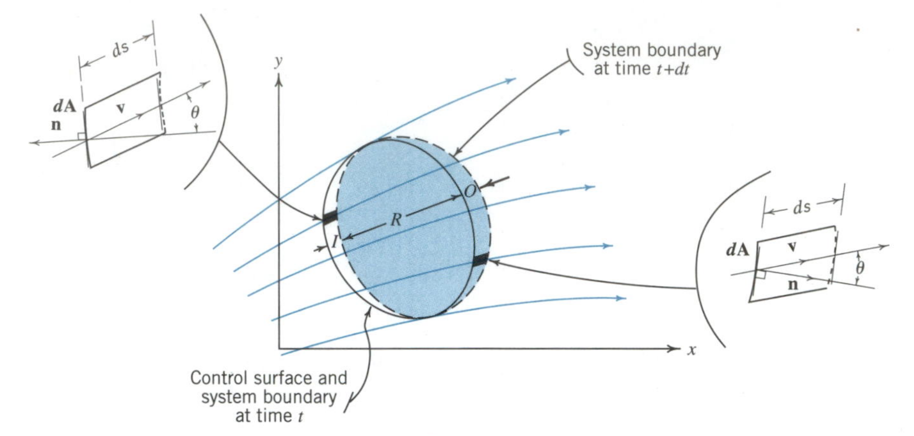

Fig. 4.3

control volume in zone R. From the conservation of mass

$$(m_I + m_R)_t = (m_R + m_O)_{t+dt}$$

and for a steady flow $(m_R)_t = (m_R)_{t+dt}$. Hence, as before,

$$(m_I)_t = (m_O)_{t+dt}$$

The mass in O is the integral of the masses moving *out* through all the incremental areas dA of the control surface in time dt (see the right-hand insert in Fig. 4.3)

$$(m_O)_{t+dt} = \iint_{C.S._{out}} \rho(ds \cos\theta)\, dA$$

because the volume of the small prism is the product of the area[1] of its base and its height (perpendicular to the base). The distance that the system boundary moves along a streamline $ds = v\,dt$, so that

$$(m_O)_{t+dt} = \iint_{C.S._{out}} \rho(v \cos\theta)\, dA\, dt$$

Note that $V \cos\theta$ is the magnitude of the velocity component *normal* to the control surface at dA; thus, the net flow out of the control volume is determined, *not* by the total velocity, but by the *normal velocity component*. The tangential component does *not* contribute to flow through the surface. In vector terms, if **n** is the *outward unit normal vector* at dA, from Appendix 6,

$$\mathbf{v} \cdot \mathbf{n} = v \cos\theta$$

Thus,

$$(m_O)_{t+dt} = dt \iint_{C.S._{out}} \rho\mathbf{v} \cdot \mathbf{n}\, dA = dt \iint_{C.S._{out}} \rho\mathbf{v} \cdot d\mathbf{A}$$

[1]For plane (two-dimensional) motion $dA = 1 \cdot dl$ where dl is the length of the differential segment of the control surface and a slice of the flow one-unit thick and perpendicular to the plane of flow is used.

when $d\mathbf{A} = \mathbf{n}\, dA$ is defined as the *directed area element*. It follows that the mass flow into I is

$$\begin{aligned}(m_I)_t &= \iint_{C.S._{in}} \rho(ds \cos \theta)\, dA \\ &= \iint_{C.S._{in}} \rho(v \cos \theta)\, dA\, dt = dt \iint_{C.S._{in}} \rho\mathbf{v} \cdot (-\mathbf{n})\, dA \\ &= dt \left\{ -\iint_{C.S._{in}} \rho\mathbf{v} \cdot \mathbf{n}\, dA \right\} = dt \left\{ -\iint_{C.S._{in}} \rho\mathbf{v} \cdot d\mathbf{A} \right\}\end{aligned}$$

because $\mathbf{n}$ is the outward normal and points ''against'' the IN flow (see the left-hand insert in Fig. 4.3). As the IN and OUT masses are equal, dividing by dt produces

$$-\iint_{C.S._{in}} \rho\mathbf{v} \cdot \mathbf{n}\, dA = \iint_{C.S._{out}} \rho\mathbf{v} \cdot \mathbf{n}\, dA; \; -\iint_{C.S._{in}} \rho\mathbf{v} \cdot d\mathbf{A} = \iint_{C.S._{out}} \rho\mathbf{v} \cdot d\mathbf{A}$$

Clearly,

$$\iint_{C.S._{out}} \rho\mathbf{v} \cdot dA + \iint_{C.S._{in}} \rho\mathbf{v} \cdot d\mathbf{A} = \oint\!\!\oint_{C.S.} \rho\mathbf{v} \cdot d\mathbf{A} = \oint\!\!\oint_{C.S.} \rho\mathbf{v} \cdot \mathbf{n}\, dA = 0 \tag{4.9}$$

in which the circle and arrow symbol on the integral sign indicates that the integral is to be taken once around the control surface in the counterclockwise direction. Thus, over the entire control surface, the sum of all mass flowrates normal to the surface is zero; that is, the mass flowrate IN must exactly equal the mass flowrate OUT in *steady flow*.

The integral continuity equation 4.9 is valid for finite or infinitesimal control volumes. The differential forms are easily obtained from Eq. 4.9. For the infinitesimal control volume of Fig. 4.4*a*, the density ρ and the velocity components u and v are defined at the center (x, y) of the element. Therefore, to first-order accuracy,

$$\rho_{AB} \cong \rho - \frac{d\rho}{dy}\frac{dy}{2} \quad \text{and} \quad v_{AB} \cong v - \frac{dv}{dy}\frac{dy}{2}, \qquad \text{while} \quad \mathbf{v} \cdot \mathbf{n}\, dA = -v_{AB}\, dx, \text{ etc.}$$

So the mass flowrates on the sides are

$$\begin{aligned}\int_{AB} \rho\mathbf{v} \cdot \mathbf{n}\, dA &\cong -\left(\rho - \frac{\partial \rho}{\partial y}\frac{dy}{2}\right)\left(v - \frac{\partial v}{\partial y}\frac{dy}{2}\right) dx \\ \int_{BC} \rho\mathbf{v} \cdot \mathbf{n}\, dA &\cong \left(\rho + \frac{\partial \rho}{\partial x}\frac{dx}{2}\right)\left(u + \frac{\partial u}{\partial x}\frac{dx}{2}\right) dy \\ \int_{CD} \rho\mathbf{v} \cdot \mathbf{n}\, dA &\cong \left(\rho + \frac{\partial \rho}{\partial y}\frac{dy}{2}\right)\left(v + \frac{\partial v}{\partial y}\frac{dy}{2}\right) dx \\ \int_{DA} \rho\mathbf{v} \cdot \mathbf{n}\, dA &\cong -\left(\rho - \frac{\partial \rho}{\partial x}\frac{dx}{2}\right)\left(u - \frac{\partial u}{\partial x}\frac{dx}{2}\right) dy\end{aligned}$$

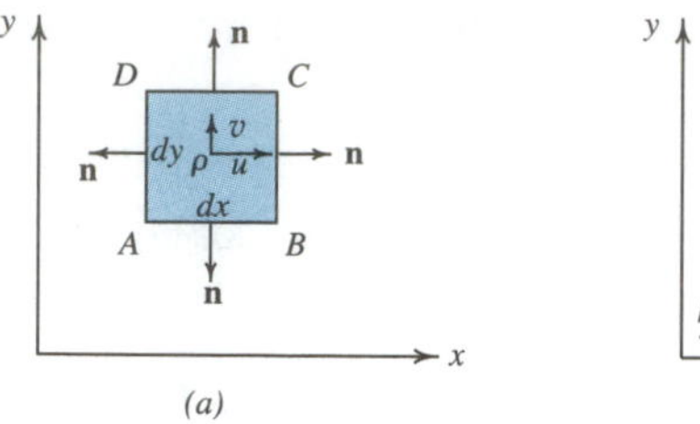

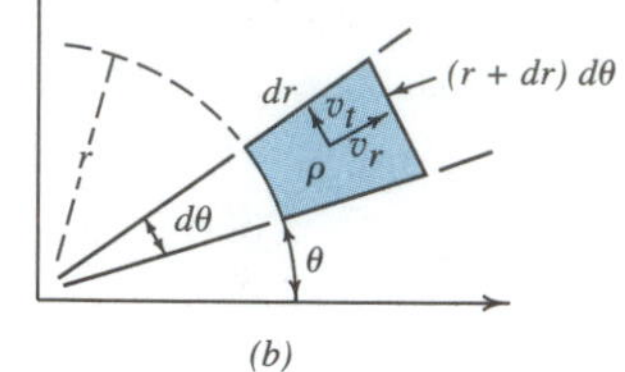

Fig. 4.4

Equation 4.9 now yields

$$\int_{AB} \rho \mathbf{v} \cdot \mathbf{n}\, dA + \int_{BC} \rho \mathbf{v} \cdot \mathbf{n}\, dA + \int_{CD} \rho \mathbf{v} \cdot \mathbf{n}\, dA + \int_{CA} \rho \mathbf{v} \cdot \mathbf{n}\, dA = 0$$

By substituting the flowrate values in differential terms, expanding the products, and retaining only terms of lowest order (largest order of magnitude), we obtain

$$\rho \frac{\partial v}{\partial y} + v \frac{\partial \rho}{\partial y} + \rho \frac{\partial u}{\partial x} + u \frac{\partial \rho}{\partial x} = 0$$

The above continuity equation for compressible, steady, two-dimensional flow can also be written as

$$\frac{\partial}{\partial x} (\rho u) + \frac{\partial}{\partial y} (\rho v) = 0 \tag{4.10}$$

For steady flow the density ρ is constant at any point in space. If it is also uniform in space, Eq. 4.10 reduces to

$$\boxed{\frac{\partial u}{\partial x} + \frac{\partial v}{\partial y} = 0} \tag{4.11}$$

It is noteworthy that a flow may be incompressible and steady, i.e., the density at any point is constant, and still the density may vary in space, e.g., stratified flows of various kinds may be of this type. Thus, Eq. 4.11 is valid for steady incompressible flows only if the density is also uniform in space.

Application of the same principles to the polar element of Fig. 4.4*b* yields the two-dimensional continuity equation in polar coordinates. For the compressible fluid this is

$$\frac{1}{r} (\rho v_r) + \frac{\partial}{\partial r} (\rho v_r) + \frac{\partial}{r \partial \theta} (\rho v_t) = 0 \tag{4.12}$$

which, for the constant density fluid, reduces to

$$\frac{v_r}{r} + \frac{\partial v_r}{\partial r} + \frac{\partial v_t}{r \partial \theta} = 0 \tag{4.13}$$

ILLUSTRATIVE PROBLEM 4.4

A mixture of ethanol (grain or ethyl alcohol) and gasoline, called "gasohol," is created here by pumping the two liquids into the "wye" pipe junction shown. The mixture is 10% alcohol and is observed to have a density of 691.1 kg/m^3 and a velocity of 1.08 m/s when the gasoline input is 30 l/s. Find the (volume) flowrate and average velocity of the incoming ethanol.

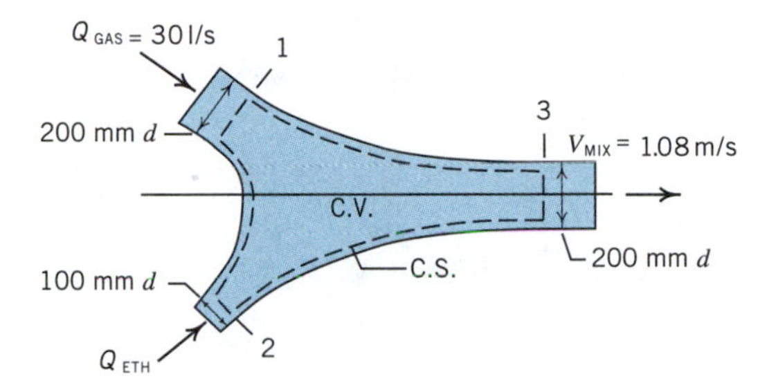

SOLUTION

In this case we will need both Eqs. 4.4 and 4.9, which are

$$\iint_{C.S._{out}} \rho \mathbf{v} \cdot d\mathbf{A} + \iint_{C.S._{in}} \rho \mathbf{v} \cdot d\mathbf{A} = \oiint_{C.S.} \rho \mathbf{v} \cdot d\mathbf{A} = \oiint_{C.S.} \rho \mathbf{v} \cdot \mathbf{n}\, dA = 0 \quad (4.9)$$

$$Q = AV \quad (4.4)$$

We define a control volume and three boundary areas through which flow passes on the diagram above and we assume that the flow in each pipe is one dimensional. The data show that

$$Q_{\text{GAS}} = 30 \text{ l/s} \qquad \rho_{\text{GAS}} = 680.3 \text{ kg/m}^3 \qquad \rho_{\text{ETH}} = 788.6 \text{ kg/m}^3$$

$$V_{\text{MIX}} = 1.08 \text{ m/s} \qquad \rho_{\text{MIX}} = 691.1 \text{ kg/m}^3$$

Then we have from the control volume and by assuming that the flow in each pipe is one-dimensional

$$\rho_1 = 680.3 \text{ kg/m}^3 \qquad \rho_2 = 788.6 \text{ kg/m}^3 \qquad \rho_3 = 691.1 \text{ kg/m}^3$$

$$V_3 = 1.08 \text{ m/s}$$

$$A_1 = \frac{\pi}{4}(0.2)^2 = 0.031 \text{ m}^2 \qquad A_2 = 0.007\,9 \text{ m}^2 \qquad A_3 = 0.031 \text{ m}^2$$

$$V_1 = 30 \times 10^{-3}/0.031 = 0.97 \text{ m/s} \quad (4.4)$$

From Eq. 4.9

$$\iint_1 \rho \mathbf{v} \cdot \mathbf{n}\, dA + \iint_2 \rho \mathbf{v} \cdot \mathbf{n}\, dA + \iint_3 \rho \mathbf{v} \cdot \mathbf{n}\, dA = 0 \quad (4.9)$$

Using the assumed uniform (i.e., average) velocity at each cross section,

$$\iint_1 \rho \mathbf{v} \cdot \mathbf{n}\, dA = -680.3 \times 0.97 \times 0.031 = -20.4 \text{ kg/s}$$

The minus sign arises because the outward normal **n** and **v** have opposite directions so $(\mathbf{v} \cdot \mathbf{n}\, dA)_1 = -V_1\, dA$. Similarly, for section 2,

$$\iint_2 \rho \mathbf{v} \cdot \mathbf{n}\, dA = -788.6 \times V_2 \times 0.0079 = -6.23V_2$$

At section 3 **v** and **n** have the same direction so

$$\iint_3 \rho \mathbf{v} \cdot \mathbf{n}\, dA = 691.1 \times 1.08 \times 0.031 = 23.1 \text{ kg/s}$$

Therefore,

$$\oiint_{\text{C.S.}} \rho \mathbf{v} \cdot \mathbf{n}\, dA = -20.4 - 6.23V_2 + 23.1 = 0 \tag{4.9}$$

$$V_2 = 0.43 \text{ m/s} \bullet$$

$$Q_{\text{ETH}} = Q_2 = A_2V_2 = 0.0079 \times 0.43 = 3.4 \times 10^{-3} \text{ m}^3/\text{s} = 3.4 \text{ l/s} \bullet \tag{4.4}$$

ILLUSTRATIVE PROBLEM 4.5

A two-dimensional incompressible flowfield is described by the equations $v_t = \omega r$ and $v_r = 0$ in which ω is a constant. Sketch this flow and show that it satisfies the continuity equation.

SOLUTION

From the problem statement, $v_t = \omega r$, $v_r = 0$, and $\omega =$ Constant. In addition, Eq. 4.13 applies, i.e.,

$$\frac{v_r}{r} + \frac{\partial v_r}{\partial r} + \frac{\partial v_t}{r \partial \theta} = 0 \tag{4.13}$$

Evidently the streamlines of this flow must be concentric circles because the radial component of velocity is everywhere zero. This flow is known as a *forced vortex.* We sketch it below.

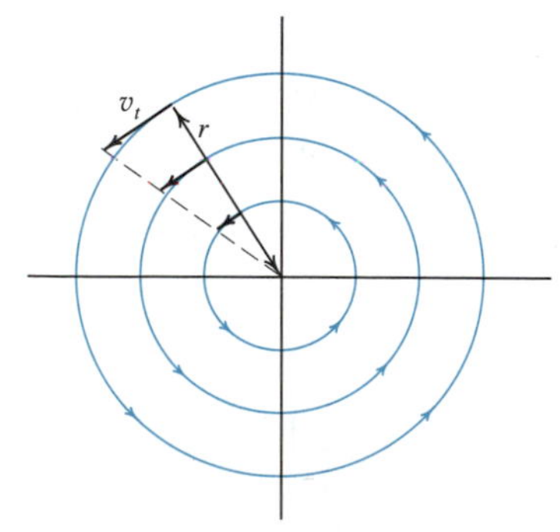

Some of its features have been examined in Section 2.7 and others are discussed in Illustrative Problem 3.4. By substituting

$v_r = 0$ and $v_t = \omega r$ in Eq. 4.13, we find $\frac{0}{r} + \frac{\partial}{\partial r}(0) + \frac{\partial}{r\partial\theta}(\omega r) = 0 + 0 + 0 = 0$

The equation of continuity is satisfied; this flow is physically possible. ●

4.4 THE REYNOLDS TRANSPORT THEOREM

The previous two sections have shown us how to relate the system and control volume concepts in simple situations and we derived interesting integral relations for the conservation of mass. In fact, we showed how to derive the conservation law for a control volume from knowledge of the conservation law for the system's mass. That is the underlying idea behind the *Reynolds Transport Theorem.* Given that laws such as mass conservation and Newton's second law are system's laws, we can make good use of a general relationship that converts the laws from one view to the other, i.e., from the system to the control volume. The elements are all present in the derivation in the previous section, requiring only some definitions and extension to unsteady flow. The essential change in procedure is that we will have to take integrals over the system or control volume to account for the changes within them as a function of time. All the vector relationships developed there are used here without re-derivation, but Fig. 4.3 is reproduced here for convenience as Fig. 4.5.

To begin, let us define two types of properties of the fluid in a system. These are *extensive properties*, which are the *total* system mass, momentum, energy, etc., and the *intensive properties*, which are the system mass, momentum, energy, etc., *per unit mass.* We can denote the extensive properties by E and intensive properties by i. If the system mass is m, then we have as we look at Fig. 4.5,

$$E = \iiint_{System} i\, dm = \iiint_{System} i\rho\, d\forall \tag{4.14}$$

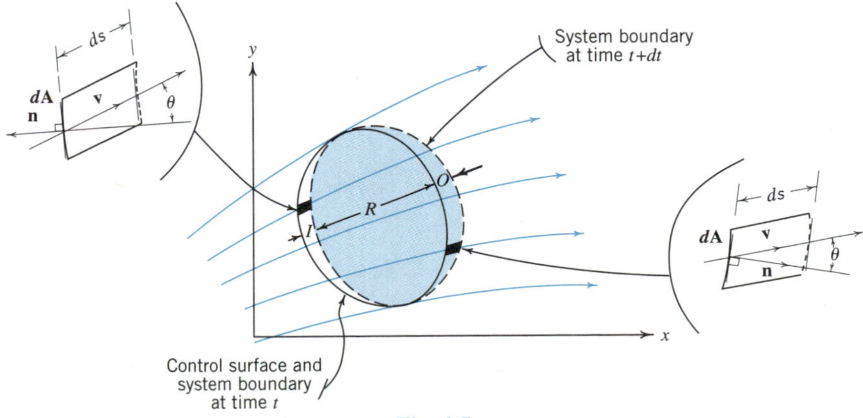

Fig. 4.5

Here, the integral is a volume integral over the entire system at a particular time. It follows that, if E is system mass m, then i is unity, if E is system momentum, then i is $\mathbf{V}$, and so forth. It appears that the most interesting thing to know would be the time rate of change of a system property, i.e., to know the rate of mass, momentum or energy increase (or decrease). That result is obtained by examining the system of Fig. 4.5 at time t and at time $t + dt$. For any extensive system property E,

$$E_{t+dt} - E_t = (E_R + E_O)_{t+dt} - (E_R + E_I)_t$$

From the previous section, it is possible to deduce these terms as follows:

$$(E_O)_{t+dt} = dt \iint_{C.S._{out}} i\rho\mathbf{v} \cdot d\mathbf{A}$$

$$(E_I)_t = dt \left(-\iint_{C.S._{in}} i\rho\mathbf{v} \cdot d\mathbf{A} \right)$$

Then from Eq. 4.14,

$$(E_R)_{t+dt} = \left(\iiint_R i\rho \, d\forall \right)_{t+dt}$$

$$(E_R)_t = \left(\iiint_R i\rho \, d\forall \right)_t$$

The result is

$$\begin{aligned} E_{t+dt} - E_t &= (E_R + E_O)_{t+dt} - (E_R + E_I)_t \\ &= \left(\iiint_R i\rho \, d\forall \right)_{t+dt} + dt \iint_{C.S._{out}} i\rho\mathbf{v} \cdot d\mathbf{A} \\ &\quad - \left(\iiint_R i\rho \, d\forall \right)_t - dt \left(-\iint_{C.S._{in}} i\rho\mathbf{v} \cdot d\mathbf{A} \right) \end{aligned}$$

Dividing both sides of this equation by dt and rearranging yields

$$\begin{aligned} \frac{E_{t+dt} - E_t}{dt} &= \frac{1}{dt} \left(\left(\iiint_R i\rho \, d\forall \right)_{t+dt} - \left(\iiint_R i\rho \, d\forall \right)_t \right) \\ &\quad + \iint_{C.S._{out}} i\rho\mathbf{v} \cdot d\mathbf{A} + \iint_{C.S._{in}} i\rho\mathbf{v} \cdot d\mathbf{A} \end{aligned}$$

In the limit as dt goes to zero, the left side of the above equation becomes the total derivative of E, while the first term on the right becomes the difference between an integral over the control volume at times $t + dt$ and t divided by dt, leading to a partial derivative with respect to time. (As dt approaches zero, the integral over the R volume is the same as the integral over the control volume.) We obtain the final result by use of the formalism of the left section of Eq. 4.9:

$$\begin{aligned}\frac{dE}{dt} &= \frac{\partial}{\partial t}\left(\iiint_{C.V.} i\rho\, d\forall\right) + \oiint_{C.S.} i\rho\mathbf{v}\cdot d\mathbf{A} \\ &= \frac{\partial}{\partial t}\left(\iiint_{C.V.} i\rho\, d\forall\right) + \iint_{C.S._{out}} i\rho\mathbf{v}\cdot d\mathbf{A} + \iint_{C.S._{in}} i\rho\mathbf{v}\cdot d\mathbf{A}\end{aligned} \tag{4.15}$$

Accordingly, for any extensive property E of the system, its time rate of change in the system is equal to its time rate of change within the control volume plus the fluxes of it across the control surface.

An easy application of Eq. 4.15 is to the conservation of mass. For the system, $E = m$, the change of mass m with time is zero according to the law of conservation of mass, and $i = 1$ yielding

$$\begin{aligned}\frac{\partial}{\partial t}\left(\iiint_{C.V.} \rho\, d\forall\right) &= -\oiint_{C.S.} \rho\mathbf{v}\cdot d\mathbf{A} \\ &= -\left(\iint_{C.S._{out}} \rho\mathbf{v}\cdot d\mathbf{A} + \iint_{C.S._{in}} \rho\mathbf{v}\cdot d\mathbf{A}\right)\end{aligned} \tag{4.16}$$

Compare this result with Eq. 4.9. Clearly, in an unsteady flow it is possible for the mass within the control volume to change if the density changes. If the flow is of uniform density or is steady, then

$$\frac{\partial}{\partial t}\left(\iiint_{C.V.} \rho\, d\forall\right) = 0$$

and so

$$-\oiint_{C.S.} \rho\mathbf{v}\cdot d\mathbf{A} = -\left(\iint_{C.S._{out}} \rho\mathbf{v}\cdot d\mathbf{A} + \iint_{C.S._{in}} \rho\mathbf{v}\cdot d\mathbf{A}\right) = 0 \tag{4.9}$$

which is Eq. 4.9 as expected. To close the loop, for the steady, one-dimensional flow of Fig. 4.1, we have, from this last result,

$$-\oiint_{C.S.} \rho\mathbf{v}\cdot d\mathbf{A} = -\left(\iint_{C.S._{out}} \rho\mathbf{v}\cdot d\mathbf{A} + \iint_{C.S._{in}} \rho\mathbf{v}\cdot d\mathbf{A}\right) = 0,$$

$$\iint_{C.S._{out}} \rho\mathbf{v}\cdot d\mathbf{A} = \rho_2 V_2 A_2 \text{ and}$$

$$\iint_{C.S._{in}} \rho\mathbf{v}\cdot d\mathbf{A} = -\rho_1 V_1 A_1$$

Accordingly,

$$\left(\iint_{C.S._{out}} \rho \mathbf{v} \cdot d\mathbf{A} + \iint_{C.S._{in}} \rho \mathbf{v} \cdot d\mathbf{A}\right) = \rho_2 V_2 A_2 + (-\rho_1 V_1 A_1) \\ = \rho_2 V_2 A_2 - \rho_1 V_1 A_1 = 0 \tag{4.1}$$

and we retrieve Eq. 4.1, viz., $\rho_2 V_2 A_2 = \rho_1 V_1 A_1$

In Chapters 5 and 6, the Reynolds Transport Theorem is used to derive the work-energy, impulse-momentum, and moment of momentum principles. The Reynolds Transport Theorem can be applied, of course, for any fluid flow property which can be defined in terms of a quantity per unit mass.

PROBLEMS

4.1. The mean velocity of water in a 100 mm pipeline is 2 m/s. Calculate the rate of flow in cubic metres per second, newtons per second, and kilograms per second.

4.2. One hundred pounds of water per minute flow through a 6 in. pipeline. Calculate the mean velocity.

4.3. Four hundred litres per minute of glycerin flow in a 75 mm pipeline. Calculate the mean velocity.

4.4. A 0.3 m by 0.5 m rectangular air duct carries a flow of 0.45 m^3/s at a density of 2 kg/m^3. Calculate the mean velocity in the duct. If the duct tapers to 0.15 m by 0.5 m size, what is the mean velocity in this section if the density is 1.5 kg/m^3 there?

4.5. Across a shock wave in a gas flow there is a great change in gas density ρ. If a shock wave occurs in a duct such that $V = 660$ m/s and $\rho = 1.0$ kg/m^3 before the shock and $V = 250$ m/s after the shock, what is ρ after the shock?

4.6. Water flows in a pipeline composed of 75 mm and 150 mm pipe. Calculate the mean velocity in the 75 mm pipe when that in the 150 mm pipe is 2.5 m/s. What is its ratio to the mean velocity in the 150 mm pipe?

4.7. A smooth nozzle with a tip diameter of 2 in. terminates a 6 in. waterline. Calculate the mean velocity of efflux from the nozzle when the velocity in the line is 10 ft/s.

4.8. Hydrogen is being pumped through a pipe system whose temperature is held at 273 K. At a section where the pipe diameter is 10 mm, the pressure and average velocity are 200 kPa and 30 m/s. Find all possible velocities and pressures at a downstream section whose diameter is 20 mm.

4.9. At a point in a two-dimensional fluid flow, two streamlines are parallel and 75 mm (3 in.) apart. At another point these streamlines are parallel but only 25 mm (1 in.) apart. If the velocity at the first point is 3 m/s (9.8 fps), calculate the velocity at the second.

4.10. A 300 mm pipeline leaves a large tank through a square-edged hole in its vertical wall. The mean velocity in the pipe is 4.5 m/s. Assuming that the fluid in the tank approaches the center of the pipe entrance radially, what is the velocity of the fluid 0.5, 1, and 2 m from the pipe entrance?

4.11. Five hundred cubic feet per second of water flow in a rectangular open channel 20 ft wide and 8 ft deep. After passing through a transition structure into a trapezoidal canal of 5 ft base width with sides sloping at 30°, the velocity is 6 fps. Calculate the depth of the water in the canal.

4.12. Calculate the mean velocities for these two-dimensional velocity profiles if $v_c = 3$ m/s or 10 ft/s.

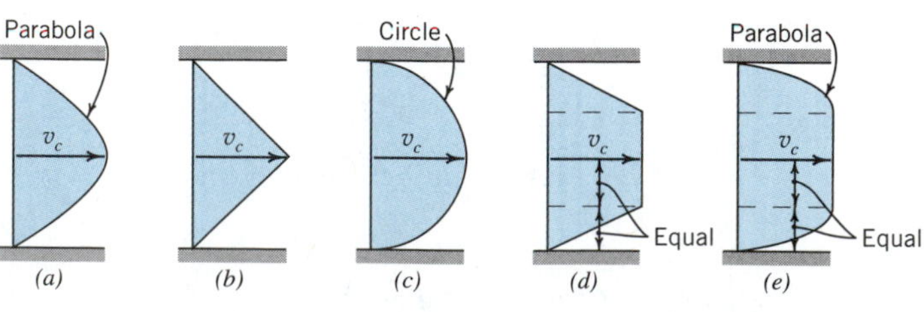

Problem 4.12

4.13. Calculate the mean velocities in the preceding problem, assuming the velocity profiles to be axisymmetric in a cylindrical passage.

4.14. If the velocity profile in a passage of width $2R$ is given by the equation $v/v_c = (y/R)^{1/n}$, derive an expression for V/v_c in terms of n: (*a*) for a two-dimensional passage, and (*b*) for a cylindrical passage.

4.15. Calculate the mean velocities at C and D assuming them to be radial.

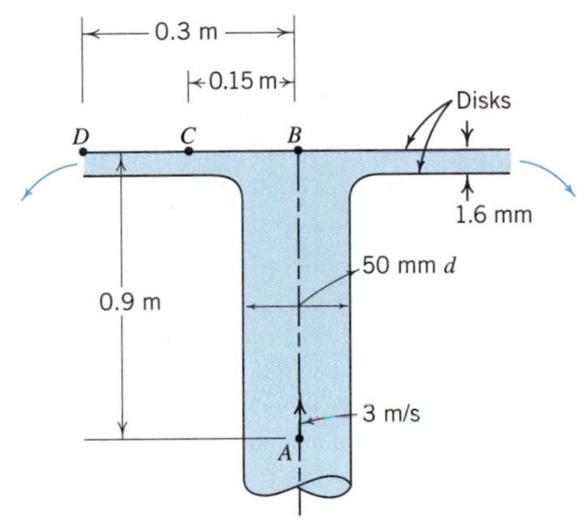

Problem 4.15

4.16. Fluid passes through this set of thin closely spaced blades. What flowrate q is required for the velocity V to be 10 ft/s (or equivalently, 3.0 m/s)?

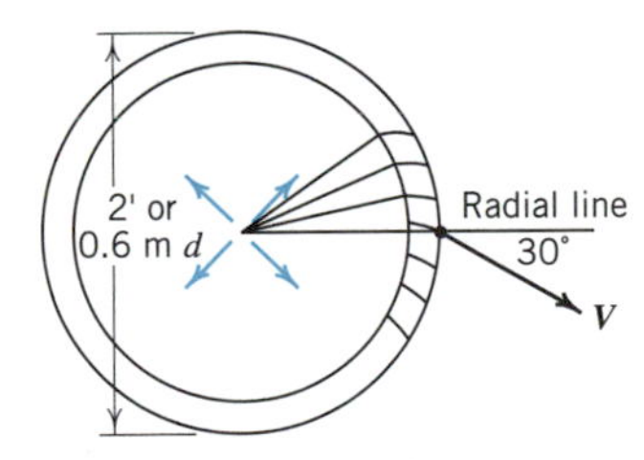

Problem 4.16

4.17. A pipeline 0.3 m in diameter divides at a Y into two branches 200 mm and 150 mm in diameter. If the flowrate in the main line is 0.3 m^3/s and the mean velocity in the 200 mm pipe is 2.5 m/s, what is the flowrate in the 150 mm pipe?

4.18. A *manifold pipe* of 3 in. diameter has four openings in its walls spaced equally along the pipe and is closed at the downstream end. If the discharge from each opening is 0.50 cfs, what are the mean velocities in the pipe between the openings?

4.19. Find the average efflux velocity V if the flow exits from a hole of area 1 m^2 in the side of the duct as shown.

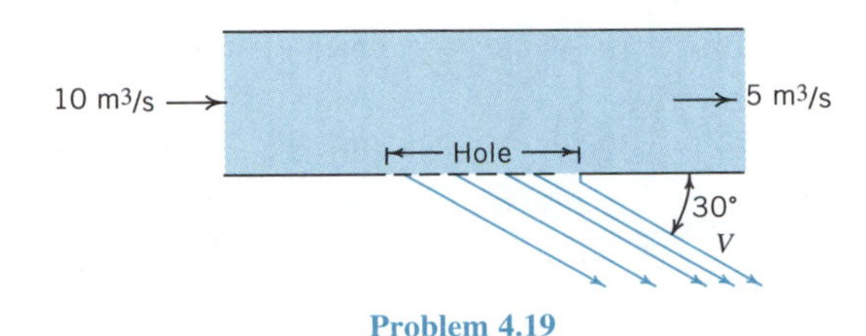

Problem 4.19

4.20. Find V for this mushroom cap on a pipeline.

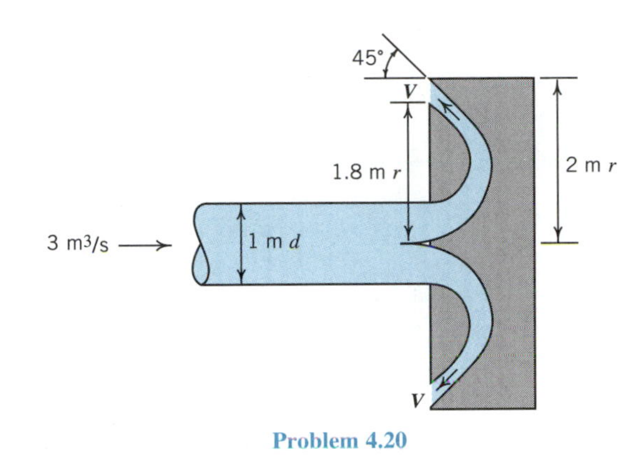

Problem 4.20

4.21. Using the wye and control volume of Illustrative Problem 4.4, find the mixture velocity and density if freshwater ($\rho_1 = 1$ Mg/m^3) enters section 1 at 50 l/s, while saltwater ($\rho_2 = 1.03$ Mg/m^3) enters section 2 at 25 l/s.

4.22. Derive an expression for the flowrate q between two streamlines of radii r_1 and r_2 for the flowfields (*a*) $v_r = 0$, $v_t = c/r$, and (*b*) $v_r = 0$, $v_t = \omega r$.

4.23. For the flowfield $v_r = c/r$, $v_t = 0$, derive an expression for the flowrate q between any two streamlines.

4.24. The flow of a uniform, incompressible fluid is described by the equations $v_r = (2\pi r)^{-1}$ and $v_t = -(2\pi r)^{-1}$. Sketch this flow, show that it satisfies the differential continuity equation, find the magnitude and direction of the velocity, and determine if the continuity equation for a finite control volume $r \leq r_0$ is satisfied.

4.25. Derive the equation of continuity in polar coordinates for the incompressible fluid.

4.26. Investigate the flowfields of problem 3.6 to see if they are physically possible (i.e., satisfy the equation of continuity).

4.27. Suppose the two-dimensional flowfield for flow against a plate (problem 3.7) is given by the equations: $u = 4x$, $v = -4y$. Show that this flow satisfies the equation of continuity.

4.28. Does the flow of problem 3.1 satisfy the equation of continuity?

4.29. Using Eq. 4.16, derive the unsteady equivalent of Eq. 4.10.

4.30. Using Eq. 4.16, derive the unsteady equivalent of Eq. 4.12.

4.31. For a differential control volume, show that Eq. 4.16 reduces to Eq. 4.11 for a steady flow with a uniform constant density.

4.32. Derive Eq. 4.1 by use of Eq. 4.16.

4.33. If $E = m\mathbf{V}_c$, which is the linear momentum of a fluid system of mass m whose center of mass moves at the velocity $\mathbf{V}_c$, then $i = \mathbf{v}$, which is the velocity of fluid elements of mass dm. Newton's second law for such a system states that the sum of the forces acting on the system must equal the change in the system's momentum [this is equivalent in a particle sense to mass times acceleration] or

$$\sum \mathbf{F} = \frac{d}{dt}(m\mathbf{V}_c).$$

Use this information and Eq. 4.15 to derive the impulse-momentum equation for a control volume.

CHAPTER

5 FLOW OF AN INCOMPRESSIBLE IDEAL FLUID

Significant insight into the basic laws of fluid flow can be obtained from a study of the flow of a hypothetical *ideal fluid*. An ideal fluid is a fluid assumed to be inviscid, or devoid of viscosity.[1] In such a fluid there are no frictional effects between moving fluid layers or between these layers and boundary walls, and thus no cause for eddy formation or energy dissipation due to friction. The assumption that a fluid is ideal allows it to be treated as an aggregation of small particles that will support pressure forces normal to their surfaces but will slide over one another without resistance. Thus the motion of these ideal fluid particles is analogous to the motion of a solid body on a frictionless plane, and the unbalanced forces existing on them cause the acceleration of these particles according to Newton's second law.

Under the assumption of frictionless motion, equations are considerably simplified and more easily assimilated by the beginner in the field. In many cases these simplified equations allow solution of engineering problems to an accuracy entirely adequate for practical use. The beginner should not jump to the conclusion that the assumption of frictionless flow leads to a useless abstraction which is always far from reality. In those real situations, where the actual effects of friction are small, the frictionless assumption will give good results; where friction is large, it obviously will not. The identification of these situations is part of the art of fluid mechanics; for example, the lift on a wing can often be predicted accurately by an inviscid analysis while the drag rarely can.

The further assumption of an incompressible (i.e., constant-density) fluid restricts the

[1]Compare with the definition of a perfect gas, Section 1.4.

present chapter to the flow of liquids and of gases that undergo negligibly small changes of pressure and temperature. The flow of gases with large density changes is discussed in Chapter 13, but there are numerous practical engineering problems that involve fluids whose densities may safely be considered constant; thus, this assumption proves to be not only a practical one but also a useful simplification in the introduction to fluid flow, because it usually permits thermodynamic effects to be disregarded.

This chapter introduces the important Bernoulli and work-energy equations, which permit us to relate and to predict pressures and velocities in a flowfield. The concepts are introduced from several points of view. Thus, the presentation begins with a simple one-dimensional analysis of an elemental fluid system moving along a streamline, then moves next to a work-energy relation that allows inclusion of pumps and turbines in the computations, and concludes with an introduction to two-dimensional flows. Significantly, while the simple analysis of a fluid system yields the important Bernoulli equation (Section 5.2), precisely the same equation is obtained *as a simplification* of the more general result of a *control volume analysis* which employs the work-energy principle (Section 5.5).

One-Dimensional Flow

5.1 EULER'S EQUATION

In 1750, Leonhard Euler first applied Newton's second law to the motion of fluid particles and thus laid the groundwork for an analytical approach to fluid dynamics.

Consider a streamline and select a small cylindrical fluid system for analysis, as shown in Fig. 5.1. The forces tending to accelerate the cylindrical fluid system are: pressure forces on the ends of the system, $p\,dA - (p + dp)\,dA = -dp\,dA$ (the pressures on the sides of the system have no effect on its acceleration), and the component of weight in the direction of motion, $-\rho g_n\,ds\,dA(dz/ds) = -\rho g_n\,dA\,dz$. The differential mass being

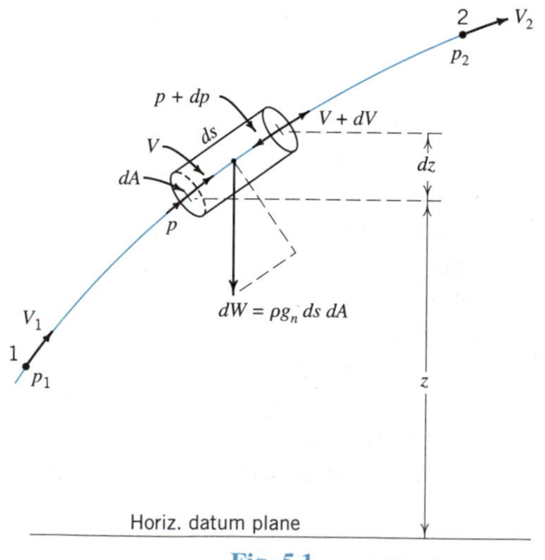

Fig. 5.1

accelerated by the action of these differential forces is $dM = \rho\, ds\, dA$. Applying Newton's second law $dF = (dM)a$ along the streamline and using the one-dimensional expression for acceleration (equation 3.2) gives (recall only steady flow is being considered)

$$-dp\, dA - \rho g_n\, dA\, dz = (\rho\, ds\, dA)V\frac{dV}{ds}$$

Dividing by $\rho\, dA$ produces the one-dimensional Euler equation

$$\frac{dp}{\rho} + V\, dV + g_n\, dz = 0$$

For incompressible flow this equation is usually divided by g_n and written

$$\frac{dp}{\gamma} + d\left(\frac{V^2}{2g_n}\right) + dz = 0$$

or for uniform density flows

$$d\left(\frac{p}{\gamma} + \frac{V^2}{2g_n} + z\right) = 0$$

5.2 BERNOULLI'S EQUATION WITH ENERGY AND HYDRAULIC GRADE LINES

For *incompressible flow* of uniform density fluid, the one-dimensional Euler equation can be easily integrated between any two points (because γ and g_n are both constant) to obtain

$$\frac{p_1}{\gamma} + \frac{V_1^2}{2g_n} + z_1 = \frac{p_2}{\gamma} + \frac{V_2^2}{2g_n} + z_2$$

As points 1 and 2 are any two arbitrary points on the streamline, the quantity

$$\frac{p}{\gamma} + \frac{V^2}{2g_n} + z = H = \text{Constant} \tag{5.1}$$

applies to all points on the streamline and thus provides a useful relationship between pressure p, the magnitude V of the velocity, and the height z above datum. Equation 5.1 is known as the *Bernoulli equation* and the *Bernoulli constant H* is also termed the *total head*.

Examination of the *Bernoulli terms* of Eq. 5.1 reveals that p/γ and z are, respectively, the pressure (either gage or absolute) and potential heads encountered in Section 2.1 and hence may be visualized as vertical distances. Pitot's experiments (1732) showed that the sum of velocity head $V^2/2g_n$ and pressure head p/γ could be measured by placing a tiny open tube (now known as a *pitot tube*) in the flow with its open end upstream. Thus, the Bernoulli equation may be visualized for liquids as, in Fig. 5.2, the sum of the terms (total head) being the constant distance between the horizontal datum plane and the *total head line* or *energy line* (EL). The *piezometric head line* or *hydraulic grade line* (HGL) drawn through the tops of the piezometer columns gives a picture of the pressure variation in the flow; evidently (1) its distance from the streamtube is a direct measure of the static pressure

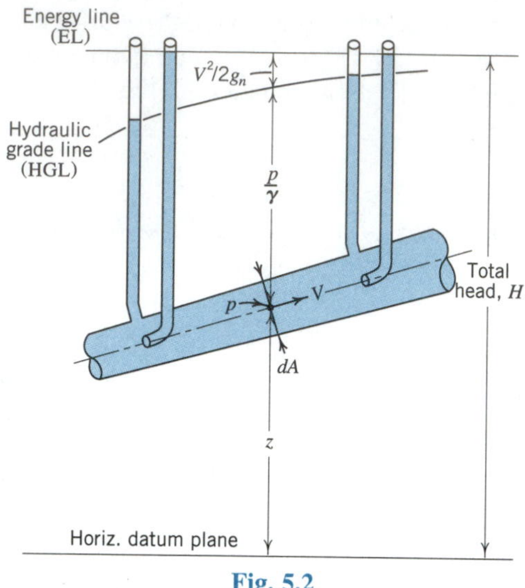

Fig. 5.2

in the flow, and (2) its distance below the energy line is proportional to the square of the velocity. Complete familiarity with these lines is essential because of their wide use in engineering practice and their great utility in problem solutions.

5.3 THE ONE-DIMENSIONAL ASSUMPTION FOR STREAMTUBES OF FINITE CROSS SECTION

The foregoing development of the Bernoulli equation has been carried out for a single streamline or infinitesimal streamtube across which the variation of p, V, and z is negligible because of the differential size of the cross section of the element. However, the engineer may apply this equation easily and fruitfully to large streamtubes such as pipes and canals once its limitations are understood. Consider a cross section of a large flow (open or closed but not a free jet) through which *all streamlines are precisely straight and parallel* (Fig. 5.3). The forces, normal to the streamlines, on the element of fluid of unit width

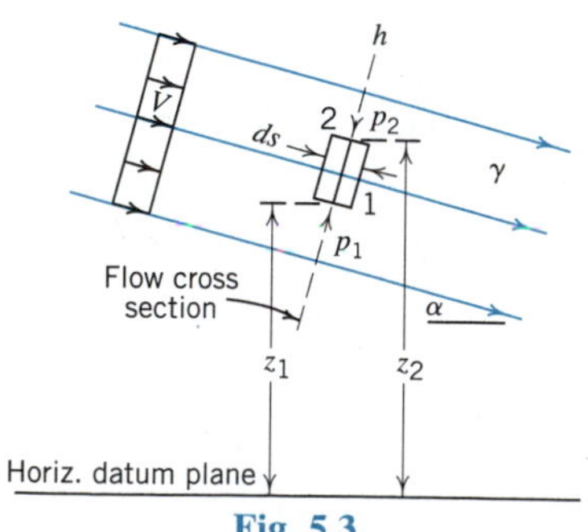

Fig. 5.3

are $(p_1 - p_2)\,ds$ and the component of the weight of the element, $\gamma h\, ds \cos \alpha$, in which $\cos \alpha = (z_2 - z_1)/h$. It is apparent that if (and only if) the streamlines are straight and parallel the acceleration toward the boundary is zero. This means that the forces defined above are in equilibrium:

$$(p_1 - p_2)\,ds = \gamma(z_2 - z_1)\,ds$$

yielding a result identical to that in Chapter 2,

$$\frac{p_1}{\gamma} + z_1 = \frac{p_2}{\gamma} + z_2 \tag{2.6}$$

and demonstrating that the quantity $(z + p/\gamma)$ is constant over the flow cross section normal to the streamlines when they are straight and parallel. This is often called a *hydrostatic pressure distribution* because the relation between z and p/γ is the same as that for a fluid at rest. This means that the Bernoulli equation of the single streamline may be extended to apply to two- and three-dimensional flows because at a given flow cross section $z + p/\gamma$ is the same for all streamlines as it is for the central streamline; stated in another way, it means that a single hydraulic grade line applies to all the streamlines of a flow, *provided these streamlines are straight and parallel.* In practice the engineer seldom if ever encounters flows containing precisely straight and parallel streamlines; however, because in pipes, ducts, and prismatic channels such lines are *essentially* straight and parallel, the approximation may be used for many practical calculations.

In ideal fluid flows, because of the absence of friction, the distribution of velocity over a cross section of a flow containing straight and parallel streamlines is uniform; that is, all fluid particles pass a given cross section at the same velocity, which is equal to the average velocity V. Accordingly, no adjustment of the $V^2/2g_n$ term of the Bernoulli equation is to be expected in extending the equation from the infinitesimal to the finite streamtube because $V^2/2g_n$ is constant across the streamtube and H is the same for every streamline in the streamtube. Later when friction is considered it will be shown that there usually is a nonuniform profile of velocity across a finite streamtube even though the streamlines are straight and parallel. A correction must be applied then to the value of $V^2/2g_n$ in the streamtube before the Bernoulli equation may be used to represent flow along a streamtube of finite cross section.

ILLUSTRATIVE PROBLEM 5.1

Water is flowing through a section of cylindrical pipe. If the static pressure at point C is 35 kPa, what are the static pressures at A and B, and where is the hydraulic grade line at this flow cross section?

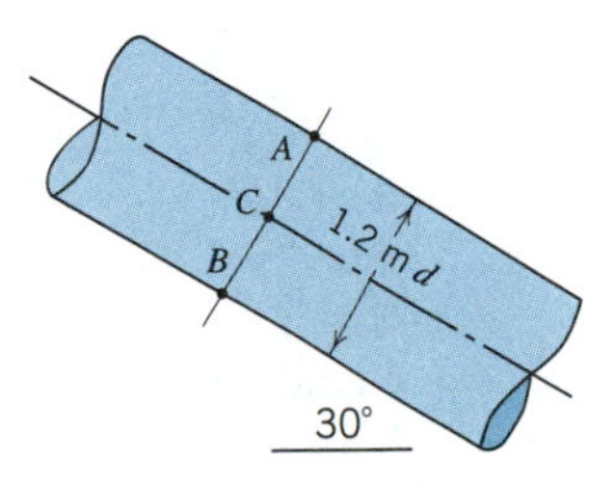

SOLUTION

This is obviously a situation where the streamlines of the flow are straight and parallel so we know that the pressure varies hydrostatically in a direction normal to the streamlines. The pressures at A and B will differ from that at C by an amount equivalent to their difference in elevation. That is, along the line A-C-B, $p/\gamma + z$ is a constant. So, we can write the following equations

$$\frac{p_A}{\gamma} + z_A = \frac{p_C}{\gamma} + z_C = \frac{p_B}{\gamma} + z_B$$

For point B,

$$p_B = \gamma(z_C - z_B) + p_C = 9.8 \times 10^3\ \text{N/m}^3 \times (0.6\ \text{m} \times \cos 30°) + 35 \times 10^3\ \text{Pa}$$

$$p_B = 40.1 \times 10^3\ \text{Pa} \qquad \text{or} \qquad 40.1\ \text{kPa}\ \bullet$$

For point A,

$$p_A = \gamma(z_C - z_A) + p_C = 9.8 \times 10^3\ \text{N/m}^3 \times (-0.6\ \text{m} \times \cos 30°) + 35.0 \times 10^3$$

$$p_A = 29.9 \times 10^3\ \text{Pa} \qquad \text{or} \qquad 29.9\ \text{kPa}\ \bullet$$

The hydraulic grade line is always positioned above the centerline of the pipe by an amount equal to the pressure head at the center of the pipe. Hence, the hydraulic grade line is

$$\frac{p_C}{\gamma} = \frac{35 \times 10^3\ \text{Pa}}{9.8 \times 10^3\ \text{N/m}^3} = 3.57\ \text{m vertically above } C.\ \bullet$$

5.4 APPLICATIONS OF BERNOULLI'S EQUATION

Before proceeding to some engineering applications of Bernoulli's equation, it should first be noted that this equation gives further aid in the interpretation of streamline pictures, Eq. 5.1 indicating that, when velocity increases, the sum $(p/\gamma + z)$ of pressure and potential head must decrease. In many flow problems, the potential head z varies little, allowing the approximate general statement: where velocity is high, pressure is low. Regions of closely spaced streamlines have been shown (Section 4.2) to indicate regions of relatively high velocity, and now from the Bernoulli equation these are seen also to be regions of relatively low pressure.

In 1643, Torricelli showed that the velocity of efflux of an ideal fluid from a small orifice under a static head varies with the square root of the head. Today, Torricelli's theorem is written

$$V = \sqrt{2g_n h}$$

the velocity being (ideally) equal to that attained by a solid body falling from rest through a height h. Torricelli's theorem is now recognized as a special case of the Bernoulli equation involving certain conditions appearing in many engineering problems. Torricelli's equation can be easily derived by applying Bernoulli's equation from the reservoir to the tip of the nozzle in Fig. 5.4. The reservoir is assumed to be very large (compared to the nozzle). Thus, the small flow from the nozzle produces negligible velocities in the reservoir

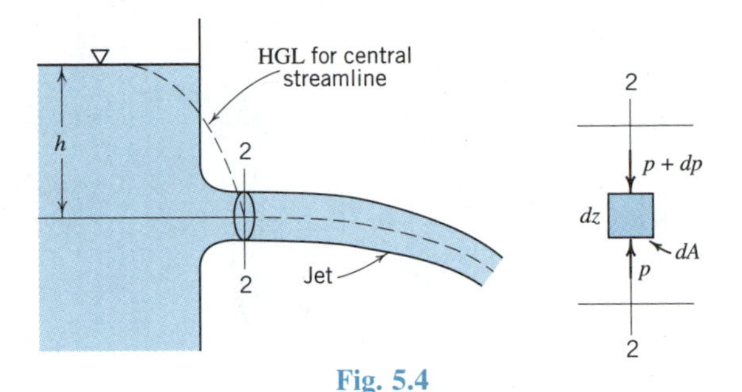

Fig. 5.4

except near the nozzle. Taking the datum plane at the center of the nozzle and choosing (arbitrarily) the center streamline give $h = z + p/\gamma$ in the reservoir where velocities are negligible. Writing Bernoulli's equation for a streamline between the reservoir and the tip of the nozzle,

$$\frac{p_1}{\gamma} + \frac{V_1^2}{2g_n} + z_1 = h = \frac{p_2}{\gamma} + \frac{V_2^2}{2g_n} + 0$$

Torricelli's equation results if $p_2 = 0$. From the validity of the Bernoulli and Torricelli equations it may be deduced that the pressure throughout the jet in the plane of the nozzle must be zero; however, analytical proof of this from Newton's law may be more convincing.

Consider the streamlines through section 2 of Fig. 5.4 to be essentially straight and parallel.[2] The pressure surrounding the jet is zero (gage) and the vertical acceleration of an elemental fluid mass $p\ dA\ dz$ is equal to g_n. The force causing the acceleration can result only from the pressure difference between top and bottom of the element, and the weight of the element. By writing Newton's second law in the vertical direction, we obtain

$$-(p + dp)\ dA + p\ dA - \gamma\ dA\ dz = -(\rho\ dA\ dz)g_n$$

from which it may be concluded that $dp = 0$. Thus there can be no pressure gradient across the jet at section 2, and with the pressures zero at the jet boundaries it follows that the pressure throughout the jet must be zero. This also shows that V cannot be constant across the jet because $p = 0$ and z varies. Downstream from section 2 it is customary to *assume* that the pressures throughout the free jet are also zero; this is equivalent to assuming that each fluid element follows a free trajectory streamline unaffected by adjacent fluid elements. This is an adequate approximation in many engineering problems, but it is not exact because the curvature and convergence of streamlines and effects such as surface tension were neglected.

ILLUSTRATIVE PROBLEM 5.2

Water flows in the pipeline shown below from the reservoir through the constriction, exiting from a nozzle at the downstream end. Calculate the pressures at sections 1, 2, 3, and 4 in the pipe and determine the elevation of the top of the free jet's trajectory.

[2]Actually, the streamlines are curved and convergent (recall the trajectory theory of particle dynamics), but such convergence and curvature may be neglected for large h and high jet velocity.

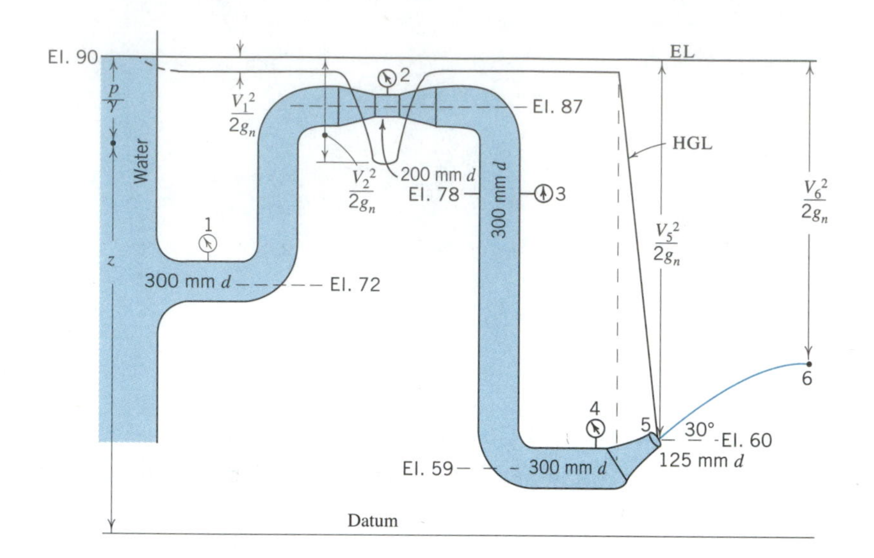

SOLUTION

First sketch the energy line; at all points in the reservoir where the velocity is negligible, $(z + p/\gamma)$ is the same. Thus, the energy line has the same elevation as the water surface. Next sketch the hydraulic grade line; this is coincident with the energy line in the reservoir where velocity is negligible but drops below the energy line over the pipe entrance where velocity is gained. The velocity in the 300 mm pipe is everywhere the same, so the hydraulic grade line must be horizontal until the flow encounters the constriction upstream from section 2. Here, as velocity increases, the hydraulic grade line must fall (possibly to a level below the constriction). Downstream from the constriction, the hydraulic grade line must rise to the original level over the 300 mm pipe and continue at this level to a point over the *base of the nozzle* at section 4. Over the nozzle, the hydraulic grade line must fall to the nozzle tip and after that follow the jet, because the pressure in the jet is everywhere zero.

Since the vertical distance between the energy line and the hydraulic grade line at any section is the velocity head at that section, it is evident that

$$\frac{V_5^2}{2g_n} = 90 \text{ m} - 60 \text{ m} = 30 \text{ m} \qquad \text{and, therefore} \qquad V_5 = 24.3 \text{ m/s}$$

The flowrate in the system can now be computed as

$$Q = VA = 24.3 \text{ m/s} \times \frac{\pi}{4}(0.125 \text{ m})^2 = 0.298 \text{ m}^3/\text{s} \bullet \tag{4.4}$$

We can now find the velocity in the 300-mm pipe using the continuity equation.

$$V_1 A_1 = V_5 A_5 \qquad \text{and so} \qquad V_1 = \frac{A_5}{A_1} V_5 = \frac{d_5^2}{d_1^2} V_5 \tag{4.5}$$

$$V_1 = \frac{(0.125 \text{ m})^2}{(0.300 \text{ m})^2} \times 24.3 \text{ m/s} = 4.22 \text{ m/s}$$

So, the velocity heads in all the reaches of 300 mm pipe are

$$\frac{V_1^2}{2g_n} = \frac{V_3^2}{2g_n} = \frac{V_4^2}{2g_n} = \frac{(4.22\text{ m/s})^2}{2g_n} = 0.91\text{ m}$$

In the constriction, the velocity head may be computed by recognizing that velocity head varies as the ratio of the diameter to the fourth power. That is,

$$\frac{V_2^2}{2g_n} = \left(\frac{d_1}{d_2}\right)^4 \frac{V_1^2}{2g_n} = \left(\frac{0.300\text{ m}}{0.200\text{ m}}\right)^4 \times 0.91\text{ m} = 4.61\text{ m}$$

Since the hydraulic grade line is above the pipe centerline by an amount equal to the pressure head, the pressures in the pipe and the constriction may be computed as follows:

$$\frac{p_1}{\gamma} = \left(90\text{ m} - \frac{V_1^2}{2g_n} - 72\text{ m}\right) = 18 - 0.91 = 71.09\text{ m} \qquad (5.1)$$

$$p_1 = 17.09 \times 9.8\text{ kN/m}^3 = 167.5\text{ kPa} \bullet$$

$$\frac{p_2}{\gamma} = \left(90\text{ m} - \frac{V_2^2}{2g_n} - 87\text{ m}\right) = 3 - 4.61\text{ m} = -1.61\text{ m} \qquad (5.1)$$

$$p_2 = -1.61 \times 9.8\text{ kN/m}^3 = -15.8\text{ kPa} \bullet$$

$$\frac{p_3}{\gamma} = \left(90\text{ m} - \frac{V_3^2}{2g_n} - 78\text{ m}\right) = 12 - 0.91\text{ m} = 11.09\text{ m} \qquad (5.1)$$

$$p_3 = 11.09 \times 9.8\text{ kN/m}^3 = 108.7\text{ kPa} \bullet$$

$$\frac{p_4}{\gamma} = \left(90\text{ m} - \frac{V_4^2}{2g_n} - 59\text{ m}\right) = 31 - 0.91\text{ m} = 30.09\text{ m} \qquad (5.1)$$

$$p_4 = 30.09 \times 9.8\text{ kN/m}^3 = 294.9\text{ kPa} \bullet$$

The velocity of the water jet at the top of its trajectory where there is no vertical component of velocity is calculated by treating the water as a collection of individual particles. In this case, classical mechanics tells us that, while the acceleration of gravity causes the vertical component of the velocity to go to zero between points 5 and 6, the horizontal velocity is unchanged. Therefore,

$$V_6 = V_5 \cos 30° = 24.3\text{ m/s} \times 0.867 = 21.0\text{ m/s}$$

Because the pressure is zero in the jet, the hydraulic grade line lies in the jet as noted above. Accordingly, z_6 is given by

$$90\text{ m} = \frac{V_6^2}{2g_n} + z_6 \qquad \text{or} \qquad z_6 = 90\text{ m} - \frac{(21.0\text{ m/s})^2}{2g_n} = 67.5\text{ m} \bullet$$

With increasing velocity or potential head, the pressure within a flowing fluid drops. However, this pressure does not drop below the absolute zero of pressure as it has been found experimentally that fluids will not sustain the tension implied by pressures less than absolute zero. Thus, a practical physical restriction is placed on the Bernoulli equation. Such a restriction is, in fact, not appropriate for gases because they expand with reduction in pressure, but it frequently assumes great importance in the flow of liquids. Actually, in liquids the absolute pressure can drop only to the vapor pressure of the liquid, whereupon spontaneous vaporization (boiling) takes place. This vaporization or formation of vapor cavities is called cavitation.[3] The formation, translation with the fluid motion, and subsequent rapid collapse of these cavities produces vibration, destructive pitting, and other deleterious effects on hydraulic machinery, hydrofoils, and ship propellers, for example.

ILLUSTRATIVE PROBLEM 5.3

For the situation described below, what diameter of constriction can be expected to produce incipient cavitation at the throat of the constriction? Water at 100°F is flowing and the barometric pressure is 14.0 psia.

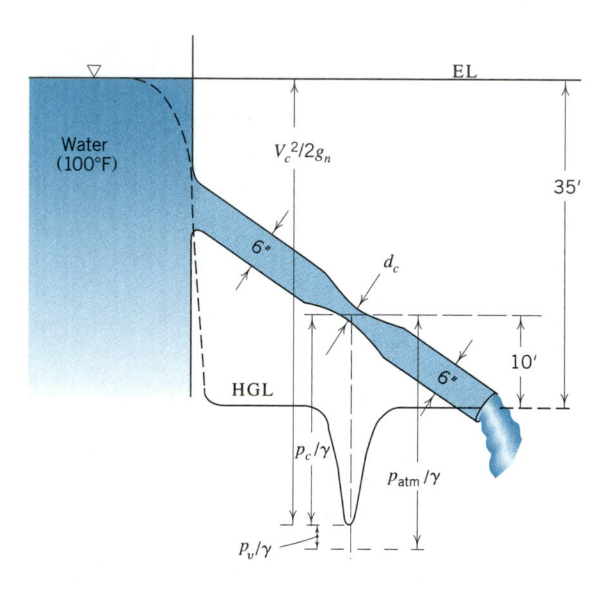

SOLUTION

First, we need to determine what gage pressure will produce cavitation. From Appendix 2, for water at 100°F, we find $\gamma = 62.0\ \text{lb/ft}^3$ and $p_v = 0.95$ psia. The atmospheric pressure head is

$$\frac{p_{atm}}{\gamma} = \frac{14.0\ \text{lb/in}^2 \times 144\ \text{in}^2/\text{ft}^2}{62.0\ \text{lb/ft}^3} = 32.5\ \text{ft}$$

We can now compute the critical gage pressure head at the constriction,

$$\frac{p_c}{\gamma} = -\left(\frac{p_{atm}}{\gamma} - \frac{p_v}{\gamma}\right) = -\left(32.5\ \text{ft} - \frac{0.95\ \text{lb/in}^2 \times 144\ \text{in}^2}{62.0\ \text{lb/ft}^3}\right) = -30.3\ \text{ft}$$

The negative sign indicates that the pressure at the section is below atmospheric.

From the above sketch, it can be seen that the velocity head at the constriction (distance from the hydraulic grade line to the energy line) is equal to (35 ft − 10 ft) + 30.3 ft = 55.3 ft.

[3]For a description of cavitation phenomena, see Appendix 4.

$$\frac{V_c^2}{2g_n} = 55.3 \text{ ft} \qquad \text{so} \qquad V_c = 59.7 \text{ ft/s}$$

From the sketch, we can also see that the velocity head at the exit from the pipe is 35 ft. This permits us to calculate the velocity at the exit and the flowrate.

$$\frac{V_{exit}^2}{2g_n} = 35 \text{ ft} \qquad \text{so} \qquad V_{\text{exit}} = 47.5 \text{ ft/s}$$

and

$$Q = (AV)_{\text{exit}} = \frac{\pi}{4}\left(\frac{6 \text{ in}}{12}\right)^2 \times 47.5 \text{ ft/s} = 9.33 \text{ ft}^3\text{/s}$$

The flow rate at the constriction must be the same, so

$$Q = A_c V_c \qquad \text{giving} \qquad A_c = \frac{Q}{V_c} = \frac{9.33 \text{ ft}^3\text{/s}}{59.7 \text{ ft/s}} = 0.156 \text{ ft}^2$$

Solving for the diameter at the constriction,

$$d_c = \sqrt{\frac{A_c}{\pi/4}} = \sqrt{\frac{0.156 \text{ ft}^2}{\pi/4}} = 0.446 \text{ ft} \qquad \text{or} \qquad 5.35 \text{ in} \bullet$$

Incipient cavitation must be assumed in ideal fluid flow problems of this nature in order for head losses to be considered negligible. Also, with little cavitation there is more likelihood of the pipe flowing full at its exit.

The Bernoulli equation is frequently written in terms of pressure rather than head and may be obtained in this form by multiplying Eq. 5.1 by γ and substituting ρ for γ/g_n; this results in

$$p_1 + \tfrac{1}{2}\rho V_1^2 + \rho g_n z_1 = p_2 + \tfrac{1}{2}\rho V_2^2 + \rho g_n z_2 \tag{5.2}$$

Here the Bernoulli terms p, $\rho V^2/2$, and $\rho g_n z$ are called static pressure, dynamic pressure, and potential pressure, respectively. The *stagnation (or total) pressure*, p_s is defined by

$$p_s = p_o + \tfrac{1}{2}\rho V_o^2 \tag{5.3}$$

and from Fig. 5.5 it can be seen that this is the local pressure at the tip of a pitot tube or,

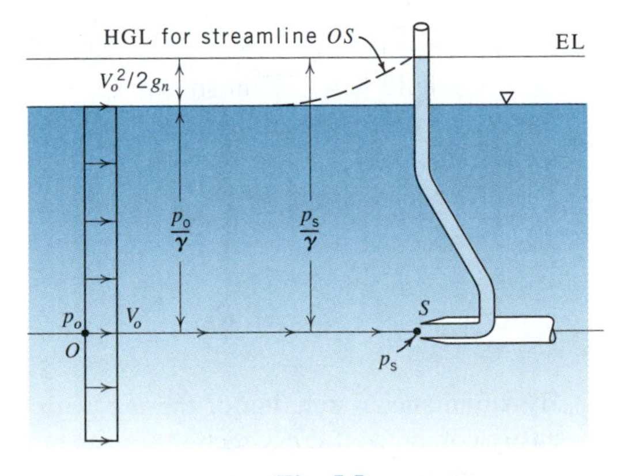

Fig. 5.5

more generally, the pressure at the zero-velocity point on the nose of any solid object in a flow. This point is called appropriately a *stagnation point* because here the flow momentarily stops, or *stagnates*. The variation of pressure and velocity along the central streamline to the nose of a solid object is shown in Fig. 5.5. Note that the pressure rises rather abruptly (but not discontinuously) from p_o to p_s just in front of the object, while the velocity decreases from V_o to zero. With stagnation pressure at s easily measurable, and pressure p_o known or measurable, the velocity V_o of the undisturbed stream can be computed from Eq. 5.3; this is the essence of the pitot tube principle which finds wide application in many velocity-measuring devices.

ILLUSTRATIVE PROBLEM 5.4

The pitot-static tube is carefully aligned with an airstream of density 1.23 kg/m^3. If the attached differential manometer shows a reading of 150 mm of water, what is the velocity of the airstream?

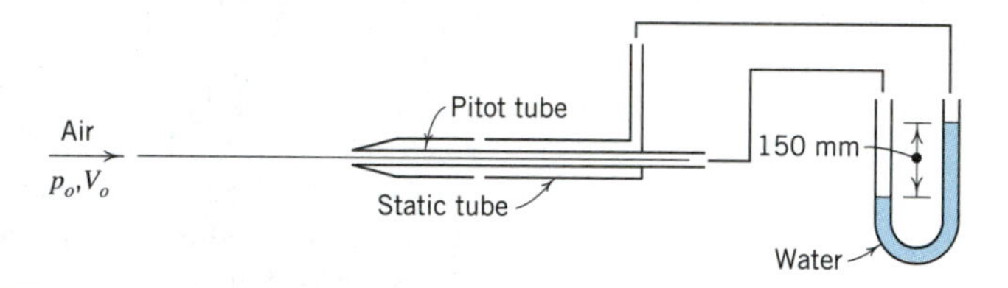

SOLUTION

The stagnation pressure will be found at the tip of the pitot-static tube. Assuming that the holes in the barrel of the pitot-static tube will sense the static pressure p_o in the undisturbed airstream, the manometer will measure $(p_s - p_o)$. Applying Eq. 5.3,

$$p_s - p_o = \tfrac{1}{2}\rho V_o^2 \tag{5.3}$$

As a result,

$$V_o = \sqrt{\frac{2(p_s - p_o)}{\rho}} = \sqrt{\frac{2(0.150 \text{ m} \times 9\,810 \text{ N/m}^3)}{1.23 \text{ kg/m}^3}} = 48.9 \text{ m/s} \bullet$$

A constriction (Fig. 5.6) in a streamtube or pipeline is frequently used as a device for metering fluid flow. Simultaneous application of the continuity and Bernoulli principles to such a constriction allows direct calculation of the flowrate when certain variables are measured.

The equations are

$$Q = A_1V_1 = A_2V_2 \tag{3.12}$$

$$\frac{p_1}{\gamma} + \frac{V_1^2}{2g_n} + z_1 = \frac{p_2}{\gamma} + \frac{V_2^2}{2g_n} + z_2 \tag{5.1}$$

By simultaneous solution of these equations the flowrate through a constriction can be easily computed if cross-sectional areas of pipe and constriction are known and pressures and heights above datum are measured, either separately or in the combination indicated.

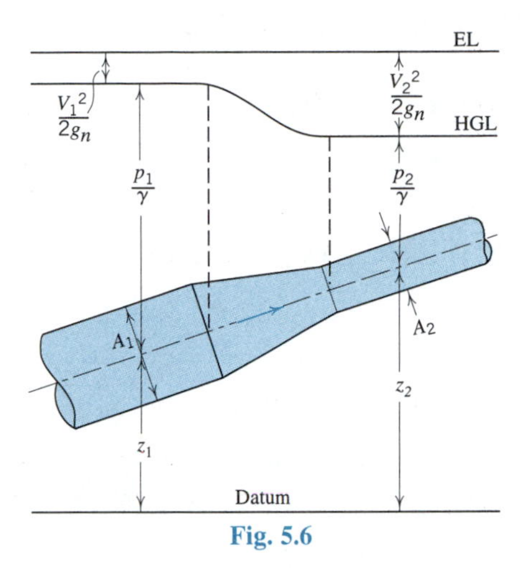

Fig. 5.6

ILLUSTRATIVE PROBLEM 5.5

Gasoline (s.g. = 0.82) flows through the pipeline shown below. Calculate the flow rate by two methods—first using the gage readings and then the differential manometer.

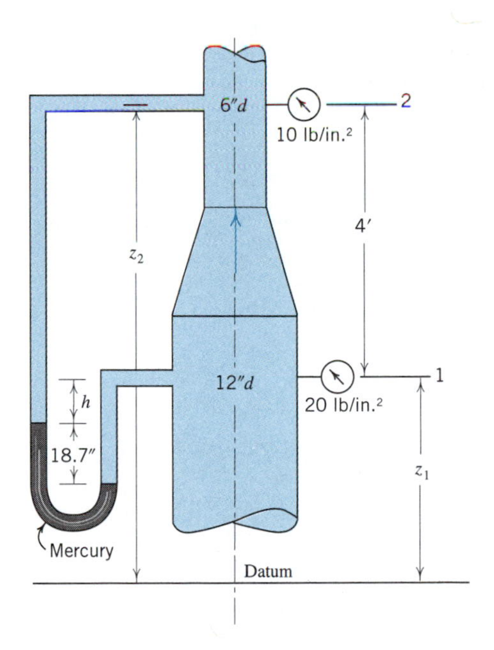

SOLUTION

The problem is solved by use of the Bernoulli equation and, in the second case, the hydrostatic relationships in the manometer tube.

Solution using the gages

We go directly to the Bernoulli equation 5.1.

$$z + \frac{p}{\gamma} + \frac{V^2}{2g_n} = H = \text{Constant} \tag{5.1}$$

This equation can be written

$$z_1 + \frac{p_1}{\gamma} + \frac{V_1^2}{2g_n} = z_2 + \frac{p_2}{\gamma} + \frac{V_2^2}{2g_n}$$

First calculating the p/γ terms,

$$\frac{p_1}{\gamma} = \frac{20 \text{ lb/in}^2 \times 144 \text{ in}^2/\text{ft}^2}{(0.82 \times 62.4 \text{ lb/ft}^3)} = 56.3 \text{ ft of gasoline}$$

$$\frac{p_2}{\gamma} = \frac{10 \text{ lb/in}^2 \times 144 \text{ in}^2/\text{ft}^2}{(0.82 \times 62.4 \text{ lb/ft}^3)} = 28.1 \text{ ft of gasoline}$$

Now, substituting into the Bernoulli equation,

$$z_1 + 56.3 \text{ ft} + \frac{V_1^2}{2g_n} = (z_1 + 4 \text{ ft}) + 28.1 \text{ ft} + \frac{V_2^2}{2g_n}$$

$$\frac{V_1^2}{2g_n} = \frac{V_2^2}{2g_n} - 24.2$$

We know that the velocity heads in a situation like this vary as the diameter ratio to the fourth power, so

$$\left(\frac{d_2}{d_1}\right)^4 \frac{V_2^2}{2g_n} = \frac{V_2^2}{2g_n} - 24.2$$

$$V_2 = \sqrt{\frac{2g_n \times 24.2}{1 - (6/12)^4}} = 40.8 \text{ ft/s}$$

The flow rate can now be found.

$$Q = A_2V_2 = \pi/4 \times (6/12)^2 \times 40.8 = 8.01 \text{ ft}^3/\text{s} \bullet \tag{4.4}$$

Solution using manometer

We first write a hydrostatic pressure equation through the manometer.

$$p_1 + \gamma h + (18.7/12)\gamma - (18.7/12)(13.57\gamma_{water}) - \gamma h - 4\gamma = p_2$$

Dividing through by the specific weight of the gasoline γ,

$$\frac{p_1}{\gamma} + (18.7/12)\left(1 - 13.57\frac{\gamma_{water}}{\gamma}\right) - 4 = \frac{p_2}{\gamma}$$

$$\frac{p_1}{\gamma} - 28.2 = \frac{p_2}{\gamma}$$

Now, substituting this into the Bernoulli equation,

$$z_1 + \frac{p_1}{\gamma} + \frac{V_1^2}{2g_n} = (z_1 + 4) + \left(\frac{p_1}{\gamma} - 28.2\right) + \frac{V_2^2}{2g_n} \quad (5.1)$$

Subtracting out identical terms leaves

$$\frac{V_1^2}{2g_n} = \frac{V_2^2}{2g_n} - 24.2$$

This equation is identical to the one developed under the previous approach using the pressure gages; hence, this analysis will lead to the same value for the flowrate. ●

The Bernoulli principle may be, of course, applied to problems of *open flow* such as the overflow structure of Fig. 5.7. Such problems feature a moving liquid surface in contact with the atmosphere and flow pictures dominated by gravitational action. A short distance upstream from the structure, the streamlines will be straight and parallel and the velocity distribution will be uniform. In this region, the quantity $z + p/\gamma$ will be constant, the pressure distribution hydrostatic, and the hydraulic grade line (for all streamtubes) located in the liquid surface; the energy line will be horizontal and located $V_1^2/2g_n$ above the liquid surface. With atmospheric pressure on the liquid surface the streamtube in the liquid surface behaves as a free jet, allowing all surface velocities to be computed from the positions of liquid surface and energy line. The prediction of velocities elsewhere in the flowfield where streamtubes are severely convergent or curved is outside the province of one-dimensional flow; suffice it to say that all such velocities are interdependent and the whole flowfield must be established before any single velocity can be computed. At section 2, however (if the streamlines there are assumed straight and parallel), the pressures and velocities may be computed from the one-dimensional assumption.

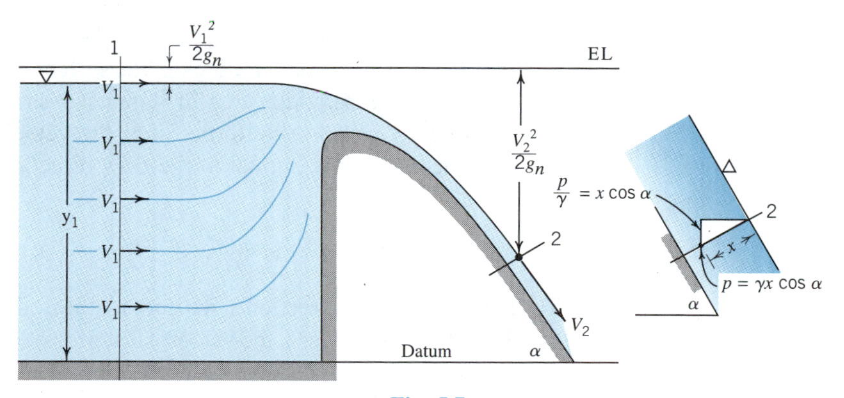

Fig. 5.7

ILLUSTRATIVE PROBLEM 5.6

At section 2 in Fig. 5.7, the water surface is at elevation 30.5 m and the 60° spillway face is at elevation 30.0 m. The velocity at the water surface at section 2 is 6.11 m/s. Calculate the pressure and velocity on the spillway face at section 2.

If the bottom of the approach channel is at elevation 29.0 m, calculate the depth and velocity in the approach channel.

SOLUTION

First, addressing the calculation of the pressure at section 2, we assume the streamlines are essentially straight and parallel at this section and the pressure varies hydrostatically normal to the streamlines. As a consequence, the difference in pressure between the surface ($p = 0$) and the spillway face is determined by the difference in elevation of the two points. Since the elevation of the water surface is 30.5 m and that of the spillway face 30.0 m, the difference in pressure is

$$\Delta p = \gamma \, \Delta_{elev} = 9\,800 \text{ N/m}^3 \times 0.5 \text{ m} = 4\,900 \text{ Pa or } 4.90 \text{ kPa}$$

Because the pressure is zero on the water surface, the pressure at the spillway face is 4.90 kPa. •

Now to calculate the velocity at the spillway face, we use the data at the water surface to establish the elevation of the energy line through use of Eq. 5.1.

$$\text{Elevation of energy line} = z + \frac{p}{\gamma} + \frac{V^2}{2g_n} = 30.5 \text{ m} + 0 + \frac{(6.11 \text{ m/s})^2}{2g_n} = 32.4 \text{ m}$$

Now, for the flow at the spillway face, the three Bernoulli terms must add up to 32.4 m. Referring to the velocity at the spillway face as V_{2_f}, we get

$$32.4 = 30.0 \text{ m} + \frac{4\,900 \text{ Pa}}{9\,800 \text{ N/m}^3} + \frac{V_{2_f}^2}{2g_n}$$

$$V_{2_f} = \sqrt{2g_n(2.4 \text{ m} - 0.5 \text{ m})} = 6.11 \text{ m/s} \;\bullet$$

The fact that the velocity is the same at both the surface and the bottom of the flow validates the one-dimensional assumption that in regions where the streamlines are straight and parallel, the velocity profile is uniform. Since the velocities at section 2 are all 6.11 m/s and since the depth of flow normal to the flow direction is 1.0 m, the flow rate per unit width is

$$q = V_2 y_2 = 6.11 \text{ m/s} \times 1.0 \text{ m} = 6.11 \text{ m}^3\text{/s/m of width} \qquad (4.7)$$

To find the velocity at section 1 we note that it is also a section where the streamlines are straight and parallel. Hence, the velocity is uniform over section 1 and continuity shows that $V_1 y_1 = V_2 y_2$. Writing the Bernoulli equation along the streamline which follows the bottom of the channel,

$$z_1 + \frac{p_1}{\gamma} + \frac{V_1^2}{2g_n} = z_2 + \frac{p_2}{\gamma} + \frac{V_2^2}{2g_n}$$

$$29.0 \text{ m} + y_1 + \frac{V_1^2}{2g_n} = 30 \text{ m} + 0.5 \text{ m} + \frac{(6.11 \text{ m/s})^2}{2g_n}$$

$$y_1 + \frac{V_1^2}{2g_n} = 3.40$$

From continuity,

$$V_1 y_1 = q = 6.11 \text{ m}^3/\text{s/m}$$

Solving this equation for V_1 and substituting into the Bernoulli equation gives

$$y_1 + \frac{6.11^2}{2g_n y_1^2} = 3.40$$

This equation is cubic and has three roots, $y_1 = 3.22$ m, $y_1 = 0.85$ m, and $y_1 = -0.69$ m. The third root is obviously invalid. The second root is also invalid because the depth at section 1 may not be less than that at section 2. That leaves us with the first root which gives the depth at section 1 as 3.22 m. ●

Returning to continuity, we can calculate the velocity at section 1 as

$$V_1 = \frac{q}{y_1} = \frac{6.11 \text{ m}^3/\text{s/m}}{3.22 \text{ m}} = 1.90 \text{ m/s} \quad \bullet$$

5.5 THE WORK-ENERGY EQUATION

The previous analysis of a fluid system yielded a useful and practical equation, the Bernoulli equation 5.1, which was applied to establish the relationships between elevation, pressure, velocity, and total head in a variety of fluid flows. However, many pipelines, for example, contain pumps or turbines which, respectively, add energy to or extract it from the fluid. There is no satisfactory method for incorporating these effects in the derivation of the Bernoulli equation, which was based on the application of Newton's second law to an infinitesimal fluid system. On the other hand, a physically meaningful derivation and additional insight into the physical meanings of the various terms in the Bernoulli equation can be obtained via a control volume analysis.

This control volume analysis utilizes the mechanical work-energy principle and applies it to fluid flow, resulting in a powerful relationship between fluid properties, work done, and energy transported. The Bernoulli equation is then seen to be equivalent to the work-energy equation for ideal fluid flow.

To develop the work-energy equation, we will use a control volume which coincides with the walls of a streamtube or full-sized conduit (see Fig. 5.8). Fluid enters and leaves the control volume at locations where the streamlines are straight and parallel and, hence, the velocity is uniform over the cross section under these conditions. We will employ a statement of the mechanical work-energy principle specific to fluid flow which states that the work done on a fluid system equals the change in the potential and kinetic energy of the system. Heat transfer and internal energy are neglected. If they were included, the result would be the First Law of Thermodynamics (see Section 7.11 and Chapter 13). The mechanical work-energy principle is expressed by the equation

$$dW = dE$$

where dW is the increment of work done and dE is the resulting incremental change in energy. If we now divide by dt we have the *rate* at which work is done and energy is changed. This is a more natural configuration for the mechanical work-energy principle for flow processes. The applicable form of the mechanical work-energy principle for flow

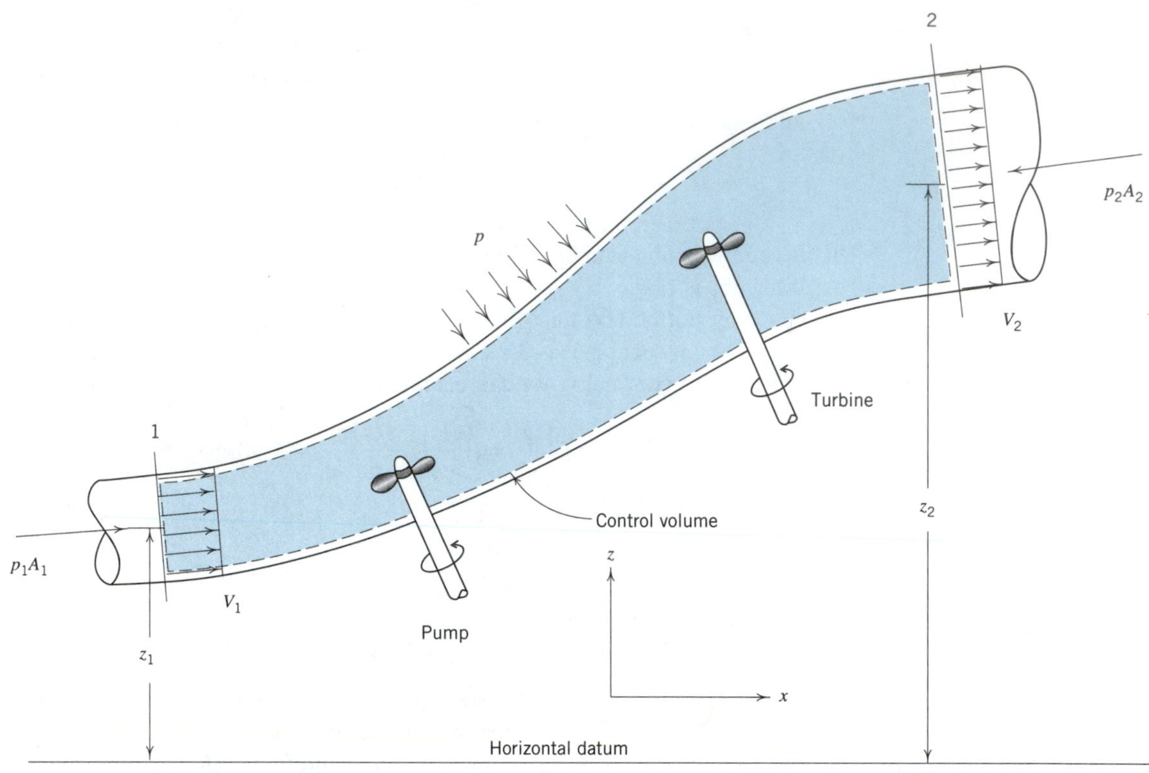

Fig. 5.8

processes becomes

$$\frac{dW}{dt} = \frac{dE}{dt}$$

The Reynolds Transport Theorem (Eq. 4.15) provides an equation for evaluating the rate of change of an extensive property (in this case, energy) of a fluid system which occupies the control volume at a given instant (dE/dt). The rate of net work done (dW/dt) will be the result of pressure forces acting at the control surface on the fluid system in the control volume and any work done by flow machines such as pumps and turbines.

To evaluate the rate of change of energy of the fluid system, we write the steady-state form of the Reynolds Transport Theorem (Eq. 4.15).

$$\frac{dE}{dt} = \iint_{c.s.out} i\rho\mathbf{v} \cdot d\mathbf{A} + \iint_{c.s.in} i\rho\mathbf{v} \cdot d\mathbf{A} \tag{4.15}$$

The *energy per unit mass* in the control volume is potential energy gz and kinetic energy $V^2/2$. For a constant density flow, the above equation becomes

$$\frac{dE}{dt} = \rho \iint_{c.s.out} \left(g_n z + \frac{V^2}{2}\right) \mathbf{v} \cdot d\mathbf{A} + \rho \iint_{c.s.in} \left(g_n z + \frac{V^2}{2}\right) \mathbf{v} \cdot d\mathbf{A}$$

where dE/dt is the rate of energy increase for the fluid system. Note that even in steady flow, the fluid system energy can change with time because the system moves through the control volume where both velocity and elevation can change. Because the velocity vector is normal to the area and because the velocity is constant over the two cross sections through which the flow enters and leaves the control volume, the above integration yields

$$\frac{dE}{dt} = \rho\left(g_n z_2 + \frac{V_2^2}{2}\right)V_2A_2 - \rho\left(g_n z_1 + \frac{V_1^2}{2}\right)V_1A_1$$

$$= \rho g_n\left(z_2 + \frac{V_2^2}{2g_n}\right)V_2A_2 - \rho g_n\left(z_1 + \frac{V_1^2}{2g_n}\right)V_1A_1$$

Recognizing from conservation of mass that $\rho g_n V_2 A_2 = \rho g_n V_1 A_1 = Q\gamma$,

$$\frac{dE}{dt} = Q\gamma\left[\left(z_2 + \frac{V_2^2}{2g_n}\right) - \left(z_1 + \frac{V_1^2}{2g_n}\right)\right] \tag{5.4}$$

Now, we proceed to evaluate the work done on the fluid system. The work done takes on three forms:

1. *Flow work* done by pressure forces via fluid entering or leaving the control volume.
2. *Machine work* done by pumps and turbines.
3. *Shear work* done by shearing forces acting on the system at the control surface.

Because we are working with an ideal fluid, there will be no shear forces. Consequently, shear work will not be considered; the effects of shear will be dealt with in Chapter 7.

First, addressing *flow work*, we note that all internal pressure forces between fluid particles cancel, hence we are left with only pressure forces at the control surfaces doing work. Further, only the pressure forces acting at cross sections 1 and 2 will do work. Those pressure forces acting at the remainder of the control surfaces will do no work because they are normal to the fluid motion. Recognizing that the pressure variation is hydrostatic over sections 1 and 2, we calculate the pressure force as the product of the pressure at the centroid of the cross section and the area of the cross section. To get the rate at which work is being done, we multiply by the velocity. Thus, the rate of net flow work done by pressure forces on the fluid system within the control volume is

$$\text{Net flow work rate} = p_1A_1V_1 - p_2A_2V_2$$

Now, considering *machine work*, we will define the work done per unit weight of fluid flowing to be E_P if the machine is a pump and $-E_T$ if it is a turbine. Then the rate of work done by the machine on the fluid system will be

$$\text{Net machine work rate} = Q\gamma E_P - Q\gamma E_T$$

In preparation for equating to Eq. 5.4, we multiply numerator and denominator of the flow work equation by γ and rearrange to get

$$\text{Net flow work rate} = \frac{p_1}{\gamma}\gamma A_1V_1 - \frac{p_2}{\gamma}\gamma A_2V_2 = Q\gamma\left(\frac{p_1}{\gamma} - \frac{p_2}{\gamma}\right)$$

Now, combining the two net-work-rate equations,

$$\text{Net work rate} = Q\gamma\left(\frac{p_1}{\gamma} - \frac{p_2}{\gamma} + E_P - E_T\right) \tag{5.5}$$

Equating Eqs. 5.4 and 5.5, we get

$$Q\gamma\left[\left(z_2 + \frac{V_2^2}{2g_n}\right) - \left(z_1 + \frac{V_1^2}{2g_n}\right)\right] = Q\gamma\left(\frac{p_1}{\gamma} - \frac{p_2}{\gamma} + E_P - E_T\right) \tag{5.6}$$

Dividing both sides by $Q\gamma$ and collecting terms with like subscripts,

$$z_1 + \frac{p_1}{\gamma} + \frac{V_1^2}{2g_n} + E_P = z_2 + \frac{p_2}{\gamma} + \frac{V_2^2}{2g_n} + E_T \tag{5.7}$$

where E_P is the energy added by the pump per unit weight of fluid flowing and E_T is the energy extracted by the turbine per unit weight of fluid flowing.

The above equation will be referred to as the *work-energy equation.* Even though it appears to be identical to the Bernoulli equation, there is a very important distinction. Each term in Eq. 5.7 represents energy per unit weight or work done per unit weight (ft-lb/lb or N·m/N = Joules/N). For example, this would permit us to add terms to the equation such as internal energy or heat transfer which would lead to the First Law of Thermodynamics (see Chapters 7 and 13). While the work-energy equation 5.7 without pumps and turbines and the Bernoulli equation 5.1 are identical and give similar results for ideal fluid flow, only the work-energy equation will be used in real fluid flow situations.

The addition or extraction of mechanical energy by a pump or turbine will appear as an abrupt rise or fall of the energy line over the respective machines. Because the engineer is usually interested in the power of such machines, it is important to be able to evaluate the added or extracted power. Recalling that E_P and E_T were energy per unit weight, one need only multiply by the weight flow rate to obtain power.

$$\text{Power} = (E_P \text{ or } E_T)\,\frac{\text{ft·lb}}{\text{lb}} \times Q\gamma\,\frac{\text{lb}}{\text{s}} = Q\gamma(E_P \text{ or } E_T)\,\frac{\text{ft·lb}}{\text{s}}$$

$$\text{Power} = (\text{E}_P \text{ or } E_T)\,\frac{\text{J}}{\text{N}} \times Q\gamma\,\frac{\text{N}}{\text{s}} = Q\gamma(E_P \text{ or } E_T)\,\frac{\text{J}}{\text{s}} = Q\gamma(E_P \text{ or } E_T)\text{ watts}$$

However, power is generally expressed as horsepower in the U.S. Customary units and as kilowatts in SI units. The corresponding equations are

$$\text{Horsepower of machine} = \frac{Q\gamma(E_P \text{ or } E_T)}{550} \tag{5.8a}$$

$$\text{Kilowatts of machine} = \frac{Q\gamma(E_P \text{ or } E_T)}{1\,000} \tag{5.8b}$$

Note that 1 hp = 0.746 kW. These equations may also be used to convert any unit energy, e.g., kinetic energy per unit weight $V^2/2g_n$, to the corresponding power.

ILLUSTRATIVE PROBLEM 5.7

The pump shown below delivers a flowrate of 0.15 m^3/s of water. How much power must the pump supply to the water to maintain gage readings of 250 mm of mercury vacuum on the suction side of the pump and 275 kPa of pressure on the discharge side?

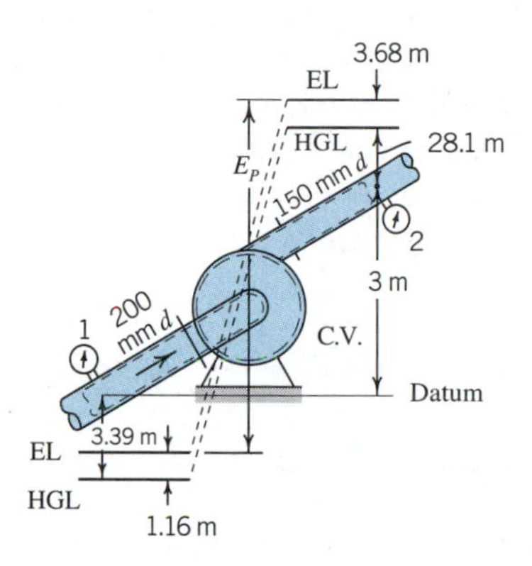

SOLUTION

Before implementing the work-energy equation, we will convert the two pressure readings to the proper units for substitution into the equation. Addressing the discharge pressure at 2 first, we need to convert this pressure to metres of water.

$$\frac{p_2}{\gamma} = \frac{275\ 000 \text{ Pa}}{9\ 800 \text{ N/m}^3} = 28.1 \text{ m}$$

Next, at section 1, we must convert millimetres of mercury to metres of water. We do this using the specific gravity of mercury from Appendix 2, Table 1.

$$\frac{p_1}{\gamma} = -250 \text{ mm} \times \frac{1}{1\ 000 \text{ mm/m}} \times 13.57 \text{ s.g. Hg} = -3.4 \text{ m}$$

To compute the velocity heads in the work-energy equation, we need to find the velocities at the two sections.

$$V_1 = \frac{Q}{A} = \frac{0.15 \text{ m}^3/\text{s}}{(\pi/4)(0.200 \text{ m})^2} = 4.77 \text{ m/s}$$

$$V_2 = V_1 \frac{A_1}{A_2} = V_1 \left(\frac{d_1}{d_2}\right)^2 = 4.77 \left(\frac{200 \text{ mm}}{150 \text{ mm}}\right)^2 = 8.48 \text{ m/s}$$

We now substitute these values into the work-energy equation for lines with a pump.

$$z_1 + \frac{p_1}{\gamma} + \frac{V_1^2}{2g_n} + E_P = z_2 + \frac{p_2}{\gamma} + \frac{V_2^2}{2g_n} \tag{5.7}$$

$$0 + (-3.4 \text{ m}) + \frac{(4.77 \text{ m/s})^2}{2g_n} + E_P = 3 \text{ m} + 28.1 \text{ m} + \frac{(8.48 \text{ m/s})^2}{2g_n}$$

$$-3.4 + 1.16 + E_P = 3.0 + 28.1 + 3.67$$

$$E_P = 37.0 \text{ J/N}$$

The power imparted to the water can be computed from Eq. 5.8b as

$$\text{Power} = \frac{Q\gamma E_P}{1\,000} = \frac{0.15 \text{ m}^3/\text{s} \times 9\,800 \text{ N/m}^3 \times 37.0 \text{ J/N}}{1\,000} = 54.4 \text{ kW} \bullet \quad (5.8b)$$

The rise in the energy line through the pump represents the energy supplied by the pump to each newton of water passing through the pump. It should be noted that the hydraulic grade lines are for the pipes only and do not include the pump passages, where flow is not one-dimensional. The positions of these HGLs give no assurance that the pump will run cavitation-free, since local velocities in the pump passages will be considerably larger than the average velocities in the pipes.

Two-Dimensional Flow

The solution of flowfield problems is much more complex than the solution of one-dimensional flow. Partial differential equations are invariably required for a formal mathematical approach to such problems. In many cases of two-dimensional ideal flow, the theory of complex variables provides exact solutions. In more general cases of real or ideal flow (see Sections 7.15 and 7.16), computer-based numerical solutions of the equations of motion enjoy wide success and engineering applicability. Many approximate techniques also exist for problem solution. The objective of the remainder of this chapter is to present an introduction to certain essentials and practical problems of importance to engineering students. To stress the similarities and differences of one- and two-dimensional flows, the subject is developed in parallel with the preceding treatment of one-dimensional flow. Coverage of advanced mathematical operations or broad generalizations is not attempted at this point; the emphasis is on giving the beginner some appreciation of the intricacies of flowfield problems as compared to the relative simplicity of those of one-dimensional flow.

5.6 EULER'S EQUATIONS

Euler's equations for a vertical two-dimensional flowfield may be derived by applying Newton's second law to a basic differential *system* of fluid of dimensions dx by dz (Fig. 5.9). The forces dF_x and dF_z on such an elemental system have been identified in Eqs. 2.1 and 2.2 of Section 2.1, and with substitution of appropriate pressures they reduce to

$$dF_x = -\frac{\partial p}{\partial x}\, dx\, dz$$

$$dF_z = -\frac{\partial p}{\partial z}\, dx\, dz - \rho g_n\, dx\, dz$$

The accelerations of the system have been derived in Eq. 3.6 of Section 3.3 for unsteady

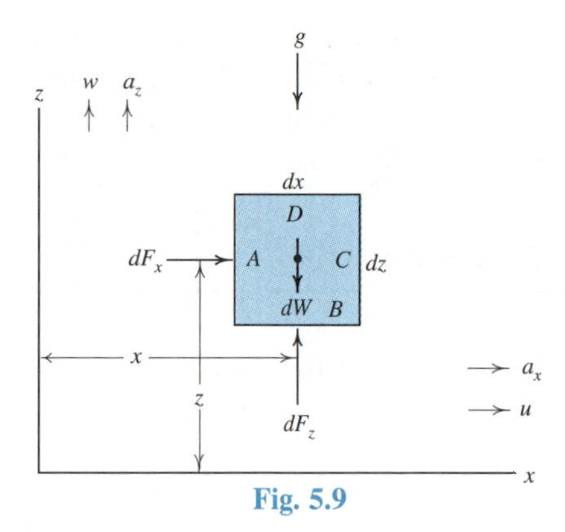

Fig. 5.9

flow; for steady flow in two dimensions, they reduce to

$$a_x = u\frac{\partial u}{\partial x} + w\frac{\partial u}{\partial z}$$

$$a_z = u\frac{\partial w}{\partial x} + w\frac{\partial w}{\partial z}$$

Applying Newton's second law by equating the differential forces to the products of the mass of the system and respective accelerations gives

$$-\frac{\partial p}{\partial x}\,dx\,dz = \rho\,dx\,dz\left(u\frac{\partial u}{\partial x} + w\frac{\partial u}{\partial z}\right)$$

$$-\frac{\partial p}{\partial z}\,dx\,dz - \rho g_n\,dx\,dz = \rho\,dx\,dz\left(u\frac{\partial w}{\partial x} + w\frac{\partial w}{\partial z}\right)$$

and by cancellation of $dx\,dz$ and slight rearrangement, the Euler equations of two-dimensional flow in a vertical plane are

$$-\frac{1}{\rho}\frac{\partial p}{\partial x} = u\frac{\partial u}{\partial x} + w\frac{\partial u}{\partial z} \tag{5.9a}$$

$$-\frac{1}{\rho}\frac{\partial p}{\partial z} = u\frac{\partial w}{\partial x} + w\frac{\partial w}{\partial z} + g_n \tag{5.9b}$$

With the equation of continuity,

$$\frac{\partial u}{\partial x} + \frac{\partial w}{\partial z} = 0 \tag{4.11}$$

we have a set of three simultaneous partial differential equations that are basic to the solution of two-dimensional flowfield problems; complete solution of these equations yields p, u, and w as functions of x and z, allowing prediction of pressure and velocity at any point in the flowfield. It is intriguing that application of a Lagrangian approach, that is, studying the dynamics of a small individual fluid system, has yielded the Eulerian equations of motion. However, recall that for Lagrangian and Eulerian systems, the velocity and acceleration at a point are the same.

5.7 BERNOULLI'S EQUATION

Bernoulli's equation can be derived by integrating the Euler equations for a uniform density flow, as was done for one-dimensional flow (Section 5.2). By multiplying the first of Eqs. 5.9 by dx and the second by dz and adding them, we find that

$$-\frac{1}{\rho}\left(\frac{\partial p}{\partial x}\,dx + \frac{\partial p}{\partial z}\,dz\right) = u\,\frac{\partial u}{\partial x}\,dx + w\,\frac{\partial u}{\partial z}\,dx + u\,\frac{\partial w}{\partial x}\,dz + w\,\frac{\partial w}{\partial z}\,dz + g_n\,dz$$

The terms $w(\partial w/\partial x)\,dx$ and $u(\partial u/\partial z)\,dz$ are added to and subtracted from the right-hand side of the equation, and terms are then collected in the following pattern:

$$-\frac{1}{\rho}\left(\frac{\partial p}{\partial x}\,dx + \frac{\partial p}{\partial z}\,dz\right) = \left(u\,\frac{\partial u}{\partial x}\,dx + w\,\frac{\partial w}{\partial x}\,dx\right) + \left(u\,\frac{\partial u}{\partial z}\,dz + w\,\frac{\partial w}{\partial z}\,dz\right) + (u\,dz - w\,dx)\left(\frac{\partial w}{\partial x} - \frac{\partial u}{\partial z}\right) + g_n\,dz$$

The bracket on the left-hand side of this equation is (see Appendix 6) the total differential dp. The sum of the first two brackets on the right-hand side is easily shown to be $d(u^2 + w^2)/2$, and the third bracket is the vorticity, ξ (Section 3.4). Reducing the equation accordingly and dividing it by g_n lead to

$$-\frac{dp}{\gamma} = \frac{d(u^2 + w^2)}{2g_n} + \frac{1}{g_n}(u\,dz - w\,dx)\xi + dz$$

and after integration to

$$\frac{p}{\gamma} + \frac{u^2 + w^2}{2g_n} + z = H - \frac{1}{g_n}\int \xi(u\,dz - w\,dx)$$

in which H is the constant of integration. Because the magnitude of the resultant velocity V at any point in the flowfield is related to its components u and w by $V^2 = u^2 + w$, the equation further simplifies to

$$\frac{p}{\gamma} + \frac{V^2}{2g_n} + z = H - \frac{1}{g_n}\int \xi(u\,dz - w\,dx) \tag{5.10}$$

This equation thus shows that the sum of the Bernoulli terms at *any* point in a steady flowfield is a constant H if the vorticity ξ is zero, that is, *if the flowfield is an irrotational* (or potential) one. Thus *for irrotational flow the same constant applies to all the streamlines of the flowfield* or, in terms of the energy line, all fluid masses in an irrotational flowfield possess the same unit energy. In a rotational flowfield the integral in Eq. 5.10 must be evaluated. However, along a streamline in any steady flow $dz/dx = w/u$ by definition. Thus, $u\,dz = w\,dx$; that is, $u\,dz - w\,dx = 0$ along a streamline. In a rotational flow, the Bernoulli terms are constant along any streamline, but the constant is different for each streamline.

5.8 APPLICATIONS OF BERNOULLI'S EQUATION

For irrotational flow of an ideal incompressible fluid, the Bernoulli equation may be applied over the flowfield with a single (horizontal) energy line completely describing the energy situation. Figure 5.10 depicts this for two representative points A and B. From the position of the points (above datum) the quantity $(p/\gamma + V^2/2g_n)$ may be determined from the position of the energy line, but the pressures p_A, p_B, and the like, cannot be calculated until the corresponding velocities V_A, V_B, and so forth are known. However, in a flowfield all the velocities are interdependent and are determined by the streamline definition and by the differential equation (4.11) of continuity; until methods are described for solving these equations, the pressures in the flowfield cannot be accurately predicted. However, the lack of formal mathematical solutions describing the entire velocity field need not deter the engineer from making a semiquantitative approach to such problems; indeed, when no formal solutions exist (as frequently happens) this is the only alternative.

One very useful tool in such approaches to flowfield problems is the effect of flow curvature on the pressure variation across the flow. For any element of streamline (Fig. 5.11) having radius of curvature r, the normal component of acceleration a_r is directed toward the center of curvature and is equal to V^2/r (Eq. 3.3). Newton's second law, applied to such an elemental fluid system on the streamline, yields a general and useful result in the analysis and interpretation of flowfields. In the radial direction the components of force on the system are

$$(p + dp)\,ds - p\,ds + dW \cos\theta = \frac{(\rho\,dr\,ds)V^2}{r}$$

From the geometry of the system and streamline, $\cos\theta = dz/dr$ and $dW = \gamma\,dr\,ds$.

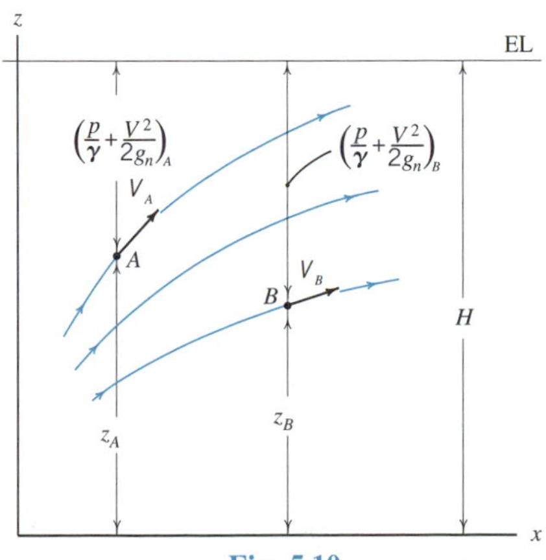

Fig. 5.10

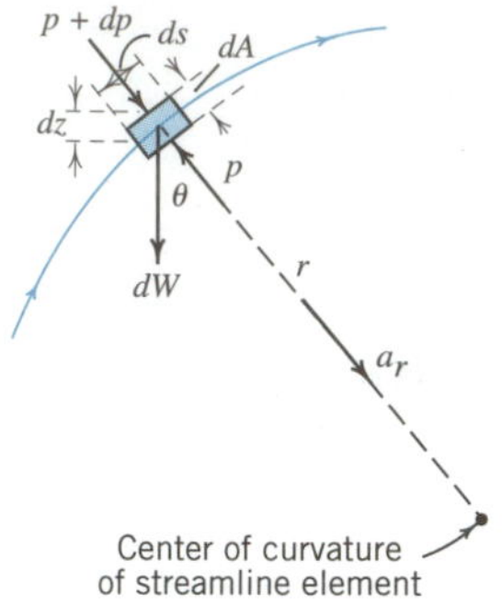

Fig. 5.11

Substituting these values in the equation above, dividing by $\gamma\, ds$, and rearranging produce

$$\frac{d}{dr}\left(\frac{p}{\gamma}+z\right)=\frac{V^2}{g_n r} \tag{5.11}$$

from which it is seen that the gradient of $(p/\gamma + z)$ along r is always positive, or that an increase of $(p/\gamma + z)$ is to be expected along a direction outward from the center of curvature; conversely, a drop of $(p/\gamma + z)$ is expected along a direction toward the center of curvature. Although this equation is formally integrable only for vortex motion (where there is a single center of curvature), it is nevertheless of great value to the engineer in the ''reading'' of streamline pictures to distinguish regions of high and low $(p/\gamma + z)$. The reader will sense that this development, which predicts the variation of $(p/\gamma + z)$ in the radial direction, will also determine (through the Bernoulli equation) the variation of velocity with radial distance. This variation may be discovered by taking the derivative (with respect to r) of the Bernoulli equation of irrotational flow:

$$\frac{d}{dr}\left(\frac{p}{\gamma}+\frac{V^2}{2g_n}+z=H\right)=\frac{d}{dr}\left(\frac{p}{\gamma}+z\right)+\frac{2V}{2g_n}\frac{dV}{dr}=0$$

or

$$\frac{d}{dr}\left(\frac{p}{\gamma}+z\right)=-\frac{V}{g_n}\frac{dV}{dr}$$

which may be equated to the expression (Eq. 5.11) obtained from Newton's second law. Thus

$$\frac{V^2}{g_n r}=-\frac{V}{g_n}\frac{dV}{dr} \qquad \text{or} \qquad V\,dr + r\,dV = 0$$

which when integrated (for a flowfield having a single center of curvature) yields

$$Vr = \text{Constant} \tag{5.12}$$

Again it is emphasized that the equation cannot be generally integrated over a whole flowfield because of the numerous centers of curvature of the streamline elements, but it may nevertheless be used to conclude that generally a *decrease*[4] of velocity is to be ex-

[4]The beginner, schooled in the mechanics of solid-body rotation where tangential velocity varies linearly with radius, is frequently surprised to discover that this law does not apply to fluid flow. One should not be surprised, however, because the mobility of fluid particles in a curved flow is infinitely greater than that of solid particles which are ''locked'' together in a rotating object.

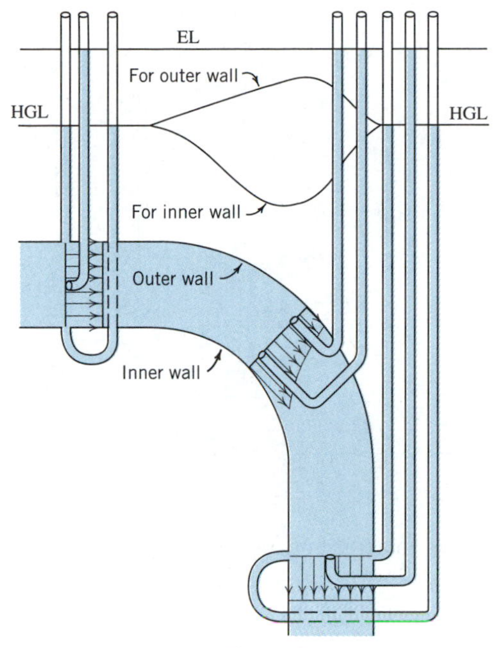

Fig. 5.12

pected with *increase* of distance from center of curvature in irrotational flowfields. The reader may be familiar with this type of velocity distribution from casual experience with vortices in the atmosphere (tornadoes) or the ''bathtub vortex'' which frequently develops when a tank is drained through an orifice in the bottom.

Figure 5.12 depicts a curved flow occurring through a passage in a vertical plane. Distortion of the velocity profile from a uniform distribution is shown; pitot tube and piezometer columns also indicate the variation of pressure and velocity throughout the flow. If these facts are supplemented by a streamline picture, streamlines that are uniformly spaced in the straight passages will be crowded together toward the inner wall of the curved passage and widely spaced toward its outer wall; *thus they cannot be concentric circular arcs.*

Another example of the effect of streamline curvature in a flowfield is seen in the convergent-divergent passage of Fig. 5.13. The streamlines *AA* along the walls are most sharply curved, whereas the central streamline possesses no curvature, and the streamlines in the region between walls and centerline are of intermediate curvature. Accordingly, it can be deduced that the velocity profile at section 2 features higher velocity at the walls than on the centerline, and the relative position of the hydraulic grade lines for these streamlines is as shown. In passing, it is of interest to note that, if incipient cavitation occurred at section 2, it would be expected to appear on the *upper wall* of the passage; for both walls at that cross section, $(p/\gamma + z)$ is the same but, with z larger for the upper wall, the pressure will be less there. This problem has been treated by one-dimensional methods in Section 5.4; it is clearly a one-dimensional problem at sections 1 and 3, but between these sections there is a flowfield which can be treated as one-dimensional only

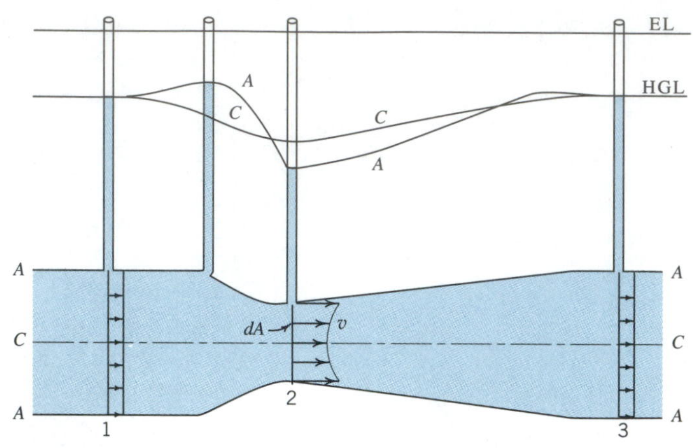

Fig. 5.13

as an approximation. The application of the continuity equation to such problems is basic and instructive. Clearly

$$Q = A_1V_1 = A_2V_2 = A_3V_3 \tag{4.5}$$

in which V_1 and V_3 are the respective mean velocities at sections 1 and 3 and also the velocities of individual particles as they pass these sections. At section 2, however, the velocities of fluid particles are very different from the mean velocity, and a relation between these is to be sought. From Eq. 4.8 the mean velocity V_2 is given by

$$V_2 = \frac{1}{A_2}\iint_{A_2} v\, dA$$

It is noted that, although V_2 can be easily calculated from Q, this yields no information on the distribution of velocity, which is determined by the shape and curvature of the passage walls.

Rigorous application of the Bernoulli equation to the flow between sections 1 and 2 is complicated by the nonuniform velocity profile at section 2. Although the same energy line applies to all the streamlines of the flowfield, the engineer is concerned with the separate terms p, z, and v, all of which vary across the flow at section 2. Here the Bernoulli equation must be written in terms of total power rather than unit energy; following the pattern of Eqs. 5.8a and 5.8b,

$$\begin{bmatrix}\text{Total power}\\ \text{at section 1}\end{bmatrix} = Q\gamma\left(\frac{p_1}{\gamma} + z_1 + \frac{V_1^2}{2g_n}\right)$$

the quantities $(p/\gamma + z)$ and $V^2/2g_n$ being constant across section 1. At section 2 the total power of the flow must be expressed as an integral, following the same pattern. Using $dQ = v\, dA$,

$$\begin{bmatrix}\text{Total power}\\ \text{at section 2}\end{bmatrix} = \iint_{A_2} (v\, dA)\gamma\left(\frac{p}{\gamma} + z + \frac{v^2}{2g_n}\right)$$

Equating the expressions for total power, the "Bernoulli equation" becomes

$$Q\left(\frac{p_1}{\gamma} + z_1 + \frac{V_1^2}{2g_n}\right) = \iint_{A_2} v\left(\frac{p}{\gamma} + z\right) dA + \iint_{A_2} \frac{v^3}{2g_n} dA \tag{5.13}$$

for which a method of solution is suggested in the following illustrative problem.

ILLUSTRATIVE PROBLEM 5.8

Ideal fluid flows through a symmetrical constriction in a two-dimensional passage. From the information given on the sketch, show how the flowrate may be calculated.

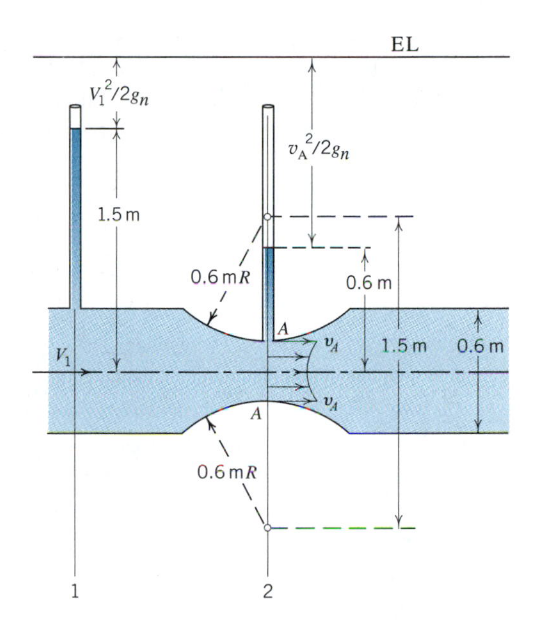

SOLUTION

The solution involves solving Eq. 5.13 by a trial-and-error process. We begin by drawing the energy line and recognizing that it applies to all points in the flow. Next, we use the one-dimensional work-energy equation and the continuity equation to find an average velocity at the constriction by assuming the velocity at point A is the average velocity over the cross section. These average velocities can be used to calculate an approximate flowrate Q. Because the one-dimensional assumption will give an average velocity at the constriction greater than the true average value, the approximate flowrate will be larger than the true flowrate.

Now, the trial solution process begins by assuming a flowrate less than the approximate flowrate first calculated. With the assumed flowrate, the velocity V_1 can be found, establishing the position of the energy line. Then, v_A can be calculated from the position of the energy line. The left-hand side of Eq. 5.13 can then be determined.

$$Q\left(\frac{p_1}{\gamma} + z_1 + \frac{V_1^2}{2g_n}\right) = \iint_{A_2} v(p/\gamma + z)\, dA + \iint_{A_2} (v^3/2g_n)\, dA \tag{5.13}$$

Next, assume a velocity profile similar to the shape indicated that satisfies continuity for the assumed Q. From this profile, calculate the distribution of $(p/\gamma + z)$ across the flow. Finally, carry out the integrations (graphically, if necessary) on the right-hand side of Eq. 5.13 to see if the equation is satisfied. On the first trial it will likely not be satisfied so assume another Q and repeat the process until Eq. 5.13 is satisfied. As a guide in the process, $d/dr(p/\gamma + z)$ at the walls may be computed from $v_A^2/g_n r$ for assistance in establishing a suitable distribution of $(p/\gamma + z)$.

The identification of stagnation points in flowfields is another useful adjunct to an understanding of the flow process. In ideal flow these points may be expected wherever a streamline is forced to turn a sharp corner; such points will be expected only on solid boundaries. Two stagnation points are shown in the external and internal flowfields of Fig. 5.14. At each stagnation point the velocity is locally zero, so the vertical distance between stagnation point and energy line is a direct measure of the pressure at the stagnation point. With this knowledge, the velocities at other points in the flowfield (if irrotational) may be calculated from static pressure measurements.

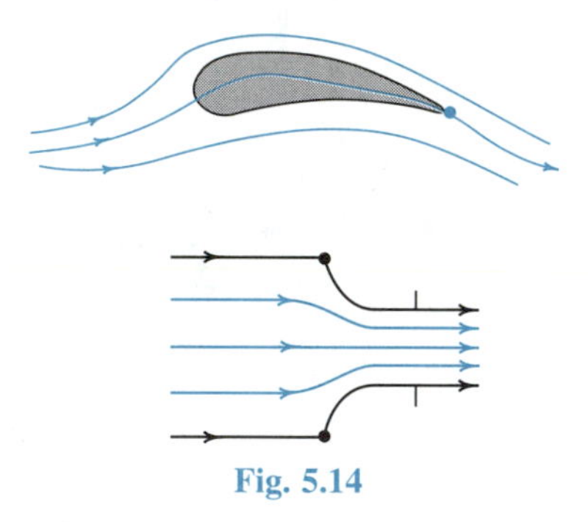

Fig. 5.14

ILLUSTRATIVE PROBLEM 5.9

Ideal fluid of specific weight 50 lb/ft^3 flows down the pipe and discharges into the atmosphere through an end-cap orifice. The pressure gage at B reads 6.0 psi and at A, 2.0 psi. Calculate the mean flow velocity in the pipe, assuming the flow is irrotational.

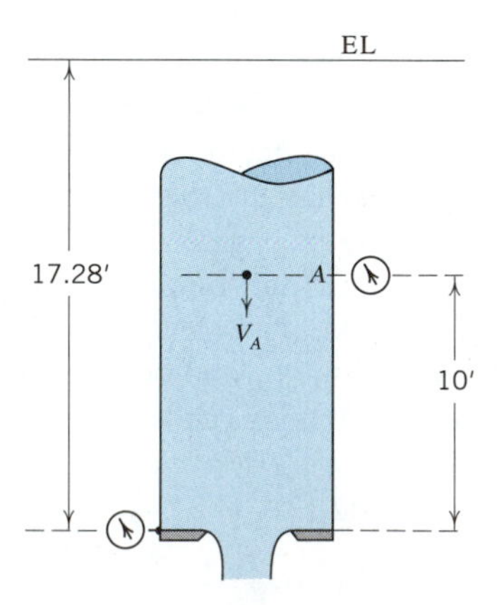

SOLUTION

Point B is easily identified as a stagnation point and can be used to establish the position of the energy line. The energy line will be a distance above the stagnation point equal to the pressure head at the point.

$$\text{Elevation}_{\text{EL}} = \frac{p_B}{\gamma} = \frac{6.0\ \text{lb/in}^2 \times 144\ \text{in}^2/\text{ft}^2}{50\ \text{lb/ft}^3} = 17.28\ \text{ft}$$

From Eq. 5.10, we know that the total head H is given by

$$z + \frac{p}{\gamma} + \frac{V^2}{2g_n} = H - \frac{1}{g_n}\int \xi(u\,dz - w\,dx) \tag{5.10}$$

where the integral term vanishes for irrotational flow. Now, at cross-section A, using elevation B as a datum, we know that

$$z_A + \frac{p_A}{\gamma} + \frac{V_A^2}{2g_n} = 10\ \text{ft} + \frac{2.0\ \text{lb/in}^2 \times 144\ \text{in}^2/\text{ft}^2}{50\ \text{lb/ft}^3} + \frac{V_A^2}{2g_n} = H = 17.28\ \text{ft}$$

$$\frac{V_A^2}{2g_n} = 17.28 - 10 - 5.76 = 1.52$$

$$V_A = 9.89\ \text{ft/s} \bullet$$

A typical problem of two-dimensional irrotational open flow is that of the sharp-crested weir, as shown in Fig. 5.15. A short distance upstream from such a structure the streamlines will be essentially straight and parallel, and a one-dimensional situation will therefore exist. Between this section and some point in the falling sheet of liquid where free fall begins, a flowfield occurs about as shown. Because of the velocity (V) of approach to the weir, the energy line may be visualized above the flow picture as indicated. The boundary streamlines BB and AA (downstream from the weir crest) are called *free stream-*

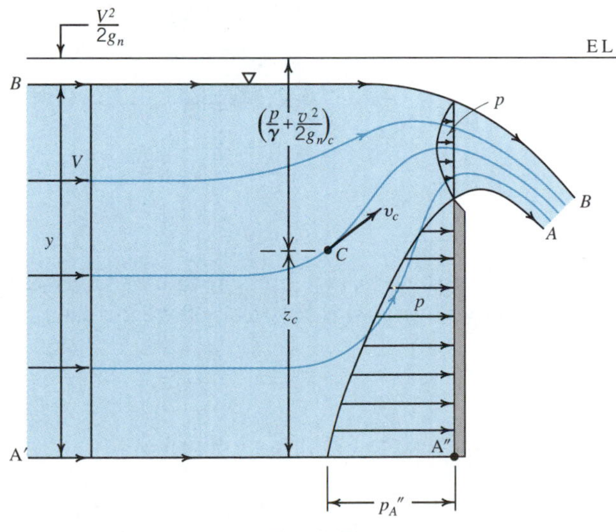

Fig. 5.15

lines, their precise position in space being unknown, but the pressure on them is everywhere constant, in this case zero (gage); once their position is established (by analysis or experiment), the velocity at any point on them may be calculated because the vertical distance between any point and the energy line is the velocity head ($v^2/2g_n$) at the point. The pressure distribution in the flow at section 1 is hydrostatic, the bottom pressure at A' being simply γy. The only stagnation point in the flow is noted at A'', at which the pressure will be $\gamma(y + V^2/2g_n)$. At any other point (C) in the flowfield, $(v_c^2/2g_n + p_c/\gamma)$ can be computed from the positions of point and energy line; however, the pressure there is not calculable without the velocity v_c, which is interrelated with all other velocities and not generally predictable until complete details of the whole flowfield are known. The pressure distribution in the plane of the weir plate is qualitatively predictable because the pressures at both boundaries (free streamlines) are zero; with the streamlines of sharpest curvature nearest to the weir crest, $d/dr\ (p/\gamma + z)$ is largest here, producing a positive pressure in the flow as shown. Downstream from this cross section the pressure within the falling sheet diminishes, becoming essentially zero as the streamlines become straighter and more parallel.

ILLUSTRATIVE PROBLEM 5.10

Calculate the flowrate through this two-dimensional nozzle discharging into the atmosphere and identify any stagnation points in the flow. Assume irrotational flow.

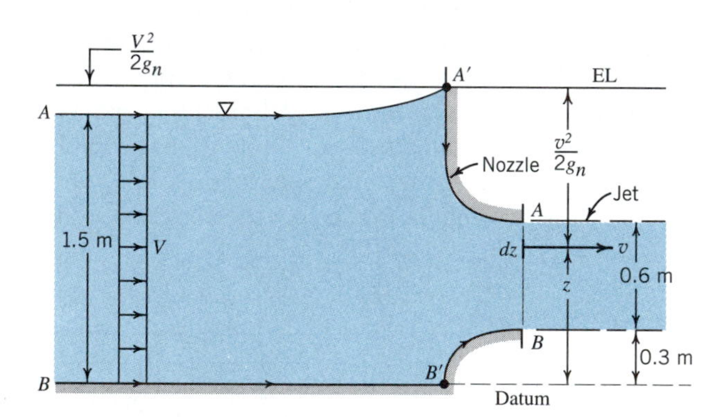

SOLUTION

We assume that the pressure in the free jet is everywhere zero, hence, the velocity in the jet will vary with distance z above the datum. The total head H for the flow is given by

$$H = z + \frac{p}{\gamma} + \frac{V^2}{2g_n} = 1.5\ \text{m} + \frac{V^2}{2g_n}$$

In the jet, the total head is given by

$$H = z + \frac{v^2}{2g_n}$$

Solving the two equations for v gives

$$v = \sqrt{2g_n\left(1.5 + \frac{V^2}{2g_n} - z\right)}$$

The flowrate q can be determined by integrating $v\,dA$ over the cross section of the jet.

$$1.5V = q = \int_{0.3}^{0.9} v\,dz = \int_{0.3}^{0.9} \sqrt{2g_n\left(1.5 + \frac{V^2}{2g_n} - z\right)}\,dz$$

The above equation must be solved by trial by assuming a value of V, performing the integration to determine if the right-hand side of the equation equals $1.5V$, and then repeating the procedure until satisfactory agreement is reached. In this case, the result is $q = 2.81\ \text{m}^3/\text{s/m}$. ●

Stagnation points on the boundary streamlines AA and BB are to be expected. The one at B' needs no comment. At some point on the plane $A'B'$ there must be a stagnation point on the top boundary streamline; assume that this is somewhere below A'. Such a point could not be a stagnation point since its distance below the energy line would indicate a velocity head and thus a velocity there. Accordingly it is concluded that the stagnation point must be at A' and the liquid surface must rise to this point.

Frequently, it is possible to obtain the complete kinematics of a flowfield by mathematical methods that yield specific equations for the streamlines and velocities, from which accelerations and pressure variations may be predicted. A classic and useful example of this is the (irrotational) flowfield about a cylinder of radius R in a rectilinear flow of velocity U, which is shown in Fig. 5.16 (see also Section 11.4). The radial and tangential components of velocity anywhere in the flowfield may be shown to be

$$v_r = U\left(1 - \frac{R^2}{r^2}\right)\cos\theta \qquad \text{and} \qquad v_t = -U\left(1 + \frac{R^2}{r^2}\right)\sin\theta$$

In such problems the velocity along the surface of the body is of greatest interest; here $r = R$, so $v_r = 0$ and $v_t = -2U\sin\theta$. For $\theta = 0$ and $\theta = \pi$, v_t will be zero, thus confirming the expected stagnation points at the head and tail of the body. Applying the

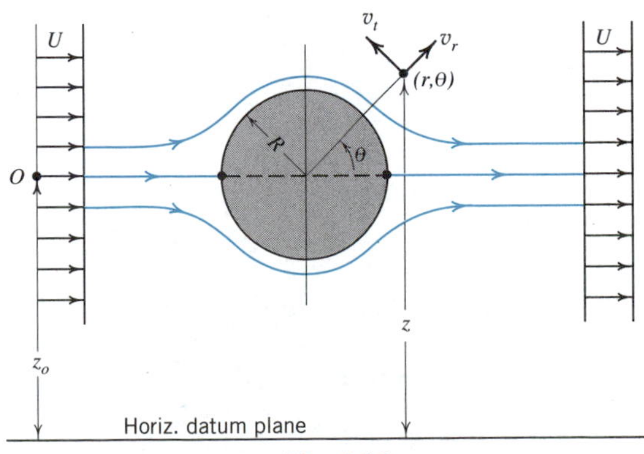

Fig. 5.16

Bernoulli equation between the undisturbed flow and any point on the body contour,

$$\frac{p_o}{\gamma} + \frac{U^2}{2g_n} + z_o = \left(\frac{p}{\gamma} + z\right) + \frac{(-2U \sin \theta)^2}{2g_n}$$

it is seen that $(p/\gamma + z)$ at any point on the cylinder may be predicted from the properties (p_o, z_o, and U) of the undisturbed flow; from the cylinder size, z is determined and thus computation of the pressures on the cylindrical surface may be accomplished. By similar methods any pressure in the flowfield may be computed. Unfortunately, this ideal fluid motion is quite different from the observed real fluid motion (see Chapter 11), and the results of the present analysis are approximately valid for real flow only on a portion of the front face of the cylinder.

ILLUSTRATIVE PROBLEM 5.11

A cylinder 6 inches in diameter extends horizontally across the test section of a large open-throat wind tunnel through which air flows at a velocity of 100 ft/s, at a pressure of 0 psi, and with a specific weight of 0.08 lb/ft^3. Calculate the theoretical values of the velocity and pressure in the flow field at the point ($\theta = 120°$, $r = 6$ in). Assume the flow is irrotational and refer to Fig. 5.16 and the previous discussion for expressions for the velocity in the flowfield.

SOLUTION

For an irrotational flow, we know the total head H is given by

$$H = z + \frac{p}{\gamma} + \frac{V^2}{2g_n} = 0 + 0 + \frac{(100 \text{ ft/s})^2}{2g_n} = 155.3 \text{ ft}$$

where the datum has been set at the centerline of the cylinder.

The expressions for velocity are

$$v_r = U\left(1 - \frac{R^2}{r^2}\right) \cos \theta \qquad v_t = -U\left(1 + \frac{R^2}{r^2}\right) \sin \theta$$

We can now compute the velocities at the point in question.

$$v_r = 100 \text{ ft/s}\left(1 - \frac{(3 \text{ in})^2}{(6 \text{ in})^2}\right) \cos 120° = -37.5 \text{ ft/s}$$

$$v_t = -100 \text{ ft/s}\left(1 + \frac{(3 \text{ in})^2}{(6 \text{ in})^2}\right) \sin 120° = -108.3 \text{ ft/s}$$

Solving for the total velocity,

$$V = \sqrt{v_r^2 + v_t^2} = \sqrt{(-37.5)^2 + (-108.3)^2} = 114.6 \text{ ft/s} \bullet$$

To calculate the pressures, we substitute the velocity into the total head expression.

$$H = 155.3 \text{ ft} = z + \frac{p}{\gamma} + \frac{V^2}{2g_n} = \left(\frac{6 \text{ in}}{12}\right) \sin 120° + \frac{p}{\gamma} + \frac{(114.6 \text{ ft/s})^2}{2g_n}$$

$$= 0.43 + \frac{p}{\gamma} + 203.9$$

Solving for p/γ,

$$\frac{p}{\gamma} = 155.3 - 203.9 - 0.43 = -49.1 \text{ ft}$$

The pressure is

$$p = -49.1\gamma = 49.1 \times 0.08 \text{ lb/ft}^3 = -3.93 \text{ lb/ft}^2$$

In gas flow problems, it is customary to neglect the z-terms in the Bernoulli equation. Had this been done here, the 0.43 would not have appeared and the pressure would have been -3.89 lb/ft^2, about 1% from that calculated. In problems with larger velocities, the z-terms are of even less importance.

ILLUSTRATIVE PROBLEM 5.12

Referring to Fig. 5.16, find a general relationship for the pressure difference between the undisturbed airflow and any point on the surface of the cylinder and plot the distribution of this pressure.

SOLUTION

We will develop an expression for the pressure difference from the Bernoulli equation using the expressions for velocity from Illustrative Problem 5.11 and the previous text material. First, we write the Bernoulli equation,

$$z_o + \frac{p_o}{\gamma} + \frac{V_o^2}{2g_n} = z_2 + \frac{p_2}{\gamma} + \frac{V_2^2}{2g_n}$$

where the zero subscript indicates the undisturbed flow. As justified in Illustrative Problem 5.11, we will neglect changes in z, drop the 2 subscript and solve for the difference in pressure.

$$p - p_o = \frac{\gamma}{2g_n}(U^2 - V^2) = \frac{\rho}{2}(U^2 - V^2)$$

where p is the pressure at any point on the surface of the cylinder and U is the undisturbed flow velocity. Now, we recognize that

$$V^2 = v_r^2 + v_t^2 = U^2\left[\left(1 - \frac{R^2}{R^2}\right)^2 \cos^2\theta + \left(-1 - \frac{R^2}{R^2}\right)^2 \sin^2\theta\right]$$

$$V^2 = U^2[0 + (-2)^2 \sin^2\theta] = 4U^2 \sin^2\theta$$

Now, replacing V^2 in the Bernoulli equation with the above,

$$p - p_o = \frac{\rho}{2}(U^2 - 4U^2 \sin^2\theta) = \frac{\rho U^2}{2}(1 - 4\sin^2\theta)$$

In order to display this pressure difference in a dimensionless fashion, we define a *pressure coefficient* C_p,

$$C_p = \frac{p - p_o}{\rho U^2/2} = \frac{\rho U^2/2(1 - 4 \sin^2 \theta)}{\rho U^2/2} = 1 - 4 \sin^2 \theta \bullet$$

The result is displayed in polar coordinate fashion on the following diagram.

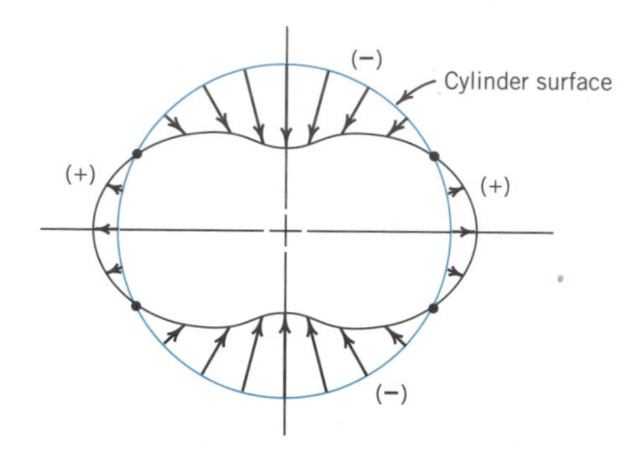

The flow through a sharp-edged opening (Fig. 5.17) produces a flowfield, which has many engineering applications. Jet contraction and flow curvature are produced by the (approximately) radial approach of fluid to the orifice, the streamlines becoming essentially straight and parallel at a section (termed the vena contracta)[5] a short distance downstream from the opening. Here and at other sections downstream, the pressure through the jet is essentially zero, as explained in Section 5.4. Elsewhere, the pressure is zero only on the free streamlines which bound the jet, but, with the centers of curvature of streamlines A, B, and C in the vicinity of O and the pressure increasing away from the center of curvature, it is apparent that, in the plane of the opening, pressures increase and velocities decrease toward the centerline; thus within the curved portion of the jet the pressures are expected

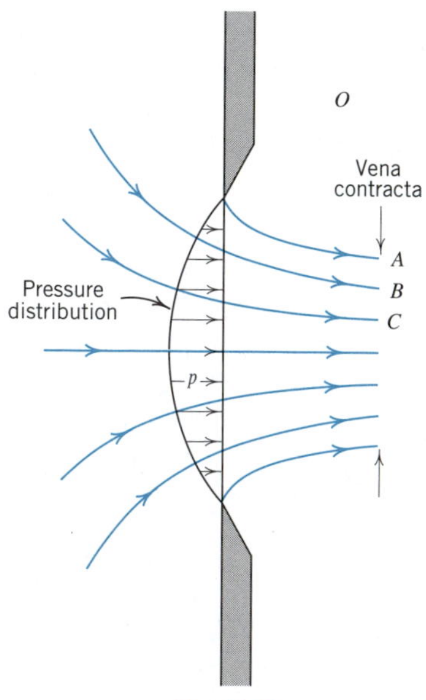

Fig. 5.17

[5]Using advanced analytical methods, Kirchhoff showed the width of the vena contracta to be $(\pi/(\pi + 2)) \times$ (width of opening) for discharge at high velocity.

to be larger than zero. In engineering practice, this problem is usually treated by one-dimensional methods, which are entirely adequate at the vena contracta; applied to the flow cross section in the plane of the opening, they are quite meaningless and lead only to contradictions.

ILLUSTRATIVE PROBLEM 5.13

A two-dimensional flow of liquid discharges from a large reservoir through the sharp-edged opening. A pitot tube at the center of the vena contracta produces the reading indicated. Calculate the velocities at points *A*, *B*, *C*, and *D* and also find the flowrate.

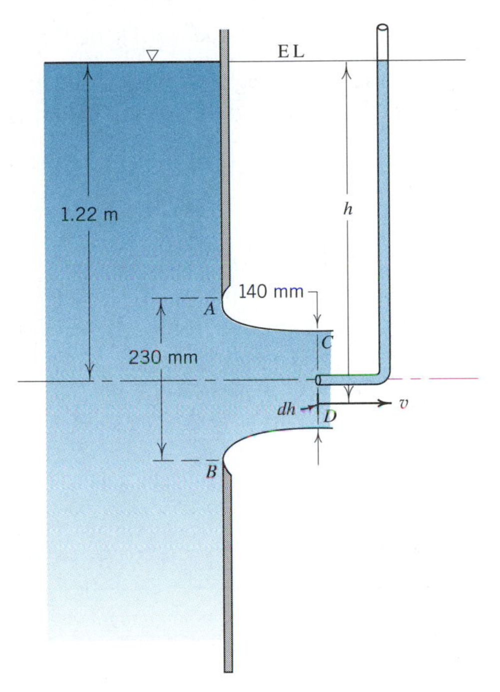

SOLUTION

The pitot tube reading determines the position of the energy line (and also that of the free surface in the reservoir). Since the pressures at points *A*, *B*, *C*, and *D* are all zero, the respective velocity heads are determined by the vertical distances between the points and the energy line. The results are as follows:

$$\frac{V_A^2}{2g_n} = 1.22 \text{ m} - 0.230 \text{ m}/2 = 1.11 \text{ m} \qquad V_A = 4.66 \text{ m/s} \bullet$$

$$\frac{V_B^2}{2g_n} = 1.22 \text{ m} + 0.230 \text{ m}/2 = 1.34 \text{ m} \qquad V_B = 5.12 \text{ m/s} \bullet$$

$$\frac{V_C^2}{2g_n} = 1.22 \text{ m} - 0.140 \text{ m}/2 = 1.15 \text{ m} \qquad V_C = 4.75 \text{ m/s} \bullet$$

$$\frac{V_D^2}{2g_n} = 1.22 \text{ m} + 0.140 \text{ m}/2 = 1.29 \text{ m} \qquad V_D = 5.03 \text{ m/s} \bullet$$

The flowrate may be found by integrating the product $v\,dA$ over the flow cross section CD. The value of velocity over this section, assuming the pressure is zero over the entire cross section of the jet, is

$$\frac{v^2}{2g_n} = h \qquad \text{or} \qquad v = \sqrt{2g_n h}$$

Recognizing that $dA = dh$ for a two-dimensional flow, and noting that the limits of integration are the vertical distances from the energy line to points C and D, respectively,

$$q = \int_C^D v\,dA = \int_{1.15}^{1.29} \sqrt{2g_n h} \times dh = \sqrt{2g_n}\,[2/3h^{3/2}]_{1.15}^{1.29}$$

$$q = 0.685\ \text{m}^3/\text{s/m} \bullet$$

It is interesting to see the result if we consider the velocity at the jet centerline to be the mean velocity and compute the flowrate under this assumption.

$$V \approx \sqrt{2g_n \times 1.22\ \text{m}} = 4.89\ \text{m/s}$$

$$q = V \times \text{width of jet} = 4.89 \times 0.140\ \text{m} = 0.685\ \text{m/s}$$

While the two results are equal, it is true only when the head on the orifice is relatively large compared with the orifice opening. Actually, the center velocity is greater than the mean, but for large ratios of head to orifice opening, there is a negligible difference between them. For a small ratio of head to orifice opening, the flow field is greatly distorted by the drooping of the jet, producing sharply curved streamlines and an ill-defined vena contracta. For such conditions, the foregoing methods of calculation may be used only as crude approximations.

5.9 STREAM FUNCTION AND VELOCITY POTENTIAL

One of the features that limits the application of the Bernoulli equation in two-dimensional flows is the need to formulate a complete set of differential equations and the related boundary conditions which can be solved to yield flow fields and the resulting velocities and pressures at all points. One method for developing these equations for two-dimensional flow relies on the concepts of the stream function and the velocity potential.

Definition of the stream function, a concept based on the continuity principle and the properties of the streamline, provides a mathematical means of solving for, as well as the plotting and interpreting of, two-dimensional and steady flowfields. Consider the streamline A of Fig. 5.18. By definition, no flow crosses it and thus the flowrate ψ across all lines OA is the same.[6] Accordingly, ψ is a constant of the streamline and, if ψ can be found as a function of x and y, the streamline can be plotted. Similarly, the flowrate between O and a closely adjacent streamline (B) will be $\psi + d\psi$, and the flowrate between the streamlines (i.e., in the streamtube) will be $d\psi$. As the flowrates into and out of the elemental triangle are equal from continuity considerations,

$$d\psi = -v\,dx + u\,dy$$

[6]Point O need not be the origin of coordinates.

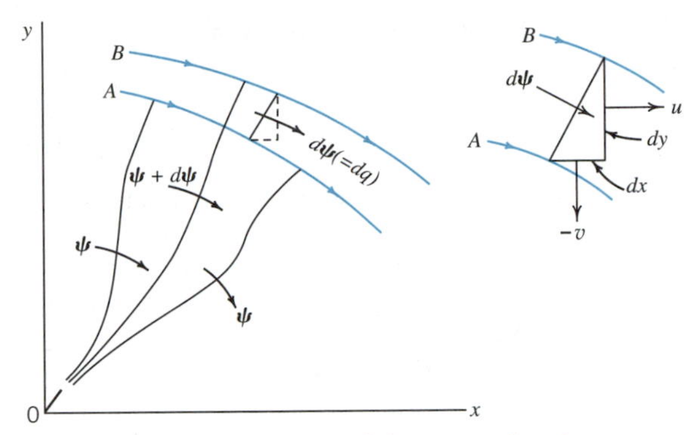

Fig. 5.18 Definition of the stream function.

However, if $\psi = \psi(x, y)$, the total derivative is (see Appendix 6)

$$d\psi = \frac{\partial \psi}{\partial x} dx + \frac{\partial \psi}{\partial y} dy \tag{5.14}$$

and comparison of these equations yields[7]

$$u = \frac{\partial \psi}{\partial y} \quad \text{and} \quad v = -\frac{\partial \psi}{\partial x} \tag{5.15}$$

Thus, if ψ is a known function of x and y, the components of velocity at any point may be obtained by taking the appropriate partial derivative of ψ. Conversely, if u and v are known as functions of x and y, ψ may be obtained by integration of Eq. 5.14, yielding

$$\psi = \int \left(\frac{\partial \psi}{\partial x}\right) dx + \int \left(\frac{\partial \psi}{\partial y}\right) dy + C$$

The equation of continuity,

$$\frac{\partial u}{\partial x} + \frac{\partial v}{\partial y} = 0 \tag{4.11}$$

can be easily (and usefully) expressed in terms of ψ by substituting the relations of Eqs. 5.15; this leads to the identity

$$\frac{\partial}{\partial x}\left(\frac{\partial \psi}{\partial y}\right) = \frac{\partial}{\partial y}\left(\frac{\partial \psi}{\partial x}\right) \quad \text{or} \quad \frac{\partial^2 \psi}{\partial x\, \partial y} = \frac{\partial^2 \psi}{\partial y\, \partial x}$$

which expresses the fact that if $\psi = \psi(x, y)$ the derivatives taken in either order yield the same result and that a flow described by a stream function satisfies the continuity equation automatically.

[7]Analogous relations for polar coordinates are

$$v_r = \frac{\partial \psi}{r \partial \theta} \quad \text{and} \quad v_t = -\frac{\partial \psi}{\partial r}$$

in which v_r is positive radially outward from the origin; v_t and θ are positive counterclockwise.

The equation for vorticity,

$$\xi = \frac{\partial v}{\partial x} - \frac{\partial u}{\partial y} \tag{3.10}$$

can also be expressed in terms of ψ by similar substitutions:

$$\xi = -\frac{\partial^2 \psi}{\partial x^2} - \frac{\partial^2 \psi}{\partial y^2}$$

However, for irrotational flows, $\xi = 0$, and the classic Laplace equation,

$$\frac{\partial^2 \psi}{\partial x^2} + \frac{\partial^2 \psi}{\partial y^2} = \nabla^2 \psi = 0$$

results. This means that the stream functions of all irrotational flows must satisfy the Laplace equation and that such flows may be identified in this manner; conversely, flows whose ψ does not satisfy the Laplace equation are rotational ones. Since both rotational and irrotational flowfields are physically possible, the satisfaction of the Laplace equation is no criterion of the physical existence of a flowfield.

ILLUSTRATIVE PROBLEM 5.14

A flowfield is described by the equation $\psi = y - x^2$. Sketch the streamlines $\psi = 0$, $\psi = 1$, $\psi = 2$. Derive an expression for the velocity V at any point in the flowfield. Calculate the vorticity of this flow.

SOLUTION

From the equation for ψ, it appears the flowfield is a family of parabolas symmetrical about the y-axis with the streamline $\psi = 0$ passing through the origin of the coordinate system. A sketch of the flowfield is shown below.

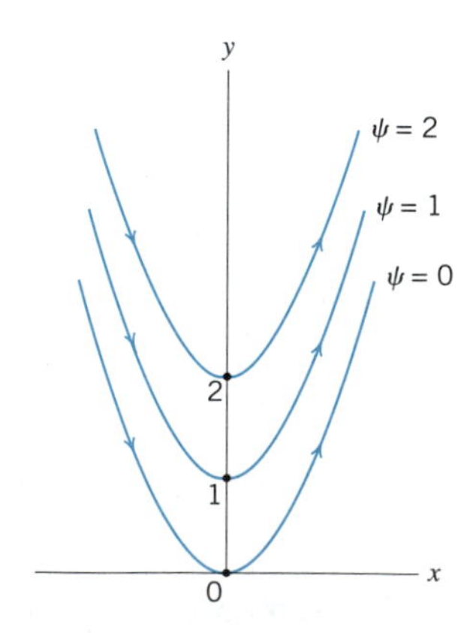

To compute the velocity at any point in the flow, we proceed to find the x- and y-components of velocity and calculate the total velocity in the conventional manner.

$$u = \frac{\partial \psi}{\partial y} = \frac{\partial}{\partial y}(y - x^2) = 1 - 0 = 1 \tag{5.15}$$

$$v = -\frac{\partial \psi}{\partial x} = -\frac{\partial}{\partial y}(y - x^2) = -(-2x) = 2x \tag{5.15}$$

These equations allow us to place the directional arrows on the streamlines depicted above. The magnitude of the velocity is calculated as

$$V = \sqrt{u^2 + v^2} = \sqrt{(2x)^2 + 1^2} = \sqrt{4x^2 + 1} \bullet$$

The vorticity of the flow is calculated from Eq. 3.10.

$$\xi = \frac{\partial v}{\partial x} - \frac{\partial u}{\partial y} = \frac{\partial}{\partial x}(2x) - \frac{\partial}{\partial y}(1) = 2 - 0 = 2 \text{ s}^{-1} \bullet \tag{3.10}$$

Since $\xi \neq 0$, this flowfield is rotational.

Suppose now that another function $\phi(x, y)$ is defined such that the negative of its derivative with respect to distance in any direction yields the velocity in that direction. For example,

$$\mathbf{V} = -\text{grad } \phi = -\left[\frac{\partial \phi}{\partial x}\mathbf{e}_x + \frac{\partial \phi}{\partial y}\mathbf{e}_y\right] = -\nabla \phi$$

so in Cartesian coordinates

$$u = -\frac{\partial \phi}{\partial x} \quad \text{and} \quad v = -\frac{\partial \phi}{\partial y} \tag{5.16}$$

or, in polar coordinates,

$$v_r = -\frac{\partial \phi}{\partial r} \quad \text{and} \quad v_t = -\frac{\partial \phi}{r \partial \theta} \tag{5.17}$$

The function ϕ is known as the *velocity potential*, and it has some significant properties which may now be examined in general terms.

The continuity equation

$$\frac{\partial u}{\partial x} + \frac{\partial v}{\partial y} = 0 \tag{4.11}$$

may be written in terms of ϕ by substitution of the above definitions, to yield the Laplace differential equation,

$$\frac{\partial^2 \phi}{\partial x^2} + \frac{\partial^2 \phi}{\partial y^2} = 0 \tag{5.18}$$

Thus all practical flows (which must conform to the continuity principle) must satisfy the Laplace equation in terms of ϕ.

Similarly, the equation for vorticity,

$$\xi = \frac{\partial v}{\partial x} - \frac{\partial u}{\partial y} \tag{3.10}$$

may be put in terms of ϕ to give

$$\xi = \frac{\partial}{\partial x}\left(-\frac{\partial \phi}{\partial y}\right) - \frac{\partial}{\partial y}\left(-\frac{\partial \phi}{\partial x}\right) = -\frac{\partial^2 \phi}{\partial x \partial y} + \frac{\partial^2 \phi}{\partial y \partial x}$$

from which a valuable conclusion may be drawn: Since $\partial^2\phi/\partial x\partial y = \partial^2\phi/\partial y\partial x$, *the vorticity must be zero for the existence of a velocity potential.* From this it may be deduced that only irrotational ($\xi = 0$) flowfields can be characterized by a velocity potential ϕ; for this reason *irrotational* flows are also known as *potential* flows.

ILLUSTRATIVE PROBLEM 5.15

Calculate the velocity potential ϕ and sketch the equipotential lines of the flowfield produced by a cylinder (often represented mathematically by a ''doublet'') of radius R in a rectilinear flow (see Fig. 5.16).

SOLUTION

For the flowfield under investigation, it can be shown that the expressions for velocity are as follows (see Sections 5.8 and 11.4):

$$v_r = U\left(1 - \frac{R^2}{r^2}\right)\cos\theta \qquad v_t = -U\left(1 + \frac{R^2}{r^2}\right)\sin\theta$$

Velocity can be expressed in terms of the velocity potential as

$$v_r = -\frac{\partial \phi}{\partial r} \qquad v_t = -\frac{1}{r}\frac{\partial \phi}{\partial \theta} \tag{5.17}$$

Combining the above sets of equations gives

$$\frac{\partial \phi}{\partial r} = -U\left(1 - \frac{R^2}{r^2}\right)\cos\theta \qquad -\frac{1}{r}\frac{\partial \phi}{\partial \theta} = -U\left(1 + \frac{R^2}{r^2}\right)\sin\theta$$

From Appendix 6, the total differential of ϕ can be written

$$d\phi = \frac{\partial \phi}{\partial r}dr + \frac{\partial \phi}{\partial \theta}d\theta = \frac{\partial \phi}{\partial r}dr + \frac{1}{r}\frac{\partial \phi}{\partial \theta}r\,d\theta$$

We can integrate this equation to obtain

$$\phi = \int\left(\frac{\partial \phi}{\partial r}\right)dr + \int\frac{1}{r}\left(\frac{\partial \phi}{\partial \theta}\right)r\,d\theta + C$$

Substituting the expressions for $\partial\phi/\partial r$ and $(1/r)\partial\phi/\partial\theta$ into the previous equation,

$$\phi = \int -U\left(1 - \frac{R^2}{r^2}\right)\cos\theta\,dr + \int U\left(1 + \frac{R^2}{r^2}\right)\sin\theta\,r\,d\theta + C$$

Expanding and rearranging,

$$\phi = -U\left[\int\cos\theta\,dr - \int r\sin\theta\,d\theta\right] + UR^2\left[\int\frac{\cos\theta}{r^2}dr + \int\frac{\sin\theta}{r}d\theta\right] + C$$

The bracketed quantities may be shown to be (respectively) equivalent to the integrals shown in

$$\phi = -U \int d(r \cos \theta) + UR^2 \int d\left(\frac{-\cos \theta}{r}\right) + C$$

and so,

$$\phi = -U\left(r + \frac{R^2}{r}\right) \cos \theta + C \bullet$$

Assuming for convenience that the constant of integration is zero, it is seen that $\phi = 0$ when $\theta = \pi/2$ or $3\pi/2$; thus, the line $\phi = 0$ coincides with y-axis. For $\theta \rightarrow \pi$ and $r \rightarrow \infty$, $\phi \rightarrow +\infty$, and, for $\theta \rightarrow 0$ and $r \rightarrow \infty$, $\phi \rightarrow -\infty$. For $\theta = \pi/3$ and $r = R$, $\phi = -RU$, and when plotted or sketched, the equipotential lines must appear as shown.

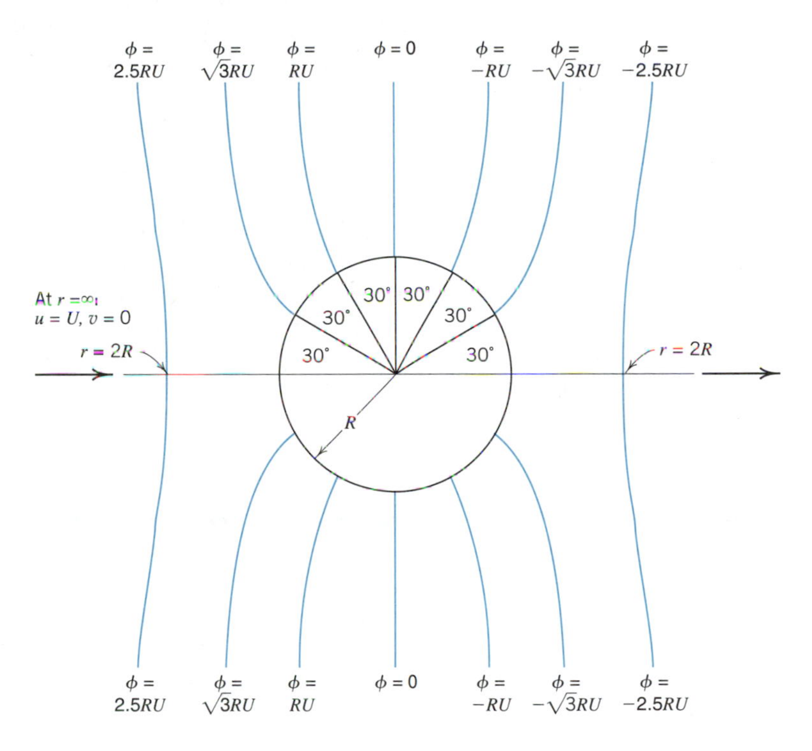

A geometric relationship between streamlines and equipotential lines may be derived from the foregoing equations and restatement of certain mathematical definitions; the latter are (with definitions of u and v inserted)

$$d\psi = \frac{\partial \psi}{\partial x} dx + \frac{\partial \psi}{\partial y} dy = -v\, dx + u\, dy$$

$$d\phi = \frac{\partial \phi}{\partial x} dx + \frac{\partial \phi}{\partial y} dy = -u\, dx - v\, dy$$

However, along any streamline ψ is constant and $d\psi = 0$, so $dy/dx = v/u$ along a streamline; also along any equipotential line ϕ is constant and $d\phi = 0$, so $dy/dx = -u/v$ along an equipotential line. The geometric significance of this is seen in Fig. 5.19 (and

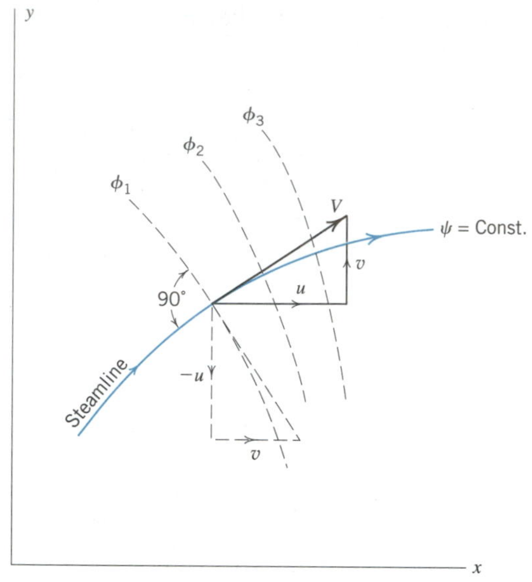

Fig. 5.19

has been suggested in the preceding Illustrative Problem): *The equipotential lines are normal to the streamlines.* Thus, the streamlines and equipotential lines (for an irrotational flow)[8] form a net, called a *flownet*, of mutually perpendicular families of lines, a fact of great significance for the study of flowfields where formal mathematical expressions of ϕ and ψ are unobtainable. Another feature of the velocity potential which may be deduced from Fig. 5.19 and Eqs. 5.16 is that the value of ϕ drops *along the direction of the flow*; that is, $\phi_3 < \phi_2 < \phi_1$.

REFERENCES

Batchelor, G. K. 1967. *An introduction to fluid dynamics.* Cambridge: Cambridge University Press.

Karamcheti, K. 1966. *Principles of ideal-fluid aerodynamics.* New York: Wiley.

Lamb, H. 1945. *Hydrodynamics.* 6th ed. New York: Dover Publications.

Milne-Thomson, L. M. 1968. *Theoretical hydrodynamics.* 5th ed. New York: Macmillan.

Robertson, J. M. 1965. *Hydrodynamics in theory and application.* Englewood Cliffs, N. J.: Prentice Hall.

Vallentine, H. R. 1967. *Applied hydrodynamics.* 2nd ed. London: Butterworths Scientific Publications.

FILMS

Rouse, H. Mechanics of fluids: fluid motion in a gravitational field. Film No. U45961, Media Library, Audiovisual Center, Univ. of Iowa.

Shapiro, A. H. Pressure fields and fluid acceleration. NCFMF/EDC Film No. 21609, Encyclopaedia Britannica Educ. Corp.

[8]There can be no equipotential lines for rotational flows because we showed that the velocity potential does not exist for rotational flows.

FILM LOOPS

Hele-Shaw analog to potential flows. Part I. Sources and Sinks in Uniform Flow, Loop No. S-FM080. Hele-Shaw analog to potential flows. Part II. Sources and Sinks, Loop No. S-FM081. Encyclopaedia Britannica Educ. Corp.

PROBLEMS

5.1. Integrate the one-dimensional Euler equation for a perfect gas at constant temperature.

5.2. Integrate the one-dimensional Euler equation for an isentropic, gas process.

5.3. On an overflow structure of 50° slope the water depth (measured normal to the surface of the structure) is 1.2 m (4 ft) and the streamlines are essentially straight and parallel. Calculate the pressure on the surface of the structure.

5.4. Water flows in a 1 m (3.3 ft) diameter horizontal supply pipe. The flow is turbulent with a maximum centerline velocity of 2 m/s (6.6 ft/s). What is the difference between the pressures at the top and the bottom of the pipe?

5.5. Water flows in a pipeline. At a point in the line where the diameter is 7 in., the velocity is 12 fps and the pressure is 50 psi. At a point 40 ft away the diameter reduces to 3 in. Calculate the pressure here when the pipe is (*a*) horizontal, and (*b*) vertical with flow downward.

5.6. In a pipe 0.3 m in diameter, 0.3 m^3/s of water are pumped up a hill. On the hilltop (elevation 48), the line reduces to 0.2 m diameter. If the pump maintains a pressure of 690 kPa at elevation 21, calculate the pressure in the pipe on the hilltop.

5.7. In a 3-in. horizontal pipeline containing a pressure of 8 psi, 100 gal/min of liquid hydrogen flow. If the pipeline reduces to 1 in. diameter, calculate the pressure in the 1 in. section.

5.8. If crude oil flows through this pipeline and its velocity at *A* is 2.4 m/s, where is the oil level in the open tube *C*?

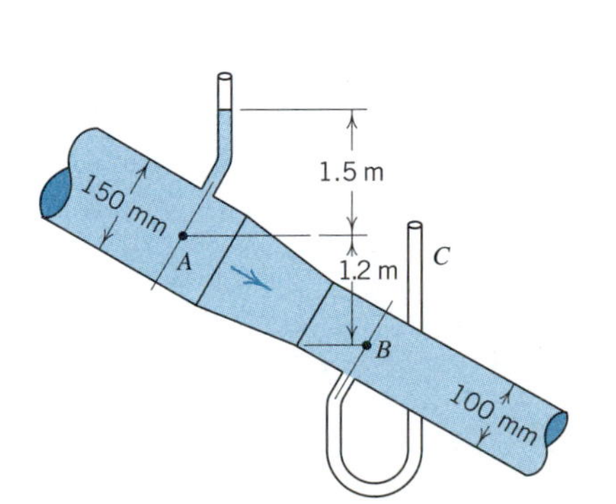

Problem 5.8

5.9. Air of specific weight 12.6 N/m^3 flows through a 100 mm constriction in a 200 mm pipeline. When the flowrate is 11.1 N/s, the pressure in the constriction is 100 mm of mercury vacuum. Calculate the pressure in the 200 mm section, neglecting compressibility.

5.10. A pump draws water from a reservoir through a 12 in. pipe. When 12 cfs are being pumped, what is the pressure in the pipe at a point 8 ft above the reservoir surface, in psi? In feet of water?

5.11. If the pressure in the 0.3 m pipe of problem 4.17 is 70 kPa, what pressures exist in the branches, assuming all pipes are in the same horizontal plane? Water is flowing.

5.12. Water is flowing. Calculate *H* (m) and *p*(kPa).

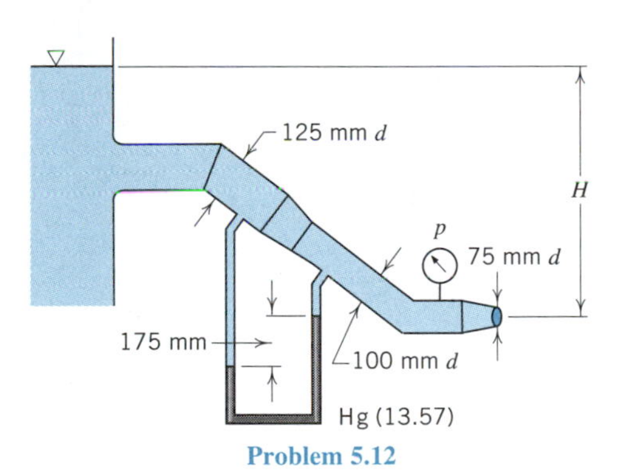

Problem 5.12

5.13. If each gage shows the same reading for a flowrate of 1.00 cfs, what is the diameter of the constriction?

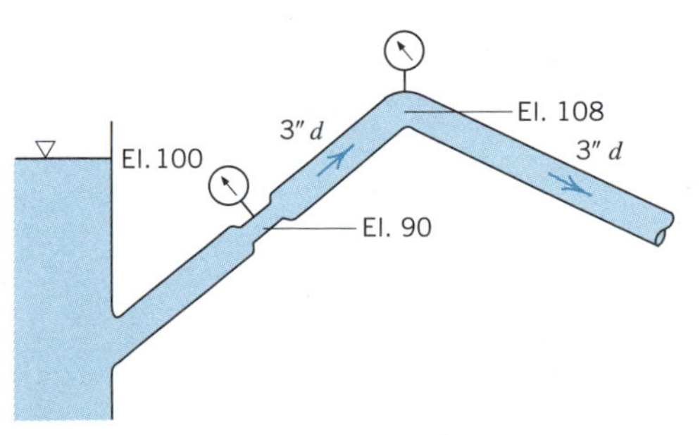

Problem 5.13

5.14. What pressure in psi is required at the base of the nozzle (point 1) to produce a flow of 75 gpm of gasoline (s.g. = 0.68)?

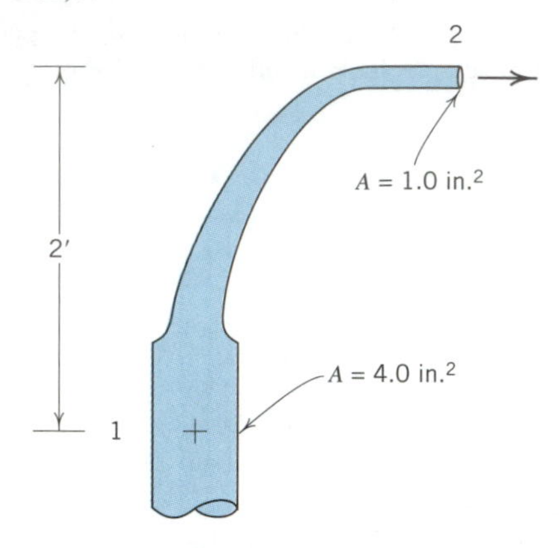

Problem 5.14

5.15. Derive a relation between A_1 and A_2 so that for a flowrate of 0.28 m^3/s the static pressure will be the same at sections 1 and 2. Also calculate the manometer reading for this condition.

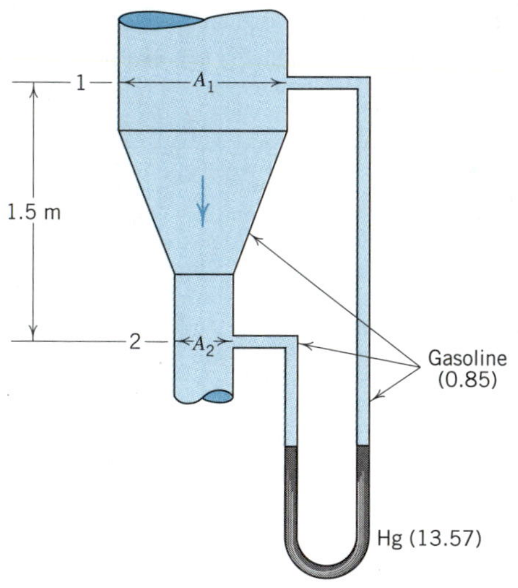

Problem 5.15

5.16. Water is flowing. Calculate the required pipe diameter, d, for the two gages to read the same.

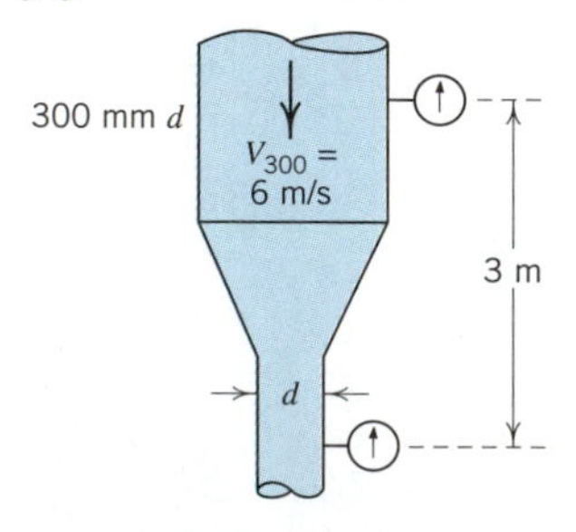

Problem 5.16

5.17. If the pipe of problem 5.16 is part of a water supply system in a moon station, find d. Recall $g_m \approx g_n/6$.

5.18. For a flowrate of 75 cfs of air ($\gamma = 0.076\,3\ \text{lb/ft}^3$) what is the largest A_2 that will cause water to be drawn up to the piezometer opening? Neglect compressibility effects.

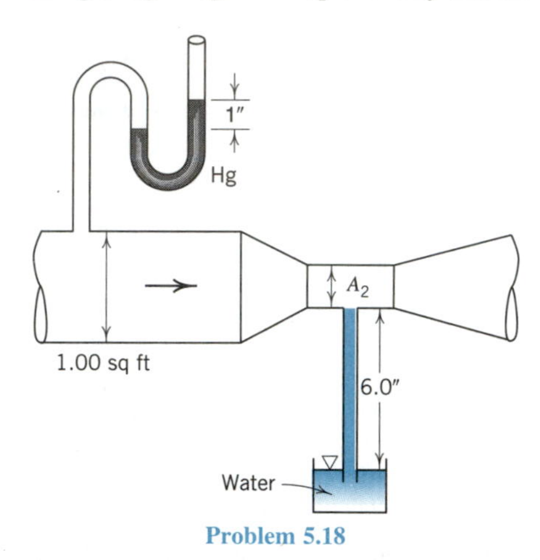

Problem 5.18

5.19. A smooth nozzle of 50 mm (2 in.) diameter is connected to a water tank. Connected to the tank at the same elevation is an open U-tube manometer containing mercury and showing a reading of 625 mm (25 in.). The lower mercury surface is 500 mm (20 in.) below the tank connection. What flowrate will be obtained from the nozzle? Water fills the tube between tank and mercury.

5.20. Water discharges from a tank through a nozzle of 50 mm (2 in.) diameter into an air tank where the pressure is 35 kPa (5 psi). If the nozzle is 4.5 m (14.8 ft) below the water surface, calculate the flowrate.

5.21. Water discharges through a nozzle of 25 mm (1 in.) diameter under a 6 m (20 ft) head into a tank of air in which a vacuum of 250 mm (10 in.) of mercury is maintained. Calculate the rate of flow.

5.22. Air is pumped through the tank as shown. Neglecting effects of compressibility, compute the velocity of air in the 100 mm pipe. Atmospheric pressure is 91 kPa and the specific weight of air 11 N/m^3.

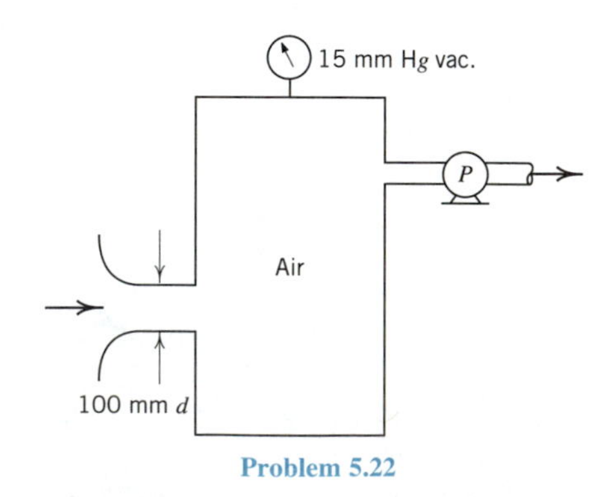

Problem 5.22

5.23. A closed tank contains water with air above it. The air is maintained at a pressure of 103 kPa (14.9 psi) and 3 m (9.8 ft) below the water surface a nozzle discharges into the atmosphere. At what velocity will water emerge from the nozzle?

5.24. Water is flowing. The flow picture is axisymmetric. Calculate the flowrate and manometer reading.

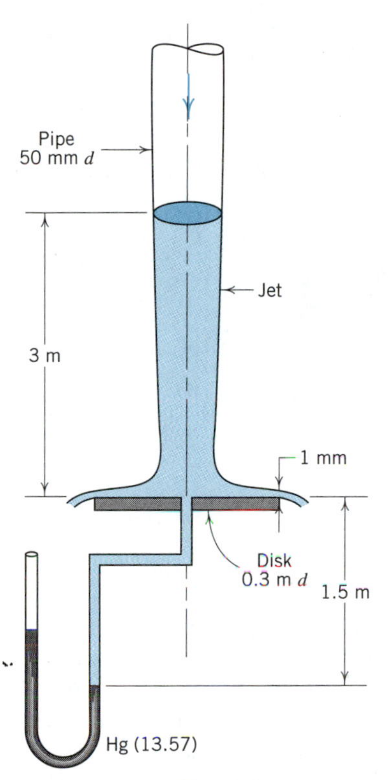

Problem 5.24

5.25. An airplane flies at altitude 30 000 ft where the air density is 0.000 9 slug/ft^3 and the air pressure 4.37 psia. A pitot tube inside the entrance of its jet engine is connected to a conventional Bourdon pressure gage inside the airplane. This gage reads 13.00 in. of mercury vacuum when the cabin is pressurized to 12.24 psia. If the streamline configuration is as shown, how much air (cfs) is passing through the engine? Assume the air is incompressible.

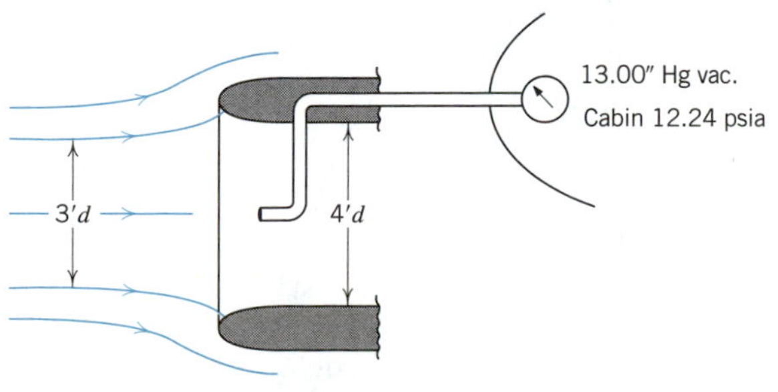

Problem 5.25

5.26. A section of a freely falling water jet is observed to taper from 1 300 mm^2 to 975 mm^2 in a vertical distance of 1.2 m. Calculate the flowrate in the jet.

5.27. Estimate the flowrate Q of water discharging from the nozzle. If the fluid discharging were oil at s.g. = 0.90 and the pitot tube still registered 8 in, what would be the flowrate?

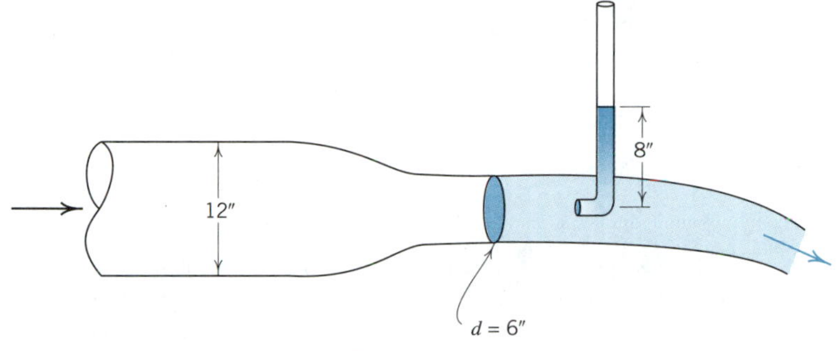

Problem 5.27

5.28. Water jets upward through a 3 in. diameter nozzle under a head of 10 ft. At what height h will the liquid stand in the pitot tube? What is the cross-sectional area of the jet at section B?

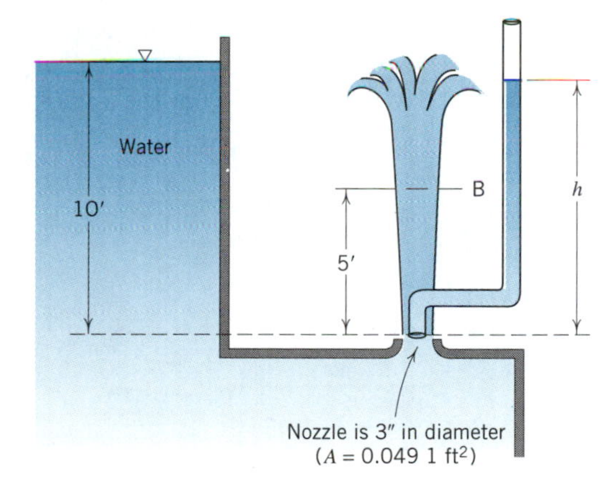

Problem 5.28

5.29. Calculate the rate of flow through this pipeline and the pressures at A, B, C, and D. Sketch the EL and HGL showing vertical distances.

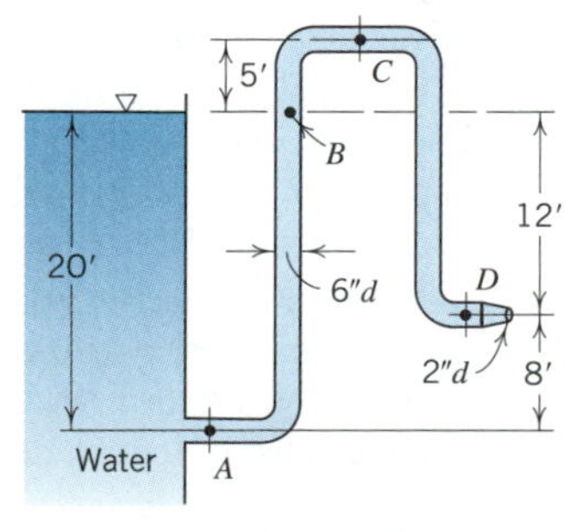

Problem 5.29

5.30. Calculate the pressure in the flow at A: (*a*) for the system shown, and (*b*) for the pipe without the nozzle. For both cases, sketch the EL and HGL.

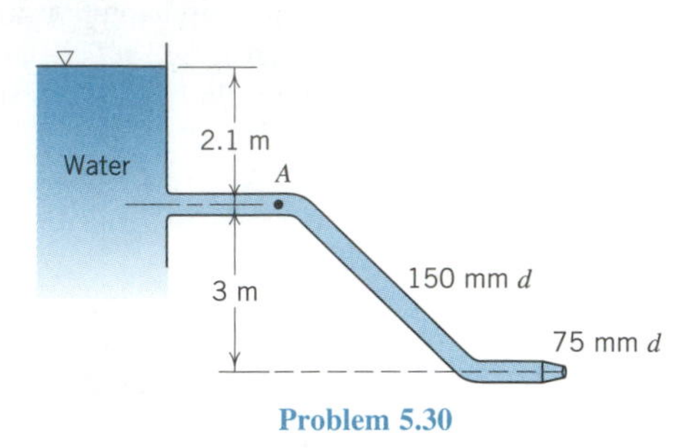

Problem 5.30

5.31. A siphon consisting of a 25 mm (1 in.) hose is used to drain water from a tank. The outlet end of the hose is 2.4 m (7.9 ft) below the water surface, and the bend in the hose is 0.9 m (3 ft) above the water surface. Calculate the pressure in the bend and flowrate.

5.32. A 75 mm horizontal pipe is connected to a tank of water 1.5 m below the water surface. The pipe is gradually enlarged to 88 mm diameter and discharges freely into the atmosphere. Calculate the flowrate and the pressure in the 75 mm pipe.

5.33. Calculate the pressure head in the 20 mm section when $h = 0.16$ m. Calculate the largest h at which the divergent tube can be expected to flow full.

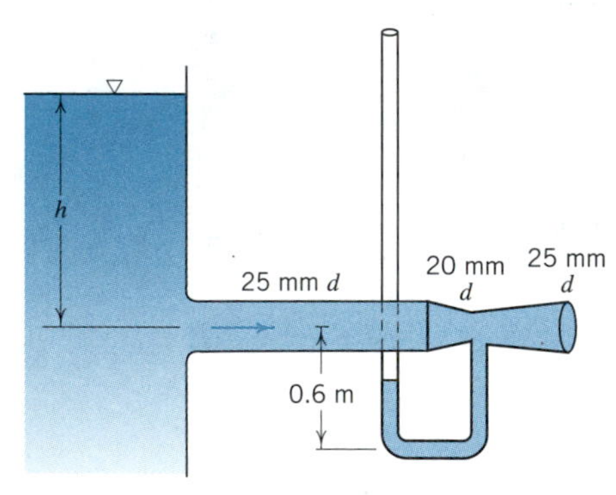

Problem 5.33

5.34. The head of water on a 50 mm diameter smooth nozzle is 3 m. If the nozzle is directed upward at angles of (*a*) 30°, (*b*) 45°, (*c*) 60°, and (*d*) 90°, how high above the nozzle will the jet rise, and how far from the nozzle will the jet pass through the horizontal plane in which the nozzle lies? What is the diameter of the jet at the top of the trajectory?

5.35. A fire hose nozzle which discharges water at 46 m/s (150 ft/s) is to throw a stream through a window 30 m (100 ft) above and 30 m (100 ft) horizontally from the nozzle. What is the minimum angle of elevation of the nozzle that will accomplish this?

5.36. The jet passes through a point A. Calculate the flowrate.

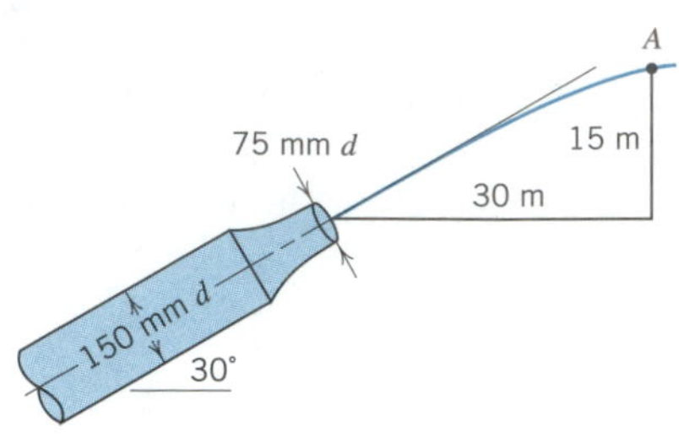

Problem 5.36

5.37. Calculate the minimum flowrate that will pass over the wall.

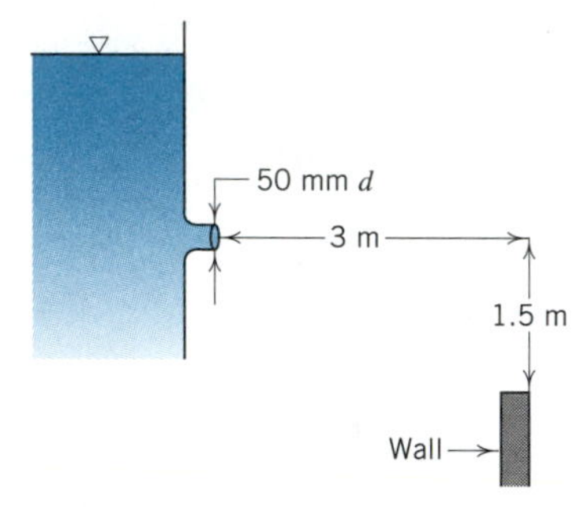

Problem 5.37

5.38. From Bernoulli's equation derive a relation between the velocities at any two points on the trajectory of a free jet in terms of the vertical distance between the points.

5.39. A jet of water falling vertically has a velocity of 24 m/s (79 ft/s) and a diameter of 50 mm (2 in.) at elevation 12 m (40 ft); calculate its velocity at elevation 4.5 m (15 ft).

5.40. Applying free trajectory theory to the centerline of a free jet discharging from a horizontal nozzle under a head h, show that the radius of curvature of the jet at the tip of the nozzle is $2h$.

5.41. For the two orifices discharging as shown, prove that $h_1 y_1 = h_2 y_2$.

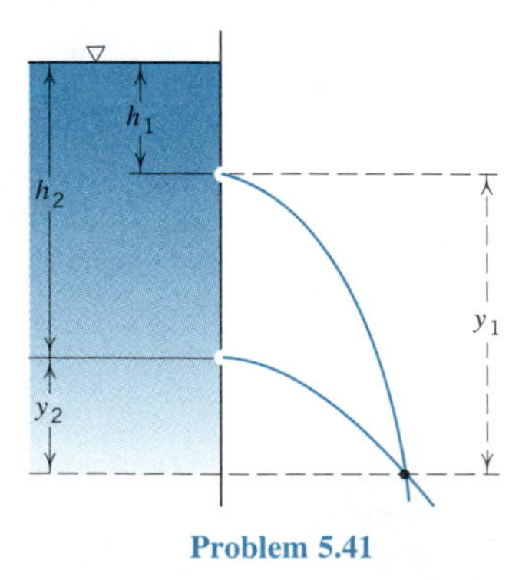

Problem 5.41

5.42. Water flows from one reservoir in a 200 mm (8 in.) pipe, while water flows from a second reservoir in a 150 mm (6 in.) pipe. The two pipes meet in a "tee" junction with a 300 mm (12 in.) pipe that discharges to the atmosphere at an elevation of 20 m (66 ft). If the water surface in the reservoirs is at 30 m (98 ft) elevation, what is the total flowrate?

5.43. What size must one set for the pipe from the second reservoir of problem 5.42 so that the flowrate from the second reservoir is twice that from the first? What is the total flowrate in this case?

5.44. A constriction of 150 mm diameter occurs in a horizontal 250 mm water line. It is stated that the pressure in the 250 mm pipe is 125 mm of mercury vacuum when the flowrate is 170 l/s. Is this possible? Why or why not?

5.45. Water flows through a 1 in. constriction in a horizontal 3 in. pipeline. If the water temperature is 150°F and the pressure in the line is maintained at 40 psi, what is the maximum flowrate that may occur? Barometric pressure is 14.7 psia.

5.46. The liquid has a specific gravity of 1.60 and negligible vapor pressure. Calculate the flowrate for incipient cavitation in the 75 mm section, assuming that the tube flows full. Barometric pressure is 100 kPa.

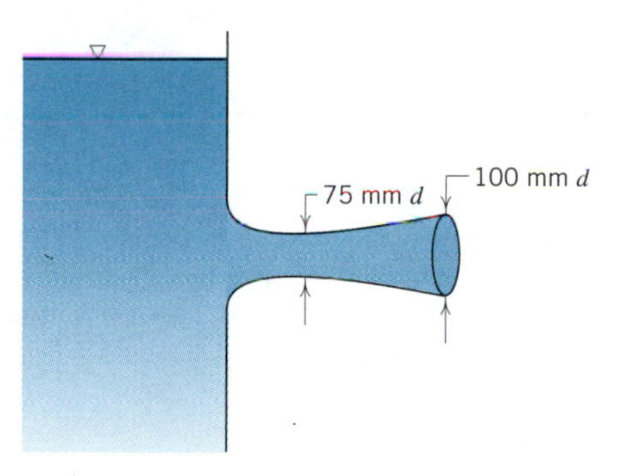

Problem 5.46

5.47. Water flows between two open reservoirs. Barometric pressure is 13.4 psia. Vapor pressure is 6.4 psia. Above what value of h will cavitation be expected in the 2 in. constriction?

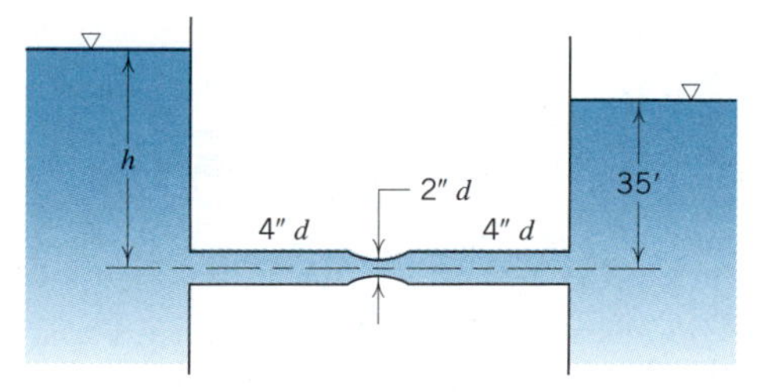

Problem 5.47

5.48. Barometric pressure is 101.3 kPa. For $h > 0.6$ m, cavitation is observed at the 50 mm section. If the pipe is horizontal and flows full throughout, what is the vapor pressure of the water?

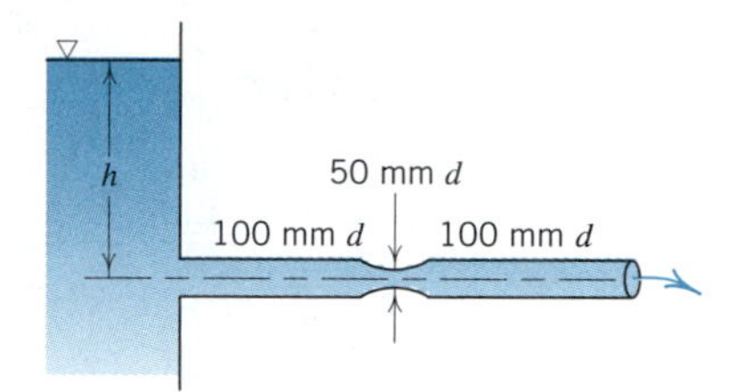

Problem 5.48

5.49. If cavitation is observed in the 50 mm section, what is the flowrate? Barometric pressure is 100 kPa.

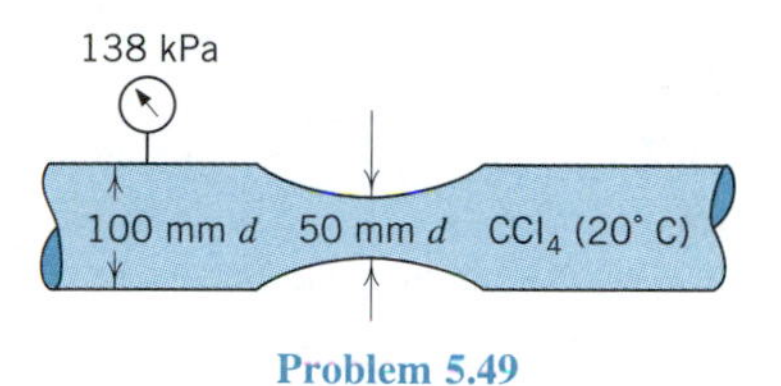

Problem 5.49

5.50. Barometric pressure is 14.0 psia. What is the maximum flowrate that can be obtained by opening the valve?

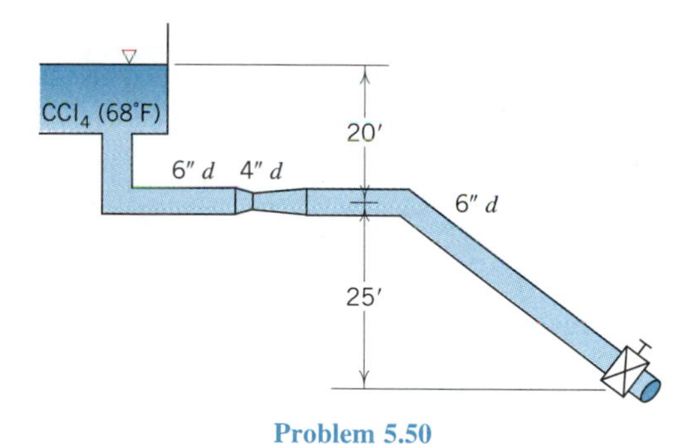

Problem 5.50

5.51. A variety of combinations of d and h will allow maximum possible flowrate to occur through this system. Derive a relationship between d and h that will always produce this. Barometric pressure is 101 kPa.

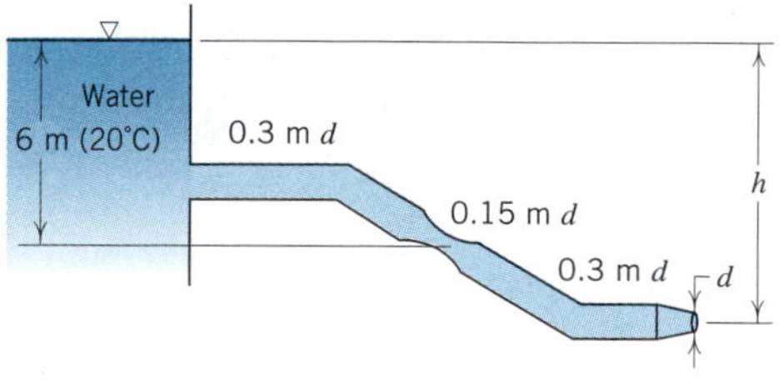

Problem 5.51

5.52. Calculate the maximum h and the minimum d that will permit cavitation-free flow through this frictionless pipe system. Atmospheric pressure = 14.5 psia; vapor pressure = 1.5 psia. Water is flowing.

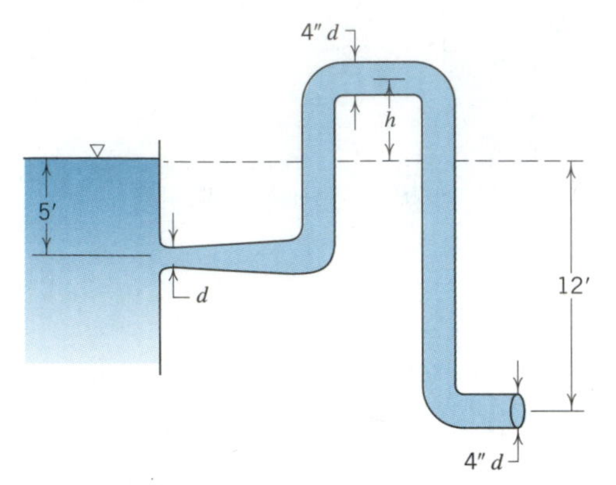

Problem 5.52

5.53. For a flowrate of 0.28 m^3/s derive a relation between h and d that will indicate incipient cavitation in the top of this siphon pipe. Over what range of h is this relation applicable?

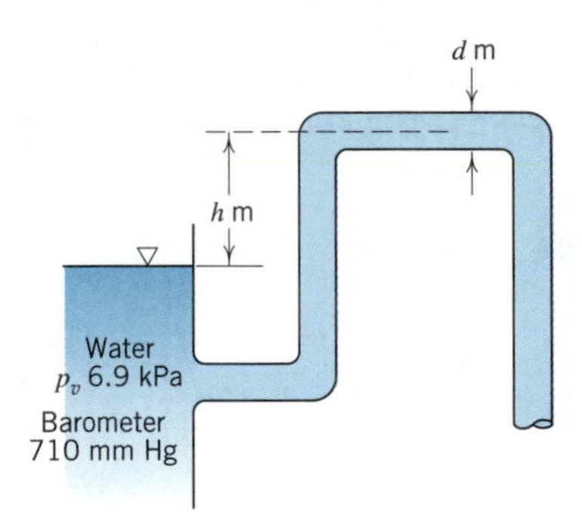

Problem 5.53

5.54. At what h will cavitation be incipient in the 2 in. section? Assume the same water in system and barometer.

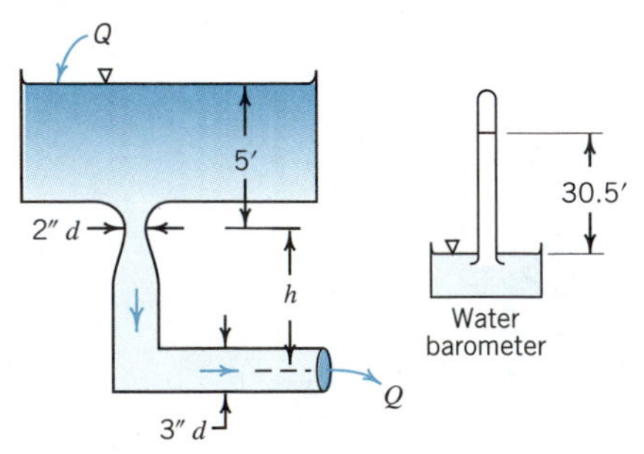

Problem 5.54

5.55. The pressure in this closed tank is gradually increased by pumping. Calculate the gage reading at which cavitation will appear in the 25 mm constriction. The barometric pressure is 94 kPa.

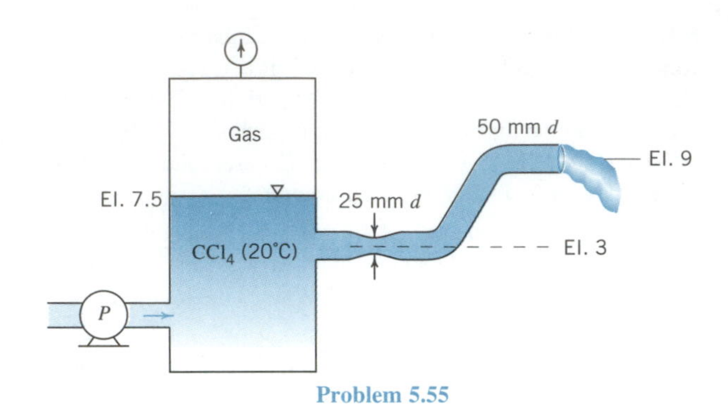

Problem 5.55

5.56. If cavitation anywhere in this pipe system is to be avoided, what is the diameter of the largest nozzle which may be used? Atmospheric pressure = 100 kPa; vapor pressure = 10.3 kPa.

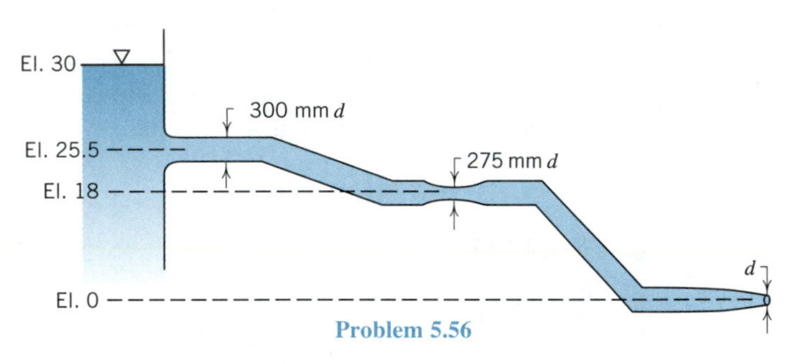

Problem 5.56

5.57. What is the smallest nozzle diameter (d) that will produce the maximum possible flowrate through this frictionless pipe system? Atmospheric pressure is 14.3 psia; vapor pressure of the water is 1.3 psia.

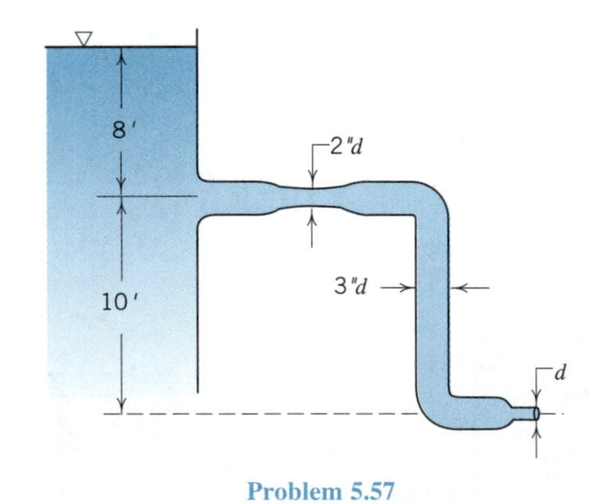

Problem 5.57

5.58. A convergent-divergent tube is connected to a large tank as in problem 5.46. It is to be proportioned so that the negative pressure head at the throat is half the height of the water surface above the centerline of the tube. For no cavitation and assuming that the tube flows full, what ratio of exit diameter to throat diameter is required?

5.59. Cavitation occurs in this convergent–divergent tube as shown. The right-hand side of the manometer is connected to the cavitation zone. The water in the right-hand tube has all evaporated leaving only vapor. Assuming an ideal fluid at 40°C, calculate the gage reading if the local atmospheric pressure is 750 mm of mercury.

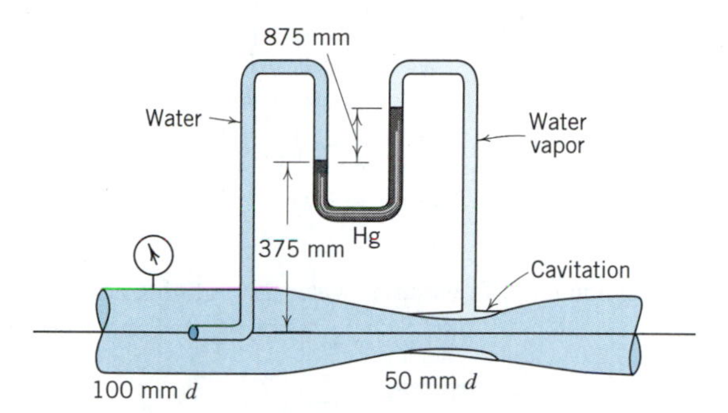

Problem 5.59

5.60. Water at 70°C (158°F) is being siphoned from a tank through a hose in which the velocity is 4.5 m/s (15 ft/s). What is the maximum theoretical height of the high point ("crown") of the siphon above the water surface that will allow this flow to occur? Assume standard barometric pressure.

5.61. Barometric pressure is 101.3 kPa and vapor pressure of the water is 32.4 kPa. Calculate the elevation of end *B* for maximum flowrate through this pipeline. Sketch the EL and HGL for this situation.

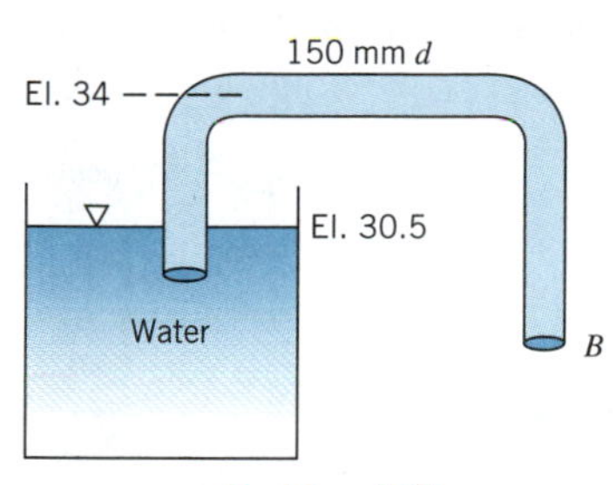

Problem 5.61

5.62. The pressure in the test section of a wind tunnel is −27 mm (−1.1 in.) of water when the velocity is 97 km/h (60 mph). Calculate the pressure on the nose of an object when placed in the test section. Assume the specific weight for air 12.0 N/m³ (0.076 3 lb/ft³).

5.63. The pressure in a 100 mm pipeline carrying 4 500 l/min of fluid weighing 11 kN/m³ is 138 kPa. Calculate the pressure of the upstream end of a small object placed in this pipeline.

5.64. A submarine moves at 46 km/h (25 knots) through salt-water (s.g. = 1.025) at a depth of 15 m (50 ft). Calculate the pressure on the nose of the submarine.

5.65. Calculate the flowrate of water in this pipeline.

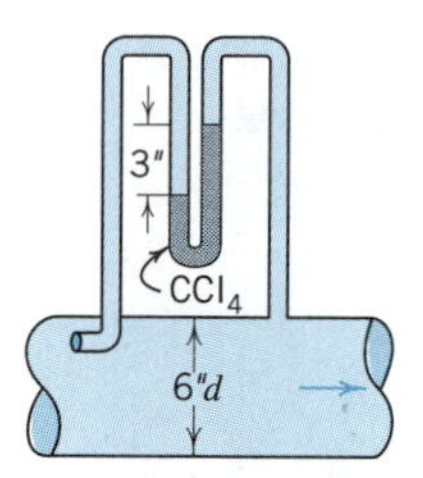

Problem 5.65

5.66. Three tenths of a cubic metre per second of water are flowing. Calculate the manometer reading (*a*) using the sketch as shown, and (*b*) when the pitot tube is at section 2 and the static pressure connection is at section 1.

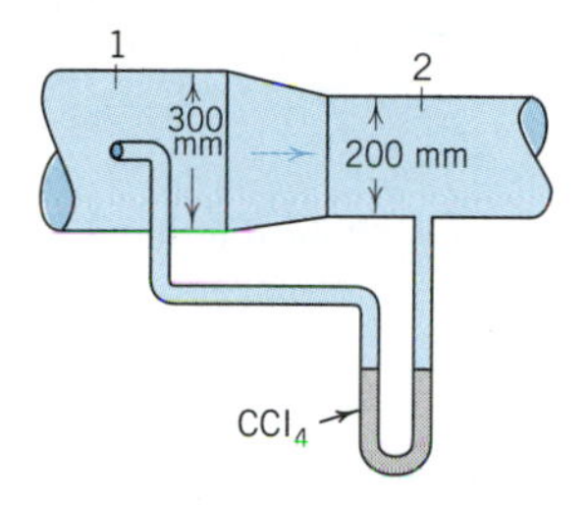

Problem 5.66

5.67. Calculate the flowrate through this pipeline.

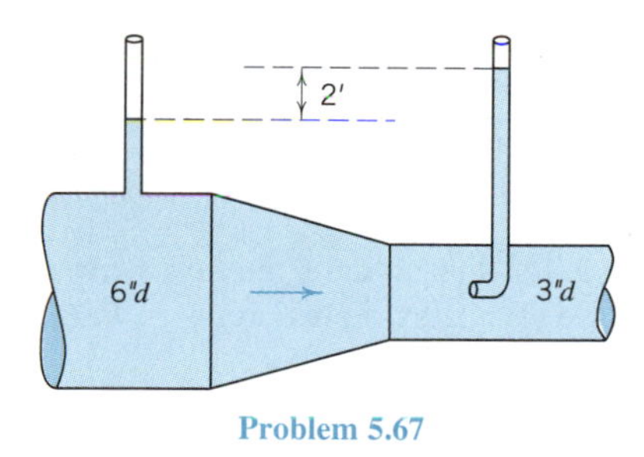

Problem 5.67

5.68. Water is flowing. Calculate the flowrate.

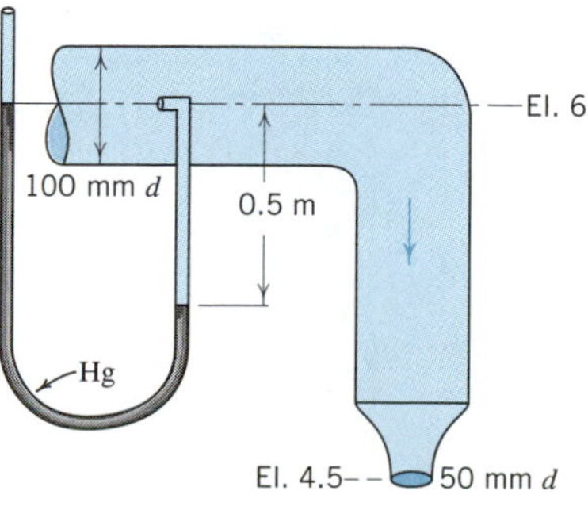

Problem 5.68

5.69. Gasoline (s.g. 0.85) is flowing. Calculate gage readings and flowrate.

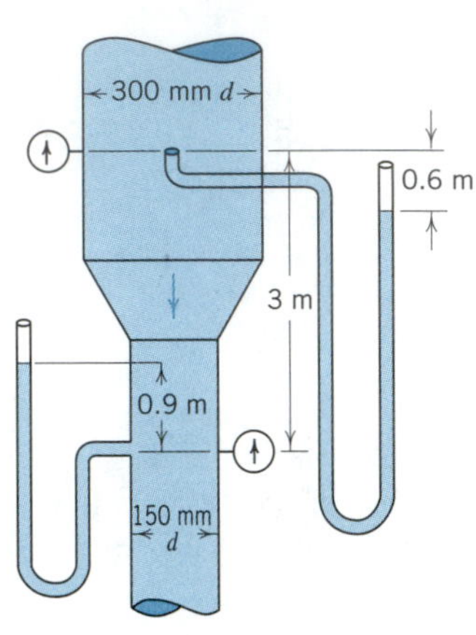

Problem 5.69

5.70. Calculate the velocity head in the 3 in. constriction.

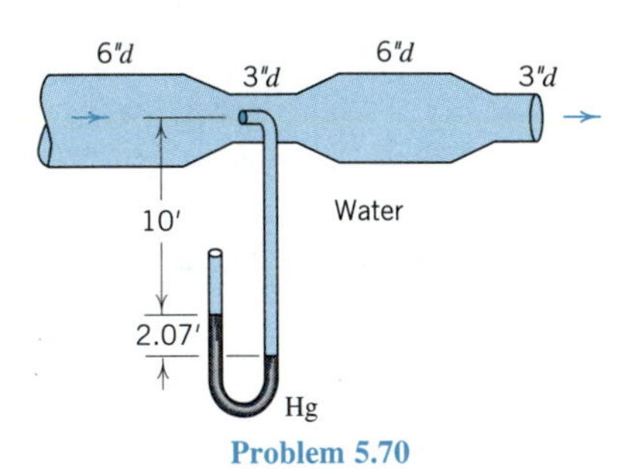

Problem 5.70

5.71. Water is flowing. Assume the flow between the disks to be radial and calculate the pressures at A, B, C, and D. The flow discharges to the atmosphere.

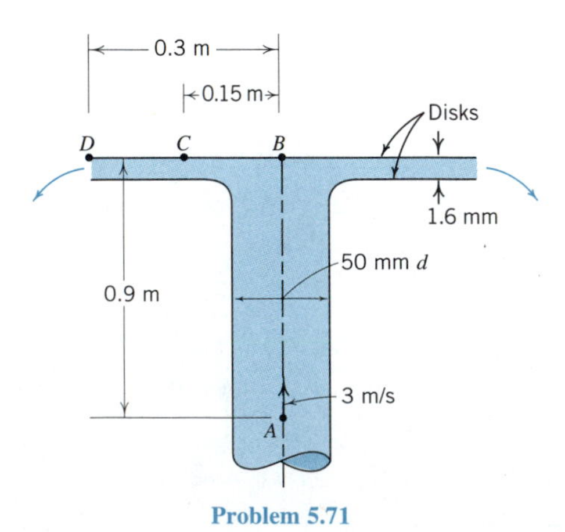

Problem 5.71

5.72. Calculate the gage reading.

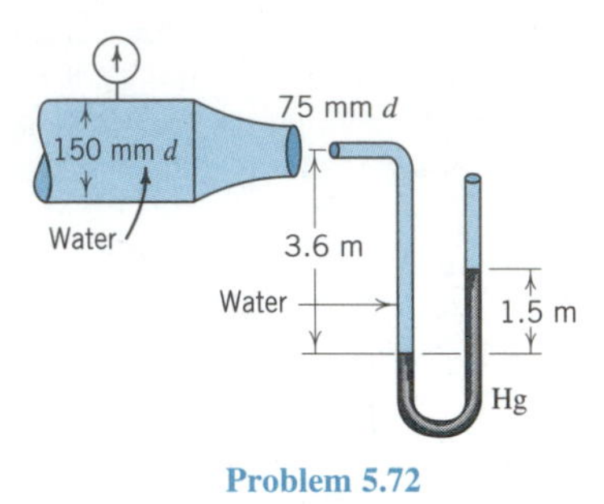

Problem 5.72

5.73. Calculate the flowrate of water through this nozzle.

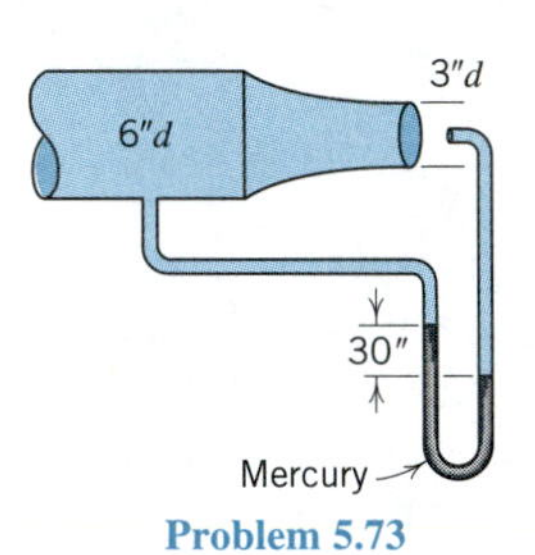

Problem 5.73

5.74. A flowrate of 1.54 m^3/s of water (specific weight 9.81 kN/m^3; vapor pressure 6.9 kPa) passes through this water tunnel. The cock is now closed. Calculate magnitude and direction of the manometer reading after the cock is opened. Atmospheric pressure is 100 kPa.

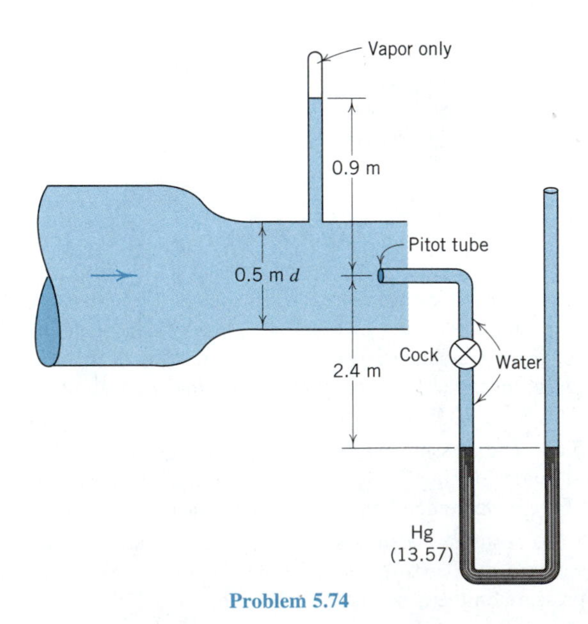

Problem 5.74

5.75. Water is flowing. Calculate the flowrate and gage reading.

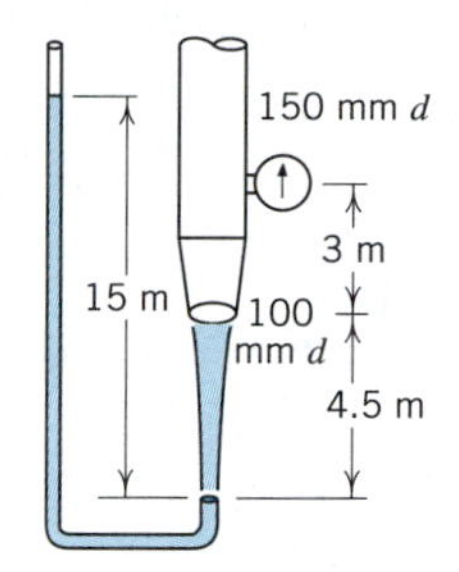

Problem 5.75

5.76. The tip of the pitot tube is at the top of the jet. Calculate the flowrate and the angle θ.

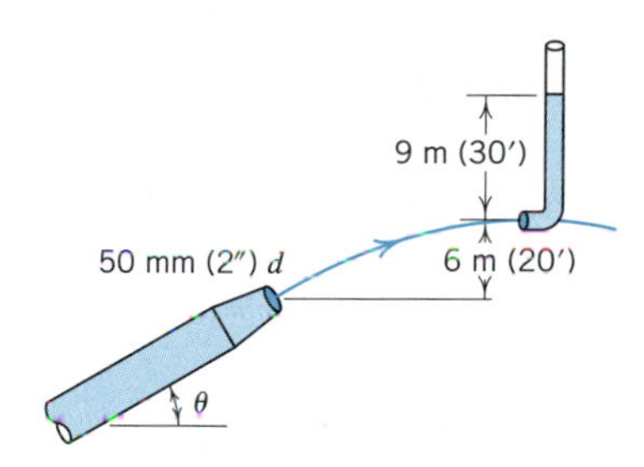

Problem 5.76

5.77. Air discharges from a duct of 300 mm diameter through a 100 mm nozzle into the atmosphere. The pressure in the duct is found by a draft gage to be 25 mm of water. Assuming the specific weight of air to be constant at 11.8 N/m^3, what is the flowrate?

5.78. If in problem 5.8 the vertical distance between the liquid surfaces of the piezometer tubes is 0.6 m, what is the flowrate?

5.79. Water flows through a 75 mm constriction in a horizontal 150 mm pipe. If the pressure in the 150 mm section is 345 kPa and that in the constriction 207 kPa, calculate the velocity in the constriction and the flowrate.

5.80. Ten cubic feet per second of liquid of specific weight 50 lb/ft^3 are flowing. Calculate the manometer reading.

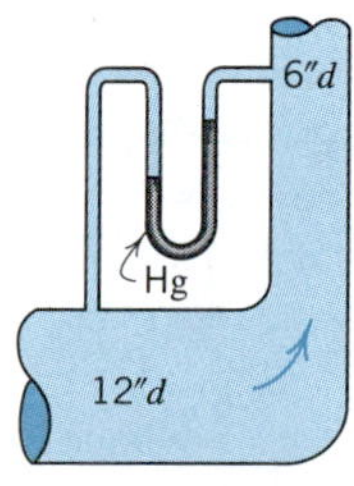

Problem 5.80

5.81. In the preceding problem calculate the flowrate when the manometer reading is 10 in.

5.82. Carbon tetrachloride flows downward through a 50 mm constriction in a 75 mm vertical pipeline. If a differential manometer containing mercury is connected to the constriction and to a point in the pipe 1.2 m above the constriction and this manometer reads 350 mm, calculate the flowrate. (Carbon tetrachloride fills manometer tubes to the mercury surfaces.)

5.83. Calculate the flowrate.

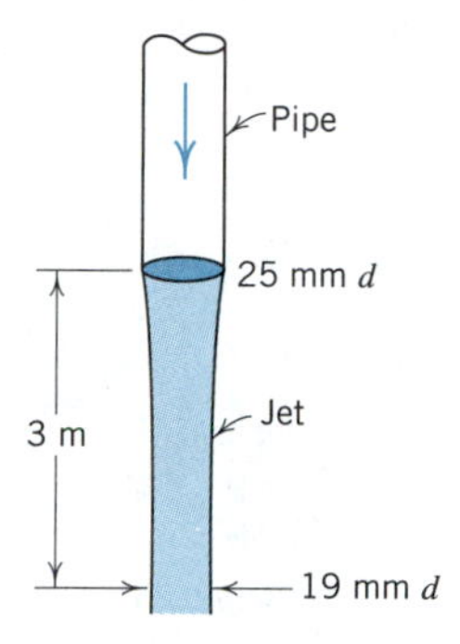

Problem 5.83

5.84. If a free jet of fluid strikes a circular disk and produces the flow picture shown, what is the flowrate?

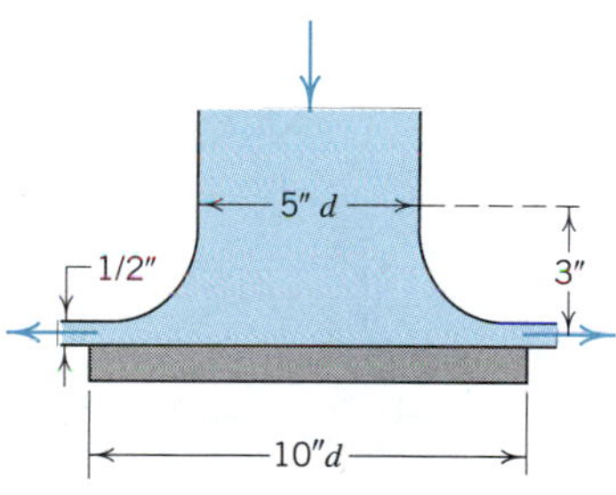

Problem 5.84

5.85. Through a transition structure between two rectangular open channels the width narrows from 2.4 m to 2.1 m, the depth decreases from 1.5 m to 1.05 m, and the bottom rises 0.3 m. Calculate the flowrate.

5.86. This "Venturi flume" is installed in a horizontal frictionless open channel of 10 ft width and water depth 10 ft. In the "throat" of the flume where the width has been narrowed to 8 ft, the water depth is observed to be 8 ft. Calculate the flowrate in the channel.

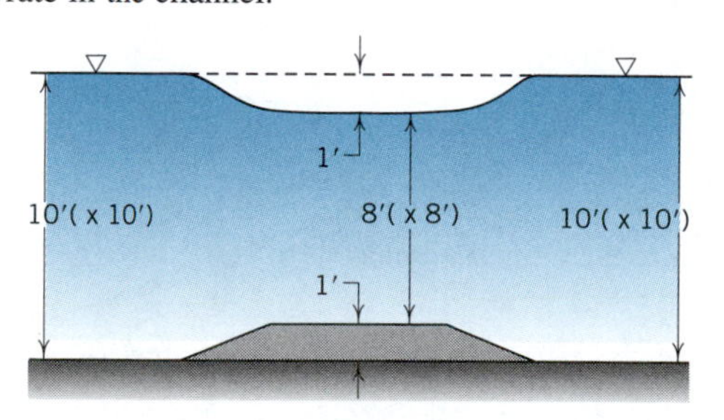

Problem 5.86

5.87. Through a transition structure between two rectangular open channels the width increases from 3 m to 3.6 m while the water surface remains horizontal. If the depth in the upstream channel is 1.5 m, what is the depth in the downstream channel?

5.88. If the two-dimensional flowrate over this sharp crested weir is 10 cfs/ft, what is the thickness of the sheet of falling water at a point 3 ft below the weir crest?

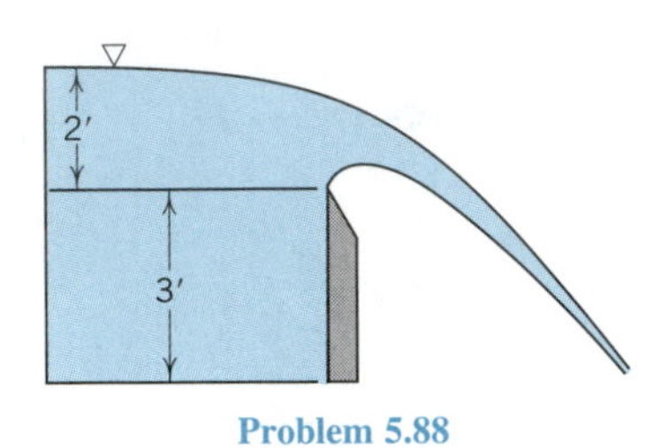

Problem 5.88

5.89. Calculate the two-dimensional flowrate through this frictionless sluice gate when the depth h is 1.5 m. Also calculate the depth h for a flowrate of 3.25 $m^3/s{\cdot}m$.

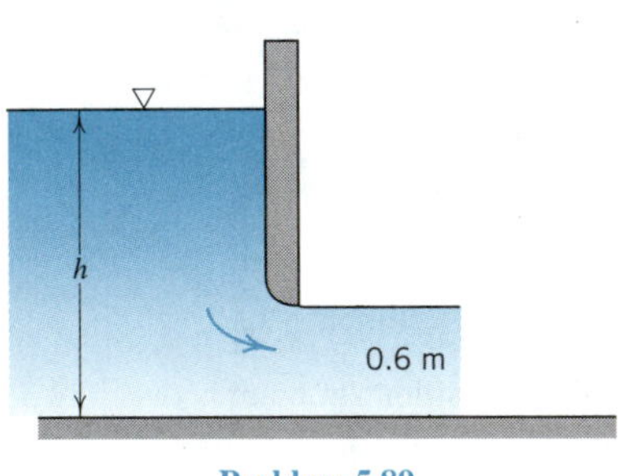

Problem 5.89

5.90. The flow in an open drainage channel passes into a pipe that carries it through a highway embankment. If the channel flows full, what is the flowrate?

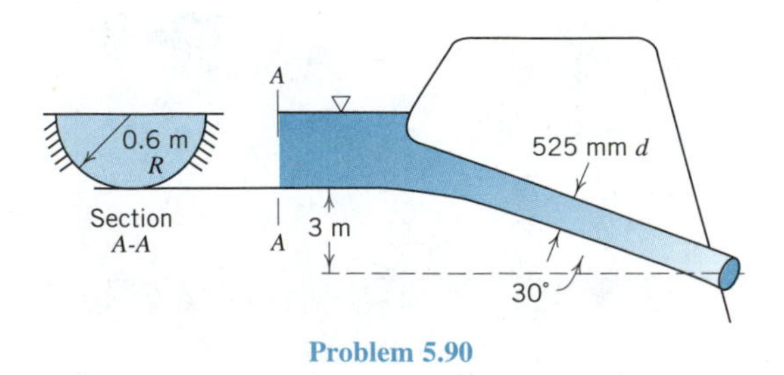

Problem 5.90

5.91. The two-dimensional gate contains a two-dimensional nozzle (slot) as shown. For a flowrate of 80 cfs/ft in the channel, predict the flowrates through the slot and under the gate.

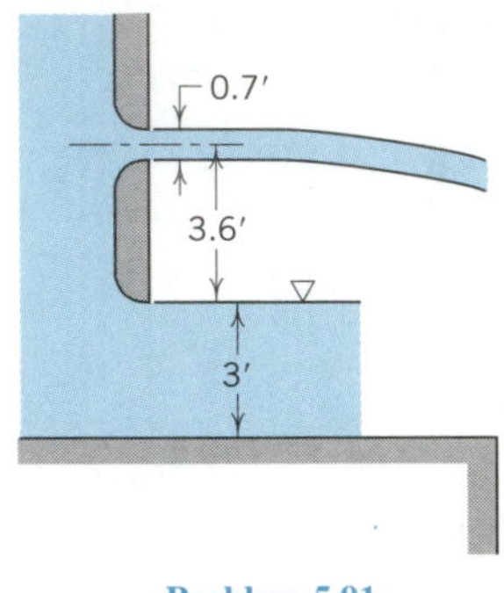

Problem 5.91

5.92. Channel and gate are 1 m wide (normal to the plane of the paper). Calculate q_1, q_2, and Q_3.

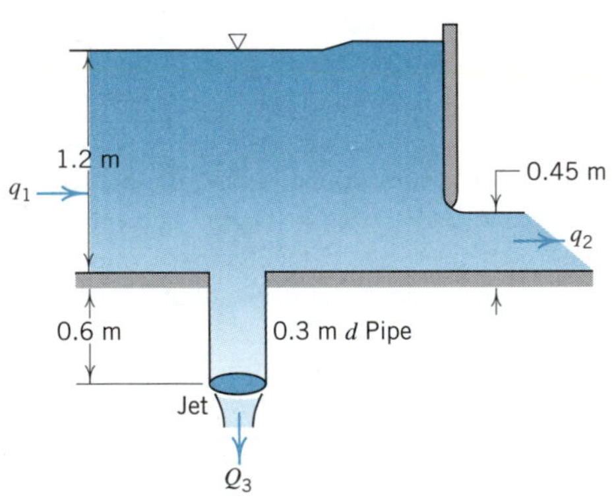

Problem 5.92

5.93. A pump having a 100 mm suction pipe and 75 mm discharge pipe pumps 32 l/s of water. At a point on the suction pipe a vacuum gage reads 150 mm of mercury; on the discharge pipe 3.6 m above this point a pressure gage reads 331 kPa. Calculate the power supplied by the pump.

5.94. A pump of what power is theoretically required to raise 900 l/min (240 gpm) of water from a reservoir of surface elevation 30 m (98 ft) to one of surface elevation 75 m (250 ft)?

5.95. If 340 l/s of water are pumped over a hill through a 450 mm pipeline, and the hilltop is 60 m above the surface of the reservoir from which the water is being taken, calculate the pump power required to maintain a pressure of 175 kPa on the hilltop.

5.96. Water is pumped from a large lake into an irrigation canal of rectangular cross section 3 m wide, producing the flow situation shown. Calculate the required pump power assuming ideal flow.

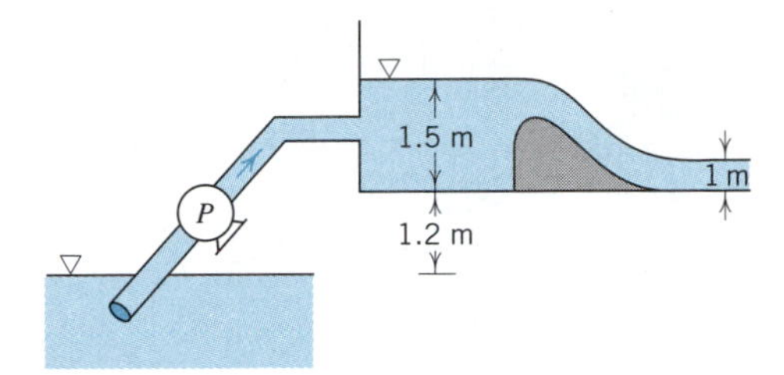

Problem 5.96

5.97. A pump has 8 in. inlet and 6 in. outlet. If the hydraulic grade line rises 60 ft across the pump when the flowrate is 5.0 cfs of sodium at 1 000°F, how much horsepower is the pump delivering to the fluid? For the same rise in hydraulic grade line, what flowrate will be maintained by an expenditure of 60 hp?

5.98. Water is flowing. Calculate the pump power for a flowrate of 28 l/s. Draw the EL and HGL.

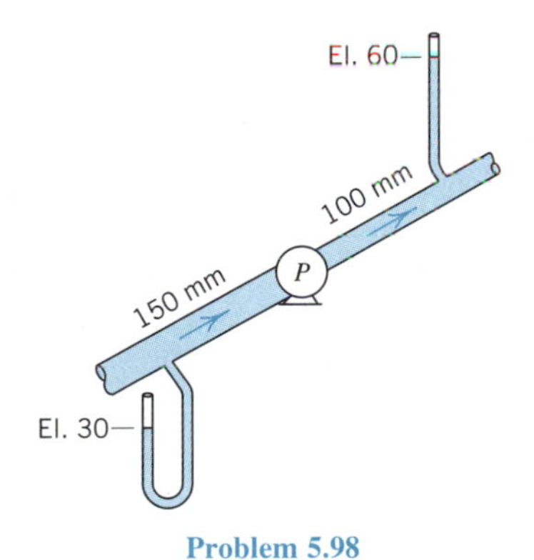

Problem 5.98

5.99. One hundred and twenty litres per second of jet fuel (JP-4) are flowing. Calculate the pump power.

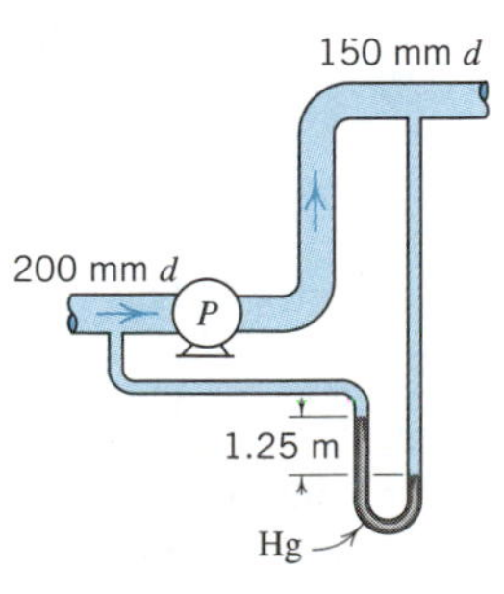

Problem 5.99

5.100. If pitot tubes replace the static pressure connections of the preceding problem and give the same manometer reading, what is the flowrate if the pump is supplying 7.5 kW to the fluid?

5.101. A pump takes water from a tank and discharges it into the atmosphere through a horizontal 50 mm nozzle. The nozzle is 4.5 m above the water surface and is connected to the pump's 100 mm discharge pipe. What power must the pump have to maintain a pressure of 276 kPa just upstream from the nozzle?

5.102. Calculate the pump power, assuming that the diverging tube flows full.

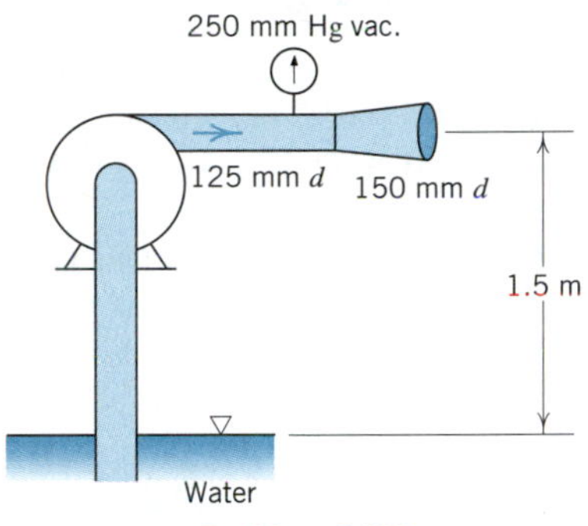

Problem 5.102

5.103. Calculate the minimum pump horsepower that will send the jet over the wall.

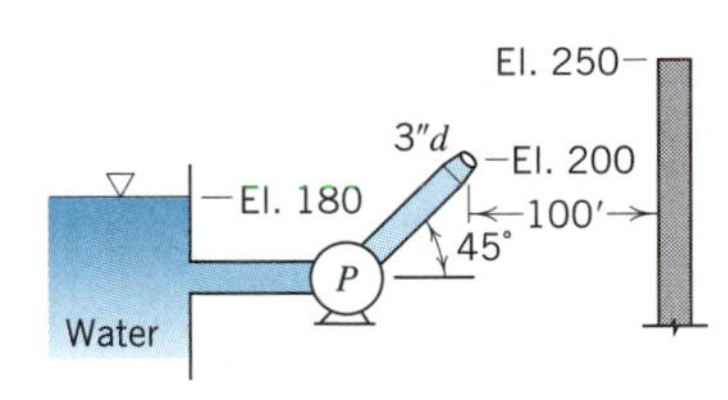

Problem 5.103

5.104. Calculate the pump power.

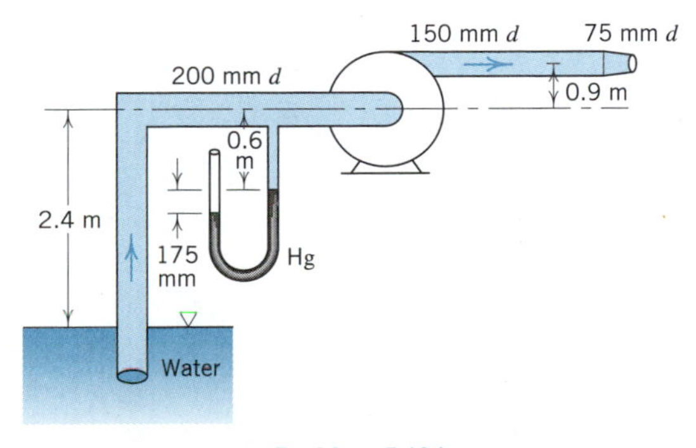

Problem 5.104

5.105. Compute the pump horsepower required to maintain a flowrate of 4.0 cfs and draw the EL and HGL if the barometric pressure is 14.3 psia and the vapor pressure is 1.0 psia. Calculate the maximum possible x for reliable operation.

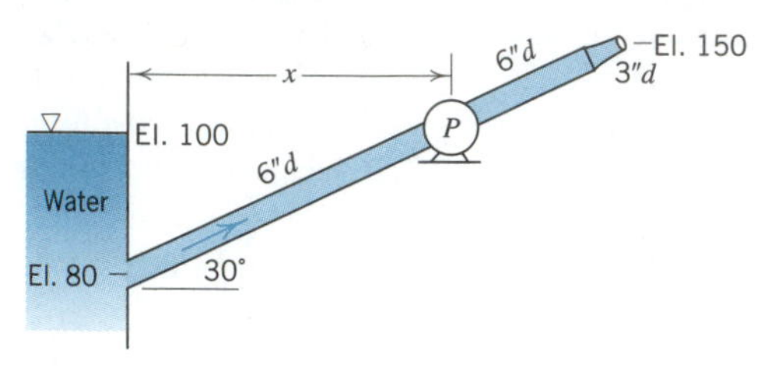

Problem 5.105

5.106. Referring to the sketch of problem 5.105 above, what flowrate will be expected when 50 horsepower are supplied to the pump?

5.107. Two cubic feet per second of ethyl alcohol are flowing in a moon station supply system. Calculate the pump horsepower.

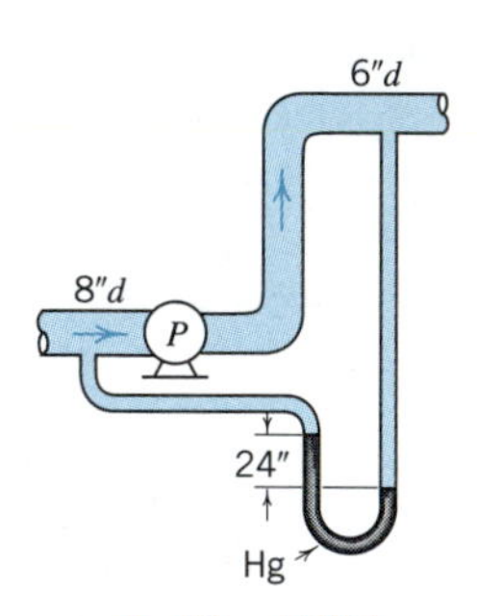

Problem 5.107

5.108. Water is flowing in this moon station supply pipeline. Calculate the pump power for a flowrate of 3 l/s.

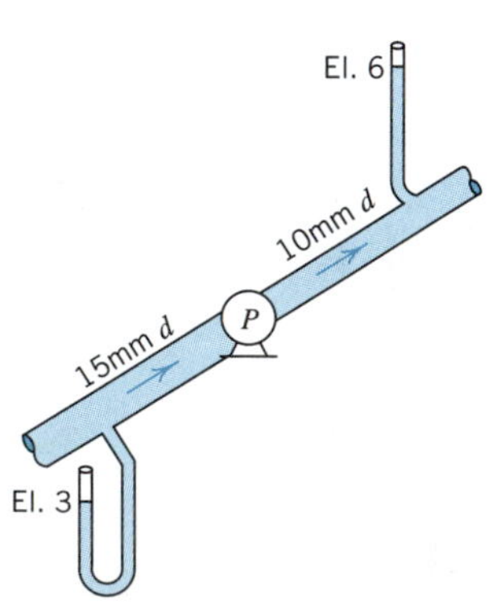

Problem 5.108

5.109. A pump is to be installed to increase the flowrate through this pipe and nozzle by 20%. Calculate the required power.

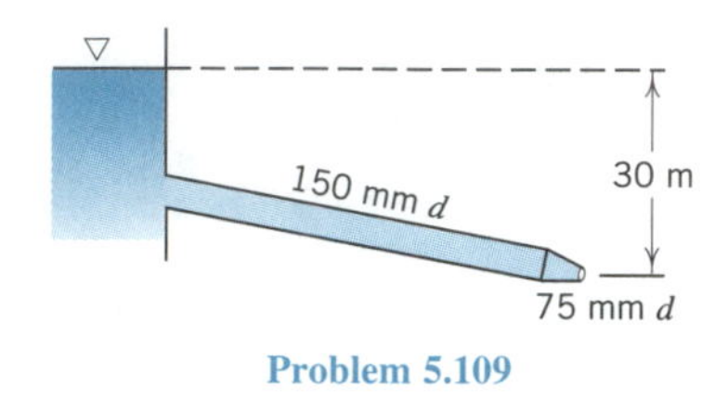

Problem 5.109

5.110. Through a 4 in. pipe, 1.0 cfs of water enters a small hydraulic motor and discharges through a 6 in. pipe. The inlet pipe is lower than the discharge pipe, and at a point on the inlet pipe a pressure gage reads 70 psi; 14 ft above this on the discharge pipe a pressure gage reads 40 psi. What horsepower is developed by the motor? For the same gage readings, what flowrate would be required to develop 10 hp?

5.111. A 7.5 kW pump is used to draw air from the atmosphere (pressure = 101.3 kPa) through a smooth bell-mouth nozzle with a 250 mm diameter throat and into a very large tank where the absolute pressure is 1.5 bar. If the air density is essentially constant at 1.25 kg/m^3, what is the velocity in the nozzle throat?

5.112. Find the flowrate in the system below and draw the energy line and the hydraulic grade line. If a pump is installed at A to increase the flowrate 20%, find the horsepower the pump must add to the water.

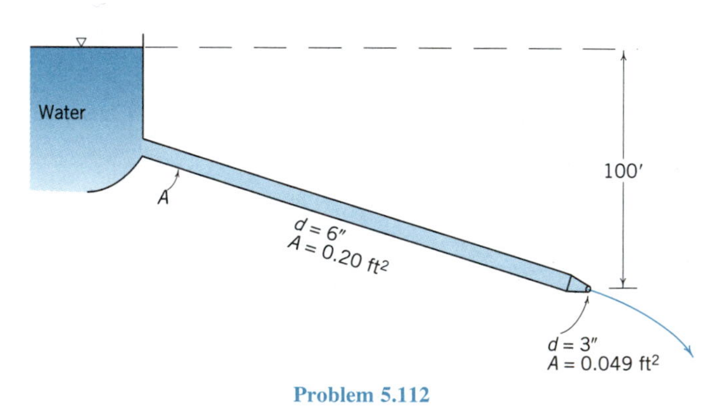

Problem 5.112

5.113. For the fire pump installation shown, compute the horsepower added to the water necessary to give a nozzle ve-

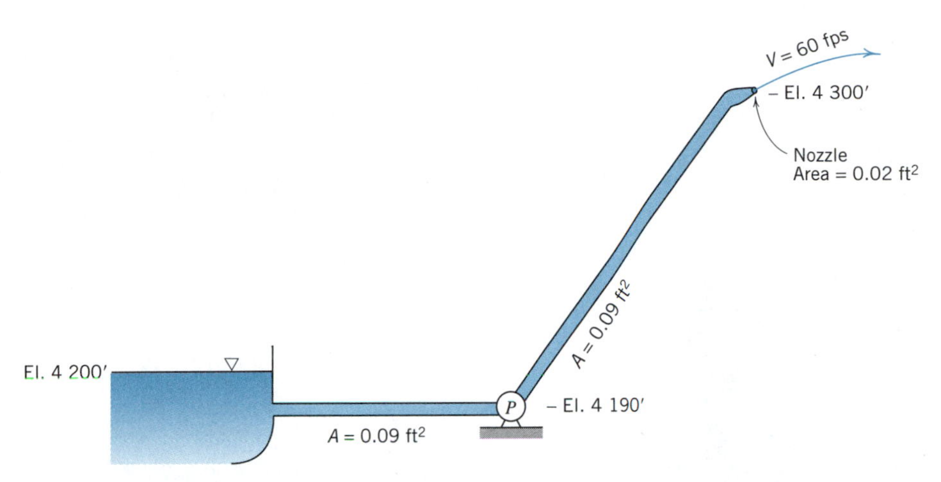

Problem 5.113

locity of 60 ft/s to the exiting stream. Also draw the energy line and the hydraulic grade line.

5.114. Water is being pumped from the lower reservoir through a nozzle into the upper reservoir. If the vacuum gage at A reads 2.4 psi vacuum,

a) find the flow velocity through the nozzle,

b) find the horsepower the pump must add to the water,

c) draw the energy line and the hydraulic grade line.

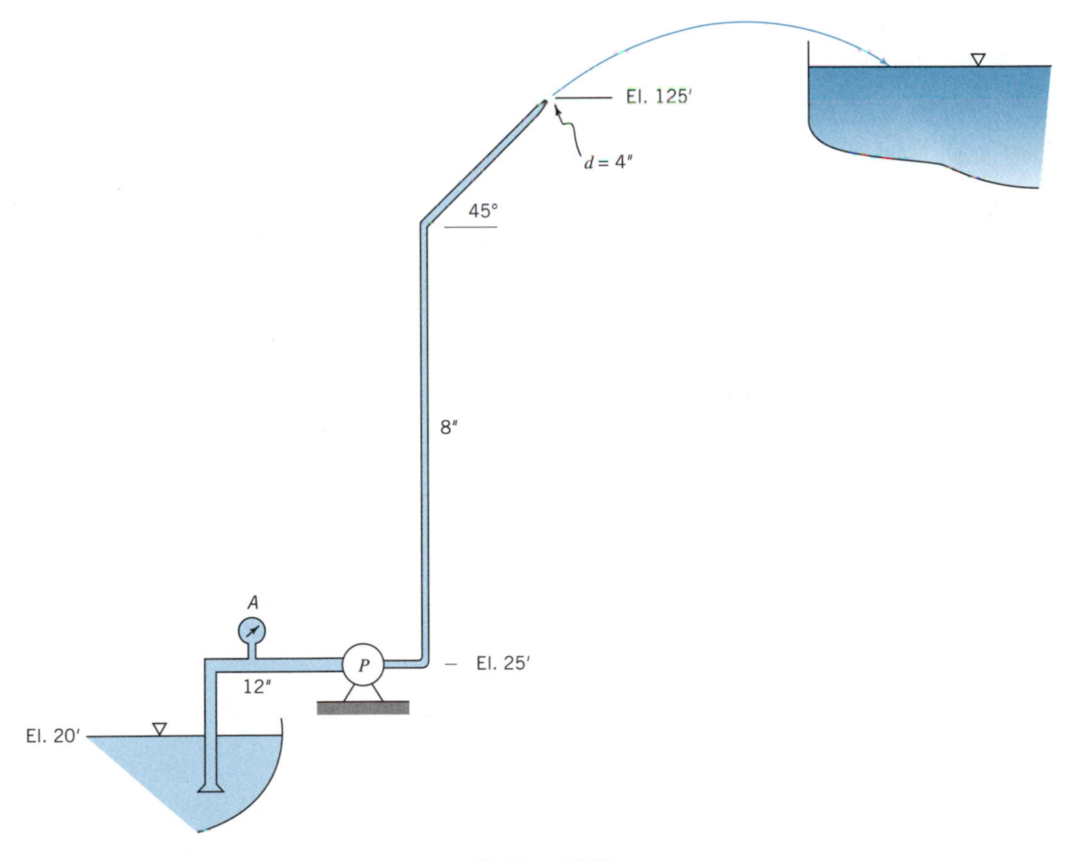

Problem 5.114

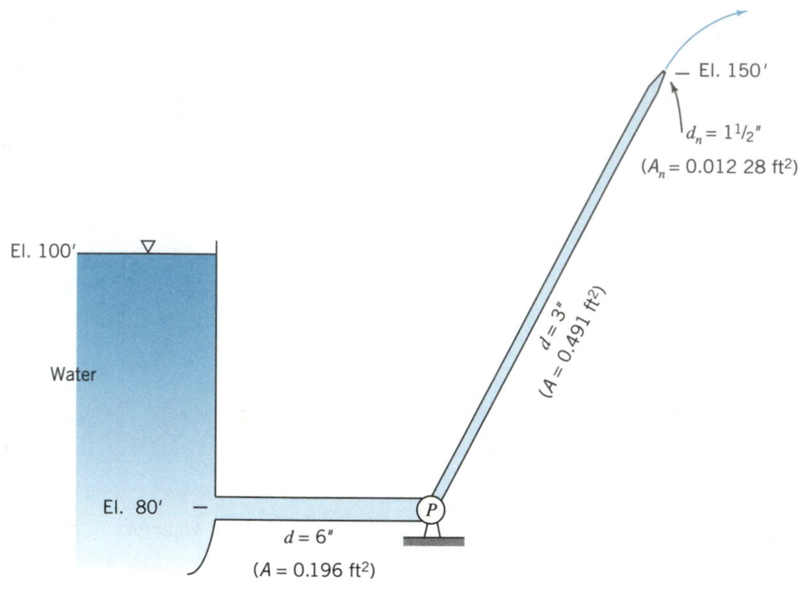

Problem 5.115

5.115. In the pumping system shown, the velocity discharging from the 1.50-inch diameter nozzle must be 50 ft/s. If water is flowing,

a) calculate Q,

b) calculate the energy added by the pump to each pound of water,

c) calculate the horsepower the pump must add to the water,

d) draw the energy line and the hydraulic grade line approximately to scale.

5.116. A hydraulic turbine in a power plant takes 3 m³/s (106 ft³/s) of water from a reservoir of surface elevation 70 m (230 ft) and discharges it into a river of surface elevation 20 m (66 ft). What theoretical power is available in this flow?

5.117. Calculate the h that will produce a flowrate of 85 l/s and a turbine output of 15 kW.

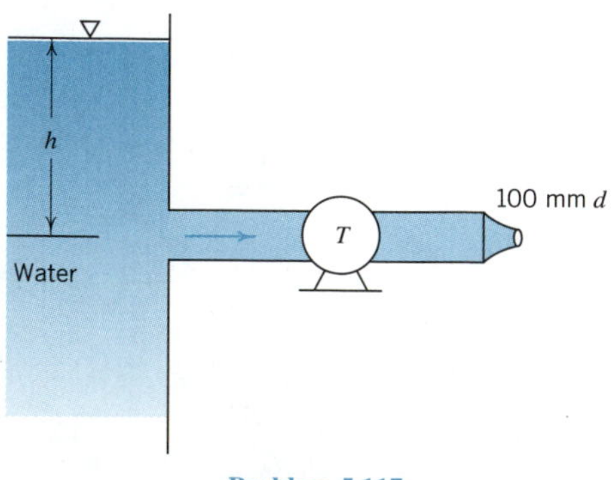

Problem 5.117

5.118. Water flows from the rectangular channel into a 2 ft diameter pipe where the turbine extracts 10 ft·lb of energy from every pound of fluid. How far above the pipe outlet will the water surface be in the column connected to the pitot tube?

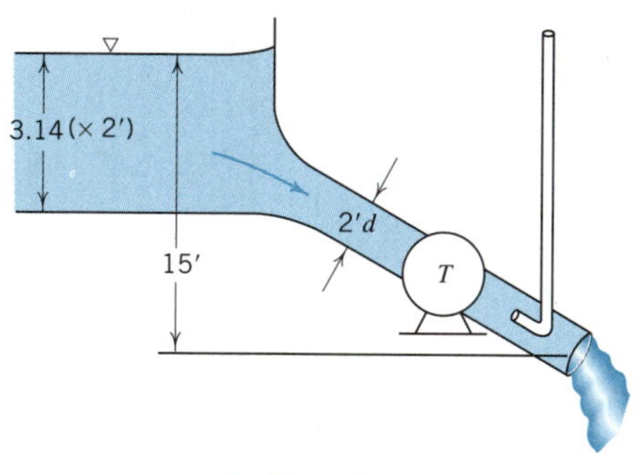

Problem 5.118

5.119. The turbine extracts from the flowing water half as much energy as remains in the jet at the nozzle exit. Calculate the power of the turbine.

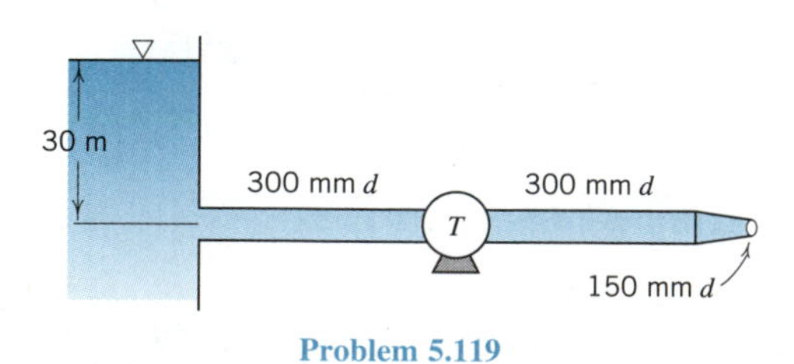

Problem 5.119

5.120. This turbine develops 100 horsepower when the flowrate is 20 cfs. What flowrate may be expected if the turbine is removed?

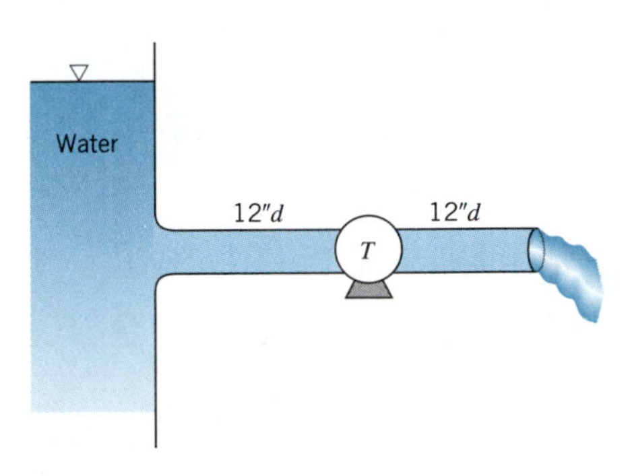

Problem 5.120

5.121. The turbine extracts power from the water flowing from the reservoir. Find the horsepower extracted if the flow through the system is 1 000 cfs. Draw the energy line and the hydraulic grade line.

5.122. Calculate the power output of this turbine.

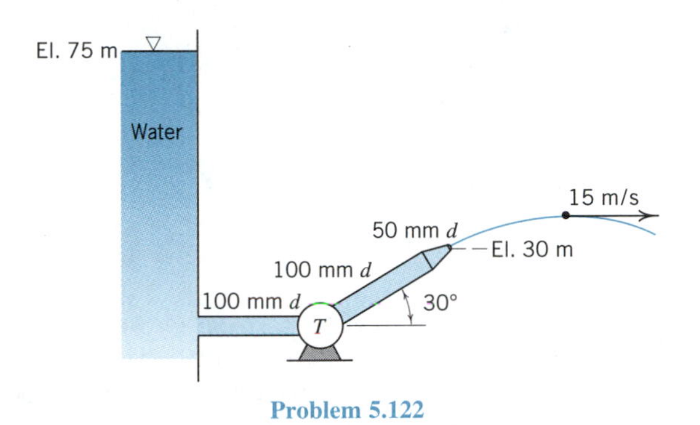

Problem 5.122

5.123. What is the maximum power the turbine can extract from the flow before cavitation will occur at some point in the system? Barometric pressure is 102 kPa, and vapor pressure of the water is 3.5 kPa.

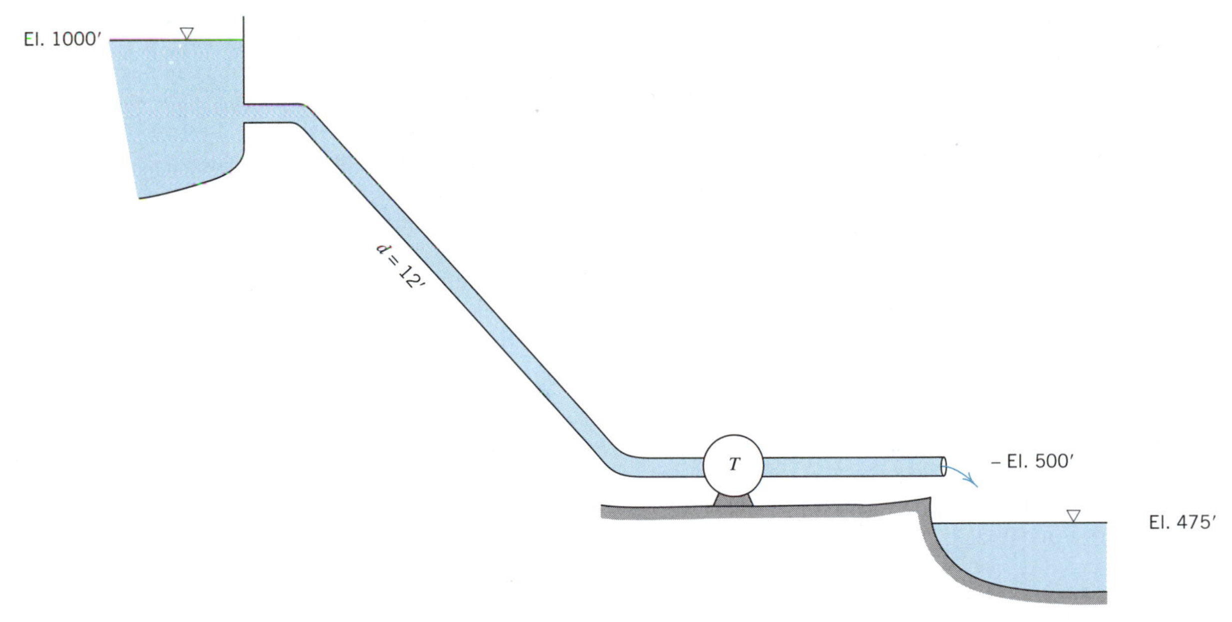

Problem 5.121

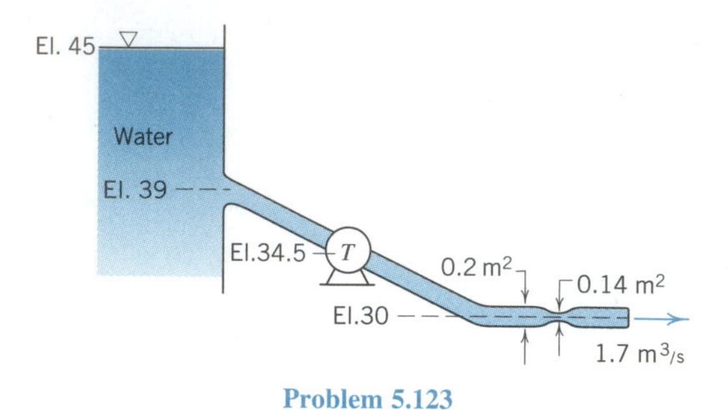

Problem 5.123

5.124. A flowrate of 7 m^3/s of water passes through this hydraulic turbine. The static pressure at the *top* of the inlet pipe is 345 kPa and across a 1.5 m diameter of the outlet pipe (''draft tube'') the stagnation pressure is 250 mm of mercury vacuum. How much power may be expected from the machine?

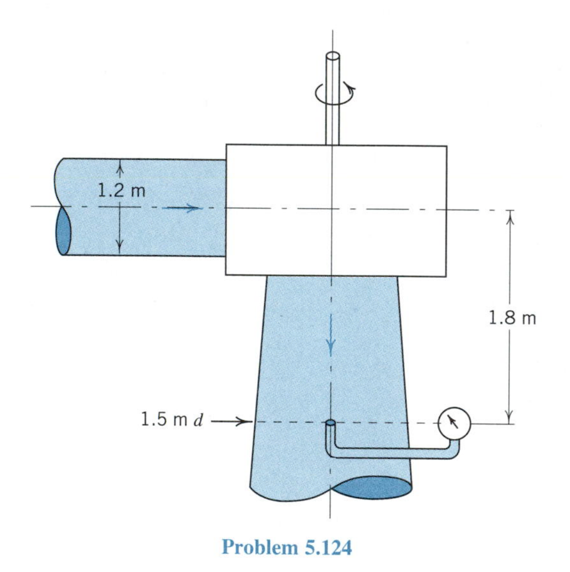

Problem 5.124

5.125. Assuming a very large tank and density of air constant at 0.002 5 slug/ft^3, calculate (*a*) pressure just upstream from the turbine, (*b*) the turbine horsepower, and (*c*) the jet velocity when the turbine is removed.

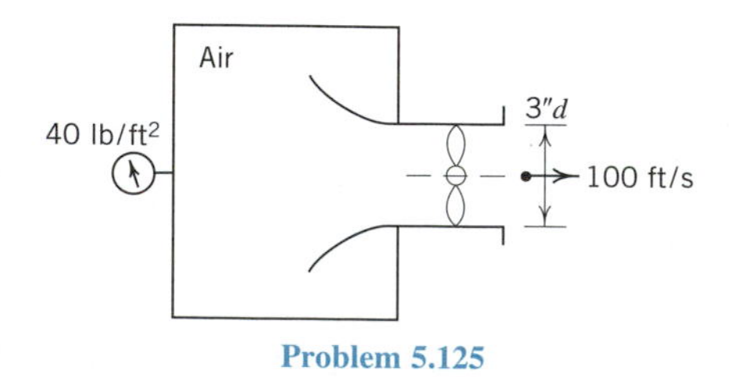

Problem 5.125

5.126. Calculate the power available in the jet of water issuing from this nozzle.

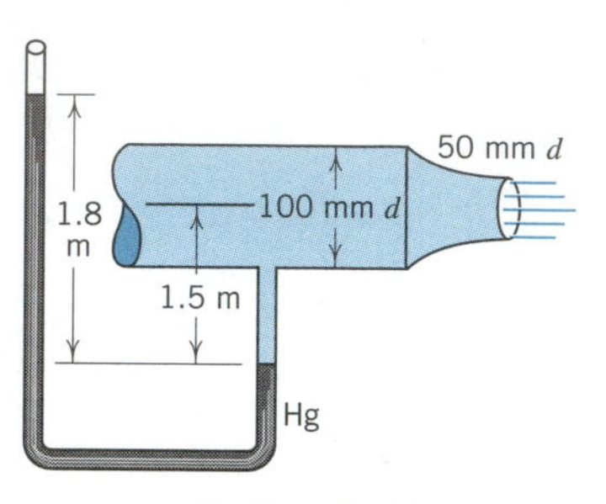

Problem 5.126

5.127. Water flows in a 0.1 m (0.33 ft) wide gap between two very large flat plates; assume the flow is two-dimensional. The flow is laminar and so the velocity profile is parabolic with a centerline velocity of 0.2 m/s (0.67 ft/s). Under the assumption that the streamlines of the flow are straight and parallel, find the change in the sum of the Bernoulli terms in Eq. 5.10 with distance from the flow centerline. Use two different methods, one of which uses the vorticity. Show that the methods are equivalent.

5.128. If the irrotational flow of problem 3.6*f* is in a horizontal plane and the pressure head at the origin of coordinates is 3 m (10 ft), what is the pressure head at point (2,2)? What is the pressure head when the flowfield is in a vertical plane?

5.129. If the irrotational flowfields of problems 3.6*l* and 3.6*m* are in a horizontal plane and the velocity and pressure head at a radius of 0.6 m are 4.5 m/s and 3 m, respectively, what pressure head exists at a radius of 0.9 m? When these flowfields are in vertical planes and the pressure head at ($\theta = 90°$, $r = 0.6$) is 3 m, what is the pressure head at ($\theta = 0°$, $r = 0.9$)?

5.130. If the irrotational flowfield of a free vortex is described by the equations $v_t = c/r$ and $v_r = 0$, derive an equation for the profile of the liquid surface.

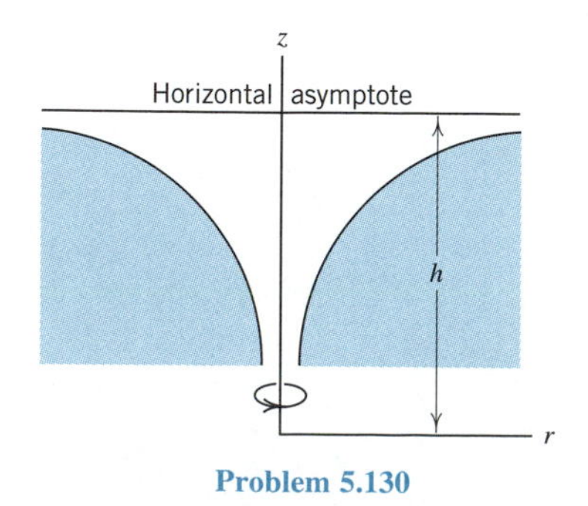

Problem 5.130

5.131. A 180° circular bend occurs in a horizontal plane in a passage 1 m (3 ft) wide. The radius of the inner wall is 0.3 m (1 ft) and that of the outer wall 1.3 m (4 ft). If the velocity (of the water) halfway around the bend on the outer wall is 3 m/s (10 ft/s) and the pressure there is 35 kPa (5 psi), what will be the velocity and pressure at the corresponding point on the inner wall (using the relation of Eq. 5.12)? Note that this equation is an approximation because the centers of curvature of all streamlines are not coincident with the center of curvature of the bend. Also calculate the approximate flowrate in the passage.

5.132. If the bend of the preceding problem is the top of a siphon, and thus in a vertical plane, what are the velocity and pressure? What is the approximate flowrate?

5.133. Estimate the peripheral velocity 30 m from the center of a tornado if the peripheral velocity 120 m from the center is 80 km/h. If the barometer reads 710 mm of mercury at the latter point, what reading can be expected at the former? Assume air density 1.23 kg/m^3.

5.134. Predict the flowrate through this two-dimensional outflow structure for a water depth of 1.5 m in the canal. Also calculate the water depth on the upstream face of the structure and the pressure at A.

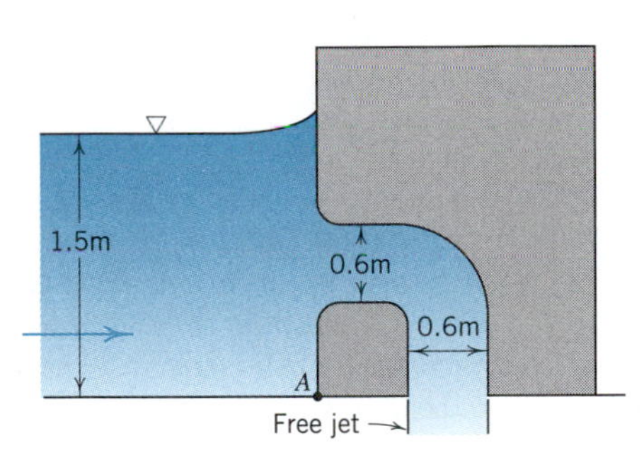

Problem 5.134

5.135. Complete the solution of Illustrative Problem 5.8.

5.136. Assuming an ideal fluid, this flowfield would occur when deep open flow passes over a submerged semicylindrical weir. If the local velocities through section 2 may be expressed by

$$v = V_1\left[1 + \left(\frac{R}{y}\right)^n\right]$$

determine the two-dimensional flowrate and the pressure at the top of the weir.

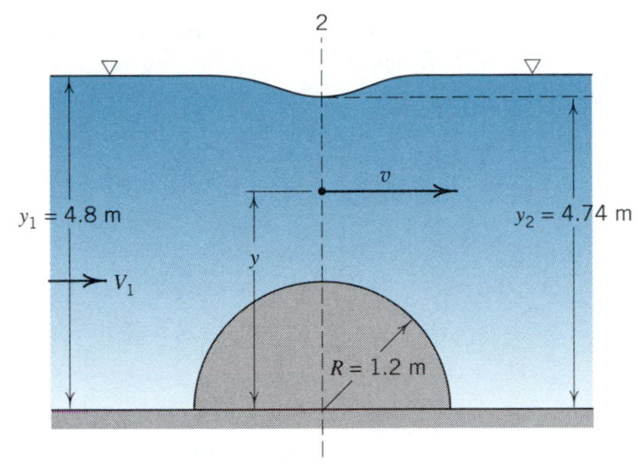

Problem 5.136

5.137. The two-dimensional flowrate over this submerged semicylinder is 11.3 m^3/s·m. The ideal fluid flowing has the density of water. The velocity over the surface of the semicylinder may be closely approximated by $v = 6 \sin \theta$. Derive an expression for the pressure on the semicylinder. Integrate this pressure distribution in such ways that it will yield the vertical and horizontal components of force exerted by the water on the cylinder. Assume the flowfield perfectly symmetrical about the line AA.

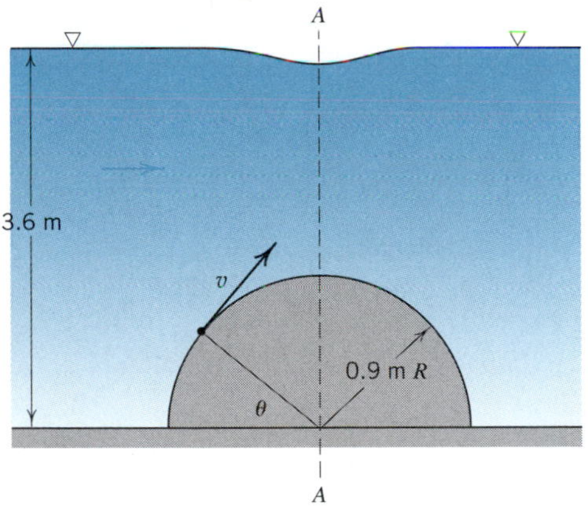

Problem 5.137

5.138. If a gently curved overflow structure placed in a high velocity flow produces the flowfield shown, and if the indicated velocity distribution is assumed at section 2, what flowrate is indicated?

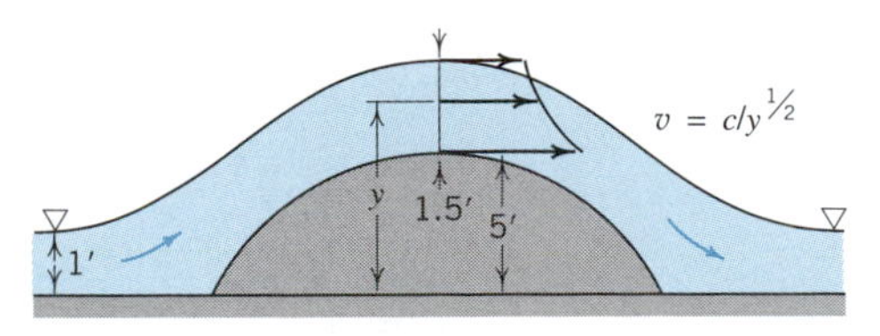

Problem 5.138

5.139. The flowrate is 0.75 m^3/s. The gage pressure in the corner at A is 83 kPa. Calculate the gage pressure at B and the power of the turbine.

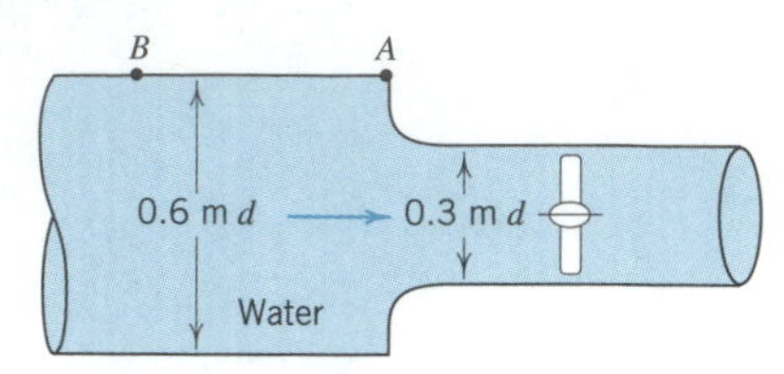

Problem 5.139

5.140. A small circular cylinder spans the test section of a large water tunnel. A water and mercury differential manometer is connected to two small openings in the cylinder, one facing directly into the approaching flow the other at 90° to this point. For a manometer reading of 10 in. (250 mm) calculate the velocity approaching the cylinder.

5.141. For the flowfield described by the velocities given in Illustrative Problem 5.15, derive the equation of a line at all points on which the pressure is p_o. See Fig. 5.16.

5.142. If the depth at the vena contracta caused by this sluice gate is 0.6 m, calculate the two-dimensional flowrate and the water depth on the upstream face of the gate.

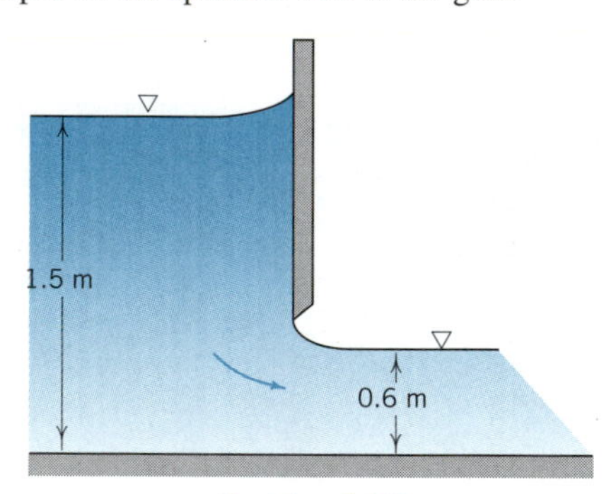

Problem 5.142

5.143. Water is flowing. The pressure at A is 1.75 psi. Calculate the flowrate, diameter of vena contracta, and pressures at points B, C, and D.

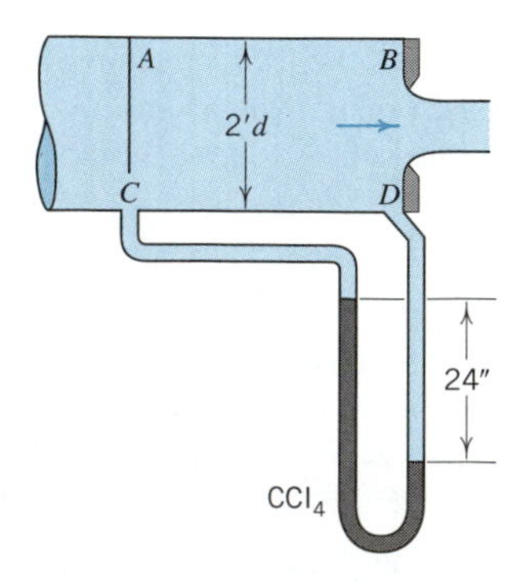

Problem 5.143

5.144. Liquid of specific weight γ flows through a *two-dimensional* ''elbow-meter'' similar to that of Fig. 14.34. A differential manometer is connected between the two pressure taps and contains liquid of specific weight γ_1. Assuming the velocity distribution at the section containing the pressure taps is given by Eq. 5.12, derive an expression for flowrate in terms of specific weights, R_i, R_o, and manometer reading, h.

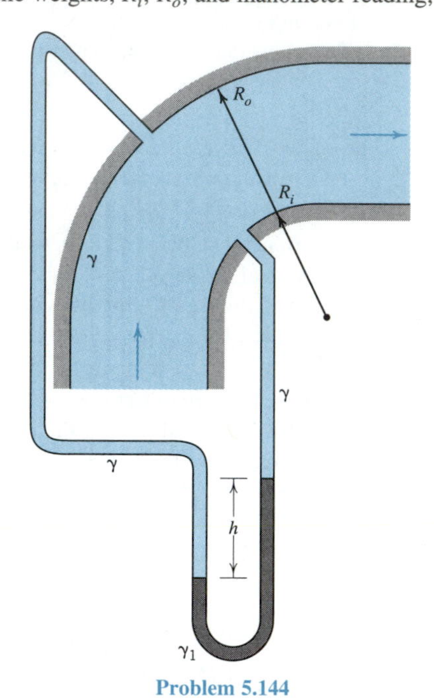

Problem 5.144

5.145. In the throat of a passage with curved walls the distribution of velocity will be axisymmetric and somewhat as shown. If this velocity distribution is characterized mathematically by $v = 3 + 13.3r^2$, what is the flowrate? If the flow is irrotational, what pressure difference is to be expected between the top and the center of the passage?

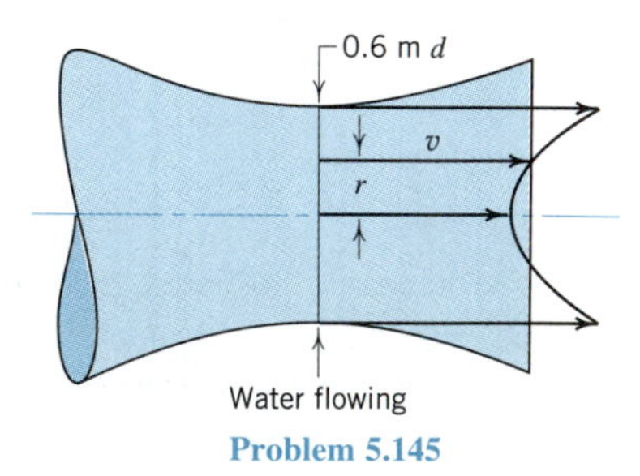

Problem 5.145

5.146. Referring to the sketch of Fig. 5.15, show that the force exerted by the water on the weir can be calculated (approximately) by

$$\gamma P \left[\frac{P}{2} + H + \frac{q^2}{2g_n(P + H)^2} \right]$$

in which P is the weir height and h the head on the weir. Will this approximate force be expected to be more or less than the exact force on the weir?

5.147. A circular disk of diameter d is placed (normal to the flow) in a free stream of pressure p_o, velocity V_o, and density ρ. If the pressure is assumed to fall parabolically from stagnation pressure at the center of the disk to p_2 at the edges and the pressure p_2 is assumed to be distributed uniformly over the downstream side of the disk, derive an expression for the drag force on the disk including only p_o, p_2, and d. Note that the drag force on such a disk may be predicted more reliably using Eq. 11.1 and $C_D = 1.12$ from Fig. 11.9.

5.148. The study of the flowfield around the elliptic cylinder shows that the local velocity at the midsection is $3V_o/2$ and that this is the maximum velocity in the whole flowfield. This cylinder is placed (with its axis vertical) in a large water tunnel as shown on the sketch. For a gage reading of 70 kPa (10.0 psi), calculate the V_o at which cavitation will be incipient. Atmospheric pressure = 96 kPa (14.0 psi). Vapor pressure = 6.9 kPa (1.0 psi).

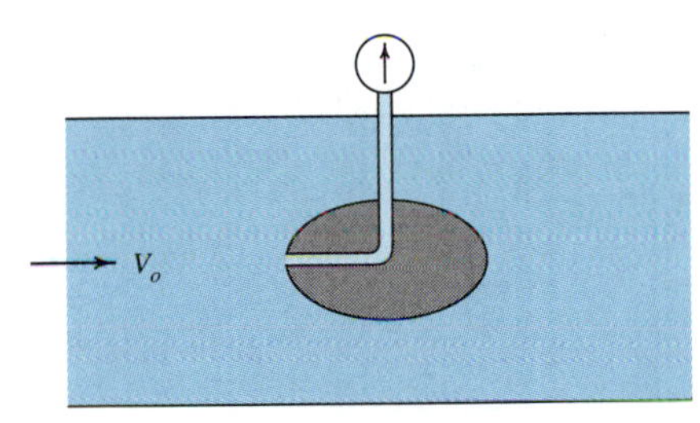

Problem 5.148

5.149. Determine the stream functions for the flowfields of problem 3.6 and plot the streamline $\psi = 2$.

5.150. Determine the stream function for a parabolic velocity profile (as in laminar flow) between parallel plates separated by a distance b. Take the origin of coordinates midway between the plates.

5.151. A flowfield is characterized by the stream function $\psi = 3x^2y - y^3$. Is this flow irrotational? If not, calculate the vorticity. Show that the magnitude of velocity at any point in the flowfield depends only on its distance from the origin of coordinates. Plot the streamline $\psi = 2$.

5.152. A flowfield is characterized by the stream function $\psi = xy$. Is this flow irrotational? Plot sufficient streamlines to determine the flow pattern. Identify at least two possible physical interpretations of this flow.

5.153. Determine the stream function for a uniform rectilinear flow at an arbitrary angle α to the x-axis and with velocity magnitude U.

5.154. A flowfield is defined by the stream function

$$\psi = Uy + \frac{q}{2\pi}\left(\tan^{-1}\frac{y-a}{x} - \tan^{-1}\frac{y+a}{x}\right)$$

Derive a general expression for the locations of all stagnation points. Sketch the flowfields implied by this stream function.

5.155. For the stream function

$$\psi = \frac{q}{2\pi}\left(\tan^{-1}\frac{y-a}{x} - \tan^{-1}\frac{y+a}{x}\right) - \frac{\Gamma}{2\pi}\ln\sqrt{x^2+y^2}$$

locate any stagnation points and sketch the flowfield. Derive an expression for the velocity at $(a, 0)$.

5.156. A flowfield is defined by the stream function

$$\psi = 12\left(\tan^{-1}\frac{y-2}{x-2} + \tan^{-1}\frac{y+2}{x+2}\right) + \frac{3}{4\sqrt{2}}(x^2+y^2)$$

Identify the components of this flowfield. Is the flowfield rotational or irrotational? Why? Are there any stagnation points? How many? Where are they located? What is the velocity at (4, 4)? Sketch the flowfield.

5.157. Determine the velocity potential ϕ for (a) the flow in problem 5.151 and (b) the flow in problem 5.152.

5.158. Plot representative ϕ and ψ lines for the flow of problem 5.152.

CHAPTER 6

THE IMPULSE-MOMENTUM PRINCIPLE

This chapter develops the third of the three basic equations of fluid mechanics—conservation of mass (continuity), work-energy, and, now, impulse-momentum. These three equations form the basis for the solution of a wide variety of fluids problems, generally in conjunction with other empirical relationships to account for frictional effects. The concept of impulse-momentum derives from Newton's second law, a vector relationship, which can be applied to a fluid particle or system and written as

$$\left(\sum \mathbf{F}\right) dt = d(m\mathbf{v}_c) \qquad \text{or} \qquad \left(\sum \mathbf{F}\right) = \frac{d}{dt}(m\mathbf{v}_c)$$

where $\mathbf{v}_c$ is the velocity of the center of mass of the system of mass m, $m\mathbf{v}_c$ is the *linear momentum*, and $(\Sigma\mathbf{F})\, dt$ is the *impulse* in the time dt from the sum of all external forces acting on the system. Although in developing the Euler equation, we used only one small fluid system, we now define the fluid system to include all of the fluid in a specified control volume at a particular time. We further restrict the analysis to steady flow to develop the principle and demonstrate its application. Although shear stress is not explicitly included in the development, the equations we derive will apply equally well to real fluids as well as ideal fluids. However, some adjustments must be made to accommodate real fluids because of the effect of friction on the velocity profiles (see Section 7.12). While we have yet to discuss in detail the mechanics of compressible flow (that is reserved until Chapter 13), the basic equations we develop in this chapter will also apply to both incompressible and compressible flow.

The first development will focus on the *linear impulse-momentum* equation and its application to some common flow situations. The *angular impulse-momentum* equation (also known as the moment of momentum equation) will then be developed and applied to specific flow problems. Whereas the linear impulse-momentum equation provides a means for calculating directly the magnitude and direction of resultant forces exerted on the flowing fluid by structures (and vice versa), the angular impulse-momentum equation provides a means of locating the line of action of the resultant forces. The angular impulse momentum equations are also particularly useful in applications to rotating fluid machinery such as pumps and turbines (see Chapter 12).

Principles and Elementary Applications

6.1 THE LINEAR IMPULSE-MOMENTUM EQUATION

We begin our development by once again turning to the general control volume previously employed to find the basic equations for conservation of mass and work-energy. Figure 6.1 recreates this diagram and shows the small individual fluid system previously used to develop the Euler equation and the resulting Bernoulli equation. We will first look at this fluid system again.

For the individual fluid system shown in the control volume of Fig. 6.1,

$$\sum \mathbf{F} = m\mathbf{a} = \frac{d}{dt}m\mathbf{v} = \frac{d}{dt}\rho\mathbf{v}\,d\forall$$

If we write this equation for every individual fluid system in the control volume at a given time, and sum them all, we get[1]

$$\sum \mathbf{F}_{ext} = \iiint_{sys} \frac{d}{dt}(\rho\mathbf{v}\,d\forall) = \frac{d}{dt}\iiint_{sys} \rho\mathbf{v}\,d\forall$$

where all the internal forces (except gravity) cancel leaving for the left-hand side of the equation only the weight of the fluid system in the control volume and the forces applied at the control surface *on* the integrated fluid system in the control volume. The forces applied at the control surface result from pressure and, as we will see in Chapter 7, shear; thus, a force at the surface of the control volume can have both normal and tangential components.

The right-hand side of the equation can be evaluated with the Reynolds Transport Theorem. Modifying Eq. 4.15 to limit the development to steady flow results in

$$\frac{d}{dt}\iiint_{sys} \rho\mathbf{v}\,d\forall = \frac{dE}{dt} = \iint_{c.s.} i\rho\mathbf{v}\cdot d\mathbf{A} = \iint_{c.s.out} \rho\mathbf{v}(\mathbf{v}\cdot d\mathbf{A}) - \iint_{c.s.in} \rho\mathbf{v}(\mathbf{v}\cdot d\mathbf{A})$$

where E is the momentum of the fluid system in the control volume at a given time and i is $\mathbf{v}$, which is the momentum per unit mass. Because the streamlines are assumed straight

[1]The integrals over all the systems and the control volume are equivalent at a given instant; however, to exchange the integral and the time derivative requires us to formally integrate over all of the systems.

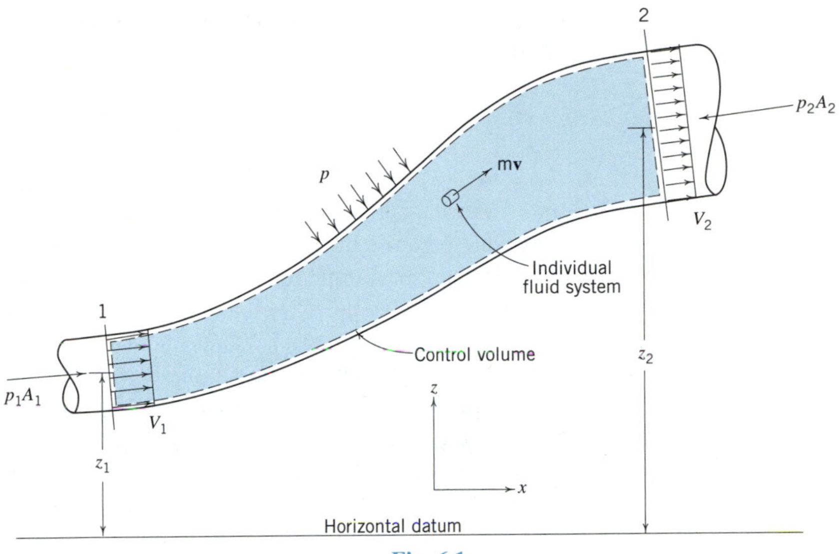

Fig. 6.1

and parallel at sections 1 and 2, velocity is constant (uniform) over the cross sections. Also, the cross-sectional area is normal to the velocity vector over the entire cross section which leads to the following result for the integration of the terms on the right:

$$\frac{d}{dt}\iiint_{sys} \rho\mathbf{v}\, d\forall = \rho_2\mathbf{V}_2Q_2 - \rho_1\mathbf{V}_1Q_1$$

But we know from conservation of mass that $Q_1\rho_1 = Q_2\rho_2 = Q\rho$ for a steady flow,[2] hence

$$\frac{d}{dt}\iiint_{sys} \rho\mathbf{v}\, d\forall = Q\rho(\mathbf{V}_2 - \mathbf{V}_1)$$

Substituting this result back into the original equation for the system yields

$$\Sigma\mathbf{F}_{ext} = Q\rho(\mathbf{V}_2 - \mathbf{V}_1) \tag{6.1}$$

In two dimensions it is common to write this result as

$$(\Sigma F_{ext})_x = Q\rho(V_{2_x} - V_{1_x}) \tag{6.2a}$$

$$(\Sigma F_{ext})_z = Q\rho(V_{2_z} - V_{1_z}) \tag{6.2b}$$

In the case where momentum enters and leaves the control volume at more than one

[2]Even if the flow is compressible, continuity requires that ρQ = constant in steady flow in a (finite) streamtube (see Section 4.2).

location (also where the streamlines are straight and parallel), a more general form of the equation is

$$\Sigma \mathbf{F}_{ext} = (\Sigma Q\rho \mathbf{V})_{out} - (\Sigma Q\rho \mathbf{V})_{in} \tag{6.3}$$

To demonstrate the application of this principle, we will address a variety of common flow problems which require the use of impulse-momentum to enable solutions. We also need to remember that the external forces may include both normal and tangential forces on the fluid in the control volume, as well as the weight of the fluid inside the control volume at a given time.

In the next sections, we discuss a number of applications of the impulse-momentum principle to situations where the flows across the control surface are one-dimensional, that is, the streamlines are essentially straight and parallel. Recall that the inherent advantage of the impulse-momentum principle is that only flow conditions at inlets and exits of the control volume are needed for successful application; detailed (and often complex) flow processes within the control volume need not be known to apply the principle. It should be emphasized that efficient application of the principle depends greatly on the judicious selection of a convenient control volume, with streamlines essentially straight and parallel at inlet and exit.

6.2 PIPE FLOW APPLICATIONS

The force exerted by a flowing fluid on a bend, enlargement, or contraction in a pipeline may be readily computed by application of the impulse-momentum principle without detailed information on shape of passage or pressure distribution therein.

The reducing pipe bend of Fig. 6.2 is typical of many of the problems encountered in fluid mechanics. The object is to calculate the force exerted by the fluid on the bend between sections 1 and 2. The control volume *ABCD* encloses the fluid within the bend and the flow at sections 1 and 2 is assumed one-dimensional. **F** is the force exerted *by the bend on the fluid* (see the free-body diagram in Fig. 6.2). The flowrate, pressures and

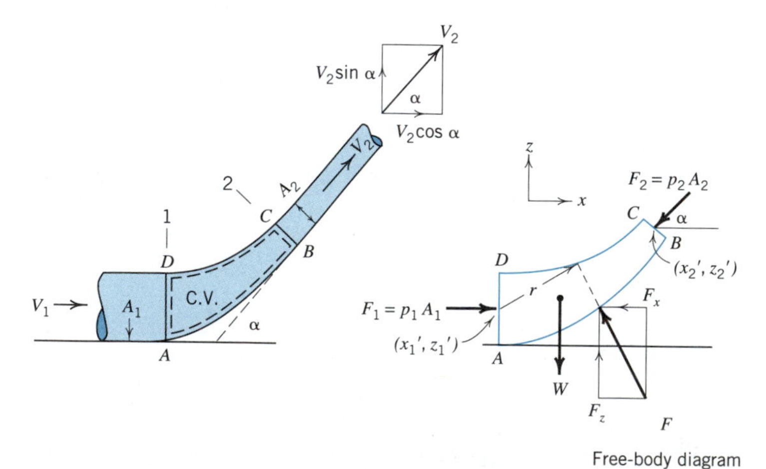

Fig. 6.2

velocities at sections 1 and 2, and the superficial flow geometry are known; **F** is to be found.

For streamlines essentially straight and parallel at sections 1 and 2, the forces F_1 and F_2 result from hydrostatic pressure distributions in the flowing fluid. If the mean pressures p_1 and p_2 are large and the pipe areas are relatively small, $F_1 = p_1A_1$ and $F_2 = p_2A_2$ can be assumed with little error to act at and along the centerline of the pipe.[3] The summation of body forces acting on the fluid in *ABCD* will, of course, yield the total weight of fluid (W). Thus, W is dependent only on the specific weight of the fluid and the control volume geometry.[4] The force **F** exerted by the bend on the fluid is the resultant of the pressure distribution over the entire interior of the bend between sections 1 and 2; although this distribution is usually unknown in detail, its resultant can be predicted by use of Eqs. 6.2a and 6.2b. The force exerted *by the fluid on the bend* is then the equal and opposite of this resultant.

When Eqs. 6.2a and 6.2b are applied to control volume *ABCD*, the results are

$$(\Sigma F_{ext})_x = p_1A_1 - p_2A_2 \cos\alpha - F_x$$

and

$$Q\rho(V_{2_x} - V_{1_x}) = Q\rho(V_2 \cos\alpha - V_1)$$

Combining the two equations to develop an expression for F_x,

$$F_x = p_1A_1 - p_2A_2 \cos\alpha + Q\rho(V_1 - V_2 \cos\alpha) \tag{6.4a}$$

Applying the same procedure in the z-direction,

$$(\Sigma F_{ext})_z = -W - p_2A_2 \sin\alpha + F_z$$

and

$$Q\rho(V_{2_z} - V_{1_z}) = Q\rho(V_2 \sin\alpha - 0)$$

In the z-direction,

$$F_z = W + p_2A_2 \sin\alpha + Q\rho V_2 \sin\alpha \tag{6.4b}$$

Of course, the resultant force F is the vector combination of F_x and F_z. Note that if the bend is relatively sharp, the weight may be negligible, depending on the magnitudes of the pressure and velocities. Of course, if the bend lies in a horizontal plane, the weight has no influence on the forces in that plane.

ILLUSTRATIVE PROBLEM 6.1

When 300 l/s of water flow through this vertical 300 mm by 200 mm reducing pipe bend, the pressure at the entrance to the bend is 70 kPa. Calculate the force exerted by the fluid on the bend if the volume of the bend is 0.085 m^3.

[3]In Fig 6.2, the points (x_1', z_1') and (x_2', z_2') identify the *centers of pressure* (not the centroids) of sections 1 and 2. However, these will be coincident for uniform pressure distributions. See Section 2.4.

[4]Frequently in engineering problems this force is small compared to the forces on the surfaces of the control volume, but the decision to neglect it requires either preliminary calculations or considerable experience. It obviously plays no part if flow is in a horizontal plane, since its direction is perpendicular to this plane.

SOLUTION

The sketch below shows the control volume selected and the freebody diagram of the fluid system with which we will work.

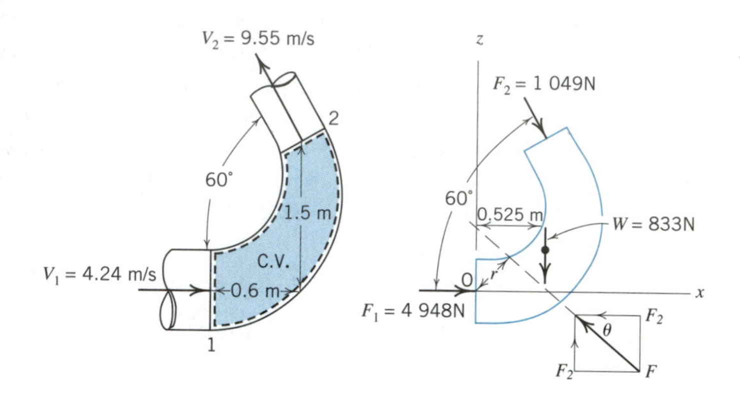

Before applying Eqs. 6.4a and 6.4b, we need to calculate some preliminary values. The flowrate of 300 l/s is equal to 0.300 m^3/s (see Appendix 1). The velocities at the entrance (section 1) and the exit (section 2) of the bend are

$$V_1 = \frac{Q}{A_1} = \frac{0.300 \text{ m}^3/\text{s}}{(\pi/4)(0.300 \text{ m})^2} = 4.24 \text{ m/s}$$

$$V_2 = V_1 \left(\frac{d_1}{d_2}\right)^2 = 4.24 \text{ m/s} \times \left(\frac{300 \text{ mm}}{200 \text{ mm}}\right)^2 = 9.55 \text{ m/s}$$

Knowing that the pressure at section 1 is 70 kPa, we need to use the work-energy equation to find the pressure at section 2.

$$z_1 + \frac{p_1}{\gamma} + \frac{V_1^2}{2g_n} = z_2 + \frac{p_2}{\gamma} + \frac{V_2^2}{2g_n} \tag{5.7}$$

Substituting the known values into the above equation yields

$$0 + \frac{70 \text{ kPa}}{9.79 \text{ kN/m}^3} + \frac{(4.24 \text{ m/s})^2}{2g_n} = 1.5 + \frac{p_2}{9.79 \text{ kN/m}^3} + \frac{(9.55 \text{ m/s})^2}{2g_n}$$

$$7.15 + 0.92 = 1.5 + \frac{p_2}{9.79} + 4.65$$

$$p_2 = 18.8 \text{ kPa}$$

The weight of the water in the control volume is

$$W = \gamma \forall = 9.79 \text{ kN/m}^3 \times 0.085 \text{ m}^3 = 0.833 \text{ kN}$$

Now we can apply Eqs. 6.4a and 6.4b.

$$F_x = p_1 A_1 - p_2 A_2 \cos \alpha + Q\rho(V_1 - V_2 \cos \alpha) \tag{6.4a}$$

$$F_z = W + p_2 A_2 \sin \alpha + Q\rho V_2 \sin \alpha \tag{6.4b}$$

Substituting our computed values into Eq. 6.4a,

$$F_x = 70\,000 \text{ Pa} \times (\pi/4)(0.300 \text{ m})^2 - 18\,800 \text{ Pa} \times (\pi/4)(0.200 \text{ m})^2 \cos 120^\circ$$
$$+ 0.300 \text{ m}^3/\text{s} \times 998 \text{ kg/m}^3 (4.24 \text{ m/s} - 9.55 \text{ m/s} \times \cos 120^\circ)$$

$$F_x = 4\,948 - (-295) + 2\,699 = 7\,942 \text{ N}$$

Now, following the same procedure for Eq. 6.4b,

$$F_z = 833 \text{ N} + 18\,800 \text{ Pa} \times (\pi/4)(0.200 \text{ m})^2 \sin 120^\circ$$
$$+ 0.300 \text{ m}^3/\text{s} \times 998 \text{ kg/m}^3 \times 9.55 \text{ m/s} \times \sin 120^\circ$$

$$F_z = 833 + 511 + 2\,476 = 3\,820 \text{ N}$$

$$F = \sqrt{F_x^2 + F_z^2} = \sqrt{7\,942^2 + 3\,820^2} = 8\,813 \text{ N} \bullet$$

The direction of the force is given by the angle θ,

$$\theta = \tan^{-1}\frac{F_z}{F_x} = \tan^{-1}\frac{3\,820}{7\,942} = 25.7^\circ \bullet$$

The resultant *force exerted on the fluid by the bend* is 8 813 N at an angle θ of 25.7°.

The plus sign on the two force components confirms the correctness of the assumed direction of the forces. Therefore, the *force exerted by the fluid on the bend* is 8.8 kN downward to the right at an angle of 25.7° with the horizontal. •

The impulse-momentum principle can be employed to predict the fall of the energy line (that is, the energy loss due to a rise in the internal energy of the fluid caused by viscous dissipation) at an abrupt axisymmetric enlargement in a passage (Fig. 6.3). Consider the control surface *ABCD*, drawn to enclose the zone of momentum change. Assume a one-dimensional flow. The flowrate Q through the control volume is

$$Q = A_1V_1 = A_2V_2 \tag{4.5}$$

The rate of change of momentum is

$$Q\rho(V_{2_x} - V_{1_x}) = (V_2 - V_1)\frac{Q\gamma}{g_n} = (V_2 - V_1)\frac{A_2\gamma V_2}{g_n}$$

The forces producing this change of momentum act on the surfaces *AB* and *CD*; if these are assumed to result from hydrostatic[5] pressure distributions over the areas, they may be calculated as

$$\Sigma F_x = p_1A_2 - p_2A_2 = (p_1 - p_2)A_2$$

By applying Eq. 6.2a, we obtain

$$\frac{p_1 - p_2}{\gamma} = \frac{V_2(V_2 - V_1)}{g_n}$$

[5]This is a good assumption for area *CD* because the conditions of one-dimensional flow are satisfied. For area *AB* it is an approximation because of the dynamics of the eddies in the "dead water" zone. Accordingly the result of the analysis will be approximate and will require experimental verification. See Section 9.9.

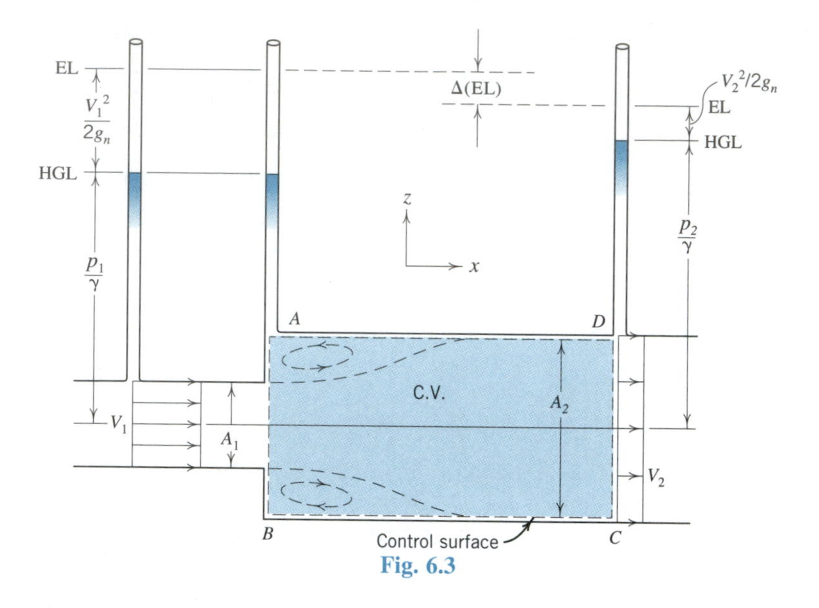

Fig. 6.3

However, from the energy line,

$$\frac{p_1 - p_2}{\gamma} = \frac{V_2^2}{2g_n} - \frac{V_1^2}{2g_n} + \Delta \text{ (EL)}$$

Equating these expressions for $(p_1 - p_2)/\gamma$ and solving for Δ (EL),

$$\Delta \text{ (EL)} = \frac{(V_1 - V_2)^2}{2g_n} \tag{6.5}$$

This analysis is a classic one of early analytic hydraulics; Δ (EL) is frequently termed *Borda-Carnot head loss*, after those who contributed to its original development.

6.3 OPEN CHANNEL FLOW APPLICATIONS

There are many applications of the impulse-momentum equations to open channel flow situations. We will look at the computation of forces on open channel structures such as sluice gates and weirs. We will analyze the hydraulic jump. And we will look at wave propagation. In some cases, the analysis yields good results; in others, only rough approximations.

The first problem we will address is the flow under a sluice gate as shown in Fig. 6.4. A control volume is selected which has uniform flow and straight and parallel streamlines at the entrance and exit. The work-energy equation and the continuity equation are applied to the flow resulting in known values of depths y_1 and y_2 and the flowrate per unit width q.[6] We can now productively apply the impulse-momentum equation to find the force the water exerts on the sluice gate. From Eq. 6.2a,

$$(\Sigma F_{ext})_x = F_1 - F_2 - F_x = Q\rho(V_{2_x} - V_{1_x}) = q\rho(V_2 - V_1)$$

[6]Note that in open channel flows, the water depth is denoted by y.

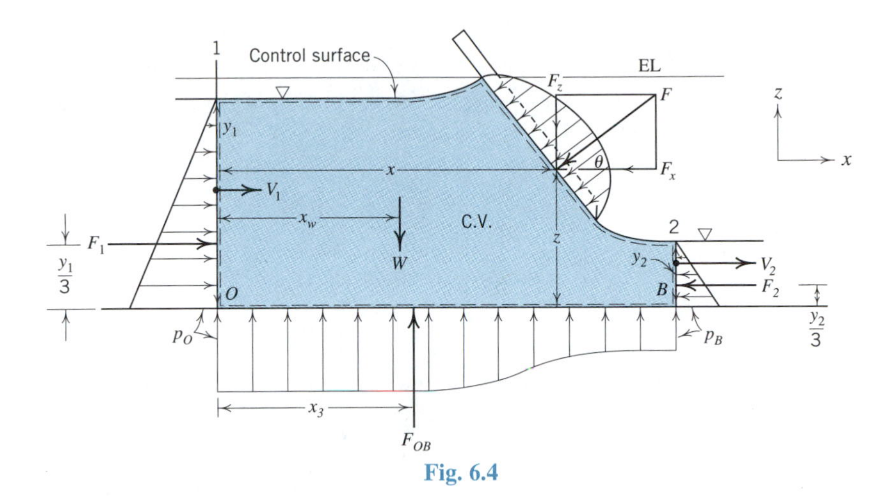

Fig. 6.4

Recognizing that the pressure distribution is hydrostatic at sections 1 and 2, and replacing V with q/A,

$$\frac{\gamma y_1^2}{2} - \frac{\gamma y_2^2}{2} - F_x = q^2\rho\left(\frac{1}{y_2} - \frac{1}{y_1}\right)$$

The x-component of the force by the sluice gate on the water is given by

$$F_x = \frac{\gamma}{2}(y_1^2 - y_2^2) + q^2\rho\left(\frac{1}{y_1} - \frac{1}{y_2}\right) \tag{6.6}$$

For an ideal fluid (and to a good approximation, for a real fluid), the force on the gate tangent to the gate is zero (no shear). Hence, the resultant force is normal to the gate and it may be computed as

$$F = F_x/\cos\theta$$

This realization eliminates the need to use the impulse-momentum equation in the z-direction.

ILLUSTRATIVE PROBLEM 6.2

This two-dimensional overflow structure (shape and size unknown) produces the flowfield shown. Calculate the magnitude and direction of the horizontal component of the resultant force the fluid exerts on the structure.

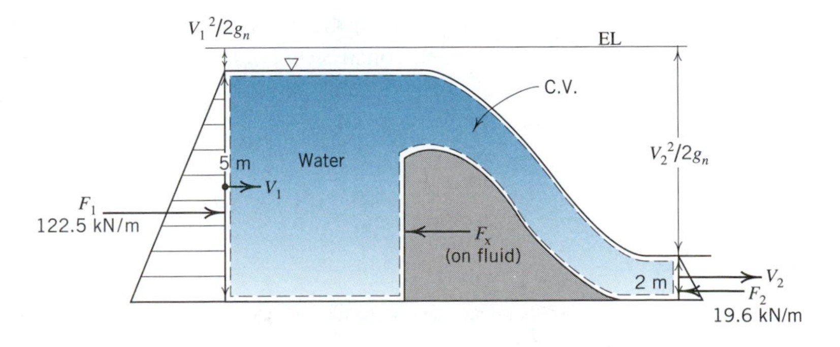

SOLUTION

Assuming an ideal fluid, we construct the energy line above the liquid surface and write the work-energy equation

$$0 + 5\text{ m} + \frac{V_1^2}{2g_n} = 0 + 2\text{ m} + \frac{V_2^2}{2g_n} \tag{5.7}$$

and the continuity equation

$$q = 5\text{ m} \times V_1 = 2\text{ m} \times V_2 \tag{4.7}$$

Substituting the continuity equation into the work-energy equation and solving gives

$$V_1 = 3.33\text{ m/s} \qquad V_2 = 8.33\text{ m/s} \qquad q = 16.65\text{ m}^3\text{/s/m}$$

For the control volume shown above, the streamlines are essentially straight and parallel at sections 1 and 2. The forces F_1 and F_2 can be calculated as follows with $\gamma = 9.8\text{ kN/m}^3$.

$$F_1 = \frac{9.8\text{ kN/m}^3 \times (5\text{ m})^2}{2} = 122.5\text{ kN} \qquad F_2 = \frac{9.8\text{ kN/m}^3 \times (2\text{ m})^2}{2} = 19.6\text{ kN}$$

Now applying the impulse-momentum equation 6.2a with $\rho = 1\,000\text{ kg/m}^3$, we get

$$(\Sigma F_{ext})_x = F_1 - F_2 - F_x = 122\,500\text{ N} - 19\,600\text{ N} - F_x$$

$$Q\rho(V_{2_x} - V_{1_x}) = 16.65\text{ m}^3\text{/s/m} \times 1\,000\text{ kg/m}^3 \times (8.33\text{ m/s} - 3.33\text{ m/s})$$

Equating,

$$122\,500 - 19\,600 - F_x = 16\,650 \times 5.00$$

$$F_x = 19\,650\text{ N} \quad \text{or} \quad 19.65\text{ kN} \quad \text{to the left}$$

Because the above force is that exerted by the structure on the fluid, the force exerted by the fluid on the structure is 19.65 kN to the right •

Frequently in open channel flow, when liquid at high velocity discharges into a zone of lower velocity, a rather abrupt rise (a standing wave) occurs in the liquid surface and is accompanied by violent turbulence, eddying, air entrainment, and surface undulations; such a wave is known as a *hydraulic jump*.[7] In spite of the foregoing complications and the consequential large head loss, application of the impulse-momentum principle gives results very close to those observed in field and laboratory; the engineering problem is to find the relation between the depths for a given flowrate. Assume a plane flow situation with flowrate q and construct a control volume (between two sections 1 and 2 where the streamlines are straight and parallel; Fig. 6.5) enclosing the jump. The only horizontal forces are seen to be (neglecting friction) the hydrostatic ones of Eq. 2.12. Applying Eq. 6.2a gives

$$(\Sigma F_{ext})_x = F_1 - F_2 = \frac{\gamma y_1^2}{2} - \frac{\gamma y_2^2}{2} = q\rho(V_2 - V_1)$$

[7]Further discussion of the hydraulic jump is found in Section 10.9.

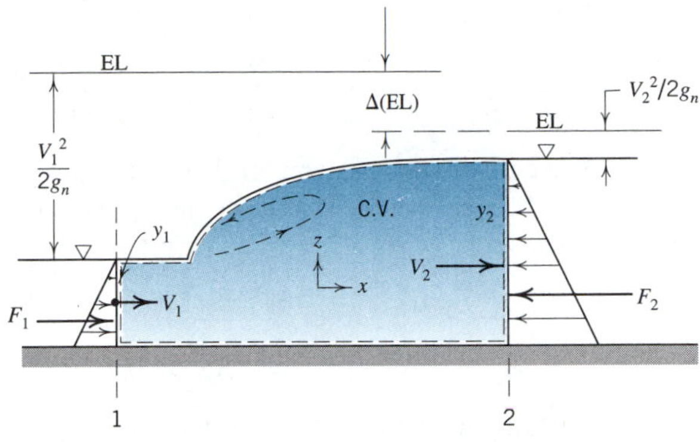

Fig. 6.5

Substituting the continuity relations $V_2 = q/y_2$ and $V_1 = q/y_1$ and rearranging give the desired relationship among y_1, y_2, and q:

$$\frac{q^2}{g_n y_1} + \frac{y_1^2}{2} = \frac{q^2}{g_n y_2} + \frac{y_2^2}{2}$$

and solving for y_2/y_1 yields

$$\frac{y_2}{y_1} = \frac{1}{2}\left[-1 + \sqrt{1 + \frac{8q^2}{g_n y_1^3}}\right] = \frac{1}{2}\left[-1 + \sqrt{1 + \frac{8V_1^2}{g_n y_1}}\right] \tag{6.7}$$

From this equation it will be seen that: (*a*) for $V_1^2/g_n y_1 = 1$, $y_2/y_1 = 1$; (*b*) for $V_1^2/g_n y_1 > 1$, $y_2/y_1 > 1$; and (*c*) $V_1^2/g_n y_1 < 1$, $y_2/y_1 < 1$.

Condition (*c*), that of a *fall of the liquid surface*, although satisfying the impulse-momentum and continuity equations, will be found to produce a rise of the energy line through the jump, and is thus physically impossible. Accordingly it may be concluded that, for a hydraulic jump to occur, the upstream conditions must be such that $V_1^2/g_n y_1 > 1$. Later (Section 8.1) $V_1^2/g_n y_1$ will be shown to be the square of the *Froude number*, the ratio of flow velocity to surface wave velocity, and thus analogous to the Mach number of compressible flow.

ILLUSTRATIVE PROBLEM 6.3

Water flows in a horizontal open channel at a depth of 0.6 m at a flowrate of 3.7 $m^3/s/m$ of width. If a hydraulic jump is possible, calculate the depth just downstream from the jump and the power dissipated in the jump.

SOLUTION

To determine if a jump is possible, we must calculate the term $V_1^2/g_n y_1$ and compare the result with 1.0 (see text above). We first calculate the velocity from continuity,

$$V_1 = \frac{q}{y_1} = \frac{3.7\ m^3/s/m}{0.6\ m} = 6.17\ m/s \tag{4.7}$$

Now,

$$\frac{V_1^2}{g_n y_1} = \frac{(6.17 \text{ m/s})^2}{9.81 \times 0.6 \text{ m}} = 6.46$$

Since this number is greater than 1.0, a jump can form.

To find the depth after the jump, we employ Eq. 6.7,

$$\frac{y_2}{y_1} = \frac{1}{2}\left[-1 + \sqrt{1 + \frac{8V_1^2}{g_n y_1}}\right] = \frac{1}{2}\left[-1 + \sqrt{1 + 8 \times 6.46}\right] = 3.13 \qquad (6.7)$$

$$y_2 = 3.13 \times y_1 = 3.13 \times 0.6 = 1.88 \text{ m} \bullet$$

To compute the power lost in the jump, we need to locate the energy line downstream of the jump. To do this, we need the velocity at section 2.

$$V_2 = \frac{q}{y_2} = \frac{3.7 \text{ m}^3\text{/s/m}}{1.88 \text{ m}} = 1.97 \text{ m/s} \qquad (4.7)$$

The drop in the energy line across the jump is the power dissipated in the jump per newton of fluid flowing.

$$\Delta \text{ EL} = \left(y_1 + \frac{V_1^2}{2g_n}\right) - \left(y_2 + \frac{V_2^2}{2g_n}\right)$$

$$= \left(0.6 + \frac{6.17^2}{2 \times 9.81}\right) - \left(1.88 - \frac{1.97^2}{2 \times 9.81}\right)$$

$$\Delta \text{ EL} = 2.54 - 2.08 = 0.46 \text{ m} \quad \text{or} \quad 0.46 \text{ N·m/N}$$

The power dissipated in the jump is, from Eq. 5.8b,

$$P = \frac{q\gamma \, \Delta \text{ EL}}{1\,000} = \frac{3.7 \text{ m}^3\text{/s/m} \times 9\,800 \text{ N/m}^3 \times 0.46 \text{ m}}{1\,000} \qquad (5.8b)$$

$$= 16.7 \text{ kW/metre of width} \bullet$$

This result indicates that the hydraulic jump is an excellent energy dissipator and it is frequently used for this purpose in engineering projects.

The velocity (or celerity) of small gravity waves in a body of water can be found using the impulse-momentum principle. The wave appears as a small localized rise in the liquid surface which propagates at a velocity a. It is important to distinguish between this type of wave where liquid movement extends over the full depth of the flow and the small surface disturbance or ripple where liquid movement is restricted to a region near the surface.

This type of wave is depicted in Fig. 6.6 moving at velocity a through a fluid at rest. To a stationary observer, the fluid motion is unsteady as the wave passes. However, if the observer moves with the wave at velocity a, the flow appears steady and can be analyzed with elementary principles.

For the "steady flow" configuration, we will assign the velocity under the wave as V. From continuity,

$$ay = V(y + dy)$$

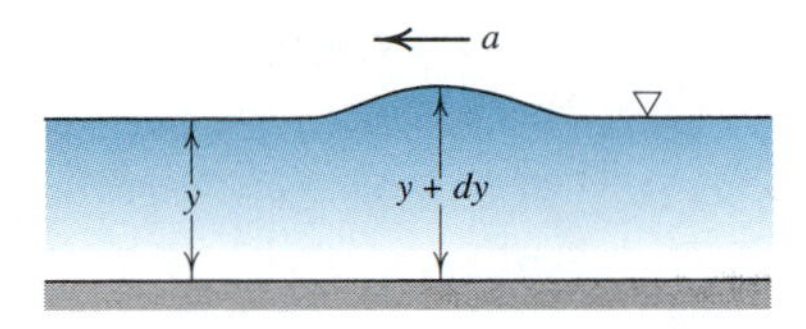

As seen by a stationary observer

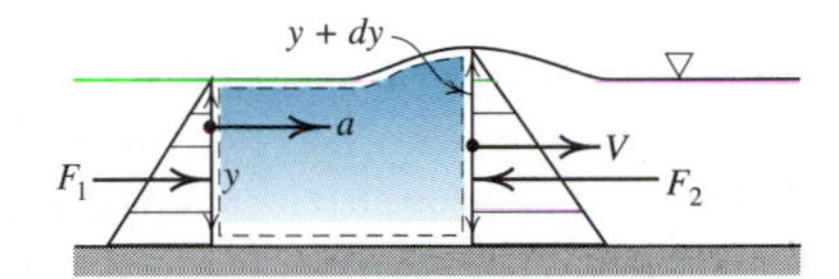

As seen by an observer moving with the wave

Fig. 6.6

And from impulse-momentum (as was done for the hydraulic jump), we get

$$\frac{\gamma y^2}{2} - \frac{\gamma(y + dy)^2}{2} = (ay)\rho(V - a) \tag{6.2a}$$

Combining these two equations gives

$$a^2 = g_n(y + dy)$$

Letting dy approach zero results in

$$a = \sqrt{g_n y} \tag{6.8}$$

For the small gravity wave, the *velocity depends only on the depth of flow* when the length of the wave is large compared to the depth of the water.

ILLUSTRATIVE PROBLEM 6.4

Earthquakes on the ocean floor often cause tsunamis (seismic sea waves) which radiate from the disturbance carrying energy much like the small gravity waves we have just analyzed. An earthquake of this nature occurred in Anchorage, Alaska in 1964, creating a tsunami which traveled throughout the Pacific ocean. If the ocean depth averages 12 000 ft, calculate the wave velocity and the time required for the wave to reach a port city 8 450 miles away, assuming the small gravity wave theory applies.

SOLUTION

Equation 6.8 will give us the wave velocity for an average depth of 12 000 ft.

$$a = \sqrt{g_n y} = \sqrt{32.2 \times 12\,000 \text{ ft}} = 622 \text{ ft/s} \quad \text{or} \quad 424 \text{ mph} \bullet \tag{6.8}$$

At this velocity, the wave will reach the port city in

$$t = \frac{8\,450 \text{ mi}}{424 \text{ mph}} = 19.9 \text{ hr} \bullet$$

While the wave velocity may seem unexpectedly high, the wave length is very long and the energy is distributed over the full 12 000 ft of depth. In the open ocean, one would not notice the wave passing. However, when the wave approaches shore and the energy becomes concentrated in the shallow water, the wave takes on dramatic form and can wash far inland and create massive damage.

Under certain circumstances in open channel flow with $V_1 > \sqrt{g_n y_1}$, a diagonal standing wave will form whose properties can be analyzed by the impulse-momentum principle. A wedge-shaped bridge pier (in a river) producing such a wave is shown in Fig. 6.7. Upstream from the wave front the liquid depth is y_1, and downstream it is y_2. By isolating a control volume $ABCD$ of length 1 m parallel to the wave front, application of the impulse-momentum and continuity principles is readily made.

Apply the continuity principle by noting that, because (from symmetry) the flowrate across AB is equal to that across CD, the flowrates across AD and BC are also equal. Accordingly,

$$q = y_1 V_1 \sin \beta = y_2 V_2 \sin (\beta - \theta) \tag{6.9}$$

The impulse-momentum principle applied along the tangential t-direction shows (from symmetry and the absence of body forces in the horizontal plane) that the forces on the surfaces AB and CD are equal and thus the t-component of velocity upstream and downstream from the wave must also be equal; therefore

$$V_1 \cos \beta = V_2 \cos (\beta - \theta)$$

Apply the impulse-momentum principle along the normal direction n; the difference in the hydrostatic forces must account for the rate of change of momentum per unit time. This

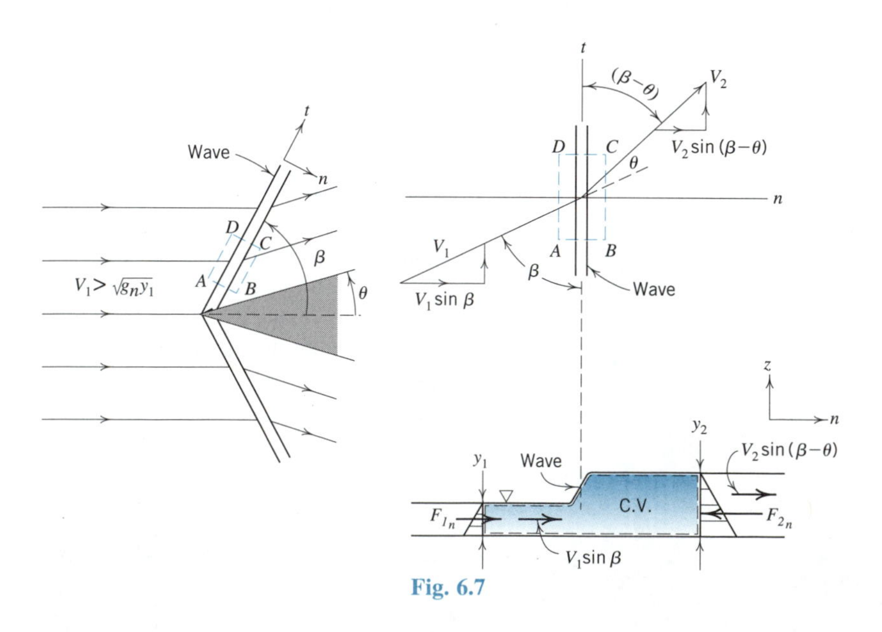

Fig. 6.7

is written (see Eq. 6.2a)

$$(\Sigma F_{ext})_n = (F_1 - F_2)_n = \frac{\gamma y_1^2}{2} - \frac{\gamma y_2^2}{2} \qquad (6.10)$$
$$= q\rho[V_2 \sin(\beta - \theta) - V_1 \sin \beta]$$

and, because $F_{2_n} > F_{1_n}$, $V_2 \sin(\beta - \theta) < V_1 \sin \beta$, and the indicated direction of deflection of the streamlines through the wave is therefore justified.

Combination of these equations yields

$$\frac{\tan^2(\beta - \theta)}{\tan^2 \beta} = \frac{1 + 2(V_2^2/g_n y_2)\sin^2(\beta - \theta)}{1 + 2(V_1^2/g_n y_1)\sin^2 \beta} \qquad (6.11)$$

which shows that (for $V_1^2/g_n y_1 > 1$): for small values of β, $V_2^2/g_n y_2 > 1$ and, for large values of β, $V_2^2/g_n y_2 < 1$. Thus the oblique standing wave resembles the hydraulic jump for large values of β but is quite unlike it for small values of β.

A relationship[8] analogous to that of Eq. 6.7 for the hydraulic jump may be set down directly by comparing the development leading to Eq. 6.7 with that leading to Eqs. 6.9 and 6.10. These are identical except for the terms $\sin \beta$ and $\sin(\beta - \theta)$ which accompany the velocities V_1 and V_2, respectively. Hence, for the oblique standing wave, Eq. 6.7 may be adjusted to read

$$\frac{y_2}{y_1} = \frac{1}{2}\left[-1 + \sqrt{1 + \frac{8V_1^2 \sin^2 \beta}{g_n y_1}}\right] \qquad (6.12)$$

ILLUSTRATIVE PROBLEM 6.5

Water flows in an open channel at a depth of 0.6 m and velocity of 6.0 m/s. A wedge-shaped pier is to be placed in the flow to produce a standing wave of angle $\beta = 40°$. Calculate the required wedge angle 2θ and the depth just downstream from the wave.

SOLUTION

We will first compute y_2 from Eq. 6.12.

$$y_2 = \frac{y_1}{2}\left[-1 + \sqrt{1 + \frac{8V_1^2 \sin^2 \beta}{g_n y_1}}\right] \qquad (6.12)$$
$$= \frac{0.6 \text{ m}}{2}\left[-1 + \sqrt{1 + \frac{8 \times (6.0 \text{ m/s})^2 \sin^2 40°}{9.81 \times 0.6 \text{ m}}}\right]$$
$$y_2 = 1.08 \text{ m} \bullet$$

From continuity, we can find $V_2 \sin(\beta - \theta)$.

$$y_1 V_1 \sin \beta = y_2 V_2 \sin(\beta - \theta) \qquad (6.9)$$

$$V_2 \sin(\beta - \theta) = \frac{y_1}{y_2} V_1 \sin \beta = \frac{0.6 \text{ m}}{1.08 \text{ m}} 6.0 \text{ m/s} \sin 40° = 2.14 \text{ m/s}$$

[8] For other relationships, see Sec. 15-17 of V. T. Chow, *Open-Channel Hydraulics*, McGraw-Hill, 1959.

Now, moving to Eq. 6.11, we can find θ.

$$\frac{\tan^2(\beta - \theta)}{\tan^2 \beta} = \frac{1 + 2(V_2^2/g_n y_2)\sin^2(\beta - \theta)}{1 + 2(V_1^2/g_n y_1)\sin^2 \beta} \tag{6.11}$$

$$\tan^2(\beta - \theta) = \tan^2 40° \frac{1 + 2[(2.14 \text{ m/s})^2/(9.81 \times 1.08 \text{ m})]}{1 + 2[(6.0 \text{ m/s})^2/(9.81 \times 0.6 \text{ m})]\sin^2 40°}$$

$$\tan(\beta - \theta) = 0.466 \qquad (\beta - \theta) = 25° \qquad \theta = 15° \qquad 2\theta = 30° \bullet$$

6.4 THE ANGULAR IMPULSE-MOMENTUM PRINCIPLE

In Section 6.1, we identified a small individual fluid system in a control volume, wrote Newton's second law for the system, and summed for every fluid system in the control volume. The result was the linear impulse-momentum equation which is a vector equation relating the sum of the external forces acting on the fluid in the control volume and the momentum transport into and out of the control volume. We now develop the *angular impulse-momentum* equation by a similar technique only this time, we will use the *moments* of the force and momentum vectors.

We begin again with the small individual fluid system (see Fig. 6.8) and establish a point O about which the moments of the forces and momentum vectors will be taken. For this system, the moments will be

$$\sum \mathbf{r} \times \mathbf{F} = \frac{d}{dt}(\mathbf{r} \times m\mathbf{v}) = \frac{d}{dt}(\mathbf{r} \times \rho \, d\forall \mathbf{v})$$

Writing this equation for every fluid system in the control volume and summing, we get

$$\sum (\mathbf{r} \times \mathbf{F}_{ext}) = \frac{d}{dt}\iiint_{sys} (\mathbf{r} \times \mathbf{v})\rho \, d\forall$$

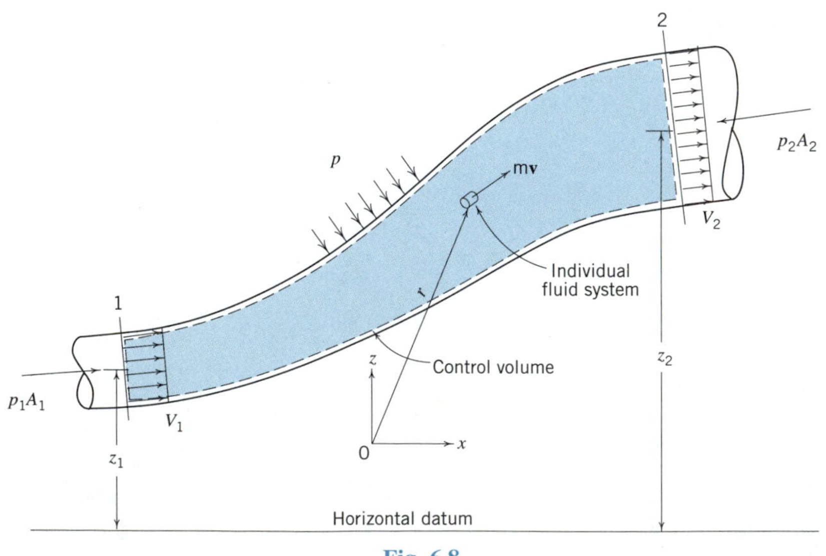

Fig. 6.8

where the integral over all the systems and the control volume are equivalent at a given instant; however, to exchange the integral and the time derivative requires us to formally integrate over all of the systems. Note that the moments of all internal forces cancel leaving only the moments of the weight of the fluid in the control volume and the moments of the external forces applied at the control surface.

Drawing on the Reynolds Transport Theorem to evaluate the integral,

$$\frac{dE}{dt} = \frac{d}{dt}\iiint_{sys} (\mathbf{r} \times \mathbf{v})\rho \, d\forall = \iint_{C.S.} i\rho\mathbf{v} \cdot d\mathbf{A}$$

$$= \iint_{C.S.out} (\mathbf{r} \times \mathbf{v})\rho\mathbf{v} \cdot d\mathbf{A} + \iint_{C.S.in} (\mathbf{r} \times \mathbf{v})\rho\mathbf{v} \cdot d\mathbf{A}$$

where E = the moment of the momentum of the fluid system and $i = \mathbf{r} \times \mathbf{v}$, the moment of momentum per unit mass.

As before, we will restrict ourselves to control volumes where the fluid enters and leaves at sections where the streamlines are straight and parallel and with the velocity normal to the cross-sectional area. Under these conditions we can simplify the above integral to

$$\frac{d}{dt}\iiint_{sys} (\mathbf{r} \times \mathbf{v})\rho \, d\forall = \iint_{C.S.out} (\mathbf{r} \times \mathbf{v})\rho \, dQ - \iint_{C.S.in} (\mathbf{r} \times \mathbf{v})\rho \, dQ$$

Because the velocity is uniform over the flow cross sections we can replace the integrals on the right with

$$\frac{d}{dt}\iiint_{sys} (\mathbf{r} \times \mathbf{v})\rho \, d\forall = Q\rho(\mathbf{r}_{out} \times \mathbf{V}_{out}) - Q\rho(\mathbf{r}_{in} \times \mathbf{V}_{in})$$

$$= Q\rho[(\mathbf{r} \times \mathbf{V})_{out} - (\mathbf{r} \times \mathbf{V})_{in}]$$

where $\mathbf{r}$ is the position vector from the moment center to the centroid of the entering or leaving flow cross section of the control volume.

Substituting this expression back into the original equation,

$$\Sigma(\mathbf{r} \times \mathbf{F}_{ext}) = \Sigma\mathbf{M}_0 = Q\rho[(\mathbf{r} \times \mathbf{V})_{out} - (\mathbf{r} \times \mathbf{V})_{in}] \tag{6.13}$$

If motion is restricted to two dimensions, the above equation can be written

$$\Sigma M_0 = Q\rho(r_2 V_{2_t} - r_1 V_{1_t}) \tag{6.14}$$

where V_t is the component of velocity normal to the moment arm r.

In rectangular components, assuming V is directed with positive components in both the x and z-direction, and with the moment center at the origin of the x-z coordinate system, for clockwise positive moments,

$$\Sigma M_0 = Q\rho[(z_2 V_{2_x} - x_2 V_{2_z}) - (z_1 V_{1_x} - x_1 V_{1_z})] \tag{6.15}$$

where x, z are the coordinates of the centroids of the flow cross-sections where the fluid enters and leaves the control volume.

Should the fluid enter or leave the control volume at more than one cross-section, the following form of Eq. 6.14 should be used.

$$\Sigma M_0 = (\Sigma Q\rho r V_t)_{out} - (\Sigma Q\rho r V_t)_{in} \tag{6.16}$$

To demonstrate the application of the principle, we will locate the position of the resultant force acting on the pipe bend in Illustrative Problem 6.1.

ILLUSTRATIVE PROBLEM 6.6

For the problem analyzed in Illustrative Problem 6.1, compute the location of the resultant force exerted by the water on the pipe bend.

SOLUTION

The results of the analysis in Illustrative Problem 6.1 are repeated below with the resultant force F = 8 813 N.

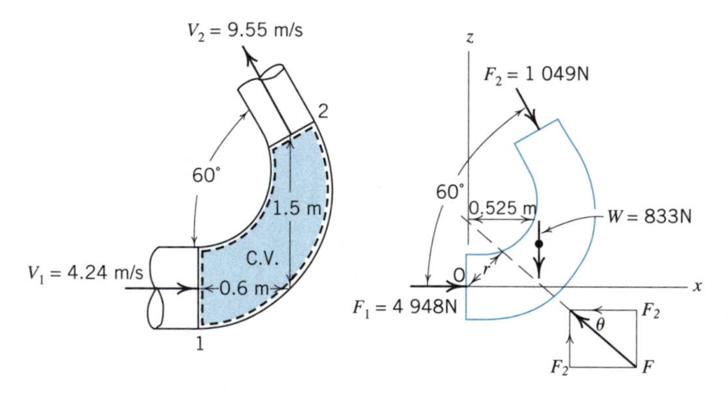

With the coordinate system established at the centroid of section 1, we can take moments of the force and momentum vectors about this origin. Note that we will assume that the forces F_1 and F_2 act at the centroid of the sections rather than at the center of pressure. While this approach introduces a small error, for pressures of this magnitude the effect is negligible.

We will use Eq. 6.15 to find the location of the resultant force. Note from the diagram that we identify the location of the resultant by a perpendicular distance r from the origin.

$$\Sigma M_0 = Q\rho[(z_2 V_{2_x} - x_2 V_{2_z}) - (z_1 V_{1_x} - x_1 V_{1_z})] \tag{6.15}$$

$$- F_2 \cos 60° \times 1.5 \text{ m} - F_2 \sin 60° \times 0.6 \text{ m} - W \times 0.525 \text{ m} + F \times r$$
$$= Q\rho[(1.5 \text{ m} \times V_2 \cos 60° - 0.6 \text{ m} \times (-V_2) \sin 60°) - (0)]$$
$$-1\,049 \text{ N} \times \cos 60° \times 1.5 - 1\,049 \text{ N} \times \sin 60° \times 0.6 - 833 \text{ N} \times 0.525$$
$$+ 8\,813 \text{ N} \times r = 0.3 \text{ m}^3/\text{s} \times 998 \text{ kg/m}^3 [(1.5 \times 9.55 \text{ m/s} \times \cos 60°$$
$$+ 0.6 \times 9.55 \text{ m/s} \times \sin 60°)]$$
$$- 787 - 545 - 437 + 8\,813r = 299.4 \times (7.16 + 4.96)$$
$$r = 0.61 \text{ m} \bullet$$

While the force exerted by the water on the bend is opposite in direction from the force F, the line of action is the same.

Flow Machines

6.5 JET PROPULSION

In a modern aircraft jet-propulsion system (Fig. 6.9), air is drawn in at the upstream end and its pressure, density, and temperature are raised by a compressor (usually axial flow). Just downstream from the compressor, fuel is injected and burned, adding energy to the fluid. The mixture then expands and passes through a gas turbine (which drives the compressor) on its way to the nozzle; from here it emerges into the atmosphere at high velocity and at a pressure equal to or greater than that surrounding the nozzle. A cross section of a modern jet engine is shown in Fig. 6.10.

An impulse-momentum analysis of such a unit allows its propulsive force to be computed. Here a control volume of indefinite horizontal and vertical extent that is fixed to the jet-propulsion unit[9] can be used, as shown in Fig. 6.9, section 1 being located well

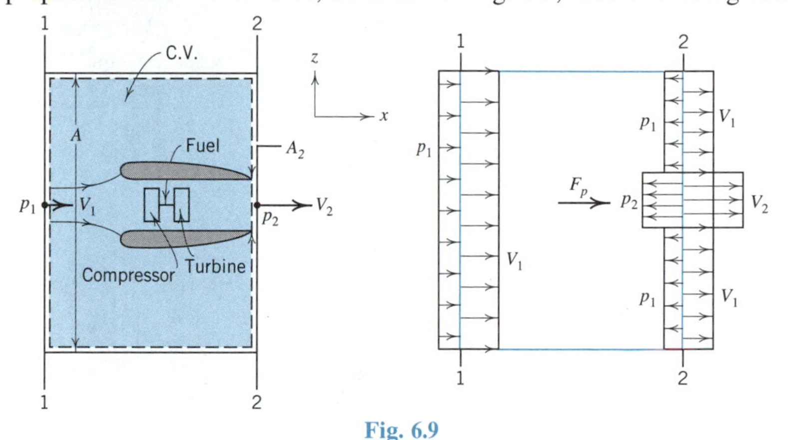

Fig. 6.9

[9]In this and several other applications, it is desirable to fix the control volume to the object being analyzed rather than to hold the control volume fixed in space as has been assumed to this point. From particle dynamics it is recalled that Newton's second law is applicable in an *inertial reference frame*. As defined in Section 4.1, a reference frame that is fixed in space is an inertial frame; but it can be shown that a reference frame which is moving at a *constant velocity* relative to a fixed frame is also an inertial frame. An inspection of the derivations of all the equations based on Newton's second law, which have been carried out to this point, reveals that, if all velocities and accelerations, and so on, are referenced to an inertial frame, then all the results remain valid and the equations are, indeed, unchanged. Accordingly, when it is useful to do so, control volumes will be fixed on the object being analyzed, but only if it is moving with a constant velocity.

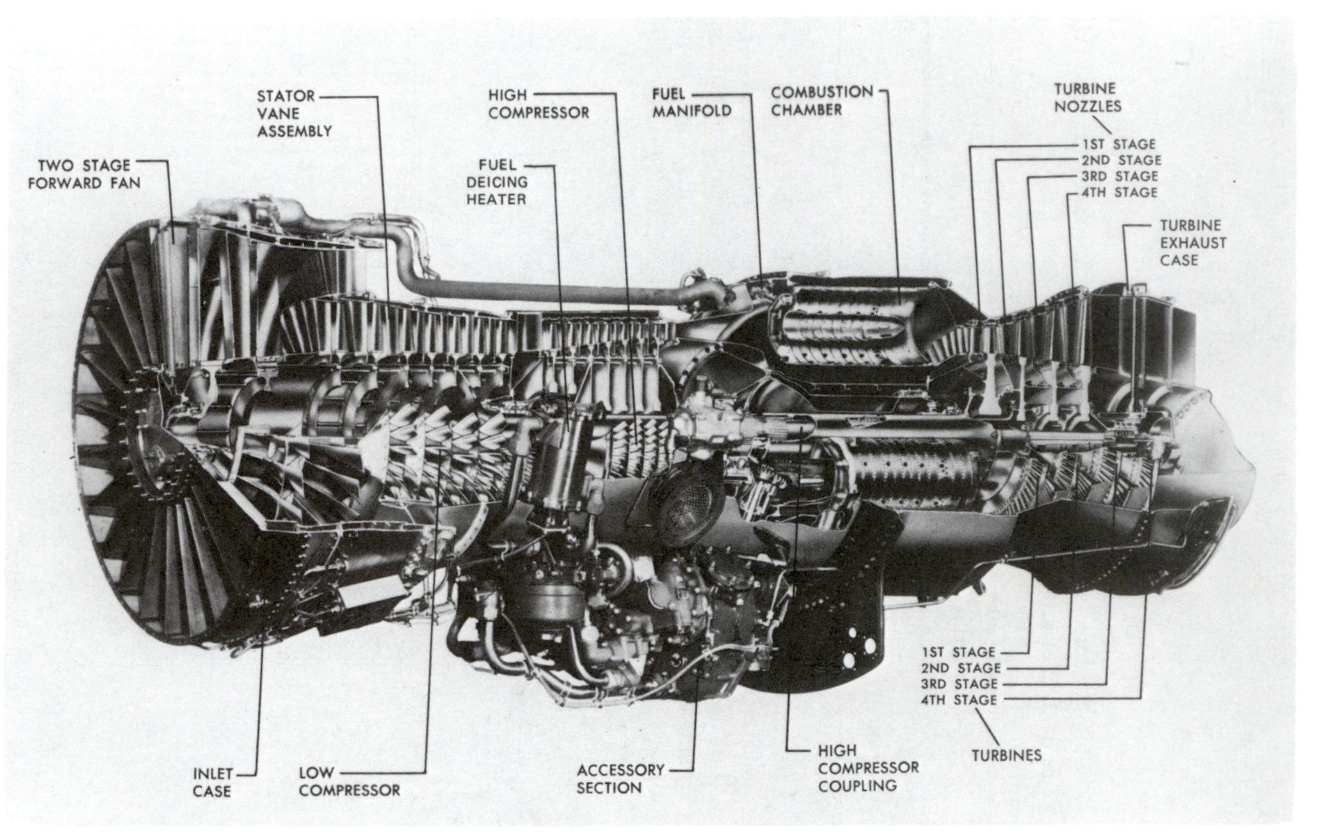

Fig. 6.10 Modern jet engine cutaway view—model JT3D. (Courtesy Pratt & Whitney Aircraft, Division of United Aircraft Corporation.)

upstream of the unit and section 2 in the plane of the nozzle exit. The speed of the jet unit relative to still air is V_1. Thus, in a coordinate frame that moves with the jet unit, the incoming air velocity far from the unit is V_1. If it is assumed that air which does not pass through the unit experiences no net change of velocity or pressure, the pressures and velocities at all points of sections 1 and 2 will be p_1 and V_1, respectively, except in the nozzle exit, where they are p_2 and V_2. Taking the summation of forces on the fluid within the control volume gives

$$\Sigma F_x = p_1 A - p_1(A - A_2) - p_2 A_2 + F_P = (p_1 - p_2)A_2 + F_P$$

in which F_P is the propulsive force exerted *by the unit on the fluid*. Now the fuel is added through the control surface and so must be considered in the fluid momentum analysis. Between the point of fuel injection and nozzle exit the velocity of the mass flowrate ($\dot{m}_f$) of fuel increases from zero to V_2; between sections 1 and 2 the velocity of the mass flowrate ($\dot{m}_a$) of air increases from V_1 to V_2. Accordingly

$$\oint_{C.S.} V_x \, d\dot{m} = V_2(\dot{m}_f) + (V_2 - V_1)(\dot{m}_a)$$

Using Eq. 6.2a yields the propulsive force, F_P, obtainable from the unit; it is

$$F_P = (\dot{m}_a)(V_2 - V_1) + (\dot{m}_f)V_2 + (p_2 - p_1)A_2 \qquad (6.17)$$

Although this force is fundamental to preliminary design calculations, it yields no information about the optimum design of passages, compressor, turbine, and so forth, for which other principles must be used. It does, however, demonstrate again the strength of the impulse-momentum principle in the bypassing of internal complexities and obtaining useful results from upstream and downstream flow conditions alone.

ILLUSTRATIVE PROBLEM 6.7

A jet-propulsion unit is to develop a propulsive force of 200 kN when moving through the US Standard Atmosphere (Appendix 2) at an altitude of 10 km and at a speed of 250 m/s. The velocity, absolute pressure, and the area at the nozzle exit are 1.2 km/s, 35 kPa, and 1.4 m^2, respectively. How much air must be drawn through the unit? Fuel may be neglected.

SOLUTION

From Appendix 2, we obtain the pressure of 26.5 kPa at a 10 km altitude in the Standard Atmosphere. We employ Eq. 6.17 to compute the thrust.

$$F_P = (\dot{m}_a)(V_2 - V_1) + (\dot{m}_f)V_2 + (p_2 - p_1)A_2 \qquad (6.17)$$

$$200 \times 10^3 \text{ N} = (\dot{m}_a)(1\,200 \text{ m/s} - 250 \text{ m/s}) + 0$$
$$+ (35.0 - 26.5) \times 10^3 \text{ Pa} \times 1.4 \text{ m}^2$$
$$\dot{m}_a = 198 \text{ kg/s} \bullet$$

A suitable compressor must then be designed to provide this flowrate.

Fig. 6.11 Controllable pitch ship propellers (4.2 m d.). (Courtesy Lips N. V. Propeller Works, Drunen, Holland.)

6.6 PROPELLERS AND WINDMILLS

Although the screw propellers of ships or aircraft cannot be analyzed with the impulse-momentum and energy principles alone, application of these principles to the problem will lead to some of the laws which characterize their design.

A pair of controllable pitch ship propellers is shown in Fig. 6.11. An idealized screw propeller with its slipstream is shown in Fig. 6.12. For such a propeller operating in an unconfined fluid, the pressures p_1 and p_4 at some distance ahead of and behind the propeller, and the pressures over the slipstream boundary, are the same. However, from the shape of the slipstream (using the continuity and Bernoulli principles) the mean pressure p_2 just upstream from the propeller is smaller than p_1, and the pressure p_3 just downstream

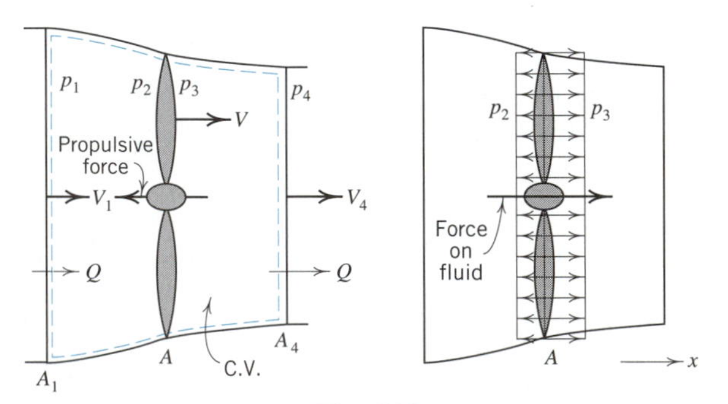

Fig. 6.12

from the propeller is larger than p_4.[10] When the fluid in the slipstream between sections 1 and 4 is isolated, it is observed that the only force acting is that exerted by the propeller on the fluid.[11] This may be computed either from the pressure difference $(p_3 - p_2)$ or from the gain in momentum flux between sections 1 and 4. Therefore

$$(p_3 - p_2)A = F = (V_4 - V_1)\rho Q = (V_4 - V_1)A\rho V \tag{6.18}$$

in which V is the mean velocity through the propeller disk; cancelling A's in this equation gives

$$p_3 - p_2 = (V_4 - V_1)\rho V \tag{6.19}$$

Now, by applying the Bernoulli principle between sections 1 and 2,

$$p_1 + \tfrac{1}{2}\rho V_1^2 = p_2 + \tfrac{1}{2}\rho V_2^2 \tag{5.2}$$

and between sections 3 and 4,

$$p_3 + \tfrac{1}{2}\rho V_3^2 = p_4 + \tfrac{1}{2}\rho V_4^2 \tag{5.2}$$

and by using $p_1 = p_4$, another expression may be derived for $(p_3 - p_2)$; it is

$$p_3 - p_2 = \tfrac{1}{2}\rho(V_4^2 - V_1^2) \tag{6.20}$$

By equating Eqs. 6.19 and 6.20,

$$V = (V_1 + V_4)/2 \tag{6.21}$$

which shows that the velocity through the propeller disk is the numerical average of the velocities at some distance ahead of and behind the propeller; in other words, there is the same increase of velocity ahead of the propeller as behind it. This result, modified slightly to allow for friction, rotational effects, and so forth, is one of the basic assumptions of propeller design.

The useful power output, P_o, derived from a propeller, is the thrust, F, multiplied by the velocity, V_1, at which the propeller is moving forward.

$$P_o = FV_1 = (V_4 - V_1)\rho Q V_1 \tag{6.22}$$

The power input, P_i, is that required to maintain continual increase of velocity of the slipstream from V_1 to V_4. From Eqs. 5.8,

$$P_i = \frac{\rho Q}{2}(V_4^2 - V_1^2) = \rho Q(V_4 - V_1)\left(\frac{V_4 + V_1}{2}\right) = \rho Q(V_4 - V_1)V \tag{6.23}$$

The ideal efficiency, η, of the propeller is then

$$\eta = \frac{P_o}{P_i} = \frac{V_1}{V} \tag{6.24}$$

and, because V is always greater than V_1, it may be concluded that the efficiency of a propeller even in an ideal fluid can never be 100%.[12]

[10] Applying this, and the fact of the preceding sentence, at the propeller tip apparently results in two different pressures there. This is due to the assumption of a uniform pressure distribution; in reality the pressure is not uniform over the propeller disk because at the propeller tip it must equal that surrounding the slipstream.

[11] The same pressure distributed over sections 1, 4, and the slipstream boundary can produce no net force.

[12] When $V = V_1$, a propeller of 100% efficiency results, but such a propeller produces no propulsive force! (See Eq. 6.18.) Practical ship and airplane propellers may have efficiencies of 80%.

ILLUSTRATIVE PROBLEM 6.8

The engine of an airplane flying through still air (specific weight $12.0\ \text{N/m}^3$) at 320 km/h (88.9 m/s) delivers 1 120 kW to an ideal propeller 3 m in diameter. Calculate the slipstream velocity, the velocity through the propeller disk, and the diameter of the slipstream ahead of and behind the propeller. Also calculate the thrust and efficiency.

SOLUTION

We begin the solution by using the equation for power, Eq. 6.23, to find the velocity V_4.

$$P_i = \frac{\rho Q}{2}(V_4^2 - V_1^2) = \frac{\rho}{2}(\pi/4)(d^2)\frac{V_1 + V_4}{2}(V_4^2 - V_1^2) \qquad (6.23)$$

where the average velocity at the propeller disk is

$$V = \frac{V_1 + V_4}{2} \qquad (6.21)$$

Substituting the data into Eq. 6.23,

$$1\,120 \times 10^3\ \text{W} = \frac{12.0\ \text{N/m}^3}{2 \times 9.81}(\pi/4)(3\ \text{m})^2\frac{88.9\ \text{m/s} + V_4}{2}[V_4^2 - (88.9\ \text{m/s})^2]$$

Solving this equation for V_4 gives $V_4 = 103$ m/s •

Now, the average velocity is

$$V = \frac{V_1 + V_4}{2} = \frac{88.9\ \text{m/s} + 103\ \text{m/s}}{2} = 95.95\ \text{m/s} \quad \bullet$$

The discharge Q is

$$Q = AV = (\pi/4)(3.0\ \text{m})^2 \times 95.95\ \text{m/s} = 678\ \text{m}^3/\text{s}$$

We can now calculate the diameter of the slipstream ahead and behind the propeller.

$$A_1 = \frac{Q}{V_1} = \frac{678\ \text{m}^3/\text{s}}{88.9\ \text{m/s}} = 7.63\ \text{m}^2 \qquad d_1 = 3.12\ \text{m} \quad \bullet$$

$$A_4 = \frac{Q}{V_4} = \frac{678\ \text{m}^3/\text{s}}{103\ \text{m/s}} = 6.58\ \text{m}^2 \qquad d_4 = 2.89\ \text{m} \quad \bullet$$

The thrust can be calculated from Eq. 6.18.

$$F = (V_4 - V_1)\rho Q = (103\ \text{m/s} - 88.9\ \text{m/s}) \times \frac{12.0\ \text{N/m}^3}{9.81} \times 678\ \text{m}^3/\text{s} \qquad (6.18)$$

$$F = 11\,700\ \text{N} \quad \text{or} \quad 11.7\ \text{kN} \quad \bullet$$

The efficiency is obtained from Eq. 6.24.

$$\eta = \frac{V_1}{V} = \frac{88.9\ \text{m/s}}{95.95\ \text{m/s}} = 0.928 \quad \text{or} \quad 92.8\% \quad \bullet \qquad (6.24)$$

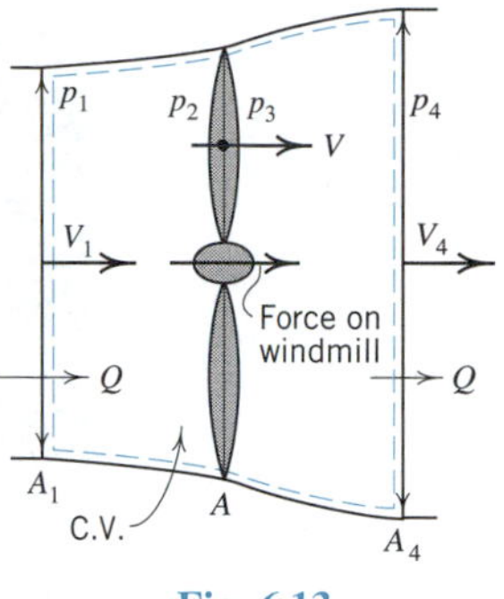

Fig. 6.13

There are many similarities between propeller and windmill, but their purposes are quite different. The propeller is designed primarily to create a propulsive force or thrust and acts as a pump. The windmill is designed to extract energy from the wind and, hence, is a turbine. Because of the different objectives of windmill and propeller, their efficiencies are calculated differently. However, comparison of Figs. 6.12 and 6.13 shows the windmill to be, as far as the flow picture is concerned, the inverse of the propeller. In the windmill the "slipstream" widens as it passes the machine and the pressure p_2 is greater than pressure p_3. However, by applying the Bernoulli and impulse-momentum principles as before, the velocity through the windmill disk, as through the propeller, may be shown to be the numerical average of V_1 and V_4.

In a frictionless machine, the power delivered to the windmill must be exactly that extracted from the air, which in turn is represented by the decrease of the kinetic energy of the slipstream between sections 1 and 4. This is the *output* of the machine and is given by

$$P_o = \frac{\rho Q}{2}(V_1^2 - V_4^2) \tag{6.25}$$

It is customary to define windmill efficiencies as the ratio of this power output to the total power *available* in a streamtube of cross-sectional area A and wind velocity V_1. Thus the efficiency of an ideal windmill is

$$\eta = \frac{P_o}{P_a} = \frac{(V_1^2 - V_4^2)AV\rho/2}{AV_1\rho V_1^2/2} = \frac{(V_1 + V_4)(V_1^2 - V_4^2)}{2V_1^3} \tag{6.26}$$

$(V_1 + V_4)/2$ having been substituted for V. The maximum efficiency is found by differentiating η with respect to V_4/V_1 and setting the result equal to zero. This gives a value of $V_4/V_1 = \frac{1}{3}$, which when substituted in Eq. 6.26 produces a maximum efficiency of $\frac{16}{27}$, or 59.3%. Because of friction and other losses this efficiency is, of course, not realized in practice; the highest possible efficiency for a real windmill appears to be around 50%. While the traditional "Dutch windmill" with large sail-like blades operates at an efficiency around 15%, the world fuel crisis has revived interest in wind energy. Since the mid-1970s major research and development efforts have focused on developing special windmills called wind turbines, with airfoil-shaped (propellerlike) blades (see Fig. 6.14), with acceptable efficiencies (up to 48%).

Stimulated by the environmental movement and tax incentives in the 1970s and 1980s, a whole technology has developed around the design and operation of wind turbines. Early efforts utilized existing airfoil technology for turbine blade design but the unique nature

Fig. 6.14 AOC 15/50 wind turbine generator [15 m rotor with hub height of 25 m; rated at 50kW at 11 m/s] Photo courtesy of Atlantic Orient Wind Systems, Inc., Norwich, VT 05055.

of wind turbines (low wind speed, high aspect ratio, and incredibly long hours of operation) led to failures from fatigue and vibration. In addition, because the turbine operates in the earth's wind boundary layer, the blades are subjected to unsymmetrical loading; that is, the blades experience higher velocities at their upper positions than at their lower. This leads to large unbalanced forces on the drive mechanism as well as speed regulation problems. Traditionally, speed control has been achieved by designing the blades to ''stall out'' at high wind speeds to help regulate the speed without excessive braking. However, new technology permitting the turbine to rotate freely, converting the electric current to 60 cycles electronically, may solve the speed regulation problem. Research and development in recent years has produced blades of high efficiency from composite materials, giving longer life and leading to competitive costs for generating power from this environmentally attractive source of energy.

ILLUSTRATIVE PROBLEM 6.9

A three-bladed wind turbine is undergoing preliminary design with the objective of producing 100 kW of electricity under an average wind speed of 36 km/h. If the efficiency of the turbine is estimated at 48%, calculate the length of blade required.

SOLUTION

First, we convert the wind speed to metres per second.

$$V_1 = 36 \text{ km/h} \times 1\,000 \text{ m/km} \div 3\,600 \text{ s/h} = 10 \text{ m/s}$$

With the required power output of 100 kW, the wind power must be

$$\text{Wind power} = \frac{P_a}{\eta} = \frac{100 \text{ kW}}{0.48} = 208.33 \text{ kW}$$

The available wind power is given as

$$\text{Available wind power} = \rho Q \frac{V_1^2}{2} = \rho A \frac{V_1^3}{2} = \rho A \frac{(10 \text{ m/s})^3}{2} = 500\rho A \text{ W}$$

Now, obtaining the value for density at sea level from Appendix 2 and matching the available wind power with the power needed,

$$500 \times 1.225 \text{ kg/m}^3 \times A = 208.33 \times 10^3 \text{ W}$$

Solving for A,

$$A = 340.1 \text{ m}^2$$

$$\text{Blade diameter} = 20.8 \text{ m}$$

The required blade length (neglecting hub diameter) will be one-half the blade diameter.

$$\text{Blade length} = 10.4 \text{ m.}$$ •

6.7 ROCKET PROPULSION

In the flow machines described so far, the energy transfers occurred to or from a stream of fluid moving through the machine. Thus, in the jet engine, a small amount of fuel, added to the stream and ignited, produces a large thrust. A rocket does not depend on the use of an external fluid stream, but carries a mass of fuel (or *propellant*) that is burned and exhausted at high speed to create thrust.

Normally, a rocket starts from rest or a uniform motion state and, while the engine is on and burning fuel, accelerates continuously. As the fuel mass is exhausted from the engine, the mass of the rocket decreases. The process and motion are unsteady and not amenable to treatment by the steady-state, impulse-momentum equations derived earlier. However, a very fundamental and useful analysis can be made by returning to the system concept.

Consider the rocket shown at time t in Fig. 6.15. It has a mass m and velocity v. For simplicity let the rocket be in the vacuum of deep space where air resistance and gravity

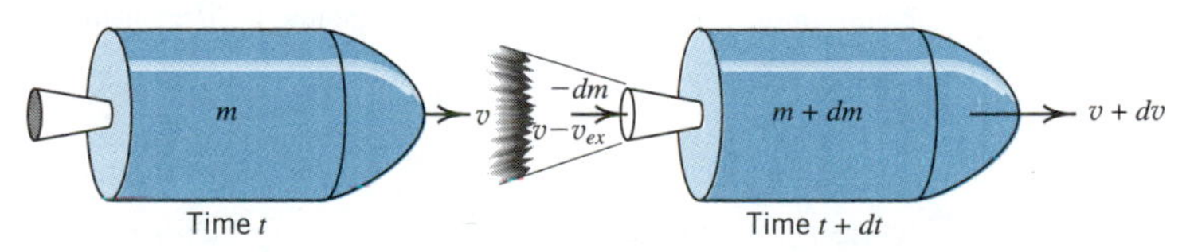

Fig. 6.15

attraction forces are negligible. There are, therefore, no external forces on the rocket. Suppose that the rocket engine is running in the interval t to $t + dt$ and exhausts a mass $(-dm)$ with an effective velocity[13] v_{ex} relative to the main rocket (as $m > 0$ and decreases with time, $dm < 0$).

Because there are no external forces acting on the system composed of the rocket frame, payload, and fuel (burned and unburned), conservation of momentum must hold for the system. Thus,

$$\Sigma(mv)_t = \Sigma(mv)_{t+dt} \tag{6.27}$$

$$mv = (m + dm)(v + dv) + (-dm)(v - v_{ex})$$

$$= mv + m\,dv + dmv_{ex} + dm\,dv$$

Canceling mv and neglecting the higher order term $dm\,dv$ yield

$$v_{ex}\,dm = -m\,dv \tag{6.28}$$

or

$$a = \frac{dv}{dt} = -v_{ex}\frac{1}{m}\frac{dm}{dt} \tag{6.29}$$

This relation can be integrated from an initial state (v_i, m_i) to obtain

$$\Delta v = v - v_i = v_{ex} \ln \frac{m_i}{m}$$

The greatest Δv obtainable is

$$\Delta v = v_{ex} \ln \frac{m_i}{m_{emp}} \tag{6.30}$$

where m_i is the initial mass of the rocket frame, fuel, and payload and m_{emp} is the mass of the empty frame and payload alone.

Because the main rocket has a mass m at time t, the thrust Th exerted by the engine is easily deduced to be

$$Th = ma = -v_{ex}\frac{dm}{dt} = v_{ex}\dot{m} \tag{6.31}$$

Here $\dot{m} = -dm/dt$ is the mass flowrate from the engine. Notice that the thrust is dependent, not on the rocket speed, but only on the effective *relative* speed of the exhaust and on the mass flowrate.

[13]Generally, to obtain the greatest exhaust velocity, rockets employ a convergent-divergent (De Laval) nozzle (Section 13.9). Thus, the exhaust pressure at the nozzle end is not always equal to the ambient or back pressure $p_a (p_a = 0$ in space). This results in a net pressure force (due to the pressure differential) that acts on the rocket over the nozzle outlet area and, accordingly, the exhaust jet must either expand or contract to reach equilibrium with the surroundings. Commonly, v_{ex} is defined, to account for this, as the effective (equilibrium) exhaust velocity

$$v_{ex} = v_o + \frac{(p_o - p_a)A_o}{\dot{m}}$$

where v_o, p_o, A_o are, respectively, the nozzle outlet velocity, pressure, and area. The term $(p_o - p_a)A_o/\dot{m}$ accounts for the effect of the pressure force on the nozzle.

REFERENCES

Rocket and Jet Propulsion

Hill, P. G., and Peterson, C. R. 1965. *Mechanics and thermodynamics of propulsion.* New York: Addison-Wesley.

Lancaster, O. E., Ed. 1959. *Jet propulsion engines.* Vol. XII in *High speed aerodynamics and jet propulsion.* Princeton: Princeton Univ. Press.

Loh, W. H. T. 1968. *Jet, rocket, nuclear, ion and electric propulsion: theory and design.* New York: Springer-Verlag.

Smith, G. G. 1950. *Gas turbines and jet propulsion for aircraft.* 5th ed. London: Illiffe.

Sutton, G. P., and Ross, D. M. 1976. *Rocket propulsion elements.* 4th ed. New York: Wiley.

Zucrow, M. J. 1958. *Aircraft and missile propulsion.* New York: Wiley.

Propellers and Windmills

Ashley, S. 1992. Turbines catch their second wind. ASME. *Mechanical Engineering.*

Eldridge, F. R. 1980. *Wind machines.* 2nd ed. New York: Van Nostrand Reinhold Co.

Glauert, H. 1947. *Elements of airfoil and airscrew theory.* 2nd ed. New York: Macmillan.

Hansen, B. C. and Butterfield, C. P. 1993. Aerodynamics of horizontal-axis wind turbines. *Ann. Rev. Fluid Mech. 25:* 115–49. Palo Alto: Annual Reviews.

Meacock, F. T. 1947. *Elements of aircraft propeller design.* London: E. and F. N. Spon.

Simmons, D. M. 1975. *Wind power.* New Jersey: Noyes Data Corp.

Taylor, D. W. 1933. *The speed and power of ships.* Rev. ed. Washington, D.C.: Ramsdell.

Theodorsen, T. 1948. *Theory of propellers.* New York: McGraw-Hill.

Torrey, W. 1976. *Wind-catchers.* Brattleboro, Vt.: S. Greene Press.

PROBLEMS

6.1. A horizontal 150 mm pipe, in which 62 l/s of water are flowing, contracts to a 75 mm diameter. If the pressure in the 150 mm pipe is 275 kPa, calculate the magnitude and direction of the horizontal force exerted on the contraction.

6.2. A horizontal 50 mm pipe, in which 1 820 l/min of water are flowing, enlarges to a 100 mm diameter. If the pressure in the smaller pipe is 138 kPa, calculate magnitude and direction of the horizontal force on the enlargement.

6.3. A conical enlargement in a vertical pipeline is 5 ft long and enlarges the pipe from 12 in. to 24 in. diameter. Calculate the magnitude and direction of the vertical force on this enlargement when 10 cfs of water flow upward through the line and the pressure at the smaller end of the enlargement is 30 psi.

6.4. A conical diverging tube is horizontal, 0.3 m long, has 75 mm throat diameter, 100 mm exit diameter, and discharges 28.3 l/s of water into the atmosphere. Calculate the magnitude and direction of the force components exerted by the water on the tube.

6.5. A 100 mm nozzle is bolted (with 6 bolts) to the flange of a 300 mm horizontal pipeline and discharges water into the atmosphere. Calculate the tension load on each bolt when the pressure in the pipe is 600 kPa. Neglect vertical forces.

6.6. For the nozzle of problem 6.5 what flowrate will produce a tension force of 7 kN in each bolt?

6.7. Calculate the force exerted by the water on this orifice plate. Assume that water in the jet between orifice plate and vena contracta weighs 4.0 lb.

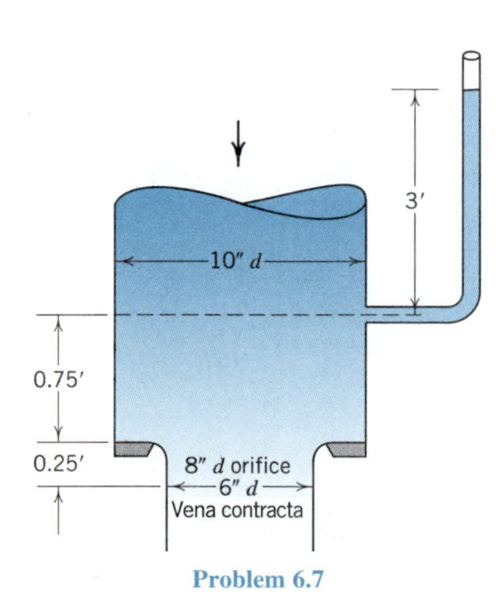

Problem 6.7

6.8. The projectile partially fills the end of the 0.3 m pipe. Calculate the force required to hold the projectile in position when the mean velocity in the pipe is 6 m/s.

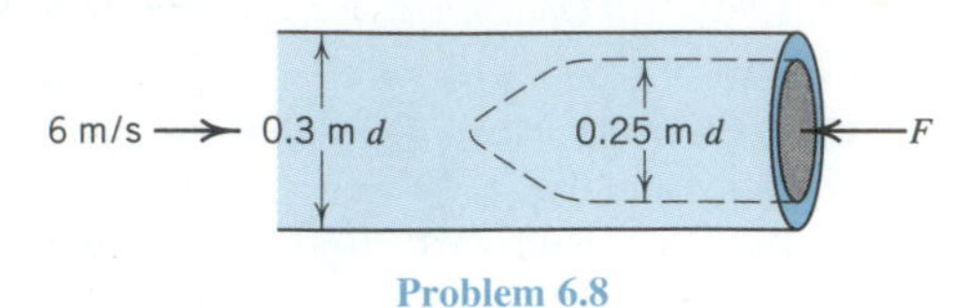

Problem 6.8

6.9. This ''needle nozzle'' discharges a free jet of water at a velocity of 30 m/s. The tension force in the stem is measured experimentally and found to be 4 448 N. Predict the horizontal force on the bolts.

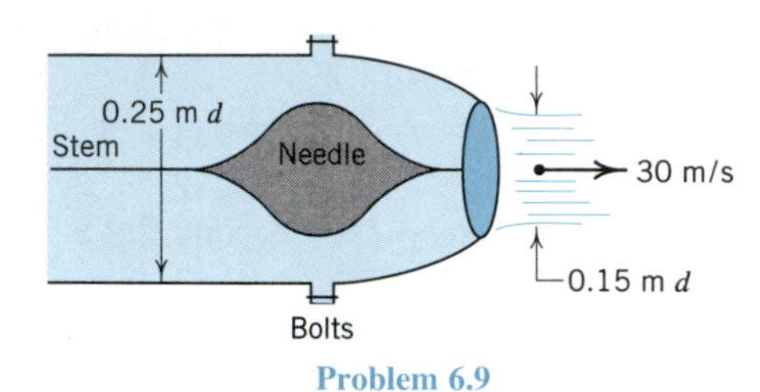

Problem 6.9

6.10. A 90° bend occurs in a 0.3 m horizontal pipe in which the pressure is 276 kPa. Calculate the magnitude and direction of the horizontal force on the bend when 0.28 m^3/s of water flow therein.

6.11. A 6 in. horizontal pipeline bends through 90° and while bending changes its diameter to 3 in. The pressure in the 6 in. pipe is 30 psi. Calculate the magnitude and direction of the horizontal force on the bend when 2.0 cfs of water flow therein. Both pipes are in the same horizontal plane.

6.12. A 100 mm by 50 mm 180° pipe bend lies in a horizontal plane. Find the horizontal force of the water on the bend when the pressures in the 100 mm and 50 mm pipes are 105 kPa and 35 kPa, respectively.

6.13. Calculate the force on the bolts. Water is flowing.

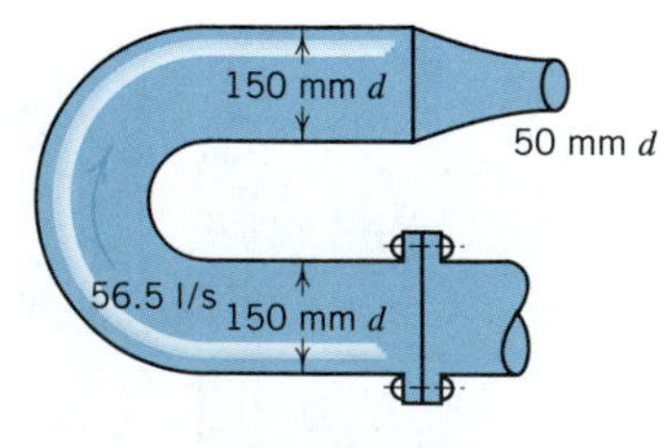

Problem 6.13

6.14. The axes of the pipes are in a vertical plane. The flowrate is 2.83 m^3/s of water. Calculate the magnitude, direction, and location of the resultant force of the water on the pipe bend.

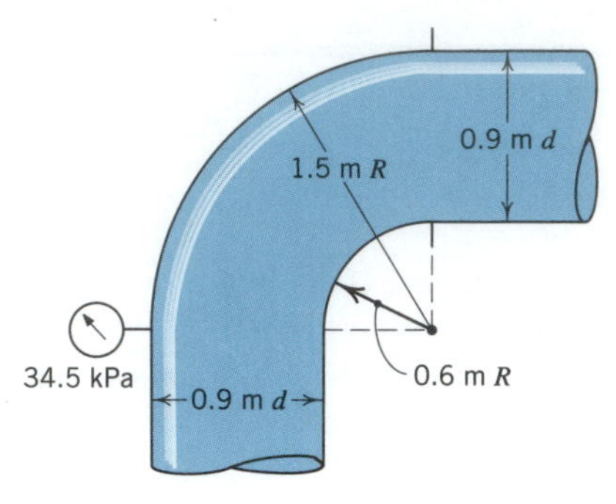

Problem 6.14

6.15. Water flows through a tee in a horizontal pipe system. The velocity in the stem of the tee is 15 ft/s, and the diameter is 12 in. Each branch is of 6 in. diameter. If the pressure in the stem is 20 psi, calculate magnitude and direction of the force of the water on the tee if the flowrates in the branches are the same.

6.16. Two types of gasoline are blended by passing them through a horizontal ''wye'' as shown. Calculate the magnitude and direction of the force exerted on the ''wye'' by the gasoline. The pressure p_3 = 145 kPa.

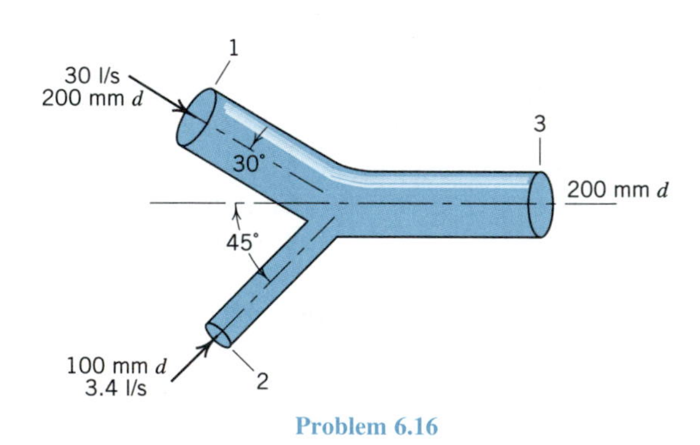

Problem 6.16

6.17. If the two pipes from the reservoirs of problem 5.42 join through the (unequally sized) branches of a horizontal tee and the discharge pipe leaves along the stem of the tee, find the magnitude and direction of the force of the water on the tee. The tee is at an elevation of 25 m (82 ft).

6.18. A nozzle of 50 mm tip diameter discharges 0.018 7 m^3/s of water vertically upward. Calculate the volume of water in the jet between the nozzle tip and a section 3.6 m above this point.

6.19. The block weighs 1 lb and is held up by the water jet issuing from the nozzle. Calculate H for a flowrate of 0.054 54 cfs. Ignore the small quantity of water above plane AA.

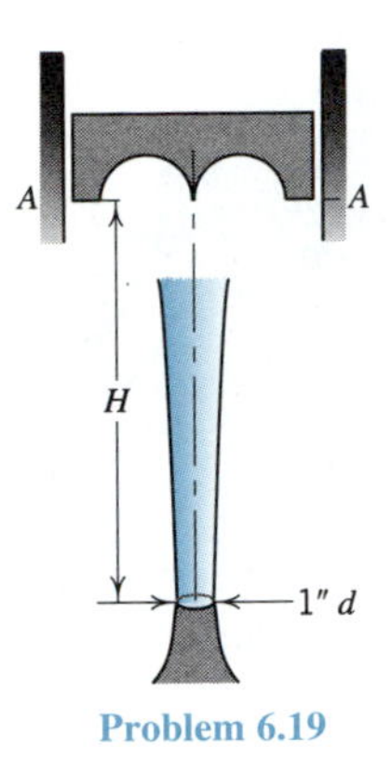

Problem 6.19

6.20. When round jets of the same velocity meet head on this flow picture results. Derive (ignoring gravity effects) an expression for θ in terms of d_1 and d_2.

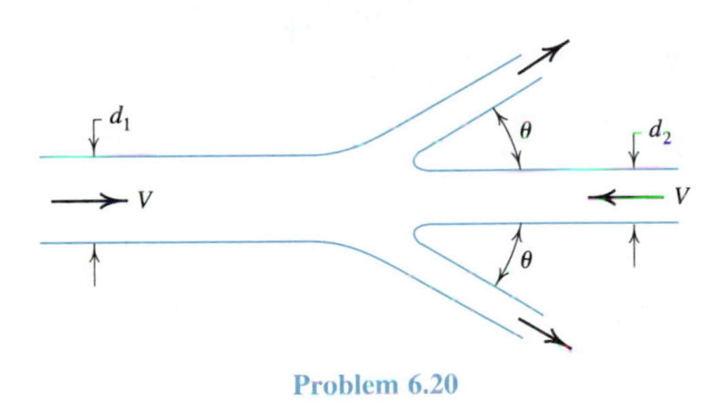

Problem 6.20

6.21. The lower tank weighs 224 N, and the water in it weighs 897 N. If this tank is on a platform scale, what weight will register on the scale beam?

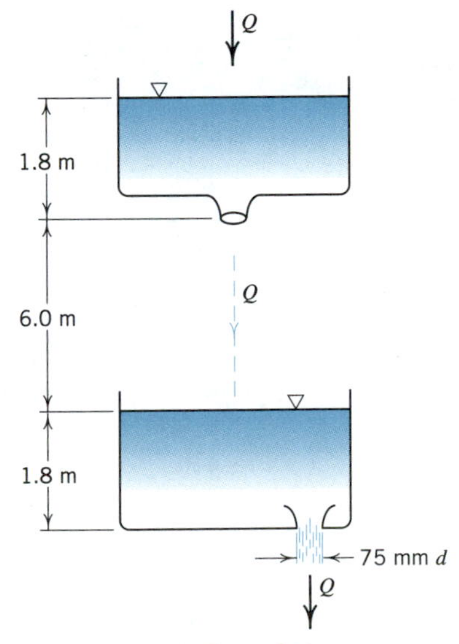

Problem 6.21

6.22. A free jet issues form this ''Borda orifice'' into the atmosphere. Calculate d/D.

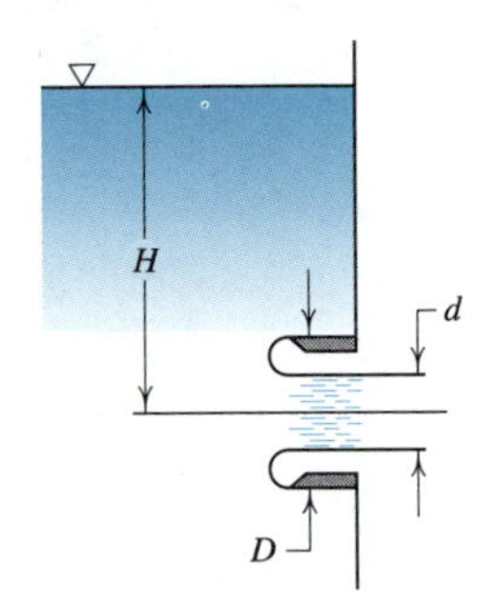

Problem 6.22

6.23. The pressure difference results from head loss caused by eddies downstream from the orifice plate. Wall friction is negligible. Calculate the force exerted by the water on the orifice plate. The flowrate is 7.86 cfs.

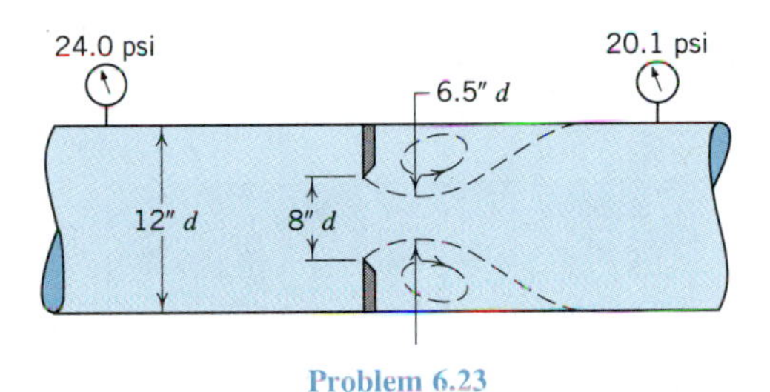

Problem 6.23

6.24. The pump, suction pipe, discharge pipe, and nozzle are all welded together as a single unit. Calculate the horizontal component of force (magnitude and direction) exerted by the water on the unit when the pump is developing a head of 22.5 m.

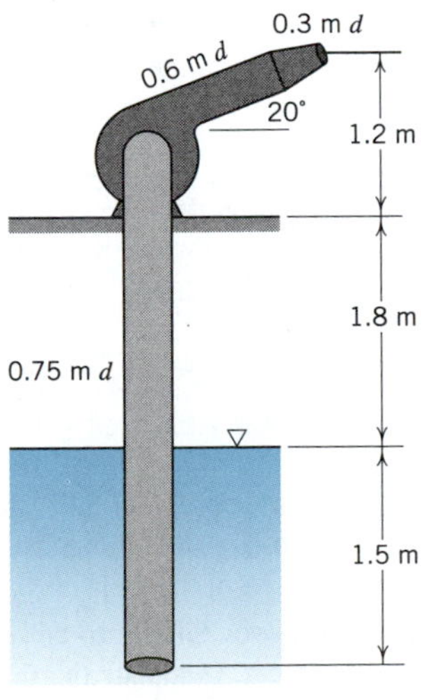

Problem 6.24

6.25. When the pump is started, strain gages at A and B indicate longitudinal tension forces in the pipes of 23 and 100 lb, respectively. Assuming a frictionless system, calculate flowrate and pump horsepower.

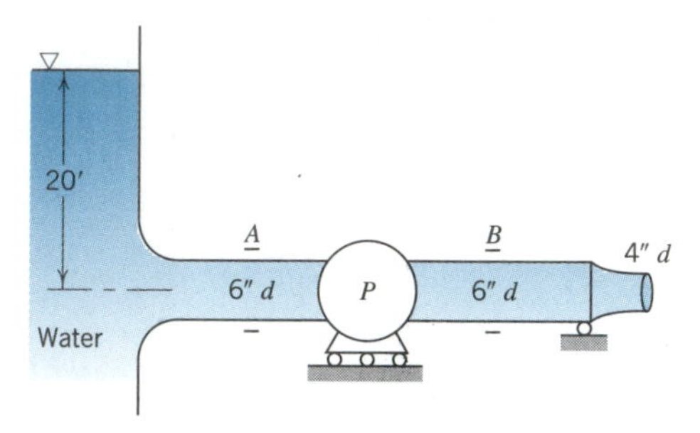

Problem 6.25

6.26. A sphere of 50 mm diameter placed on the centerline of a water (40°C) tunnel produces a vapor cavity as shown. The cavity occurs in such high speed flows because the pressure behind the body is reduced to the vapor pressure. Calculate the force exerted by water and vapor on the sphere if the velocity at section 1 is 30 m/s.

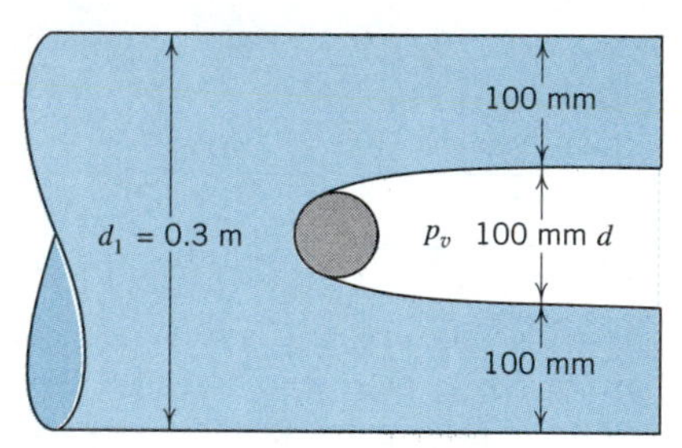

Problem 6.26

6.27. Calculate the drag on the streamlined axisymmetric body. The velocity defect downstream of the body marks the *wake* caused by friction effects on the body.

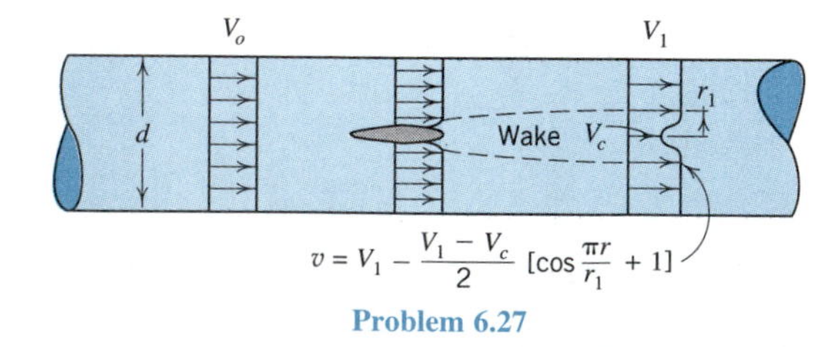

Problem 6.27

6.28. Referring to the sketch of problem 5.88, prove that the thickness (measured *vertically*) of the falling sheet of water is constant for all locations far from the weir crest.

6.29. The flowrate passing over this sharp-crested weir in a channel of 1 ft width is 3.5 cfs. Calculate the magnitude and direction of the force exerted by the water on the weir plate.

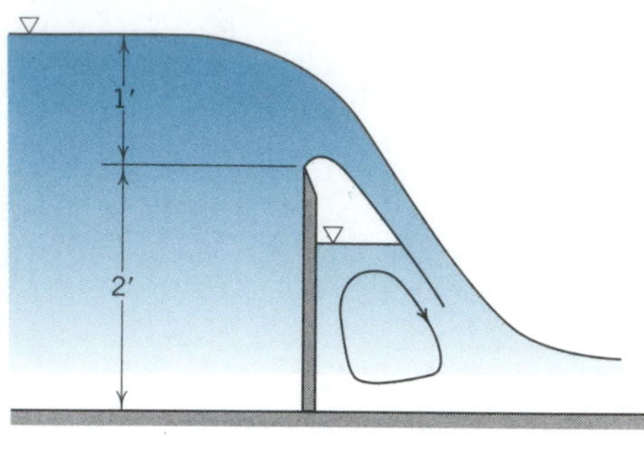

Problem 6.29

6.30. The passage is 1.2 m wide normal to the paper. What will be the horizontal component of force exerted by the water on the structure?

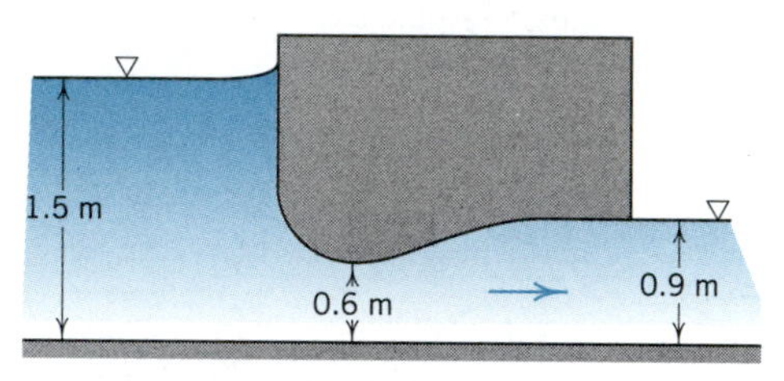

Problem 6.30

6.31. Find the force exerted by the flowing water on the 10-ft-wide gate. Can you calculate the force R necessary to keep the gate in equilibrium? Why or why not? Neglect friction.

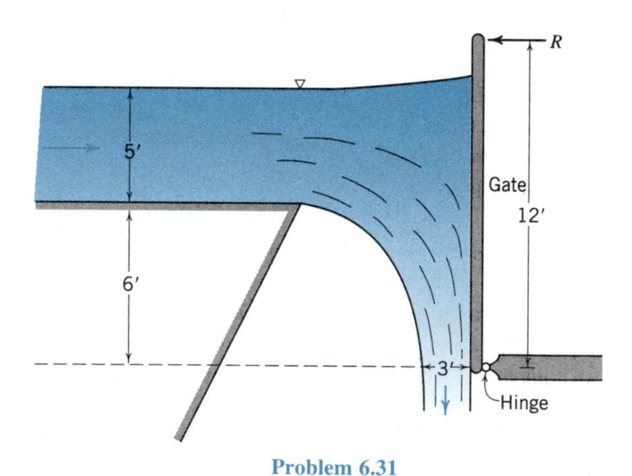

Problem 6.31

6.32. Flow occurs over the spillway shown. Determine the magnitude and direction of the horizontal force exerted by the water on the spillway. Assume the water is an ideal fluid. The spillway is 10 ft wide.

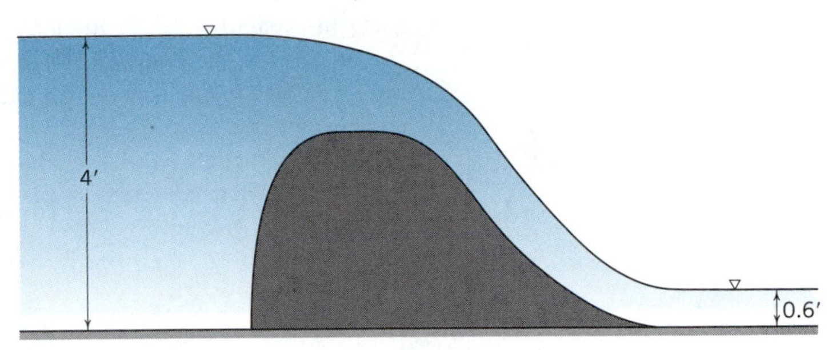

Problem 6.32

6.33. If the two-dimensional flowrate through this sluice gate is 50 cfs/ft, calculate the horizontal and vertical components of force on the gate, neglecting wall friction.

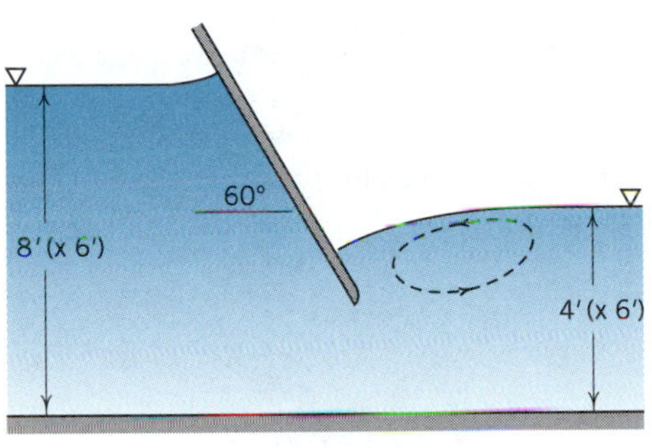

Problem 6.33

6.34. Calculate the magnitude and direction of the horizontal component of force exerted by the flowing water on this (hatched) outflow structure. Assume velocity distribution uniform where streamlines are straight and parallel.

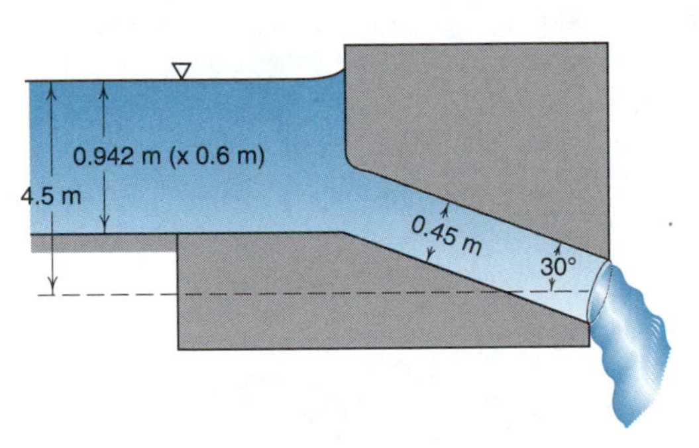

Problem 6.34

6.35. Calculate the horizontal component of force exerted by the water on this "submerged sluice gate." The pressure distribution at section 2 *may be assumed hydrostatic*. Between sections 2 and 3 head losses are large because of diffusion and roller. All wall and bottom friction may be neglected. Consider the flow field two-dimensional.

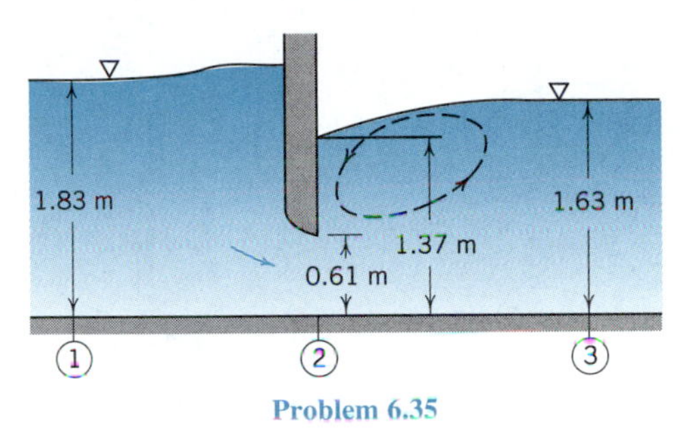

Problem 6.35

6.36. Flow from the end of a two-dimensional open channel is deflected vertically downward by the gate *AB*. Calculate the force exerted by the water on the gate. At (and downstream from) *B* the flow may be considered a free jet.

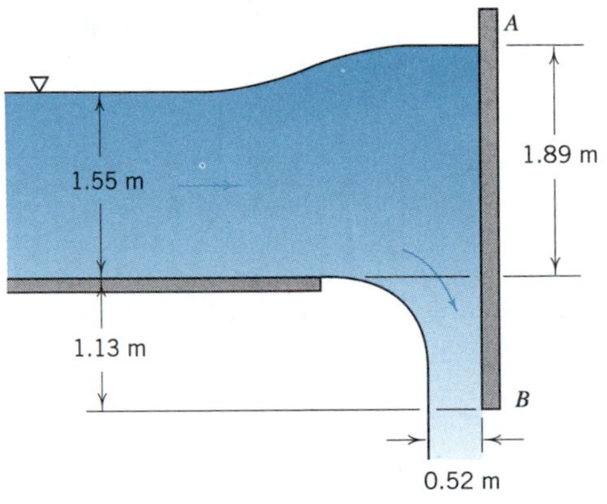

Problem 6.36

6.37. Calculate the magnitude and direction of the horizontal force exerted by the water on the frictionless "drop structure" *AB*. Assume the structure to be 1 ft wide normal to the paper.

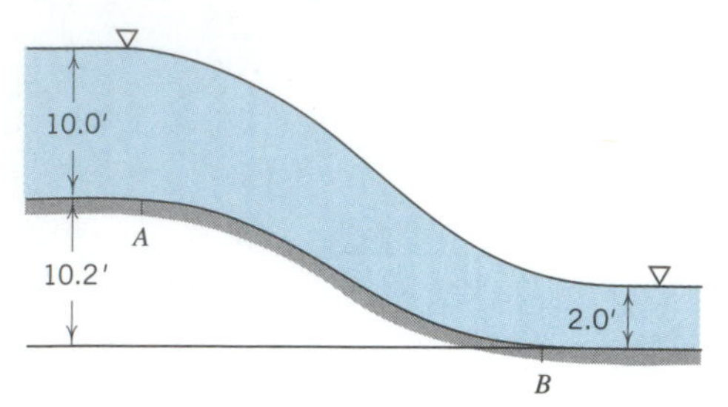

Problem 6.37

6.38. Calculate the magnitude and direction of the horizontal and vertical components of resultant force exerted by the flowing water on the "flip bucket" *AB*. Assume that the water between sections *A* and *B* weighs 2.69 kN and that downstream from *B* the moving fluid may be considered to be a free jet.

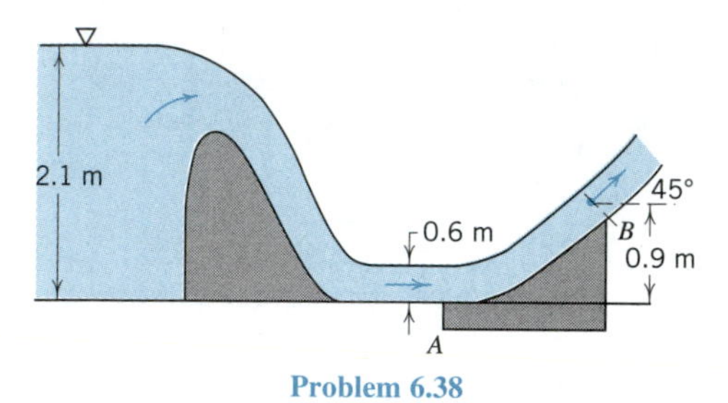

Problem 6.38

6.39. Calculate F_x exerted by the water on the block which has been placed at the end of this horizontal open channel. The channel and block are 4 ft wide normal to the paper.

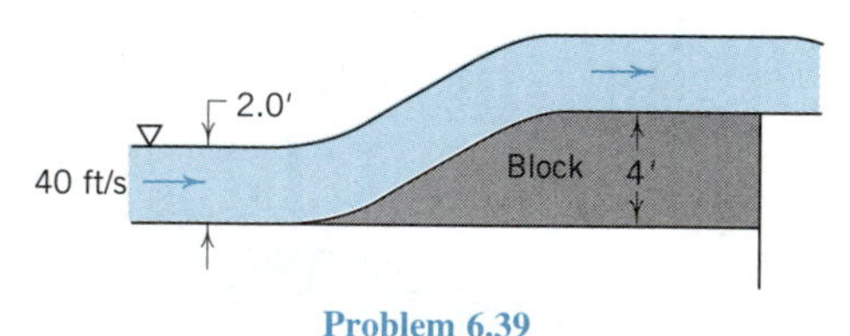

Problem 6.39

6.40. This two-dimensional overflow structure (gray) at the end of an open channel produces a free jet as shown. The water depth in the channel is 1.5 m and the thickness of the jet at the top of its trajectory is 0.6 m. Predict the horizontal component of force by water on structure if the structure is 3 m wide normal to the plane of the paper.

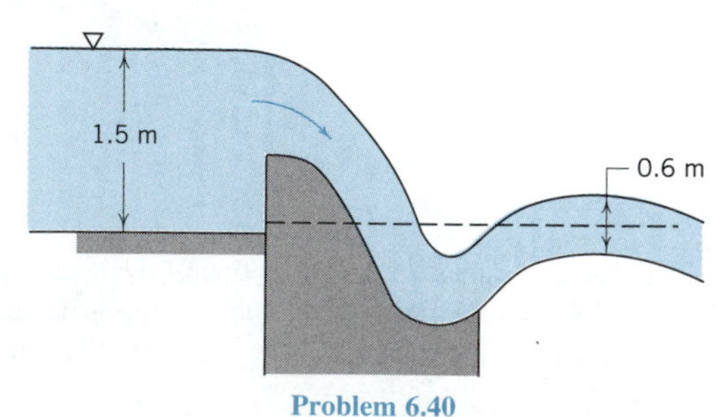

Problem 6.40

6.41. In sketch (*a*) the "cylinder gate" is closed. In sketch (*b*) the gate has been raised 2 ft and the upstream water depth increased 2 ft. Calculate the magnitudes of the horizontal force components by water on gate. Compare the magnitudes of the vertical components. Which of these latter will be the larger? Why? Will both resultant forces pass through the center of the gate? Why or why not?

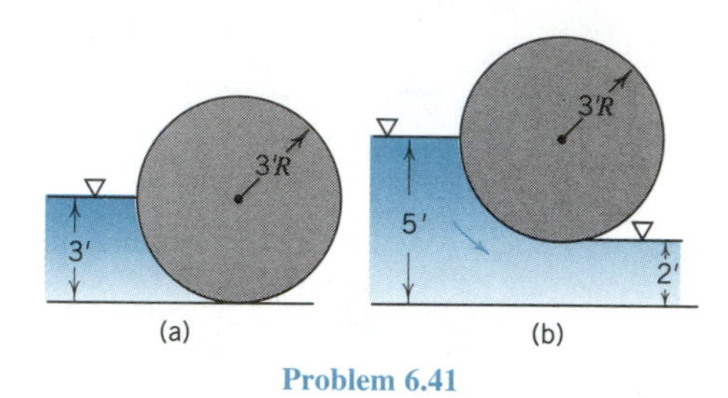

Problem 6.41

6.42. Calculate the force, *F*, required to drive the scoop (gray) at such velocity that the top of the jet centerline is 3 m above the channel bottom. Assume the scoop and channel 1.5 m wide normal to the paper and that the scoop extracts all of the water from the channel.

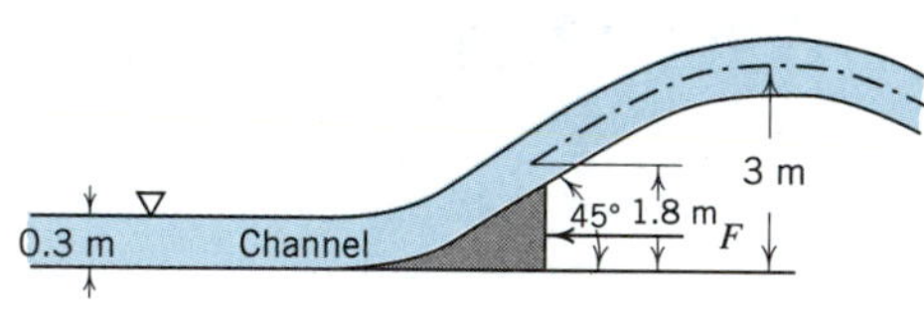

Problem 6.42

6.43. Upstream from an axisymmetric abrupt enlargement in a horizontal passage the mean pressure is 140 kPa (20 psi) and the diameter is 150 mm (6 in.). Downstream from the enlargement the mean pressure is 210 kPa (30 psi) and the diameter is 300 mm (12 in.). Estimate the flowrate through the passage and the force on the enlargement.

6.44. Find the pressure change in and force on the abrupt contraction. $Q = 0.2\ \text{m}^3/\text{s}$, $A_1 = 0.1\ \text{m}^2$, $A_2/A_1 = 0.4$, $p_1 = 200$ kPa, $\Delta(\text{EL}) = 0.3\ V_2^2/2g_n$ (see Section 9.9), and water (5°C) is flowing.

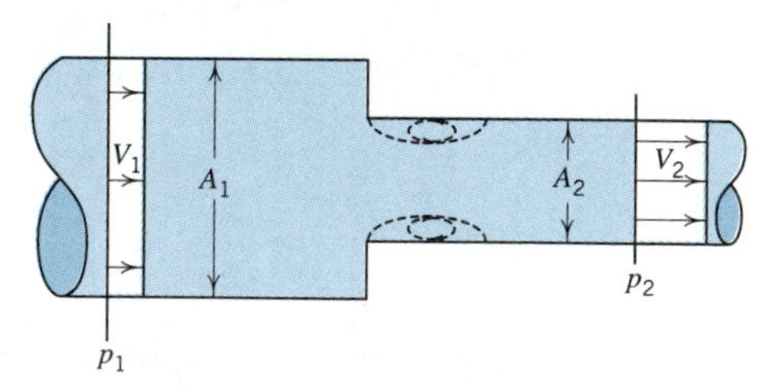

Problem 6.44

6.45. Find the horizontal force exerted on the overflow-diversion structure shown between sections 1, 2, and 3. The

velocity at section 1 is 10 ft/s and the structure is 10 ft wide. Neglect friction.

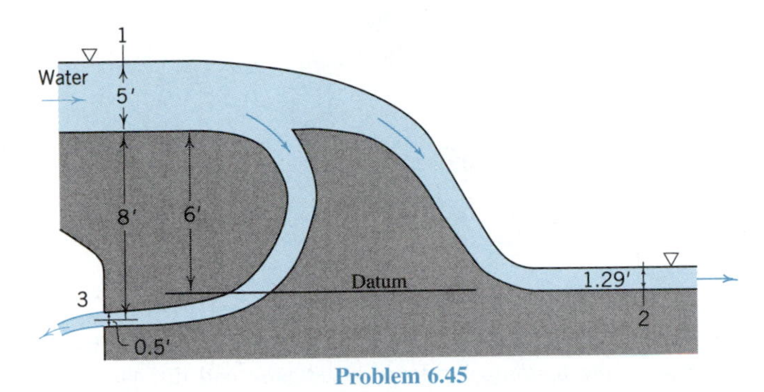

Problem 6.45

6.46. Derive equations for the pressure change and energy loss in the abrupt enlargement. Compare the results with those of Section 6.2. The velocity profiles are parabolas, representing a laminar flow.

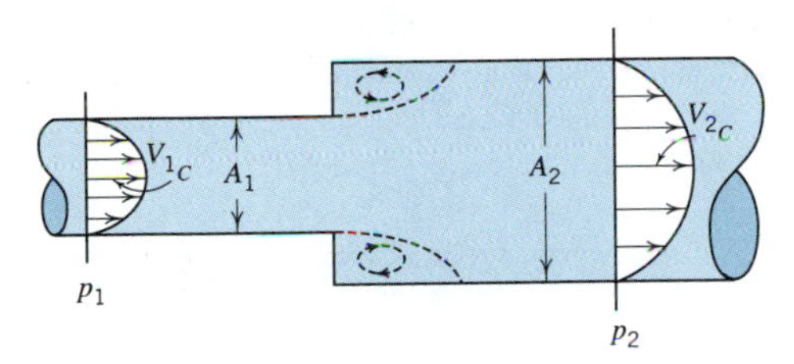

Problem 6.46

6.47. At a point in a rectangular channel 6 m (20 ft) wide just downstream from a hydraulic jump, the depth is observed to be 3.6 m (12 ft) when the flowrate is 60 m^3/s (2 000 ft^2/s). What were the depth and velocity just upstream from the jump?

6.48. A hydraulic jump is observed in an open channel 3 m (10 ft) wide. The approximate depths (such depths are very difficult to measure accurately) upstream and downstream from the jump are 0.6 m (2 ft) and 1.5 m (5 ft), respectively. Estimate the flowrate in the channel.

6.49. This corrugated ramp is used as an energy dissipator in a two-dimensional open channel flow. For a flowrate of 5.4 $m^3/s{\cdot}m$ calculate the head lost, the power dissipated, and the horizontal component of force exerted by the water on the ramp.

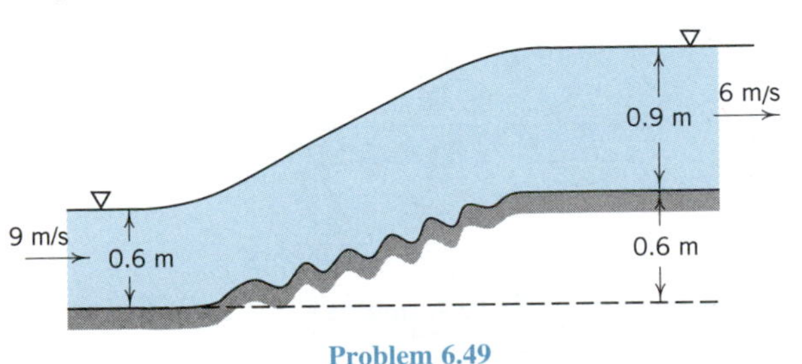

Problem 6.49

6.50. In an open channel 1.5 m (5 ft) wide, the depth of water is 0.3 m (1 ft). Small standing waves (caused by imperfections in the sidewalls) are observed on the water surface at an angle of 30° with the walls. Estimate the flowrate in the channel.

6.51. Calculate the drop in the energy line (the head loss) produced by the oblique wave described in Illustrative Problem 6.5.

6.52. In an open rectangular channel 3 m (10 ft) wide the water depth is 0.3 m (1 ft). A standing wave 0.15 m (0.5 ft) high is produced by narrowing the channel with a straight diagonal wall. The angle between the wave and the original channel wall is observed to be 60°. Estimate the flowrate in the channel. What must be the angle of the diagonal wall to produce such a wave?

6.53. For the configuration of problem 6.13, calculate the torque about the pipe's centerline in the plane of the bolted flange that is caused by the flow through the nozzle. The nozzle centerline is 0.3 m above the flange centerline. What is the effect of this torque on the force on the bolts calculated in problem 6.13? Neglect the effects of the weights of the pipe and the fluid in the pipe.

6.54. A fire truck is equipped with a 20 m (66 ft) long extension ladder which is attached at a pivot and raised to an angle of 45°. A 100 mm (4 in.) diameter fire hose is laid up the ladder and a 50 mm (2 in.) diameter nozzle is attached to the top of the ladder so that the nozzle directs the stream horizontally into the window of a burning building. If the flowrate is 30 l/s (1 ft^3/s), compute the torque exerted about the ladder pivot point. The ladder, hose, and the water in the hose weigh about 150 N/m (10 lb/ft).

6.55. Calculate the torque exerted on the flange joint by the fluid flow as a function of the pump flowrate. Neglect the weight of the 100 mm diameter pipe and the fluid in the pipe.

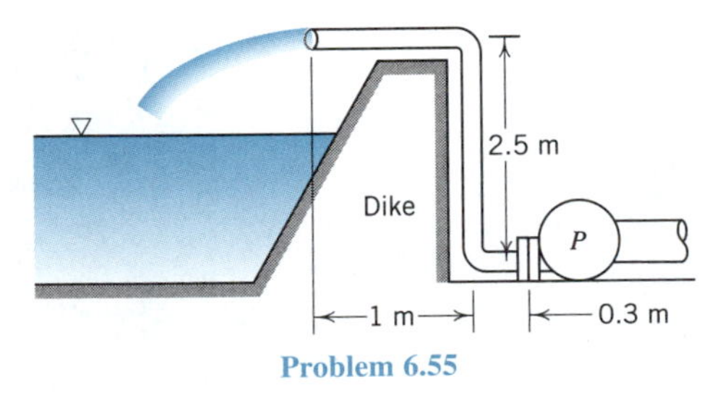

Problem 6.55

6.56. A 16 l/s horizontal jet of water of 25 mm diameter strikes a stationary blade which deflects it 60° from its original direction. Calculate the vertical and horizontal components of force exerted by the liquid on the blade.

6.57. The jet shown strikes the semicylindrical vane, which is in the vertical plane. Calculate the horizontal component of force on the vane. Neglect all friction.

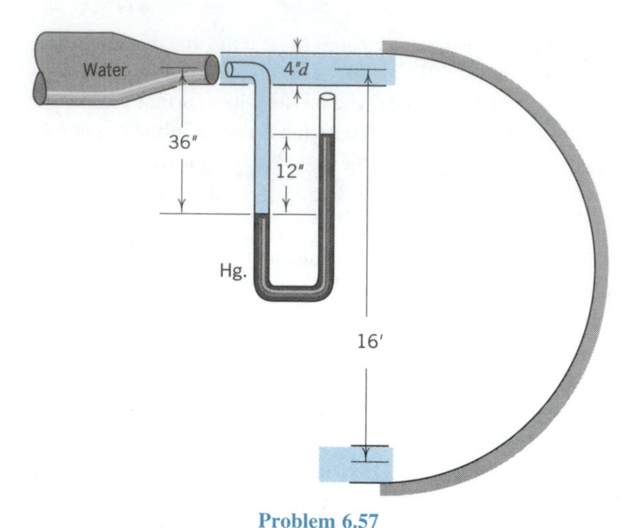

Problem 6.57

6.58. Calculate the resultant force of the water on this plate. The flowrate is 42.5 l/s.

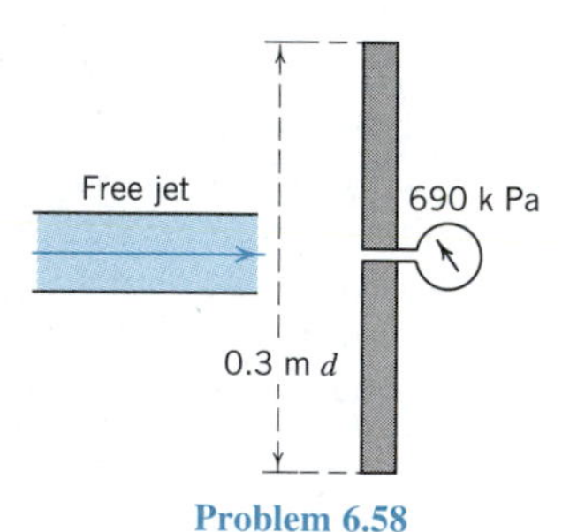

Problem 6.58

6.59. The plate covers the 125 mm diameter hole. What is the maximum H that can be maintained without leaking?

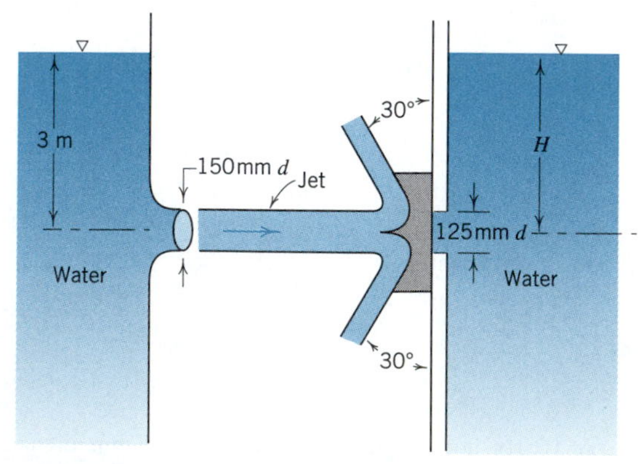

Problem 6.59

6.60. Calculate the magnitude and direction of the vertical and horizontal components and the total force exerted on this stationary blade by a 50 mm jet of water moving at 15 m/s.

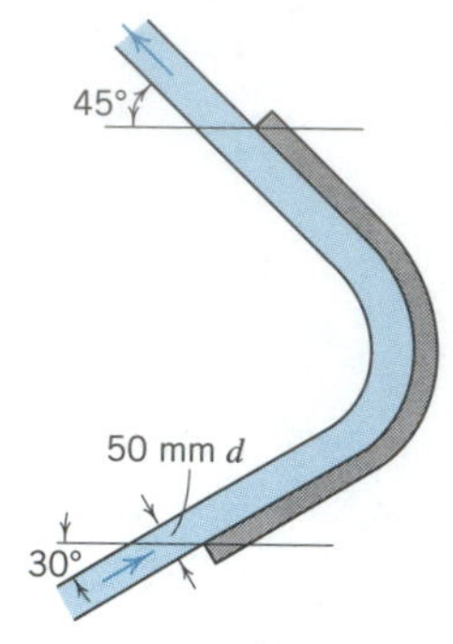

Problem 6.60

6.61. A smooth nozzle of 50 mm (2 in.) tip diameter is connected to the bottom of a large water tank and discharges vertically downward. The water surface in the tank is 1.5 m (5 ft) above the tip of the nozzle. Three metres (10 ft) below the nozzle tip there is a horizontal disk of 100 mm (4 in.) diameter surface. Calculate the force exerted by the water jet on the disk.

6.62. A jet falls from a reservoir into a weightless dish that floats on the surface of a second reservoir. If the submerged volume of the dish is 0.12 m^3, calculate the weight of the water within the dish (i.e., below the line AB).

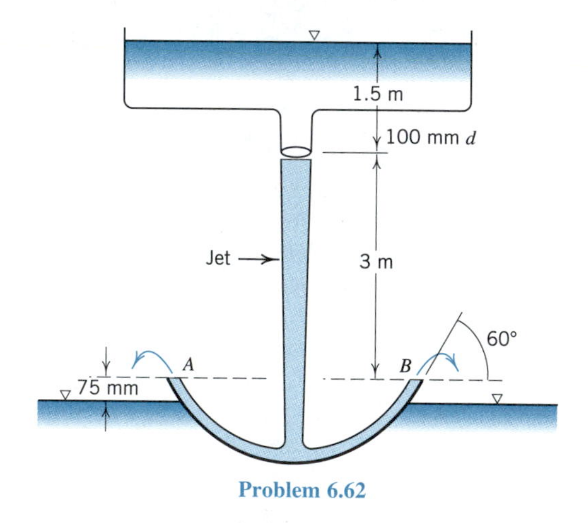

Problem 6.62

6.63. This water jet of 50 mm diameter moving at 30 m/s is divided in half by a "splitter" on the stationary flat plate. Calculate the magnitude and direction of the force on the plate. Assume that flow is in a horizontal plane.

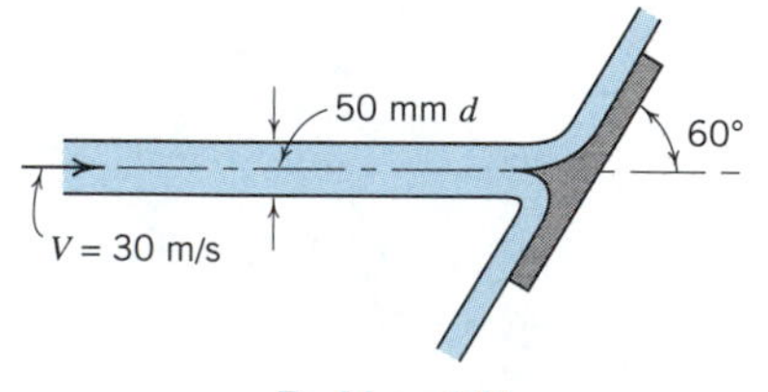

Problem 6.63

6.64. If the splitter is removed from the plate of the preceding problem and sidewalls are provided on the plate to keep the flow two-dimensional, how will the jet divide after striking the plate?

6.65. Calculate the gage reading when the plate is pushed horizontally to the left at a constant speed of 25 ft/s. Also calculate the force and power required to push the plate at this speed.

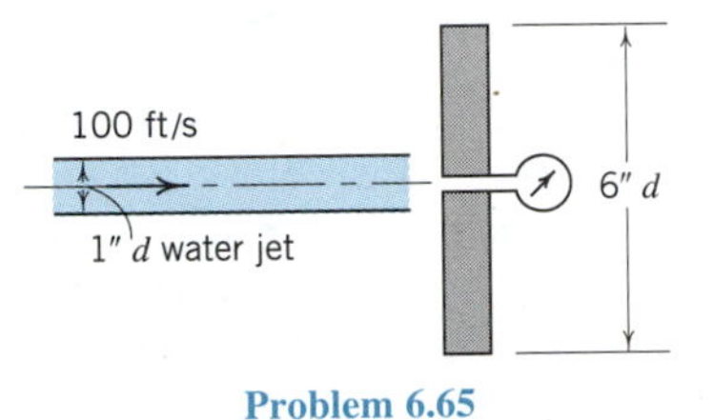

Problem 6.65

6.66. The pressure in the pipe at section 1 is 30 psi and the cross-sectional areas of the pipe at sections 1 and 2 are 0.20 ft^2 and 0.05 ft^2, respectively. Compute the vertical component of the force exerted by the water on the deflecting vane. Neglect viscous forces and the weight of the water passing over the vane.

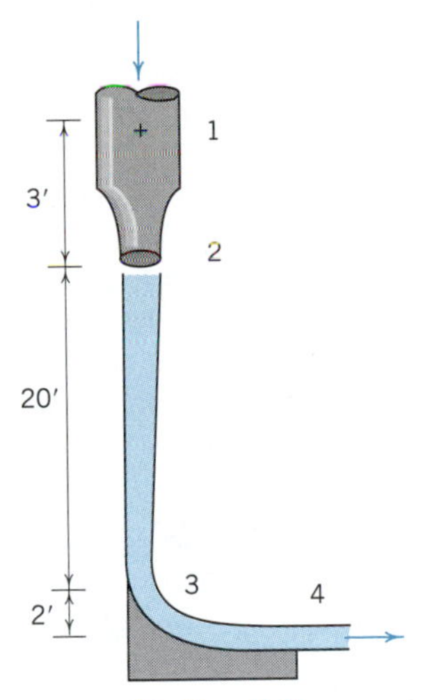

Problem 6.66

6.67. The jet of oil (s.g. = 0.90) falls 20 ft from the end of a vertical pipe and strikes a splitter plate. One half of the jet is deflected horizontally and the other half continues to fall vertically. What is the vertical component of the force necessary to hold the splitter plate in position? Neglect friction, the weight of the fluid in contact with the splitter, and show the control volume with all forces and velocities labeled.

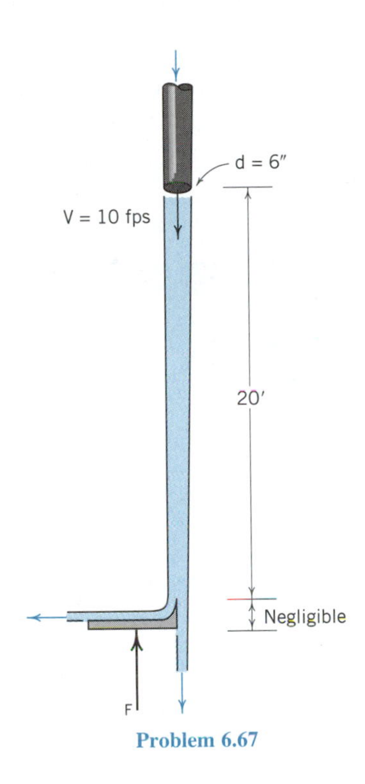

Problem 6.67

6.68. The gray structure deflects a jet of water as shown. The pressure at *a* is 80 psi. Find the magnitude and direction of the horizontal component of the force exerted *on the structure* by the deflected jet.

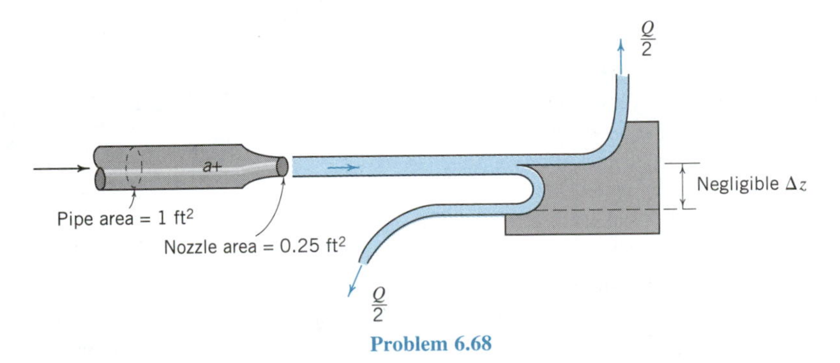

Problem 6.68

6.69. Calculate the thrust and power of the jet engine of Illustrative Problem 6.7, as functions of air speed for the given nozzle conditions. Plot the results and find the point of maximum power delivery.

6.70. At high speeds the compressor and turbine of the jet engine may be eliminated entirely. The result is called a *ramjet* (a subsonic configuration is shown). Here the incoming air is slowed and the pressure increases; the air is heated in the widest part by the burning of injected fuel. The heated air exhausts at high velocity from the converging nozzle. What nozzle area A_2 is needed to deliver a 90 kN thrust at an air speed of 270 m/s if the exhaust velocity is the sonic velocity for the heated air, which is at 1 000 K. Assume that the jet operates at an altitude of 12 km and neglect the fuel mass and pressure differentials.

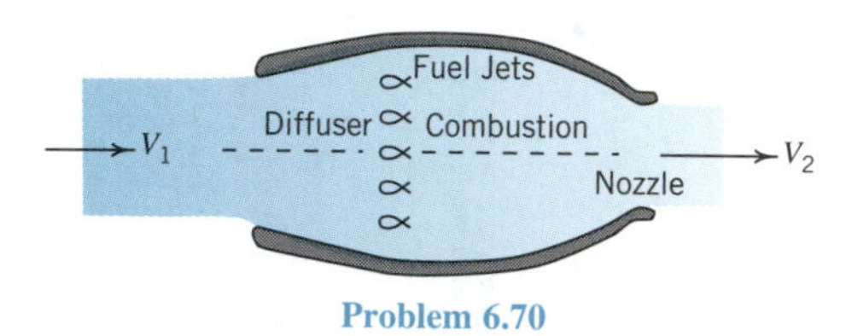

Problem 6.70

6.71. A horizontal convergent-divergent nozzle is bolted to a water tank from which it discharges under a 3.6 m head. The throat and tip diameters of the nozzle are 125 mm and 150 mm, respectively. Determine the net horizontal force exerted on the tank-and-nozzle combination by the flowing fluid.

6.72. The pump maintains a pressure of 10 psi at the gage. Calculate the tension force in the cable.

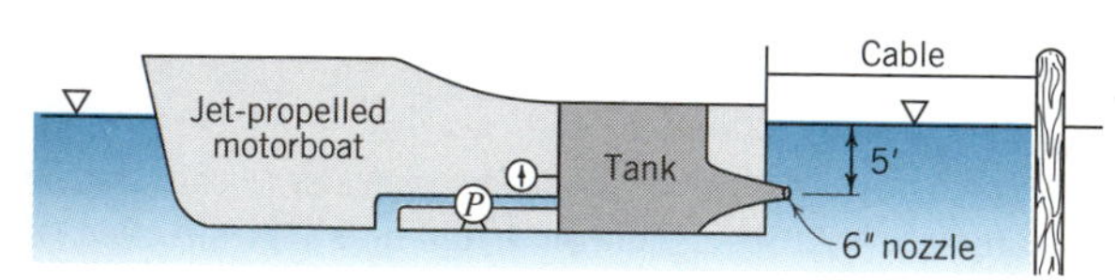

Problem 6.72

6.73. A motorboat moves up a river at a speed of 9 m/s (30 ft/s) (relative to the land). The river flows at a velocity of 1.5 m/s (5 ft/s). The boat is powered by a jet-propulsion unit which takes in water at the bow and discharges it (beneath the surface) at the stern. Measurements in the jet show its velocity (relative to the boat) to be 18 m/s (60 ft/s). For a flowrate through the unit of 0.15 m³/s (5.3 ft³/s), calculate the propulsive force produced.

6.74. A head of 1.8 m is maintained on the nozzle. What is the propulsive force: (*a*) on the car, (*b*) on the blade, (*c*) on the tank and nozzle?

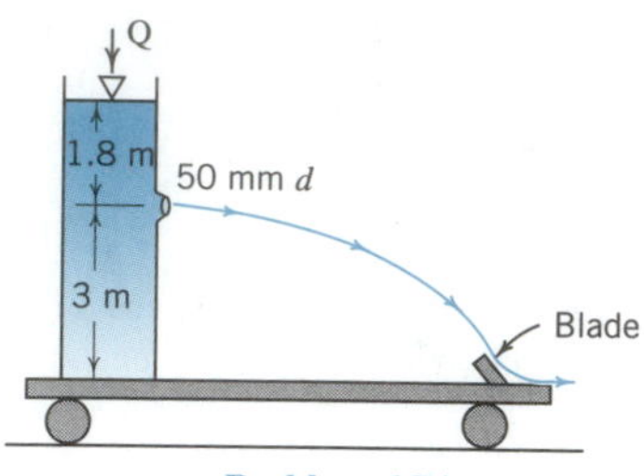

Problem 6.74

6.75. A ship moves up a river at 32 km/h (20 mph) relative to the shore. The river current has a velocity of 8 km/h (5 mph). The velocity of the water a short distance behind the propellers is 64 km/h (40 mph) relative to the ship. If the velocity of 2.8 m³/s (100 cfs) of water is changed by the propeller, calculate its thrust.

6.76. An airplane flies at 200 km/h through still air of specific weight 12 N/m³. The propeller is 2.4 m in diameter, and its slipstream has a velocity of 290 km/h relative to the fuselage. Calculate: (*a*) the propeller efficiency, (*b*) the velocity through the plane of the propeller, (*c*) the power input, (*d*) the power output, (*e*) the thrust of the propeller, and (*f*) the pressure difference across the propeller disk.

6.77. A propeller must produce a thrust of 9 kN to drive an airplane at 280 km/h. An ideal propeller of what size must be provided if it is to operate at 90% efficiency? Assume air density 1.23 kg/m³.

6.78. A "flying platform" is supported by four downward-directed jets, one at each corner of the platform for stability. The platform is to support a gross load of 10 kN (2 250 lb). What flowrate of standard sea-level air must be delivered by each fan? How much power must be supplied by each motor? Consider the air to be frictionless and incompressible. Assume the platform to be stationary and well off the ground.

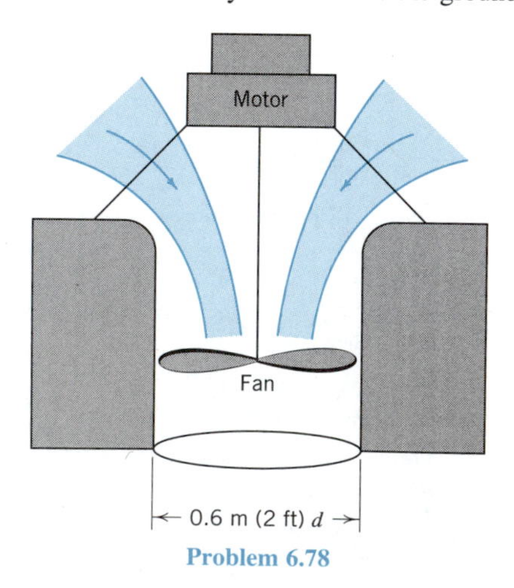

Problem 6.78

6.79. This ducted propeller unit drives a ship through still water at a speed of 4.5 m/s. Within the duct the mean velocity of the water (relative to the unit) is 15 m/s. Calculate the propulsive force produced by the unit. Calculate the force exerted on the fluid by the propeller. Account for the difference between these forces.

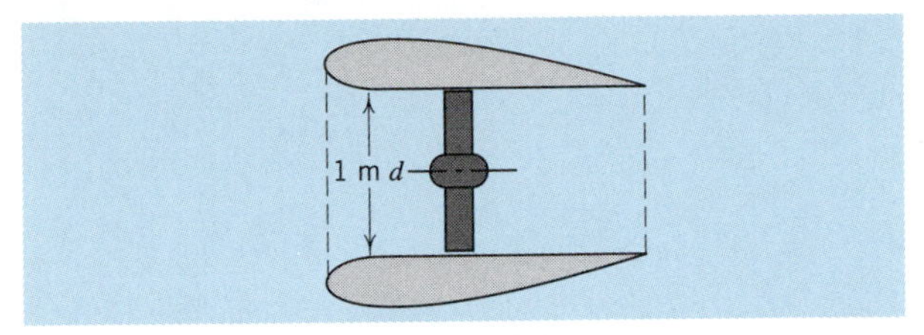

Problem 6.79

6.80. Show that this ducted propeller system when moving forward at velocity V_1 will have an efficiency given by $2V_1/(V_4 + V_1)$. If for a specific design and point of operation, $V_2/V_1 = 9/4$ and $V_4/V_2 = 5/4$, what fraction of the propulsive force will be contributed: (*a*) by the propeller, and (*b*) by the duct?

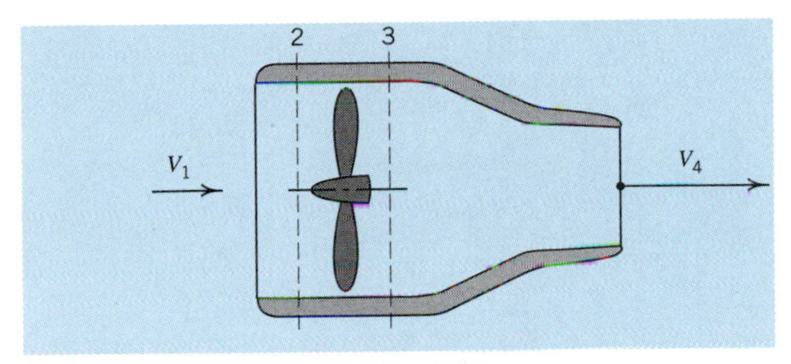

Problem 6.80

6.81. This ducted propeller unit (now operating as a turbine) is towed through still water at a speed of 7.5 m/s. Calculate the maximum power that the propeller can develop. Neglect all friction effects.

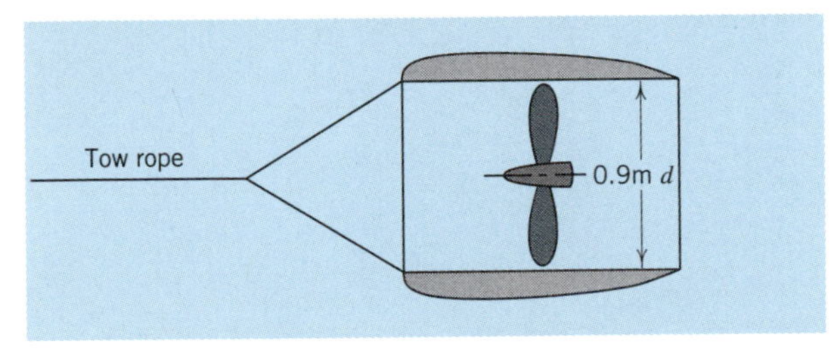

Problem 6.81

6.82. A propeller type of turbine wheel is installed near the end of a horizontal pipeline discharging water to the atmosphere. Derive an expression for the power to be expected from the turbine in terms of the gage pressure upstream from the turbine disk, and the flowrate.

6.83. What is the maximum power that can be expected from a windmill 30 m (100 ft) in diameter in a wind of 50 km/h (30 mph)? Assume air density 1.225 kg/m^3 (0.002 38 slug/ft^3).

6.84. If an ideal windmill is operating at best efficiency in a wind of 48 km/h, what is the velocity through the disk and at some distance behind the windmill? What is the thrust on this windmill, assuming a diameter of 60 m and an air density of 1.23 kg/m^3? What are the mean pressures just ahead of and directly behind the windmill disk?

6.85. The flowrate through a 3 m (10 ft) diameter windmill is 170 m^3/s (6 000 ft^3/s). The mean pressures just upstream and downstream from the windmill plane are 240 Pa (5 lb/ft^2) and −190 Pa (−3.9 lb/ft^2), respectively. Assuming the air density 1.3 kg/m^3 (0.002 5 slug/ft^3), calculate the wind velocity, the axial force on the windmill, and the power of the machine.

6.86. Transverse thrusters are used to make large ships fully maneuverable at low speeds without tugboat assistance. A transverse thruster consists of a propeller mounted in a duct; the unit is then mounted below the waterline in the bow or stern of the ship. The duct runs completely across the ship. Calculate the thrust developed by a 1 865 kW unit (supplied to the propeller) if the duct is 2.8 m in diameter and the ship is stationary.

6.87. Derive an expression for the thrust against the supports of a stationary rocket if fuel is supplied at a mass rate $\dot{m}_f$ and the exit velocity is v_{ex}.

6.88. A vehicle in deep space has a mass of 1 000 kg (69 slug); if v_{ex} is 1 km/s (0.62 mi/s), how much fuel must be burned to increase the speed by 500 m/s (1 600 ft/s)?

6.89. A space vehicle is designed so that the fuel to payload ratio is 25 to 1. What is the ratio of the maximum vehicle velocity to the engine exhaust velocity if the vehicle is initially at rest, in deep space (no external or body forces act on it)?

CHAPTER 7

FLOW OF A REAL FLUID

One of the major objectives of Chapter 1 was to characterize the fluid state, and an important observed fluid property mentioned there was its viscosity which permits the development of shear stresses and resistance to flow. When the fluid is static (no flow), only the normal stress or pressure exists, and Chapter 2 focused on these cases of fluid statics. Of course, viscosity played no role. In Chapter 3, kinematics was studied. Although not excluded, viscous effects again played no role because kinematics is concerned with motions without regard for the forces which cause them. The examination of the conservation of mass, control volumes and systems, which led to the Reynolds Transport Theorem, in Chapter 4 did not require the explicit consideration of viscous effects; but their impact was implicitly included in setting the shape of velocity profiles and in establishing some of the forces that can cause the time rate of change of the extensive fluid properties defined in Chapter 4.

The flow of an ideal incompressible fluid was considered in Chapter 5, while the compressible ideal fluid is dealt with in Chapter 13, along with other aspects of compressible flow. The ideal fluid was defined to be inviscid, that is, devoid of viscosity. Thus, there were no frictional effects between moving fluid layers or between the fluid and bounding walls. The observed *no-slip* condition for a real fluid at a solid wall did not apply, and the ideal fluid slipped freely along bounding walls. The Euler, Bernoulli, and work-energy equations as they were derived in Chapter 5 are strictly applicable only to ideal fluid flows. However, in Chapter 13 (Section 13.1), the analysis of the First Law of Thermodynamics leads to an energy equation that includes heat transfer and the internal energy of the fluid as additions to the terms in the work-energy equation. Later in this chapter, we will use these concepts and see that the energy equation derived from the First Law can be applied directly to real fluid flows because, although shear work is neglected in the derivation, the frictional forces at fixed walls (e.g., in bounded flows, such as those in pipes, etc.) do no work, and so make no contribution in the energy equation.

On the other hand, forces acting on a fluid do cause momentum changes even though no work is done. Therefore, when the impulse-momentum principle was applied in Chapter 6 to derive Eqs. 6.2 and 6.15 for ideal fluid flows, a crucial set of forces, namely, shear forces at solid boundaries, was not included. This neglect is trivial to fix, however; for real fluids one merely includes the appropriate frictional forces in the left-hand side of these equations when the forces acting on the control volume are being summed in any problem solution. In this chapter, we make specific use of this ''correction'' in Sections 7.4, 7.10, and 7.15.

In a real fluid flow then, viscosity introduces resistance to motion by causing shear or friction forces between fluid particles and between these and boundary walls. For flow to take place, work must be done against these resistance forces, and in the process energy is converted into heat. The inclusion of viscosity also allows the existence of two physically distinct flow regimes (as described in Section 1.6) and, in addition (by causing separation and secondary flows), it frequently produces flow situations entirely different from those of the ideal fluid. The effects of viscosity on the velocity profile also render invalid the assumption of a uniform velocity distribution. The derivation of the Euler equations (e.g., see Section 5.6 for a two-dimensional analysis) can be altered to include the shear stress in a real fluid in addition to the normal stress or pressure already included there. This is accomplished by adding the appropriate frictional forces as suggested above. The result is a set of nonlinear, second-order partial differential equations, called the Navier-Stokes equations. These equations applied to both steady and unsteady, incompressible and compressible fluids have a major role in the numerical simulation of complex geophysical flows in the atmosphere, the ocean, estuaries, lakes, and reservoirs. For that reason, a brief introduction to the Navier-Stokes equations is included at the end of this chapter. Unfortunately, few useful analytical solutions to these equations are known and engineers must resort to numerical simulations which employ the equations to gain useful results. Of course, many important results can be obtained by use of approximate equations, reduced numbers of dimensions (e.g., one-dimensional flows), and steady flows. This requires a good basic understanding of a variety of physical phenomena, which are described in this chapter and which are basic manifestations of fluid friction.

This chapter is divided into four functional sections. First, the concepts of laminar and turbulent flow and the conditions under which they occur are examined and the influence of solid boundaries is introduced in a section on qualitative views (more detail on pipe flows is provided in Chapter 9). Next, using these concepts as a base, a range of physical situations which occur frequently is surveyed. For the survey the flows are collected into two useful classes: *external flows* and *internal flows* (fully bounded). Finally, the Navier-Stokes equations are derived and some applications are discussed. This chapter provides the key for what follows in Chapters 8, 9, 10, 11, and 13.

Qualitative Views

7.1 LAMINAR AND TURBULENT FLOW

In laminar flow, agitation of fluid particles is of a molecular nature only (and, hence, at a length scale of the order of the mean free path of the molecules). On the usual macroscopic scale of observation, these particles appear then to be constrained to motion in essentially parallel paths by the action of viscosity. The shearing stress between adjacent layers in a simple parallel flow is determined in laminar flow by the viscosity and is completely defined by the differential equation (Section 1.6)

$$\tau = \mu \frac{dv}{dy} \tag{1.12}$$

where the stress is the product of viscosity and the velocity gradient (Fig. 7.1). Later (Section 7.15), a more general stress relationship is derived for general flows with curving streamlines. If the laminar flow is disturbed by wall roughness or some other obstacle, the disturbances are rapidly damped by viscous action, and downstream the flow is smooth again. A laminar flow is stable against such disturbances, but a turbulent flow is not. In turbulent flow, fluid particles do not remain in layers, but move in a heterogeneous fashion through the flow, sliding past other particles and colliding with some in an entirely haphazard manner that results in a rapid and continuous macroscopic mixing of the flowing fluid, with length scales which are very much greater than the molecular scales in laminar flow.

The random motion and the observed eddies in a turbulent flow suggest that both the inertia forces, associated with the accelerations during the motion, and the viscous forces, induced by the action of the viscosity, may be important. When the viscous forces are dominant, the flow might be expected to be laminar. When the inertia forces are dominant, the flow might well be turbulent. These characteristics were demonstrated by Reynolds,[1] with an apparatus similar to that of Fig. 7.2. Water flows from a tank through a bell-mouthed glass pipe, the flow being controlled by the valve A. A thin tube, B, leading from a reservoir of dye, C, has its opening within the entrance of the glass pipe. With low velocities in the glass pipe, a thin filament of dye issuing from the tube did not diffuse but formed a thin (essentially) straight line parallel to the axis of the pipe. As the valve was opened and greater velocities were attained, the dye filament wavered and broke, eventually diffusing through the water flowing in the pipe. Reynolds found that the mean velocity at

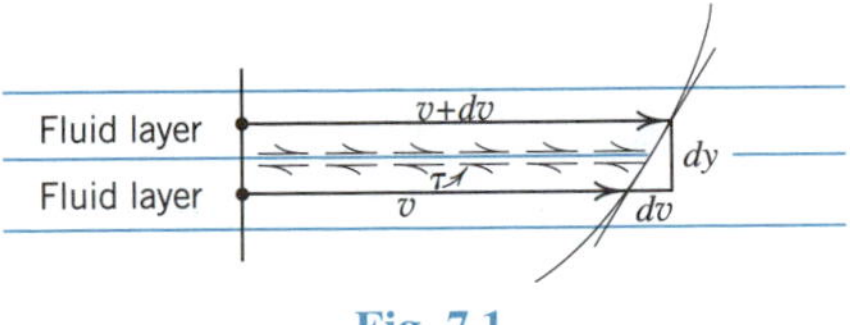

Fig. 7.1

[1]O. Reynolds, "An Experimental Investigation of the Circumstances Which Determine Whether the Motion of Water Shall Be Direct or Sinuous and of the Law of Resistance in Parallel Channels," *Phil. Trans. Roy. Soc.* vol. 174, part III, p. 935, 1883. See also *Osborne Reynolds and Engineering Science Today*, Manchester University Press (Barnes & Noble, Inc.), 1970.

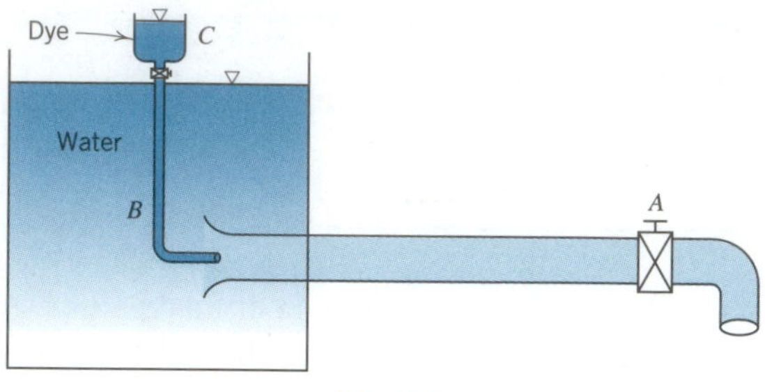

Fig. 7.2

which the filament of dye began to break up (termed the *critical velocity*) was dependent on the degree of quiescence of the water in the tank, higher critical velocities being obtainable with increased quiescence. He also discovered that if the dye filament had once diffused it became necessary to decrease the velocity in order to restore it, but that the restoration always occurred at approximately the same mean velocity in the pipe.

Since rapid mixing of fluid particles during flow would cause diffusion of the dye filament, Reynolds deduced that at low velocities this mixing was absent and that the fluid particles moved in parallel layers, or laminae, sliding past adjacent laminae but not mixing with them; this is the regime of *laminar flow*. Of course, molecular diffusion does occur, but it is too slow to have any noticeable effect in this experiment. Since at higher velocities the dye filament diffused through the pipe, it was apparent that rapid intermingling of fluid particles was occurring and the flow was *turbulent*. Laminar flow broke down into turbulent flow at some critical velocity above that at which turbulent flow was restored to the laminar condition, the former velocity being an *upper critical velocity*, and the latter, a *lower critical velocity*.

Other evidence of the existence of two flow regimes can be obtained from the simple experiment illustrated in Fig. 7.3. Here the fall of pressure between two points in a section of a long straight pipe is measured by the manometer reading h and correlated with the mean velocity V. For the small values of V a plot of h against V yields a straight line ($h \propto V$), but at higher values of V a nearly parabolic curve ($h \propto V^2$) results. Evidently the flow is laminar in the first case, but turbulent in the second. Between the two flow regimes lies an interesting *transition region*; as V is increased, the data follow the line OABCD but with diminishing V follow *DCAO*. From these results and Reynolds' observations it may be deduced that points A and B define the lower and upper critical velocities, respectively.

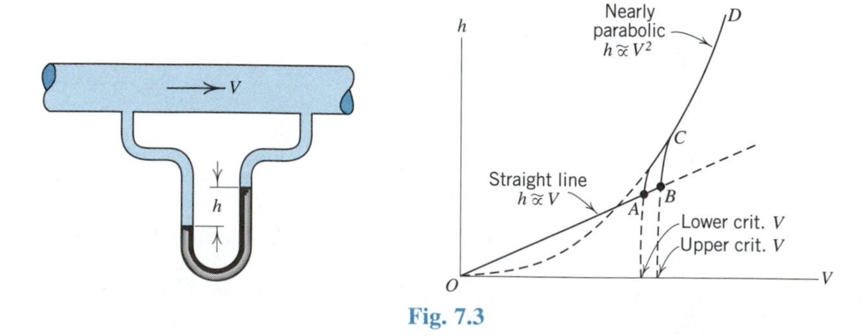

Fig. 7.3

Reynolds was able to generalize his conclusions from his dye stream experiments by the introduction of a dimensionless term **R**, later called the Reynolds number, which was defined by

$$\mathbf{R} = \frac{Vd\rho}{\mu} \quad \text{or} \quad \frac{Vd}{\nu} \tag{7.1}$$

in which V is the mean velocity in the pipe, d the pipe diameter, and ρ and μ the density and viscosity of the flowing fluid. Reynolds found that certain critical values of the Reynolds number, $\mathbf{R}_c$, defined the upper and lower critical velocities for all fluids flowing in all sizes of pipes, and thus deduced that single numbers define the limits of laminar and turbulent pipe flow *for all fluids*.

The upper limit of laminar flow is indefinite, being dependent on several incidental conditions such as: (1) initial quiescence of the fluid, (2) shape of pipe entrance, and (3) roughness of pipe, and these values are of little practical interest. The lower limit of turbulent flow, defined by the lower critical Reynolds number, is of greater engineering importance; it defines a condition below which all turbulence entering the flow from any source will eventually be damped out by viscosity. This lower critical Reynolds number thus sets a limit below which laminar flow will always occur; many experiments have indicated the lower critical Reynolds number to have a value of approximately 2 100.

The concept of a critical Reynolds number delineating the regimes of laminar and turbulent flow is indeed a useful one in promoting concise generalization of certain flow phenomena. Applying this concept to the flow of *any fluid in cylindrical pipes*, the engineer can predict that the flow will be laminar in $\mathbf{R} < 2\,100$ and turbulent if $\mathbf{R} >> 2\,100$. In fact, as seen in Chapter 9, when $2\,100 < \mathbf{R} < 4\,000$, a transition occurs and the flow can be either laminar or turbulent. However, *critical Reynolds number is very much a function of boundary geometry*. For flow between parallel walls (using mean velocity V, and spacing d) $\mathbf{R}_c \cong 1\,000$; for flow in a wide open channel (using mean velocity V and water depth d) $\mathbf{R}_c \cong 500$; for flow about a sphere (using approach velocity V and sphere diameter d) $\mathbf{R}_c \cong 1$. Such critical Reynolds numbers must be determined experimentally; because of the complex origins of turbulence, analytical methods for predicting critical Reynolds numbers have yet to be developed.

It is shown in Chapter 8 that **R**, in fact, is the ratio of the inertia forces to the viscous forces in the flow. When **R** is small, viscous forces dominate. When **R** is large, inertia forces dominate. More important, from the form of the expression for **R** and the above discussion, it is clear that the laminar or turbulent regimes are not determined just by flow velocity. Laminar flows (small **R**) are characterized by low velocities, small length scales (e.g., small diameter pipes), and fluids with high kinematic viscosity. Turbulent flows (large **R**) are characterized by high velocities, large length scales, and fluids of low kinematic viscosity.

ILLUSTRATIVE PROBLEM 7.1

Water at 15°C has a kinematic viscosity of 1.139×10^{-6} m^2/s and flows in a cylindrical pipe of 30 mm diameter. Calculate the largest flowrate for which laminar flow can be expected. What is the equivalent flowrate for air?

SOLUTION

Assume that the flow is one-dimensional and the flowrate

$$Q = AV \tag{4.4}$$

while the Reynolds number

$$\mathbf{R} = \frac{Vd}{\nu} \tag{7.1}$$

For this calculation, the kinematic viscosity of water and the pipe diameter are given as $\nu = 1.139 \times 10^{-6}$ m²/s and $d = 0.03$ m. Taking $\mathbf{R}_c = 2\,100$ as the conservative upper limit for laminar flow,

$$2\,100 = \frac{V(0.03)}{0.000\,001\,139}; \qquad V = 0.080 \text{ m/s} \tag{7.1}$$

$$Q_{\text{water}} = (0.080) \times \left(\frac{\pi}{4}\right)(0.03)^2 = 5.64 \times 10^{-5} \text{ m}^3\text{/s} \bullet \tag{4.4}$$

The water has the given viscosity at about 15°C so from Appendix 2 we have for air at zero altitude, which is at that temperature,

$$\nu_{\text{air}} = (\mu/\rho)_{\text{air}} = (1.789 \times 10^{-5} \text{ Pa·s})/(1.225 \text{ kg/m}^3) = 1.46 \times 10^{-5} \text{ m}^2\text{/s}$$

and

$$2\,100 = \frac{V(0.03)}{0.000\,014\,6}; \qquad V = 1.022 \text{ m/s} \tag{7.1}$$

$$Q_{\text{air}} = (1.022) \times \left(\frac{\pi}{4}\right)(0.03)^2 = 7.22 \times 10^{-4} \text{ m}^3\text{/s} = 13Q_{\text{water}} \bullet \tag{4.4}$$

7.2 TURBULENT FLOW AND EDDY VISCOSITY

If the Reynolds number is high enough, virtually every type of flow will be turbulent. Turbulence is found in the atmosphere, in the ocean, in the flow about aircraft and missiles, in most pipe flows, in estuaries and rivers, and in the wakes of moving vehicles. This turbulence is generated primarily by friction effects at solid boundaries or by the interaction of fluid streams that are moving past each other with different velocities. Despite its ubiquitous appearance, turbulence is difficult to define. Tennekes and Lumley[2] suggest that the best one can do is to list characteristics of turbulent flow. They list the following:

1. Irregularity or randomness in time and space.
2. Diffusivity or rapid mixing.
3. High Reynolds number.
4. Three-dimensional vorticity fluctuations (see Section 3.4).
5. Dissipation of the kinetic energy of the turbulence by viscous shear stresses.

[2]H. Tennekes and J. L. Lumley, *A First Course in Turbulence*, The MIT Press, 4th Printing, 1977.

6. Turbulence is a continuum phenomenon even at the smallest scales.
7. Turbulence is a feature of fluid flows, not a property of the fluids themselves.

The first three characteristics were mentioned earlier in the chapter, the rapid mixing being significant because it implies high rates (compared to laminar flow) of momentum and heat transfer through the flow. The vorticity fluctuations symbolize the essential three-dimensional nature of turbulence. The characteristic of dissipation makes it clear that there must be a continuous energy supply, for example, from the mean flow to the turbulence, or it will decay. In contrast, the essentially irrotational motion in the long water waves on the ocean involves little dissipation, explaining why great storm waves can travel over whole oceans without losing their identity.

To analyze turbulence it is useful to focus on the fluid particles. These particles are observed to travel in *randomly* moving fluid masses of varying sizes called *eddies*; these cause at any point in the flow, a rapid and irregular pulsation of velocity about a well-defined mean value. This may be visualized as in Fig. 7.4, where v is the time mean velocity and v' is the instantaneous velocity, necessarily a function of time. The instantaneous velocity v' may be considered to be composed of the vector sum of the mean velocity and the components of the pulsations v_x and v_y, both functions of time; by defining and isolating v_x and v_y in this manner, certain essentials of turbulence can be fruitfully studied. Measurements of v_x and v_y by hot-wire anemometer[3] (Section 14.9) yield a record similar to that of Fig. 7.4, which because of the random nature of turbulence discloses no regular period or amplitude; nevertheless such records allow the definition of certain turbulence characteristics. The root-mean-square (rms) values $(\overline{v_x^2})^{1/2}$ or $(\overline{v_y^2})^{1/2}$ are a measure of the violence of turbulent fluctuations, that is, the magnitude of departure of v' from v; the value $(\overline{v_x^2})^{1/2}/v$ is the *relative intensity* of turbulence. The mean time interval[4] between reversals in the sign of v_x (or v_y) is a measure of the *scale* of the turbulence because it is a measure of the size of the turbulent eddies passing the point. In general, the intensity of turbulence increases with velocity, and scale increases with boundary dimensions. The former is easily imagined from the more rapid diffusion of the dye filament in a Reynolds apparatus with increased velocity; the latter can be visualized from the expectation that turbulent eddies will be larger in a large canal than in a small pipe for the same mean

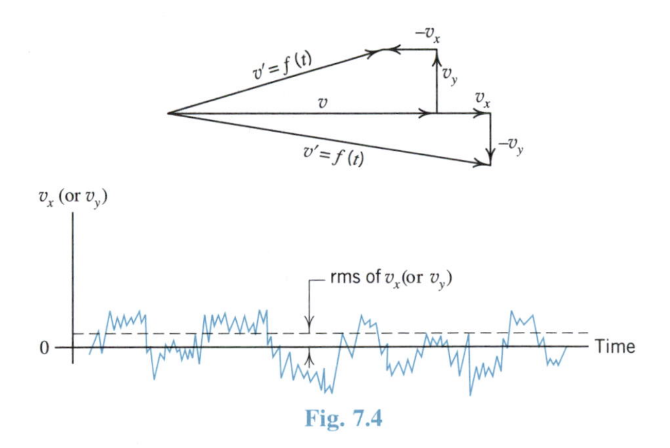

Fig. 7.4

[3]The time mean velocity $v = \overline{v'}$, where the overbar denotes a time average value, can be measured directly with a small pitot tube.

[4]This would correspond to the half-period of a simple harmonic vibration.

velocity. Indeed, as a rule of thumb, the largest eddy size expected is equal to a characteristic length of the flow, that is, the pipe radius, a channel width or depth, a boundary layer thickness, and so forth.

Because turbulence is an entirely chaotic motion of small fluid masses through short distances in every direction as flow takes place, the motion of individual fluid particles is impossible to trace and characterize mathematically. However, mathematical relationships may be obtained by considering the average motion of aggregations of fluid particles or by statistical methods.

Shearing stress in a simple parallel turbulent flow may be visualized by considering two adjacent points in a flow cross section (Fig. 7.5); at one of these points the mean velocity is v, at the other, $v + \Delta v$. If the small transverse distance between the points is l, a velocity gradient dv/dy (of the mean velocity) is implied. If v and $v + \Delta v$ are now taken to be the mean velocities of (fictitious) fluid layers, the turbulence velocity v_y represents the observed transverse motion of small fluid masses between layers, such mass being transferred in one direction or in the other. However, before transfer these fluid masses have velocities (v and $v + \Delta v$), which after transfer become $v + \Delta v$ and v, respectively; this means that their *momentum is changed* during the transfer process—tending to speed up the slower layer and slow down the faster one—just as if there were a shearing stress between them. Thus, the existence of shearing stress in turbulent flow is deducible from momentum considerations.

For a century, engineers have grappled with the problem of developing useful and accurate expressions (in terms of mean velocity gradients and other flow properties) for turbulent shear stress. Some progress has been made, but perfection has not been attained (indeed, it does not appear attainable at this time). The following is only a brief discussion of a few essentials to provide perspective.[5]

The first attempt to express turbulent shear stress in mathematical form was made by Boussinesq,[6] who followed the pattern of the laminar flow equation and wrote

$$\tau = \varepsilon \frac{dv}{dy} \tag{7.2}$$

where the *eddy viscosity*, ε, was a property of the flow (not the fluid alone) which depended primarily on the structure of the turbulence. From its definition the eddy viscosity can be seen to have the disadvantageous feature of varying from from point to point throughout the flow. Nevertheless this first expression for turbulent shear is used frequently today because of the comparison between μ and ε and because it has been possible to develop useful, if not theoretically satisfying, expressions for ε in many engineering problems.

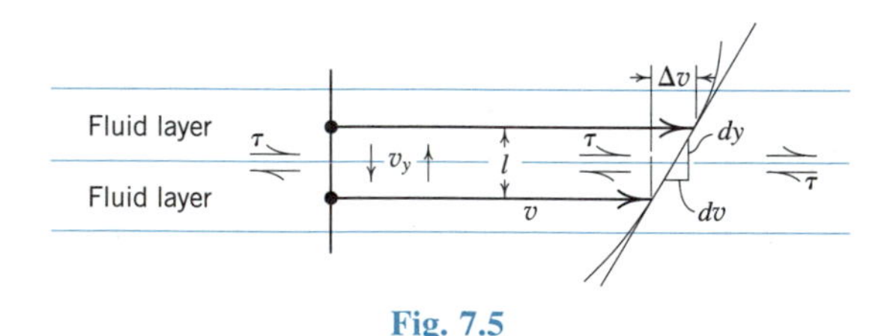

Fig. 7.5

[5]H. Tennekes and J. L. Lumley, *A First Course in Turbulence*, The MIT Press, 4th Printing, 1977, provides a complete guide for the novice.

[6]J. Boussinesq, "Essay on the Theory of Flowing Water," French Academy of Sciences, 1877.-

Now the equation is usually written

$$\tau = (\mu + \varepsilon)\frac{dv}{dy}$$

to cover the combined situation where both viscous action and turbulent action are present in a flow. For the limiting conditions where the flow is entirely laminar or entirely turbulent, ε or μ is taken to be zero, respectively, and the foregoing equation reverts to Eq. 1.12 or 7.2.

Taking the fluctuating velocities of fluid particles due to turbulence as v_y and v_x, respectively, normal to and along the direction of general mean motion, Reynolds[7] confirmed that these turbulence velocities caused an effective mean shearing stress in turbulent flow and he showed that the stress could be written as

$$\tau = -\rho\overline{v_x v_y}$$

in which $\overline{v_x v_y}$ is the mean value of the product of $v_x v_y$. Terms of the form $-\rho\overline{v_x v_y}$ are now called *Reynolds stresses.* Prandtl[8] succeeded in relating the velocities of turbulence to the general flow characteristics by proposing that small aggregations of fluid particles are transported by turbulence a certain mean distance,[9] l, from regions of one velocity to regions of another and in so doing suffer changes in their general velocities of motion (see Fig. 7.5). Prandtl termed the distance l the *mixing length* and suggested that the change in velocity, Δv, incurred by a fluid particle moving through the distance l was proportional to v_x and to v_y, that is, $l\, dv/dy \propto v_x$ and $l\, dv/dy \propto v_y$.

From this he suggested

$$\tau = -\rho\overline{v_x v_y} = \rho l^2 \left(\frac{dv}{dy}\right)^2 \tag{7.3}$$

as a valid equation for shearing stress in turbulent flow. Then, from Eq. 7.2, the eddy viscosity in a turbulent flow is

$$\varepsilon = \rho l^2 (dv/dy) \tag{7.4}$$

These expressions, although satisfactory in many respects, have the disadvantage that l is an unknown function of y that, in case of flows near a boundary or wall, presumably becomes smaller as the boundary or wall is approached. Of course, l must be chosen so that the right-hand side of Eq. 7.3 agrees with the actual stress.

In the case of flow near a bounding wall, the turbulence is strongly influenced by the wall, and v_x and v_y must be zero at the wall. It is intuitive then to let the mixing length l vary directly with the distance from the wall y. This gives

$$l = \kappa y \tag{7.5}$$

The constant κ is the so-called von Kármán constant which has been determined by use of experimental data. The nominal value of the Kármán constant is 0.4. Incorporating

[7]O. Reynolds, "On the Dynamical Theory of Incompressible Viscous Fluids and the Determination of the Criterion," *Phil. Trans. Roy. Soc.*, A1, vol. 186, p. 123, 1895.

[8]L. Prandtl, "Ueber die ausgebildete Turbulenz," *Proc. 2nd Intern. Congr., Appl. Mech.*, Zurich, p. 62, 1926.

[9]This concept is analogous to that of the mean free path in molecular theory.

assumption Eq. 7.5 into Eq. 7.3 yields

$$\tau = \rho \kappa^2 y^2 \left(\frac{dv}{dy}\right)^2 \tag{7.6}$$

The linear variation of l near a wall is found to give a mean velocity profile that agrees closely with experiments (see Chapter 9).

ILLUSTRATIVE PROBLEM 7.2

Show that, if the velocity profile in laminar flow is parabolic, the shear stress profile must be a straight line.

SOLUTION

For a parabolic relation between v and y, $v = c_1 y^2 + c_2$. From Eq. 1.12,

$$\tau = \mu \frac{dv}{dy} \tag{1.12}$$

Therefore, because $dv/dy = 2c_1 y$, $\tau = \mu(2c_1 y)$ and clearly, $\tau \propto y$ •

ILLUSTRATIVE PROBLEM 7.3

A turbulent flow of water occurs in a pipe of 2 m diameter. The velocity profile is measured experimentally and found to be closely approximated by the equation $v = 10 + 0.8 \ln y$, in which v is in metres per second and y (the distance from the pipe wall) is in metres. The shearing stress in the fluid at a point $\frac{1}{3}$ m from the wall is calculated from measurements of pressure drop (see Section 7.10) to be 103 Pa. Calculate the eddy viscosity, mixing length, and turbulence constant at this point.

SOLUTION

The velocity gradient is needed and from the given profile

$$v = 10 + 0.8 \ln y, \qquad \frac{dv}{dy} = \frac{0.8}{y} = \frac{0.8}{(0.33)} = 2.4 \text{ s}^{-1}$$

at a distance of 0.33 m from the wall.

From Eq. 7.2, $\tau = \varepsilon \dfrac{dv}{dy}$ yields $103 \text{ Pa} = 103 \text{ N/m}^2 = \varepsilon(2.4 \text{ s}^{-1})$,

$$\varepsilon = 42.9 \text{ Pa·s (Note } \mu_{\text{water}} \approx 10^{-3} \text{ Pa·s)} \; \bullet$$

From Eq. 7.3, $\tau = -\rho\overline{v_x v_y} = \rho l^2 \left(\frac{dv}{dy}\right)^2$ yields (if $\rho = 1\,000$ kg/m^3)

$$103 \text{ N/m}^2 = 1\,000 l^2 (2.4)^2, \qquad l = 0.134 \text{ m} \bullet$$

From Eq. 7.6, $\tau = \rho\kappa^2 y^2 \left(\frac{dv}{dy}\right)^2$ yields

$$103 = 1\,000\kappa^2 \left(\frac{1}{3}\right)^2 (2.4)^2, \qquad \kappa = 0.401 \bullet$$

The magnitude of the eddy viscosity ε when compared with the viscosity μ (approximately 0.001 Pa·s) is of special interest in that it provides a direct comparison between the (large) turbulent shear and (small) laminar shear for the same velocity gradient. The mixing length, l, when compared with the pipe radius is found to be about 10% of the latter dimension; this is a nominal value of correct order of magnitude, as is the turbulence constant, κ.

7.3 FLUID FLOW PAST SOLID BOUNDARIES

A knowledge of flow phenomena near a solid boundary is of great value in engineering problems because, in practice, flow is always affected to some extent by the solid boundaries over which it passes: for example, the classic aeronautical problem is the external flow of fluid over the surfaces of an object such as a wing or fuselage, and in many other branches of engineering the problem of internal flow *between* solid boundaries, as in pipes and channels, is of paramount importance.

For a real fluid, experimental evidence shows that the velocity of the layer adjacent to the surface is zero (relative to the surface). This means that a velocity profile must show a velocity of zero at the boundary. In visualizing the flow over a boundary surface it is well to imagine a very thin layer of fluid, possibly having the thickness of but a few molecules, adhering to the surface with a continuous increase of velocity of the fluid as one moves farther away from the surface, the magnitude of the velocity being dependent on the shear in the fluid. For rough surfaces this simple picture is somewhat compromised because small eddies tend to form between the roughness projections, causing local unsteadiness of the flow.

Laminar flow occurring over smooth[10] or rough boundaries (Fig. 7.6) possesses essentially the same properties, the velocity being zero at the boundary surface and the shear stress throughout the flow being given by Eq. 1.12. Thus, in laminar flow, *surface roughness has no effect* on the flow picture as long as the roughness projections are small relative to the flow cross-section size.

In turbulent flow, however, the roughness of the boundary surface will affect the physical properties of the fluid motion. When turbulent flow occurs over *smooth* solid boundaries, it is always separated from the boundary by a *sublayer* of viscosity-dominated flow (Fig. 7.7). This sublayer has been observed experimentally and its existence may be

[10]In the dynamics of solids, a ''smooth'' surface is often assumed to be a frictionless one; this concept is irrelevant here.

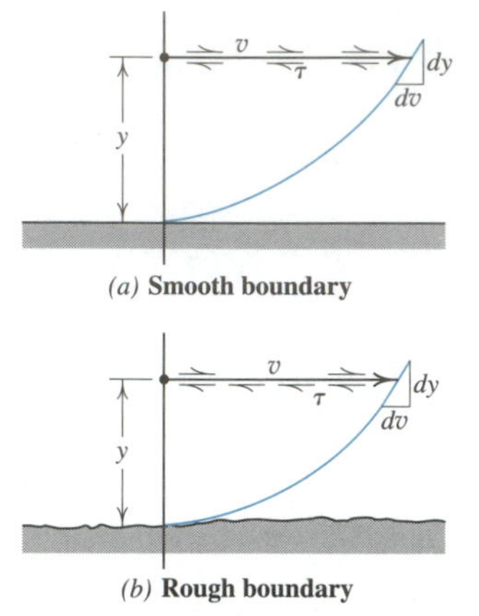

Fig. 7.6 Laminar flow over boundaries that *act* smooth.

justified theoretically by the following simple reasoning: The presence of a boundary in a turbulent flow will curtail the freedom of the turbulent mixing process by reducing the available mixing length, and, in a region very close to the boundary, the available mixing length is reduced to zero (i.e., the turbulence is completely extinguished) and a film of viscous flow over the boundary results.

In the viscous sublayer the shear stress, τ, is given by the viscous or laminar flow Eq. 1.12, and at a distance from the boundary, where turbulence is fully developed, by Eqs. 7.3 or 7.6. Between the latter region and the viscous sublayer lies a transition zone in which shear stress results from a complex combination of both turbulent and viscous action, turbulent mixing being inhibited by the viscous effects due to the proximity of the wall. The fact that there is a transition from fully developed turbulence to no turbulence at the boundary surface shows that the viscous sublayer, although given (for convenience) an arbitrary thickness, δ_v (Fig. 7.7), does *not* imply a sharp line of demarcation between the laminar and turbulent regions. Intensive research on the viscous sublayer has shown

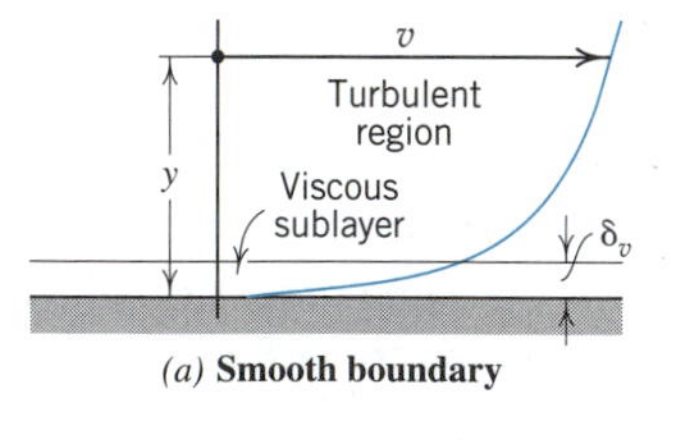

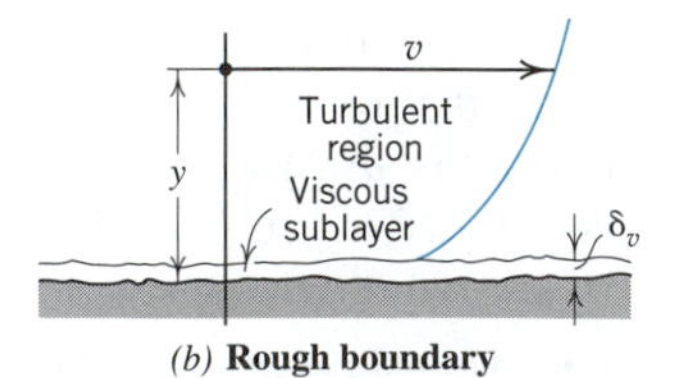

Fig. 7.7 Turbulent flow over boundaries that *act* smooth.

that its thickness varies with time, showing the sublayer flow to be unsteady, this unsteadiness being associated with eddy formation adjacent to the surface. As these eddies or spots of turbulence are carried away from the wall, their presence accounts for the increased turbulence observed in the transition zone. See J. T. Davies, *Turbulence Phenomena*, Academic Press, 1972.

As stated above, the roughness of boundary surfaces will affect the physical properties of turbulent flow, and the effect of this roughness is dependent on the relative size of roughness and viscous sublayer. A boundary surface is said to be *smooth* if its projections or protuberances are so completely submerged in the viscous sublayer (Fig. 7.7*b*) that they have no effect on the structure of the turbulence. However, experiments have shown that roughness heights larger than about one-third of sublayer thickness will augment the turbulence and have some effect on the flow. Thus the thickness of the viscous sublayer is the criterion of effective roughness, and, because the thickness of this sublayer also depends on certain properties of the flow, it is quite possible for the same boundary surface to behave as a smooth one or a rough one, depending on the size of the Reynolds number and of the viscous sublayer which tends to form over it.[11]

External Flows

Each of us has a mental image of internal and external fluid flows. Beyond the obvious physical differences, there are marked differences in the results sought in analysis of these flows. In *external flows*, one seeks the flow pattern around an object immersed in the fluid (over a wing or a flat plate, etc.), the lift and drag (resistance to motion) on the object, and perhaps the patterns of viscous action in the fluid as it passes around a body. For *internal flows*, the focus is often not on lift or drag, but on energy or head losses, pressure drops, and cavitation where energy is dissipated, because in internal flow, energy or work is used to move fluid through passages. In external flow cases, energy or work is used typically to move the object through the fluid.

7.4 CHARACTERISTICS OF THE BOUNDARY LAYER

Analyses made on the basis of ideal (inviscid) flow over streamlined bodies produce two results that are contrary to observation. First, the calculated drag on the body is negligible, while the observed drag is not. Second, the ideal fluid slips smoothly by the body; real fluid does not. For fluids with small viscosity (e.g., air and water), Prandtl[12] first suggested the boundary layer concept to explain the resistance of streamlined bodies, flat plates parallel to the flow, and so forth. The essential point is that the *frictional aspects of the flow are confined to the boundary layer* and perhaps a wake behind the body (where the flow is *rotational*), but outside the boundary layer the viscosity of the fluid is essentially inoperative, that is, the flow is effectively frictionless and *irrotational*. The boundary layer

[11]It will be shown later that sublayer thickness decreases with increasing Reynolds number. Usually, in practice, the change of a smooth surface to a rough one results from an increase of Reynolds number brought about by an increase of velocity.

[12]*Proc. Third Intern. Math. Congress*, Heidelberg, 1904.

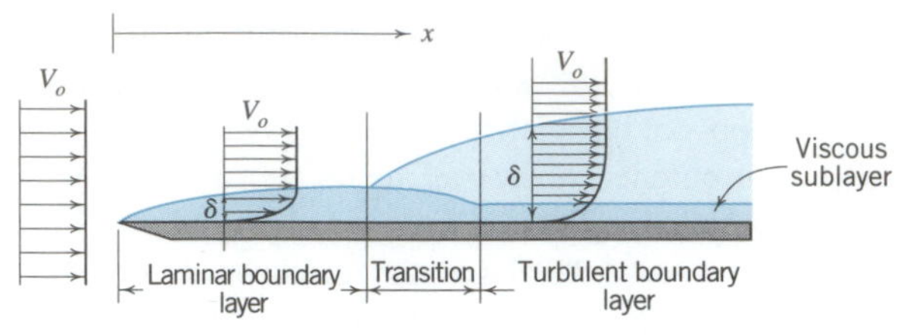

Fig. 7.8 Boundary layers on a flat plate.

idea has found wide application in numerous problems of fluid dynamics and has provided a powerful tool for the analysis of problems of fluid resistance; it has probably contributed more to progress in modern fluid mechanics than any other single idea.

Boundary layer phenomena may be most easily visualized on a smooth flat plate parallel to the oncoming flow (Fig. 7.8). On this plate the boundary layer may be either laminar or turbulent; however, a rational approach can tell us which type of flow prevails. Assuming the leading edge of the plate to be smooth and even, a *laminar boundary layer* is to be expected adjacent to the upstream portions of the plate since the boundary layer is thin, viscous action intense, and turbulence is inhibited. A similar argument was advanced in the previous section to justify the laminar character of the sublayer over a smooth surface. Two Reynolds numbers are widely used to define the character of the boundary layer; they are defined in terms of the undisturbed velocity V_o, boundary layer thickness δ, and distance from the leading edge of the plate x, as shown in Fig. 7.8:

$$\mathbf{R}_x = \frac{V_o x}{\nu} \quad \text{and} \quad \mathbf{R}_\delta = \frac{V_o \delta}{\nu} \tag{7.7}$$

and experiments on flat plates have shown that the typical or nominal critical values for these Reynolds numbers are 500 000 and 3 900, respectively. Below these values, laminar boundary layers are to be expected, whereas Reynolds numbers above these values suggest that the flow is transitioning to turbulent boundary layers.

The boundary layer starts with zero thickness at the leading edge of the plate where viscous action begins and steadily increases in thickness as increased viscous action extends into the flow and slows down more and more fluid; in terms of Reynolds numbers, $\mathbf{R}_\delta$ varies over the length of the laminar boundary layer from zero at the leading edge of the plate to the critical value of about 3 900 at the downstream point in the layer where it begins to turn turbulent. The critical value of 3 900 is a nominal one suitable for use in rough calculations and textbook problems. For excessive turbulence in the oncoming flow or a rough leading edge on the plate, the critical number will be considerably less than 3 900; indeed, if these upstream conditions are very poor, it may be essentially zero, implying a turbulent boundary layer beginning at the leading edge of the plate. The critical value of $\mathbf{R}_\delta$ is also very sensitive to pressure gradient; a favorable gradient (pressure decreasing in the direction of the flow; see Fig. 7.19) stabilizes the laminar boundary layer and produces a critical value of $\mathbf{R}_\delta$ greater than 3 900. Conversely, an unfavorable pressure gradient (pressure increasing in the direction of the flow, as in a diffuser) causes early breakdown of the laminar flow and a critical value of considerably less than 3 900.

At the point where the end of the laminar boundary layer is reached, instabilities in the boundary layer produce breakdown of the laminar structure of the flow and cause

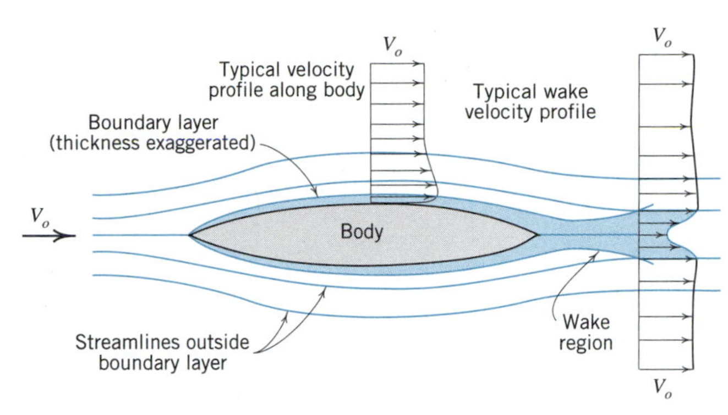

Fig. 7.9

turbulence to begin. After the onset of turbulence, the (transition) boundary layer thickens rapidly, developing into a turbulent boundary layer having many of the characteristics of turbulent pipe flow (Chapter 9); as in the pipe flow, a thin *viscous sublayer* (Fig. 7.8) will exist between the solid surface and the turbulence of the boundary layer (cf., Section 7.3).

Comparison of flat plate boundary layers with those of the streamlined body (Fig. 7.9) reveals certain differences between them. The superficial similarity between Figs. 7.8 and 7.9 is noted immediately. However, for the streamlined body: (a) its surface has a curvature that may affect the boundary layer development due to either inertial effects or induced separation (See Section 7.7) if the body is particularly blunt, and (b) the velocity in the irrotational flow just outside the boundary layer changes continuously along the body because of the disturbance to the overall flow offered by the body of finite width (this is verified by solution of the problem of ideal flow past a body).

Outside the boundary layer the fluid motion is accurately described by ideal fluid theory. This main ideal flow acts as an ''outer'' flow which establishes both the velocity at the edge of the ''inner'' flow or boundary layer and the pressure distribution along the body.[13] An engineering approach to solution of such a flow problem is to solve first the ''outer'' problem of ideal fluid motion about the body, ignoring viscous effects entirely. Then, using the ''outer'' solution values of velocity and pressure at the surface of the body as approximate values for the edge of the boundary layer, the ''inner'' viscous flow problem is solved. Experiments have demonstrated that this is often an effective and accurate process. For streamlined shapes, this procedure gives, from the ''outer'' solution, the pressure distribution (including an accurate estimate of the lift force, if any) and, from the ''inner'' solution, an estimate of the friction force or drag on the shape.

Another important characteristic of boundary layers can be deduced[14] from their relative thinness; that is, the tangential component of velocity u is much greater than the normal component v and the gradients $\partial u/\partial x$, $\partial v/\partial x$, and $\partial v/\partial y$ are negligible compared to $\partial u/\partial y$. In the general case noted here, the coordinate directions x and y are taken to be locally tangent and normal to each point along the body on which the boundary layer is growing. As a consequence of the velocity behavior in the boundary layer,[14] the pressure change across it is negligible, so $\partial p/\partial y \approx 0$ and $\partial p/\partial x \approx dp/dx$.

While solution of the flow in a boundary layer is difficult, often requiring sophisticated mathematical and/or computer methods, the so-called *von Kármán momentum integral*

[13] K. Karamcheti, *Principles of Ideal-Fluid Aerodynamics*, Wiley, 1966, pp. 52–55.

[14] H. Schlichting, *Boundary-Layer Theory*, 7th ed., McGraw-Hill, 1979, Chap. VII.

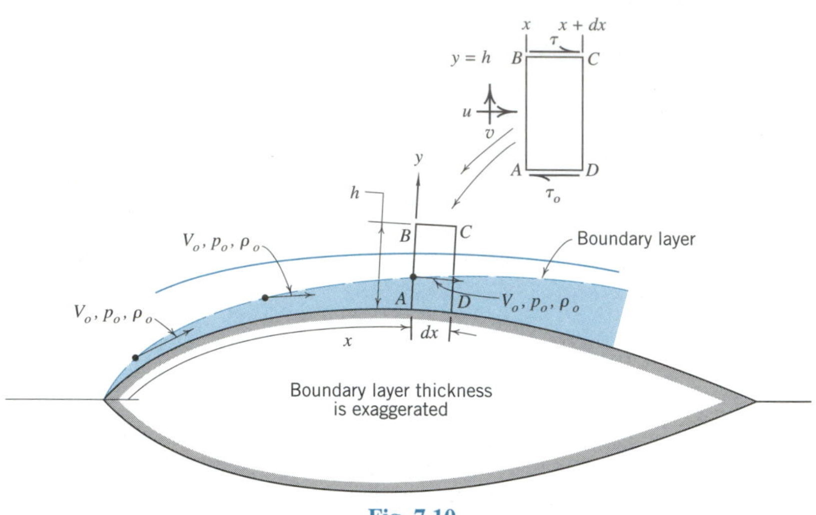

Fig. 7.10

equation of the boundary layer is simple to derive and very useful in the analysis of boundary layer behavior. The derivation leads in turn to natural definitions of useful quantities such as the boundary layer displacement and momentum thicknesses, as well as to determination of the surface shear stress, τ_o, by analysis or experiment.

The development of a boundary layer in a variable-density, two-dimensional flow over a streamlined shape is shown in Fig. 7.10. (This derivation does not depend on, but can be related to compressible flow concepts which are presented in Chapter 13.) The integral equation is developed by analysis of a control volume *ABCD* of height h (that is greater than the boundary layer thickness) and differential length dx. The velocity, $V_o(x)$, at the edge of the boundary layer, the pressure, $p_o(x)$, and density, $\rho_o(x)$, are presumed known from analysis of the outer ideal flow. Neither the pressure nor the velocity of the ideal flow is expected to change significantly over y-distances of the order of the boundary-layer thickness, δ. In addition the curvature of the body has a negligible influence on the dynamics of the boundary layer, provided that the curvature is not appreciable, that is, that the local radius of curvature is very large compared to δ. Accordingly the continuity Eq. 4.9 and the momentum equation from Section 6.1 can be applied directly to the control volume *ABCD*.

From Eq. 4.9

$$\oint_{ABCD} \rho \mathbf{v} \cdot \mathbf{n}\, dA = \left[\int_0^h \rho u\, dy\right]_{at\, x+dx} + [\rho v]_{y=h}\, dx - \left[\int_0^h \rho u\, dy\right]_{at\, x} = 0 \quad (4.9)$$

Thus, by rewriting the terms evaluated at $x + dx$ and x as a derivative, we find that

$$\frac{d}{dx}\left(\int_0^h \rho u\, dy\right) dx + [\rho v]_{y=h}\, dx = 0$$

The second term represents the mass transport through the top of the control volume resulting from the fact that the streamlines are not parallel to the body surface because of the growing boundary layer. From Section 6.1,

$$\Sigma F_x = [-p_o]_{at\, x+dx} h + [p_o]_{at\, x} h - \tau_o\, dx + [\tau]_{y=h}\, dx$$

and

$$\oint_{ABCD} \mathbf{v}_x(\rho \mathbf{v} \cdot \mathbf{n}\, dA) = \left[\int_0^h \rho u^2\, dy\right]_{at\, x+dx} + [\rho u v]_{y=h}\, dx - \left[\int_0^h \rho u^2\, dy\right]_{at\, x}$$

Hence,

$$-\frac{dp_o}{dx} h\, dx - \tau_o\, dx + [\tau]_{y=h}\, dx = \frac{d}{dx}\left(\int_0^h \rho u^2\, dy\right) dx + [\rho u v]_{y=h}\, dx$$

Because $h > \delta$, $[\tau]_{y=h} = 0$ and other terms evaluated at $y = h$ have free stream or ideal flow values. Thus, the above equations become

$$\rho_o v_o = -\frac{d}{dx}\left(\int_0^h \rho u\, dy\right) \tag{7.8}$$

where v_o is the small vertical velocity component at the edge of the boundary layer. Letting[15] $u_o = u_{y=h} = V_o$,

$$-\frac{dp_o}{dx} h - \tau_o = \frac{d}{dx}\left(\int_0^h \rho u^2\, dy\right) + \rho_o V_o v_o \tag{7.9}$$

Introducing Eq. 7.8 into Eq. 7.9,

$$-\frac{dp_o}{dx} h - \tau_o = \frac{d}{dx}\left(\int_0^h \rho u^2\, dy\right) - V_o \frac{d}{dx}\left(\int_0^h \rho u\, dy\right) \tag{7.10}$$

However, p_o and V_o are related in the ideal, outer flow by the Euler equation (Section 5.1). Neglecting gravity effects,

$$\frac{dp_o}{\rho_o} + V_o\, dV_o = 0$$

Hence,

$$\frac{dp_o}{dx} = -\rho_o V_o \frac{dV_o}{dx}$$

and Eq. 7.10 becomes

$$-\tau_o = \frac{d}{dx}\left(\int_0^h \rho u^2\, dy\right) - V_o \frac{d}{dx}\left(\int_0^h \rho u\, dy\right) - \rho_o V_o \frac{dV_o}{dx} h \tag{7.11}$$

If the density variation and velocity profile in the boundary layer are known or measured, along with p_o, ρ_o, and V_o, Eq. 7.11 allows direct evaluation of the shear stress along the body. In using Eq. 7.11 it is convenient to define certain physically based terms to simplify the equation.

First, δ must be defined; this is difficult because there is no sharp demarcation between the boundary layer and the ideal fluid zones. Usually δ is defined as the distance from the solid boundary at which the velocity u reaches 99% of the free stream value V_o.

[15]Strictly speaking, the magnitude of the velocity at the edge of the boundary layer $V_o = (u_o^2 + v_o^2)^{1/2}$. However, usually $u_o \gg v_o$, so $u_o \approx V_o$.

Second, the term $\int_0^\delta \rho u \, dy$ is the mass flowrate through the boundary layer. In the absence of the boundary layer, the mass flowrate would be $\rho_o V_o \delta > \int_0^\delta \rho u \, dy$ because there is a *flow deficit* in the boundary layer caused by retardation of the fluid. A *displacement thickness*, δ_1, is defined to represent the thickness of an imaginary layer needed to carry the deficit flow; then,

$$\rho_o V_o \, \delta_1 = \int_0^\delta (\rho_o V_o - \rho u) \, dy = \int_0^h (\rho_o V_o - \rho u) \, dy$$

because $\rho_o V_o - \rho u \equiv 0$ for $h \geq y \geq \delta$. Thus,

$$\delta_1 = \int_0^{\delta \text{ or } h} \left(1 - \frac{\rho u}{\rho_o V_o}\right) dy \tag{7.12}$$

and δ_1 is a measure of the outward displacement of the ideal flow streamlines caused by the boundary layer.

Third, the retardation of the flow in the boundary layer also causes a momentum flux deficit in comparison to the ideal fluid flow momentum flux. Let δ_2 be the *momentum thickness* and represent the thickness of an imaginary layer of fluid of velocity V_o that carries a momentum flux equal to the deficit caused by the boundary-layer profile. It follows that (remember to account for flux into the top of the control volume)

$$\rho_o V_o^2 \, \delta_2 = \int_0^\delta \rho u (V_o - u) \, dy = \int_0^h \rho u (V_o - u) \, dy$$

or

$$\delta_2 = \int_0^{\delta \text{ or } h} \frac{\rho u}{\rho_o V_o} \left(1 - \frac{u}{V_o}\right) dy \tag{7.13}$$

Equation 7.11 can now be recast in terms of δ_1 and δ_2 and values in the free stream only. From Eqs. 7.12 and 7.13,

$$\int_0^h \rho u \, dy = -\rho_o V_o \, \delta_1 + \rho_o V_o h$$

$$\int_0^h \rho u^2 \, dy = V_o(-\rho_o V_o \, \delta_1 + \rho_o V_o h) - \rho_o V_o^2 \, \delta_2$$

Introducing these results and making a number of rearrangements in Eq. 7.11 leads to

$$\frac{\tau_o}{\rho_o V_o^2} = \frac{d\delta_2}{dx} + \delta_2 \left[\left(2 + \frac{\delta_1}{\delta_2}\right) \frac{dV_o/dx}{V_o} + \frac{d\rho_o/dx}{\rho_o} \right] \tag{7.14}$$

It is customary to express τ_o in terms of a dimensionless *local friction coefficient* c_f in the form $\tau_o = c_f\rho_o V_o^2/2$. Thus,

$$\frac{c_f}{2} = \frac{d\delta_2}{dx} + \delta_2\left[\left(2 + \frac{\delta_1}{\delta_2}\right)\frac{dV_o/dx}{V_o} + \frac{d\rho_o/dx}{\rho_o}\right] \tag{7.15}$$

For a constant density flow, $d\rho_o/dx = 0$ and

$$\frac{c_f}{2} = \frac{d\delta_2}{dx} + \delta_2\left[\left(2 + \frac{\delta_1}{\delta_2}\right)\frac{dV_o/dx}{V_o}\right] \tag{7.16}$$

For ρ_o = constant and in the absence of a pressure gradient,

$$\frac{c_f}{2} = \frac{d\delta_2}{dx} \tag{7.17}$$

that is, the shear stress is proportional to the rate of change of the momentum deficit along the surface.

7.5 THE LAMINAR BOUNDARY LAYER—INCOMPRESSIBLE FLOW

An approximate analysis of the laminar boundary layer developing on a flat plate in a zero pressure gradient (Fig. 7.8) can be carried out by assuming the velocity profile in the layer to be parabolic.[16] For a parabolic profile

$$u = \frac{V_o(2\,\delta y - y^2)}{\delta^2}$$

Hence, from Eq. 7.13,

$$\delta_2 \cong \int_0^\delta \frac{1}{\delta^2}(2\,\delta y - y_2)\left(1 - \frac{2\,\delta y - y^2}{\delta^2}\right)dy = \frac{2}{15}\,\delta \tag{7.13}$$

From Eq. 7.17 it follows that (with $\rho_o = \rho$ = constant here)

$$\frac{\tau_o}{\rho V_o^2} = \frac{c_f}{2} \cong \frac{2}{15}\frac{d\delta}{dx} \tag{7.18}$$

and

$$\tau_o \cong \frac{2}{15}\rho V_o^2\frac{d\delta}{dx} \cong u\left(\frac{2V_o}{\delta}\right) \tag{7.19^{17}}$$

[16]The actual velocity profile is nearly parabolic but not precisely so, and this fact will necessitate subsequent adjustments in the derived equations. A comprehensive analytical solution of this boundary-layer problem was accomplished by H. Blasius (*Zeit. Math. Physik*, vol. 56, p. 1, 1908), and his predicted velocity distributions have been confirmed experimentally many times. The advanced mathematical methods used by Blasius preclude development of his exact solution in an elementary text.

because $\tau_o = \mu(du/dy)_{y=0}$ and $(du/dy)_{y=0} = 2V_o/\delta$. The variables of this equation may be separated and the equation integrated:

$$\int_0^{\delta} \delta\, d\delta \cong \frac{15\mu}{\rho V_o}\int_0^x dx; \qquad \frac{\delta^2}{2} \cong \frac{15\mu x}{\rho V_o} \tag{7.20}$$[17]

giving the desired relationship between δ and x, which is conveniently expressed in terms of another Reynolds number, $\mathbf{R}_x = V_o x/\nu$.

$$\frac{\delta}{x} \cong \sqrt{\frac{30}{V_o x/\nu}} \cong \sqrt{\frac{30}{\mathbf{R}_x}} \tag{7.21}$$[17]

Information on shear stress and its variation along the plate can now be obtained from Eq. 7.19, with $d\delta/dx$ obtained from Eq. 7.21. The result is

$$\tau_o \cong \frac{\rho V_o^2}{2}\sqrt{\frac{8}{15\mathbf{R}_x}} \tag{7.22}$$[17]

This equation shows τ_o to vary *inversely* with $\sqrt{x}$; this means that the frictional stress on the upstream portions of the plate will be larger than that on the portions downstream. From Eq. 7.18,

$$c_f \cong \sqrt{\frac{8}{15\mathbf{R}_x}} \tag{7.23}$$[17]

If Eq. 7.23 is integrated along the plate, the drag coefficient,[18] C_f, is obtained, where D is the total drag force

$$C_f = \frac{D}{\frac{1}{2}\rho A V_o^2} = \frac{1}{A}\int_0^x c_f\, dx = \frac{1}{x}\int_0^x c_f\, dx = \sqrt{\frac{32}{15\mathbf{R}_x}} \tag{7.24}$$[17]

which shows the drag coefficient to be a function of Reynolds number (for incompressible flow) as predicted by the dimensional analysis of Section 11.2.

Other useful relationships may be derived from the foregoing equations. The relation between $\mathbf{R}_x$ and $\mathbf{R}_\delta$ may be obtained by mere rearrangement of Eq. 7.21; the result is

$$\mathbf{R}_x \cong \frac{\mathbf{R}_\delta^2}{30} \tag{7.25}$$[17]

[17]For agreement with the exact results of Blasius' analysis and also with experimental values, the coefficients of the following equations should be adjusted as indicated:

Equation Numbers	Change from	Change to
7.19	$2/15$	0.1278
	2	1.732
7.20	2×15	27
7.21 & 7.25	30	27
7.22 & 7.23	$\sqrt{8/15}$	0.664
7.24	$\sqrt{32/15}$	1.328

[18]The subscript f is used here to emphasize the wholly frictional nature of the coefficient. Because the flow is two-dimensional, C_f is the coefficient per unit width; that is, $A = 1 \cdot x$.

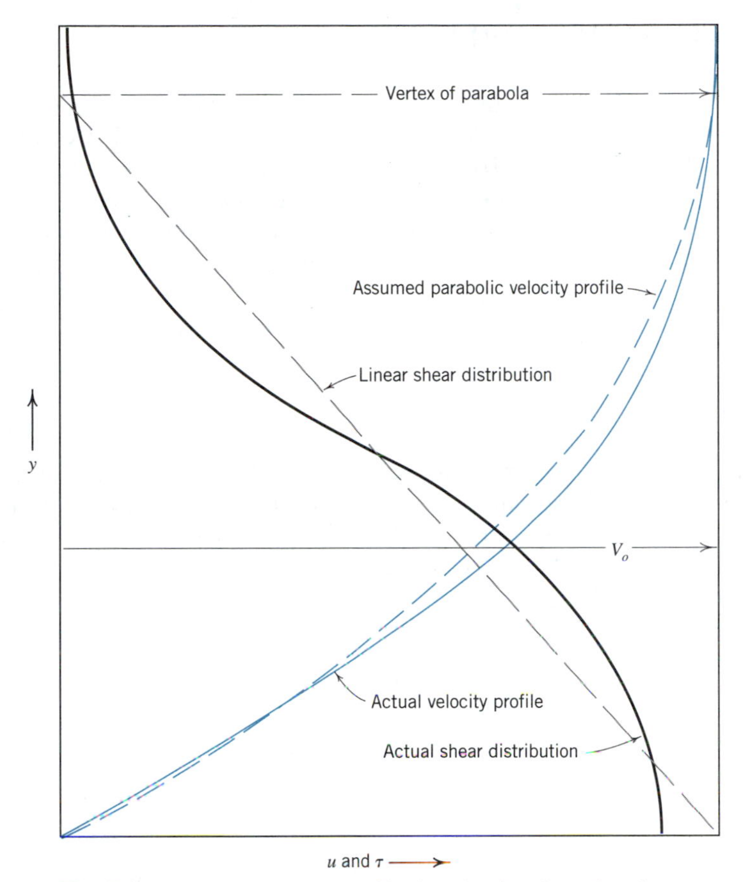

Fig. 7.11 Shear and velocity profiles in a laminar boundary layer.

When the critical value of 3 900 is substituted for $\mathbf{R}_\delta$, $\mathbf{R}_x \cong 500\,000$, showing that, in terms of plate length x, a laminar boundary layer is not to be expected beyond this critical value of $\mathbf{R}_x$.

Further insight into boundary layer properties may be obtained by comparing (see Fig. 7.11) the assumed distributions of velocity and shear with those derived by the exact analysis of Blasius (for which the velocity profile has been accurately confirmed by experiment). Experience with pipe flow has shown (Section 9.2) that a parabolic velocity profile in laminar flow is accompanied by a linear shear stress distribution. However, the exact analysis shows that the velocity profile is not precisely parabolic nor the stress profile precisely linear, the most important property of both of these profiles being their asymptotic character at the outer edge of the boundary layer. From Eq. 7.12 and the parabolic profile, the displacement thickness $\delta_1 = \delta/3$, which is very close to the value obtained for the exact analysis. This quantity is important in precise calculations since it in effect augments the thickness of plate or body and thus alters the pressure distributions on them; in the case of the flat plate, for example, the deflection of the streamlines causes a slight increase of free stream velocity along the plate accompanied by a slight drop of pressure, and a small favorable pressure gradient. Even though these features have been ignored in Blasius' exact analysis of the laminar boundary layer, there is no appreciable effect on the final results, all of which have been confirmed by experiment.

ILLUSTRATIVE PROBLEM 7.4

According to "An enthusiast's guide to the technology of the America's Cup," Advertising Supplement, *Scientific American*, 1992, the ocean-going International America's Cup Class (IACC) 1992 racing yachts were typically about 23 metres long with a waterline length of 17 metres and a draft of 4 metres. About 45% of the total drag on the yacht arises from skin friction, the remaining components being form drag on the hull (10%; see Section 11.1), wave-making drag (25%), and drag on the keel and other subsurface appendages (20%). If the yacht is operating at speeds of 1 and 10 knots, what fraction of the boundary layer on the hull is laminar and what is the drag on that portion?

SOLUTION

Assume that the flow over the wetted surface of the yacht can be approximated by the boundary layers on both sides of a 4 m deep and 17 m long flat plate. From Appendix 1, the yacht speeds of 1 and 10 knots are 0.52 m/s and 5.16 m/s, respectively, and, from Appendix 2, the density and viscosity of seawater are 1 025 kg/m^3 and 1.07×10^{-3} Pa·s, respectively.

According to experiments (Section 7.4), the flow will remain laminar for $\mathbf{R}_x$ less than or equal to 500 000. Equations 7.7 give then

$$\mathbf{R}_x = \frac{V_o x}{\nu} \text{ and so, for 1 knot, } x_{\max} = \mathbf{R}_{x_{\max}} \left(\frac{\mu}{\rho}\right) / V_o$$

$$= 500\,000 \left(\frac{1.07 \times 10^{-3}}{1\,025}\right) / 0.52 = 1.0 \text{ m} \bullet \quad (7.7)$$

Thus, 1/17th or less than 6% is laminar at 1 knot; it follows that for 10 knots, $x_{\max} = 0.1$ m and less than 1% is laminar at 10 knots. •

From Eq. 7.24, the drag coefficient is

$$C_f = \frac{D}{\frac{1}{2} A \rho V_o^2} = \sqrt{\frac{32}{15 \mathbf{R}_x}} = \sqrt{\frac{32}{15 \times 500\,000}} = 0.002 \quad (7.24)$$

Because the draft is 4 m, $A = 4x_{max}$ here so

$$D = 0.002(\tfrac{1}{2} \times 4x_{\max})(1\,025)V_o^2 = 4.1 x_{\max} V_o^2$$

For 1 knot, $D_1 = 4.1(1.0)(0.52)^2 = 1.1$ N; therefore the total drag in the laminar boundary layers on both sides of the hull is 2.2 N •

For 10 knots, $D_{10} = 4.1(0.1)(5.16)^2 = 11$ N; therefore the total drag in the laminar boundary layers on both sides of the hull is 22 N •

The effects of the turbulent boundary layer are examined in Illustrative Problem 7.5.

7.6 THE TURBULENT BOUNDARY LAYER—INCOMPRESSIBLE FLOW

An approximate analysis of the turbulent boundary layer developing on a flat plate in a zero pressure gradient (Fig. 7.8) can be made by assuming that the seventh-root velocity profile and the accompanying shear stress expression developed by Blasius for established turbulent pipe flow are applicable. For the boundary layer, Eqs. 9.27 and 9.28 (Section 9.3) are (with $\rho_o = \rho$ here)

$$u = V_o \left(\frac{y}{\delta}\right)^{1/7} \tag{9.27}$$

$$\tau_o = 0.046\,4 \left(\frac{\nu}{V_o\,\delta}\right)^{1/4} \frac{\rho V_o^2}{2} \tag{9.28}$$

Hence, from equation 7.13,

$$\delta_2 \cong \int_0^\delta \left(\frac{y}{\delta}\right)^{1/7} \left[1 - \left(\frac{y}{\delta}\right)^{1/7}\right] dy = \frac{7}{72}\,\delta \tag{7.26}$$

From Eq. 7.17, it follows that

$$\frac{\tau_o}{\rho V_o^2} = \frac{c_f}{2} \cong \frac{7}{72}\frac{d\delta}{dx} \tag{7.27}$$

and using Eq. 9.28

$$\tau_o \cong \frac{7}{72}\,\rho V_o^2 \frac{d\delta}{dx} \cong 0.046\,4 \left(\frac{\nu}{V_o\,\delta}\right)^{1/4} \frac{\rho V_o^2}{2}$$

Separating variables and integrating,[19]

$$\int_0^\delta \delta^{1/4}\, d\delta \cong 0.238\,5 \left(\frac{\nu}{V_o}\right)^{1/4} \int_0^\delta dx$$

Performing the integration and substituting $\mathbf{R}_x$ for $V_o x/\nu$ gives

$$\frac{\delta}{x} \cong \left(\frac{0.007\,9}{\mathbf{R}_x}\right)^{1/5} = \frac{0.38}{\mathbf{R}_x^{0.2}} \tag{7.28}$$

allowing the approximate shape and size of the turbulent boundary layer to be predicted. Substitution of δ from Eq. 7.28 into Eq. 7.27 yields

$$\frac{\tau_o}{\rho V_o^2} = \frac{c_f}{2} \cong \frac{0.03}{\mathbf{R}_x^{0.2}} \tag{7.29}$$

It follows that the drag coefficient is

$$C_f \cong \frac{0.074}{\mathbf{R}_x^{0.2}} \tag{7.30}$$

[19]It is assumed for this integration and the definition of $\mathbf{R}_x$ that the boundary layer is turbulent all the way to the leading edge of the plate.

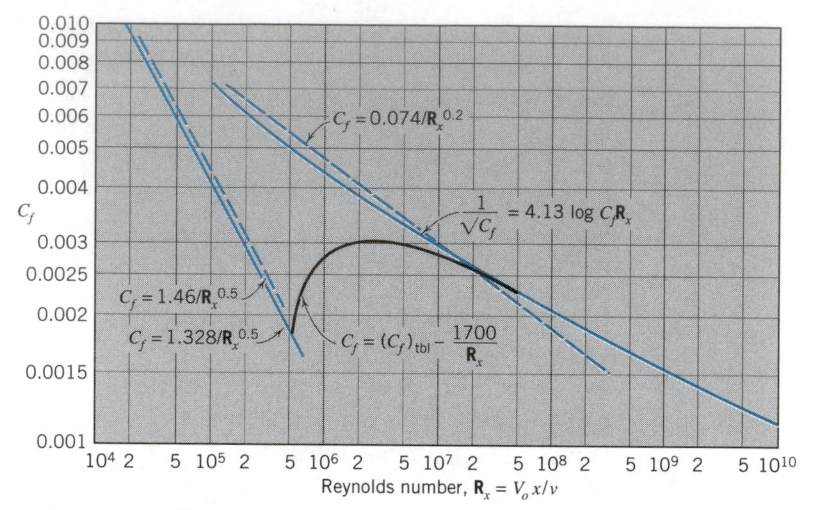

Fig. 7.12 Drag coefficients for smooth, flat plates.

and again the drag coefficient is seen to be a function of Reynolds number only. The seventh-root law in boundary layers, as in pipe flow, can be expected to be adequate only over a limited range of Reynolds numbers which must be determined by experiment; comparison of these on Fig. 7.12 shows this range of $\mathbf{R}_x$ to be from 10^5 to about 10^8 and implies that computations for boundary-layer thickness may be made with Eq. 7.28 in this range.

A more refined analysis of the turbulent boundary layer has been made by von Kármán,[20] who used the logarithmic law of velocity distribution which leads, with some adjustment of coefficients, to

$$\frac{1}{\sqrt{C_f}} = 1.70 + 4.15 \log C_f\mathbf{R}_x$$

However, Schoenherr[21] found that further adjustment of the constants gave an equation which more accurately represented the results of a wide range of experiments; his equation is

$$\frac{1}{\sqrt{C_f}} = 4.13 \log C_f\mathbf{R}_x \tag{7.31}$$

A plot of C_f against $\mathbf{R}_x$ for smooth flat plates (Fig. 7.12) with laminar and turbulent boundary layers bears a striking resemblance to the Blasius-Stanton diagram for smooth circular pipes (Fig. 9.9). However, the critical Reynolds number is not so well defined as its counterpart in pipe flow because of flow conditions which are not as well controlled. With increased initial turbulence in the approaching flow, earlier breakdown of the laminar boundary layer occurs, thus reducing the critical Reynolds number; roughening the leading

[20]Th. von Kármán, "Turbulence and Skin Friction," *Jour. Aero. Sci.*, vol. 1, p. 1, 1934.

[21]K. E. Schoenherr, "Resistance of Flat Plates Moving through a Fluid," *Trans. Soc. Nav. Arch. and Marine Engrs.*, vol. 40, p. 279, 1932. See also Schlichting (footnote 14).

edge of the plate has also been found to decrease the critical Reynolds number by decreasing the flow stability and causing earlier breakdown of the laminar layer. On the other hand, a decrease in pressure along the flow (*favorable pressure gradient*) produced by curvature of the boundary surface or at a pipe inlet (Fig. 7.19) has been found to delay the breakdown of the laminar boundary layer; application of this principle has produced the so-called *laminar flow wing* for aircraft which exhibits a drag well below that of ordinary wings. The sizable reduction in drag obtained by maintenance of a laminar boundary layer is obvious at once from Fig. 7.12; in aerodynamic practice, suction slots and porous materials, along with smooth leading edges, and appropriately shaped wing profiles are used to accomplish this reduction.

When smooth plates feature a laminar boundary layer followed by a turbulent one (Fig. 7.8), experimental drag coefficients fall between the laminar and turbulent lines of Fig. 7.12, leaving the former abruptly and approaching the latter asymptotically. Prandtl expressed this fact mathematically by

$$C_f = (C_f)_{\text{tbl}} - \frac{1\,700}{\mathbf{R}_x} \tag{7.32}$$

in which $(C_f)_{\text{tbl}}$ would be obtained from the Schoenherr equation 7.31; Eq. 7.32 is also plotted on Fig. 7.12, giving a critical Reynolds number of 500 000. This value, however, should be taken only as a typical or nominal one; the range of critical Reynolds numbers determined experimentally without a favorable pressure gradient range from 100 000 to 600 000. For favorable pressure gradients they may be considerably larger than 600 000.

ILLUSTRATIVE PROBLEM 7.5

For the IACC yacht of Illustrative Problem 7.4, calculate the total friction drag and the boundary layer thickness at the stern of the yacht when it is moving at a speed of 10 knots; ignore the laminar boundary layer and assume that sea conditions are sufficiently rough so as to induce a turbulent boundary layer from the bow of the yacht.

SOLUTION

Again, assume that the flow over the wetted surface of the yacht can be approximated by the boundary layers on both sides of a 4 m deep and 17 m long flat plate. From Appendix 1, the yacht speed of 10 knots is equivalent to 5.16 m/s, and, from Appendix 2, the density and viscosity of seawater are 1 025 kg/m^3 and 1.07×10^{-3} Pa·s, respectively.

From Eq. 7.7, at the stern $$\mathbf{R}_x = \frac{V_o x}{\nu} = \frac{5.16 \times 17}{(1.07 \times 10^{-3}/1\,025)} = 84.0 \times 10^6 \tag{7.7}$$

Equation 7.28 for the shape and Eq. 7.30 for the drag coefficient of the turbulent boundary layer can be used; they are

$$\frac{\delta}{x} = \frac{0.38}{\mathbf{R}_x^{0.2}} \quad \text{and} \quad C_f \cong \frac{0.074}{\mathbf{R}_x^{0.2}} \tag{7.28 \& 7.30}$$

Rearranging and inserting the appropriate values produces

$$\delta_{stern} = 17 \times \left(\frac{0.38}{(84.0 \times 10^6)^{0.2}}\right) = 0.17 \text{ m} = 170 \text{ mm} \bullet$$

$$C_f \cong \frac{0.074}{(84.0 \times 10^6)^{0.2}} = 0.001\,9$$

so $$D_{total} = 2 \times D = 2 \times \left(\frac{1}{2} A\rho V_o^2 C_f\right) = (4 \times 17)(1\,025)(5.16)^2(0.001\,9)$$

$$= 3\,570 \text{ N} = 3.57 \text{ kN} \bullet$$

Clearly, by comparison with the results of Illustrative Problem 7.4, the laminar layer's contribution to the drag is negligible.

7.7 SEPARATION

Separation of moving fluid from boundary surfaces is another important difference between the flow of ideal and real fluids. The mathematical theory of the ideal fluid yields no information about the expectation of separation even in simple cases where intuition alone would predict separation with complete certainty. Examples are shown in Fig. 7.13 for a sharp projection on a wall and a flat plate normal to a rectilinear flow. For the ideal fluid the flowfields will be found to be symmetrical upstream and downstream from such obstructions, the fluid rapidly accelerating toward the obstruction and decelerating in the same pattern downstream from it. However, the engineer would reason that the inertia of the moving fluid would prevent its following the sharp corners of such obstructions and that consequently separation of fluid from boundary surface is to be expected there, resulting in asymmetric flowfields featured by eddies and wakes downstream from the obstructions. Motion pictures of such eddies disclose that they are basically unsteady—forming, being swept away, and re-forming—thus absorbing energy from the flow and

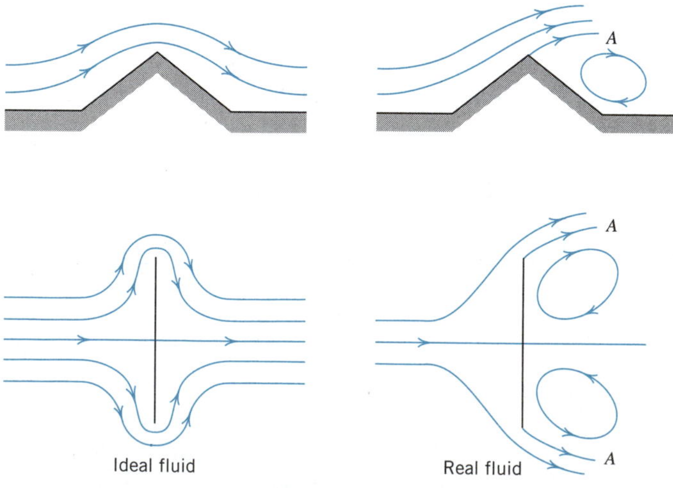

Fig. 7.13

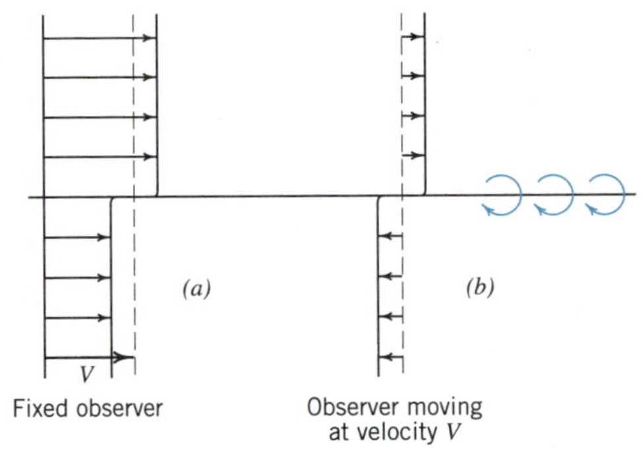

Fig. 7.14

dissipating it in heat as they decay in an extensive zone downstream from the obstruction; thus the sketches of Fig. 7.13 are to be taken as time-average flow pictures which are intended to convey the essentials but not the complete details of flow separation.

Surfaces of discontinuity (indicated by A on Fig. 7.13) divide the live stream from the adjacent and more sluggishly moving eddies. Across such surfaces there will be high velocity gradients and accompanying high shear stress, but no discontinuity of pressure. The tendency for surfaces of discontinuity to break up into smaller eddies may be seen from the simplified velocity profile of Fig. 7.14*a*. An observer moving at velocity V would see the relative velocity profile of Fig. 7.14*b* from which the tendency for eddy formation is immediately evident.

Under special circumstances the streamlines of a surface of discontinuity become *free streamlines*,[22] which are streamlines along which the pressure is constant. It is apparent that the surfaces of discontinuity of Fig. 7.13 do not quite satisfy this requirement, and thus their streamlines may be considered free streamlines only as a crude approximation. Another example of interest is the cavitation zone behind a disk (Fig. 7.15) or other sharp-cornered object in a liquid flowfield; here the constant pressure imposed on the free streamline is the vapor pressure of the liquid. Bernoulli's equation applied along any free streamline shows that, if pressure is constant, $z + v^2/2g_n$ is also constant; if variations in z are negligible or zero, free streamlines are also lines of constant velocity. Free streamline theory is concerned with the prediction of form and positions of such lines from their unique properties and the hydrodynamical equations of flowfield theory.

Although the prediction of separation may be quite simple for sharp-cornered obstructions, this is a considerably more complex matter for gently curved (streamlined) objects

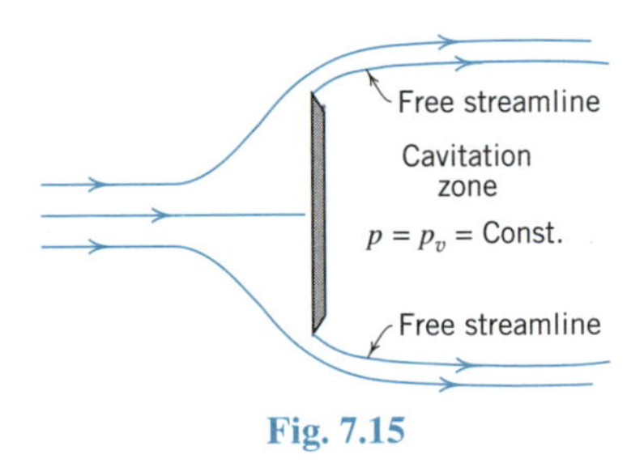

Fig. 7.15

[22]Some examples of free streamlines were cited in Section 5.8.

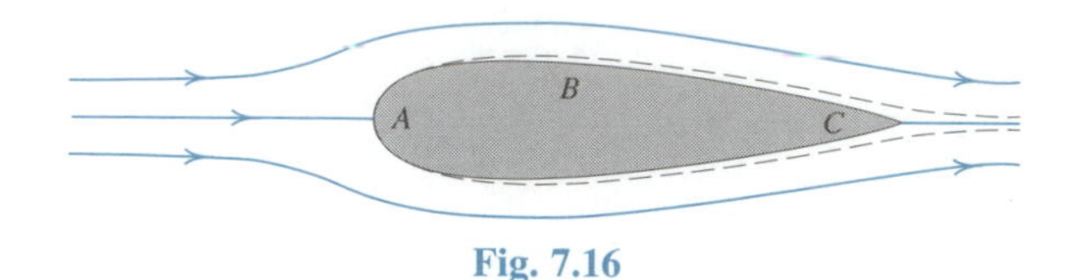

Fig. 7.16

or surfaces. Suppose that separation of flow from the streamlined strut of Fig. 7.16 does not occur. This is a very reasonable assumption for small ratios of thickness to length but much less reasonable for high values of this ratio. For no separation, the flowfield is virtually identical with that of the ideal fluid except for the growth of thin boundary layers between the strut and the remainder of the flowfield (cf., Fig. 7.9). The boundary layers coalesce at the trailing edge of the strut, producing a narrow wake of fine-grained eddies. Proceeding along the surface of the strut from (the stagnation point) A to B, the pressure falls because the flow is accelerating, producing a favorable pressure gradient which "strengthens" the boundary layer. From B to C, however, the pressure rises as the flow decelerates because the body is thinning and its effect on the flow diminishes, producing an adverse (unfavorable) pressure gradient which may "weaken" the boundary layer sufficiently to cause separation. The likelihood of separation is enhanced with increase of thickness-to-length ratio of the strut, which from the increased divergence of the streamlines adjacent to BC will produce a larger unfavorable pressure gradient. This adverse pressure gradient penetrates the boundary layer and serves to produce a force opposing the motion of its fluid; if the gradient is large enough, the slowly moving fluid near the wall will be brought to rest and begin to accumulate, diverting the live flow outward from the surface and producing an eddy accompanied by separation of the flow from the body surface. After separation has occurred, the flowfield near the separation point will appear about as shown in Fig. 7.17. Obviously the analytical prediction of separation point location is an exceedingly difficult problem requiring accurate quantitative information on the phenomena cited above; for this reason the prediction of separation and location of separation points on gently curved bodies or obstructions are usually obtained more reliably from experiment than from analysis.

The foregoing may now be generalized into the following simple axiom: *Acceleration of real fluids tends to be an efficient process, deceleration an inefficient one.* Accelerated motion, as it occurs along the surface of the front end of a submerged object, is accompanied by a favorable pressure gradient which serves to stabilize the boundary layer and thus minimize energy dissipation. Decelerated motion is accompanied by an adverse pressure gradient which tends to promote separation, instability, eddy formation, and large energy dissipation.

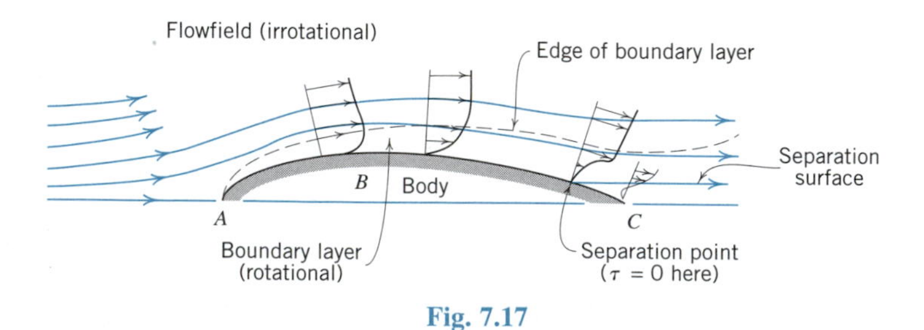

Fig. 7.17

7.8 SECONDARY FLOW

Another consequence of wall friction is the creation of a flow within a flow—a *secondary flow* superposed on the main *primary* flow. Figure 7.18*a* shows a cross section through a river bend where the secondary motion serves to deposit material at the inside of the bend and to assist in scouring the outer side, thus producing the well-known meandering characteristic of natural streams. In Fig. 7.18*b* is shown the horseshoe-shaped vortex that is produced by projections from a boundary surface. Its origin can be easily seen from the velocity profile along the wall which leads to the stagnation pressure at A being larger than that at B. This pressure difference maintains a downward secondary flow from A to B, thus inducing a vortex type of motion, the core of the vortex being swept downstream around the sides of the projection. This principle is used on the wings of some jet aircraft, the devices (called vortex generators) being used to draw higher energy fluid down to the wing surface to forestall large-scale separation.

Because of the complex origin and geometry of secondary flows they have (so far) defied rigorous analysis. They are cited in this elementary text only so that the reader will become aware of them and make due allowance for their possible existence in analyzing new problems. A conservative attitude is to expect secondary flow phenomena wherever there are severe irregularities in boundary geometry.

Internal Flows

For the flow of a real fluid in ducts, channels, and pipes many of the concepts developed in earlier chapters can be applied with only minor, but important, changes needed to account for viscous effects. Interestingly, boundary layers, separation, and secondary flows are as important in internal flows as they were in the external flow cases.

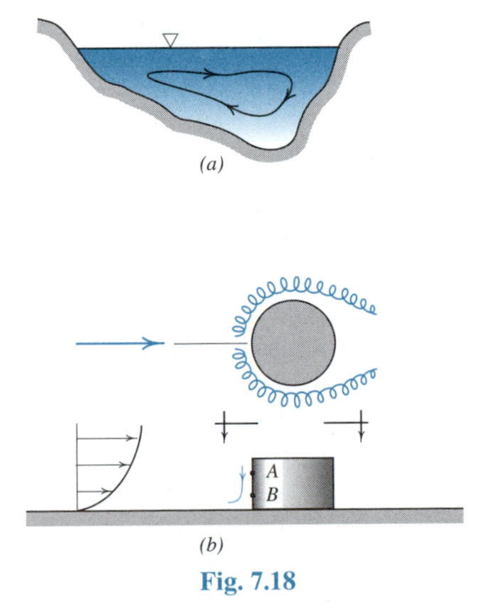

Fig. 7.18

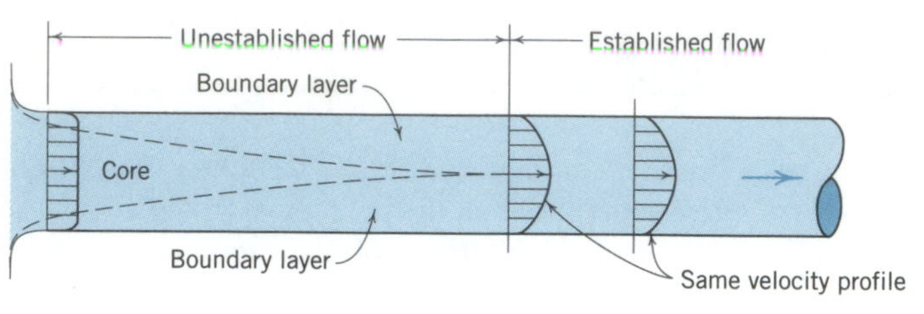

Fig. 7.19

7.9 FLOW ESTABLISHMENT—BOUNDARY LAYERS

Just as a boundary layer beginning at the edge of a flat plate in an external flow grows in the downstream direction, viscous effects begin their influence at the entrance to a pipe and establish themselves in internal flows by a process of boundary layer growth (Fig. 7.19). The zone of this growth is the zone of *unestablished* flow and it may be of importance in some engineering problems; it is described here for the simplest case—flow from a large reservoir into a cylindrical passage.

The unestablished flow zone may contain many different flow phenomena. With an unrounded entrance, separation (Section 7.13) will dominate the flow picture—featured by localized eddy formation close to the entrance and followed by decay of the eddies in the final established flow; measurements show that this process extends over a distance of some 50 pipe diameters. However, a rounded (streamlined) entrance is of more engineering interest. For this case, the unestablished flow zone will be dominated by the growth of boundary layers along the walls (accompanied by a diminishing *core* of irrotational fluid at the center of the passage); established flow results from the merging of these boundary layers, a phenomenon that occurs in most internal flows.

The mechanism of boundary layer growth in a pipe entrance can be described as follows. As the fluid flows into the entrance, high velocity gradients (dv/dy) develop in the vicinity of the boundary. These gradients are associated with large frictional stresses in the boundary layer which, as in the case of flow over a flat plate, ''eat their way'' into the flow by slowing down successive fluid elements. Thus the boundary layers steadily thicken until they meet and so envelop the whole flow; downstream from this point the influence of wall friction is felt throughout the flowfield. The flow is, therefore *established* (there is no further change in the velocity profiles) and is everywhere *rotational.*

Of course, the flow in a boundary layer may be either laminar or turbulent. If the Reynolds number $\mathbf{R} = Vd/\nu$ for the established flow is less than 2 100, it may be safely inferred that the established laminar flow has resulted from the growth of laminar boundary layers. In this case the zone of establishment has a length x given[23] by $x/d \approx \mathbf{R}/20$. For slightly higher $\mathbf{R}$, the flow may be laminar up to and past $x/d \approx \mathbf{R}/20$, with subsequent transition to turbulent flow before the flow is truly established. If $\mathbf{R} \gg 2\,100$, the boundary layers are ultimately turbulent. For a well-shaped and smooth entrance the boundary layers may be laminar at the upstream end of the zone of unestablished flow, followed by turbulent boundary layers in the downstream portion of this region. In practical cases at high $\mathbf{R}$ the unestablished flow region may extend for 100 diameters or so, but the flow is effectively established beyond $x/d \approx 20$ and the details of the flow are of minor interest.[23] In Section 9.9 the net effect at entrances is treated as a head loss, which is a function of the velocity head in the established flow region.

[23]W. M. Kays and M. E. Crawford, *Convective Heat and Mass Transfer*, 2nd ed., McGraw-Hill, 1980.

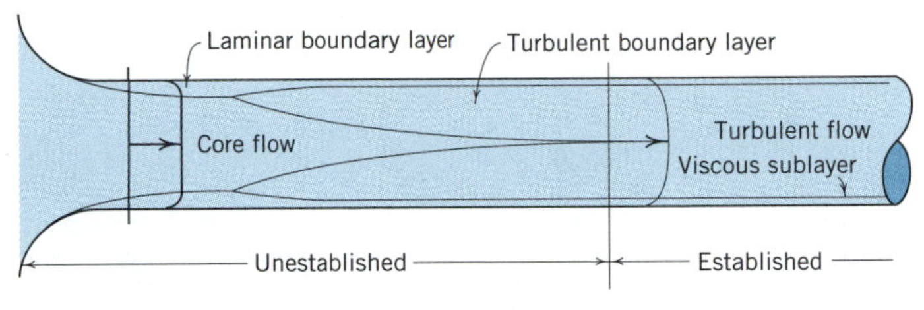

Fig. 7.20

Comparison of flat plate boundary layers with those of the pipe entrance (Fig. 7.19) will reveal certain subtle but critical differences between them. The superficial similarity between Figs. 7.8 and 7.19 is noted immediately; however, for the pipe entrance: (*a*) the plate has been rolled into a cylinder so it is not flat, (*b*) the core velocity steadily increases downstream whereas the corresponding free stream velocity of Fig. 7.8 remains essentially constant, and (*c*) the pressure in the fluid diminishes[24] in a downstream direction whereas for the flat plate there is no such pressure variation.

Although this is not the place for a detailed analysis of the pipe entrance boundary layer problem, certain qualitative facts should be noted (Fig. 7.20). Between the turbulence (of boundary layer or established flow) and solid boundary there exists the viscous sublayer cited in Section 7.3. The velocity of the core flow increases over the region of establishment from a value slightly more than the mean velocity, Q/A, to the centerline velocity of the established flow. The thickening of the laminar boundary layer causes a decrease in velocity gradient and thus a decrease in wall shearing stress in a downstream direction. With the change of velocity profile in the turbulent boundary layer a rather sudden increase of wall shearing stress can be expected after the boundary layer has changed from laminar to turbulent; thereafter this shear stress continues to decrease in a downstream direction. Since the core flow may be treated as frictionless, the fall of pressure may be predicted from the simple Bernoulli equation (without head loss) once the core velocities are known; this has been proved by experiment to be reliable and accurate in the upstream portion of the region of establishment (where the boundary layer is thin) but less accurate in the downstream portion.

7.10 SHEAR STRESS AND HEAD LOSS

A crucial question to be raised is "What is the effect of the friction forces on the boundary of a control volume, such as the inside of a pipe?" The impulse-momentum principle of Section 6.1 (or we could use the Reynolds Transport theorem of Section 4.4 directly) provides a clear answer. A simple analysis will do, so we consider a *one-dimensional* analysis of a steady compressible flow (see Section 4.2) in an element of a cylindrical passage (Fig. 7.21); this use of a circular cylindrical pipe provides a concrete example and allows us to isolate the effect of the shear stresses from other effects such as those due to a change in flow cross section (cf., Section 6.2). The wall stress τ_o is the basic resistance (shear) stress to be investigated and will produce a force on the solid periphery of passage opposing the direction of fluid motion. The impulse-momentum analysis is applied *along*

[24]Recall that this is known as a *favorable pressure gradient*, since it extends the length of the laminar boundary layer and thus reduces frictional effects.

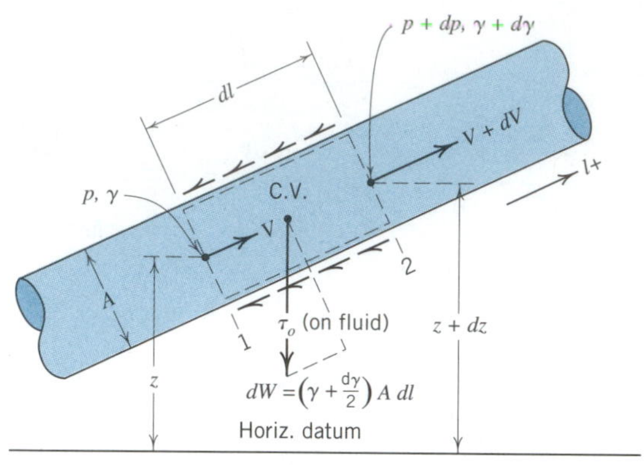

Fig. 7.21

the direction of this streamtube to the control volume bounded by sections 1, 2, and the streamtube boundary. The result is (cf., Eqs. 6.2)

$$pA - (p + dp)A - \tau_o P\, dl - \left(\gamma + \frac{d\gamma}{2}\right) A\, dl\, \frac{dz}{dl} = (V + dV)^2 A(\rho + d\rho) - V^2 A\rho$$

in which P is the perimeter of the streamtube and $dW(dz/dl)$ is the component of the fluid's weight along the streamtube. The effects of pumps and turbines are neglected here and the flow is assumed to be one-dimensional. Dividing the equation by $A\gamma$ (recalling that $\gamma = \rho g_n$) and neglecting the smaller terms that contain products of differential quantities, one can obtain

$$\frac{dp}{\gamma} + d\left(\frac{V^2}{2g_n}\right) + dz = -\frac{\tau_o\, dl}{\gamma R_h} \qquad (7.33)$$

where here the ratio A/P is known as the hydraulic radius R_h; further discussion of it is found in Section 9.7.

For an established *incompressible flow* in a tube of constant cross section, τ_o is not a function of l, γ is constant, and $d\left(\frac{1}{\gamma}\right) = 0$. Hence, Eq. 7.33 may be written as

$$d\left(\frac{p}{\gamma} + \frac{V^2}{2g_n} + z\right) = -\frac{\tau_o\, dl}{\gamma R_h}$$

This equation can be integrated between points 1 and 2 to yield

$$\left(\frac{p_1}{\gamma} + \frac{V_1^2}{2g_n} + z_1\right) - \left(\frac{p_2}{\gamma} + \frac{V_2^2}{2g_n} + z_2\right) = \frac{\tau_o(l_2 - l_1)}{\gamma R_h} \qquad (7.34)$$

Following the notation used in Section 6.2 (e.g., in Eq. 6.5) we note that the difference between the bracketed terms in Eq. 7.34 is the drop in the energy line between points 1 and 2. Accordingly, Eq. 7.34 can be rewritten as

$$\left(\frac{p_1}{\gamma} + \frac{V_1^2}{2g_n} + z_1\right) - \left(\frac{p_2}{\gamma} + \frac{V_2^2}{2g_n} + z_2\right) = \Delta(\text{EL}) = \frac{\tau_o(l_2 - l_1)}{\gamma R_h}$$

Given that the energy line is also the total head line, the common definition that is applied is to call the drop in the energy line the *head loss* h_L. Then, between points 1 and 2 along a streamtube

$$\left(\frac{p_1}{\gamma} + \frac{V_1^2}{2g_n} + z_1\right) - \left(\frac{p_2}{\gamma} + \frac{V_2^2}{2g_n} + z_2\right) = h_{L_{1-2}} \tag{7.35}$$

Accordingly, it is clear that the effect of the shear stress on the control volume (i.e., the force exerted on the fluid in the control volume by the bounding solid surfaces of, say, a pipe) is to cause a loss of energy in the flow that is proportional to the stress and the length of the section over which the stress acts and inversely proportional to the effective radius of the control volume. For this simple case of incompressible flow

$$h_{L_{1-2}} = \frac{\tau_o(l_2 - l_1)}{\gamma R_h} \tag{7.36}$$

giving a simple relation between resistance stress τ_o and the head loss caused by the action of resistance. Comparing Eq. 5.7, which expresses the work energy principle for an ideal fluid, and Eq. 7.35 shows that along a stream tube the energy and its changes can be tracked by simply using the appropriate equation between any two points; most importantly, the effects of pumps, turbines and friction are ''additive'' terms in the equation.

In previous applications of the impulse-momentum principle to flows such as the hydraulic jump and the abrupt expansion (Section 6.2), the drop in the energy line or head loss has been attributed to a ''rise in the internal energy of the fluid caused by viscous dissipation.'' That is a correct interpretation of what happens in the present case as a result of the viscous shear stresses. By borrowing some ideas from the elementary thermodynamics presented later in Chapter 13, we can prove this and close the loop between the work energy equation and Eq. 7.35. This is done in the next section.

Finally, it should be apparent that the foregoing analysis may be similarly applied to any streamtube of the flow; it is useful to do this for a streamtube of radius r and concentric with the axis of a cylindrical pipe (Fig. 7.22). For such a streamtube the frictional stress, τ, will be that exerted on the outermost fluid layer of the streamtube by the adjacent (more slowly moving) fluid. Without repeating the foregoing development, the following sub-

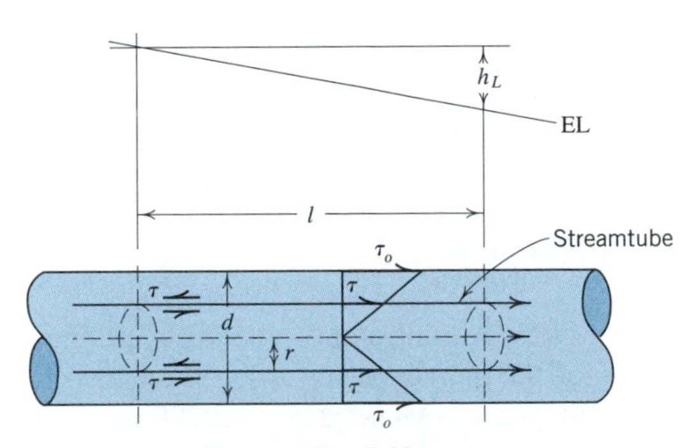

Fig. 7.22

stitutions may be made in Eq. 7.36: τ for τ_o, $r/2$ for R_h, h_L for $h_{L_{1-2}}$, and l for $l_1 - l_2$; this yields

$$\tau = \left(\frac{\gamma h_L}{2l}\right) r \tag{7.37}$$

and shows that, in established pipe flow, the shear stress τ in the fluid varies linearly with distance from the centerline of the pipe. The relationships of Eq. 7.36 and 7.37 have been developed without regard to the flow regime; *thus, it follows that they are equally applicable to both laminar and turbulent flow in pipes.* These equations are used later as the first step in the analytical treatment of these problems.

ILLUSTRATIVE PROBLEM 7.6

Water flows in a 0.9 m by 0.6 m rectangular conduit. The head lost in 60 m of this conduit is determined (experimentally) to be 10 m. Calculate the resistance stress exerted between fluid and conduit walls.

SOLUTION

Since

$$h_{L_{1-2}} = \frac{\tau_o(l_2 - l_1)}{\gamma R_h} \tag{7.36}$$

$$R_h = A/P$$

and the data (given or obtained from Appendix 2) are

$$\gamma = 9.8 \text{ kN/m}^3 \qquad h_{L_{1-2}} = 10 \text{ m} \qquad l_2 - l_1 = 60 \text{ m}$$

$$A = 0.9 \times 0.6 = 0.54 \text{ m}^2 \qquad P = 2(0.9 + 0.6) = 3 \text{ m}$$

we have

$$\tau_o = \frac{10 \times 9.8 \times 10^3 \times 0.54/3}{60} = 0.29 \text{ kPa} \bullet \tag{7.36}$$

Since this flow is *not* axisymmetric, τ_o must be presumed to be the *mean* shear stress on the perimeter of the conduit.

ILLUSTRATIVE PROBLEM 7.7

If the results cited in the preceding problem are obtained for water flowing in a cylindrical pipe 0.6 m in diameter, what shear stress is to be expected (*a*) between fluid and pipe wall, and (*b*) in the fluid at a point 200 mm from the wall?

SOLUTION

The additional relevant equation is

$$\tau = \left(\frac{\gamma h_L}{2l}\right) r \tag{7.37}$$

and

$$d = 0.6 \text{ m} \qquad r = 0.1 \text{ m}$$

See the previous problem for Eq. 7.36, γ, $h_{L_{1-2}} = h_L$, $l = l_2 - l_1$, R_h, etc.

Clearly,

$$R_h = d/4; \qquad \text{then}$$

$$\tau_o = \left(\frac{10 \times 9.8 \times 10^3}{60}\right) 0.15 = 0.25 \text{ kPa} \bullet \tag{7.36}$$

From the linear variation of τ with r (Eq. 7.37),

$$\tau = \tfrac{100}{300}(0.25 \times 10^3) = 83.3 \text{ Pa} = 0.083\,3 \text{ kPa} \bullet$$

7.11 THE FIRST LAW OF THERMODYNAMICS AND SHEAR STRESS EFFECTS

While the thermodynamics of fluid flow is an important focus of Chapter 13 on compressible flow, the First Law of Thermodynamics is a simple empirical law that can help shed some light here on the relationship between shear stresses and energy dissipation. Some terminology is necessary. A *system* remains some fixed, identifiable, quantity of matter. A system *property* is an observable or measurable characteristic of the system, for example, temperature, density, pressure, etc. When, over time, transfers across the system boundary occur or work is done on or by the system, it is said to have undergone a *process*. The First Law of Thermodynamics is an empirical conservation law for processes; the law expresses the conservation of energy in any process (barring here nuclear mass-energy conversions and electromagnetic effects). From Section 5.5, it is known that a fluid system possesses both kinetic and potential energy and can do work on its surroundings. From thermodynamics, it is known that energy transfers occur across system boundaries when there is a temperature difference there; this energy in transition as a result of a temperature difference is called *heat*. Furthermore, a fluid system possesses energy as a result of the kinetic energy of its molecules and the forces between them; this property of a system is known as *internal energy* and manifests itself in temperature, high or low temperature implying high or low internal energy, respectively. The First Law states that in a process for a system

$$dQ + dW = dE \tag{7.38}$$

where during some time increment dt, dQ is the heat transferred to the system, dW is the work done on the system, and dE is the change in the *total* energy of the system. If we now divide Eq. 7.38 by dt, we have a rate equation for the heat transferred, the work done and the energy change per unit time:

$$\frac{dQ}{dt} + \frac{dW}{dt} = \frac{dE}{dt} \tag{7.39}$$

This is a natural configuration for the application of the First Law and conforms to the form of the Reynolds Transport Theorem Eq. 4.15 which provides an equation for evaluating the rate of change of an extensive property (in this case the total energy) of a fluid system which occupies a control volume at a given instant.

Just as the work-energy principle and a control volume analysis were used in Section 5.5 to derive the work-energy equation 5.7, the First Law of Thermodynamics and a control volume analysis are now used to derive an energy equation (which, in fact, reduces to Eq. 5.7 in the absence of heat transfer and of density variations). Consider the control volume which coincides with the walls of a streamtube or full-sized conduit (Fig. 7.23); this is a steady flow and is the same configuration used in Section 5.5. We can use some of the results obtained there. Fluid enters and leaves the control volume at locations where the streamlines are straight, parallel, and perpendicular to the control volume surface; for convenience (not necessity) we also assume that the velocity is uniform over the cross section under these conditions. We need to evaluate three elements. First, we need to find the rate of heat transfer to the system of fluid found in the control volume at some instant. Second, we need to determine the rate of work done on this system of fluid. Third, we must find the rate of change of total energy and relate it to the energy of the fluid in the control volume. These three elements are the components of Eq. 7.39.

In thermodynamics, it is common to define the heat transfer rate dQ/dt in terms of q_H, which is the heat added to the fluid in the control volume per unit of mass passing through the control volume. According to the conservation of mass, the mass flow rate

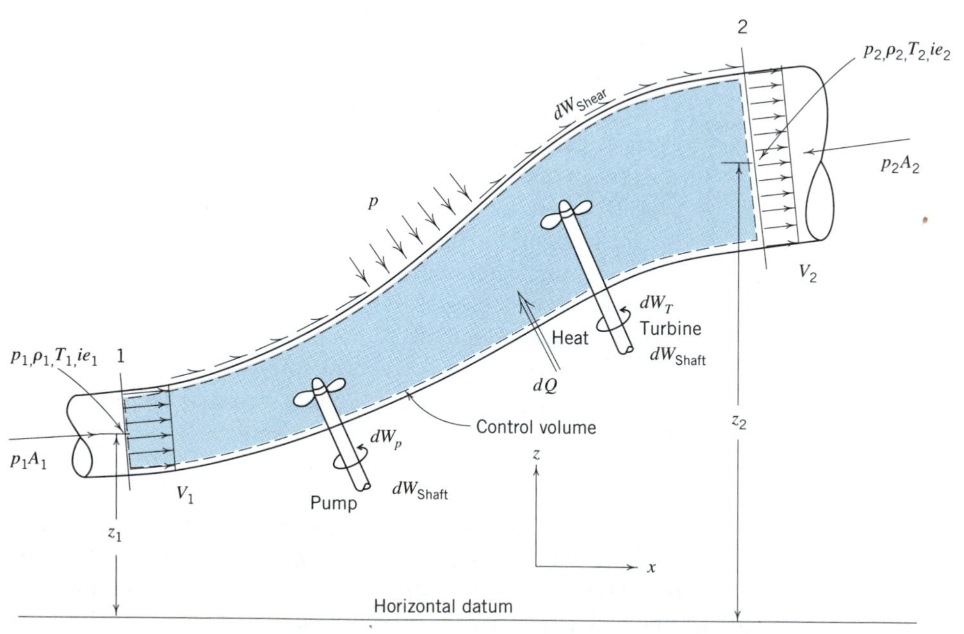

Fig. 7.23

$\dot{m} = \rho_1 A_1 V_1 = \rho_2 A_2 V_2$ in this steady flow. Therefore,

$$\frac{dQ}{dt} = \dot{m} q_H \quad \text{or} \quad q_H = \frac{1}{\dot{m}} \frac{dQ}{dt} = \left[\frac{\text{N·m}}{\text{kg}}\right] = \left[\frac{\text{joule}}{\text{kg}}\right] = \left[\frac{\text{ft·lb}}{\text{slug}}\right] \tag{7.40}$$

The work done on the system can take three forms: (a) flow or pressure work (see Section 5.5), (b) shaft (or machine) work (see Section 5.5), and (c) shear work. In ideal fluid flow the shear work is zero because the fluid is inviscid and cannot support shear stress. For real fluids, the shear work can be nonzero only where the system boundary cuts through a fluid zone so that the system can exert a shear force across the boundary on fluid that is in motion; recall that work is done by a force moving through a distance. However, since we are dealing with a fixed control volume, the shear work is zero because the boundary is fixed, the real fluid does not actually move at a solid boundary, and no work can be done. In addition, there is no shear work done at the entrance or exit of the control volume when the flow is perpendicular to the control volume surface. [This occurs because the shear stress force is parallel to the control volume surface and so perpendicular to the velocity; hence, there is no work done.] Thus, even for real fluids the shear work is zero under these conditions.

From Section 5.5, we see that the shaft or machine work done to the system can be given by

$$\frac{dW_{\text{shaft}}}{dt} = Q\gamma(E_P - E_T) = \dot{m} g_n (E_P - E_T)$$

or

$$\frac{1}{\dot{m}} \frac{dW_{\text{shaft}}}{dt} = g_n E_P - g_n E_T = \left[\frac{\text{J}}{\text{kg}}\right] = \left[\frac{\text{ft·lb}}{\text{slug}}\right] \tag{7.41}$$

where E_T and E_P are energies withdrawn by turbines or added by pumps *per unit weight* of fluid passing through the control volume.

Again referring to Section 5.5, we have the flow work rate

$$\frac{dW_{\text{flow}}}{dt} = p_1 A_1 V_1 - p_2 A_2 V_2 \quad \text{or} \quad \frac{1}{\dot{m}} \frac{dW_{\text{flow}}}{dt} = \frac{p_1}{\rho_1} - \frac{p_2}{\rho_2} = \left[\frac{\text{J}}{\text{kg}}\right] = \left[\frac{\text{ft·lb}}{\text{slug}}\right] \tag{7.42}$$

The total energy of the system is given by

$$E = \iiint_{\text{System}} i \cdot dm = \iiint_{\text{System}} (\text{kinetic} + \text{potential} + \text{internal energies}) \cdot dm$$

$$= \iiint_{\text{System}} \left(\frac{1}{2} V^2 + g_n z + ie\right) \cdot \rho \, d\forall$$

and ie is the internal energy per unit mass. [As an example, for perfect gases, thermodynamics shows ie to be a function of temperature only and the change of internal energy related to the change of temperature by $ie_1 - ie_2 = c_v(T_1 - T_2)$. For liquids, $c_p = c_v = c$ and $\Delta ie = c\,\Delta T$, also.] The system in this case can be considered the fluid within the control volume in Fig. 7.23 at some initial instant. At a later instant that system will have moved through the control volume and perhaps gained energy due to heat transfer or work done by the surroundings on the system. Thus, although the flow in the control volume is steady and unchanging at any point (the Eulerian view), the system state is changing (the Lagrangian view) and the time variation of E for the system is entirely appropriate.

The Reynolds Transport Theorem Equation 4.15 says, for this case of a steady flow with one-dimensional flows at the entrance and exit of the control volume,

$$\frac{dE}{dt} = \iint_{C.S._{out}} i \cdot (\rho \mathbf{v} \cdot d\mathbf{A}) + \iint_{C.S._{in}} i \cdot (\rho \mathbf{v} \cdot d\mathbf{A}) \tag{4.15}$$

Thus, from the definition of the total energy above

$$\frac{dE}{dt} = \iint_{C.S._{out}} \left(\frac{1}{2}V^2 + g_n z + ie\right)(\rho \mathbf{v} \cdot d\mathbf{A}) + \iint_{C.S._{in}} \left(\frac{1}{2}V^2 + g_n z + ie\right)(\rho \mathbf{v} \cdot d\mathbf{A})$$

Because the velocity vector is normal to the area and because the velocity has been assumed constant over the two cross sections where the flow enters and leaves the control volume, the above integration yields

$$\frac{dE}{dt} = \left(\frac{1}{2}V^2 + g_n z + ie\right)_2 \cdot (\rho_2 V_2 A_2) - \left(\frac{1}{2}V^2 + g_n z + ie\right)_1 \cdot (\rho_1 V_1 A_1)$$

However, $\dot{m} = \rho_2 V_2 A_2 = \rho_1 V_1 A_1$; thus,

$$\frac{1}{\dot{m}}\frac{dE}{dt} = \left(\frac{1}{2}V^2 + g_n z + ie\right)_2 - \left(\frac{1}{2}V^2 + g_n z + ie\right)_1 \tag{7.43}$$

Now, we divide Eq. 7.39 by $\dot{m}$ and introduce the results from Eqs. 7.40, 7.41, 7.42, and 7.43 to achieve

$$q_H + (g_n E_P - g_n E_T) + \left(\frac{p_1}{\rho_1} - \frac{p_2}{\rho_2}\right) = \left(\frac{1}{2}V^2 + g_n z + ie\right)_2 - \left(\frac{1}{2}V^2 + g_n z + ie\right)_1$$

This equation can be divided by g_n and reorganized to a more conventional form as the First Law of Thermodynamics for a control volume

$$\left(\frac{p_1}{\gamma_1} + \frac{V_1^2}{2g_n} + z_1\right) + E_P = \left(\frac{p_2}{\gamma_2} + \frac{V_2^2}{2g_n} + z_2\right) + E_T + \frac{1}{g_n}(ie_2 - ie_1 - q_H) \tag{7.44}$$

When Eq. 7.44 is written for the flow of liquids and gases in the numerous engineering situations in which there is negligible change of fluid density and the effects of pumps and turbines are neglected, Eq. 7.44 becomes a direct analogy to Eq. 7.35 and both apply to precisely the same conditions. Equation 7.44 becomes

$$\left(\frac{p_1}{\gamma} + \frac{V_1^2}{2g_n} + z_1\right) - \left(\frac{p_2}{\gamma} + \frac{V_2^2}{2g_n} + z_2\right) = \frac{1}{g_n}(ie_2 - ie_1 - q_H)$$

Noting also Eq. 7.36, it is clear then that

$$h_{L_{1-2}} = \frac{\tau_o(l_2 - l_1)}{\gamma R_h} = \frac{1}{g_n}(ie_2 - ie_1 - q_H) \tag{7.45}$$

In such problems the separate values of ie_2, ie_1, and q_H are usually not required, and the "packaging" of them into a single term, $h_{L_{1-2}}$, proves highly effective for engineering use. The equation offers proof that head loss is not a loss of total energy but rather a conversion of energy into heat, part of which leaves the fluid, the remainder serving to increase its internal energy. This is the practical case of incompressible flow as it appears in many engineering applications; here head loss is a permissible and useful concept because heat energy leaving the flow and energy converted into internal energy are seldom recoverable and are in effect lost from the useful total of pressure, velocity, and potential energies. For compressible flow this is not generally true, since the useful total of energies will include the internal energy (see Section 13.14).

Introducing Eq. 7.45 into Eq. 7.44 under conditions of constant density generates a reasonably (but not completely) general energy equation for steady incompressible flow

$$\left(\frac{p_1}{\gamma} + \frac{V_1^2}{2g_n} + z_1\right) + E_P = \left(\frac{p_2}{\gamma} + \frac{V_2^2}{2g_n} + z_2\right) + E_T + h_{L_{1-2}} \tag{7.46}$$

For the record, chemical, electrical, and atomic energies and the kinetic energy of turbulence are excluded from this analysis and result. Also, because of the use of mean velocity, pressure, density, and temperature (implicit in the internal energy), correction factors should be applied to the terms on the left side of Eq. 7.46, as well as to the head loss term, to allow for the nonuniform distribution of these quantities across the flow at sections 1 and 2. These are discussed in Section 7.12, which follows. However, these correction factors are usually close to unity for practical and turbulent flows, and such refinement of the equation is generally not necessary for engineering use; these correction factors have been omitted here.

ILLUSTRATIVE PROBLEM 7.8

A flowrate of 1.42 m^3/s of water occurs in a streamtube, say a pipe of variable cross section, containing a pump, but no turbine. The pump delivers 300 kW to the flowing fluid. Measurements at two points, i.e., 1 and 2 along the streamtube, show that $A_1 = 0.4$ m^2, $A_2 = 0.2$ m^2, $z_1 = 9$ m, $z_2 = 24$ m, $p_1 = 138$ kPa, $p_2 = 69$ kPa. Calculate the head lost between sections 1 and 2.

SOLUTION

Taking $\gamma = 9.81$ kN/m^3 and using Eqs. 4.4 and 5.8 produces

$$Q = AV,\ 1.42 = 0.4V_1 = 0.2V_2 \qquad V_1 = 3.55\ \text{m/s} \qquad V_2 = 7.1\ \text{m/s} \tag{4.4}$$

$$P = Q\gamma E_P \qquad \text{and so} \qquad E_P = (300 \times 10^3)/(9\,810 \times 1.42)$$

$$= 21.54\ \text{J/N} = 21.54\ \text{N·m/N} = 21.54\ \text{m} \tag{5.8}$$

Apply Eq. 7.46 with $E_T = 0$ to find (recall $g_n = 9.81 \text{ m/s}^2$)

$$\left(\frac{p_1}{\gamma_1} + \frac{V_1^2}{2g_n} + z_1\right) + E_P = \left(\frac{p_2}{\gamma_2} + \frac{V_2^2}{2g_n} + z_2\right) + E_T + h_{L_{1-2}} \tag{7.46}$$

$$\left(\frac{138\,000}{9\,810} + \frac{(3.55)^2}{2g_n} + 9\right) + 21.54 = \left(\frac{69\,000}{9\,810} + \frac{(7.1)^2}{2g_n} + 24\right) + h_{L_{1-2}} \tag{7.46}$$

$$h_{L_{1-2}} = 11.7 \text{ m} \bullet$$

7.12 VELOCITY DISTRIBUTION AND ITS SIGNIFICANCE

The shearing stresses of laminar and turbulent flow produce velocity distributions characterized by reduced velocities near boundary surfaces. These deviations from the uniform velocity distribution of ideal fluid flow necessitate alterations in the methods for calculation of velocity head and momentum flux. In many practical problems, these alterations are so small that they may be neglected. The effect of nonuniform velocity distribution on the computation of flowrate has been indicated in Section 4.2, which might usefully be restudied at this point.

The kinetic energy of fluid moving in the differentially small streamtube of Fig. 7.24 is (from Section 5.5) $(dQ)\gamma v^2/2g_n$ or $\rho v^3 dA/2$. The momentum flux (Section 6.1) is $(dQ)\rho v$ or $\rho v^2\, dA$. The total kinetic energy and momentum flux are the respective integrals of these differential quantities; thus

$$\text{Total kinetic energy (J/s or ft·lb/s)} = \frac{\rho}{2}\iint_A v^3\, dA \tag{7.47}$$

$$\text{Momentum flux (N or lb)} = \rho \iint_A v^2\, dA \tag{7.48}$$

Although these quantities are computed as indicated in many engineering situations, they are also usefully expressed in terms of mean velocity V and total flowrate Q, using the same forms of the equations; thus

$$\text{Total kinetic energy} = \alpha Q\gamma\frac{V^2}{2g_n} = Q\gamma\left(\alpha\frac{V^2}{2g_n}\right) \tag{7.49}$$

$$\text{Momentum flux} = \beta Q\rho V \tag{7.50}$$

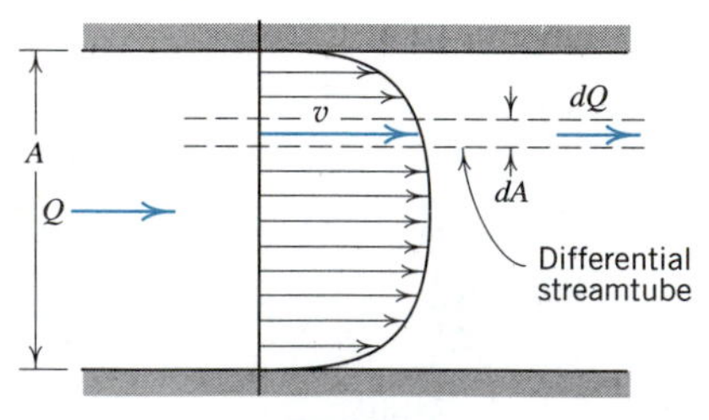

Fig. 7.24

in which α and β are dimensionless and represent correction factors to the conventional velocity head $V^2/2g_n$ and momentum flux $Q\rho V$, respectively. For a uniform velocity distribution $\alpha = \beta = 1$; for a nonuniform velocity profile $\alpha > \beta > 1$. Expressions for α and β may be derived by equating Eqs. 7.47 and 7.49 and Eqs. 7.48 and 7.50, substituting $\int^A v\,dA$ for Q, and obtaining

$$\alpha = \frac{1}{V^2}\frac{\iint_A v^3\,dA}{\iint_A v\,dA} \quad \text{and} \quad \beta = \frac{1}{V}\frac{\iint_A v^2\,dA}{\iint_A v\,dA} \tag{7.51}$$

from which it is easily seen that $\alpha = \beta = 1$ for uniform velocity distributions. Comparing the rather pointed (far from uniform) velocity distributions of laminar flow (Fig. 7.6) with the rather flattened (nearer to uniform) ones of turbulent flow (Fig. 7.7), it may be concluded directly that high values of α and β are to be expected for the former, lower ones for the latter.

For the established flow of a real fluid in a prismatic passage the analysis of Section 5.3 applies,[25] and $(p/\gamma + z)$ is a constant throughout any flow cross section. The hydrostatic pressure distribution is not affected by viscosity, but only by the streamline curvature. From this it is immediately seen that $(z + p/\gamma + v^2/2g_n)$ cannot be constant throughout the flow cross section, and this in turn requires some re-examination of the energy line concept. Taking A and C as typical streamlines, it is noted from Fig. 7.25 that each streamline is associated with a different energy line; in other words, the flow along different

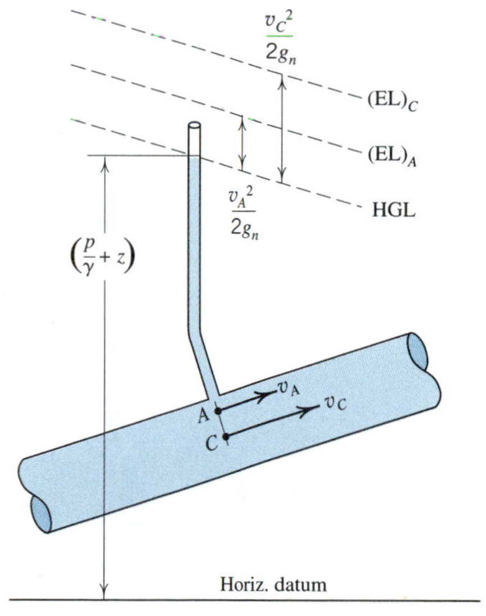

Fig. 7.25

[25]Provided that the shearing stresses in the fluid perpendicular to the direction of motion may be neglected. These are safely negligible in engineering problems except for flows dominated by viscous action. Such flows occur at very low Reynolds numbers and low velocity; they are frequently termed *creeping motions*.

streamlines possesses different amounts of total energy, and a comprehensive energy line picture would be a "bundle of energy lines"—one for each streamline. However, for flow in a parallel-walled passage, such as a pipe, duct, or open channel, the properties of individual streamlines are seldom of interest, and the whole aggregation of streamlines (i.e., the whole flow) is characterized by a single effective energy line a distance $\alpha V^2/2g_n$ above the hydraulic grade line; thus alterations in the work-energy equation due to nonuniform velocity distribution are concentrated in the coefficient α alone. However, in problems involving flowfields each streamline will in general be associated with different amounts of total energy, a fact essential to the analysis and interpretation of such problems.

Another consequence of nonuniform velocity distribution can be shown in its simplest aspects by considering a flow through a short constriction (Fig. 7.26) in a passage where the coefficient α changes from section to section. The trend of change of α from section 1 to section 2 may be established by noting that all fluid particles in passing from section 1 to section 2 experience the same change in $(p/\gamma + z)$ or the same drop in the hydraulic grade line. The velocities at section 1 are relatively small, and those at section 2 relatively large; with frictional effects in the constriction relatively small, the increase in the velocity of all particles from section 1 to section 2 will tend to be about the same; the velocity profile at section 2 will therefore be flatter than that at section 1, and α_2 will accordingly be less than α_1. The *exact head loss* between sections 1 and 2 is given by the drop in the energy line between these sections; from Fig. 7.26 (cf., Eq. 7.46),

$$\text{Exact } h_{L_{1-2}} = \left(\alpha_1 \frac{V_1^2}{2g_n} + \frac{p_1}{\gamma} + z_1\right) - \left(\alpha_2 \frac{V_2^2}{2g_n} + \frac{p_2}{\gamma} + z_2\right) \tag{7.52}$$

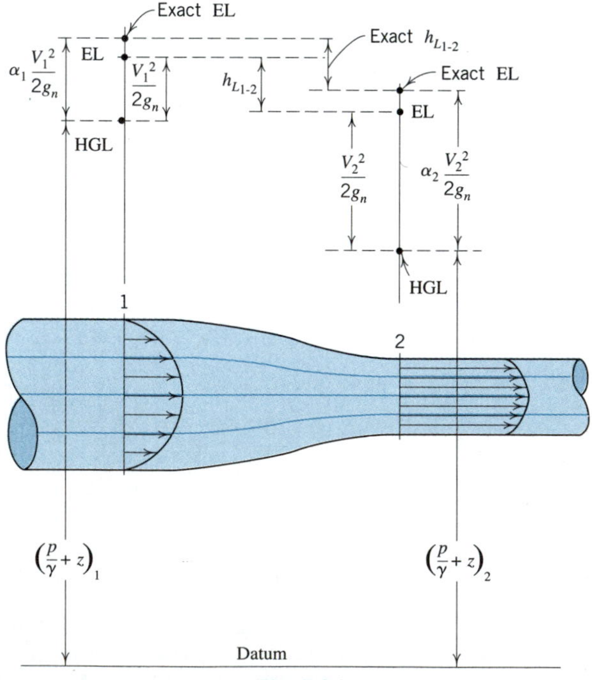

Fig. 7.26

However, the conventional head loss used in engineering computations is defined by ignoring the α terms in the work-energy equation and writing

$$h_{L_{1-2}} = \left(\frac{V_1^2}{2g_n} + \frac{p_1}{\gamma} + z_1\right) - \left(\frac{V_2^2}{2g_n} + \frac{p_2}{\gamma} + z_2\right) \tag{7.53}$$

Comparison of the two equations leads to

$$h_{L_{1-2}} = \text{Exact } h_{L_{1-2}} + (\alpha_2 - 1)\frac{V_2^2}{2g_n} - (\alpha_1 - 1)\frac{V_1^2}{2g_n} \tag{7.54}$$

from which it is seen that the conventional head loss does not equal exact head loss unless $(\alpha_2 - 1)V_2^2/2g_n = (\alpha_1 - 1)V_1^2/2g_n$. However, constriction of the passage causes V_2 to be greater than V_1, while the flattening of the velocity profile causes $(\alpha_2 - 1)$ to be less than $(\alpha_1 - 1)$; thus there are some compensating features in Eq. 7.54 which tend to make exact head loss not very different from conventional head loss in this simple example. Although in other cases these conventional and exact head losses may differ considerably, it should be noted that this is no serious obstacle in most engineering problems; no matter how these losses may be defined or related to each other, there can be (for a given flowrate) only one value for the change of $(p/\gamma + z)$ between sections 1 and 2; if calculations are made with this fact in mind, reliable predictions can be made from conventional head losses even though these losses are never precisely equal to the exact ones.

ILLUSTRATIVE PROBLEM 7.9

Assuming Fig. 7.24 to represent a parabolic velocity profile in a passage bounded by two infinite planes of spacing $2R$ and maximum velocity v_c, calculate q, α, and β.

SOLUTION

The necessary equations are Eqs. 4.8 and 7.51, namely,

$$q = VA = V(2R) = \iint_A v\,dA \tag{4.8}$$

$$\alpha = \frac{1}{V^2}\frac{\displaystyle\iint_A v^3\,dA}{\displaystyle\iint_A v\,dA} \quad \text{and} \quad \beta = \frac{1}{V}\frac{\displaystyle\iint_A v^2\,dA}{\displaystyle\iint_A v\,dA} \tag{7.51}$$

Taking r as the distance from centerline of the passage to any local velocity, v, and element of area dA, $dA = dr$. The equation of the parabola is $v = v_c(1 - r^2/R^2)$.

$$q = 2\int_0^R v_c\left(1 - \frac{r^2}{R^2}\right)\text{dr} = \frac{2}{3}(2Rv_c) \tag{4.8}$$

Since q also equals $2RV$, $V = 2v_c/3$.

$$\alpha = \frac{2\int_0^R v_c^3\left(1 - \frac{r^2}{R^2}\right)^3 \mathrm{dr}}{\left(\frac{2v_c}{3}\right)^2 \frac{2}{3}(2Rv_c)} = \frac{54}{35} = 1.54 \tag{7.51}$$

$$\beta = \frac{2\int_0^R v_c^2\left(1 - \frac{r^2}{R^2}\right)^2 dr}{\left(\frac{2v_c}{3}\right) \frac{2}{3}(2Rv_c)} = \frac{6}{5} = 1.20 \tag{7.51}$$

The meaning of these figures is that the exact velocity head is more than 54% greater than $V^2/2g_n$, and the exact momentum flux 20% greater than $Q\rho V$. Differences of this magnitude warn the engineer that α and β should be considered when applying the energy and momentum equations to one-dimensional laminar flow problems—unless their effects can be shown to have negligible consequence in the results desired.

7.13 SEPARATION

Separation has great impact on design and performance of certain familiar internal flow systems. For example, separation of the flow from the blades of a pump impeller or turbine runner can cause serious degradation of efficiency and often damage to the system. Careful design can prevent separation in these cases; but for flow through a valve or metering orifice or a sudden expansion, separation is inevitable. Figure 7.27 shows both classical ideal fluid and real fluid flows for three cases (see the discussion in Section 7.7 also).

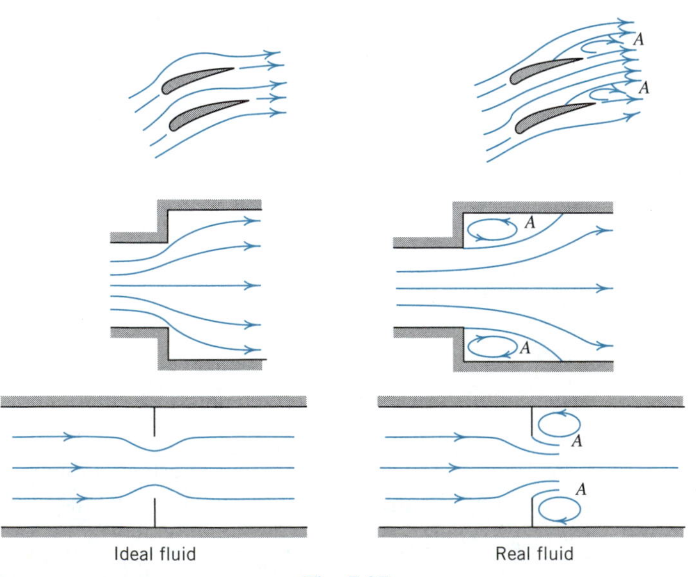

Fig. 7.27

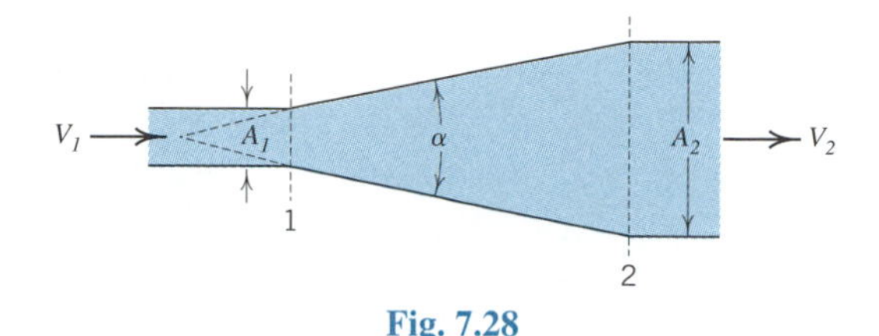

Fig. 7.28

A classic internal flow in which separation is often a major problem is that in a *diffuser* (Fig. 7.28). Here the engineering objective is to provide an expanding passage of proper shape and minimum length that will yield minimum head loss or maximum pressure rise for reduction of mean velocity from V_1 to V_2. A short passage with sharp wall curvatures will prove inefficient because of flow separation and large energy dissipation: a longer one will be efficient but too space-consuming and will produce too much boundary resistance. The optimum design lies between the extremes but (as yet) cannot be determined wholly by analytical methods; again experiments must be used to supplement and confirm analytical attacks on problems where separation is involved. The optimum included angle, α, in simple diffusers lies between 6° and 8° and depends on the velocity distribution in section 1 at the entrance to the diffuser. In cases in which wide-angle diffusers must be used, their performance can often be markedly improved by use of fixed flow-guide-vanes mounted in the diffuser body or by suction along the diffuser walls to remove slow moving fluid. Both efforts serve to suppress separation in the diffuser.

To reiterate a point made earlier: *Acceleration of real fluids tends to be an efficient process, deceleration an inefficient one.* Accelerating motion through a convergent nozzle, for example, is accompanied by a favorable pressure gradient which stabilizes the boundary layer and thus minimizes energy dissipation. Decelerating motion is accompanied by an adverse pressure gradient which promotes separation, instability, eddy formation, and large energy dissipation. This axiom may be extended to explain the difference between the maximum efficiencies obtained in comparable complex machines such as hydraulic turbines and centrifugal pumps. In turbines the flow passages are predominantly convergent and the flow is accelerated; in the pump the opposite situation obtains. Hydraulic turbine efficiencies have been obtained up to 94%, but maximum centrifugal pump efficiencies are around 87%. The striking difference between these figures can be attributed primarily to the inherent efficiency and inefficiency of the acceleration and deceleration flow processes, respectively.

7.14 SECONDARY FLOW

The secondary flow that develops in internal flow of real fluids is typified by the double spiral motion produced by a gradual bend in a closed passage. Consider the circular pipe bend of Fig. 7.29 and assume that its curves are gentle enough that separation is not to be expected. For an ideal fluid flowing under these conditions it has been shown (Section 5.8) that a pressure gradient develops across the bend due to the centrifugal forces of fluid particles as they move through the bend. Stability occurs in the ideal fluid when this pressure gradient brings about a balance between centrifugal and centripetal forces on the fluid particles. In a real fluid this stability is disrupted by the velocity reduction toward

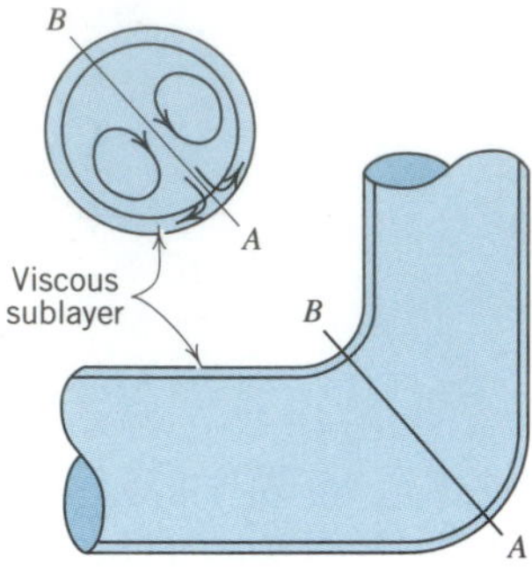

Fig. 7.29

the walls. The reduction of velocity at the outer part, A, of the bend reduces the centrifugal force of particles moving near the wall, causing the pressure at the wall to be below that which would be maintained in an ideal fluid. However, the velocities of fluid particles toward the center of the bend are about the same as those of the ideal fluid, and the pressure gradient developed by their centrifugal forces is about the same. The ''weakening'' of the pressure gradient at the outer wall will cause a flow to be set up from the center of the pipe toward the wall which will develop into the twin eddy motion shown, and this secondary motion added to the main flow will cause a double spiral motion. As shown in Section 9.9, a combination of the double spiral flow and separation leads to increased head losses in bends, a large part of which may be caused by increased wall shear stress in and downstream from the bend because of the redistribution of the flow streamlines by the secondary flow.

The Navier-Stokes Equations for Incompressible Flow

The equations of motion can be derived, as we have seen, either by application of Newton's second law to an element of fluid or by application of the impulse-momentum principle for control volumes. The derived equations are known to accurately represent the flow physics for Newtonian fluids (cf., Section 1.6) in very general circumstances, including three-dimensional unsteady flows with variable density. However, the equations are nonlinear and so their use has not blossomed until the advent of powerful computers made numerical simulations feasible and efficient. In this section, our attention is confined to two-dimensional, unsteady, but incompressible flows of fluids with a uniform and constant density fluid.[26] There is little difficulty in extending the derivation to three dimensions, but it is left as an exercise (Problem 7.80). The so-called Navier-Stokes equations derived here are applicable to both laminar and turbulent flow and underlie much of the practice of modern fluid mechanics (see, for example, virtually any issue of the *Journal of Fluid Mechanics*, *The International Journal for Numerical Methods in Fluids*, or the *Journal of Fluids Engineering*). These equations are introduced here for these reasons.

[26]More general and complete derivations may be found in H. Schlichting, *Boundary-Layer Theory*, 7th ed., New York: McGraw-Hill, 1979, and J. O. Hinze, *Turbulence*, 2nd ed., New York: McGraw-Hill, 1975; some historical context is given by H. Rouse and S. Ince, *History of hydraulics*, New York: Dover, 1963.

7.15 DERIVATION OF THE NAVIER-STOKES EQUATIONS FOR TWO-DIMENSIONAL FLOW

The derivation is carried out for an infinitesimal control volume and an incompressible, constant density fluid. The key elements of the derivation are the Reynolds Transport Theorem (Section 4.4, Eq. 4.15) and the impulse-momentum principle (Section 6.1) in which $E = m\mathbf{V}_c$ and $i = \mathbf{V}$. In the derivation in Section 6.1, it was learned that there are two types of forces that can act on the control volume, namely, body and surface forces. Body forces act on each element of mass in the control volume, e.g., the gravitational attractive force; surface forces act at the control surface, e.g., frictional forces. In Section 5.6, the Euler equations were derived for two-dimensional, steady flows by examination of the forces acting on an infinitesimal mass element; the result was Eqs. 5.9. In the case of these Euler equations, the forces acting were the gravitational attractive force (a body force) and the pressure (or normal stress) force (a surface force). For a real fluid, however, the viscosity leads to the presence of surface forces, namely, shear forces or shear stresses in the moving fluid (cf., Chapter 1, Eq. 1.12 and the introduction to Chapter 5) in addition to the pressure force or normal stress. Thus, it is necessary to determine what forces can and do act on an infinitesimal control volume in a real fluid flow and to generalize the simple relationship between shear stress and rate of strain in the flow given by Eq. 1.12.

Before applying the Reynolds Transport Theorem, it is convenient to generate the expressions for the surface forces. To describe the two-dimensional, unsteady motion in the vertical plane we employ the Cartesian coordinates (x, z). However, consider for a moment the three-dimensional sketch of Fig. 7.30, which visualizes the surface forces. The fluid element has dimensions of dx and dz in the vertical plane, but is of unit depth in the y-direction, i.e., perpendicular to the x-z plane. To facilitate the work of readers who seek to delve into the details and want more depth than given here, the pictured stresses follow the notation of Schlichting,[27] who gives a full description of the origin of the stresses

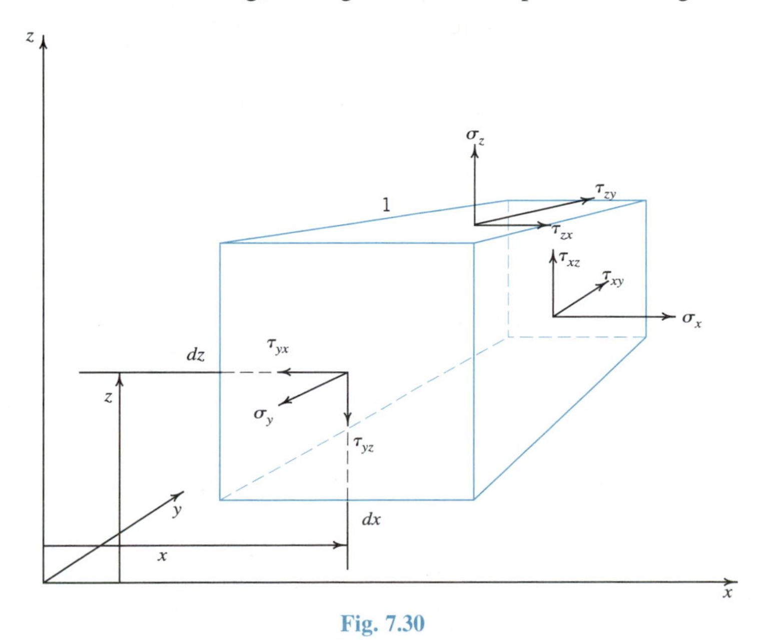

Fig. 7.30

[27]H. Schlichting, *Boundary-Layer Theory*, 7th ed., New York: McGraw-Hill, 1979.

and their form. The stresses are designated to show whether they are normal (σ) or shear stresses (τ); they act on the surface areas of the element and have the dimensions of force per unit area. Only the stresses on the visible faces are shown in Fig. 7.30; appropriate and similar forces act on the other faces, but have the opposite orientation. Normal stresses are always assumed to act in the direction of the outward normal on the element, which is a control volume here. The shear stresses, caused by the relative movement of fluid inside and outside the control surface, are defined to be positive in the direction of the coordinates on the forward faces of the element (i.e., those faces farthest from the coordinate origin). Accordingly, the stress τ_{xy} lies on the face of the element lying in a y-z plane (whose location is thus a function of x) and points in the positive y-direction.; the stress τ_{zx} lies on the face of the element lying in x-y plane (whose location is thus a function of z) and points in the positive x-direction, but the stress τ_{yx} lies in the face of the element lying in a z-x plane (whose location is a function of y) and points in the negative x-direction, etc.

With these preliminaries in hand, we recall that Newton's second law can be written as

$$\sum \mathbf{F} = \frac{d}{dt}(E) = \frac{d}{dt}(m\mathbf{v}_c) \tag{7.55}$$

for a system and so we need to define the forces acting on the control volume and then find the rate of change of the extensive property E. We can use the two-dimensional picture of Fig. 7.31 to derive the active forces on the control volume ABCD in the form of their x and z components; note that

$$\sum \mathbf{F} = \sum F_x\mathbf{e}_x + \sum F_z\mathbf{e}_z$$

(Check Appendix 6.) If everything is implicitly multiplied by the unit depth of the element in the direction perpendicular to the paper, the y-direction, then the force summations are

$$\sum F_x = \left(\sigma_x + \frac{\partial \sigma_x}{\partial x}\frac{dx}{2}\right) dz - \left(\sigma_x - \frac{\partial \sigma_x}{\partial x}\frac{dx}{2}\right) dz + \left(\tau_{zx} + \frac{\partial \tau_{zx}}{\partial z}\frac{dz}{2}\right) dx - \left(\tau_{zx} - \frac{\partial \tau_{zx}}{\partial z}\frac{dz}{2}\right) dx \tag{7.56}$$

$$= \frac{\partial \sigma_x}{\partial x} dx\, dz + \frac{\partial \tau_{zx}}{\partial z} dx\, dz$$

$$\sum F_z = \left(\sigma_z + \frac{\partial \sigma_z}{\partial z}\frac{dz}{2}\right) dx - \left(\sigma_z - \frac{\partial \sigma_z}{\partial z}\frac{dz}{2}\right) dx + \left(\tau_{xz} + \frac{\partial \tau_{xz}}{\partial x}\frac{dx}{2}\right) dz - \left(\tau_{xz} - \frac{\partial \tau_{xz}}{\partial x}\frac{dx}{2}\right) dz - \rho g_n\, dz\, dx \tag{7.57}$$

$$= \frac{\partial \sigma_z}{\partial z} dz\, dx + \frac{\partial \tau_{xz}}{\partial x} dz\, dx - \rho g_n\, dz\, dx$$

The Reynolds Transport Theorem equation 4.15 states

$$\frac{dE}{dt} = \frac{\partial}{\partial t}\left(\iiint_{C.V.} i\rho\, d\forall\right) + \iint_{C.S.} i\rho\mathbf{v}\cdot d\mathbf{A} \tag{4.15}$$

We must now evaluate the terms in this equation for the infinitesimal control volume in Fig. 7.31; something similar was done in Section 4.3 and we can follow that example. To begin, with $i = \mathbf{v}$,

$$\frac{\partial}{\partial t}\left(\iiint_{C.V.} i\rho \, d\text{V}\right) = \frac{\partial}{\partial t}\left(\iiint_{C.V.} \mathbf{v}\rho \, d\text{V}\right) = \frac{\partial}{\partial t}\left(\iiint_{C.V.} (u\mathbf{e}_x + w\mathbf{e}_z)\rho \, d\text{V}\right)$$

$$= \frac{\partial}{\partial t}(u\mathbf{e}_x + w\mathbf{e}_z)\rho \, dz \, dx = \left(\frac{\partial u}{\partial t}\mathbf{e}_x + \frac{\partial w}{\partial t}\mathbf{e}_z\right)\rho \, dz \, dx \qquad (7.58)$$

because the density is constant. Next, we have

$$\oiint_{C.S.} i\rho\mathbf{v} \cdot d\mathbf{A} = \oiint_{C.S.} \mathbf{v}(\rho\mathbf{v} \cdot \mathbf{n} \, dA) = \left(\iint_{AB} + \iint_{BC} + \iint_{CD} + \iint_{DA} (\mathbf{v}(\rho\mathbf{v} \cdot \mathbf{n} \, dA))\right)$$

For the control volume of Fig. 7.31,

$$\iint_{BC} (\mathbf{v}(\rho\mathbf{v} \cdot \mathbf{n} \, dA)) \cong \mathbf{e}_x\left[\left(u + \frac{\partial u}{\partial x}\frac{dx}{2}\right) \times \rho\left(u + \frac{\partial u}{\partial x}\frac{dx}{2}\right) dz\right]$$
$$+ \mathbf{e}_z\left[\left(w + \frac{\partial w}{\partial x}\frac{dx}{2}\right) \times \rho\left(u + \frac{\partial u}{\partial x}\frac{dx}{2}\right) dz\right]$$

$$\iint_{CD} (\mathbf{v}(\rho\mathbf{v} \cdot \mathbf{n} \, dA)) \cong \mathbf{e}_x\left[\left(u + \frac{\partial u}{\partial z}\frac{dz}{2}\right) \times \rho\left(w + \frac{\partial w}{\partial z}\frac{dz}{2}\right) dx\right]$$
$$+ \mathbf{e}_z\left[\left(w + \frac{\partial w}{\partial z}\frac{dz}{2}\right) \times \rho\left(w + \frac{\partial w}{\partial z}\frac{dz}{2}\right) dx\right]$$

$$\iint_{DA} (\mathbf{v}(\rho\mathbf{v} \cdot \mathbf{n} \, dA)) \cong \mathbf{e}_x\left[\left(u - \frac{\partial u}{\partial x}\frac{dx}{2}\right) \times \left(-\rho\left(u - \frac{\partial u}{\partial x}\frac{dx}{2}\right) dz\right)\right]$$
$$+ \mathbf{e}_z\left[\left(w - \frac{\partial w}{\partial x}\frac{dx}{2}\right) \times \left(-\rho\left(u - \frac{\partial u}{\partial x}\frac{dx}{2}\right) dz\right)\right]$$

$$\iint_{AB} (\mathbf{v}(\rho\mathbf{v} \cdot \mathbf{n} \, dA)) \cong \mathbf{e}_x\left[\left(u - \frac{\partial u}{\partial z}\frac{dz}{2}\right) \times \left(-\rho\left(w - \frac{\partial w}{\partial z}\frac{dz}{2}\right) dx\right)\right]$$
$$+ \mathbf{e}_z\left[\left(w - \frac{\partial w}{\partial z}\frac{dz}{2}\right) \times \left(-\rho\left(w - \frac{\partial w}{\partial z}\frac{dz}{2}\right) dx\right)\right]$$

Simplifying by combining terms yields

$$\oiint_{C.S.} \mathbf{v}(\rho\mathbf{v} \cdot \mathbf{n} \, dA) = \mathbf{e}_x(\rho \, dx \, dz)\left(2u\frac{\partial u}{\partial x} + u\frac{\partial w}{\partial z} + w\frac{\partial u}{\partial z}\right)$$
$$+ \mathbf{e}_z(\rho \, dx \, dz)\left(2w\frac{\partial w}{\partial z} + u\frac{\partial w}{\partial x} + w\frac{\partial u}{\partial x}\right)$$

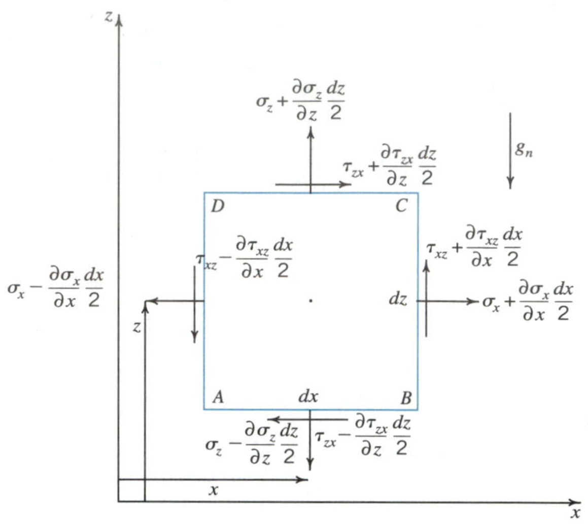

Fig. 7.31

Next, we need to be reminded (from our derivation of the Euler equations) that the kinematic equation of continuity (conservation of mass) needs to accompany the dynamic equations in order to fully describe the flow. Thus, for the current case, Eq. 4.11

$$\frac{\partial u}{\partial x} + \frac{\partial w}{\partial z} = 0 \tag{4.11}$$

will be included in the final set of equations and, indeed, it can be used now to bring the above equations to their final form, viz.,

$$\oiint_{C.S.} \mathbf{v}(\rho\mathbf{v} \cdot \mathbf{n}\, dA) = \mathbf{e}_x(\rho\, dx\, dz)\left(u\frac{\partial u}{\partial x} + w\frac{\partial u}{\partial z}\right) + \mathbf{e}_z(\rho\, dx\, dz)\left(w\frac{\partial w}{\partial z} + u\frac{\partial w}{\partial x}\right) \tag{7.59}$$

Now, joining Eq. 7.55 with Eq. 4.15 and inserting the values found in Eqs. 7.56 through 7.59 produces

$$\mathbf{e}_x\left(\frac{\partial \sigma_x}{\partial x} + \frac{\partial \tau_{zx}}{\partial z}\right) dx\, dz + \mathbf{e}_z\left(\frac{\partial \sigma_z}{\partial z} + \frac{\partial \tau_{xz}}{\partial x} - \rho g_n\right) dx\, dz$$
$$= \mathbf{e}_x\left(\frac{\partial u}{\partial t} + u\frac{\partial u}{\partial x} + w\frac{\partial u}{\partial z}\right)\rho\, dx\, dz + \mathbf{e}_z\left(\frac{\partial w}{\partial t} + u\frac{\partial w}{\partial x} + w\frac{\partial w}{\partial z}\right)\rho\, dx\, dz$$

It follows that the components in each direction must separately satisfy this equation and so upon canceling the common terms we have

$$\rho\left(\frac{\partial u}{\partial t} + u\frac{\partial u}{\partial x} + w\frac{\partial u}{\partial z}\right) = \frac{\partial \sigma_x}{\partial x} + \frac{\partial \tau_{zx}}{\partial z}$$

and (7.60)

$$\rho\left(\frac{\partial w}{\partial t} + u\frac{\partial w}{\partial x} + w\frac{\partial w}{\partial z}\right) = \frac{\partial \sigma_z}{\partial z} + \frac{\partial \tau_{xz}}{\partial x} - \rho g_n$$

It remains to find a relationship between the stresses in these equations, the flow variables, and the viscosity. This relationship was embodied in the *Stokes' hypothesis*; we shall only state it here for the constant density, incompressible, and two-dimensional flow that is being considered. However, it is a generalization of the ideas put forth in Section 1.6, namely, that the shear stress is proportional to the rate of strain of the fluid. The Stokes' hypothesis or *constitutive relation* depends on a result that Schlichting shows by application of the moment of momentum principle about the center of control volume ABCD. Importantly, he finds that the shear stresses are symmetric, that is, $\tau_{xz} = \tau_{zx}$. This is a general result that applies to the shear stresses in three dimensions as well. A statement of the Stokes' hypothesis is

$$\begin{aligned} \sigma_x &= -p + 2\mu \frac{\partial u}{\partial x} \\ \sigma_z &= -p + 2\mu \frac{\partial w}{\partial z} \\ \tau_{xz} &= \tau_{zx} = \mu \left(\frac{\partial u}{\partial z} + \frac{\partial w}{\partial x} \right) \end{aligned} \tag{7.61}$$

Here, the rate of strain is $\frac{\partial u}{\partial z} + \frac{\partial w}{\partial x}$ and the pressure is defined to be

$$p = -(\sigma_x + \sigma_z). \tag{7.62}$$

This pressure is known as the *thermodynamic pressure* and its definition as the average of the normal stresses is also valid in three dimensions. Inserting Eqs. 7.61 and 7.62 into Eq. 7.60 produces

$$\rho \left(\frac{\partial u}{\partial t} + u \frac{\partial u}{\partial x} + w \frac{\partial u}{\partial z} \right) = \frac{\partial}{\partial x} \left(-p + 2\mu \frac{\partial u}{\partial x} \right) + \frac{\partial}{\partial z} \left(\mu \left(\frac{\partial w}{\partial x} + \frac{\partial u}{\partial z} \right) \right)$$

and

$$\rho \left(\frac{\partial w}{\partial t} + u \frac{\partial w}{\partial x} + w \frac{\partial w}{\partial z} \right) = \frac{\partial}{\partial z} \left(-p + 2\mu \frac{\partial w}{\partial w} \right) + \frac{\partial}{\partial x} \left(\mu \left(\frac{\partial w}{\partial x} + \frac{\partial u}{\partial z} \right) \right) - \rho g_n$$

Recalling that the density and the viscosity are constant in this flow situation and using the continuity equation 4.11 cited above to simplify the equations one obtains the Navier-Stokes equations, which apply to laminar and turbulent flow:

$$a_x = \frac{du}{dt} = \frac{\partial u}{\partial t} + u \frac{\partial u}{\partial x} + w \frac{\partial u}{\partial z} = -\frac{1}{\rho} \frac{\partial p}{\partial x} + \nu \left(\frac{\partial^2 u}{\partial x^2} + \frac{\partial^2 u}{\partial z^2} \right)$$

and (7.63)

$$a_y = \frac{dw}{dt} = \frac{\partial w}{\partial t} + u \frac{\partial w}{\partial x} + w \frac{\partial w}{\partial z} = -\frac{1}{\rho} \frac{\partial p}{\partial z} + \nu \left(\frac{\partial^2 w}{\partial x^2} + \frac{\partial^2 w}{\partial z^2} \right) - g_n$$

The total or substantial derivatives were defined in Eqs. 3.5 and 3.6.

The Navier-Stokes equations can be written in cylindrical coordinates for axisymmetric flows (r, z) in the form

$$a_r = \frac{du_r}{dt} = \frac{\partial u_r}{\partial t} + u_r\frac{\partial u_r}{\partial r} + u_z\frac{\partial u_r}{\partial z} = -\frac{1}{\rho}\frac{\partial p}{\partial r} + \nu\left(\frac{\partial^2 u_r}{\partial r^2} + \frac{1}{r}\frac{\partial u_r}{\partial r} - \frac{u_r}{r^2} + \frac{\partial^2 u_r}{\partial z^2}\right)$$

and

$$a_z = \frac{du_z}{dt} = \frac{\partial u_z}{\partial t} + u_r\frac{\partial u_z}{\partial x} + u_z\frac{\partial u_z}{\partial z} = -\frac{1}{\rho}\frac{\partial p}{\partial z} + \nu\left(\frac{\partial^2 u_z}{\partial r^2} + \frac{1}{r}\frac{\partial u_z}{\partial r} + \frac{\partial^2 u_z}{\partial z^2}\right) \tag{7.64}$$

where the velocities in the r- and the z-directions are given by u_r and u_z, respectively, and the gravitational force is neglected.

To make a comparison with Eqs. 5.9, we assume steady inviscid flow, so that the viscous and time derivative terms vanish, yielding as expected the Euler equations

$$u\frac{\partial u}{\partial x} + w\frac{\partial u}{\partial z} = -\frac{1}{\rho}\frac{\partial p}{\partial x}$$

and

$$u\frac{\partial w}{\partial x} + w\frac{\partial w}{\partial z} = -\frac{1}{\rho}\frac{\partial p}{\partial z} - g_n \tag{5.9}$$

7.16 APPLICATIONS OF THE NAVIER-STOKES EQUATIONS

As noted earlier, there are only a few useful analytic exact solutions to the Navier-Stokes equations. In this section, we introduce, in the form of illustrative problems, two analytic solutions that relate to laminar steady flow problems, which we have encountered or will encounter in this book. These are (1) fluid flow between large parallel plates in which one is moving (the Couette flow of Section 1.6) and (2) flow through a pipe (Section 9.2). The flow between rotating cylinders (Section 1.6) is left as an exercise. Then, the time-averaged Navier-Stokes equations for turbulent flow are derived and the Boussinesq eddy viscosity of Section 7.2 is generalized.

ILLUSTRATIVE PROBLEM 7.10

Derive the velocity profile for the steady unidirectional flow between two large flat plates which are parallel to each other and separated by a distance h (cf., Fig. 1.7). Let the top plate move with a velocity V as shown in the figure below.

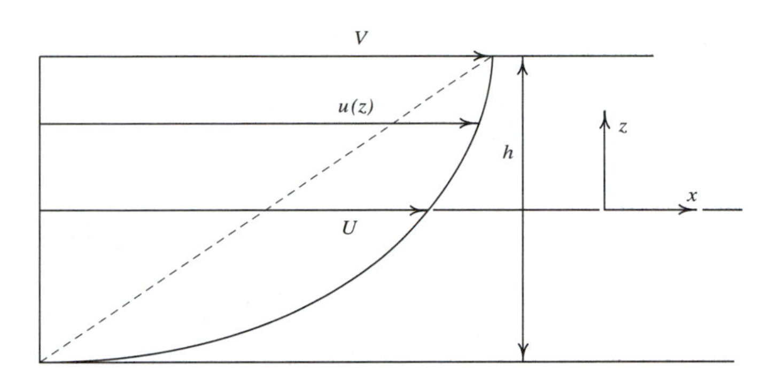

SOLUTION

From the problem specification it is possible to draw several conclusions:

1. The no-slip boundary conditions at the plates produce: $u = 0$ and $w = 0$ at $z = -h/2$; but $u = V$ and $w = 0$ at $z = h/2$.
2. Because the flow is parallel and steady, the vertical velocities are all zero.
3. It follows from the continuity equation 4.11 that the derivatives in the x-direction of the horizontal velocities are zero.
4. The temporal acceleration is zero.

Note also that since there are no velocity variations in the x-direction, the partial derivatives of the velocity can be replaced by total derivatives because u is a function of z alone. The remaining Navier-Stokes terms are then

$$0 = -\frac{1}{\rho}\frac{\partial p}{\partial x} + \nu\frac{d^2u}{dz^2} \quad \text{and} \quad 0 = -\frac{1}{\rho}\frac{\partial p}{\partial z} - g_n \tag{7.63}$$

The second equation yields the hydrostatic relationship that the vertical variation of $p \propto -\rho g_n$, while the first equation can be integrated with respect to z. If the pressure gradient in the x-direction is specified, then the velocity profile is known. The integration yields $u(z) = \frac{1}{\nu}\left(\frac{1}{\rho}\frac{\partial p}{\partial x}\right)\frac{z^2}{2} + C_1 z + C_2$, where C_1 and C_2 are unknown constants; since $u(-h/2) = 0$, $C_2 = -\frac{1}{\nu}\left(\frac{1}{\rho}\frac{\partial p}{\partial x}\right)\frac{h^2}{8} - C_1\left(\frac{-h}{2}\right)$. Because $u(h/2) = V$, $V = \frac{1}{\nu}\left(\frac{1}{\rho}\frac{\partial p}{\partial x}\right)\frac{h^2}{8} + C_1\left(\frac{h}{2}\right) + C_2$. Solving for C_1 and C_2 yields $C_1 = \frac{V}{h}$ and $C_2 = \frac{V}{2} - \frac{1}{\nu}\left(\frac{1}{\rho}\frac{\partial p}{\partial x}\right)\frac{h^2}{8}$.

Upon rearrangement and simplification (e.g., $\nu\rho = \mu$) the velocity profile is

$$u(z) = \frac{1}{2\mu}\left(\frac{\partial p}{\partial x}\right)\left(z^2 - \left(\frac{h}{2}\right)^2\right) + V\left(\frac{1}{2} + \frac{z}{h}\right) \bullet$$

This profile consists of two distinct parts; if the pressure gradient is zero, then the profile is linear as we found in Chapter 1. If the upper plate is still, then the profile is parabolic and the maximum velocity depends on the pressure gradient as $u(0) = U = -\frac{h^2}{8\mu}\left(\frac{\partial p}{\partial x}\right)$. Other variations are left for the problems.

ILLUSTRATIVE PROBLEM 7.11

Find the velocity profile for steady incompressible laminar flow in a pipe of diameter d as shown below. This is known as the *Hagen-Poiseuille Flow*.

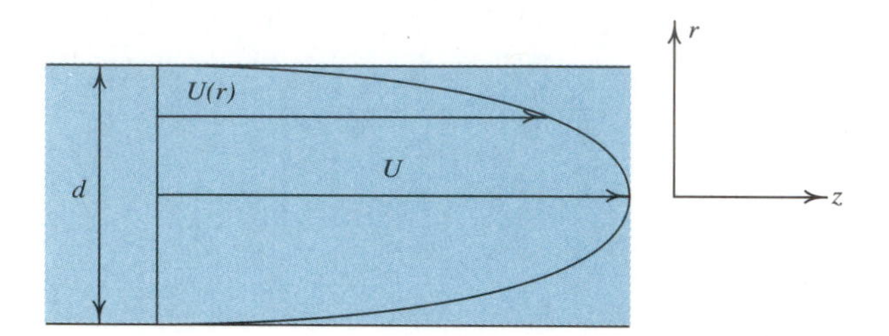

SOLUTION

Making use of the insights from the previous problem, we can simplify Eq. 7.64 by remembering that there is no radial velocity, no gradients with z (so remaining partial derivatives become total derivatives), and the temporal terms are zero. The results are

$$0 = -\frac{1}{\rho}\frac{\partial p}{\partial r} \rightarrow \frac{dp}{dr} = 0$$

$$0 = -\frac{1}{\rho}\frac{\partial p}{\partial z} + \nu\left(\frac{\partial^2 u_z}{\partial r^2} + \frac{1}{r}\frac{\partial u_z}{\partial r}\right) = -\frac{1}{\rho}\frac{dp}{dz} + \nu\left(\frac{d^2 u_z}{dr^2} + \frac{1}{r}\frac{du_z}{dr}\right)$$

Assuming that the pressure gradient is set by some external set of conditions and noting that

1. $u_z\left(\frac{d}{2}, z\right) = u_z\left(\frac{d}{2}\right) = 0$ because of the no-slip condition at the pipe wall.
2. $\frac{\partial u_z}{\partial r}(0, z) = \frac{du_z}{dr}(0)$ because the profile is rotationally symmetric; otherwise, the shear stress across the centerline is infinite where the velocity profile slope is discontinuous. Also, the above governing equation is singular at $r = 0$, so clearly for a well-behaved solution the derivative must be zero.

The resulting equation and boundary conditions can be integrated to produce the velocity profile

$$u_z = -\frac{1}{4\mu}\left(\frac{dp}{dz}\right)\left(\left(\frac{d}{2}\right)^2 - r^2\right) \bullet$$

Clearly, the profile is parabolic (see Section 9.2) and the velocity on the centerline is

$$U = \frac{d^2}{16\mu}\left(-\frac{dp}{dz}\right) \bullet$$

Early in this chapter, the difference between the instantaneous and mean velocities was examined briefly, leading us to the definition of the Reynolds stress and some attempts to represent it. The Navier-Stokes equations, being able to represent the full range of the fluid motion—laminar or turbulent, provide a useful vehicle for further examination of these points. The velocity components can be considered to be composed of two parts, i.e., a mean and a fluctuation. In Section 7.2, an intuitive notation is used of the velocity components of the simple flow considered there; here the standard notation for these components is introduced. The mean value is the value at a point found by taking the time average of the instantaneous velocities (that is, the average of the instantaneous values over all times as $t \to \infty$); the fluctuation is the difference between the instantaneous and the mean values. Thus, the instantaneous velocity may be *decomposed* as

$$u(x, z, t) = \overline{u}(x, z) + u' \quad \text{and} \quad w(x, z, t) = \overline{w}(x, z) + w',$$

where $\overline{\varphi} = \lim\limits_{T\to\infty} \dfrac{1}{T}\displaystyle\int \varphi\, dt$

In all other sections of this book the overbar is not used to designate the mean velocity because turbulent fluctuations are not discussed and it is unnecessary to distinguish between the mean and the fluctuations. Later, in discussion of unsteady flows the nomenclature will be made clear as the discussion proceeds. If these definitions are introduced into the Navier-Stokes equations 7.63 and the result is time averaged, one obtains the equations for the mean flow as follows:

1. Introduce velocity decomposition into the Navier-Stokes equations

$$\frac{\partial(\overline{u} + u')}{\partial t} + (\overline{u} + u')\frac{\partial(\overline{u} + u')}{\partial x} + (\overline{w} + w')\frac{\partial(\overline{u} + u')}{\partial z} = -\frac{1}{\rho}\frac{\partial p}{\partial x} + \nu\left(\frac{\partial^2(\overline{u} + u')}{\partial x^2} + \frac{\partial^2(\overline{u} + u')}{\partial z^2}\right)$$

and

$$\frac{\partial(\overline{w} + w')}{\partial t} + (\overline{u} + u')\frac{\partial(\overline{w} + w')}{\partial x} + (\overline{w} + w')\frac{\partial(\overline{w} + w')}{\partial z} = -\frac{1}{\rho}\frac{\partial p}{\partial z} + \nu\left(\frac{\partial^2(\overline{w} + w')}{\partial x^2} + \frac{\partial^2(\overline{w} + w')}{\partial z^2}\right) - g_n$$

2. Time average, noting that $\overline{u'} = \overline{w'} = 0$, $\overline{\overline{u}} = \overline{u}$, and $\overline{\overline{u}u'} = 0$ by definition of the time average

$$\overline{u}\frac{\partial\overline{u}}{\partial x} + \overline{w}\frac{\partial\overline{u}}{\partial z} + \frac{\partial}{\partial x}(\overline{u'^2}) + \frac{\partial}{\partial z}(\overline{u'w'}) = -\frac{1}{\rho}\frac{\partial\overline{p}}{\partial x} + \nu\left(\frac{\partial^2\overline{u}}{\partial x^2} + \frac{\partial^2\overline{u}}{\partial z^2}\right)$$

and

$$\overline{u}\frac{\partial\overline{w}}{\partial x} + \overline{w}\frac{\partial\overline{w}}{\partial z} + \frac{\partial}{\partial x}(\overline{u'w'}) + \frac{\partial}{\partial z}(\overline{w'^2}) = -\frac{1}{\rho}\frac{\partial\overline{p}}{\partial z} + \nu\left(\frac{\partial^2\overline{w}}{\partial x^2} + \frac{\partial^2\overline{w}}{\partial z^2}\right) - g_n$$

where use has been made of the relationships

$$\overline{\frac{\partial}{\partial x}u'^2} = \frac{\partial}{\partial x}\overline{u'^2} \quad \text{and} \quad \overline{u'\frac{\partial u'}{\partial x} + w'\frac{\partial u'}{\partial z}} = \frac{\partial}{\partial x}(\overline{u'^2}) + \frac{\partial}{\partial z}(\overline{u'w'})$$

3. Rearrange the equations to show the parallel between the viscous terms and the now unknown correlations of the fluctuating terms; this introduction of unknown correlations by averaging is called the *closure problem*; it is resolved and the equations are *closed*, for example, as in Section 7.2 by defining the correlations in terms of mean flow quantities. Then the equations may be solved, but the closures are approximate and the subject of ongoing research to this day. Otherwise, one must derive equations for the fluctuating quantities; unfortunately, that results in more unknown correlations that have to be dealt with approximately as before.

The rearranged equations are

$$\overline{u}\frac{\partial \overline{u}}{\partial x} + \overline{w}\frac{\partial \overline{u}}{\partial z} = -\frac{1}{\rho}\frac{\partial \overline{p}}{\partial x} + \frac{\partial}{\partial x}\left(\nu\frac{\partial \overline{u}}{\partial x} - \overline{u'^2}\right) + \frac{\partial}{\partial z}\left(\nu\frac{\partial \overline{u}}{\partial z} - \overline{u'w'}\right)$$

and

$$\overline{u}\frac{\partial \overline{w}}{\partial x} + \overline{w}\frac{\partial \overline{w}}{\partial z} = -\frac{1}{\rho}\frac{\partial \overline{p}}{\partial z} + \frac{\partial}{\partial x}\left(\nu\frac{\partial \overline{w}}{\partial x} - \overline{u'w'}\right) + \frac{\partial}{\partial z}\left(\nu\frac{\partial \overline{w}}{\partial z} - \overline{w'^2}\right) - g_n$$

While many variations are possible and most have been tried, perhaps the simplest way to deal with the unknown correlations (the Reynolds stresses; cf., Section 7.2) is to imitate the Stokes' hypothesis of Eq. 7.61 and define the Reynolds stresses in terms of an eddy viscosity as (the Boussinesq approach, see Section 7.2)

$$\overline{u'^2} = -2\varepsilon\frac{\partial \overline{u}}{\partial x}; \qquad \overline{w'^2} = -2\varepsilon\frac{\partial \overline{w}}{\partial z}; \qquad \overline{u'w'} = \varepsilon\left(\frac{\partial \overline{u}}{\partial z} + \frac{\partial \overline{w}}{\partial x}\right)$$

The result, as the reader may confirm, is

$$\overline{u}\frac{\partial \overline{u}}{\partial x} + \overline{w}\frac{\partial \overline{u}}{\partial z} = -\frac{1}{\rho}\frac{\partial \overline{p}}{\partial x} + \frac{\partial}{\partial x}\left((\nu + \varepsilon)\frac{\partial \overline{u}}{\partial x}\right) + \frac{\partial}{\partial z}\left((\nu + \varepsilon)\frac{\partial \overline{u}}{\partial z}\right)$$

and (7.65)

$$\overline{u}\frac{\partial \overline{w}}{\partial x} + \overline{w}\frac{\partial \overline{w}}{\partial z} = -\frac{1}{\rho}\frac{\partial \overline{p}}{\partial z} + \frac{\partial}{\partial x}\left((\nu + \varepsilon)\frac{\partial \overline{w}}{\partial x}\right) + \frac{\partial}{\partial z}\left((\nu + \varepsilon)\frac{\partial \overline{w}}{\partial z}\right) - g_n$$

In the event that one takes both the kinematic and eddy viscosities to be constant, the result appears exactly as the original Navier-Stokes equations would for steady flow, viz.,

$$\begin{aligned} \overline{u}\frac{\partial \overline{u}}{\partial x} + \overline{w}\frac{\partial \overline{u}}{\partial z} &= -\frac{1}{\rho}\frac{\partial \overline{p}}{\partial x} + \nu_T\left(\frac{\partial^2 \overline{u}}{\partial x^2} + \frac{\partial^2 \overline{u}}{\partial z^2}\right) \\ \overline{u}\frac{\partial \overline{w}}{\partial x} + \overline{w}\frac{\partial \overline{w}}{\partial z} &= -\frac{1}{\rho}\frac{\partial \overline{p}}{\partial z} + \nu_T\left(\frac{\partial^2 \overline{w}}{\partial x^2} + \frac{\partial^2 \overline{w}}{\partial z^2}\right) - g_n \end{aligned} \tag{7.66}$$

where $\nu_T = \nu + \varepsilon$.

However, most numerical simulations use a more complex definition for the eddy viscosity, e.g., as in Section 7.2, it is allowed to be a function of velocity gradients and length scales of the mean flow.

REFERENCES

Batchelor, G. K. 1953. *The theory of homogeneous turbulence.* Cambridge: Cambridge Univ. Press.

Hinze, J. O. 1975. *Turbulence.* 2nd ed. New York: McGraw-Hill.

Kovasznay, L. S. G. 1970. The turbulent boundary layer. *Annual Review of Fluid Mechanics* 2: 95–112. Palo Alto: Annual Reviews.

Loitsianskii, L. G. 1970. The development of boundary-layer theory in the U.S.S.R. *Annual Review of Fluid Mechanics* 2: 1–14. Palo Alto: Annual Reviews.

McDowell, D. M., and Jackson, J. D., Eds. 1970. *Osborne Reynolds and engineering science today.* New York: Manchester University Press (Barnes & Noble).

Rott, N. 1990. Note on the history of the Reynolds number. *Annual Review of Fluid Mechanics* 22: 1–11. Palo Alto: Annual Reviews.

Schlichting, H. 1979. *Boundary layer theory.* 7th ed. New York: McGraw-Hill.

Tennekes, H., and Lumley, J. L. 1977. *A first course in turbulence* 4th Printing Cambridge: The MIT Press.

Townsend, A. A. 1976. *The structure of turbulent shear flow.* 2nd ed. Cambridge: Cambridge Univ. Press.

FILMS

Abernathy, F. H. Fundamentals of boundary layers. NCFMF/EDC Film No. 21623, Encyclopaedia Britannica Educ. Corp.

Rouse, H. Mechanics of Fluids: characteristics of laminar and turbulent flow. Film No. U56159, Media Library, Audiovisual Center, Univ. of Iowa.

Stewart, R. W. Turbulence. NCFMF/EDC Film No. 21626, Encyclopaedia Britannica Educ. Corp.

Taylor, E. S. Secondary flow. NCFMF/EDC Film No. 21612, Encyclopaedia Britannica Educ. Corp.

PROBLEMS

7.1. When 0.001 9 m^3/s of water flow in a 76 mm pipeline at 21°C, is the flow laminar or turbulent?

7.2. Glycerin flows in a 25 mm (1 in.) pipe at a mean velocity of 0.3 m/s (1 ft/s) and temperature 25°C (77°F). Is the flow laminar or turbulent?

7.3. Carbon dioxide flows in a 50 mm pipe at a velocity of 1.5 m/s, temperature 66°C, and absolute pressure 380 kPa. Is the flow laminar or turbulent?

7.4. What is the maximum flowrate of air that may occur at laminar condition in a 100 mm (4 in.) pipe at an absolute pressure of 200 kPa (30 psia) and 40°C (104°F)?

7.5. What is the smallest diameter of pipeline that may be used to carry 6.3 l/s (100 gpm) of jet fuel (JP-4) at 15°C (59°F) if the flow is to be laminar?

7.6. Derive an expression for pipeline Reynolds number in terms of Q, d, and ν.

7.7. A fluid flows in a 75 mm pipe which discharges into a 150 mm line. What is the Reynolds number in the 150 mm pipe if that in the 75 mm pipe is 20 000?

7.8. If water (20°C or 68°F) flows at constant depth (i.e., uniformly) in a wide open channel, below what flowrate q may the regime be expected to be laminar?

7.9. What is the maximum speed at which a spherical sand grain of diameter 0.254 mm (0.01 in.) may move through water (20°C or 68°F) and the flow regime be laminar?

7.10. In the laminar flow of an oil of viscosity 1 Pa·s the velocity at the center of a 0.3 m pipe in 4.5 m/s and the velocity distribution is parabolic. Calculate the shear stress at the pipe wall and within the fluid 75 mm from the pipe wall.

7.11. If the turbulent velocity profile in a pipe 0.6 m in diameter may be approximated by $v = 3.56y^{1/7}$(v m/s, y m) and the shearing stress in the fluid 0.15 m from the pipe wall is 23.0 Pa, calculate the eddy viscosity, mixing length, and turbulence constant at this point. Specific gravity of the fluid is 0.90.

7.12. A turbulent flow in a boundary layer has a velocity profile $v = (v_*/\kappa) \ln y + C$, where κ is the Kármán constant and the *friction velocity* v_* is defined as $\sqrt{\tau_o/\rho}$ and τ_o is the wall shear stress. Find expressions for the eddy viscosity ε and the shear stress $\tau(y)$ if the mixing length relationship $l = \kappa y$ is assumed valid.

7.13 Given that $\tau = \tau_o$ across the entire boundary layer in a turbulent flow, where τ_o is the wall shear stress, use equation 7.6 to derive the expected velocity profile. *Hint:* express the result in terms of the *friction velocity* $v_* = \sqrt{\tau_o/\rho}$ and (*a*) $y_o > 0$, which is the value of y at which the velocity goes to zero near the wall, or (*b*) δ which is the boundary layer thickness, that is, the y-value at which the velocity is equal to the undisturbed (or free stream) velocity V_o.

7.14. In a pipe flow the shear stress varies linearly with distance from the wall so $\tau = \tau_o(1 - y/R)$, where R is the pipe radius. Also, if v_c is the centerline velocity in the pipe, the typical observed velocity profile is accurately approximated by $(v - v_c)/v_* = (1/\kappa)\ln(y/R)$. Find and plot the variations of the eddy viscosity and the mixing length. Show that $l \sim \kappa y$ for small y/R.

7.15. The thickness of the viscous sublayer (Fig. 7.7) is usually defined by the intersection of a laminar velocity profile (adjacent to the wall) with a turbulent velocity profile. If the former may be described by $v = c_1 y$ and the latter by $v = c_2 y^{1/7}$, derive an expression for the thickness of the film in terms of c_1 and c_2.

7.16. For flow over smooth surfaces, a large number of experimental data can be represented by $v/v_* = v_* y/\nu$ in the viscous sublayer and $v/v_* = 2.5 \ln(v_* y/\nu) + 5$ in the fully turbulent region. Plot this composite "law of the wall" as v/v_* versus $y_+ = v_* y/\nu$ and locate the nominal thickness δ_v of the viscous sublayer at the intersection of the given curves. Recall $v_* = \sqrt{\tau_o/\rho}$. Discuss how δ_v depends on τ_o, ρ, and μ.

7.17. When oil (kinematic viscosity 1×10^{-4} m^2/s, specific gravity 0.92) flows at a mean velocity of 1.5 m/s through a 50 mm pipeline, the head lost in 30 m of pipe is 5.4 m. What will be the head loss when the velocity is increased to 3 m/s?

7.18. If in the preceding problem fluid with a kinematic viscosity 1×10^{-5} m^2/s is flowing the head lost will be 15 m if the roughness is very great. What head loss will occur when the mean velocity is increased to 3 m/s?

7.19. For flow over a smooth plate, what approximately is the maximum length of the laminar boundary layer if $V_o = 9.0$ m/s (30 ft/s) in the irrotational uniform flow and the fluid is air? Water?

7.20. Estimate the maximum laminar boundary layer thicknesses in the preceding problem.

7.21. A model of a thin streamlined body is placed in a flow for testing. The body is 0.9 m (3 ft) long and the flow velocity is 0.6 m/s (2 ft/s). What ν is needed to ensure that the boundary layer on the body is laminar?

7.22. Separation is found to occur 0.76 m (2.5 ft) from the nose of the body in the preceding problem when the fluid is water. Is the boundary layer flow laminar or turbulent at separation?

7.23. In an incompressible boundary-layer flow, the velocity profile is found to be $u/V_o = (y/\delta)^{1/5}$. Find δ_1 and δ_2 as functions of δ.

7.24. Prove that, if $u/V_o = (y/\delta)^{1/m}$ in incompressible flow in a boundary layer, then $\delta_2/\delta = m/[(m+1)(m+2)]$; find δ_1/δ also.

7.25. For flow of problem 7.24, find an expression for c_f in the zero pressure-gradient case.

7.26. Use the momentum integral equation to derive expressions for δ_1, δ_2, and c_f for flow in an incompressible boundary layer with the linear velocity profile $u/V_o = y/\delta$. How do the results differ from those obtained in Section 7.5?

7.27. Repeat problem 7.26 for a velocity profile $u/V_o = \sin(\pi y/2\,\delta)$.

7.28. Plot the local friction coefficient c_f, the boundary layer thickness ratio δ/x, and the drag coefficient C_f for both laminar and turbulent boundary layers on a flat plate for $\mathbf{R}_x$ from 0 to 500 000, assuming in the turbulent case that the layer is tripped at the leading edge and so is fully turbulent along the length of the plate. Discuss the ratio of drag forces as a function of $\mathbf{R}_x$.

7.29. What is the relationship between the boundary layer thickness in laminar and turbulent flows as a function of $\mathbf{R}_x$? Between the drag coefficients?

7.30. A smooth plate 3 m long and 0.9 m wide moves through still sea level air (Appendix 2) at 4.5 m/s. Assuming the boundary layer to be wholly laminar, calculate (*a*) the thickness of the layer at 0.5, 1.0, 1.5, 2.0, 2.5, and 3.0 m from the leading edge of the plate; (*b*) the shear stress, τ_o, at those points; and (*c*) the total drag force on one side of the plate. (*d*) Calculate the thickness at the above points if the layer is turbulent. (*e*) Calculate the total drag for the turbulent boundary layer. (*f*) What percentage saving in drag is effected by a laminar boundary layer?

7.31. A smooth flat plate 2.4 m long and 0.6 m wide is placed in an airstream (101.3 kPa and 15°C) of velocity 9 m/s. Calculate the total drag force on this plate (2 sides) if the boundary layer at the trailing edge is (*a*) laminar, (*b*) transition, and (*c*) turbulent.

7.32. A flat-bottomed barge having a 150 ft by 20 ft bottom is towed through still water (60°F) at 10 mph. What is the frictional drag force exerted by the water on the bottom of the barge? How long could the laminar portion of the boundary layer be, using a critical Reynolds number of 537 000? What is the thickness of the laminar layer at its downstream end? What is the approximate thickness of the boundary layer at the rear end of the bottom of the barge?

7.33. European InterCity Express trains operate at speeds of up to 280 km/hr. Suppose that a train is 120 m long. Treat the sides and top of the train as a smooth flat plate 9 m wide. When the train moves through still air at sea level (Appendix 2), calculate the possible length of the laminar boundary layer and the thickness of this layer at its downstream end. What is the thickness of the boundary layer at the rear end of the train? What is the viscous drag force on the train and what power must be expended to overcome this resistance at maximum speed? At 50% of maximum?

7.34. Grumman Corp. has proposed (*Mechanical Engineering, 115*, 8, August 1993, p. 74f) to build a magnetic levitation train to operate at a top speed of 300 mph. The vehicle is 114 ft long. Assuming that the sides and top can be treated approximately as a smooth flat plate of 30 ft width, with a turbulent boundary layer on it, calculate the drag force and the power expended to overcome the drag at the maximum speed.

7.35. The U.S. Navy has built the *Sea Shadow* (*Mechanical Engineering, 115*, 6, June 1993, p. 68f), which is a *small waterplane twin-hull* (SWATH) ship whose object is to achieve the same reduced radar profile as the STEALTH aircraft. This catamaran is 160 ft long and its twin hulls have a draft of 14 ft. Assume that the ocean turbulence triggers a fully turbulent boundary layer on the sides of each hull. Treat these as flat plate boundary layers and calculate the drag on the ship and the power required to overcome it as a function of speed. Plot the results for speeds from 5 to 13 knots.

7.36. Estimate the frictional drag encountered by the hull of a submarine when traveling deeply submerged at a speed of 33 km/h. The length of the hull is 60 m and its surface area 1 700 m^2. Assume water density 1 026 kg/m^3 and kinematic viscosity 1.2×10^{-6} m^2/s.

7.37. The two rectangular smooth flat plates are to have the same drag in the same fluid stream. Calculate the required x. If the two plates are combined into the T-shape indicated, what ratio exists between the drag of the combination and that of either one? Assume laminar boundary layers in all calculations.

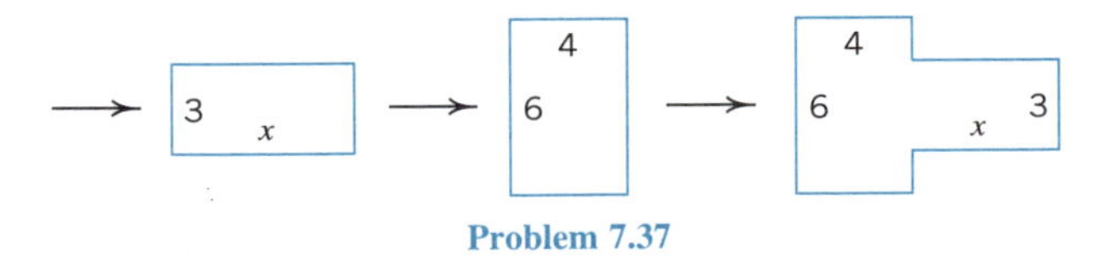

Problem 7.37

7.38. If at the trailing edge of a smooth flat plate 10 ft (3 m) long, covered by a laminar boundary layer, the shearing stress is 0.000 01 lb/ft^2 (0.000 48 Pa), what is the total drag force on one side of the plate? What is the shearing stress at a point halfway between leading and trailing edges? What are the drag forces on the front and rear halves of one side of the plate?

7.39. A smooth flat plate 2.5 ft (0.76 m) long is immersed in water at 68°F (20°C) flowing at an undisturbed velocity of 2 ft/s (0.61 m/s). Calculate the shear stress at the center of the plate, assuming a laminar boundary layer over the whole plate.

7.40. A fluid stream of uniform velocity 3 m/s approaches a flat plate 30 m long which is parallel to the oncoming flow. The Reynolds number calculated with the full length of the plate is 400 000. How much space (m^2) is occupied by the boundary layer?

7.41. A boundary layer on a flat plate immersed in water increases in thickness from 0.152 m to 0.155 m in a distance of 0.3 m. The velocity of the undisturbed flow is 7.6 m/s. The flow in the boundary layer is not laminar. If there is a (fictitious) linear velocity distribution in the boundary layer, calculate the mean local shearing stress on this section of the plate.

7.42. The velocity profile through the boundary layer at the downstream end of a flat plate is found to conform to the equation $v/V_o = (y/\delta)^{1/8}$ in which V_o is 20 ft/s (6.1 m/s) and δ is 1 ft (0.3 m). Calculate the drag force exerted on (one side of) this plate if the fluid density is 2.0 slugs/ft^3 (1 031 kg/m^3).

7.43. The velocity distributions upstream and downstream from the transition region of a flat plate boundary layer are as indicated. Calculate the total drag force on the plate length AB.

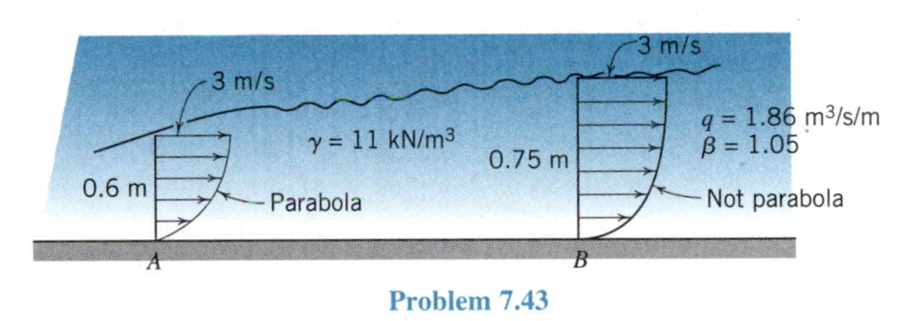

Problem 7.43

7.44. Show that for a laminar boundary layer on a flat plate the kinetic energy lost between free stream and any point in the boundary layer is 31.5% of the total kinetic energy of a portion of the free stream containing the flowrate in the boundary layer at the above point. How may this loss of kinetic energy be accounted for?

7.45. Accurate values of C_f from Fig. 7.12 are 0.001 85, 0.001 81, and 0.001 99 for Reynolds numbers 514 000, 537 000, and 560 000, respectively, which seem to indicate that the local wall shear is larger at the beginning of transition than at the end of the laminar boundary layer. From these figures what percentage increase in mean wall shear is indicated for the reaches implied by the Reynolds numbers?

7.46. A smooth flat plate 20 ft long moves through a fluid of kinematic viscosity 0.000 1 ft^2/s and s.g. 0.80. When $\mathbf{R}_x$ is

500 000 the shear stress on the plate is 0.067 lb/ft^2. Estimate the total drag (per foot of width) on one side of the plate.

7.47. A javelin is about 2.6 m long and 50 mm in diameter. By treating its surface as a flat plate of length 2.6 m and 50π mm width, estimate the friction drag as a function of speed in sea level air. Plot the result. Up to what speed does the boundary layer remain laminar? What is the drag at 15 m/s?

7.48. From Section 7.9 information, calculate the allowable maximum velocity at the centerline of the established flow of water in a 10 mm or 0.4 in. diameter smooth pipe if the boundary layer must remain laminar. How many pipe diameters from the smooth entrance does established flow begin?

7.49. Carry out the formal derivation of Eq. 7.37 by application of the impulse-momentum principle.

7.50. Use Eq. 7.37 to derive the velocity profile and calculate the flowrate for laminar pipe flow.

7.51. When fluid of specific weight 50 lb/ft^3 flows in a 6 in. pipeline, the frictional stress between fluid and pipe is 0.5 psf. Calculate the head lost per foot of pipe. If the flow rate is 2.0 cfs, how much power is lost per foot of pipe?

7.52. If the head lost in 30 m (100 ft) of 75 mm (3 in.) pipe is 7.6 m (25 ft) when a certain quantity of water flows therein, what is the total dragging force exerted by the water on this reach of pipe?

7.53. Air ($\gamma = 12.6$ N/m^3) flows through a horizontal 0.3 m by 0.6 m rectangular duct at a rate of 15 N/s. Find the mean shear stress at the wall of the duct if the pressure drop in a 100 m length is 160 Pa. Compute the power lost per metre of duct length.

7.54. If the velocity near a pipe wall has the form $v = A \ln y + B$, while Eq. 7.37 is valid in a pipe of radius R, find how the mixing length l in Eq. 7.3 varies with y. Show that this variation agrees with the discussion of Section 7.2 for small y and propose a method to determine κ.

7.55. In a reach of cylindrical pipe in which air is flowing the mean velocity is observed to increase from 400 to 500 ft/s and the mean temperature to rise from 100°F to 120°F. How much heat is being added to each pound of air in this reach of pipe?

7.56. In a reach of cylindrical pipe wrapped with perfect insulation ($q_H = 0$) the absolute pressure is observed to drop from 690 to 345 kPa and the mean velocity to increase from 91 to 175 m/s. Predict the temperatures in the pipe at each end of the reach if air is flowing in the pipe.

7.57. When water flows at a mean velocity of 3 m/s in a 300 mm pipe, the head loss in 100 m of pipe is 3 m. Estimate the rise of temperature of the water if the pipe is wrapped with perfect insulation so ($q_H = 0$). If the pipe is not insulated, how much heat must be extracted from the water to hold its temperature constant? The change in the internal energy of water is equal to the product of its specific heat and the temperature difference (see Section 7.11; specific heat for water is given in Appendix 2).

7.58. If a zone of unestablished flow may be idealized to the extent shown and the centerline may be treated as a streamline in an ideal fluid, calculate the drag force exerted by the sidewalls (between sections 1 and 2) on the fluid if the flow is: (*a*) two-dimensional and 0.3 m wide normal to the paper and (*b*) axisymmetric. The fluid flowing has specific gravity 0.90.

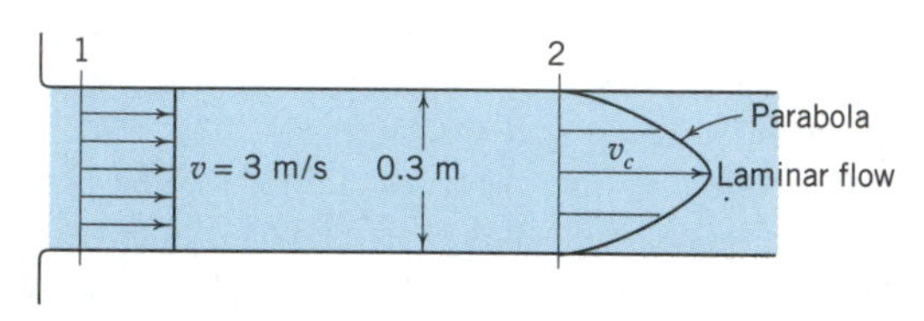

Problem 7.58

7.59. Calculate α, β, and the momentum flux for the velocity profiles of problem 4.12. Assume that the passage is 0.6 m or 2 ft high, 0.3 m or 1 ft wide normal to the paper, and that water is flowing.

7.60. Calculate α, β, and the momentum flux for the velocity profiles of problem 4.13. Assume that the passage is 1 m or 3 ft in diameter and that water is flowing.

7.61. Calculate α and β for the velocity profile of problem 4.14.

7.62. Just downstream from the nozzle tip the velocity distribution is as shown. Calculate the flowrate past section 1, α, β, and the momentum flux. Assume water is flowing.

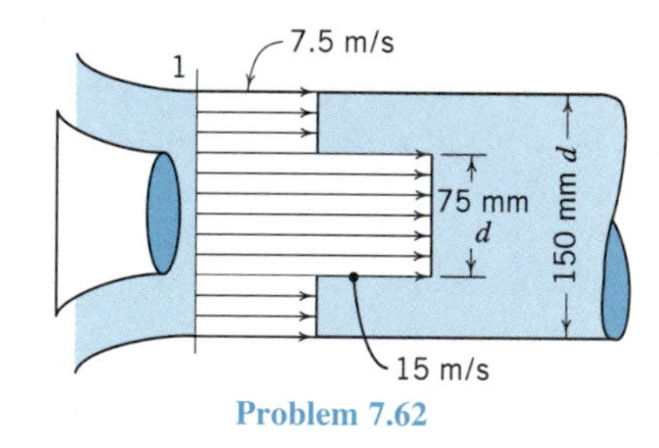

Problem 7.62

7.63. Calculate α and β for the flow in this two-dimensional passage if q is 1.5 m^3/s·m.

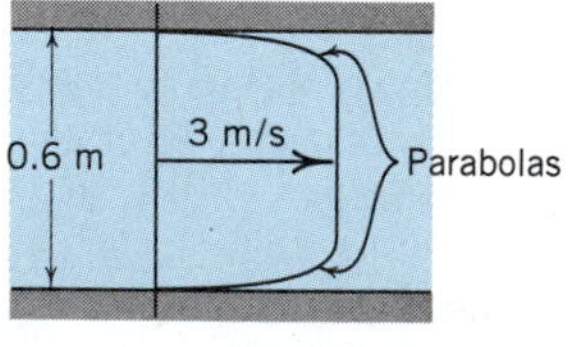

Problem 7.63

7.64. If the velocity profile in a two-dimensional open channel may be approximated by the *parabola* shown, calculate the flowrate and the coefficients α and β.

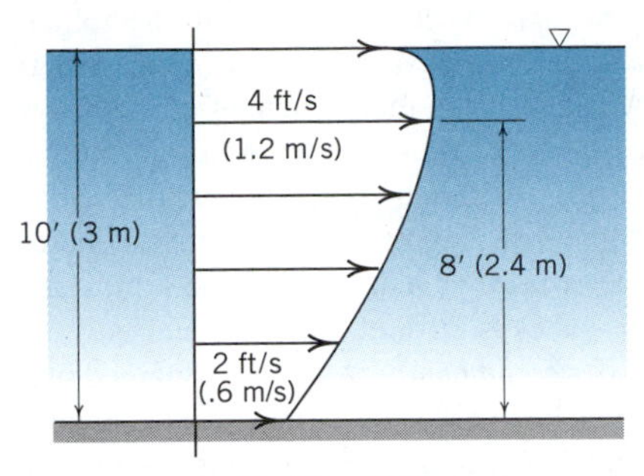

Problem 7.64

7.65. A horizontal nozzle having a cylindrical tip of 75 mm diameter attached to a 150 mm water pipe discharges 0.05 m^3/s. In the pipe just upstream from the nozzle the pressure is 62.6 kPa and α is 1.05. In the issuing jet α is 1.01. Calculate the conventional and exact head losses in the nozzle.

7.66. At a section just downstream from a sluice gate in a wide horizontal open channel the water depth is observed to be 0.20 ft, $q = 2$ cfs/ft, and the velocity distribution uniform. At another section further downstream the depth is 0.30 ft and the velocity profile follows the "seventh-root law" $v/v_s = (y/0.30)^{1/7}$ in which v_s is the surface velocity and y the distance from the channel bottom. Calculate the total drag force (per foot of width) exerted by the water on the length of channel bottom between the two sections.

7.67. These phenomena are observed in a laboratory channel 1 m wide, and the indicated measurements taken. Many velocity measurements at sections 1 and 2 allow calculation of β_1 and β_2, which turn out to be 1.02 and 1.07, respectively. Calculate the total drag force exerted by the water on the walls and bottom of the length of channel between sections 1 and 2.

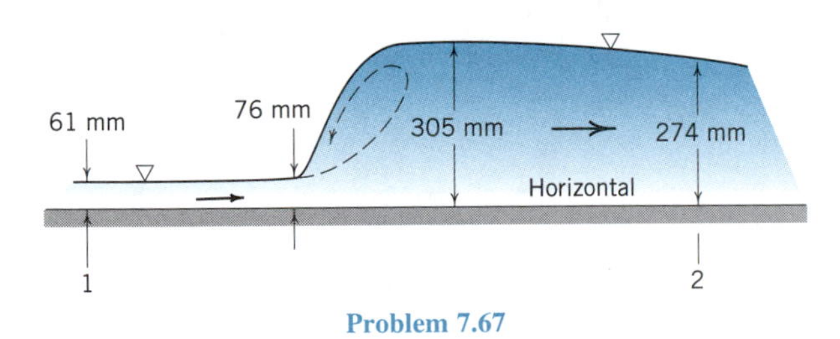

Problem 7.67

7.68. Calculate the horizontal component of force exerted by the fluid on the upstream half of the semicylinder of problem 5.136.

7.69. If the velocity profiles at the upstream and downstream ends of the mixing zone of a jet pump may be approximated as shown, and wall friction may be neglected, calculate the rise of pressure from section 1 to section 2, and the power lost in the mixing process. Water is flowing.

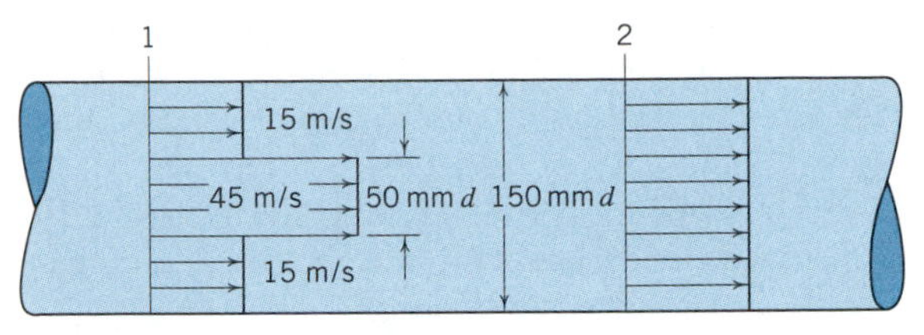

Problem 7.69

7.70. Calculate the pressure rise in the preceding problem if the velocity profile at the downstream end of the mixing zone is defined by $v = C(R - r)^{1/7}$.

7.71. The sketch depicts a simplified flow situation in a two-dimensional wind- or water-tunnel in which the velocity distributions have been measured upstream and downstream from the strut. If wall friction is negligible, determine the drag of the strut in terms of the density of the fluid.

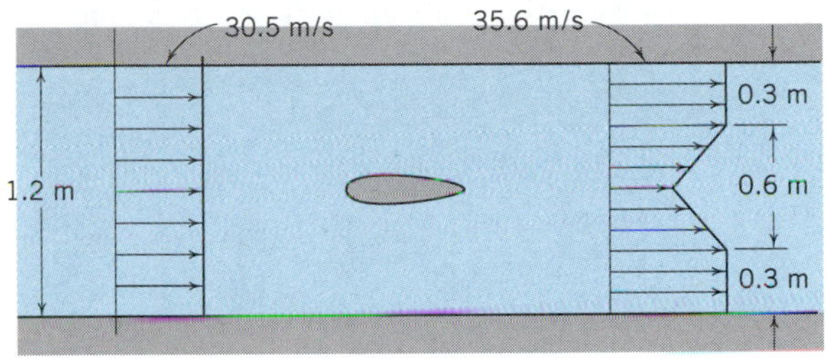

Problem 7.71

7.72. A long straight thin-walled tube of 25 mm or 1 in. diameter is towed through water (20°C or 68°F) at a speed of 0.15 m/s or 0.5 ft/s. The mean velocity of the water *through* the tube is observed to be 0.06 m/s or 0.2 ft/s (relative to the tube) with a parabolic velocity distribution in the downstream region of the tube. Calculate the total drag force exerted on the *inside* of the tube.

7.73. This nozzle discharges ideal liquid of specific weight 9.80 kN/m^3 to the atmosphere. The velocity in the pipe is 3.0 m/s. The velocity through the jet may be assumed to vary linearly from top to bottom. The nozzle is of such shape that the weight of liquid contained therein is 135.0 N, with center of gravity 100 mm to the right of section 1. Calculate accurately the magnitude, direction, and location of the resultant force exerted by the liquid on the nozzle.

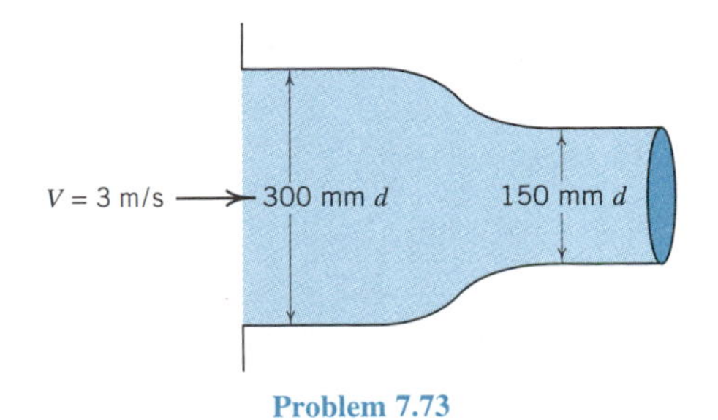

Problem 7.73

7.74. Calculate the minimum length of an optimum vaneless diffuser connecting the 1 m or 3.3 ft diameter test section of a wind tunnel to the 3 m or 10 ft diameter main circulation pipe.

7.75. Using the profile derived in Illustrative Problem 7.10, plot the nondimensional velocity profile u/V as a function of the nondimensional pressure gradient

$$\left(\left(\frac{h^2}{8\mu V}\right)\frac{\partial p}{\partial x}\right)$$

Hint: try the coordinate $Z = \left(\frac{z}{h/2}\right)$.

7.76. For the flow in Illustrative Problem 7.10, what is the pressure gradient for which the flowrate is zero? Plot that profile.

7.77. Apply the Navier-Stokes equations to the steady state flow of Illustrative Problem 1.9. Given the speed of the inner cylinder, find the velocity profile in the flow between the cylinders.

7.78. What is the relationship between the shear stress at the wall of the pipe and the pressure gradient in the Hagen-Poiseuille flow of Illustrative Problem 7.11?

7.79. Assume that the flow in Illustrative Problem 7.10 is turbulent and that the averaged Navier-Stokes equations for a steady flow apply. If the mixing length of the eddy viscosity varies linearly with distance from each solid boundary according to Eq. 7.5 with $\kappa = 0.4$, compute the velocity profile for a small number of pressure gradients, including zero, if the top plate is moving with a speed of 1 m/s, the spacing between the plates is 30 mm, and the fluid is water. Do the solutions differ significantly from ones obtained by assuming that the flow is laminar? Is the flow expected to be turbulent? Hint: This problem is not as easy at it might seem; the velocity gradient at the lower wall has to be guessed and the flow integrated to the upper wall. If the velocity from the profile is not 1 m/s, then another guess must be made, etc.

7.80. Extend the derivation of the unsteady Navier-Stokes equations 7.63 to three-dimensions.

CHAPTER

8 SIMILITUDE, DIMENSIONAL ANALYSIS, AND NORMALIZATION OF EQUATIONS OF MOTION

This chapter is different from those that preceded it. In those we were concerned with learning new principles of fluid mechanics and with new physics. Our goal here is to learn how to begin to interpret fluid flows. Most real fluid flows are complex and can be solved, at best, only approximately by analytical methods. Experiments, both physical and numerical, play a crucial role in actually understanding the phenomena occurring, in verifying analytical solutions, in suggesting what approximations are valid, and ultimately in providing results that cannot be obtained by theoretical analysis alone.

Unfortunately, as soon as we try to plan a set of physical experiments or numerical experiments, often called *simulations*, we are faced with three dilemmas. First, the number of possible and relevant variables or physical parameters is huge and so the potential number of experiments is beyond our resources. This applies to physical and to numerical experiments. Second, many, if not most, real flow situations are either too large or far too small for convenient experiment at their true size, involve situations where instruments cannot be placed, or present field conditions which are so uncontrolled as to make systematic study tedious or indeed impossible. When testing the real thing (which is called the *prototype*) is not feasible, a physical *model* (that is, a scaled version of the prototype) can be constructed and the performance of the prototype simulated in the physical model.

If on the other hand, the equations that govern the motion are thought to be known, then we may construct a numerical model of the real situation and carry out a numerical experiment or numerical simulation of the prototype situation. Such numerical models are widely used and, often, very large—so-called—supercomputers are used to solve the equations. These results of atmospheric simulations, often seen on television as hurricanes, are tracked across the Caribbean and predictions made about landfalls and necessary evacuations; likewise, numerical simulations play a role in tornado and thunderstorm warnings. Unfortunately, we can say with certainty that the mathematical equations represent all of the real physics in only a few cases, e.g., in the case of the Navier-Stokes equations that we first met in the previous chapter. Thus, our third dilemma is that the solutions produced by numerical simulations often must be verified or the numerical models calibrated by use of physical models or measurements in the prototype.

Historically, physical models have been used for over a hundred years and it was near the latter part of the nineteenth century that models began to be used extensively to study flow phenomena that could not be solved by analytical methods or by means of available experimental results from prototype situations. Over the era of modern engineering, the use of models and our confidence in model studies have steadily increased: aeronautical engineers obtain important design information for model tests in wind tunnels—occasionally using some parts of or all of a prototype; naval architects test ship models in large towing basins and water tunnels (see Appendix 4 for an example); mechanical engineers test models of turbines and pumps and predict the performance of the full-scale machines from these tests; and civil and environmental engineers work with models of hydraulic structures, river sections, estuaries and coastal bays and seas to obtain more reliable solutions to design and/or pollution problems. The justifications for models include *economics*—a model, being small compared to the prototype typically, costs little compared to the prototype which it is built to model, and its results may lead to savings of many times its costs—and *practicality*—in a model test environmental and flow conditions can be rigorously controlled. On the other hand, the flow in a river or the wind that carries pollution over a city can be neither controlled nor (often) predicted in detail. Of course, models are not always smaller than the prototype. Flow models of the human circulation systems, e.g., models of the blood flow in the human brain, whose goal was to assess the impact of clots on the flow, are examples of models that are typically much larger than the prototype and use fluids different from the prototype fluid!

In this chapter, the laws of similitude are described first and provide a basis for interpretation of physical and numerical model results and for crafting both physical and numerical experiments. Then, the method of dimensional analysis is shown to lead us to further insights. Finally, the concept of nondimensionalization or normalization of the governing equations is introduced. In this book, only the concept can be introduced, but this method is very powerful, particularly for numerical simulation analysis. Kline[1] terms this collection of methods *fractional analysis* and defines them as procedures for finding some information about the solutions, usually short of a complete answer.

The following gives an example and insight to the rest of the chapter. As noted above, when one is faced with planning or interpreting a set of experiments in a model or prototype situation, it is rapidly discovered that to carry each pertinent flow variable or physical parameter through its appropriate range involves obtaining a prohibitive number of individual experimental data points. Indeed, for a simple incompressible flow where the goal is to find the drag force on a body, Fox and McDonald[2] and White[3] both estimate that 10^4 separate data points are needed to describe the drag force D as a function of the characteristic body size d and the fluid velocity V, density ρ, and viscosity μ. The second major topic of this chapter, namely, dimensional analysis, can be applied to this problem and the

[1]S. J. Kline, *Similitude and Approximation Theory*, Springer-Verlag, 1986.

[2]R. W. Fox and A. T. McDonald, *Introduction to Fluid Mechanics*, 3rd ed. Wiley, 1985.

[3]F. M. White, *Fluid Mechanics*, 2nd ed. McGraw-Hill, 1986.

results leavened with the knowledge gained from the laws of similitude, to learn that the drag can be characterized completely by a relationship between only *two groups* of variables. In fact, the drag coefficient $C_D = D/\rho V^2 d^2$ is functionally related to the Reynolds number $\mathbf{R} = Vd\rho/\mu$ so

$$C_D = f(\mathbf{R}) \tag{8.1}$$

Just as Reynolds found (see Section 7.1) that the single number **R** defined the limits of laminar and turbulent flow for all fluids in all pipes, it is now found (without recourse to any experiment) that, for an incompressible flow and a given set of bodies with the same *shape* (say, a prototype and any models without regard for their size), their drag is characterized by **R** for all body sizes and all fluids. Therefore, using one conveniently sized body and one fluid, one needs to obtain only 10 to 20 data points at different velocities (e.g., see Section 11.3) to define the desired relationship among D, V, d, ρ, and μ. This assistance provided by the laws of similitude and by dimensional analysis in finding the important groups of variables in each flow situation is an indispensable aid in planning and interpreting experiments. In addition, an important understanding is often gained about the physical phenomena being studied, and clues on what to look for in the experiment are often revealed.

8.1 SIMILITUDE AND PHYSICAL MODELS

Although the basic theory for the interpretation of model tests is quite simple, it is seldom possible to design and operate a physical model of a fluid phenomenon from theory alone; here the *art* of engineering must be practiced with experience, judgment, ingenuity, and patience if useful results are to be obtained, correctly interpreted, and prototype performance predicted from results.

Similitude of flow phenomena not only occurs between a prototype and its model but also may exist between various natural phenomena if certain laws of similarity are satisfied. Similarity thus becomes a means of correlating the apparently divergent results obtained from similar fluid phenomena and as such becomes a valuable tool of modern fluid mechanics; the application of the laws of similitude will be found to lead to more comprehensive solutions and, therefore, to a better understanding of fluid phenomena in general.

There are three basic types of similitude; all three must be obtained if complete similarity is to exist between fluid phenomena. However, in some special cases effective and useful approximate similarity is obtained without satisfying this condition, perhaps the most well known case being a relaxation of the similarity of friction forces in models of ships. The first and simplest type of similarity is *geometric similarity*, which states that the flowfield and boundary geometry of model and of the prototype have the same shape and, therefore, that the ratios between corresponding lengths in model and prototype are the same. While most models are geometrically similar to their prototypes, departure from geometric similarity (resulting in a *distorted* model) has been frequently used in the past (for economic and physical reasons) in models of rivers, harbors, estuaries, and so forth. In this case, the flows are not similar and the models have to be calibrated and adjusted to make them perform properly. Now, numerical models are regarded as superior to distorted models and are usually used in their place.

Now, consider the prototype and model bodies in Fig. 8.1, which could represent cross sections of struts in the inlet of a turbine or struts supporting a hydrofoil on a ship. The

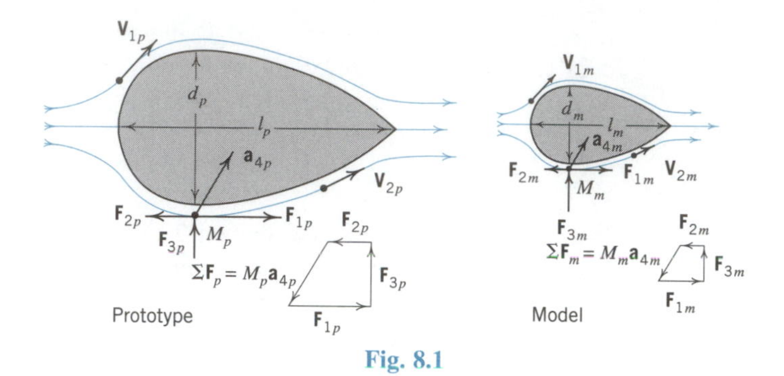

Fig. 8.1

model and prototype must have the same shape to geometrically similar; thus, for the characteristic lengths we have

$$\frac{d_p}{d_m} = \frac{l_p}{l_m}$$

Corollaries of geometric similarity are that corresponding areas vary with the squares of their linear dimensions, so that

$$\frac{A_p}{A_m} = \left(\frac{d_p}{d_m}\right)^2 = \left(\frac{l_p}{l_m}\right)^2$$

and that volumes vary with the cubes of their linear dimensions.

The second type of similarity is *kinematic similarity*; this requires that, in addition to the flowfields having the same shape, the ratios of *corresponding* velocities and accelerations must be the same throughout the flow. For example, returning to the geometrically similar objects of Fig. 8.1, for the typical velocities and accelerations indicated we obtain

$$\mathbf{V}_{1p}/\mathbf{V}_{1m} = \mathbf{V}_{2p}/\mathbf{V}_{2m} \qquad \text{and} \qquad \mathbf{a}_{3p}/\mathbf{a}_{3m} = \mathbf{a}_{4p}/\mathbf{a}_{4m}$$

Thus, flows with geometrically similar streamlines are kinematically similar.

The third type of similarity is *dynamic similarity*. This means that in order to maintain the geometric and kinematic similarity between flowfields, the forces acting on *corresponding* fluid masses must be related by ratios similar to those for kinematic similarity. Suppose now that three types of force can act at any point in the model and the prototype; these might include, for example, gravity, viscous and pressure forces. With these forces shown acting on the corresponding fluid masses M_p and M_m of Fig. 8.1, the vector polygons may be drawn, and from the geometric similarity of these polygons and from Newton's second law (which, of course, is operative in both model and prototype)

$$\frac{\mathbf{F}_{1p}}{\mathbf{F}_{1m}} = \frac{\mathbf{F}_{2p}}{\mathbf{F}_{2m}} = \frac{\mathbf{F}_{3p}}{\mathbf{F}_{3m}} = \frac{M_p \mathbf{a}_{4p}}{M_m \mathbf{a}_{4m}} \tag{8.2}$$

Here, to facilitate our work, we define the product of the mass and the acceleration, $M\mathbf{a}$, to be the "inertia force, $\mathbf{F}_I$." For dynamic similarity these force ratios must be maintained on all the corresponding fluid masses throughout the flowfields; thus it is evident that they can be governed only by relations between the dynamic and kinematic properties of the flows and by the physical properties of the fluids involved.

Complete similarity requires simultaneous satisfaction of geometric, kinematic, and dynamic similarity. Kinematically similar flows must be geometrically similar, of course. But, if the *mass distributions* in flows are similar (i.e., if the density ratios for corresponding fluid masses are the same for all masses), then from Eq. 8.2 it is seen that kinematically similar flows are automatically completely similar. As an example, for constant-density flows, kinematic similarity guarantees complete similarity, while dynamic plus geometric similarity guarantees kinematic similarity.

Referring again to Fig. 8.1, it is apparent that

$$\mathbf{F}_{1p} + \mathbf{F}_{2p} + \mathbf{F}_{3p} = M_p\mathbf{a}_{4p}$$

and

$$\mathbf{F}_{1m} + \mathbf{F}_{2m} + \mathbf{F}_{3m} = M_m\mathbf{a}_{4m}$$

and from this it may be concluded that, if the ratios between three of the four corresponding terms in these equations are the same, the ratio between the corresponding fourth terms must be the same as that between the other three. Thus, one of the ratios of Eq. 8.2 is redundant, and dynamic similarity is characterized by an equality of force ratios numbering one less than the forces. Any force ratio may be selected for elimination, depending on the quantities that are desired in the equations. When the first force ratio is eliminated from Eq. 8.2, the equation may be rewritten in various ways. It is usually written in the form of simultaneous equations,

$$\frac{M_p\mathbf{a}_{4p}}{\mathbf{F}_{2p}} = \frac{M_m\mathbf{a}_{4m}}{\mathbf{F}_{2m}} \tag{8.3}$$

$$\frac{M_p\mathbf{a}_{4p}}{\mathbf{F}_{3p}} = \frac{M_m\mathbf{a}_{4m}}{\mathbf{F}_{3m}} \tag{8.4}$$

The scalar magnitudes of forces that *may* affect a flowfield include: pressure, F_P; inertia, F_I; gravity, F_G; viscosity, F_V; elasticity, F_E; and surface tension, F_T. This discussion excludes, for example, the Coriolis force of rotating systems (yields a Rossby number) the buoyancy forces in stratified flow (yields a Richardson number), and the forces in an oscillating flow (yields a Strouhal number). Since the considered forces are taken to be those on or of any fluid mass, they may be generalized by the following fundamental relationships.

$$F_P = (\Delta p)A = (\Delta p)l^2$$

$$F_I = Ma = \rho l^3\left(\frac{V^2}{l}\right) = \rho V^2 l^2$$

$$F_G = Mg_n = \rho l^3 g_n$$

$$F_V = \mu\left(\frac{dv}{dy}\right)A = \mu\left(\frac{V}{l}\right)l^2 = \mu V l$$

$$F_E = EA = El^2$$

$$F_T = \sigma l$$

In each case these fundamental forces are obtained by replacing derivatives and other terms in the actual force expressions obtained in earlier chapters by difference expressions and characteristic lengths, velocities, pressures, and so forth, of the flow. Here l and V are a characteristic or typical length and velocity for the system, while ρ, μ, E, and σ are fluid properties and g_n is the acceleration due to gravitational acceleration.

To obtain dynamic similarity between two flowfields when all these forces act, all corresponding force ratios must be the same in model and prototype; thus dynamic similarity between two flowfields when all possible forces are acting may be expressed (after the pattern of Eqs. 8.3 and 8.4) by the following five simultaneous equations:

$$\left(\frac{F_I}{F_P}\right)_p = \left(\frac{F_I}{F_P}\right)_m = \left(\frac{\rho V^2}{\Delta p}\right)_p = \left(\frac{\rho V^2}{\Delta p}\right)_m \qquad \mathbf{E}_p = \mathbf{E}_m \tag{8.5}$$

$$\left(\frac{F_I}{F_V}\right)_p = \left(\frac{F_I}{F_V}\right)_m = \left(\frac{Vl\rho}{\mu}\right)_p = \left(\frac{Vl\rho}{\mu}\right)_m \qquad \mathbf{R}_p = \mathbf{R}_m \tag{8.6}$$

$$\left(\frac{F_I}{F_G}\right)_p = \left(\frac{F_I}{F_G}\right)_m = \left(\frac{V^2}{lg_n}\right)_p = \left(\frac{V^2}{lg_n}\right)_m \qquad \mathbf{F}_p = \mathbf{F}_m \tag{8.7}$$

$$\left(\frac{F_I}{F_E}\right)_p = \left(\frac{F_I}{F_E}\right)_m = \left(\frac{\rho V^2}{E}\right)_p = \left(\frac{\rho V^2}{E}\right)_m \qquad \mathbf{M}_p = \mathbf{M}_m \tag{8.8}$$

$$\left(\frac{F_I}{F_T}\right)_p = \left(\frac{F_I}{F_T}\right)_m = \left(\frac{\rho l V^2}{\sigma}\right)_p = \left(\frac{\rho l V^2}{\sigma}\right)_m \qquad \mathbf{W}_p = \mathbf{W}_m \tag{8.9}$$

Over the history of fluid mechanics a set of dimensionless numbers of dynamic similarity have come to be well known, and each is associated (typically) with the name of the person who introduced it; these ubiquitous numbers are:

$$\text{Euler number, } \mathbf{E} = V\sqrt{\frac{\rho}{2\Delta p}} \tag{8.10}$$

$$\text{Reynolds number, } \mathbf{R} = \frac{Vl}{\nu} \tag{8.11}$$

$$\text{Froude number, } \mathbf{F} = \frac{V}{\sqrt{lg_n}} \tag{8.12}$$

$$\text{Cauchy[4] number, } \mathbf{C} = \mathbf{M}^2 = \frac{\rho V^2}{E} \tag{8.13}$$

$$\text{Weber number, } \mathbf{W} = \frac{\rho l V^2}{\sigma} \tag{8.14}$$

Each number has a physical interpretation as a force ratio, being the ratio of the inertia force in a flow to the particular force represented by the number; for example, **R** is the ratio of the inertia to the viscous force. It is apparent now that the foregoing force-ratio equations may be written in terms of the dimensionless numbers as indicated above. Following the argument leading to Eqs. 8.3 and 8.4, it will be noted that only four of these

[4]Because the sonic velocity $a = \sqrt{E/\rho}$ (Section 1.5), the Mach number, **M**, and Cauchy number, **C**, are related by $\mathbf{C} = \mathbf{M}^2$.

equations are independent; thus, if any four of them are simultaneously satisfied (e.g., 8.6 to 8.9), dynamic similarity will be ensured and the fifth equation (8.5) will be satisfied automatically.

Fortunately, in most engineering problems four simultaneous equations are not necessary, since some of the forces stated above (1) may not act, (2) may be of negligible magnitude, or (3) may oppose other forces in such a way that the effect of both is reduced. In each new problem of similitude a good understanding of fluid phenomena is necessary to determine how the problem may be satisfactorily simplified by the elimination of the irrelevant, negligible, or compensating forces. It is particularly important to remember that Eqs. 8.5 through 8.9 express ratios of forces. For example, the Froude number **F** is the ratio of inertia to gravity forces. Accordingly, a large Froude number **F** implies that the gravity forces are weak compared to the inertial forces and might possibly be neglected or have little influence on the flow. The reasoning involved in such analyses is best illustrated by citing certain simple and recurring engineering examples.

Reynolds Similarity

In the classical low-speed submerged body problem typified by the conventional airfoil of Fig. 8.2 there are no surface tension phenomena, negligible compressibility (elastic) effects, and gravity does not affect the flowfield. Thus three of the four equations (8.6 to 8.9), which are typically selected to be satisfied simultaneously, are not relevant to the problem, and dynamic similarity is obtained between model and prototype when the Reynolds numbers or ratio of inertia to viscous forces are the same, that is, when

$$\left(\frac{Vl}{\nu}\right)_p = \mathbf{R}_p = \mathbf{R}_m = \left(\frac{Vl}{\nu}\right)_m \tag{8.6}$$

providing that model and prototype are geometrically similar and are similarly oriented to their oncoming flows. Since the equation places no restriction on the fluids of the model and prototype, the latter could move through air and the former be tested in water; if the Reynolds numbers of model and prototype could be made the same, dynamic similitude would result. If for practical reasons the same fluid is used in model and prototype and the kinematic viscosities are equal, the product (Vl) must be the same in both; this means that the velocities around the model will be *larger* than the corresponding ones around the prototype. In aeronautical research the model is frequently tested with compressed air in a *variable density* (or *pressure*) *wind tunnel*; here $\nu_m < \nu_p$ and large velocities past the model are not required. Once equality of Reynolds numbers is obtained in model and prototype, it follows that the ratio of any corresponding forces (such as lift or drag) will be equal to the ratio of any other relevant corresponding forces. Thus (for drag force)

$$\left(\frac{D}{\rho V^2 l^2}\right)_p = \left(\frac{D}{\rho V^2 l^2}\right)_m \tag{8.15}$$

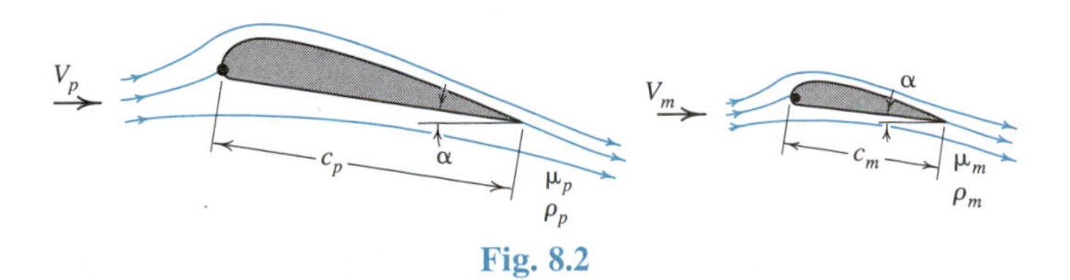

Fig. 8.2

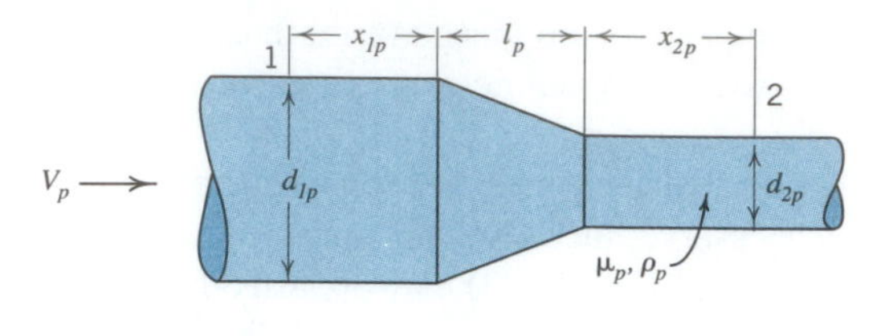

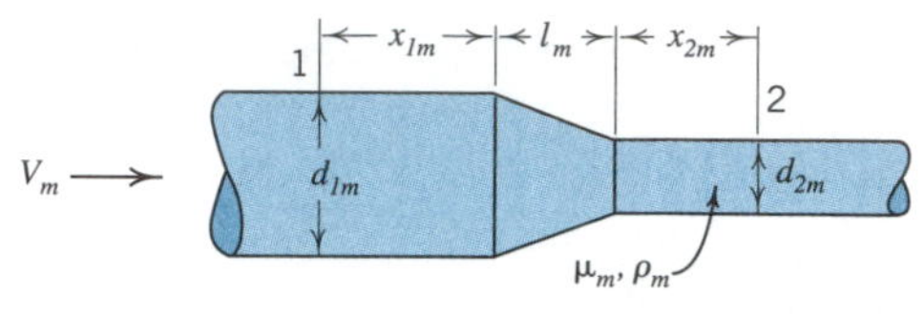

Fig. 8.3

from which the drag of the prototype may be predicted directly from drag and velocity measurements in the model; no corrections for "scale effect" are needed if the Reynolds numbers are the same in model and prototype.

Much of the foregoing reasoning may also be applied to the flow of incompressible fluids through closed passages. Consider, for example, the flows through prototype and model of the contraction of Fig. 8.3. For geometric similarity $(d_2/d_1)_p = (d_2/d_1)_m$, $(l/d_1)_p = (l/d_1)_m$, $(x_1/d_1)_p = (x_1/d_1)_m$, $(x_2/d_2)_p = (x_2/d_2)_m$; *and the roughness pattern of the two passages must be similar in every detail.* Surface tension and elastic effects are nonexistent and gravity does not affect the flowfields. Accordingly dynamic similarity results when equation 8.6 is satisfied, that is, when

$$\left(\frac{V\,d_1}{\nu}\right)_p = \mathbf{R}_p = \mathbf{R}_m = \left(\frac{V\,d_1}{\nu}\right)_m \tag{8.6}$$

from which it follows that Eq. 8.5 is satisfied automatically, so

$$\left(\frac{p_1 - p_2}{\rho V^2}\right)_p = \left(\frac{p_1 - p_2}{\rho V^2}\right)_m \tag{8.5}$$

allowing (for example) prediction of prototype pressure drop $(p_1 - p_2)$ from model measurements. In addition, the ratio between the model and prototype flowrates can be deduced because d_{1p}, d_{1m}, V_p, and V_m are known. Clearly, $Q_p = (\pi/4)d_{1p}^2 V_p$ and $Q_m = (\pi/4)d_{1m}^2 V_m$; therefore, $Q_m/Q_p = (d_{1m}^2/d_{1p}^2)(V_m/V_p)$. If $d_{1m}/d_{1p} = 1/5$ and water is used in both model and prototype, then by virtue of Eq. 8.6, $V_m/V_p = 5$ and $Q_m/Q_p = 1/5$. Here again it is immaterial whether the fluids are the same, dynamic similarity being ensured by the equality[5] of the Reynolds numbers in model and prototype.

ILLUSTRATIVE PROBLEM 8.1

Water (0°C) flows in a 75 mm horizontal pipeline at a mean velocity of 3 m/s. The pressure drop in 10 m of this pipe is 14 kPa. With what velocity must gasoline (20°C) flow in a

[5]Equality of the Reynolds numbers will not produce dynamic similarity in this example unless established flow exists upstream of the contraction; this has been assumed.

geometrically similar 25 mm pipeline for the flows to be dynamically similar, and what pressure drop is to be expected in $3\frac{1}{3}$ m of the 25 mm pipe?

SOLUTION

For this flow dynamic similarity is achieved by making the model and prototype Reynolds numbers equal

$$\left(\frac{V\,d}{\nu}\right)_p = \mathbf{R}_p = \mathbf{R}_m = \left(\frac{V\,d}{\nu}\right)_m \tag{8.6}$$

and it follows then that the Euler numbers for model and prototype will be the same, i.e.,

$$\left(\frac{\Delta p}{\rho V_o^2}\right)_p = \left(\frac{\Delta p}{\rho V_o^2}\right)_m \tag{8.5}$$

The data for this problem are

$$V_p = 3\ \text{m/s} \qquad \Delta p_p = 14\ \text{kPa} \qquad d_p = 75\ \text{mm} \qquad \mu_p = 1.781 \times 10^{-3}\ \text{Pa·s}$$

$$\rho_p = 998.8\ \text{kg/m}^3 \qquad d_m = 25\ \text{mm} \qquad \rho_m = 998.2 \times 0.68 = 680.3\ \text{kg/m}^3$$

$$\mu_m = 2.9 \times 10^{-4}\ \text{Pa·s}$$

Thus, Eq. 8.6 yields

$$\frac{3 \times (0.075) \times 999.8}{0.001\,781} = \frac{V_m \times (0.025) \times 680.3}{0.000\,29} \qquad V_m = 2.16\ \text{m/s} \bullet \tag{8.6}$$

and so Eq. 8.5 gives

$$\frac{14}{[999.8 \times (3)^2]} = \frac{\Delta p_m}{[680.3 \times (2.16)^2]} \qquad \Delta p_m = 4.94\ \text{kPa} \bullet \tag{8.5}$$

Froude Similarity

Another example of wide engineering interest is the modelling of the flowfield about an object (such as the ship of Fig. 8.4) moving on the surface of a liquid. Here geometric similarity is obtained by a carefully dimensioned model suitably weighted so that

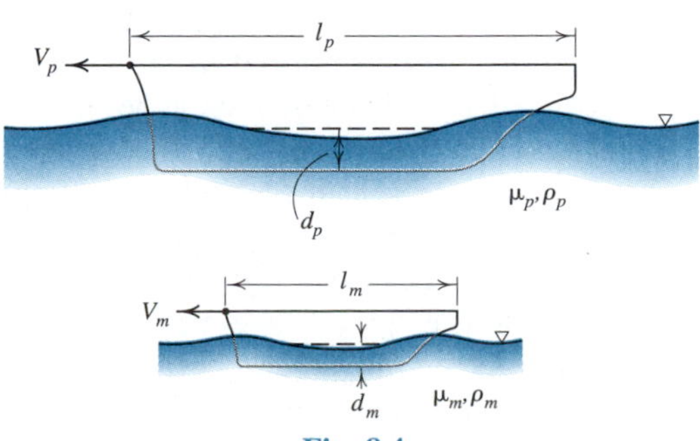

Fig. 8.4

$(d/l)_p = (d/l)_m$. Compressibility of the liquid is of no consequence in such problems, and surface tension may also be ignored if the model is not too small. However, the motion of the ship through the liquid generates water waves on the liquid surface and these waves move under the influence of gravity. Thus, the drag of the ship is the sum of the effects of the energy dissipated in wave generation (which results in a clearly visible and characteristic wave pattern) and of the frictional action of the liquid on the hull. *If frictional effects are assumed to be negligible* [as they could be for an unstreamlined object where most of the resistance is associated with the wave pattern], then dynamic similitude is characterized by

$$\left(\frac{V}{\sqrt{lg_n}}\right)_p = \mathbf{F}_p = \mathbf{F}_m = \left(\frac{V}{\sqrt{lg_n}}\right)_m \tag{8.7}$$

From this it follows (as for the submerged object of Fig. 8.2) that

$$\left(\frac{D}{\rho V^2 l^2}\right)_p = \left(\frac{D}{\rho V^2 l^2}\right)_m \tag{8.15}$$

which means that, if the model is tested with $\mathbf{F}_p = \mathbf{F}_m$, and the drag measured, the drag of the prototype may be predicted for the corresponding speed; the latter is (from Eq. 8.7) $V_p = V_m\sqrt{l_p/l_m}$.

For ship hulls of good design the contribution of wave pattern and frictional action to the drag are of the same order, and neither can be ignored. Here the phenomena will be associated with both the Froude and the Reynolds numbers, requiring for dynamic similitude the two simultaneous equations

$$\left(\frac{V}{\sqrt{lg_n}}\right)_p = \mathbf{F}_p = \mathbf{F}_m = \left(\frac{V}{\sqrt{lg_n}}\right)_m \tag{8.7}$$

$$\left(\frac{Vl}{\nu}\right)_p = \mathbf{R}_p = \mathbf{R}_m = \left(\frac{Vl}{\nu}\right)_m \tag{8.6}$$

Taking g_n the same for model and prototype and solving these equations (by eliminating V) yields $\nu_p/\nu_m = (l_p/l_m)^{3/2}$, indicating that a specific relation between the viscosities of the liquids is required once the model scale is selected. This means (1) that a liquid of appropriate viscosity must be found for the model test, or (2) if the same liquid is used for model and prototype, the model must be as large as the prototype! Since liquids of appropriate viscosity may not exist and full-scale models are obviously impractical, the engineer is forced to choose between the two equations, since both cannot be satisfied simultaneously with the same liquid for model and prototype. This is done by operating the model so that $\mathbf{F}_p = \mathbf{F}_m$ (resulting in $\mathbf{R}_p \gg \mathbf{R}_m$) and then correcting the test results by experimental data dependent on Reynolds number. This is known as correcting for the *scale effect* and is a correction necessitated by incomplete similitude; there would be no scale effect if Eqs. 8.6 and 8.7 could both be satisfied. William Froude originated this technique for ship model testing in England around 1870, but the same equations and principles apply to any flowfield controlled by the combined action of gravity and viscous action; models of rivers, harbors, hydraulic structures, and open-flow problems in general are good examples. However, such models are considerably more difficult to operate and interpret than ship models because of the less well-defined frictional resistance caused by variations of surface roughness and complex boundary geometry.

ILLUSTRATIVE PROBLEM 8.2

A ship of 400 ft length is to be tested by a model 10 ft long. If the ship travels at 30 knots, at what speed must the model be towed for dynamic similitude between model and prototype? If the drag of the model is 2 lb, what prototype drag is to be expected?

SOLUTION

We shall determine the model speed from the equality of Froude numbers, which are given by

$$\left(\frac{V}{\sqrt{lg_n}}\right)_p = \mathbf{F}_p = \mathbf{F}_m = \left(\frac{V}{\sqrt{lg_n}}\right)_m \tag{8.7}$$

A knot is a nautical mile (6 080 ft) per hour; therefore, V_p = 30 knots × 6080 ft/hr/(3600 s/hr) = 50.7 ft/s; also known are l_p = 400 ft, l_m = 10 ft, g_n = 32.2 ft/s^2, and $\rho_m = \rho_p$. Using Eq. 8.7 produces

$$\frac{(50.7)^2}{400 \times 32.2} = \frac{V_m^2}{10 \times 32.2}; \qquad \text{so} \qquad V_m = 8.0 \text{ ft/s} \bullet \tag{8.7}$$

Given the equality of Froude numbers and assuming that the viscous effects are negligible, Eq. 8.15 can be used since the drag force ratios will be equal due to the similarity. Accordingly,

$$\left(\frac{D}{\rho V^2 l^2}\right)_p = \left(\frac{D}{\rho V^2 l^2}\right)_m \qquad \text{and so with } D_m = 2 \text{ lb,}$$

$$\frac{D_p}{(50.7)^2(400)^2} = \frac{2}{(8)^2(10)^2} \qquad D_p = 128\,525 \text{ lb} \bullet \tag{8.15}$$

With no information on hull form, frictional effects cannot be included in this problem; however, if they were included the predicted drag of the prototype would be somewhat *less* than 128 525 lb. See Illustrative Problem 7.5 in Section 7.6.

ILLUSTRATIVE PROBLEM 8.3

A short smooth hydraulic overflow structure passes a flowrate of 600 m^3/s. What flowrate should be used with a 1 : 15 model of this structure to obtain dynamic similitude if friction may be neglected?

SOLUTION

Because of the predominance of gravitational action, similitude will result when Eq. 8.7 is satisfied. Writing V as Q/A or Q/l^2, the equality of Froude numbers may be expressed as

$$\left(\frac{Q}{l^2\sqrt{lg_n}}\right)_p = \left(\frac{Q}{l^2\sqrt{lg_n}}\right)_m \tag{8.7}$$

whence with $\left(\frac{l_m}{l_p}\right) = \frac{1}{15}$, and $Q_p = 600 \text{ m}^3/\text{s}$, we have

$$Q_m = 600 \times \left(\frac{1}{15}\right)^{5/2} = 0.69 \text{ m}^3/\text{s} \bullet$$

Mach (Cauchy) Similarity

An example of similitude in compressible fluid flow is that of the projectile of Fig. 8.5. Here gravity and surface tension do not affect the flowfield, and similitude will result from the actions of resistance and elasticity (compressibility) characterized by

$$\left(\frac{Vl}{\nu}\right)_p = \mathbf{R}_p = \mathbf{R}_m = \left(\frac{Vl}{\nu}\right)_m \tag{8.6}$$

$$\left(\frac{V}{a}\right)_p = \mathbf{M}_p = \mathbf{M}_m = \left(\frac{V}{a}\right)_m \tag{8.8}$$

which may be solved to yield $l_p/l_m = (\nu_p/\nu_m)(a_m/a_p)$, showing that a relation must exist between model scale and the viscosities and sonic speeds in the gases used in model and prototype if dynamic similarity is to be complete. In this situation (unlike the analogous one for the ship model), gases are available that will allow the equation to be satisfied.

ILLUSTRATIVE PROBLEM 8.4

The forces acting on a jet plane traveling at a velocity of 900 m/s at an altitude of 6 km are to be estimated by using a 1 : 10 model that is tested by firing the model like a projectile in a tank of carbon dioxide at 20°C. Calculate the required velocity of the model and the pressure required in the tank for dynamic similitude between the model and the prototype.

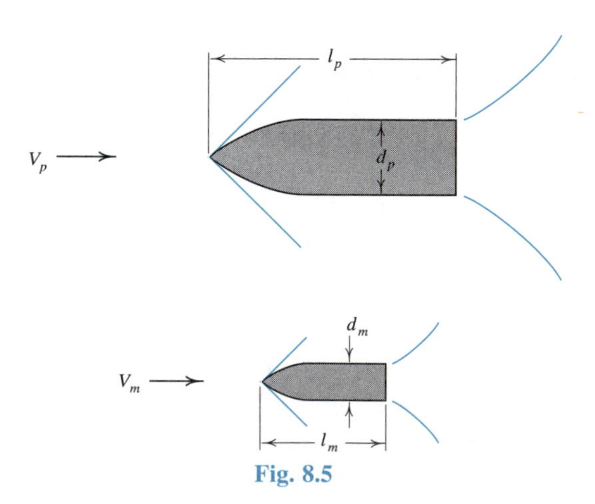

Fig. 8.5

SOLUTION

Assume that the plane flies in a U.S. Standard Atmosphere (Appendix 2). We need to use two equations from Chapter 1, namely,

$$\rho = p/RT \tag{1.3}$$

$$a = \sqrt{kRT} \tag{1.11}$$

and we know that $V_p = 900$ m/s $l_m/l_p = 1/10$ alt = 6 km $T_m = 293$ K
For dynamic similarity $\mathbf{R}_p = \mathbf{R}_m$ and $\mathbf{M}_p = \mathbf{M}_m$. Using

$$\left(\frac{V}{a}\right)_p = \mathbf{M}_p = \mathbf{M}_m = \left(\frac{V}{a}\right)_m \tag{8.8}$$

gives $V_m = 900(a_m/a_p)$. From Appendix 2, $T_p = 249$ K, $R_p = 286.8$ J/kg·K, $k_p = 1.4$, $R_m = 187.8$ J/kg·K, and $k_m = 1.28$. Calculating a_m and a_p from Eq. 1.11, $a_m = 265.4$ m/s and $a_p = 316.2$ m/s, yielding $V_m = 755.4$ m/s. ●
Now, using

$$\left(\frac{Vl}{\nu}\right)_p = \mathbf{R}_p = \mathbf{R}_m = \left(\frac{Vl}{\nu}\right)_m \tag{8.6}$$

(with numerical values from Fig. 1.6 or Appendix 2),

$$\frac{900 \times 10 \times 0.660}{1.595} = \frac{755.4 \times 1 \times \rho_m}{1.47} \qquad \rho_m = 7.25 \text{ kg/m}^3 \tag{8.6}$$

Finally, from Eq. 1.3,

$$7.25 = \frac{p_m}{187.8 \times 293} \qquad p_m(\text{abs}) = 399 \text{ kPa} \;\bullet \tag{1.3}$$

Euler Similarity—The Cavitation Number

When prototype cavitation is to be modeled, the foregoing equations of similitude are inadequate, since they do not include vapor pressure p_v, the attainment of which is the unique feature of the cavitation.[6] Consider the cavitating hydrofoil prototype and model of Fig. 8.6, the model fixed in a water tunnel where p_o and V_o can be controlled. Geometric similarity and the same orientation to the oncoming flow are assumed, and complete dynamic similitude is desired. Here gravity will have small effect on the flowfield, compressibility of the fluid will be insignificant, and surface tension may be neglected.[7] Accordingly, dynamic similitude is obtained when the Reynolds number and the appropriate form of the Euler number (i.e., the inverse of **E**) are used, that is, when

$$\left(\frac{V_o l}{\nu}\right)_p = \mathbf{R}_p = \mathbf{R}_m = \left(\frac{V_o l}{\nu}\right)_m \tag{8.6}$$

$$\left(\frac{p_o - p_v}{\rho V_o^2}\right)_p = \boldsymbol{\sigma}_p = \boldsymbol{\sigma}_m = \left(\frac{p_o - p_v}{\rho V_o^2}\right)_m \tag{8.16}$$

[6]See Appendix 4.

[7]Surface tension may have critical influence in such problems if the cavitation is incipient; the gas content of the liquid is also important. Both of these points are ignored here to simplify the problem.

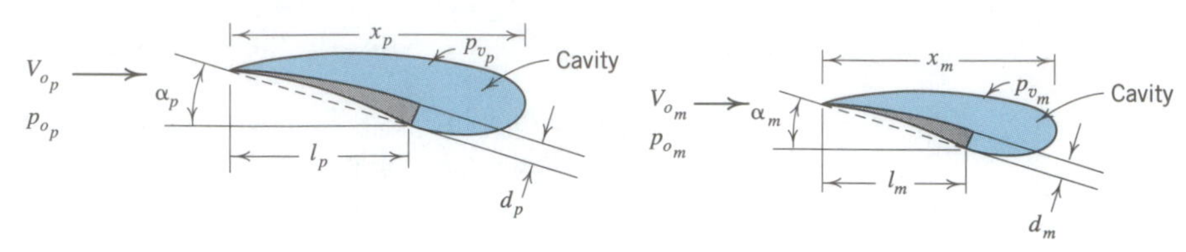

Fig. 8.6

When these equations are satisfied the cavity shapes in model and prototype will be the same; that is, $(x/l)_p = (x/l)_m$. The second of the two simultaneous equations represents an equality of special Euler numbers (known as the *cavitation number* $\boldsymbol{\sigma}$) containing the (absolute) pressures, p_o and p_v, peculiar to cavitation. If these equations can be simultaneously satisfied by the use of appropriate liquid or adjustment of tunnel velocities and pressures or both, then (as for the submerged body without cavitation)

$$\left(\frac{D}{\rho V_o^2 l^2}\right)_p = \left(\frac{D}{\rho V_o^2 l^2}\right)_m \tag{8.15}$$

and

$$\left(\frac{\Delta p}{\rho V_o^2}\right)_p = \left(\frac{\Delta p}{\rho V_o^2}\right)_m \tag{8.5}$$

in which Δp in model and prototype represents the pressure changes between any two *corresponding* points in the flowfields of model or prototype. Here (as for the ship model) it is very difficult to satisfy both of the simultaneous equations (8.6 & 8.16) with essentially the same liquid in model and prototype. Again the engineer must select and satisfy the more important of the two equations and correct the test results to allow for the unsatisfied equation. In this problem it is clear that the existence and extent of the cavitation zone dominate the flowfield and frictional aspects are of comparatively small importance; accordingly the tests would be carried out with the cavitation numbers the same in model and prototype and empirical adjustments made for the frictional phenomena.

ILLUSTRATIVE PROBLEM 8.5

Consider the flow in a venturi meter as shown here, which is used to measure the flowrate in a pipeline. In this case a large meter (12″ × 6″) is installed in a production line; here the dimensions are the diameters at sections 1 and 2, respectively. There is a suspicion that cavitation is occurring in the venturi's 6″ section, but carrying out tests in the production line is not possible. However, the local university fluids laboratory is found to own a geometrically similar 4″ × 2″ venturi, which is installed in the laboratory water supply line. If the production and laboratory venturis are run under dynamically similar conditions, then it should be possible to ascertain if cavitation is occurring in the production meter by examining conditions in the laboratory meter. Under what conditions should the laboratory (i.e., the model) venturi be operated if the 12″ × 6″ (prototype) venturi data show that $p_1 - p_2 = 10$ psi and $Q = 8.0$ ft^3/s. Assume that the pressure at section 2 is the vapor pressure since cavitation is suspected and that both the laboratory and production venturis operate with water at 68°F.

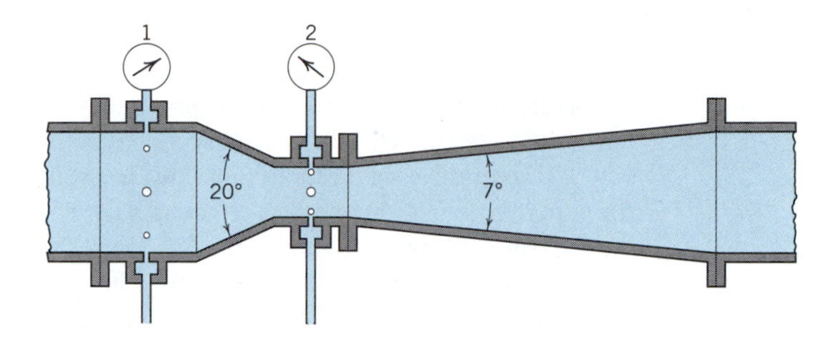

SOLUTION

In this case, geometric similarity exists and dynamic similarity will be guaranteed by equality of Reynolds and cavitation numbers in the production and laboratory (model) venturi meters. Therefore, Eqs. 8.6 and 8.7 are appropriate and recalling that the pressure at section 2 is presumed to be the vapor pressure, we have

$$\left(\frac{V_1 d_1}{\nu}\right)_p = \mathbf{R}_p = \mathbf{R}_m = \left(\frac{V_1 d_1}{\nu}\right)_m \tag{8.6}$$

$$\left(\frac{p_1 - p_v}{\rho V_1^2}\right)_p = \boldsymbol{\sigma}_p = \boldsymbol{\sigma}_m = \left(\frac{p_1 - p_v}{\rho V_1^2}\right)_m \tag{8.16}$$

Since the kinematic viscosity of the fluid is the same in the model and in the prototype and because $Q = \frac{\pi}{4} d_1^2 V_1$, we rewrite Eq. 8.6 and solve for the relation between the flowrates; the result is

$$Q_m = \left(\frac{d_{1m}}{d_{1p}}\right) Q_p; \qquad \text{thus,} \qquad Q_m = \left(\frac{4}{12}\right) \times 8.0 = 2.7 \text{ ft}^3/\text{s} \bullet$$

Using the relation between flowrate and velocity and noting that because the model and prototype use the same fluid at the same temperature, the density drops out of Eq. 8.16, we have

$$\left(\frac{p_1 - p_v}{Q^2/d_1^4}\right)_m = \left(\frac{p_1 - p_v}{Q^2/d_1^4}\right)_p$$

$$(p_1 - p_v)_m = \left(\frac{Q_m^2 d_p^4}{Q_p^2 d_m^4}\right)(p_1 - p_v)_p = \left(\frac{Q_m}{Q_p}\right)^2 \left(\frac{d_p}{d_m}\right)^4 (p_1 - p_v)_p \tag{8.16}$$

Thus, by using the relationship between flowrates and diameters from above,

$$(p_1 - p_v)_m = \left(\frac{d_{1m}}{d_{1p}}\right)^2 \left(\frac{d_{1p}}{d_{1m}}\right)^4 (p_1 - p_v)_p = \left(\frac{d_{1p}}{d_{1m}}\right)^2 (p_1 - p_v)_p$$

$$= \left(\frac{12}{4}\right)^2 (10 \text{ psi}) = 90 \text{ psi}$$

From Appendix 2, $p_{v_m} = 0.34$ psia, so $p_{1_m} = 90.34$ psia •

If the laboratory model venturi is operated then at the calculated flowrate and pressure, the conditions at section 2 should be similar to those at section 2 in the production venturi and an examination made to see if cavitation is actually occurring (See Appendix 4).

Comment

It is a rare occurrence in engineering problems when pure theory alone leads to a complete answer. This is particularly true in the construction, operation, and interpretation of models where compromises of the type above are almost invariably necessary; there is no substitute for experience and judgment in effecting such compromises. A major aid to the engineer are the equations which govern a fluid process. They serve as a guide to key dimensionless parameters. This topic has been developed in depth by Kline.[8] Section 8.3 herein gives an introduction to the topic; however, simple examples to whet the appetite are seen from the analyses of the hydraulic jump and shock waves in Sections 6.3 and 13.12. The hydraulic jump equation 6.7 shows that the ratio of the depths before and after the jump y_1/y_2 is a function of only the Froude number $\mathbf{F}^2 = V_1^2/g_n y_1$. In the normal shock wave case, the results in Eqs. 13.31 and 13.32 demonstrate that the proper pressure ratio to be analyzed is $(p_2 - p_1)/p_1$ not p_2/p_1, and that both the pressure ratio and the velocity ratio of values ahead of and behind the shock are functions of only the upstream Mach number $\mathbf{M}$ and the ratio of specific heats k.

The foregoing treatment of similitude is intended as an introduction to the theory and to the numerous practical applications cited in the literature at the end of the chapter. Also it is now clear that certain dimensionless groups play a key and recurring role in fluid mechanics problems.

8.2 DIMENSIONAL ANALYSIS

Dimensional analysis is the mathematics of the dimensions of quantities and is closely related to the laws of similitude. Dimensional analysis is a very useful tool and in addition to the examples cited in this section, there is a very significant application of this method in Section 9.6. There, dimensional analysis is used to deduce the relationship between the shear stress and other parameters of a pipe flow. The analysis leads to a definition of a flow resistance factor, called the *friction factor*, in terms of only two dimensionless groups and the experimental data bear out the result in a convincing fashion.

The methods of dimensional analysis are built on Fourier's *principle of dimensional homogeneity* (1882), which states that an equation expressing a physical relationship between quantities must be dimensionally homogeneous; that is, the dimensions of each side of the equation must be the same. This principle was utilized in Chapter 1 in obtaining the dimensions of density and kinematic viscosity, and it was recommended as a valuable means of checking engineering calculations. Further investigation of the principle will reveal that it affords a means of ascertaining the forms of physical equations from knowledge of relevant variables and their dimensions. Although dimensional manipulations cannot be expected to produce analytical solutions to physical problems, dimensional analysis proves a powerful tool in formulating problems which defy analytical solution and must be solved experimentally. Here dimensional analysis comes into its own by pointing the way toward a maximum of information from a minimum of experiment. It accomplishes this by the formation of dimensionless groups, some of which are identical with the force ratios developed with the laws of similitude.

In the SI and U.S. Customary unit systems, there are four basic dimensions that are

[8]S. J. Kline, *Similitude and Approximation Theory*, Springer-Verlag, 1986.

directly relevant to fluid mechanics, namely, length (L), mass (M), time (t), and thermodynamic temperature (T). Only the first three have been included in the following discussion, but should temperature be needed in an analysis, its presence does not change the principles or process. A summary of the fundamental quantities of fluid mechanics and their dimensions and units is given in Appendix 1, the conventional system of capital letters being followed to indicate the dimensions of quantities. An exception is made with respect to the time dimension (t), to avoid confusion with thermodynamic temperature. In problems of mechanics, the basic relation between force and mass is given by the Newtonian law: force (or weight) = mass × acceleration; and, therefore, dimensionally,

$$F = \frac{ML}{t^2}$$

through which the dimension of force can be expressed in terms of the basic dimensions. Thus, there are only three *independent* fundamental dimensions here.

To illustrate the mathematical steps in a simple dimensional problem, suppose that it is known that the power, P, which can be derived from a hydraulic turbine is dependent on the rate of flow through the machine, Q, the specific weight of the fluid, γ, and the unit mechanical energy, E_T, which is given up by every unit weight of fluid as it passes through the machine. Suppose that the relation between these four variables is unknown but it is known that these are the only variables involved in the problem. (This suggests, correctly, that experience and analytical ability in determining the relevant variables are necessary before the methods of dimensional analysis can be successfully applied; contrast this with the situation in the case of normalization in Section 8.3.) With the meager knowledge of the appropriate variables, the following mathematical statement can be made (it formally posits that the power is a function of the selected variables):

$$P = f(Q,\gamma,E_T)$$

From the principle of dimensional homogeneity it is apparent that the quantities involved cannot be added or subtracted since their dimensions are different. This principle limits the equation to a combination of products of powers of the quantities involved, which may be expressed in the general form

$$P = CQ^a\gamma^b E_T^c$$

in which a, b, c are unknowns that we will determine by dimensional methods; unfortunately, the dimensionless constant C cannot be determined, except by running a numerical or physical experiment. The equation can be written dimensionally as

$$(\text{Dimensions of } P) = (\text{Dimensions of } Q)^a(\text{Dimensions of } \gamma)^b(\text{Dimensions of } E_T)^c$$

or

$$\frac{ML^2}{t^3} = \left(\frac{L^3}{t}\right)^a\left(\frac{M}{L^2t^2}\right)^b(L)^c$$

The principle of dimensional homogeneity being applied, the exponent of *each* of the fundamental dimensions is the same on each side of the equation, giving the following equations in the exponents of the dimensions:

$$M: \quad 1 = b$$

$$L: \quad 2 = 3a - 2b + c$$

$$t: \quad -3 = -a - 2b$$

Solving for a, b, and c yields

$$a = 1 \qquad b = 1 \qquad c = 1$$

and resubstituting these values in the equation above for P gives

$$P = CQ\gamma E_T$$

The form of the equation (confirmed by Eqs. 5.8) has, therefore, been derived, without physical analysis, solely from consideration of the dimensions of the quantities which were known to enter the problem. The magnitude of C must be obtained either (1) from a physical analysis of the problem or (2) from experimental measurements of P, Q, γ, and E_T.

From the foregoing problem it appears that in dimensional analysis (of problems in mechanics) only three equations can be written since there are only three independent fundamental dimensions: M, L, and t. This fact limits the completeness with which a problem with more than three unknowns may be solved, but it does not limit the utility of dimensional analysis in obtaining the form of the terms of the equation. This point may be fruitfully illustrated and a more general analysis procedure developed by reexamining the ship model problem previously treated (Section 8.1) by the laws of similitude. Consider the ship of Fig. 8.4 having a certain *shape* and draft, d. *Shape* is emphasized here so that it cannot be confused with *size*. The shape of a ship is *not* fixed by *algebraic* dimensions such as l, d, and so forth. With shape constant and size variable, a series of geometrically similar objects is implied. The drag of the ship will depend on the size of the ship [characterized by its length (l)], the viscosity (μ) and density (ρ) of the fluid, the velocity (V) of the ship, and the acceleration (g_n) due to gravity [which dominates the surface wave pattern set up as the ship moves through the water]. (We note again that the selection of relevant variables is the crucial first step in any problem of dimensional analysis, and obviously this cannot be done without some experience and ''feel'' for the problem; the mathematical processes of dimensional analysis cannot overcome the selection of irrelevant variables! On the other hand, experiments will show that irrelevant variables are not correlated to the main features; having too many variables is less of a sin than having too few!)

While Lord Rayleigh put forth the early bases of a dimensional analysis method, Buckingham[9] provided a broad generalization known as the Π-theorem. The modern version of the Buckingham theorem can be stated as follows (see, e.g., Kline[10]):

1. If n variables are functions of each other (such as D, l, ρ, μ, V, and g_n of the present example), then k equations of their exponents (a, b, c, etc.) can be written (where k is the largest number of variables among the n variables which *cannot* be combined into a dimensionless group).
2. In most cases, k is equal to the number m of independent dimensions (equal to three here, that is, M, L, and t). Generally, $k \leqq m$.
3. Application of dimensional analysis allows expression now of the functional relationship in terms of $(n - k)$ distinct dimensionless groups. In the present case, $n = 6$, $k = m = 3$, so three groups are expected. Indeed, the previous similitude analysis yielded $D/\rho l^2 V^2$, **F**, and **R**!

[9]E. Buckingham, ''Model experiments and the forms of empirical equations,'' *Trans. A. S. M. E.* 37, 1915. pp. 263ff.

[10]S. J. Kline, *Similitude and approximation theory*, Springer-Verlag, 1986.

Buckingham designated the dimensionless groups by the Greek (capital) letter Π. The important advantage of the Π-theorem is that it shows in advance of the analysis how many groups are to be expected and allows the engineer flexibility in formulating them (particularly as it is already known that certain groups, e.g., force ratios, are relevant).

For the present problem of drag on a ship, the functional relationship among the variables can be written in the form

$$f(D, l, \rho, \mu, V, g_n) = 0$$

by transferring all terms to the left side of the equation. The number of Π's to be expected is $n - m = 6 - 3 = 3$, so the relation between dimensionless groups should be expressible as

$$f'(\Pi_1, \Pi_2, \Pi_3) = 0$$

Two crucial points are, first, that one should check to see if $k < m$ and, second, that there are a sizable number of ways to combine six variables into three dimensionless groups. A rational approach is needed. Experience with similitude suggests that V, l, and ρ appear in most dimensionless groups because they all appear in the inertia force of a similitude analysis; they can be called the *repeating variables* since we will assume for many problems that one of their number will appear in every Π-group. Variables such as D, μ, σ, and E appear only in the unique group describing the ratio of the inertia force to the force related to the variable. Therefore, a useful procedure might be:

1. Find the largest number of variables which do not form a dimensionless Π-group. In the present case the number of independent dimensions is $m = 3$ and V, ρ, and l cannot be formed into a Π-group, so $k = m = 3$.
2. Determine the number of Π-groups to be formed. Here, $n = 6$, $k = m = 3$, and thus three Π's are required.
3. Combine sequentially the variables that cannot be formed into a dimensionless group (usually the repeating variables noted above), with each of the remaining variables to form the requisite Π-groups. Here then the three Π's would be

$$\Pi_1 = f_1(D, \rho, V, l)$$

$$\Pi_2 = f_2(\mu, \rho, V, l)$$

$$\Pi_3 = f_3(g_n, \rho, V, l)$$

Thus, it remains only to determine the detailed form of these dimensionless groups. Using Π_1 as an example to illustrate the method,

$$\Pi_1 = D^a \rho^b V^c l^d$$

Writing the equation dimensionally,

$$M^0 L^0 t^0 = \left(\frac{ML}{t^2}\right)^a \left(\frac{M}{L^3}\right)^b \left(\frac{L}{t}\right)^c (L)^d$$

The following equations in the exponents of the dimensions are obtained:

$$M\text{: } 0 = a + b$$

$$L\text{: } 0 = a - 3b + c + d$$

$$t\text{: } 0 = 2a + c$$

Solving these equations in terms of a, b, c, or d (a will be used here),

$$b = -a \qquad c = -2a \qquad d = -2a$$

and substituting them in the equation above for Π_1,

$$\Pi_1 = \left(\frac{D}{\rho l^2 V^2}\right)^a$$

Since a dimensionless group raised to a power is of no more significance than the group itself, the exponent may be taken as any convenient number other than zero, and Π_1 as $D/\rho V^2 l^2$. Similarly Π_2 is obtained as $Vl\rho/\mu$ and Π_3 as $V/\sqrt{lg_n}$. But

$$\mathbf{F} = \frac{V}{\sqrt{lg_n}} \qquad \text{and} \qquad \mathbf{R} = \frac{Vl\rho}{\mu}$$

allowing the equation to be written in the more general form

$$f'\left(\frac{D}{\rho l^2 V^2}, \mathbf{R}, \mathbf{F}\right) = 0$$

or

$$\frac{D}{\rho l^2 V^2} = f''(\mathbf{F}, \mathbf{R})$$

showing without experiment, but from dimensional analysis alone, that $D/\rho l^2 V^2$ depends only on the Froude and Reynolds numbers. This sort of result is the main objective of dimensional analysis in engineering problems and may always be obtained by application of the technique used here. Although it gives no clue to the functional relationship among $D/\rho l^2 V^2$, $\mathbf{F}$, and $\mathbf{R}$, it has arranged the numerous original variables into a relation between a smaller number of dimensionless groups of variables and has thus indicated how test results should be processed for concise presentation. Use of the Π-theorem in dimensional analysis is thus seen to be highly efficient in formulating dimensionless groups which may be easily interpreted in terms of those of geometric, kinematic, and dynamic similitude.

The laws of similitude showed that the flowfield about a prototype surface ship is dynamically similar to that of its model if (with geometric similarity) $\mathbf{F}_p = \mathbf{F}_m$ and $\mathbf{R}_p = \mathbf{R}_m$—and that satisfaction of these conditions leads to $(D/\rho l^2 V^2)_p = (D/\rho l^2 V^2)_m$. The result of dimensional analysis is the same since it has demonstrated the existence of a unique functional relationship among $D/\rho l^2 V^2$, $\mathbf{F}$, and $\mathbf{R}$ for a ship of one shape. Thus (whatever this relationship) for the same $\mathbf{F}$ and same $\mathbf{R}$ in model and prototype, $(D/\rho l^2 V^2)_p = (D/\rho l^2 V^2)_m$.

ILLUSTRATIVE PROBLEM 8.6

A hydraulic jump occurs in an open channel as shown. Clearly for this flow there are two relevant length scales, namely, y_1 and y_2, as well as the usual variables such as the two-dimensional flowrate q, the acceleration g_n due to gravity, the fluid density ρ and the viscosity μ. Find the functional relationship between these variables.

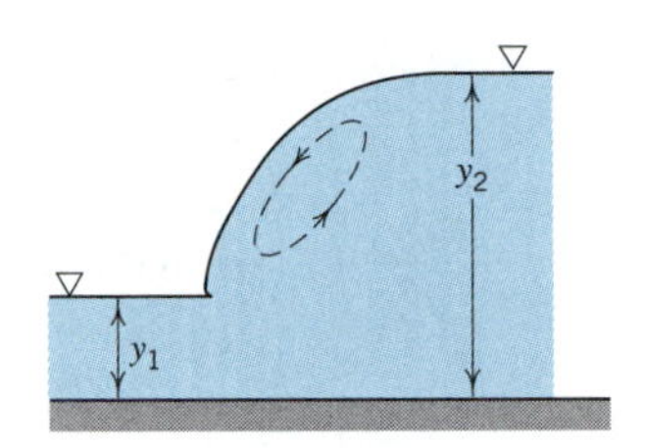

SOLUTION

The relation to be analyzed is

$$f(y_1,y_2,q,\rho,\mu,g_n) = 0$$

Follow the steps outlined above.

1. q, y_1, and ρ cannot be formed into a dimensionless group and $m = 3$ (M,L,t) so $k = m = 3$.
2. There are $n = 6$ variables, so $n - k = 3$.
3. The Π-groups are formed as (note the repeating variables)

$$\Pi_1 = f_1(y_2,\ q,\ y_1,\ \rho)$$

$$\Pi_2 = f_2(g_n,\ q,\ y_1,\ \rho)$$

$$\Pi_3 = f_3(\mu,\ q,\ y_1,\ \rho)$$

For Π_1,

$$M^0L^0t^0 = L^a\left(\frac{L^2}{t}\right)^b L^c\left(\frac{M}{L^3}\right)^d$$

$$M\text{: } 0 = 0 + 0 + 0 + d$$

$$L\text{: } 0 = a + 2b + c - 3d$$

$$t\text{: } 0 = 0 - b + 0 + 0$$

Clearly, $d = 0$ and $b = 0$; thus, $a = -c$;

$$\Pi_1 = \left(\frac{y_2}{y_1}\right)^c \qquad \text{or (making the arbitrary, but useful choice } c = 1) \qquad \Pi_1 = \frac{y_2}{y_1}$$

For Π_2,

$$M^0L^0t^0 = \left(\frac{L}{t^2}\right)^a\left(\frac{L^2}{t}\right)^b L^c\left(\frac{M}{L^3}\right)^d$$

$$M\text{: } 0 = 0 + 0 + 0 + d$$

$$L\text{: } 0 = a + 2b + c - 3d$$

$$t\text{: } 0 = -2a - b + 0 + 0$$

Here, $d = 0$ and $b = -2a$ so $-3a + c = 0$; it follows that $c = 3a$ and $b = -2a$ and so

$$\Pi_2 = \left(\frac{g_n y_1^3}{q^2}\right)^a$$ or making again an arbitrary, but insightful choice of $a = -1$

$$\Pi_2 = \left(\frac{q^2}{g_n y_1^3}\right)$$

For Π_3,

$$M^0 L^0 t^0 = \left(\frac{M}{Lt}\right)^a \left(\frac{L^2}{t}\right)^b L^c \left(\frac{M}{L^3}\right)^d$$

$$M\colon 0 = a + 0 + 0 + d$$

$$L\colon 0 = -a + 2b + c - 3d$$

$$t\colon 0 = -a - b + 0 + 0$$

Now, $b = -a$, $d = -a$, and $c = a + 3d - 2b = -a$ so

$$\Pi_3 = \left(\frac{\mu}{q y_1 \rho}\right)^a; \qquad \text{if } a = -1, \text{ then} \qquad \Pi_3 = \frac{q y_1}{\nu}$$

If we recognize the two-dimensional flowrate q as the product of the mean velocity V_1 times y_1, then it is clear that $\Pi_2 = \mathbf{F}^2$, where **F** is a Froude number, and $\Pi_3 = \mathbf{R}$, where **R** is a Reynolds number. Thus,

$$\frac{y_2}{y_1} = f\left(\frac{q^2}{g_n y_1^3}, \frac{q y_1}{\nu}\right) = f(\mathbf{F}, \mathbf{R}) \; \bullet$$

It appears that the depth ratio in a hydraulic jump depends on the Froude and Reynolds numbers. The remarkable and informative form of this result should be compared to the precise result found in Section 6.3. Experiments have shown that the friction effects in the jump itself are negligible and the depth ratio depends on the Froude number.

ILLUSTRATIVE PROBLEM 8.7

A smooth, symmetric body of cross-sectional area A moves through a compressible fluid of density ρ, viscosity μ, and modulus of elasticity E (recall that the sonic or acoustic velocity $a = \sqrt{E/\rho}$) and with a velocity V_o. If the drag force exerted on the body is D, find the functional dependence of D on the other variables.

SOLUTION

The relation to be analyzed is

$$f(D, A, \rho, \mu, V_o, E) = 0$$

In the compressible flow about an immersed body, the effect of gravity is generally negligible, so g_n is not included here.

Follow the steps outlined above.

1. Here A, ρ, and V_o cannot be formed into a dimensionless group and $m = 3(M, L, t)$ so $k = m = 3$.
2. There are $n = 6$ variables, so $n - k = 3$.
3. The Π-groups are formed as

$$\Pi_1 = f_1(D, A, \rho, V_o)$$

$$\Pi_2 = f_2(\mu, A, \rho, V_o)$$

$$\Pi_3 = f_3(E, A, \rho, V_o)$$

For Π_1,

$$\frac{M^0L^0}{t^0} = \left(\frac{ML}{t^2}\right)^a (L^2)^b \left(\frac{M}{L^3}\right)^c \left(\frac{L}{t}\right)^d$$

$$M: 0 = a + c$$

$$L: 0 = a + 2b - 3c + d$$

$$t: 0 = -2a - d$$

Solving in terms of c: $a = -c$, $b = c$, and $d = 2c$, so

$$\Pi_1 = \left(\frac{A\rho V_o^2}{D}\right)^c \quad \text{or} \quad (\text{letting } c = -1)\ \Pi_1 = \frac{D}{A\rho V_o^2}$$

For Π_2,

$$\frac{M^0L^0}{t^0} = \left(\frac{M}{Lt}\right)^a (L^2)^b \left(\frac{M}{L^3}\right)^c \left(\frac{L}{t}\right)^d$$

which yields in terms of d: $a = -d$, $b = \frac{1}{2}d$, and $c = d$, so

$$\Pi_2 = \left(\frac{\sqrt{A}\rho V_o}{\mu}\right)^d \quad \text{or} \quad (\text{for } d = 1)\ \Pi_2 = \frac{\sqrt{A}\rho V_o}{\mu}$$

For Π_3,

$$\frac{M^0L^0}{t^0} = \left(\frac{M}{t^2L}\right)^a (L^2)^b \left(\frac{M}{L^3}\right)^c \left(\frac{L}{t}\right)^d$$

which yields in terms of a: $b = 0$, $c = -a$, and $d = -2a$, so

$$\Pi_3 = \left(\frac{E}{\rho V_o^2}\right)^a \quad \text{or} \quad \left(\text{for } a = -\frac{1}{2}\right)\ \Pi_3 = \frac{V_o}{\sqrt{E/\rho}}$$

As $\sqrt{E/\rho} = a$, $\Pi_3 = V_o/a$. Accordingly,

$$f'\left(\frac{D}{A\rho V_o^2}, \frac{\sqrt{A}\rho V_o}{\mu}, \frac{V_o}{a}\right) = 0$$

or defining $\mathbf{R} = \sqrt{A}\rho V_o/\mu$ and $\mathbf{M} = V_o/a$,

$$\frac{D}{A\rho V_o^2} = f''(\mathbf{R}, \mathbf{M}) \bullet$$

that is, the drag depends on the Reynolds and Mach numbers (see Section 11.2).

8.3 NORMALIZATION OF EQUATIONS

Normalization is the process of nondimensionalizing the governing equations of a mathematical model in terms of the significant scales of the flowfield under consideration. In general, the boundary and initial conditions for the flow are also normalized, but in this introduction, we shall not do that. For a full treatment,[11] the excellent book by Kline (*Similitude and Approximation Theory*, Springer-Verlag, 1986) is recommended; the discussion here is similar to his approach. This process reduces classes of models to a standard form so that only a few numerical computations or physical experiments need be made to obtain answers for a range of conditions. In essence, if one knows the equations that govern the flow, then one can obtain from normalization of the equations *all* of the variables that need to be considered and *all* of the similitude force ratios or Π terms that are applicable.

Under the assumption that the governing equations are known, e.g., the Navier-Stokes equations, the Euler equations, Bernoulli's equation, etc., there are three key steps to be performed. First, the significant scales of the problem must be selected, with the object of selecting scales so that the maximum values of the nondimensional dependent and independent variables have values that are of the order of unity everywhere. Second, the variables of the equations are made nondimensional by defining new variables using the selected scales. Third, the equations are made nondimensional and simplified so that the governing similitude terms are obvious. This process is now illustrated by application to the Navier-Stokes equations, which were derived in the previous chapter (Section 7.15).

For simplicity, we use the two-dimensional equations for an incompressible flow with constant density in the x- and z-plane, namely,

$$\frac{\partial u}{\partial t} + u\frac{\partial u}{\partial x} + w\frac{\partial u}{\partial z} = -\frac{1}{\rho}\frac{\partial p}{\partial x} + \frac{\mu}{\rho}\left(\frac{\partial^2 u}{\partial x^2} + \frac{\partial^2 u}{\partial z^2}\right) \tag{8.17}$$

$$\frac{\partial w}{\partial t} + u\frac{\partial w}{\partial x} + w\frac{\partial w}{\partial z} = -\frac{1}{\rho}\frac{\partial p}{\partial z} - g_n + \frac{\mu}{\rho}\left(\frac{\partial^2 w}{\partial x^2} + \frac{\partial^2 w}{\partial z^2}\right) \tag{8.18}$$

Step 1: In order to select the significant scales, it is necessary to know what physical problem or class of problems is to be solved or modeled. Suppose that here it is the flow past the shape or body first used in Section 8.1. A natural length scale is the maximum dimension of the shape l. Notice that we have chosen only one length scale; for a boundary layer where the cross-stream thickness of the layer is much less than its streamwise length, it would be better to chose different length scales in the x- and z-directions. A natural velocity scale and one likely to be known to us would be the velocity V far upstream of the shape. From the applications of the Bernoulli equation in Section 5.4, we know that the maximum pressure is likely to occur at the nose of the shape, where there is a stagnation point and that the pressure difference between there and far upstream is $\rho V^2/2$. Finally, we need to select a time scale: there are two choices. One can wait and see if a useful scale emerges from the normalization process; alternatively, we observe that the time that it takes a fluid particle to traverse the length of the shape is a useful measure and select a time scale of l/V.

[11]See also R. Street, *The Analysis and Solution of Partial Differential Equations*, Chap. 10, Monterey: Brooks-Cole, 1973.

Step 2: Now it is possible to define a new set of nondimensional variables as follows:

$$X = \frac{x}{l}; \qquad Z = \frac{z}{l}; \qquad T = \frac{t}{l/V}$$

$$U = \frac{u}{V}; \qquad W = \frac{w}{V}; \qquad P = \frac{p}{\frac{1}{2}\rho V^2}$$

To use these, we write them as $u = VU$, $w = VW$, $x = lX$, etc. and insert them into the governing equations.

Step 3: According to the chain rule of partial differentiation,

$$\frac{\partial u}{\partial t} = \frac{\partial}{\partial T}(VU)\frac{dT}{dt} = \frac{V}{l}\frac{\partial}{\partial T}(VU) = \frac{V^2}{l}\frac{\partial U}{\partial T}; \qquad \frac{\partial u}{\partial x} = \frac{\partial}{\partial X}(VU)\frac{dX}{dx} = \frac{V}{l}\frac{\partial U}{\partial X}$$

and $\dfrac{\partial^2 u}{\partial x^2} = \dfrac{V}{l^2}\dfrac{\partial^2 U}{\partial X^2}$, while

$$\frac{\partial w}{\partial z} = \frac{\partial}{\partial Z}(VW)\frac{dZ}{dz} = \frac{V}{l}\frac{\partial W}{\partial Z} \quad \text{and} \quad \frac{\partial p}{\partial x} = \frac{\partial}{\partial X}\left(\frac{1}{2}\rho V^2 P\right)\frac{dX}{dx} = \frac{\frac{1}{2}\rho V^2}{l}\frac{\partial P}{\partial X}, \quad \text{etc.}$$

Introducing these results plus the appropriate expressions for the remaining terms into Eqs. 8.17 and 8.18 produces

$$\frac{V^2}{l}\left(\frac{\partial U}{\partial T} + U\frac{\partial U}{\partial X} + W\frac{\partial U}{\partial Z}\right) = -\frac{\frac{1}{2}V^2}{l}\frac{\partial P}{\partial X} + \frac{\mu V}{\rho l^2}\left(\frac{\partial^2 U}{\partial X^2} + \frac{\partial^2 U}{\partial Z^2}\right)$$

$$\frac{V^2}{l}\left(\frac{\partial W}{\partial T} + U\frac{\partial W}{\partial X} + W\frac{\partial W}{\partial Z}\right) = -\frac{\frac{1}{2}V^2}{l}\frac{\partial P}{\partial Z} - g_n + \frac{\mu V}{\rho l^2}\left(\frac{\partial^2 W}{\partial X^2} + \frac{\partial^2 W}{\partial Z^2}\right)$$

These equations can now be simplified by dividing both sides by V^2/l to achieve

$$\frac{\partial U}{\partial T} + U\frac{\partial U}{\partial X} + W\frac{\partial U}{\partial Z} = -\frac{1}{2}\frac{\partial P}{\partial X} + \frac{\mu}{\rho V l}\left(\frac{\partial^2 U}{\partial X^2} + \frac{\partial^2 U}{\partial Z^2}\right)$$

$$\frac{\partial W}{\partial T} + U\frac{\partial W}{\partial X} + W\frac{\partial W}{\partial Z} = -\frac{1}{2}\frac{\partial P}{\partial Z} - \frac{g_n l}{V^2} + \frac{\mu}{\rho V l}\left(\frac{\partial^2 W}{\partial X^2} + \frac{\partial^2 W}{\partial Z^2}\right)$$

We see immediately that the key similitude parameters are the Froude number $\mathbf{F} = \sqrt{V/lg_n}$ and the Reynolds number $\mathbf{R} = \rho V l/\mu = Vl/\nu$. The normalized equations also give us the specific forms of these force ratios, as well as the form of the pressure coefficient which has been defined by the nondimensionalization of the pressure (compare with Eq. 8.5). If one were to run a numerical simulation using these nonlinear equations, it would be clear that the normalized form is preferable since the solution for given Froude and Reynolds numbers is valid for *all* flows with those same values irrespective of the actual physical scales of the flows. A similar conclusion would be reached for physical experiments. Of course, we have not normalized the boundary conditions and were we to do that, we might find additional parameters, such as the Weber number which arises in connection with effects of surface tension on, say, the water surface in a channel.

ILLUSTRATIVE PROBLEM 8.8

Consider Eq. 5.1, which gives the Bernoulli equation for points along a streamline in an ideal incompressible flow of uniform density:

$$\frac{p}{\gamma} + \frac{V^2}{2g_n} + z = H \tag{5.1}$$

Here H is the total head (see Section 5.2). What can be learned here by normalization of this equation?

SOLUTION

Select the scales for length, H, and velocity, V_T, but let us not choose the exact value of the velocity scale now. We have then $z = ZH$; $V = \overline{V}V_T$; and we let $p = P\left(\frac{1}{2}\frac{\gamma}{g_n}V_T^2\right)$. Because there are no derivatives here, direct substitution of these relationships into Eq. 5.1 yields

$$\frac{\frac{1}{2}\frac{\gamma}{g_n}V_T^2 P}{\gamma} + \frac{V_T^2\overline{V}^2}{2g_n} + HZ = H$$

Dividing by H and rearranging gives

$$\frac{1}{2}\frac{V_T^2}{g_n H}(P + \overline{V}^2) + Z = 1$$

Clearly, $V_T^2/2g_nH$ is the square of a Froude number which is the only applicable force ratio. Also we see that choosing the scaling velocity $V_T = \sqrt{2g_nH}$ produces a very simple equation

$$P + \overline{V}^2 + Z = 1$$

which governs all flows irrespective of the Froude number or the total head H! It is worth noting that V_T in this case is precisely equal to the velocity given by Torricelli's theorem (see Section 5.4). Can you imagine why?

REFERENCES

Allen, J. 1947. *Scale models in hydraulic engineering.* London: Longmans, Green.

A.S.C.E. 1942. Hydraulic models. *A.S.C.E. Manual of Engineering Practice*, No. 25.

Bridgman, P. W. 1931. *Dimensional analysis.* Rev. ed. New Haven: Yale University Press.

Buckingham, E. 1915. Model experiments and the forms of empirical equations. *Trans. A.S.M.E.* 37: 263.

Duncan, W. J. 1953. *Physical similarity and dimensional analysis.* London: Edward Arnold.

Hudson, R. Y., Herrmann, F. A., Sager, R. A., Whalin, R. W., Keulegan, G. H., Chatham, C. E., Jr., and Hales, L. Z. 1979. ''Coastal hydraulic models.'' U.S. Army Corps of Engineers *Coastal Engineering Research Center Spec. Rep. No.* 5, May.

Ippen, A. T. 1970. Hydraulic scale models. In *Osborne Reynolds and engineering science today*, Eds. McDowell and Jackson. New York: Barnes and Noble.

Ipsen, D. C. 1960. *Units, dimensions, and dimensionless numbers.* New York: McGraw-Hill.

Kline, S. J. 1986. *Similitude and approximation theory.* New York: Springer-Verlag.

Langhaar, H. L. 1951. *Dimensional analysis and theory of models.* New York: Wiley.

Rott, N. 1985. Jakob Ackeret and the history of the Mach number. *Annual Review of Fluid Mechanics* 17: 1–9. Palo Alto: Annual Reviews.

Rott, N. 1990. Note on the history of the Reynolds number. *Annual Review of Fluid Mechanics* 22: 1–11. Palo Alto: Annual Reviews.

Sedov, L. I. 1959. *Similarity and dimensional methods in mechanics.* New York: Academic Press.

Taylor, E. S. 1974. *Dimensional analysis for engineers.* London: Oxford University Press.

FILMS

Eisenberg, P. Cavitation. NCFMF/EDC Film No. 21620, Encyclopaedia Britannica Educ. Corp.

Shapiro, A. H. The fluid dynamics of drag. NCFMF/EDC Films No. 21601-21604, Encyclopaedia Britannica Educ. Corp.

PROBLEMS

8.1. An airplane wing of 3 m chord length moves through still air at 15°C and 101.3 kPa at a speed of 320 km/h. A 1 : 20 scale model of this wing is placed in a wind tunnel, and dynamic similarity between model and prototype is desired. (*a*) What velocity is necessary in a tunnel where the air has the same pressure and temperature as that in flight? (*b*) What velocity is necessary in a variable-density wind tunnel where absolute pressure is 1 400 kPa and temperature is 15°C? (*c*) At what speed must the model move through water (15°C) for dynamic similarity?

8.2. A body of 1 m (3.3 ft) length moving through air at 60 m/s (200 ft/s) is to be studied by use of a 1 : 6 model in a water tunnel in which the pressure level is maintained high enough to prevent cavitation. Using air (altitude zero) and water properties from Appendix 2, what velocity should be provided in the test section of the water tunnel for dynamic similarity between model and prototype? What would be the ratio of the prototype drag force to the model drag force? Assume the water temperature 15°C (59°F).

8.3. In a water (20°C) tunnel test the velocity approaching the model is 24 m/s. The pressure head difference between this point and a point on the object is 60-m of water. Calculate the corresponding pressure difference (kPa) on a 12 : 1 prototype in an airstream (101.3 kPa and 15°C) if model and prototype are tested under conditions of dynamic similarity.

8.4. A flat plate 1.5 m long and 0.3 m wide is towed at 3 m/s in a towing basin containing water at 20°C, and the drag force is observed to be 14 N. Calculate the dimensions of a similar plate which will yield dynamically similar conditions in an airstream (101.4 kPa and 15°C) having a velocity of 18 m/s. What drag force may be expected on this plate?

8.5. It is desired to obtain dynamic similarity between 2 cfs of water at 50°F flowing in a 6 in. pipe and crude oil flowing at a velocity of 30 ft/s at 90°F. What size of pipe is necessary for the crude oil?

8.6. A large Venturi meter (Section 14.12) for air flow measurement has $d_1 = 1.5$ m and $d_2 = 0.9$ m. It is to be calibrated using a 1 : 12 model with water the flowing fluid. When 0.07 m^3/s pass through the model, the drop in pressure from section 1 to section 2 is 172 kPa. Calculate the corresponding flowrate and pressure drop in the prototype. Use densities and viscosities for air and water given in Appendix 2 (altitude zero). Assume that these properties do not change. Assume the water temperature 15°C.

8.7. This 1 : 12 pump model (using water at 15°C) simulates a prototype for pumping oil of specific gravity 0.90. The input to the model is 0.522 kW. Calculate the viscosity of the oil and the prototype power for complete dynamic similarity between model and prototype.

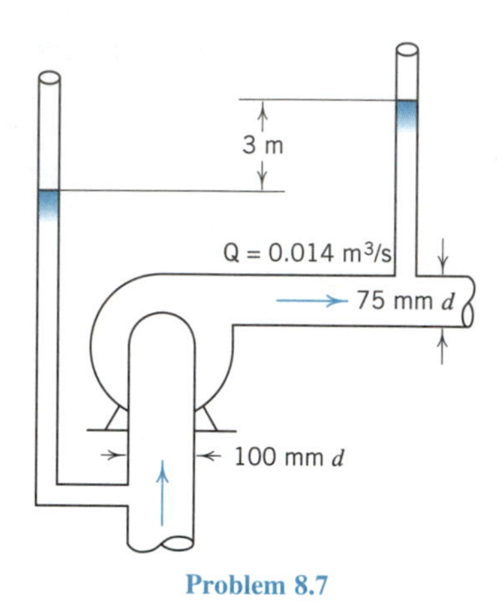

Problem 8.7

8.8. When crude oil flows at 60°F in a 2 in. horizontal pipeline at 10 ft/s, a pressure drop of 480 psi occurs in 200 ft of pipe. Calculate the pressure drop in the corresponding length of 1 in. pipe when gasoline at 100°F flows therein at the same Reynolds number.

8.9. A flowrate of 0.18 m^3/s of water (20°C) discharges from a 0.3 m pipe through a 0.15 m nozzle into the atmosphere. The axial force component exerted by water on the nozzle is 3 kN. If frictional effects may be ignored, what corresponding force will be exerted on a 4 : 1 prototype of nozzle and pipe discharging 1.13 m^3/s of air (101.4 kPa and 15°C) to the atmosphere? If frictional effects are included, the axial force component is 3.56 kN. What flowrate of air is then required for dynamic similarity? What is the corresponding force on the nozzle discharging air?

8.10. The pressure drop in a certain length of 0.3 m horizontal waterline is 68.95 kPa when the mean velocity is 4.5 m/s and the water temperature 20°C. If a 1 : 6 model of this pipeline using air as the working fluid is to produce a pressure drop of 55.2 kPa in the corresponding length when the mean velocity is 30 m/s, calculate the air pressure and temperature required for dynamic similarity between model and prototype.

8.11. A model of this disk rotating in a casing is to be constructed with air as the working fluid. What should be the corresponding dimensions and speed of the model for the torques of model and prototype to be the same? The air is at 59°F and 5.0 psia.

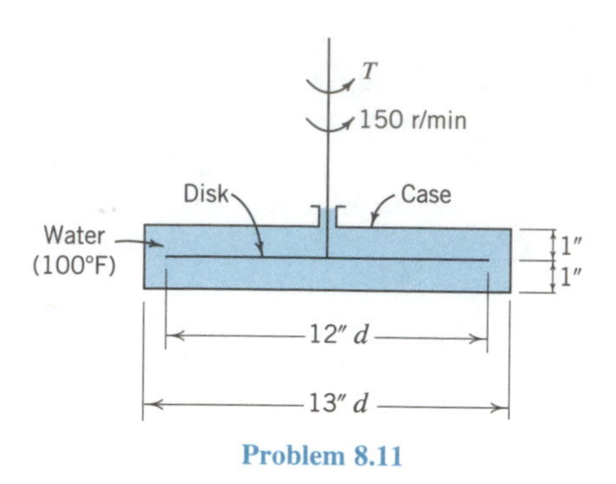

Problem 8.11

8.12. A force of 9 N is required to tow a 1 : 50 ship model at 4.8 km/h. Assuming the same water in towing basin and sea, calculate the corresponding speed and force in the prototype if the flow is dominated by: (*a*) density and gravity, (*b*) density and surface tension, and (*c*) density and viscosity.

8.13. A tanker 300 m (980 ft) long is to be tested by a 1 : 50 scale model. If the ship is to travel at 46 km/h (25 knots), at what speed must the model be towed to obtain dynamic similarity (neglecting friction) with its prototype?

8.14. A ship model 1 m long (with negligible skin friction) is tested in a towing basin at a speed of 0.6 m/s. To what ship velocity does this correspond if the ship is 60 m long? A force of 4.45 N is required to tow the model; what propulsive force does this represent in the prototype?

8.15. A seaplane is to take off at 130 km/h (80 mph). If the maximum speed available for testing its model is 4.5 m/s (15 ft/s), what is the largest model scale which can be used?

8.16. A ship 120 m (400 ft) long moves through freshwater at 15°C (59°F) at 32 km/h (17.4 knots). A 1 : 100 model of this ship is to be tested in a towing basin containing a liquid of specific gravity 0.92. What viscosity must this liquid have for both Reynolds' and Froude's laws to be satisfied? At what velocity must the model be towed? What propulsive force on the ship corresponds to a towing force of 9 N (2 lb) in the model?

8.17. A perfect fluid discharges from an orifice under a static head. For this orifice and a geometrically similar model of the same, what are the ratios between velocities, flowrates, and jet powers in model and prototype in terms of the model scale?

8.18. The flowrate from a 1.2 in. diameter orifice under a 12 ft static head is 0.133 cfs of water at 68°F. A 1 : 12 model of this setup is to be operated at conditions completely dynamically similar to those of the prototype. The liquid used in the model has a specific gravity of 1.40 and surface-tension effects may be ignored. What flowrate is required in the model? What viscosity should the liquid have?

8.19. The discharge of oil from a tank through an orifice is to be modeled using water as the flowing fluid. The kinematic viscosity of the oil is eight times that of the water. The specific gravity of the oil is 0.90. What oil flowrate is represented by 0.002 2 m^3/s in the model? If the force exerted on the model tank bottom is 210 N, what is the corresponding force in the prototype?

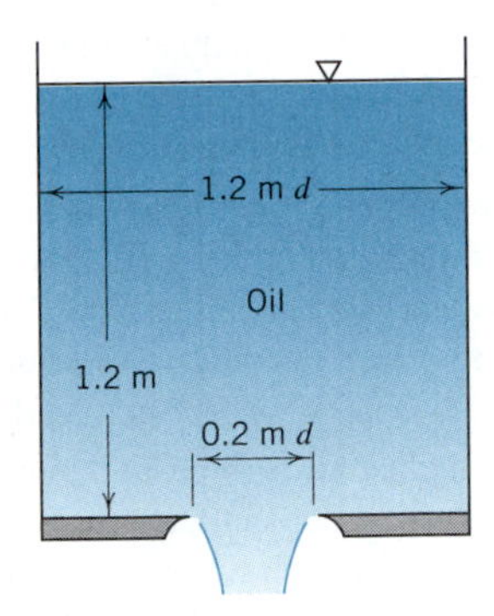

Problem 8.19

8.20. An overflow structure 480 m long is designed to pass a flood flow of 3 400 m^3/s. A 1 : 20 model of the *cross section* of the structure is built in a laboratory channel 0.3 m wide. Calculate the required laboratory flowrate if the actions of viscosity and surface tension may be neglected. When the model is tested at this flowrate, the pressure at a point on the model is observed to be 50 mm of mercury vacuum; how should this be interpreted for the prototype?

8.21. Water at 15°C or 59°F (see Appendix 2) flows over a model of a spillway structure 0.3 m or 1 ft high. If a second model is built one-half the size of the first, what should be the corresponding properties of the liquid (for use in the smaller model) to secure complete dynamic similarity between the two models?

8.22. A 1 : 30 scale model of a cavitating overflow structure is to be tested in a vacuum tank wherein the pressure is maintained at 2.0 psia. The prototype liquid is water at 70°F. The barometric pressure on the prototype is 14.5 psia. If the liquid to be used in the model has a vapor pressure of 1.50 psia, what values of density, viscosity, and surface tension must it have for complete dynamic similarity between model and prototype?

8.23. The plot shows the calibration curve for a sharp-crested circular weir of 150 mm diameter discharging from an infinite reservoir. Show how a calibration curve for a similar weir of 225 mm diameter could be prepared from these data alone. Sketch this new curve on the plot.

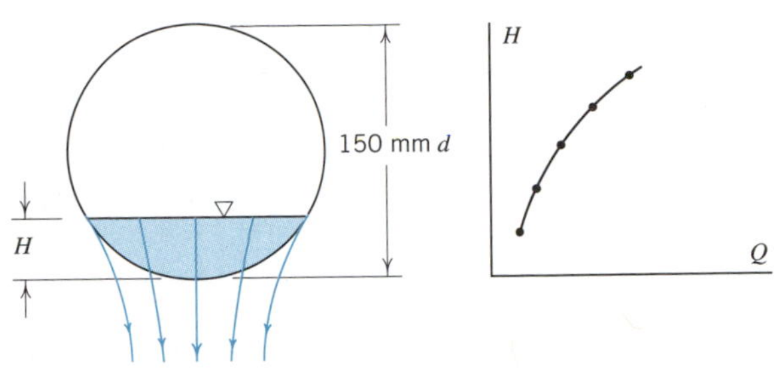

Problem 8.23

8.24. A hydraulic jump from 0.6 m to 1.5 m (2 ft to 5 ft) is to be modeled in a laboratory channel at a scale of 1 : 10. What (two-dimensional) flowrate should be used in the laboratory channel? What are the Froude numbers upstream and downstream from the jump in model and prototype?

8.25. If a 1 : 2 000 tidal model is operated to satisfy Froude's law, what length of time in the model represents a day in the prototype?

8.26. An open cylindrical tank of 1.2 m (4 ft) diameter contains water to a depth of 1.2 m (4 ft). The tank is rotated at 100 r/min about its axis (which is vertical). A half-size model of this tank is to be made with mercury as the fluid. At what speed must the model be rotated for similarity? What is the ratio of the pressures at corresponding points in the liquids? What model law is operative in this situation?

8.27. A laboratory model test is to be made of an ocean breakwater. The ocean wave periods average about 10 seconds (between the arrival of wave crests); however, the laboratory wave generator can only make waves with a 1 second period. What must be the model scale?

8.28. When a sphere of 0.25 mm diameter and specific gravity 5.54 is dropped in water at 25°C it will attain a constant velocity of 0.07 m/s. What specific gravity must a 2.5 mm sphere have so that when it is dropped in crude oil (25°C) the two flows will be dynamically similar when the terminal velocity is attained?

8.29. The flow about a 150 mm artillery projectile which travels at 600 m/s through still air at 30°C and absolute pressure 101.4 kPa is to be modeled in a high-speed wind tunnel with a 1 : 6 model. If the wind tunnel air has a temperature of −18°C and absolute pressure of 68.9 kPa, what velocity is required? If the drag force on the model is 35 N, what is the drag force on the prototype if skin friction may be neglected?

8.30. A cavitation zone is expected on an overflow structure when the flowrate is 140 m^3/s, atmospheric pressure 101.3 kPa, and water temperature 5°C. The cavitation is to be reproduced on a 1 : 20 model of the structure operating in a vacuum tank with water at 50°C. Disregarding frictional and surface-tension effects, determine the flowrate and absolute pressure (kPa) to be used in the tank for dynamic similarity.

8.31. A model is to be built of a flow phenomenon which is dominated by the action of gravity and surface tension forces. Derive an expression for the model scale in terms of the physical properties of the fluid.

8.32. A liquid rises a certain distance in a capillary tube. If this phenomenon is to be modeled, derive a force-ratio expression which must be equal in model and prototype. Compare this with Eq. 1.16.

8.33. For small models and (small) prototypes of surface ships and overflow structures, the actions of gravity, viscosity, and surface tension may be of equal importance. For dynamic similarity between model and prototype, what relation must exist between viscosity, surface tension, and model scale?

8.34. Prove by dimensional analysis that centrifugal force $= CMV^2/r$.

8.35. Prove by dimensional analysis that $\dot{m} = CA\rho V$.

8.36. Assume that the velocity acquired by a body falling from rest (without resistance) depends on weight of body, acceleration due to gravity, and distance of fall. Prove by dimensional analysis that $V = C\sqrt{g_n h}$ and is thus independent of the weight of the body.

8.37. If $\mathbf{C} = f(V, \rho, E)$, prove the expression for Cauchy number by dimensional analysis.

8.38. If $\mathbf{R} = f(V, l, \rho, \mu)$, prove the expression for Reynolds number by dimensional analysis.

8.39. If $\mathbf{W} = f(V, l, \rho, \sigma)$, prove the expression for Weber number by dimensional analysis.

8.40. A physical problem is characterized by a relationship among length, velocity, density, viscosity, and surface tension. Derive all possible dimensionless groups significant to this problem.

8.41. Derive by dimensional analysis an expression for the local velocity in established pipe flow through a smooth pipe if this velocity depends only on mean velocity, pipe diameter, distance from pipe wall, and density and viscosity of the fluid.

8.42. Derive by dimensional analysis a general expression for the capillary rise of liquids in small tubes if this depends on tube diameter, and the specific weight and surface tension of the liquid. Determine the final functional form by comparison with Eq. 1.16.

8.43. The speed of shallow water waves in the ocean (e.g., seismic sea waves or *tsunamis*) depends only on the still water depth and the acceleration due to gravity; derive an expression for wave speed.

8.44. If the velocity of deep water waves depends only on wave length and acceleration due to gravity, derive an expression for wave velocity.

8.45. Derive an expression for the velocity of very small ripples on the surface of a liquid if this velocity depends only on ripple length and density and surface tension of the liquid.

8.46. Derive an expression for the axial thrust exerted by a propeller if the thrust depends only on forward speed, angular speed, size, and viscosity and density of the fluid. How would the expression change if g_n were a relevant variable in the case of a ship propeller?

8.47. Derive an expression for drag force on a smooth submerged object moving through incompressible fluid if this force depends only on speed and size of object and viscosity and density of the fluid. Discuss results in terms of Fig. 11.9.

8.48. Derive an expression for the head lost in an established incompressible flow in a smooth pipe if this loss of head depends only on diameter and length of pipe, density, viscosity, and mean velocity of the fluid, and acceleration due to gravity.

8.49. Derive an expression for the drag force on a smooth object moving through compressible fluid if this force depends only on speed and size of object, and viscosity, density, and modulus of elasticity of the fluid. Discuss in terms of Fig. 11.5 by converting the result to relevant dimensions for a sphere.

8.50. Derive an expression for the velocity of a jet of viscous liquid issuing from an orifice under static head if this velocity depends only on head, orifice size, acceleration due to gravity, and viscosity and density of the fluid.

8.51. Derive an expression for the flowrate over an overflow structure if this flowrate depends only on size of structure, head on the structure, acceleration due to gravity, and viscosity, density, and surface tension of the liquid flowing.

8.52. Derive an expression for terminal velocity of smooth solid spheres falling through incompressible fluids if this velocity depends only on size and density of sphere, acceleration due to gravity, and density and viscosity of the fluid.

8.53. A circular disk of diameter d and of negligible thickness is rotated at a constant angular speed, ω, in a cylindrical casing filled with a liquid of viscosity μ and density ρ. The casing has an internal diameter D, and there is a clearance y between the surfaces of disk and casing. Derive an expression for the torque required to maintain this speed if it depends only on the foregoing variables.

8.54. Two cylinders are concentric, the outer one fixed and the inner one movable. A viscous incompressible fluid fills the gap between them. Derive an expression for the torque required to maintain constant-speed rotation of the inner cylinder if this torque depends only on the diameters and lengths of the cylinders, the viscosity and density of the fluid, and the angular speed of the inner cylinder.

8.55. Derive an expression for the frictional torque exerted on the journal of a bearing if this torque depends only on the diameters of journal and bearing, their axial lengths (these are the same), viscosity of the lubricant, angular speed of the journal, and the transverse load (force) on the bearing.

8.56. For a hydraulic jump (Fig. 10.23), derive by dimensional analysis an expression for y_2 if y_2 depends only on q, y_1, g_n, μ, and ρ. Compare the resulting expression with Eq. 10.24.

8.57. Derive by dimensional analysis an expression for the pressure drop, Δp, over the length, X, of unestablished flow if Δp depends only on X, pipe diameter, flowrate, and density and viscosity of the fluid . See Fig. 7.19.

8.58. Fluid flows horizontally through a bed of uniform spherical sand grains. Derive by dimensional analysis an expression for the pressure drop along the flow in terms of velocity, distance, size of grains, and density and viscosity of the fluid.

8.59. The force, F, exerted by the flowing liquid on this two-dimensional sluice gate is to be studied by dimensional analysis. Assuming the flow frictionless, derive an expression for this force in terms of the other variables relevant to the problem.

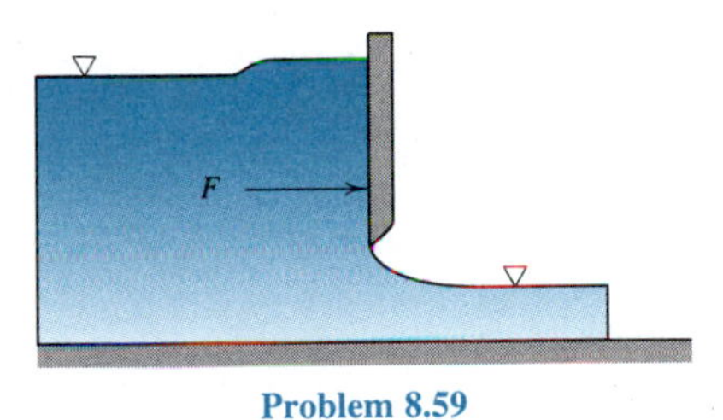

Problem 8.59

8.60. The time of formation of liquid drops is to be studied experimentally with the apparatus shown. The variables indicated are thought to be the relevant and independent ones. Make a dimensional analysis of the problem assuming "formation time" to be the dependent variable. Show the result in terms of dimensionless groups.

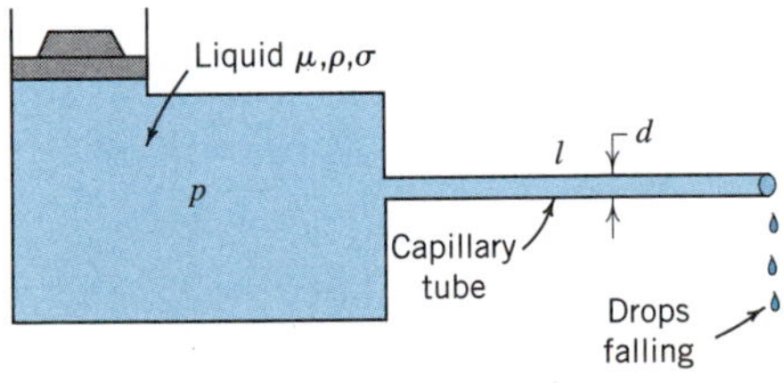

Problem 8.60

8.61. Tests on the established flow of six different liquids in smooth pipes of various sizes yield the following data:

SI Units

Diameter mm	Velocity m/s	Viscosity mPa·s	Density kg/m^3	Wall Shear Pa
300	2.26	862.0	1 247	51.2
250	2.47	431.0	1 031	33.5
150	1.22	84.3	907	5.41
100	1.39	44.0	938	9.67
50	0.20	1.5	861	0.162
25	0.36	1.0	1 000	0.517

U.S. Customary Units

Diameter in.	Velocity ft/s	Viscosity $\times$ 10^5 lb·s/ft^2	Density slug/ft^3	Wall Shear lb/ft^2
12	7.43	1 800	2.42	1.070
10	8.10	900	2.00	0.700
6	4.00	176	1.76	0.113
4	4.56	92	1.82	0.202
2	0.67	3.11	1.67	0.003 38
1	1.17	2.10	1.94	0.010 8

Make a dimensional analysis of this problem and a plot of the resulting dimensionless numbers as ordinate and abscissa. What conclusions may be drawn from the plot?

8.62. A long cylinder of diameter d rotates coaxially at angular speed ω within another long cylinder of diameter D. Between the cylinders is a fluid of density ρ and viscosity μ. Make a dimensional analysis of this problem taking as the dependent variable the torque T (per metre of cylinder length) to maintain steady state conditions. *For laminar flow* a physical analysis of this problem yields the equation

$$T = \pi\mu\omega\, d^3/2(D - d),$$

if $(D - d)$ is small. With this equation as a guide, show on a sketch plot the general trend of the family of curves to be expected in the general solution of this problem.

8.63. In a hypothetical problem the following variables are thought to be significant: linear velocity, length, pressure difference, modulus of elasticity, and absolute viscosity. How many different dimensionless numbers may be derived from these? How many will be needed for a general solution of the problem?

8.64. In a hypothetical problem for dimensional analysis, flowrate (m^3/s or ft^3/s) is thought to be a function of size, acceleration due to gravity, and kinematic viscosity. Derive (or write down from experience) all of the dimensionless numbers obtainable from this aggregation of four variables. How many of these will be significant for a complete solution of the problem?

8.65. Derive an expression for modeling the thrust of geometrically similar screw propellers moving through incompressible fluids if this thrust depends only on propeller diameter, velocity of propeller through the fluid, rotative speed, the density and viscosity of the fluid, and the acceleration due to gravity.

8.66. Normalize the Navier-Stokes equations for steady flow, but do not choose the velocity scale until the end. Hint: Can you see why $2g_n l$ is a good choice?

CHAPTER

9 FLOW IN PIPES

The problem of fluid flow in pipelines—the prediction of flowrate through pipes of given characteristics, the calculation of energy conversions therein, and so forth—is encountered in many areas of engineering practice; they afford an opportunity of applying many of the foregoing principles to (essentially one-dimensional) fluid flows of a comparatively simple and controlled nature. The subject of pipe flow embraces only those problems in which pipes flow completely full; pipes that flow partially full, such as sewer lines and culverts, are treated as open channels and are discussed in the next chapter.

The solution of practical pipe flow problems results from application of the work-energy principle, the equation of continuity, and the principles and equations of fluid resistance. Resistance to flow in pipes is offered not only by long reaches of pipe but also by pipe fittings, such as bends and valves, which dissipate energy by producing relatively large-scale turbulence.

Steady Flow

9.1 FUNDAMENTAL EQUATIONS

The work-energy equation for incompressible fluid motion in pipes is, recalling Section 7.12, (if h_L replaces $h_{L_{1-2}}$)

$$z_1 + \frac{p_1}{\gamma} + \alpha_1 \frac{V_1^2}{2g_n} = z_2 + \frac{p_2}{\gamma} + \alpha_2 \frac{V_2^2}{2g_n} + h_L \tag{9.1}$$

However, in most problems of pipe flow, the α term may be omitted for several reasons. (1) Most engineering pipe flow problems involve turbulent flow in which α is only slightly more than unity. (2) In laminar flow where α is large, velocity heads are usually negligible when compared to the other terms. (3) The velocity heads in most pipe flows are usually so small compared to other terms that inclusion of α has little effect on the final result. (4) Engineering answers are not usually required to an accuracy which would justify the inclusion of α in the equation. Application of Eq. 9.1 to practical problems thus depends primarily on an understanding of the factors which affect the head loss, h_L, and the methods available for calculating this quantity.

Early experiments (circa 1850) on the flow of water in long, straight, cylindrical pipes (Fig. 9.1) indicated that head loss varied (approximately) directly with velocity head and pipe length, and inversely with pipe diameter. Using a dimensionless coefficient of proportionality, f, called the *friction factor*, Darcy, Weisbach, and others proposed equations of the form

$$h_L = f \frac{l}{d} \frac{V^2}{2g_n} \tag{9.2}$$

Observations indicated that the friction factor f depended primarily on pipe roughness but also on velocity and pipe diameter; more recently it was observed that the friction factor also depended on the viscosity of the fluid flowing. This equation, usually called the *Darcy-Weisbach equation*, is still the basic equation for head loss caused by established pipe friction (not pipe fittings) in long, straight, uniform pipes.

Equations 9.2 and 7.36 may now be combined to give a basic relation between frictional stress, τ_o, and friction factor, f; this is

$$\tau_o = \frac{f\rho V^2}{8} \tag{9.3}$$

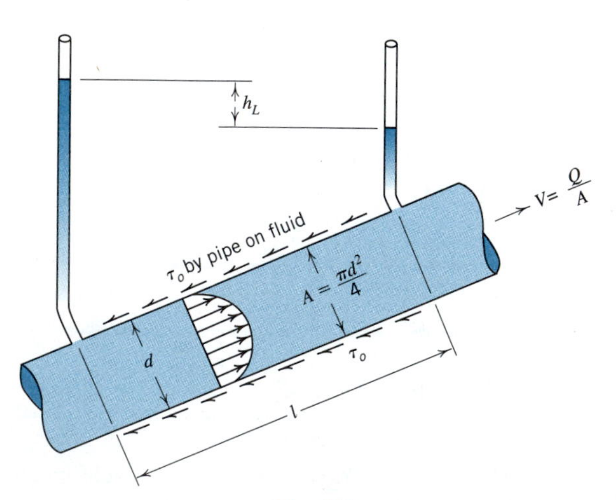

Fig. 9.1

In this fundamental equation relating wall shear to friction factor, density, and mean velocity, it is apparent that, with f dimensionless, $\sqrt{\tau_o/\rho}$ must have the dimensions of velocity; this is known as the *friction velocity*, v_*, which (from Eq. 9.3) is related to the friction factor and the mean velocity by

$$v_* = \sqrt{\frac{\tau_o}{\rho}} = V\sqrt{\frac{f}{8}} \tag{9.4}$$

However, the physical meaning of the friction velocity is not revealed by this algebraic definition; since it is a velocity which embodies only wall shear and fluid density, it is defined by the same equation whatever the flow regime (laminar or turbulent) or whatever the boundary texture (rough or smooth). For this reason it is a useful generalization that finds wide application in further developments.

ILLUSTRATIVE PROBLEM 9.1

Water flows in a 150 mm diameter pipeline at a mean velocity of 4.5 m/s. The head lost in 30 m of this pipe is measured experimentally and found to be 5.33 m. Calculate the friction velocity in the pipe.

SOLUTION

We will use Eq. 9.4 to calculate the friction velocity but first we must find the Darcy-Weisbach friction factor f. We resort to Eq. 9.2 because all the terms in that equation are known except f.

$$h_L = f\frac{l}{d}\frac{V^2}{2g_n} \tag{9.2}$$

Solving Eq. 9.2 for f gives

$$f = \frac{2g_n}{V^2}\frac{d}{l}h_L = \frac{2 \times 9.81}{(4.5\text{ m/s})^2}\frac{0.150\text{ m}}{30\text{ m}}5.33\text{ m} = 0.026$$

Now, bringing in Eq. 9.4,

$$v_* = V\sqrt{\frac{f}{8}} = 4.5\text{ m/s}\sqrt{\frac{0.026}{8}} = 0.26\text{ m/s} \bullet \tag{9.4}$$

9.2 LAMINAR FLOW

Although the facts of fluid flow can usually be established by experiment, an analytical approach to the problem is also necessary to an understanding of the mechanics of the flow. For the mechanics of a real fluid, this consists of the application of basic physical

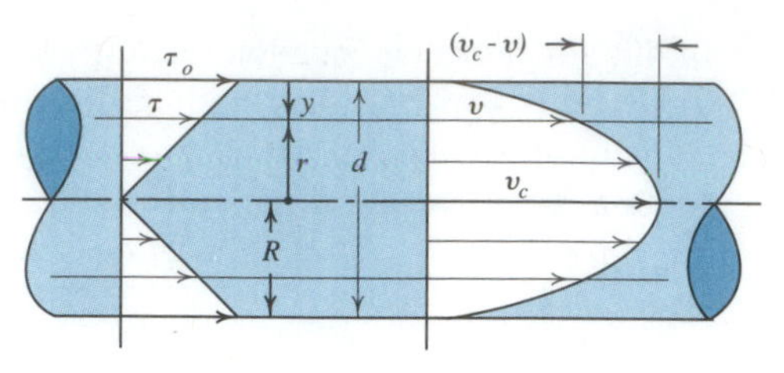

Fig. 9.2

laws which have themselves been verified by experiment; the use of ''pure theory'' alone is seldom possible in this field.

Analysis of laminar flow in a pipeline (Fig. 9.2) may be begun with the following established facts: (1) symmetrical distribution of shear stress and velocity, (2) maximum velocity at the center of the pipe and no velocity at the wall (the no-slip condition), (3) linear shear stress distribution in the fluid given by Eq. 7.37 (from application of the impulse-momentum principle), and (4) shear stress in the fluid given by Eq. 1.12 (appropriate for laminar flow). In Fig. 9.2 it is clear that $r = R - y$ and $dr = -dy$; thus, equating the expressions for τ yields

$$\tau = \left(\frac{\gamma h_L}{2l}\right) r = \mu \frac{dv}{dy} = -\mu \frac{dv}{dr}$$

We now proceed to develop expressions for the velocity profile and head loss in laminar flow by integrating the above equation. Before integrating, we utilize conditions which prevail at the wall to develop an expression for τ as a function of r. At the wall ($r = R$ and $y = 0$), we know from Eq. 7.37 that $\tau_o = \gamma h_L R/2l$, or in another form, $\gamma h_L/2l = \tau_o/R$. Replacing $\gamma h_L/2l$ in the above equation with τ_o/R,

$$\frac{dv}{dr} = -\frac{1}{\mu}\tau = -\frac{1}{\mu}\left(\frac{\gamma h_L}{2l}\right) r = -\frac{1}{\mu}\frac{\tau_o}{R} r = -\frac{\tau_o}{\mu R} r$$

Integrating this expression gives

$$v = -\frac{\tau_o}{\mu R}\left(\frac{r^2}{2}\right) + C$$

Implementing the no-slip boundary condition at $r = R$,

$$0 = -\frac{\tau_o}{\mu R}\left(\frac{R^2}{2}\right) + C$$

So $C = \tau_o R^2/2\mu R$ and the velocity profile is found to be parabolic and of the form

$$v = \frac{\tau_o}{2\mu R}(R^2 - r^2) \tag{9.5}$$

or, recognizing that when $r = 0$, $v = v_c = \tau_o R^2/2\mu R$, and the above equation becomes

$$v = v_c\left(1 - \frac{r^2}{R^2}\right) \tag{9.5}$$

Recalling that $v_*^2 = \tau_o/\rho$, Eq. 9.5 can be written (with $\nu = \mu/\rho$) as

$$v = \frac{v_*^2}{2\nu R}(R^2 - r^2) \quad \text{or} \quad \frac{v}{v_*} = \frac{v_*}{2\nu R}(R^2 - r^2) \tag{9.6}$$

or with $r^2 = (R - y)^2$,

$$\frac{v}{v_*} = \frac{v_*}{\nu}\left(y - \frac{y^2}{2R}\right) \tag{9.6}$$

Thus, the velocity profile can be expressed in terms of distance from the wall and v_* is seen to be a characteristic "velocity" of the laminar profile. We introduce this concept because v_* will appear when we examine turbulent velocity profiles.

It is interesting to note that near the pipe wall, the term $y^2/2R$ becomes negligible compared to y and the *velocity profile near the wall is essentially linear.* In dimensionless form, Eq. 9.6 becomes

$$\frac{v}{v_*} = \frac{v_* y}{\nu} \quad \text{where } y \ll R \tag{9.7}$$

We will approach development of a head loss equation by finding the flowrate Q by integrating the velocity profile given by Eq. 9.5.

$$Q = \int_0^R v(2\pi r\, dr) = \frac{\pi\tau_o}{\mu R}\int_0^R (R^2 - r^2)\, dr = \frac{\pi\tau_o R^3}{4\mu}$$

Once again bringing in Eq. 7.37 where $\tau_o = \gamma h_L R/2l$, the above equation becomes

$$Q = \frac{\pi R^4 \gamma h_L}{8\mu l} = \frac{\pi\, d^4 \gamma h_L}{128\mu l} \tag{9.8}$$

Since $Q = \pi R^2 V$, Eq. 9.8 becomes

$$V = \frac{\gamma R^2 h_L}{8\mu l} = \frac{\gamma\, d^2 h_L}{32\mu l}$$

Solving for h_L,

$$h_L = \frac{32\mu l V}{\gamma\, d^2} \tag{9.9}$$

Equation 9.8 shows that in laminar flow the flowrate, Q, which will occur in a circular pipe, varies directly with the head loss and with the fourth power of the diameter but inversely with the length of pipe and viscosity of the fluid flowing. These facts of laminar flow were established experimentally, independently, and almost simultaneously by Hagen (1839) and Poiseuille (1840), and thus the law of laminar flow expressed by the equations above is termed the *Hagen-Poiseuille* law. The experimental verification (by Hagen, Poiseuille, and later investigators) of the derivations above serves to confirm the assumptions (1) that there is no velocity adjacent to a solid boundary (i.e., no "slip" between fluid and pipe wall), and (2) that in laminar flow the shear stress is given by $\tau = \mu\, dv/dy$, which were taken for granted in the foregoing derivations.

Equation 9.9 shows that, *in a laminar flow, head loss varies with the first power of the velocity*. Equating the Darcy-Weisbach equation 9.2 for head loss to Eq. 9.9 yields an expression for the friction factor; it is

$$f = \frac{64\mu}{Vd\rho} = \frac{64}{\mathbf{R}} \tag{9.10}$$

Thus, in laminar flow the friction factor depends *only* on the Reynolds number.

ILLUSTRATIVE PROBLEM 9.2

One hundred gallons per minute of oil (s.g. = 0.90 and $\mu = 0.001\,2\ \text{lb}\cdot\text{s/ft}^2$) flow through a pipeline 3 inches in diameter. Calculate the centerline velocity, the head loss in 1 000 ft of pipe, and the shear stress and velocity at a point 1 inch from the centerline.

SOLUTION

Our first move is to calculate the average velocity V and the Reynolds number to insure that the flow is laminar.

$$V = \frac{Q}{A} = \frac{100\ \text{gal/min}/(7.48\ \text{gal/ft}^3 \times 60\ \text{s/min})}{(\pi/4)(3\ \text{in}/12)^2} = 4.54\ \text{ft/s}$$

$$\mathbf{R} = \frac{Vd\rho}{\mu} = \frac{4.54 \times 3/12 \times (0.90 \times 1.936)}{0.001\,2} = 1\,648$$

Since $\mathbf{R} < 2\,100$, we know that laminar flow exists. From Illustrative Problem 4.3 in Chapter 4, we know that $v_c = 2V$ for circular pipes with parabolic velocity distributions.

$$v_c = 2V = 2 \times 4.54 = 9.08\ \text{ft/s} \bullet$$

To compute the head loss, we use Eq. 9.9.

$$h_L = \frac{32\mu l V}{\gamma\, d^2} = \frac{32 \times 0.001\,2 \times 1\,000 \times 4.54}{(0.90 \times 62.4) \times (3/12)^2} = 49.7\ \text{ft of oil} \bullet \tag{9.9}$$

To compute the shear stress 1.0 inch from the centerline, we use Eq. 7.37 expressing shear stress as a linear function of radius.

$$\tau = \left(\frac{\gamma h_L}{2l}\right) r = \frac{(0.90 \times 62.4) \times 49.7}{2 \times 1\,000} \times \frac{1}{12} = 0.116\ \text{lb/ft}^2 \bullet \tag{7.37}$$

For the velocity 1.0 inch from the centerline, we use Eq. 9.5.

$$v = v_c\left(1 - \frac{r^2}{R^2}\right) = 9.08 \times \left[1 - \left(\frac{1.0}{1.5}\right)^2\right] = 5.04\ \text{ft/s} \bullet \tag{9.5}$$

ILLUSTRATIVE PROBLEM 9.3

A fluid flows from a large pressurized tank through a 100 mm long, 4 mm diameter tube. In a 600 sec time period, 1 300 cm^3 of fluid are collected in a measuring cup. If the head loss in the tube is 1 m, calculate the kinematic viscosity ν. Check to verify that the flow is laminar.

SOLUTION

First, we will calculate the flowrate by the fundamental relationship Q = volume/time. The volume is 1 300 $\text{cm}^3/(10^6\ \text{cm}^3/\text{m}^3) = 13 \times 10^{-4}\ \text{m}^3$. The time is 600 s.

$$Q = \frac{\text{Volume}}{\text{Time}} = \frac{13 \times 10^{-4}}{600} = 2.17 \times 10^{-6}\ \text{m}^3/s$$

We will use a modified version of Eq. 9.8 to find ν.

$$Q = \frac{\pi\, d^4 \gamma h_L}{128 \mu l} = \frac{\pi\, d^4 (\rho g_n) h_L}{128 \mu l} = \frac{\pi\, d^4 g_n h_L}{128 (\mu/\rho) l} = \frac{\pi\, d^4 g_n h_L}{128 \nu l} \tag{9.8}$$

Now, solving the above equation for ν,

$$\nu = \frac{\pi\, d^4 g_n h_L}{128 Q l} = \frac{3.14 \times (4/1\,000)^4 \times 9.81 \times 1}{128 \times (2.17 \times 10^{-6}) \times (100/1\,000)} = 0.000\,28\ \text{m}^2/s \bullet$$

To calculate the Reynolds number, we need the average velocity V.

$$V = \frac{Q}{A} = \frac{2.17 \times 10^{-6}}{(\pi/4) \times (4/1\,000)^2} = 0.17\ \text{m/s}$$

Using Eq. 8.11 to obtain Reynolds number, we get

$$\mathbf{R} = \frac{Vd}{\nu} = \frac{0.17 \times (4/1\,000)}{0.000\,28} = 2.4 \tag{8.11}$$

Since $\mathbf{R} = 2.4$ is well below 2 100, we have a strongly laminar flow and the analysis is correct. •

9.3 TURBULENT FLOW—SMOOTH PIPES

Pipe flow with friction effects has been seen to be a viscosity-inertia phenomenon and, thus, to be characterized by a Reynolds number (Sections 7.1 and 8.1). For smooth pipes, the discussion about flow past solid boundaries (Section 7.3) is relevant and strongly suggests the existence of a viscous sublayer near the pipe walls.

Drawing from the analysis of Section 7.2, the total shear stress can be written as

$$\tau = \left(\mu \frac{dv}{dy} - \rho \overline{v_x v_y} \right) \tag{9.11}$$

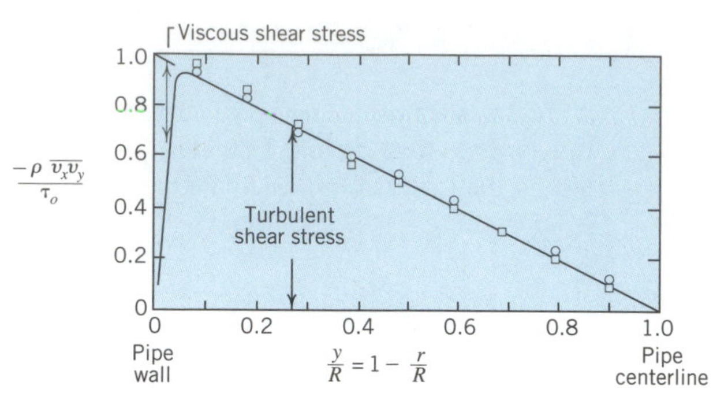

Fig. 9.3 Typical variation of shear stress in a turbulent pipe flow (to scale for $\mathbf{R} \gg 2\,100$). Data ($\mathbf{R} \sim 5 \times 10^4$ – ○; $\mathbf{R} \sim 5 \times 10^5$ – □) from J. Laufer, N.A.C.A. Report 1174, 1954.

where $\mu\, dv/dy$ is the viscous stress and $-\rho\overline{v_x v_y}$ is the turbulent (Reynolds) stress. It has already been shown (Section 7.10) that the total stress is a linear function of radius r in a pipe. Experiments show (see the typical result sketched in Fig. 9.3) that over most of the flow, the turbulent stress dominates, but the maximum stress equals the viscous stress at the wall, where the turbulent shear stress is zero. Attempts to represent the Reynolds stress have only been partly successful; this area represents one of the classic unsolved problems of fluid flow. As derived in Section 7.2, Prandtl's mixing length theory is a plausible and practical, although not theoretically rigorous, approach. It is used here to derive the velocity distribution and other quantities because the results obtained are close to experimental data and the process gives insight to the physics of the flow.

The analytical treatment is begun by assuming that the viscous stress in Eq. 9.11 is negligible over most of the flow, employing the Prandtl relationship (Eq. 7.3) for the turbulent shear stress, and equating this to the linear total shear stress relation (Eq. 7.37) for pipes. The result is (recalling from Fig. 9.2 that $r = R - y$ and $dr = -dy$)

$$\tau_o\left(1 - \frac{y}{R}\right) = \rho l^2 (dv/dy)^2 \tag{9.12}$$

where y is measured from the pipe wall. Now in Section 7.2 it was argued that near the wall ($y \ll R$), $l = \kappa y$. However, here on the centerline ($y = R$), $\tau = 0$, so either $l = 0$ or $dv/dy = 0$ there (or both equal zero). Thus, some insight is needed to proceed further; it comes from experiment. Nikuradse's systematic and comprehensive measurements[1] of velocity profiles in smooth pipes ($5 \times 10^5 < \mathbf{R} < 3 \times 10^6$) and in pipes with a uniform sand grain roughness showed that *all* velocity profiles (not just in smooth pipes) could be characterized by the single equation

$$\frac{v_c - v}{v_*} = -2.5 \ln \frac{y}{R} \qquad \text{(All pipes)} \tag{9.13}$$

[1]J. Nikuradse, "Strömungsgesetze in rauhen Rohren," *VDI-Forschungsheft*, 361, 1933. Translation available N.A.C.A., *Techn Mem.* 1292.

Therefore, it must be that $dv/dy \propto 1/y$ and, from Eq. 9.12, $l \propto y\sqrt{1 - (y/R)}$. Near the wall, $l = \kappa y$; it follows that, in a pipe flow,

$$l = \kappa y \left(1 - \frac{y}{R}\right)^{1/2} \tag{9.14}$$

Equation 9.12 now becomes

$$\tau_o/\rho\kappa^2 y^2 = (dv/dy)^2$$

or

$$dv/dy = \frac{\sqrt{\tau_o/\rho}}{\kappa y} = \frac{v_*}{\kappa y} \tag{9.15}$$

which illustrates again the pervasive presence of the friction velocity v_*. Integrating Eq. 9.15 produces

$$v = \frac{v_*}{\kappa} \ln y + C \tag{9.16}$$

There are two ways to approach evaluation of this unknown constant C.

First, if $v = v_c$ at $y = R$ (on the centerline), then

$$v_c = \frac{v_*}{\kappa} \ln R + C$$

and $C = v_c - (v_*/\kappa) \ln R$. The result is

$$v = \frac{v_*}{\kappa} \ln y - \frac{v_*}{\kappa} \ln R + v_c$$

or

$$\frac{v_c - v}{v_*} = -\frac{1}{\kappa} \ln \frac{y}{R} \tag{9.13}$$

For the usual value $\kappa = 0.40$ this is the experimentally derived Eq. 9.13.

Second, one can try to find C in terms of the no-slip condition ($v = 0$) at $y = 0$. But at $y = 0$, $v = -\infty$, according to Eq. 9.16. This is at least unrealistic (and not unexpected because the neglected viscous stresses dominate at the boundary)! It must be that the "turbulent" profile is replaced by a viscous-dominated profile near the wall (see Section 7.3) and that there is some appropriate distance from the wall which marks the onset of this process.

Now, very near the wall in a laminar flow, the velocity profile is linear (Section 9.2). If we then match a linear laminar flow velocity profile with the turbulent profile at some distance y' from the boundary, and if we include additional experimental data from Nikuradse's work, we come up with a different form of the turbulent velocity profile in *smooth pipes*

$$\frac{v}{v_*} = 2.5 \ln \frac{v_* y}{\nu} + 5.5 \tag{9.17}$$

or in terms of common logarithms,

$$\frac{v}{v_*} = 5.75 \log \frac{v_* y}{\nu} + 5.5 \quad \text{(Smooth pipes)} \tag{9.17}$$

which is the general equation of the velocity profile for turbulent flow in *smooth pipes.*

Using Eq. 9.17 and the fact that a viscous sublayer must exist near the smooth wall, it is possible to describe the structure of the flow in some detail. In the viscous sublayer, the laminar shear stress relation holds and the velocity profile is given by

$$\frac{v}{v_*} = \frac{v_* y}{\nu} \tag{9.7}$$

The nominal extent of the viscous sublayer is obtained by finding the intersection of the viscous profile of Eq. 9.7 with the turbulent profile given by Eq. 9.17. At the intersection y', the velocities are the same, so

$$\frac{v_* y'}{\nu} = 5.75 \log \frac{v_* y'}{\nu} + 5.5$$

This equation is satisfied by $v_* y'/\nu = 11.6$ and the nominal sublayer thickness[2] δ_v is equal to y'; therefore,

$$\frac{v_* \delta_v}{\nu} = 11.6 \qquad \text{or} \qquad \delta_v = 11.6\,(\nu/v_*) \tag{9.18}$$

These results are illustrated, together with some experimental data, in Fig. 9.4.

We will use continuity to develop another useful equation for turbulent flow in smooth pipes. The flowrate Q in a turbulent pipe flow is given by

$$Q = \int_0^R v(2\pi r\, dr) = 2\pi v_* \int_0^R \left(5.75 \log \frac{v_*(R - r)}{\nu} + 5.5\right) r\, dr$$

if the contribution of the viscous sublayer and the imperfection of the profile there and at the centerline are ignored. (At the centerline the derivative dv/dy must be zero, but the present Eq. 9.17 does not yield this result.) After integration

$$\frac{Q}{\pi R^2 v_*} = 5.75 \log \frac{v_* R}{\nu} + 1.75$$

If the mean velocity $V = Q/\pi R^2$, then

$$\frac{V}{v_*} = 5.75 \log \frac{v_* R}{\nu} + 1.75 \quad \text{(Smooth pipes)} \tag{9.19}$$

[2] There is no precise boundary of the sublayer and no actual intersection of the two velocity distribution curves; a gradual transition from laminar to turbulent action occurs through a ''buffer zone'' extending from $3.5 < v_* y/\nu < 30$. The velocity profile of Fig. 9.4 is taken to be a universal one for the turbulent flow of fluids over smooth surfaces, providing there is no transfer of heat between fluid and surface.

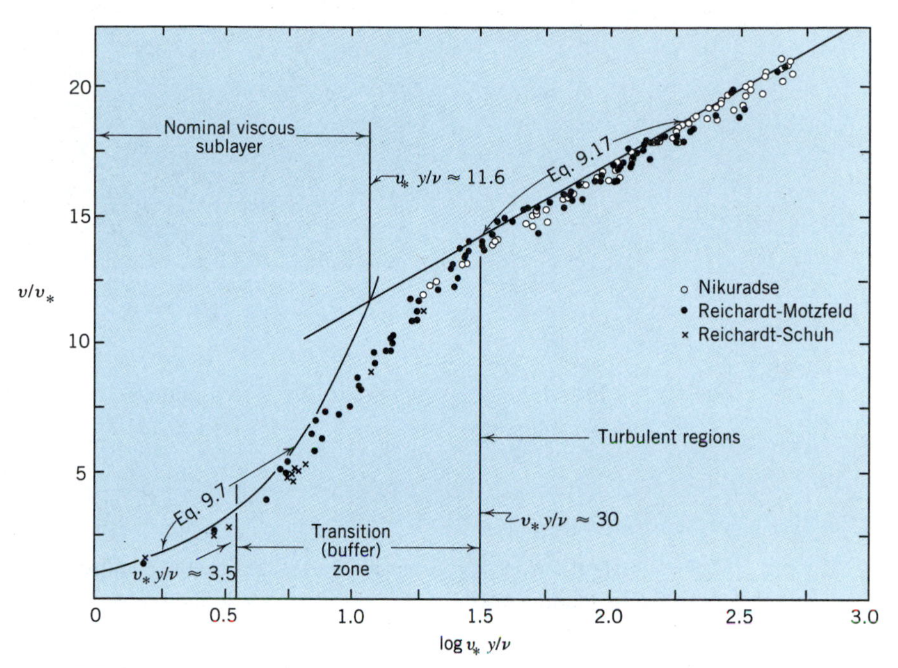

Fig. 9.4 Velocity distribution near a smooth wall. (Adapted from N.A.C.A. Technical Memorandum 1047.)

From Eq. 9.17 evaluated at $y = R$, where $v = v_c$,

$$\frac{v_c}{v_*} = 5.75 \log \frac{v_* R}{\nu} + 5.5$$

Subtracting Eq. 9.19 from this equation, recalling that $v_* = V\sqrt{f/8}$ (Eq. 9.4), and adjusting the result to conform to experiment, we get

$$\frac{v_c}{V} = 1 + 4.07\sqrt{f/8} \qquad \text{(Smooth pipes)} \qquad (9.20)$$

Figure 9.5 gives a comparison (for the same flowrate and mean velocity) of typical velocity profiles for laminar and turbulent pipe flow.

Next, we can develop a relationship between the friction factor f and the Reynolds number $\mathbf{R}$. This is accomplished by introducing Eq. 9.4 into Eq. 9.19 (replacing R with $d/2$) and again adjusting the result to conform with experiment to give

$$\frac{1}{\sqrt{f}} = 2.0 \log(\mathbf{R}\sqrt{f}) - 0.8 \qquad \text{(Smooth pipes)} \qquad (9.21)$$

where $\mathbf{R} = Vd/\nu$ is the pipe Reynolds number.

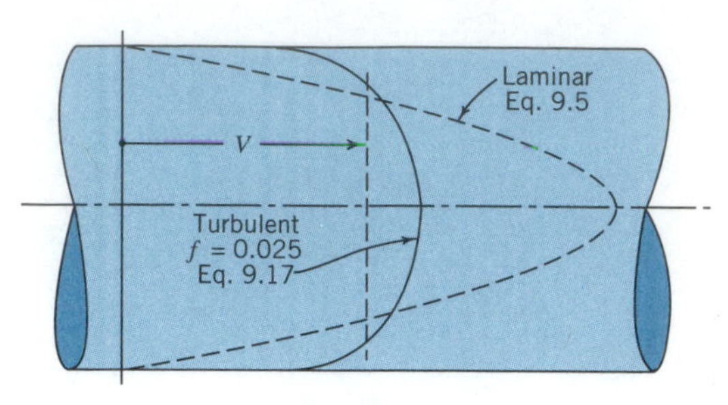

Fig. 9.5

Introducing Eq. 9.4 into the Eq. 9.18 produces a more useful expression for the laminar sublayer thickness

$$\frac{\delta_v}{d} = \frac{11.6\nu}{v_* d} = \frac{11.6\nu}{V\sqrt{f/8}d} = \frac{32.8}{\mathbf{R}\sqrt{f}} \tag{9.22}$$

where $v_* = V\sqrt{f/8}$ has been used. From this equation the decrease of sublayer thickness with increasing Reynolds number is readily seen. Furthermore, writing Eq. 9.22 in the form

$$\mathbf{R}\sqrt{f} = 32.8/(\delta_v/d)$$

and substituting it in the right-hand side of Eq. 9.21 yields the surprising result that

$$\frac{1}{\sqrt{f}} = 2.0 \log[32.8/(\delta_v/d)] - 0.8 \tag{9.23}$$

that is, for turbulent flow over smooth walls, the friction factor is a function *only* of the ratio of the sublayer thickness to the pipe diameter.

ILLUSTRATIVE PROBLEM 9.4

Water at 20°C flows in a 75-mm diameter smooth pipeline. According to a wall shear meter (Section 14.11), $\tau_o = 3.68 \text{ N/m}^2$. Calculate the thickness of the viscous sublayer, the friction factor, the mean velocity and flowrate, the centerline velocity, the shear stress and velocity 25 mm from the pipe centerline, and the head lost in 1 000 m of this pipeline.

SOLUTION

Because the friction velocity appears in so many of the relevant equations, we will calculate that quantity first. By definition,

$$v_* = \sqrt{\frac{\tau_o}{\rho}} = \sqrt{\frac{3.68 \text{ N/m}^2}{998 \text{ kg/m}^3}} = 0.061 \text{ m/s}$$

Now we employ Eq. 9.18 to compute sublayer thickness.

$$\delta_v = 11.6(\nu/v_*) = 11.6(1 \times 10^{-6} \text{ m}^2/\text{s})/(0.061 \text{ m/s}) \tag{9.18}$$

$$= 1.9 \times 10^{-4} \text{ m} = 0.19 \text{ mm} \bullet$$

We turn to Eq. 9.23 to calculate the friction factor.

$$\frac{1}{\sqrt{f}} = 2.0 \log[32.8/(\delta_v/d)] - 0.8 \tag{9.23}$$

$$= 2.0 \log[32.8/(0.19 \text{ mm}/75 \text{ mm})] - 0.8 = 7.42$$

$$f = 0.018 \bullet$$

Equation 9.19 will be used to compute the average velocity although Eq. 9.4 would do as well.

$$\frac{V}{v_*} = 5.75 \log \frac{v_* R}{\nu} + 1.75 \tag{9.19}$$

$$= 5.75 \log \frac{0.061 \text{ m/s} \times 0.037\,5 \text{ m}}{1 \times 10^{-6} \text{ m}^2/\text{s}} + 1.75 = 21.1$$

$$V = 1.29 \text{ m/s} \bullet$$

$$Q = VA = 1.29 \text{ m/s} \times (\pi/4) \times (0.075 \text{ m})^2 \doteq 0.005\,7 \text{ m}^3/\text{s} \bullet$$

To compute the centerline velocity, we utilize Eq. 9.20.

$$\frac{v_c}{V} = 1 + 4.07\sqrt{f/8} = 1 + 4.07\sqrt{.018/8} = 1.193 \tag{9.20}$$

$$v_c = 1.54 \text{ m/s} \bullet$$

We know that shear stress varies linearly with radius. The point in question is 25 mm from the pipe centerline or 2/3 of the way to the wall. As a consequence, the shear stress at this point is 2/3 of the value at the wall or 2.45 N/m^2. •

Equation 9.17 can be used to find the velocity at this point (remember, y is measured outward from the wall).

$$\frac{v}{v_*} = 5.75 \log \frac{v_* y}{\nu} + 5.5 = 5.75 \log \frac{0.061 \text{ m/s} \times .012\,5 \text{ m}}{1 \times 10^{-6} \text{ m}^2/\text{s}} + 5.5 = 22.1 \tag{9.17}$$

$$v = 1.35 \text{ m/s} \bullet$$

Finally, we calculate the head loss in 1 000 m of pipe using the Darcy-Weisbach formula.

$$h_L = f \frac{l}{d} \frac{V^2}{2g_n} = 0.018 \frac{1\,000 \text{ m}}{0.075 \text{ m}} \frac{(1.29 \text{ m/s})^2}{2 \times 9.81} = 20.4 \text{ m} \bullet \tag{9.2}$$

Before the development of the foregoing generalizations by Prandtl, von Kármán, and Nikuradse, a pioneering effort was made by Blasius to relate velocity profile, wall shear, and friction factor for turbulent flow in smooth pipes. Although Blasius's work has been superseded by these generalizations, it is still of some importance in engineering (in spite of its limited scope and empiricism) because of a mathematical simplicity which allows easy visualization and leads directly to useful (but approximate) results.

First, Blasius[3] showed that the curve representing the friction factor (for $3\,000 < \mathbf{R} < 100\,000$) could be closely approximated by the equation

[3]H. Blasius, *Forschungsarbeiten auf dem Gebiete des Ingenieurwesens*, 131, 1913.

$$f = \frac{0.316}{\mathbf{R}^{0.25}} \quad \text{(Blasius)} \tag{9.24}$$

When this is substituted in the Darcy-Weisbach Eq. 9.2, it is noted that $h_L \propto V^{1.75}$ for the turbulent flow in smooth pipes with $\mathbf{R} < 10^5$.

Substituting Eq. 9.24 into Eq. 9.3 produces

$$\tau_o = \frac{0.316}{(2RV\rho/\mu)^{0.25}} \frac{\rho V^2}{8} = 0.033\,2\mu^{1/4}R^{-1/4}V^{7/4}\rho^{3/4} \tag{9.25}$$

Blasius then assumed that the turbulent velocity profile could be approximated (see Fig. 9.6) by a power relationship

$$\frac{v}{v_c} = \left(\frac{y}{R}\right)^m$$

For this equation the mean velocity V may be related to the center velocity v_c by applying Eq. 4.8:

$$V\pi R^2 = \int_R^0 v_c \left(\frac{y}{R}\right)^m 2\pi(R - y)(-dy)$$

which gives

$$\frac{V}{v_c} = \frac{2}{(m + 1)(m + 2)} \tag{9.26}$$

from which, by substitution into the profile equation, we derive

$$V = \frac{2}{(m + 1)(m + 2)} v \left(\frac{R}{y}\right)^m$$

Substituting this expression for V into Eq. 9.25,

$$\tau_o = 0.033\,2 \left[\frac{2}{(m + 1)(m + 2)}\right]^{7/4} \mu^{1/4}R^{-1/4+(7m/4)}v^{7/4}y^{-(7m/4)}\rho^{3/4}$$

However (Blasius reasoned), wall shear τ_o could depend only on the form of the velocity profile and the physical properties of the fluid but could not be affected by pipe size R; thus the exponent of R must be zero, and from this the exponent m must be equal to $1/7$. Validation of this depended on experimental measurements of turbulent velocity profiles

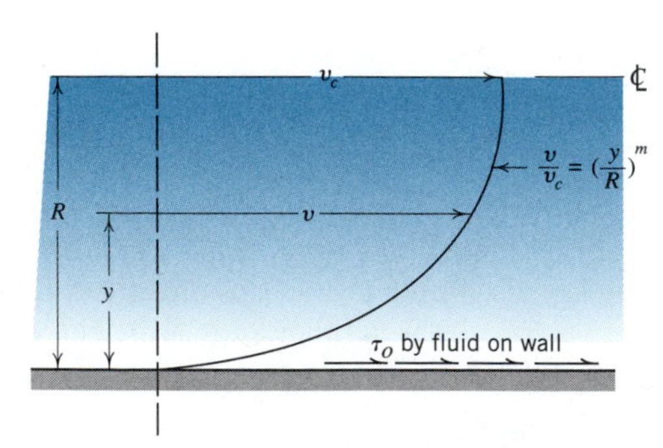

Fig. 9.6

which were found to agree quite well with the hypothesis above; thus the so-called *seventh-root law* for turbulent velocity distribution has been widely accepted. It is written

$$\frac{v}{v_c} = \left(\frac{y}{R}\right)^{1/7} \tag{9.27}$$

A useful corollary of this law is an equation for wall shear τ_o in terms of v_c and R. Using $m = 1/7$ in Eq. 9.26 yields $V/v_c = 49/60$, and, substituting $49v_c/60$ into Eq. 9.25 for V and rearranging,

$$\tau_o = 0.046\,4 \left(\frac{\mu}{v_c \rho R}\right)^{1/4} \frac{\rho v_c^2}{2} \tag{9.28}$$

which will be used later in the approximate analysis of turbulent boundary layers.

ILLUSTRATIVE PROBLEM 9.5

For the conditions of Illustrative Problem 9.4, calculate the friction factor, wall shear stress, centerline velocity, and the velocity 25 mm from the pipe centerline using the seventh-root law.

SOLUTION

The friction factor can be calculated from Eq. 9.24. However, we will first calculate the Reynolds number **R**.

$$\mathbf{R} = \frac{Vd}{\nu} = \frac{1.29 \text{ m/s} \times 0.075 \text{ m}}{1 \times 10^{-6} \text{ m}^2\text{/s}} = 96\,750$$

Checking to see if we are in the range of applicability of the Blasius approximation, we note that 96 750 is greater than 3 000 and less than 100 000; so, the Blasius approach should yield good results.

Now, calculating the friction factor,

$$f = \frac{0.316}{\mathbf{R}^{0.25}} = \frac{0.316}{(96\,750)^{0.25}} = 0.018 \bullet \tag{9.24}$$

Before calculating wall shear stress, we must find the centerline velocity. We go to Eq. 9.26.

$$\frac{V}{v_c} = \frac{2}{\left(\frac{1}{7} + 1\right)\left(\frac{1}{7} + 2\right)} = \frac{49}{60} \quad \text{so} \tag{9.26}$$

$$v_c = \frac{60}{49} V = \frac{60}{49} \times 1.29 \text{ m/s} = 1.58 \text{ m/s} \bullet$$

Now, employing Eq. 9.28, we can calculate the wall shear stress.

$$\tau_o = 0.046\,4 \left(\frac{\nu}{v_c R}\right)^{0.25} \frac{\rho v_c^2}{2} \quad \text{with} \quad \nu = \mu/\rho \tag{9.28}$$

$$= 0.046\,4 \left(\frac{1 \times 10^{-6}\ \text{m}^2/\text{s}}{1.58\ \text{m/s} \times 0.037\,5\ \text{m}}\right)^{0.25} \frac{998\ \text{kg/m}^3(1.58\ \text{m/s})^2}{2}$$

$$= 3.70\ \text{Pa} \bullet$$

Finally, to calculate the velocity 25 mm from the centerline, we use Eq. 9.27.

$$\frac{v_{25}}{v_c} = \left(\frac{y}{R}\right)^{1/7} = \left(\frac{12.5}{37.5}\right)^{1/7} = 0.855 \tag{9.27}$$

$$v_{25} = 1.35\ \text{m/s} \bullet$$

Compare these results with those of Illustrative Problem 9.4.

9.4 TURBULENT FLOW—ROUGH PIPES

Pipe friction in rough pipes at high Reynolds numbers will be governed primarily by the size and pattern of the roughness, since disruption of the viscous sublayer will render any laminar flow action in that region ineffective. However, experiments show, as noted above, that the logarithmic velocity profile given by Eq. 9.13 is applicable for both smooth and rough pipes. A development similar to that used for smooth pipe flow, except with the average roughness height e used instead of the viscosity, and tempered by the experimental work of Nikuradse on pipes with sand grain roughness, yields the following equation for the velocity profile.

$$\frac{v}{v_*} = 5.75 \log \frac{y}{e} + 8.5 \quad \text{(Rough pipes)} \tag{9.29}$$

where, for example, e is the sand grain diameter.

As before the flowrate Q is found by integration

$$Q = \int_0^R v(2\pi r\, dr) = 2\pi v_* \int_0^R \left[5.75 \log \left(\frac{R - r}{e}\right) + 8.5\right] r\, dr$$

yielding

$$V = Q/\pi R^2 = v_* \left[5.75 \log \frac{R}{e} + 4.75\right]$$

Thus,

$$\frac{V}{v_*} = 5.75 \log \frac{R}{e} + 4.75 \quad \text{(Rough pipes)} \tag{9.30}$$

(The reader can use this equation and Eq. 9.29 to verify that v_c/V is again given by Eq. 9.20, which is, clearly, then valid for all turbulent flows.) Using $v_* = V\sqrt{f/8}$ in Eq. 9.30,

$$\frac{1}{\sqrt{f}} = 2.03 \log \frac{R}{e} + 1.68$$

While the form of this result is correct, comparison with Nikuradse's experimental results suggests that the constants must be adjusted; the final equation is

$$\frac{1}{\sqrt{f}} = 2.0 \log \frac{R}{e} + 1.75 \qquad (9.31)$$

or

$$\frac{1}{\sqrt{f}} = 2.0 \log \frac{d}{e} + 1.14 \qquad \text{(Rough pipes)} \qquad (9.31)$$

For turbulent flow in rough pipes the friction factor is a function *only* of the relative roughness e/R or e/d and is *not* a function of Reynolds number (compare Eq. 9.31 to Eq. 9.21). Friction effects are produced in fully rough flow by roughness alone, without dependence on viscous action.

ILLUSTRATIVE PROBLEM 9.6

The mean velocity in a 300-mm pipeline is 3 m/s. The relative roughness of the pipe is 0.002 and the kinematic viscosity of the water is 9×10^{-7} m^2/s. Determine the friction factor, the centerline velocity, the velocity 50 mm from the pipe wall, and the head lost in 300 m of this pipe under the assumption that the pipe is rough.

SOLUTION

We have all the information to calculate the friction factor using Eq. 9.31, noting that the relative roughness $e/d = 0.002$.

$$\frac{1}{\sqrt{f}} = 2.0 \log \frac{d}{e} + 1.14 = 2.0 \log \frac{1}{0.002} + 1.14 = 6.54 \qquad (9.31)$$

$$f = 0.023\,4 \bullet$$

Equation 9.29 can be employed to find the centerline velocity but first we must calculate the friction velocity.

$$v_* = V\sqrt{\frac{f}{8}} = 3 \text{ m/s} \times \sqrt{\frac{0.023\,4}{8}} = 0.162 \text{ m/s}$$

Now, with $e/d = 0.002$, $e = 0.002 \times 300 \text{ mm} = 0.6 \text{ mm}$. With $y = 150 \text{ mm}$, $y/e = R/e = 150 \text{ mm}/0.6 \text{ mm} = 250$. We now substitute these values in Eq. 9.29 with $v = v_c$ and $y = R$.

$$\frac{v_c}{v_*} = 5.75 \log \frac{R}{e} + 8.5 = 5.75 \log 250 + 8.5 = 22.3 \qquad (9.29)$$

$$v_c = 22.3 \times v_* = 22.3 \times 0.162 = 3.61 \text{ m/s} \bullet$$

Once again we turn to Eq. 9.29 to find the velocity 50 mm from the wall. In this instance, $y/e = 50 \text{ mm}/0.6 \text{ mm} = 83.3$.

$$\frac{v_{50}}{v_*} = 5.75 \log \frac{y_{50}}{e} + 8.5 = 5.75 \log 83.3 + 8.5 = 19.54 \qquad (9.29)$$

$$v_{50} = 19.54 \times v_* = 19.54 \times 0.162 = 3.17 \text{ m/s} \bullet$$

Finally, we use the Darcy-Weisbach equation to compute the head loss in 300 m of the pipe.

$$h_L = f \frac{l}{d} \frac{V^2}{2g_n} = 0.023\,4 \frac{300 \text{ m}}{0.300 \text{ m}} \frac{(3 \text{ m/s})^2}{2 \times 9.81} = 10.7 \text{ m of water} \bullet \qquad (9.2)$$

9.5 CLASSIFICATION OF SMOOTHNESS AND ROUGHNESS—IMPACT ON FRICTION FACTOR

Turbulent flow in a smooth pipe was found (e.g., Eq. 9.23) to involve the sublayer thickness δ_v as a characteristic length. Turbulent flow in a rough pipe (e.g., Eq. 9.31) had the absolute roughness e as a characteristic length. In cases of transition where the pipe acts as neither smooth nor fully rough, e/δ, must be a significant parameter.

In laminar flow $\delta_v = R$ because the viscous effects dominate the whole flow. In any reasonable pipe e/R is not large and experiments confirm that, for example, the effects of roughness on laminar flow are negligible. Thus, $f = 64/\mathbf{R}$ and the other results derived in Section 9.2 are valid for smooth or rough pipes as long as the flow is laminar.

The significant relationship for classification of pipe surfaces as smooth or rough in turbulent flow is derived by expanding e/δ_v in terms of the sublayer thickness definition Eq. 9.22, namely, by writing

$$\frac{e}{\delta_v} = \frac{e/d}{\delta_v/d} = \frac{e/d}{32.8/\mathbf{R}\sqrt{f}} = \left(\frac{e}{d}\right) \frac{\mathbf{R}\sqrt{f}}{32.8}$$

so

$$\frac{e}{d} \mathbf{R}\sqrt{f} = 32.8 \frac{e}{\delta_v} \qquad (9.32)$$

Now the friction factor results obtained in the previous sections can be plotted versus $(e/d)\mathbf{R}\sqrt{f}$ and analyzed in terms of experimental results.

For fully rough flow,

$$\frac{1}{\sqrt{f}} - 2.0 \log \frac{d}{e} = 1.14 \tag{9.31}$$

so it is convenient to plot $(1/\sqrt{f}) - 2.0 \log (d/e)$ versus $(e/d)\mathbf{R}\sqrt{f}$ because the rough flow region is a horizontal line. For smooth flow,

$$\frac{1}{\sqrt{f}} = 2.0 \log(\mathbf{R}\sqrt{f}) - 0.8 \tag{9.21}$$

Adding $-2.0 \log(d/e)$ to both sides,

$$\frac{1}{\sqrt{f}} - 2.0 \log \frac{d}{e} = 2.0 \log \left(\frac{e}{d}\mathbf{R}\sqrt{f}\right) - 0.8 \tag{9.33}$$

Equations 9.31 and 9.33 (i.e., modified 9.21) are plotted on Fig. 9.7. The shaded band represents the uniform sand-grain roughness data of Nikuradse. The flow is easily classified now. In particular,

For smooth flow: $\frac{e}{d}\mathbf{R}\sqrt{f} \leq 10$

For transition flow: $10 < \frac{e}{d}\mathbf{R}\sqrt{f} < 200$

For rough flow: $200 \leq \frac{e}{d}\mathbf{R}\sqrt{f}$

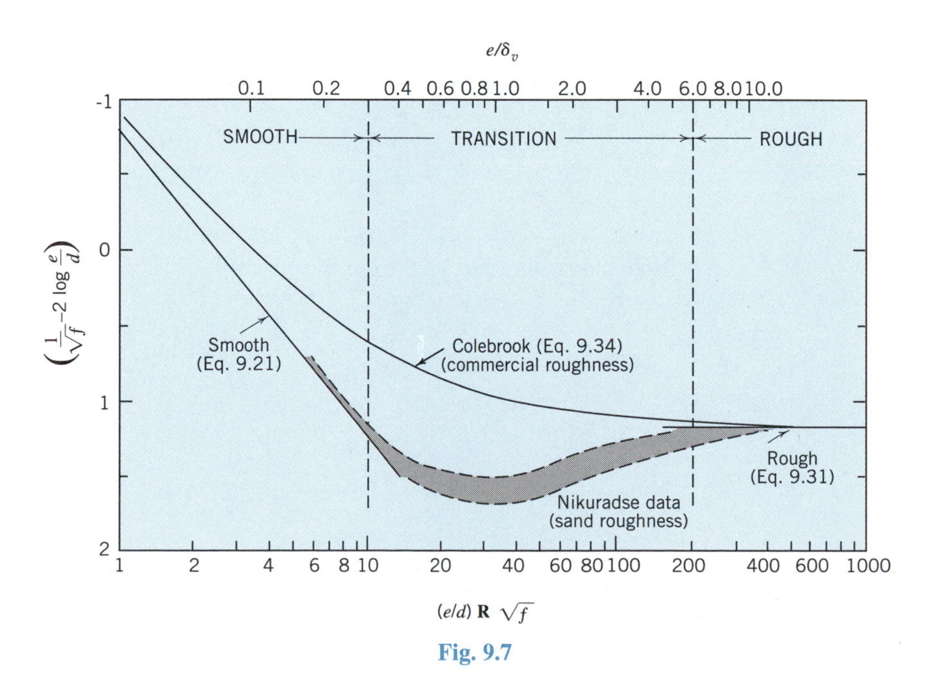

Fig. 9.7

From Eq. 9.32, it follows that the pipe acts as a smooth pipe when $e/\delta_v \leq 0.3$, while the pipe acts rough when $6 \leq e/\delta_v$. The e/δ_v scale is plotted along the upper boundary of Fig. 9.7. The general conclusion to be drawn is that the absolute size is not a measure of the effect of surface roughness on fluid flow; the effect is dependent on the size of the roughness *relative to the thickness of the viscous sublayer.*

Unfortunately, the excellent results of Nikuradse cannot be applied directly to engineering problems because the roughness patterns of commercial pipes are entirely different, much more variable, and much less definable than the artificial roughnesses used by Nikuradse. However, Colebrook[4] has shown how these results may be applied toward a quantitative measure of commercial pipe roughness. He found that any pipe of commercial roughness when tested to a high enough Reynolds number gave a friction factor which no longer varied with Reynolds number. This allowed comparison with Eq. 9.31 and, from measurement of the friction factor, permitted an *equivalent* sand-grain size, e, to be computed. Obtaining e in this manner and plotting test results on many pipes, Colebrook found results from all pipes to be closely clustered about a single line having the equation

$$\frac{1}{\sqrt{f}} - 2\log\frac{d}{e} = 1.14 - 2\log\left[1 + \frac{9.28}{\mathbf{R}(e/d)\sqrt{f}}\right] \tag{9.34}$$

This result is also shown on Fig. 9.7. Because of the variability of commercial roughness patterns, the previously observed distinctions between smooth, transition, and rough flow are not present for commercial pipes (see Section 9.6).

Another means of classifying roughness effects is to use the velocity profiles directly. For smooth flow, Eq. 9.13 can be written (for $\kappa = 0.40$) as

$$\frac{v_c - v}{v_*} = 5.75\log\frac{R}{y} \tag{9.35}$$

On the other hand, using Eq. 9.29 for rough flow with $v = v_c$

$$\frac{v_c}{v_*} = 5.75\log\frac{R}{e} + 8.5$$

and subtracting Eq. 9.29 from this gives

$$\frac{v_c - v}{v_*} = 5.75\left(\log\frac{R}{e} - \log\frac{y}{e}\right)$$

or

$$\frac{v_c - v}{v_*} = 5.75\log\frac{R}{y}$$

[4]C. F. Colebrook, "Turbulent Flow in Pipes, with Particular Reference to the Transition Region between the Smooth and Rough Pipe Laws," *Jour. Inst. Civil Engrs.*, London, p. 133, February 1939.

That is, Eq. 9.35 is valid for both smooth and rough flow (see Eq. 9.13). Writing Eq. 9.35 as

$$\frac{v}{v_*} = \frac{v_c}{v_*} + 5.75 \log \frac{y}{R}$$

$$= \frac{v_c}{v_*} + 5.75 \log \frac{e}{R} + 5.75 \log \frac{y}{e}$$

$$= A + 5.75 \log \frac{y}{e}$$

suggests that the term

$$A = \frac{v_c}{v_*} + 5.75 \log \frac{e}{R} \tag{9.36}$$

is a constant in a fully rough flow. In fact, $A = 8.5$ according to experimental data. For smooth flow (Eq. 9.17)

$$\frac{v}{v_*} = 5.5 + 5.75 \log \frac{v_* y}{\nu} \tag{9.17}$$

$$= 5.5 + 5.75 \log \frac{v_* y}{\nu}\left(\frac{e}{e}\right)$$

$$= 5.5 + 5.75 \log \frac{v_* e}{\nu} + 5.75 \log \frac{y}{e}$$

$$= A + 5.75 \log \frac{y}{e}$$

but now

$$A = 5.50 + \log \frac{v_* e}{\nu} \tag{9.37}$$

and $v_* e/\nu$ is the *Roughness Reynolds Number*. Figure 9.8 is a plot of A versus $v_* e/\nu$ for Nikuradse's sand roughness data.[5] The results suggest that

For smooth flow: $v_* e/\nu \leq 3.5$

For transition flow: $3.5 < v_* e/\nu < 70$

For wholly rough flow: $70 \leq v_* e/\nu$

If for smooth flow $v_* e/\nu \leq 3.5$ and, from Eq. 9.18, $v_* \delta_v/\nu = 11.6$, it follows that $(11.6/\delta_v)e \leq 3.5$ or $e/\delta_v \leq 0.3$, which was obtained from the friction factor analysis. Likewise, if $70 \leq v_* e/\nu$, $70 \leq (11.6/\delta_v)e$ or $6 \leq e/\delta_v$ for rough flow. Thus, both means of classifying roughness effects are consistent.

[5]Figure 9.8 is adapted from N.A.C.A. *Tech Mem.* 1292, which is a translation of J. Nikuradse, "Strömungsgesetze in rauhen Rohren," *VDI-Forschungsheft*, 361, 1933.

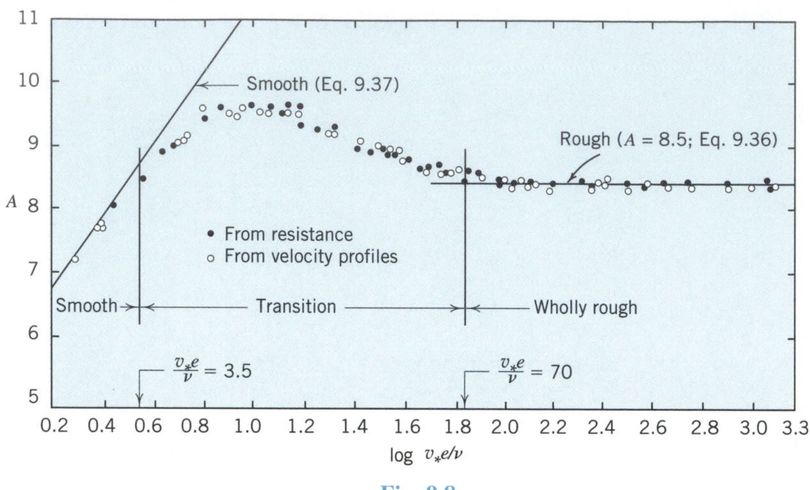

Fig. 9.8

ILLUSTRATIVE PROBLEM 9.7

Check Illustrative Problem 9.6 to establish whether or not the flow is truly rough as assumed.

SOLUTION

To determine if the flow is rough, we refer to Fig. 9.8 and the associated equations. For wholly rough flow, $v_* e/\nu \geq 70$. So, we can calculate the value of this quantity for the flow in question and compare with the above inequality. From the previous illustrative problem,

$$v_* = 0.162 \text{ m/s}, \; e = 0.6 \text{ mm or } 0.000\,6 \text{ m}, \qquad \text{and} \qquad \nu = 9 \times 10^{-7} \text{ m}^2/\text{s}$$

$$\frac{v_* e}{\nu} = \frac{0.162 \times 0.000\,6}{9 \times 10^{-7}} = 108$$

Since $108 > 70$, we conclude the flow is rough and the assumption used in Illustrative Problem 9.6 is valid. •

9.6 PIPE FRICTION FACTORS

The evaluation of pipe friction factors is extensively based on experimental work so it is informative to approach the representation of the friction factor from the standpoint of dimensional analysis and similitude. From the analyses presented in previous sections, it is known that the wall shear stress τ_o depends on the mean velocity V, the pipe diameter d, the mean height of the wall roughness projections e, the fluid density ρ, and the fluid viscosity μ. Thus, writing this relationship in the same form used in Chapter 8, we have

$$\phi(\tau_o, V, d, e, \rho, \mu) = 0$$

Using the Π-theorem (Section 8.2), with V, d, and ρ as the repeating variables, gives

$$\Pi_1 = \frac{\tau_o}{\rho V^2} \qquad \Pi_2 = \frac{Vd\rho}{\mu} \qquad \Pi_3 = \frac{e}{d}$$

Setting Π_1 as a function of Π_2 and Π_3, we obtain

$$\tau_o = \rho V^2 \phi' \left(\frac{Vd\rho}{\mu}, \frac{e}{d} \right) = \rho V^2 \phi' \left(\mathbf{R}, \frac{e}{d} \right) = \frac{\rho V^2}{8} \phi'' \left(\mathbf{R}, \frac{e}{d} \right)$$

Comparison with equation 9.3 shows that

$$f = \phi'' \left(\mathbf{R}, \frac{e}{d} \right) \tag{9.38}$$

Thus, the dimensional analysis approach confirms the experimental results and previous analyses that the friction factor depends only on the Reynolds number of the flow and the pipe relative roughness e/d. The physical significance of Eq. 9.38 may be stated briefly as *the friction factors of a number of pipes will be the same if their Reynolds numbers, roughness patterns, and relative roughnesses are the same.* When this is interpreted by the laws of similitude, its basic meaning is *the friction factors of the pipes are the same if their flow pictures in every detail are geometrically and dynamically similar.*

Equation 9.38 suggests a convenient means of presenting the experimental data on the friction factor. This was first used by Blasius in 1913 and by Stanton[6] in 1914 and consists of a logarithmic plot of friction factor versus Reynolds number, with relative roughness as the parameter. Figure 9.9 shows Nikuradse's[7] data plotted in the Blasius-Stanton format, revealing the relationship between f, $\mathbf{R}$, and e/d. Recall that Nikuradse's experiments employed fixing sand grains on the inside of the pipe to establish an easily

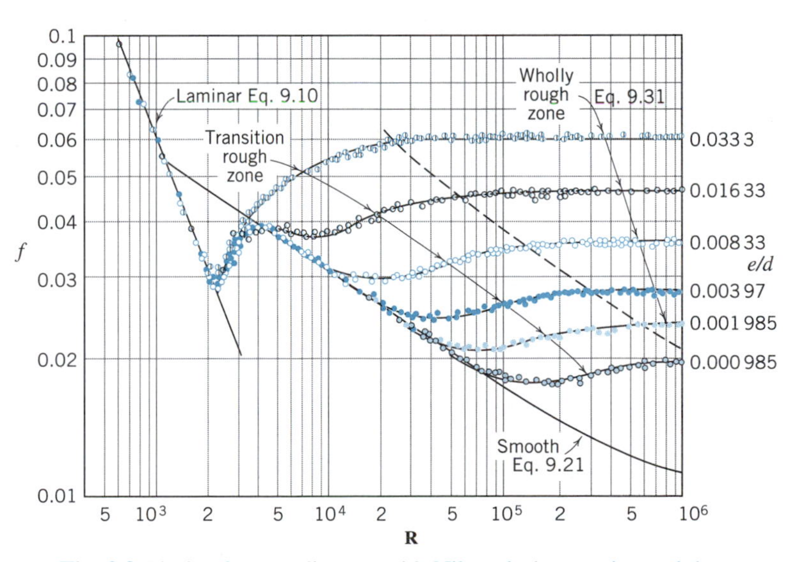

Fig. 9.9 Blasius-Stanton diagram with Nikuradse's experimental data.

[6] This plot will be referred to as the Blasius-Stanton diagram

[7] J. Nikuradse, "Strömungsgesetze in rauhen Rohren," *VDI-Forschungsheft*, 361, 1933. Translation available in N.A.C.A., *Tech. Mem.* 1292.

measurable index of roughness. The purpose was to demonstrate conclusively the interrelationship among f, **R**, and e/d. In that regard, the effort was successful and the following important fundamentals were clearly established:

1. The physical difference between the laminar and turbulent flow regimes is indicated by the change in the relationship of f to **R** near the critical Reynolds number of 2 100.
2. The laminar regime is characterized by a single curve, given by the equation $f = 64/\mathbf{R}$ for all surface roughnesses. This confirms that head loss in laminar flow is independent of surface roughness and that $h_L \propto V$.
3. In turbulent flow a curve of f versus **R** exists for every relative roughness, e/d, and the horizontal aspect of the curves confirms that for rough pipes the roughness is more important than the Reynolds number in determining the magnitude of the friction factor.
4. At high Reynolds numbers the friction factors of rough pipes become constant, dependent wholly on the roughness of the pipe, and thus independent of the Reynolds number. From the Darcy-Weisbach equation it may be concluded that $h_L \propto V^2$ for completely turbulent flow over rough surfaces.
5. Although the lowest curve was obtained from tests on hydraulically smooth pipes, many of Nikuradse's rough pipe test results coincide with it for $5\,000 < \mathbf{R} < 50\,000$. Here the roughness is submerged in the viscous sublayer (Sections 9.3 and 9.5) and can have no effect on friction factor and head loss, which depend on viscosity effects alone. From the Darcy-Weisbach equation $h \propto V^{1.75}$ for turbulent flow in smooth pipes where the friction factor is given by the Blasius expression (Eq. 9.24).
6. The series of curves for the rough pipes diverges from the smooth pipe curve as the Reynolds number increases. In other words, pipes that are smooth at low values of **R** become rough at high values of **R**. This is explained by the thickness of the viscous sublayer decreasing (Sections 9.3 and 9.5) as the Reynolds number increases, thus exposing smaller roughness protuberances to the turbulent region and causing the pipe to exhibit the properties of a rough pipe.

ILLUSTRATIVE PROBLEM 9.8

Water at 100°F flows in a 3 inch pipe at a Reynolds number of 80 000. If the pipe is lined with uniform sand grains 0.006 inches in diameter, how much head loss is to be expected in 1 000 ft of the pipe? How much head loss would be expected if the pipe were smooth?

SOLUTION

The relative roughness of the pipe is 0.006 in./3 in. = 0.002. From Fig. 9.9, with a Reynolds number of 80 000 and $e/d = 0.002$,

$$f \cong 0.021$$

In order to calculate the head loss from the Darcy-Weisbach formula, we must find the average velocity. This is accomplished by use of the Reynolds number,

$$\mathbf{R} = \frac{Vd}{\nu}$$

where, from Appendix 2, $\nu = 0.739 \times 10^{-5}\ \text{ft}^2/\text{s}$. Solving the above equation for V,

$$V = \frac{\mathbf{R}\nu}{d} = \frac{80\,000 \times 0.739 \times 10^{-5}}{3/12} = 2.36\ \text{ft/s}$$

Now, the Darcy-Weisbach equation can be used to compute the head loss.

$$h_L = f\frac{l}{d}\frac{V^2}{2g_n} \cong 0.021\ \frac{1\,000}{3/12}\ \frac{(2.36)^2}{2 \times 32.2} = 7.3\ \text{ft}\ \bullet$$

To calculate the head loss in a smooth pipe of the same size and length, we must again find the friction factor. While we could easily read the approximate value from Fig. 9.9, we note that there is an alternate method available in this situation. Because **R** is in the range of applicability of the Blasius power relationship ($3\,000 < \mathbf{R} < 100\,000$), we can use Eq. 9.24 to compute the friction factor.

$$f = \frac{0.316}{(80\,000)^{0.25}} = 0.018\,8 \tag{9.24}$$

The head loss in the smooth pipe is

$$h_L = f\frac{l}{d}\frac{V^2}{2g_n} = 0.018\,8\ \frac{1\,000}{3/12}\ \frac{(2.36)^2}{2 \times 32.2} = 6.5\ \text{ft}\ \bullet$$

Earlier (see Section 9.5), Colebrook showed that Nikuradse's results were not representative of commercial pipes. Roughness patterns and variations in the height of roughness projections in commercial pipes resulted in friction factors which were considerably different than Nikuradse's in the transition zone between smooth and wholly rough turbulent flow. Benedict[8] gives a particularly lucid discussion of this apparent discrepancy. He notes that the work of Colebrook and others on commercial pipe rather conclusively demonstrates that the friction factor varies as described by Eq. 9.34 rather than as shown in Fig. 9.9. Later, Moody[9] presented the equations of Colebrook in graphical form using the Blasius-Stanton format. The result of this work is shown in Fig. 9.10. This plot, known as the *Moody diagram*, along with the e-values for commercial pipes given in Fig. 9.11, can be used directly in the solution of engineering problems. A computer program, DARCY, which calculates a friction factor from the Moody diagram, is given in Appendix 7. However, a word of caution is necessary. The roughness of commercial pipe materials varies widely with the manufacturer, with years in service, and with liquid conveyed. Corrosion of pipe wall material and deposition of scale, slime, and such can drastically increase the roughness of the pipe and the resulting friction factor. One of the practical considerations of pipe flow calculations, which poses a difficult problem for the engineer, is the prediction of friction factors which will actually be realized after a pipeline is constructed. Extensive practical experience is a valuable asset in his situation.

[8]Benedict, R. P. 1980. *Fundamentals of Pipe Flow.* New York: John Wiley & Sons.

[9]L. F. Moody, "Friction Factors for Pipe Flow," *Trans. A.S.M.E.*, vol. 66, 1944. Figures 9.10, 9.11, and 9.12 are adapted from this publication with permission.

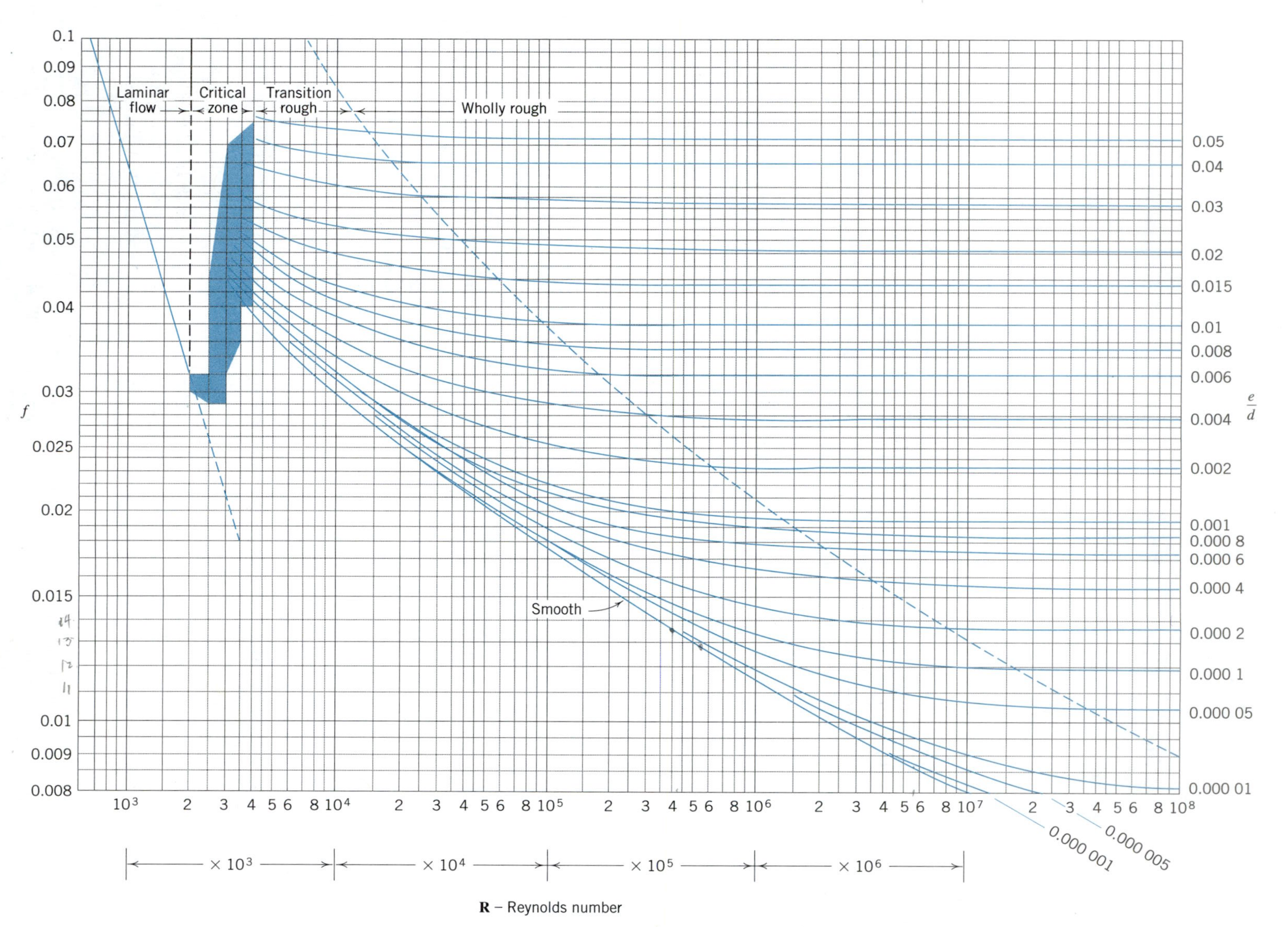

Fig. 9.10 Relation of friction factor, Reynolds number, and roughness for commercial pipes (see footnote 9).

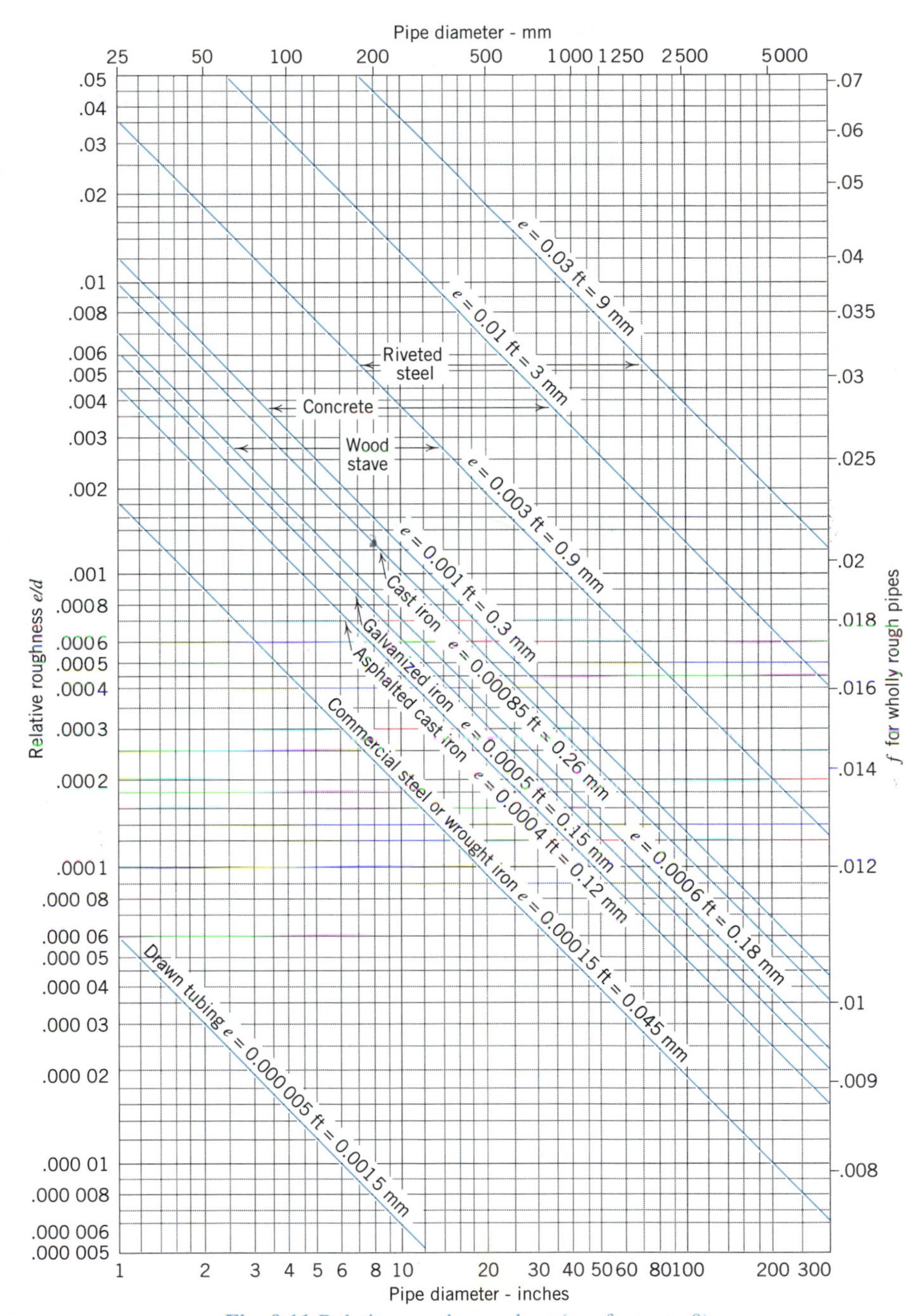

Fig. 9.11 Relative roughness chart (see footnote 9).

ILLUSTRATIVE PROBLEM 9.9

Water at 100°F flows in a 3 inch pipe at a Reynolds number of 80 000. This is a *commercial pipe* with an equivalent sand grain roughness of 0.006 in. What head loss is to be expected in 1 000 ft of this pipe?

Note that this is the same problem as Illustrative Problem 9.8 except we are using commercial pipe.

SOLUTION

We will use the Moody diagram to find the friction factor but first we must calculate the relative roughness e/d.

$$\frac{e}{d} = \frac{0.006 \text{ in.}}{3.00 \text{ in.}} = 0.002$$

With $\mathbf{R} = 80\,000$ and $e/d = 0.002$, we go to the Moody diagram of Fig. 9.10 and find

$$f \cong 0.025\,5$$

From Illustrative Problem 9.8, $V = 2.36$ ft/s. We can now compute the head loss using the Darcy-Weisbach equation.

$$h_L = f\frac{l}{d}\frac{V^2}{2g_n} = 0.025\,5\,\frac{1\,000}{3/12}\,\frac{2.36^2}{2 \times 32.2} = 8.8 \text{ ft} \bullet \qquad (9.2)$$

Note that the commercial pipe with an *equivalent* sand grain roughness equal to that of a pipe lined with real sand grains 0.006 inches in diameter has a 30% higher head loss at this Reynolds number. However, both of these pipes would have the same head loss under smooth pipe and wholly rough conditions.

ILLUSTRATIVE PROBLEM 9.10

Crude oil at 20°C flows in a riveted steel pipe 1.00 m in diameter at a mean velocity of 2.0 m/s. What range of head loss is to be expected in 1 km of pipeline?

SOLUTION

The question posed above suggests there may be some uncertainty in the evaluation of the head loss. We will begin our investigation by calculating the Reynolds number. From Appendix 2, crude oil at 20°C has a density $\rho = 855.6$ kg/m^3 and an absolute viscosity $\mu = 71.8 \times 10^{-4}$ Pa·s.

$$\mathbf{R} = \frac{Vd\rho}{\mu} = \frac{2.0 \text{ m/s} \times 1.00 \text{ m} \times 855.6 \text{ kg/m}^3}{71.8 \times 10^{-4} \text{ Pa·s}} = 2.4 \times 10^5$$

Now, to find the relative roughness, we go to Fig. 9.11 for riveted steel pipe 1 m in diameter and read

$$\frac{e}{d} = 0.000\,9 \text{ to } 0.009$$

So now we know what the uncertainty is about. Apparently there are several ways to construct riveted steel pipe with the number, type and spacing of rivets open to question. We will take the two extreme values and find the head loss for each. From the Moody diagram of Fig. 9.10,

$$\frac{e}{d} = 0.000\,9 \qquad f = 0.020\,5 \quad \text{(By program DARCY, App. 7, } f = 0.020\,4\text{)}$$

$$\frac{e}{d} = 0.009 \qquad f = 0.036\,5 \quad \text{(By program DARCY, App. 7, } f = 0.036\,5\text{)}$$

The Darcy-Weisbach equation 9.2 yields

$$h_L = f\frac{l}{d}\frac{V^2}{2g_n} = 0.020\,5\,\frac{1\,000}{1}\,\frac{2^2}{2 \times 9.81} = 4.2 \text{ m} \bullet$$

$$h_L = f\frac{l}{d}\frac{V^2}{2g_n} = 0.036\,5\,\frac{1\,000}{1}\,\frac{2^2}{2 \times 9.81} = 7.4 \text{ m} \bullet$$

The range of head loss is 4.2 m to 7.4 m per km.

9.7 PIPE FRICTION IN NONCIRCULAR PIPES—THE HYDRAULIC RADIUS

Although the majority of pipes used in engineering practice are of circular cross section, occasions arise when calculations must be carried out on head loss in rectangular ducts and other conduits of noncircular form. The foregoing equations for circular pipes may be adapted to these special problems through use of the *hydraulic radius* concept.

The hydraulic radius, R_h, is defined as the area, A, of the flow cross section divided by its "wetted perimeter," P (see Section 7.10). In a circular pipe of diameter d,

$$R_h = \frac{d}{4} \qquad \text{or} \qquad d = 4R_h \tag{9.39}$$

This value may be substituted into the Darcy-Weisbach equation for head loss and into the expression for the Reynolds number with the following results:

$$h_L = \frac{f}{4}\frac{l}{R_h}\frac{V^2}{2g_n} \tag{9.40}$$

and

$$\mathbf{R} = \frac{V(4R_h)\rho}{\mu} \tag{9.41}$$

from which the head loss in many conduits of noncircular cross section may be calculated with the aid of the Moody diagram of Fig. 9.10.

The calculation of lost head in noncircular conduits thus involves the calculation of the hydraulic radius of the flow cross section and the use of the friction factor obtained for an *equivalent* circular pipe having a diameter d equal to $4R_h$. In view of the complexities of viscous sublayers, turbulence, roughness, shear stress, and so forth, it seems surprising at first that a circular pipe equivalent to a noncircular conduit may be obtained so easily, and it would, therefore, be expected that the method might be subject to certain limitations.

The method gives satisfactory results when the problem is one of turbulent flow but, if it is used for laminar flow, large errors are introduced.

The foregoing facts may be justified analytically by examining further the structure of Eq. 9.40 in which h_L varies with $1/R_h$. From the definition of the hydraulic radius, its reciprocal is the *wetted perimeter per unit of flow cross section* and is, therefore, an index of the extent of the boundary surface in contact with the flowing fluid. The hydraulic radius may be safely used in the equation above when resistance to flow and head loss are primarily dependent on the extent of the boundary surface, as for turbulent flow in which pipe friction phenomena are confined to a thin region adjacent to the boundary surface and thus vary with the size of this surface. However, severe deviation from a circular flow cross section will prevent the hydraulic radius from accounting for changes in head loss, even for cases of turbulent flow. Tests on turbulent flow through annular passages,[10] for example, show a large increase in friction factor with increase of the ratio of core diameter to pipe diameter.

In laminar flow, friction phenomena result from the action of viscosity throughout the whole body of the flow, are independent of roughness, and are not primarily associated with the region close to the boundary walls. In view of these facts the hydraulic radius technique cannot be expected to give reliable conversions from circular to noncircular passages in laminar flow. This expectation is borne out both by experiment and by analytical solutions of laminar flow in noncircular passages.

ILLUSTRATIVE PROBLEM 9.11

Calculate the loss of head and the pressure drop when air at an absolute pressure of 101.3 kPa and 15°C flows through 600 m of 450 mm by 300 mm smooth rectangular duct with a mean velocity of 3 m/s.

SOLUTION

We will treat this problem in the conventional fashion substituting $4R_h$ for d in the formulas. Since the duct is smooth, we will only need the Reynolds number to find the friction factor. From Appendix 2, we find values of $\rho = 1.225 \text{ kg/m}^3$ and $\mu = 1.789 \times 10^{-5}$ Pa·s for air at the specified pressure and temperature. The hydraulic radius for the duct is

$$R_h = \frac{\text{Area}}{\text{Wetted perimeter}} = \frac{0.45 \text{ m} \times 0.30 \text{ m}}{2 \times 0.45 \text{ m} + 2 \times 0.30 \text{ m}} = 0.090 \text{ m}$$

Now, we can compute the Reynolds number from Eq. 9.41.

$$\mathbf{R} = \frac{V(4R_h)\rho}{\mu} = \frac{3 \text{ m/s} \times (4 \times 0.090 \text{ m}) \times 1.225 \text{ kg/m}^3}{1.789 \times 10^{-5}} = 73\,950 \qquad (9.41)$$

From the smooth pipe line on the Moody diagram of Fig. 9.10,

$$f \cong 0.019$$

[10]See W. M. Owen, "Experimental Study of Water Flow in Annular Pipes," *Trans. A.S.C.E.*, vol. 117, 1952; and J. E. Walker, G. A. Whan, and R. R. Rothfus, "Fluid Friction in Noncircular Ducts," *Jl.A.I. Ch.E.*, vol. 3, 1957.

Equation 9.40 can be used to calculate the head loss in 600 m of the pipe.

$$h_L = f \frac{l}{4R_h} \frac{V^2}{2g_n} = 0.019 \frac{600}{4 \times 0.090} \frac{3^2}{2 \times 9.81} = 14.5 \text{ m of air} \bullet \qquad (9.40)$$

The pressure drop is found from the equation

$$\Delta p = \gamma h_L = \rho g h_L = 1.225 \times 9.81 \times 14.5 = 174 \text{ Pa} \bullet$$

9.8 PIPE FRICTION—EMPIRICAL FORMULAS

The Darcy-Weisbach equation 9.2 has provided a rational basis for the analysis and computation of head loss. However, historically, a number of empirical formulas have been and still are being used for pipe friction calculations in engineering practice. A treatment of pipe friction would be incomplete without mentioning at least two of these formulas and how they relate to the Darcy-Weisbach equation.

By far the most widely used of these empirical formulations is the Hazen-Williams[11] formula. It was developed to permit calculating the capacity of pipes to convey water.

$$\text{(U.S. Customary units)} \quad V = 1.318 C_{hw} R_h^{0.63} S^{0.54} \qquad (9.42a)$$

$$\text{(SI units)} \quad V = 0.849 C_{hw} R_h^{0.63} S^{0.54} \qquad (9.42b)$$

where R_h is the hydraulic radius, S is the head loss per unit length h_L/l, and C_{hw} is a roughness coefficient associated with the pipe material. Table 1 gives some typical values of the Hazen-Williams coefficient.

TABLE 1 Hazen-Williams Coefficient C_{hw} and Manning n-values[a]

	C_{hw}	n
Extremely smooth pipes—PVC	150–160	0.009
Copper, aluminum tubing	150	0.010
Asbestos cement	140	0.011
New cast iron	130	0.013
Welded steel	130–140	0.012
Concrete	120–140	0.011–0.014
Ductile iron (cement lined)	140	0.011
Vitrified clay pipe	—	0.011–0.013
Riveted steel	110	0.013–0.017
Old cast iron	100	0.015–0.035

[a]These are typical values but, because of variabilities in fabrication, the user should consult the pipe manufacturer for recommended values of roughness coefficients.

[11]G. S. Williams and A. H. Hazen, *Hydraulic Tables*, 3rd ed., John Wiley & Sons, 1933.

Unlike the Darcy-Wesibach equation which can be applied to both laminar and turbulent flow over a wide range of fluids and temperatures, the Hazen-Williams formula is restricted to turbulent flow of water at normal temperature in a limited size range of relatively smooth pipes. However, the formula is so constructed that it permits direct calculation of the velocity of flow (or discharge) for a known allowable head loss. This feature is attractive in the design of water pipelines. Conversely, use of the Darcy-Weisbach equation in the same circumstances often requires a trial solution as one estimates the Reynolds number to find the friction factor and makes successive adjustments as one closes in on the answer.

In view of the formula configuration and the rather indefinite descriptions of the roughness coefficients, it is difficult to judge the range of validity of the Hazen-Williams formula without wide experience in its application. It is not clear whether C_{hw} is a measure of absolute or relative roughness, whether there is any effect of Reynolds number in the formula, whether it applies only in smooth or rough pipe situations, etc. These questions can be answered fairly conclusively if the formula is rewritten in the form of the Darcy-Weisbach equation. Working with the U.S. Customary version, replacing R_h with $d/4$, S with h_L/l, and solving for h_L gives

$$h_L = \left[\frac{194}{C_{hw}^{1.85}(Vd)^{0.15}\ d^{0.015}}\right]\frac{l}{d}\frac{V^2}{2g_n}$$

If $(V\ d)^{0.15}$ is multiplied by $(\nu/\nu)^{0.15}$, a Reynolds number appears and the equation may be written as

$$h_L = \left[\frac{194}{\nu^{0.15}C_{hw}^{1.85}\ d^{0.015}\mathbf{R}^{0.15}}\right]\frac{l}{d}\frac{V^2}{2g_n}$$

Comparing this equation with the Darcy-Weisbach equation 9.2 and taking a nominal ν-value of 0.000 01 ft^2/s for water, the "equivalent f-value" for the Hazen-Williams formula is

$$\text{“}f\text{”} = \frac{1\,090}{C_{hw}^{1.85}\ d^{0.015}\mathbf{R}^{0.15}} \tag{9.43}$$

The most effective means of visualizing the meaning of this "f-value" is to plot the equation on the Moody diagram for several C_{hw} values. This is done on Fig. 9.12. The following conclusions can now be drawn in regard to the range of applicability of the Hazen-Williams formula.

1. The Hazen-Williams formula is a "transition" formula working best as a smooth pipe formula or in the "early" transition zone between smooth and rough pipe flow. Note the $C_{hw} = 150$ smooth-pipe value applies fairly well over a range of Reynolds numbers from 100 000 on but the plot suggests a value higher than 150, say 160, should be used for Reynolds numbers above 1 000 000. Also note that the formula grossly underestimates friction factors in smooth pipes for Reynolds numbers below 10 000 (typical of trickle irrigation systems) and does not apply at all for laminar flow. Because the lines for the various C_{hw}-values on Fig. 9.12 all have a constant slope, the Hazen-Williams formula should never be used for wholly rough pipes. More specifically, the Hazen-Williams formula should be used only

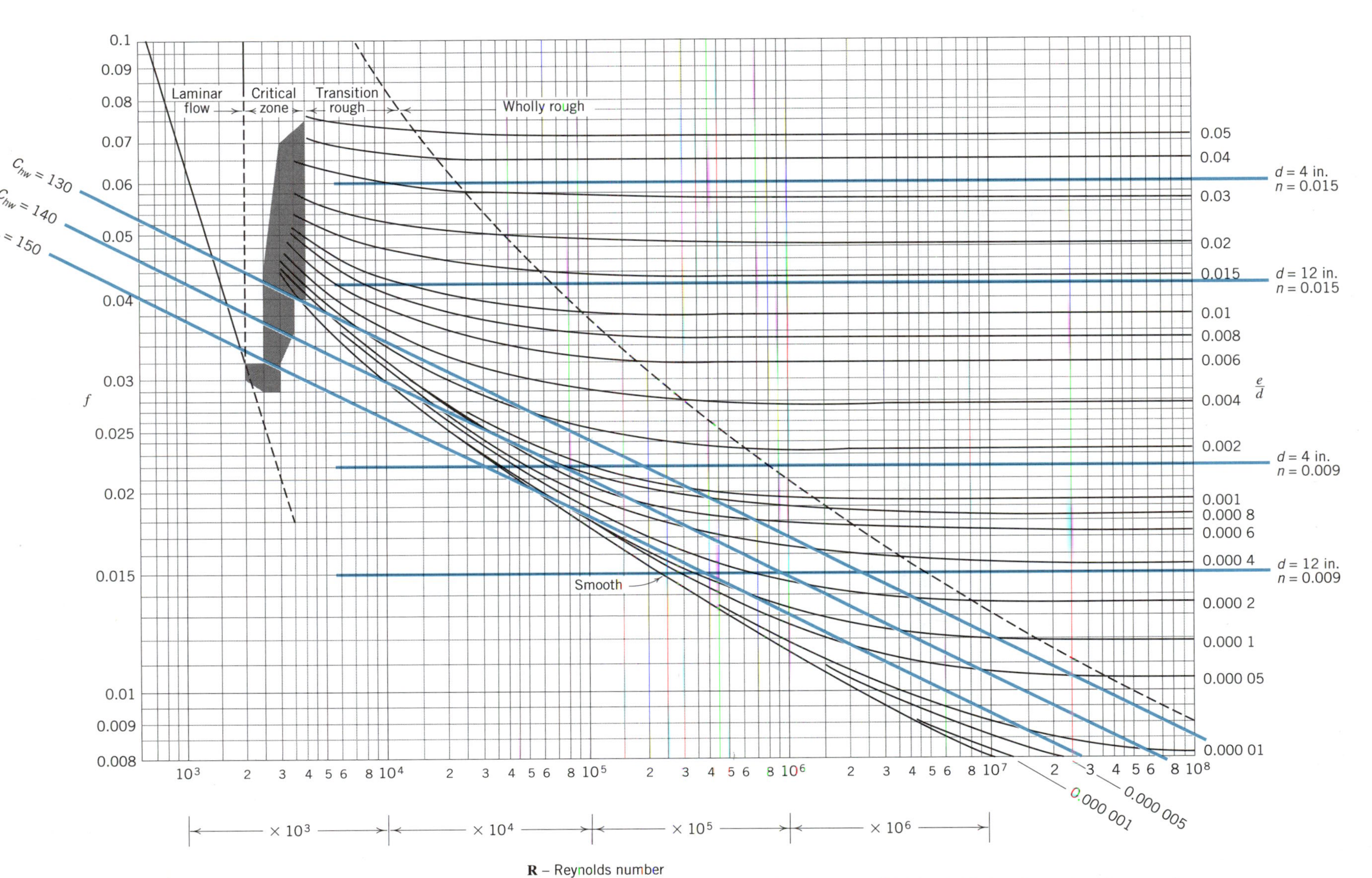

Fig. 9.12 Comparison of Hazen-Williams and Manning roughness coefficients with the Darcy-Weisbach friction factor (see footnote 9).

in zones where the pipe is relatively smooth and in the "early" part of its transition to rough flow. If there is a question as to applicability, it is always possible to estimate the Reynolds number of the flow and, with the proposed C_{hw} value, check Fig. 9.12 to verify the applicability of the formula.

2. Addressing the issue of relative roughness, we see it is clear from the above equation for "f" that there is a small relative roughness effect built into the equation although the 0.015 exponent dampens the impact considerably. However, in the "near-smooth" zone where the Hazen-Williams formula applies, the effect of relative roughness is small, thereby minimizing the error.
3. There is definitely a strong Reynolds number effect built into the Hazen-Williams formula which renders it accurate in the "near-smooth" transition range for Reynolds numbers greater than 100 000 or with the appropriate value of C_{hw}, e.g., $C_{hw} = 130$ for $10\,000 < \mathbf{R} < 100\,000$ on Fig. 9.12.

The second formula selected for examination is the Manning equation, noted for its application to open channel flow but also used for pipe flow. The Manning formulas for pipe flow are given as

$$\text{(U.S. Customary units)} \quad V = \frac{1.49}{n} R_h^{2/3} S^{1/2} \tag{9.44a}$$

$$\text{(SI units)} \quad V = \frac{1}{n} R_h^{2/3} S^{1/2} \tag{9.44b}$$

where R_h and S are as described for the Hazen-Williams formula and n is a roughness coefficient. Some typical values are given in Table 1.

We will investigate the range of applicability of the Manning formula in the same manner as was done with the Hazen-Williams formula. Again using the U.S. Customary version of the formula, and arranging in the Darcy-Weisbach configuration, we obtain

$$\text{``}f\text{''} = \frac{185n^2}{d^{1/3}} \tag{9.45}$$

It seems clear from this equation that there is no Reynolds number effect but there is a fairly strong relative roughness effect. This equation is plotted in Fig. 9.12 for two different n-values and two different diameters to show the effect of relative roughness. The following conclusions can be drawn regarding the applicability of the Manning formula.

1. There is no Reynolds number effect so the formula must be used only in the wholly rough flow zone where its horizontal slope can accurately match Darcy-Weisbach values provided the proper n-value is selected.
2. The relative roughness effect is correct in the sense that, for a given roughness, a larger pipe will have a smaller friction factor.
3. In a general sense, because the formula is valid only for rough pipes, the rougher the pipe, the more likely the Manning formula will apply.

Any other formula can be compared with the Darcy-Weisbach equation in the manner used above. For empirically-based formulas, this approach is strongly recommended unless the user is confident the formula is applicable to the case at hand.

ILLUSTRATIVE PROBLEM 9.12

A pipeline is to convey 20 ft^3/s of water between two reservoirs 5 mi apart and 200 ft different in elevation. Considering only pipe friction losses, use the Darcy-Weisbach, Hazen-Williams, and Manning formulas to select a diameter of welded steel pipe that will meet this requirement.

SOLUTION

Hazen-Williams Solution

From Table 1, $C_{hw} = 135$. We will modify Eq. 9.42a to create a formula for the flowrate Q.

$$Q = AV = \frac{\pi}{4} d^2 \times 1.318 C_{hw} R_h^{0.63} S^{0.54} \tag{9.42a}$$

$$20 = \frac{\pi}{4} d^2 \times 1.318 \times 135 \times \left(\frac{d}{4}\right)^{0.63} \left(\frac{200 \text{ ft}}{5 \text{ mi} \times 5\,280 \text{ ft/mi}}\right)^{0.54}$$

$$d^{2.63} = 4.79$$

$$d = 1.82 \text{ ft} \qquad \text{or} \qquad 21.8 \text{ in} \bullet$$

The design diameter would be the next nominal size, e.g., 22-in pipe. The equivalent f-value from Eq. 9.43 is $f = 0.016$.

Manning Solution

From Table 1, $n = 0.012$. Again we modify Eq. 9.44a to create a formula for the discharge Q.

$$Q = AV = \frac{\pi}{4} d^2 \times \frac{1.49}{n} R_h^{2/3} S^{1/2} \tag{9.44a}$$

$$20 = \frac{\pi}{4} d^2 \times \frac{1.49}{0.012} \left(\frac{d}{4}\right)^{2/3} \left(\frac{200 \text{ ft}}{5 \text{ mi} \times 5\,280 \text{ ft/mi}}\right)^{1/2}$$

$$d^{8/3} = 5.94$$

$$d = 1.95 \text{ ft} \qquad \text{or} \qquad 23.4 \text{ in} \bullet$$

The design diameter would be 24 inches. The equivalent f-value from Eq. 9.45 is $f = 0.021$.

Darcy-Weisbach Solution

A trial-and-correct solution technique is required because the Reynolds number and the relative roughness are unknown because the diameter is initially unknown. The normal procedure would be to assume a diameter, use it to compute **R** and e/d, solve for d, then recompute **R** and e/d, iterating until a satisfactory accuracy is reached. Since we are aware of the approximate diameter from the previous solutions, we will use that as a first "guess." From Fig. 9.11, $e = 0.000\ 15$ ft so

$$e/d = 0.000\ 15\ \text{ft}/2.0\ \text{ft} = 0.000\ 075$$

Using water at a normal temperature to compute Reynolds number, $\nu = 1.2 \times 10^{-5}$ ft²/s.

$$\mathbf{R} = \frac{Vd}{\nu} = \frac{Qd}{A\nu} = \frac{Qd}{(\pi/4)\ d^2\nu} = \frac{4Q}{\pi d\nu}$$

$$= \frac{4 \times 20\ \text{ft}^3/\text{s}}{\pi \times 2\ \text{ft} \times 1.2 \times 10^{-5}\ \text{ft}^2/\text{s}} = 1.1 \times 10^6$$

From Fig. 9.10, $f = 0.013\ 5$. Substituting this value into the Darcy-Weisbach equation gives

$$h_L = f\frac{l}{d}\frac{V^2}{2g_n} = f\frac{l}{d}\frac{Q^2}{A^2 2g_n} = f\frac{l}{d}\frac{Q^2}{(\pi/4 \times d^2)^2 \times 2g_n} = \frac{16}{\pi^2}f\frac{l}{d^5}\frac{Q^2}{2g_n}$$

$$d^5 = \frac{16}{\pi^2}f\frac{l}{h_L}\frac{Q^2}{2g_n} = \frac{16}{\pi^2}0.013\ 5\ \frac{5\ \text{mi} \times 5\ 280\ \text{ft/mi}}{200\ \text{ft}}\frac{(20\ \text{ft}^3/\text{s})^2}{2 \times 32.2} = 17.94$$

$$d = 1.78\ \text{ft} \qquad \text{or} \qquad 21.4\ \text{in} \bullet$$

The design diameter in this case would be 22 inches if that size pipe were available.

Even though there are rather dramatic differences in f-values (almost 50% variation), the resulting impact on design pipe diameter is substantially less. This is a consequence of the fact that the diameter varies at about the 1/5th power of the f-value, so that dramatic differences in f-value have a greatly reduced impact.

9.9 LOCAL LOSSES IN PIPELINES

Into the category of local losses in pipelines fall those losses incurred by change of cross section, bends, elbows, valves, and fittings of all types. Although in long pipelines these are distinctly minor losses and can often be neglected without serious error, in shorter pipelines an accurate knowledge of their effects must be known for correct engineering calculations.

The general aspects of local losses in pipelines may be obtained from a study of the flow phenomena about an abrupt obstruction placed in a pipeline (Fig. 9.13), which creates flow conditions typical of those which dissipate energy and cause local losses. Local losses usually result from rather abrupt changes (in magnitude or direction) of velocity; in general, increase of velocity (acceleration) is associated with small head loss but decrease of velocity (deceleration) causes large head loss because of the production of large-scale turbulence (Section 7.13). In Fig. 9.13, useful energy is extracted to create eddies as the fluid decelerates between sections 2 and 3, and this energy is dissipated in heat as the eddies decay between sections 3 and 4. Local losses in pipe flow are, therefore, accomplished in the pipe downstream from the source of the eddies, and the pipe friction processes in this length of pipe are complicated by the superposition of large-scale turbulence on the normal turbulence pattern. To make local loss calculations possible it is necessary to assume

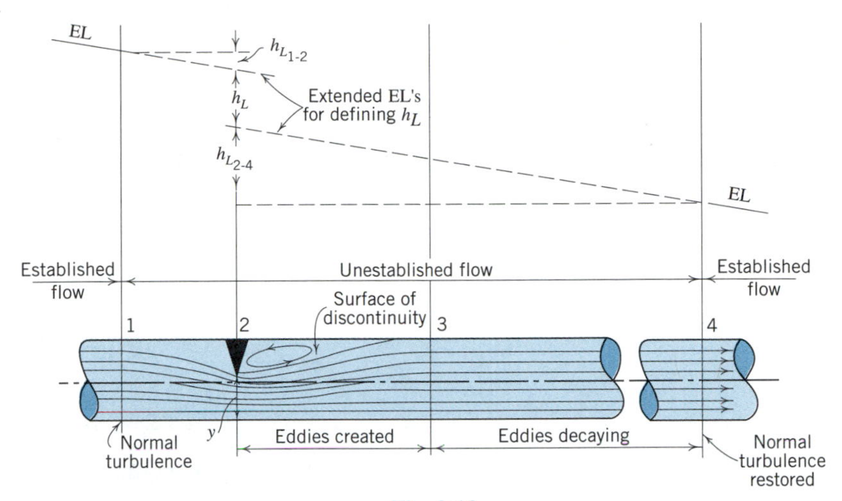

Fig. 9.13

separate action of the normal turbulence and large-scale turbulence, although in reality a complex combination of the two processes exists. Assuming the processes independent allows calculation of the losses due to established pipe friction, $h_{L_{1-2}}$ and $h_{L_{2-4}}$, and also permits the loss h_L, due to the obstruction alone, to be assumed concentrated at section 2. This is a great convenience for engineering calculations since the total lost head in a pipeline may be obtained by simple addition of established pipe friction and local losses without detailed consideration of the above-mentioned complications.

Early experiments with water (at high Reynolds number) indicated that local losses vary approximately with the square of velocity and led to the proposal of the basic equation

$$h_L = K_L \frac{V^2}{2g_n} \tag{9.46}$$

in which K_L, the *loss coefficient*, is, for a given flow geometry, practically constant at high Reynolds number; the loss coefficient tends to increase with increasing roughness and decreasing Reynolds number,[12] but these variations are usually of minor importance in turbulent flow. The magnitude of the loss coefficient is determined primarily by the flow geometry, that is, by the shape of the obstruction or pipe fitting.

When an *abrupt enlargement* of section (Fig. 9.14) occurs in a pipeline, a rapid deceleration takes place, accompanied by characteristic large-scale turbulence, which may persist in the larger pipe for a distance of 50 diameters or more downstream before the normal turbulence pattern of established flow is restored. Simultaneous application of the continuity, Bernoulli, and momentum principles to this problem has shown (Section 6.2) that (with certain simplifying assumptions)

$$h_L = K_L \frac{(V_1 - V_2)^2}{2g_n} \tag{9.47}$$

[12]Note that this is the same trend followed by the friction factor, f.

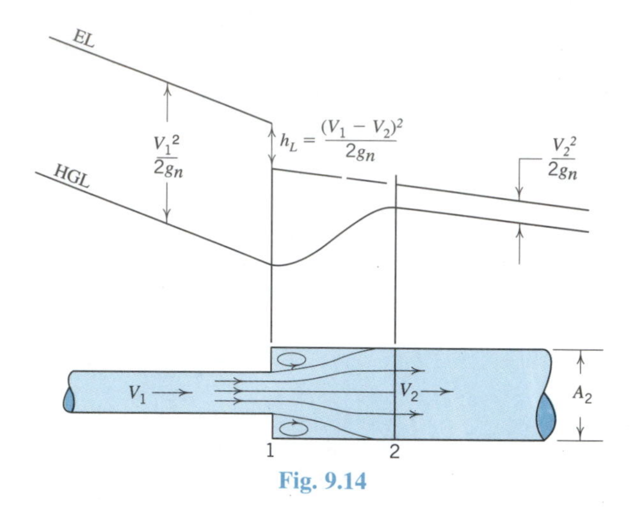

Fig. 9.14

in which $K_L \cong 1$. Experimental determinations of K_L confirm this value within a few percent, making it quite adequate for engineering use. A special case of an abrupt enlargement exists when a (relatively small) pipe discharges into a (relatively large) tank or reservoir. Here the velocity downstream from the enlargement may be taken to be zero, and when the lost head (called the exit loss) is calculated from Eq. 9.47 it is found to be the velocity head in the pipe.

The loss of head due to *gradual enlargement* is, of course, dependent on the shape of the enlargement. Tests have been carried out by Gibson[13] on the losses in conical enlargements, and the results are expressed by Eq. 9.47, in which K_L is primarily dependent on the cone angle but is also a function of the area ratio, as shown in Fig. 9.15. Because of the large surface of the conical enlargement which contacts the fluid, the coefficient K_L

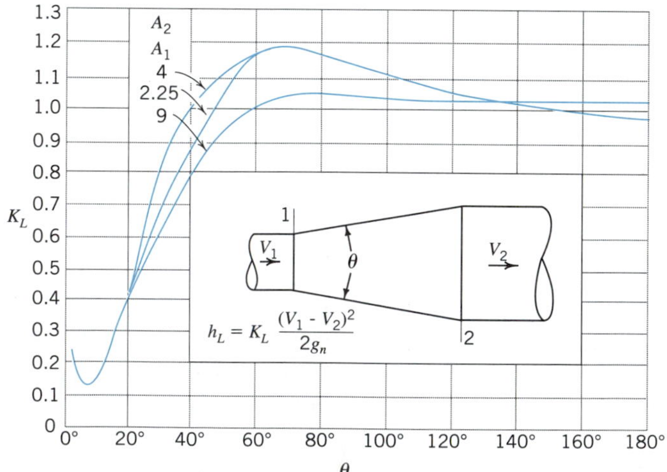

Fig. 9.15 Loss coefficients for conical enlargements.[13] (Source: A. H. Gibson, Hydraulics and its Applications, 4th ed., 1930.)

[13]A. H. Gibson, *Hydraulics and Its Applications*, 4th ed., p. 93, D. Van Nostrand Co., 1930. See also *Engineering Data Book*, 2nd Ed. Hydraulic Institute, Parsippany, NJ, 1990.

embodies the effects of wall friction as well as those of large-scale turbulence. In an enlargement of small central angle, K_L will result almost wholly from surface friction; but, as the angle increases and the enlargement becomes more abrupt, not only is the surface area reduced but also separation occurs, producing large eddies, and here the energy dissipated in the eddies determines the magnitude of K_L. From the plot it may be observed that (1) there is an optimum cone angle of about 7° where the combination of the effects of surface friction and eddying turbulence is a minimum; (2) it is better to use a sudden enlargement than one of cone angle around 60°, since K_L is smaller for the former.

Gradual enlargements in passages (termed *diffusers*) of various forms are widely used in engineering practice for *pressure recovery*, that is, pressure rise in the direction of flow. No attempt is made here to review the extensive literature[14] on diffusers, but it should be realized that Gibson's results cannot give a reliable solution to this problem. His tests were made, as are all such tests for local losses, with long straight lengths of pipe upstream and downstream from the enlargement (see Fig. 9.13). The designer, however, is frequently more interested in the pressure rise through the diffuser and substitution of a short nozzle for the upstream pipe length. The tests of Gibson and others have shown that the pressure will continue to rise for a few pipe diameters downstream from section 2, owing primarily to readjustment of velocity distribution from a rather pointed one caused by deceleration through the diffuser to the flatter one of turbulent flow.[15] From this it may be concluded that pressure rise through the diffuser, computed from application of the Bernoulli equation with data from Fig. 9.15, will be larger than that which will actually be realized. Substitution of a nozzle for the upstream pipe length will alter the inlet velocity distribution from the standard one of turbulent flow to a practically uniform one with a thin boundary layer (see Fig. 7.19). The effect is to reduce the losses by stabilizing the flow and delaying separation; not only do smaller loss coefficients result, but the cone angle for minimum losses is larger, thus allowing a shorter diffuser for the same area ratio.

ILLUSTRATIVE PROBLEM 9.13

A 300 mm horizontal water line enlarges to a 600 mm line through a 20° conical enlargement. When 0.30 m^3/s flow through this line, the pressure in the smaller pipe is 140 kPa. Calculate the pressure in the larger pipe, neglecting pipe friction.

SOLUTION

In determining the head loss in a conical enlargement, Eq. 9.47 applies with the K_L-value obtained from Fig. 9.15. First, we will calculate the velocities in each pipe.

$$V_{300} = \frac{Q}{A_{300}} = \frac{0.30\ \text{m}^3/\text{s}}{(\pi/4)(0.300\ \text{m})^2} = 4.24\ \text{m/s}$$

$$V_{600} = \frac{Q}{A_{600}} = \frac{0.30\ \text{m}^3/\text{s}}{(\pi/4)(0.600\ \text{m})^2} = 1.06\ \text{m/s}$$

[14]See, for example, E. G. Reid, "Performance Characteristics of Plane-Wall Two-Dimensional Diffusers," *N.A.C.A., Tech. Note* 2888, 1953; S. J. Kline, D. E. Abbott, and R. W. Fox, "Optimum Design of Straight-Walled Diffusers," *Trans. A.S.M.E.*, vol. 81, 1959; and J. M. Robertson and H. R. Fraser, "Separation Prediction in Conical Diffusers," *Trans. A.S.M.E. (Series D)*, vol. 82, no. 1, 1960.

[15]Compare this with the opposite situation discussed in Section 7.12.

From Fig. 9.15, $K_L = 0.43$. To compute the pressure in the large pipe, we turn to the work-energy equation 7.46 without pumps or turbines.

$$z_{300} + \frac{p_{300}}{\gamma} + \frac{V_{300}^2}{2g_n} = z_{600} + \frac{p_{600}}{\gamma} + \frac{V_{600}^2}{2g_n} + h_L \tag{7.53}$$

where $h_L = K_L \dfrac{(V_{300} - V_{600})^2}{2g_n}$

Taking the datum as the pipe centerline eliminates z from the calculations leaving

$$\frac{140 \times 10^3 \text{ Pa}}{9\,800 \text{ N/m}^3} + \frac{(4.24 \text{ m/s})^2}{2 \times 9.81} = \frac{p_{600}}{\gamma} + \frac{(1.06 \text{ m/s})^2}{2 \times 9.81} + 0.43\,\frac{(4.24 - 1.06)^2}{2 \times 9.81}$$

$$\frac{p_{600}}{\gamma} = 14.6 \text{ m}$$

$$p_{600} = 14.6 \times 9\,800 = 143\,000 \text{ Pa} = 143 \text{ kPa} \bullet$$

This is the pressure to be expected a metre or so downstream from the end of the enlargement.

Flow through an *abrupt contraction* is shown in Fig. 9.16 and is featured by the formation of a vena contracta (Section 5.8) and subsequent deceleration and reexpansion of the live stream of flowing fluid.

Experimental measurements of K_L are somewhat conflicting in magnitude although they exhibit a well-established trend from 0.5 for $A_2/A_1 = 0$ to 0 for $A_2/A_1 = 1$. In view of this it is entirely adequate for engineering practice to use a synthesis of analytical approaches and generally accepted experimental information.[16] The result is given in Table 2 where $C_c = A_c/A_2$.

A square-edged pipe entrance (Fig. 9.17*a*) from a large body of fluid is the limiting case of the abrupt contraction, with $A_2/A_1 = 0$ (A_1 is virtually infinite here). The head loss is expressed by Eq. 9.46 in which K_L is close to 0.5 for highly turbulent flow, as mentioned above.

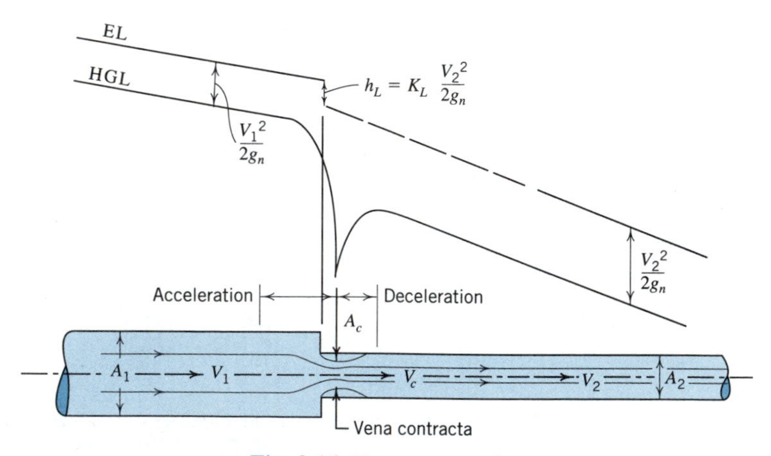

Fig. 9.16 Abrupt contraction.

[16]J. Weisbach, *Die Experimental Hydraulik*, 1855.

TABLE 2 Contraction, C_c, and Loss, K_L, Coefficients for Abrupt Contractions

A_2/A_1	0	0.1	0.2	0.3	0.4	0.5	0.6	0.7	0.8	0.9	1.0
C_c	0.617	0.624	0.632	0.643	0.659	0.681	0.712	0.755	0.813	0.892	1.00
K_L	0.50	0.46	0.41	0.36	0.30	0.24	0.18	0.12	0.06	0.02	0

The entrance of Fig. 9.17*b* is known as a re-entrant one. If the pipe wall is very thin and if the plane of the opening is more than one pipe diameter upstream from the reservoir wall, the loss coefficient will be close to 0.8, this high value resulting mainly from the small vena contracta and consequent large deceleration loss. For thick-walled pipes the vena contracta can be expected to be larger and the loss coefficient less than 0.8; Harris[17] has shown that pipes of wall thicknesses greater than 0.05*d* (if square-edged) will give a loss coefficient equal to that of the square-edged entrance.

If the edges of a pipe entrance are rounded to produce a streamlined *bell-mouth* (Fig. 9.18) the loss coefficient can be materially reduced. Hamilton[18] has shown that any radius of rounding greater than 0.14*d* will prevent the formation of a vena contracta and thus eliminate the head loss due to flow deceleration. The nominal value of K_L for such an entrance is about 0.1, but its exact magnitude will depend on the detailed geometry of the entrance and structure of the boundary layer (see Fig. 7.19 of Section 7.9).

The head loss caused by short well-streamlined gradual contractions (Fig. 9.19) is so small that it may usually be neglected in engineering problems. However, an appreciable fall of the hydraulic grade line over such contractions is to be expected; in the pipe downstream from the contraction the hydraulic grade line will be found to slope more steeply

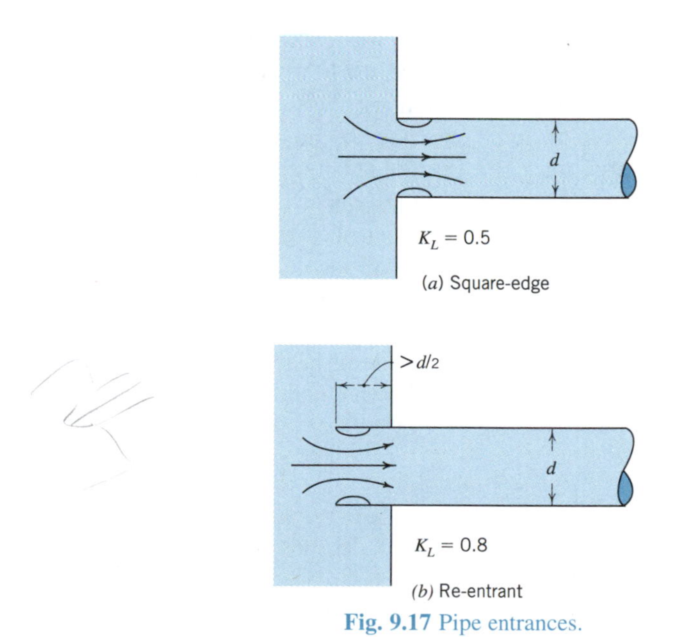

Fig. 9.17 Pipe entrances.

[17] C. W. Harris, "The Influence of Pipe Thickness on Reentrant Intake Losses," *Univ. Wash. Eng. Expt. Sta., Bull.* 48, 1928.

[18] J. B. Hamilton, "The Suppression of Intake Losses by Various Degrees of Rounding," *Univ. Wash. Eng. Expt. Sta., Bull.* 51, 1929.

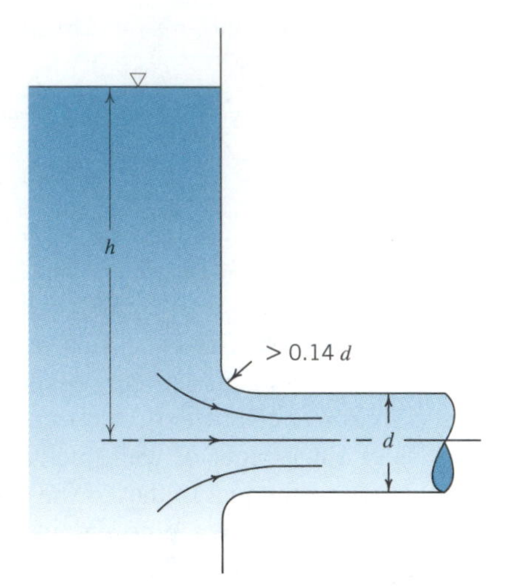

Fig. 9.18 Bell-mouth entrance.

than that for established flow because of the change of velocity distribution and boundary layer growth, which cause an increase of α (see Fig. 7.19 of Section 7.9). These effects are known to be small and may be ignored in many problems. A nominal value of K_L (for use in Eq. 9.46 with $V = V_2$) for short well-streamlined contractions is 0.04; by careful design this figure may be lowered to 0.02, but for long contractions values much larger than 0.04 are to be expected because of extensive wall friction.

Losses of head in *smooth pipe bends* are caused by the combined effects of separation, wall friction, and the twin-eddy secondary flow described in Section 7.14; for bends of large radius of curvature, the last two effects will predominate, whereas, for small radius of curvature, separation and the secondary flow will be the more significant. The loss of head in a bend is expressed by Eq. 9.46 in which the head loss is the drop of the extended energy lines (see Fig. 9.13) between the entrance and exit of the bend. Reliable and extensive information on this subject will be found in the work of Itō,[19] a small but typical portion of which is shown in Fig. 9.20. Head loss coefficients for smooth pipe bends provide another example of the dependence of K_L on shape of passage (determined by θ and R/d) and Reynolds number; the research of Hofmann[20] provides information on the (expected) dependence of loss coefficient on relative roughness as well. To the engineer the most significant feature of head loss in bends is the minimum value of K_L occurring at certain values of R/d which allows selection of bend shapes for maximum efficiency in pipeline design.

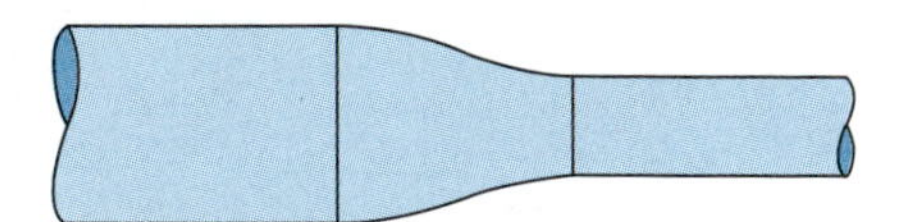

Fig. 9.19 Gradual contraction.

[19]H. Itō, "Pressure Losses in Smooth Pipe Bends," *Trans. A.S.M.E. (Series D)*, vol. 82, no. 1, 1960.

[20]A. Hofmann, "Loss in 90° Pipe Bends of Constant Circular Cross Section," *Trans. Hydr. Inst. of Munich Techn. Univ., Bull.* 3, A.S.M.E., 1935.

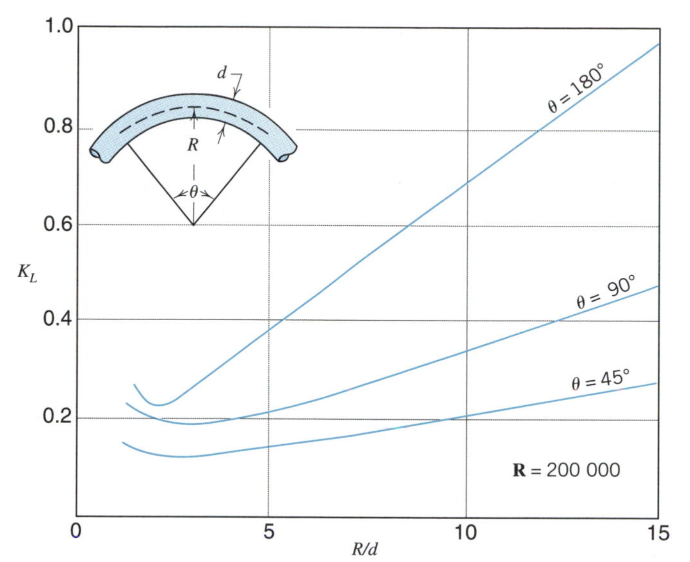

Fig. 9.20 Itō's loss coefficients for smooth bends (**R** = 200 000).

Tests made on bends with $R/d = 0$ have given values of K_L around 1.1. Such bends are known as *miter bends* (Fig. 9.21) and are used widely in large ducts such as wind and water tunnels, where space does not permit a bend of large radius. In these bends, installation of guide vanes materially reduces the head loss and at the same time breaks up the spiral motion and improves the velocity distribution downstream.

The losses of head caused by commercial pipe fittings occur because of their rough and irregular shapes which produce excessive large-scale turbulence. The shapes of commercial pipe fittings are determined more by structural properties, ease in handling, and production methods than by head loss considerations, and it is, therefore, not feasible or economically justifiable to build pipe fittings having completely streamlined interiors in order to minimize head loss. The loss of head in commercial pipe fittings is usually expressed by Eq. 9.46 with V the mean velocity in the pipe and K_L a constant (at high Reynolds numbers), the magnitude of which depends on the shape of the fitting. Values of K_L for various common fittings are available in the *Engineering Data Book* of the Hydraulic Institute; typical values are presented in Table 3.

It is generally recognized that when fittings are placed in close proximity the total head loss caused by them is less than their numerical sum obtained by the foregoing methods. Systematic tests have not been made on this subject because a simple numerical sum of losses gives a result in excess of the actual losses, and thus, produces an error on the conservative side when design calculations of pressures and flowrates are to be made.

Fig. 9.21 Miter bends.

TABLE 3 Approximate Loss Coefficient, K_L, for Commercial Pipe Fittings[21]

Valves, wide open	Screwed		Flanged
Globe	10		5
Gate	0.2		0.1
Swing-check		2	
Angle		2	
Foot		0.8	
Return bend	1.5		0.2
Elbows			
90°—regular	1.5		0.3
—long radius	0.7		0.2
45°—regular	0.4		—
—long radius	—		0.2
Tees			
Line flow	0.9		0.2
Branch flow	2		1

ILLUSTRATIVE PROBLEM 9.14

A 3 inch 10 000 ft pipeline carries water at a velocity of 2.36 ft/s and has a pipe friction loss of 88 ft, neglecting local losses. If the line contains a sharp-edged reservoir entrance, a wide-open, screwed globe valve, and four 90° regular, screwed elbows and exits directly into a reservoir, compute the local losses in the line and the percent error incurred by neglecting them.

SOLUTION

We will use Eq. 9.46 to compute the local losses but first we need to find the loss coefficients for each of the local losses.

Sharp-edged entrance	From Fig. 9.17, $K_L = 0.5$
Wide-open, screwed globe valve	From Table 3, $K_L = 10$
90° regular, screwed elbows	From Table 3, $K_L = 1.5$ ea.
Exit into reservoir	The entire velocity head is lost, so $K_L = 1.0$

Now, we place all of these loss coefficients into Eq. 9.46 to get

$$\text{Total local loss} = \sum K_L \frac{V^2}{2g_n} = (0.5 + 10 + 4 \times 1.5 + 1.0)\frac{(2.36 \text{ ft/s})^2}{2 \times 32.2}$$

$$\text{Total local loss} = 1.51 \text{ ft} \bullet$$

$$\text{Percent error} = 100\frac{\text{local loss}}{\text{pipe friction loss}} = 100\frac{1.51}{88} = 1.7\% \bullet$$

[21]Adapted from the *Engineering Data Book*, 2nd ed, 1990, with permission of the Hydraulic Institute, Parsippany, NJ.

9.10 PIPELINE PROBLEMS—SINGLE PIPES

All steady-flow pipe problems may be solved by application of the work-energy and continuity equations, and the most effective method of doing this is the construction of energy and hydraulic grade lines. From such lines the variations of pressure, velocity, and unit energy can be clearly seen for the whole problem; thus, the construction of these lines becomes equivalent to writing numerous equations, but the lines lend a clarity to the solution of the problem which equations alone never can.

In engineering offices, tables, charts, nomograms, computer software, and so forth, are employed where numerous pipe-flow problems are to be solved. Although all these methods are different, they have their foundations in the work-energy principle, usually with certain approximations; no attempt is made here to cover these many methods—the following discussion will be confined to the application of the work-energy and continuity principles and the use of certain approximations.

Engineering pipe-flow problems usually consist of (1) calculation of head loss and pressure variation from flowrate and pipeline characteristics, (2) calculation of flowrate from pipeline characteristics and the head which produces flow, and (3) calculation of required pipe diameter to pass a given flowrate between two regions of known pressure difference. The first of these problems can be solved directly, but solution by trial is required for the other two.[22] Trial-and-error solutions are necessitated by the fact that the friction factor, f, and loss coefficients, K_L, depend on the Reynolds number, which in turn depends on flowrate and pipe diameter, the unknowns of problem types 2 and 3, respectively. However, many engineering pipeline problems involve flow in rough pipes at high Reynolds numbers. Here trial solutions are seldom required (1) because of the tendency of f and K_L toward constancy in this region, (2) because of the inevitable error[23] in selecting f from Fig. 9.10, and (3) because engineering answers are usually not needed to a precision which warrants trial-and-error solution in the light of the foregoing facts. Construction of energy and hydraulic grade lines for some typical pipeline problems will indicate further approximations which may frequently be used in the solution of engineering problems.

Consider the calculation of flowrate in a pipeline laid between two tanks or reservoirs having a difference of surface elevation H (Fig. 9.22). The energy line must start in one reservoir surface and end in the other; using a gradual drop to represent head loss due to

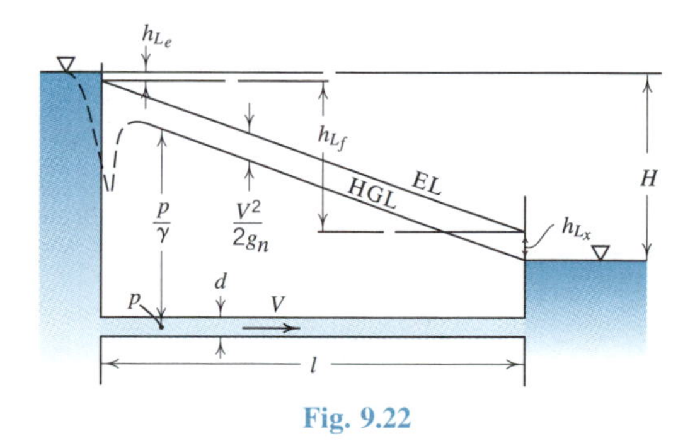

Fig. 9.22

[22]Unless special plots are devised for circumventing this.

[23]Due to the inexactness of definition of the roughness.

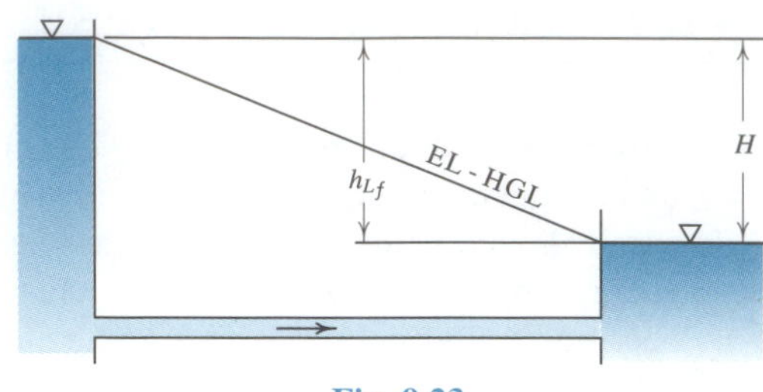

Fig. 9.23

pipe friction, h_{L_f}, and abrupt drops to represent entrance and exit losses, h_{L_e} and h_{L_x}, the energy line is constructed as shown. It is apparent from the energy line that

$$h_{L_e} + h_{L_f} + h_{L_x} = H$$

which is the work-energy equation written between the reservoir surfaces. When the appropriate expressions for the head losses are substituted (Sections 9.1 and 9.9),

$$\left(0.5 + f\frac{l}{d} + 1\right)\frac{V^2}{2g_n} = H$$

If turbulent flow is assumed and a nominal value for f of 0.03 selected (Fig. 9.10), the quantity in parentheses becomes 4.5, 31.5, and 301.5 for l/d-values of 100, 1 000, and 10 000, respectively. The quantities 0.5 and 1.0, which result from inclusion of local losses, have a decreasing effect on the solution with increasing l/d; if these terms were omitted entirely, errors of about 18, 2, and 0.3%, respectively, would be produced in the velocity and flowrate. Evidently, the effect of local losses in pipelines of common length is so small that they may often be neglected entirely, appreciably simplifying calculations. Another convenient approximation accompanies the above; increasing l/d also decreases $V^2/2g_n$ and thus brings the energy line and the hydraulic grade line closer together; since $V^2/2g_n$ is of the order of the local losses, it is consistent to neglect this also, thus making energy and hydraulic grade lines coincident (except near the entrance) and necessitating the construction of only one line. When the single line is drawn for this pipeline problem (Fig. 9.23), the equation

$$h_L = f\frac{l}{d}\frac{V^2}{2g_n} = H$$

may be written, and the velocity and flowrate may be obtained by trial-and-error procedure. The foregoing approximations are convenient in engineering problems but, of course, cannot be applied blindly and without some experience; preliminary calculations similar to those above will usually indicate the effect of such approximations on the accuracy of the result.

ILLUSTRATIVE PROBLEM 9.15

A clean cast iron pipeline 0.30 m in diameter and 300 m long connects two reservoirs having surface elevations 60 m and 75 m. Calculate the flowrate through this line, assuming water at 10°C and a square-edged entrance.

SOLUTION

Before using the work-energy equation to calculate the flowrate, we need to make some preliminary calculations. The Reynolds number depends on the unknown velocity but we will introduce all the known quantities into the expression for **R**. From Appendix 2, for water at 20°C, $\nu = 1.306 \times 10^{-6}\ \text{m}^2/\text{s}$.

$$\mathbf{R} = \frac{Vd}{\nu} = \frac{V \times 0.30\ \text{m}}{1.306 \times 10^{-6}\ \text{m}^2/\text{s}} = 229\,000\ V$$

From Fig. 9.11, $e/d = 0.000\,83$ for clean cast iron pipe. To arrive at a reasonable estimate of the Reynolds number so that a good first estimate of the f-value can be found, we assume $V \approx 2$ m/s. Inserting this number into the expression for **R**,

$$\mathbf{R} = 229\,000 \times 2 = 458\,000$$

From Fig. 9.10 with $e/d = 0.000\,83$ and $\mathbf{R} = 458\,000$, we find $f = 0.020$. Now, we use this value along with the local loss coefficients in the work-energy equation. Those local loss coefficients are:

$$K_L = 0.5 \text{ for a square-edged entrance}$$

$$K_L = 1.0 \text{ for exit into a reservoir}$$

We are now prepared to utilize the work-energy Eq. 7.35.

$$z_1 + \frac{p_1}{\gamma} + \frac{V_1^2}{2g_n} = z_2 + \frac{p_2}{\gamma} + \frac{V_2^2}{2g_n} + h_L \tag{7.35}$$

Note that the choice of points 1 and 2 at the surface of the two reservoirs means that $p_1 = p_2 = 0$ and $V_1 = V_2 = 0$.

$$75 + 0 + 0 = 60 + 0 + 0 + \left(0.5 + f\frac{l}{d} + 1.0\right)\frac{V^2}{2g_n}$$

$$15 = \left(0.5 + \frac{0.020 \times 300\ \text{m}}{0.30\ \text{m}} + 1.0\right)\frac{V^2}{2 \times 9.81} = 1.096\ V^2$$

$$V = 3.70\ \text{m/s}$$

This is considerably above our estimate so we must re-compute **R**, find a new f-value and find V again through the work-energy equation. With $V = 3.70$ m/s, $\mathbf{R} = 847\,250$. From Fig. 9.10, $f = 0.019\,3$. Substituting this value into the work-energy equation yields

$$V = 3.76\ \text{m/s}$$

This is clearly close enough to the first calculated value that we can be confident another iteration is unnecessary. The flowrate is

$$Q = VA = 3.76\ \text{m/s} \times (\pi/4)(0.30\ \text{m})^2 = 0.266\text{m}^3/\text{s} \ \bullet$$

ILLUSTRATIVE PROBLEM 9.16

A smooth PVC pipeline 200 ft long is to carry a flowrate of 0.1 ft^3/s between two water tanks whose difference in surface elevation is 5 ft. If a square-edged entrance and water at 50°F are assumed, what diameter of pipe is required.

SOLUTION

This type of problem represents the most difficult to solve because the unknown diameter renders both the Reynolds number and the relative roughness as unknown. The only simplification in the case of smooth pipes is that the relative roughness is irrelevant, i.e., we are on the smooth pipe line on the Moody diagram.

We again turn to Appendix 2 to obtain the ν-value of 1.41×10^{-5} ft^2/s for water at 50°F. We can also use the technique of representing the Reynolds number in terms of the flowrate as

$$\mathbf{R} = \frac{Vd}{\nu} = \frac{Qd}{A\nu} = \frac{Q}{(\pi/4)\, d \times \nu} = \frac{0.1\ \text{ft}^3/\text{s}}{(\pi/4)\, d \times 1.41 \times 10^{-5}\ \text{ft}^2/\text{s}} = \frac{9\,020}{d}$$

Following the previous illustrative problem and applying the work-energy equation, we get

$$5 = \left(0.5 + f\,\frac{200}{d} + 1.0\right)\frac{V^2}{2g_n}$$

We are now involved in a trial-and-error solution for V. The best procedure is to set up a table for the solution process.

Assume d	**R**	f	V	$V^2/2g_n$	(.....)	**Right-Hand Side**
0.25	36 000	0.022	2.04	0.064 4	19.1	1.23
0.20	45 100	0.021 2	3.18	0.157	22.7	3.56
0.18	50 100	0.020 8	3.93	0.240	24.6	5.90
0.187	48 200	0.021 0	3.64	0.206	24.0	4.94

The value of $d = 0.187$ ft produces a reasonable match to the left-hand side of the equation ($\approx$1% error). Hence, we will have a diameter of $d = 2.24$ inches, which means we will specify the next larger nominal diameter $d = 2.5$ inches. •

Another interesting application of the work-energy principle occurs when a pipeline extending from a reservoir terminates in a nozzle. This situation for cases where the main line velocity head is both significant and negligible is shown in Fig. 9.24. However, in either case, the velocity head at the nozzle exit cannot be neglected. In the most general case where we will consider local losses, the work-energy equation is

$$z_o + \frac{p_o}{\gamma} + \frac{V_o^2}{2g_n} = z_2 + \frac{p_2}{\gamma} + \frac{V_2^2}{2g_n} + h_L$$

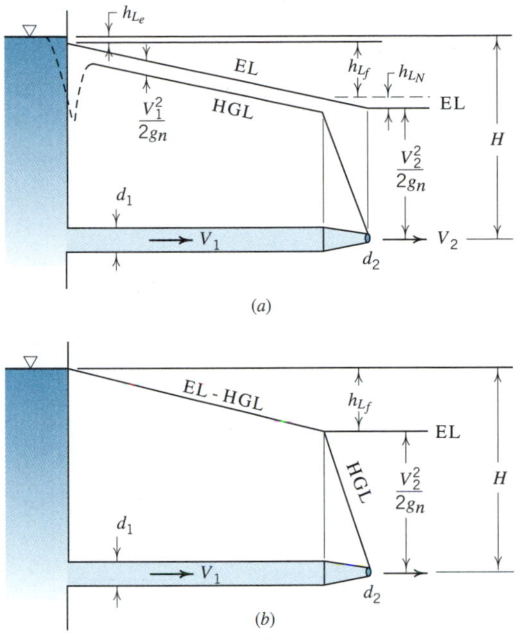

Fig. 9.24

where z_o, p_o, and V_o are values in the reservoir and

$$h_L = \left(h_{L_e} + f\frac{l}{d_1}\frac{V_1^2}{2g_n}\right)$$

Now, from Fig. 9.24,

$$H + 0 + 0 = 0 + 0 + \frac{V_2^2}{2g_n} + \left(K_{L_e}\frac{V_1^2}{2g_n} + f\frac{l}{d_1}\frac{V_1^2}{2g_n}\right)$$

$$H = \frac{V_2^2}{2g_n} + \left(K_{L_e} + f\frac{l}{d_1}\right)\frac{V_1^2}{2g_n}$$

Recognizing from continuity that $A_1V_1 = A_2V_2$ or $V_2 = (d_1/d_2)^2V_1$,

$$H = \left(\frac{d_1^2}{d_2^2} + K_{L_e} + f\frac{l}{d_1}\right)\frac{V_1^2}{2g_n}$$

The above equation can be solved directly if flowrate is known or by trial, if H is known.

The above analysis applies to a number of applications such as sprinkler design, fire suppression hose nozzles, etc. One application in the power industry is to the impulse turbine where the power in the jet is converted to electrical energy. Specifically, we would be concerned with maximizing the power in the jet for a given H. The expression for jet power, in U.S. Customary units, is

$$\text{Power} = \left(Q\gamma\frac{\text{lb}}{\text{s}}\right)\left(\frac{V^2}{2g_n}\frac{\text{ft}\cdot\text{lb}}{\text{lb}}\right) = Q\gamma\frac{V^2}{2g_n}\frac{\text{ft}\cdot\text{lb}}{\text{s}}$$

The first equation for H can be written, neglecting local losses, as

$$\frac{V_2^2}{2g_n} = H - f\frac{l}{d_1}\frac{V_1^2}{2g_n} = H - f\frac{l}{d_1}\frac{Q^2}{2g_n A_1^2} = H - \frac{flQ^2}{2g_n\, d_1 A_1^2}$$

Substituting this equation into the power equation gives

$$\text{Power} = Q\gamma\left(H - \frac{flQ^2}{2g_n\, d_1 A_1^2}\right)$$

Taking $dP/dQ = 0$ for maximization of the jet power,

$$\frac{flQ^2}{2g_n\, d_1 A_1^2} = \frac{H}{3}$$

which shows that the maximum jet power may be expected when[24]

$$\frac{flV_1^2}{2g_n\, d_1} = \frac{H}{3} \quad \text{and} \quad \frac{V_2^2}{2g_n} = \frac{2H}{3}$$

from which pipe and nozzle may be sized for any available flowrate.

When a pipeline runs above its hydraulic grade line, negative pressure in the pipe is indicated (Section 5.4). Sketching pipe and hydraulic grade line to scale indicates regions of negative pressures and critical points which may place limitations on the flowrate. Theoretically the absolute pressure in a pipeline may fall to the vapor pressure of the liquid, at which point cavitation (Appendix 4) sets in; however, this extreme condition is to be avoided in pipelines, and much trouble can be expected before such low pressures are attained. Most engineering liquids contain dissolved gases which will come out of solution well before the cavitation point is reached; since such gases go back into solution very slowly, they move with the liquid as large bubbles, collect in the high points of the line, reduce the flow cross section, and tend to disrupt the flow. In practice, large negative pressures in pipes should be avoided if possible by improvements of design; where such negative pressures cannot be avoided they should be prevented from exceeding about two thirds of the difference between barometric and vapor pressures.

ILLUSTRATIVE PROBLEM 9.17

A pipeline is being designed to convey water between two reservoirs whose elevations are shown below. The pipeline is 20 km long and the preliminary pipeline profile has the line passing over a ridge where the pipeline elevation is 1 313 m at a distance of 4 km from the upstream reservoir.

There is concern that the ridge is too high and will create an unacceptably low pressure in the pipeline. What is your recommendation as to the feasibility of the proposed location of the pipeline?

[24] In hydropower practice, $V_2^2/2g_n$ will be found to be considerably larger than $\frac{2}{3}H$, depending on the economic value of the water. The derived result could be used for design only in a situation where the liquid was of no value.

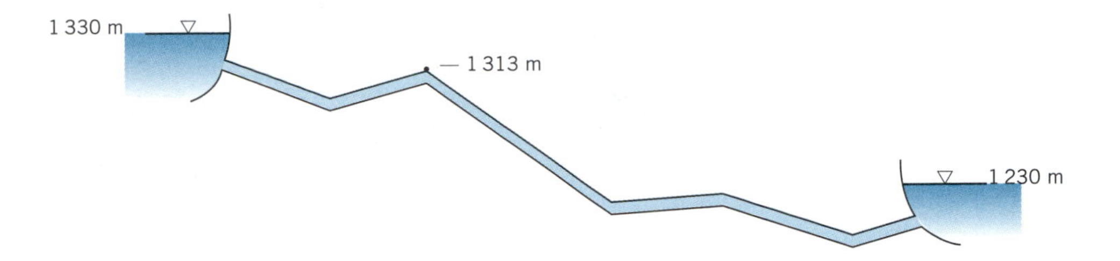

SOLUTION

In a pipeline this long we will neglect local losses and consider the energy line and the hydraulic grade line to be coincident. Therefore, the EL-HGL will fall uniformly 100 m over the 20 km line. Since the ridge is 4 km or 1/5 of the length of the pipeline downstream from the reservoir, the EL-HGL elevation at the ridge will be

$$\text{Elev EL-HGL}_{\text{ridge}} = 1\,330 \text{ m} - 1/5 \times 100 \text{ m} = 1\,310 \text{ m}$$

Since the elevation of the ridge is 1 313 m, this means the pressure head at the ridge is approximately -3 m.

While the pressure is negative, it is only about 1/3 of the negative pressure head which would be required to reach the vapor pressure head of water (approximately -10 m of water) and does not violate the 2/3 rule suggested in the previous reading material. So, in that sense, the design is acceptable. However, there are some practical considerations which must be addressed. Entrained air and dissolved gases will come out of solution at negative pressures and accumulate at the summit of the pipeline. These bubbles will constrict the flow and cause local losses, which will reduce the capacity of the pipeline. They cannot be removed by traditional means such as air release valves because these devices require positive pressure to force out the gases. With these practical considerations in mind, it seems prudent to recommend that an alternative route be sought for the pipeline.

Although to date we have been working with gravity-flow pipelines, a more common occurrence is the pumped pipeline. There are two general types of pumped pipelines. The first is where the pumps (or pump) are located at the upstream end of the pipeline. These are known as source pumps and they draw liquid from wells, reservoirs, tanks, wet and dry wells in plants, etc. The second type is where the pumps (or pump) are located at some intermediate point in the pipeline. These are known as booster pumps.

In either pumping situation, the engineer is interested in determining the power required to meet flowrate and pressure demands specified for the system. This usually requires the use of the work-energy and the fluid power equation. From Section 7.11 (Eq. 7.46),

$$z_1 + \frac{p_1}{\gamma} + \frac{V_1^2}{2g_n} + E_P = z_2 + \frac{p_2}{\gamma} + \frac{V_2^2}{2g_n} + h_L \tag{9.48}$$

where E_P is the work per unit weight added to the fluid by the pump. And for power,

$$WHP = \frac{Q\gamma E_P}{550} \quad \text{(U.S. Customary)} \tag{9.49a}$$

$$WKW = \frac{Q\gamma E_P}{1\,000} \quad \text{(SI)} \tag{9.49b}$$

where *WHP* and *WKW* represent power added to the fluid by the pump. To determine power requirements for the engine or motor driving the pump, the efficiency of the pump must be known. Illustrative Problem 9.18 addresses the typical booster pump situation where one must be concerned about maintaining a minimum pressure on the suction side of the pump.

In the situation where a specific pump is being employed in a pipeline system, a different approach is generally used. This approach is dictated by two factors. First, we must recognize that a pump is a reactive element in a pipe system; that is, the flowrate through the pump and the head increase across the pump depend on the pipeline system in which the pump is installed. The pump will respond to the demand placed on it by the system. Second, the demand the system makes on the pump depends on the friction losses in the system as functions of flowrate and the vertical lift (often called static lift) required between the two ends of the pipeline. This situation is best understood by examining the graphical representation of both the system demand and the pump performance.

Figure 9.25 shows a commercial turbine pump characteristic diagram, which graphically depicts the amount of head increase (total head) the pump will supply for a given flowrate (U.S. gallons per minute) for the four different impeller sizes (curves A, B, C, and D). This amount of head may be reduced by some local losses in the pump discharge

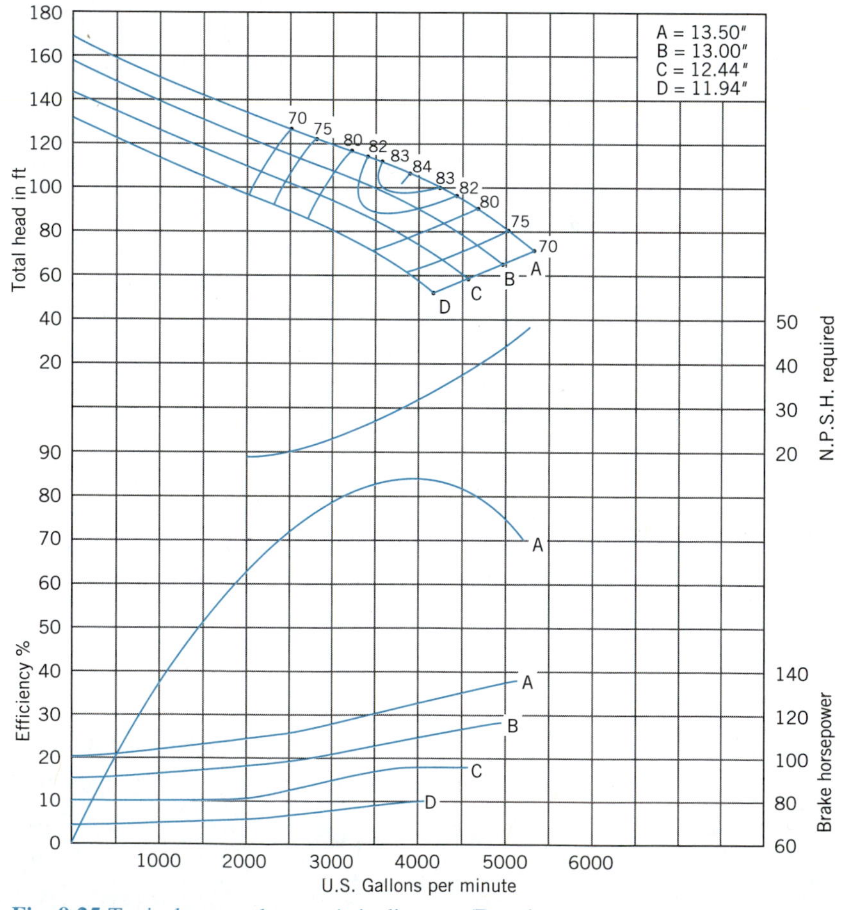

Fig. 9.25 Typical pump characteristic diagram (Data by courtesy of ITT A-C Pump).

column or other fixtures but for our purposes, we will assume the Total Head is the E_P-value found in Eq. 9.48. The exception would be in the case of multistage pumps where the flow is directed through two or more pump impellers within the unit. In this case, the value of head from the pump characteristic diagram would have to be multiplied by the number of stages before it becomes E_P. Also depicted are the power requirements necessary to drive the various-sized impellers (curves A, B, C, and D) along with efficiency contours. For example, we can see from Fig. 9.25 that the 13.50 inch impeller (curve A) is the most efficient impeller at 84%. It pumps 3 800 gal/m at this efficiency and requires 124 hp to accomplish this.

Figure 9.26 is a plot of the head required by the system for a wide range of flowrate and different vertical lifts, commonly occurring as the result of reservoir or well-level fluctuations. The head versus discharge data from the pump characteristic diagram is also plotted on Fig. 9.26 for one, two, and three pumps in parallel because flow demands may vary resulting in the need for flexibility in efficiently meeting demand. The intersection of the system demand and pump supply lines on the graph represents a solution to the mutual needs of the two systems and provides a visual interpretation of their requirements. For example, Fig. 9.26 shows that with the reservoir at its lowest level, one pump will produce 4 000 gal/min at best efficiency. If we operate two pumps, we will produce 6 900 gal/min and with three pumps, 8 500 gal/min. Note that if we were pumping 8 200 to 8 500 gal/min, we would need three pumps if the reservoir were at the lowest level but only two pumps if it were at the highest level. From this demonstration, it should be clear that graphical representations are very helpful in establishing operational rules for the pipeline system. Illustrative Problem 9.19 provides additional experience in the use of system demand and pump supply curves.

ILLUSTRATIVE PROBLEM 9.18

Calculate the horsepower that the pump must supply to the water (50°F) in order to pump 2.5 ft^3/s through a clean cast iron pipe from the lower reservoir to the upper reservoir. Neglect local losses and velocity heads.

Using the criteria for minimum allowable pressure suggested earlier in this section, compute the maximum dependable flow which can be pumped through this system.

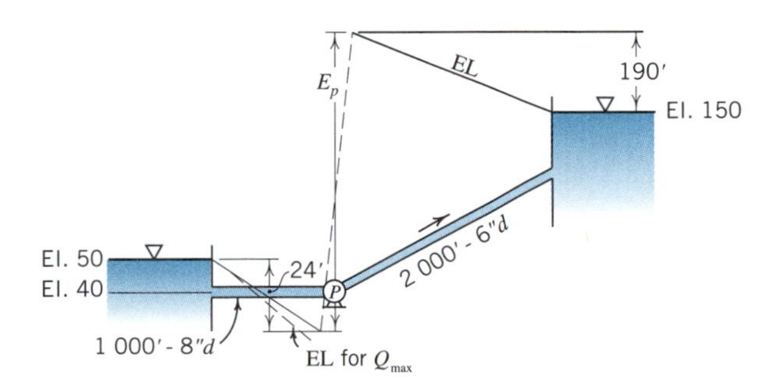

SOLUTION

The strategy for the solution is to find E_P from the work-energy Eq. 9.48 and use Eq. 9.49a to compute the power required. First, however, we will need to calculate the quantities

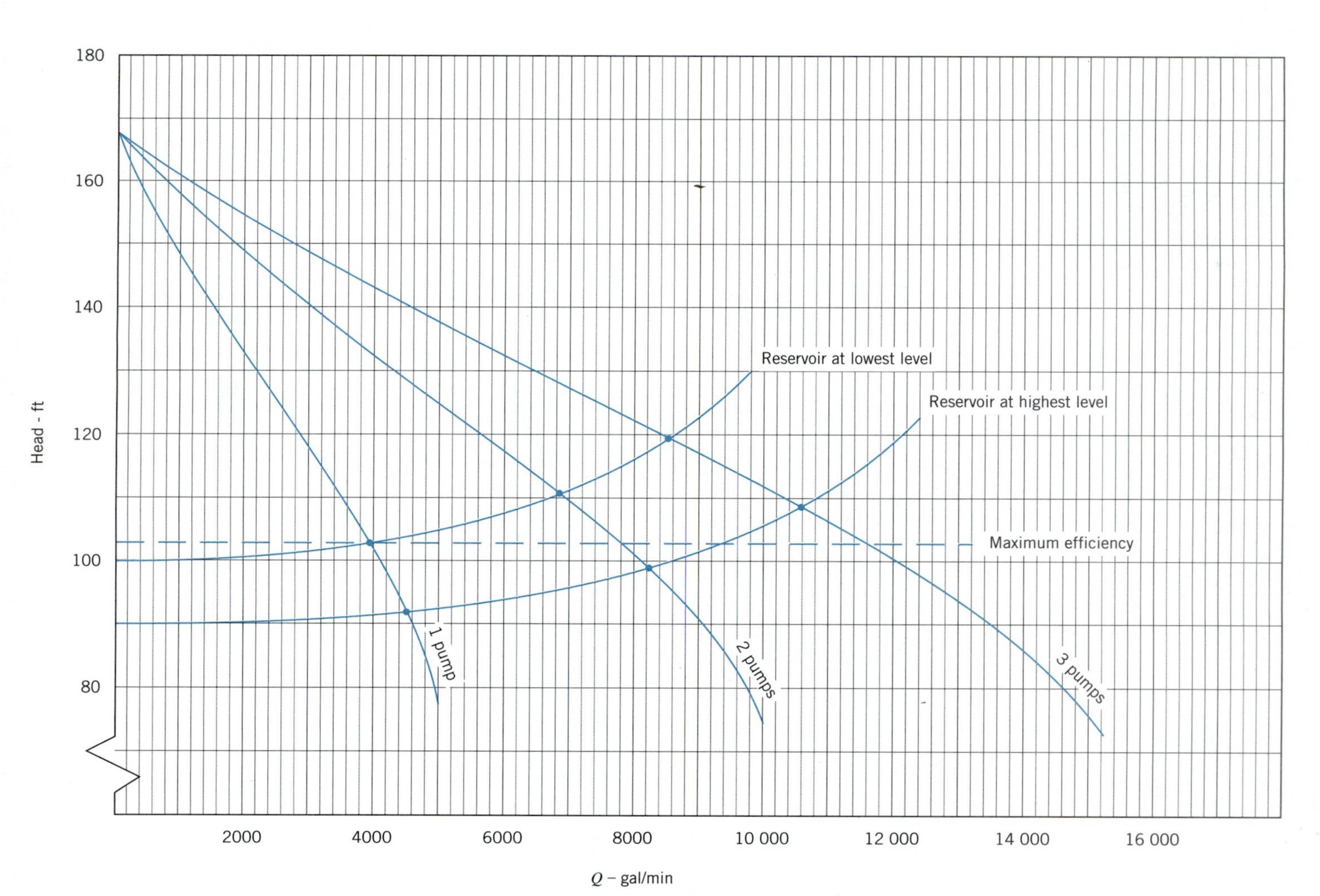

Fig. 9.26 System demand and pump supply curves.

that are required in the work-energy equation. The velocities of flow are:

$$V = \frac{Q}{A} \qquad V_8 = \frac{2.5\ \text{ft}^3/\text{s}}{(\pi/4)(8\ \text{in}/12)^2} = 7.16\ \text{ft/s} \qquad V_6 = \frac{2.5\ \text{ft}^3/\text{s}}{(\pi/4)(6\ \text{in}/12)^2} = 12.72\ \text{ft/s}$$

Now, to find the Darcy-Weisbach friction factor we need the Reynolds number and the relative roughness for both size pipes. Appendix 2 provides the ν-value.

8 inch pipe

$$\mathbf{R}_8 = \frac{Vd}{\nu} = \frac{7.16\ \text{ft/s} \times (8\ \text{in}/12)}{1.41 \times 10^{-5}\ \text{ft}^2/\text{s}} = 338\,500$$

$$\left(\frac{e}{d}\right)_8 = 0.001\,28\ \text{(from Fig. 9.11)}$$

$$f \cong 0.021\ \text{(from Fig. 9.10)}$$

6 inch pipe

$$\mathbf{R}_6 = \frac{Vd}{\nu} = \frac{12.72\ \text{ft/s} \times (6\ \text{in}/12)}{1.41 \times 10^{-5}\ \text{ft}^2/\text{s}} = 451\,000$$

$$\left(\frac{e}{d}\right)_6 = 0.001\,71\ \text{(from Fig. 9.11)}$$

$$f \cong 0.022\ \text{(from Fig. 9.10)}$$

The head loss in each of the pipes can now be calculated.

$$h_{L_8} = f\frac{L}{d}\frac{V^2}{2g_n} = 0.021\,\frac{1\,000\ \text{ft}}{8\ \text{in}/12}\,\frac{(7.16\ \text{ft/s})^2}{2 \times 32.2} = 25\ \text{ft}$$

$$h_{L_6} = f\frac{L}{d}\frac{V^2}{2g_n} = 0.022\,\frac{2\,000\ \text{ft}}{6\ \text{in.}/12}\,\frac{(12.72\ \text{ft/s})^2}{2 \times 32.2} = 221\ \text{ft}$$

We now insert all of these quantities into Eq. 9.48 using 1 as the upstream reservoir and 2 as the downstream reservoir.

$$z_1 + \frac{p_1}{\gamma} + \frac{V_1^2}{2g_n} + E_P = z_2 + \frac{p_2}{\gamma} + \frac{V_2^2}{2g_n} + h_L$$

$$50\ \text{ft} + 0 + 0 + E_P = 150\ \text{ft} + 0 + 0 + 25\ \text{ft} + 221\ \text{ft} \tag{9.48}$$

$$E_P = 346\ \text{ft-lb/lb}$$

The power delivered to the water can now be calculated from Eq. 9.49a.

$$WHP = \frac{Q\gamma E_P}{550} = \frac{2.5\ \text{ft}^3/\text{s} \times 62.4\ \text{lb/ft}^3 \times 346\ \text{ft-lb/lb}}{550} = 98\ \text{hp}\ \bullet \tag{9.49a}$$

From the previous discussion on allowable pressures in pipelines, we recall that the pressure head should not fall below -20 ft of water. As can be seen from the diagram above, this would establish a pressure head of -20 ft at the suction side of the pump. Writing the work-energy equation between the upstream reservoir and the suction side of the pump gives, neglecting velocity head,

$$z_1 + \frac{p_1}{\gamma} + \frac{V_1^2}{2g_n} = z_s + \frac{p_s}{\gamma} + \frac{V_s^2}{2g_n} + h_{L_8}$$

$$50 \text{ ft} + 0 + 0 = 40 \text{ ft} + (-20 \text{ ft}) + 0 + h_{L_8}$$

$$h_{L_8} = 30 \text{ ft}$$

Now, we go to the Darcy-Weisbach equation, assuming that the f-value remains the same, and calculate V_8.

$$h_{L_8} = 30 \text{ ft} = 0.021 \frac{1\,000 \text{ ft}}{8 \text{ in}/12} \frac{V_8^2}{2 \times 32.2}$$

$$V_8 = 7.8 \text{ ft/s}$$

The maximum dependable flowrate that can be pumped through the system is

$$Q_{\text{max}} = V_8 A = 7.8 \text{ ft/s} \times (\pi/4)(8 \text{ in}/12)^2 = 2.7 \text{ ft}^3\text{/s} \bullet$$

A quick check on the change in Reynolds number shows that it has not increased enough to alter our original value; hence the value for Q_{max} above is correct.

It must be recognized that the "rule of thumb" used above should be used only if more precise information is not available. Because velocities within the pump may be considerably higher than in the pipeline and because there are flow curvatures in the pump which additionally decrease pressure, it is entirely possible that the pressure at the suction side of the pump may have to be considerably higher than the "rule of thumb" suggests. Pump vendor catalogs and pump manufacturers can be of assistance in determining these requirements.

ILLUSTRATIVE PROBLEM 9.19

The pump whose characteristics are given in Fig. 9.25 is proposed for use in the pipe system shown below. In order to provide the head necessary to deliver the required flowrate to the upper reservoir, a four-stage pump is planned. Using the graphical technique of Fig. 9.26, determine the flowrate produced by the proposed pumping configuration and estimate the efficiency of the pump. Neglect local losses and use a Darcy-Weisbach f-value of 0.018.

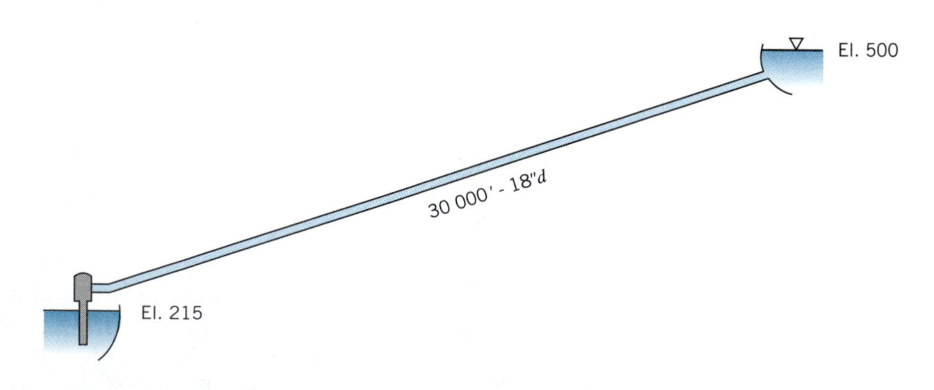

SOLUTION

System Demand Curve

To construct the system demand curve, we use the work-energy Eq. 9.48 with the points 1 and 2 in the lower and upper reservoir, respectively.

$$z_1 + \frac{p_1}{\gamma} + \frac{V_1^2}{2g_n} + E_P = z_2 + \frac{p_2}{\gamma} + \frac{V_2^2}{2g_n} + f\frac{L}{d}\frac{V^2}{2g_n} \qquad (9.48)$$

$$215 \text{ ft} + 0 + 0 + E_P = 500 \text{ ft} + 0 + 0 + .018\frac{30\,000 \text{ ft}}{18 \text{ in}/12}\frac{V^2}{2 \times 32.2}$$

$$E_P = 285 + 5.59\,V^2$$

Q – ft³/s	V – ft/s	5.59 V^2	285 + 5.59 V^2
0	0	0	285
2	1.13	7.1	292
4	2.26	28.6	314
6	3.40	64.6	350
8	4.53	114.7	400
10	5.66	179.1	464

Pump Supply Curve

From Fig. 9.25,

Q – gal/min	Q[a] – ft³/s	Head/stage – ft	Head/Four stages – ft
0	0	168	672
1 000	2.23	150	600
2 000	4.45	133	532
3 000	6.68	119	476
4 000	8.91	103	412
5 000	11.14	78	312

[a]Flowrate in ft³/s is found by Q (gal/min) ÷ 449.

Now, we can plot the two curves and find the flowrate from their intersection. The proposed configuration will produce 8.7 ft³/s (3900 gal/min) at nearly the maximum efficiency of 84%. From these results, we conclude that the proper pumping configuration has been selected.

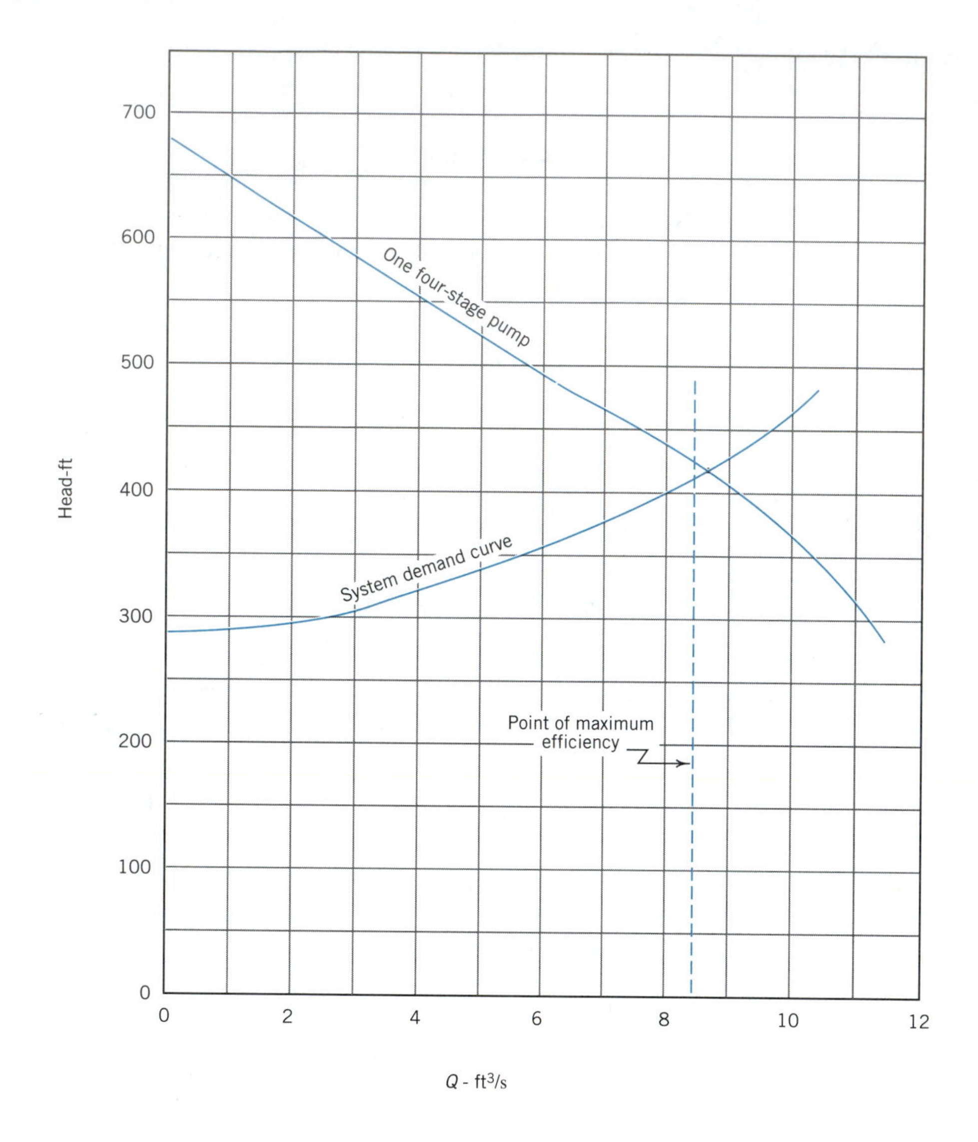

9.11 PIPELINE PROBLEMS—MULTIPLE PIPES

Some of the more challenging problems of pipeline analysis and design involve the flow of fluids in multiple-pipe systems. The system may be as small as two pipes separating and rejoining (Fig. 9.27) or as complex as several hundred pipes interconnected in a massive network. In either case, the basic principles of analysis are the same although the techniques of analysis vary depending on the system complexity. In virtually all cases, velocity heads and local losses are neglected with the EL and HGL considered coincident. As a consequence, the EL-HGLs of the pipes in the system form a continuous network above the pipes, joining at the pipe junctions. Variations of the friction factor f with Reynolds number are often neglected unless handled transparently by a computer program. In our treatment, we will begin with the simplest application in order to develop an understanding of the analysis concepts. Our first case will be the one-loop network.

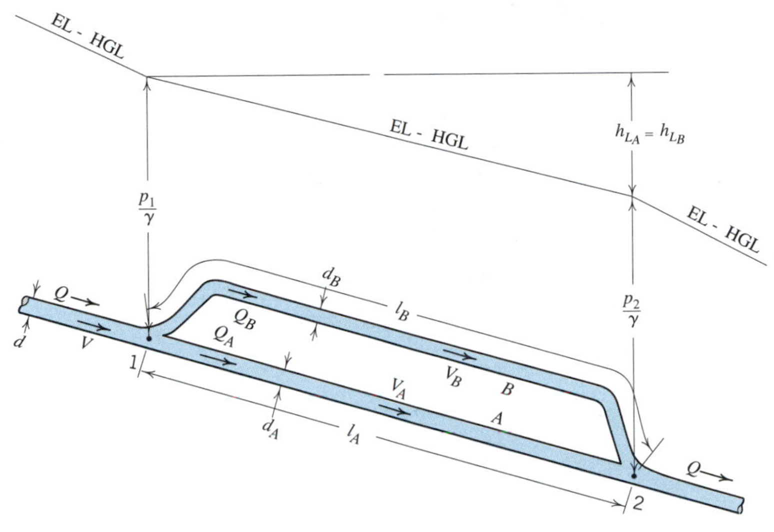

Fig. 9.27

In pipeline practice, laying a pipeline parallel to an existing pipeline for a fraction of its length is a standard method of increasing capacity without having to replace an entire line or construct a parallel line of equal length (see Fig. 9.27). Analysis takes into account the fact that the *head loss through both the branches of the loop must be the same* if the EL-HGL network above the pipes is to be continuous throughout the system.

Application of the continuity principle shows that the flowrate in the main line is equal to the sum of the flowrates in the branches. Thus the following simultaneous equations may be written:

$$h_{L_A} = h_{L_B}$$
$$Q = Q_A + Q_B$$

Head losses are expressed in terms of flowrate through the Darcy-Weisbach equation (9.2),

$$h_L = f\frac{l}{d}\frac{V^2}{2g_n} = \frac{fl}{2g_n\,d}\frac{16Q^2}{\pi^2\,d^4} = \left(\frac{16\,fl}{2\pi^2 g_n\,d^5}\right)Q^2 \tag{9.50}$$

This equation may be generalized by writing it as

$$h_L = KQ^n \tag{9.51}$$

Substituting this in the first of the simultaneous equations above,

$$K_A Q_A^n = K_B Q_B^n$$
$$Q = Q_A + Q_B$$

Solution of these simultaneous equations allows prediction of the division of a flowrate Q into flowrates Q_A and Q_B when the pipe characteristics are known. Application of these principles also allows prediction of the increased flowrate obtainable by looping an existing pipeline.

ILLUSTRATIVE PROBLEM 9.20

A 300 mm pipeline 1 500 m long is laid between two reservoirs having a difference in surface elevation of 24 m. The maximum flowrate obtainable through this line (with all valves wide open) is 0.15 m^3/s. When this pipe is looped with a 600 m pipe of the same size and material laid parallel and connected to it, what percent increase in maximum flowrate may be expected?

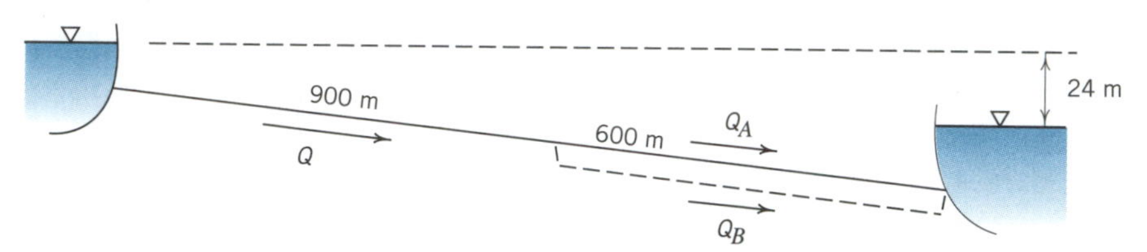

SOLUTION

Using the standard procedure of neglecting local losses and velocity heads, we can compute the K-factor in the equation $h_L = KQ^2$ (Eq. 9.51).

$$K_{1\,500} = \frac{h_L}{Q^2} = \frac{24 \text{ m}}{(0.15 \text{ m}^3/\text{s})^2} = 1\,067$$

for the original 1 500 m line. Recognizing from Eq. 9.50 that K is a linear function of length, we conclude that

$$K_{600} = K_{1\,500}\frac{600}{1\,500} = 427$$

for the looped section and

$$K_{900} = K_{1\,500}\frac{900}{1\,500} = 640$$

for the unlooped portion of the original line.

For the original 300-mm pipeline, the head loss in the looped plus unlooped portion gives

$$h_L = 24 \text{ m} = K_{900}Q^2 + K_{600}Q_A^2 = 640\,Q^2 + 427\,Q_A^2$$

while the head loss in the new pipe in the looped portion plus the loss in the 900 ft of the old pipe is

$$h_L = 24 \text{ m} = K_{900}Q^2 + K_{600}Q_B^2 = 640\,Q^2 + 427\,Q_B^2$$

in which Q_A and Q_B are the flowrates in the parallel branches. Solving these equations by eliminating Q shows that $Q_A = Q_B$ (which is to be expected from the symmetry in this problem).

Since from continuity, $Q = Q_A + Q_B$, then $Q_A = Q/2$. Substituting this into the first equation yields

$$Q = 0.18 \text{ m}^3/\text{s}$$

Thus, the gain in capacity is 0.03 m^3/s or $100\,\frac{0.03 \text{ m}^3/\text{s}}{0.15 \text{ m}^3/\text{s}} = 20\%$ •

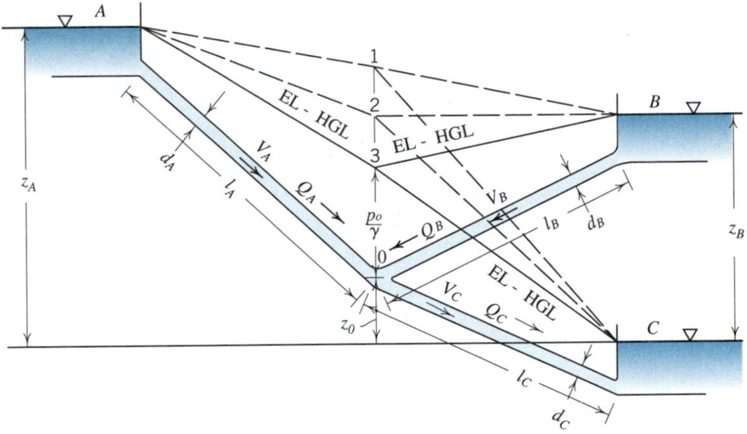

Fig. 9.28

Another engineering example of a multiple pipe system is the classic *three-reservoir problem* of Fig. 9.28 in which pipes lead from (three or more) reservoirs to a common point; this problem may be solved advantageously by the use of the energy line. Here flow may take place (1) from reservoir A into reservoirs B and C, or (2) from reservoir A to C without inflow or outflow from reservoir B, or (3) from reservoirs A and B into reservoir C. For situation (1), writing head losses in the form of KQ^n,

$$z_A - K_A Q_A^n = K_C Q_C^n$$
$$z_A - K_A Q_A^n = z_B + K_B Q_B^n$$
$$Q_A = Q_B + Q_C$$

For situation (3),

$$z_A - K_A Q_A^n = K_C Q_C^n$$
$$z_A - K_A Q_A^n = z_B - K_B Q_B^n$$
$$Q_A = Q_C - Q_B$$

For situation (2), $Q_B = 0$ and the sets of equations above become identical. In view of the physical flow picture only one of these sets of equations can be satisfied; this set may be discovered by a preliminary calculation using situation (2) in which Q_A and Q_C may be computed from the first two equations. If $Q_A > Q_C$, it can be seen from the continuity equation that the first set of equations should be used; if $Q_A < Q_C$, the second set will yield a solution. Having identified the set of equations valid for the problem, these may then be solved (by trial) to yield the flowrates Q_A, Q_B, and Q_C from which the pressures at all points in the lines may be predicted.

Multiple pipe systems reach their greatest complexity in the problems of distribution of flow in pipe networks such as those of a city water distribution system. Space does not permit comprehensive treatment of this professional engineering problem here, but one method of analysis is presented to illustrate the basic principles (the reader should consult the references listed at the end of the chapter for a more comprehensive treatment of this subject).

A pipe network of a water system is the aggregation of connected pipes used to distribute water to users in a specified area, such as a city or subdivision. The network consists of pipes of various sizes, geometric orientations, and hydraulic characteristics plus pumps, valves and fittings, and so forth. Figure 9.29 shows a simple network in plan view. The pipe junctions are indicated by the capital letters A–H, the individual pipes by the numbers 1–10, and loops (closed circuits of pipes) by the Roman numerals I–III. *Flows are assumed positive in a clockwise direction around each loop.* Pipes 1, 3, 4, and 2 comprise loop I. Pipes 4, 8, 10, and 7 comprise loop II. Pipe 4 is common to both loops.

The solution of any network problem must satisfy the continuity and the work-energy principles throughout the network. The continuity principle states that the net flowrate into any pipe junction must be zero. The work-energy principle requires that at any junction there be only one position of the EL-HGL, that is, that the net head loss around any single loop (see Fig. 9.29) of the network must be zero. To determine the net head loss, one moves around the loop in a clockwise direction adding head losses in the pipes algebraically with minus signs assigned to those losses in pipes where the flow opposes the clockwise motion, e.g., for loop II, the head loss in pipe 4 would be negative. Applying these principles to each junction and loop of the network in Fig. 9.29 yields a set of simultaneous equations. The equations for loop I are

$$\sum_A Q = -Q_A + Q_2 - Q_1 = 0$$

$$\sum_F Q = Q_1 + Q_F - Q_3 = 0$$

$$\sum_E Q = Q_3 - Q_4 - Q_8 = 0$$

$$\sum_B Q = -Q_2 + Q_4 + Q_7 + Q_5 = 0$$

$$\sum_{\text{I}} h_L = K_1 Q_1^n + K_3 Q_3^n + K_4 Q_4^n + K_2 Q_2^n = 0$$

There are similar equations for the other loops. In constructing the equations, flow directions have been *assumed* (these may not turn out to be the *actual* flow directions). It is also assumed that the pipe size, length, and hydraulic characteristics are known as well as the network inflows and outflows (Q_A, etc.), pump locations and pump characteristics, and network layout and elevations (needed if pressures in the network are to be calculated). The ten unknown flowrates, Q_i, for $i = 1, 10$ are to be found.

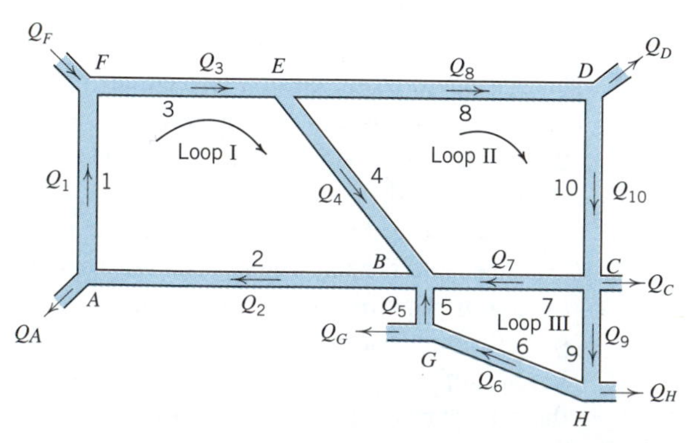

Fig. 9.29

The solution for the ten unknown flowrates is obtained by a trial-and-correction or iterative process. There are several ways to solve for these flowrates, the simplest and easiest to understand being the Hardy Cross method,[25] which is described here. It is important to point out that there are several desktop computer equation solvers which can be used to solve these equations, as well. More powerful methods, far beyond the scope of this work, are widely used in the solution of complex pipe network problems.

The essence of the method is to start with a best estimate of a set of initial values, Q_{0_i}, that satisfy continuity at each junction and then to systematically adjust these values, keeping continuity satisfied, until the head loss equations around each loop are satisfied to a desired level of accuracy. *All the equations of continuity at the pipe junctions are automatically and continuously satisfied by this approach.* Hence, only the head loss equations remain and *the number of simultaneous equations to be solved is reduced to the number of loops.*

If the first estimates are reasonably accurate, the true flowrates, Q_i, should only be a small increment, Δ_L, different from the original, Q_{0_i} in each loop; that is, for a pipe that is in only one loop, the first iteration gives

$$Q_i = Q_{0_i} \pm \Delta_L \tag{9.52}$$

where the sign ($\pm$) depends on the directions assumed for Q_{0_i}. To maintain continuity, the correction, Δ_L, for a loop is applied to every pipe in the loop with proper attention to the direction of flow. For example,

$$Q_3 = Q_{0_3} + \Delta_I,\ Q_8 = Q_{0_8} + \Delta_{II}, \qquad \text{but} \qquad Q_4 = Q_{0_4} + \Delta_I - \Delta_{II}$$

because pipe 4 belongs to two loops and the Δ_L correction is established with a clockwise positive sign convention around each loop. In general a head loss equation takes the form

$$\sum_L h_{L_i} = \sum_L \pm K_i Q_i^n = 0 \tag{9.53}$$

where the $\pm$ sign depends on the direction of flow (as explained earlier) and the Q_i are taken to be the *magnitudes* of the flowrate. For the assumed flow directions and loops in Fig. 9.29, the head loss Eqs. 9.53 are (the K_i are all positive)

$$K_1 Q_1^n + K_3 Q_3^n + K_4 Q_4^n + K_2 Q_2^n = 0$$

$$-K_4 Q_4^n + K_8 Q_8^n + K_{10} Q_{10}^n + K_7 Q_7^n = 0$$

$$K_5 Q_5^n - K_7 Q_7^n + K_9 Q_9^n + K_6 Q_6^n = 0$$

Introducing Eq. 9.52 into 9.53,

$$\sum_L h_{L_i} = \sum_L \pm K_i (Q_{0_i} \pm \Delta_L)^n = 0$$

Expanding this equation by the binomial theorem and neglecting all terms containing Δ_L raised to a higher power, e.g., Δ_L^2, Δ_L^3, and so forth, because Δ_L is presumed to be small, we get for the first iteration,

$$\sum_L h_{L_i} = \sum_L \pm K_i (Q_{0_i}^n \pm n Q_{0_i}^{n-1}\, \Delta_L + \text{negligible}) = 0$$

[25] H. Cross, "Analysis of Flow in Networks of Conduits and Conductors," *Univ. Illinois Eng. Expt. Sta., Bull.* 286, 1936. See R. W. Jeppson, *Analysis of Flow in Pipe Networks*, Ann Arbor Science, 1976, for the Hardy Cross and other methods plus computer programs.

Solving for Δ_L, the first correction for loop L becomes

$$\Delta_L = -\frac{\sum_L \pm K_i Q_{0_i}^n}{\sum_L |nK_i Q_{0_i}^{n-1}|} \tag{9.54}$$

The absolute value must be used in the denominator to insure the proper sign for Δ_L. This equation is used to compute the flowrate correction Δ_L for each loop in the network.

Because some pipes share loops and because higher level terms in the binomial expansion were neglected, Eq. 9.54 does not produce a set of Δ_L's that precisely corrects all the Q_{0_i} to the final Q_i which satisfy the head loss equations. Hence, the iterative process must be continued with successive sets of Δ_L-values computed and flowrates corrected until the values of Q_i converge with suitable accuracy to final values. The iteration equation expressing this process is

$$\Delta_L^{(j+1)} = -\frac{\sum_L \pm K_i (Q_i^{(j)})^n}{\sum_L |nK_i (Q_i^{(j)})^{n-1}|}$$

where j indicates the result of the jth trial-and-correction for loop L. The end point of the iterative process is generally reached when all the Δ_L have dropped below a preestablished accuracy limit.

To add a pump to a pipe in the network, an expression representing the head increase versus the capacity curve is required. One method to accomplish this is to fit a polynomial curve to the pump characteristics to form an equation of the form

$$E_{P_i} = a_0 + a_1 Q_i + a_2 Q_i^2 + a_3 Q_i^3 + \cdots\cdots$$

with as many coefficients, a_i, as necessary to provide a good representation of the pump curve. For example, if a pump is added to the line 8 in loop II of Fig. 9.29, the head loss equation for loop II becomes

$$-K_4 Q_4^n + K_8 Q_8^n - (a_0 + a_1 Q_8 + a_2 Q_8^2 + \cdots\cdots) + K_{10} Q_{10}^n + K_7 Q_7^n = 0$$

and the analysis proceeds as before.

The analysis procedure as demonstrated in Illustrative Problem 9.21 requires a consistent way to determine whether the head loss is positive or negative and whether a Δ_L should be added to or subtracted from the flowrate in a given pipe. Drawing on the work of Jeppson, we accomplish this in the following way. We assign an algebraic sign to each flowrate in the network, giving a positive sign to those flows which move in a clockwise direction around the loop and a negative sign to those that flow counterclockwise. This determines the sign of the head loss consistent with our previously established sign convention. The Δ_L's can then be computed and added algebraically to the positive or negative flows in the pipes of the loop with confidence that continuity at each junction will be preserved and with no further effort needed to insure the adjustments are correct. This technique is straightforward for hand calculation and easy to program on a computer, an important consideration in network analysis. A basic Hardy Cross program named HARDY is provided in Appendix 7 to assist the student in solving simple network problems.

ILLUSTRATIVE PROBLEM 9.21

A parallel commercial steel pipe network was built in two parts. As shown below, section ACD is the original line; the parallel section ABC was then added; then section BD was added to complete the job. By accident, a valve is left open in the short pipe BC. What are the resulting flowrates in all the pipes, neglecting local losses and assuming the flows are wholly rough.

The pipe table below constructed using the Darcy-Weisbach equation with $K_i = 16 f_i l_i / 2\pi^2 g_n d_i^5$ gives all the pertinent pipe characteristics.

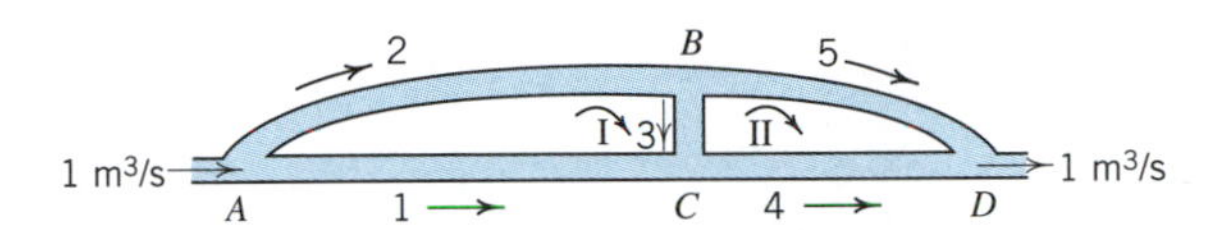

Pipe No.	Length (m)	Diameter (m)	e/d	f	K_i (Eq. 9.50)
1	1 000	0.5	9×10^{-5}	0.012	31.7
2	1 000	0.4	1×10^{-4}	0.012	96.8
3	100	0.4	1×10^{-4}	0.012	9.7
4	1 000	0.5	9×10^{-5}	0.012	31.7
5	1 000	0.3	1.4×10^{-4}	0.013	442.0

SOLUTION

We begin the analysis by writing out the equations for Δ for each loop.

$$\Delta_I = -\frac{-K_1 Q_{0_1}^2 + K_2 Q_{0_2}^2 + K_3 Q_{0_3}^2}{2(|K_1 Q_{0_1}| + |K_2 Q_{0_2}| + |K_3 Q_{0_3}|)} \quad \text{(Zero subscript used for first iteration)}$$

$$\Delta_{II} = -\frac{-K_3 Q_{0_3}^2 + K_5 Q_{0_5}^2 - K_4 Q_{0_4}^2}{2(|K_3 Q_{0_3}| + |K_5 Q_{0_5}| + |K_4 Q_{0_4}|)} \quad \text{(Zero subscript used for first iteration)}$$

The adjustment equations are

Initial Calculation	**Subsequent Calculations**
$Q_1 = Q_{0_1} + \Delta_I$	$Q_1^{(j+1)} = Q_1^{(j)} + \Delta_I^{(j)}$
$Q_2 = Q_{0_2} + \Delta_I$	$Q_2^{(j+1)} = Q_2^{(j)} + \Delta_I^{(j)}$
$Q_3 = Q_{0_3} + \Delta_I - \Delta_{II}$ (Loop I)	$Q_3^{(j+1)} = Q_3^{(j)} + \Delta_I^{(j)} - \Delta_{II}^{(j)}$ (Loop I)
$Q_3 = Q_{0_3} + \Delta_{II} - \Delta_I$ (Loop II)	$Q_3^{(j+1)} = Q_3^{(j)} + \Delta_{II}^{(j)} - \Delta_I^{(j)}$ (Loop II)
$Q_4 = Q_{0_4} + \Delta_{II}$	$Q_4^{(j+1)} = Q_4^{(j)} + \Delta_{II}^{(j)}$
$Q_5 = Q_{0_5} + \Delta_{II}$.	$Q_5^{(j+1)} = Q_5^{(j)} + \Delta_{II}^{(j)}$

where j is the iteration index.

The iteration process is carried out by setting up a table for systematically calculating the Δ-values and correcting the flowrates in the various pipes. The following table illustrates this for the first two iterations.

Loop I

			First Iteration			Second Iteration		
Pipe	K	Q_0	KQ_0^2	KQ_0	$Q^{(1)}$	KQ^2	KQ	$Q^{(2)}$
1	31.7	−0.5	− 7.93	15.85	−0.63	−12.58	19.97	−0.64
2	96.8	0.5	24.20	48.40	0.37	13.25	35.82	0.36
3	9.7	0.4	1.55	3.88	0.12	0.14	1.16	0.15
			17.82	68.13		0.81	56.95	
			$\Delta_I^{(1)} = -0.13$			$\Delta_I^{(2)} = -0.01$		

Loop II

			First Iteration			Second Iteration		
Pipe	K	Q_0	KQ_0^2	KQ_0	$Q^{(1)}$	KQ^2	KQ	$Q^{(2)}$
3	9.7	−0.4	− 1.55	3.88	−0.12	− 0.14	1.16	−0.15
4	31.7	−0.9	−25.68	28.53	−0.75	−17.83	23.78	−0.79
5	442.0	0.1	4.42	44.20	0.25	27.63	110.50	0.21
			−22.81	76.61		9.66	135.44	
			$\Delta_{II}^{(1)} = 0.15$			$\Delta_{II}^{(2)} = -0.04$		

If one additional iteration were performed, it would show that both Δ's were zero. This confirms that the iterative process has converged and the results given under column $Q^{(2)}$ are the final flowrates, accurate to 0.01 m^3/s. ●

Unsteady Flow

Unsteady flow in piping systems is a common occurrence. Indeed, steady flow is so rare that one might question the advisability of devoting so much time to a study of its behavior. However, in many cases, the unsteadiness occurring in a pipeline system is of little consequence because of its transient nature and its small magnitude of change, hence, virtually all hydraulic design is based on steady flow analysis. It is with those few cases wherein significant changes in velocity cause large changes in pressure that we are concerned.

Unsteady flow problems in engineering practice are of significant importance because they can cause excessive pressures, vibration, cavitation, and noise far beyond that indicated by steady flow analysis. In fact, the problems created by hydraulic transients may be so severe as to cause physical or performance failure of a system. Further, any one cause of a hydraulic transient may create different effects, benign in one situation and destructive in another, depending on the physical configuration of the system, the mechanical components in the system, the physical properties of the pipe and liquid, and the existence of free air in the system. Often, piping systems are far too complex and the flow situations therein too uncertain to permit accurate simulation by mathematical means. It requires considerable judgment and experience to simplify the problem to one that can be analyzed and still provide useful information regarding the performance of the original system.

9.12 UNSTEADY FLOW AND WATER HAMMER IN PIPELINES

The analysis of unsteady flow in pipeline systems can be divided into two broad categories. The first, called *surge* or *rigid water column* theory treats the fluid as an inelastic substance wherein pressure changes propagate instantaneously throughout the system and elastic properties of the pipe walls are of no consequence. The equations describing this type of flow are generally ordinary differential equations which can be solved in closed form or with relatively straightforward numerical techniques. Where applicable, this approach is the easiest to apply and should always be considered as a possibility to adequately approximate problems under consideration.

The second category of problems is classified under *elastic* or *water hammer* theory wherein the elasticity of both the fluid and the pipe walls is taken into account in the calculations. Pressure waves created by velocity changes depend on these elastic properties and they propagate throughout the pipeline system at speeds depending directly on these elastic properties. While the elastic theory more accurately reflects the behavior of the unsteady flow system, successful analysis hinges on the ability to solve two nonlinear partial differential equations. As a consequence, the analysis is more complex and difficult to manage than for inelastic theory. However, in the 1960s Streeter and Wylie[26] demonstrated that, with the assistance of a high-speed digital computer, the method of characteristics can be applied to solve the equations in a relatively general and easily understood manner. Their text represents a compilation of computer analysis techniques and was the most significant book in the area of water hammer analysis to be published in years.

Before computer analysis, the general equations describing water hammer in pipeline systems were simplified in some manner to permit solution by arithmetic, graphical, or algebraic means. Nonlinear terms were neglected, friction was included by lumping or approximating, or it was left out altogether. Matching boundary conditions at pumps and turbines was, at best, difficult and understood by relatively few engineers. Today, modern analysis techniques, including numerical methods for solving partial differential equations, have brought within reach of most engineers the capability of solving accurately a wide range of water hammer problems. Although digital computers are needed, the desktop PC, commonly available to all practicing engineers, is more than adequate to accomplish most of these analyses.

In approaching the solution of transient problems, one of the tasks is the determination of which method (rigid water column theory or elastic theory) to employ. To assist in this determination, we will examine the action of water hammer in a simple pipeline situation. The observations we make will help us determine when to apply elastic theory and will facilitate grasping the significance of the sequence of events occurring later in more complicated problems.

Because including friction in unsteady flow analysis is important and because it may be necessary to apply the analysis to a wide variety of problems including other liquids

[26]Streeter, V. L. and E. B. Wylie, *Hydraulic Transients*, McGraw-Hill, 1967.

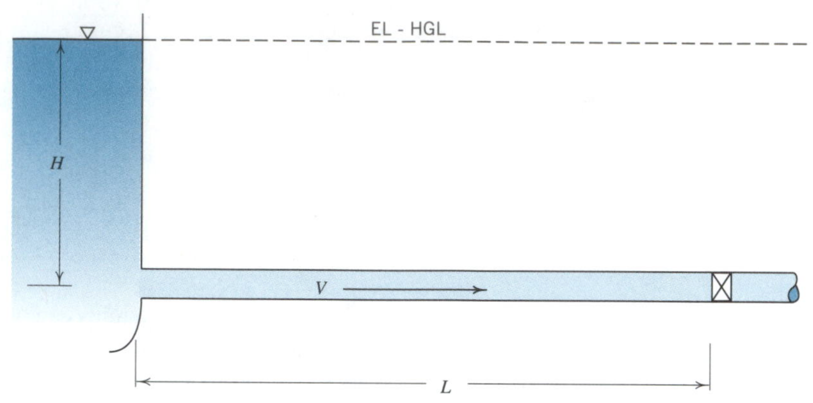

Fig. 9.30 Steady-state flow situation for simple water hammer (friction negligible).

as well as water, a pipe friction formula must be used which is sufficiently versatile to encompass these needs. Accordingly, the Darcy-Weisbach formula, Eq. 9.2 will be used in conjunction with the Moody diagram in Fig. 9.10.

To grasp a basic understanding of the action of a pipe system carrying liquid under the action of water hammer waves, it is easiest to consider as simple a system as possible. The system we will examine is shown in Fig. 9.30 as a horizontal, constant-diameter pipe leading from a reservoir to some unknown destination far downstream. A valve is placed a distance L from the reservoir. Friction in the line is assumed negligible to simplify the analysis; and because velocity heads are generally quite small in relation to water hammer pressures, the difference between the energy gradeline EL and the hydraulic gradeline HGL will be neglected.

Water hammer will be introduced into the system by suddenly closing the valve. The activity will occur both upstream and downstream of the valve but, for our purposes, we will observe only what occurs upstream of the valve.

Upon sudden closure of the valve, the velocity of water at the valve is forced suddenly to zero. As a consequence, the pressure head at the valve increases suddenly by an amount ΔH (see Fig. 9.31). The magnitude of ΔH is just the amount of pressure head necessary to change the momentum of the liquid initially flowing at velocity V at the valve to zero.

The increase in pressure at the valve results in a stretching of the pipe and an increase in the density of the liquid. The amount of pipe stretching and liquid density increase depends on the pipe material and size and the liquid elasticity. Generally, for common pipe materials and liquids, the percentage change is less than 0.5%. The deformation has been greatly exaggerated in Fig. 9.31 for purposes of illustration.

The pressure increase propagates upstream at a wave speed a, which is determined by the elastic properties of the system and the liquid and the system geometry. The wave speed will remain constant so long as they remain constant. Traveling at a speed a, the wave will reach the reservoir in a time L/a. At this time, the velocity in the pipe is everywhere zero, the pressure head is everywhere $H + \Delta H$, the pipe is stretched, and the fluid is compressed.

Under these conditions the liquid in the pipe is not in equilibrium because the pressure head in the reservoir is only H. As a result, flow begins to occur toward the reservoir as the distended pipe ejects liquid in that direction. The reverse velocity is equal in magnitude to the initial steady velocity (as a result of neglecting friction) and the source of liquid for

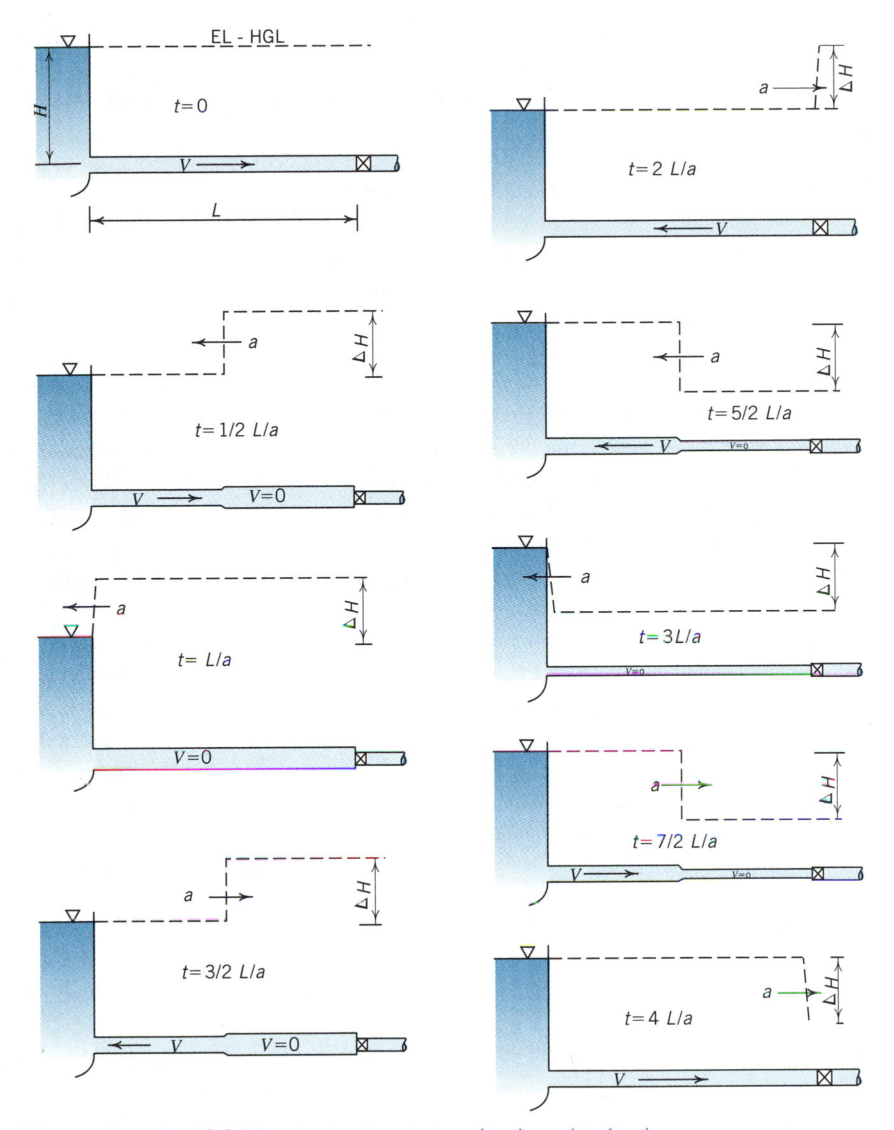

Fig. 9.31 Pressure wave propagation in a simple pipe system.

the reverse flow is the liquid previously stored in the stretched pipe walls as compressed liquid.

This process continues and at time 2 L/a, the pressure has returned to normal (but with reverse flow occurring) throughout the pipe. However, there is no source of liquid at the valve to supply the upstream flow hence the pressure head drops an additional ΔH to force the reverse velocity to zero. This drop in pressure causes the pipe to shrink and the liquid to expand.

At time 3 L/a this effect has propagated to the reservoir and the velocity of flow is everywhere zero. However, the pipe pressure head is ΔH below that of the reservoir. Consequently, the pipe sucks in liquid from the reservoir creating a velocity of flow equal to and in the same direction as the original steady flow. While this is occurring, the pressure in the pipe is also returning to its original value.

After time 4 L/a, this wave has reached the valve and at this instant the flow is identical to its original steady state configuration. This elapsed time constitutes one wave period. As time goes on, this cycle of events will continue without abatement (in the absence of friction).

Some fundamental concepts can be gained from examining more closely what occurs in this system. For example, it is clear that the time parameter that best describes the sequence of events in a meaningful fashion is not time alone but the ratio L/a. It is informative to plot the pressure head at various points in the pipeline as a function of time as shown in Fig. 9.32. Note the pressure head at the valve fluctuates between $H \pm \Delta H$, whereas the pressure head at other locations also experiences periods of time when its value is H.

One basic point can be made from Fig. 9.32*b*. Note that the pressure does not increase at a point until enough time has occurred for the wave to travel from the closed valve. Once the pressure head has increased, it remains at that level only long enough for ''relief'' to arrive back from the reservoir. This idea of ''time of communication'' or ''message propagation time'' is fundamental to a good understanding of the happenings in a system undergoing water hammer.

A second important point can be seen by examining Fig. 9.32*a* more closely. Suppose that instead of closing the valve suddenly, we were to close it in 10 steps, each increasing the pressure head at the valve by $\Delta H/10$. A further requirement would be that the complete closure of the valve would be accomplished before 2 L/a seconds had elapsed. It is clear that the pressure head at the valve would still build up to the full ΔH value because ''relief'' from the reservoir could not arrive before 2 L/a seconds. The point to be made is that a valve need not be closed suddenly to create the maximum water hammer pressure. Indeed, any closure time less than the time necessary for relief to return from a reservoir (a larger pipe may also act much like a reservoir) will result in full water hammer pressures. This time of 2 L/a is known as the *critical time of closure*. In fact, because of the manner in which a valve shuts off flow in a pipeline by creating large head losses, it may be necessary to close the valve in a time much greater than 2 L/a to prevent high pressures from occurring.

To summarize, if the action which causes unsteady flow, e.g., a closing valve, takes place over a period of many 2 L/a time intervals, then it would be more appropriate to use rigid water column theory. If, on the other hand, the action was completed in only a few L/a time intervals or less, then elastic theory should be utilized. Unfortunately, there are no firm rules which indicate the proper analysis technique in those cases where it is not reasonably clear which method to use. Experience in transient analysis is the key to making the proper decision in these cases. To help generate some experience in the two methods of analysis, we will develop the basic theory and examine some elementary applications.

9.13 RIGID WATER COLUMN THEORY

To analyze unsteady flow problems in pipe systems by rigid water column theory, we begin by developing an equation describing flow in a single pipe. The approach is to apply Newton's second law to a small cylindrical fluid system at the pipe centerline as was done in Chapter 5 on Fig. 5.1, but with the addition of a shear stress on the system surface.

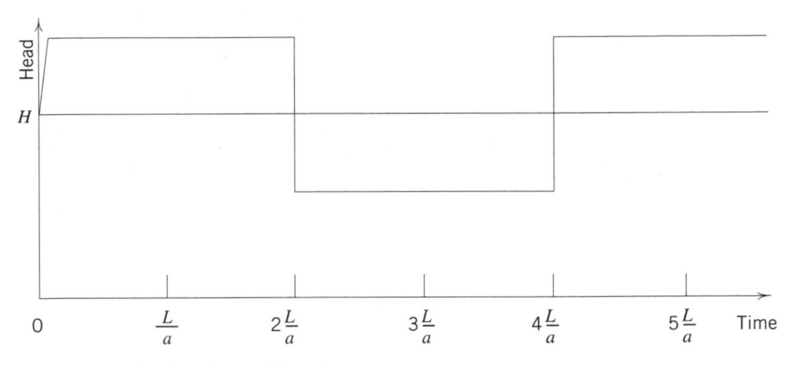

a) Pressure head vs. time at the valve.

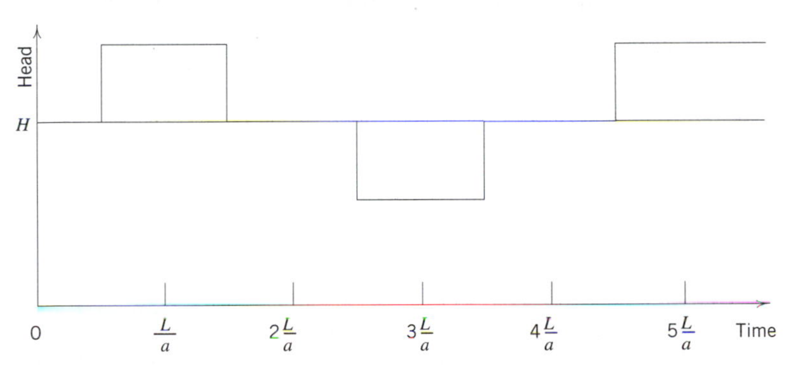

b) Pressure head vs. time at the midpoint.

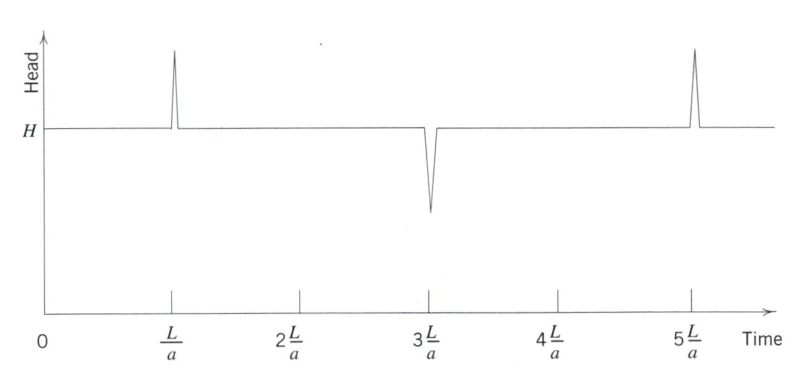

c) Pressure head vs. time at the reservoir.

Fig. 9.32 Pressure head versus time at three locations along the pipe.

Because of the differential size of the fluid system, the resulting differential equation is equally valid for compressible and incompressible flow and can be used in both rigid water column and elastic analyses. The Euler equation modified to include shear stress τ is

$$-\frac{1}{\gamma}\frac{\partial p}{\partial s} - \frac{\partial z}{\partial s} - \frac{4\tau}{\gamma d} = \frac{1}{g_n}\frac{dV}{dt} \tag{9.55}$$

When the system diameter is expanded to the size of the pipe diameter, Eq. 9.55 becomes

$$-\frac{1}{\gamma}\frac{\partial p}{\partial s} - \frac{\partial z}{\partial s} - \frac{4\tau_o}{\gamma d} = \frac{1}{g_n}\frac{dV}{dt}$$

where τ_o is the wall shear stress and V is the average velocity in the pipe.

Because the above form with the shear stress τ_o is not directly useful, we substitute the relation between τ_o and the Darcy-Weisbach friction factor f. The result of this substitution is

$$-\frac{1}{\gamma}\frac{\partial p}{\partial s} - \frac{\partial z}{\partial s} - \frac{f}{d}\frac{V^2}{2g_n} = \frac{1}{g_n}\frac{dV}{dt}$$

Recognizing that z is a function only of s and represents the elevation of the pipe centerline above some datum, we can change the partial derivative to a total derivative. This cannot be done for p and V which are functions of both s and t. Finally, the equation has the form

$$-\frac{1}{\gamma}\frac{\partial p}{\partial s} - \frac{dz}{ds} - \frac{f}{d}\frac{V^2}{2g_n} = \frac{1}{g_n}\frac{dV}{dt} \tag{9.56}$$

The unsteady flow equation can be used to solve a wide range of pipeline problems *which fall within the domain of rigid water column theory*. We will address some of the basic problems.

If the discharge in the pipeline shown in Fig. 9.33 is controlled by the valve at the downstream end, the pressure in the pipe is everywhere equal to H_o when the valve is closed. When the valve is suddenly opened, the pressure at the valve drops instantly to zero and the fluid begins to accelerate.

The equation describing this flow is obtained by integrating Eq. 9.56 with respect to s from point 1 to point 2.

$$-\int \frac{1}{\gamma}\frac{\partial p}{\partial s}\,ds - \int \frac{dz}{ds}\,ds - \int \frac{fV^2}{2g_n d}\,ds = \int \frac{1}{g_n}\frac{dV}{dt}\,ds$$

In a horizontal constant-diameter pipe, the integration is made quite easy because $(dz/ds) = 0$ and V is a function of time only. We also assume the f-value in unsteady

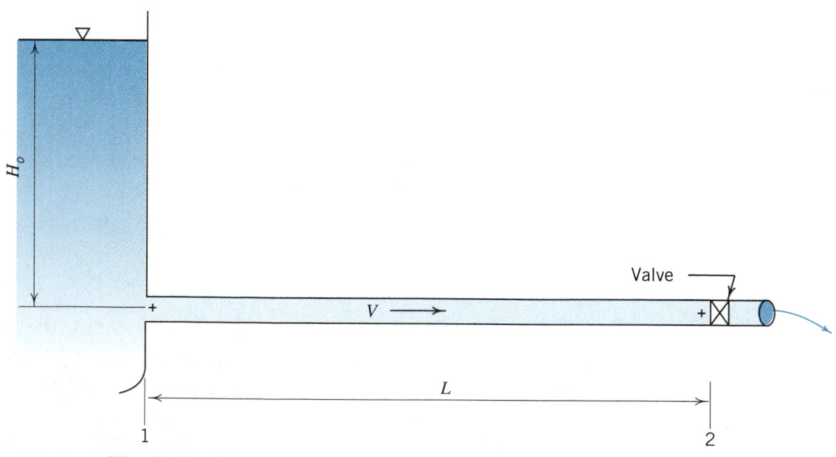

Fig. 9.33 Simple system for applying rigid water column theory.

flow is the same as for a steady flow at a velocity equal to the instantaneous value. The result is

$$\frac{p_1}{\gamma} - \frac{p_2}{\gamma} - \frac{fL}{2g_n\, d} V^2 = \frac{L}{g_n}\frac{dV}{dt} \tag{9.57}$$

Because the pressure head p_1/γ = constant = H_o and because $p_2/\gamma = 0$ for $t > 0$, the equation is reduced to

$$H_o - \frac{fL}{2g_n\, d} V^2 = \frac{L}{g_n}\frac{dV}{dt} \tag{9.58}$$

Integration is performed by separating the variables to form

$$\int dt = \frac{L}{g_n}\int \frac{dV}{H_o - (fL/2g_n\, d)V^2}$$

The integration gives the following equation for the time necessary to accelerate the flow to a given velocity V.

$$t = \sqrt{\frac{L\, d}{2g_n f H_o}}\ \ln\left[\frac{\sqrt{(2g_n H_o\, d/fL)} + V}{\sqrt{2g_n H_o\, d/fL)} - V}\right] \tag{9.59}$$

Recognizing that, neglecting local losses, $\sqrt{(2g_n H_o\, d/fL)}$ = steady-state velocity V_o, the equation for t becomes

$$t = \frac{LV_o}{2g_n H_o}\ln\left[\frac{V_o + V}{V_o - V}\right] \tag{9.60}$$

It is important to note that as steady flow is approached, $V \rightarrow V_o$ and as a consequence $t \rightarrow \infty$. Of course, this answer is unacceptable so we propose that when $V = 0.99\ V_o$, we have essentially steady flow. With this interpretation,

$$t_{99} = 2.65\frac{LV_o}{g_n H_o} \tag{9.61}$$

ILLUSTRATIVE PROBLEM 9.22

A horizontal 24 inch pipe 10 000 ft long leaves a reservoir 100 ft below the surface and terminates in a valve. The steady-state friction factor for the pipe is 0.018 and it is assumed to remain constant during the acceleration process. If the valve is opened suddenly, calculate how long it will take for the velocity to reach 99% of its final value. Neglect local losses.

SOLUTION

This solution is covered in the previous reading material and requires the application of Eq. 9.61. However, we must first calculate the steady state velocity from the Darcy-Weisbach equation.

$$h_L = 100 \text{ ft} = f \frac{l}{d} \frac{V_o^2}{2g_n} = 0.018 \frac{10\,000 \text{ ft}}{24 \text{ in}/12} \frac{V_o^2}{2 \times 32.2} \tag{9.2}$$

$$V_o = 8.46 \text{ ft/s}$$

Now, from Eq. 9.61,

$$t_{99} = 2.65 \frac{LV_o}{g_n H_o} = \frac{2.65 \times 10\,000 \text{ ft} \times 8.46 \text{ ft/s}}{32.2 \times 100 \text{ ft}} = 70 \text{ sec } \bullet \tag{9.61}$$

Valve closure can cause some analysis problems beyond those of instantaneous valve openings. The difficulty occurring in this problem is precipitated by the fact that the pressure just upstream of the valve is no longer zero, but is determined by loss characteristics of the flow through the valve.

Figure 9.33 can still be used to represent the problem. At $t = 0$, the velocity is V_o and the EL-HGL is approximately a straight line between the reservoir surface and the pipe outlet (neglecting local losses) under steady flow conditions.

The differential equation representing this problem is the same as Eq. 9.57, except that $p_1/\gamma = H_o$ here; thus

$$H_o - \frac{p_2}{\gamma} - \frac{fL}{2g_n d} V^2 = \frac{L}{g_n} \frac{dV}{dt} \tag{9.62}$$

Unfortunately, there are two dependent variables (viz, p_2 and V) so we need another equation.

The second equation results from an energy equation written across the valve

$$\frac{p_2}{\gamma} = K_L \frac{V^2}{2g_n} \tag{9.63}$$

where K_L is the valve loss coefficient. Substituting this equation into Eq. 9.62 gives

$$H_o - \left(K_L + \frac{fL}{d}\right) \frac{V^2}{2g_n} = \frac{L}{g_n} \frac{dV}{dt} \tag{9.64}$$

If K_L were a constant, integration would proceed as with the flow establishment case. However, K_L is a function of the amount the valve is open. Further complicating the problem is the fact that there is not an equation directly relating K_L to either time or velocity. Hence, the solution to the differential equation must be a numerical one.

The approach would be to write the equation in finite difference form. With a valve closing schedule specified, the value of K_L would be known at any time and would be averaged over each Δt time interval. One form of the equation would be the implicit relationship.

$$V(t + \Delta t) = V(t) + \frac{g_n \Delta t}{L} \left[H_o - \left(\overline{K}_L + f \frac{L}{d} \right) \frac{1}{2g_n} \left(\frac{V(t) + V(t + \Delta t)}{2} \right)^2 \right]$$

where

$$\overline{K}_L = 0.5[K_L(t) + K_L(t + \Delta t)]$$

That there is indeed a limit of applicability to this approach can be seen with Eq. 9.62. As faster and faster valve closure times are used, dV/dt becomes quite large and, in the limit, goes to infinity. According to Eq. 9.62, in the limit $p_2/\gamma \to \infty$ also. The point at which rigid water column theory fails to give acceptable results and a move to elastic theory is necessary is hard to establish, because it depends on the individual problem and the accuracy in analysis required.

If local losses occur in the pipe system to the extent that they have a noticeable effect on the results, then they must be incorporated into the analysis. For a single pipe this can be done in two ways. In the first method, the pipe is broken into two pieces (as in Illustrative Problem 9.23) and each portion is set up separately with the two solutions coupled at the local loss location via a work-energy equation. The second method includes the local loss in the differential equation along with the pipe friction term. This can be done by absorbing it into the pipe friction term by increasing the friction factor or simply by adding it in as a separate term.

Assuming that the local loss be represented as $h_L = K_L(V^2/2g_n)$, these two latter techniques result in the following modifications of Eq. 9.57, respectively,

$$\frac{p_1}{\gamma} - \frac{p_2}{\gamma} - \frac{f'L}{2g_n\, d} V^2 = \frac{L}{g_n}\frac{dV}{dt}$$

where

$$f' = f + K_L \frac{d}{L}$$

and

$$\frac{p_1}{\gamma} - \frac{p_2}{\gamma} - \left(f\frac{L}{d} + K_L\right)\frac{V^2}{2g_n} = \frac{L}{g_n}\frac{dV}{dt} \tag{9.65}$$

It is *important not to use the traditional equivalent length method* to represent the local loss. This technique adds length to the pipe and the subsequent increase in liquid mass will distort the true dynamic behavior of the system.

ILLUSTRATIVE PROBLEM 9.23

Water flows from one reservoir to another through the horizontal pipe at a velocity of 10 ft/s. The shutdown plan calls for a valve closure schedule, which will cause the velocity to decrease linearly to zero in 100 sec. The valve is located at the center of the 6 440 ft long pipeline. Estimate the maximum and minimum pressures which will occur in the system, locate them, and give the time at which they will occur.

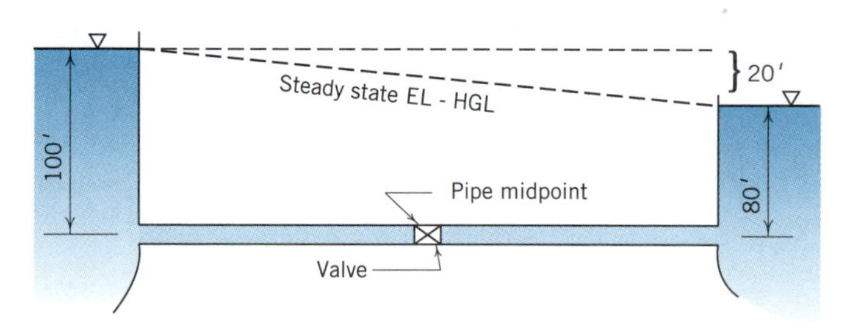

SOLUTION

The version of the unsteady flow equation which applies to this case is Eq. 9.57.

$$\frac{p_1}{\gamma} - \frac{p_2}{\gamma} - \frac{fl}{2\,d} V^2 = \frac{l}{d}\frac{dV}{dt} \tag{9.57}$$

Given that the velocity will decrease linearly with time,

$$\frac{dV}{dt} = \frac{-10 \text{ ft/s}}{100 \text{ s}} = -0.10 \text{ ft/s}^2$$

We will now solve the problem in two sections.

Upstream Section

Substituting into Eq. 9.57 for the upstream section,

$$\frac{p_1}{\gamma} - \frac{p_2}{\gamma} - \frac{fl}{2\,d} V^2 = \frac{l}{d}\frac{dV}{dt} \tag{9.57}$$

$$100 \text{ ft} - \frac{p_2}{\gamma} - f\frac{l}{d}\frac{V^2}{2g_n} = \frac{3\,220 \text{ ft}}{32.2}(-0.10 \text{ ft/s}^2)$$

$$\frac{p_2}{\gamma} = 110 - \frac{fl}{2g_n\,d} V^2$$

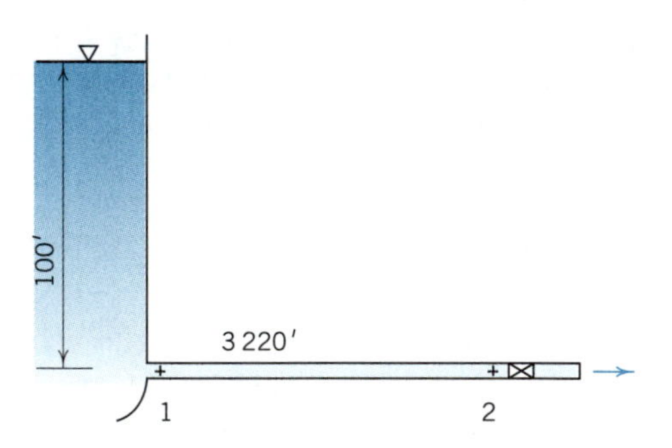

Because we are looking for extreme values of pressure, it is clear that the maximum pressure will occur when $V = 0$, the instant of complete valve closure. Conversely, the minimum pressure occurs under steady flow just before the valve begins to close.

In summary,

$$\left(\frac{p_2}{\gamma}\right)_{\max} = 110 \text{ ft at } t = 100 \text{ sec} \bullet$$

$$\left(\frac{p_2}{\gamma}\right)_{\min} = 90 \text{ ft just before valve closure begins} \bullet$$

Downstream Section

Now, using Eq. 9.57 for the downstream section,

$$\frac{p_3}{\gamma} - \frac{p_4}{\gamma} - \frac{fl}{2\,d}V^2 = \frac{l}{d}\frac{dV}{dt}$$

$$\frac{p_3}{\gamma} - 80\text{ ft} - f\frac{l}{d}\frac{V^2}{2g_n} = \frac{3\,220\text{ ft}}{32.2}(-0.10\text{ ft/s}^2) \tag{9.57}$$

$$\frac{p_3}{\gamma} = 70 + \frac{fl}{2g_n\,d}V^2$$

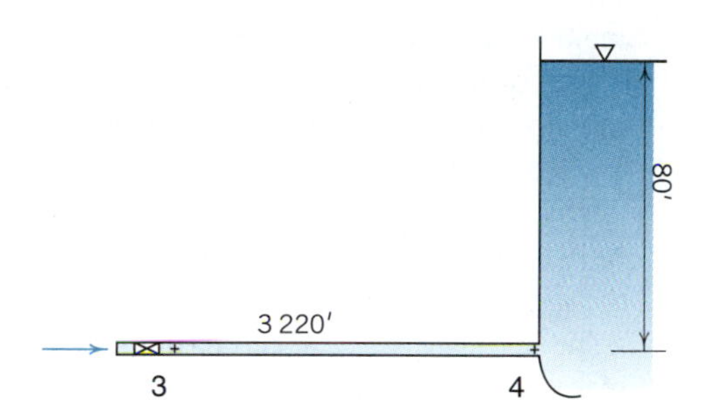

Under steady flow conditions, the pressure head just downstream of the valve is 90 ft. The instant the valve begins to move, it suddenly drops to 80 ft. At the instant the valve reaches complete closure, it has reduced to 70 ft.

In summary,

$$\left(\frac{p_3}{\gamma}\right)_{max} = 90\text{ ft at steady flow just before valve closure begins} \bullet$$

$$\left(\frac{p_3}{\gamma}\right)_{min} = 70\text{ ft at } t = 100\text{ sec} \bullet$$

Considering the entire system, the extreme pressure heads are as follows:

$$\left(\frac{p}{\gamma}\right)_{max} = 110\text{ ft at the upstream side of the value at } t = 100\text{ sec} \bullet$$

$$\left(\frac{p}{\gamma}\right)_{min} = 70\text{ ft at the downstream side of the valve at } t = 100\text{ sec} \bullet$$

If the pipe system is complex, the previous techniques must be applied to the individual components of the system. The second method discussed earlier is recommended with the technique of computing an f' to distribute the local loss along the entire pipe. After this has been accomplished, the analysis proceeds as before.

9.14 ELASTIC THEORY (WATER HAMMER)

For situations in which the velocity changes suddenly and the pipeline is relatively long, the elastic properties of the pipe and liquid enter into the analysis. Earlier we saw how a pipeline behaves under the action of a suddenly closed valve. The suddenly closed valve caused an increase in pressure head ΔH to occur, which propagated at a speed a. It remains now to develop means to calculate ΔH and a and broaden the range of applications from that of the simple example.

Impulse-Momentum

The previously integrated unsteady flow equations cannot be used because they have not included elastic effects. We will employ the impulse-momentum equation and the conservation of mass principle to develop an appropriate set of equations for an impulsive change in velocity. The impulse-momentum equation will be used to develop an equation for ΔH. We know that a change in velocity ΔV will cause a pressure head change ΔH to propagate upstream at some speed a. To begin, we will use a piece of pipe δL long (Fig. 9.34), where δL is arbitrarily small but not differentially small as dL would be. The pressure wave and the pipe bulge (which is caused by the pressure head change ΔH) propagate upstream at a speed a. The wave speed in this work is defined as the speed relative to the observer at rest with respect to the pipe rather than the speed relative to the flowing water. In the case of relatively rigid pipes, either approach used gives essentially the same result. Because this is an unsteady flow situation, the impulse-momentum equation for steady flow cannot be used. However, in this case it is possible to use a translating ''inertial'' coordinate system to transform the unsteady flow into a ''steady'' flow (see Section 4.1). Recall that the transitional velocity must be constant which requires that a be constant. If we move our reference system to the left at speed a we have, for all appearances, a steady flow (Fig. 9.35). (This is an important technique that is widely used in the analysis of unsteady flows.)

From Chapter 6, we have the one-dimensional impulse-momentum equation.

$$\Sigma \mathbf{F}_{ext} = (\Sigma Q\rho\mathbf{V})_{out} - (\Sigma Q\rho\mathbf{V})_{in} \tag{6.3}$$

where Q is the discharge, ρ is the liquid density, and $\Sigma \mathbf{F}_{ext}$ is the sum of the external forces acting. The momentum correction factor for nonuniform velocity profiles has been assigned the value of 1.0.

Considering only the component of this vector equation parallel to the pipe and noting

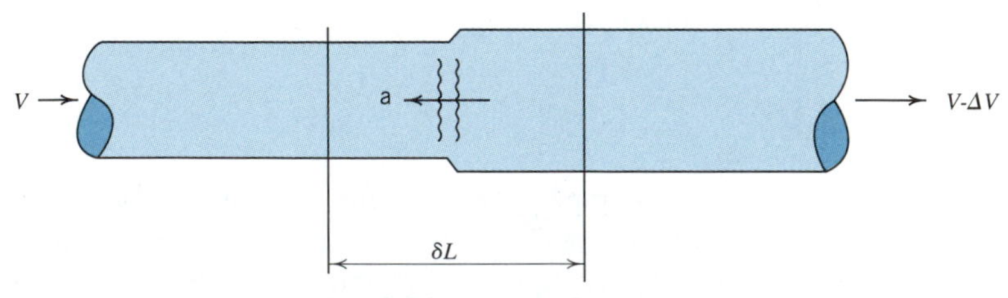

Fig. 9.34 Unsteady flow case.

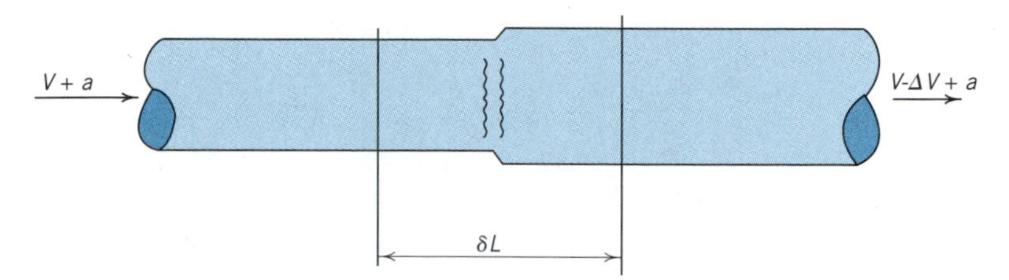

Fig. 9.35 Steady flow case.

that momentum enters and leaves the section of pipe δL long at only one section each, we can write

$$(\Sigma F_{ext}) = \dot{m}(V_{out} - V_{in}) \tag{9.67}$$

where $\dot{m} = Q\rho =$ constant. To apply the impulse-momentum equation we must specify a control volume and take into account all forces acting on the fluid in the control volume at a particular instant and at that same instant evaluate the momentum fluxes into and out of the control volume. We will choose a control volume coinciding with the inside of the pipe walls over the length δL and including the flow cross section at each end of the pipe section δL long. This control volume, the fluid in it, and the external forces acting are shown in Fig. 9.36.

The wall shear force caused by friction will be neglected because its size is limited by a very small δL. Also, because we are considering only relatively rigid pipe (steel, concrete, etc.), the pipe bulge will be very small and F_3 will also be negligible.

Application of Eq. 9.67 gives

$$F_1 - F_2 = \dot{m}(V - \Delta V + a - V - a) = \dot{m}(-\Delta V)$$

where $\dot{m} = (V + a)A\rho$ and ΔV is the *reduction* in velocity.

If the pressure at (1) were p then the pressure at (2) would be $p + \Delta p$.

$$pA - (p + \Delta p)(A + \delta A) = (V + a)A\rho(-\Delta V)$$

Expanding this equation and recognizing that $\Delta p = \gamma\,\Delta H$ and δA is very small compared to ΔH, A, and γ, we can neglect the small terms with the result

$$-\Delta H\gamma A = (V + a)A\rho(-\Delta V)$$

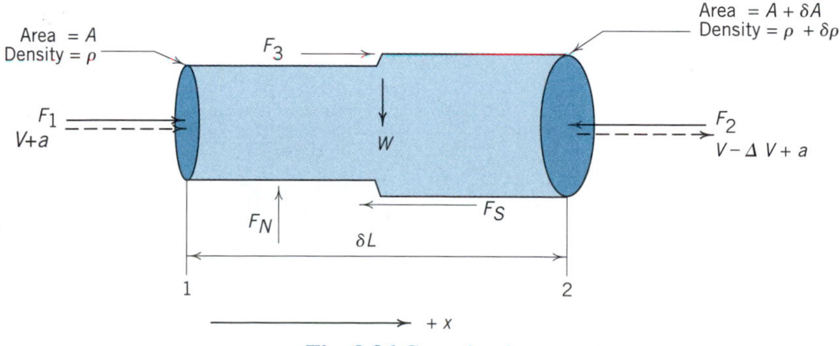

Fig. 9.36 Control volume.

In slightly different form, this equation can be written

$$\Delta H = \frac{\rho}{\gamma} \Delta V(V + a)$$

or

$$\Delta H = \frac{a \, \Delta V}{g_n} \left(1 + \frac{V}{a}\right) \tag{9.68}$$

In most cases involving rigid pipes (even PVC with a wave speed of only 1 200 fps), the value of V/a is less than 0.01. Accordingly, Eq. 9.68 is generally used (and is always used in this text) as

$$\Delta H = \frac{a}{g_n} \Delta V \tag{9.69}$$

It is clear from Eq. 9.69 that ΔH depends on the wave speed a and cannot be determined until a value of a is established.

To develop an equation for the wave speed a we will consider conservation of mass in the section of pipe δL long, which was used in the previous section to find an equation for ΔH. The procedure used will be to examine the mass flow into and out of the portion of pipe of length δL over the time period required for the wave to pass through that portion of the pipe. The net inflow of mass will be equated to the increased mass storage in δL to yield an equation for a.

To begin, we will look at the situation when the wave first reaches the δL section and then at the time the wave has just passed through the section δt later (Fig. 9.37). It is clear that δL and δt are related via the wave speed by $\delta L = a \, \delta t$

Conservation of Mass

During the time period δt an amount of liquid has accumulated in the section of pipe given by the amount

$$\delta M = \text{mass accumulated} = VA\rho \, \delta t - (V - \Delta V)(\rho + \delta\rho)(A + \delta A) \, \delta t$$

Expanding parentheses and neglecting small terms gives

$$\delta M = A\rho \, \Delta V \, \delta t$$

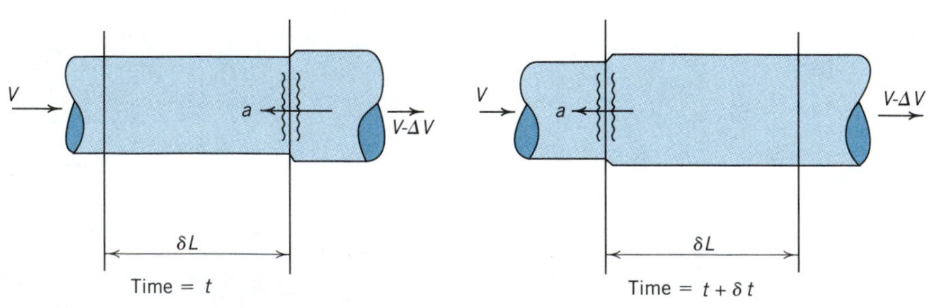

Fig. 9.37 Wave passage through section of pipe.

or writing in terms of wave speed and δL

$$\delta M = A\rho \, \Delta V \frac{\delta L}{a} \tag{9.70}$$

This amount of extra liquid is accumulated in section δL by being compressed slightly and by stretching the pipe slightly to provide storage room.

Because the pressure has increased during the passage of the wave, the volume of the liquid in the section will compress slightly to a higher density. The equation describing this relationship is that defining the bulk modulus of elasticity which can be found as Eq. 1.5.

$$E = -\frac{dp}{d\forall / \forall} \tag{1.5}$$

where E = the bulk modulus of elasticity of the liquid and p, $\forall$ are the pressure and volume, respectively. Recognizing that $dp \approx \Delta p$ (E is relatively constant over a wide range of pressure in the absence of free or entrained air), Eq. 1.5 becomes

$$\delta \forall = -\Delta p \frac{\delta L A}{E} \tag{9.71}$$

where $\delta \forall$ is the change in volume of the liquid in the pipe section δL long as the result of a pressure change Δp.

Because the increased pressure stretches the pipe, there is more room made available to store the net mass inflow of liquid. When the pipe stretches circumferentially it may also stretch longitudinally so both contributions to change in pipe volume should be evaluated. The result of this stretching gives

$$\delta \forall = \frac{\pi}{4} d^2 \, \delta L(\Delta \varepsilon_1 + 2 \, \Delta \varepsilon_2)$$

where ε_1 and ε_2 represent unit strains in the longitudinal and radial directions, respectively. If the pipe is restrained from longitudinal stretching, then $\delta \forall$ can be expressed in terms of pressure as

$$\delta \forall = \frac{\pi}{4} d^2 \, \delta L \left(\frac{1 - \mu_p^2}{E_p} \right) \left(\frac{\Delta p \, d}{e_p} \right) \tag{9.72}$$

where e_p is the pipe wall thickness and E_p is the modulus of elasticity and μ_p is the Poisson's ratio of the pipe material (see Table 4 for typical values). Now considering conservation of mass, we already have Eq. 9.70 expressing the amount of mass which has accumulated in the δL pipe section in δt seconds. We can write a different expression for the mass change in the δL pipe section after wave passage. The mass change in the section is

$$\delta M = (\rho + \delta\rho)(A \, \delta L + \delta \forall) - \rho A \, \delta L \tag{9.73}$$

Combining the expressions for δM in Eqs. 9.70 and 9.73, and inserting Eqs. 9.71 and 9.72

TABLE 4 Moduli of Elasticity and Poisson's Ratios for Common Pipe Materials

Material	Modulus of Elasticity E_p (psi)	Poisson's Ratio μ_p
Steel	30×10^6	0.30
Ductile iron	24×10^6	0.28
Copper	16×10^6	0.36
Brass	15×10^6	0.34
Aluminum	10.5×10^6	0.33
PVC	4×10^5	0.45
Fiberglass reinforced plastic (FRP)	4.0×10^6 (radial) 1.3×10^6 (long.)	0.27–0.30 (radial) 0.20–0.24 (long.)
Asbestos cement	3.4×10^6	0.30
Concrete[a]	$57\,000\sqrt{f_c'}$	0.24 (dynamically)

[a] f_c' = 28-day strength

in place of $\delta \forall$ and simplifying we find

$$a = \frac{\sqrt{E/\rho}}{\sqrt{1 + \frac{E}{E_p}\frac{d}{e_p}(1 - \mu_p^2)}} \tag{9.74}$$

Although Eq. 9.74 applies only to thin-walled pipes, restrained from axial deformation (most common situation), it gives results similar to other types of restraint. For a more thorough treatment of this subject, refer to Watters,[27] Wylie and Streeter,[28] or Chaudhry.[29]

To assist in calculating wave speeds in pipes constructed of common materials, see Table 4 which includes E_p-values and μ_p-values. The value of E for water can be taken as approximately 318 000 psi (see Appendix 2). It should be noted that a small amount of free air suspended as bubbles in the water can drastically reduce the E-value. However, evaluating the amount of air, its distribution, its pervasiveness, and the exact effect on E is most difficult. Consequently, in the design situation, the larger conservative value of E without free air is commonly used because it generally predicts the most severe water hammer pressures. The presence of any free air in the system can be considered an unforeseen, but fortuitous, occurrence, at least in the sense that it reduces E.

In the limit the pipe can become completely rigid without causing the wave speed to become infinite. This limiting value is obtained by passing E_p to ∞ in Eq. 9.74. With the nominal value of E = 318 000 psi, the resulting wave speed is approximately 4 860 fps. This number has no practical value in design because it is far too high to serve as even an approximate wave speed for preliminary design. With even a limited amount of experience, the designer can make far better estimates for wave speed in the pipe on which he or she is working.

[27]Watters, G. Z., *Analysis and Control of Unsteady Flow in Pipelines*, Butterworth, 1984.
[28]Wylie, E. B. and V. L. Streeter, *Fluid Transients in Systems*, Prentice-Hall, 1993.
[29]Chaudhry, M. H., *Applied Hydraulic Transients*, 2nd ed., Van Nostrand Reinhold, 1987.

ILLUSTRATIVE PROBLEM 9.24

Steady flow in a 24 inch pipeline 10 000 ft long occurs at a velocity of 6 ft/s. The pipe is fabricated of steel and has a wall thickness of 0.25 in. Calculate the wave speed in the pipe and the head increase resulting from sudden valve closure.

What is the longest valve closure time that will produce the same maximum pressure at the valve?

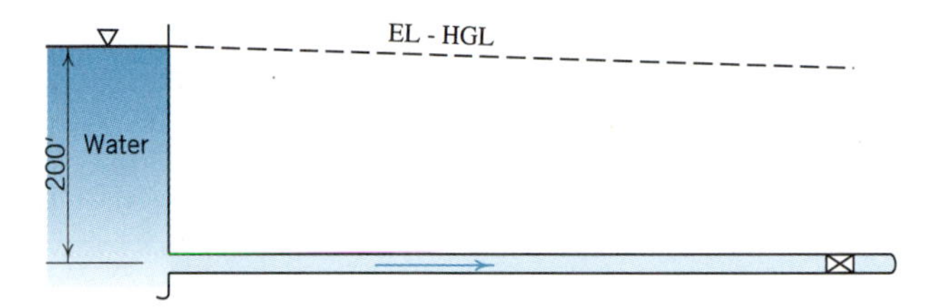

SOLUTION

We go to Eq. 9.74 to compute the wave speed using the applicable properties from Appendix 2 and Table 4.

$$a = \frac{4\,860 \text{ ft/s}}{\sqrt{1 + \dfrac{Ed}{e_p E_p}(1 - \mu_p^2)}} \tag{9.74}$$

$$a = \frac{4\,860 \text{ ft/s}}{\sqrt{1 + \dfrac{3.18 \times 10^5 \text{ lb/in}^2 \times 24 \text{ in}}{0.25 \text{ in} \times 3 \times 10^7 \text{ lb/in}^2}(1 - 0.3^2)}}$$

$$a = 3\,502 \text{ ft/s} \bullet$$

To calculate the head increment, we use Eq. 9.69.

$$\Delta H = \frac{a}{g_n}\Delta V \tag{9.69}$$

$$\Delta H = \frac{3\,502 \text{ ft/s}}{32.2} \times (6 \text{ ft/s})$$

$$\Delta H = 653 \text{ ft } (282 \text{ lb/in}^2) \bullet$$

To find the longest valve closure time that will produce the same high pressure, we recall that the critical valve closure time is $2L/a$. Any valve closure in a time less than $2L/a$ will produce the same pressure as sudden valve closure. So, the critical valve closure time is

$$2L/a = 2 \times 10\,000 \text{ ft}/3\,502 \text{ ft/s} = 5.71 \text{ s}$$

Any valve closure time less than 5.6 s will produce the maximum pressure head of 661 ft. •

Air Entrainment

When free air occurs in a pipeline, either as small bubbles or in discrete lumps, the wave speed in the pipeline is decreased dramatically. As a consequence, the wage propagation patterns and the pressures resulting from water hammer are substantially affected. If the air-water mixture is assumed to be uniformly distributed throughout a portion of the pipeline, the wave speed in that portion of the line can be computed using Eq. 9.74. However, special care must be taken to include the elastic properties of both air and water in determining E and ρ. The value of the modulus of elasticity for the mixture is developed from Eq. 1.5 by replacing the relative change in overall volume by the sum of the relative changes in volume of the air and the water. The result is

$$E_{mix} = \frac{E}{1 + \alpha_{mix}\left(\dfrac{E}{E_{air}} - 1\right)} \tag{9.75}$$

where E, E_{air} = modulus of elasticity of the liquid and air, respectively, and α_{mix} = void fraction (volume of air per total volume of mixture). For density, the same approach is used resulting in the equation

$$\rho_{mix} = (1 - \alpha_{mix})\rho \tag{9.76}$$

where ρ = density of the liquid.

Substituting Eqs. 9.75 and 9.76 into Eq. 9.74 and recognizing $E/E_{air} \gg 1$, the result is

$$a = \frac{\sqrt{E/\rho_{mix}}}{\sqrt{1 + \dfrac{E}{E_p}\dfrac{d}{e_p}(1 - \mu_p^2) + \alpha_{mix}\dfrac{E}{E_{air}}}} \tag{9.77}$$

This same equation and a detailed description of the difficulties encountered in solving water hammer problems with entrained air is given by Tullis, Streeter, and Wylie.[30]

It is clear that the wave speed depends on the pressure in the pipeline because the values of α_{mix} and E_{air} depend on the pressure. As a consequence, the wave speed varies with the passage of a pressure wave. This fact greatly complicates the analysis procedure.

An example problem is presented to demonstrate the dramatic effect of small fractions of air on the wave speed. However, before the wave speed can be calculated, the thermodynamic process followed by the air as it is compressed must be determined. Wylie and Streeter[31] suggest using an isothermal process with $E_{air} = p$. The other extreme would be to use an isentropic process with $E_{air} = 1.4p$. If some provision for heat transfer is made, then the polytropic process with $E_{air} = 1.2p$ may be appropriate. The effect on the wave speed of these various assumptions is demonstrated in Illustrative Problem 9.25.

[30]Tullis, J. P., V. L. Streeter and E. B. Wylie, "Water Hammer Analysis with Air Release", *Second International Conference on Pressure Surges*, Paper C3, BHRA Fluid Engineering, Cranfield, Bedford, England, Sept. 1976.

[31]Wylie, E. B. and V. L. Streeter, *Fluid Transients*, McGraw-Hill, 1978.

ILLUSTRATIVE PROBLEM 9.25

For the pipeline of Illustrative Problem 9.24, compute the wave speed for entrained air percentages of 0.10, 0.50, 1.0, and 2.0. Assume a polytropic thermodynamic process for the air with an exponent of 1.2.

To test the effect on wave speed of the thermodynamic process, calculate the wave speed for 1% entrained air for both isothermal and isentropic processes and compare with the results using the polytropic process.

SOLUTION

To solve the problem we use Eq. 9.77, which includes the effect on the wave speed of entrained air.

$$a = \frac{\sqrt{E/\rho_{mix}}}{\sqrt{1 + \frac{Ed}{e_p E_p}(1 - \mu_p^2) + \alpha_{mix}\frac{E}{E_{air}}}} \tag{9.77}$$

The term in the numerator will remain approximately 4 860 ft/s because for low percentages of air, the density of the mixture changes negligibly. The modulus of elasticity for a gas is $1.2p$ for a polytropic process. For air under an *absolute* pressure head of 200 ft of water,

$$E_{air} = 1.2 \times \frac{200 \text{ ft}}{2.31 \text{ ft per lb/in}^2} = 104 \text{ lb/in}^2$$

Now, turning to Eq. 9.77, using a nominal value of $E = 3 \times 10^5$ lb/in^2,

$$a = \frac{4\,860 \text{ ft/s}}{\sqrt{1 + \frac{3 \times 10^5 \text{ lb/in}^2 \times 24 \text{ in}}{0.25 \text{ in} \times 3 \times 10^7 \text{ lb/in}^2}(1 - 0.3^2) + \alpha_{mix}\frac{3 \times 10^5 \text{ lb/in}^2}{104 \text{ lb/in}^2}}}$$

Substituting the four different values of α_{mix} into the above equation gives

$$\alpha_{mix} = 0.10\% \qquad a = 2\,227 \text{ ft/s} \bullet$$

$$\alpha_{mix} = 0.50\% \qquad a = 1\,206 \text{ ft/s} \bullet$$

$$\alpha_{mix} = 1.00\% \qquad a = 880 \text{ ft/s} \bullet$$

$$\alpha_{mix} = 2.00\% \qquad a = 635 \text{ ft/s} \bullet$$

It is clear from these results that even the smallest amount of entrained air in the liquid causes a drastic reduction in wave speed, which translates into a substantial reduction in the head increase below that for water with no entrained air.

For the isothermal and isentropic processes, the only term to change in Eq. 9.77 is E_{air}.

$$(E_{air})_{isothermal} = 1.0\,\frac{200 \text{ ft}}{2.31 \text{ ft per lb/in}^2} = 87 \text{ lb/in}^2$$

$$(E_{air})_{isentropic} = 1.4\,\frac{200 \text{ ft}}{2.31 \text{ ft per lb/in}^2} = 121 \text{ lb/in}^2$$

The results for 1% air entrainment are:

(Isothermal) $a = 808$ ft/s 8% below polytropic
(Isentropic) $a = 946$ ft/s 7.5% above polytropic

Regardless of the process, the error is not great. Other uncertainties are likely to make this difference insignificant.

In many instances, it is desirable to be able to estimate portions of pressure waves reflected and transmitted at pipe junctions. We already know that at reservoirs, none of the positive (or negative) pressure wave is transmitted into the reservoir. We will look first at series pipe junctions and then at tee junctions, assuming no head loss occurs at the junctions.

Series Pipe Junctions

The equations of momentum and continuity are applied to a pressure head increase of ΔH approaching a junction. After the wave reaches the junction, ΔH_1 passes through (is transmitted) and $(\Delta H\text{-}\Delta H_1)$ is reflected. Figure 9.38 shows the configuration of the *EL-HGL* before and after the occurrence. The results of the analysis show that

$$\Delta H_1 = \frac{2a_1A_2}{a_2A_1 + a_1A_2}\Delta H \tag{9.78}$$

where A is the pipe cross-sectional area. For equal a-values,

$$\Delta H_1 = \frac{2A_2}{A_1 + A_2}\Delta H \tag{9.79}$$

Tee Junctions

The situation for tee junctions is shown in Fig. 9.39. Using the same analysis technique as before leads to the following equations

$$\Delta H_1 = \Delta H_2 = \frac{2a_1a_2A_3}{a_2a_3A_1 + a_1a_3A_2 + a_1a_2A_3}\Delta H \tag{9.80}$$

or for similar a-values in all pipes,

$$\Delta H_1 = \Delta H_2 = \frac{2A_3}{A_1 + A_2 + A_3}\Delta H \tag{9.81}$$

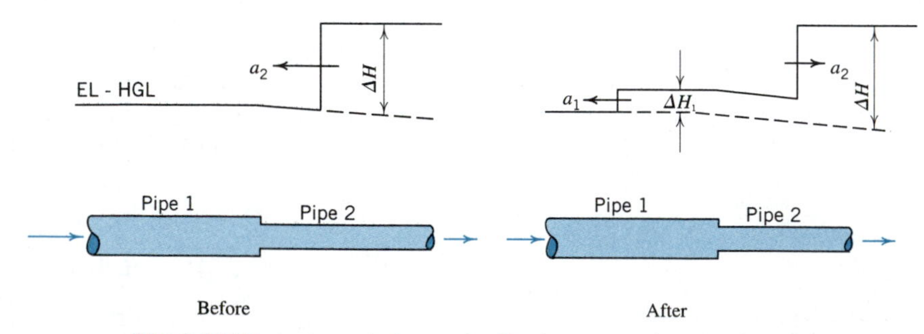

Fig. 9.38 Wave transmission and reflection at a series pipe junction.

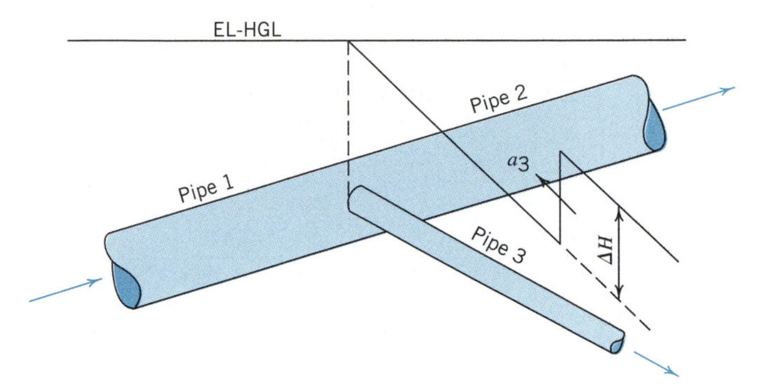

Fig. 9.39 Wave transmission and reflection at a tee junction.

ILLUSTRATIVE PROBLEM 9.26

A 24 inch pipe with a wave speed of 3 300 ft/s and a flow velocity of 1.0 ft/s reduces to a 6 inch pipe with a wave speed of 3 700 ft/s and a velocity of 16 ft/s. The ΔH-increment for sudden flow stoppage in the 6 inch pipe would be 1 838 ft. Find the portion of this increment that would be transmitted into the 24 inch pipe.

SOLUTION

In this application, Eq. 7.78 is used. But first we will calculate the areas of the two pipes.

$$A_{24} = \frac{\pi}{4} d^2 = \frac{\pi}{4}\left(\frac{24 \text{ in}}{12}\right)^2 = 3.14 \text{ ft}^2 \qquad A_6 = \frac{\pi}{4}\left(\frac{6 \text{ in}}{12}\right)^2 = 0.196 \text{ ft}^2$$

Now, we calculate the head increment.

$$\Delta H_1 = \frac{2a_1A_2}{a_2A_1 + a_1A_2}\Delta H \tag{9.78}$$

$$= \frac{2 \times 3\,300 \text{ ft/s} \times 0.196 \text{ ft}^2}{3\,700 \text{ ft/s} \times 3.14 \text{ ft}^2 + 3\,300 \text{ ft/s} \times 0.196 \text{ ft}^2} \times 1\,838 \text{ ft}$$

$$\Delta H_1 = 194 \text{ ft or } 10.6\% \text{ of the original head increment} \; \bullet$$

In many respects, the large pipe acts much like a reservoir.

ILLUSTRATIVE PROBLEM 9.27

If the 24 inch main in the previous illustrative problem had the 6 inch line teeing into it, how much of the head increment would pass into the 24 inch pipe for sudden flow stoppage in the 6 inch pipe? In this example, refer to Fig. 9.39 and use the values of $V_1 = 4.0$ ft/s, $V_2 = 3.376$ ft/s, and $V_3 = 10$ ft/s. The ΔH-increment in the 6 inch pipe would be 1 150 ft for sudden flow stoppage.

SOLUTION

In this application, Eq. 9.80 is used.

$$\Delta H_1 = \Delta H_2 = \frac{2a_1a_2A_3}{a_2a_3A_1 + a_1a_3A_2 + a_1a_2A_3}\Delta H \tag{9.80}$$

$$= \frac{2 \times 3\,300 \text{ ft/s} \times 3\,300 \text{ ft/s} \times 0.196 \text{ ft}^2}{3\,300 \times 3\,700 \times 3.14 + 3\,300 \times 3\,700 \times 3.14 + 3\,300 \times 3\,300 \times 0.196} \times 1\,150$$

$\Delta H_1 = \Delta H_2 = 62$ ft or 5.4% of the original head increment •

Dead-End Pipes

If a pipe system contains a member which carries no flow and terminates in a dead end, for example, a closed valve, then a unique situation exists, which could cause unexpected high pressures. This is actually a special case of the tee junction situation discussed previously. As a high-pressure wave passes the junction from which the dead-end pipe extends, a high pressure wave moves into the pipe and induces a flow velocity toward the dead end. When the pressure wave and the induced velocity reach the dead end, the flow is abruptly stopped, thereby increasing the head at the dead end two increments in pressure.

While pipe system geometry, pipe sizes, and hydraulic friction losses all affect the total pressure increase to some extent, the maximum pressure increase occurs when the dead-end pipe is very small in relation to the main pipe. Under this condition and with small frictional effects, the pressure increase in the dead-end pipe is, at most, equal to twice the value of the pressure head increase of the wave initially passing the junction. The following illustrative problem demonstrates the dead-end pipe effect for two extreme cases.

ILLUSTRATIVE PROBLEM 9.28

A dead-end pipe (Pipe 3) 3 000 ft long extends from a 12 inch line as shown below. If the dead-end pipe is 1 inch in diameter, find the maximum pressure head increase in this line if the mainline flow velocity of 5 ft/s is suddenly stopped. Use a wave speed of 3 000 ft/s for all pipes and neglect any frictional effects.

What is the result if all three pipes are 12 inches in diameter?

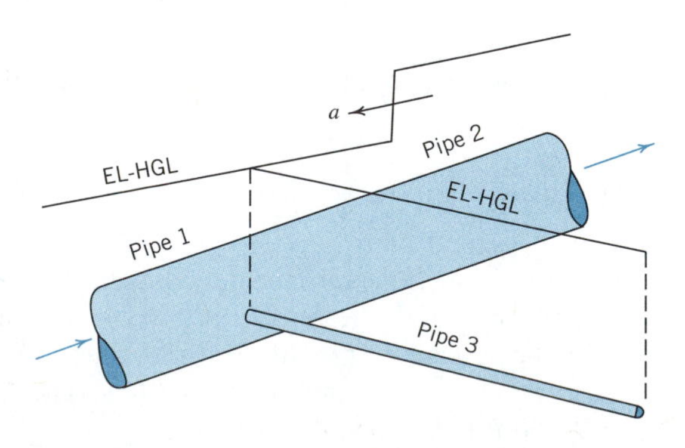

SOLUTION

First, we must calculate the head increment resulting from sudden flow stoppage. We use Eq. 9.69 to find ΔH.

$$\Delta H = \frac{a}{g_n} \Delta V = \frac{3\,000 \text{ ft/s}}{32.2} 5 \text{ ft/s} = 466 \text{ ft} \tag{9.69}$$

With all the wave speeds the same, we can use Eq. 9.81.

$$\begin{aligned}\Delta H_3 &= \frac{2A_2}{A_1 + A_2 + A_3} \Delta H = \frac{2d_2^2}{d_1^2 + d_2^2 + d_3^2} \Delta H \qquad (9.81)\\ &= \frac{2 \times (12 \text{ in.})^2}{(12 \text{ in.})^2 + (12 \text{ in.})^2 + (1 \text{ in.})^2} \times 466 \text{ ft}\\ &= 464 \text{ ft}\end{aligned}$$

Wave reflection at the dead end will double this value resulting in a *maximum head increase of 928 ft.* •

If all three pipes are 12 inches in diameter, Eq. 9.81 becomes

$$\Delta H_3 = \frac{2 \times (12 \text{ in.})^2}{3 \times (12 \text{ in.})^2} \times 466 \text{ ft} = \frac{2}{3} \times 466 = 311 \text{ ft}$$

Doubling this to find the effect of wave reflection gives a *maximum head increase of 622 ft.* •

It is interesting to note that a computer analysis of these two situations verifies the results above. In addition, if friction is considered, the maximum head increases are less than calculated above. Specifically, a friction factor of 0.020 in the 1 inch pipe would reduce the maximum head increase from 928 ft to 770 ft. So, the doubling method of estimating maximum head increases gives conservative results.

Computer Analysis

By far, the solution of hydraulic transient problems is accomplished most conveniently with the assistance of a desktop computer. Storage requirements are modest and computational times are well within the capability of this type of machine. While the development of the theory and analysis techniques associated with computer analysis are beyond the scope of this work, a basic computer program HAMMER has been included in Appendix 7 to permit the interested student to explore to a limited degree the behavior of a single pipe under the action of a rapid velocity change.

REFERENCES

Benedict, R. P. 1980. *Fundamentals of pipe flow.* New York: John Wiley & Sons.

Chaudhry, M. H. 1987. *Applied hydraulic transients.* 2nd. ed. New York: Van Nostrand Reinhold.

Colebrook, C. F. and White, C. M. 1937. The reduction of carrying capacity of pipes with age. *Jour. Inst. Civil Engrs., London* 7:99.

Hydraulic Institute. 1990. *Engineering data book.* 2nd ed. Cleveland: Hydraulic Institute.

Jeppson, R. W. 1976. *Analysis of flow in pipe networks.* Ann Arbor: Ann Arbor Science Pub., Inc.

Schlichting, H. 1979. *Boundary-layer theory.* 7th ed. New York: McGraw-Hill, Chapter XX.

Streeter, V. L. and Wylie, E. B. 1967. *Hydraulic transients.* New York: McGraw-Hill.

Watters, G. Z. 1984. *Analysis and control of unsteady flow in pipelines.* Boston: Butterworth Pub.

Wylie, E. B. and Streeter, V. L. 1978. *Fluid transients.* New York: McGraw-Hill.

Wylie, E. B. and Streeter, V. L. 1993. *Fluid transients in systems.* New Jersey: Prentice-Hall.

FILMS

NCFMF Film Loops. Encyclopaedia Britannica Educ. Corp.

FM-15 Incompressible flow through area contractions and expansions.

FM-16 Flow from a reservoir to a duct.

FM-17 Flow patterns in venturis, nozzle and orifices.

FM-69 Flow through tee-elbow.

FM-134 Laminar and turbulent pipe flow.

PROBLEMS

9.1. When 0.3 m^3/s of water flows through a 150 mm constriction in a 300 mm horizontal pipeline, the pressure at a point in the pipe is 345 kPa, and the head lost between this point and the constriction is 3 m. Calculate the pressure in the constriction.

9.2. A 50 mm nozzle terminates a vertical 150 mm pipeline in which water flows downward. At a point on the pipeline a pressure gage reads 276 kPa. If this point is 3.6 m above the nozzle tip and the head lost between point and tip is 1.5 m, calculate the flowrate.

9.3. A 12 in. pipe leaves a reservoir of surface elevation 300 at elevation 250 and drops to elevation 150, where it terminates in a 3 in. nozzle. If the head lost through line and nozzle is 30 ft, calculate the flowrate.

9.4. A vertical 150 mm pipe leaves a water tank of surface elevation 24. Between the tank and elevation 12, on the line, 2.4 m of head are lost when 56 l/s flow through the line. If an open piezometer tube is attached to the pipe at elevation 12, what will be the elevation of the water surface in this tube?

9.5. A water pipe gradually changes from 6 in. to 8 in. diameter accompanied by an increase of elevation of 10 ft. If the pressures at the 6 in. and 8 in. sections are 9 psi and 6 psi, respectively, what is the direction of flow: (*a*) for 3 cfs and (*b*) for 4 cfs?

9.6. A pump of what power is required to pump 0.56 m^3/s of water from a reservoir of surface elevation 30 to one of surface elevation 75, if in the pump and pipeline 12 metres of head are lost?

9.7. Through a hydraulic turbine flow 2.8 m^3/s of water. On the 1 m inlet pipe at elevation 43.5, a pressure gage reads 345 kPa. On the 1.5 m discharge pipe at elevation 39, a vacuum gage reads 150 mm of mercury. If the total head lost through pipes and turbines between elevations 43.5 and 39 is 9 m, what power may be expected from the machine?

9.8. In a 225 mm pipeline 0.14 m^3/s of water are pumped from a reservoir of surface elevation 30 over a hill of elevation 50. A pump of what power is required to maintain a pressure of 345 kPa on the hilltop if the head lost between reservoir and hilltop is 6 m?

9.9. The pressure drop, Δp, in a pipe is known to be a function of l, d, V, ρ, and a friction factor, f. Recalling that $h_L = \Delta p/\gamma$, use dimensional analysis to derive Eq. 9.2.

9.10. Derive expressions for wall shearing stress, wall velocity gradient, and friction velocity in terms of V, d, ρ, and μ for laminar flow in a pipe.

9.11. When a horizontal laminar flow occurs between two parallel plates of infinite extent 0.3 m apart, the velocity at the midpoint between the plates is 2.7 m/s. Calculate (*a*) the flowrate through a cross section 0.9 m wide, (*b*) the velocity gradient at the surface of the plate, (*c*) the wall shearing stress if the fluid has viscosity 1.44 Pa·s, (*d*) the pressure drop in each 30 m along the flow.

9.12. Glycerin (10°C or 50°F) flows in a 50 mm (or 2 in.) pipeline. The center velocity is 2.4 m/s (or 8 ft/s). Calculate the flowrate and the head loss in 3 m (or 10 ft) of pipe.

9.13. In a laminar flow of 0.007 m^3/s in a 75 mm pipeline the shearing stress at the pipe wall is known to be 47.9 Pa. Calculate the viscosity of the fluid.

9.14. Oil of viscosity 0.48 Pa·s and specific gravity 0.90 flows with a mean velocity of 1.5 m/s in a 0.3 m pipeline. Calculate shearing stress and velocity 75 mm from the pipe centerline.

9.15. A flowrate of 1.0 l/min of oil of specific gravity 0.92 exists in this pipeline. Is this flow laminar? What is the viscosity of the oil? For the same flow in the opposite direction, what manometer reading is to be expected?

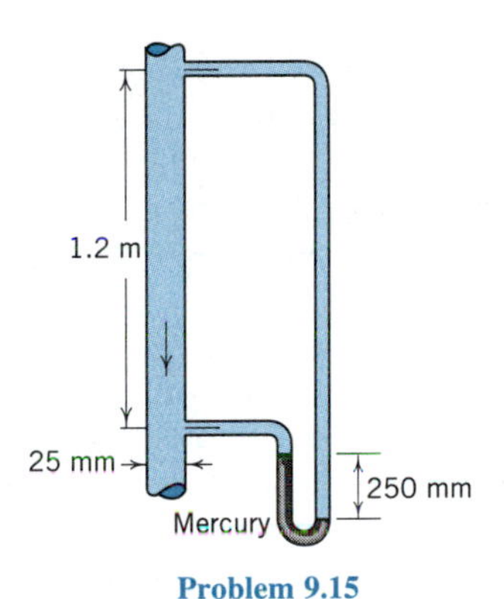

Problem 9.15

9.16. In a laminar flow in a 12 in. pipe the shear stress at the wall is 1.0 psf and the fluid viscosity 0.002 $lb{\cdot}s/ft^2$. Calculate the velocity gradient 1 in. from the centerline.

9.17. A fluid of specific gravity 0.90 flows at a Reynolds number of 1 500 in a 0.3 m pipeline. The velocity 50 mm from the wall is 3 m/s. Calculate the flowrate and the velocity gradient at the wall.

9.18. Plot the dimensionless mixing length l/R (Eq. 9.14) versus y/R for pipe flow with $\kappa = 0.40$. Over what range of y/R is $l/R = 0.4(y/R)$ within 10% of the "correct" value?

9.19. In a turbulent flow in a 0.3 m pipe the centerline velocity is 6 m/s, and that 50 mm from the pipe wall 5.2 m/s. Calculate the friction factor and flowrate.

9.20. If the velocity past a smooth surface in turbulent flow depends only on the distance from the surface, viscosity and density of fluid, and wall shear, show by dimensional analysis that $v/v_* = F(v_* y/\nu)$. See Eq. 9.17.

9.21. To determine the frictional stress exerted by fluid on a smooth wall, velocities v_1 and v_2 are measured in the turbulent zone at distances y_1 and y_2 from the wall. Derive an expression for the frictional stress in terms of the four measured quantities and the fluid density.

9.22. Solve problem 9.11 with turbulent flow, smooth plates, and fluid density and viscosity 1 000 kg/m^3 and 0.001 4 Pa·s, respectively.

9.23. Solve problem 9.12 assuming water (10° or 50°F) flowing in a smooth pipe.

9.24. Solve problem 9.13 assuming turbulent flow and a smooth pipe with fluid density 1 000 kg/m^3 and wall shear 4.8 Pa.

9.25. Solve problem 9.14 assuming a viscosity of 0.004 8 Pa·s and the pipe a smooth one.

9.26. Solve problem 9.16 assuming a wall shear of 0.10 psf, viscosity 0.000 02 $lb{\cdot}s/ft^2$, and density 1.94 $slugs/ft^3$. Assume turbulent flow.

9.27. Solve problem 9.17 for a Reynolds number of 150 000 and a smooth pipe.

9.28. Three-tenths of a cubic metre per second of liquid (s.g. 1.25) flows in a smooth pipe of 300 mm diameter at Reynolds number 10 000. Predict the velocity where $y = \delta_v$.

9.29. Fluid flows in a very smooth cylindrical pipe at Reynolds number 50 000. Using the velocity distribution of Eq. 9.27 and the methods of Section 9.3, determine the thickness of the viscous sublayer as a percentage of the pipe radius.

9.30. Fluid of density 1 030 kg/m^3 and kinematic viscosity 1.86×10^{-5} m^2/s flows parallel to a very smooth plane surface. The velocities at 75 mm and 4.4 mm from the wall are measured and found to be 0.3 m/s and 0.08 m/s, respectively. Using *both* of these measurements calculate the shearing stress on the surface.

9.31. Show that the seventh-root law velocity profile gives velocity gradients of $v_c/7R$ at the center of the pipe and infinity at the wall.

9.32. Fluid flows in a 6 in. or 150 mm smooth pipe at a Reynolds number of 25 000. Compare values of V/v_c computed from Eq. 9.20 and from the seventh-root law.

9.33. If the $f - \mathbf{R}$ relationship for $10^5 < \mathbf{R} < 10^6$ may be approximated by a straight line (on log-log plot) between $\mathbf{R} = 10^5, f = 0.018\,0$ and $\mathbf{R} = 10^6, f = 0.011\,5$, what value of m should be used in the equation $v/v_c = (y/R)^m$ for the velocity profile?

9.34. Calculate a value of $v_* \, \delta_v/\nu$ that defines the viscous sublayer thickness when the seventh-root law is used for the turbulent velocity profile.

9.35. Solve problem 9.11 for turbulent flow, rough plates with $e = 0.5$ mm, and fluid density and viscosity 1 000 kg/m^3 and 0.001 4 Pa·s, respectively.

9.36. Solve problem 9.12 assuming water (10°C) flowing in a rough pipe with $e = \frac{1}{2}$ mm.

9.37. Solve problem 9.14 assuming turbulent flow and a rough pipe having $e = 0.5$ mm with fluid viscosity 0.000 48 Pa·s.

9.38. Solve problem 9.16 assuming a wall shear of 0.20 psf, rough surface with $e = 0.03$ in., and fluid viscosity and density 0.000 020 lb·s/ft² and 1.94 slugs/ft³, respectively. Assume turbulent flow.

9.39. Solve problem 9.17 for a Reynolds number of 1.5×10^6 and rough surface having $e = 0.5$ mm.

9.40. Water flows in a smooth pipeline at a Reynolds number of 10^6. After many years of use it is observed that half the original flowrate produces the same head loss as for the original flow. Estimate the size of the relative roughness of the deteriorated pipe.

9.41. A horizontal rough pipe of 150 mm diameter carries water at 20°C. It is observed that the fall of pressure along this pipe is 184 kPa per 100 m when the flowrate is 60 l/s. What size of smooth pipe would produce the same pressure drop for the same flowrate?

9.42. In an established flow in a pipe of 0.3 m diameter the centerline velocity is 3.05 m/s and the velocity 75 mm from the centerline 2.73 m/s. Identify: (*a*) the flow as laminar or turbulent and (*b*) the pipe wall as smooth or rough.

9.43. If the size of uniform sand-grain roughness in a 10 in. or 254 mm pipe is 0.02 in. or 0.5 mm, below what approximate Reynolds number will the pipe behave as a hydraulically smooth one? What will be the thickness of the viscous sublayer at this Reynolds number?

9.44. A single layer of steel spheres is stuck to the glass-smooth floor of a two-dimensional open channel. Water of kinematic viscosity 9.3×10^{-7} m²/s flows in the channel at a depth of 0.3 m and surface velocity of $\frac{1}{4}$ m/s. Show that for spheres of 7.2 mm and 0.3 mm diameter that the channel bottom should be classified *rough* and *smooth*, respectively.

9.45. For the velocity measurements shown what is the largest roughness (e) which would allow the surface to be classified as smooth?

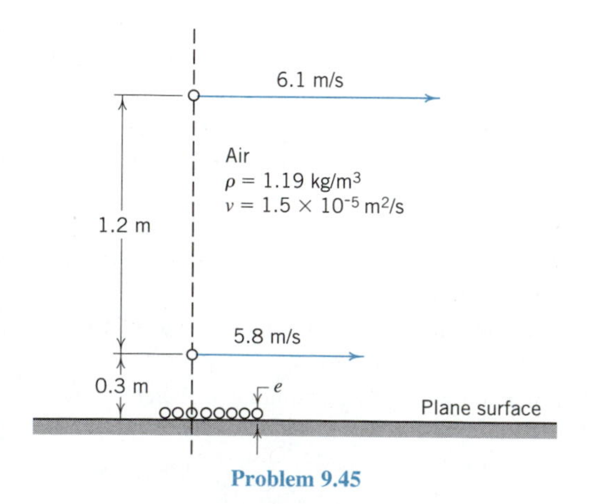

Problem 9.45

9.46. Water (20°C or 68°F) flows in a 3 m (10 ft) smooth pipe of $6\frac{1}{4}$ mm (0.25 in.) diameter. Plot head loss against velocity to substantiate the sketch-plot of Fig. 7.3.

9.47. Water flows through a section of 300 mm pipeline 300 m long running from elevation 90 to elevation 75. A pressure gage at elevation 90 reads 275 kPa, and one at elevation 75 reads 345 kPa. Calculate head loss, direction of flow, and shear stress at the pipe wall and 75 mm from the pipe wall. If the flowrate is 0.14 m³/s, calculate the friction factor and friction velocity.

9.48. When a liquid flows in a horizontal 150 mm (6 in.) pipe, the shear stress at the walls is 100 Pa (2.0 lb/ft²). Calculate the pressure drop in 30 m (100 ft) of this pipeline. What is the shearing stress in the liquid 25 mm (1 in.) from the pipe centerline?

9.49. If the friction factor for a 250 mm (10 in.) waterline is 0.030, and the shearing stress in the water 50 mm (2 in.) from the pipe centerline is 14 Pa (0.3 lb/ft²), what is the flowrate?

9.50. When 0.28 m³/s of water flow in a 0.3 m pipeline, 63 kW are lost in friction in 300 m of pipe. Calculate head loss, friction factor, friction velocity, and shear stress at the pipe wall.

9.51. In a 12 in. pipe, 15 cfs of water flow upward. At a point on the line at elevation 100, a pressure gage reads 130 psi. Calculate the pressure at elevation 150, 2 000 ft up the line, assuming the friction factor to be 0.02.

9.52. A 50 mm water line possesses two pressure connections 15 m apart (along the pipe) having a difference of elevation of 3 m. When water flows upward through the line at a velocity of 3 m/s, a differential manometer attached to the pressure connections and containing mercury and water shows a reading of 254 mm. Calculate the friction factor of the pipe.

9.53. If 680 l/min of water at 20°C flow in a 150 mm pipeline having roughness protuberances of average height 0.75 mm, and if similar roughness having height 0.375 mm exists in a 75 mm pipe, what flowrate of crude oil (40°C) must take place therein for the friction factors of the two pipes to be the same?

9.54. If two cylindrical horizontal pipes are geometrically similar and the flows in them dynamically similar, how will the pressure drops in corresponding lengths vary with density and mean velocity of the fluid?

9.55. Calculate the loss of head in 300 m (1 000 ft) of 75 mm (3 in.) PVC pipe when water at 27°C (80°F) flows therein at a mean velocity of 3 m/s (10 ft/s).

9.56. In a 50 mm (2 in.) pipeline, 95 l/min (25 gpm) of glycerin flow at 20°C (68°F). Calculate the loss of head in 50 m (160 ft) of this pipe.

9.57. If 45 kg/min of air flow in a 75 mm galvanized iron pipeline at 1 200 kPa abs. and 27°C, calculate the pressure drop in 90 m of this pipe. Assume the air to be of constant density.

9.58. If 0.34 m^3/s of water flows in a 0.3 m riveted steel pipe at 21°C, calculate the smallest loss of head to be expected in 150 m of this pipe.

9.59. A 3 in. smooth pipeline 100 ft long carries 100 gpm of crude oil. Calculate the head loss when the oil is at (*a*) 80°F, (*b*) 110°F.

9.60. In a laboratory test, 222 kg/min of water at 15°C flow through a section of 50 mm pipe 9 m long. A differential manometer connected to the ends of this section shows a reading of 480 mm. If the fluid in the bottom of the manometer has a specific gravity of 3.20, calculate the friction factor and the Reynolds number.

9.61. Carbon dioxide flows in a horizontal 100 mm wrought iron pipeline at a velocity of 3 m/s. At a point in the line a pressure gage reads 690 kPa and the temperature is 40°C. What pressure is lost as the result of friction in 30 m of this pipe? Barometric pressure is 101.3 kPa. Assume the fluid is of constant density.

9.62. When water at 20°C (68°F) flows through 6 m (20 ft) of 50 mm (2 in.) smooth pipe, the head lost is 0.3 m (1 ft). Calculate the flowrate.

9.63. When glycerin (25°C or 77°F) flows through a 30 m (100 ft) length of 75 mm (3 in.) pipe, the head loss is 36 m (120 ft). Calculate the flowrate.

9.64. A pump of what power is required to pump 40 l/min of crude oil from a tank of surface elevation 12 to one of elevation 18 through 450 m of 75 mm pipe, if the oil is at (*a*) 25°C, (*b*) 40°C?

9.65. If the head lost in 150 m of 75 mm smooth pipe is 21 m when the flowrate is 8.5 l/s, is the flow laminar or turbulent?

9.66. When the flowrate in a certain smooth pipe is 0.14 m^3/s, the friction factor is 0.06. What friction factor can be expected for a flowrate of 0.71 m^3/s of the same fluid in the same pipe?

9.67. A fluid of kinematic viscosity 4.6×10^{-4} m^2/s flows in a certain length of cast iron pipe of 0.3 m diameter with a mean velocity of 3 m/s and head loss of 4.5 m. Predict the head loss in this length of pipe when the velocity is increased to 6 m/s.

9.68. When 0.14 m^3/s of water (20°C) flow in a smooth 150 mm pipeline, the head lost in a certain length is 4.5 m. What head loss can be expected in the same length for a flowrate of 0.28 m^3/s? How is the answer changed if the pipe is concrete?

9.69. Warm oil (s.g. 0.92) flows in a 2 in. or 50 mm smooth pipeline at a mean velocity of 8 ft/s or 2.4 m/s and Reynolds number 7 500. Calculate the wall shear stress. As the oil cools, its viscosity increases; what higher viscosity will produce the same shear stress? Neglect variation in specific gravity. The flowrate does not change.

9.70. The same fluid flows through 300 m (1 000 ft) of 75 mm (3 in.) and 300 m (1 000 ft) of 100 mm (4 in.) smooth pipe. The two flows are adjusted so that their Reynolds numbers are the same. What is the ratio between their head losses?

9.71. When 57 l/s of liquid (s.g. 1.27, viscosity 0.012 Pa·s) flow in 150 m of 150 mm pipe, the head lost is 11.2 m. Is this pipe rough or smooth?

9.72. Fluid of specific gravity 0.92 and viscosity 0.096 Pa·s flows in a 50 mm smooth pipeline. If $\mathbf{R}$ = 2 100, calculate the head lost in 30 m of pipe if the flow is (*a*) laminar, (*b*) turbulent.

9.73. The head lost in 150 m of 0.3 m pipe having sand-grain roughness projections 2.5 mm high is 12 m for a flowrate of 0.28 m^3/s. What head loss can be expected when the flowrate is 0.56 m^3/s?

9.74. A liquid of specific gravity 0.85 flows in a 100 mm (4 in.) diameter commercial steel pipe. The flowrate is 4.25 l/s (65 gpm). If the pressure drop over a 60 m (200 ft) length of horizontal pipe is 1.75 kPa (0.25 psi), determine the viscosity of the liquid.

9.75. Water flows in a 100 mm (4 in.) commercial steel pipe at 15°C (59°F). If the center velocity is 1 m/s (3.3 ft/s) what is the flowrate?

9.76. Carbon dioxide flows in a 75 mm wrought iron pipe at an absolute pressure of 345 kPa and 10°C. If the center velocity is 0.6 m/s, calculate the mass flowrate.

9.77. Air at 101.3 kPa and 15.6°C flows in a horizontal triangular smooth duct, having 200 mm sides, at a mean velocity of 3.6 m/s. Calculate the pressure drop per metre of duct. Assume the air has constant density.

9.78. Three-tenths of a cubic metre per second of water flows in a smooth 230 mm square duct at 10°C. Calculate the head lost in 30 m of this duct.

9.79. A concrete conduit of cross-sectional area 10 ft^2 and wetted perimeter 12 ft carries water at 50°F at a mean velocity of 8 ft/s. Calculate the smallest head loss to be expected in 200 ft of this conduit.

9.80. A semicircular concrete conduit of 1.5 m (5 ft) diameter carries water at 20°C (68°F) at a velocity of 3 m/s (10 ft/s). Calculate the smallest loss of head to be expected per metre (foot) of conduit.

9.81. What relative roughness is equivalent to a Hazen-Williams coefficient of 140 for $10^5 < \mathbf{R} < 10^6$?

9.82. A new 12 in. riveted steel pipeline carries a flowrate of 2.5 cfs of water. Calculate the head loss of 1 000 ft of this pipe using the Hazen-Williams method. Calculate f and an approximate value of e.

9.83. An 18 in. new riveted steel pipeline 1 000 ft long runs from elevation 150 to elevation 200. If the pressure at elevation 150 is 100 psi and at elevation 200 is 72 psi, what flowrate can be expected through the line?

9.84. Smooth masonry pipe of what diameter is necessary to carry 50 cfs between two reservoirs of surface elevations 250 and 100 if the pipeline is to be 2 miles long?

9.85. What Hazen-Williams coefficient will yield the same head loss as the Darcy-Weisbach equation for a 2 in. smooth pipe with flow at Reynolds number 10^5? Compare the result with the values of Table 1.

9.86. Laboratory tests on cylindrical pipe yield the empirical formula $h_L = 0.002\,583\; lV^{2.14}\, d^{-0.86}$ with head loss in m, length in m, diameter in m, and velocity in m/s. Water of kinematic viscosity 9.3×10^{-7} m^2/s was used in the tests and ranges of d and V were: $0.03 < d < 0.06$ and $0.6 < V < 1.5$. Analyze the formula and comment on its possible validity.

9.87. In the early hydraulic literature there are many empirical head loss formulas of the form $h_L/l = CV^x/d^y$. Show from the shape of the curves on the Moody diagram that $x \leq 2$, $y \geq 1$, and $x + y = 3$.

9.88. If 0.14 m^3/s of water flow through a 150 mm horizontal pipe which enlarges abruptly to 300 mm diameter, and if the pressure in the smaller pipe is 138 kPa, calculate the pressure in the 300 mm pipe, neglecting pipe friction.

9.89. The fluid flowing has specific gravity 0.90; $V_{75} = 6$ m/s; $\mathbf{R} = 10^5$. Calculate the gage reading.

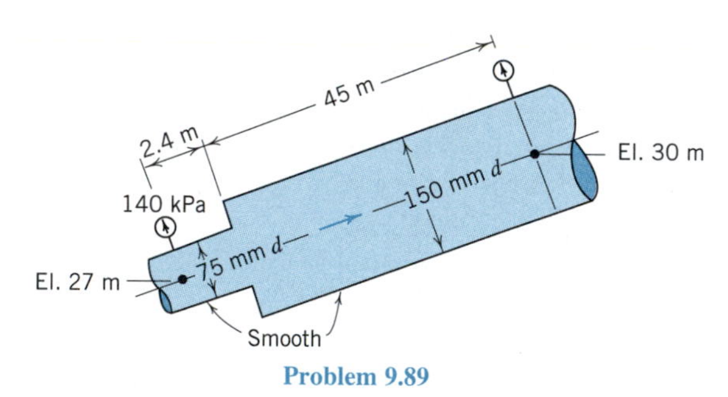

Problem 9.89

9.90. Water is flowing. Calculate the direction and approximate magnitude of the manometer reading.

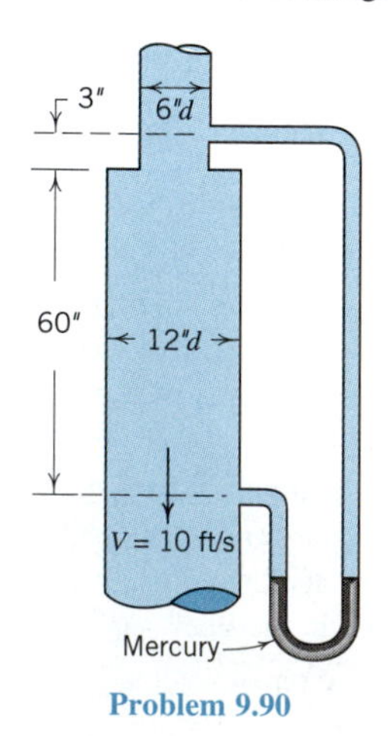

Problem 9.90

9.91. Solve problem 9.90 assuming conical enlargements of 70° and 7°.

9.92. Calculate the magnitude and direction of the manometer reading. Water is flowing.

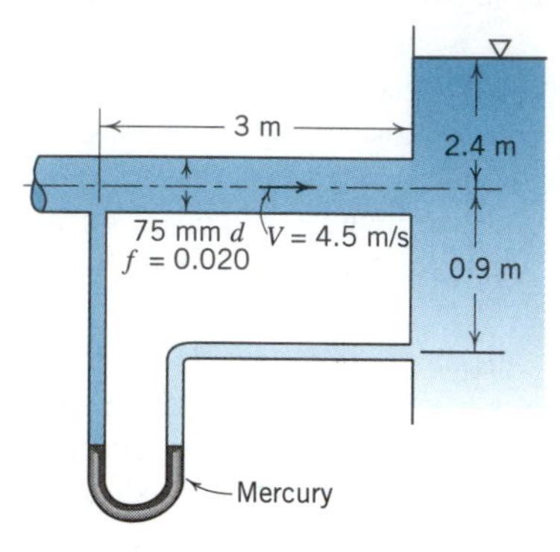

Problem 9.92

9.93. Calculate the approximate loss coefficient for this gradual enlargement.

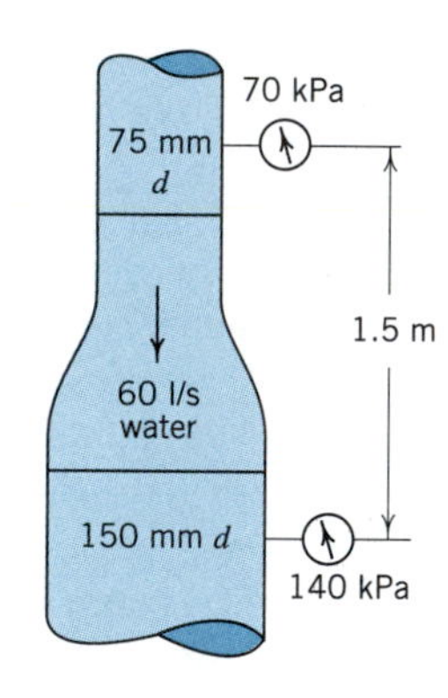

Problem 9.93

9.94. Experimental determination of local losses and loss coefficients are made from measurements of the hydraulic grade lines in zones of established flow. Calculate the head loss and loss coefficients for this gradual enlargement from the data given.

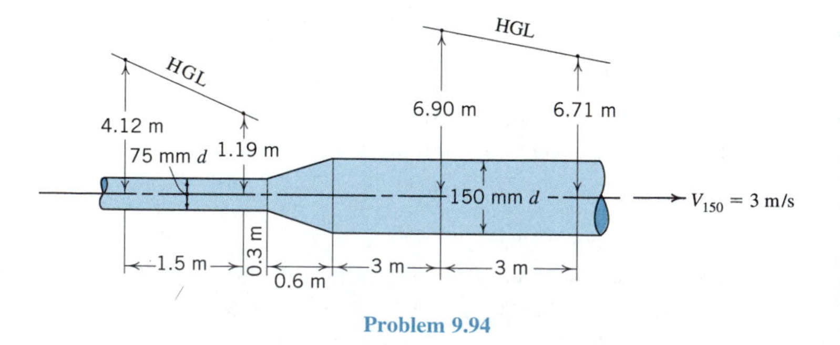

Problem 9.94

9.95. The mean velocity of water in a 150 mm horizontal pipe is 0.9 m/s. Calculate the loss of head through an abrupt contraction to 50 mm diameter. If the pressure in the 150 mm pipe is 345 kPa, what is the pressure in the 50 mm pipe, neglecting pipe friction?

9.96. A 150 mm horizontal waterline contracts abruptly to 75 mm diameter. A pressure gage 150 mm upstream from the contraction reads 34.5 kPa when the mean velocity in the 150 mm pipe is 1.5 m/s. What will pressure gages read 0.6 m downstream and just downstream from the contraction if the diameter of the vena contracta is 61 mm? Neglect pipe friction.

9.97. Water is flowing. Calculate the gage reading when V_{12} is 8 ft/s.

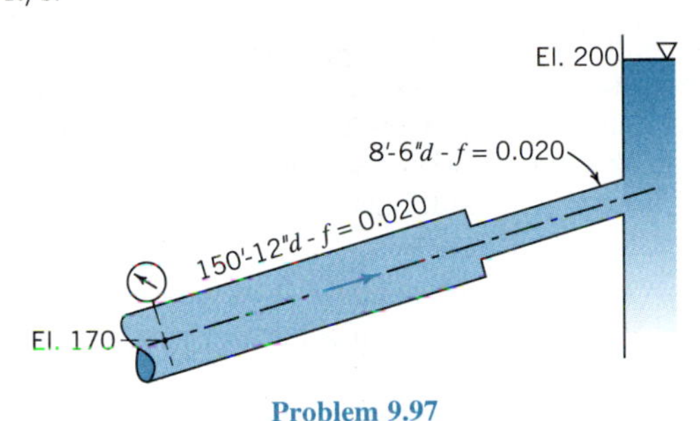

Problem 9.97

9.98. Calculate the head loss and loss coefficient caused by this restricted contraction.

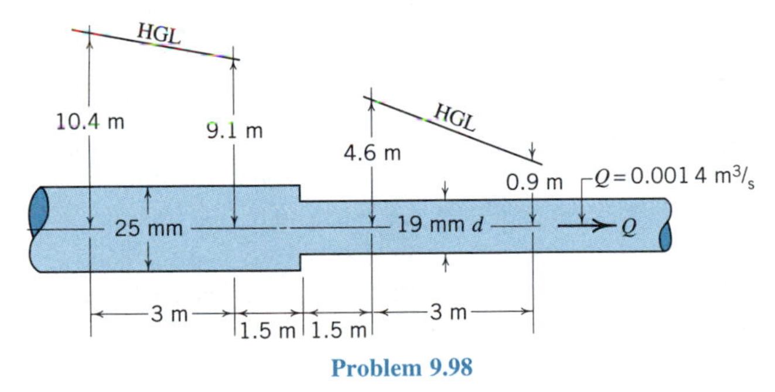

Problem 9.98

9.99. Calculate the magnitude and direction of the manometer reading.

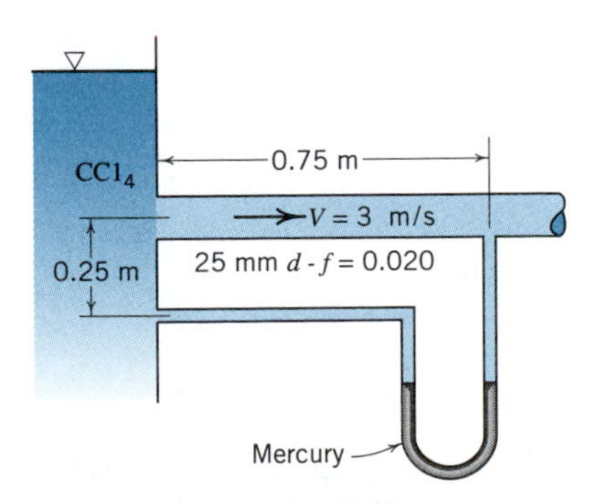

Problem 9.99

9.100. Determine the head loss and loss coefficient for this restricted pipe entrance.

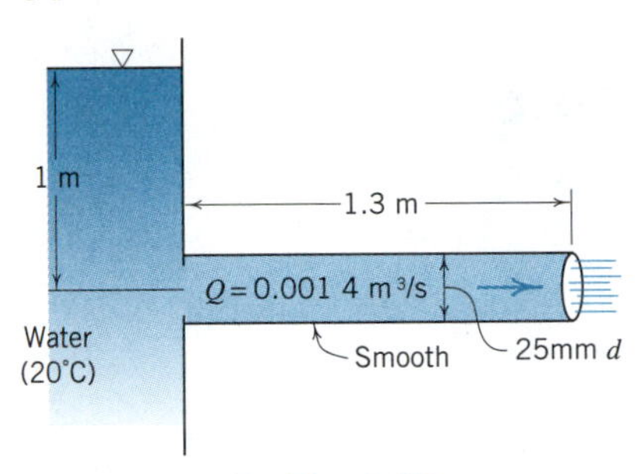

Problem 9.100

9.101. The 6 in. suction pipe for a pump extends 10 ft vertically below the free surface of water in a tank. If the mean velocity in the pipe is 20 ft/s, what is the pressure in the pipe at the level of the liquid surface if the entrance is (*a*) rounded, (*b*) re-entrant? Assume the friction factor of the pipe to be 0.020.

9.102. A horizontal pipeline of 150 mm diameter leaves a tank (square-edged entrance) 15 m below its water surface and enters another tank 6 m below its water surface. If the flowrate in the line is 0.11 m^3/s, what will gages read on the pipeline a short distance (say 0.6 m) from the tanks? Neglect pipe friction.

9.103. If the length of the unestablished flow zone downstream from a rounded pipe entrance (see Fig. 7.19 of Section 7.9) is 50 pipe diameters when the Reynolds number (Vd/ν) is 1 800, what is the loss coefficient of the entrance if the total head may be assumed to remain constant along the central streamline?

9.104. Solve the preceding problem for a turbulent flow in a smooth pipe at Reynolds number 100 000 if the length of the unestablished flow zone is 25 diameters. Assume that the seventh-root law (Section 9.3) is applicable.

9.105. A 90° smooth bend in a 6 in. or 150 mm pipeline has a radius of 5 ft or 1.5 m. If the mean velocity through the bend is 10 ft/s or 3 m/s and the Reynolds number 200 000, what head loss is caused by the bend? If the bend were unrolled and established flow assumed to exist in the length of pipe, what percent of the total head loss could be considered due to wall friction?

9.106. A 90° screwed elbow is installed in a 50 mm (2 in.) pipeline having a friction factor of 0.03. The head lost at the elbow is equivalent to that lost in how many metres (feet) of the pipe? Repeat the calculation for a 25 mm (1 in.) pipe.

9.107. A 50 mm pipeline 1.5 m long leaves a tank of water and discharges into the atmosphere at a point 3.6 m below the water surface. In the line close to the tank is a valve. What flowrate can be expected when the valve is a (*a*) gate valve, (*b*) globe valve? Assume a square-edged entrance, a friction factor of 0.020, and screwed valves.

9.108. Calculate the total tension in the bolts. Neglect entrance loss.

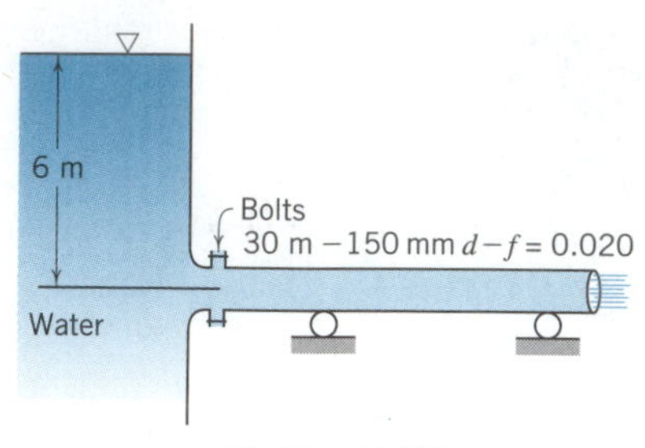

Problem 9.108

9.109. Water flows at 10°C from a reservoir through a 25 mm pipeline 600 m long which discharges into the atmosphere at a point 0.3 m below the reservoir surface. Calculate the flowrate, assuming it to be laminar and neglecting local losses and velocity head in the pipeline. Check the assumption of laminar flow.

9.110. Glycerin flows through a 2 in. horizontal pipeline leading from a tank and discharging into the atmosphere. If the pipeline leaves the tank 20 ft below the liquid surface and is 100 ft long, calculate the flowrate when the glycerin has a temperature of (*a*) 50°F, (*b*) 70°F. Neglect local losses and velocity head.

9.111. A horizontal 50 mm PVC pipeline leaves (square-edged entrance) a water tank 3 m below its free surface. At 15 m from the tank, it enlarges abruptly to a 100 mm pipe which runs 30 m horizontally to another tank, entering it 0.6 m below its surface. Calculate the flowrate through the line (water temperature 20°C), including all head losses.

9.112. Water flows from a tank through 60 m of horizontal 50 mm PVC pipe and discharges into the atmosphere. If the water surface in the tank is 1.2 m above the pipe, calculate the flowrate, considering losses due to pipe friction only, when the water temperature is (*a*) 10°C, (*b*) 40°C.

9.113. A smooth 12 in. pipeline leaves a reservoir of surface elevation 500 at elevation 460. A pressure gage is located on this line at elevation 400 and 1 000 ft from the reservoir (measured along the line). Calculate the gage reading when 10 cfs of water (68°F) flow in the line. Neglect local losses.

9.114. One-quarter of a cubic metre per second of liquid (20°C) is to be carried between two tanks having a difference of surface elevation of 9 m. If the pipeline is smooth and 90 m long, what pipe size is required if the liquid is (*a*) crude oil, (*b*) water? Neglect local losses.

9.115. A horizontal 50 mm pipeline leaves a water tank 6 m below the water surface. If this line has a square-edged entrance and discharges into the atmosphere, calculate the flowrate, neglecting and considering the entrance loss, if the pipe length is (*a*) 4.5 m, (*b*) 45 m. Assume a friction factor of 0.025.

9.116. Calculate the flowrate from this water tank if the 6 in. pipeline has a friction factor of 0.020 and is 50 ft long. Is cavitation to be expected in the pipe entrance? The water in the tank is 5 ft deep.

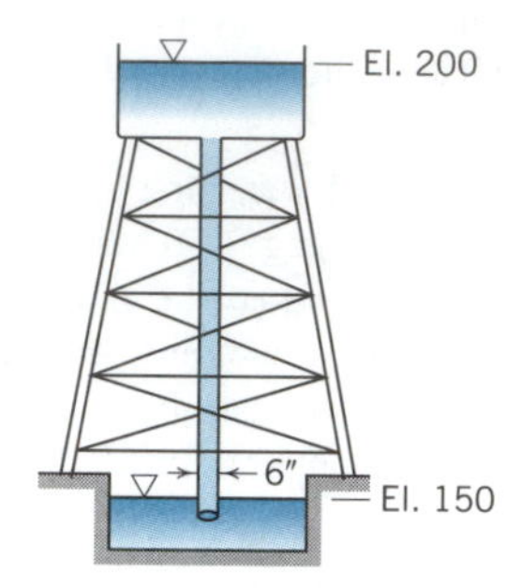

Problem 9.116

9.117. A 300 mm horizontal pipe 300 m long leaves a reservoir of surface elevation 60 at elevation 54. This line connects (abrupt contraction) to a 150 mm pipe 300 m long running to elevation 30, where it enters a reservoir of surface elevation 39. Assuming friction factors of 0.02, calculate the flowrate through the line.

9.118. What is the maximum flow which may be theoretically obtained in problem 9.117 when the 150 mm and 300 mm pipes are interchanged?

9.119. A long 0.3 m pipeline laid between two reservoirs carries a flowrate of 0.14 m^3/s of water. A parallel pipe of the same friction factor is laid beside this one. Calculate the approximate diameter of the second pipe if it is to carry 0.28 m^3/s.

9.120. A 6 in. horizontal smooth pipe 1 000 ft long takes oil from a large tank and discharges it into the atmosphere. At the midpoint of the pipe the pressure is 10.0 psi. If the specific gravity and viscosity of the oil are 0.88 and 0.000 5 $lb{\cdot}s/ft^2$, respectively, calculate (*a*) the flowrate and (*b*) the pressure in the tank on the same level as the pipe.

9.121. There is a leak in a horizontal 0.3 m pipeline having a friction factor of 0.025. Upstream from the leak two gages 600 m apart on the line show a difference of 138 kPa. Downstream from the leak two gages 600 m apart show a difference of 124 kPa. How much water is being lost from the pipe per second?

9.122. The pipe is filled and the plug then removed. Estimate the steady flowrate.

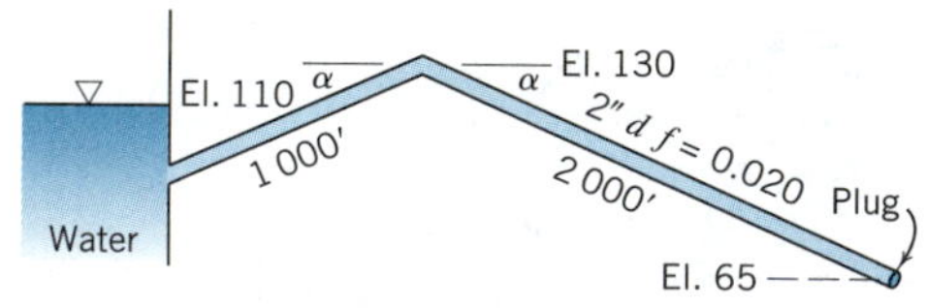

Problem 9.122

9.123. An irrigation siphon has the dimensions shown and is placed over a dike. Estimate the flowrate to be expected under a head of 0.3 m. Assume a re-entrant entrance, a friction factor of 0.020, and bend loss coefficients of 0.20.

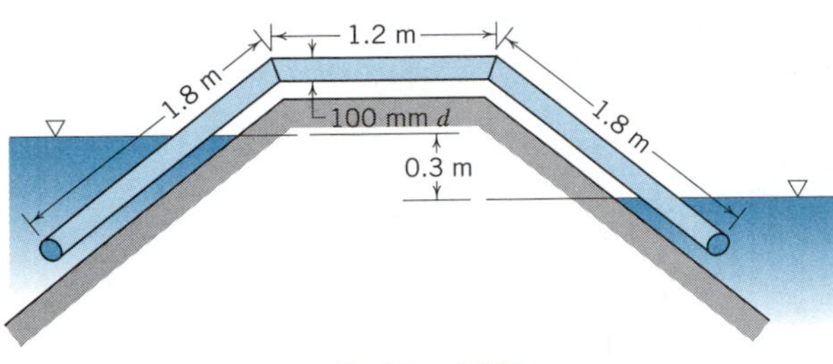

Problem 9.123

9.124. Calculate the flowrate and the gage reading, neglecting local losses and velocity heads.

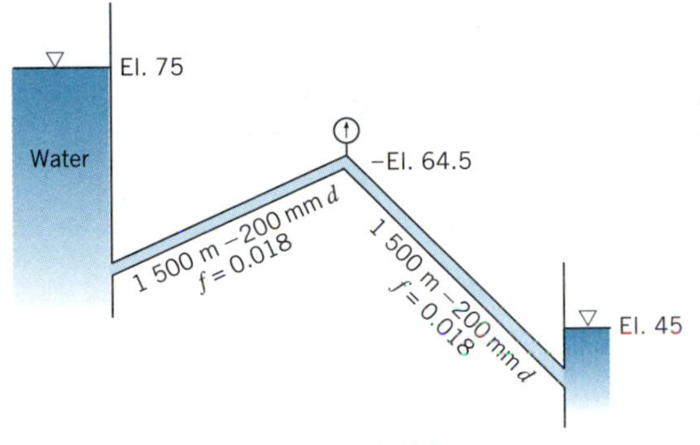

Problem 9.124

9.125. At least 0.08 m^3/s of oil ($\nu = 1.7 \times 10^{-5}\ m^2/s$) are to flow between two reservoirs having a 15 m difference of free surface elevation. Three hundred millimetre rough steel pipe of equivalent sand-grain size 3 mm and 250 mm smooth pipe are available at the same cost. Which pipe should be used? Provide calculations to justify the choice.

9.126. A 380 mm pipeline having equivalent sand-grain roughness of 6 mm carries 1/3 m^3/s of water (20°C). If a smooth liner is installed in the pipe, thereby reducing the diameter to 350 mm,·what (percent) reduction of head loss can be expected in the latter pipe for the same flowrate?

9.127. A 6 ft diameter pipeline 4 miles long between two reservoirs of surface elevations 500 and 300 ft carries a flowrate of 250 cfs of water (68°F). It is proposed to increase the flowrate through the line by installing a glass-smooth liner. Above what liner diameter may an increase of flowrate be expected? What is the maximum increase to be expected? Assume the 6 ft diameter to be measured to the midpoint of the roughness projections. Neglect all local head losses.

9.128. Compute the flowrate in the system neglecting all local losses except the loss in the partially open valve, which has a K_L of 10.0. The liquid is water at 60°F ($\nu = 1.22 \times 10^{-5}\ ft^2/s$). The roughness value for the pipe is 0.06 in. Draw the energy line carefully, labeling all changes in slope.

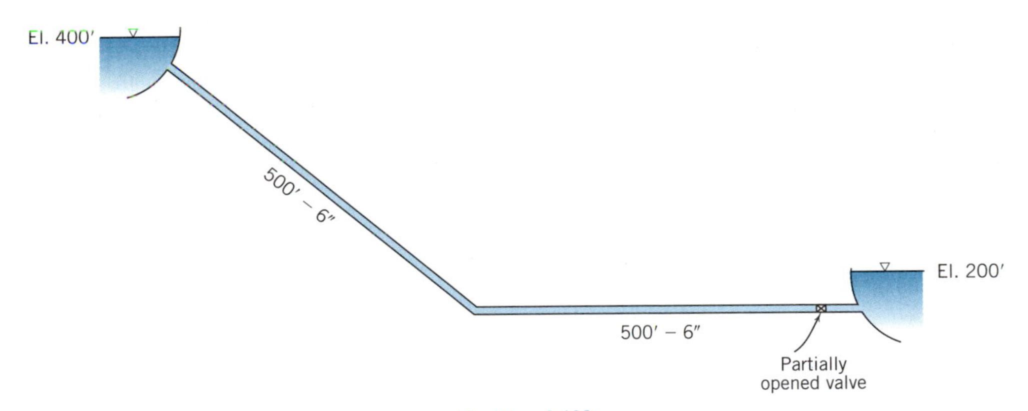

Problem 9.128

9.129. A straight concrete pipe (e = 0.12 in.) 12 in. inside diameter and 1 000 ft long discharges into the atmosphere. Compute the flowrate of water at 60°F if the local loss at the entrance is neglected. Draw the energy line, labeling the changes in slope.

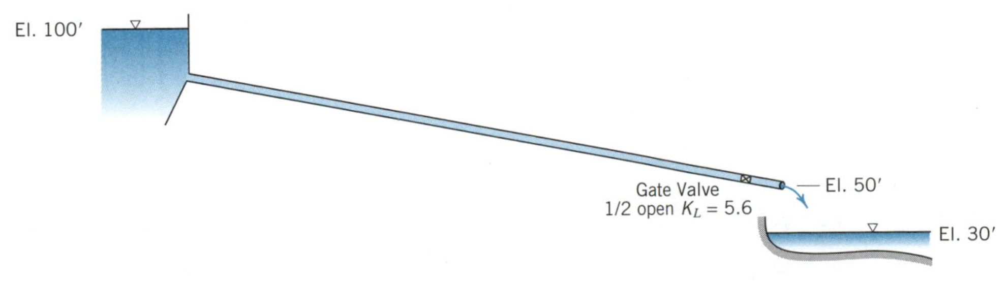

Problem 9.129

9.130. A concrete pipeline ($e = 0.12$ in.) 3 000 ft long conveys water ($\nu = 1.1 \times 10^{-5}$ ft^2/s between two reservoirs. If the discharge is 4.0 cfs, find the elevation of the lower reservoir. A partially open valve near the lower reservoir has a loss coefficient of 8.0. Draw the energy line carefully, labeling all changes in slope.

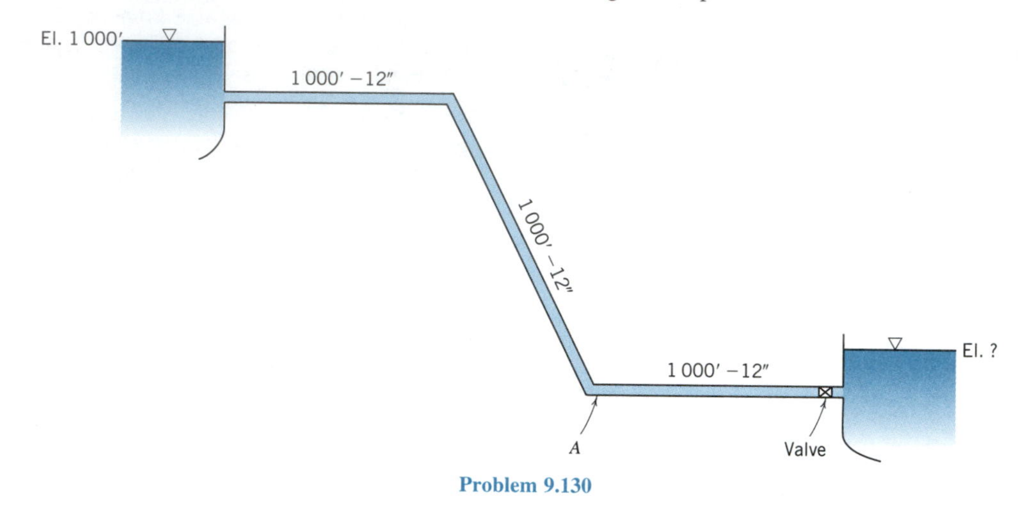

Problem 9.130

9.131. It is desired to double the flowrate between the two reservoirs of Problem 9.130 by installing a pump at location *A*. At the same time, the valve will be opened and its loss coefficient will become negligible. What horsepower must be supplied to the water by the pump to accomplish this objective?

9.132. A 0.3 m pipeline 450 m long leaves (square-edged entrance) a reservoir of surface elevation 150 at elevation 138 and runs to elevation 117, where it discharges into the atmosphere. Calculate the flowrate and sketch the energy and hydraulic grade lines (assuming that $f = 0.022$) (*a*) for these conditions, and (*b*) when a 75 mm nozzle is attached to the end of the line, assuming the lost head caused by the nozzle to be 1.5 m. How much power is available in the jet?

9.133. A 300 mm pipeline ($f = 0.020$) is horizontal and 60 m long and runs between two reservoirs. It leaves the high-level reservoir at a point 36 m below its surface and enters the low-level reservoir 6 m below its surface. (*a*) Assume standard square-edged entrance and exit and compute the flowrate to be expected. (*b*) A nozzle of 225 mm tip diameter is now attached to the downstream end of the pipe; neglecting head losses *in* the nozzle, what reduction of flowrate will be expected? (*c*) The nozzle is now replaced with a diffuser tube of 375 mm exit diameter. Assuming that the diffuser tube flows full and the losses therein may be neglected, what increase of flowrate is to be expected?

9.134. A pipeline of 0.3 m diameter ($f = 0.020$) runs 600 m from a reservoir of surface elevation 150 to a point at elevation 105 where it terminates in a 150 mm diameter nozzle. (*a*) Neglecting entrance and nozzle losses calculate the flowrate through this pipeline. (*b*) If a turbine (100% efficiency) is now installed toward the middle of the pipeline, what is the maximum power that may be expected from it? (*c*) Could a larger power than that of (*b*) be obtained by changing the nozzle size? If so, calculate the required nozzle diameter. Assume the setting of the turbine low enough so that there are no cavitation considerations in this problem.

9.135. Water flows from a large reservoir through 2 500 ft of 12 in. pipe of constant friction factor 0.030. The pipe terminates in a frictionless nozzle discharging to the atmosphere. The nozzle is at elevation 100, the reservoir surface at elevation 300. What size (diameter) of nozzle should be provided to maximize the jet horsepower? Calculate this maximum horsepower.

9.136. A 0.6 m pipeline 900 m long leaves (square-edged entrance) a reservoir of surface elevation 150 at elevation 135 and runs to a turbine at elevation 60. Water flows from the turbine through a 0.9 m vertical pipe ("draft tube") 6 m long to tail water of surface elevation 56. When 0.85 m^3/s flow through pipe and turbine, what power is developed? Take $f = 0.020$; include exit loss; neglect other local losses and those within the turbine. How much power may be saved by replacing the above draft tube with a 7° conical diffuser of the same length?

9.137. A pump close to a reservoir of surface elevation 100 pumps water through a 6 in. pipeline 1 500 ft long and discharges it at elevation 200 through a 2 in. nozzle. Calculate the pump horsepower necessary to maintain a pressure of 50 psi behind the nozzle, and sketch accurately the energy line, taking $f = 0.020$.

9.138. The horizontal 200 mm suction pipe of a pump is 150 m long and is connected to a reservoir of surface elevation 90 m, 3 m below the water surface. From the pump, the 150 mm discharge pipe runs 600 m to a reservoir of surface elevation 126, which it enters 10 m below the water surface. Taking f to be 0.020 for both pipes, calculate the power required to pump 0.085 m^3/s from the lower reservoir. What is the maximum dependable flowrate that may be pumped through this system (*a*) with the 200 mm suction pipe, and (*b*) with a 150 mm suction pipe?

9.139. A 0.3 m pipeline 3.2 km long runs on an even grade between reservoirs of surface elevations 150 and 120, entering the reservoirs 10 m below their surfaces. The flowrate through the line is inadequate, and a pump is installed at elevation 125 to increase the capacity of the line. Assuming f to be 0.020, what pump power is required to pump 0.17 m^3/s downhill through the line? Sketch accurately the energy line before and after the pump is installed. What is the maximum dependable flowrate that may be obtained through the line?

9.140. Assuming disruption of the flow when the negative pressure head reaches 6 m, to what elevation may the water surface in the tank be lowered by this siphon? Calculate the flowrates when the water surface is at elevation 29.4 and at the point of disruption.

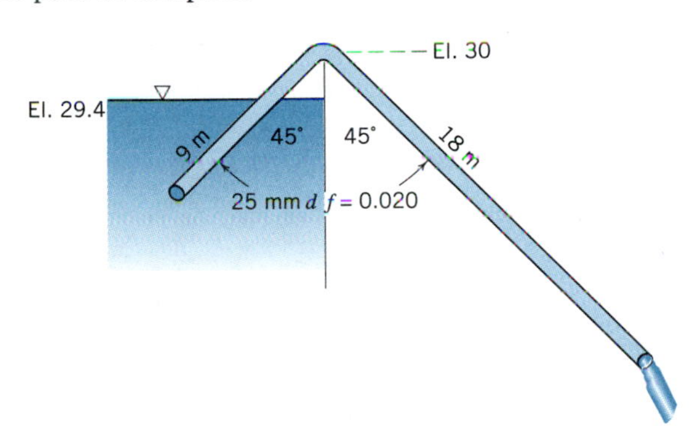

Problem 9.140

9.141. Calculate the smallest reliable flowrate that can be pumped through this pipeline. Assume atmospheric pressure 14.7 psia.

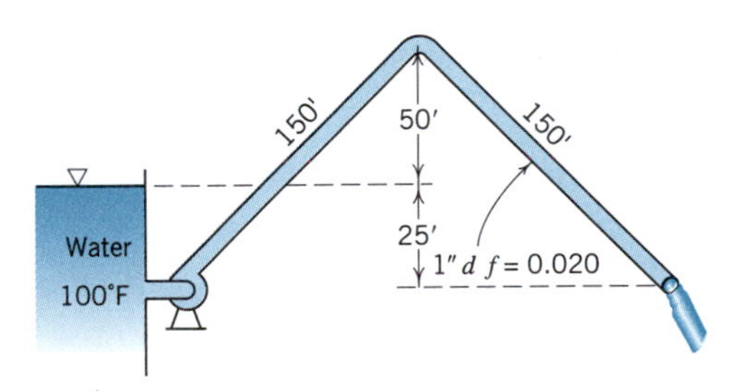

Problem 9.141

9.142. The suction side of a pump is connected directly to a tank of water at elevation 30. The water surface in the tank is at elevation 39. The discharge pipe from the pump is horizontal, 300 m long, of 150 mm diameter, and has a friction factor of 0.025. The pipe runs from west to east and terminates in a 75 mm frictionless nozzle. The nozzle stream is to discharge into a tank whose western upper edge is horizontal and is located at elevation 15 and 30 m east of the tip of the nozzle. What is the minimum power that the pump may supply for the stream to pass into the tank?

9.143. The pump is required to maintain the flowrate which would have occurred without any friction. What power pump is needed? Neglect local losses.

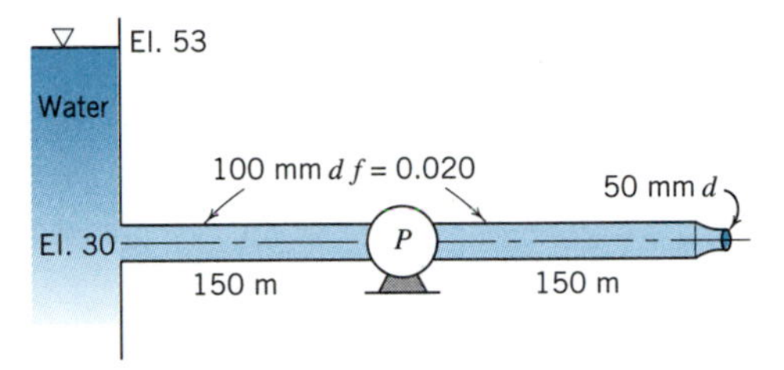

Problem 9.143

9.144. The flowrate through this pipeline and nozzle when the pump is not running (and assumed not to impede the flow) is 0.28 m^3/s. How much power must be supplied by the pump to produce the same flowrate with a 100 mm nozzle at the end of the line?

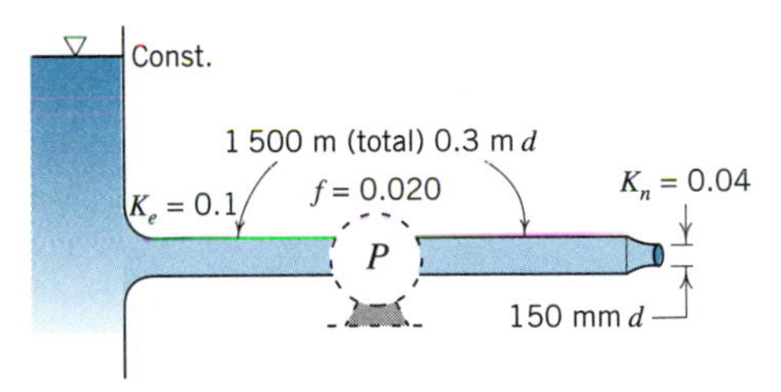

Problem 9.144

9.145. If the turbine extracts 530 hp from the flow, what flowrate must be passing through the system? What is the maximum power obtainable from the turbine?

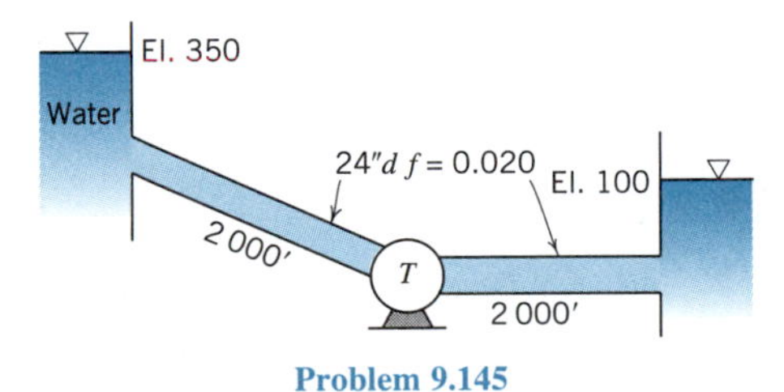

Problem 9.145

9.146. If there were no pump, 0.14 m^3/s of water would flow through this pipe system. Calculate the pump power required to maintain the same flowrate in the opposite direction.

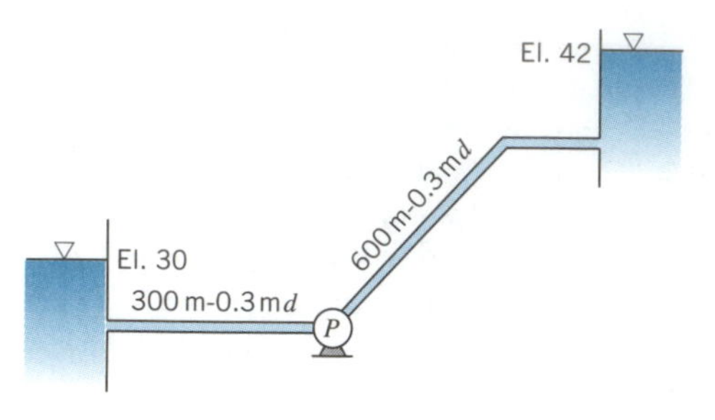

Problem 9.146

9.147. Water ($\nu = 1.0 \times 15^{-5}$ ft^2/s, $\rho = 1.94$ slug/ft^3) is pumped through the asphalted cast iron pipeline from the lower to the upper reservoir at a rate of 10.0 cfs. Find the horsepower that the pump must add to the water to accomplish this. Include local losses. Sketch the energy line approximately to scale and label the changes in slope. In addition, calculate the average wall shear stress in psi and the centerline velocity in the pipe away from the interference of bends and valves.

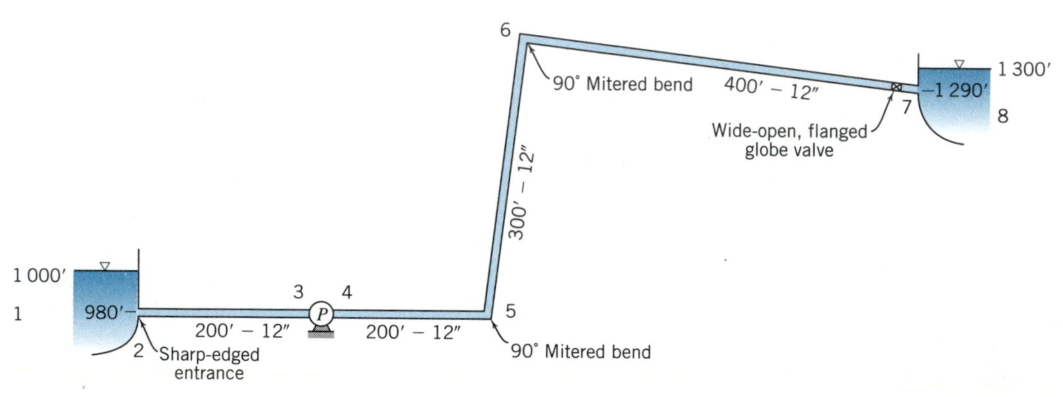

Problem 9.147

9.148. What horsepower must the pump add to the water to pump 9.0 cfs to the upper reservoir? Use a kinematic viscosity of 1.0×10^{-5} ft^2/s. Include local losses except at bends. Draw the energy line labeling all breaks in slope. Also find the centerline velocity and the shear stress 3 in. from the wall. Is the pipe smooth, rough, or transition?

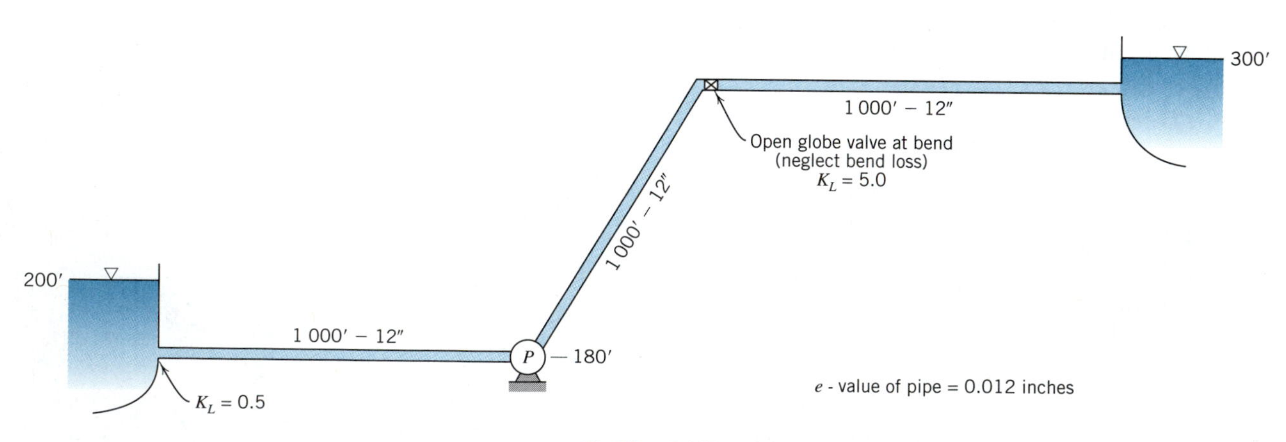

Problem 9.148

9.149. The flowrate is 0.13 m^3/s without the pump. Calculate the approximate pump power required to maintain a flowrate of 0.17 m^3/s.

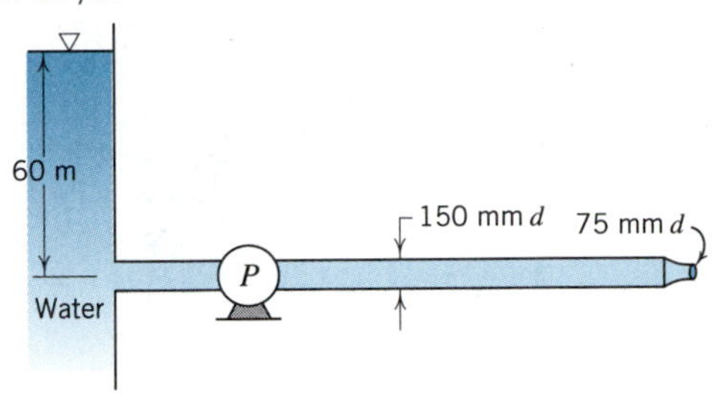

Problem 9.149

9.150. The pitot tube is on the centerline of the pipe. The flowrate is 2 cfs of water. Calculate the horsepower transferred from pump to fluid.

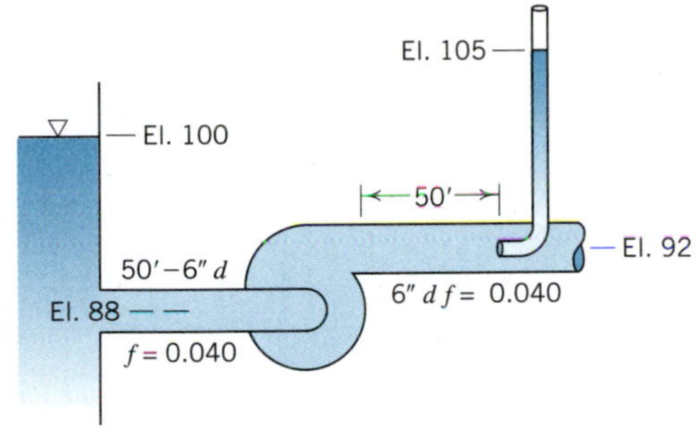

Problem 9.150

9.151. The pump of Fig. 9.25 (curve *A*) is to be installed between two reservoirs that are 8 000 m apart and have an elevation difference of 80 m. Select a pipe material and size so that the pump operates at maximum efficiency.

9.152. The pump of problem 9.151 is mounted in a 0.3 m diameter steel pipe of variable length that discharges to the atmosphere. The pipe is horizontal and its entrance (smooth bellmouth) is located at elevation 50 in a reservoir whose surface elevation is 100. For the range of flow for which the relation between head and flowrate are known for the pump, calculate and plot a relationship between flowrate and the length of the pipe.

9.153. For the situation of problem 9.152 find the length of pipe at which the pump operates at maximum efficiency.

9.154. A two-stage pump (impeller diameter 12 1/6 in.) whose pump characteristics are given below will pump what flowrate of water through the pipe system shown? Draw the system demand curve and the pump supply curve to obtain your solution. (See page 424)

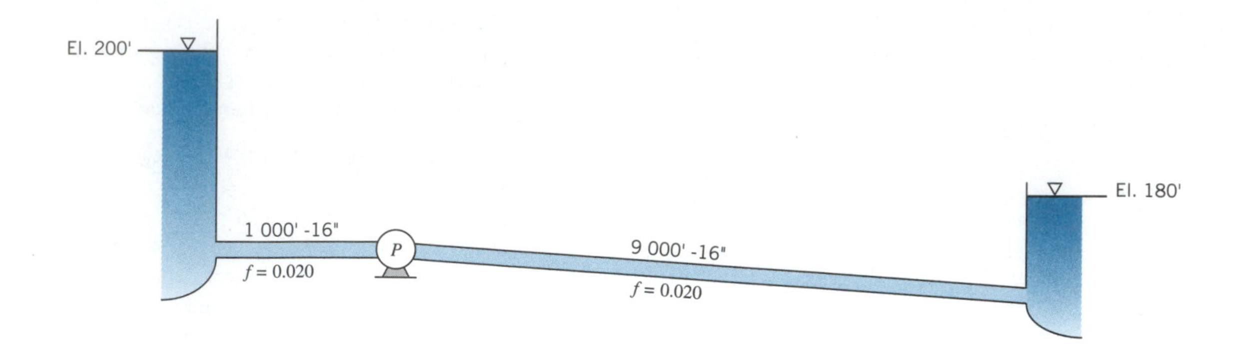

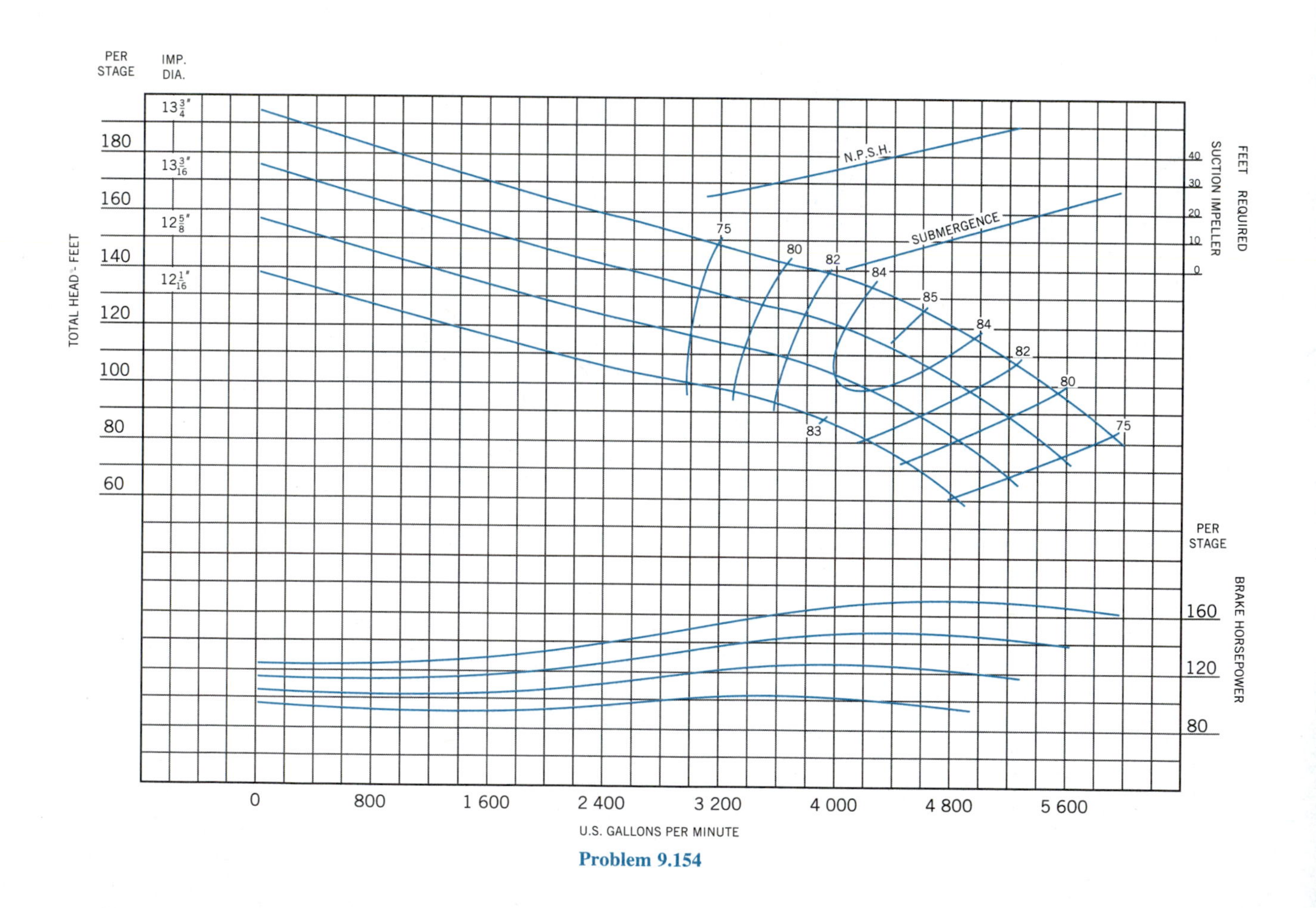

Problem 9.154

9.155. Two two-stage pumps whose characteristics are shown below (13 3/8 in. impeller) act in parallel to pump water to the upper reservoir through the 30 inch pipe. Neglecting local losses, compute the flowrate through each pump and the flowrate in the pipeline. Draw the system demand curve and the pump supply curve to obtain your solution.

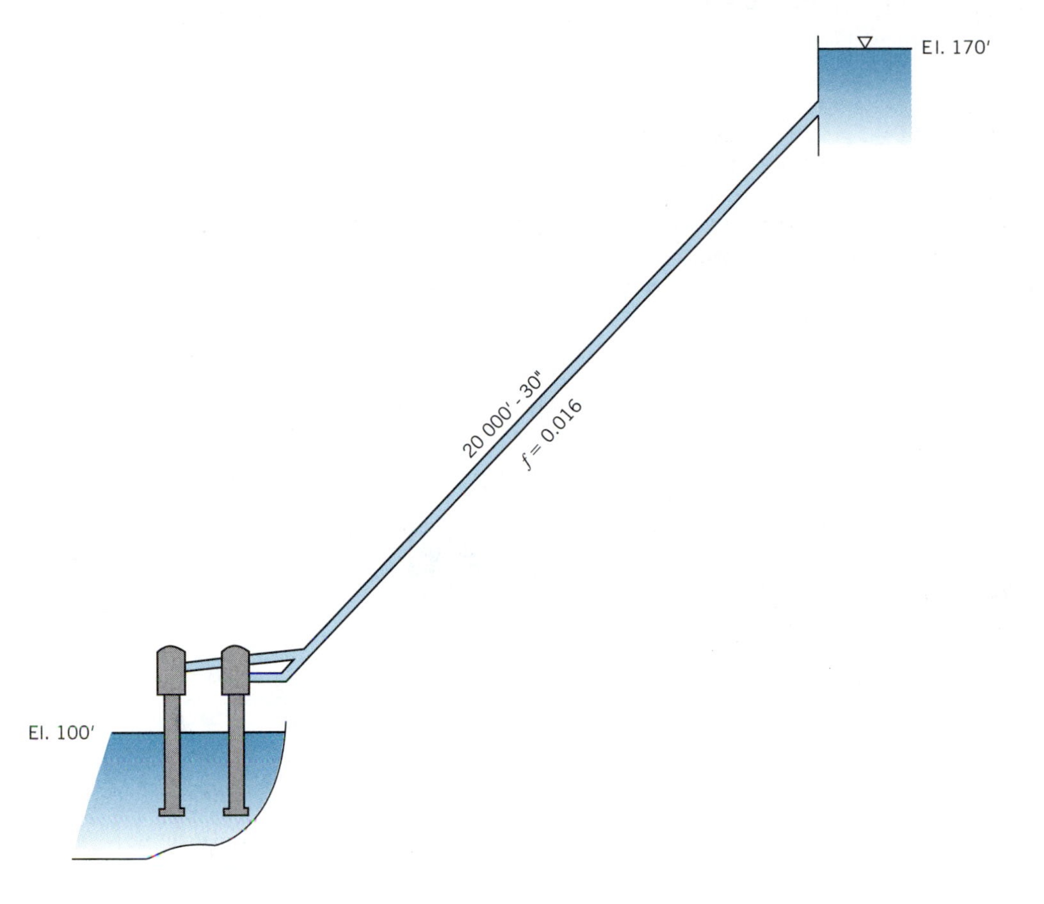

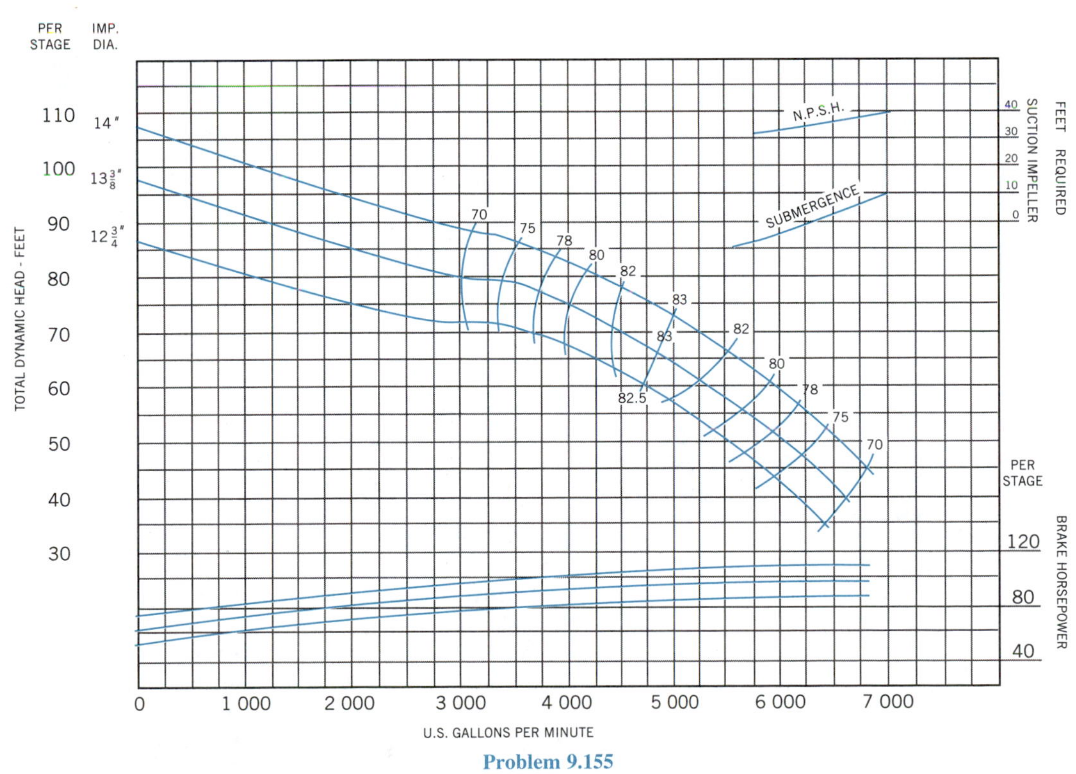

Problem 9.155

9.156. Five 25 mm diameter PVC pipes will conduct 2.4 l/s of water (20°C) between two reservoirs whose respective surface levels are constant. Find the diameter of a single PVC pipe that will carry the same flowrate between these reservoirs.

9.157. A 24 in. pipeline branches into a 12 in. and an 18 in. pipe, each of which is 1 mile long, and they rejoin to form a 24 in. pipe. If 30 cfs flow in the main pipe, how will the flow divide? Assume that $f = 0.018$ for both branches.

9.158. For the configuration shown, derive an equation for the velocity in the service line in terms of its diameter, friction factor, and length and of the velocities in the two sections of the venturi meter. Neglect minor losses. Note that in an Arctic setting, flow in the main line produces flow in the smaller home service line, thus keeping the lines from freezing up.

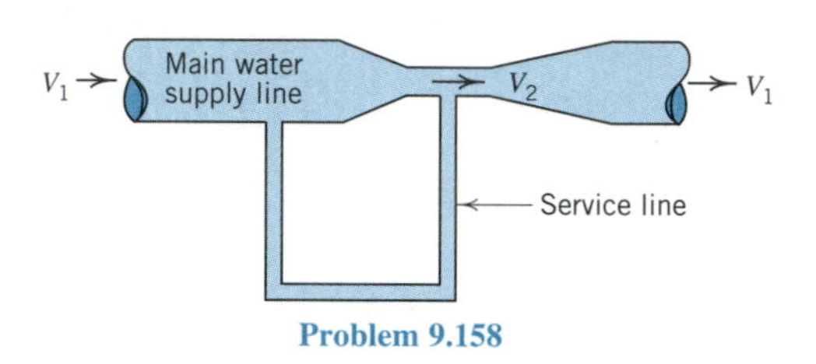

Problem 9.158

9.159. A 0.6 m pipeline carrying 0.85 m^3/s divides into 150 mm, 200 mm, and 300 mm branches, all of which are the same length and enter the same reservoir below its surface. Assuming that $f = 0.020$ for all pipes, how will the flow divide?

9.160. A straight 12 in. pipeline 3 miles long is laid between two reservoirs of surface elevations 500 and 350 entering these reservoirs 30 ft beneath their free surfaces. To increase the capacity of the line, a 12 in. line 1.5 miles long is laid from the original line's midpoint to the lower reservoir. What increase in flowrate is gained by installing the new line? Assume that $f = 0.020$ for all pipes.

9.161. Calculate the nozzle diameter which will maximize the jet power. Calculate this (maximum) jet power. Neglect local losses. Assume all friction factors 0.020.

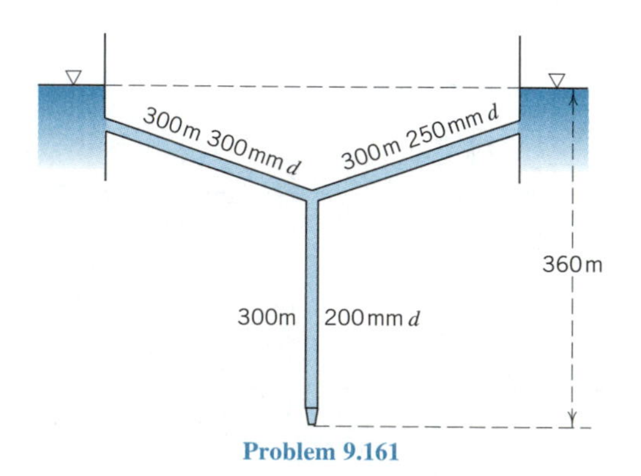

Problem 9.161

9.162. Solve problem 5.42 for three 1 000 m (3 300 ft) commercial steel pipes, neglecting entrance losses, but not losses in the tee. State assumptions made.

9.163. Solve problem 5.43 under the conditions set in problem 9.162.

9.164. Both of the pipelines are composed of 1 200 m of 0.3 m pipe terminating in a 0.1 m nozzle. The friction factors for both pipes are 0.020. The head losses in the nozzles are negligible. Points A and B are the midpoints of the lines. There is a short length of pipe between A and B containing a closed gate valve. When this valve is opened, will the total flowrate from the nozzles increase or decrease?

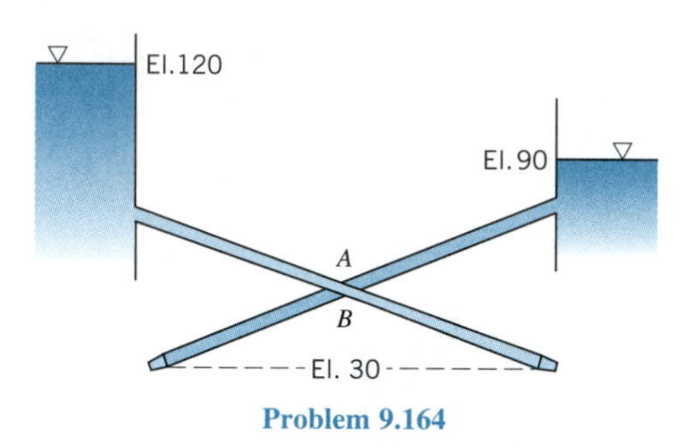

Problem 9.164

9.165. If the pump supplies 300 hp to the water, what flowrate will occur in the pipes? Assume a friction factor of 0.020 for both pipes.

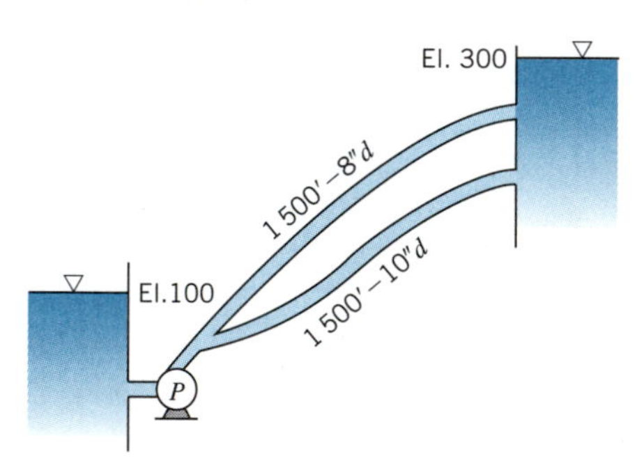

Problem 9.165

9.166. A 0.9 m pipe divides into three 0.45 m pipes at elevation 120. The 0.45 m pipes run to reservoirs which have surface elevations 90, 60, and 30, these pipes having respective lengths of 3.2, 4.8, and 6.8 kilometres. When 1.4 m^3/s flows in the 0.9 m line, how will the flow divide? Assume that $f = 0.017$ for all pipes.

9.167. Reservoirs A, B, and C have surface elevations 500, 400, and 300, respectively. A 12 in. pipe 1 mile long leaves reservoir A and runs to point O at elevation 450. Here the pipe divides and an 8 in. pipe, 1 mile long, runs from O to B and a 6 in pipe, 1.5 miles long, runs from O to C. Assuming that $f = 0.020$, calculate the flowrates in the lines.

9.168. Three pipes join at a common point at elevation 105. One, a 0.3 m line 600 m long, goes to a reservoir of surface elevation 120; another, a 150 mm line 900 m long, goes to a reservoir of surface elevation 150; the third (150 mm) runs 300 m to elevation 75, where it discharges into the atmosphere. Assuming that $f = 0.020$, calculate the flowrate in each line. Calculate these flowrates when a 50 mm nozzle is attached to the end of the third pipe.

9.169. A 0.3 m pipeline 600 m long leaves a reservoir of surface elevation 150 and runs to elevation 120, where it divides into two 150 mm lines each 300 m long, both of which discharge into the atmosphere, one at elevation 135, the other at elevation 105. Calculate the flowrates in the three pipes if all friction factors are 0.025.

9.170. The pump is to deliver 110 l/s to the outlet at elevation 165 and 220 l/s to the upper reservoir. Calculate pump power and the required diameter of the 300 m pipe.

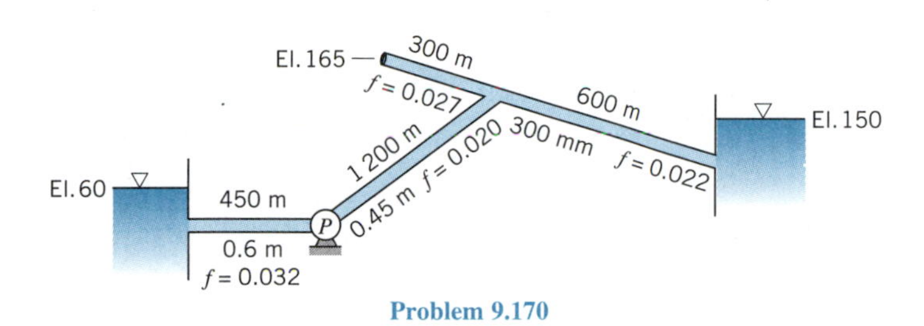

Problem 9.170

9.171. A two-stage pump whose characteristics are shown in Problem 9.154 is pumping water to the two reservoirs. The 13 3/4 in. impeller is being used. Calculate the flowrate through the pump and in each line. Draw the energy line for each pipe.

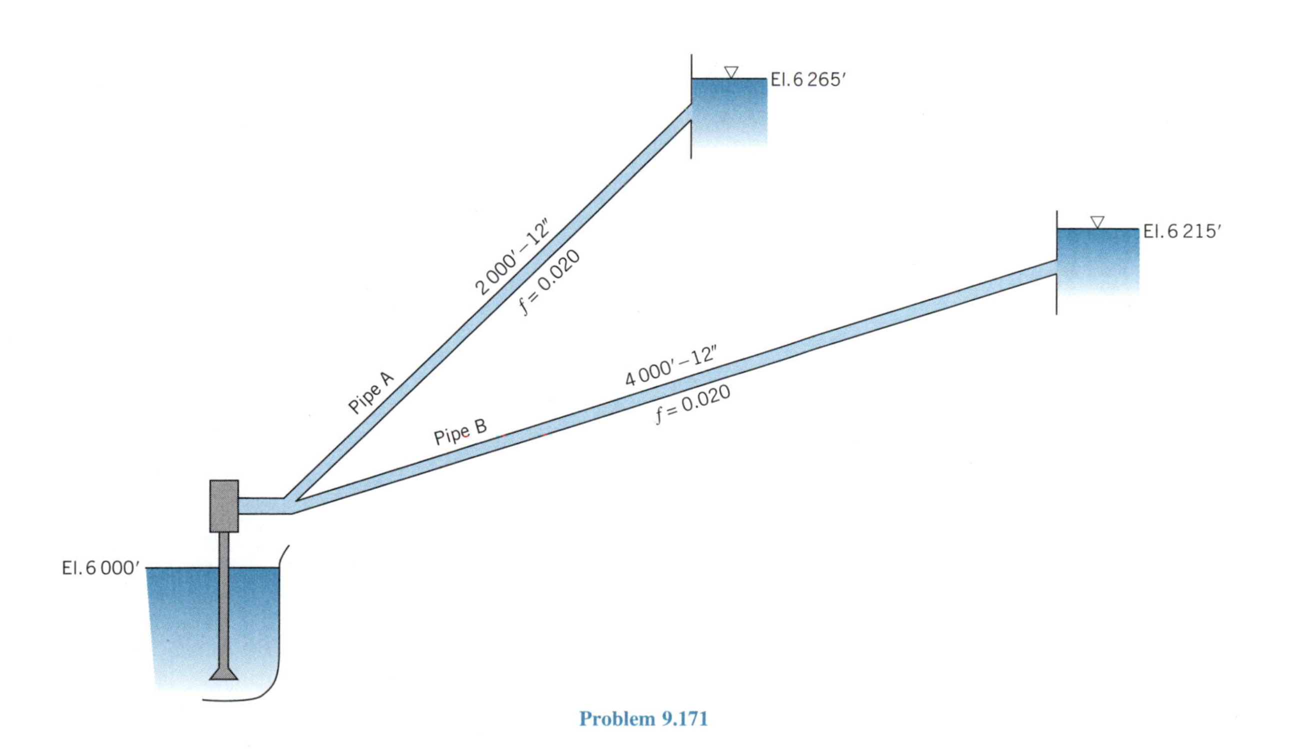

Problem 9.171

9.172. Water is flowing. For $Q_8 = 4.0$ cfs, calculate Q_6, Q_{12}, and pump horsepower.

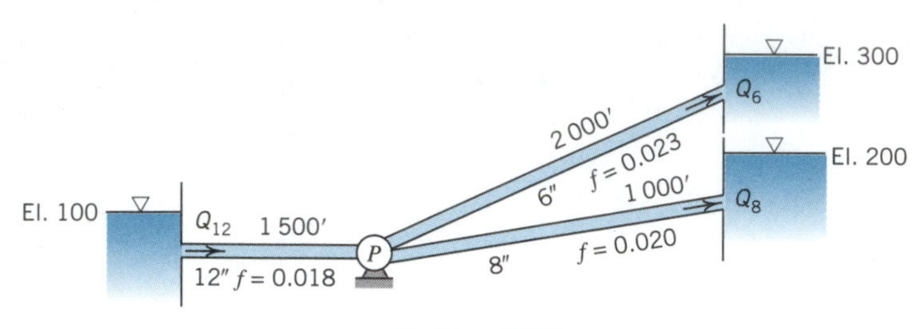

Problem 9.172

9.173. If a turbine replaces the pump in the pipe system of problem 9.172, what is the maximum horsepower it can produce?

9.174. An analogy to the laminar flow of fluids in pipes may be made with the flow of electric current through a resistance. If head loss is analogous to voltage drop, and flowrate to current, what combination of quantities in the fluid problem is analogous to electric resistance? Applying this idea to problem 9.175, determine the flowrates in the pipes if the kinematic viscosity of the fluid flowing is 2.8×10^{-4} m^2/s.

9.175. Calculate the flowrates in the pipes of this loop if all friction factors are 0.020.

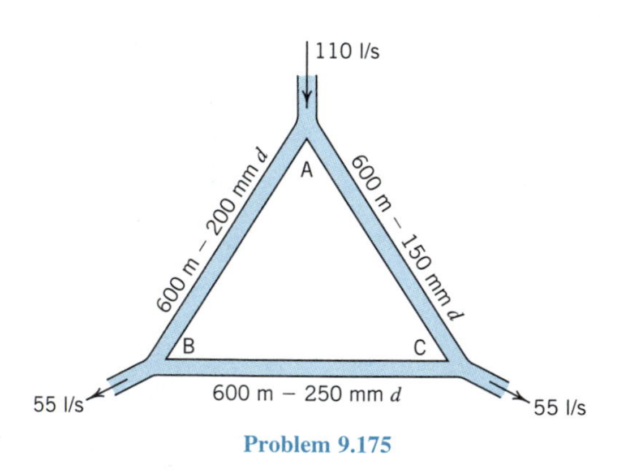

Problem 9.175

9.176. The head loss in each of the pipes shown is given by the formula $h_L = KQ^2$ where $K_{AB} = 10$, $K_{AC} = 5$, $K_{BC} = 2$, $K_{BD} = 10$, $K_{CD} = 5$. Compute the ΔQ-value for each loop and the new value for Q_0, which would be used in the second trial. The assumed flowrates for the first trial are shown on the sketch.

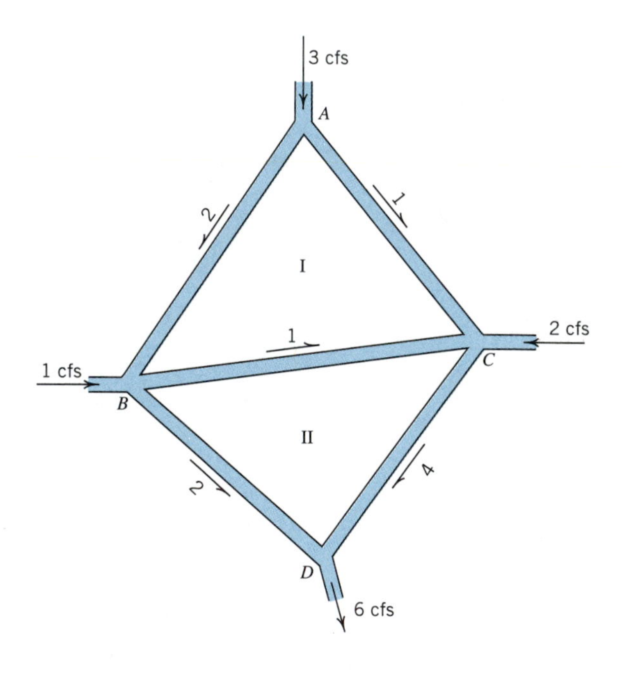

Problem 9.176

9.177. In the network shown, pipe 2 is common to three network loops. The K-values in the formula $h_L = KQ^2$, where Q is in cfs are $K_1 = K_2 = K_3 = 10$, $K_4 = K_5 = K_6 = K_7 = 5$. Compute the first Δ calculation for each loop and adjust the original flowrate estimates in preparation for the second iteration. The original estimates are shown on the sketch.

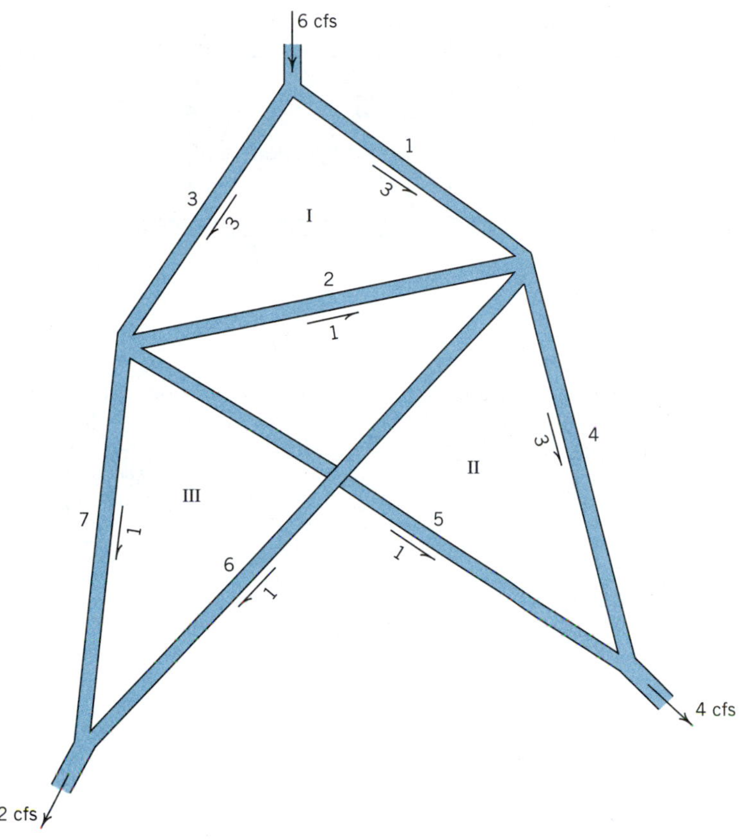

Problem 9.177

9.178. Calculate the flowrates in the pipes and pressures at the junctions of this network. Assume rough flow and concrete pipes. Neglect local losses and losses in the short pipe from the reservoir. The network is at elevation 10; the reservoir at 70.

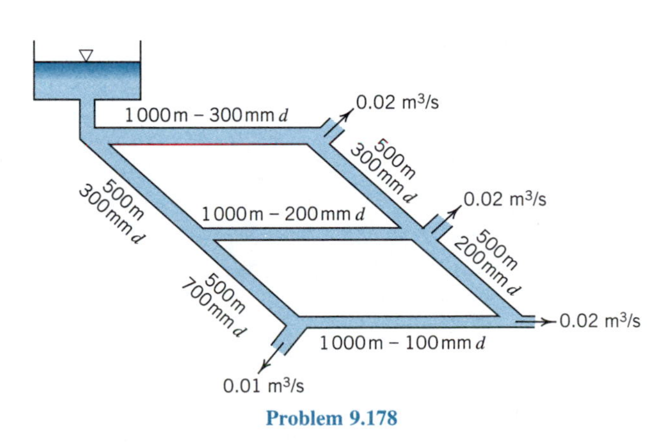

Problem 9.178

9.179. Calculate the flowrates in the pipes of this network. The length, diameter, and f-values are shown for each pipe.

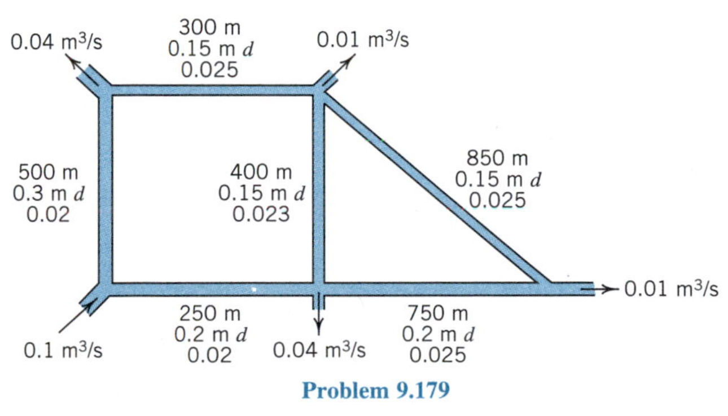

Problem 9.179

9.180. Calculate the flowrates in the pipes of this network. The length, diameter, and f-values are shown for each pipe.

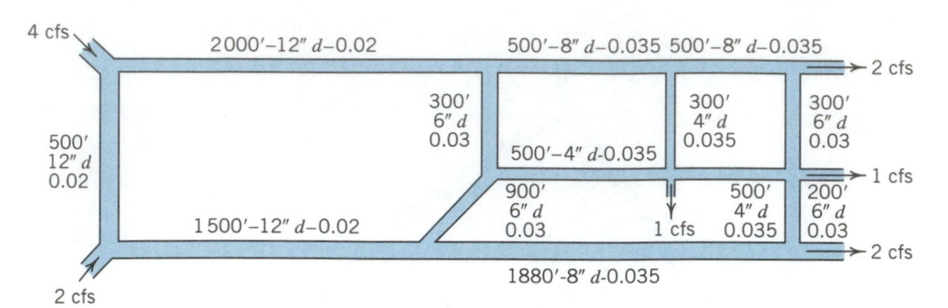

Problem 9.180

9.181. Suppose that the head versus flowrate characteristic of a pump is approximately $E_P = 70(1 - 2Q)$. Calculate the flowrates in the pipes in this network with and without the pump. Let $f = 0.015$ in all pipes.

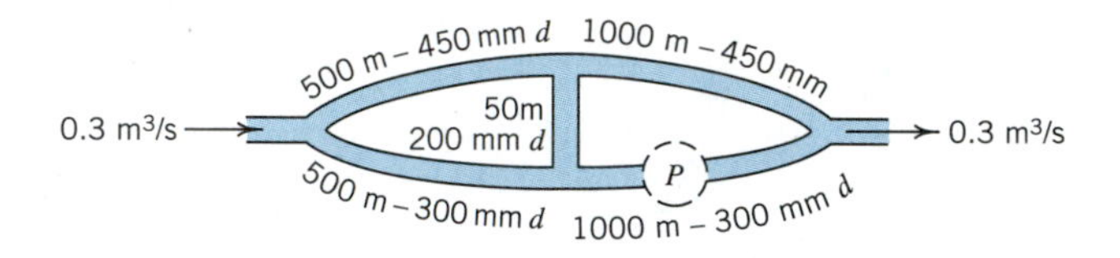

Problem 9.181

9.182. The pipe is initially full of water with the valve completely closed. Compute the time required for the flow velocity to reach 99% of its final value for sudden valve opening. When completely open, the valve head loss is negligible.

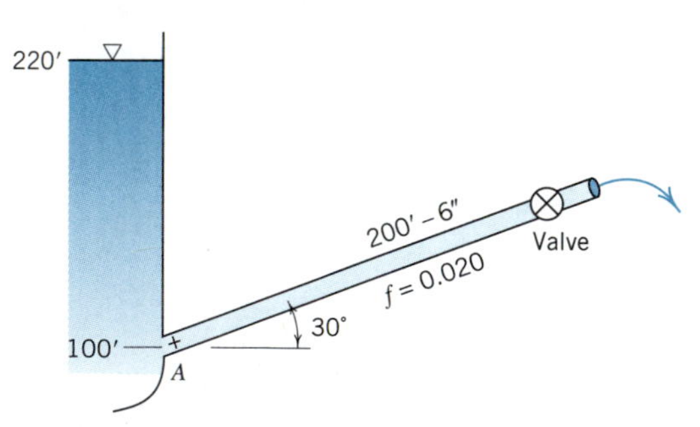

Problem 9.182

9.183. For the same conditions, solve the preceding problem if the valve were located at section A instead.

9.184. The pressure head in the horizontal pipe line of length L is H_2 before the valve is opened.

a. If the valve is opened suddenly, find an equation for the time required for the flow to reach 99% of its final value.

b. When it is desired to shut off the flow in the pipeline, the velocity of 10 ft/s is decreased linearly to zero in 100 s. What is the minimum head in the system, where does it occur and at what time? For this part let $L =$ 3 220 ft, $H_2 = 100$ ft, $H_1 = 200$ ft, $D = 1.0$ ft, $f =$ 0.020.

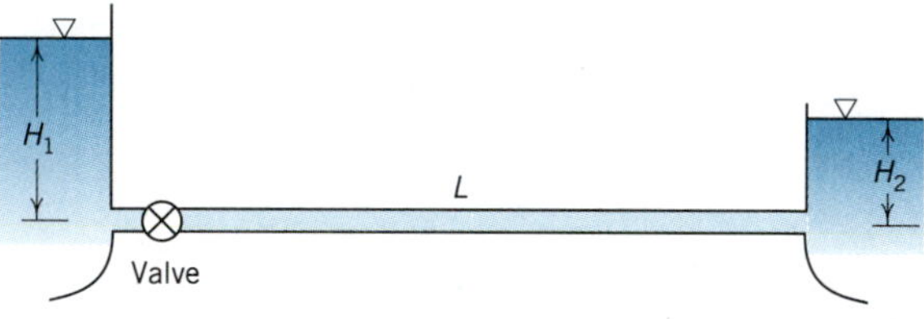

Problem 9.184

9.185. A horizontal pipe 10 000 ft long and 8 ft in diameter leaves a reservoir under a head of 60 ft and terminates in a valve. If the Darcy-Weisbach friction factor is 0.030, how long after the sudden opening of the valve will the velocity of flow reach a value 99% of its final value? Neglect local losses.

9.186. The globe valve in the pipe below is opened instantaneously. If the loss coefficient for the wide-open valve is 3.0, how many seconds will it take for the flow velocity to reach 99% of its final value?

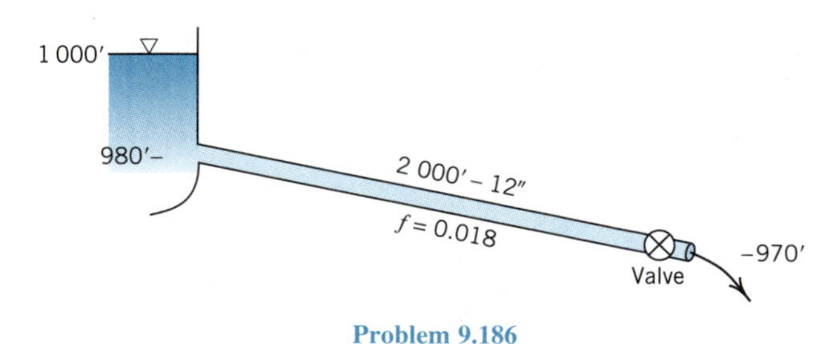

Problem 9.186

9.187. The globe valve in the pipe of the previous problem is opened instantaneously to establish flow in the pipe. If the valve's wide-open loss coefficient is 6.3, how many seconds will it take for the flow velocity to reach 99.9% of its final value?

9.188. With the valve in the pipeline wide open, a steady flow of velocity 10 ft/s occurs. Under these conditions the head loss through the wide-open valve and the local losses are negligible. The valve is partially closed creating a head loss which causes the flow velocity to decrease linearly with time to 5 ft/s in 10 s. What, when, and where is the maximum and minimum pressure head occurring in the pipeline? Find the loss coefficient for the partially open valve after valve movement ceases.

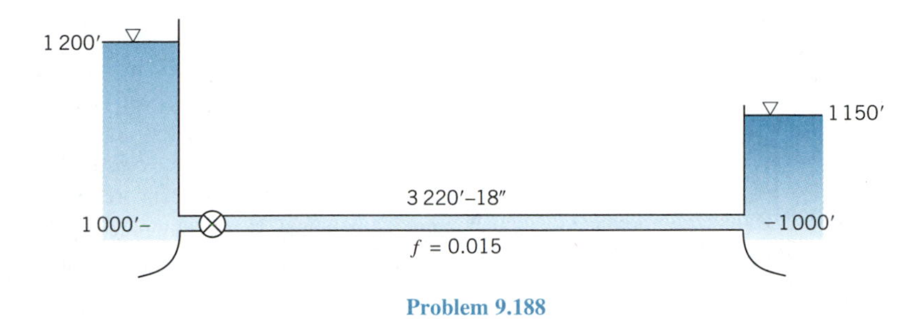

Problem 9.188

9.189. The valve is closed in a manner which causes the velocity of flow to decrease quadratically to zero according to the relation

$$V = V_o \left[1 - \frac{t}{t_c}\right]^2$$

where the initial velocity, V_o, is 5 ft/s and the time of valve closure, t_c, is 30 s. Find the minimum pressure head in the pipe, its location and, the time after valve closure begins at which it occurs.

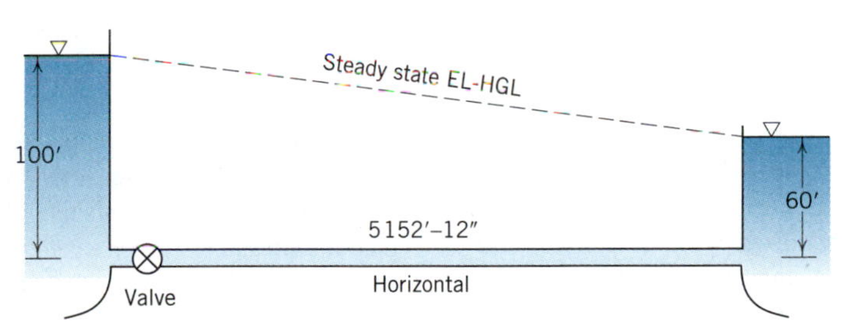

Problem 9.189

9.190. At time $t = 0$ the valve in the pipe is closed. It is *proposed* to open the valve in such a manner that the velocity in the pipe will increase linearly with time to its steady state value of 10 ft/s in 100 s. Find the maximum and minimum pressure head occurring in the system for the proposed scheme. Can this operating scheme work? Explain.

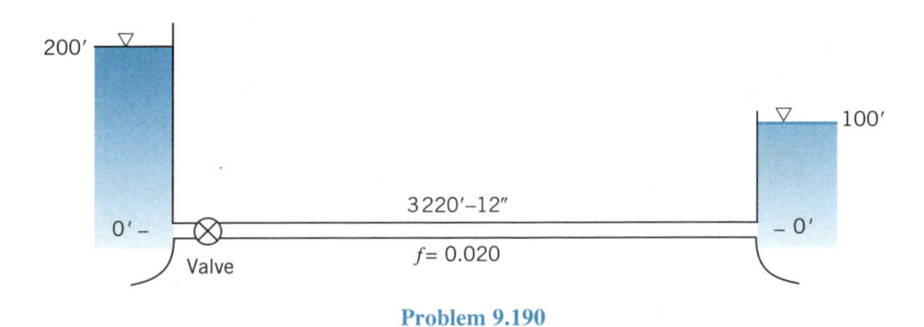

Problem 9.190

9.191. A high-pressure water system is being designed for use in a lumber mill to remove bark from logs. The main portion of the pipe system is 6 in. steel pipe with walls 0.219 in. thick. The inside diameter of the pipe is 6.187 in., and the working pressure in the pipe is 1 094 psi (working stress of the steel is 15 000 psi). Under steady flow conditions, the pressure is about 750 psi with a flow velocity of 10 fps. Because of the operation there is a potential for rapid valve closure. Compute the wave speed and make a recommendation as to whether the water hammer pressures being developed could overstress the pipe material.

9.192. Calculate the wave speed in the following situations.

a. Steel pipe—36 in. diam. ID, 0.375 in. wall thickness

b. Ductile iron—18 in. diam. ID, 0.50 in. wall thickness

c. Aluminum—4 in. diam. ID, 0.10 in. wall thickness

d. Asbestos cement—11.56 in. diam. ID, 1.26 in. wall thickness

e. PVC (Class 125)—6.22 in. diam. ID, 0.20 in. wall thickness

Note the difference in a-values for the wide range of pipe materials and sizes above.

9.193. A 14 in. Class 51 ductile iron pipe is used to convey water between two reservoirs. The outside diameter is 15.30 in. and, for this class of pipe, the wall thickness is 0.36 in. Assuming the pipe can be considered thin-walled, compute the wave speed.

9.194. For the pipe in Problem 9.193, compute the percent change in pipe volume, which occurs as the result of a sudden stoppage of a 10 fps flow. What is the percent change in the density of the water?

9.195. A plastic supply pipe in a building water system is anchored at both ends with glued joints along its length. If $\mu_p = 0.5$ for this material, what is the wave speed? The line is 800 ft long, 6.00 in. inside diameter with a 0.200 in. wall thickness. The modulus of elasticity for the material is 500 000 psi. The water in the pipe normally flows at a velocity of 10 ft/s. The system valves are so designed that they can be closed very quickly. If the steady state-pressure in the pipe is about 100 psi, what is the maximum pressure you would estimate to occur in the system under the worst water hammer conditions?

9.196. A steel pipe 8 000 ft long, 6 ft inside diameter, and with a wall thickness of 0.50 in. is stressed at 7 000 psi. It has water flowing at 5 ft/s. For sudden valve closure, how much flow enters the pipe after closure?

9.197. Calculate the wave speed in a water-filled copper tube. The tube is 0.375 in. inside diameter and 75 ft long with a wall thickness of 0.03 in. The steady state pressure in the tube is 73 psi.

9.198. For the pipes in problem 9.192, compute the pressure increase in psi generated by the sudden stoppage of a flow velocity of 7 ft/s. If the operating pressure in each case is 100 psi, calculate the percent increase in pressure caused by the sudden stoppage.

9.199. A 12 in. PVC line 2 000 ft long is laid below ground with concrete anchor blocks at each bend. If the inside diameter of the pipe is 12.091 in. and the wall thickness is 0.311 in., compute the wave speed. What is the change in pipe volume caused by a velocity of 10 ft/s brought suddenly to rest? What is the critical valve closure time which would still generate the full water hammer pressure at the valve?

9.200. In problem 9.195, what is the critical valve closure time which would still generate the full water hammer pressure at the valve?

9.201. In problem 9.191, it is proposed to reduce the water hammer pressure by injecting air into the flow. Using the polytropic thermodynamic process, calculate the percent air, which must be injected into the flow to reduce the water hammer pressure by 50%.

9.202. In a forced main where wastewater is being pumped at 100 psi from a treatment facility, the free air in the wastewater is estimated at 1.0%. If the wave speed in the pipe is 3 600 ft/s with no free air, what will be its reduced value in the presence of free air? Use a polytropic thermodynamic process and assume the E-value of wastewater is the same as pure water.

9.203. A residence is being served by a 1 inch service line teed into an 8 inch water line. Assuming the wave speeds in all the lines are identical at 3 500 ft/s, calculate the water hammer pressure wave transmitted into the 8 inch line resulting from the sudden stoppage of a 10 ft/s flow in the service connection.

9.204. A fire hydrant at the end of a 12 inch water main is normally maintained at a pressure of 75 psi. Assume that the flow is negligible at the time that the hydrant is suddenly opened to induce a velocity of 20 ft/s in a 2 inch fire hose. What reduced pressure would occur in the water main? Use a wave speed of 3 000 ft/s in the water main.

9.205. During a fire, a fire hydrant draws enough water from an 8 inch main to induce a velocity of 8 ft/s in the main at a pressure of 50 psi. If the wave speed in the main is 3 600 ft/s, what maximum pressures will be imposed in the homes along the main supplied by 1 inch service connections if the hydrant is suddenly closed bringing the main velocity to zero?

CHAPTER

10 FLOW IN OPEN CHANNELS

Open-channel flow embraces a variety of problems that arise when water flows in natural water courses, canals, irrigation ditches, sewer lines, flumes, and so forth—a province of paramount importance to the civil engineer. Although open-channel problems practically always involve the flow of water, and although the experimental results used in these problems were obtained by hydraulic tests, modern fluid mechanics indicates the extent to which these results may be applied to the flow of other liquids in open channels.

10.1 FUNDAMENTALS

In the problems of pipe flow, as may be seen from the hydraulic grade line, the pressure in the pipe can vary along the pipe, depending on the conditions imposed on the end of the line. Open-channel flow, however, is characterized by the fact that pressure conditions are determined by the constant pressure, usually atmospheric, existing on the entire surface of the flowing liquid. Usually, pressure variations within an open-channel flow can be determined by the principles of hydrostatics since the streamlines are ordinarily straight, close to parallel, and approximately horizontal (compare Fig. 5.3, Section 5.3); when these conditions are satisfied, the hydraulic grade lines for all the streamtubes are the same and coincide with the liquid surface. There are, of course, many exceptions to the foregoing situation and, in flowfields where streamlines are convergent, divergent, curved, or steeply sloping, each streamline has, in general, its own hydraulic grade line and these do not lie in the liquid surface.

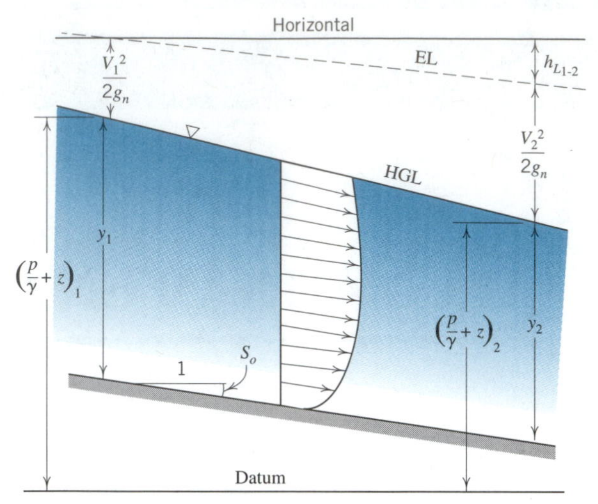

Fig. 10.1

A typical reach of open-channel flow is shown in Fig. 10.1.[1] The individual streamlines of the flow are very slightly convergent but may be safely assumed to be close enough to parallel so that the hydraulic grade line coincides with the liquid surface. The energy line is located one velocity head[2] above the hydraulic grade line, and the head loss between two points is defined, as usual, as the drop in the energy line between these two points. Thus, the energy line-hydraulic grade line approach to one-dimensional flow problems can be expected to find wide application in many problems of open-channel flow.

Open-channel flow may be laminar or turbulent, steady or unsteady, *uniform* or *varied*, *subcritical* or *supercritical.* The complexity of unsteady open-flow problems forbids their treatment in an elementary text, but the other categories are examined herein; the emphasis, however, is on steady turbulent flow, which is the problem generally encountered in practice. The definitions of subcritical and supercritical flow are presented in Section 10.6, but the significance, causes, and limits of uniform and varied flow must be examined first. The meaning of these terms and also the fundamentals of open-channel flow may be seen from a comparison of ideal-fluid flow and real-fluid flow in identical prismatic channels[3] leading from reservoirs of the same surface elevation (Fig. 10.2). No resistance is encountered by the ideal fluid as it flows down the channel. Because of this lack of resistance, the ideal fluid continually accelerates under the influence of gravity. Thus, the mean velocity continually increases, and with this increase of velocity a reduction in flow cross section is required by the continuity principle. Reduction in flow cross section is characterized by a decrease in depth of flow; since the depth of flow continually *varies* (from section to section, not with time) this type of fluid motion is termed *varied flow*.[4]

[1]The channel slopes in many of the illustrations in this chapter have been exaggerated to emphasize their existence. In open-channel practice, slopes are very seldom encountered which are greater than 1 (vertical) in 100 (horizontal), or 0.01.

[2]Strictly, this distance should be $\alpha V^2/2g_n$ but, in view of the relatively flat velocity profile, α is in most cases only slightly greater than unity.

[3]A prismatic channel has an unvarying cross-sectional shape and a constant bottom slope.

[4]The term *nonuniform flow* is also used, but *varied flow* is preferred.

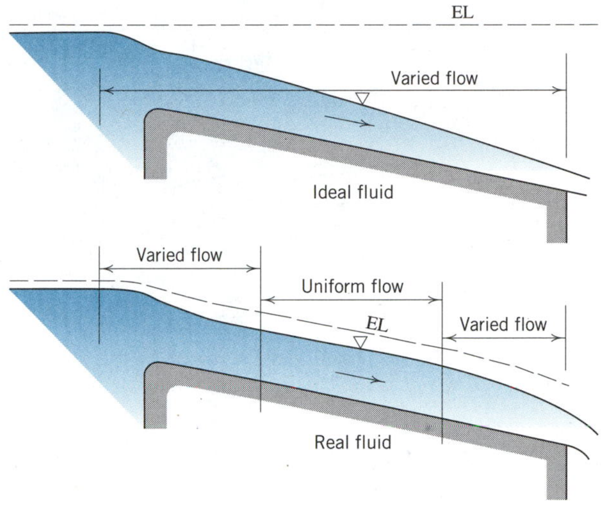

Fig. 10.2

When real fluid flows in the same channel, there are resistance forces due to fluid viscosity and channel roughness. Analysis of the resistance forces originating from these same properties in pipes has shown that such forces depend on the velocity of flow (Eq. 9.3). Thus, in the upper end of the channel where motion is slow, resistance forces are small, but the components of gravity forces in the direction of motion are about the same as for the ideal fluid. The resulting unbalanced forces in the direction of motion bring about acceleration and varied flow in the upper reaches of the channel. However, as the velocity increases down the channel, the forces of resistance build until they finally balance those caused by gravity. When this force balance occurs, constant-velocity motion is attained; it is characterized by no change of flow cross section and thus no change in depth of flow; this is *uniform flow*. Near the lower end of the channel as the free overfall is approached, pressure and gravity forces again exceed resistance forces and varied flow results.

Obviously, an inequality between the above-mentioned forces is more probable than a balance of these forces, and, hence, varied flow occurs in practice to a far greater extent than uniform flow. In short channels, for example, uniform-flow conditions may never be attained because of the long reach of channel necessary for the establishment of uniform flow; nevertheless, solution of the uniform-flow problem forms the basis for open-channel flow calculations.

10.2 UNIFORM FLOW—THE CHEZY EQUATION

The fundamental equation for uniform open-channel flow may be derived readily by equating (since there is no change in momentum) the equal and opposite force components of gravity and flow resistance and applying some of the fundamental notions of fluid mechanics encountered in the analysis of pipe flow. Consider the uniform flow of a liquid at a depth y_o between sections 1 and 2 of the open channel of Fig. 10.3. The forces acting on the liquid in the control volume *ABCD* are (1) the forces of static pressure, F_1 and F_2,

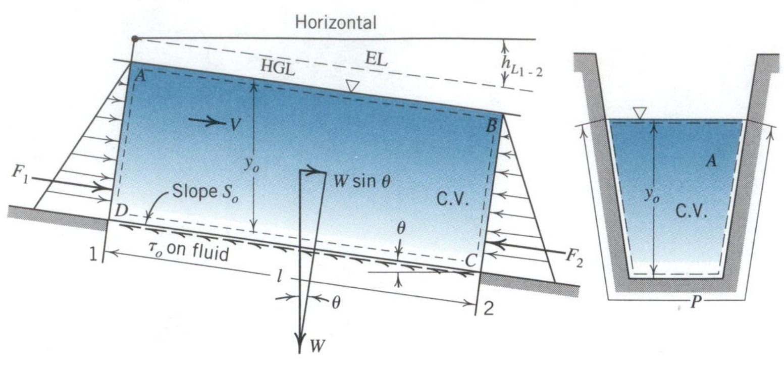

Fig. 10.3

acting on the ends of the liquid mass; (2) the weight, W, of the liquid mass, which has a component $W \sin \theta$ in the direction of motion; (3) the pressure forces exerted on the liquid mass by the bottom and sides of the channel;[5] and (4) the force of resistance exerted on the liquid mass by the bottom and sides of the channel $Pl\overline{\tau}_o$ where $\overline{\tau}_o$ is the average shear stress around the perimeter of the channel. A summation of forces in the direction of motion gives

$$F_1 + W \sin \theta - F_2 - Pl\overline{\tau}_o = 0$$

since there is no change in momentum between sections 1 and 2. Obviously, $F_1 = F_2$, $W = A\gamma l$, and $\sin \theta = h_L/l$ because the EL, HGL, and the channel bed are all parallel in uniform flow. The slope, S_o, of both channel bed and liquid surface is $\tan \theta$, and for the small slopes encountered in open-channel practice the approximation $\tan \theta = \sin \theta$ may be made. The term A/P is recognized (Sections 7.10 and 9.7) as the hydraulic radius, R_h. When these substitutions are made in the above equation, an expression for the mean shear stress, $\overline{\tau}_o$, results; it is

$$\overline{\tau}_o = \gamma R_h S_o \tag{10.1}$$

In pipe flow, τ_o was shown (Section 9.1) to be given by $f\rho V^2/8$ in which f, the friction factor, is dependent on surface roughness and Reynolds number, but more dependent on the magnitude of the roughness for highly turbulent flow. The mechanism for real fluid motion is similar in pipes and open channels, and if it is assumed[6] that the hydraulic radius concept will account adequately for the differences in the cross-sectional shapes of circular pipes and open channels, these expressions for τ_o and $\overline{\tau}_o$ may be equated. Solving for V and replacing γ/ρ by g_n,

$$V = \sqrt{\frac{8g_n}{f}}\sqrt{R_h S_o}$$

or, defining $C = \sqrt{8g_n/f}$,

$$V = C\sqrt{R_h S_o} \tag{10.2}$$

[5]These are not shown since they do not enter the following equations.

[6]Experience has shown this to be a valid assumption for regular prismatic channels (canals, flumes, etc.) with turbulent flow. It cannot be expected to apply to laminar flow (Section 10.4) or to channels of irregular or distorted cross section (such as natural streams at flood stage).

This is called the *Chezy equation* after the French hydraulician who established this relationship experimentally in 1775. By applying the continuity principle, the equation may be represented in terms of flowrate as

$$Q = CA\sqrt{R_h S_o} \tag{10.3}$$

which is the fundamental equation for uniform flow in open channels.

10.3 THE CHEZY COEFFICIENT AND THE MANNING *n*

From the time of Chezy to the present, many experiments in field and laboratory have been performed to determine the magnitude of the Chezy coefficient, C, and its dependence on other variables. The simplest relation and the one most widely used is the result of the work of Manning[7] and others and may be derived from an analysis of data obtained from his own experiments and those of others. The results may be summarized by the empirical relation

$$\text{(SI units): } C = \frac{R_h^{1/6}}{n}$$

which may be combined with the Chezy equation to give the so-called Chezy-Manning equation

$$\text{(SI units): } Q = \left(\frac{1}{n}\right) A R_h^{2/3} S_o^{1/2}$$

These last two equations were derived from the metric data; hence, the unit of length is the metre. Although n is often supposed to be a characteristic of the channel roughness, it is convenient to consider n to be dimensionless.[8] Then, while the U.S. Customary and metric (SI) equations differ, the values of n are the same in both systems. Unfortunately, the coefficient $(1/n)$ in the equations has the dimensions $t^{-1}L^{1/3}$; as a consequence, conversion to U.S. Customary units requires inclusion of the factor $(3.28)^{1/3} = 1.49$, where 3.28 is the number of feet in a metre. The result is

$$\text{(U.S. Customary units): } C = \frac{1.49 R_h^{1/6}}{n}$$

and

$$\text{(U.S. Customary units): } Q = \left(\frac{1.49}{n}\right) A R_h^{2/3} S_o^{1/2}$$

[7] *Trans. Inst. Civil Engrs. Ireland*, vol. 20, p. 161, 1890.

[8] See V. T. Chow, *Open-Channel Hydraulics*, McGraw-Hill, 1959, p. 98, for a complete discussion of the dimensional aspects of n.

Accordingly, the results can be summarized as

$$C = \frac{uR_h^{1/6}}{n} \tag{10.4}$$

and

$$Q = \left(\frac{u}{n}\right) AR_h^{2/3}S_o^{1/2} \tag{10.5}$$

where $u = 1$ for SI units and $u = 1.49$ for U.S. Customary units.

The Manning n is obtained from a descriptive statement of the channel character; some typical values are given in Table 5. There is no substitute for experience and judgment in the interpretation and selection of values for n. Systematic experiments and analyses, similar to those of Nikuradse and Colebrook in the field of pipe flow, have yet to produce a clear definition of open-channel roughness or a scientific interpretation of the Chezy C or Manning n.

It is possible, however, to learn something of the character of n by comparing with the friction factor f. From the relationship $C = \sqrt{8g_n/f}$ and Eq. 10.4,

$$n = uR_h^{1/6}\sqrt{\frac{f}{8g_n}}$$

Introducing numerical values for g_n in the respective systems,

$$\text{(U.S. Customary units): } n = 0.093f^{1/2}R_h^{1/6}$$

$$\text{(SI units): } n = 0.113f^{1/2}R_h^{1/6}$$

Because $(3.28)^{1/6} = 1.22$ and $0.133/0.093 = 1.22$, the n-values obtained are equal irrespective of units. Clearly, n is not an absolute roughness coefficient because it depends on

TABLE 5 Manning *n*-values[9]

Type of Conduit	Minimum	Normal	Maximum
Pipes flowing partly full			
Welded steel	0.010	0.012	0.014
Coated cast iron	0.010	0.013	0.014
Corrugated metal	0.021	0.024	0.030
Cement mortar lined (neat)	0.010	0.011	0.013
Concrete culvert (finished)	0.011	0.012	0.014
Concrete pipe (steel form)	0.012	0.013	0.014
Vitrified clay	0.011	0.014	0.017
Drainage tile	0.011	0.013	0.017

[9]Table 5 was adapted, with permission of McGraw-Hill Book Company, from V. T. Chow, *Open Channel Hydraulics*, Chapter 5, copyright © 1959 McGraw-Hill Book Co. H. H. Barnes, "Roughness Characteristics of Natural Channels," *U.S. Geol. Survey Wat. Supply Pap.* 1849, 1967, gives a set of typical n values together with matching descriptive data and color photographs of natural channels.

TABLE 5 (Continued)

Type of Conduit	Minimum	Normal	Maximum
Lined open channels			
Smooth steel	0.011	0.012	0.014
Wood (planed)	0.010	0.012	0.014
Wood (unplaned)	0.011	0.013	0.015
Cement (neat)	0.010	0.011	0.013
Concrete (troweled)	0.011	0.013	0.015
Concrete (float finish)	0.013	0.015	0.016
Concrete (unfinished)	0.014	0.017	0.020
Gunite	0.016	0.019	0.023
Brick	0.012	0.015	0.018
Rubble masonry	0.017	0.025	0.030
Asphalt (smooth)	—	0.013	—
Asphalt (rough)	—	0.016	—
Unlined open channels			
Earth, straight and uniform, clean	0.016	0.018	0.020
Earth, straight and uniform, short vegetation	0.022	0.027	0.033
Earth, winding and sluggish, clean	0.023	0.025	0.030
Earth, winding, sluggish, short vegetation	0.025	0.030	0.033
Gravel, straight and uniform, clean	0.022	0.025	0.030
Dredged, clean	0.025	0.028	0.033
Rock cuts, smooth and uniform	0.025	0.035	0.040
Rock cuts, jagged and irregular	0.035	0.040	0.050
Natural channels			
Clean, straight, no riffles or pools	0.025	0.030	0.033
Clean, winding, some pools and shoals	0.033	0.040	0.045
Sluggish, weedy, deep pools	0.050	0.070	0.080
Mountain streams, gravel, cobbles	0.030	0.040	0.050
Mountain streams, cobbles, large boulders	0.040	0.050	0.070
Flood plains, pasture	0.025	0.030	0.035
Flood plains, light brush and trees	0.035	0.050	0.060
Flood plains, heavy stand of timber	0.080	0.100	0.120

the hydraulic radius and must exhibit the same characteristics as the friction factor, which in turn depends on relative roughness and Reynolds number. Interestingly, increases in R_h tend to be offset by related decreases in f so that n may change only gradually for a given boundary surface for increasing depths and flowrates.

On the basis of experience and by careful use of judgment, it is possible to estimate n and use the Chezy-Manning equation successfully. When used with care it is simple and reliable; however, it is a dimensionally nonhomogeneous (see Section 8.2), empirical equation. This nonhomogeneity leads to the requirement that n or the constants in the equation have dimensions and makes it impossible to establish a fundamentally sound basis for determining n.

ILLUSTRATIVE PROBLEM 10.1

A rectangular channel lined with asphalt is 20 ft wide and laid on a slope of 0.000 1. Calculate the depth of uniform flow in this channel when the flowrate is 400 ft^3/s.

SOLUTION

The units given are U.S. Customary, so we will use Eq. 10.5 with $u = 1.49$.

$$Q = \left(\frac{u}{n}\right) AR_h^{2/3}S_o^{1/2} \tag{10.5}$$

The values to be substituted into this equation are:

$$B = 20 \text{ ft} \qquad S_o = 0.000\,1 \text{ ft/ft} \qquad Q = 400 \text{ ft}^3/\text{s}$$

We need to calculate the hydraulic radius R_h, so it is necessary to obtain expressions for P and A.

$$P = 20 + 2y_o \text{ ft} \qquad A = 20y_o \text{ ft}^2$$

$$R_h = \left(\frac{A}{P}\right) = \left(\frac{20y_o}{20 + 2y_o}\right) \text{ ft}$$

From Table 5, $n = 0.013$. Substituting all these values into Eq. 10.5,

$$400 = \left(\frac{1.49}{0.013}\right)(20y_o)\left(\frac{20y_o}{20 + 2y_o}\right)^{2/3}(0.000\,1)^{1/2}$$

Reducing the equation as much as possible, we obtain

$$y_o\left(\frac{20y_o}{20 + 2y_o}\right)^{2/3} = 17.45$$

Solving by trial,

$$y_o = 6.85 \text{ ft} \bullet$$

For rough asphalt the n-value would be 0.016 (23% greater than 0.013), the depth would have been 7.95 ft (16% greater than 6.85 ft).

ILLUSTRATIVE PROBLEM 10.2

A concrete-lined canal with an n-value of 0.014 is constructed on a slope of 0.33 m/km and conveys 23 m^3/s of water. Find the uniform flow depth if the canal is trapezoidal in cross section with a bottom width of $B = 6$ m and side slopes of $z = 1.5$.

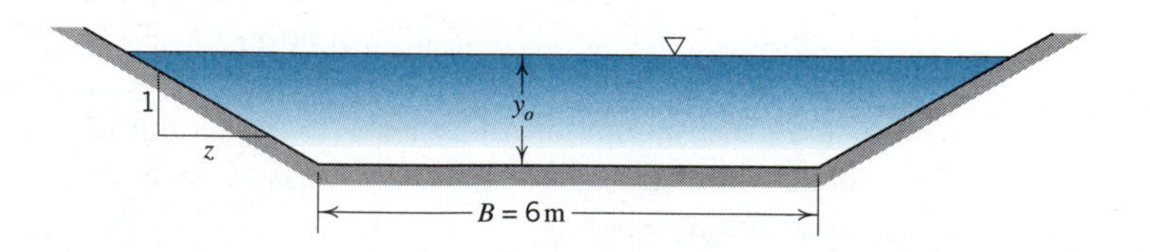

SOLUTION

Once again, Eq. 10.5 applies with $u = 1.00$.

$$Q = \left(\frac{u}{n}\right) A R_h^{2/3} S_o^{1/2} \tag{10.5}$$

The values to be substituted into this equation are

$B = 6$ m $\qquad S_o = (0.33 \text{ m/km})/(1\,000 \text{ m/km}) = 0.000\,33$ m/m $\qquad Q = 23 \text{ m}^3/\text{s}$

We need to calculate the hydraulic radius R_h, so it is necessary to obtain expressions for P and A.

$$P = 6 + 2\sqrt{1.5^2 + 1.0^2}\, y_o = 6 + 3.61 y_o \text{ m}$$

$$A = 6y_o + 2\left[\frac{1}{2}(y_o \times 1.5y_o)\right] = 6y_o + 1.5y_o^2 \text{ m}^2$$

$$R_h = \left(\frac{A}{P}\right) = \left(\frac{6y_o + 1.5y_o^2}{6 + 3.61y_o}\right) \text{ m}$$

Substituting into Eq. 10.5,

$$23 = \left(\frac{1.00}{0.014}\right)(6y_o + 1.5y_o^2)\left(\frac{6y_o + 1.5y_o^2}{6 + 3.61y_o}\right)^{2/3} (0.000\,33)^{1/2}$$

Solving by trial,

$$y_o = 1.77 \text{ m} \bullet$$

10.4 UNIFORM LAMINAR FLOW[10]

Laminar flow ($Vy_o/\nu \lesssim 500$) in open channels occurs in drainage from streets, airport runways, parking areas, and so forth. Here the flow is in thin sheets of virtually infinite width (i.e., without sidewalls) with resistance only at the bottom of the sheet; thus the flow is essentially a two-dimensional one. A definition sketch for uniform laminar flow is shown in Fig. 10.4. From the pipe flow analysis of Section 9.2, a parabolic velocity profile and

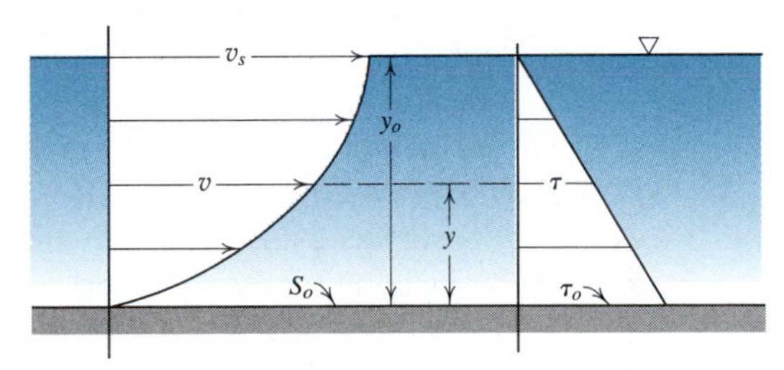

Fig. 10.4

[10]See W. M. Owen, "Laminar to Turbulent Flow in a Wide Open Channel," *Trans. A.S.C.E.*, vol. 119, 1954, and L. G. Straub, E. Silberman, and H. C. Nelson, "Open Channel Flow at Small Reynolds Number," *Trans. A.S.C.E.*, Vol. 123, 1957.

linear shear stress profile are to be expected. The hydraulic radius for such a flow may be deduced by considering a section of width b normal to the flow; the flow cross section is thus a rectangle of depth y_o and width b. However, the wetted perimeter (i.e., the boundary offering resistance to the flow) is only the channel bottom, of width b; therefore $R_h = A/P = by_o/b = y_o$. Inserting this in Eq. 10.1,

$$\tau_o = \gamma y_o S_o$$

However, in laminar flow, τ_o is also given by $\mu(dv/dy)_o$ in which $(dv/dy)_o$ is the velocity gradient at the channel bottom. From the properties of the parabolic velocity profile, $(dv/dy)_o$ is $2v_s/y_o$ and the mean velocity V is $2v_s/3$. Combining these, equating the two expressions for τ_o, and substituting ρg_n for γ, and ν for μ/ρ produce

$$V = \frac{g_n y_o^2 S_o}{3\nu} \tag{10.6}$$

which relates mean velocity, slope, and depth when the flow is laminar. As in pipes, the limit of laminar flow is defined by an experimentally determined critical Reynolds number, in this case having a value of around 500 if the Reynolds number is defined as Vy_o/ν. From this a specific relationship between Chezy coefficient and Reynolds number may be derived for this laminar flow. Substitute V from equation 10.6 into the definiton for Reynolds number to obtain

$$\mathbf{R} = g_n y_o^3 S_o/3\nu^2 \tag{10.7}$$

By rearranging Eq. 10.6 to the Chezy form,

$$V = \frac{g_n \sqrt{S_o}\, y_o^{3/2}}{3\nu} \sqrt{y_o S_o} \tag{10.8}$$

and by comparing Eqs. 10.7 and 10.8 we find that

$$C = \sqrt{g_n \mathbf{R}/3} \tag{10.9}$$

10.5 HYDRAULIC RADIUS CONSIDERATIONS

With the hydraulic radius R_h playing a prominent role in the equations of open-channel flow, and with depth variation a basic characteristic of such flows, the variation of hydraulic radius with depth (and other variables) becomes an important consideration. Clearly this is a problem of section geometry that requires no principles of mechanics for solution.

Consider first the variation of hydraulic radius with depth in a rectangular channel (Fig. 10.5*a*) of width B. Here $R_h = By/(B + 2y)$ and it is immediately evident that: for $y = 0$, $R_h = 0$ and, for $y \to \infty$, $R_h \to B/2$; therefore the variation of R_h with y must be as shown. From this comes a useful engineering approximation: *for narrow deep sections* $R_h \cong B/2$; since any (nonrectangular) section when deep and narrow approaches a rectangle this approximation may be used for any deep and narrow section—for which the hydraulic radius may be taken to be one half of the mean width.

Another useful engineering approximation may be discovered by examining the variation of hydraulic radius with channel width (Fig. 10.5*b*) for a constant depth y. With $R_h = By/(B + 2y)$ it is noted that: for $B = 0$, $R_h = 0$ and, for $B \to \infty$, $R_h \to y$; thus the variation of R_h with B is as shown. From this it may be concluded that for *wide shallow*

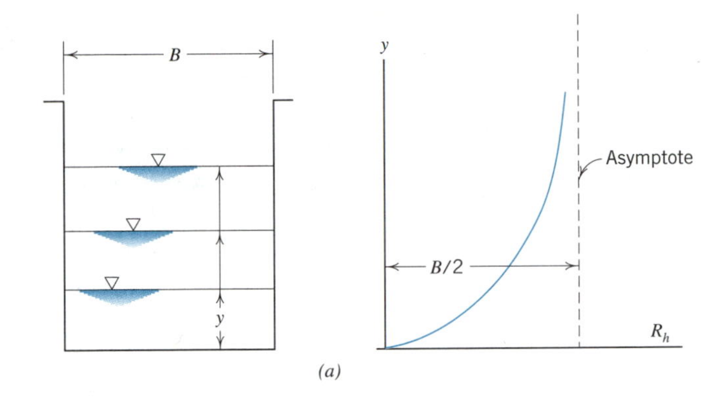

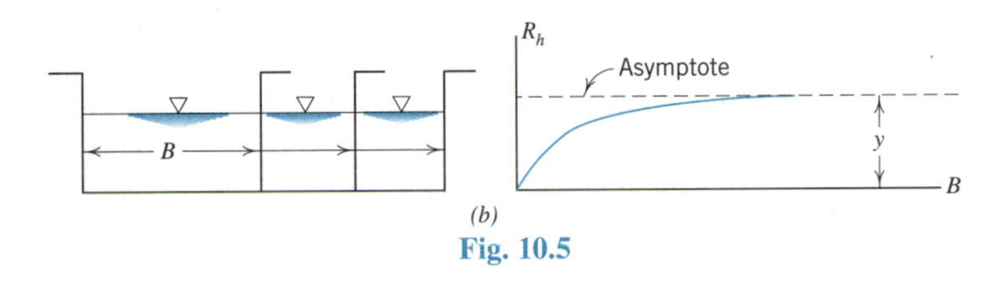

Fig. 10.5

rectangular sections $R_h \cong y$; for nonrectangular sections the approximation is also valid if the section is wide and shallow—here the hydraulic radius is approximately the mean depth. It is informative to investigate the hydraulic radius of cross sections other than rectangular. A typical nonrectangular section is the trapezoidal one of Fig. 10.6 for which $R_h = A/P = (By + zy^2)/(B + 2\sqrt{1 + z^2}y)$. The derivative of R_h with respect to y will allow investigation of the form of the relationship between R_h and y. Performing this operation yields

$$\frac{dR_h}{dy} = \frac{1 + 2z(y/B)\left[1 + (y/B)\sqrt{1 + z^2}\right]}{1 + 4\sqrt{1 + z^2}(y/B)\left[1 + (y/B)\sqrt{1 + z^2}\right]} \tag{10.10}$$

For $0 \leq \alpha \leq 90°$ $(0 \leq z \leq \infty)$ and for all values of y, the denominator of Eq. 10.10 is larger than the numerator. Therefore, $dR_h/dy < 1$; and, also, dR_h/dy diminishes with increasing y. Thus the form of the variation of R_h with y is as shown on Fig. 10.7. It

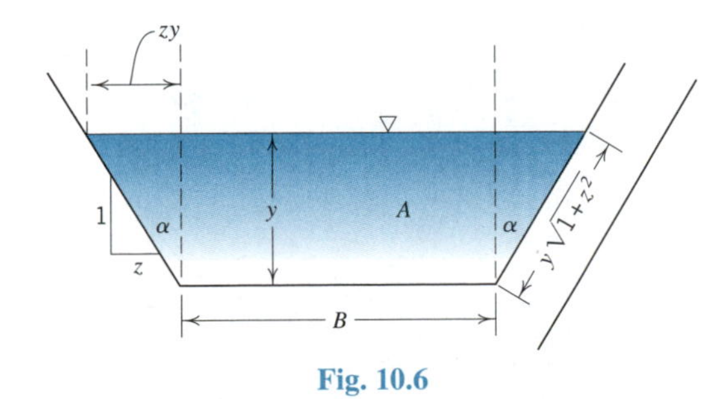

Fig. 10.6

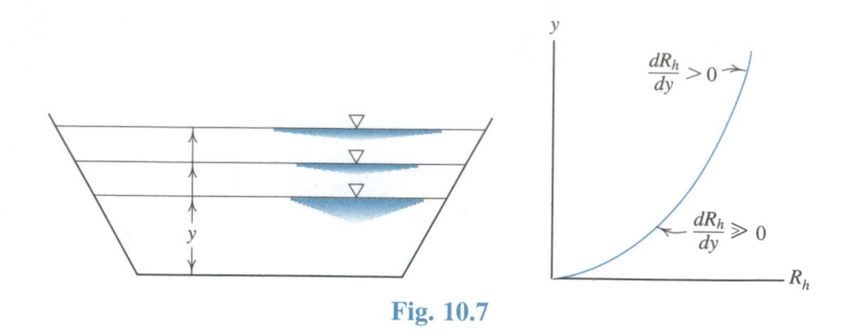

Fig. 10.7

superficially resembles the curve of Fig. 10.5*a* but has no vertical asymptote since dR_h/dy does not approach zero as y approaches infinity. These are also general properties of most (nontrapezoidal) channel sections with divergent side walls.

Another common nonrectangular cross section is the circular pipe flowing partially full. It is of wide use in engineering practice for underground open channels such as sewer lines, storm drains, and culverts. Here, for channels flowing half full or completely full, $R_h = d/4$; and, for $y = 0$, $R_h = 0$. Thus, a continuous variation of R_h with y may be expected to feature a maximum value as shown in Fig. 10.8. Of more engineering significance in such problems is the variation of $AR^{2/3}$ with y, since this combination of A and R_h appears in the Chezy-Manning equation 10.5. With $AR^{2/3}$ plotted against y (see Fig. 10.8), another maximum point is noted; from this it may be concluded that for given S_o and n there is a point at which the (uniform) flowrate is also maximum. Engineers usually disregard this in channel design, but it sometimes helps to explain phenomena which occur when channels with convergent walls flow nearly full.

Another important engineering problem of section geometry is the reduction of boundary resistance by minimization of the wetted perimeter for a given area of flow cross section. With A fixed, $R_h = A/P$, and, with P to be minimized, it is apparent that this may be considered a problem of maximization of the hydraulic radius. The desirability of this is evident from the above, but it is to be noted that reduction of wetted perimeter also tends to reduce the cost of lining material, grading, and construction. Hence a channel section of maximum hydraulic radius not only results in optimum hydraulic design but also tends toward a section of minimum cost. For this reason a section of maximum hydraulic radius is known as the *most efficient*, or *best, hydraulic cross section.* Since there is little hope for a generalized solution for all possible section shapes (the circular section

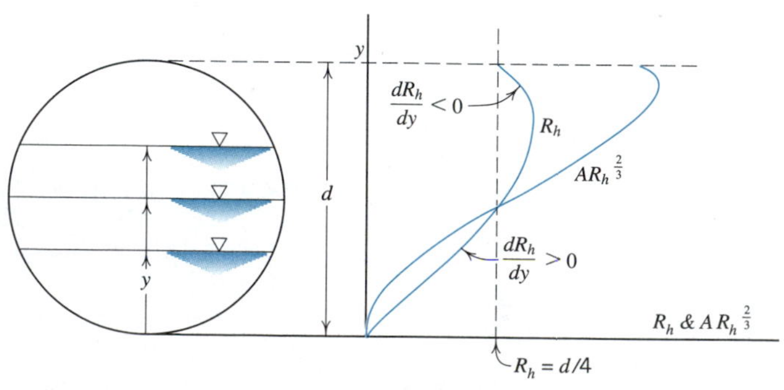

Fig. 10.8

is the best), the trapezoidal one (Fig. 10.6) is again studied because of its wide use and good approximation to other sections. As before, $A = By + zy^2$ and $P = B + 2y\sqrt{1 + z^2}$, but here[11] A and z are fixed and the relation between B and y for maximum R_h is sought. Eliminating B from the expression for P by using the equation for A, R_h may be written as a function of y:

$$R_h = \frac{A}{P} = \frac{A}{A/y - zy + 2y\sqrt{1 + z^2}}$$

Differentiating R_h with respect to y and equating the result to zero gives

$$A = y^2(2\sqrt{1 + z^2} - z) \qquad B = 2y(\sqrt{1 + z^2} - z)$$

which, when the expression for A is substituted into the foregoing expression for R_h, yields

$$R_{h_{\max}} = y/2 \tag{10.11}$$

Thus, for *best hydraulic design conditions, a trapezoidal open channel should be proportioned so that its hydraulic radius is close to one-half of the depth of flow*; this may also be used as a rough guide for the design of other channel sections that approach trapezoidal shape.

Since a rectangle is a special form of trapezoid, the foregoing results may be applied directly to the rectangular channel. Here again $R_{h_{\max}} = y/2$ and, as $z = 0$, $A = 2y^2$. Since A also equals By, it is evident that $B = 2y$ and that the *best proportions for a rectangular channel exist when the depth of flow is one-half the width of the channel.*

ILLUSTRATIVE PROBLEM 10.3

Find the dimensions of the most-efficient cross section for a rectangular channel that is to convey a uniform flow of 10 m³/s if the channel is lined with gunite concrete and is laid on a slope of 0.000 1.

SOLUTION

For the most efficient cross section, the hydraulic radius R_h is one-half the depth. And, as we have seen in the previous development for most-efficient cross-sections for rectangular channels, the area is given by $2y_o^2$. Substituting this information into Eq. 10.5 gives

$$Q = \left(\frac{u}{n}\right) AR_h^{2/3}S_o^{1/2} = \left(\frac{u}{n}\right)(2y_o^2)\left(\frac{y_o}{2}\right)^{2/3} S_o^{1/2} \tag{10.5}$$

From Table 5, the normal n-value for a gunite-lined channel is 0.019 and for SI units, $u = 1.0$.

[11]The slope of the side walls z is limited in an earth canal by the angle of repose of the soil. If the canal is appropriately lined, z may have any value.

Substituting all the values into Eq. 10.5,

$$10 = \left(\frac{1.0}{0.019}\right)(2y_o^2)\left(\frac{y_o}{2}\right)^{2/3}(0.000\ 1)^{1/2}$$

Reducing the equation to its simplest form,

$$y_o^{8/3} = 15.1$$

$$y_o = 2.77 \text{ m} \qquad B = 2y_o = 5.54 \text{ m} \bullet$$

Note that the *solution for most efficient cross section is not a trial solution* as in Illustrative Problem 10.1.

ILLUSTRATIVE PROBLEM 10.4

Solve Illustrative Problem 10.3 for a trapezoidal channel with side slopes $z = 2$.

SOLUTION

Once again, for the most efficient cross section, the hydraulic radius R_h is one half of the depth. From the previous development for most efficient cross section for trapezoidal channels,

$$A = y^2(2\sqrt{1 + z^2} - z)$$

Substituting into Eq. 10.5,

$$Q = \left(\frac{u}{n}\right) AR_h^{2/3}S_o^{1/2} = \left(\frac{u}{n}\right)\left[y_o^2(2\sqrt{1 + z^2} - z)\right]\left(\frac{y_o}{2}\right)^{2/3} S_o^{1/2} \qquad (10.5)$$

This equation can be solved for y_o to yield a direct solution for the uniform flow depth.

$$y_o = \left[\frac{1.59\ Qn}{uS_o^{1/2}(2\sqrt{1 + z^2} - z)}\right]^{3/8}$$

Substituting the numerical values into the above equation yields

$$y_o = \left[\frac{1.59 \times 10 \times 0.019}{1.0 \times 0.000\ 1^{1/2}(2\sqrt{1 + 2^2} - 2)}\right]^{3/8}$$

$$y_o = 2.56 \text{ m} \qquad B = 2y_o(\sqrt{1 + z^2} - z) = 2 \times 2.56(\sqrt{1 + 4} - 2) = 1.21 \text{ m} \bullet$$

Note that, once again, the use of the *most efficient cross section has resulted in a direct solution* for the channel shape.

For an extended or distorted channel cross section such as that of Fig. 10.9, which represents a simplification of a river section in flood, routine calculation of hydraulic radius from A/P will lead to large errors; such a calculation would imply that the effect of boundary resistance is uniformly distributed through the flow cross section, which is clearly not the case. Furthermore, accurate estimation of the effective value of n is virtually im-

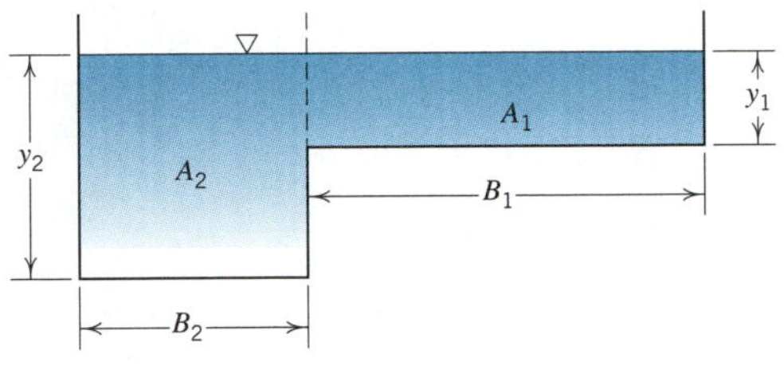

Fig. 10.9

possible since n for the narrow deep portion of the section may be very different from that of the wide shallow one. A logical (but not necessarily precise) means of treating such problems is to consider them composed of parallel channels separated by the vertical dashed line. The total flowrate is then expressed as $Q_1 + Q_2$, and the separate Q's by the Chezy-Manning equation,

$$Q = \left(\frac{u}{n_1}\right) A_1 R_{h_1}^{2/3} S_o^{1/2} + \left(\frac{u}{n_2}\right) A_2 R_{h_1}^{2/3} S_o^{1/2}$$

in which $A_1 = B_1 y_{o_1}$, $A_2 = B_2 y_{o_2}$, $P_1 = B_1 + y_{o1}$, $P_2 = B_2 + 2y_{o_2} - y_{o_1}$, and for SI units $u = 1$, while for US Customary units $u = 1.49$ as before.

10.6 SPECIFIC ENERGY, CRITICAL DEPTH, AND CRITICAL SLOPE—WIDE RECTANGULAR CHANNELS

Many problems of open-channel flow are solved by a special application of the work-energy principle using the channel bottom as the datum plane. The concept and use of *specific energy* (the distance between the channel bottom and the energy line) were introduced by Bakhmeteff in 1912 and have proved fruitful in the explanation and analysis of new and old problems of open-channel flow. Specific energy concepts permit determination of how water surface and velocity vary with change in flow cross-section much as the Bernoulli and work-energy equations did in pipe flow. Today, a knowledge of the fundamentals of specific energy is absolutely necessary in coping with the advanced problems of open-channel flow. These fundamentals and a few of their applications are developed in the following paragraphs.

With the specific energy E defined as the vertical distance from the channel bottom to the energy line (Fig. 10.10),

$$E = y + \frac{V^2}{2g_n} \tag{10.12}$$

or in terms of flowrate, which in steady flow is the same through each cross section,

$$E = y + \frac{1}{2g_n}\left(\frac{Q}{A}\right)^2 \tag{10.13}$$

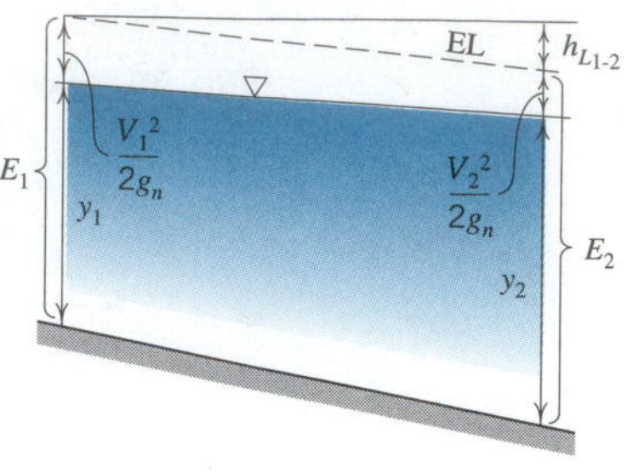

Fig. 10.10

It is convenient at this point to deal with flow in a wide channel of rectangular cross section to simplify the equations for better illustration of fundamentals; the principles may be applied, of course, to channels of other shapes (Section 10.7), but the resulting equations are considerably more unwieldy. The assumption of a wide rectangular channel allows the use of the two-dimensional approximation and the use of the two-dimensional flowrate, q. Substituting q/y for V in Eq. 10.12 gives

$$E = y + \frac{1}{2g_n}\left(\frac{q}{y}\right)^2 \qquad \text{or} \qquad q = \sqrt{2g_n(y^2E - y^3)} \tag{10.14}$$

This equation gives clear and simple relationships between specific energy, flowrate, and depth. Although a three-dimensional plot of these variables may be visualized, a better understanding of the equation may more easily be acquired by (1) holding q constant and studying the relation between E and y and (2) holding E constant and examining the relation between q and y. Plotting these relations yields, respectively, the *specific energy diagram* and *q-curve* of Fig. 10.11. Computer programs TRAP1, TRAP2, QCURV1, and QCURV2 are listed in Appendix 7 to assist in generating the data needed to plot specific energy diagrams and q-curves. Since these curves are merely different plots of the same equation, it may be expected (proof will be offered later) that the points of minimum E and maximum q are entirely equivalent. The depth associated with these points is known as the critical depth, y_c, and it is a boundary line between zones of open-channel flow, which are very different in character. Flows at depths greater than the critical (and velocities less than the

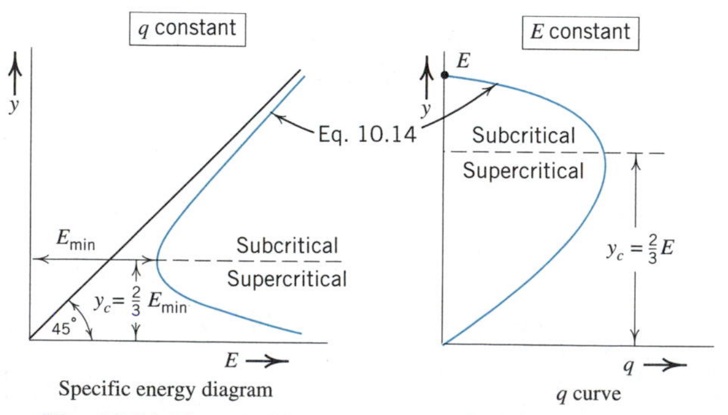

Fig. 10.11 The specific energy diagram and the q-curve.

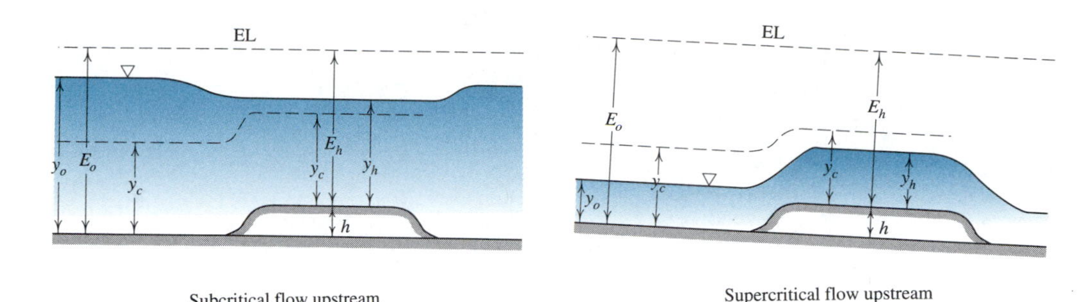

Fig. 10.12 Flow over a smooth hump.

critical) are known as *subcritical*[12] flows; flows at depths less than the critical (and velocities greater than the critical) are known as *supercritical* flows.

To develop an understanding of how these diagrams can be used, we will consider two types of local changes in flow cross section—the channel hump and the channel constriction. Working first with the channel hump (see Fig. 10.12), we assume that the structure is sufficiently smooth and streamlined so that no additional friction losses are introduced. As a consequence, the energy line remains parallel to the channel bottom. As flow moves from the upstream channel over the hump, the distance from the channel bottom to the energy line (E) decreases even though the total energy remains the same. From the specific energy diagram in Fig. 10.11, it appears that the depth of flow will either increase or decrease as flow moves onto the hump depending on whether the upstream flow is supercritical or subcritical. If the upstream flow is subcritical, then the decrease in E as the flow passes over the hump will cause a decrease in depth of flow of sufficient magnitude to result in a drop in the water surface over the hump. This result can be verified by considering the velocity head over the hump. Since the depth decreases over the hump, the flow velocity and the velocity head must increase. Because the water surface is below the energy line a distance equal to the velocity head, the water surface over the hump must be lower than the water surface upstream of the hump; hence, a drop in the water surface occurs as the flow moves onto the hump. Conversely, if the upstream flow is supercritical, an increase in depth over the hump would occur.

Problems of this nature can be readily visualized and solved using the specific energy diagram. Figure 10.13 demonstrates how the hump height influences the flow not only over the hump but, possibly, upstream as well. Note from Fig. 10.13 that once the value of E_o is found the value of h can be subtracted from it and the value of y_h can be read directly from the diagram. If the hump were made high enough ($h_{max} = E_o - E_{min}$), the critical depth would occur over the hump. Further, if the hump were made even higher, the flow would not have enough specific energy to flow over the hump and the hump would essentially become a "dam" or "choke." In this case, if the upstream flow is subcritical, water would be "backed up," gaining elevation until it obtained just enough specific energy to flow over the "dam" at critical depth and minimum energy. If the upstream flow is supercritical, a hydraulic jump would form (see Sections 6.3 and 10.9)

[12]The terms *subcritical* and *supercritical* (analogous to the *subsonic* and *supersonic* of compressible fluid motion) refer to the velocity; depths greater than critical give velocities less than that occurring at critical depth, and vice versa.

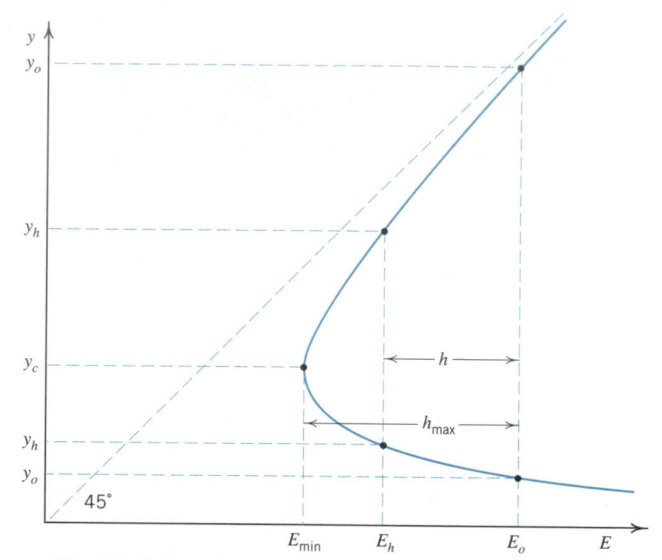

Fig. 10.13 Specific energy diagram for flow over a hump.

and flow would again back up, gaining enough elevation to flow over the hump at minimum energy and critical depth.

The situation for channel constrictions is very similar (see Fig. 10.14). If we again assume the constriction is so smooth that no additional friction losses are incurred, then E = constant through the constriction and the q-curve applies. With total Q a constant, the flowrate per unit width q increases as the flow passes through the constriction. From Fig. 10.11 it is clear that if the upstream flow is subcritical, a decrease in flow depth occurs in the constriction; if the upstream flow is supercritical, a depth increase occurs in the constriction. Fig. 10.14 depicts this result.

Problems of this type can be visualized and solved using the q-curve. Figure 10.15 shows how the width of the constriction affects the flow through the constriction and possibly upstream as well. If the width of the constriction is known, then the q-value can computed as $q_{con} = (B_o/B_{con})q_o$ and the value of the depth y_{con} in the constriction can be read from the q-curve. Again, if the channel constriction is made sufficiently narrow ($q_{max} = (B_o/B_{min})q_o$, critical flow conditions will exist in the constriction. And if the constriction is narrowed further, the flow will be unable to pass through with the specific energy available and the constriction becomes a choke. For subcritical flow upstream, flow

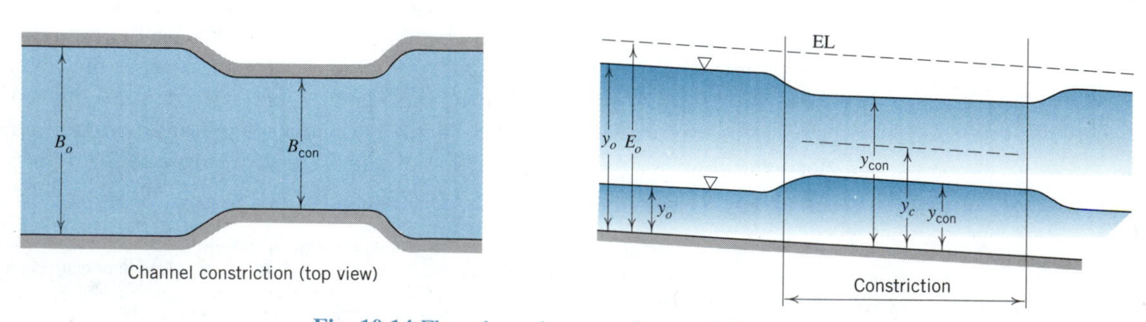

Fig. 10.14 Flow through a smooth constriction.

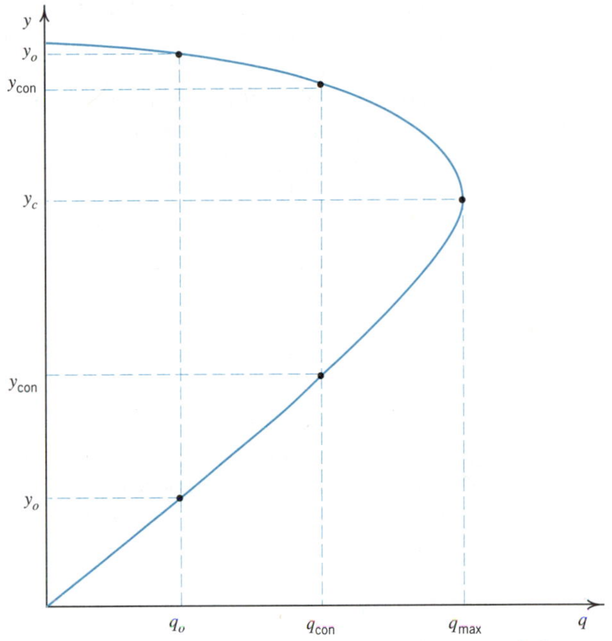

Fig. 10.15 The q-curve for flow through a constriction.

will then back up much as occurred with the hump, effectively creating an unconventional dam, until sufficient specific energy is built up to pass the flow through the constriction at the critical depth. If supercritical flow exists upstream, a hydraulic jump will form and the flow will back up, gaining enough elevation to flow through the constriction at minimum energy conditions.

It is clear from the situations described above that specific energy E and q can increase or decrease along a channel and the depths of flow will change in a manner determined by the specific energy diagram or q-curve. The direction of change depends on whether or not the flow is sub- or supercritical. Fig. 10.11 also shows that for a given specific energy, there are two depths, one less than the critical depth and one greater. These two depths are known as *alternate depths*.

The importance of the calculation of critical depth as a means of identifying the type of flow is apparent from the previous discussion, and equations for this may be obtained from the specific energy diagram where $dE/dy = 0$, and from the q-curve where $dq/dy = 0$. Performing the differentiations of Eqs. 10.14 yields (from the first equation)

$$q = \sqrt{g_n y_c^3} \qquad \text{or} \qquad y_c = \sqrt[3]{\frac{q^2}{g_n}} \tag{10.15}$$

and (from the second equation)

$$y_c = \frac{2E_{\min}}{3} \qquad \text{or} \qquad E_{\min} = \frac{3y_c}{2} \tag{10.16}$$

Substituting the latter relation into Eq. 10.14 for q gives $q = \sqrt{g_n y_c^3}$ as before, thus proving that the points of minimum E and maximum q have the same properties. Of great significance is the fact that *critical depth is dependent on flowrate only* (Eq. 10.15); the many other variables of open flow are not relevant to the computation of this important parameter. Equation 10.15 also suggests the use of critical flow as a means of flowrate measurement; if critical flow can be created or identified in a channel, the depth may be measured and flowrate determined.

Critical velocity V_c (velocity at the critical depth) may be obtained from combining the two equations $q = V_c y_c$ and $q = \sqrt{g_n y_c^3}$ to eliminate q. The result is

$$V_c = \sqrt{g_n y_c}$$

This equation is similar to another equation of open channel flow (see Section 6.3),

$$a = \sqrt{g_n y}$$

which gives the speed of propagation of a small gravity wave on the surface of a liquid of depth y. Therefore in supercritical flow, small gravity waves will not propagate upstream but rather will be swept downstream because the flow velocity is greater than the wave propagation speed. Conversely, the waves will propagate upstream in a subcritical flow.

Since wave phenomena are characterized by the interaction of inertia and gravity effects, it is possible to express critical flow considerations in terms of the Froude number (see Section 8.1). Defining Froude number by $\mathbf{F} = V/\sqrt{g_n y}$, we find that $\mathbf{F} < 1$ for subcritical flow and $\mathbf{F} > 1$ for supercritical flow. And of course for critical flow, $\mathbf{F} = 1$.

ILLUSTRATIVE PROBLEM 10.5

Uniform subcritical flow at a depth of 5 ft occurs in a long rectangular channel of width 10 ft, having a Manning n-value of 0.015 and laid on a slope of 0.001 ft/ft. Calculate

(a) the minimum height of hump which can be built in the floor of the channel to produce critical depth, and

(b) the maximum width of constriction which will produce critical depth.

SOLUTION

(a) In addressing the solution to part (a), we recognize that it is a situation where the specific energy diagram applies. Referring to Figs. 10.11 and 10.12, it is clear that building a hump in the channel will cause a decrease in specific energy over the hump which in turn leads to a decrease in depth of flow over the hump for subcritical flow. If we build the hump high enough, the depth of flow will decrease to the critical depth.

The first step is to calculate the specific energy corresponding to uniform flow. But this requires knowing the discharge so we must use the Chezy-Manning Eq. 10.5 to solve for the flowrate.

$$Q = \left(\frac{u}{n}\right) A R_h^{2/3} S_o^{1/2} \tag{10.5}$$

$$= \left(\frac{1.49}{0.015}\right)(5 \times 10)\left(\frac{5 \times 10}{10 + 2 \times 5}\right)^{2/3} (0.001)^{1/2} = 289 \text{ ft}^3/\text{s}$$

Now we can use Eq. 10.13 to find the specific energy for uniform flow.

$$E_o = y_o + \frac{1}{2g_n}\left(\frac{Q}{A_o}\right)^2 = 5 + \frac{1}{2 \times 32.2}\left(\frac{289}{5 \times 10}\right)^2 = 5.52 \text{ ft} \qquad (10.13)$$

Because we want critical flow to occur over the hump, we recognize that this condition represents minimum specific energy E_{min}. In order to find E_{min} we need to calculate the critical depth y_c from Eq. 10.15.

$$y_c = \sqrt[3]{\frac{q^2}{g_n}} = \sqrt[3]{\frac{(289/10)^2}{32.2}} = 2.96 \text{ ft} \qquad (10.15)$$

Now E_{min} can be found from Eq. 10.16.

$$E_{\min} = \frac{3}{2}y_c = \frac{3}{2} \times 2.96 = 4.44 \text{ ft} \qquad (10.16)$$

The hump height required to just cause critical depth to occur over the hump is obtained by subtracting E_{min} from E_o as seen in the sketch below.

$$\text{Hump height } x = E_o - E_{min} = 5.52 - 4.44 = 1.08 \text{ ft} \bullet$$

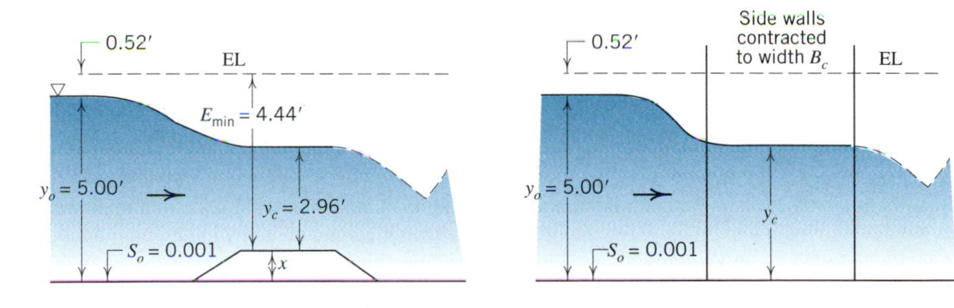

A hump height smaller than 1.08 ft will lower the water surface over the hump but will not cause critical depth; a hump higher than 1.08 ft will produce critical depth over the hump and also cause the water upstream of the hump to ''back up,'' creating a dam. Regardless of how high the hump becomes, the depth of flow over the hump will remain the critical depth.

(b) In arriving at a solution to part (b) we recognize that the q-curve applies and that a narrowing of the channel width will cause the flow depth to decrease and approach the critical. When the width has been reduced to the point where critical depth has just been reached, we know that $y_c = 2/3E_{min}$ (Eq. 10.16). Because there has been no change in specific energy as the flow moves through the constriction, E_o becomes E_{min} in the constriction. Eq. 10.16 gives

$$y_c = \frac{2}{3}E_{\min} = \frac{2}{3}E_o = \frac{2}{3}5.52 = 3.68 \text{ ft} \qquad (10.16)$$

Now, knowing the critical depth, we can go back to Eq. 10.15 and reconstruct the width B_c required to produce this condition.

$$y_c = 3.68 = \sqrt[3]{\frac{q^2}{g_n}} = \sqrt[3]{\frac{(289 \text{ ft}^3 s/B_c \text{ ft})^2}{32.2}} \qquad (10.15)$$

Solving the above equation for B_c gives

$$B_c = 7.21 \text{ ft} \bullet$$

If the constriction is made any narrower than this, the flow will back up increasing the specific energy in the constriction. However, under these conditions, the depth of flow in the constriction will increase while still remaining at the critical and be computed using Eq. 10.16.

If uniform flow were to occur at critical depth, the slope of the channel bottom under these conditions would be called the *critical slope* S_c. For a rectangular channel of great width ($R_h = y_c$), a simple expression for S_c is obtained by equating the flowrates of Eqs. 10.5 and 10.15.

$$q = \sqrt{g_n y_c^3} = \frac{u}{n} y_c^{5/3} S_c^{1/2}$$

Solving for S_c,

$$S_c = \frac{g_n n^2}{u^2 y_c^{1/3}} \tag{10.17}$$

This equation shows that critical slope is a function of critical depth for a given n-value. Because critical depth varies with q, the critical slope of a channel is also a function of q. However, because y_c appears to the 1/3 power in Eq. 10.17, S_c does not vary a great deal over a wide range of flow. Fig. 10.16 shows, in general, the variation of critical slope with critical depth for a wide rectangular channel. Although the form of this function is different for other cross sections, this feature is retained for practically all sections except those which are narrow and deep; it justifies the useful rough approximation that over a relatively large range of depths (and flowrates) S_c may be assumed constant. Uniform critical flow is, of course, a rare borderline situation between subcritical and supercritical flows. On the specific energy diagram it is noted that, in the vicinity of the critical depth, the depth may change considerably with little variation of specific energy; physically this means that, since many depths may occur for practically the same specific energy, flow near the critical depth will possess a certain instability (which manifests itself by undulations in the liquid surface). Uniform flow near the critical depth has been observed in both field and laboratory to possess these characteristics, and because of this the designer seeks to prevent uniform flow situations close to the critical.

Slopes greater than and less than the critical slope, S_c, are known, respectively, as *steep* and *mild* slopes. Evidently, channels of steep slope (if long enough) will produce supercritical uniform flows, and channels of mild slope (if long enough) will produce

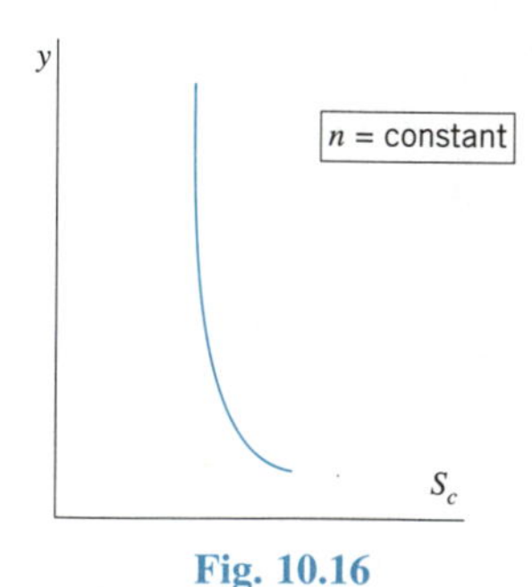

Fig. 10.16

subcritical uniform flows. Whether or not uniformity is actually realized, the depth of uniform flow may always be computed from flowrate, slope, channel shape, and roughness through the Chezy-Manning equation. Depths computed in this manner are called *normal* or *neutral* depths; although such depths may not occur in short channels, they are, nevertheless, useful parameters of the flow and essential to an understanding of problems of varied flow.

ILLUSTRATIVE PROBLEM 10.6

For a flowrate of 500 cfs in a rectangular channel 40 ft wide, the water depth is 4 ft. Is the flow subcritical or supercritical? If $n = 0.017$, what is the critical slope of this channel for this flowrate?

What channel slope would be required to produce uniform flow at a depth of 4 ft?

SOLUTION

To determine whether or not the flow is subcritical or supercritical, we calculate the critical depth and compare it with the existing depth. First we compute the flowrate per unit width q

$$q = Q/B = \frac{500 \text{ ft}^3/\text{s}}{40 \text{ ft}} = 12.5 \text{ ft}^3/\text{s/ft}$$

Now we use Eq. 10.15 to compute the critical depth.

$$y_c = \sqrt[3]{\frac{q^2}{g_n}} = \sqrt[3]{\frac{12.5^2}{32.2}} = 1.69 \text{ ft} \tag{10.15}$$

Since the existing flow depth of 4 ft is greater than the critical depth of 1.69 ft, the flow is classified as subcritical. •

With the critical depth of 1.69 ft and a width of channel of 40 ft, it is reasonable to use Eq. 10.17 for wide rectangular channels to compute the critical slope.

$$S_c = \frac{g_n n^2}{u^2 y_c^{1/3}} = \frac{32.2 \times 0.017^2}{1.49^2 \times 1.69^{1/3}} = 0.003\,5 \text{ ft/ft} \quad \bullet \tag{10.17}$$

To compute the channel slope which would convey 500 ft^3/s at a uniform flow depth of 4 ft, we use the Chezy-Manning equation.

$$Q = \left(\frac{u}{n}\right) A R_h^{2/3} S_o^{1/2} \tag{10.5}$$

$$S_o^{1/2} = Q\left(\frac{n}{u}\right)\left(\frac{1}{AR_h^{2/3}}\right)$$

$$= 500\left(\frac{0.017}{1.49}\right)\left(\frac{1}{(4 \times 40)\left(\dfrac{4 \times 40}{40 + 2 \times 4}\right)^{2/3}}\right) = 0.015\,98$$

$$S_o = 0.000\,255 \text{ ft/ft} \quad \bullet$$

The slope S_o is a *mild* slope since it is less than S_c.

10.7 SPECIFIC ENERGY, CRITICAL DEPTH, AND CRITICAL SLOPE—NONRECTANGULAR CHANNELS

The application of the principles of Section 10.6 to channels of nonrectangular cross section (Fig. 10.17) leads to more generalized and complicated mathematical expressions, owing to the more difficult geometrical aspects of such problems. The specific energy equation is written as before,

$$E = y + \frac{1}{2g_n}\left(\frac{Q}{A}\right)^2 \tag{10.13}$$

Here A is some function of y, depending on the form of the channel cross section, and the equation becomes

$$E = y + \frac{1}{2g_n}\left(\frac{Q}{f(y)}\right)^2 \tag{10.18}$$

which is analogous to Eq. 10.14 and leads to a specific energy diagram and a Q-curve (Fig. 10.18) having the same superficial appearance as those of Fig. 10.11 but for which there are other critical depth relationships. These may be worked out for the generalized channel cross section by differentiating Eq. 10.13 in respect to y and setting the result equal to zero:

$$\frac{dE}{dy} = 1 + \frac{Q^2}{2g_n}\left(-\frac{2}{A^3}\frac{dA}{dy}\right) = 0$$

From Fig. 10.17, $dA = b\,dy$ in which b is the channel width *at the liquid surface* because dA/dy refers to the increase in area with depth, which always occurs at the surface; dA/dy is thus equal to b, and substitution gives

$$\frac{Q^2}{g_n} = \frac{A^3}{b} \qquad \text{or} \qquad \frac{Q^2 b}{g_n A^3} = 1 \tag{10.19}$$

as the equation which allows calculation of critical depth in nonrectangular channels. Substituting V for Q/A in Eq. 10.19 and defining a Froude number by

$$\mathbf{F} = \sqrt{\frac{Q^2 b}{g_n A^3}} = \frac{V}{\sqrt{g_n(A/b)}} \tag{10.20}$$

it is evident that (as for the wide rectangular channel): for subcritical flow $\mathbf{F} < 1$, for critical flow $\mathbf{F} = 1$, and for supercritical flow $\mathbf{F} > 1$.

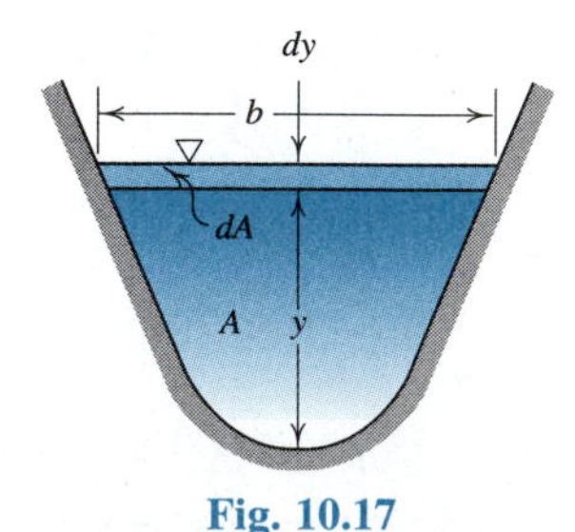

Fig. 10.17

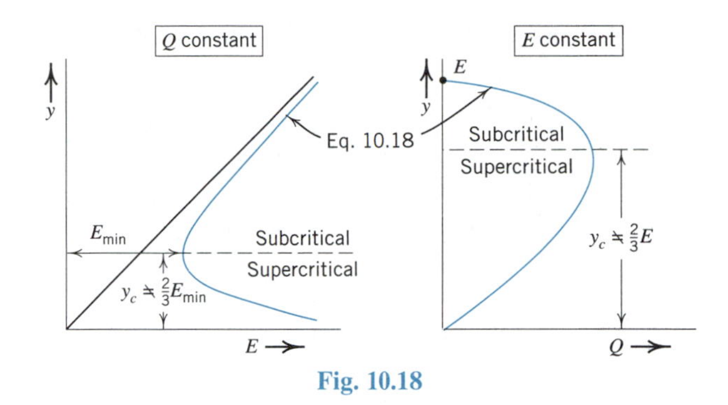

Fig. 10.18

Critical slope for a nonrectangular channel may be derived by following the same procedure as for the wide rectangular one:

$$Q = \sqrt{\frac{g_n A^3}{b}} = \left(\frac{u}{n}\right) A R_h^{2/3} S_c^{1/2}$$

from which we obtain

$$S_c = \frac{g_n n^2}{u^2} \left(\frac{A}{b R_h^{4/3}}\right) \tag{10.21}$$

In this expression A, b, and R_h are all functions of depth and dependent on the section shape. Critical slope is thus a function of section shape also, and the form of the variation of S_c and y will usually be similar to that of Fig. 10.16. Computer programs TRAP1 and TRAP2 are listed in Appendix 7 to assist in generating the data needed to plot specific energy diagrams for trapezoidal channels.

ILLUSTRATIVE PROBLEM 10.7

A flow of 28 m³/s occurs in an earth-lined canal having a base width of 3 m, side slopes $z = 2$ and with a Manning n-value of 0.022. Calculate the critical depth and critical slope.

SOLUTION

To calculate the critical depth, we employ Eq. 10.19.

$$\frac{Q^2 b}{g_n A^3} = 1 = \frac{(28\ \text{m}^3/\text{s})^2 \times (3 + 4y_c\ \text{m})}{9.81 \times (3y_c + 2y_c^2\ \text{m}^2)^3} \tag{10.19}$$

This equation has only one unknown y_c whose value must be determined by trial solution. The result is

$$y_c = 1.50\ \text{m} \bullet$$

Now in order to compute the critical slope, we need only to substitute the critical depth of 1.50 ft into Eq. 10.21. But first we will compute the values of area and hydraulic radius corresponding to the critical flow conditions.

$$A = 3y_c + 2y_c^2 = 3 \times 1.50 + 2 \times 1.50^2 = 9.00 \text{ m}^2$$

$$R_h = \frac{A}{P} = \frac{A}{3 + 2\sqrt{5} \times 1.50} = \frac{9.00}{9.70} = 0.93 \text{ m}$$

Placing these values into Eq. 10.21,

$$S_c = \frac{g_n n^2}{u^2}\left(\frac{A}{bR_h^{4/3}}\right) = \frac{9.81 \times 0.022^2}{1^2}\left(\frac{9.00}{9.00 \times 0.93^{4/3}}\right) \qquad (10.21)$$

$$= 0.005\,23 \text{ m/m} \;\bullet$$

10.8 CONTROLS AND THE OCCURRENCE OF CRITICAL DEPTH

The analysis of open-channel flow problems usually begins with the location of points in the channel at which the relationship between water depth and flowrate is known or controllable. These points are known as *controls* since their existence governs, or controls, the liquid depths in the reach of a channel either upstream or downstream from such points. In a broad sense control points usually feature a change from subcritical to supercritical flow. In this context, controls occur at physical barriers, for example, at sluice gates, dams, weirs, drop structures, etc., where critical flow is forced to occur, and at changes in channel slope, for example, from mild to steep, which causes the flow to pass through the critical depth. It is also possible that, where the channel slope changes (e.g., from steep to less steep or to mild), the flow will move from a normal depth upstream to another normal depth or into a hydraulic jump downstream. In this case the slope change point is a control for the downstream reach and the depth at the control is the normal depth for the flow upstream of the slope change. The prediction of control point location is, thus, a powerful tool for ''blocking in'' the variety of flow phenomena to be expected before making detailed calculations.

The most obvious place where critical depth can be expected is in the situation pictured in Fig. 10.19, where a long channel of mild slope ($S_o < S_c$) is connected to a long channel of steep slope ($S_o > S_c$). Far up the former channel uniform subcritical flow at normal depth, y_{o1}, will occur, and far down the latter a uniform supercritical flow at a smaller normal depth, y_{o2}, can be expected. These two uniform flows will be connected by a reach of varied flow in which at some point the depth must pass through the critical. Experience shows that this point is close to (actually slightly upstream from) the break in slope but

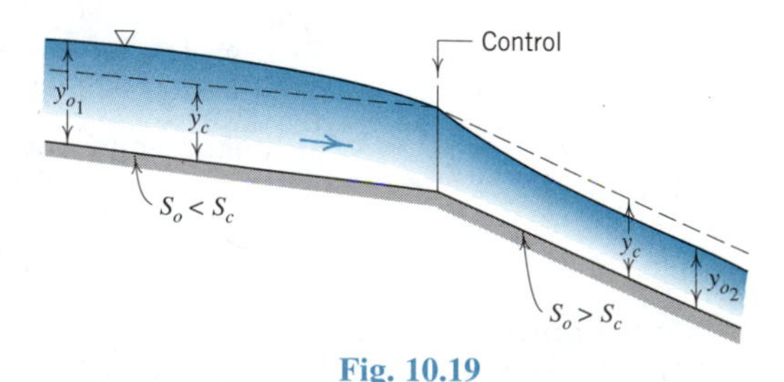

Fig. 10.19

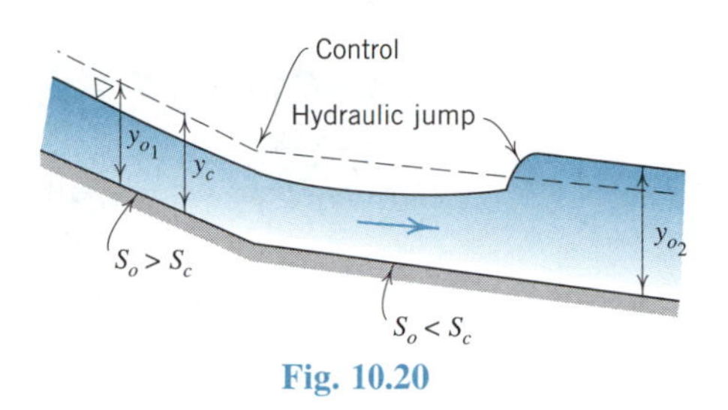

Fig. 10.20

may be assumed to be at the break for most purposes. It is to be noted, however, that in the immediate vicinity of the change in bottom slope, the streamlines are both curved and convergent and the problem cannot be treated as one-dimensional flow with hydrostatic pressure distribution. The fact that the critical depth is not found precisely at the point of slope change prevents the use of such points in the field for obtaining the exact flowrate by depth measurement and application of Eqs. 10.15 or 10.19; however, the possibility of this method as an approximate means of flowrate measurement should not be overlooked.

When a long channel of steep slope discharges into one of mild slope (Fig. 10.20), normal depths will occur upstream and downstream from the point of slope change, but usually the critical depth will not be found near this point. Under these conditions a *hydraulic jump* (Sections 6.3 and 10.9) will form whose location will be dictated (through varied flow calculations) by the details of slopes, roughness, channel shapes, and so forth, but the critical depth will be found within the hydraulic jump. (See Section 10.9).

The occurrence of critical depth on overflow structures can be proved by examining the flow over the top of a high frictionless broad-crested weir (Fig. 10.21) equipped with a movable sluice gate at the downstream end and discharging from a large reservoir of constant surface elevation. With gate closed (position A), the depth of liquid on the crest will be y_A, and the flowrate will obviously be zero, giving point A on the q-curve. With the gate raised to position B, a flowrate q_B will occur, with a decrease in depth from y_A to y_B. This process will continue until the gate is lifted clear of the flow (C) and can therefore no longer affect it. With the energy line fixed in position at the reservoir surface level and, therefore, giving constant specific energy, it follows that points A, B, and C have outlined the upper portion of the q-curve, that the flow occurring without gates is maximum, and that the depth on the crest is, therefore, the critical depth. For flow over weirs, a relation between head and flowrate is desired; this may be obtained by substituting (for a very high

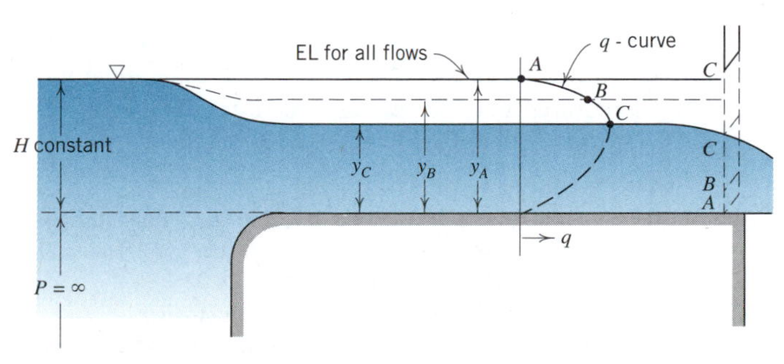

Fig. 10.21

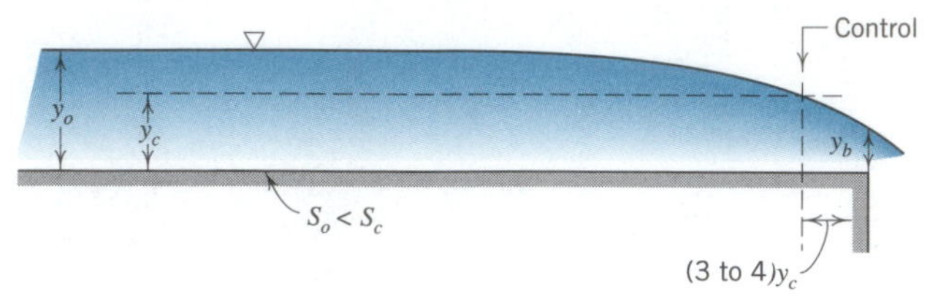

Fig. 10.22

weir) $y_c = 2H/3$ in Eq. 10.15, which yields

$$q = \sqrt{g_n y_c^3} = \sqrt{g_n \left(\frac{2H}{3}\right)^3} = 0.577 \times \frac{2}{3}\sqrt{2g_n}H^{3/2} \tag{10.22}$$

The last form of this equation is readily compared with the standard weir equation (Section 14.16); here 0.577 is the *weir coefficient.* Because of neglect of friction in the analysis, this coefficient of 0.577 is higher than those obtained in experiments; tests on high, broad-crested weirs give coefficients between 0.50 and 0.57 (depending on the details of weir shape).

The reasoning of the foregoing paragraph may be extended to a *free outfall* (Fig. 10.22) from a long channel of mild slope, to conclude that the depth must pass through the critical in the vicinity of the brink. Rouse[13] has found that for such rectangular channels the critical depth occurs a short distance (3 to $4y_c$) upstream from the brink and that the brink depth (y_b) is 71.5% of the critical depth.[14] Using this figure and Eq. 10.15, Rouse proposes the free outfall as a simple device for metering the flowrate, which requires only measurement of the brink depth. Free outfalls also provide an opportunity for recognition of the limitations of specific energy theory. Upstream from the point of critical depth, the streamlines are essentially straight and parallel and the pressure distribution hydrostatic; between this point and the brink neither of these conditions obtains, so the one-dimensional theory is invalid in this region. For example, it *cannot* be concluded that the flow at the brink is supercritical merely because the brink depth is less than the critical depth.

Critical depth may be obtained at points in an open channel where the channel bottom is raised by the construction of a low hump or the channel is constricted by moving in the sidewalls (see Section 10.6). Such contractions (usually containing a rise in the channel bottom) are known generally as Venturi flumes, and specific designs[15] of these flumes are widely used in the measurement of irrigation water. Preliminary analysis of such contractions (which feature accelerated flow) may be made assuming one-dimensional flow and neglecting head losses, but final designs require more refined information either from other designs or from laboratory experiments. The advantages of such flumes are their ability to pass sediment-laden water without deposition and the small net change of water level required between entrance and exit channels.

[13]H. Rouse, "Discharge Characteristics of the Free Overfall," *Civil Engr.* vol. 6, no. 4, p. 257, 1936. See also T. Strelkoff and M. S. Moayeri, "Pattern of Potential Flow in a Free Overfall," *Jour. Hydr. Div., A.S.C.E.*, vol. 96, no. HY4, pp. 879–901, April, 1970.

[14]Experiments on free outfalls from *circular conduits* of mild slope show brink depth to be about 75% of critical depth when the brink depth is less than 60% of the diameter of the conduit.

[15]See, for example, P. Ackers, W. R. White, J. A. Perkins, and A. J. M. Harrison, *Weirs and Flumes for Flow Measurement*, John Wiley & Sons, 1978.

10.9 THE HYDRAULIC JUMP

When a change from supercritical to subcritical flow occurs in open flow a *hydraulic jump* appears, through which the depth increases abruptly in the direction of flow. In spite of the complex appearance of a hydraulic jump with its turbulence[16] and air entrainment, it may be successfully analyzed by application of the impulse-momentum principle (see Section 6.3) to yield results and relationships which conform closely with experimental observations.

A hydraulic jump in an open channel of small slope is shown in Fig. 10.23.[17] In engineering practice the hydraulic jump frequently appears downstream from overflow structures (spillways) or underflow structures (sluice gates) where velocities are high. It may be used as an effective dissipator of kinetic energy (and thus prevent scour of the channel bottom) or as a mixing device in water or sewage treatment designs where chemicals are added to the flow. In design calculations the engineer is concerned mainly with prediction of existence, size, and location of the jump.

In Section 6.3, the basic equation for the jump in a rectangular[18] open channel was derived; it is

$$\frac{q^2}{g_n y_1} + \frac{y_1^2}{2} = \frac{q^2}{g_n y_2} + \frac{y_2^2}{2} \tag{10.23}$$

the solution of which may be written

$$\frac{y_2}{y_1} = \frac{1}{2}\left[-1 + \sqrt{1 + \frac{8V_1^2}{g_n y_1}}\right] \tag{10.24}$$

or

$$\frac{y_1}{y_2} = \frac{1}{2}\left[-1 + \sqrt{1 + \frac{8V_2^2}{g_n y_2}}\right] \tag{10.25}$$

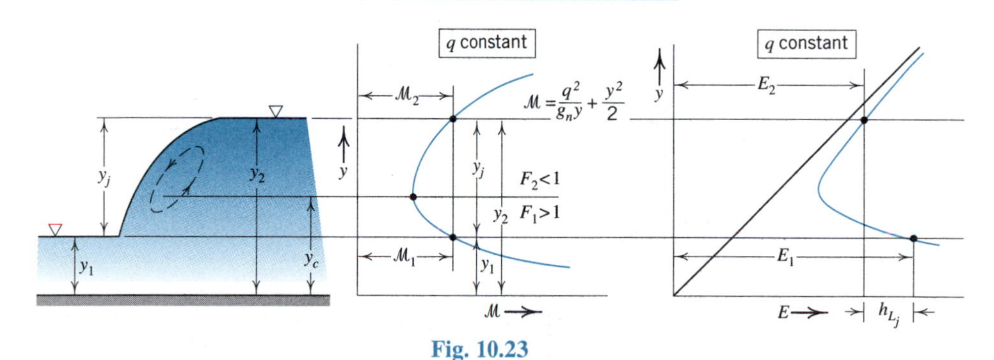

Fig. 10.23

[16]See H. Rouse, T. T. Siao, and S. Nageratnam, "Turbulence Characteristics of the Hydraulic Jump," *Trans. A.S.C.E.*, vol. 124, 1959.

[17]The jump is not so steep as shown in the figure; the length of the jump is approximately $6y_2$. See U. S. Bureau of Reclamation, "Research studies on stilling basins, energy dissipators, and associated appurtenances," *Hydr. Lab. Rep.*, Hydr 399, 1 June 1955, and V. T. Chow, *Open Channel Hydraulics*, McGraw-Hill, 1959, Chapter 15.

[18]Here the analysis is confined to the rectangular channel for both mathematical simplicity and practical application. The same methods may be applied to channels of nonrectangular cross section.

Fig. 10.24

in which $V_1^2/g_n y_1$ is recognized as $\mathbf{F}_1^2$ and $\mathbf{F}_2^2 = V_2^2/g_n y_2$; these equations show that $y_2/y_1 > 1$ only when $\mathbf{F}_1 > 1$ and $\mathbf{F}_2 < 1$, thus proving the necessity of supercritical flow for jump formation. Another way of visualizing this is by defining a quantity $\mathcal{M}$ (called force + momentum) by

$$\mathcal{M} = \frac{q^2}{g_n y} + \frac{y^2}{2}$$

and plotting $\mathcal{M}$ as a function of y (Fig. 10.23) for constant flowrate, whereupon the solution of the equation occurs when $\mathcal{M}_1 = \mathcal{M}_2$; the curve obtained features a minimum value of $\mathcal{M}$ at the critical depth,[19] and thus, superficially resembles (but must not be confused with) the specific energy diagram. After construction of this curve and with one depth known, the corresponding, or *conjugate*, depth may be found by passing a vertical line through the point of known depth. Since a vertical line is a line of constant $\mathcal{M}$, the intersection of this line and the other portion of the curve gives a point where $\mathcal{M}_1$ is equal to $\mathcal{M}_2$, and allows the conjugate depth and height of jump y_j to be taken directly from the plot. The shape of the $\mathcal{M}$-curve also shows clearly and convincingly that hydraulic jumps can take place only across the critical depth (i.e., from supercritical to subcritical flow). Computer programs TRAP1 and TRAP2 are listed in Appendix 7 to assist in generating the data needed to plot specific energy diagrams and force + momentum ($\mathcal{M}$) diagrams.

Because of eddies (rollers), air entrainment, and flow decelerations in the hydraulic jump, large head losses are to be expected; they may be calculated (as usual) from the fall of the energy line, h_{L_j}, as can be seen on the specific energy diagram in Fig. 10.23.

$$h_{L_j} = \left(y_1 + \frac{V_1^2}{2g_n} + z_1\right) - \left(y_2 + \frac{V_1^2}{2g_n} + z_2\right)$$

in which $z_1 = z_2$ if the channel bottom is horizontal. For very small jumps, the eddies and air entrainment disappear, the form of the jump changes to that of a smooth standing wave, known as an *undular* jump (Fig. 10.24), with very small head loss. Laboratory tests show that undular jumps are to be expected for $\mathbf{F}_1 \lesssim 1.7$.

10.10 VARIED FLOW

For the design of open channels and the analysis of their performance, the engineer must be able to predict the forms and calculate the locations of the various types of water surface profiles which may occur under nonuniform flow conditions. The first of these objectives can be obtained by studying the differential equation of varied flow. It will be demonstrated that just by observing the character of this differential equation, we will be able to predict how depth varies with distance along the channel; we will learn where uniform flow may

[19]This may be easily proved by setting $d\mathcal{M}/dy$ equal to zero.

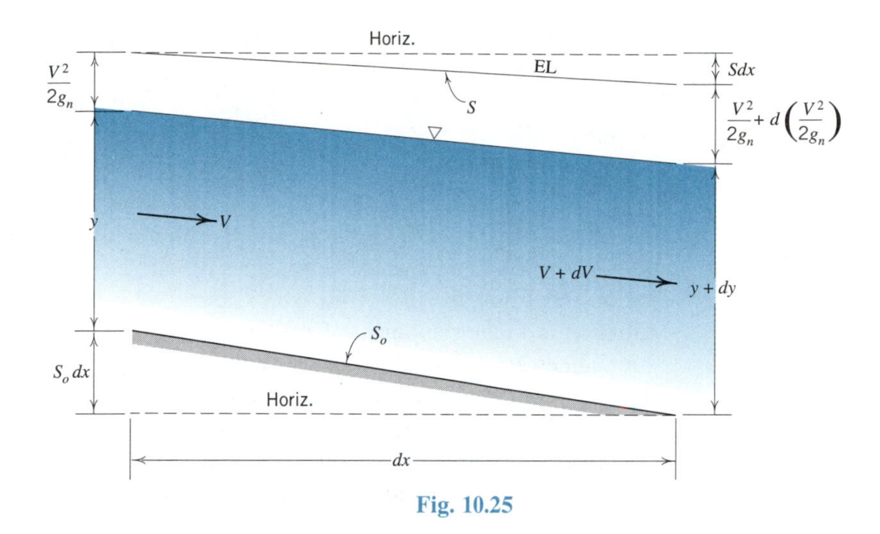

Fig. 10.25

be expected; and we will be able to predict the curvature of the water surface profile. To achieve the second objective, we will construct a finite difference model of the varied flow equation and show how it can be integrated numerically to give values of depth vs distance along the channel.

Directing our attention to the first objective, we will develop the differential equation of varied flow by considering the differential length of channel shown in Fig. 10.25. It should be noted that the channel bottom slope S_o is so small that distance along the channel is virtually identical to horizontal distance x. Defining the slope of the energy line as S, the drops in the energy line and the channel bottom over the distance dx are $S\,dx$ and $S_o\,dx$, respectively. Noting from Fig. 10.25 that the distance between the two parallel horizontal lines is constant, we can write the following equation.

$$S_o\,dx + y + \frac{V^2}{2g_n} = y + dy + \frac{V^2}{2g_n} + d\left(\frac{V^2}{2g_n}\right) + S\,dx \tag{10.26}$$

Dividing Eq. 10.26 by dx and cancelling equal terms produces

$$\frac{dy}{dx} + \frac{d}{dx}\left(\frac{V^2}{2g_n}\right) = S_o - S$$

Multiplying the second term by dy/dy, factoring out dy/dx, and solving for dy/dx yields

$$\frac{dy}{dx} = \frac{S_o - S}{1 + \dfrac{d}{dy}\left(\dfrac{V^2}{2g_n}\right)}$$

The derivative in the denominator of the fraction must be put into a more useful form before the meaning of the equation can be fully explored. Referring to Section 10.7, one can see that

$$\frac{d}{dy}\left(\frac{V^2}{2g_n}\right) = \frac{1}{2g_n}\frac{d}{dy}\left(\frac{Q^2}{A^2}\right) = -\frac{Q^2 b}{g_n A^3} = -\mathbf{F}^2$$

Making this substitution into the varied flow equation gives

$$\frac{dy}{dx} = \frac{S_o - S}{1 - \mathbf{F}^2} \tag{10.27}$$

The forms of all possible water surface profiles can be deduced from this equation.

Before addressing the various types of surface profiles, we should note two clarifications regarding the use of Eq. 10.27. First, we have assumed hydrostatic pressure distribution in the development of the equation so application is limited to situations where streamlines are essentially straight and parallel with small channel slopes S_o. Second, it must be recognized that dy/dx is the variation of depth of flow along the channel, not the slope of the water surface.

One additional observation must be made before an understanding of water surface profiles can be attempted. While the Chezy-Manning equation is generally considered a uniform flow equation, it is our intention to use it to evaluate energy loss (slope of the energy line, S) for varied flow so long as the streamlines of the flow are essentially straight and parallel. Recall that in uniform flow, the slope of the energy gradient is parallel to the channel bottom ($S = S_o$). For varied flow, the slope of the energy gradient will be taken as (compare with Eq. 10.5)

$$S = \left(\frac{Qn}{uAR_h^{2/3}}\right)^2 \tag{10.28}$$

This plausible assumption used in the calculation of S has never been precisely confirmed by experiment, but errors arising from it are small compared to the uncertainty in the selection of an n-value, and over the years, it has proved to be a reliable basis for design calculations.

For uniform flow conditions, Eq. 10.28 permits calculation of S_o for values of A_o and R_{h_o}. Recognizing that $AR_h^{2/3}$ increases with depth in almost all channels (except circular pipes), we see that if depth increases above y_o, then $AR_h^{2/3}$ increases causing S to decrease below S_o. As a consequence, when $y > y_o$, then $S < S_o$ and when $y < y_o$, then $S > S_o$. This knowledge will be essential in determining the sign of the numerator in Eq. 10.27.

We will classify water surface profiles according to the type of channel slope on which they occur. The classification proceeds as follows:

$S_o < S_c$	$y_o > y_c$	Mild slope	M-profile
$S_o > S_c$	$y_o < y_c$	Steep slope	S-profile
$S_o = S_c$	$y_o = y_c$	Critical slope	C-profile
$S_o = 0$	$y_o = \infty$	Horizontal slope	H-profile
$S_o < 0$	y_o undefined	Adverse slope	A-profile

We will break down the classification further within any given profile type by using the current depth in relation to the critical and normal depth. For example, consider the mild-slope channel shown in Fig. 10.26. We have shown the critical and normal depths for a given flowrate as dashed lines effectively dividing the possible flow profiles into three zones, labeled Zone 1, Zone 2, and Zone 3. A flow profile occurring in Zone 1 would be labeled an M_1 profile; one in Zone 2, and M_2 profile and so on. As we will soon see, knowledge of which zone a specific flow profile is in will permit us to predict the shape

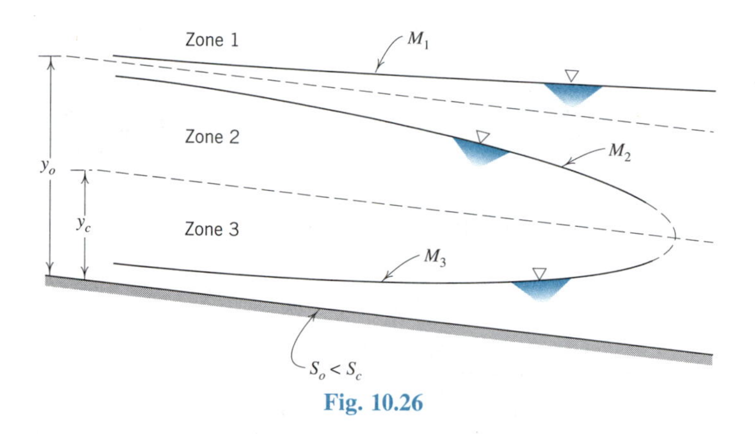

Fig. 10.26

of the water surface, the variation of depth along the channel, and where the critical or normal depth will occur.

Let us continue the example by using Eq. 10.27 to confirm the form of the three M-profiles shown on Fig. 10.26. Working in Zone 1 first, we see that if the depth of flow at some point is greater than the normal depth of flow, then $S < S_o$ and the numerator of Eq. 10.27 is positive. Further, the depth is also greater than the critical, so subcritical flow is occurring ($\mathbf{F} < 1$) causing the denominator of Eq. 10.27 to be positive. As a consequence, the sign of dy/dx is positive, which means the depth of flow is increasing in the downstream direction. Conversely, as we move upstream the depth decreases, approaching the normal depth. As we get closer and closer to the normal depth, the numerator of Eq. 10.27 becomes vanishingly small while the denominator remains finite. The result is that the depth does not reach the normal until we have moved an infinite distance upstream. All of this information permits us to conclude that an M_1 profile is a water surface with increasing depth in the downstream direction, concave upward, with normal depth being approached asymtotically in the upstream direction.

Now, turning our attention to Zone 2, we note that the flow depth is less than the normal depth so that $S > S_o$. This means that the numerator of Eq. 10.27 is now negative in sign. However, the denominator remains positive because we are still in subcritical flow where $\mathbf{F} < 1$. Thus, in Zone 2, Eq. 10.27 shows that dy/dx is negative, meaning the flow depth decreases in the downstream direction. To further describe the shape of the flow profile, we note that as the flow depth decreases in the downstream direction, we approach the critical depth ($\mathbf{F} = 1$). Eq. 10.27 shows us that the denominator approaches zero while the numerator remains finite. The apparent consequence is that the water surface becomes vertical; however, in the practical sense, the hydrostatic pressure distribution assumption is violated and the flow passes through the critical depth at the downstream end of the mild-sloped channel. Looking in the upstream direction, we once again see the flow depth approaching the normal but, as before, it will reach normal depth at an infinite distance upstream. We can now characterize the M_2 profile as a water surface with decreasing depth in the downstream direction, concave downward and approaching critical depth, while in the upstream direction, approaching the normal depth asymptotically.

If we were to pursue the above procedure for Zone 3 for the M-profile water surfaces and do the same for the other four types of profiles, we would get the results shown in Table 6. It should be noted in Table 6 that there are no H_1 or A_1 curves because there is no zone above the normal depth of infinity. Also, there is no C_2 profile because Zone 2 has been squeezed out when $y_o = y_c$ for critical slopes.

TABLE 6 Surface Profiles of Varied Flow

Mild $S_o < S_c$	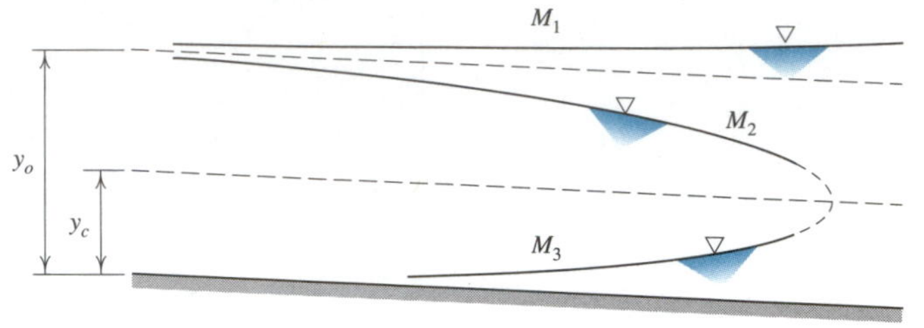 M_1, M_2, M_3, y_o, y_c
Steep $S_o > S_c$	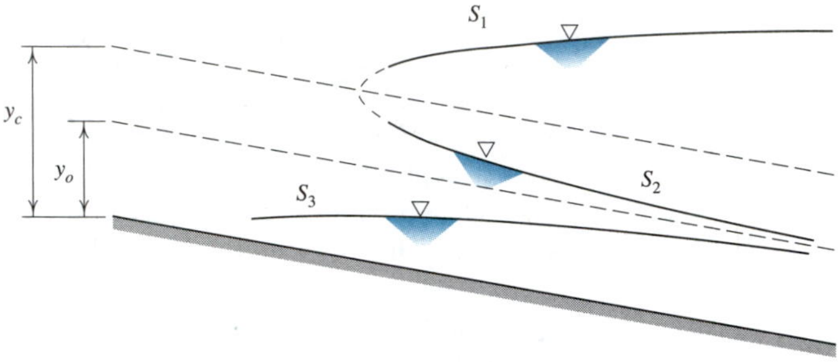 S_1, S_2, S_3, y_c, y_o
Critical $S_o = S_c$	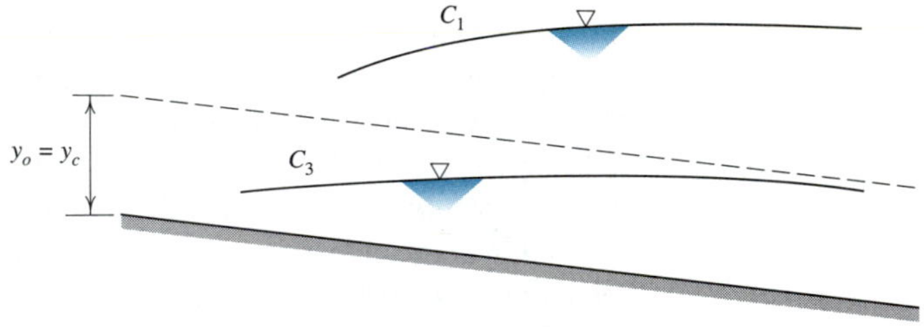 C_1, C_3, $y_o = y_c$
Horizontal $S_o = 0$	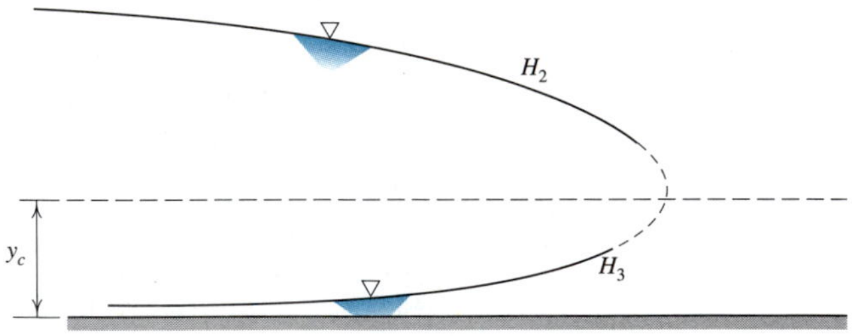 H_2, H_3, y_c
Adverse $S_o < 0$	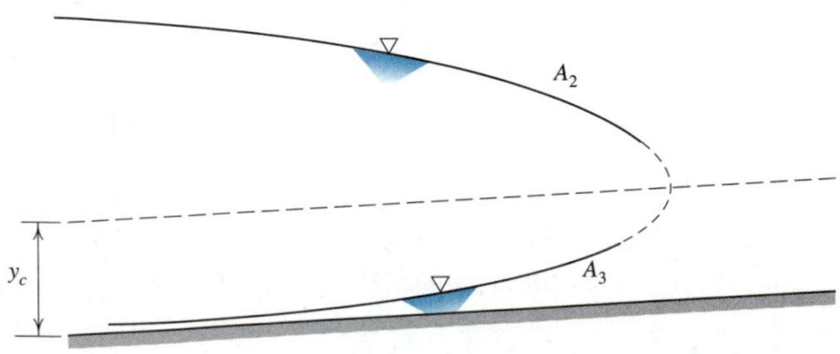 A_2, A_3, y_c

ILLUSTRATIVE PROBLEM 10.8

Using Table 6, identify the water surface profiles for the following cases.

(a) Illustrative Problem 10.5, upstream of the hump if the hump were higher than 1.08 ft.
(b) Upstream and downstream of the break in slope shown in Fig. 10.19.
(c) On the mild slope just upstream of the hydraulic jump in Fig. 10.20.
(d) On the portion of the channel upstream of the free overfall in Fig. 10.22.

SOLUTION

(a) For hump heights up to 1.08 ft, the flow upstream remains at the normal depth. As indicated in the problem statement, the uniform flow is subcritical hence the slope is mild. An increase in the hump height above 1.08 ft will cause the depth of flow upstream of the hump to increase. From Table 6, a water surface profile on a mild slope at a depth greater than the normal is classified as an *M_1 profile*.

(b) Fig. 10.19 is reproduced below in order to facilitate the solution. Note that the upstream slope is mild whereas the downstream slope is steep. Table 6 shows that on a mild slope where the depth of flow is greater than the critical but less than the normal, we will have an *M_2 profile* concave downward with flow approaching the critical depth downstream. On the downstream slope, the depth is less than the critical but greater than the normal. According to Table 6, the water surface follows an *S_2 profile* with flow approaching the normal depth downstream.

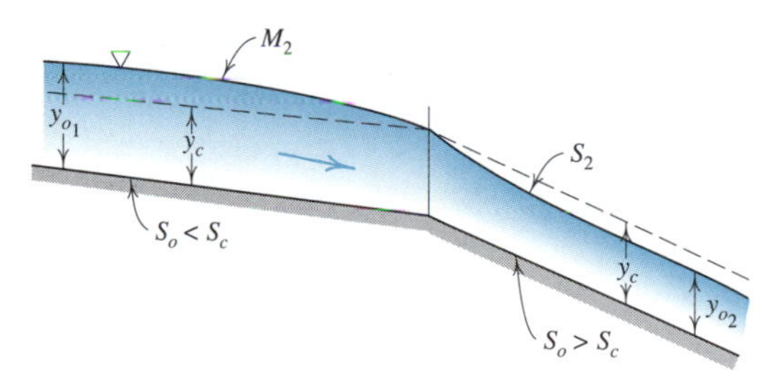

(c) On Fig. 10.20, we note that the channel downstream of the break has a mild slope and that the depth is less than the critical. Of course, on a mild slope this means the depth is also less than the normal. From Table 6, it can be seen that the water surface forms an *M_3 profile* with flow depth increasing in the downstream direction until it forms an hydraulic jump.

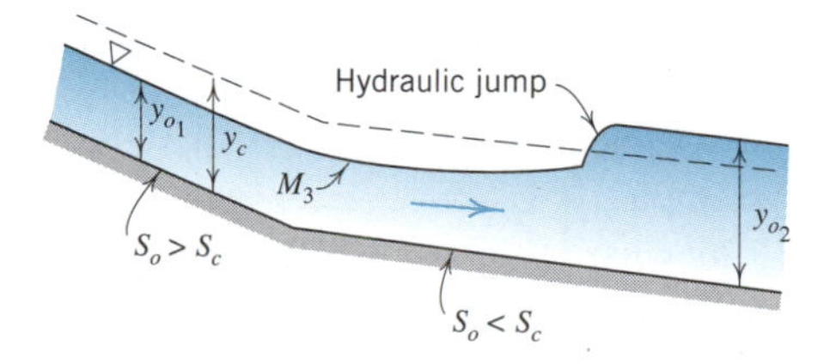

(d) Fig. 10.22 reproduced below shows flow over a free overfall from a channel with a mild slope. It is clear from the figure that the flow depth is greater than the critical and less than the normal. From Table 6, we see that the water surface must be an M_2 *profile* with the depth approaching the critical at the end of the channel.

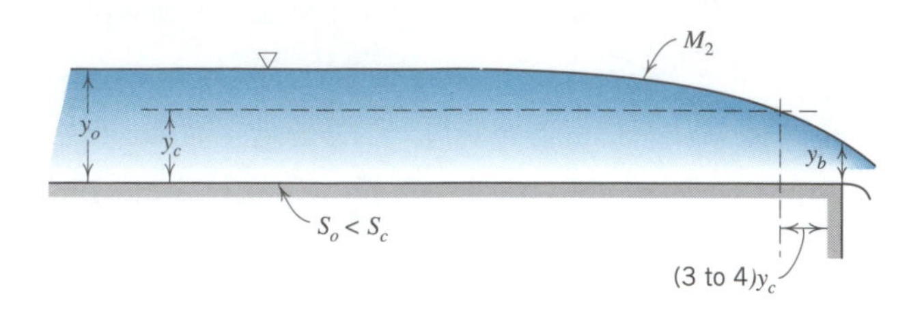

Knowing, in a qualitative sense, how depth varies along a channel for a given set of conditions, we must now develop a means of computing the water surface profile along the channel. Over the last century, numerous attempts have been made to achieve this objective by integrating the differential equation of varied flow. These efforts have met with varying degrees of success but have declined to historical interest with the advent of the digital computer. Currently, virtually all approaches to computing water surface profiles rely on the techniques of numerical analysis and digital computing (see Chaudhry[20]). This is due to the conceptual simplicity of application of numerical methods to solving the differential equation for varied flow, particularly in prismatic channels. However, the numerical approach is virtually the only practical method to use in natural channels. In this work we will address only the elementary application to flows in prismatic channels.

The most straightforward technique to use in numerically solving the varied flow equation is referred to as the *direct step method*. To set up this approach, we return to Eq. 10.26 and write it in a slightly different form recognizing that $y + V^2/2g_n = E$ (specific energy).

$$E + S_o\,dx = E + dE + S\,dx$$

so

$$dE = (S_o - S)\,dx$$

Now, writing this equation in finite difference form,

$$\Delta x = \frac{E_{i+1} - E_i}{S_o - \bar{S}} \tag{10.29}$$

where i and $i + 1$ represent locations along the channel separated by a distance Δx. The only uncertainty in applying this equation is the evaluation of $\bar{S}$, the average value of S in the interval. Experience has shown that $\bar{S}$ varies very gradually along the channel so if the values of Δx are kept reasonably small, Eq. 10.29 can produce good results.

The standard computational procedure is to determine a point in the channel where the depth is known and use that point as the starting point for the computations. (For example, it is often convenient to start at or near a control). A depth is then chosen somewhat larger (or smaller) than the initial depth and the distance upstream (or downstream) to the chosen depth is computed from Eq. 10.29. However, this calculation cannot

[20]Chaudhry, M. H., *Open Channel Flow*, Prentice Hall, 1993.

proceed until a means of calculating $\overline{S}$ is determined. While there are several possibilities for evaluating $\overline{S}$, using the average of the S-values calculated at each end of the channel reach is simple, yet is about as accurate as any other technique (according to Chaudhry[20]). Recalling Eq. 10.28,

$$\overline{S} = \frac{S_i + S_{i+1}}{2} = \frac{1}{2}\left(\frac{Qn}{u\mathrm{A}_i R_{h_i}^{2/3}}\right)^2 + \frac{1}{2}\left(\frac{Qn}{u\mathrm{A}_{i+1} R_{h_{i+1}}^{2/3}}\right)^2 \tag{10.30}$$

Subsequent Δx's can be computed by specifying additional depths and computing the associated Δx's until the required water surface profile has been generated. Program DIRSTEP is listed in Appendix 7 to assist in calculating surface profiles using this method.

A note of caution is in order. The normal depth cannot be used as a starting point because the differential equation of varied flow is undefined at that point (remember the depth of flow reaches the normal depth only at infinity). Also, the depth increment must be monitored to insure that the depth never crosses the normal depth or the critical depth or erroneous results will be generated. Determination of the incremental changes in depth to use in Eq. 10.29 can only be gained with experience but as a guideline, do not exceed 10% of the difference between the starting depth and the normal or critical depth as an initial estimate of the depth increment.

ILLUSTRATIVE PROBLEM 10.9

A flowrate of 10 m^3/s occurs in a rectangular channel 6 m wide, lined with concrete (troweled) and laid on a slope of 0.000 1 m/m. If the depth at a point in the channel is 1.50 m, how far (upstream or downstream) from this point will the depth be 1.65 m?

SOLUTION

The first step in the solution of this problem is to classify the flow profile. To do this we need to calculate the normal and critical depths. To obtain the normal depth, we use Eq. 10.5 with $u = 1$. From Table 5, we select an n-value of 0.013.

$$Q = \left(\frac{u}{n}\right) AR_h^{2/3} S_o^{1/2} = \left(\frac{1}{0.013}\right)(6y_o)\left(\frac{6y_o}{6 + 2y_o}\right)^{2/3} 0.000\,1^{1/2} \tag{10.5}$$

$$10 = 0.769(6y_o)\left(\frac{6y_o}{6 + 2y_o}\right)^{2/3}$$

Solving this equation by trial,

$$y_o = 1.94 \text{ m}$$

To calculate the critical depth, we use Eq. 10.15.

$$y_c = \sqrt[3]{\frac{q^2}{g_n}} = \sqrt[3]{\frac{(10/6)^2}{9.81}} = 0.66 \text{ m} \tag{10.15}$$

Because $y_o > y_c$, the slope is mild and because $y_c < y < y_o$, we are in Zone 2. From Table 6, it is clear we have an M_2 profle where depth of flow decreases in the downstream direction. From this result, we now can conclude that the depth of 1.65 m will occur *upstream* of the 1.50 m depth.

To perform the varied flow computations using Eq. 10.29, we will set up a computation table. We will use depth increments of 0.05 m.

$$\Delta x = \frac{E_{i+1} - E_i}{S_o - \bar{S}} \tag{10.29}$$

y	A	V	$V^2/2g_n$	E	P	R_h	S	$\bar{S}$	Δx	Σx
1.50	9.00	1.11	0.062 8	1.562 8	9.00	1.000	.000 209			0
								.000 199	471	
1.55	9.30	1.08	0.059 4	1.609 4	9.10	1.022	.000 190			471
								.000 182	557	
1.60	9.60	1.04	0.055 1	1.655 1	9.20	1.043	.000 173			1 028
								.000 166	711	
1.65	9.90	1.10	0.052 0	1.702 0	9.30	1.065	.000 159			1 739

Total distance = 1 739 m ●

A disadvantage of the previous method is that the depth of flow at a particular distance cannot be found directly. That is, the selection of y-values yields Δx's. We cannot select a Δx and solve directly for a y-value. This difficulty has been overcome by Prasad,[21] who produced an efficient technique for determining depth for a given value of Δx. The essence of his method is outlined in the following paragraphs for a prismatic channel. In his paper, Prasad generalizes this analysis to natural channels with lateral inflow.

The integration of the varied flow equation is based on Eq. 10.27.

$$y' = \frac{dy}{dx} = \frac{S_o - S}{1 - \mathbf{F}^2} = \frac{S_o - \left(\dfrac{Qn}{uAR_h^{2/3}}\right)^2}{1 - \dfrac{Q^2 b}{g_n A^3}} \tag{10.31}$$

Equation 10.31 is a first-order, nonlinear, ordinary differential equation for y as a function of x. It may be integrated by any number of means.[22] Here, an iterative technique known as the trapezoidal method is used.

Consider Fig. 10.27, which is essentially the same as the definition sketch of Fig. 10.25. Because y is a function of x,

$$y_{i+1} = y_i + \frac{dy}{dx}\Delta x = y_i + y'\,\Delta x \tag{10.32}$$

where the subscripts indicate the location along the channel. For the numerical integration Δx is presumed to be relatively very small so that to a good approximation, in the interval i to $i + 1$, $y' = 0.5(y'_i + y'_{i+1})$. Eq. 10.32 becomes

$$y_{i+1} = y_i + (y'_i + y'_{i+1})\,\Delta x \tag{10.33}$$

Equations 10.31 and 10.33 contain two unknowns, y_{i+1} and y'_{i+1}, provided the channel characteristics A, n, P, b, Q, and y_i are presumed known. The solution for y_{i+1} and y'_{i+1} proceeds as follows:

[21]R. Prasad, "Numerical Method of Computing Flow Profiles," *Jour. Hydr. Div., A.S.C.E.*, vol. 96, no. HY1, pp. 75–86, January, 1970.

[22]C. F. Gerald and P. O. Wheatley, *Applied Numerical Analysis*, Addison Wesley, 1989.

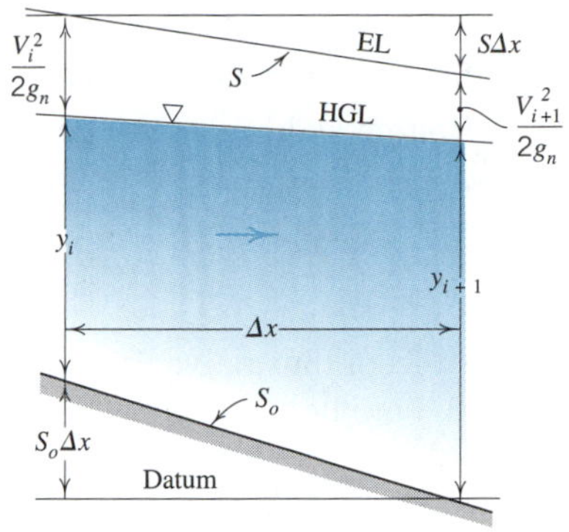

Fig. 10.27

(a) Use Eq. 10.31 to calculate y'_i where y_i is known either as the initial point or from a previous cycle of this calculation.

(b) Set $y'_{i+1} = y'_i$ as a first estimate.

(c) Use current values of y'_i and y'_{i+1} to calculate y_{i+1} from Eq. 10.33 for a selected Δx.

(d) Find a revised estimate of y'_{i+1} from Eq. 10.31 using the y_{i+1}-value from step (c).

(e) If the new y'_{i+1} value is not close enough to the previously calculated value, repeat steps (c), (d), and (e) using the latest estimate of y'_{i+1} found in step (d).

(f) Once the iteration procedure has yielded successive estimates of y'_{i+1} and y_{i+1} within acceptable limits of accuracy, proceed to the next section of channel and repeat the process.

(g) Terminate the process when the desired reach has been covered.

The question of accuracy always comes up when choosing a Δx for each reach of channel. A good check on the accuracy is to try a solution with the Δx-value halved. If the profile is essentially the same as before, the results should be good. Because this approach involves a great deal of iterative calculations, a computer program adapted from Prasad[23] is included for the convenience of the reader in solving problems of this type. The program PRASAD with the definition of input variables is shown in Appendix 7.

ILLUSTRATIVE PROBLEM 10.10

Solve Illustrative Problem 10.9 using the Prasad technique and the computer program PRASAD in Appendix 7.

SOLUTION

With the computer program at hand, the immediate task is to prepare the data needed for input to the program. Much of the necessary data is available from the problem statement. Following the comment statements in the program source listing which define the input, we have the following information:

[23]Adapted from R. Prasad, ''Numerical Method of Computing Flow Profiles,'' *Jour. Hydr. Div., A.S.C.E.*, vol. 96, no. HY1, p. 77 and 85, January, 1970.

Unit indicator = U = 1.0	Total length = L = 1 800 m
Slope = SZERO = 0.000 1	n-value = N = 0.013
Bottom width = B = 6 m	Side slope = Z = 0
Reach length = DX = 100 m	Flow rate = Q = 10 m^3/s
Initial depth = YI = 1.50 m	Direction indicator = DIR = −1

Note that from Illustrative Problem 10.9, we had discovered that the water surface profile was an M_2 profile so we know that depth increases in the upstream direction. This permits us to set DIR = −1 because we want calculations to proceed upstream. The selection of DX = 100 m is just an arbitrary choice. And finally, we would not normally know what value to set for L because our problem was posed in terms of a depth change rather than a distance. From the results of Illustrative Problem 10.9, we know that 1 800 m is an adequate distance.

Upon execution of PRASAD, the user will be prompted to key in the data listed above. The results of the analysis will be saved on a file for subsequent viewing and printing.

The results of the analysis for two separate DX-values are shown below:

```
** INPUT DATA **

TOTAL LENGTH = 1800.
       SLOPE = .000100
     N-VALUE = .0130
BOTTOM WIDTH = 6.0
  SIDE SLOPE = .00
REACH LENGTH = 100.
   FLOW RATE = 10.0
INITIAL DEPTH = 1.50

** ANALYSIS RESULTS **

      X           Y(X)
     .0          1.500
 -100.0          1.512
 -200.0          1.523
 -300.0          1.533
 -400.0          1.543
 -500.0          1.553
 -600.0          1.563
 -700.0          1.572
 -800.0          1.580
 -900.0          1.589
-1000.0          1.597
-1100.0          1.605
-1200.0          1.612
-1300.0          1.620
-1400.0          1.627
-1500.0          1.634
-1600.0          1.640
-1700.0          1.647
-1800.0          1.653
```

```
** INPUT DATA **

TOTAL LENGTH = 1800.
       SLOPE = .000100
     N-VALUE = .0130
BOTTOM WIDTH = 6.0
  SIDE SLOPE = .00
REACH LENGTH = 580.
   FLOW RATE = 10.0
INITIAL DEPTH = 1.50

** ANALYSIS RESULTS **

      X           Y(X)
     .0          1.500
 -580.0          1.561
-1160.0          1.610
-1740.0          1.650
```

There are three points to make as the result of this analysis. First, both approaches, as given in the two illustrative problems, give the same results. When three steps are used by both techniques, the distance upstream to a depth of 1.65 m is the same for both techniques. Second, a decrease in the step size does not necessarily result in a significant increase in the accuracy of the result; however, having a computer program available to explore the consequences of various step sizes permits one to make a thorough investigation of the problem. Third, which approach to use depends on whether you are seeking depth of flow at a given distance (Prasad) or distance to a given depth (direct step).

Another application of the varied flow equation involves the location of a hydraulic jump. In this particular case, we must utilize Eqs. 10.24 and 10.25 from Section 10.9 which describe relationships between depths on the two sides of the hydraulic jump. The problem is posed as shown on Fig. 10.28 where a steep channel slope changes to a mild slope. Both channels are of great length so uniform flow will eventually occur on both extremes of the channel. Initially, we do not know whether the jump occurs on the steep slope or the mild slope. If the jump occurs on the mild slope (Fig. 10.28), then the flow remains at normal depth on the steep slope and decelerates on the mild slope until a jump forms. Because the depth of flow on the mild-sloped channel is less than both the normal and critical depth, we are in Zone 3 and an M_3 profile occurs (see Table 6). The depth on the downstream side of the jump is the normal depth y_{o_2} in the mild-sloped channel. However, if the jump occurs on the steep slope (Fig. 10.29), the flow jumps from the normal depth on the steep slope to the conjugate depth y_2, decelerates until it reaches the normal depth y_{o_2} at the beginning of the mild-sloped channel. Because the depth of flow downstream of the jump is greater than both the normal and critical depths, we are in Zone 1 and an S_1 profile occurs.

In order to position the jump, we need to determine which scenario applies. The key to the determination is the conjugate depths across the jump corresponding to the two normal depths. For a given flow rate in the channel, the conjugate depth for the normal depth in the downstream channel can be found from Eq. 10.25 (y_1 in Fig. 10.26). If the depth y_1 is greater than the normal depth in the steep channel, and M_3 profile will be needed to decelerate the flow and increase the depth so that the jump can occur. If, on the other hand, the conjugate depth computed from Eq. 10.25 is less than the normal depth in the steep channel, then we have an impossible situation; for an M_3 profile, the depth cannot decrease in the downstream direction. Consequently, the jump must occur in the steep channel. Having discovered in which channel the jump occurs, we can now work with the

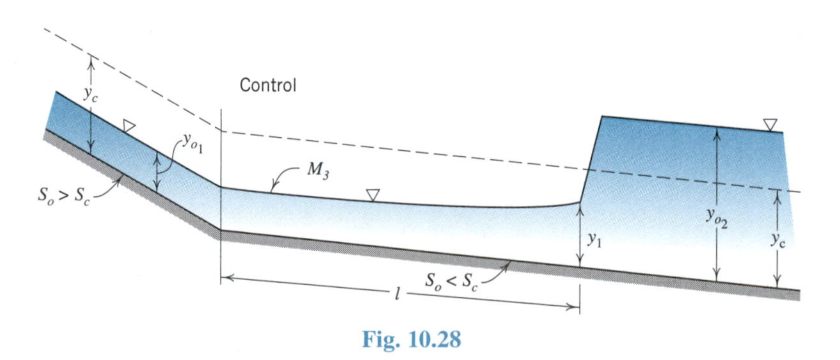

Fig. 10.28

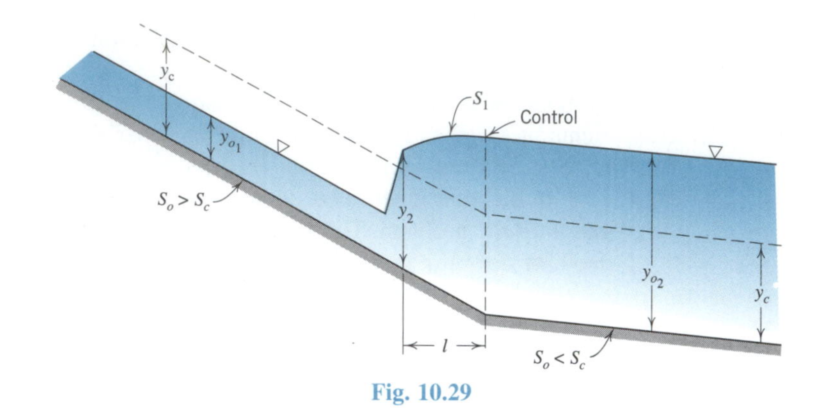

Fig. 10.29

varied flow equation to locate it. It is interesting to note that an increase in the n-value in the downstream channel could increase the normal depth y_{o_2} (and decrease the conjugate depth y_1) to the extent that it would be impossible for the jump to form on the downstream slope ($y_1 < y_{o_1}$). So, channel roughness does play a role in the positioning of hydraulic jumps.

ILLUSTRATIVE PROBLEM 10.11

A flow of 500 ft^3/s occurs in a long rectangular channel 12 ft wide with an n-value of 0.014. The channel is shown below and has a break in slope. The channel slope upstream of the break in slope is 0.012 and downstream of the break is 0.001 5. Uniform flow calculations determine that the normal depth upstream of the break is 2.46 ft and downstream of the break is 5.13 ft. The critical depth for this flow rate is 3.78 ft.

Verify that a hydraulic jump must occur and locate its position in the channel.

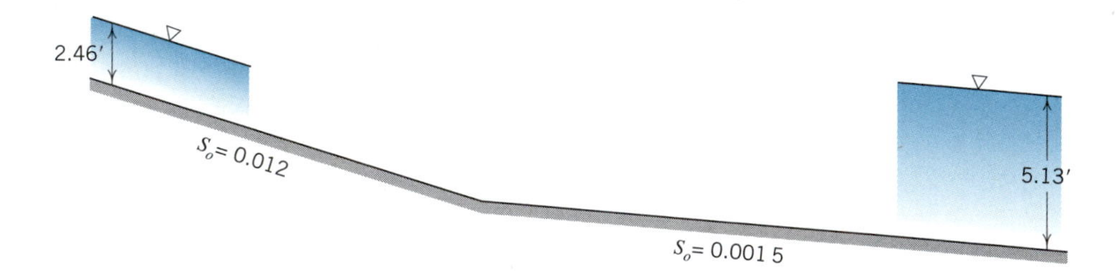

SOLUTION

The channel slope upstream of the break in slope is steep because the normal depth is less than the critical. Downstream of the break in slope, the normal depth is greater than the critical, hence this slope is mild. Under these conditions, for a long channel downstream of the break, a hydraulic jump will occur.

To determine whether the jump is upstream or downstream of the channel break, we will assume it occurs downstream and check to see if the assumption is valid. To do this, we use a slightly modified form of Eq. 10.25 to calculate the conjugate depth to the downstream normal depth.

$$y_1 = \frac{y_2}{2}\left[-1 + \sqrt{1 + \frac{8V_2^2}{g_n y_2}}\right] \tag{10.25}$$

where

$$V_2 = \frac{Q}{A_2} = \frac{500}{12 \times 5.13} = 8.12 \text{ ft/s}$$

Substituting this value into Eq. 10.25 gives

$$y_1 = \frac{5.13}{2}\left[-1 + \sqrt{1 + \frac{8 \times 8.12^2}{32.2 \times 5.13}}\right] = 2.69 \text{ ft}$$

Because the conjugate depth of 2.69 ft is greater than the normal depth of 2.46 ft on the steep slope, the conditions are right for an M_3 curve to form. Consequently, the jump will occur downstream of the break in channel slope.

To locate the jump, we will compute the distance downstream from the break in slope to the point where the depth of 2.69 ft occurs. We will use Eq. 10.29 and only one computational step to simplify the calculation. However, before using Eq. 10.29, we must compute the E-values at both depths and the average energy gradient slope $\bar{S}$.

$$E_{2.69} = y + \frac{V^2}{2g_n} = 2.69 + \frac{\left(\dfrac{500}{12 \times 2.69}\right)^2}{2 \times 32.2} = 6.420 \text{ ft}$$

$$E_{2.46} = y + \frac{V^2}{2g_n} = 2.46 + \frac{\left(\dfrac{500}{12 \times 2.46}\right)^2}{2 \times 32.2} = 6.900 \text{ ft}$$

$$S_{2.69} = \left(\frac{Qn}{uAR_h^{2/3}}\right)^2 = \left(\frac{500 \times 0.014}{1.49 \times (12 \times 2.69)\left(\dfrac{12 \times 2.69}{12 + 2 \times 2.69}\right)^{2/3}}\right)^2 = 0.009\,28$$

$$S_{2.46} = 0.012 \qquad \text{(same as channel slope upstream of the break in slope)}$$

Now we can use Eq. 10.29.

$$\Delta x = \frac{E_{i+1} - E_i}{S_o - \bar{S}} \tag{10.29}$$

$$\Delta x = \frac{E_{2.69} - E_{2.46}}{S_o - \left(\dfrac{S_{2.69} + S_{2.46}}{2}\right)} = \frac{6.420 - 6.900}{0.0015 - \left(\dfrac{0.00928 + 0.012}{2}\right)} = 53 \text{ ft}$$

The hydraulic jump is located 53 ft *downstream* of the break in the channel slope. •

REFERENCES

Ackers, P., White, W. R., Perkins, J. A., and Harrison, A. J. M. 1978. *Weirs and flumes for flow measurement.* New York: Wiley.

Chaudhry, M. H. 1993. *Open-channel flow.* New Jersey: Prentice Hall.

Chow, V. T. 1959. *Open-channel hydraulics.* New York: McGraw-Hill.

Elevatorski, E. A. 1959. *Hydraulic energy dissipators.* New York: McGraw-Hill.

Henderson, F. M. 1966. *Open channel flow.* New York: Macmillan.

Ippen, A. T. 1950. Channel transitions and controls. Chapter VIII in *Engineering hydraulics*, Ed. H. Rouse. New York: Wiley.

McBean, E., and Perkins, F. 1975. Numerical errors in water profile computation. *Jour. Hydr. Div., A.S.C.E.* 101, HY11: 1389–1403 (see also 1977: 103, HY6: 665–666).

Minton, P. and Sobey, R. J. 1973. Unified nondimensional formulation for open channel flow. *Jour. Hydr. Div., A.S.C.E.* 99, HY1: 1–12.

Posey, C. J. 1950. Gradually varied open channel flow. Chapter IX in *Engineering hydraulics*, Ed. H. Rouse. New York: Wiley.

Prasad, R. 1970. Numerical method of computing flow profiles. *Jour. Hydr. Div., A.S.C.E.* 96, HY1: 75–86.

FILMS

Rouse, H. Mechanics of fluids: fluid motion in a gravitational field. Film No. U45961, Media Library, Audiovisual Center, Univ. of Iowa.

St. Anthony Falls Hydraulic Laboratory. Some phenomena of open channel flow. Film No. 3, St. Anthony Falls Hydr. Lab., Minneapolis, Minn.

PROBLEMS

10.1. Liquid of specific weight γ flows uniformly down an inclined plane of slope angle θ. Derive an expression for the pressure on the plane in terms of γ, θ, and the depth y (measured *vertically* from the plane to the surface of the liquid).

10.2. Water flows uniformly at a depth of 1.2 m (4 ft) in a rectangular canal 3 m (10 ft) wide, laid on a slope of 1 m per 1 000 m (1 ft per 1 000 ft). What is the mean shear stress on the sides and bottom of the canal?

10.3. Calculate the mean shear stress over the wetted perimeter of a circular sewer 3 m in diameter in which the depth of uniform flow is 1 m and whose slope is 0.000 1.

10.4. What is the mean shear stress over the wetted perimeter of a triangular flume 2.4 m (8 ft) deep and 3 m (10 ft) wide at the top, when the depth of uniform flow is 1.8 m (6 ft)? The slope of the flume is 1 in 200, and water is flowing.

10.5. Calculate the Chezy coefficient that corresponds to a friction factor of 0.030.

10.6. Uniform flow occurs in a rectangular channel 10 ft wide at a depth of 6 ft. If the Chezy coefficient is 120 $\text{ft}^{1/2}/\text{s}$, calculate friction factor, Manning n, and approximate height of the roughness projections.

10.7. What uniform flowrate will occur in a rectangular timber flume 1.5 m wide and having a slope of 0.001 when the depth therein is 0.9 m?

10.8. Calculate the uniform flowrate in an earth-lined (n = 0.020) trapezoidal canal having bottom width 3 m (10 ft), sides sloping 1 (vert.) on 2 (horiz.), laid on a slope of 0.000 1, and having a depth of 1.8 m (6 ft).

10.9. A steel flume in the form of an equilateral triangle (apex down) of 1.2 m sides is laid on a slope of 0.01. Calculate the uniform flowrate that occurs at a depth of 0.9 m.

10.10. This large (uniform) open channel flow is to be modeled (without geometric distortion) in the hydraulic laboratory at a scale of 1 to 9. What flowrate, bottom slope, and Manning n will be required in the model?

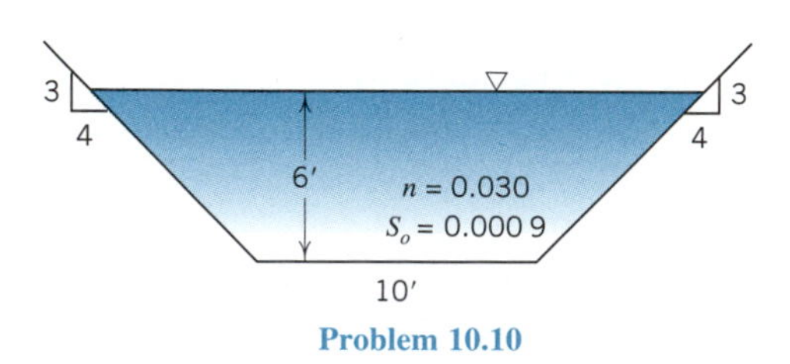

Problem 10.10

10.11. A rectangular open channel of 2.4 m width has a homogeneous roughness estimated to have an effective height of 6.4 mm. To visualize the variation of the various resistance coefficients with depth calculate and tabulate: relative roughness, friction factor, Chezy coefficient, and Manning n for depths of 0.15, 0.3, 0.6, 1.2, 1.8, and 2.4 m.

10.12. What uniform flowrate will occur in this canal cross section if it is laid on a slope of 1 in 2 000 and has $n = 0.017$?

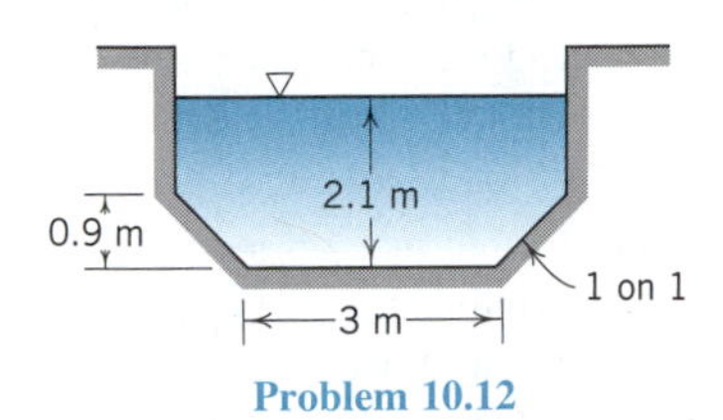

Problem 10.12

10.13. A semicircular canal of 1.2 m (4 ft) radius is laid on a slope of 0.002. If n is 0.015, what uniform flowrate will exist when the canal is brim full?

10.14. A flume of timber has as its cross section an isosceles triangle (apex down) of 2.4 m base and 1.8 m altitude. At what depth will 5 m^3/s flow uniformly in this flume if it is laid on a slope of 0.01?

10.15. At what depth will 4.25 m^3/s flow uniformly in a rectangular channel 3.6 m wide lined with rubble masonry and laid on a slope of 1 in 4 000?

10.16. At what depth will 400 cfs flow uniformly in an earth-lined ($n = 0.025$) trapezoidal canal of base width 15 ft, having side slopes 1 on 3, if the canal is laid on a slope of 1 in 10 000?

10.17. Calculate the depth of uniform flow for a flowrate of 11 m^3/s in the open channel of problem 10.12.

10.18. An earth-lined trapezoidal canal of base width 10 ft and side slopes 1 (vert.) on 3 (horiz.) is to carry 100 cfs uniformly at a mean velocity of 2 ft/s. What slope should it have?

10.19. What slope is necessary to carry 11 m^3/s uniformly at a depth of 1.5 m in a rectangular channel 3.6 m wide, having $n = 0.017$?

10.20. A trapezoidal canal of side slopes 1 (vert.) or 2 (horiz.) and having $n = 0.017$ is to carry a uniform flow of 37 m^3/s (1 300 ft^3/s) on a slope of 0.005 at a depth of 1.5 m (5 ft). What base width is required?

10.21. A rectangular channel 1.5 m wide has uniform sand grain roughness of diameter 6 mm. To observe the variation of n with depth, calculate values of n for depths of 0.03, 0.3, 0.6, and 0.9 m.

10.22. Compute the discharge in the unfinished concrete highway gutter shown if $z = 24$, $y_o = 0.22$ ft, and the slope is 3 ft per 100 ft.

10.23. A winding natural channel has a uniform flow of 30 m^3/s at a particular depth in the winter. What will be the uniform flowrate in the summer at the same depth when the channel is very weedy?

10.24. A flowrate of 0.1 cfs of oil (per foot of width) is to flow uniformly down an inclined glass plate at a depth of 0.05 ft. Calculate the required slope. The viscosity and specific gravity of the oil are 0.009 $lb \cdot s/ft^2$ and 0.90, respectively.

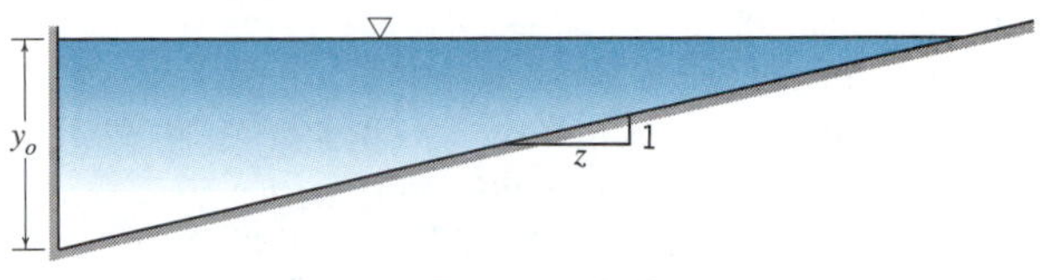

Problem 10.22

10.25. Water (20°C) flows uniformly in a channel at a depth of 0.009 m. Assuming a critical Reynolds number of 500, what is the largest slope on which laminar flow can be maintained? What mean velocity will occur on this slope?

10.26. What flowrate (per foot of width) may be expected for water (68°F) flowing in a wide rectangular channel at a depth of 0.02 ft if the channel slope is 0.000 1? Assume laminar flow and confirm by calculating Reynolds number.

10.27. At what depth will a flowrate of water (20°C) of 0.28 $l/s \cdot m$ occur in a wide open channel of slope 0.000 15? Assume laminar flow and confirm by calculating Reynolds number.

10.28. Plot curves similar to those of Fig. 10.8 for an equilateral triangle (apex up). Find the maximum points of the curves mathematically.

10.29. Plot curves similar to those of Fig. 10.8 for a square, laid diagonal vertical. Find the maximum points of the curves mathematically.

10.30. Derive an equation for the shape of the walls of an open channel that features no variation of hydraulic radius with water depth.

10.31. The cross section of an open channel is 6 m wide at the water surface and 3 m deep at the center. If its shape may be closely approximated by a parabola calculate the hydraulic radius.

10.32. This canal cross section is proposed for carrying a flowrate of 1 760 cfs at a velocity of 5 ft/s. With the same area dimension, propose another (trapezoidal) one which will be hydraulically *better*. What dimensions should be used (for the same area) for the section to be hydraulically *best*? The side slopes, bottom width, and depth may all be changed.

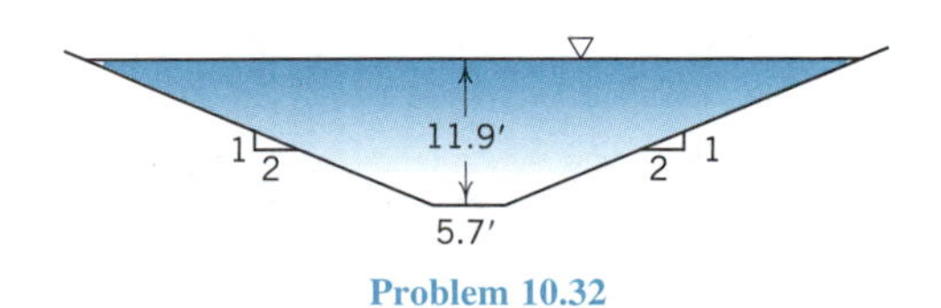

Problem 10.32

10.33. What uniform flowrate occurs in a 1.5 m circular brick conduit laid on a slope of 0.001 when the depth of flow is 1.05 m? What is the mean velocity of this flow?

10.34. Calculate the depth at which 0.7 m^3/s (25 ft^3/s) will flow uniformly in a smooth cement-lined circular conduit 1.8 m (6 ft) in diameter, laid on a slope of 1 in 7 000.

10.35. Rectangular channels of flow cross section 50 ft^2 have dimensions (width × depth) of (*a*) 25 ft by 2 ft, (*b*) 12.5 ft by 4 ft, (*c*) 10 ft by 5 ft, and (*d*) 5 ft by 10 ft. Calculate the hydraulic radii of these sections.

10.36. A channel flow cross section has an area of 18 m^2. Calculate its best dimensions if (*a*) rectangular, (*b*) trapezoidal with 1 (vert.) or 2 (horiz.) side slopes, and (*c*) V-shaped.

10.37. Calculate the required width of a rectangular channel to carry 45 m^3/s uniformly at best hydraulic conditions on a slope of 0.001 if n is 0.035.

10.38. What are the best dimensions for a trapezoidal canal having side slopes 1 (vert.) on 3 (horiz.) and n of 0.020 if it is to carry 40 m^3/s (1 400 ft^3/s) uniformly on a slope of 0.009?

10.39. A canal is being built to carry 1 000 cfs of water. The canal is lined with float-finished concrete, and its slope is 2.5 ft per mile. It is trapezoidal in cross section with side slopes of $z = 2$. Calculate the normal depth and bottom width of the canal under uniform flow conditions for the most efficient cross section.

10.40. What is the minimum slope at which 200 cfs may be carried uniformly in a rectangular channel having an n-value of 0.014 and a mean velocity of 3 ft/s?

10.41. A canal lined with concrete ($n = 0.014$) carries 2 500 cfs of water. The slope of the canal is 10 ft in 10 miles. It is trapezoidal in cross-section with side slopes of $z = 2$. Calculate the normal depth of flow and the bottom width for the most efficient cross-section.

10.42. What is the minimum slope at which 5.67 m^3/s may be carried uniformly in a rectangular channel (having a value of n of 0.014) at a mean velocity of 0.9 m^3/s.

10.43. What is the minimum slope at which 28 m^3/s (1 000 ft^3/s) may be carried uniformly at a mean velocity of 0.6 m/s (2 ft/s) in a trapezoidal canal having $n = 0.025$ and sides sloping 1 (vert.) on 4 (horiz.)?

10.44. Prove that the best form for a V-shaped open-channel section is one of vertex angle 90°.

10.45. This flood channel has a Manning n of 0.017 and a slope of 0.000 9. Estimate the depth of uniform flow for a flowrate of 1 200 cfs.

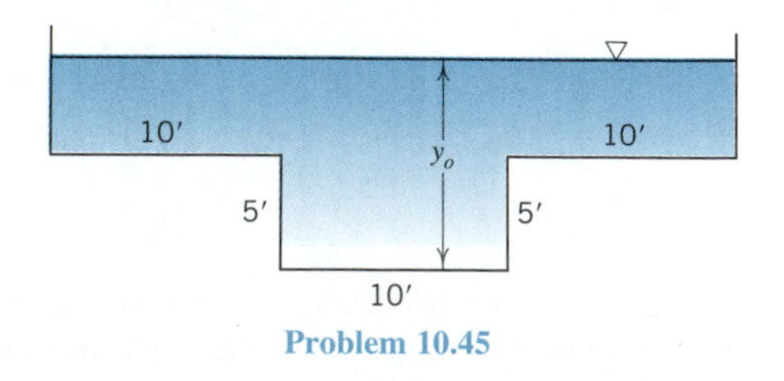

Problem 10.45

10.46. Calculate the specific energies when 225 cfs flow in a rectangular channel of 10 ft width at depths of (*a*) 1.5 ft, (*b*) 3 ft, and (*c*) 6 ft.

10.47. At what depths may 0.85 m^3/s flow in a rectangular channel 1.8 m wide if the specific energy is 1.2 m?

10.48. In a two-dimensional open-channel flow the water depth is 0.9 m (3 ft) and the specific energy 2.4 m (8 ft). For the same flowrate what depths may be expected for a specific energy of 3 m (10 ft)?

10.49. Eight hundred cubic feet per second flow in a rectangular channel of 20 ft width having $n = 0.017$. Plot accurately the specific energy diagram for depths from 0 to 10 ft, using the same scales for y and E. Determine from the diagram (*a*) the critical depth, (*b*) the minimum specific energy, (*c*) the specific energy when the depth of flow is 7 ft, and (*d*) the depths when the specific energy is 8 ft. What type of flow exists when the depth is (*e*) 2 ft, (*f*) 6 ft; what are the channel slopes necessary to maintain these depths? What type of slopes are these, and (*g*) what is the critical slope, assuming the channel to be of great width?

You may use the program in Appendix 7.

10.50. Flow occurs in a rectangular channel of 6 m width and has a specific energy of 3 m. Plot accurately the q-curve. Determine from the curve (*a*) the critical depth, (*b*) the maximum flowrate, (*c*) the flowrate at a depth of 2.4 m, and (*d*) the depths at which a flowrate of 28.3 m^3/s may exist, and the flow condition at these depths.

You may use the program in Appendix 7.

10.51. Five hundred cubic feet per second flow in a rectangular channel 15 ft wide at depth of 4 ft. Is the flow subcritical or supercritical?

10.52. If 11 m^3/s flow in a rectangular channel 5.4 m wide with a velocity of 1.5 m/s, is the flow subcritical or supercritical?

10.53. If 8.5 m^3/s flow uniformly in a rectangular channel 3.6 m wide having $n = 0.015$ and laid on a slope of 0.005, is the flow subcritical or supercritical? What is the critical slope for this flowrate, assuming the channel to be of great width?

10.54. A uniform flow in a rectangular channel 12 ft wide has a specific energy of 8 ft; the slope of the channel is 0.005 and the Chezy coefficient 120 $ft^{1/2}/s$. Predict all possible depths and flowrates.

10.55. What is the maximum flowrate which may occur in a rectangular channel 2.4 m wide for a specific energy of 1.5 m?

10.56. Thirty cubic metres per second (1 000 ft^3/s) flow uniformly in a rectangular channel, 4.5 m (15 ft) wide (n is 0.018, S_o is 0.002), at best hydraulic conditions. Is this flow subcritical or supercritical? What is the critical slope for the flowrate, assuming the channel to be of great width?

10.57. In a wide rectangular channel, n is 0.017 and does not vary over a depth range from 0.3 to 1.5 m. To confirm the trend of Fig. 10.16, calculate values of the critical slope for depths of 0.3, 0.9, and 1.5 m. What flowrates are associated with these depths? What is the average value of the critical slope for this depth range? What is the maximum (percent) deviation from this average value?

10.58. Assuming the canal entrance frictionless, calculate the two-dimensional flowrate and determine the water depths at the lettered sections.

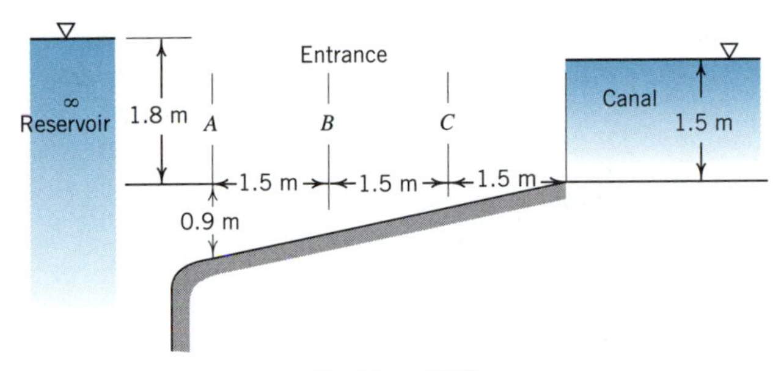

Problem 10.58

10.59. This "Venturi flume" has straight walls, horizontal floor, and may be assumed frictionless. For a flowrate of 300 cfs and energy line 10 ft above the floor determine the water depths at the lettered sections. Plot and use a q-curve in the solution of this problem. You may use the program in Appendix 7.

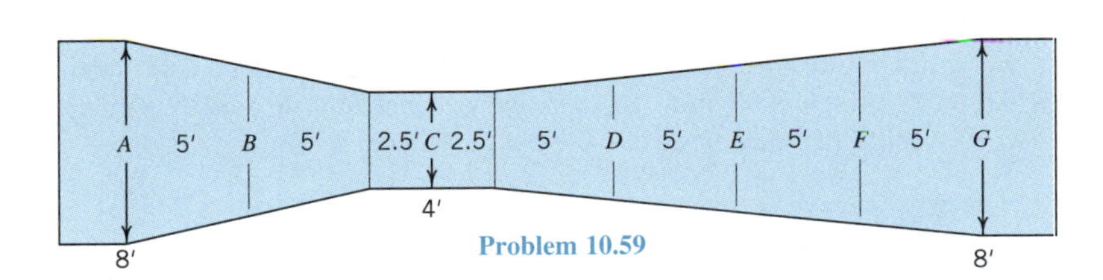

Problem 10.59

10.60. Eleven cubic metres per second are diverted through ports in the bottom of the channel between sections 1 and 2. Neglecting head losses and assuming a horizontal channel, what depth of water is to be expected at section 2? What channel width at section 2 would be required to produce a depth of 2.5 m?

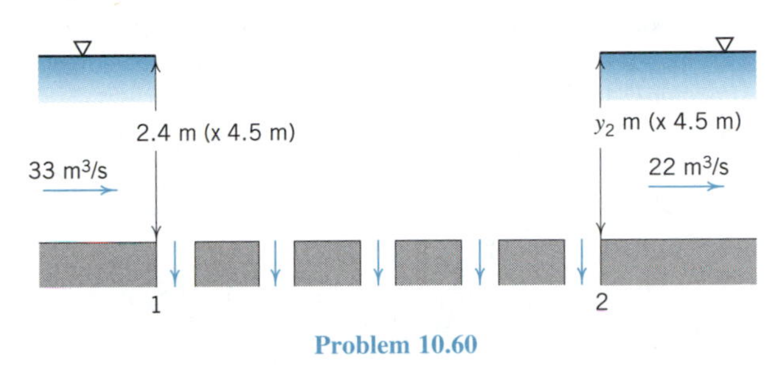

Problem 10.60

10.61. Derive a general expression for critical slope in terms of b, y, and n for a rectangular channel.

10.62. Calculate the specific energy when 300 cfs flow at a depth of 4 ft in a trapezoidal channel having base width 8 ft and sides sloping at 45°.

10.63. Calculate the specific energy when 2.8 m^3/s flow at a depth of 0.9 m in a V-shaped flume if the width at the water surface is 1.2 m.

10.64. Determine the critical depth for a flowrate of 2 m^3/s in this square open channel.

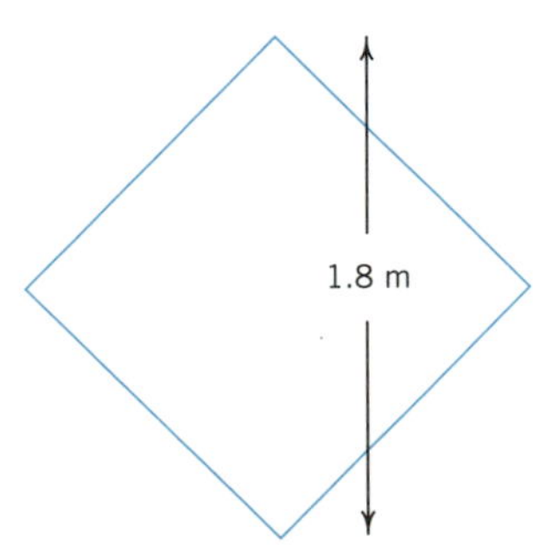

Problem 10.64

10.65. At what depths may 800 cfs flow in a trapezoidal channel of base width 12 ft and side slopes 1 (vert.) on 3 (horiz.) if the specific energy is 7 ft?

10.66. Eleven cubic metres per second flow in a trapezoidal channel of base width 4.5 m and side slopes 1 (vert.) on 3 (horiz.). Calculate the critical depth and the ratio of critical depth to minimum specific energy. If $n = 0.020$, what is the critical slope?

10.67. Uniform flow occurs in a trapezoidal canal of base width 1.5 m and side slopes 1 (vert.) and 2 (horiz.) laid on a slope of 0.002. The depth is 1.8 m. Is this flow subcritical or supercritical?

10.68. A sewer line of elliptical cross section 10 ft high by 7 ft wide carries 200 cfs of water at a depth of 5 ft. Is this flow subcritical or supercritical?

10.69. What is the critical depth for a flowrate of 14 m^3/s in a channel having the cross section of problem 10.12?

10.70. For best hydraulic cross section in a trapezoidal channel of side slopes 2 (vert.) on 1 (horiz.), a uniform flowrate of 1 000 cfs occurs at critical depth. If n is 0.017, calculate the critical slope.

10.71. Plot the variation of critical slope with depth for an open channel whose cross section is an isosceles triangle (apex up) of 2.4 m base and 2.4 m altitude.

10.72. Derive an expression for critical depth for an open channel of V-shaped cross section with side slopes of 45°. What is the ratio between critical depth and minimum specific energy for this channel?

10.73. Solve the preceding problem for a channel of parabolic cross section defined by $y = x^2$.

10.74. Calculate more exact values of critical slopes for problems (*a*) 10.49, (*b*) 10.53, (*c*) 10.56 by including the effect of the sidewalls.

10.75. Solve problem 10.57 for a rectangular channel of 4.5 m width.

10.76. What theoretical flowrate will occur over a high broad-crested weir 30 ft long when the head thereon is 2 ft?

10.77. The elevation of the crest of a high broad-crested weir is 100.00 ft. If the length of this weir is 12 ft and the flowrate over it 200 cfs, what is the elevation of the water surface upstream from the weir?

10.78. If the energy line is 0.9 m (3 ft) above the crest of a frictionless broad-crested weir of 1.2 m (4 ft) height, what is the water depth just upstream from the weir?

10.79. When the depth of water just upstream from a frictionless broad-crested weir 0.6 m (2 ft) high is 0.9 m (3 ft), what flowrate per metre (foot) of crest length can be expected?

10.80. A dam 1.2 m high and having a broad horizontal crest is built in a rectangular channel 4.5 m wide. For a depth of water *on* the crest of 0.6 m, calculate the flowrate and the depth of water just upstream from the dam.

10.81. The velocity in a rectangular channel of 3 m width is to be reduced by installing a smooth broad-crested rectangular weir. Before installation the mean velocity is 1.5 m/s and the water depth 0.3 m; after installation these quantities are to be 0.3 m/s and 1.5 m, respectively. What height of weir is required?

10.82. Flow occurs over and under this control gate as shown. Calculate the flowrate (per foot of gate width), assuming streamlines straight and parallel at section 2.

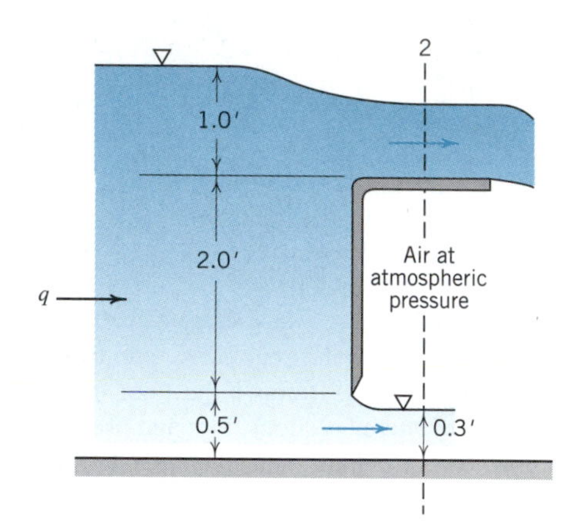

Problem 10.82

10.83. Assuming frictionless flow, calculate the two-dimensional flowrate over the weir crest and the water depths at the lettered sections.

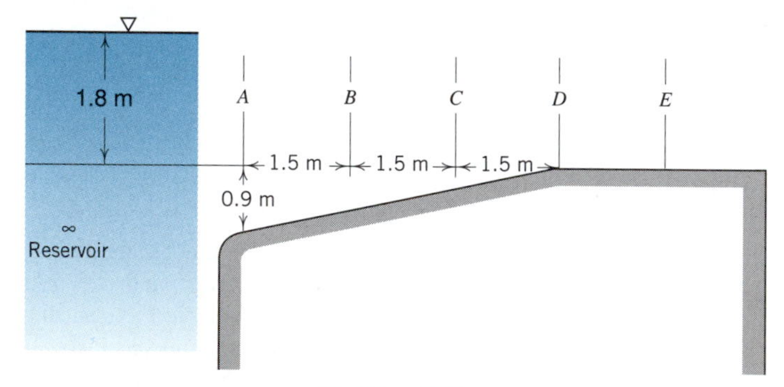

Problem 10.83

10.84. Four hundred cubic feet per second flow in a long rectangular open channel 12 ft wide which ends in a free outfall. The slope of the channel is 0.000 9 and Manning's n is 0.017. A frictionless broad-crested weir is to be installed near the end of the channel to produce uniform flow throughout the length of the channel. What weir height is required?

10.85. In the preceding problem, a sluice gate installed near the end of the channel is used to accomplish the same purpose. If the water depth just downstream from the gate is equal to the gate opening, what gate opening is required to produce uniform flow?

10.86. Show that the freely falling sheet of water well downstream from the brink of Fig. 10.22 must have a thickness (measured vertically) of $2y_c/3$.

10.87. An open, rectangular channel 1.5 m wide and laid on a mild slope ends in a free outfall. If the brink depth is measured as 0.264 m, what flowrate exists in the channel?

10.88. Fifty cubic feet per second flow in a rectangular channel 8 ft wide. If this channel is laid on a mild slope and ends in a free outfall, what depth at the brink is to be expected?

10.89. A horizontal pipe of 0.6 m diameter discharges as a free outfall. The depth of water at the brink of the outfall is 0.15 m. Calculate the flowrate.

10.90. This rectangular laboratory channel is 1 m wide and has a movable section 30 m long set on frictionless rollers and connected to the fixed channel with material so flexible that it transmits no force. When the indicated steady flow is established in the channel the reading on the spring scales is observed to increase by 75 N. Calculate Manning's n for the channel.

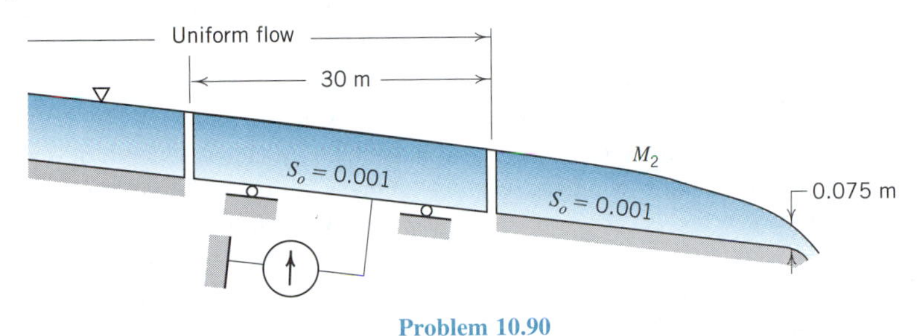

Problem 10.90

10.91. Is this water surface profile possible? If not, what profile is to be expected? The channel and constriction are both of rectangular cross section. The flowrate is 7.2 m^3/s.

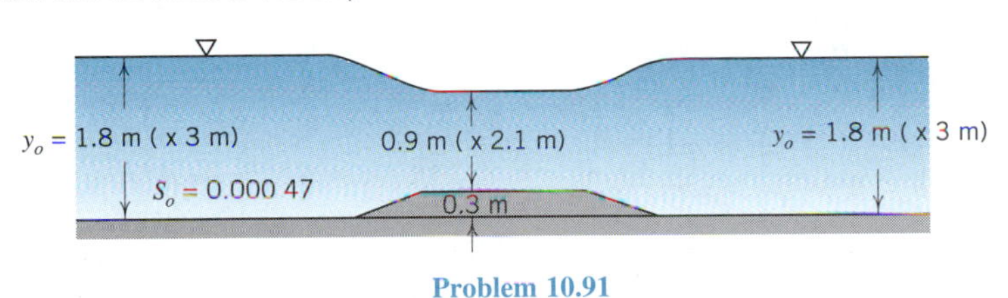

Problem 10.91

10.92. Up to what flowrate can critical depth be expected to exist in this constriction? For convenience assume Chezy C for the channel 86 $ft^{1/2}/s$ and neglect head losses caused by hump and constriction. Assume the channel very long upstream and downstream from the constriction.

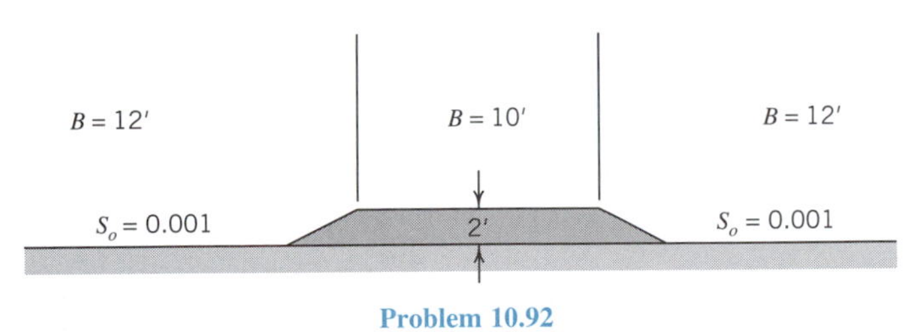

Problem 10.92

10.93. The critical depth is maintained at a point in a rectangular channel 1.8 m wide by building a gentle hump 0.3 m high in the bottom of the channel. When the depth over the hump is 0.66 m, what water depths are possible just upstream from the hump?

10.94. If 150 cfs flow uniformly in a rectangular channel 10 ft wide, laid on a slope of 0.000 4, and having $n = 0.014$, what is the minimum height of the hump that may be built across this channel to create critical depth over the hump? Sketch the energy line and water surface, showing all vertical dimensions. Neglect head losses caused by the hump.

10.95. Uniform flow in a rectangular channel occurs at a depth of 0.45 m and velocity of 7.5 m/s. When a smooth frictionless hump 0.6 m high is built in the floor of the channel, what depth can be expected on the hump? What hump height would be required to produce critical depth on the hump? What would happen if a taller hump than this were installed.

10.96. Solve the preceding problem for a depth of 3 m and velocity of 1.125 m/s.

10.97. Uniform flow at 6.00 ft depth occurs in a rectangular open channel 10 ft wide having Manning n of 0.017 and laid on a slope of 0.003 6. There is a flexible hump in the floor of the channel which can be raised or lowered. Neglecting losses caused by the hump: (*a*) How large may the hump height be made without changing the water depth just upstream from the hump? (*b*) How large should the hump height be to make the water depth 7.00 ft just upstream from the hump?

10.98. The water depth over a hump of 0.3 m height in a rectangular open channel is 1.05 m. Just upstream and just downstream from the hump the water depths are 1.5 m. What depths would be expected on and just upstream from the hump when the channel downstream from the hump is removed? There will be no change of flowrate.

10.99. A long, rectangular channel 10 ft wide carries 375 cfs of water at a normal depth of 7.0 ft. Calculate the critical depth and minimum energy. Find the depth of flow over and just upstream of a smooth hump 1.0 ft high and one 3.0 ft high. What is the minimum hump height that will cause critical flow over the hump?

10.100. A rectangular channel 12 ft wide is laid on a slope of 2.22 ft per 10 000 ft. The channel has an n-value of 0.016 and carries 352 cfs at a normal depth of 9.00 ft. A smooth hump is placed in the bottom of the channel. What would be the depth of flow over a 3-ft high hump? How high can the hump be made without backing flow up? What would be the depth of flow over and just upstream of a 6-ft high hump?

10.101. A very long, rectangular channel 13 ft wide carries 926 cfs at a normal depth of 8.0 ft. The n-value of the channel is 0.016, and the slope of the channel is 0.001 12. Calculate the critical depth and the minimum energy. Find the depth of flow over humps of 1.0, 2.0, and 3.0 ft heights.

10.102. A long rectangular channel 10 ft wide carries a flowrate of 372 cfs uniformly at a depth of 2.00 ft and ends in a free outfall. When a smooth hump 1.50 ft high is installed at the very end of this channel, what depth is to be expected on the hump? How large a hump would cause the depth there to be critical?

10.103. A discharge of 363.2 cfs occurs at a uniform depth of 4.00 ft in a rectangular channel 20 ft wide. This channel is to be narrowed to cause critical flow in the constricted section. What is the maximum width of constricted section that will accomplish this? For this situation, neglect head losses and sketch the energy line and water surface, showing the vertical dimensions. If the constriction is narrowed to an 8 ft width, what depth is to be expected in the constriction?

10.104. A rectangular channel 3.6 m wide is narrowed to a 1.8 m width to cause critical flow in the contracted section. If the depth in this section is 0.9 m, calculate the flowrate and the depth in the 3.6 m section, neglecting head losses in the transition. Sketch energy line and water surface, showing all pertinent vertical dimensions.

10.105. Four and one-quarter cubic metres per second flow uniformly in a rectangular channel 3 m wide having n = 0.014, and laid on a slope of 0.000 4. This channel is to be narrowed to cause critical flow in the contracted section. What is the maximum width of contracted section which will accomplish this? For this width and neglecting head losses, sketch the energy line and water surface, showing vertical dimensions. If the contraction is narrowed to 1.2 m width, what depths are to be expected in and just upstream from the contraction?

10.106. A uniform flow of 21.2 m^3/s occurs in a rectangular channel 4.5 m wide at a depth of 3 m. A hump 0.6 m high is built in the bottom of the channel, and at the same point width is reduced to 3.6 m. When friction is neglected, what is the water depth over the hump?

10.107. A flow of 50 cfs per foot of width occurs in a wide rectangular channel at a depth of 6 ft. A bridge to be constructed across this channel requires piers spaced 20 ft on centers. Assuming that the noses of the piers are well streamlined and frictionless, how thick may they be made without causing "backwater effects" (i.e., deepening of the water) upstream from the bridge?

10.108. A flow of 8.5 m^3/s occurs in a long rectangular channel 3 m wide, laid on a slope of 0.001 6, and having n = 0.018. There is a smooth gradual constriction in the channel to 1.8 m width. What depths may be expected in and just upstream from the constriction?

10.109. A long, rectangular channel 15 ft wide has $n = 0.014$ and is laid on a slope of 0.000 90. In this channel, there is a smooth constriction to 13.5 ft width containing a hump 1.50 ft high. Predict the water depths in and just upstream from the constriction for a flowrate of 1 000 cfs.

10.110. If the channel in problem 10.99 were narrowed to cause critical flow to occur in the constriction, what is the maximum width of the constriction that would accomplish this? What would be the depth of flow in the constriction and just upstream?

10.111. In a trapezoidal canal of 3 m base width and sides sloping at 45°, 23 m^3/s flow uniformly at a depth of 3 m. The channel is constricted by raising the sides to a vertical position. Calculate the depth of water in the constriction, neglecting local head losses. What is the minimum height of hump which may be installed in the constriction to produce critical depth there?

10.112. A long, rectangular channel of 4.5 m width, slope of 0.001, and n of 0.015 reduces to 3.3 m width as it passes through a culvert in a highway embankment. Below what flowrate can depths greater than normal depth be expected upstream from the embankment?

10.113. A uniform flow of 21.25 m^3/s occurs in a rectangular channel 4.6 m wide at a depth of 3 m. A hump 0.6 m high is built into the channel and at the same location, the width is reduced to 3.7 m. Neglecting friction, what is the water depth in the constriction?

10.114. A long, rectangular channel 3 m wide flows at a depth of 3 m with a discharge of 14 m^3/s. The channel is constricted to 2.4 m and a smooth hump is built into the channel in the constriction. How high can the hump be built without causing the flow to be backed up?

10.115. If the channel of problem 10.101 were constricted to the point where critical flow conditions just occurred, what would be the width of the constriction? If a 1.0 ft hump were installed in the channel and the width were reduced in that part of the channel over the hump, what constriction width would just cause critical flow to occur?

10.116. For a rectangular open channel of 8 ft width, n = 0.015, and slope = 0.003 5, the following data are given. Approximately how far apart are sections 1 and 2?

Section	Depth (ft)	Velocity (ft/s)	Hydraulic radius (ft)	Specific energy (ft)
1	3.00	15.00	1.715	6.49
2	3.20	14.06	1.775	6.26

10.117. The channels of Fig. 10.19 are of 3.6 m width and have n = 0.020. Their slopes are 0.005 00 and 0.010 0 and the depth of water at the break in slope is 1.5 m. Considering this depth to be critical depth, how far upstream and downstream from this point will depths of 1.62 and 1.38 m be expected? If the respective slopes were slightly decreased, would the calculated distances be lengthened or shortened?

10.118. This uniform flow occurs in a very long, rectangular channel. The sluice gate is lowered to a position so that the opening is 0.6 m. Sketch the new water surface profiles to be expected and identify them by letter and number. Calculate and show all significant depths.

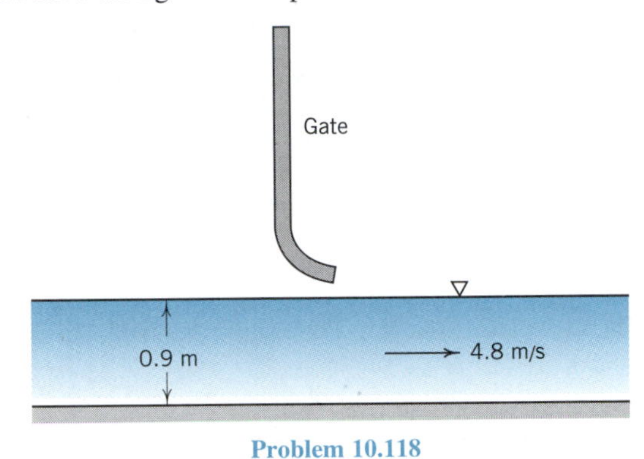

Problem 10.118

10.119. Solve the preceding problem for the same flowrate but with 1.8 m depth of uniform flow.

10.120. The depth of uniform flow for a flowrate of 300 cfs in an infinitely long rectangular open channel of 10 ft width is 5.00 ft. A single short smooth structure exists in the channel. This structure consists of a hump 2 ft high and a constriction to 8 ft width; hump and constriction are at the same flow cross section. Sketch the expected water surface profiles in the channel upstream and downstream from the structure. Identify any gradually varied flow profiles by letter and number. Calculate and show on the sketch all dimensions.

10.121. Solve the preceding problem for depths of uniform flow of (*a*) 2.50 ft and (*b*) 7.50 ft.

10.122. In each case, label the water surface profile with the appropriate letter and number symbol. Use an "*x*" if any one doesn't fit into a category. In each sketch, the channels are presumed to extend indefinitely in the upstream and downstream directions.

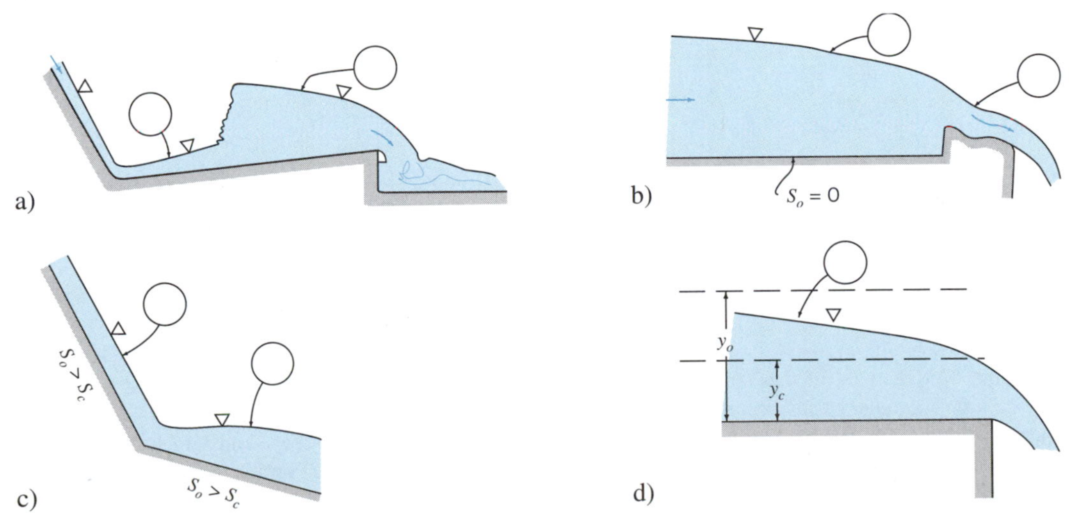

Problem 10.122 (continued on next page)

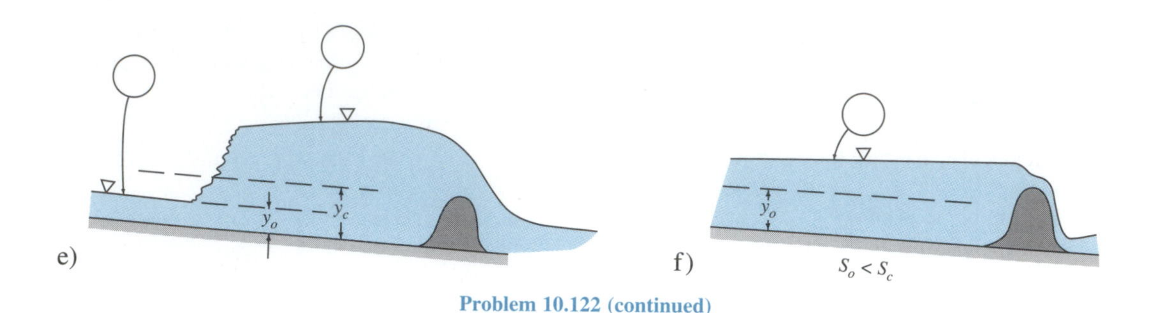

Problem 10.122 (continued)

10.123. Identify the following surface profiles as M1, S2, etc., if possible.

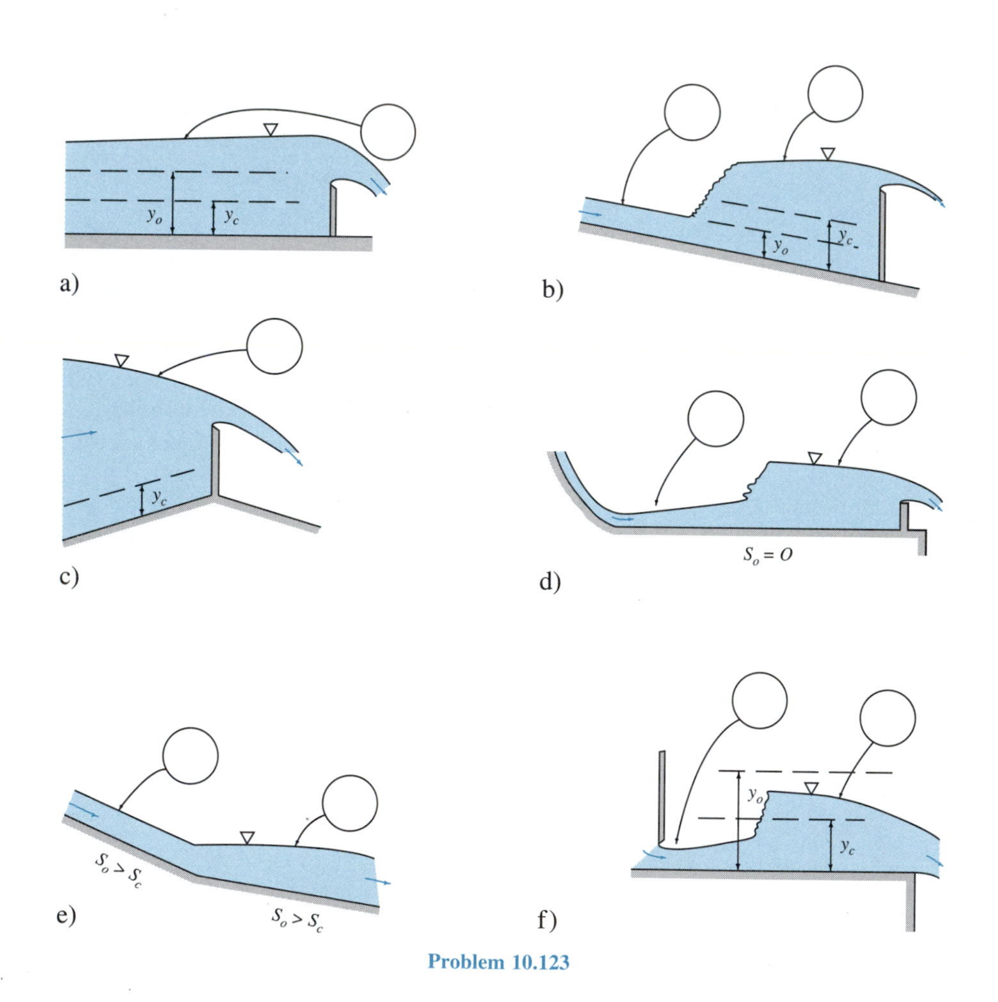

Problem 10.123

10.124. The length of ''Reach 1'' of the San Luis Canal in California is 25.25 km. The canal has a bottom width of 33 m and side slopes 7 (vert) on 15 (horiz). The bottom slope of this reach of canal is zero. With no flow in the canal, the depth of water will be everywhere 10.0 m, and the water surface coincident with that of the large reservoir (forebay) at the upstream end of the canal. Flow in this canal is produced by pumps at the downstream end which extract a (maximum) flowrate of 371 m^3/s. For this flowrate what drop in the water surface is to be expected at the downstream end of the canal?

Assume that the water surface in the forebay remains constant and that n is 0.014 9.

10.125. A uniform flow at 5 ft depth occurs in a long, rectangular channel 15 ft wide, having $n = 0.017$, and laid on a slope of 0.001 0. A hump is built across the floor of the channel of such height that the water depth just upstream from the hump is raised to 6.00 ft. Identify the surface profile upstream from the hump and calculate the distance from the hump to where a depth of 5.60 ft is to be expected. Use depth intervals of 0.20 ft for noncomputer solutions.

10.126. Water discharges through a sluice gate into a long, rectangular open channel 2.4 m wide having $n = 0.017$. The depth and velocity at the vena contracta are 0.6 m and 7.5 m/s, respectively. Identify the surface profiles if the channel slope is (*a*) favorable, (*b*) zero, (*c*) adverse. If these slopes are (*a*) 0.002, (*b*) 0.000, and (*c*) -0.002, how far downstream from the sluice will a depth of 0.67 m be expected?

10.127. A rectangular channel 3 m wide is laid on a slope of 0.002 and has Manning n of 0.014 9. At a certain point in the channel the water depth is 0.60 m. Sixty metres downstream from this point the depth is 0.67 m. Estimate the flowrate and calculate the water surface profile.

10.128. The flowrate in a trapezoidal channel of base width 10 ft, side slope 2 on 3, and $n = 0.014\,9$ is 600 cfs. A very long reach of this channel has a slope of 0.000 3. At a point in the channel, the depth is 5.00 ft. Approximately how far (upstream or downstream) may a depth of 6.00 ft be expected?

10.129. Construct the computer solution to problem 10.125.

10.130. Compute the profile downstream from a point where $y = 1.9$ m in a trapezoidal channel in which $n = 0.02$, $z = 2$, $S_o = 0.002$, $b = 30$ and $Q = 211\ \text{m}^3/\text{s}$. How far can calculation proceed?

10.131. A very wide rectangular channel of bottom slope 0.002 has a constant $C = 100\ \text{ft}^{1/2}/\text{s}$. At a point in this channel the water depth is 5.0 ft.; 150 ft. downstream the depth is 4.7 ft. Estimate the flowrate per foot of width and identify the type of the water surface profile.

10.132. Eleven cubic metres per second flow uniformly in an infinitely long trapezoidal canal having base width 3 m, side slopes 1 on 1, Manning $n = 0.021$, and depth 0.75 m. The channel is locally constricted by raising the walls to a vertical position. This alteration is short and streamlined and may be assumed to cause no additional head loss. What water depths are to be expected in the constriction, just upstream from there, and just downstream from there? Also find any other depths that can be obtained without use of the equations of gradually varied flow. Show all results on a sketch with any gradually varied flow profiles identified by letter and number.

10.133. Compute all gradually varied flow profiles for problem 10.132.

10.134. An overflow dam is built in the channel of problem 10.100, which causes a depth of flow just upstream of the dam to be 11 ft. Classify the surface profile that will occur in the channel upstream of the dam. Calculate the distance upstream to the section where the depth is 9 ft using depth increments of 1.0 ft.

10.135. Construct the computer solution to problem 10.134 using depth increments of 0.1 ft.

10.136. Eight hundred cubic feet per second flow in a rectangular channel of 20 ft width. Plot the $\mathcal{M}$-curve of hydraulic jumps on the specific energy diagram of problem 10.49. From these curves determine (*a*) the depth after a hydraulic jump has taken place from a depth of 1.5 ft, (*b*) the height of this jump, (*c*) the specific energy before the jump, (*d*) the specific energy after the jump, (*e*) the loss of head in the jump, and (*f*) the total horsepower lost in the jump.

10.137. A hydraulic jump occurs in a rectangular open channel. The water depths before and after the jump are 0.6 m and 1.5 m, respectively. Calculate the critical depth.

10.138. For a rectangular open channel, prove that hydraulic jumps can occur only across the critical depth. Prove this for a nonrectangular channel.

10.139. If the maximum $\mathbf{F}_1$ for an undular hydraulic jump is $\sqrt{3}$, what is the maximum value of y_2/y_1 for such jumps? Show that the head lost in a hydraulic jump is given by $h_L/y_1 = (y_2/y_1 - 1)^3/4(y_2/y_1)$. Note that such head losses will be very small (usually negligible) for low values of y_2/y_1.

10.140. A supercritical flow of 100 cfs occurs at 3 ft depth in a V-shaped open channel of side slopes 45°. Calculate the depth just downstream from a hydraulic jump in this flow.

10.141. A hydraulic jump occurs in a V-shaped open channel with sides sloping at 45°. The depths upstream and downstream from the jump are 0.9 m and 1.2 m, respectively. Estimate the flowrate in the channel.

10.142. A hydraulic jump occurs in a horizontal storm sewer of circular cross section 1.2 m in diameter. Before the jump, the water depth is 0.6 m and just downstream from the jump the sewer is full with a gage pressure of 7 kPa at the top. Predict the flowrate.

10.143. A rectangular channel 3 m wide carries a flowrate of $14\ \text{m}^3/\text{s}$ uniformly at 0.6 m depth. The channel is constricted at the end to produce a hydraulic jump in the channel. Calculate the width of constriction required for the jump to be just upstream from the constriction.

10.144. The depths of water upstream and downstream from a hydraulic jump on the horizontal ''apron'' downstream from a spillway structure are observed to be approximately 3 ft and 8 ft. If the structure is 200 ft long (perpendicular to the direction of flow), about how much horsepower is being dissipated in this jump?

10.145. Calculate y_2, h, and y_3 for this two-dimensional flow picture. State any assumptions clearly.

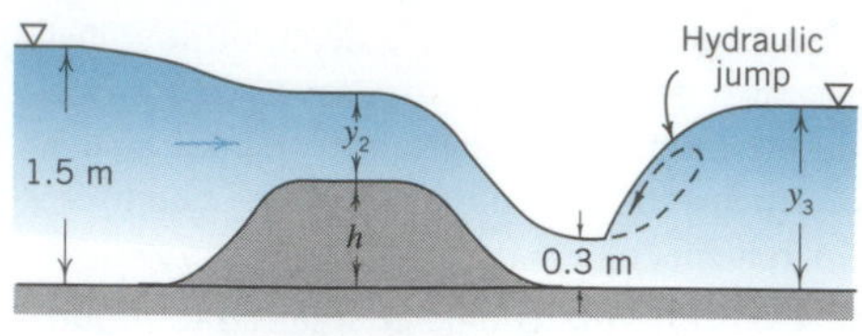

Problem 10.145

10.146. The situation shown on the sketch is observed (in the field) in a rectangular, open channel 5 ft wide and 12 ft deep. The 10 ft and 6 ft depths are easily (although not accurately) measured. The depth at the vena contracta cannot be obtained because of danger and inaccessibility. From the measurements shown, estimate the flowrate in the channel.

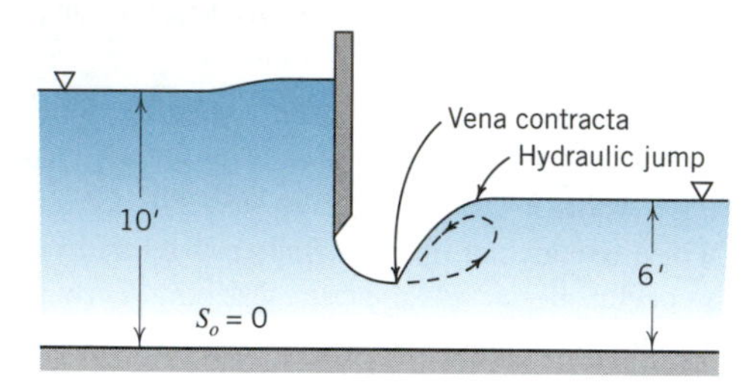

Problem 10.146

10.147. What shape (y_2/y_1) of hydraulic jump will cause dissipation of one third of the total energy of the approaching flow?

10.148. Determine the horizontal component of force exerted on the broad-crested weir of problem 14.94 using (*a*) a conventional impulse-momentum analysis and (*b*) the $\mathcal{M}$-diagram of Fig. 10.23.

10.149. Determine the force component of problem 6.32 by use of the $\mathcal{M}$-diagram of Fig. 10.23.

10.150. In which zone (*AB*, *BC*, etc.) in this flow is a hydraulic jump to be expected? Show numerical proof for selection of zone. The numbers above the H_2 profile are the water depths for that profile. The numbers below the channel floor are the depths for the H_3 profile. Upstream of section *A* the flow may be considered frictionless.

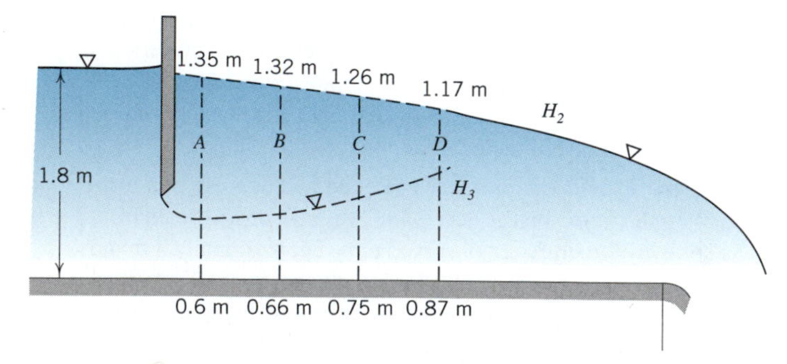

Problem 10.150

10.151. A very long, rectangular channel of 8 ft width and n of 0.017, laid on a slope of 0.03, carries a flowrate of 200 cfs and ends in a free outfall. Across the downstream end of the channel a smooth frictionless hump is constructed. Calculate the depths (*a*) on the hump, (*b*) just upstream from the hump, and (*c*) far up the channel, and sketch and identify the surface profiles. If a hydraulic jump occurs calculate the depths y_1 and y_2 and the varied flow profiles. The hump is 2.5 ft high.

10.152. Water leaves the spillway at the base of a dam with a flowrate of 113 m^3/s in a 15 m wide channel. The depth of flow is 0.67 m as it leaves the spillway and enters the channel, which has a slope of 0.001 and an n-value of 0.025. The normal depth of flow in this channel is 3.34 m for this flowrate. Design considerations require the jump to form within 25 m of the spillway. Neglecting the weight component in the hydraulic jump, determine if the jump will occur within the required distance.

10.153. In the very long channel shown, a flow of 76.5 m^3/s occurs. The normal depth is 3.3 m, the channel slope is 0.000 339, and the n-value is 0.014. Flow enters the channel at the base of a spillway at a depth of 0.90 m. Will a hydraulic jump occur downstream? If it does occur, how far downstream will it occur and how much energy will be lost in the jump per newton of water?

10.154. Water enters a rectangular channel (n = 0.014 9) of great width (assume two-dimensional flow) at a velocity of 40 ft/s and a depth of 2.0 ft. An H3 curve forms downstream and the water jumps to a depth of 8.0 ft some distance downstream. Find the distance downstream to the location of the jump.

10.155. Flow is introduced into the channel of problem 10.99 at a high velocity and a depth of 1.00 ft. At some distance down the channel, a hydraulic jump occurs. Calculate the depth of flow just upstream of the jump and the total energy lost in the jump. If the n-value for the channel is 0.013, calculate how far down the channel the jump occurs.

10.156. The slope of the channel of problem 10.100 is increased to produce a uniform flow depth of 1.5 ft. A hydraulic jump is caused to occur in the channel by obstructing the flow with a channel hump. Sketch the flow profile upstream of the hump and classify the surface profile. Calculate the conjugate depth and find the head loss in the jump.

10.157. A flowrate of 254 cfs occurs in a very long rectangular open channel 10 ft wide, having slope and Manning n such that the depth of uniform flow is 6.00 ft. In this channel there is a smooth hump 2 ft high and (at the same section) a constriction to 6 ft width. Predict the water depths; (*a*) in the constriction, and in the channel, (*b*) just upstream, and (*c*) just downstream from the constriction. Calculate and sketch the water-surface profile carefully, identifying any profiles of gradually varied flow by letter and number.

10.158. The flowrate through this Venturi flume is 28.3 m^3/s; the flume may be considered frictionless. The slope of the approach channel is such that the normal depth is 2.7 m. Over what depth range at the downstream end of the flume may hydraulic jumps be expected in the flume?

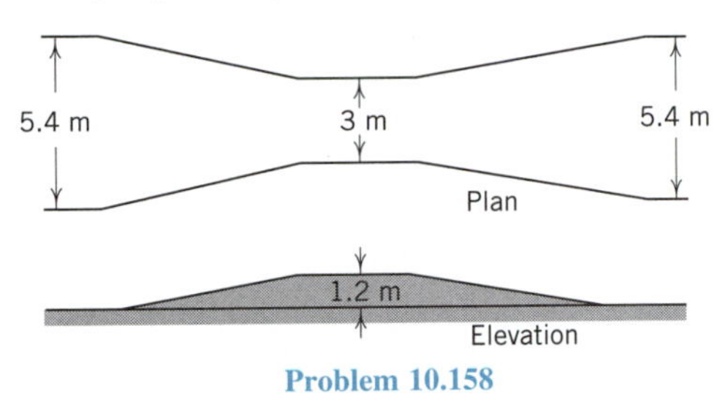

Problem 10.158

10.159. Neglecting wall, bottom, and hump friction (but not losses in the jump), what height of hump will produce this flow picture?

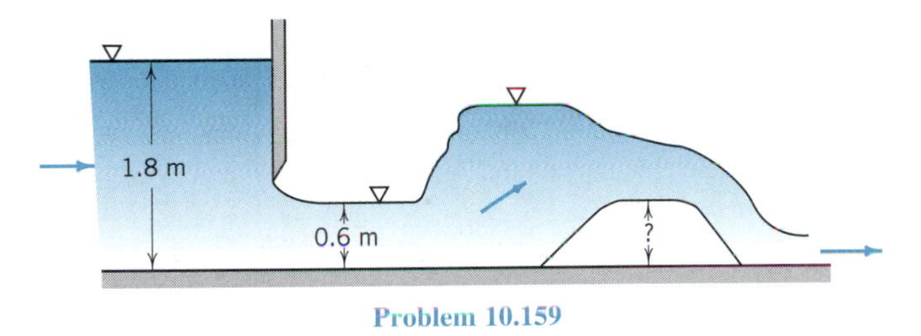

Problem 10.159

10.160. The sketch shows a plan view of a frictionless Venturi flume of rectangular cross section with horizontal floor. A hydraulic jump from 1 ft to 3 ft occurs at section 3. Calculate the depth of water at sections 1, 2, and 4. Section 1 may be assumed to be infinitely wide.

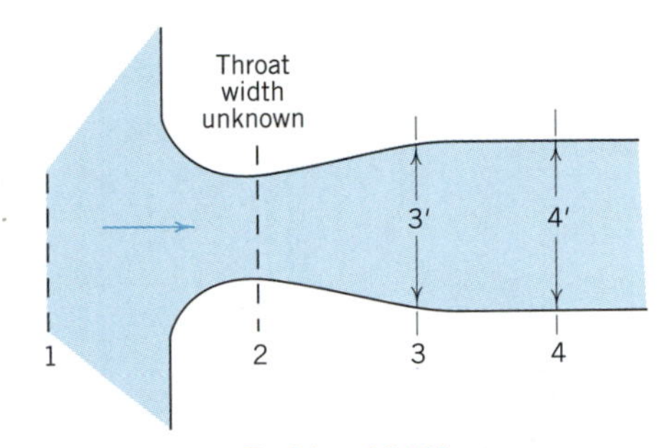

Problem 10.160

10.161. The flowrate in Fig. 10.28 is 15.4 m^3/s. If the channels are 3.6 m wide with $n = 0.017$ and the downstream channel laid on a slope of 0.002 28, what depth must exist in this channel for a hydraulic jump to occur to uniform flow? Calculate the length l if $y_{o_1} = 0.80$ m.

10.162. The flowrate in Fig. 10.29 is 15.4 m^3/s, the channels are 3.67 m wide, and $n = 0.017$. The downstream channel is laid on a slope of 0.001 5. If $y_{o_1} = 0.79$ m, calculate l.

CHAPTER

11 LIFT AND DRAG—INCOMPRESSIBLE FLOW

Fluid resistance is experienced by all bodies when they move through a real fluid, whether they be cars or planes or boats or trains, or even surfers! Thus, the problems involving the forces acting on a solid body when fluid flows by it are no longer exclusively the domain of the naval architect and aeronautical engineer, who focus on ships, propellers, wings and fuselages. The effective and safe design of buildings, bridges, automobiles, trucks, and trains demands a knowledge of fluid resistance or drag. The fluid principles describing lift are applied in the design of propellers, wind-, water-, or gas turbines, and pumps. Even a major league baseball pitcher might benefit from the knowledge of why his curve ball curves and how far! Because these principles find wide application, it is essential for all engineers to be familiar with the fundamental mechanics of the fluid motion in and about such objects. The purpose of this chapter is to outline the fundamental and elementary aspects of the external flow about immersed objects for incompressible flow; the compressible flow is treated in Chapter 13. This chapter is built in three major sections. First, we will consider some key fundamentals, including definitions and dimensional analysis. Second, the elements of lift and drag will be described and the essential relationships presented. Finally, we turn to some applications, including baseballs, offshore pilings, wind turbines, automobiles, etc.

Fundamentals

11.1 DEFINITIONS

In general, when flow occurs about an object that is either asymmetrical or whose axis is not aligned with the flow, the flowfield will be asymmetrical, the local velocities and pressures on either side of the object will be different, and a force normal to the oncoming flow will be exerted. Accompanying this, we expect (based on our study of boundary layers in Chapter 7) that the action of the frictional stress due to the boundary layers on the surface of the body will produce a net force approximately along the direction of the oncoming flow and opposing the motion of the body. These pressure and friction forces produce a pair of net forces, which are perpendicular to each other and are called (from their aeronautical ancestry) *lift* and *drag*. A classic and instructive example of such forces is seen in the flow about the airfoil (or hydrofoil) of Fig. 11.1, on which the lift L, drag D, resultant force F, angle of attack α, and chord c are shown; the length of the foil perpendicular to the plane of the paper is termed the *span*. The force F, which is seen to be the resultant of L and D, is also the resultant of all forces of pressure and friction exerted by fluid on foil. However, often the contribution of the frictional stresses to the lift may

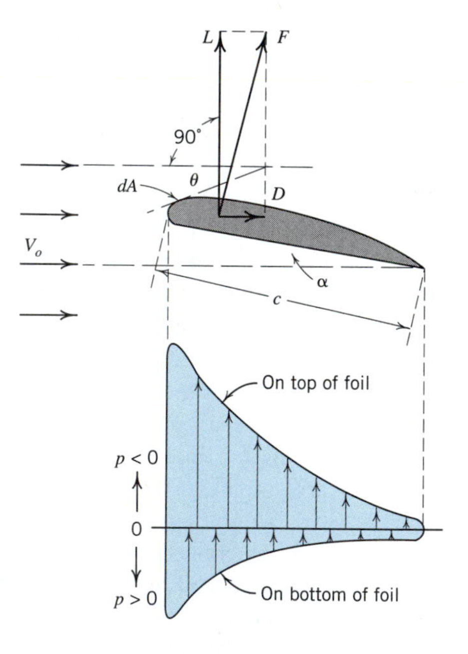

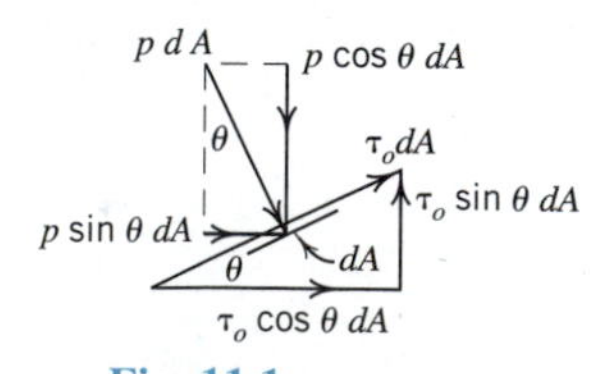

Fig. 11.1

be neglected when such stresses are small compared to the pressure and act in a direction roughly normal to L. When permissible this is an important simplification of the problem in that it allows L to be considered to be the result of pressure variation alone and thus permits the use of the ideal fluid for analytical predictions of lift.

Prediction of the drag force on immersed objects is much more difficult than that of lift, since usually no simplifications are possible and both pressure and frictional forces must be considered. However, both lift and drag may be predicted from experimental measurements on small models in wind tunnels, water tunnels, or towing basins. In addition, a combination of hydrodynamical and boundary-layer theory can often be used.

The foregoing preliminary discussion becomes more meaningful through formal application to a surface element of an immersed body, followed by appropriate integration. Consider the element dA of Fig. 11.1 on which a pressure p and frictional stress τ_o act. The differential drag and lift on this element are seen to be

$$dD = p\,dA \sin\theta + \tau_o\,dA \cos\theta$$

$$dL = -p\,dA \cos\theta + \tau_o\,dA \sin\theta$$

which may be integrated to yield

$$D = \iint_s p\,dA \sin\theta + \iint_s \tau_o\,dA \cos\theta$$

$$L = -\iint_s p\,dA \cos\theta + \iint_s \tau_o\,dA \sin\theta$$

in which $\iint_s$ designates the *integral over the surface of the object*. Under the assumption that the second integral of the lift expression is often negligible, the lift is expressed approximately by

$$L \approx -\iint_s p\,dA \cos\theta$$

The integrals in the drag equation are of equal importance; the first one is called the *pressure drag* D_p, and the second the *frictional drag* D_f. The former will depend on the form of the object and flow separation, the latter on the extent and character of the boundary layer. Although the prediction of separation points is generally a very complex problem dependent on both body form and boundary-layer properties, the breakdown of total drag into pressure drag and frictional drag proves of great value in studying these separately in further detail. Frictional drag may be isolated by considering the flow past a thin flat plate parallel to the oncoming flow (Fig. 11.2*a* and check Section 7.4); here $\sin\theta = 0$, $\cos\theta = 1$, $D_p = 0$, and $D = D_f = \iint_s \tau_o\,dA$. Pressure drag may be isolated by studying the flow about a flat plate normal to the oncoming flow (Fig. 11.2*b*); here $\cos\theta = 0$, $\sin\theta = 1$, $D_f = 0$, and $D = D_p = \iint_s p\,dA$.

Consider the circular disk, sphere, and streamlined body of Fig. 11.3, all of which have the same cross-sectional area normal to the flow (frontal area), and are immersed in the same turbulent flow; all these flows will feature a stagnation point on the upstream side of the object and a maximum local pressure there.

For the disk, separation will be expected at the edges with high local velocity and low pressure. This reduced pressure, being adjacent to the wake, will be transmitted into it, causing the downstream side of the disk to be exposed to a mean pressure considerably

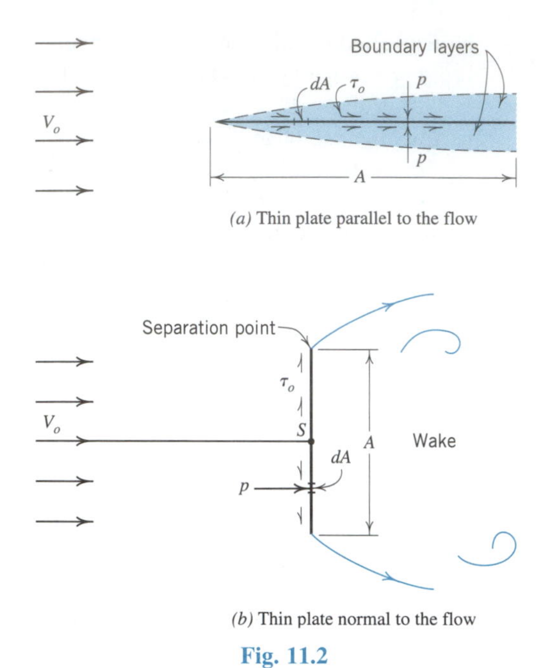

(a) Thin plate parallel to the flow

(b) Thin plate normal to the flow

Fig. 11.2

below that on the upstream side. The result is a large drag force caused wholly by pressure since none of the shear forces on the disk has components in the original direction of motion.

For the sphere, the wake is smaller than that of the disk and (from the streamline picture) will contain a somewhat higher pressure, leading to the expectation that the drag of the sphere is considerably smaller than that of the disk.[1] For the sphere, the frictional drag is not zero since all shear stresses acting on the sphere will have components parallel to the oncoming flow. These shear stresses are extremely difficult to calculate, but they are small and result in a frictional drag which is negligible compared to the pressure drag for spheres and other objects of similar (blunt) form.

For the well-streamlined body, the wake may be extremely small, being only the width of the boundary layer at the tail of the object. The pressure in such a wake is comparatively large, since the gentle contour of the body allows deceleration of the flow and consequent regain of pressure without incurring separation. Thus the pressure drag of such objects is a very small fraction of that of the disk. However, the frictional drag of streamlined bodies is considerably larger than that of the sphere, since streamlining has brought more surface

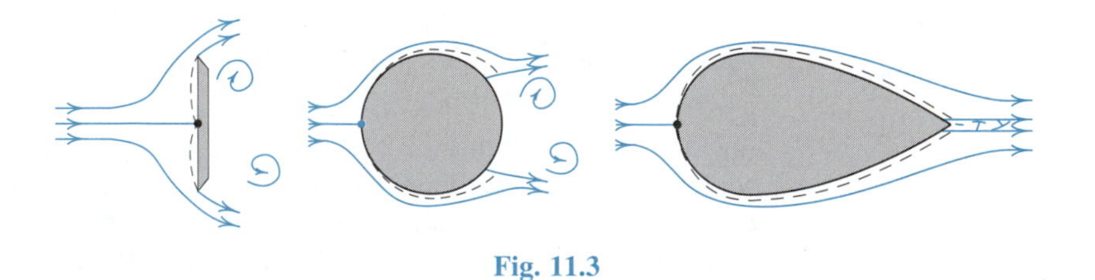

Fig. 11.3

[1]Experiments show the total drag of the sphere to be about one-third that of the disk.

area in contact with the flow. For well-streamlined objects, frictional drag is usually larger than pressure drag but both are so small that their total is only about one-fortieth that of the disk.

The foregoing examples illustrate the fact that the viscosity property of a fluid is the root of the drag problem. Viscosity has been seen to cause drag either by frictional effects on the surface of an object or through pressure drag by causing separation and the creation of a low-pressure wake behind the object. By streamlining an object, the size of its wake is decreased and a reduction in pressure drag is accomplished, but in general an increase in frictional drag is incurred.

For an ideal fluid in which there is no viscosity and thus no cause for frictional effects or wake formation regardless of the shape of the object about which flow is occurring, it is evident that the drag of the object is zero. Two centuries ago, d'Alembert's observation that all objects in an ideal fluid exhibit no drag was a fundamental and disturbing paradox; today this fact is a logical consequence of the fundamental reasoning presented above.

11.2 DIMENSIONAL ANALYSIS

The general aspects of drag and lift forces on completely[2] immersed bodies may be examined advantageously by dimensional analysis before further consideration of the physical details of the problem.

The smooth object of Fig. 11.4 having area[3] A moves through a fluid of density ρ, viscosity μ, and modulus of elasticity E, with a velocity V_o. If the drag force exerted on the body is D,

$$D = f_1(A, \rho, \mu, V_o, E)$$

and similarly

$$L = f_2(A, \rho, \mu, V_o, E)$$

The Buckingham Π-method of dimensional analysis (Section 8.2) shows that, in each case, three distinct nondimensional groups can be formed. Choosing a length ($\sqrt{A}$), ρ, and V_o

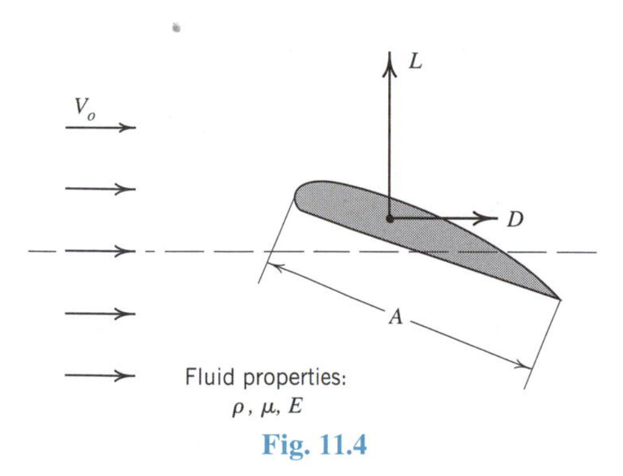

Fig. 11.4

[2]The drag of a surface vessel, a partially submerged body, was analyzed in Sections 8.1 and 8.2.

[3]A convenient significant area may be selected, for example, the product of maximum chord and span for a wing.

as the repeating variables (as in Section 8.2) leads to the following Π-terms:

$$\Pi_1 = \rho\sqrt{A}V_o/\mu = \mathbf{R}$$

$$\Pi_2 = \rho V_o^2/E = V_o^2/a^2 = \mathbf{M}^2$$

(because $a = \sqrt{E/\rho}$) which are common to both the lift and drag cases plus

$$\Pi_3 = D/A\rho V_o^2$$

and

$$\Pi_4 = L/A\rho V_o^2$$

Accordingly, the very general results are

$$D = \frac{f_3(\mathbf{R}, \mathbf{M})A\rho V_o^2}{2} \qquad \text{and} \qquad L = \frac{f_4(\mathbf{R}, \mathbf{M})A\rho V_o^2}{2}$$

If drag and lift coefficients, C_D and C_L, respectively, are defined by[4]

$$C_D = \frac{D}{\frac{1}{2}A\rho V_o^2} \qquad \textit{and} \qquad C_L = \frac{L}{\frac{1}{2}A\rho V_o^2} \tag{11.1}$$

it follows that

$$C_D = f_3(\mathbf{R}, \mathbf{M}) \qquad \text{and} \qquad C_L = f_4(\mathbf{R}, \mathbf{M})$$

These equations indicate: (1) that bodies having the same shape and the same alignment with the flow (i.e., models of each other) possess the same drag and lift coefficients if their Reynolds numbers and Mach numbers are the same, or (2) that the drag and lift coefficients of bodies of given shape and alignment may be expected to depend on their Reynolds and Mach numbers only. Thus dimensional analysis has, as in previous problems (ship resistance and pipe friction), opened the way to a comprehensive treatment of the resistance of immersed bodies by indicating the dimensionless combinations of variables on which the drag coefficient depends. An example of this is given in Fig. 11.5, where experimentally determined drag coefficients for spheres are plotted as contours on a Reynolds number-Mach number graph.

From the foregoing dimensional analysis alone no conclusion can be reached on the quantitative effects of **R** and **M** on the drag and lift coefficients, but theory and experiment both show that **R** is predominant when the fluid may be considered incompressible and **M** predominant when compressibility effects must be considered. Usually this means that over a wide range of **R**, where **M** is small and velocities subsonic, the fluid may be considered incompressible and C_D and C_L functions of **R** only. On the other hand, when **M** approaches or exceeds unity and velocities approach or exceed that of sound, C_D and C_L are functions of **M** only, whatever the magnitude of **R**. Although such a division of flow problems is useful and convenient, it is somewhat arbitrary as there is no definite point at which the effects of compressibility begin, such effects being present at all velocities; therefore it is to be expected that there will be exceptions to this convenient division of flow problems and situations encountered where the effects of **R** and **M** are of the same order and in which neither may be ignored. Such problems are highly complex and far

[4] L/A or D/A is a force per unit area, while $\frac{1}{2}\rho V_o^2$ is the dynamic pressure.

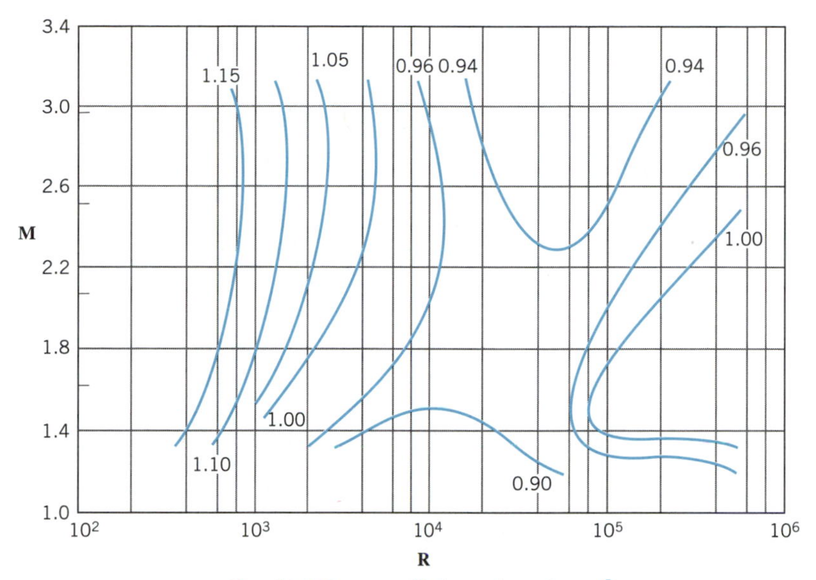

Fig. 11.5 Drag coefficients for spheres.[5]

beyond the scope of an elementary text, but the beginner should be aware of their existence and not expect them to be classified by usual and convenient methods. Reynolds and Mach numbers have been shown (Section 8.1) to be, respectively, ratios of inertia-to-viscous and inertia-to-elastic forces; from this it may be concluded directly that the flow phenomena, when governed by Reynolds number, will result from viscous action (boundary layers, etc.) but, when governed by Mach number, from elastic phenomena (shock waves, etc.). Examples of problems dealing with the first situation are the settling of solid particles through a fluid, the forces of wind on structures, and most aerodynamic problems in the field of commercial aviation. Examples of problems concerned with the second situation are the motions of missiles, rockets, propeller tips, supersonic aircraft, and elements of gas turbines and high-speed compressors.

ILLUSTRATIVE PROBLEM 11.1

The lift and drag coefficients of an approximately rectangular airfoil of 36 m span and 7.5 m chord are 1.2 and 0.02, respectively, when at an angle of attack of 7°. Calculate the power required to drive this airfoil (in horizontal flight) at 600 km/h through still, standard air at an altitude of 4 km. What lift force is obtained when this power is expended? Also calculate the Reynolds and Mach numbers.

SOLUTION

To apply Eqs. 11.1, we must first obtain the density of the air, the area of the wing, and the flow velocity in m/s. From Appendix 2, $\rho = 0.909 \text{ kg/m}^3$, $A = 7.5 \times 36 = 270 \text{ m}^2$,

[5]A. May, "Supersonic Drag of Spheres at Low Reynolds Numbers in Free Flight," *Jl. Appl. Physics*, vol. 28, 1957, pp. 910–912.

and $V_o = 600$ km/h $= 166.7$ m/s; so

$$C_D = \frac{D}{\frac{1}{2}A\rho V_o^2}: \quad D = \frac{1}{2}A\rho V_o^2 C_D \tag{11.1}$$

$$= 0.5 \times 270 \times 0.909 \times (166.7)^2 \times 0.02 = 68.2 \text{ kN}$$

From mechanics, the power is equal to the work done per unit time, i.e., the drag times the velocity:

$$\text{Power} = (68.2 \times 10^3 \text{ N}) \times (166.7 \text{ m/s}) = 11.4 \times 10^6 \text{ N·m/s}$$
$$= 11.4 \text{ MW} = 15\,250 \text{ hp} \bullet$$

Then,

$$C_L = \frac{L}{\frac{1}{2}A\rho V_o^2}: \quad L = \frac{1}{2}A\rho V_o^2 C_L \tag{11.1}$$

$$= 0.5 \times 270 \times 0.909 \times (166.7)^2 \times 1.2 = 4\,090 \text{ kN} \bullet$$

At 4 km altitude, Appendix 2 yields $T = -4.5°\text{C}$ and $\mu = 1.661 \times 10^{-5}$ Pa·s and so using a definition of the Reynolds number based on chord c,

$$\mathbf{R} = \frac{V_o c\rho}{\mu} = \frac{166.7 \times 7.5 \times 0.909}{1.661 \times 10^{-5}} = 68.4 \times 10^6 \bullet$$

Next, from Appendix 2, the gas constant $R = 286.8$ J/kg·K and the adiabatic exponent $k = 1.4$; therefore, from Eq. 1.11,

$$a = \sqrt{kRT} = \sqrt{1.4 \times 286.8(-4.5 + 273.2)} = 328.5 \text{ m/s} \tag{1.11}$$

and finally,

$$\mathbf{M} = \frac{V_o}{a} = \frac{166.7}{328.5} = 0.508 \bullet$$

For comparison, the Boeing 747B has four engines whose thrust is up to 209 kN each and it has flown at a gross weight of 3 650 kN.

Dimensional analysis can also give us some significant insight into the case of a body accelerating relative to the fluid through which it moves; this occurs when piles supporting an offshore platform are exposed to the oscillating flow field of water waves in the ocean. For this case, the question arises as to whether there is an additional drag or resistance force associated with the acceleration. Intuitively, we expect so because the body must push fluid out of the way as it moves through it. If the body is accelerating, then some mass of fluid is accelerated too and a force proportional to the product of the mass and the acceleration should be expected. Indeed, such a force is found. The effect is that the body acts as if it were more massive than before and this is known as the *added mass* effect.

Suppose now that a smooth body of characteristic dimension d moves through a fluid of density ρ and is accelerating with an acceleration dV_o/dt. Assuming for the moment that the fluid is incompressible and that viscous effects are negligible, we apply dimensional analysis to learn if there is a dimensionless grouping that can be composed from these variables and the effective ''inertial'' drag D_I on the body (related to an acceleration and the inertia force of Eq. 8.2). Using the approach outlined in Section 8.2, we posit that $D_I = f(\rho, d, dV_o/dt)$ and therefore, $D_I = C\rho^a d^b (dV_o/dt)^c$. We write that

$$(\text{Dimensions of } D_I) = (\text{Dimensions of } \rho)^a (\text{Dimensions of } d)^b (\text{Dimensions of } dV_o/dt)^c$$

or

$$\frac{ML}{t^2} = \left(\frac{M}{L^3}\right)^a (L)^b \left(\frac{L}{t^2}\right)^c$$

It follows that

$$M: \quad 1 = a$$

$$L: \quad 1 = -3a + b + c$$

$$t: \quad -2 = -2c$$

Solving for a, b, and c yields: $a = 1$, $c = 1$, and $b = 3$ and resubstituting these values in the equation above for D_I gives

$$D_I = C\rho\, d^3 (dV_o/dt)$$

By writing this result as

$$C_I = \frac{D}{\rho\, d^3 (dV_o/dt)} = \frac{D}{\rho \forall (dV_o/dt)} \tag{11.2}$$

we define a *coefficient of inertia* and we can observe that the inertial drag is proportional to the cube of the characteristic dimension of the body, i.e., to the volume $\forall$ of the body. The drag is proportional to the mass of the fluid displayed by the body times the acceleration of the body. For the cylindrical pile, it can be shown theoretically that $C_I = 2$, while for a sphere $C_I = 1.5$. In applications, the pressure and friction drag terms are added to the inertial or added mass term to form the complete drag force (see Section 11.8).

Drag and Lift

11.3 PROFILE DRAG

Although in steady flow the total drag force on any immersed object is always the sum of frictional and pressure drag, it will be seen later that this breakdown of the drag is inconvenient for objects (such as airfoils) on which a transverse (lift) force is exerted. Here the total drag is considered to be the sum of (1) that which would be developed if the airfoil had no ends (i.e., two-dimensional flow), and (2) that produced by any end effects. Since

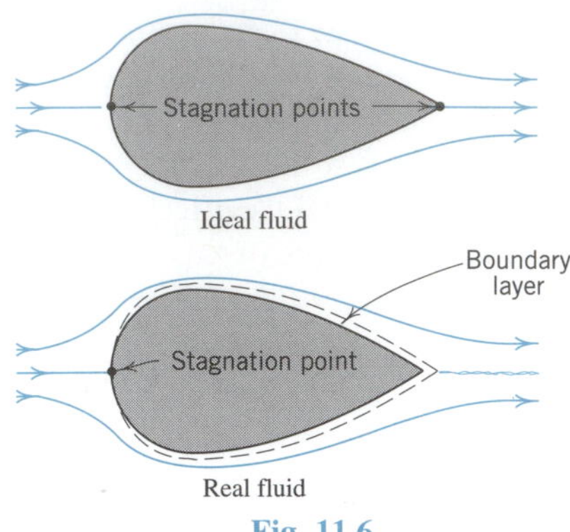

Fig. 11.6

the former depends only on the shape (profile) and orientation of the airfoil, it is called *profile drag*, whereas the latter, which depends on the airfoil plan form and is *induced* by the lift force, is termed *induced drag*. Evidently, for objects which exhibit no lift, the induced drag will be zero and the profile drag equal to the total drag.

Pressure drag has been shown (Section 11.1) to be that part of total drag resulting from pressure variation over the surface of an object and to be dependent on wake formation downstream from the object. In general, when wakes are large, pressure drag is large and, when wake width is reduced by streamlining, pressure drag is reduced also; pressure drag is thus critically dependent on the existence and position of flow separation which in turn depends on shape of object and structure of the boundary layer. Consider first the case of a well-streamlined object (Fig. 11.6 and Fig. 7.9); here there is no separation, and the flowfield for ideal or real fluid flow is essentially the same except for boundary-layer growth and lack of trailing edge stagnation point in the latter. In this case a good approximation to the pressure distribution on the object may be made by neglecting the boundary layer thickness and applying the methods of mathematical hydrodynamics; such computations may be refined (with considerable difficulty) through altering the shape of the object by the displacement thickness of the boundary layer. These methods cannot, of course, yield the pressure at the trailing edge of the body, which for ideal flow is the stagnation pressure but for real flow is considerably less than this. However, the pressure may be estimated (by extrapolating the pressure distribution over the body) with sufficient accuracy that the pressure variation over the body may be appropriately integrated to allow a reasonable estimate of the pressure drag. More accurate values of pressure drag may be obtained from experimentally determined pressure distributions in the same manner.

For a blunt object (Fig. 11.7) there is a drastic difference between the ideal and real flowfields caused by flow separation and wake formation in the latter. Here the critical feature either for analytical calculations or for understanding of experimental results is the position of the separation points. Lack of resistance and energy dissipation in the ideal flow will allow fluid particles adjacent to the object to move between stagnation points, accelerating over the upstream end of the object in a favorable pressure gradient and decelerating over the downstream end through an unfavorable pressure gradient. For the real fluid, boundary-layer growth will begin at the stagnation point and energy will be dissipated in overcoming resistance caused by shear stresses in the boundary layer. The momentum of fluid particles in the boundary layer will thus be considerably less than those at corresponding positions in the ideal flowfield; the momentum of such particles will be further reduced by the unfavorable pressure gradient until at some point they will come

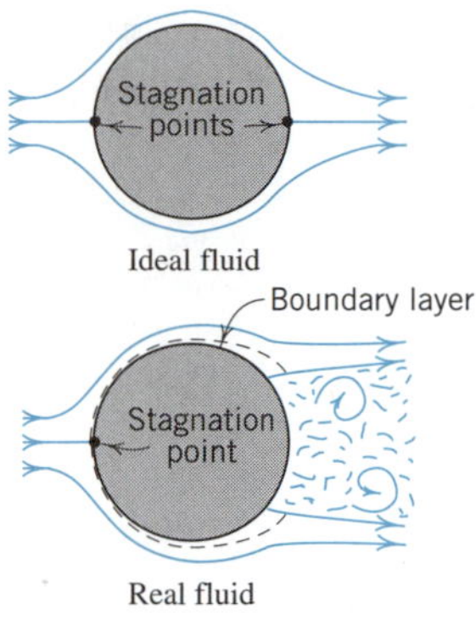

Fig. 11.7

to rest, accumulate, and be given a rotary motion by the surrounding flow; separation of the live flow from the object then results as the eddy increases in size. This description of the separation process applies at the inception of flow and at the beginning of wake formation. Once separation has occurred, a new flowfield is established and there is no reason to expect reattachment[6] of the live stream to the object.

With separation dependent on boundary-layer growth, it may be expected that the laminar or turbulent character of the boundary layer will be of critical importance in determining the position of separation. A simple comparison of laminar and turbulent layers of the same ρ, V_o, and δ will show their momentum fluxes to be $8\rho V_o^2\,\delta/15$ and $7\rho V_o^2\,\delta/9$, respectively—the momentum flux of the turbulent layer being nearly 50% greater[7] than that of the laminar one (see Sections 7.5 and 7.6). Thus, the turbulent boundary layer may be considered the "stronger" of the two and better able to survive an unfavorable pressure gradient; accordingly, it may be expected that the separation point for a turbulent boundary layer will be found farther downstream than that for a laminar boundary layer.

Quantitative aspects of profile drag may be obtained from a study of the experimentally determined drag coefficients of various objects. First, consider the basic sphere, which has been exhaustively studied. At a low Reynolds number (Fig. 11.8*a*) the flow will close behind the sphere and no wake will form; under these conditions profile drag is composed almost entirely of frictional drag. Stokes[8] has shown analytically that, in laminar flow at very low Reynolds numbers, where inertia forces may be neglected and those of viscosity alone considered, the drag of a sphere of diameter d, moving at a velocity V_o through a fluid of viscosity μ, is given by

$$D = 3\pi\mu V_o d \tag{11.3}$$

and this equation has been confirmed by many experiments. The drag coefficient, C_D, for the sphere under these conditions may be found by equating the preceding expression to 11.1:

$$\frac{C_D A\rho V_o^2}{2} = 3\pi\mu V_o d$$

[6]However, this may occur under special conditions for well-streamlined objects.

[7]A better comparison of laminar and turbulent boundary layers on the same object shows an even larger difference.

[8]G. G. Stokes, *Mathematical and Physical Papers*, vol. III, p. 55, Cambridge University Press, 1901.

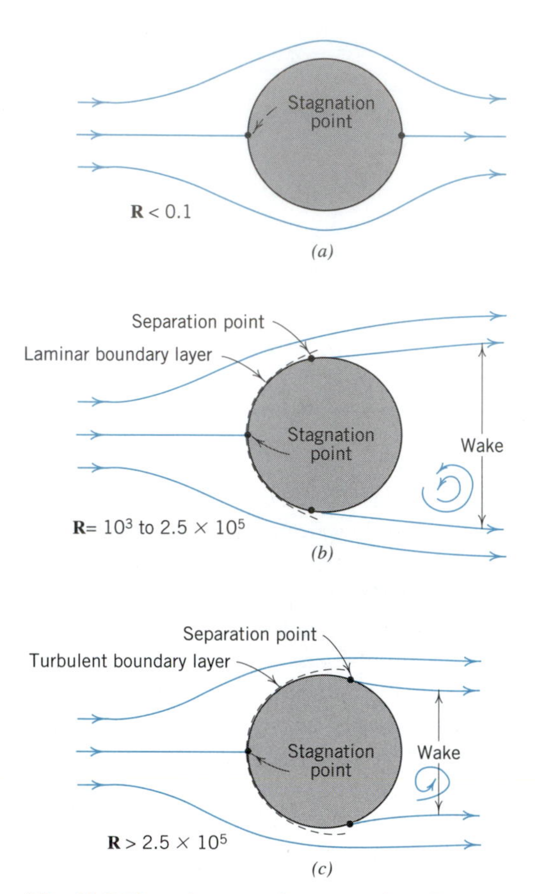

Fig. 11.8 Flow about a sphere at various Reynolds numbers.

Taking A to be the area of the projection of the object on a plane, normal to the direction of V_o, $A = \pi d^2/4$, and substituting this above gives

$$C_D = \frac{24\mu}{V_o d\rho} = \frac{24}{\mathbf{R}}$$

Thus, the drag coefficients of spheres at low velocities are dependent only on the Reynolds number—another confirmation of the results of the dimensional analysis of Section 11.2.

As the Reynolds number increases, the drag coefficients of spheres continue to depend only on the size of this number, and a plot of experimental results over a large range of Reynolds numbers for spheres of many sizes, tested in many fluids, gives the single curve of Fig. 11.9.

Up to a Reynolds number of 0.1, the Stokes equation applies accurately and the drag coefficient results from frictional effects. As the Reynolds number is increased to about 10, separation and weak eddies begin to form, enlarging into a fully developed wake near a Reynolds number of 1 000; in this range the drag coefficient results from a combination of pressure and frictional drag, the latter becoming about 5% of total drag as a Reynolds number of 1 000 is reached (see Fig. 11.8). Above this figure the effects of friction become even smaller and the drag problem becomes primarily one of pressure drag.

The drag coefficient of the sphere ranges (approximately) between 0.4 and 0.5 from $\mathbf{R} \cong 1\,000$ to $\mathbf{R} \cong 250\,000$, at which point it suddenly drops more than 50% and then

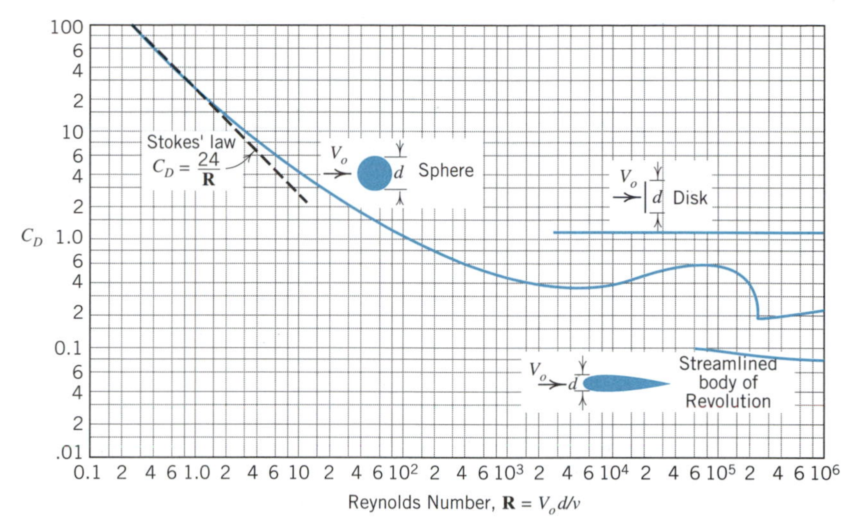

Fig. 11.9 Drag coefficients for sphere, disk, and streamlined body.[9]

increases gradually with further increase in the Reynolds number (see Figs. 11.8 and 11.9). In this range of Reynolds numbers, experiments have shown the separation point to be upstream from the midsection of the sphere, resulting in a relatively wide turbulent wake; the boundary layer on the surface of the sphere from stagnation point to separation point has been found to be laminar up to $\mathbf{R} \cong 250\,000$. With further increase in $\mathbf{R}$ the length of the laminar boundary layer decreases, the boundary layer flow past the separation point becomes turbulent, and the separation point moves to a position downstream from the center of the sphere (Fig. 11.8*c*), causing a decrease in the width of the wake and consequent decrease in the drag coefficient.

The change from laminar to turbulent boundary layer on a flat plate has been seen (Section 7.4) to occur at a critical Reynolds number dependent on the turbulence of the approaching flow. It also occurs with a sphere, and with increased turbulence in the approaching flow the sudden drop in the drag coefficient curve occurs at lower Reynolds number. Thus, a sphere may be used as a relative measure of turbulence by noting the Reynolds number at which a drag coefficient of 0.30 (see Fig. 11.9) is obtained. Before the development of the hot-wire anemometer (Section 14.9), this method was used to compare the turbulence characteristics of different wind tunnels.

The drag coefficient of a thin circular disk placed normal to the flow shows practically no variation with the Reynolds number, since the separation point is fixed at the edge of the disk and cannot shift from this point, regardless of the condition of the boundary layer. Thus, the width of the wake remains essentially constant, as does the drag coefficient. This idea may be usefully generalized and applied to all brusque or very rough objects in a fluid flow; experiments indicate that such objects have drag coefficients which are essentially constant in the range of high Reynolds numbers.[10]

[9]Data from L. Prandtl, *Ergebnisse der aerodynamischen Versuchsanstalt zu Göttingen*, vol. II, p. 29, R. Oldenbourg, 1923; and G. J. Higgens, ''Tests of the N. P. L. Airship Models in the Variable Density Wind Tunnel,'' *N.A.C.A. Tech. Note* 264, 1927.

[10]Compare this with the relation of the friction factor f, and Reynolds number, $\mathbf{R}$, for rough pipes, Figs. 9.9 and 9.10, and also the fact that the local loss coefficients of pipe flow show little variation with the Reynolds number.

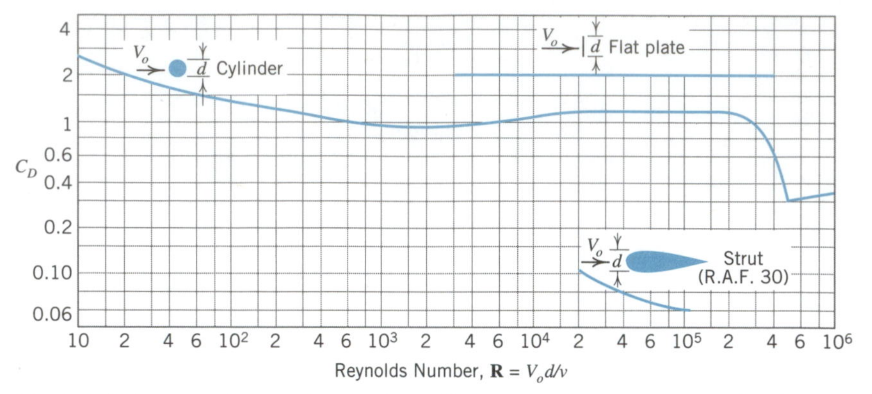

Fig. 11.10 Drag coefficients for circular cylinders, flat plates, and streamlined struts of infinite length.[12]

The drag coefficients of circular cylinders placed normal to the flow show characteristics similar to those of spheres. The coefficients shown in Fig. 11.10 are for infinitely long cylinders. The drag coefficients of streamlined struts[11] and flat plates of infinite length are also shown for comparison. The total drags of the flat plate and cylinder contain negligible frictional drag at ordinary velocities, whereas the streamlined strut, because of its small turbulent wake, possesses little pressure drag. The curves are typical of those resulting from tests of blunt and streamlined objects.

Long blunt objects such as cylinders, when placed crosswise to a fluid flow, sometimes exhibit the property of shedding large eddies regularly and alternately from opposite sides (Fig. 11.11).[13] Because of von Kármán's studies of the stability of these regular vortex patterns they are generally known as *Kármán vortex streets*. Experiments have shown that eddies will be shed regularly (with radian frequency ω) from a circular cylinder for Reynolds numbers between 60 and 5 000 (see Fig. 11.10) and that the nondimensional frequency of the phenomenon, the so-called *Strouhal number*, $S = \omega d/V_o$, is equal to approximately 0.42π over much of this range. The regular periodic nature of the eddy formation produces transverse forces on the cylinder that are also periodic, and thus tend to produce transverse oscillations. These considerations are vital to the design of elastic

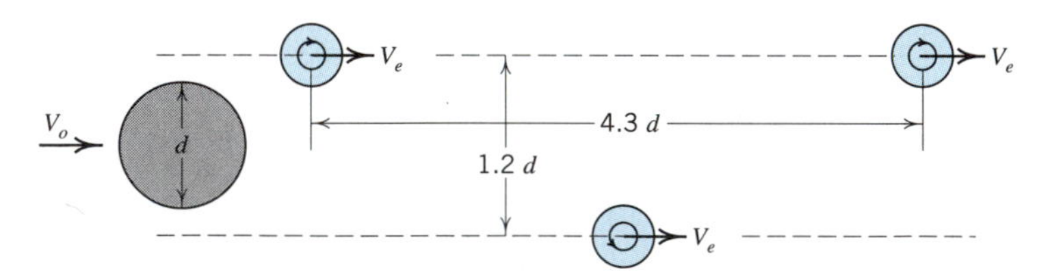

Fig. 11.11 Schematic of typical vortex street.

[11]The area to be used in the drag equation is the projection of the body on a plane normal to the direction of flow.

[12]Data from L. Prandtl, *Ergebnisse der aerodynamischen Versuchsanstalt zu Göttingen*, vol. II, p. 24, R. Oldenbourg, 1923; and B. A. Bakhmeteff, *Mechanics of Fluids*, Part II, p. 44, Columbia University Press, 1933.

[13]See Film Loop No. S-FM012, "Flow Separation and Vortex Shedding," Encyclopaedia Britannica Educational Corp.

structures such as tall chimneys and suspension bridges which are exposed to the wind. They become critical if the natural frequency of vibration of the structure is close to the frequency of the eddy formation because then resonance occurs and the energy in the fluid can be transmitted to the structure. The oscillations appear and grow in the structural frame; occasionally they grow without limit and this is catastrophic (e.g., such oscillations were a possible contributor to the Tacoma Narrows bridge failure in 1940)!

ILLUSTRATIVE PROBLEM 11.2

Calculate the ratio between the drag forces on the same sphere in the same fluid stream at Reynolds numbers 200 000 and 400 000.

SOLUTION

From Fig. 11.9, the drag coefficients at these Reynolds numbers are 0.43 and 0.20, respectively. With sphere size and fluid the same, Reynolds numbers will vary directly with velocity, and drag with $C_D\mathbf{R}^2$. Writing this as a ratio produces

$$\frac{D_{400\,000}}{D_{200\,000}} = \frac{0.2(400\,000)^2}{0.43(200\,000)^2} = 1.86 \bullet$$

Frequently, for small changes of Reynolds number such calculations are made neglecting the change of drag coefficient. Had this been done here, the error would have been 115%.

11.4 LIFT AND CIRCULATION

An important development in the understanding of lift was the *Kutta-Joukowsky theorem*, named after those who first developed it. While we shall not develop all of the underlying hydrodynamics to make the proof, we shall use the stream function defined in Section 5.9 to gain some insight. For steady, two-dimensional, inviscid flow, it is possible to construct analytic representations of the stream function for some simple flows. For example,

1. Horizontal rectilinear flow: $\psi = Uy$

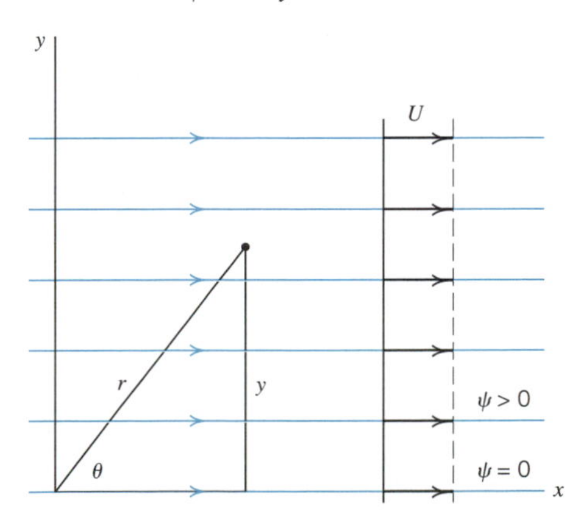

2. Source of fluid: $\psi = \dfrac{q\theta}{2\pi}$

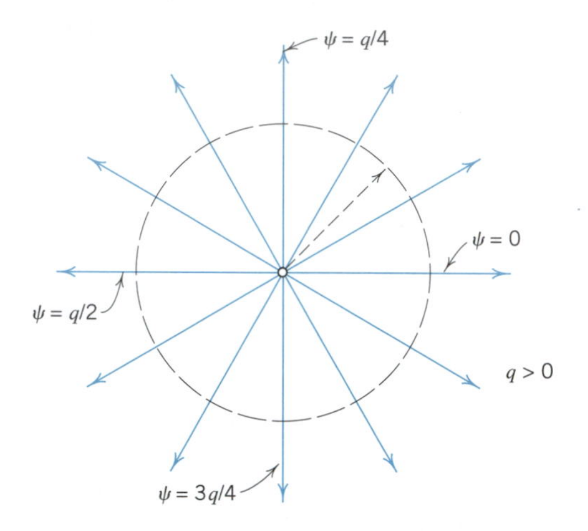

3. Free vortex (see problem 3.16): $\psi = \dfrac{\Gamma}{2\pi} \ln r$

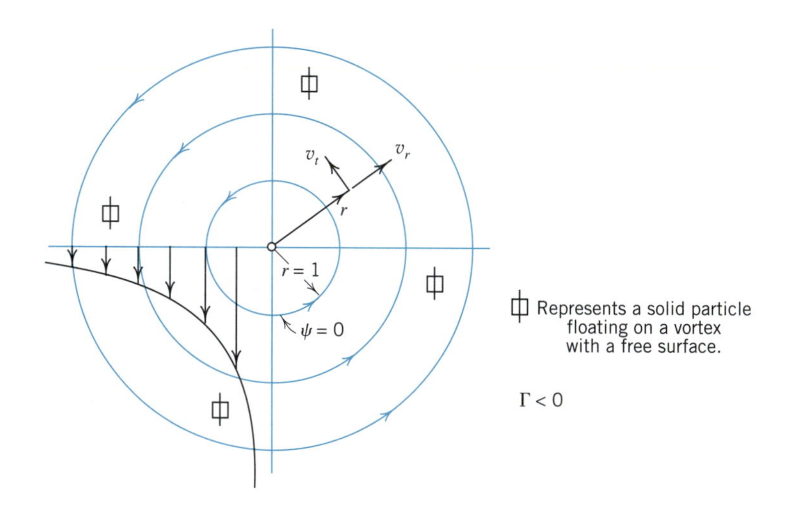

In these expressions, U is the fluid velocity in the rectilinear flow, q is the two-dimensional flowrate from the source, and Γ is the circulation (Section 3.4) or strength of the vortex. It can be shown[14] that a linear combination of the rectilinear flow, a source, a negative source (a sink), and the free vortex generate a flow about a cylinder as shown in Fig. 11.12 (in this picture, $\sin \alpha = \Gamma/4\pi UR < 1$), having a stream function

$$\psi = U\left(r - \frac{R^2}{r}\right) \sin \theta + \frac{\Gamma}{2\pi} \ln r$$

[14]K. Karamcheti, *Principles of ideal-fluid aerodynamics*. New York: Wiley, 1966.

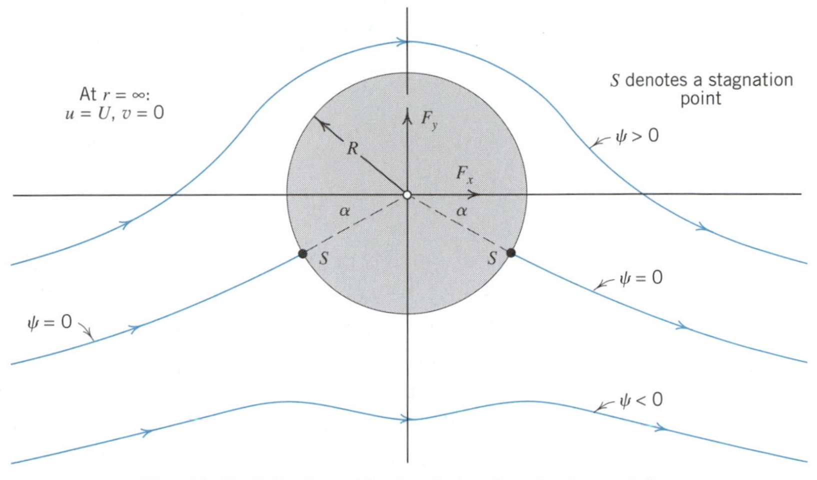

Fig. 11.12 Cylinder with circulation in a horizontal flow.

and velocities

$$v_r = \frac{\partial \psi}{r \partial \theta} = U\left(1 - \frac{R^2}{r^2}\right)\cos\theta \quad \text{and} \quad v_t = -\frac{\partial \psi}{\partial r} = -U\left(1 + \frac{R^2}{r^2}\right)\sin\theta - \frac{\Gamma}{2\pi r}$$

Application of the work-energy equation 5.5 between a point far from the cylinder and a point on the cylinder and integration of the pressure forces over the surface yields the startling results:

$$F_x = 0 \qquad \text{and} \qquad F_y = \rho U \Gamma \tag{11.4}$$

The first of these might have been anticipated from symmetry of the flowfield, and it shows that the drag force exerted by an ideal fluid on the body contour is zero. This may be generalized to apply to bodies of any shape and is known as the *d'Alembert paradox.*[15] The expression for F_y is of more far-reaching consequence in modern fluid mechanics as it shows that a circulation Γ and a velocity U are both necessary to the existence of a transverse (*lift*) force. The foregoing derivation, when generalized to apply to a body contour of any shape, is the *Kutta-Joukowsky theorem.* It serves to explain certain familiar phenomena in which bodies spinning in real fluids create their own circulation and when exposed to a rectilinear flow are acted on by a transverse force; some examples are the forces on the rotating cylinders of a ''rotorship'' and the transverse force which causes a pitched baseball to curve. However, its most important engineering application is in the theory of the lift force exerted on airfoils, hydrofoils, and the blades of turbines, pumps, propellers, wind turbines, windmills, and so forth. The theory does not, of course, explain the origin of circulation about a nonspinning object but does demonstrate that a circulation is required for the generation of a transverse force.

[15]In d'Alembert's time the discrepancy between theory and observation was considered to be a contradiction; today it is completely explained by the action of viscosity.

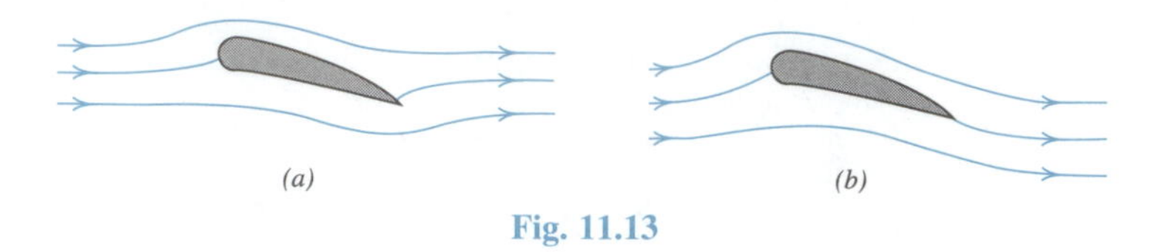

Fig. 11.13

By the use of the mathematics of complex variables, *conformal transformations* of these flows may be constructed. By using the Joukowsky transformation, the flow in Fig. 11.12 may be distorted into an airfoil or hydrofoil in a flowfield as shown in Fig. 11.13. When the circulation is zero, one obtains the flow shown in Fig. 11.13*a*, which yields no transverse force and (the unrealistic) flow around the trailing edge of the foil. However, when the circulation is nonzero and is chosen so that the flow is tangent to the surfaces at the trailing edge, then the picture is as in Fig. 11.13*b*, lift is produced, and it closely approximates experimental results. Of course, as noted earlier, viscous effects in the boundary layer must be accounted for to reconcile theoretical and experimental drag results.

The Kutta-Joukowsky theorem has demonstrated that circulation about an object is one of the requirements for the existence of a lift force on the object. Although it is not difficult to imagine a rotating body in a viscous fluid inducing its own circulation, to explain the origin of circulation about an airfoil, or an element of a propeller or turbine blade, requires knowledge of other principles.

Consider the flow conditions about a typical airfoil as it starts to move. Before motion begins, the circulation about the foil is obviously zero (Fig. 11.14*a*). As motion starts, the circulation about the airfoil tends to remain zero, and the potential flow of Fig. 11.14*b* tends to be set up, but such a flow, which includes a stagnation point near the rear of the airfoil and flow around its sharp trailing edge, cannot be maintained in a real fluid because of separation. This momentary potential flow gives way immediately to the flow of Fig. 11.14*c*, and in the process a circulation Γ develops about the airfoil and a vortex, the *starting vortex* (Fig. 11.14*d*), is shed from the airfoil. During the creation of this vortex, however, the circulation around a closed curve, including and at some distance from the airfoil, is not changed and must still be zero; thus, from the properties of circulation, the circulation about, or the strength of, the starting vortex must be equal and opposite to that about the airfoil; the existence of circulation about an airfoil is thus dependent on the creation of a starting vortex.[16]

Applications

In this applications portion of this chapter, we explore briefly several applications of the drag and lift principles covered in the earlier parts of this chapter. Sections 11.5 and 11.6 describe the effect of finite length on airfoils, the polar diagrams for lift and drag that are

[16]See Film Loop No. S-FM010, ''Generation of Circulation and Lift for an Airfoil,'' Encyclopaedia Britannica Educational Corp.

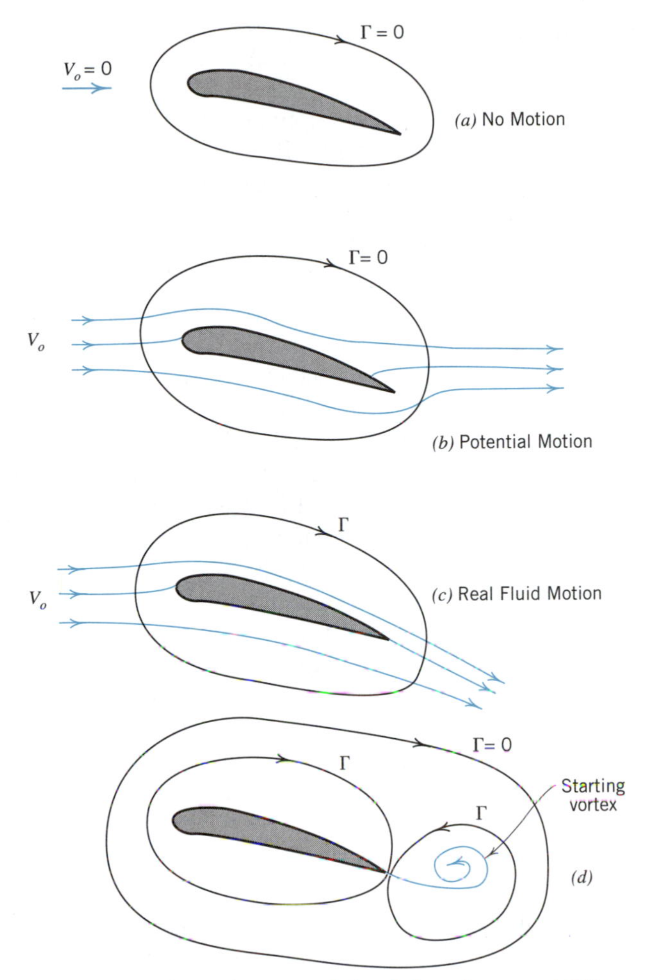

Fig. 11.14 Development of circulation about an airfoil.

used in design, and the augmentation of lift by use of camber and flaps. Section 11.7 introduces us to the realm of sports balls by examining the fascinating dynamics of the curveball in baseball. The physics of sports balls has received considerable attention in recent years and some interesting literature is cited in this section as well as at the end of the chapter. Studying the curveball exposes the reader to one of the key elements of sports ball flight, namely, the lift force induced by spinning the ball, and provides an opportunity for calculating the trajectory of the pitched ball. These elements are relevant to the motion of cricket, soccer, and golf balls as well and the references provide access to the specifics. In Section 11.8, the drag and lift forces on buildings and on offshore pilings are considered briefly, with a goal of introducing the effects on the forces of structural shape and unsteadiness of the flow. Finally, Section 11.9 provides some insight to the forces acting on road vehicles and the differences in drag and lift for various vehicles, ranging from trucks to racing cars, as well as the trade-off between ideal streamlining and the production vehicle.

11.5 AIRFOILS—FINITE LENGTH

When fluid flows about airfoils of finite length, flow phenomena result which affect both lift and drag of the airfoil; these phenomena may be understood by further investigation and application of the foregoing circulation theory of lift.

Since pressure on the bottom of an airfoil is greater than that on the top, flow will escape from below the airfoil at the ends and flow toward the top, thus distorting the general flow about the airfoil, causing fluid to move inward over the top of the airfoil and outward over the bottom (Fig. 11.15). As the fluid merges at the trailing edge of the airfoil, a surface of discontinuity is set up, and flows above and below this surface are, respectively, inward and outward as shown. The tendency for vortices to form from these velocity components is apparent, and, in fact, this surface of discontinuity is a sheet of vortices. However, such a vortex sheet is unstable, and the rotary motions contained therein combine to form two large vortices trailing from the tips of the airfoil (Fig. 11.15); these are called tip vortices and are often visible when an airfoil passes through dust-laden air or as vapor trails produced by condensation of atmospheric moisture. (See Fig. A.4 for an illustration of tip vortices made visible by cavitation on a ship propeller.)

Since the pressure difference between top and bottom of an airfoil must reduce to zero at the tips, it is evident that the lift per unit length of span varies over the span (Fig. 11.16), being maximum at the center and reducing to zero at the tips. The total lift of the airfoil is, of course, the total resulting from this lift diagram. Since lift per unit length of span varies directly with circulation ($L = \Gamma \rho V_o$), a diagram showing distribution of circulation over the span has the same shape as a diagram of lift distribution. The variation of lift and circulation over the span of an airfoil cannot, of course, be disregarded in a rigorous treatment of the subject, but such treatment leads to mathematical and physical complexities which are beyond the scope of this volume. A simple physical picture may be obtained, however, from the following analysis in which lift and circulation will be assumed to be distributed uniformly over the span (Fig. 11.16).

One of the properties of vortices is that their axes can end only at solid boundaries. Since there is no solid boundary at the end of the airfoil, the circulation Γ cannot stop here, but must continue to exist about the axes of the tip vortices (Fig. 11.17). The axes of the tip vortices extend rearward to the axis of the starting vortex; thus, according to the theory, the axis of the vortex having circulation Γ does not end, but is a closed curve composed of the axes of the airfoil, tip vortices, and starting vortex. In the real fluid the

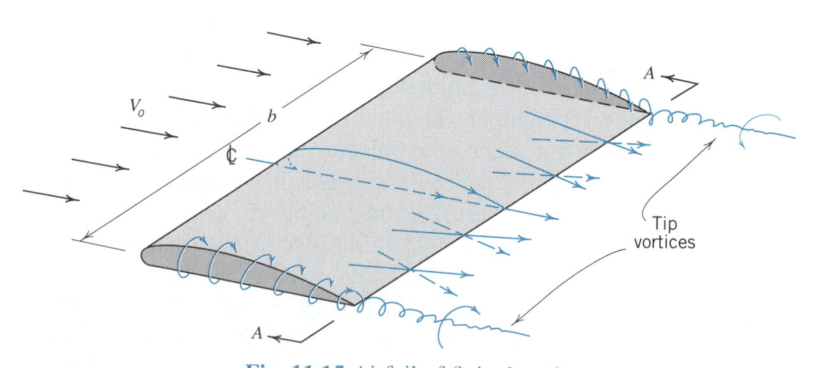

Fig. 11.15 Airfoil of finite length.

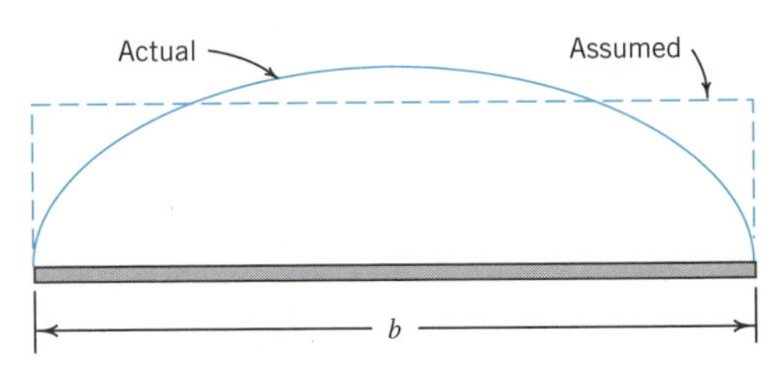

Fig. 11.16 Distribution of lift and circulation over an airfoil of finite length.

circulation persists only about the airfoil and portions of the tip vortices close to the airfoil; the starting vortex and remainder of the tip vortices are extinguished by viscous action.

The circulations about the tip vortices induce a downward motion in the fluid passing over an airfoil of finite length and in so doing affect both lift and drag *by changing the effective angle of attack*. The strength of this induced motion will, obviously, depend on the proximity of the tip vortices and, thus, upon the span of the airfoil which may be expressed in terms of the *aspect ratio* b^2/A.

An airfoil of finite span is shown at angle of attack α in the horizontal flow of Fig. 11.18. The vertical velocity induced near the wing by the tip vortices decreases the angle of attack by a small angle α_i, making the effective angle of attack $(\alpha - \alpha_i)$. This effective angle of attack is that for no induced velocity or, in other words, it is the angle of attack which would be obtained if the foil had infinite span and aspect ratio. Calling this angle of attack α_o,

$$\alpha_o = \alpha - \alpha_i \tag{11.5}$$

Now, treating the airfoil is one of infinite span at an angle of attack α_o, the lift L_o exerted on such an airfoil is by definition normal to the direction of flow in which it is placed; therefore L_o is normal to the effective velocity V_o and at an angle α_i with the vertical. The lift, L, on the airfoil of finite span is normal to the approaching horizontal velocity V and is the vertical component of L_o. But L_o also has a component in the direction of the original velocity V, which is a drag force, D_i, called the *induced drag* because its existence depends on the downward velocity induced by the tip vortices. Thus an additional drag force, D_i, must be added to profile drag in computing the total drag of a body of finite length about which a circulation exists. Calling the profile drag D_o, since it is the drag of an airfoil of infinite span (which has no end effects and, therefore, no induced drag) the total drag, D,

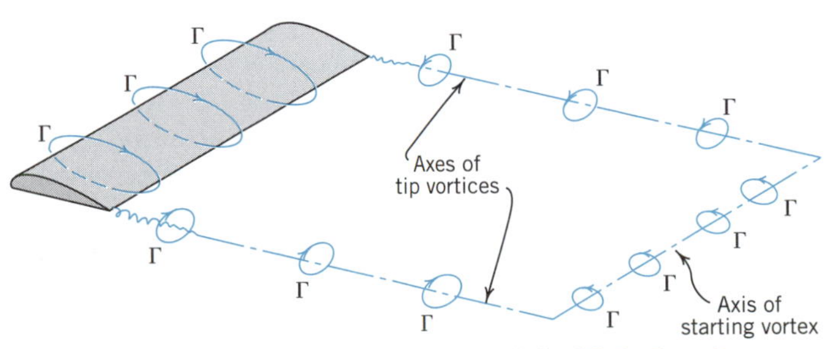

Fig. 11.17 Circulation about an airfoil of finite length.

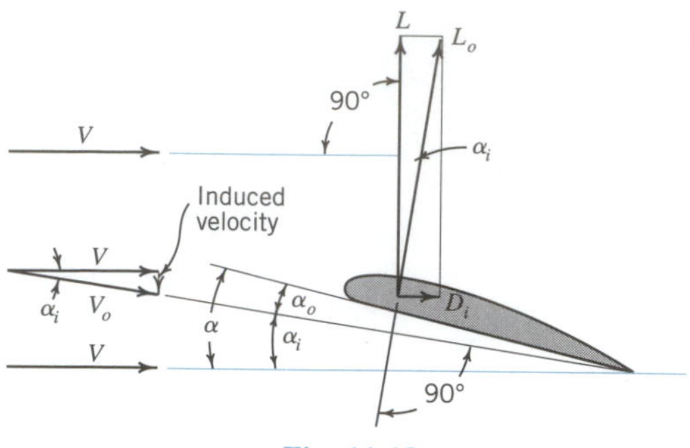

Fig. 11.18

of an airfoil of finite length is given by

$$D = D_o + D_i$$

which, by dividing by $A\rho V^2/2$, may be expressed in terms of dimensionless drag coefficients as

$$C_D = C_{D_o} + C_{D_i} \tag{11.6}$$

From the foregoing statements and Fig. 11.18, it is evident that induced drag, D_i, is related to lift L, angle α, and aspect ratio b^2/A; the equations relating these variables are of great practical importance. Since α_i is small,

$$L = L_o \qquad V = V_o \qquad D_i = L\alpha_i$$

If the distribution of lift over a wing of finite span is taken to be a semiellipse[17] (see Fig. 11.16), it may be shown that

$$\alpha_i = \frac{C_L}{\pi b^2/A} \tag{11.7}$$

With drag and lift proportional to their respective coefficients, $D_i = L\alpha_i$ may be written $\alpha_i = C_{D_i}/C_L$, and by substitution of this into Eq. 11.7 there results

$$C_{D_i} = \frac{C_L^2}{\pi b^2/A} \tag{11.8}$$

which relates lift and induced drag through their dimensionless coefficients and shows that induced drag is inversely proportional to aspect ratio, becoming zero at infinite aspect ratio (infinite span) and increasing as aspect ratio and span decrease—thus offering mathematical proof of the foregoing statements on the effect of span, aspect ratio, and proximity of tip vortices on induced downward velocity and induced drag.

The use of the derived expressions for α_i and C_{D_i} in Eqs. 11.7 and 11.8 allows airfoil data obtained at one aspect ratio to be converted into corresponding conditions at infinite aspect ratio, and these data, in turn, to be reconverted to airfoils of any aspect ratio; thus extensive testing of airfoils of the same profile at various aspect ratios becomes unnecessary.

[17]An assumption which gives minimum induced drag and conforms well (but not perfectly) with fact.

ILLUSTRATIVE PROBLEM 11.3

A rectangular airfoil of 6 ft (1.83 m) chord and 36 ft (11 m) span (aspect ratio 6) has a drag coefficient of 0.054 3 and lift coefficient of 0.960 at an angle of attack of 7.2°. What are the corresponding lift and drag coefficients and angle of attack for a wing of the same profile and aspect ratio 8?

SOLUTION

The physical dimensions of the airfoil are irrelevant to this analysis, whose strategy is to convert the given coefficients to those for infinite aspect ratio and then convert those to aspect ratio 8 by use of the following equations:

$$\alpha_o = \alpha - \alpha_i \tag{11.5}$$

$$C_D = C_{D_o} + C_{D_i} \tag{11.6}$$

$$\alpha_i = \frac{C_L}{\pi b^2/A} \tag{11.7}$$

$$C_{D_i} = \frac{C_L^2}{\pi b^2/A} \tag{11.8}$$

Thus, we have:

For aspect ratio 8:

$$C_L = 0.960 \text{ (negligible change of lift coefficient)} \bullet$$

For aspect ratio 6:

$$C_{D_i} = \frac{(0.960)^2}{6\pi} = 0.048\,9 \tag{11.8}$$

For aspect ratio ∞:

$$C_{D_o} = 0.054\,3 - 0.048\,9 = 0.005\,4 \tag{11.6}$$

For aspect ratio 8:

$$C_D = 0.005\,4 + \frac{(0.960)^2}{8\pi} = 0.042\,1 \bullet \tag{11.6}$$

For aspect ratio 6:

$$\alpha_i = \frac{0.960}{6\pi} = 0.050\,9 \text{ radian} = 2.9^\circ \tag{11.7}$$

For aspect ratio ∞:

$$\alpha_o = 7.2 - 2.9 = 4.3^\circ \tag{11.5}$$

For aspect ratio 8:

$$\alpha = 4.3 + \left(\frac{0.960}{8\pi}\right)\left(\frac{360}{2\pi}\right) = 6.5^\circ \bullet \tag{11.5 \& 11.7}$$

11.6 AIRFOILS—LIFT AND DRAG DIAGRAMS

The relation between lift and induced drag coefficients suggests plotting lift coefficient against drag coefficient and gives the so-called *polar diagram* of Fig. 11.19, which is used extensively in airplane design.[19] On this diagram Eq. 11.8 appears as a parabola passing through the origin and symmetrical about the C_D-axis, the position of the parabola depending on the aspect ratio. Since the two curves are for airfoils of the same aspect ratio, the horizontal distance between them is the profile drag coefficient C_D. However, the diagram shows much more than this. The important ratio of lift to drag is the slope of a straight line drawn between origin and the point for which this ratio is to be found; the maximum value of this ratio is the slope of a straight line tangent to the curve and passing through the origin; on the diagram are also easily seen the points of zero lift and minimum drag, and the point of maximum lift or *stall*,[20] which determines stalling angle above which lift no longer continues to increase with angle of attack; the end of the upper solid portion of the curve is the point at which the flow separates completely from the upper side of the

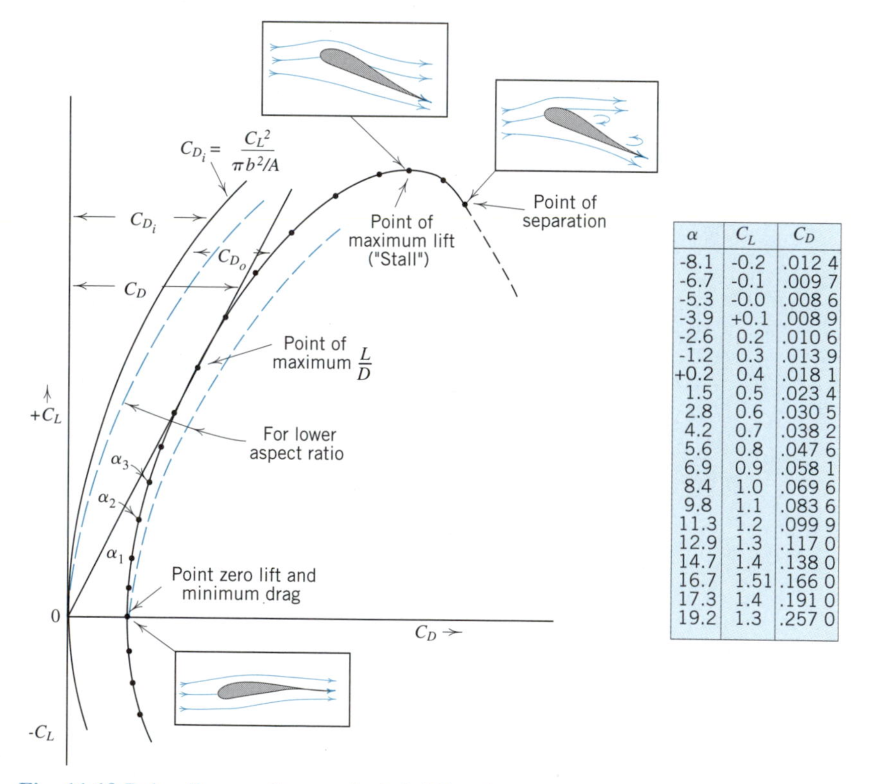

α	C_L	C_D
-8.1	-0.2	.012 4
-6.7	-0.1	.009 7
-5.3	-0.0	.008 6
-3.9	+0.1	.008 9
-2.6	0.2	.010 6
-1.2	0.3	.013 9
+0.2	0.4	.018 1
1.5	0.5	.023 4
2.8	0.6	.030 5
4.2	0.7	.038 2
5.6	0.8	.047 6
6.9	0.9	.058 1
8.4	1.0	.069 6
9.8	1.1	.083 6
11.3	1.2	.099 9
12.9	1.3	.117 0
14.7	1.4	.138 0
16.7	1.51	.166 0
17.3	1.4	.191 0
19.2	1.3	.257 0

Fig. 11.19 Polar diagram for a typical airfoil and numerical data for the Clark-Y airfoil.[18]

[18] A 14.6 m × 2.4 m (48 ft × 8 ft) rectangular airfoil tested at **R** ~ 6 000 000. Data from A. Silverstein, "Scale Effect on Clark-Y Airfoil Characteristics from N.A.C.A. Full-Scale Wind-Tunnel Tests," *N.A.C.A. Rept.* 502, 1934.

[19] See, for example, the end-of-chapter references by Stinton and Blevins.

[20] The mechanism of stall has received some intensive study. See H. W. Emmons, R. E. Kronauer, and J. A. Rockett, "A Survey of Stall Propagation—Experiment and Theory," and S. J. Kline, "On the Nature of Stall." Both papers in *Trans. A.S.M.E. (Series D)*, vol. 81, 1959.

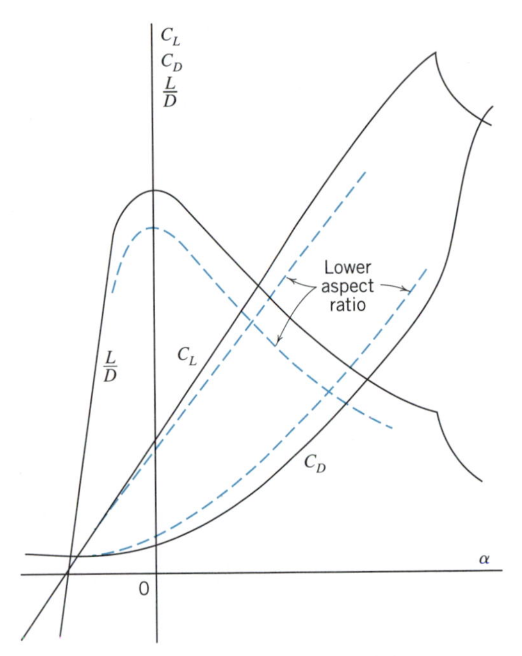

Fig. 11.20 Airfoil characteristics.

wing, forming a wake which increases the profile drag (and, therefore, the drag coefficient) and is accompanied by a large drop in lift and lift coefficient because of increased pressure on the upper side of the wing. Curves for other aspect ratios may be obtained[21] by the methods of the preceding Illustrative Problem and will be found to follow the trend indicated. Worthy of note are the equal horizontal distances C_{D_o} between corresponding curves and the decrease of L/D ratio with decreasing aspect ratio.

Another method of presenting airfoil data is to plot C_L, C_D, and L/D against angle of attack. Such a plot for the Clark-Y data of Fig. 11.19 is shown on Fig. 11.20. Of significance are the slope of the straight portion of the C_L curve, the location of the point of maximum L/D, the overall shape of the C_D curve, and the change in the position of the curves with change of aspect ratio.

The data of Fig. 11.19 were obtained at Reynolds numbers around 6×10^6, and from many foregoing statements it should be expected that the data will change with changing Reynolds number. The following trends, which are confirmed by experiment, are of some interest in the light of foregoing principles. With increasing Reynolds number the drag coefficient at low angles of attack decreases; here the drag coefficient contains predominantly frictional effects and its variation with Reynolds number is similar to that of the flat plate (Fig. 7.12). With increased turbulence, due either to increased initial turbulence or increased Reynolds number, the maximum lift coefficient usually increases; in other words, higher angles of attack can be attained without causing separation. Here the momentum of the turbulent boundary layer delays separation, allowing high-velocity flow to cling to the upper side of the airfoil, causing lower pressures and greater lift.

[21]Because of deviations from the assumed semielliptical lift distribution which led to Eqs. 11.7 and 11.8, the latter must be corrected by experimental coefficients which depend on the plan form of the wing. The size of the corrections increases with the aspect ratio, but the order of these is 5% on C_{D_i} and 15% on α_i for rectangular wings.

TABLE 7 Performance of Flaps in Wind-Tunnel Tests: Clark-Y Airfoil of Chord c = 10 Inches (Data and sketches drawn from NACA TN 422 and TRs 427 & 534)

Flap Type (size of flap and location of hinge point as a fraction of chord c)	$C_{L_{max}}$	L/D at $C_{L_{max}}$
No flap	1.3	7.6
Plain flap (0.3c and deflected at 45°)	2.0	4.0
Split flap (0.3c, hinged at 0.7c, and deflected at 45°)	2.2	4.3
Split flap (0.3c, hinged at 0.8c, and deflected at 45°)	2.3	4.4
Fowler flap (0.3c and deflected at 40°)	2.8	4.5
Fowler flap (0.4c and deflected at 40°)	3.1	4.2

No flap Plain flap Split flap Fowler flap

Two approaches are used to increase the lift of an airfoil in addition to increasing the angle of attack. These are adding *camber* to the foil shape and using *flaps*. The symmetric airfoil develops lift when placed at an angle of attack in the flow. An airfoil has camber when the center-line of the profile is curved; when the center section is elevated the camber is positive since then the curved surface yields increased circulation and lift about the airfoil. For example, at zero angle of attack, a symmetric airfoil has zero lift, while a positively cambered foil has a positive lift (see the Clark-Y foil polar diagram in Fig. 11.19). Designers then need to balance the increased lift from camber with the increased drag that it produces. One effective means of dealing with this balance is to effectively add camber on demand by the use of flaps. All airline passengers are familiar with the use of large flaps on commercial aircraft during take-off and landings when particularly high lift is needed at reduced speeds. Stinton[22] assembled data from NACA tests on an aspect ratio 6 Clark-Y airfoil operating at a Reynolds number of 609 000 to show the effectiveness of flaps. Some results from the NACA tests are reproduced in Table 7; a more extensive summary of the NACA results is given in Table 3-2, page 100, of Stinton's book. The most effective flap increases the lift by 139%, but the L/D ratio drops dramatically, indicating severely increased drag and thus necessitating maximum power on take-off for example. Among the configurations in the table below, the split flap leaves the airfoil essentially intact and drops from its lower surface. The Fowler flap is the most effective and is of a type often seen on large commercial airplanes.

Quite different challenges occur in the case of human powered flight[23] where the flow Reynolds numbers are only about 500 000 (the Reynolds number for the data in Fig. 11.19 was 6 million) and it is important that the drag be minimized because of the limitation of the human power source. Drela describes the characteristics of successful human-powered aircraft (HPA) airfoils. The polar diagrams for successful HPA airfoils show one significant similarity and some marked differences from the polar for the Clark-Y airfoil in Fig. 11.19. Both the Clark-Y and HPA airfoils have essentially the same maximum lift coefficient, i.e., about 1.5. However, the HPA airfoils show no tendency to stall (so the polar curve is

[22]D. Stinton, *The Design of the Aeroplane*, New York: Van Nostrand Reinhold, 1983, pp. 85 and 100.

[23]M. Drela, ''Aerodynamics of human-powered flight,'' *Annual Review of Fluid Mechanics*, vol. 22, 1990, pp. 93–110.

essentially flat from the point of maximum lift onward) and their lift to drag ratio is of the order of 100, while the maximum lift to drag ratio of the Clark-Y (and most common commercial airfoils) is of the order of 10.

11.7 SOME AERODYNAMICS OF BASEBALLS

Perhaps the most well-known characteristic of baseballs and golf balls is their curved flight—sometimes intentional, sometimes not. A well-thrown curve ball is useful in baseball; a slice or hook in a drive on a golf fairway is usually distinctly undesirable. Mehta[24] has written an entertaining treatise on sports balls, covering the behavior of cricket balls, baseballs and golf balls. Here we will draw on some of his insights and results and discuss both the *Magnus effect* (the development of lift on rotating bodies as described in Section 11.4) and the trajectory that can be computed for a pitched baseball. In that case (as well as in golf, tennis or soccer) a lateral deflection of the ball from a straight or only gravity-influenced line of flight is achieved.

In Section 11.4, it was shown (See Eq. 11.4) that the lift force on a *cylinder* is proportional to the product of the velocity of the body relative to the fluid and the circulation about the body. For golf and baseballs this circulation is induced by spinning the *spherical* ball about an axis that is perpendicular to its plane of flight. While the real flow is not nearly as smooth and symmetric as that in the ideal fluid case, the essential effect is the same and the result is that the ball experiences a lift (i.e., a lateral force), due to the circulation and separation pattern caused by the boundary layer development in the real fluid flow about the spinning ball. This lift is in the plane of flight, perpendicular to the oncoming flow (or the velocity of the ball), and its sign depends on the spin direction. From the results in Section 11.4 (Fig. 11.12) we can conclude that the spin must be counterclockwise (the ball top moves forward in the direction of motion) for the ball to curve downward in addition to any gravity effect. Baseball pitchers can change the ''break'' of their curveballs by adjusting the angle between their pitching arm and the ground. A ball pitched with a fully overhead motion will break down, but one pitched by a ''side-winder'' (someone who throws so that their arm is parallel to the ground at release) will break across the plate.

Mehta explains in some detail the influence of the seams on cricket balls and on baseballs, as well as defining the essence of the ''knuckle ball.'' Here we focus only on spin and the curve ball. First, let us assemble some useful data. Briggs[25] points out that the ''accepted world record for the fastest pitch'' belongs to Bob Feller at 98.6 mph (144 ft/s or 43.9 m/s) and that the maximum measured spin rate of a pitched baseball is about 1600 rpm. Thus, a pitched ball completes about 16 revolutions between the pitcher's mound and the plate. In his work Briggs examined spin rates of 1200 and 1800 rpm and ball speeds between 75 ft/s and 150 ft/s (22.9 m/s and 45.7 m/s). The distance from the pitcher's mound to home plate is 60.5 ft (about 18.4 m); the diameter d of the average baseball is about 2.90 in. or 7.37 cm and its weight W_B is about 0.32 lb or 1.42 N. Mehta

[24]R. D. Mehta, ''Aerodynamics of sports balls,'' *Annual Review of Fluid Mechanics*, vol. 17, 1985, pp. 151–189.

[25]L. J. Briggs, ''Effect of spin and speed on the lateral deflection of a baseball; and the Magnus effect for smooth spheres,'' *Am. J. Physics*, vol. 27, 1959, pp. 589–596. Reprinted in A. Armenti, Jr., Ed. *The Physics of Sports*, New York: American Institute of Physics, 1991.

interprets Briggs's data to show that the lateral force on the curveball is dependent on the square of the speed of the pitch (i.e., the lift coefficient is independent of Reynolds number). From Eq. 11.1, we have

$$C_L = \frac{L}{\frac{1}{2} A \rho V_o^2} \tag{11.3}$$

where $A = \pi d^2/4 = 0.046 \text{ ft}^2 = 0.004\,3 \text{ m}^2$ is the cross-sectional or projected area of the baseball, $\rho = 0.002\,38 \text{ slugs/ft}^3 = 1.225 \text{ kg/m}^3$ is the density of sea-level air (in Denver at Mile-High Stadium, the density is 14% less!), and V_o is the speed of the pitched ball. From Briggs's data, one can determine that C_L is independent of the speed and is proportional to the rotating speed of the ball. Average values from his experiments yield $C_{L_{1200}} = \pm 0.096$ for 1200 rpm and $C_{L_{1800}} = \pm 0.142$. Here, the sign depends on the sense of the rotation in each case.

The flight of the ball can be estimated by the method used by Briggs, namely, assume that the induced lift is constant over the flight of the ball and that the forward motion of the ball is at constant speed, i.e., the drag is negligible over the distance from the mound to the plate. Measurements of actual pitches seem to agree with these assumptions and suggest that a constant side force is involved because the ball flight path is parabolic. In fact, by using the data in Fig. 11.9 it is straightforward to show that the drag effect is smaller than 0.2% for the travel to the plate. Now, thinking of the ball as a particle and using Eqs. 3.4 and 3.5, we have (assuming that the mound is the origin of coordinates, that the x-axis runs from the mound through the center of home plate, and that the vertical coordinate z has its origin at the ground)

$$\frac{dx}{dt} = V_o \qquad \text{and} \qquad \frac{d^2 z}{dt^2} = \left(\frac{L}{m_B}\right) - g_n,$$

where V_o is the speed of the pitch, L is the lift force due to rotation, $m_B = W_B/g_n =$ the mass of the ball, and g_n represents the effect of gravity. These equations may be integrated under the assumptions made. This yields, under the assumption that the initial vertical velocity of the pitch is zero (i.e., it is released horizontally),

$$x = V_o t \qquad \text{and} \qquad z = \frac{g_n}{2}\left(\frac{L}{W_B} - 1\right) t^2 + z_R \tag{11.4}$$

where z_R is the height of release of the pitch. We can replace t in the second equation by using the first equation. We obtain

$$z = \frac{g_n}{2}\left(\frac{L}{W_B} - 1\right)\frac{x^2}{V_o^2} + z_R = \frac{g_n}{2}\left(\frac{\frac{1}{2} A \rho V_o^2 C_L}{W_B} - 1\right)\frac{x^2}{V_o^2} + z_R \tag{11.5}$$

$$= \frac{A \rho g_n C_L}{4 W_B} x^2 - \frac{g_n x^2}{2 V_o^2} + z_R$$

Here, the first term on the right represents the effect of the rotation on the deflection of the ball from a straight path. This term is independent of ball speed and dependent only on rotation rate, while the second term on the right is that due to gravity. Its effect decreases rapidly with speed.

ILLUSTRATIVE PROBLEM 11.4

Calculate the deflection of a pitched baseball due to rotation if it is rotating at 1 200 rpm or at 1 800 rpm.

SOLUTION

We will apply Eq. 11.5, ignoring the effect of gravity and assuming that the lift coefficients are positive, viz.,

$$z - z_R = \frac{A\rho g_n C_L}{4W_B} x^2$$

where now $x = 60.5$ ft or 18.4 m, $A = 0.046\ \text{ft}^2 = 0.004\,3\ \text{m}^2$, $\rho = 0.002\,38\ \text{slugs/ft}^3 = 1.225\ \text{kg/m}^3$, $g_n = 32.2\ \text{ft/s}^2 = 9.81\ \text{m/s}^2$, $W_B = 0.32\ \text{lb} = 1.42$ N, and the lift coefficients are 0.096 and 0.142 for 1 200 and 1 800 rpm, respectively. Working in U.S. Customary units, we have

$$z - z_R = \frac{0.046 \times 0.002\,38 \times 32.2C_L}{4 \times 0.32}(60.5)^2 = 10.08C_L$$

Accordingly,

$$(z - z_R)_{1\,200} = 10.08 \times 0.096 = 0.968\ \text{ft} = 11.6\ \text{in.} \bullet$$

$$(z - z_R)_{1\,800} = 10.08 \times 0.142 = 1.43\ \text{ft} = 17.2\ \text{in.} \bullet$$

The reader can easily determine that the equivalent SI equations are

$$(z - z_R)_{1\,200} = 3.08 \times 0.096 = 0.30\ \text{m} = 30\ \text{cm} \bullet$$

$$(z - z_R)_{1\,800} = 3.08 \times 0.142 = 0.43\ \text{m} = 43\ \text{cm} \bullet$$

ILLUSTRATIVE PROBLEM 11.5

A pitcher throws a ball in the fully overhead position and releases the pitch at a height of 7.5 ft above the ground; it is rotating forward at 1 500 rpm. If the vertical extent of the strike zone is 2.0 ft $< z <$ 4.5 ft, in what velocity range must the pitch be?

SOLUTION

By considering Eq. 11.5, we see that the deflection due to the spin is fixed by the information at hand and independent of the speed and that z and z_R are known. Given that the distance to the plate is known, we can obtain an equation for the allowable range of speed.

Equation 11.5 is

$$z = \frac{A\rho g_n C_L}{4W_B} x^2 - \frac{g_n x^2}{2V_o^2} + z_R \tag{11.5}$$

Using the given data and the data from Illustrative Problem 11.4, we have

$$z = 10.08C_L - \frac{32.2(60.5)^2}{2V_o^2} + 7.5 = 10.08C_L - \frac{58\,930}{V_o^2} + 7.5$$

Solving for V_o^2 yields

$$V_o^2 = \frac{58\,930}{(7.5 - z) + 10.08C_L}$$

Now, the Briggs data give only lift coefficients for rotations of 1 200 and 1 800 rpm, but his hypothesis is that the lift is proportional to the rotation speed. Under this assumption a linear interpolation of the values produces

$$C_{L1\,500} = \frac{C_{L1\,200} + C_{L1\,800}}{2} = \frac{0.096 + 0.142}{2} = 0.119$$

Because the ball is rotating forward, there is a downward lateral force relative to the positive z coordinate and we choose the minus sign for the lift. Accordingly,

$$V_o^2 = \frac{58\,930}{(7.5 - z) + 10.08(-0.119)} = \frac{58\,930}{(7.5 - z) - 1.2} = \frac{58\,930}{6.3 - z}$$

The minimum speed corresponds to the bottom of the strike zone and the maximum speed to the top; the result is

$$z = 2.0 \text{ ft:} \quad V_o^2 = \frac{58\,930}{6.3 - 2.0} = 13\,075; \quad V_o = 117.1 \text{ ft/s} \cong 80 \text{ mph} \bullet$$

$$z = 4.5 \text{ ft:} \quad V_o^2 = \frac{58\,930}{6.3 - 4.5} = 32\,739; \quad V_o = 180.9 \text{ ft/s} \cong 123 \text{ mph} \bullet$$

To place the pitch in the strike zone the pitcher must achieve a velocity greater than 80 mph; clearly with a horizontal release, it is not humanly possible (remember Feller's record noted above) to throw a pitch that rises above the strike zone in this case. Note that the ''curve-ball'' effect moves the ball downward by 1.2 ft; the remaining drop is due to the gravitational force.

11.8 FORCES ON STRUCTURES

The range of structures that could be considered is large and there are many ways to analyze and consider the forces acting on these structures. We shall examine a small subset of the possible variations here. In particular, we focus on a few key issues, namely, the drag force on the structure as a whole, integration of the drag force per unit section of structure (this allows us to compute the moment of forces acting about the base for example), the potential for lift forces to act on a structure, the concept of decomposing the structure into subunits to calculate the forces, and the effect of unsteady flow on the structural forces.

Hoerner[26] provides the drag coefficients of a number of building shapes and some of these are tabulated in Table 8. These are experimental results for buildings set on the ground in a boundary layer, where generally the building is significantly taller than the boundary layer thickness and the wind speed is typical of atmospheric values (14 m/s) far from the surface. The drag coefficient is defined according to Eq. 11.2, with the projected area A being the area perpendicular to the wind vector. Notice the distinct drop in the drag coefficient when the structures with the square planform are rotated so that the wind motion is aligned with the diagonal. In addition, if the pressure coefficient

$$C_p = \frac{p - p_o}{\rho V_o/2},$$

where p is the pressure on the surface of the building, ρ is the density of air, and V_o and p_o are the representative velocity and pressure in the undisturbed stream far from the building, then the coefficient is positive on the front face of a building (e.g., $+0.7$ on the tall rectangle in face-on orientation in the table) and negative on the rear face (e.g., -0.8 on the tall rectangle); this can lead to leakage of air from the backside of a building or pressures from inside to outside on the back-side of a building (relative to the wind) that are large enough to pop out windows when the wind is strong. The case of the large sphere is instructive as well. When the sphere is far from the ground the flow about it is symmetrical, there is no lift, and the drag is in the range of the very high Reynolds number data shown in Fig. 11.9. However, when the sphere is near the ground it is not possible for the flow to go under the sphere as easily as it goes over so that the flow is asymmetric and there is lift; in addition the drag coefficient is significantly increased.

In general, the velocity varies with height in the atmospheric boundary layer in which structures are immersed. Thus, the drag force caused by the mean wind field should be calculated by summing the drag forces on each vertical segment of the building to account for the variation of velocity. Cermak[27] presents a commonly used form of the mean wind profile in the form

$$V/V_o = (z/\delta)^{1/n} \tag{11.6}$$

Here, V_o is the velocity at the edge of the boundary layer whose thickness is δ. He shows a relationship between the boundary layer thickness δ, the actual surface roughness, and the exponent in Eq. 11.6. In particular, for frictionless flow the profile is uniform and $1/n = 0$, whereas for flow over building complexes such as city centers $1/n = 0.4$. The Blasius smooth turbulent flow profile in Section 9.3 yields $1/n = 1/7 \approx 0.14$.

The force on any element of the building will be known if the drag coefficient is known and then for the structure shown in Fig. 11.21

$$dD = \frac{1}{2}\, dA \rho V^2 C_D = \frac{1}{2}\, \rho V^2 C_D W\, dz$$

[26]S. F. Hoerner, *Fluid-dynamic drag*, 2nd ed., 1965, New Jersey: Hoerner, pp. 4-3 and 4-8.

[27]J. E. Cermak, "Aerodynamics of buildings," *Annual Review of Fluid Mechanics*, vol. 8, 1976, pp. 75–106.

TABLE 8 Drag Coefficients for Simple Buildings [adapted from Hoerner, *Fluid-Dynamic Drag*, 2nd Ed., 1965]

Configuration	Height relative to boundary layer thickness	Orientation of building relative to wind stream	Drag coefficient C_D
Cube	> 1	face on	1.05
Cube	> 1	diagonal	0.80
Rectangle - square base	about 2 times larger	face on	1.30
Rectangle - square base	about 2 times larger	diagonal	0.95
Rectangle - square base	$\gg 1$	face on	1.50
Rectangle - square base	$\gg 1$	diagonal	1.05
Pyramid - square base	about 2-3 times larger	face on	1.14
Pyramid - square base	about 2-3 times larger	diagonal	0.83
Large sphere (275 ft)	unknown	without ground effect	0.19 [no lift]
		with ground effect	0.30 [C_L= 0.03]

Here, $W\,dz$ is the projected area of the structure at any elevation z above the ground and V is given by Eq. 11.6. It follows that if the height of the structure is H, the total drag force and the moment of that force about the base are

$$D = \frac{1}{2}\rho \int_0^H (V^2 C_D W)\, dz \tag{11.7}$$

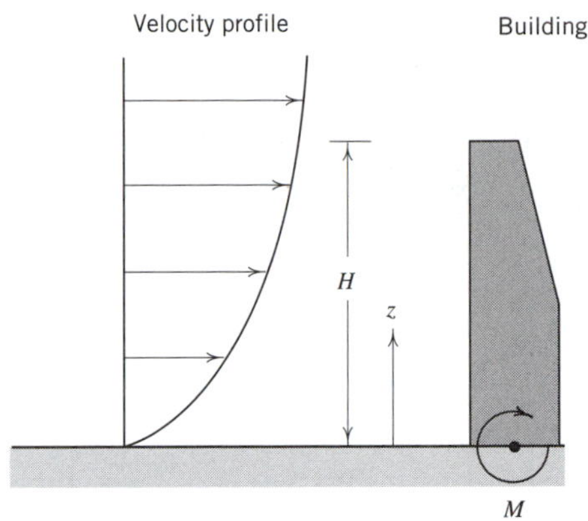

Fig. 11.21 Wind load on a building.

and

$$M = \frac{1}{2}\rho \int_0^H (V^2 C_D W) z \, dz \tag{11.8}$$

ILLUSTRATIVE PROBLEM 11.6

An architect decides to build a theme building for a fair and selects cylindrical shapes for the structure. The building is to have a restaurant at the top. The geometry is as follows:

Restaurant Segment: 50 ft diameter and 20 ft high with flat roof.
Building stem: 25 ft diameter and 200 ft high.

Compute the drag force on the total structure and the moment of that force about the center of the stem at the ground. The maximum load is expected to occur under gale force winds (up to 42 mph), the boundary layer thickness under these conditions is about 1 500 ft and the profile coefficient in Eq. 11.6 appears to be about 0.3 for the conditions at the fairgrounds by Cermak's standards.

SOLUTION

In this example, we need to decompose the structure into two parts and calculate the drag and moment on each part, summing the result. The density and viscosity of sea level air are 0.002 38 slugs/ft^3 and 1.93×10^{-5} ft^2/s, respectively. Converting the gale force winds to ft/s yields 62 ft/s and so Eq. 11.6 becomes

$$V = 62(z/1\,500)^{0.3} = 6.91z^{0.3} \tag{11.6}$$

Thus, we can estimate the Reynolds numbers for the restaurant and stem sections of the building as $\mathbf{R}_{restaurant} = 6.91(220)^{0.3} \times 50 \times 0.002\,38/1.93 \times 10^{-5} = 2.1 \times 10^5$; the Reynolds number for the stem based on the velocity at 200 ft and its 25 ft diameter will be less. However, from Fig. 11.10, the drag coefficient is relatively constant and equal to about 1.0 for $\mathbf{R} \leq 2 \times 10^5$. From Eqs. 11.7 and 11.8, writing each building segment separately, we have

$$
\begin{aligned}
D = D_{stem} + D_{restaurant} &= \frac{1}{2}\rho \int_0^{200} (V^2 C_D W)\, dz + \frac{1}{2}\rho \int_{200}^{220} (V^2 C_D W)\, dz \\
&= \frac{1}{2} 0.002\,38 \times (6.91)^2 \times 1.0 \times 25 \int_0^{200} z^{0.6}\, dz \\
&\quad + \frac{1}{2} 0.002\,38 \times (6.91)^2 \times 1.0 \times 50 \int_{200}^{220} z^{0.6}\, dz \\
&= 1.42 \int_0^{200} z^{0.6}\, dz + 2.84 \int_{200}^{220} z^{0.6}\, dz \qquad (11.7) \\
&= 1.42 \left[\frac{1}{1.6} (200^{1.6} - 0) \right] \\
&\quad + 2.84 \left[\frac{1}{1.6} (220^{1.6} - 200^{1.6}) \right] \\
&= 4\,264 + 498 = 4\,762 \text{ lb} \bullet
\end{aligned}
$$

and

$$
\begin{aligned}
M = M_{stem} + D_{restaurant} &= \frac{1}{2}\rho \int_0^{200} (V^2 C_D W) z\, dz + \frac{1}{2}\rho \int_{200}^{220} (V^2 C_D W) z\, dz \\
&= \frac{1}{2} 0.002\,38 \times (6.91)^2 \times 1.0 \times 25 \int_0^{200} z^{1.6}\, dz \\
&\quad + \frac{1}{2} 0.002\,38 \times (6.91)^2 \times 1.0 \times 50 \int_{200}^{220} z^{1.6}\, dz \\
&= 1.42 \int_0^{200} z^{1.6}\, dz + 2.84 \int_{200}^{220} z^{1.6}\, dz \qquad (11.8) \\
&= 1.42 \left[\frac{1}{2.6} (200^{2.6} - 0) \right] \\
&\quad + 2.84 \left[\frac{1}{2.6} (220^{2.6} - 200^{2.6}) \right] \\
&= 524\,800 + 295\,160 = 819\,960 \text{ ft-lb} \bullet
\end{aligned}
$$

The procedure for calculating the forces on cylindrical piles in water is not significantly different than that presented above for a building, except that in the water the wave action induces an unsteady flow. In a uniform current the forces on a pile are calculated as above for a tall, thin building. Therefore, we can focus on the wave-induced drag separately. If a current is also present then the total horizontal mean velocity must be used in the drag force calculation.

The dimensional analysis for unsteady flow in Section 11.2 yielded Eq. 11.2, which gave an expression for the inertial drag caused by the unsteady flow. Using that result and the usual drag coefficient for the profile and viscous drag one can construct the *Morrison equation*[28]

$$dD = \frac{1}{2} dA \rho V|V| C_D + d\forall \rho \frac{dV}{dt} C_I = \frac{1}{2}(d \cdot dz)\rho V|V| C_D + \left(\frac{\pi d^2}{4} dz\right) \rho \frac{dV}{dt} C_I \quad (11.9)$$

Here, ρ is the density of water, V is the instantaneous horizontal fluid velocity that would exist in the absence of the pile, dV/dt is the instantaneous acceleration of the fluid, d is the diameter of the pile, C_D is the drag for a cylinder which is affected by the maximum Reynolds number (see Fig. 11.10), and C_I is the coefficient of inertia which has a theoretical value of 2.0 for cylinders in waves. The absolute value signs in the drag expression are necessary because in a water wave the flow direction reverses and so it is necessary to make the equations direction sensitive.

To integrate Eq. 11.9 and to create an equivalent to Eq. 11.8 for the moment acting on a pile about its base, we need to know the velocity field for the water waves in Fig. 11.22. For so-called *shallow-water waves* of small height, the equations are not complex. A shallow water wave is one that is very long compared to the water depth h (e.g., storm-generated waves near the shore) and when its height is also small (i.e., very much smaller than its length L), the vertical motions and the effects of the free surface motion η on the integrals can be neglected. The resulting equation for the horizontal motion of the water is (it is assumed that the pile sits at the location $x = 0$ in the x-z coordinate system): $V = V_o \cos(2\pi t/T)$, where T is the period of the wave in seconds (i.e., the time for the wave motion to repeat) and so $dV/dt = -(2\pi/T)V_o \sin(2\pi t/T)$. Introducing these values into Eq. 11.9 produces

$$dD = \frac{1}{2}(d \cdot dz)\rho V_o^2 \cos(2\pi t/T)|\cos(2\pi t/T)| C_D - \frac{2\pi}{T}\left(\frac{\pi d^2}{4} dz\right) \rho V_o \sin(2\pi t/T) C_I$$

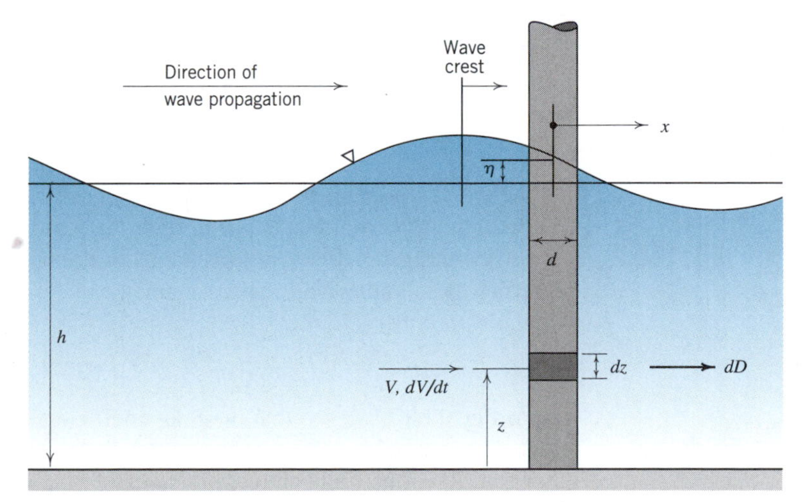

Fig. 11.22 Horizontal wave forces on a cylindrical pile.

[28]R. D. Blevins, *Applied fluid dynamics handbook*, New York: Van Nostrand Reinhold, 1984, p. 478; R. L. Wiegel, *Oceanographical engineering*, New Jersey: Prentice Hall, 1964, p. 251.

Integration of this equation from the bottom of the water surface (the effect of the small variations of the water surface from the still water line are negligible for the small-amplitude shallow-water waves considered here) yields

$$D = \int_0^h \left[\frac{1}{2}(d)\rho V_o^2 \cos(2\pi t/T)|\cos(2\pi t/T)|C_D - \frac{2\pi}{T}\left(\frac{\pi d^2}{4}\right)\rho V_o \sin(2\pi t/T)C_I \right] dz$$

$$D = \frac{1}{2}\rho V_o^2(h\,d)\cos(2\pi t/T)|\cos(2\pi t/T)|C_D - \frac{2\pi}{T}\left(\frac{\pi d^2}{4}h\right)\rho V_o \sin(2\pi t/T)C_I \tag{11.10}$$

In Eq. 11.10, the terms involving h and d are the cross-sectional area of the pile and the pile volume, respectively, as expected for drag and inertial forces. These forces are in quadrature, i.e., 90 degrees out of phase, so that one is zero when the other is at a maximum or minimum. The relative magnitude of these forces depends significantly on the pile diameter—large diameter piles or cylindrical structures experience a dominance of the inertial effect. Demonstration of this and the derivation of the moment equation are left to the Problems at the end of the chapter.

ILLUSTRATIVE PROBLEM 11.7

A circular support leg for an offshore tower sits in 25 m of water in the near-shore ocean. It is 2 m in diameter. Long waves with a height of 2 m, a period of 32 seconds, and a length of 500 metres are hitting the leg. These can be shown to be shallow-water waves and the fluid velocity that they induce is $V = V_o \cos(2\pi t/T) = 0.63 \cos(0.196t)$ m/s. What is the total drag force on the leg and what are the maximum pressure and friction drag and maximum inertial drag forces?

SOLUTION

While we may assume that the inertial drag coefficient is about 2.0, we need to calculate the maximum Reynolds number in order to estimate the pressure and friction drag coefficient from Fig. 11.10. From Appendix 2, we obtain the density and viscosity of sea water; entering these values and the leg diameter in Eq. 7.1 yields the maximum value of the Reynolds number

$$\mathbf{R} = V_o d\rho/\mu = 0.63 \times 2 \times 1\,024/(10.7 \times 10^{-4}) = 1.2 \times 10^6 \tag{7.1}$$

From Fig. 11.10, we see that the drag coefficient is about 0.35. (Experiments have been done to examine the effect of the unsteadiness on the drag and inertial coefficients; for this high Reynolds number and these long period waves the effect is negligible.) With this information in hand, we can use Eq. 11.10, which gives the total force on the leg:

$$D = \frac{1}{2}\rho V_o^2(hd)\cos(2\pi t/T)|\cos(2\pi t/T)|C_D - \frac{2\pi}{T}\left(\frac{\pi d^2}{4}h\right)\rho V_o \sin(2\pi t/T)C_I \tag{11.10}$$

Here,

$$\frac{1}{2}\rho V_o^2(hd) = \frac{1}{2}(1\,024)(0.63)^2(25 \times 2) = 10\,161;$$

$$\frac{2\pi}{T}\left(\frac{\pi d^2}{4}h\right)\rho V_o = \frac{2 \times 3.14}{32}\left(\frac{3.14 \times 4}{4}25\right)1\,024 \times 0.63 = 9\,948$$

$$D = 10\,161\cos(0.196t)|\cos(0.196t)|(0.35) - 9\,948\sin(0.196t)(2.0)$$

$$D = 3\,556\cos(0.196t)|\cos(0.196t)| - 19\,896\sin(0.196t)\text{ N} \bullet \quad \text{or}$$

$$= 3.56\cos(0.196t)|\cos(0.196t)| - 19.9\sin(0.196t)\text{ kN} \bullet$$

From this result, it is clear that the maximum drag force is 3.56 kN and the maximum inertial force is 19.9 kN. • In this case the inertial force dominates, since $(D_{inertia})_{\max}/(D_{pressure+friction})_{\max} \approx 5.6$.

11.9 SOME AERODYNAMICS OF ROAD VEHICLES

There is an old maxim that "form follows function." This is certainly the case for road vehicles. The aerodynamic shape of road vehicles is strongly "shaped" by their functional needs. The automobile has to have wheels and to contain people, who in turn must be able to drive, ride, sit comfortably, and see out of the vehicle. On the other hand, tractor-trailer rigs (trucks) that transport goods across continents must maximize the transportable volume of the trailer and the flexibility of the tractor to attach to many different trailers. The design of such vehicles is influenced by many factors, including functional issues such as air flow through radiators, visibility from the driver's seat, physical performance in handling, acceleration and braking, current fashion, economics and the competition, government regulation and policy, and the available technology. Hucho and Sovran[29] and Hucho[30] make several key points about road vehicles, namely,

a. Road vehicles are bluff bodies moving very close to the ground.

b. Their geometry is complex.

c. The flow over the vehicles is both three dimensional and turbulent (see Fig. 11.23).

d. Flow separation and trailing vortices are common (see Fig. 11.23).

e. The drag is mainly pressure drag in contrast to the drag on aircraft and ships, which are affected primarily by friction drag.

f. The air resistance is still not predictable by theory alone and model and full-scale tests are widely used.

In this section, we will look briefly at some of the issues of the drag on road vehicles. The works of Hucho and Sovran provide both more detail and broader coverage, including the

[29]W. H. Hucho and G. Sovran, "Aerodynamics of road vehicles," *Annual Review of Fluid Mechanics*, vol. 25, 1993, pp. 485–537.

[30]W. H. Hucho, Ed., *Aerodynamics of road vehicles*, London: Butterworths, 1987, Chaps. 4 & 8 in particular.

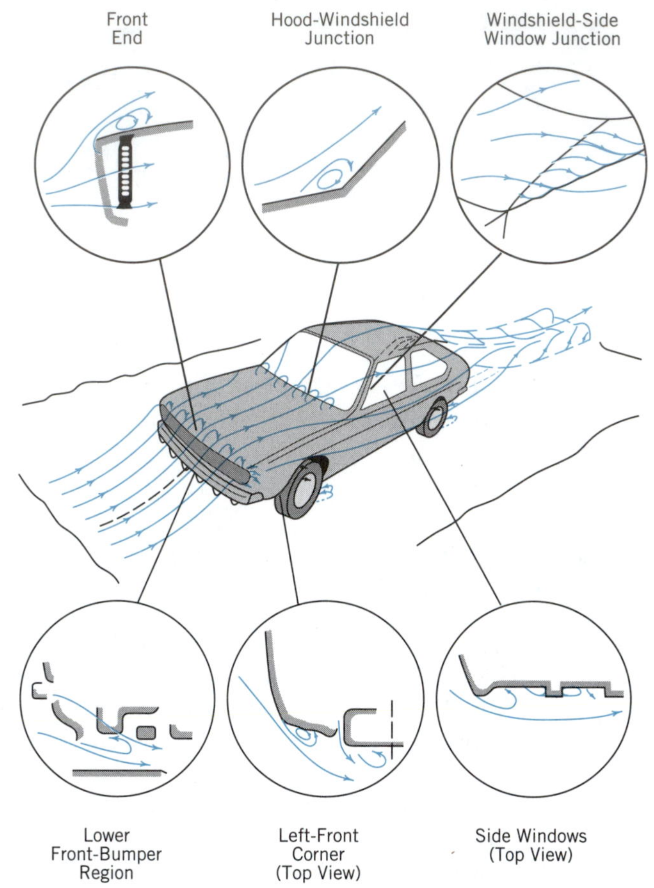

Fig. 11.23 Flow around a car, showing separation and vortices (from Hucho and Sovran, reproduced with permission, from the *Annual Review of Fluid Mechanics*, vol. 25, © 1993, by Annual Reviews Inc.).

effects of side winds, lift on the vehicle, etc. for passenger cars, high-performance vehicles, and commercial vehicles.

Figure 11.24 shows a chronological picture of the drag of automobiles. The upper band shows how the drag coefficient has dropped as the aerodynamics of the vehicles has become more sophisticated. The bottom band shows first the drag coefficient obtained by Klemperer in 1922 for an aerodynamic body with wheels; as Hucho and Sovran point out, it was more than 40 years before a ''real'' car (and even that was a research vehicle) was built with an equivalently low drag coefficient. Hucho shows as well that the development of a car body design moves through five phases. For example, for the 1982 Audi 100 III, these phases and their equivalent drag coefficients were

a. Define streamlined basic body with correct dimensions $-$ $C_D = 0.16$.
b. Define basic car shape $-$ $C_D = 0.18$.
c. Define car body model with technical characteristics of the car $-$ $C_D = 0.24$.
d. Define car body model with styling refinements $-$ $C_D = 0.29$.
f. Select final production car design $-$ $C_D = 0.30$.

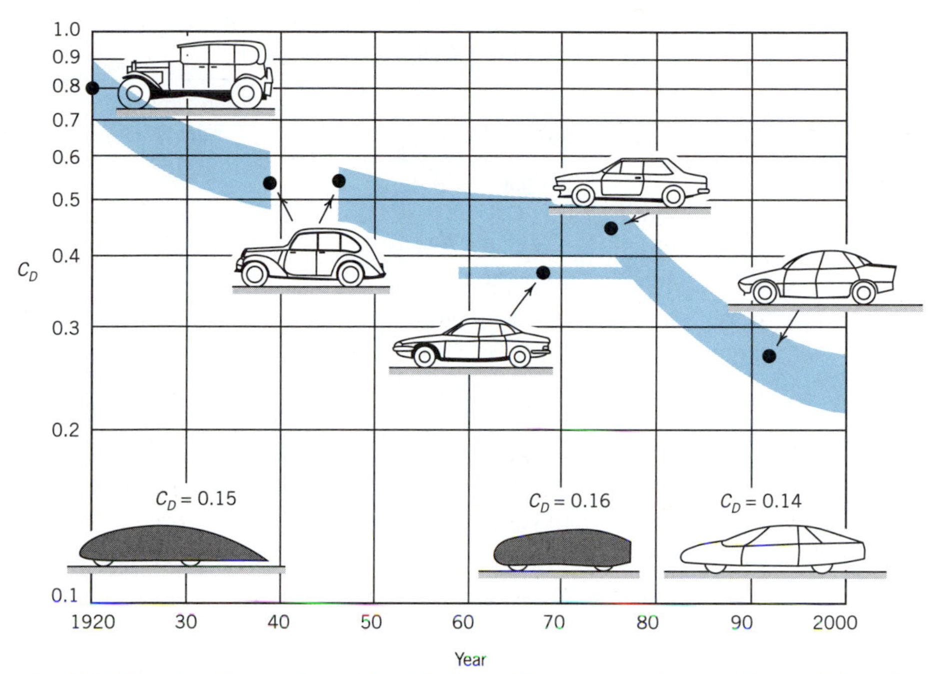

Fig. 11.24 The drag history of cars (from Hucho and Sovran, reproduced, with permission, from the *Annual Review of Fluid Mechanics*, Volume 25, © 1993, by Annual Reviews Inc.).

In this same source, Hucho provides an extensive list of the drag coefficients and frontal areas of passenger cars, from which the drag forces can be calculated as a function of speed. His data range from those for mini to medium to luxury to sports cars. A small sample is included here in Table 9. From this sample, we see that the sedan can be designed to have a drag coefficient equivalent to or less than a high performance sports car, signaling that the design of sedans has improved in recent years (cf., Fig. 11.24) and that, possibly, other factors may be more important than drag in the ultimate design of sports cars.

It is useful to note that the total resistance for a car, i.e., the force that needs to be overcome by the power train, is composed of an aerodynamic component and a rolling resistance. Emmelmann[30] shows that the ratio of the aerodynamic drag to the total resistance for cars varies with the drag coefficient and with speed; thus,

C_D	V (km/h)	Ratio $= \dfrac{\text{Aerodynamic drag}}{\text{Total resistance}}$
0.3	60	0.42
	80	0.55
	100	0.65
	120	0.69
0.4	60	0.48
	80	0.63
	100	0.72
	120	0.76

TABLE 9 Drag Coefficients of Passenger Cars (Adapted from Hucho[30], data by B. Heil)[a]

Vehicle Name	Approximate Drag Coefficient C_D	Frontal Area A (m^2)
Sedans		
VW Golf GTI	0.35	1.91
VW Golf Cabrio GL	0.48	1.86
BMW 318i	0.39	1.86
Honda Accord 1.8 EX	0.41	1.88
Volvo 360 GLT	0.40	1.95
Mercedes 200 D	0.29	2.05
BMW 323i	0.38	1.86
Volvo 740 GLE	0.41	2.16
Mercedes 300 E	0.30	2.06
BMW M 535i	0.37	2.04
Porsche 928 S	0.39	1.96
Mercedes 500 SEL	0.36	2.16
Sports cars		
Porsche 944 Turbo	0.33	1.90
Chevrolet Corvette	0.37	1.80
Toyota Celica Supra 2.8i	0.38	1.83
Chevrolet Camaro Z 28 E	0.37	1.94

[a]Often automobile models are offered in a convertible style; it is known that when the top is down on a convertible, the drag coefficient can easily increase by 30%.

In addition to consideration of drag, the car designer must be concerned with lift. As the air moves past the vehicle the flow is accelerated over the hood and roof and decelerated by the radiator and windshield. The latter two tend mainly to increase the drag, but the former two can produce pressures that are less than the free stream pressure and the net effect may be a lift on the vehicle. The flow beneath the vehicle may also accelerate and produce a *ground effect*, which in this case yields a negative lift (contrary to the positive lift which occurs for aircraft influenced by ground effect). Many readers have seen the effects of lift when racing cars lose their ground effect at high speed. Flegl and Rauser[29] show that the front axle lift on a Porsche 911 Carrera is about 50 lb at 90 mph when no aerodynamic aids are used. They say ''For driving safety and handling, high-performance cars must have negative lift, . . . There are three main ways to generate negative lift: variation of the basic vehicle configuration, mounting of negatively inclined wings, (and) built-in ground effect.'' With a front air dam and rear spoiler installed, the 911 has a front axle lift of less than 10 lbs at 150 mph. Hucho and Sovran report that for racing cars two means are used to produce a large negative lift. The use of negatively cambered wings is effective and their effect is well-defined, but the drag increase is large. The use of ground-effect in the form of what amounts to a ''venturi channel'' (cf., Chapter 9) under the vehicle to lower the pressure there has little drag penalty and produces a large downward force. However, this force can vary if the configuration of the racecar relative to the ground changes, producing the potential for extreme changes in the negative lift.

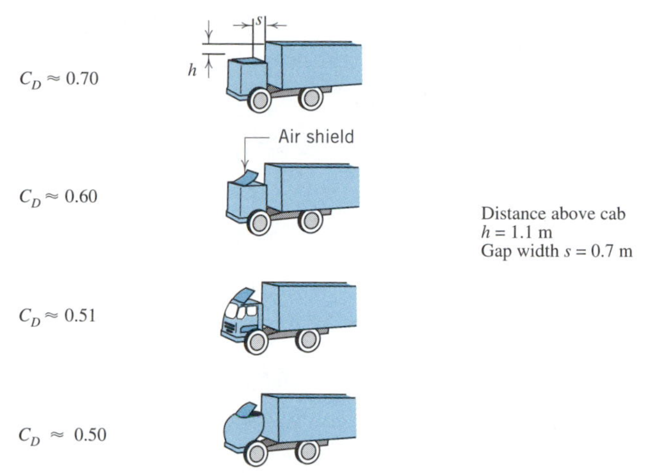

Fig. 11.25 Tractor-trailer configuration and drag reduction (after Götz[30]).

As suggested above, the productivity demand on commercial vehicles often limits the allowable shape variations. In the case of tractor-trailer truck rigs, the basic shape is a set of boxes. In addition to their larger frontal area (about 4.5 times that of a typical car or 9 m^2), the typical large tractor-trailer rig has a drag coefficient in the range of 0.7 to 1.0, depending on whether one or more trailers follow the tractor. It is also the case that the fuel consumption is driven by three factors, namely, the air drag, the rolling resistance, and the acceleration/climbing resistance. When fuel consumption is computed, it is found that these factors contribute about 18%, 50%, and 32%, respectively, of the total consumption (Götz[30]). Thus, while air drag is important, it may not be the dominant factor in a design. Accordingly, typical practice focuses on improving the air flow over the tractor cab and onto the trailer body (See Fig. 11.25). Interestingly, the results show that the addition of an air shield and moderate redesign and ''smoothing'' of the cab produce significant results (Fig. 11.25), while additional fairings, rounding of the trailer corners, etc., are not effective in producing significant additional drag reductions for head-on flow. If the truck is operating in a cross wind, devices that effectively reduce the gap between and smooth the flow between the tractor cab and the trailer can have a significant positive effect.

ILLUSTRATIVE PROBLEM 11.8

What power is needed to overcome air drag at 90 km/h for the VW Golf, the Mercedes 500 SEL, the Porsche 944 and a typical tractor-trailer rig? Assume sea level air properties.

SOLUTION

From Eq. 11.1, the drag force is

$$D = \frac{1}{2} A\rho V_o^2 C_D \tag{11.1}$$

For this case, $V_o = 90$ km/h $= 25.0$ m/s, while from Appendix 2, $\rho = 1.225$ kg/ft^3. We will tabulate the results, recalling that the power required to overcome the air resistance is $P = D \times V_o =$ work done per unit time. Thus,

$$D = \frac{1.225}{2}(25.0)^2 AC_D = 382.8AC_D \quad \text{and} \quad P = DV_o = 9\,570.3AC_D$$

Vehicle	Frontal Area, A (m^2)	C_D	Drag, D (kN)	Power, P (kW)
VW Golf	1.91	0.35	0.26	6.4
Mercedes 500 SEL	2.06	0.30	0.24	5.9
Porsche 944 Turbo	1.90	0.33	0.24	6.0
Tractor-Trailer Rig	4.5 × average car =	No air-shield-0.7	2.41	60.3
	4.5 × 2.0 = 9.0	Air-shield-0.6	2.07	51.7

REFERENCES

Anderson, J. D. 1989. *Hypersonic and high temperature gas dynamics.* New York: McGraw-Hill.

Armenti, Angelo, Jr., ed. 1992. *The physics of sports.* v.1. New York: American Institute of Physics.

Bertin, J. J., and Smith, M. L. 1989. *Aerodynamics for engineers.* 2d. ed. New Jersey: Prentice Hall.

Blevins, R. D. 1984. *Applied fluid dynamics handbook.* New York: Van Nostrand Reinhold Co. Especially Chapters 10–12.

Cermak, J. E. 1976. Aerodynamics of buildings. *Annual Review of Fluid Mechanics* 8: 75–106.

Dean, R. G., and Harleman, D. R. F. 1966. Interaction of structures and waves. *Estuary and coastline hydrodynamics.* A. T. Ippen, Ed. New York: McGraw-Hill, 341–403.

Drela, M. 1990. Aerodynamics of human-powered flight. *Annual Review of Fluid Mechanics* 22: 93–110.

Durand, W. F., ed. 1963. *Aerodynamic theory.* New York: Dover Publications. Six vols.

Faltinsen, O. M. 1990. Wave loads on offshore structures. *Annual Review of Fluid Mechanics* 22: 35–56.

Hanson, A. C., and Butterfield, C. P. 1993. Aerodynamics of horizontal-axis wind turbines. *Annual Review of Fluid Mechanics* 25: 115–149.

Hoerner, S. F. 1965. *Fluid-dynamic drag.* 2nd ed. New Jersey: Hoerner, S. F.

Hucho, W.-H., ed. 1987. *Aerodynamics of road vehicles.* London: Butterworths.

Hucho, W. H., and Sovran, G. 1993. Aerodynamics of road vehicles. *Annual Review of Fluid Mechanics* 25: 485–537.

Kuethe, A. M., and Chow, C. Y. 1986. *Foundations of aerodynamics: bases of aerodynamic design.* 4th. ed. New York: Wiley.

Labrujère, Th. E., and Sloof, J. W. 1993. Computational methods for the aerodynamic design of aircraft components. *Annual Review of Fluid Mechanics* 25: 183–214.

Liu, H. 1991. *Wind engineering: a handbook for structural engineers.* New Jersey: Prentice Hall.

Mehta, R. D. 1985. Aerodynamics of sports balls. *Annual Review of Fluid Mechanics* 17: 151–189.

Nieuwland, G. Y., and Spee, B. M. 1973. Transonic airfoils: recent developments in theory, experiment and design. *Annual Review of Fluid Mechanics* 5: 119–150.

Shapiro, A. H. 1961. *Shape and flow: the fluid dynamics of drag.* New York: Anchor Books.

Stinton, D. 1983. *The design of the airplane.* New York: Van Nostrand Reinhold.

Sutton, O. G. 1949. *The science of flight.* New York: Penguin Books.

Torenbeek, E. 1982. *Synthesis of subsonic airplane design.* Delft: Delft University Press/Martinus–Nijhoff, Pub.

Wiegel, R. L. 1964. *Oceanographical engineering.* New Jersey: Prentice Hall.

FILM

Shapiro, A. H. The fluid dynamics of drag. NCFMF/EDC Film Nos. 21601 through 21604.

PROBLEMS

11.1. A rectangular airfoil of 40 ft span and 6 ft chord has lift and drag coefficients of 0.5 and 0.04, respectively, at an angle of attack of 6°. Calculate the drag and horsepower necessary to drive this airfoil at 50, 100, and 150 mph horizontally through still air (40°F and 13.5 psia). What lift forces are obtained at these speeds?

11.2. A rectangular airfoil of 9 m span and 1.8 m chord moves horizontally at a certain angle of attack through still air at 240 km/h. Calculate the lift and drag, and the power necessary to drive the airfoil at this speed through air of (*a*) 101.3 kPa and 15°C, and (*b*) 79.3 kPa and − 18°C. $C_D = 0.035$; $C_L = 0.46$. Calculate the speed and power required for condition (*b*) to obtain the lift of condition (*a*).

11.3. If $C_L = 1.0$ and $C_D = 0.05$ for an airfoil, then find the span needed for a rectangular wing of 30 ft (10 m) chord to lift 800 000 lb (3 560 kN) at a take-off speed of 175 mph (282 km/h). What is the wing drag at take-off?

11.4. The drag coefficient of a circular disk when placed normal to the flow is 1.12. Calculate the force and power necessary to drive a 12 in. (0.3 m) disk at 30 mph (48 km/h) through (*a*) standard air at sea level (Appendix 2), and (*b*) water.

11.5. The drag coefficient of a blimp is 0.04 when the area used in the drag formula is the two-thirds power of the volume. Calculate the drag of a blimp of this shape having a volume of 14 000 m^3 when moving at 100 km/h through still standard air at sea level (Appendix 2).

11.6. A wing model of 5 in. chord and 2.5 ft span is tested at a certain angle of attack in a wind tunnel at 60 mph using air at 14.5 psia and 70°F. The lift and drag are found to be 6.0 lb and 0.4 lb, respectively. Calculate the lift and drag coefficient for the model at this angle of attack.

11.7. A wing of 46.5 m^2 plan form area is to produce a lift of 44.5 kN in level flight through standard air at sea level (Appendix 2) at 402 km/h by expending 450 kW. What C_L and C_D are required?

11.8. A steel sphere of 0.25 in. diameter is fired at a velocity of 2 000 ft/s at an altitude of 30 000 ft in the U.S. Standard Atmosphere (Appendix 2). Calculate the drag force on this sphere.

11.9. Calculate the drag on a 0.01 m diameter sphere and plot it as a function of Reynolds number if the sphere is moving through the atmosphere at an altitude of 10 000 m and at speeds greater than the speed of sound.

11.10. Repeat problem 11.9 for a 0.03 ft diameter sphere moving through sea level air at speeds greater than the speed of sound.

11.11. Calculate the force needed to overcome inertial forces and accelerate a 1 m diameter sphere at the rate of 1 m/s^2 in sea water. Assume that friction and pressure drag are negligible.

11.12. Calculate the inertial force on a 2 ft diameter pile which supports a pier if the pile is 25 ft long and subject to an onrushing wave whose fluid is accelerating at a rate of 0.5 ft/s^2.

11.13. A steel sphere (s.g. = 7.8) of 13 mm diameter falls at a constant velocity of 0.06 m/s through an oil (s.g. = 0.90). Calculate the viscosity of the oil, assuming that the fall occurs in a large tank.

11.14. What constant speed will be attained by a lead (s.g. = 11.4) sphere of 0.5 in. diameter falling freely through an oil of kinematic viscosity 0.12 ft^2/s and s.g. 0.95, if the fall occurs in a large tank?

11.15. Assuming a critical Reynolds number of 0.1, calculate the approximate diameter of the largest air bubble which will obey Stokes' law while rising through a large tank of oil of viscosity 0.004 $lb{\cdot}s/ft^2$ (0.19 Pa·s) and s.g. 0.90.

11.16. Glass spheres of 0.1 in. diameter fall at constant velocities of 0.1 and 0.05 ft/s through two different oils (of the same specific gravity) in very large tanks. If the viscosity of the first oil is 0.002 $lb{\cdot}s/ft^2$, what is the viscosity of the second?

11.17. Calculate the drag of a smooth sphere of 12 in. or 0.3 m diameter in a stream of standard sea level air (Appendix 2) at Reynolds numbers of 1, 10, 100, and 1 000.

11.18. Calculate the drag of a smooth sphere of 0.5 m diameter when placed in an airstream (15°C and 101.3 kPa) if the velocity is (*a*) 6 m/s, and (*b*) 8.4 m/s. At what velocity will the sphere attain the same drag which it had at a velocity of 6 m/s?

11.19. Estimate the drag on the streamlined body of Fig. 11.9 if it has a 0.15 m diameter and is being tested in an airstream (101.3 kPa and 15°C) at 27 m/s.

11.20. A modern submarine is 252 ft (77 m) long. It is shaped much like a blimp, with a circular cross section and a length-diameter ratio of 7.6 to 1. Compute and plot a curve of power versus speed in seawater (s.g. = 1.03) if $C_D \approx 0.08$.

11.21. A sphere of 10 in. diameter is tested in a wind tunnel with standard sea level air (Appendix 2) at 80 mph. At what speed must a 2 in. sphere be towed in water (68°F) for these spheres to have the same drag coefficients? What are the drag forces on these two spheres?

11.22. A steel sphere (s.g. 7.82) of 51 mm diameter is released in a large tank of oil (s.g. 0.82, viscosity 0.96 Pa·s). Calculate the terminal velocity of this sphere.

11.23. A cylindrical chimney 0.9 m in diameter and 22.5 m high is exposed to a 56 km/h wind (15°C and 101.3 kPa); estimate the bending moment at the bottom of the chimney. Neglect end effects.

11.24. The drag force exerted on an object (having a volume of 0.028 m^3 and s.g. of 2.80) is 445 N when moving at 6 m/s through oil (s.g. 0.90). If this object is allowed to fall through the same oil, is its terminal velocity larger or smaller than 6 m/s?

11.25. A standard marine torpedo is 0.533 m in diameter and about 7.2 m long. Make an engineering estimate of the power required to drive this torpedo at 80 km/h through freshwater at 20°C. Assume hemispherical nose, cylindrical body, and flat tail. C_D for a solid hemisphere (flat side downstream) is about 0.42.

11.26. This thin smooth wing lands at a speed of 200 km/h in air of weight density 12.0 N/m^3 and kinematic viscosity 1.4×10^{-5} m^2/s. A braking parachute is released to slow it down. Calculate the approximate diameter of the parachute required to produce an extra drag equal to the wing drag at this speed. Assume flow about the wing two-dimensional.

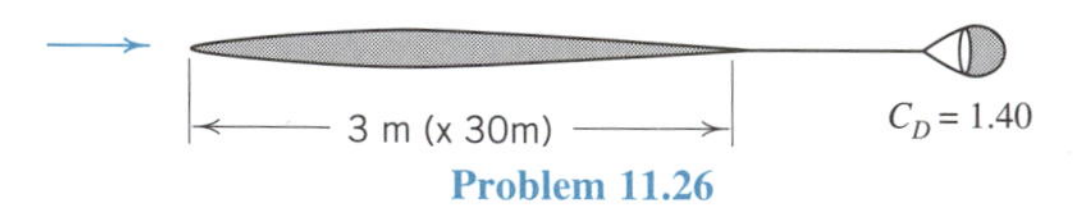

Problem 11.26

11.27. A thin circular disk is placed (normal to the flow) in an airstream of velocity 152 m/s, absolute pressure 138 kPa, and density 1.29 kg/m^3. If the pressure on the upstream side of the disk may be assumed to vary elliptically from stagnation pressure at the center to p_x at the edges, estimate the magnitude of p_x if this pressure is also exerted over the downstream side of the disk.

11.28. A large truck has an essentially boxlike body that causes flow separation at the front edges of the cab at any speed. The drag is mostly profile drag and $C_D = 0.75$. If the projected frontal area of the truck is 9 m^2 (97 ft^2), determine and plot as a function of speed between zero and the legal limit the power that must be delivered to the road to propel the truck.

11.29. The frequency ω of the shedding of eddies in a Kármán vortex street behind a cylinder is thought to depend on the diameter d of the cylinder, the flow velocity V_o, and the fluid density ρ. Use dimensional analysis to find the appropriate non-dimensional number(s) for this situation and compare your result to the Strouhal number S defined in Section 11.3.

11.30. A guy wire on a 300 m high television transmitting antenna has a 25 mm diameter. In a wind of 3 m/s, determine the drag load on the 400 m long wire and the frequency of vortex shedding. The natural frequency of vibration of a wire varies according to Mersenne's law: $n = (T/\rho)^{1/2}/2l$ where T is the wire tension, $\rho = 4$ kg/m, and l is the wire length. At what value of T is the natural vibration of the wire in resonance with the vortex shedding? Let $\nu_{\text{air}} = 1.6 \times 10^{-5}$ m^2/s. If the allowable stress in the steel wire is 575 000 kPa, can T ever be great enough to produce resonance?

11.31. Determine above what wind velocity the disturbance in the air induced by vortex shedding from telephone wires of 2 mm (0.08 in.) diameter can be heard. Humans can receive sound in the range of 20 to 20 000 Hz. What is the highest frequency tone generated within the range of **R** for regular vortex shedding?

11.32. Plot the velocity distributions along the horizontal and vertical axes for the flow in a free vortex in which the circulation is 2π m^2/s.

11.33. Plot the velocity distribution (as a function of θ) on the surface of a unit radius cylinder in an irrotational horizontal flow, where the circulation about the cylinder is 2π m^2/s and the velocity far from the cylinder is 1 m/s. How does the flow picture change if the circulation is doubled? If the circulation is zero?

11.34. Calculate the lift per unit length on the cylinder in problem 11.33 under all three conditions if the fluid is water and if the fluid is sea level air.

11.35. If Γ is the mean circulation about a wing per foot or metre of span, calculate the circulation about the wing at midpoint and quarter-points of the span, assuming a semielliptical lift distribution.

11.36. Derive a general expression for lift coefficient in terms of circulation.

11.37. An airfoil of 1.5 m chord and 9 m span develops a lift of 14 kN when moving through air of specific weight 12.0 N/m^3 at a velocity of 160 km/h. What is the mean circulation about the wing?

11.38. If the mean velocity adjacent to the top of a wing 1.8 m chord is 40 m/s and that adjacent to the bottom of the wing 31 m/s when the wing moves through still air (11.8 N/m^3) at 33.5 m/s, estimate the lift per metre of span.

11.39. A model wing of 5 in. chord and 3 ft span is tested in a wind tunnel (60°F and 14.5 psia) at 60 mph, and the lift and drag are found to be 9.00 and 0.460 lb, respectively, at an angle of attack of 6.7°. Assuming a semielliptical lift distribution, calculate (*a*) the lift and drag coefficients, (*b*) C_{D_i}, (*c*) C_{D_o}, (*d*) the corresponding angle of attack for an airfoil of infinite span, (*e*) the corresponding angle of attack for a foil of this type with aspect ratio 5, and (*f*) the lift and drag coefficients at this aspect ratio.

11.40. An airfoil of infinite span has lift and drag coefficients of 1.31 and 0.062, respectively, at an angle of attack of 7.3°. Assuming semielliptical lift distribution, what will be the corresponding coefficients for an airfoil of the same profile but aspect ratio 6? What will be the corresponding angle of attack?

11.41. From the data of Fig. 11.19, calculate the lift and drag coefficients for a Clark-Y airfoil of aspect ratio 8, and plot the polar diagram for this airfoil.

11.42. The Clark-Y airfoil of Fig. 11.19 is to move at 180 mph (290 km/h) through standard sea level air (Appendix 2). Determine the minimum drag, drag at optimum L/D, and drag at point of maximum lift. Calculate the lift at these points and the power that must be expended to obtain these lifts.

11.43. Using Fig. 7.12 and assuming a turbulent boundary layer, what approximate percentage of the total drag (at zero lift) can be attributed to skin friction for the Clark-Y airfoil of Fig. 11.19?

11.44. The take-off weight of an aircraft is 3 000 kN (393 tons). A design analysis considers the airfoil configurations of Table 7 for take-off conditions. Using the Clark-Y foil as a standard, calculate the percentage change in drag force and airfoil area needed at take off for the other foils.

11.45. The NACA 23012 airfoil has a maximum lift to drag ratio of 70 and a lift coefficient of 0.75 at an angle of attack of 5.5°. How would the required airfoil area and drag for this airfoil compare to those of the Clark-Y airfoil under the same condition of maximum L/D?

11.46. A human-powered aircraft has a gross weight of 240 lb including the pilot. Its wing has a lift coefficient of 1.5 and a lift-to-drag ratio of 70. Estimate the wing area needed and the pilot power that must be provided for this craft to cruise at 15 mph. Assume that the wing profile drag is about 40% of the total drag.

11.47. Derive the equation for the horizontal position of a pitched baseball under the assumption that the drag on the baseball is significant and so affects the time of travel to the plate. Calculate the time of travel and the speed at the plate and compare them to the result from Eq. 11.4 and the initial speed of 85 mph, respectively. Assume that the drag coefficient for the baseball can be estimated from Fig. 11.9 by use of the initial pitched speed.

11.48. If the strike zone in baseball is approximately from $2.0 < z < 4.5$ ft and 17 inches wide (the width of home plate) so $-0.71 < y < +0.71$ ft, determine which of the following pitches is a strike:

a. Released at 90 mph by a right-handed pitcher at $y = +2.5$ ft, $z = 5$ ft, and spinning at 1 800 rpm about a vertical axis.

b. Released at 80 mph by a right-handed pitcher at $y = +0$ ft, $z = 8$ ft, and spinning at 1 800 rpm about a horizontal axis.

c. Released at 95 mph by a left-handed pitcher at $y = -0.7$ ft, $z = 7$ ft, and spinning at 1 200 rpm about a horizontal axis.

d. Released at 85 mph by a right-handed pitcher at $y = +1.4$ ft, $z = 7$ ft, and spinning at 1 200 rpm about a vertical axis and at 800 rpm about a horizontal axis. This is a pitch thrown from a three-quarters overhead position and so has a spin axis not parallel to the ground. Extrapolate any data needed.

11.49. The Sears Tower in Chicago is 1 454 ft (443 m) tall. Assuming that it is a tall rectangle with a square base of 120 ft (37 m) sides, calculate the maximum drag force on the building and the force when the wind is along the diagonal of the structure as a function of wind speed from Beauford Wind Scales of strong breeze (28 mph; 12 m/s) to hurricane (75 mph; 33 m/s). Assuming that the wind field is uniform, calculate the moment about the base of the Tower also.

11.50. It is proposed to build a pyramidal building with a square base with sides of 160 ft (49 m), which has the same volume as the Sears Tower. Calculate the maximum drag force on this building and compare it to that for the Sears Tower under hurricane force conditions (75 mph; 33 m/s).

11.51. Calculate the lift and drag forces on the large sphere (diameter = 275 ft = 84 m) in Table 8. Assume that ground effect is present and a Gale wind speed on the Beauford Scale (19 m/s; 42 mph).

11.52. Suppose that the structure in Illustrative Problem 11.6 exists in an open field so that the coefficient $1/n = 1/7$ in the velocity profile equation 11.6. What are the resulting drag and moment?

11.53. Calculate the drag forces on a 1/200 scale model of the Sears Tower that is tested in a large water flume under conditions corresponding to those in problem 11.49. Ignore any free surface effects and assume dynamic similarity and that the drag coefficient is unchanged.

11.54. What are the drag and the moment to be expected in a 1/50 scale model of the structure in Illustrative Problem 11.6

if it is placed in a water flume and operated under dynamically similar conditions?

11.55. Plot the ratio of maximum inertia to maximum drag forces as a function of pile diameter for the situation of Illustrative Problem 11.7. Describe how the ratio changes.

11.56. Derive an equation for the moment about the base of the forces on a pile for small-amplitude and shallow-water waves.

11.57. Using the components of Eq. 11.10 prove (i) that, when the maximum values of the drag and inertial forces are equal, then

$$\frac{d_e}{V_o T} = \frac{1}{\pi^2}\left(\frac{C_D}{C_I}\right)$$

where d_e is the value of the pile diameter at which the forces are equal given V_o and T, and (ii) that, for given values of the velocity and period, when $d > d_e$, the inertial force dominates.

11.58. Calculate the power required to drive the VW Golf Cabrio, the Mercedes 300 E, and the Chevrolet Corvette at 80 and 120 km/h (51 and 76 mph). Account for the total resistance.

11.59. What is the ratio of the power required to drive the research car versus the 1920 sedan in Fig. 11.24 at a speed of 60 km/h (38 mph)? Account for total resistance.

11.60. Show that the power required to overcome air resistance for a car is $P = \frac{1}{2} A\rho V_o^3 C_D$.

11.61. Assume that the Mercedes 200 D is offered in a convertible model whose drag coefficient is essentially the same as the hard roof model when the convertible top is up. Calculate the power required to drive the car at 120 km/h with the top up. Up to what speed can the car operate with the top down without using more power? Account for total resistance.

11.62. Suppose that the engine in the Chevrolet Corvette can deliver a maximum of 100 hp (75 kW) towards overcoming total resistance (air resistance plus rolling resistance). What is the top speed of this car?

11.63. If the drag coefficient of a tractor-trailer rig is reduced by 50%, what percentage change in fuel consumption can be expected?

11.64. If air drag accounts for only 18% of the total resistance of a tractor-trailer with a $C_D = 0.70$, what power is required to drive this rig at 55 mph (87 km/h)? Using the same power, how much faster can the rig go with an air shield that cuts the drag coefficient to 0.50? Assume that the frontal area of the truck is 9 m^2 or 97 ft^2.

11.65. Given that an engine can deliver 200 hp (150 kW) to overcome total resistance for a tractor-trailer rig, plot the top speed as function of the air drag coefficient (between 0.45 and 1.0). Assume that air drag accounts for 20% of the total resistance. Now calculate the top speeds if the engine can deliver 400 hp (300 kW); why are the changes so small?

11.66. If the fraction of total resistance that is represented by air drag is f_D, modify the power equation of problem 11.60 so the modified formula yields the power to overcome the total resistance.

CHAPTER

12 INTRODUCTION TO FLUID MACHINERY

Many machines are characterized by the transfer of energy, forces, or torques between a moving stream of fluid and the solid elements of the machine. These are *flow machines*. They include a large and important selection of common units, such as jet engines, rocket motors, airplane and marine propellers, steam turbines for electricity generation and ship propulsion, garden insecticide sprays (jet pumps), industrial centrifugal pumps, and the massive turbines of hydroelectric plants. Some elementary principles of flow machines, such as jet propulsion, propellers and windmills, and rocket propulsion, were discussed in Sections 6.5 through 6.7.

The energy and force transfers in flow machines can be either to or from the fluid and can be accomplished in a number of configurations. Thus, it is useful to further categorize flow machines. In *turbines* energy is taken from the fluid stream; in *pumps* (*compressors*, *blowers*, or *fans*) energy is added to the fluid stream. If the transfers occur between the fluid stream and a machine element that rotates about a fixed axis (a compressor rotor, turbine runner, or pump impeller), the machine is classed as a *turbomachine*. In turbomachines, the flow may be essentially *axial* (parallel to the axis of rotation of the rotating element), *radial*, or *mixed* (a combination of radial and axial).

12.1 ANALYSIS AND CHARACTERISTICS OF TURBOMACHINES

Momentum principles applicable to turbomachines were discussed in Section 6.4. These principles provide the means for examining specific flow patterns, power transfers from or to blades, and so forth. However, because of the complexity of most turbomachines and often their great size, it is common practice to model such machines. In addition, it is useful to be able to make generalizations and predictions about common characteristics or differences between generic types of machines (e.g., centrifugal and axial flow pumps or impulse and reaction turbines). Dimensional analysis and the principles of similitude are very valuable in such matters.

The first step is to select the relevant variables. For the class of turbomachines in which no combustion or significant heat transfer occurs, the power, P, required by or delivered by geometrically similar turbomachines will depend on: the runner or impeller diameter, D (a characteristic length); the rotative speed, N; the volume flowrate, Q; the energy, H, added or subtracted, respectively, from each unit mass[1] of fluid passing through the machine; and the fluid characteristics, namely, viscosity, μ, density, ρ, and elasticity, E. From Section 8.2 and Eqs. 5.8 it is already known that $P \propto \rho Q H$; however, it is very instructive to step back and begin afresh with a new dimensional analysis. Doing this leads directly to significant design coefficients and to a deeper understanding of the relationships among the variables relevant to turbomachines. At the appropriate point in what follows, the relation $P \propto \rho Q H$ is used to coalesce the results to a final form from which the nature of the proportionality factor is explicitly seen.

Accordingly, for the variables listed above, it is expected that the general relation

$$f(P, D, N, Q, H, \mu, \rho, E) = 0$$

is valid. There are eight variables and three independent dimensions[2] so $8 - 3 = 5$ Π-terms can be constructed. A Buckingham Π-analysis enables μ, E, P, Q, and H to be successively combined with the remaining three variables ρ, D, and N to obtain the required five Π-terms (see the problem 12.2 to find the consequence of using D, ρ, Q or H, ρ, Q as a basis).

Because N has the dimension t^{-1}, ρ, D, N, and μ will form a Reynolds number $\mathbf{R}$. Set

$$\Pi_1 = \mu^a \rho^b D^c N^d$$

Solving the dimensional equations gives

$$\Pi_1 = \frac{\rho N D^2}{\mu} = \mathbf{R}$$

Similarly, using E, ρ, D, and N yields

$$\Pi_2 = \frac{\rho N^2 D^2}{E} = \frac{N^2 D^2}{a^2} = \mathbf{M}^2$$

where $\mathbf{M}$ is a Mach number and $a^2 = E/\rho$.

[1]From Section 5.5, E_P or E_T appear as abrupt rises or falls in the energy line across the machine and have the units J/N or ft·lb/lb. Often E_P or E_T are called the *head* on the unit. Because g_n is not a relevant parameter in this analysis and because it is the mass of fluid that possesses energy, here the *head*, $H = E_{(T \text{ or } P)} g_n$, is defined as the energy per unit mass; (the dimensions of H) $= L^2/t^2$.

[2]Clearly, none of the sets of terms (ρ, D, N), (D, ρ, Q) or (H, ρ, Q) can be combined into a Π-group; hence, $k = m = 3$.

Now it remains to generate three Π-terms for P, Q, and H, respectively. Taking

$$\Pi_3 = P^a \rho^b D^c N^d$$
$$\Pi_4 = Q^a \rho^b D^c N^d$$
$$\Pi_5 = H^a \rho^b D^c N^d$$

and solving the dimensional equations gives

$$\Pi_3 = \frac{P}{\rho N^3 D^5} = \text{Power Coefficient } \mathbf{C}_P$$

$$\Pi_4 = \frac{Q}{ND^3} = \text{Capacity Coefficient } \mathbf{C}_Q$$

$$\Pi_5 = \frac{H}{N^2 D^2} = \text{Head Coefficient } \mathbf{C}_H$$

In summary, it has been possible to obtain the required five Π-terms by a combination of the methods and concepts developed earlier in Chapter 8. The results can be put in the following alternative forms:

$$\mathbf{C}_P = \frac{P}{\rho N^3 D^5} = f'(\mathbf{C}_Q, \mathbf{C}_H, \mathbf{R}, \mathbf{M})$$

$$\mathbf{C}_Q = \frac{Q}{ND^3} = f''(\mathbf{C}_P, \mathbf{C}_H, \mathbf{R}, \mathbf{M})$$

$$\mathbf{C}_H = \frac{H}{N^2 D^2} = f'''(\mathbf{C}_P, \mathbf{C}_Q, \mathbf{R}, \mathbf{M})$$

However, as noted above there is a specific relationship among, P, Q, H, and ρ, that is, $P \propto \rho QH$. Accordingly, the dimensionless Π-term

$$\Pi_3' = \frac{P}{\rho QH} = \frac{\mathbf{C}_P}{\mathbf{C}_Q \cdot \mathbf{C}_H} = f^{IV}(\mathbf{C}_Q, \mathbf{C}_H, \mathbf{R}, \mathbf{M}) \tag{12.1}$$

can replace $\Pi_3 = \mathbf{C}_P$.

If now $\mathbf{C}_Q$ and $\mathbf{C}_H$ are held constant for a set of similar machines, all the hydraulic losses in head due to flow friction, eddying, separation, and so forth, are embodied by the Reynolds number effect. However, other losses in power occur in turbomachines (in bearings, etc.) that do not show up as head losses. In the particular case of incompressible flow, $\mathbf{M}$ is not a relevant parameter. Then, Eq. 12.1 becomes (with $\mathbf{C}_Q$ and $\mathbf{C}_H$ held constant)

$$\frac{P}{\rho QH} = f^{V}(\mathbf{R}) = \eta_H$$

where η_H is the hydraulic efficiency of the machine. If the mechanical efficiency η_M (which is a dimensionless number) is taken as a parameter to measure nonhydraulic losses, then, in general,

$$\frac{P}{\rho QH} = f^{VI}(\eta_M, \mathbf{R}) = \eta$$

where η is the total efficiency of the machine. Clearly, then, under the constraint that $\mathbf{C}_Q$ and $\mathbf{C}_H$ are held constant in a set of similar machines, the proportionality between P and ρQH depends only on the hydraulic and mechanical efficiencies plus Mach number effects (if any).

Shepherd[3] discusses η and indicates that differences in η between model and prototype can be represented, for example, for hydraulic reaction turbines, by an equation of the form

$$\frac{1 - \eta_m}{1 - \eta_p} = \left(\frac{D_p}{D_m}\right)^{1/5} \tag{12.2}$$

One reason for taking this approach is that the gross Reynolds number defined herein is not a good measure of the similarity of the detailed flow inside a turbomachine. In models, relative roughness effects (caused by not being able to achieve detailed geometric similarity of roughness elements) and scale differences cause transition from laminar to turbulent flow and separation or eddy formation to take place in relatively different places in model and prototype. Equation 12.2 gives a measure of these ''scale-effects.'' Thus, gross dynamic and geometric similarity *do not* guarantee detailed kinematic similarity.

Examination of $\Pi_4(\mathbf{C}_Q)$ suggests that similarity can be gaged in terms of kinematic similarlity, that is, similarity of velocity patterns, in otherwise geometrically similar machines. The product $ND \propto u$ which is the peripheral velocity of the runner or impeller, while $Q/D^2 \propto V$, which is the radial velocity through the runner or impeller. Thus,

$$\Pi_4 = \frac{Q}{ND^3} \propto \frac{V_r}{u} = \Pi_4'$$

Holding Π_4' (and Π_4) constant in a set of geometrically similar machines implies similarity of velocity triangles in the machines (see Section 12.3). Geometrically similar machines having similar velocity triangles are called *homologous machines*. The performance characteristics of any machine in a homologous set are obtained from knowledge of the characteristics of one machine in the set, for example, from data obtained in a model test.

Some variation in efficiency η among homologous machines may have to be accounted for (see Eq. 12.2) because it is sometimes not possible to keep $\mathbf{R}$ constant for all machines. Holding $\mathbf{R}$ constant for a model and prototype requires $(ND^2)_m = (ND^2)_p$; that is, $N_m = N_p(D_p/D_m)^2$. Thus, a one-fifth scale model of a turbine would have to be run at 25 times prototype speed. This may not be feasible; hence, smaller machines usually have lower $\mathbf{R}$ and proportionately larger friction effects.

Because the relationship between P and ρ, Q and H is known and η is expected to handle both $\mathbf{R}$ and mechanical effects, a new dimensional analysis for an *incompressible* fluid flowing in a *hydraulic* turbine or pump might consider only

$$f(D, Q, H, N, \eta) = 0$$

Two new Π-terms are expected because η is already dimensionless while $k = 2$ (why?). The previous analysis suggests that, for example,

$$\frac{H}{N^2D^2} = f'\left(\frac{Q}{ND^3}, \eta\right) \quad \text{or} \quad \mathbf{C}_H = f'(\mathbf{C}_Q, \eta)$$

[3]D. G. Shepherd, *Principles of Turbomachinery*, Macmillan, 1956.

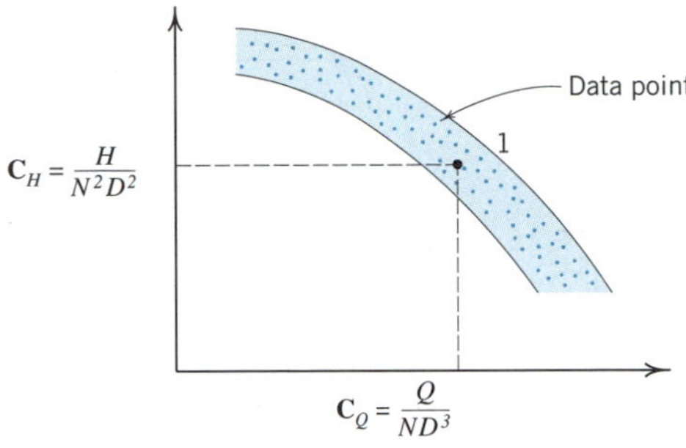

Fig. 12.1

As Shepherd notes, a plot of experimental results for a set of geometrically similar turbomachines of several sizes takes the form shown in Fig. 12.1. The data points lie in a band as shown because η varies not only with D but also with changes in **R** and in mechanical losses in the same machine as its speed changes. If η remains essentially constant, the set of geometrically similar machines represented by point 1 on Fig. 12.1 has the following characteristics

$$H \propto D^2$$

$$Q \propto D^3$$

Thus, a 10% decrease in D causes a 20% reduction in head and an almost 30% reduction in flowrate. In fact, for *homologous* machines, we have the following equations which relate size, head, flowrate, speed, and power between models and prototypes of turbomachines.

$$\left(\frac{Q}{ND^3}\right)_m = \left(\frac{Q}{ND^3}\right)_p \tag{12.3a}$$

$$\left(\frac{H}{N^2D^2}\right)_m = \left(\frac{H}{N^2D^2}\right)_p \tag{12.3b}$$

$$\left(\frac{P}{\rho N^3D^5}\right)_m = \left(\frac{P}{\rho N^3D^5}\right)_p \tag{12.3c}$$

While the previous analysis gave considerable insight to turbomachines, two other dimensionless terms are very widely used in engineering practice because through them it is possible to characterize various classes of pumps, compressors, and turbines without regard for their size (D is excluded in forming the terms). For pumps and compressors, the flowrate Q, head H, speed N, and maximum efficiency $\eta_{\max}$ are key performance parameters ($\eta_{\max}$ includes ρ and μ effects) for machines of all sizes. Two Π-terms, Π_4 and Π_5, are obtained with these variables; eliminating D by taking the ratio of $(\Pi_4)^{1/2}$ to $(\Pi_5)^{3/4}$ gives

$$\frac{\mathbf{C}_Q^{1/2}}{\mathbf{C}_H^{3/4}} = N'_S = \frac{NQ^{1/2}}{H^{3/4}} \tag{12.4a}$$

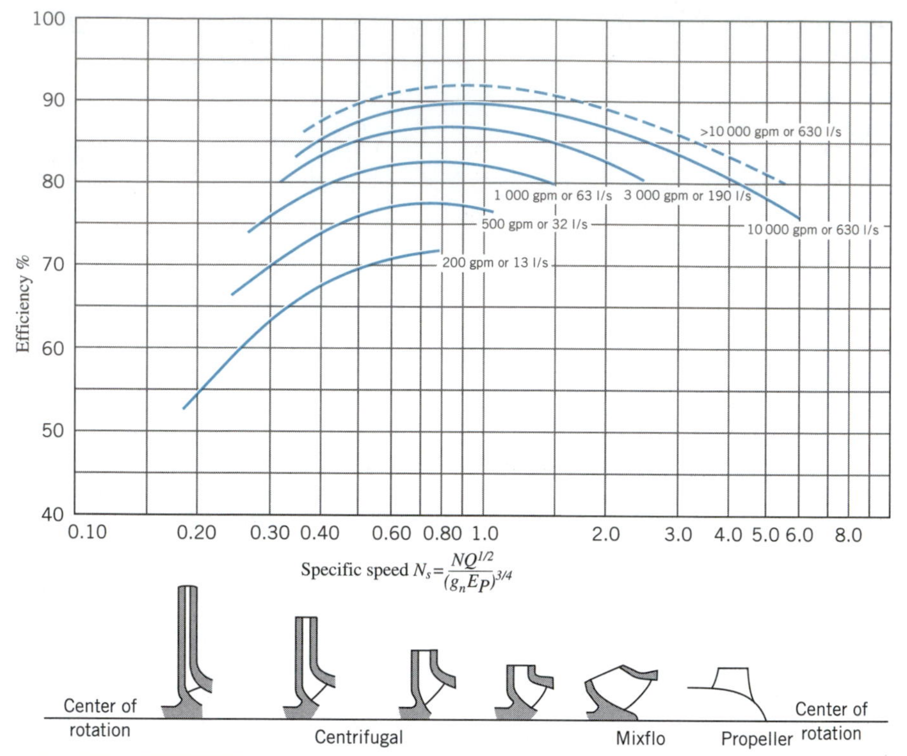

Fig. 12.2 Variation of pump impeller shapes and efficiency with specific speed (by permission of Ingersoll–Dresser Pump Company).

where N_S' is referred to as the *specific speed* and is a constant for a homologous set of machines. In its current form, however, it is not consistent with the definition of pump head set forth in Section 5.5. This is easily fixed by replacing H (head per unit mass as described in footnote 1) with $g_n E_P$ (where E_P is head per unit weight). The result is

$$N_S = \frac{NQ^{1/2}}{(g_n E_P)^{3/4}} \tag{12.4b}$$

where N_S is the *dimensionless specific speed*, N is in rad/s, Q is in ft^3/s in U.S. Customary units or m^3/s in SI units, and E_P is in feet in U.S. Customary units or metres in SI units. The value of g_n is the appropriate value for the system of units being used.

Figure 12.2 is a plot of efficiency η versus N_S for a wide range of pumps. This plot is based on data presented in a similar plot in Karassik, et al,[4] in U.S. Customary units. However, Fig. 12.2 is dimensionless, as described above, and may be used for both U.S. Customary and SI units. The lines are the locus of maximum efficiency η_{max} data points. Here η is the total efficiency of the pump and includes mechanical and hydraulic losses. Thus, the various locations of the curves (shown here to be related to flow and by implication to machine size) can be interpreted as representing variations in **R** and η_M (see above).

[4]Karassik, I. J., Krutzsch, W. C., Fraser, W. H., and Messina, J. P. *Pump Handbook.* McGraw-Hill, 1986.

Because N_S is constant for a homologous set of machines, it is often referred to as a *type characteristic*, that is, its numerical value suggests the type of pump which will be required for a given situation. For example, at a given N, we can see from Eq. 12.4b that a low specific speed will require a relatively low flow and a high head. Because $E_P \propto D^2$ for fixed N, a typical low-N_S pump has essentially a radial flow to generate the maximum pressure increase and a low Q to limit velocities within the narrow pump passages (this holds down head losses caused by frictional effects). As seen from Fig. 12.2, this result suggests a centrifugal pump. For a large N_S, a high Q and small E_P are appropriate, conditions which are satisfied by axial flow (propeller) pumps in which the high flow moves through relatively large passages at reasonable speeds without causing excessive head losses due to friction. If one attempts to operate a centrifugal pump at high specific speeds, the large Q will drive the efficiency down (due to increased friction-induced head loss) far below the efficiency of an axial flow pump whose shape better suits the flow requirements (review Fig. 12.2).

For hydraulic turbines the relevant performance and fluid variables are P, H, N, ρ, and η_{max}. Using the ratio of $(\Pi_3)^{1/2}$ to $(\Pi_5)^{5/4}$ leads to a *specific speed for turbines*

$$\frac{\mathbf{C}_P^{1/2}}{\mathbf{C}_H^{5/4}} = N_S' = \frac{NP^{1/2}}{\rho^{1/2}H^{5/4}} \tag{12.5a}$$

Once again we have an expression which is not in correspondence with our previous definition for turbine head; however, if we replace H with $g_n E_T$, we obtain

$$N_S = \frac{NP^{1/2}}{\rho^{1/2}(g_n E_T)^{5/4}} \tag{12.5b}$$

where N_S is the *dimensionless specific speed*, N is in rad/s, P is in ft·lb/s in U.S. Customary units or watts in SI units, and E_T is in feet in U.S. Customary units or metres in SI units. The value of g_n is the appropriate value for the system of units being used.

Figure 12.3 is a representative plot of η_{max} versus N_S for turbines. The plot is dimensionless and may be used for either U.S. Customary or SI units. Again, N_S is an indicator of the performance characteristics of turbines. For fixed N, Eq. 12.5b suggests that a low N_S corresponds to relatively high head and low power output (and hence low Q). In this

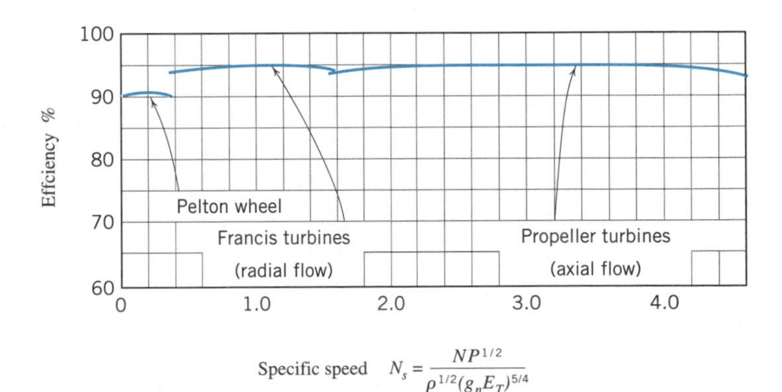

Fig. 12.3 Typical peak efficiencies as a function of specific speed for different types of hydraulic turbine.

range Fig. 12.3 indicates the Pelton wheel and the Francis reaction turbine (see Sections 12.2 and 12.3) have the best efficiencies. At high N_S, the low head and large power output (hence, high Q) are typical of axial flow turbines.

An interesting development in hydropower is the expanding use of pumped storage to provide peak power generation by storage of water pumped to an elevated reservoir during off-peak (and low power cost) hours. A typical installation uses a combination pump/turbine-motor/generator machine to pump the water up to the elevated reservoir and by reverse flow through the same machine, generate power. Figure 12.4 shows a runner for the Coo Trois Ponts profect in Belgium. The machines are Francis-type reversible pump/turbines. When generating at rated turbine capacity (145 MW), the specific speed of the pump/turbine is 0.62. When pumping at rated discharge (46 m^3/s), the specific speed is 0.60. Note where these numbers fall on Figs. 12.2 and 12.3. It should not be surprising that the two specific speeds are almost equal. If $P = \gamma QH$ (in an ideal machine where $\eta = 1$), the N_S quoted for pumps (Eq. 12.4b) and for turbines (Eq. 12.5b) are just alternate forms of the same equation. Here, the efficiencies are greater than 90% and approximately equal, so the machine has essentially the same specific speed operating as either a pump or a turbine.

Finally, the performance of hydraulic pumps and turbines can be seriously degraded by cavitation. An energy line analysis (see Section 5.4) of the flow leading from a reservoir to a pump or from a turbine to the reservoir into which the turbine discharges will show that the minimum head typically occurs in these machines at the inlet (suction side) of a pump and at the outlet (discharge side) of a turbine. In the case of pumps, the total useful head at this point is referred to as the *net positive suction head, NPSH*. Pump manufacturers specify the *NPSH* which is required if their pump is to operate without cavitating. The pump station designer must see to it that this minimum *NPSH* is provided along with any factors of safety which seem prudent.

Fig. 12.4 Coo Trois Ponts (Belgium) pumped-storage pump turbine runner (Francis-type). (Courtesy of Allis-Chalmers.)

The most useful form of the *NPSH* equation is

$$NPSH = \frac{p_0}{\gamma} - \frac{p_v}{\gamma} - z_s - h_L \tag{12.6}$$

where p_0 is generally the atmospheric pressure (unless liquid is being pumped from a pressurized tank), p_v is the vapor pressure of the liquid, z_s is the elevation of the pump impeller inlet above the water surface in the suction supply, and h_L is the head loss in the pump suction piping (generally included in the *NPSH* required by the pump unless there is additional suction piping required beyond that supplied with the pump). Once one knows the *NPSH* required by the pump, the liquid being pumped, and the atmospheric pressure, the vertical position of the pump which will eliminate cavitation can be computed.

For turbines the traditional approach is to quantify cavitation tendencies by use of the Thoma cavitation parameter $\boldsymbol{\sigma}$ which is defined as

$$\boldsymbol{\sigma} = \frac{NPSH}{E_T}$$

NPSH is defined in a manner similar to pumps as

$$NPSH = \frac{p_0}{\gamma} - \frac{p_v}{\gamma} - z_s$$

where z_s is the elevation of the runner exit above the tailwater surface. Draft tube losses are incorporated in the *NPSH* value.

As it happens, the specific speed concept, so useful in predicting efficiencies of all types of pumps and turbines, can be employed to useful purpose in establishing criteria in regard to cavitation prevention in both types of machines. In this case, because cavitation occurs at the pump impeller entrance and the turbine runner exit, Eqs. 12.4b and 12.5b are adapted using *NPSH* instead of E_P and E_T. As a result, a *suction specific speed* is defined as

$$\text{Pumps: } N_{SS} = \frac{NQ^{1/2}}{(g_n NPSH)^{3/4}} \tag{12.7a}$$

$$\text{Turbines: } N_{SS} = \frac{NP^{1/2}}{\rho^{1/2}(g_n NPSH)^{5/4}} \tag{12.7b}$$

The data of Wislicenus[5] show, according to Dixon,[6] that cavitation begins at essentially the same value of N_{SS} for all pumps and at a different, but constant, value for all turbines whose passages have been designed to minimize cavitation. Thus, to avoid cavitation in the pump inlet, $N_{SS} \leq 3$, and in the turbine exit, $N_{SS} \leq 4$. These numbers are useful in preliminary design and prior to tests of a specific unit because they provide a means of obtaining estimates of the minimum *NPSH* requirements for a given installation.

[5]G. F. Wislicenus, *Fluid Mechanics of Turbomachinery*, 2nd ed., McGraw-Hill, 1965.

[6]S. L. Dixon, *Fluid Mechanics, Thermodynamics of Turbomachinery*, 3rd ed., Pergamon, 1978.

12.2 DEFLECTORS AND BLADES—THE IMPULSE TURBINE

When a free jet is deflected by a blade surface, a change of momentum occurs and a force is exerted on the blade. If the blade is allowed to move, this force will act through a distance, and power may be derived from the moving blade; this is the basic principle of the impulse turbine.

The jet of Fig. 12.5 is deflected by a fixed blade and may be assumed to be in a horizontal plane.[7] With the control surface drawn around the region of momentum change, it is seen at once that the force exerted by the blade is the only force acting on the fluid. Therefore, from Eqs. 6.2

$$\Sigma F_x = -F_x = (V_2 \cos \beta - V_1)\rho Q \tag{12.8a}$$

$$\Sigma F_y = F_y = (V_2 \sin \beta - 0)\rho Q \tag{12.8b}$$

If this blade now moves (Fig. 12.6) with a constant velocity u in the same direction as the original jet, the jet is no longer deflected through an angle β because the leaving velocity, V_2, is now the resultant of the blade velocity and the velocity of the fluid over the blade. The velocity, v, of fluid relative to the blade is $(V_1 - u)$, and if friction is neglected, this relative velocity is the same at the entrance and exit of the blade system. Therefore, as the jet leaves the blade, it has an absolute velocity, V_2, equal to the vector sum of u and $(V_1 - u)$. Then, from Fig. 12.6 and Eqs. 6.2, with the control volume fixed to the blade,

$$\Sigma F_x = -F_x = (V_{2x} - V_{1x})\rho Q = -(V_1 - u)(1 - \cos \beta)\rho Q \tag{12.9}$$

$$\Sigma F_y = F_y = (V_{2y} - V_{1y})\rho Q = (V_1 - u) \sin \beta \rho Q \tag{12.10}$$

Engineers frequently prefer to treat such problems with the relative velocities v_1 and v_2, both of which are equal to $(V_1 - u)$, thus reducing the problem to that of the stopped blade of Fig. 12.5; the validity of this may be seen from the velocity triangles and by noting that the substitution of $(V_1 - u)$ for V_1 and V_2 in Eqs. 12.8a and 12.8b will yield Eqs. 12.9 and 12.10, respectively.

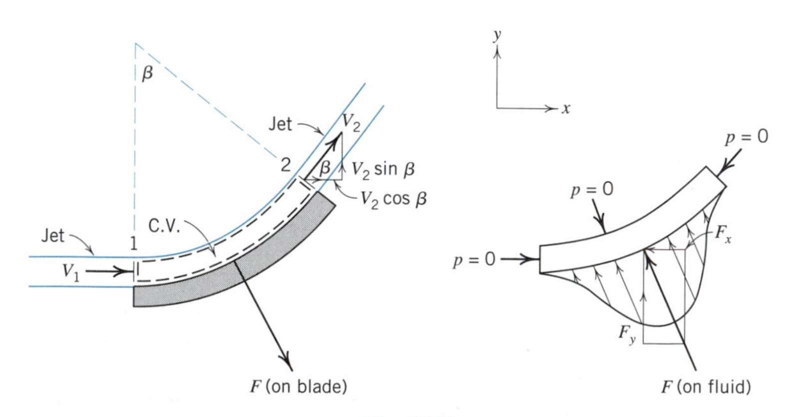

Fig. 12.5

[7]The difference in elevation between beginning and end of blade is usually negligible in practical blade problems.

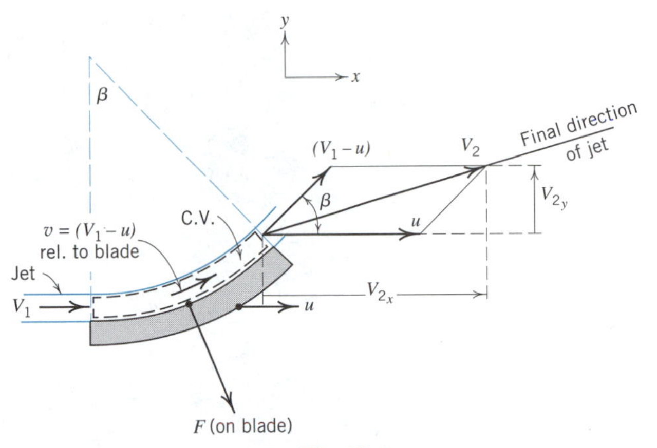

Fig. 12.6

For a single moving blade the flowrate, Q, in these equations is not the flowrate in the jet, since less[8] fluid is deflected per second by a single blade than is flowing in the jet. However, in the practical application of this theory to a turbine, where a series of blades is used, the fluid deflected *by the blade system* per second is the same as that flowing.

Problems of this type should be recognized as unsteady-flow problems for which the equations have been written for time-average conditions. As a blade moves through the jet a time-varying force will act on it, rising from zero and falling back to zero. From this it is clear that the foregoing equations can give no information about the maximum force exerted on the blade or the force on the blade at any instant of time, or the optimum spacing of blades to secure a required deflection of fluid; nevertheless, the equations may be used effectively in preliminary design and performance calculations.

In an impulse turbine (Fig. 12.7) a series of blades of the type above is mounted on the periphery of a wheel. Consider the idealized blade in Fig. 12.8; while a blade is in the jet it is moving in a direction approximately parallel to that of the jet; thus the equations above may be applied directly to find the power characteristics of the machine.[9] The force component F_y does no work, since it acts through no distance; when the component F_x is multiplied by u, the power transferred from jet to turbine is obtained. In a frictionless machine this is the output and is given by

$$P = Q\rho(V_1 - u)(1 - \cos\beta)u \tag{12.11}$$

From this it is evident (1) that no power is obtained from the machine when $u = 0$ (machine stopped) and when $u = V_1$ (runaway speed) and (2) that, with Q, ρ, β, and V_1 constant, the relationship between P and u is a parabolic one. Taking $dP/du = 0$ to obtain the properties of maximum output shows that this will occur when $u = V_1/2$ and is given by

$$P_{\max} = Q\rho \frac{V_1^2}{4}(1 - \cos\beta) \tag{12.12}$$

[8]For a blade moving in the same direction as the jet.

[9]Notice that the actual Pelton wheel blades or buckets split the jet and deflect it symmetrically to the sides. It is easy to see that the simple analysis for a single blade (Fig. 12.8) is valid for each Pelton bucket. Indeed, the control volume shown there surrounds all the moving blades and is fixed in space. As long as the jet path shown is the correct time-averaged picture, one can apply Eq. 6.2a directly to obtain F_x. The result is precisely that from Eq. 12.9, which came from a moving control volume analysis.

Fig. 12.7 Runner of the six-nozzle vertical Pelton-type turbine "Castaic" in U.S.A. (Courtesy of Escher Wyss Limited, Zurich, Switzerland.)

For a blade angle of 180°, this becomes

$$P_{\max} = \frac{Q\rho V_1^2}{2} = \frac{Q\gamma V_1^2}{2g_n} \tag{12.13}$$

As this is exactly the power of the free jet (see Eq. 5.8), it may be concluded that all the jet power may be theoretically extracted and transferred to the machine (at 100% efficiency), if (1) the peripheral speed is one-half the jet speed and if (2) the blade angle is 180°. In practice, blade angles will be found to be around 165°, operating peripheral speeds to be about 48% of jet velocity, with resulting peak efficiencies near 90% (see Section 12.1).

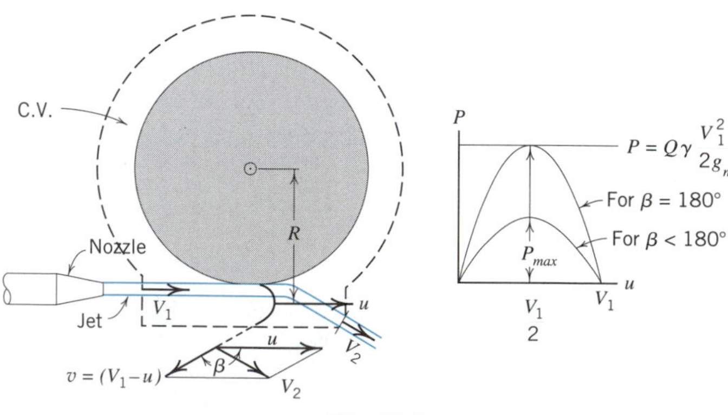

Fig. 12.8

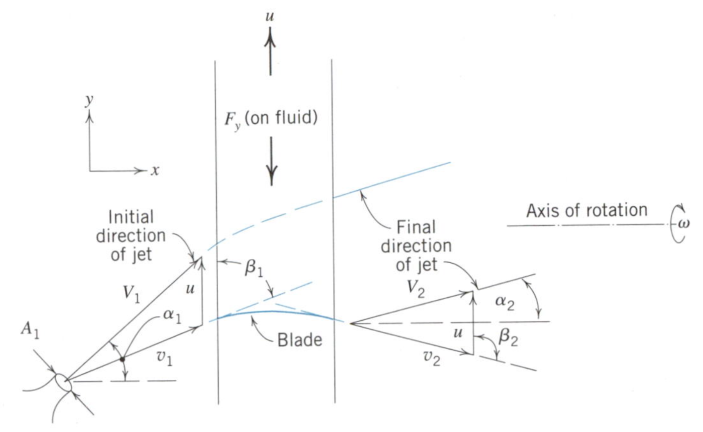

Fig. 12.9a

Fig. 12.9b Steam turbine rotor and blade cascades. (Courtesy of DELAVAL Turbine, Inc.)

Another and more general type of free-jet turbine (Fig. 12.9) may be successfully analyzed by using the energy and impulse-momentum principles. In Fig. 12.9*a* a jet plays upon a series of blades mounted on the periphery of a rotor wheel. The sketch represents, for example, one stage of an impulse steam turbine (Fig. 12.9*b*). The power P transmitted from jet to blade system, called a *cascade*, is (neglecting friction) exactly that lost from the jet (see Eq. 5.8):

$$P = \frac{Q\rho(V_1^2 - V_2^2)}{2} \tag{12.14}$$

which shows that $V_2 < V_1$. This power is calculable from $P = F_y u$, in which F_y is the working component of force exerted by the jet on the moving blade system. From the impulse-momentum principle, F_y (on the fluid) is given by

$$\Sigma F_y = -F_y = (V_2 \sin \alpha_2 - V_1 \sin \alpha_1)\rho Q \tag{12.15}$$

showing that $V_2 \sin \alpha_2 < V_1 \sin \alpha_1$. These requirements (without regard to the details of the blade geometry) justify the relative magnitudes and positions of V_1 and V_2 shown in Fig. 12.9*a*.

Suitable blades to produce these changes of velocity may now be designed by observing that the absolute velocity (V) of the fluid is the vector sum of relative velocity (v) and blade velocity (u), the relative velocity being everywhere tangent to the blade (that is, the fluid does not separate from the blade) for good design. Accordingly, the velocity triangles are as shown and the inlet and exit blade angles, β_1 and β_2, determined. A further requirement of a frictionless system is that the relative velocities v_1 and v_2 must be equal, as in Fig. 12.6.

The foregoing equations and requirements allows the preliminary design of free-jet turbines to produce a given power at a given flowrate and required speed; this is illustrated in the following problems.

ILLUSTRATIVE PROBLEM 12.1

An impulse turbine of 6 ft diameter is driven by a jet of water 2 inches in diameter moving at 200 ft/s. Calculate the force on the blades and the horsepower developed at 250 rpm. The blade angles are 150°.

SOLUTION

We calculate the flowrate first as

$$Q = AV = \frac{\pi}{4}\left(\frac{2 \text{ in}}{12}\right)^2 \times 200 \text{ ft/s} = 4.36 \text{ ft}^3\text{/s}$$

Next, we must find the blade speed u.

$$\begin{aligned} u &= \text{speed(rad/s)} \times \text{radius of rotation} \\ &= (250 \text{ rev/m}) \times (2\pi \text{ rad/rev})/(60 \text{ s/m}) \times \frac{6 \text{ ft}}{2} \\ &= 78.5 \text{ ft/s} \end{aligned}$$

From Eq. 12.9, using a ρ-value of 1.936 slug/ft^3 (see Appendix 2), we can compute the force on the blades.

$$\begin{aligned} -F_x &= -(V_1 - u)(1 - \cos \beta)\rho Q \qquad (12.9) \\ &= -(200 \text{ ft/s} - 78.5 \text{ ft/s})(1 - \cos 150°) \\ &\quad \times 1.936 \text{ slug/ft}^3 \times 4.36 \text{ ft}^3\text{/s} \\ &= -1\,914 \text{ lb} \\ F_x &= 1\,914 \text{ lb} \bullet \end{aligned}$$

The power delivered to the turbine can be computed two ways:

Method 1

Using (force) × (velocity),

$$P = \frac{F_x \times u}{550} = \frac{1\,914 \text{ lb} \times 78.5 \text{ ft/s}}{550} = 273 \text{ hp}$$

Method 2

We will use Eq. 12.14 to compute the power but first we must find the relative velocities v_1 and v_2, and the absolute velocity at the exit V_2. From the previous reading material we have

$$v_1 = v_2 = V_1 - u = 200 \text{ ft/s} - 78.5 \text{ ft/s} = 121.5 \text{ ft/s}$$

Now, solving the velocity triangle below for V_2,

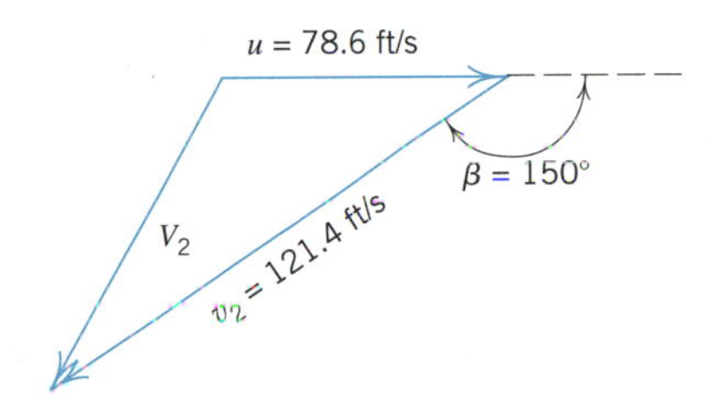

Using the law of cosines,

$$V_2^2 = u^2 + v_2^2 - 2uv_2 \cos(180 - \beta)$$

$$= (78.5 \text{ ft/s})^2 + (121.5 \text{ ft/s})^2 - 2 \times 78.5 \text{ ft/s} \times 121.5 \text{ ft/s} \times \cos 30°$$

$$= 4\,405$$

$$\text{V}_2 = 66.4 \text{ ft/s}$$

Substituting these values into Eq. 12.14,

$$P = \frac{Q\rho(V_1^2 - V_2^2)}{2 \times 550} \tag{12.14}$$

$$= \frac{4.36 \text{ ft}^3/\text{s} \times 1.936 \text{ slug/ft}^3[(200 \text{ ft/s})^2 - (66.4 \text{ ft/s})^2]}{2 \times 550}$$

$$= 273 \text{ hp}$$ ●

ILLUSTRATIVE PROBLEM 12.2

A free-jet impulse turbine is to produce 74.6 kW at a blade speed of 23 m/s. A water jet having a velocity $V_1 = 46$ m/s, a diameter = 51 mm, and an $\alpha_1 = 60°$ is used to drive the machine. Calculate the required blade angles β_1 and β_2.

SOLUTION

To see the relationship among the α and β angles, refer to Fig. 12.9*a*. The velocity triangles applicable to this case are shown below.

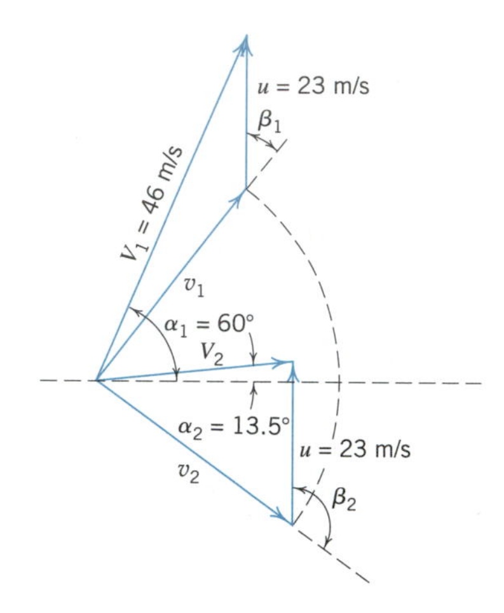

First, we calculate the flowrate.

$$Q = AV = \left(\frac{\pi}{4}\right)(0.051\ \text{m})^2 \times 46\ \text{m/s} = 0.094\ \text{m}^3/\text{s}$$

Referring to the force × velocity method for computing power, we can reverse the equation to find the force F_y on the blades.

$$P = \frac{F_y \times u}{1\,000} \quad \text{or} \quad F_y = \frac{1\,000 \times P}{u} = \frac{1\,000 \times 74.6\ \text{kW}}{23\ \text{m/s}} = 3\,240\ \text{N}$$

Using Eq. 12.14 for power, we can solve for the absolute velocity of the fluid leaving the blades.

$$P = \frac{Q\rho(V_1^2 - V_2^2)}{2} \tag{12.14}$$

$$V_2 = \sqrt{V_1^2 - \frac{2P}{Q\rho}} = \sqrt{(46\ \text{m/s})^2 - \frac{2 \times 74.6 \times 1\,000\ \text{W}}{0.094\ \text{m}^3/\text{s} \times 1\,000\ \text{kg/m}^3}}$$

$$= 23.0\ \text{m/s}$$

We now utilize Eq. 12.15 to find α_2

$$-F_y = (V_2 \sin \alpha_2 - V_1 \sin \alpha_1)\rho Q \tag{12.15}$$

$$-3\,240\ \text{N} = (23.0\ \text{m/s} \times \sin \alpha_2 - 46\ \text{m/s} \times \sin 60°) \times 0.094\ \text{m}^3/\text{s} \times 1\,000\ \text{kg/m}^3$$

Solving for α_2 gives

$$\alpha_2 = 13.5°$$

We can solve the velocity triangle at the exit of the blades to obtain β_2 and v_2 giving a value for $\beta_2 = 128.3°$. Again recognizing that $v_1 = v_2$, we can solve the velocity triangle at the entrance for β_1. The result is $\beta_1 = 54°$. ●

12.3 REACTION TURBINE AND CENTRIFUGAL PUMP

One of the most important engineering applications of the impulse-momentum principle is the design of turbines, pumps, and other turbomachines such as fluid drives and torque converters, all of which involve very complex three-dimensional flowfields beyond the scope of this book. However, the principle may be applied to an assumed two-dimensional flowfield in a reaction turbine and centrifugal pump (Figs. 12.10 and 12.11) to demonstrate the method and to gain some basic understanding of the design and operation of such machines.

Fig. 12.10 Reaction turbine runner—58 MW high pressure Francis turbine for the Rosshag power station of the Oesteneichische Tauemkroftwerke AG (operates under 672 m head). (Courtesy of Escher Wyss Limited, Zurich, Switzerland.)

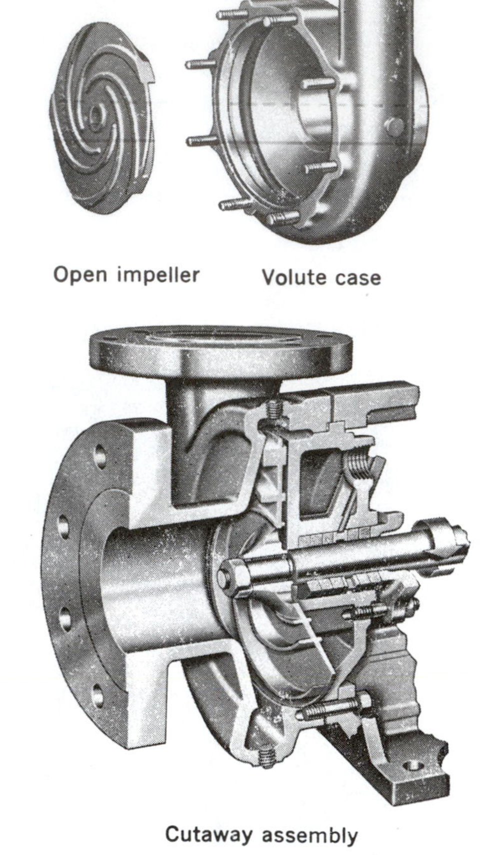

Fig. 12.11 Centrifugal pump. (Courtesy of Peerless Pump Division, FMC Corporation.)

Figure 12.12 is a definition sketch for idealizations of both of these machines, which feature a moving blade system symmetrical about the axis of rotation. This blade system is the essential part of the rotating element of the machine, called the *runner* for the turbine and the *impeller* for the pump. As fluid flows through these blade systems, the tangential component of the absolute velocity changes, decreasing through the runner of the turbine, increasing through the impeller of the pump. In constructing control surfaces (concentric circles) around the blade systems, section 1 becomes the inlet and section 2 the outlet of the control volume for both machines. The impulse-momentum principle may now be applied. But, while the other analyses above dealt with linear momentum, now the moment of momentum is the key to the analysis. By resolving the velocities (**V**) into tangential and radial components, the moment of the momentum vectors may be easily computed because the radial components of velocity pass through the center of moments; the result (for both types of machines) is

$$(\Sigma Q\rho V_t)_{out} - (\Sigma Q\rho V_t)_{in} = -Q\rho(r_2 V_{t_2} - r_1 V_{t_1}) \quad \text{(clockwise positive)} \quad (12.16)$$

The forces exerted on the fluid within the control volume are produced by the pressures on sections 1 and 2, and by the blades of the moving system; as the former are wholly

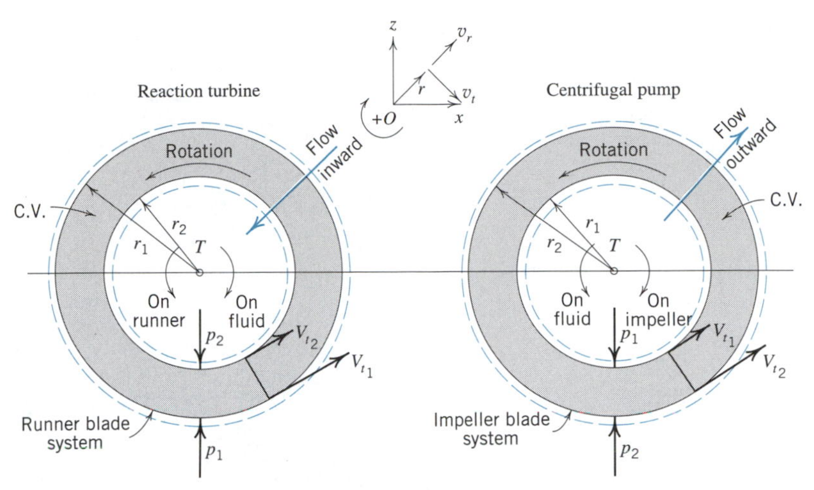

Fig. 12.12

radial, they can have no moment about O. The radial components of the latter also cancel for the same reason, leaving only the tangential components exerted by the blade system on the fluid to be considered. Although these forces are not identified as to magnitude and location, their total effect is a *torque*, T, *by the blade system on the fluid*. Applying Eq. 6.14,

$$T = -Q\rho(r_2V_{t_2} - r_1V_{t_1}) \tag{12.17}$$

From the relative magnitude of $r_1V_{t_1}$ and $r_2V_{t_2}$ it is evident that T (on the fluid) is clockwise ($T > 0$) for the turbine and counterclockwise ($T < 0$) for the pump. The *torques on runner and impeller will necessarily be in opposite directions.*

The change of unit energy, E, associated with either of these machines may be calculated from Eq. 12.17 by recalling from mechanics that $P = T\omega$ and from Eq. 5.8 that $P = Q\gamma E$. The result is

$$E = \frac{\omega}{g_n}(r_1V_{t_1} - r_2V_{t_2}) \tag{12.18}$$

which gives some insight into the relation among the internal dynamics of the machine and the head, E_P or E_T, which was emphasized in Section 5.5. For a pump, where $V_{t_2}r_2 > V_{t_1}r_1$, Eq. 12.18 is written

$$E_P = \frac{\omega}{g_n}[V_{t_2}r_2 - V_{t_1}r_1] \tag{12.19}$$

but for the turbine, where $V_{t_2}r_2 < V_{t_1}r_1$,

$$E_T = \frac{\omega}{g_n}[V_{t_1}r_1 - V_{t_2}r_2] \tag{12.20}$$

The alternative to this adjustment in the equation is to set up a special sign convention; this is convenient in more advanced analysis but hardly necessary here.

An idealized two-dimensional reaction turbine is shown in Fig. 12.13 with accompanying velocity diagrams. Fixed guide vanes exert a torque on the fluid which gives it a

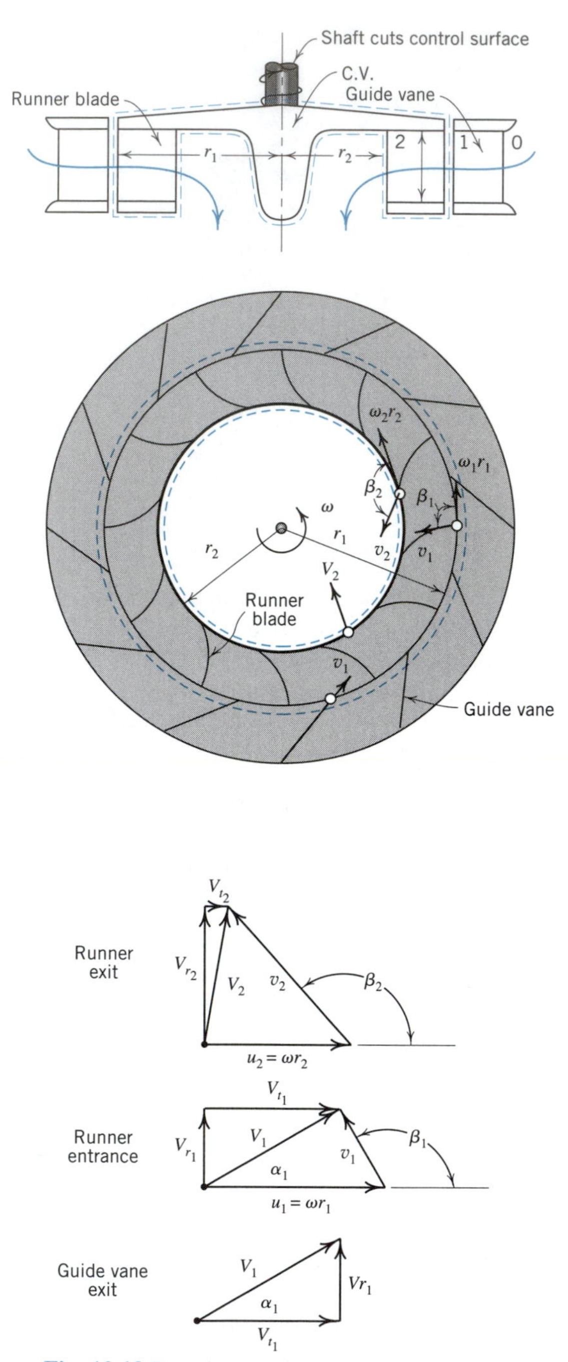

Fig. 12.13 Francis-type reaction turbine schematic.

tangential velocity component; because these guide vanes do not move, this torque does no work. The fluid then passes through the moving runner, through which the tangential component of velocity decreases, producing a torque according to Eq. 12.17. As in Section 12.2, the blades of the runner must be designed for smooth flow to accomplish this. Blade angles are determined from the velocity diagrams as before, the size of such diagrams being dictated by the equation of continuity ($q = 2\pi r_1 V_{r_1} = 2\pi r_2 V_{r_2}$), the angular speed of the runner through $u = \omega r$, and the guide vane angle, α. Important features of the velocity triangles are: (1) the increase of V_r through the runner and (2) the same V_1 at guide vane exit and runner entrance. The latter indicates no abrupt change in magnitude and direction of the velocity as it passes from guide vanes to moving runner; because this is imperative for good design, runner blades are shaped to accomplish it.

A section through an idealized two-dimensional centrifugal pump is shown in Fig. 12.14. With no guide vanes upstream from the impeller, the tangential component of absolute velocity is zero at the inlet and it increases through the impeller as torque is exerted on the fluid. This power is transferred from impeller to fluid and the energy of the fluid increases; both pressure and kinetic energies of the fluid increase through the impeller. As in the turbine, blade angles required to produce specified heads and flowrates for specified rotational speeds may be deduced from the velocity triangles. After the fluid leaves the impeller, it enters the pump (volute) casing where the flow is decelerated, converting the high kinetic energy of the fluid into pressure head, and the fluid is channeled (with as few losses as possible) to the outlet pipe.

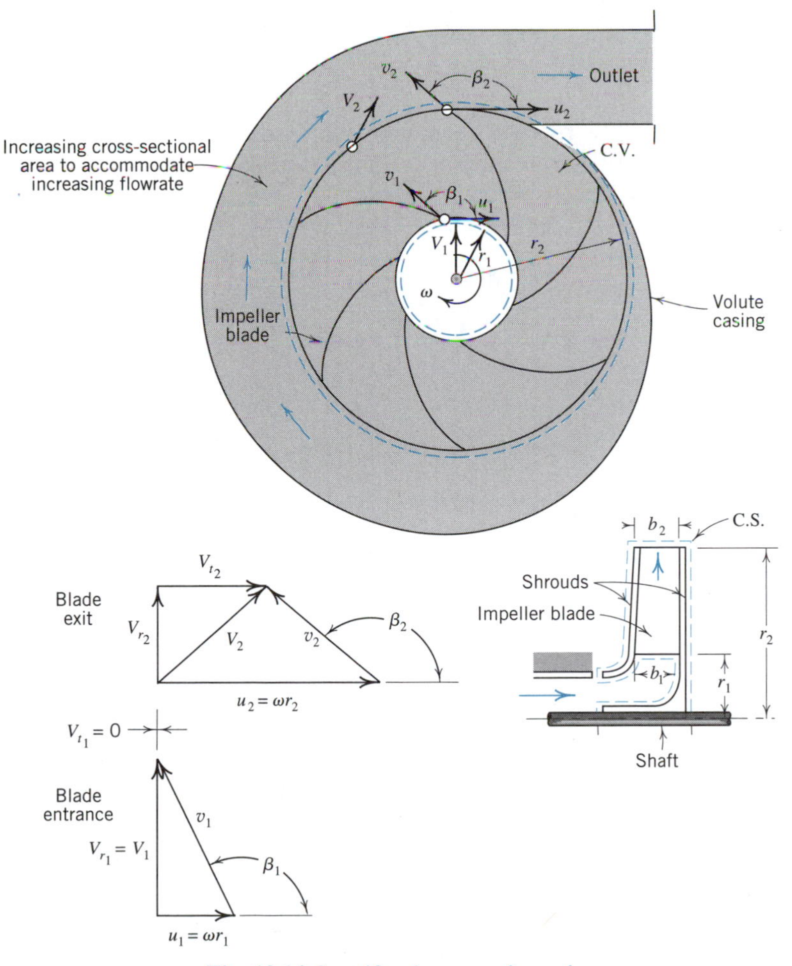

Fig. 12.14 Centrifugal pump schematic.

ILLUSTRATIVE PROBLEM 12.3

A two-dimensional reaction turbine has $r_1 = 5$ ft, $r_2 = 3.5$ ft, $\beta_1 = 60°$, $\beta_2 = 150°$, and a thickness B of 1.0 ft parallel to the axis of rotation. With a guide vane angle of 15° and a flowrate of 333 cfs of water, calculate the required speed of the runner for smooth flow at the inlet. For this condition also calculate the torque exerted on the runner, the power developed, the energy extracted from each pound of fluid, and the pressure drop through the runner.

SOLUTION

First we sketch the velocity diagrams at both the inlet and the exit of the turbine.

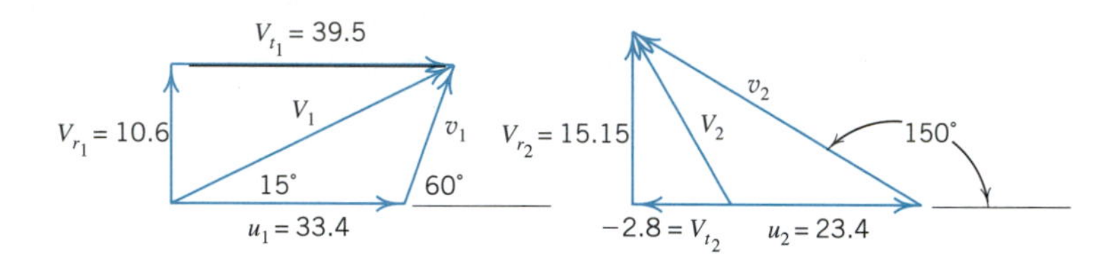

To satisfy continuity, the product of the inlet and exit flow areas and the radial component of the absolute velocity must equal the flowrate.

$$Q = 2\pi r \times B \times V_r$$

$$V_{r_1} = \frac{Q}{2\pi r_1 B} = \frac{333 \text{ cfs}}{2\pi \times 5 \text{ ft} \times 1.0 \text{ ft}} = 10.6 \text{ ft/s}$$

$$V_{r_2} = \frac{Q}{2\pi r_2 B} = \frac{333 \text{ cfs}}{2\pi \times 3.5 \text{ ft} \times 1.0 \text{ ft}} = 15.15 \text{ ft/s}$$

From the velocity triangles,

$$V_{t_1} = 10.6 \text{ ft/s} \times \cot 15° = 39.5 \text{ ft/s}$$

$$u_1 = \omega r_1 = 39.5 \text{ ft/s} - 10.6 \text{ ft/s} \times \tan 30° = 33.4 \text{ ft/s}$$

$$\omega = \frac{33.4 \text{ ft/s}}{5 \text{ ft}} = 6.68 \text{ rad/s} \quad \text{or} \quad 63.8 \text{ rpm} \bullet$$

$$u_2 = \omega r_2 = 6.68 \text{ rad/s} \times 3.5 \text{ ft} = 23.4 \text{ ft/s}$$

$$V_{t_2} = 23.4 \text{ ft/s} - 15.15 \text{ ft/s} \times \cot 30° = -2.8 \text{ ft/s}$$

$$T = -[-2.8 \text{ ft/s} \times 3.5 \text{ ft} - 39.5 \text{ ft/s} \times 5 \text{ ft}] \times 333 \text{ ft}^3\text{/s} \times 1.936 \text{ slugs/ft}^3$$

$$T = 133\,300 \text{ ft·lb} \bullet$$

The power developed is

$$\text{Horsepower} = \frac{T\omega}{550} = \frac{133\,300 \text{ ft·lb} \times 6.68 \text{ rad/s}}{550} = 1\,620 \text{ hp} \bullet$$

From the power equation,

$$P = \frac{Q\gamma E_T}{550} \tag{5.8}$$

$$E_T = \frac{550 \times P}{Q\gamma} = \frac{550 \times 1\,620 \text{ hp}}{333 \text{ ft}^3/\text{s} \times 62.4 \text{ lb/ft}^3} = 42.9 \text{ ft}\cdot\text{lb/lb} \bullet$$

From the vector diagrams above, $V_1 = 40.8$ ft/s, $V_2 = 15.4$ ft/s, and, applying the work-energy equation between sections 1 and 2,

$$z_1 + \frac{p_1}{\gamma} + \frac{V_1^2}{2g_n} = z_2 + \frac{p_2}{\gamma} + \frac{V_2^2}{2g_n} + E_T \tag{5.7}$$

$$0 + \frac{p_1}{\gamma} + \frac{(40.8 \text{ ft/s})^2}{2g_n} = \frac{p_2}{\gamma} + \frac{(15.4 \text{ ft/s})^2}{2g_n} + 42.9 \text{ ft}\cdot\text{lb/lb}$$

Solving for $p_1 - p_2$,

$$p_1 - p_2 = 9.0 \text{ psi} \bullet$$

ILLUSTRATIVE PROBLEM 12.4

A two-dimensional centrifugal pump impeller has $r_1 = 0.3$ m, $r_2 = 1.0$ m, $\beta_1 = 120°$, $\beta_2 = 135°$, and a thickness $B = 0.1$ m parallel to the axis of rotation. If it delivers 2 m³/s with no tangential velocity component at the entrance, what is the rotational speed? For this condition, calculate torque and power to the machine, the energy given to each newton of water, and the pressure rise through the impeller.

SOLUTION

First, we sketch the velocity diagrams at the entrance and exit of the impeller.

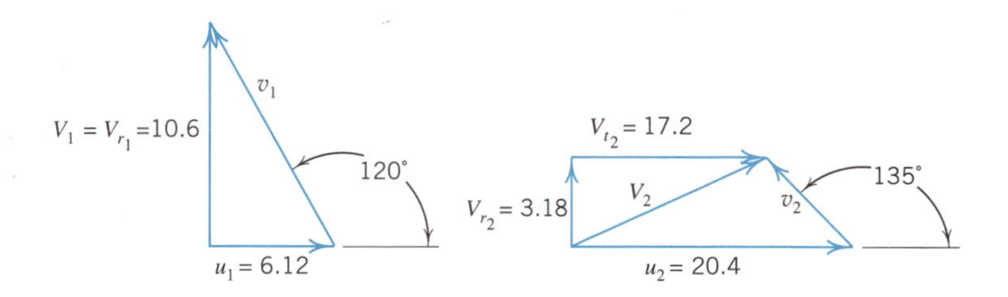

Now, we use continuity to find the radial components of the absolute velocity.

$$V_{r_1} = V_1 = \frac{Q}{2\pi r_1 B} = \frac{2 \text{ m}^3/\text{s}}{2\pi \times 0.3 \text{ m} \times 0.1 \text{ m}} = 10.6 \text{ m/s}$$

$$V_{r_2} = \frac{Q}{2\pi r_2 B} = \frac{2 \text{ m}^3/\text{s}}{2\pi \times 1.0 \text{ m} \times 0.1 \text{ m}} = 3.18 \text{ m/s}$$

From the velocity triangles above,

$$u_1 = 10.6 \text{ m/s} \times \tan 30° = 6.12 \text{ m/s}$$

$$\omega = \frac{u_1}{r_1} = \frac{6.12 \text{ m/s}}{0.3 \text{ m}} = 20.4 \text{ rad/s} = 195 \text{ rpm} \bullet$$

$$u_2 = \omega r_2 = 20.4 \text{ rad/s} \times 1.0 \text{ m} = 20.4 \text{ m/s}$$

$$V_{t_2} = 20.4 \text{ m/s} - \frac{3.18 \text{ m/s}}{\tan 45°} = 17.2 \text{ m/s}$$

The torque can be computed from Eq. 12.17.

$$T = -Q\rho(r_2 V_{t_2} - r_1 V_{t_1}) \tag{12.17}$$

$$= -2 \text{ m}^3/\text{s} \times 1\,000 \text{ kg/m}^3 \times (1.0 \text{ m} \times 17.2 \text{ m/s} - 0)$$

$$T = -34\,400 \text{ N} = -34.4 \text{ kN} \bullet$$

The energy added to each newton of fluid can be computed with the help of Eq. 5.8b.

$$E_P = \frac{P}{Q\gamma} = \frac{-T\omega}{Q\gamma} = \frac{-(-34.4 \times 10^3 \text{ N}\cdot\text{m}) \times 20.4 \text{ rad/s}}{2 \text{ m}^3/\text{s} \times 9\,800 \text{ N/m}^3}$$

$$= 35.8 \text{ m} \quad \text{or} \quad 35.8 \text{ J/N} \bullet$$

To find the power, we use Eq. 5.8b.

$$P = \frac{Q\gamma E_P}{1\,000} = \frac{2 \text{ m}^3/\text{s} \times 9\,800 \text{ N/m}^3 \times 35.8 \text{ m}}{1\,000} = 700 \text{ kW} \bullet$$

The work-energy Eq. 5.7 is used to calculate the pressure increase across the pump.

$$z_1 + \frac{p_1}{\gamma} + \frac{V_1^2}{2g_n} + E_P = z_2 + \frac{p_2}{\gamma} + \frac{V_2^2}{2g_n} \tag{5.7}$$

$$0 + \frac{p_1}{\gamma} + \frac{(10.6 \text{ m/s})^2}{2g_n} + 35.8 \text{ m} = 0 + \frac{p_2}{\gamma} + \frac{(17.5 \text{ m/s})^2}{2g_n}$$

Solving the above equation for the pressure increase gives

$$p_2 - p_1 = 254 \text{ kPa} \bullet$$

12.4 PRACTICAL APPLICATIONS

We have seen in Section 9.10 how a pump (or pumps) fits into a pipeline once the pump and pipeline have been selected. We now broaden this application to develop a process for the preliminary selection of a pumping system, given a pipeline design situation.[10] This process begins by selecting a pumping head from the system demand curve. It in-

[10]ASCE, *Pressure Pipeline Design for Water and Wastewater*, Sect. 4, 1992.

volves using specific speed to assist in deciding how many pumps are needed to work in parallel or how many stages may be required for each pump. And it will cover vertical placement of the pump to prevent cavitation over the range of operating conditions. Once the pumping configuration has been decided and a pump selected, we will address trimming the impeller to precisely match the required head increase and discharge requirements to the system. Finally, the power requirements for the motor or engine which will drive the pump will be determined. Illustrative Problems 12.5 and 12.6 demonstrate this process.

ILLUSTRATIVE PROBLEM 12.5

An engineer for a water district has determined that the pumping plant for a new pipeline will be required to produce a head increase of 360 ft while pumping 18 000 gpm of water. In addition, to provide redundancy in case of pump breakdown, routine maintenance, or substantial variations in the flow demand, a typical design would employ a minimum of four pumps operating in parallel. Using specific speed concepts as a guide, recommend a pumping configuration and a level of submergence of the pump impellers that will prevent cavitation. Also specify any impeller trimming that will be necessary to match the operating conditions and recommend the size of the electric motor required to power the pumps. The plant is situated at sea level.

SOLUTION

We first assume that the minimum number of four pumps will be acceptable and compute the specific speed on that basis.

$$N_S = \frac{NQ^{1/2}}{(g_n E_P)^{3/4}} \tag{12.4b}$$

Estimating the pump speed to be 1 750 rpm or 183.3 rad/s and noting that the flow through each pump will be 4 500 gpm,

$$N_S = \frac{183.3 \text{ rad/s} \times (4\,500 \text{ gpm}/449 \text{ gpm/cfs})^{1/2}}{(32.2 \times 360 \text{ ft})^{3/4}} = 0.52$$

From Fig. 12.2, it appears that the pump required is on the borderline between radial flow impellers (centrifugal pumps) and Francis impellers (turbine pumps). So we will investigate both possibilities.

Centrifugal Pumps

Searching the pump manufacturer's catalogs for a single centrifugal pump which will produce 360 ft of head at 4 500 gpm, we find the pump represented by the characteristic diagram below will fit the requirements reasonably well. At the required 4 500 gpm, the pump will generate 375 ft of head with the $19\frac{3}{8}$ in. impeller; so some trimming of the impeller will be necessary.

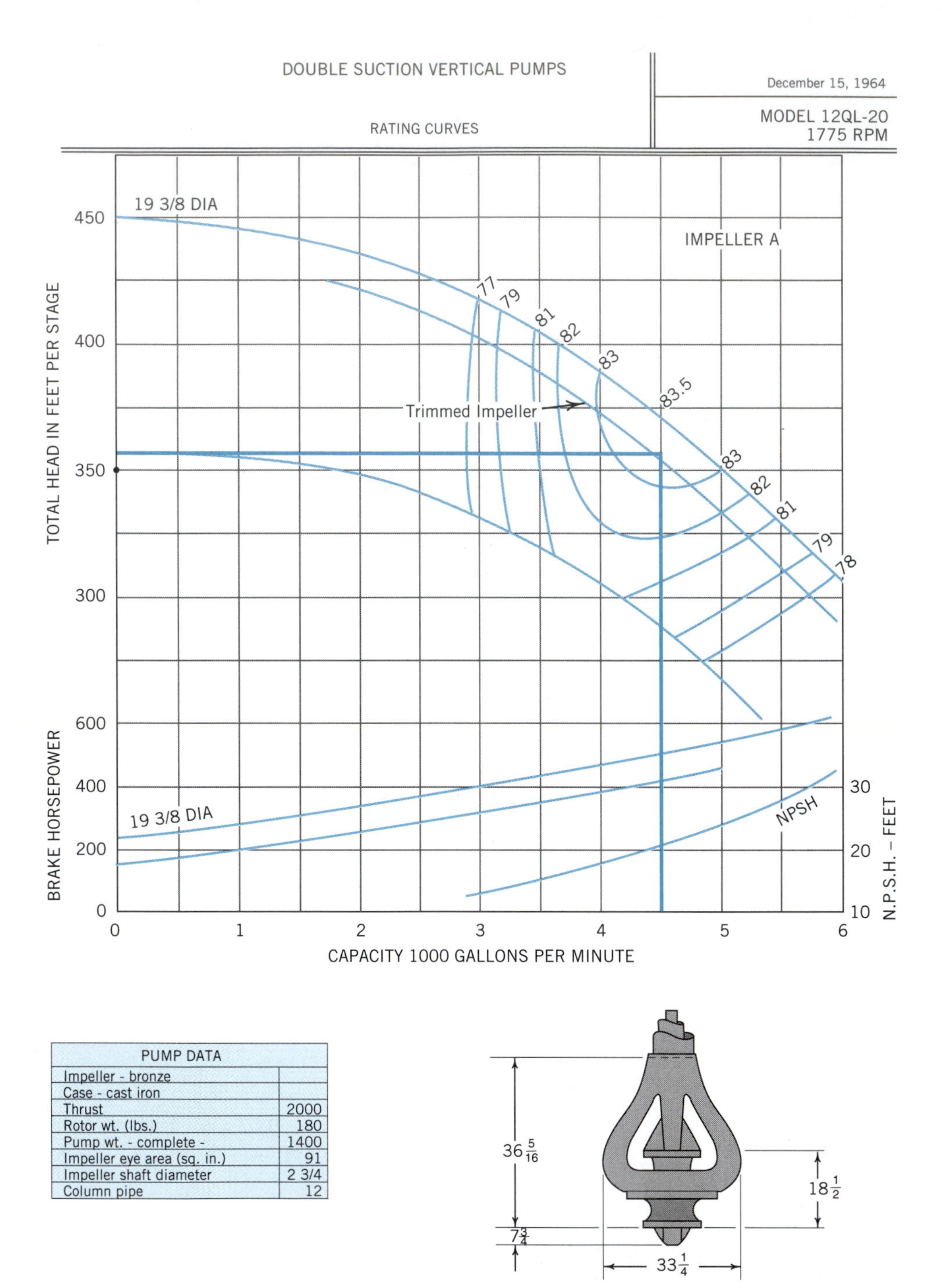

PUMP DATA	
Impeller - bronze	
Case - cast iron	
Thrust	2000
Rotor wt. (lbs.)	180
Pump wt. - complete -	1400
Impeller eye area (sq. in.)	91
Impeller shaft diameter	2 3/4
Column pipe	12

(By permission of Ingersoll–Dresser Pump Company.)

From Section 12.1, we know that $E_P(H/g_n)$ varies as the square of the diameter and Q as the cube. That is,

$$E_P \propto D^2 \qquad Q \propto D^3$$

Designating the original curve for the $19\frac{3}{8}$ in. impeller as E_{P_0} vs. Q_0, we will generate a new characteristic curve, E_{P_1} versus Q_1, which will provide the required flowrate of 4 500 gpm at a head of 360 ft. The procedure for accomplishing this is to select a trial value of D_1/D_0, then draw a revised curve of E_{P_1} versus Q_1 by selecting several values of E_{P_0} and Q_0 from the original curve and multiplying them by $(D_1/D_0)^2$ and $(D_1/D_0)^3$, respectively. After plotting and checking to see if the new curve matches the head and flowrate requirements, adjust D_1/D_0 and repeat as necessary until a satisfactory trimming is determined. In this case, $D_1/D_0 = 0.987$ produces the desired characteristics and the trimmed impeller curve is plotted on the pump diagram.

Turning to the submergence requirements, we note that the required *NPSH* is 20 ft at the design flowrate of 4 500 gpm. Because impeller trimming is done at the outlet of the impeller and cavitation occurs at the inlet to the impeller, the trimming has no effect on the *NPSH* requirements. However, there is a chance that the discharge of the pump may temporarily exceed the design discharge, so we will use a conservative value of *NPSH* of 30 ft as a safety factor.

Referring to Section 12.1, we utilize the expression for *NPSH*

$$NPSH = \frac{p_0}{\gamma} - \frac{p_v}{\gamma} - z_s - h_{L_{entr}} \tag{12.6}$$

Assuming the $h_{L_{entr}}$-value is included in the *NPSH*-value and recognizing that the atmospheric pressure head at sea level is 34 ft of water and the vapor pressure head is approximately 1 ft of water,

$$30 = 34 - 1 - z_s$$

$$z_s = 3 \text{ ft} \bullet$$

This value indicates that the pump could be placed 3 ft above the lowest anticipated water level in the pump well, sump, or forebay. Depending on whether the pump has a submersible impeller or is supplied by a suction line, appropriate attention should be exercised to prevent vortexing and air-sucking at the entrance to the impeller.

Power requirements are read from the characteristic diagram as 510 hp for the impeller size selected. Trimming the impeller does reduce the power requirements, but to cover the contingency of temporary flows above the design flowrate, we will not consider this reduction. Rather, we will specify an increased power of 575 hp or 600 hp for the motor to cover the contingency.

Turbine Pumps

Our review of the pump manuals reveals a turbine pump which will produce the required flowrate, and with multistaging, the required head without trimming. The characteristic diagram for this pump is shown on the next page. We will use the $12\frac{5}{8}$ in. impeller with four stages.

The *NPSH* requirements are 39 ft at 4 500 gpm but to be safe, we will use 45 ft. Again, turning to the expression for *NPSH*,

$$NPSH = \frac{p_0}{\gamma} - \frac{p_v}{\gamma} - z_s - h_{L_{entr}} \tag{12.6}$$

$$45 = 34 - 1 - z_s$$

$$z_s = -12 \text{ ft} \bullet$$

The bottom impeller of the multistage pump must be placed 12 ft below the sump water surface. We further note on the characteristic diagram that a curve of submergence is given. For the same flowrate that gave us an *NPSH*-value of 45 ft, we read 13 ft from the submergence curve. The calculated value of 12 ft for the submergence reasonably checks the 13-ft value found from the submergence curve. So, either approach can be used for pumping water at normal temperatures at sea level.

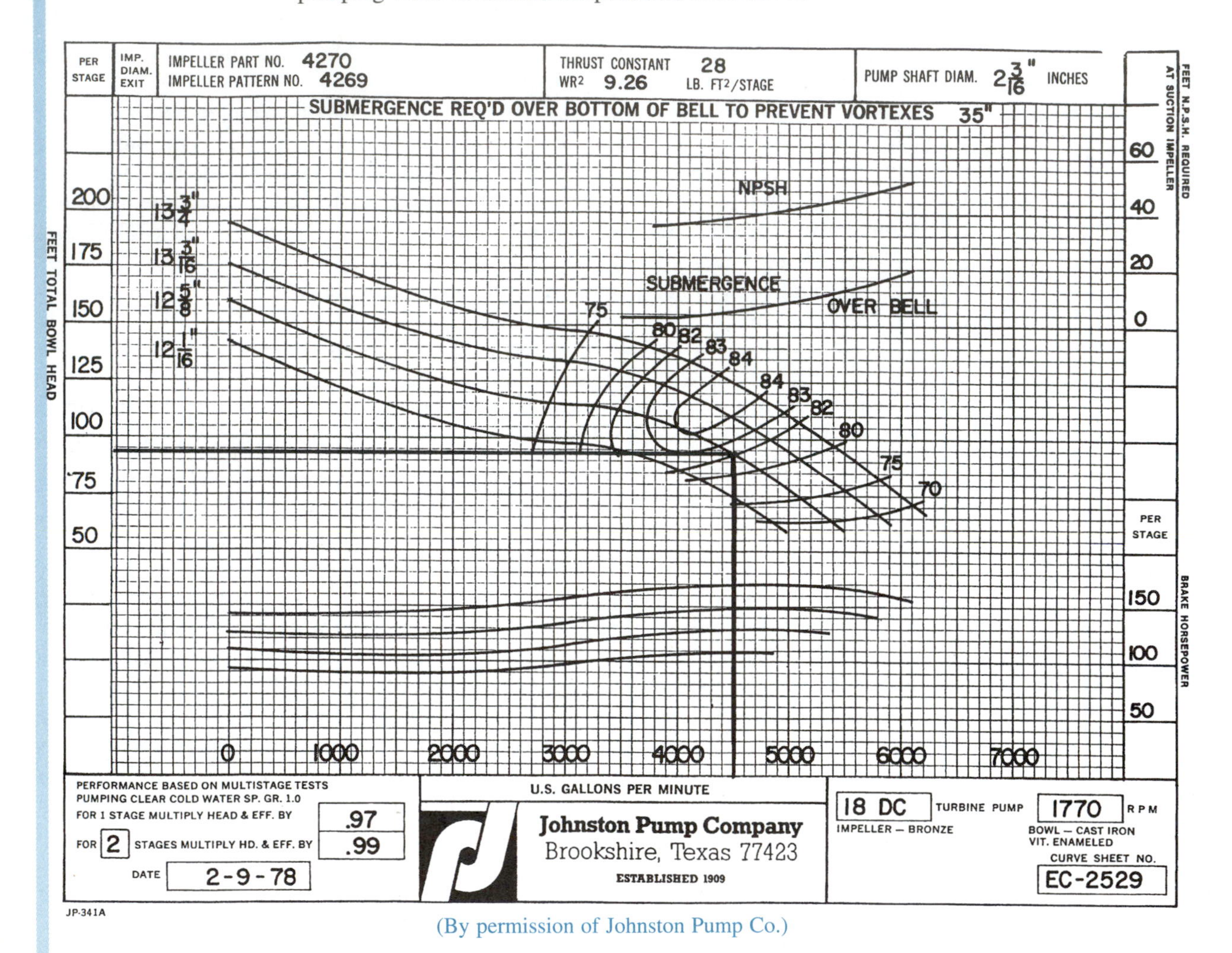

(By permission of Johnston Pump Co.)

Power requirements per stage are 126 hp giving a total power requirement of 504 hp. Recognizing that the maximum power occurs at the design flowrate, we need no safety factor for flowrate variation. A 525 hp motor would be adequate to drive the pump selected.

ILLUSTRATIVE PROBLEM 12.6

Because of unfavorable topography, a canal does not have sufficient gradient to convey the required flowrate to its destination. One solution to the problem under consideration is to terminate the canal, move a short distance uphill and begin another canal which will have the necessary gradient available to deliver the required flowrate. This solution requires pumping the water from the lower canal to the upper canal. If the canal design flowrate is 290 cfs of water and the pump head required to lift the water up to the higher canal is 13 ft, select a pumping configuration that will satisfy these requirements. The canal is located at elevation 210 ft above mean sea level.

SOLUTION

Once again we begin with the assumption that a minimum of four pumps will be necessary to provide an acceptable level of redundancy. Computing the specific speed for four pumps operating in parallel, expecting an average speed of 700 rpm (73.3 rad/s), we find

$$N_S = \frac{NQ^{1/2}}{(g_n E_P)^{3/4}} \tag{12.4b}$$

$$N_S = \frac{73.3 \text{ rad/s} \times (290 \text{ cfs/4 pumps})^{1/2}}{(32.2 \times 13 \text{ ft})^{3/4}} = 6.74$$

From Fig. 12.2, it appears our specific speed is beyond the value corresponding to high efficiency propeller (axial flow) pumps. This suggests that we should use more pumps in parallel. Recomputing specific speed with six pumps in parallel gives

$$N_S = \frac{73.3 \text{ rad/s} \times (290 \text{ cfs/6 pumps})^{1/2}}{(32.2 \times 13 \text{ ft})^{3/4}} = 5.51$$

This N_S-value is at the upper end of the propeller pump range, so we search our pump catalogs for a propeller pump which will provide 290 cfs × 449 gpm/cfs ÷ 6 pumps = 21 700 gpm at 13 ft of head. We find the pump whose characteristics are shown on the next page produces 22 000 gpm at 13 ft of head with the 18 in. impeller.

To determine the depth below the water surface which we must place the pump impeller, we turn again to the *NPSH*-value from the pump diagram. At the design flowrate of 21 700 gpm, an *NPSH* of 29 ft is required. However, we will use a value of 35 ft as a safety measure.

$$NPSH = \frac{p_0}{\gamma} - \frac{p_v}{\gamma} - z_s - h_{L_{entr}}$$

$$35 \text{ ft} = 34 \text{ ft} - 1 \text{ ft} - z_s - 0 \tag{12.6}$$

$$z_s = -2 \text{ ft} \bullet$$

The pump impeller must be kept 2 ft below the water surface to prevent cavitation. We note from the characteristic diagram that the manufacturer recommends almost 7 ft of submergence for the impellers at the flowrate corresponding to *NPSH* = 35 ft. This is to prevent vortexing and air sucking at high flowrates. With this factor controlling, we will have to place the pump impellers 7 ft below the canal water surface.

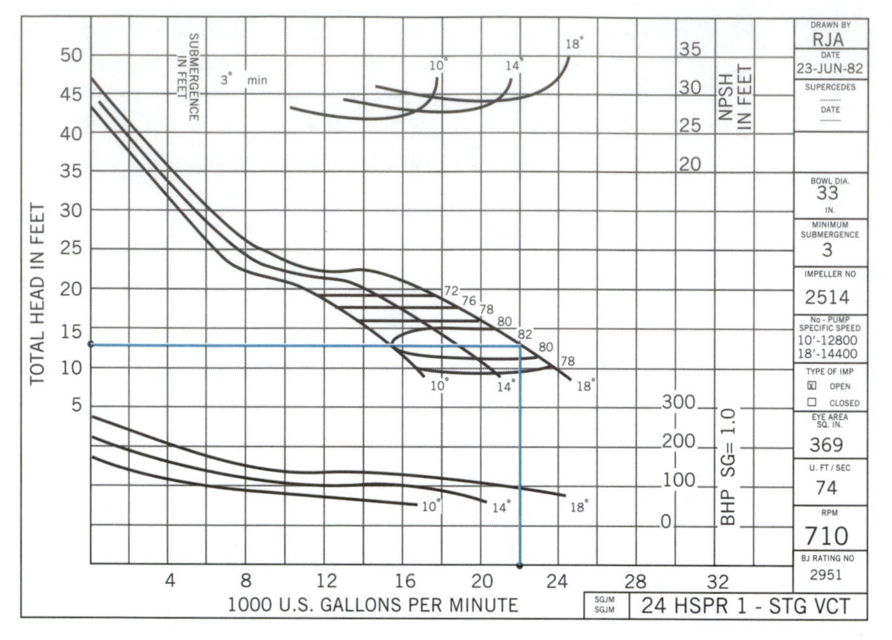

(By permission of BW/IP International, Inc.—Pump Division.)

Selection of a hydraulic turbine configuration for a power generation site follows closely that of a pump configuration determination. The specific speed is used to provide a preliminary configuration once the available head and discharge data and the anticipated power requirements are determined. Illustrative Problems 12.7 and 12.8 demonstrate this initial design process.

ILLUSTRATIVE PROBLEM 12.7

A potential hydroelectric site has an available head of 750 m and the utility company developing the site wants to generate 650 MW. At this high head, the turbines will be Pelton wheels and are anticipated to run at 360 rpm. Compute the turbine configuration and the required flowrate to accommodate this plan.

SOLUTION

We will calculate the specific speed for the site as though only one turbine were operating. But first, we will convert the rotational speed to radians per second.

$$N = 360 \text{ rpm} \times 2\pi \text{ rad/rev} \div 60 \text{ s/m} = 37.7 \text{ rad/s}$$

Now, we use Eq. 12.5b to compute the specific speed.

$$N_S = \frac{NP^{1/2}}{\rho^{1/2}(g_n E_T)^{5/4}} \tag{12.5b}$$

$$= \frac{37.7 \text{ rad/s} \times (650\,000 \times 10^3 \text{W})^{1/2}}{(1\,000 \text{ kg/m}^3)^{1/2}(9.81 \times 750 \text{ m})^{5/4}} = 0.45$$

From Fig. 12.3, this N_S-value is too high for a Pelton wheel so we must consider parallel units. Peak efficiency occurs at about $N_S = 0.2$ (see Fig. 12.3), so we will use that value to determine the number of turbines, N_{units}, in parallel.

$$N_S = 0.2 = \frac{37.7 \text{ rad/s} \times (650\,000 \times 10^3\text{W} \div N_{units})^{1/2}}{(1\,000 \text{ kg/m}^3)^{1/2}(9.81 \times 750 \text{ m})^{5/4}}$$

So, the number of units is

$$N_{units} = 4.97 \qquad \text{or} \qquad 5 \text{ units} \bullet$$

Figure 12.3 indicates that the maximum efficiency to be expected is approximately 91%. Using this figure, we can determine the flowrate required to produce this amount of power.

From Section 12.1,

$$\eta = \frac{P}{Q\gamma H}$$

$$0.91 = \frac{650\,000 \times 10^3 \text{ W}}{Q \times 9\,800 \text{ N/m}^3 \times 750 \text{ m}}$$

So, the required discharge to produce this amount of power is

$$Q = 97.2 \text{ m}^3/\text{s} \bullet$$

ILLUSTRATIVE PROBLEM 12.8

A substantial portion of the flow from a major river is diverted through a power tunnel to supply a projected hydroelectric site downstream. A flow of 4 000 cfs through the tunnel results in a head of 450 ft available at the plant site. Determine the type of turbine most suitable for this application and estimate the power which can be developed at the site. The turbine speed is initially planned as 600 rpm to match the 60 hz generators. Also, to provide some flexibility in meeting power demands, a minimum of four turbines is required.

SOLUTION

First, we will estimate the available water power at the site. From Eq. 5.8a,

$$\text{Power} = \frac{Q\gamma E_P}{550} = \frac{4\,000 \text{ cfs} \times 62.4 \text{ lb/ft}^3 \times 450 \text{ ft}}{550} = 204\,220 \text{ hp}$$

The efficiency of the turbine is estimated as 95% (see Fig. 12.3) so that the power generated will be

$$P = 0.95 \times 204\,220 \text{ hp} = 194\,000 \text{ hp}$$

Now, with four units operating, each unit must produce 48 500 hp while turning at 62.8 rad/s. The specific speed of each unit is

$$N_S = \frac{NP^{1/2}}{\rho^{1/2}(g_n E_T)^{5/4}} \tag{12.5b}$$

$$= \frac{62.8 \text{ rad/s} \times (48\,500 \text{ hp} \times 550 \text{ ft·lb/s/hp})^{1/2}}{(1.94 \text{ slug/ft}^3)^{1/2}(32.2 \times 450 \text{ ft})^{5/4}} = 1.47$$

From Fig. 12.3, it appears this value, while near the upper limit, is still acceptable for Francis type turbines. However, a slightly better efficiency may be realized by using more units. Consequently, we will select *six units*, each generating 32 300 hp and having a specific speed of

$$N_S = \frac{62.8 \text{ rad/s} \times (32\,300 \text{ hp} \times 550 \text{ ft·lb/s/hp})^{1/2}}{(1.94 \text{ slug/ft}^3)^{1/2}(32.2 \times 450 \text{ ft})^{5/4}} = 1.20$$

This specific speed appears to give the highest efficiency for a Francis-type turbine (see Fig. 12.3) and justifies our original assumption of 95% efficiency.

The power generated is expected to be approximately

$$P = \eta \frac{Q\gamma H}{550} = 0.95 \frac{4\,000 \text{ cfs} \times 62.4 \text{ lb/ft}^3 \times 450 \text{ ft}}{550}$$

$$= 194\,000 \text{ hp} \qquad \text{or} \qquad 144.7 \text{ MW} \bullet$$

REFERENCES

ASCE. 1992. *Pressure pipeline design for water and wastewater.* 2nd ed. Sect. 4.

Benaroya, A. 1978. *Centrifugal pumps.* Tulsa, Okla.: Petroleum Pub. Co.

Daugherty, R. L., Franzini, J. B., and Finnemore, E. J. 1985. *Fluid mechanics with engineering applications.* 8th ed. New York: McGraw-Hill.

Dixon, S. L. 1978. *Fluid mechanics, thermodynamics of turbomachinery.* 3rd ed. New York: Pergamon Press.

Hawthorne, W. R., Ed. 1964. *Aerodynamics of turbines and compressors.* Vol. X of *High speed aerodynamics and jet propulsion.* Princeton: Princeton Univ. Press.

Hicks, T. G., and Edwards, T. W. 1971. *Pump application engineering.* New York: McGraw-Hill.

Hydraulic Institute. 1965. *Standards of the hydraulic institute.* 11th ed. New York.

Hydraulic Institute. 1983. *Standards for centrifugal, rotary, and reciprocating pumps.* 14th ed. Cleveland.

Karassik, I. J., Krutzsch, W. C., Fraser, W. H., and Messina, J. P., Eds. 1986. *Pump handbook.* 2nd ed. New York: McGraw-Hill.

Kovalev, N. N. 1965. *Hydroturbines.* Washington, D.C.: Office of Technical Services, U.S. Dept. of Commerce.

Mosonyi, E. 1960, 1963. *Water power development.* Vols. I and II. Budapest: Akadémiai Kiado.

Sanks, R. L., Ed. 1989. *Pumping station design.* Boston: Butterworth Pub.

Shepherd, D. G. 1956. *Principles of turbomachinery.* New York: Macmillan.

PROBLEMS

12.1. Derive an expression for modeling the thrust of geometrically similar screw propellers moving through incompressible fluids if this thrust depends only on propeller diameter, velocity of propeller through the fluid, rotative speed, the density and viscosity of the fluid, and the acceleration due to gravity.

12.2. Find the resulting Π-groups when (*a*) D, ρ and Q or (*b*) H, ρ, and Q are the repeating variables in the analysis of a turbomachine where the relevant variables are P, D, N, Q, H, μ, ρ, and E (see paragraph 3 of Section 12.1). Discuss how to interpret each Π obtained.

12.3. Derive an expression for the modeling of the performance of geometrically similar centrifugal compressors. The relevant variables are flowrate, the outlet to inlet pressure ratio, rotative speed, impeller diameter, and the viscosity, density, elasticity, and specific heat ratio of the fluid.

12.4. Derive an expression for the power of hydraulic machines if this power depends only on the angular speed, size, and surface roughness of the rotating element of the machine, flowrate, and the density and viscosity of the fluid flowing.

12.5.[11] A 1 : 15 scale model of a hydraulic turbine of 4.5 m (15 ft) diameter is found to develop 2.4 kW (3.21 hp), with a flowrate of 280 l/s (10.0 ft^3/s) of water, under a head of 1.14 m (3.75 ft) while rotating at 500 r/min. Calculate the corresponding power, flowrate, head, and speed of the prototype, assuming the same efficiency in model and prototype.

12.6.[11] A centrifugal pump has a 250 mm diameter impeller, which rotates at 1 750 r/min and delivers a flowrate of 0.07 m^3/s of water at a head of 9 m. Calculate the corresponding quantities for a half-scale model of the pump if the efficiencies and fluid of model and prototype are the same.

12.7. A 0.3 m (1 ft) diameter model of a ship propeller rotates at 500 r/min. The velocity well upstream from the model is 1.5 m/s (5 ft/s). The thrust produced is observed to be 133 N (30 lb). If the prototype (of 1.8 m or 6 ft diameter) drives a ship through still water at a speed of 3 m/s (10 ft/s), what must its thrust and rotational speed be for its operation to be dynamically similar to that of its model? Assume the same efficiency and same (fresh) water in model and prototype.

12.8. In a set of geometrically similar pumps the manufacturer changes the impeller diameter by a factor of two between two models. What change in N is required to make H change by a factor of two also? What is the ratio of flowrates under these conditions?

12.9. A Francis turbine is to operate under a head of 46 m and deliver 18.6 MW while running at 150 r/min. The runner diameter is 4 m. A 1 m diameter model is operated in a laboratory under the same head. Find the model speed, power, and flowrate. If model efficiency is 90%, what is prototype efficiency?

12.10. A pump is being built to deliver 120 m (400 ft) head and 5.7 m^3/s (200 ft^3/s) at 200 r/min. A geometrically similar model is to be run. The diameter of the prototype impeller is 1.8 m (6 ft) while 0.57 m^3/s (20 ft^3/s) is the available laboratory flowrate. What should be the model size, speed, and head if the model and prototype efficiencies are the same?

[11]When head is expressed in metres, g_n becomes a relevant parameter and Froude number equality must hold; as $V \propto ND$, $(N^2D/g_n)_m = (N^2D/g_n)_p$.

12.11. A Kaplan (propeller with variable pitchblades) turbine with a rated capacity of 83 MW at a head of 24 m and 86 r/min was one of 14 units installed at the McNary project on the Columbia River. The characteristic runner diameter is 7 m. If a 6 m head is available in the laboratory, what should be the model scale, flowrate, and r/min?

12.12. A turbine model that is 0.3 m in diameter develops 150 kW under a 15 m head at 2 000 r/min. Assuming no change in efficiency, what power would a 4.5 m diameter prototype, operating at 100 r/min under a 225 m head, develop?

12.13. What type of pump should be selected for problem 12.10? What maximum efficiency is expected?

12.14. A pump is designed to operate at 1 800 r/min with 83.7% efficiency, and deliver 250 l/s (4 000 gpm) with a power consumption of 141 kW (189.5 hp). What is its specific speed and type?

12.15. A manufacturer states that it can deliver a pump operating at 690 r/min that delivers 285 l/s or 4 500 gpm at 78% efficiency with a power consumption of about 5.2 kW or 7 hp under a head of 1.5 m or 5 ft. What type of pump is it? Is the efficiency claim reasonable, too high, or too low?

12.16. Determine N_S for both model and prototype turbines in problem 12.11.

12.17. A small turbine must be run at 250 r/min to match generator characteristics and must deliver 745.7 kW from a 6 m head of water. Compute the specific speed of the unit, select a turbine type to be used, and compute the flowrate through a maximum efficiency turbine.

12.18. The Francis turbines for the Shasta plant in California are rated at 76.8 MW (103 000 hp) at 138.5 r/min. If $N_S = 0.74$, find the rated head and expected efficiency of the units.

12.19. A propeller turbine at the Bonneville plant on the Columbia River has $N_S = 2.8$ and a rated head of 18 m (60 ft); it operates at 75 r/min. How much power will the turbine deliver and at what maximum efficiency?

12.20. Reversible pump-turbines are used to provide pumped storage of water and peaking electric power. One such unit operates at 600 r/min and is rated for 125 MW at 376 m head as a turbine and for 29 m^3/s at 404 m head as a pump. Calculate the unit's specific speeds; what type of unit is it?

12.21. An impulse turbine delivers 69 000 kW from 750 m head. At what speed should the runner rotate for maximum efficiency?

12.22. Calculate the minimum required *NPSH* (in metres or feet of water) to avoid cavitation in the pumps of the following problems: (*a*) problem 12.14, (*b*) problem 12.15, and (*c*) problem 12.20 (when operated as a pump).

12.23. Calculate the minimum required *NPSH* (in metres or feet of water) to avoid cavitation in the turbines of the following problems: (*a*) problem 12.17, (*b*) problem 12.18, (*c*) problem 12.19, and (*d*) problem 12.20 (when operated as a turbine).

12.24. If the model turbine in problem 12.12 requires a *NPSH* of 18 m to avoid cavitation, what minimum *NPSH* is required in the prototype?

12.25. The blade is one of a series. Calculate the force exerted by the jet on the blade system.

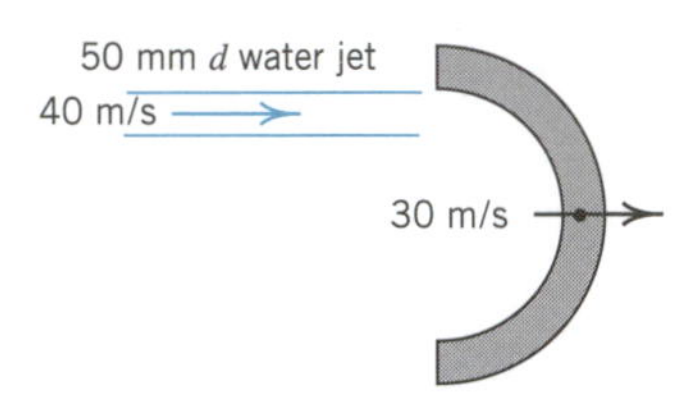

Problem 12.25

12.26. This blade is one of a series. What force is required to move the series horizontally against the direction of the jet of water at a velocity of 15 m/s (or 50 ft/s)? What power is required to accomplish this motion?

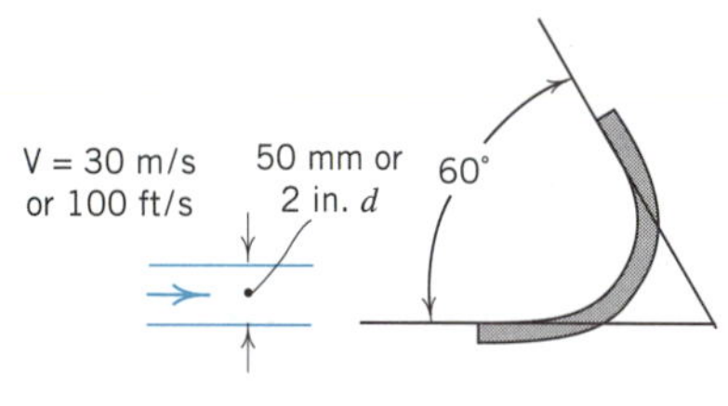

Problem 12.26

12.27. From a water jet of 2 in. (50 mm) diameter, moving at 200 ft/s (60 m/s), 180 hp (135 kW) are to be transferred to a blade system (Fig. 12.6) which is moving in the direction of the jet at 50 ft/s (15 m/s). Calculate the required blade angle.

12.28. A series of blades (Fig. 12.6), moving in the same direction as a water jet of 1 in. (25 mm) diameter and of velocity 150 ft/s (46 m/s), deflects the jet 75° from its original direction. What relation between blade velocity and blade angle must exist to satisfy this condition? What is the force on the blade system?

12.29. If a system of blades (Fig. 12.6) is free to move in a direction parallel to that of a jet, prove that the direction of the force on the blade system is the same for all blade velocities.

12.30. A crude impulse turbine has flat radial blades and is in effect a "paddle wheel." If the 25 mm (1 in.) diameter water jet has a velocity of 30 m/s (100 ft/s) and is tangent to the rim of the wheel, calculate the approximate power of the machine when the blade velocity is 12 m/s (40 ft/s).

12.31. For a flowrate of 12 l/s and turbine speed of 65 r/min, estimate the power transferred from jet to turbine wheel.

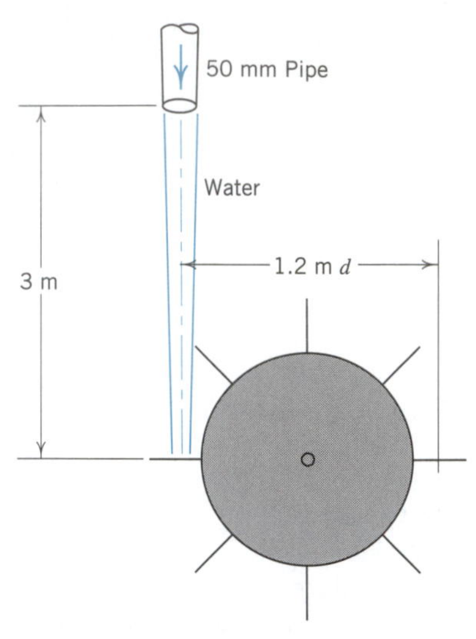

Problem 12.31

12.32. A 150 mm pipeline equipped with a 50 mm nozzle supplies water to an impulse turbine 1.8 m in diameter having blade angles of 165°. Plot a curve of theoretical power versus r/min when the pressure behind the nozzle is 690 kPa. What is the force on the blades when the maximum power is being developed? Plot a curve of force on the blades versus r/min.

12.33. The velocity of the water jet driving this impulse turbine is 45 m/s. The jet has a 75 mm diameter. After leaving the buckets the (absolute) velocity of the water is observed to be 15 m/s in a direction 60° to that of the original jet. Calculate the mean tangential force exerted by jet on turbine wheel and the speed (r/min) of the wheel.

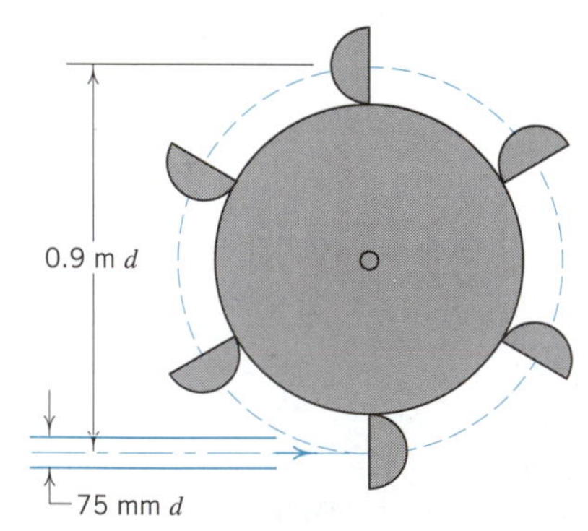

Problem 12.33

12.34. When an air jet of 1 in. diameter strikes a series of blades on a turbine rotor, the (absolute) velocities are as shown. If the air is assumed to have a constant specific weight of 0.08 lb/ft^3, what is the force on the turbine rotor? How much horsepower is transferred to the rotor? What must be the velocity of the blade system?

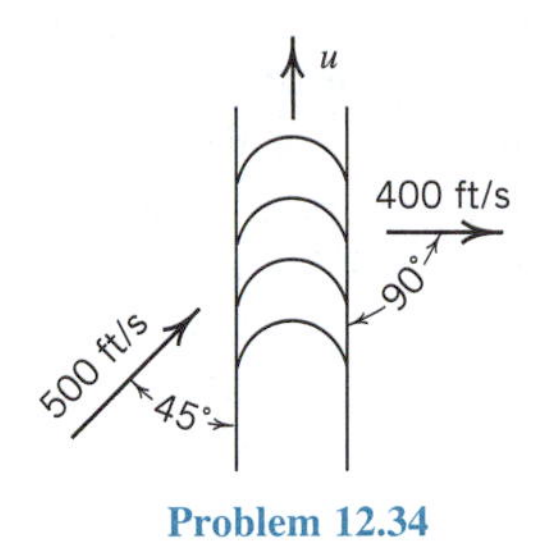

Problem 12.34

12.35. This system of blades develops 149 kW under the influence of the jet shown. Calculate the blade velocity.

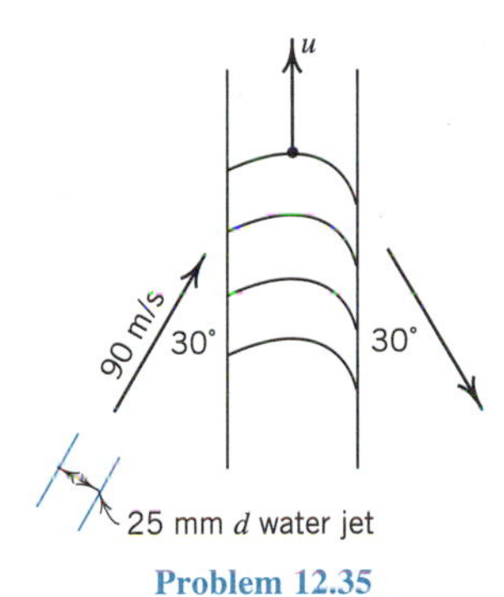

Problem 12.35

12.36. If $\alpha_1 = \alpha_2$, calculate: α, F_x on the blade system, and the power developed.

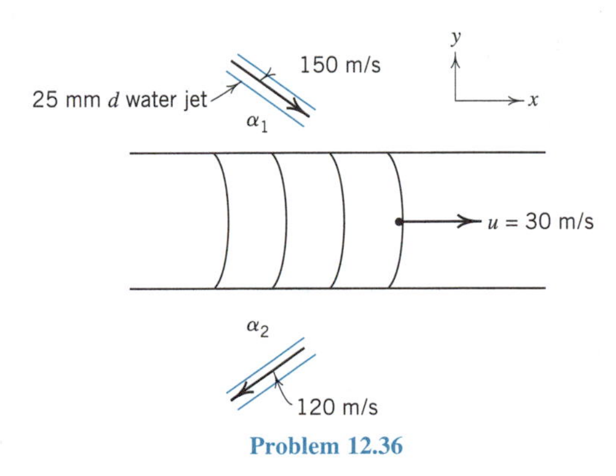

Problem 12.36

12.37. In passing through this blade system, the absolute jet velocity decreases from 136 to 73.8 ft/s (41.5 to 22.5 m/s). If the flowrate is 2.0 ft^3/s (57 l/s) of water, calculate the power transferred to the blade system and the vertical force component exerted on the blade system.

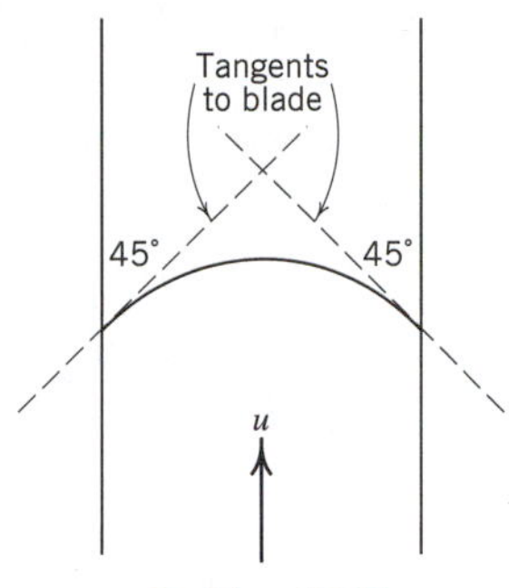

Problem 12.37

12.38. If $u = v_1 = v_2 = 100$ ft/s (30 m/s) for the blade system of the preceding problem, calculate the absolute velocities of the jet entering and leaving the blade system. If the flowrate is 5 ft^3/s (140 l/s), how much power may be expected from the machine?

12.39. The flowrate is 57 l/s (2.0 ft^3/s) of water, $u = 9$ m/s (30 ft/s), and $v_1 = v_2 = 12$ m/s (40 ft/s). Calculate: the absolute velocity of the water entering and leaving the system, F_x and F_y (magnitude and direction) on the blade system, and the power transferred from jet to blade system.

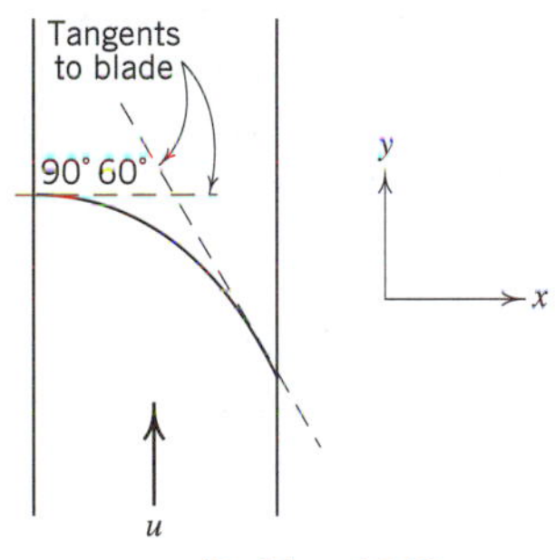

Problem 12.39

12.40. The (absolute) velocities and directions of the jets entering and leaving the blade system are as shown. Calculate the power transferred from jet to blade system and the blade angles required.

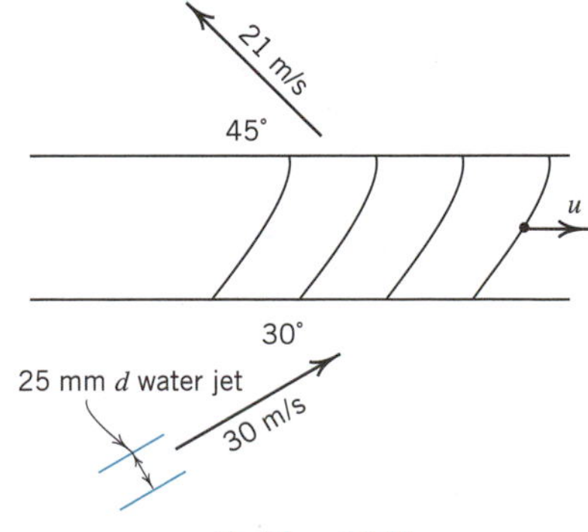

Problem 12.40

12.41. This stationary blade is pivoted at point O. Calculate the torque (about O) exerted thereon when a 2 in. water jet moving at 100 ft/s passes over it as shown.

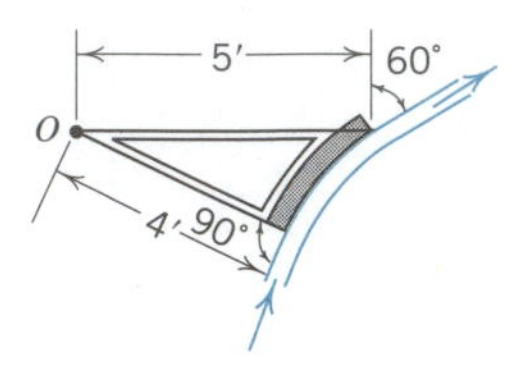

Problem 12.41

12.42. Eight thousand cubic feet (225 m^3) per second of water flow into a hydraulic turbine whose guide vanes, set at an angle of 15°, have an exit circle of 8 ft (2.4 m) radius and are 6 ft (1.8 m) high. In the draft tube the flow is observed to have no tangential component. What is the torque on the runner? If the runner is rotating at 150 r/min, how much power is being delivered to the turbine and how much energy is being extracted from each pound (newton) of water?

12.43. A simple reaction turbine has $r_1 = 0.9$ m, $r_2 = 0.6$ m, and its flow cross section is 0.3 m high. The guide vanes are set so that $\alpha_1 = 30°$. When 2.83 m^3/s of water flow through this turbine the angle α_2 is found to be 60°. Calculate the torque exerted on the turbine runner. If the angle β_2 is 150°, calculate the speed of rotation of the turbine runner and the angle β_1 necessary for smooth flow into the runner. Calculate the power developed by the turbine at the above speed.

12.44. The nozzles are all of 25 mm diameter and each nozzle discharges 7 l/s of water. If the turbine rotates at 100 r/min, calculate the power developed.

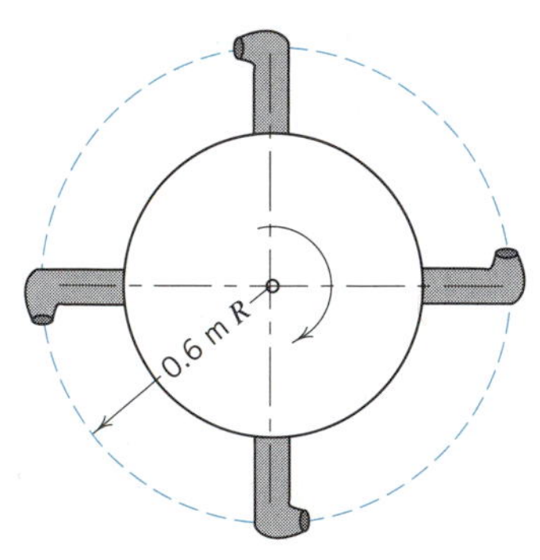

Problem 12.44

12.45. A centrifugal pump impeller having $r_1 = 50$ mm, $r_2 = 150$ mm, and width $b = 37.5$ mm is to pump 225 l/s of water and supply 12.2 J of energy to each newton of fluid. The impeller rotates at 1 000 r/min. What blade angles are required? What power is required to drive this pump? Assume smooth flow at the inlet of the impeller.

12.46. A centrifugal pump impeller having dimensions and angles as shown rotates at 500 r/min. Assuming a radial direction of velocity at the blade entrance, calculate the flowrate, the pressure difference between inlet and outlet of blades, and the torque and power required to meet these conditions.

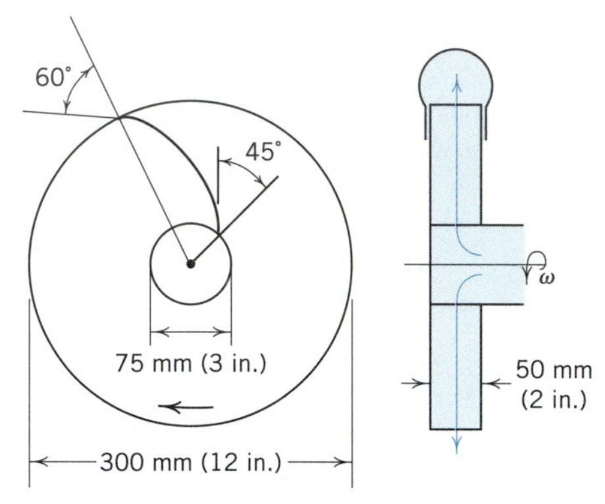

Problem 12.46

12.47. If the impeller of the preceding problem rotates between horizontal planes of infinite extent and the flowrate is 25 l/s (0.86 ft^3/s), what rise of pressure may be expected between one point having $r = 150$ mm (6 in.) and another having $r = 225$ mm (9 in.)?

12.48. At the outlet of a pump impeller of diameter 0.6 m and width 150 mm, the (absolute) velocity is observed to be 30 m/s at an angle of 60° with a radial line. Calculate the torque exerted on the impeller.

12.49. The operating characteristics of the Ingersoll-Dresser pump in the pipeline shown below are given in Illustrative Problem 12.5. However, the pump is to be driven by a geared motor which will turn the pump at 1 600 rpm.

Construct the E_P versus Q curve for this speed using the $19\frac{3}{8}$ inch impeller. Estimate the power requirement to pump 6 000 gpm and calculate the flowrate in the system, neglecting local losses.

If *NPSH* varies in the same manner as E_P under a change in speed, what will be the *NPSH*-value at the above flowrate? Estimate the submergence of the pump.

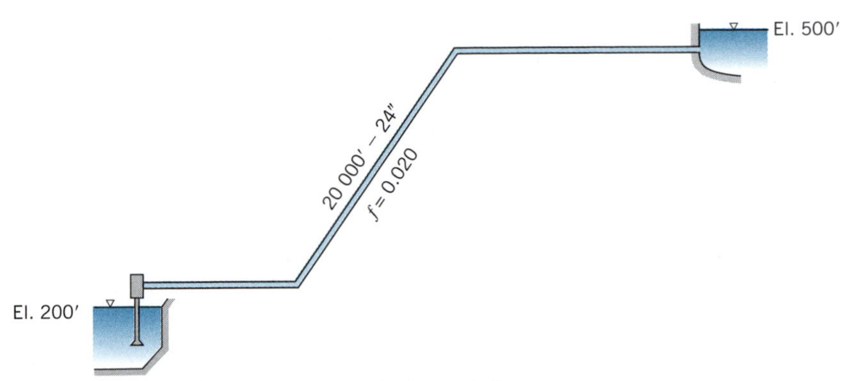

Problem 12.49

12.50. If the pump of Problem 12.49 were run at 1 160 rpm, what would be the discharge and head at the point of maximum efficiency?

What horsepower would be required to drive the pump at the point of maximum efficiency?

12.51. The pump of Illustrative Problem 12.5, running at 1 775 rpm and using the $19\frac{3}{8}$ inch impeller, supplies the pipeline below while operating at maximum efficiency. Find the pipeline loss coefficient K in the equation $h_L = KQ^2$ (Q in gpm) for this condition. Neglect local losses.

If two of these pumps operate in parallel, what is the flowrate between the two reservoirs? Assume the pipeline K-value remains unchanged.

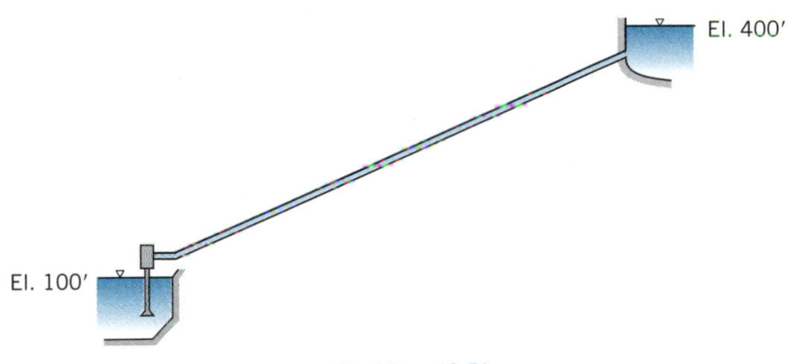

Problem 12.51

CHAPTER

13 FLOW OF COMPRESSIBLE FLUIDS

The material in this chapter is intended to impart some feeling for the subject of the flow of compressible fluids, an understanding of some of its difficulties, and an appreciation of the striking differences between compressible and incompressible flow. An understanding of the problems of compressible flow depends on the successful synthesis of fluid mechanics with thermodynamics. We assume that the readers have not yet had formal training in thermodynamics. Accordingly, some basic thermodynamics is included here with explanations which, although incomplete for comprehensive understanding, are sufficient for pursuit of the subject at hand. All the thermodynamics required for the remainder of this chapter is presented just below and in Section 13.1, including a review of the material on the First Law of Thermodynamics introduced in Section 7.11. The rest of the chapter is divided into five major segments. First, we cover one- and two-dimensional ideal fluid flows in two separate segments. Second, we apply the impulse-momentum theorem to examine shock waves. Third, we extend the pipe flow principles of Chapter 9 to the compressible flow of real fluids in pipes. Finally, the principles of Chapter 11 on lift and drag are extended to compressible flow situations.

To begin it is necessary to define or redefine certain terms in their thermodynamic context. A fluid *system* remains some fixed, identifiable quantity of matter and a system *property* remains an observable or measurable characteristic of the system, for example, temperature, density, pressure, and so forth. The *state* of a system is established by examination of the properties. If at two different times the properties of the system are all the same, the system is said to be at the same state at both times. When a system's state changes, the system has undergone a *process* which usually involves transfers across the system's boundaries and work done by or on the system. If a process leads to a final state that is the same as the initial state, this process is a *cycle*.

The two basic laws of thermodynamics are stated for processes. As we learned in Section 7.11, the First Law is a conservation law for processes and leads to definition of a new property, called the *internal energy of the fluid*. The Second Law prescribes the permitted direction in which a process may proceed and leads to the new property, called *entropy*. Implicit in these laws is also the concept of the reversible process. A process is *reversible* if and only if, upon completion of the process, the system and all of its surroundings can be returned to their initial states, that is, the process can be completely undone. All real fluid flow processes are *irreversible*, but some can be approximated by reversible processes, when the causes of the irreversibility, that is, viscous action and heat conduction across finite temperature differences, do not dominate the motion.

In Sections 13.2 through 13.11 of this chapter the assumption of an ideal (inviscid) fluid restricts the discussion to *frictionless* flow processes, as in Chapter 5 for incompressible fluid. A further restriction often made is the assumption of no heat transfer to or from the fluid, which is the definition of an *adiabatic* process. In thermodynamics, a *frictionless adiabatic* process is called an *isentropic* one, because, as seen below, it is accompanied by no change of entropy. Since this process involves no heat transfer or friction, it is, therefore, a reversible process. Such processes are closely approximated in practice if they occur with small friction and with such rapidity that there is little opportunity for heat transfer. An example is high-velocity gas flow (*gas dynamics*) over short distances (i.e., nozzles rather than pipes) where the effects of friction and heat transfer actually are small. The real fluid flow in pipes is treated in Sections 13.14 through 13.16.

High velocity gas flow is associated in general with large changes of pressure, temperature, and density, but changes in these variables are, of course, much smaller in low-velocity flow. Therefore, many of the latter problems may be treated approximately, yet satisfactorily, by the methods of Chapter 5. However, there is no precise boundary between high-velocity and low-velocity motion, and whether a gas may be treated as an incompressible fluid depends on the accuracy of the results required and the physics of the actual flow. For example, in the aerodynamics of low-speed general aviation aircraft it is a sufficiently accurate approximation to consider the air incompressible; for current commercial jet aircraft, missiles, jet engines, etc., such an assumption would obviously be unsatisfactory. Only experience with the equations and situations of fluid mechanics can provide the bases for dealing with this dilemma.

13.1 THE LAWS OF THERMODYNAMICS

Thermodynamics is concerned with the study of the interactions of work, heat, and system properties. In Section 7.11, we used the First Law of Thermodynamics to derive the energy equation 7.44 for steady flow (which reduces to the work-energy equation 5.7 in the absence of heat transfer and of density variations). We will not re-derive that result here; rather, after recalling a few definitions, we will use it and generalize it for thermodynamic applications. Recall then that the First Law of Thermodynamics is an empirical law that expresses the conservation of energy in any process (barring nuclear mass-energy conversions and electromagnetic effects). From Section 5.5, we know that a system possesses both kinetic and potential energy and can do work on its surroundings. From thermodynamics, it is known that energy transfers occur across system boundaries when there is a temperature difference; this energy in transition as a result of a temperature difference is called *heat*. Furthermore, a fluid system possesses energy as a result of the kinetic energy of its molecules and the forces between them; this property of a system is known as *internal energy* and manifests itself in temperature, high or low temperature implying high or low internal energy, respectively. It followed that application of the Reynolds Transport Theorem Eq. 4.15, where the extensive property of the system was the total energy of the system and the intensive property was the total energy per unit mass, together with the First Law written as

$$\frac{dQ}{dt} + \frac{dW}{dt} = \frac{dE}{dt}$$

yielded the energy equation

$$\left(\frac{p_1}{\gamma_1} + \frac{V_1^2}{2g_n} + z_1\right) - \left(\frac{p_2}{\gamma_2} + \frac{V_2^2}{2g_n} + z_2\right) = (E_T - E_P) + \frac{1}{g_n}(ie_2 - ie_1 - q_H) \quad (7.44)$$

Here, q_H is the heat added to the fluid in the control volume per unit of mass passing through the control volume 1221 in Fig. 13.1, dQ/dt is the rate of transfer of heat to the system, dW/dt is the rate of work done on the system, and dE/dt is the rate of change of the total energy of the system. Recall that the total energy of the system is given by

$$E = \iiint_{\text{System}} (\text{kinetic} + \text{potential} + \text{internal energies}) \cdot dm$$

$$= \iiint_{\text{System}} \left(\frac{1}{2}v^2 + g_n z + ie\right) \cdot \rho d\forall$$

and ie is the internal energy per unit mass. Note that the system in this case can be considered the fluid within the control volume 1221 of Fig. 13.1 at some initial instant. At a later instant, that system will have moved through the control volume and perhaps gained energy due to heat transfer or work done. Thus, although the flow in the control volume is steady and unchanging at any point, the system state is changing and the time variation of E for the system is entirely appropriate. Finally, we recall that E_T and E_P are the energies withdrawn by turbines or added by pumps per unit weight of fluid passing through the control volume.

By using the common combination of terms $p/\rho + ie$, called the *specific enthalpy h*, Eq. 7.44 can be converted to the steady flow energy equation, which can be written as

$$q_H = g_n(E_T - E_P) + h_2 - h_1 + \frac{1}{2}(V_2^2 - V_1^2) + g_n(z_2 - z_1) \quad (13.1)$$

or

$$\frac{h_1}{g_n} + z_1 + \frac{V_1^2}{2g_n} + E_P + \frac{q_H}{g_n} = \frac{h_2}{g_n} + z_2 + \frac{V_2^2}{2g_n} + E_T \quad (13.2)$$

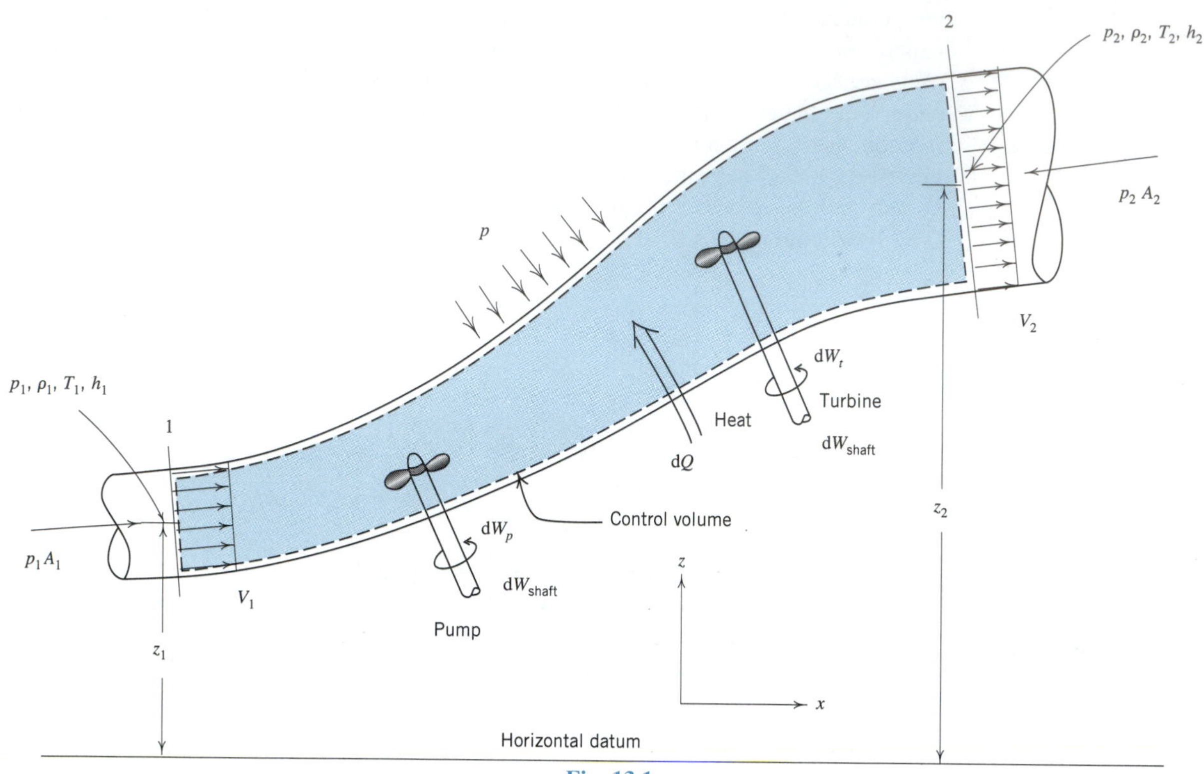

Fig. 13.1

The difference between this last equation and Eqs. 5.1 and 5.7 should be noted. The enthalpy depends only on the system properties including p, ρ, and temperature T, and so enthalpy is also a property. The units of h are J/kg or ft·lb/slug.

ILLUSTRATIVE PROBLEM 13.1

Air flows as a perfect gas in a pipe without friction between the points indicated. If heat $q_H = -1 \times 10^5$ J/kg is lost between these points from each unit mass of fluid, find V_2.

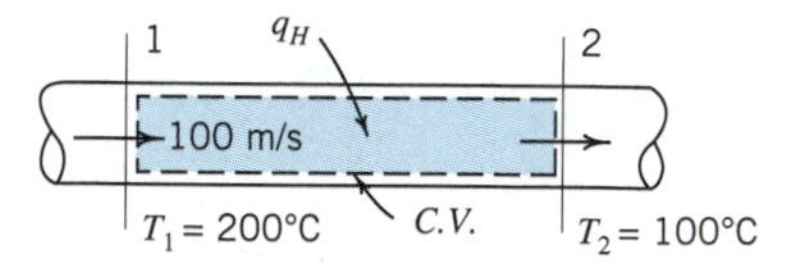

SOLUTION

Equation 13.1 is the best format to use, viz.,

$$q_H = g_n(E_T - E_P) + h_2 - h_1 + \tfrac{1}{2}(V_2^2 - V_1^2) + g_n(z_2 - z_1) \qquad (13.1)$$

Then, from Section 1.4, we have Eq. 1.3

$$p/\rho = RT \tag{1.3}$$

and from Section 1.5, we have the relationship between the specific heats

$$c_p - c_v = R$$

From Section 7.11, we recall that, for perfect gases, thermodynamics shows the internal energy per unit mass to be a function of temperature only and so $ie = c_{vT}$. Using the given data yields, since

$$h = \frac{p}{\rho} + ie \qquad \text{and} \qquad c_p = 1\,003 \text{ J/kg·K} \qquad \text{(Appendix 2)},$$

$$h_2 - h_1 = (R + c_v)(T_2 - T_1) = c_p(T_2 - T_1)$$

Here, $E_T = E_P = 0$ and $z_2 - z_1 = 0$. Thus,

$$-1 \times 10^5 = 1\,003(100 - 200) + \tfrac{1}{2}(V_2^2 - 100^2) \tag{13.1}$$

$$V_2 = 103 \text{ m/s} \bullet$$

The Second Law of Thermodynamics is interpreted in terms of a quantity called the *entropy* S defined as

$$dS = dQ_{\text{reversible}}/T$$

where T is the absolute temperature at the point of transfer and $dQ_{\text{reversible}}$ is the heat transferred to a system during a reversible process (the temperature differences across the boundary must be vanishingly small). For a *nonflow process* in which a fluid system of uniform and invariant chemical composition undergoes only a simple reversible change in volume by working against the boundaries and receiving heat in time dt, the First Law yields (only the pressure acting during the volume change $d(m/\rho)$ does work)

$$dQ = pd\left(\frac{m}{\rho}\right) + d(m\,ie)$$

or

$$T\frac{dS}{m} = pd\left(\frac{1}{\rho}\right) + d(ie) \tag{13.3}$$

where m is the system mass, ρ its density, and ie its internal energy per unit mass. If the *specific entropy* $s = S/m$ and the specific enthalpy h are introduced

$$ds = \frac{1}{T}\left(dh - \frac{1}{\rho}\,dp\right) \tag{13.4}$$

As all the variables on the right in the above Eqs. 13.3 and 13.4 are properties of the fluid, it follows that *entropy is* also a *fluid property*. In consequence the equations must hold for *any process* even though entropy is defined only in terms of heat transfer in a reversible process. By using Eq. 13.4 and an arbitrarily prescribed datum for s (only differences are important), the state of fluids can be described very usefully on entropy versus enthalpy (*Mollier*) diagrams as will be seen in the subsequent sections.

The Second Law can be stated in many useful forms. Here only an extension of the law is needed, namely, that in any process

$$dS \geq \frac{dQ}{T} \tag{13.5}$$

Equality holds for a reversible process, and hence, the *entropy cannot decrease in an adiabiatic* ($dQ = 0$) *process*. In all natural or real adiabatic processes

$$(dS)_{\text{adiabatic}} > 0 \tag{13.6}$$

A reversible, adiabatic process is, thus, properly called *isentropic* ($dS = 0$).

The entropy relations for a perfect gas ($p/\rho = RT$ and c_p and c_v are constant) are easy to derive and instructive. First, by definition the specific heats are

$$c_v = \left(\frac{\partial ie}{\partial T}\right)_v$$

$$c_p = \left(\frac{\partial h}{\partial T}\right)_p = \frac{\partial}{\partial T}\left(ie + \frac{p}{\rho}\right)_p = c_v + \frac{\partial}{\partial T}(RT)_p = c_v + R$$

so, as discovered above,

$$h_2 - h_1 = c_p(T_2 - T_1)$$

$$ie_2 - ie_1 = c_v(T_2 - T_1)$$

Then, from Eq. 13.4,

$$ds = \frac{1}{T}\left(c_p\, dT - \frac{1}{\rho}\, dp\right) = c_p \frac{dT}{T} - \frac{R}{p}\, dp$$

and

$$s_2 - s_1 = c_p \ln\left(\frac{T_2}{T_1}\right) - R \ln\left(\frac{p_2}{p_1}\right)$$

$$= c_v \left[\ln\left(\frac{T_2}{T_1}\right) + \frac{R}{c_v} \ln\left(\frac{T_2}{T_1}\right) - \frac{R}{c_v} \ln\left(\frac{p_2}{p_1}\right)\right]$$

$$= c_v \ln\left[\left(\frac{T_2}{T_1}\right)^{1+(R/c_v)} \left(\frac{p_2}{p_1}\right)^{R/c_v}\right]$$

Now, as $k = c_p/c_v = 1 + (R/c_v)$,

$$s_2 - s_1 = c_v \ln\left[\left(\frac{T_2}{T_1}\right)^k \left(\frac{p_2}{p_1}\right)^{1-k}\right] \tag{13.7}$$

and the entropy change which occurs when a system undergoes a process can be computed. Because entropy is a property, the entropy change undergone by fluid systems as they pass through the control volume of Fig. 13.1, for example, can be obtained via Eq. 13.7 by use of the property values known (the state of the fluid) at sections 1 and 2.

ILLUSTRATIVE PROBLEM 13.2

Calculate the entropy change between sections 1 and 2 in Illustrative Problem 13.1 if $\rho_1 = 0.74 \text{ kg/m}^3$ and $\rho_2 = 0.47 \text{ kg/m}^3$.

SOLUTION

Use the data and diagram from the previous problem, along with Eq. 1.3

$$p/\rho = RT,$$

$$k = c_p/c_v,$$

and from Appendix 2,

$$c_p = 1\,003 \text{ J/kg·K},$$

$$R = 286.8 \text{ J/kg·K} \quad \text{and} \quad k = 1.40.$$

From the data, $c_v = 1\,003/1.40 = 716.4$ J/kg·K. From Eq. 1.3, the absolute pressures

$$p_1 = 0.74 \times 286.8(200 + 273.2) = 100.4 \text{ kPa} \tag{1.3}$$

$$p_2 = 0.47 \times 286.8(100 + 273.2) = 50.3 \text{ kPa} \tag{1.3}$$

Thus, since Eq. 13.7 is

$$s_2 - s_1 = c_v \ln\left[\left(\frac{T_2}{T_1}\right)^k \left(\frac{p_2}{p_1}\right)^{1-k}\right] \tag{13.7}$$

we have

$$s_2 - s_1 = 716.4 \ln\left[\left(\frac{373.2}{473.2}\right)^{1.4} \left(\frac{50.3}{100.4}\right)^{-0.4}\right] = -40 \text{ J/kg·K} \bullet \tag{13.7}$$

One-Dimensional Ideal Flow

13.2 EULER'S EQUATION AND THE ENERGY EQUATION

In the previous section a control volume analysis led to the general energy equation 13.1 for a streamtube. On the other hand, a development of an Euler equation for one-dimensional flow of a compressible ideal fluid can be carried out by analysis of a system just as was done in Section 5.1 for an incompressible fluid. This leads to an energy equation for one-dimensional flow in the absence of heat transfer and shear or shaft work.

Consider a streamline and a small cylindrical fluid system for analysis, as shown in Fig. 13.2. Although fluid density will vary along the streamline, the mean density ρ of the differential fluid system shown differs negligibly from that at its ends and thus may be taken as constant throughout the system. (A similar situation is discussed in Section 2.1.)

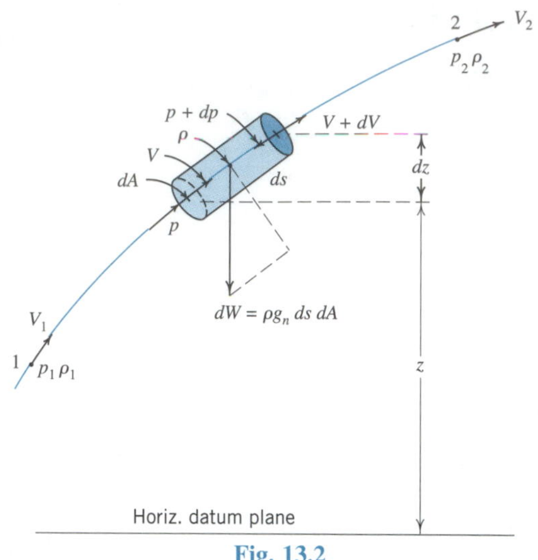

Fig. 13.2

Euler's equation is therefore (as before)

$$\frac{dp}{\rho} + V\,dV + g_n\,dz = 0$$

For compressible flow, however, the term $g_n\,dz$ is usually dropped and the Euler equation written

$$\frac{dp}{\rho} + V\,dV = 0 \qquad \text{or} \qquad \frac{dp}{\gamma} + \frac{V\,dV}{g_n} = 0 \tag{13.8}$$

This simplification is justified by the fact that compressible-flow problems are usually concerned with gases of light weight and with flows of small vertical extent in which changes of pressure and velocity are predominant and changes of elevation negligible by comparison. See Illustrative Problem 5.11 of Section 5.8.

When there is no heat transfer and no work done by pumps and turbines in the flow of an ideal fluid, the motion is isentropic and the steady flow energy Eq. 13.1 for the streamline in Fig. 13.2 becomes

$$h_1 + \tfrac{1}{2}V_1^2 = h_2 + \tfrac{1}{2}V_2^2 \qquad \text{or} \qquad ie_1 + \frac{p_1}{\rho_1} + \tfrac{1}{2}V_1^2 = ie_2 + \frac{p_2}{\rho_2} + \tfrac{1}{2}V_2^2 \tag{13.9}$$

when $z_2 - z_1$ is neglected. In differential form, the second of these energy equations can be written as

$$d(ie) + pd\left(\frac{1}{\rho}\right) + \frac{1}{\rho}\,dp + V\,dV = 0$$

and, because of Eq. 13.3 with $dS = 0$, the energy equation reduces to

$$\frac{dp}{\rho} + V\,dV = 0$$

which is recognized as the Euler equation. *Thus, the energy and Euler equations are identical for isentropic flow.*

For application to perfect gases, Eqs. 13.9 may be written in other useful terms. Recall that $ie = c_vT$, $p/\rho = RT$, $h = (c_v + R)T$, and $c_v + R = c_p$. Thus, substituting these relations into the first of Eqs. 13.9 gives

$$c_pT_1 + \frac{V_1^2}{2} = c_pT_2 + \frac{V_2^2}{2} \tag{13.10}$$

The foregoing equations actually apply equally well to frictional and frictionless flow (the derivation of the energy equation 13.1 did not exclude friction effects), but only for the adiabatic case in which no heat is added to or extracted from the fluid; also they provide only for the situation where no mechanical energy is added to or extracted from the fluid by pump or turbine. For Sections 13.3 through 13.11 their application will be confined to frictionless adiabatic (*isentropic*) flow.

13.3 INTEGRATION OF THE EULER EQUATION

When the Euler equation is integrated along the streamline (or small streamtube) for isentropic flow of perfect gases, it becomes

$$\frac{V_2^2 - V_1^2}{2} = \int_{p_2}^{p_1} \frac{dp}{\rho} = \frac{p_1}{\rho_1}\frac{k}{k-1}\left[1 - \left(\frac{p_2}{p_1}\right)^{(k-1)/k}\right]$$

$$\text{or} \quad \frac{p_2}{\rho_2}\frac{k}{k-1}\left[\left(\frac{p_1}{p_2}\right)^{(k-1)/k} - 1\right] \tag{13.11}$$

in which the integration has been performed by substitution of the isentropic relation $p_1/\rho_1^k = p_2/\rho_2^k$. The same result can also be obtained directly from the isentropic energy equation (13.10) by rearranging it and substituting $kp/\rho(k - 1)$ for c_pT and $(p_2/p_1)^{1/k}$ for ρ_2/ρ_1 [see Section 1.5]. From Eqs. 13.10 and 13.11, the variation of absolute pressure and temperature (and thus density) with velocity may be predicted along the streamline of Fig. 13.2 or the control volume of Fig. 13.1.

Equation 13.10, which shows $(c_pT + V^2/2)$ to be constant along any streamline in adiabatic flow, is frequently written

$$c_pT_1 + \frac{V_1^2}{2} = c_pT_2 + \frac{V_2^2}{2} = c_pT_s \tag{13.10}$$

At a stagnation point where V is zero, T is the *stagnation temperature* T_s, which is seen to be constant for all points on the streamline. Thus, stagnation temperature in adiabatic flow is analogous to the total head of ideal incompressible flow.

For the isentropic flow of vapors other than perfect gases, Eq. 13.9 may be used directly; the specific enthalpy h is a function of pressure and temperature and may be obtained from appropriate tables and diagrams; for isentropic flow, values of h_1 and h_2 for the same entropy must be used.

ILLUSTRATIVE PROBLEM 13.3

At one point on a streamline in an airflow the velocity, absolute pressure, and temperature are 30 m/s, 35 kPa, and 150°C, respectively. At a second point on the same streamline the velocity is 150 m/s. If the process along the streamline is assumed isentropic, calculate the pressure and temperature at the second point.

SOLUTION

From Appendix 2, $c_p = 1\,003$ J/kg·K; using Eq. 13.10

$$c_pT_1 + \frac{V_1^2}{2} = c_pT_2 + \frac{V_2^2}{2} = c_pT_s \tag{13.10}$$

and the given data, we have

$$\frac{(150)^2 - (30)^2}{2} = 1\,003(T_1 - T_2) \tag{13.10}$$

$$T_1 - T_2 = 10.8°\text{C} \qquad T_2 = 139.2°\text{C} \bullet$$

Using Eqs. 13.11

$$\frac{V_2^2 - V_1^2}{2} = \int_{p_2}^{p_1} \frac{dp}{\rho} = \frac{p_1}{\rho_1}\frac{k}{k-1}\left[1 - \left(\frac{p_2}{p_1}\right)^{(k-1)/k}\right] \tag{13.11}$$

and substituting RT_1 for p_1/ρ_1,

$$\frac{(150)^2 - (30)^2}{2} = 288(150 + 273.2)\frac{1.4}{0.4}\left[1 - \left(\frac{p_2}{p_1}\right)^{0.286}\right] \tag{13.11}$$

$$\left(\frac{p_2}{p_1}\right)^{0.286} = 0.974\,7 \qquad \frac{p_2}{p_1} = 0.914\,1$$

$$p_2(\text{absolute}) = 32 \text{ kPa} \bullet$$

ILLUSTRATIVE PROBLEM 13.4

Steam in a large tank is at an absolute pressure and temperature of 2 400 kPa and 500°C, respectively. The steam flows from the tank through a smooth nozzle and into a passage, and at a point in the passage the absolute pressure is observed to be 1 500 kPa. Determine the temperature and velocity at this point assuming an isentropic process.

SOLUTION

Refer to the *Mollier diagram* contained in any text on thermodynamics; this is a plot of specific enthalpy h against specific entropy s for various pressures and temperatures.

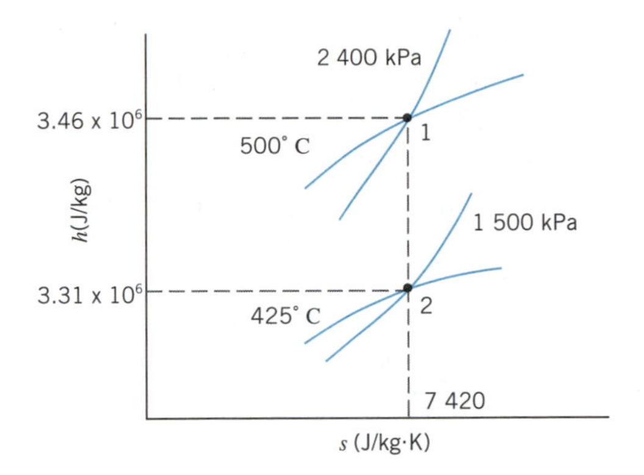

At the intersection of the 2 400 kPa and 500°C lines on the diagram, h is found to be 3.46 MJ/kg. Now drop vertically down the chart (at constant entropy) to the 1 500 kPa line. From the temperature line passing through point 2 the temperature may be read; it is 425°C. ● For this point the specific enthalpy h is found to be 3.31 MJ/kg. Substituting h_1 and h_2 into Eq. 13.9, which is

$$h_1 + \tfrac{1}{2}V_1^2 = h_2 + \tfrac{1}{2}V_2^2, \tag{13.9}$$

and noting that the velocity V_1 in the large tank will be zero, yields

$$\frac{V_2^2}{2} = 3.46 \times 10^6 - 3.31 \times 10^6 \qquad V_2 = 548 \text{ m/s} \bullet \tag{13.9}$$

13.4 THE STAGNATION POINT

In gas dynamics, Eq. 13.11 is usually expressed in terms of Mach number, **M**. Using the first form of the equation, this may be easily accomplished by recalling (Eq. 1.10) that the acoustic (sonic) velocity a is given by $\sqrt{kp/\rho}$ and thus $a_1^2 = kp_1/\rho_1$. By substituting this into Eq. 13.11 and rearranging, we find that

$$\frac{V_2^2}{a_1^2} = \mathbf{M}_1^2 + \frac{2}{k-1}\left[1 - \left(\frac{p_2}{p_1}\right)^{(k-1)/k}\right] \tag{13.12}$$

Now consider the application of this equation to a stagnation point (S) in a compressible flow (Fig. 13.3). With the fluid compressible, the rise of pressure at the stagnation point causes compression of the fluid, producing a higher density (ρ_s) and temperature

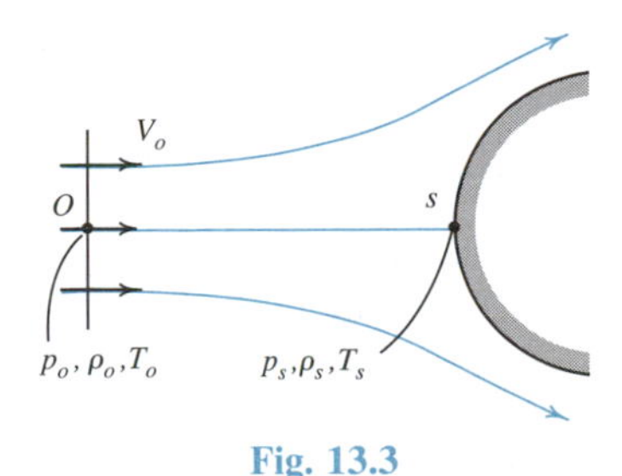

Fig. 13.3

(T_s) there. Evidently, the extent of these compression effects depends primarily on the magnitude of the stream velocity V_o; they are large at high velocities and small (often negligible) at low velocities. At the stagnation point, $V_2 = 0$ and $p_2 = p_s$; substituting these values and $p_1 = p_o$ and $\mathbf{M}_1 = \mathbf{M}_o$ into Eq. 13.12, the stagnation pressure p_s is given by

$$\frac{p_s}{p_o} = \left[1 + \mathbf{M}_o^2\,\frac{k-1}{2}\right]^{k/(k-1)} \tag{13.13}$$ [1]

If the right-hand side of this equation is expanded by the binomial theorem there results (retaining the first three terms)

$$p_s = p_o + \tfrac{1}{2}\rho_o V_o^2[1 + \tfrac{1}{4}\mathbf{M}_o^2 + \ldots] \tag{13.14}$$ [1]

Comparison of this equation with Eq. 5.3 of Section 5.4 shows that the effects of compressibility have been isolated in the bracketed quantity and that these effects depend only on the Mach number. The bracketed quantity may thus be considered a "compressibility correction factor" and the effect of compressibility on $(p_s - p_o)$ calculated with fair precision.[2]

From Eq. 13.13 it can be observed that measurements of p_s and p_o allow calculation of the "free stream" Mach number $\mathbf{M}_o$ of the undisturbed flow. However, to obtain the velocity V_o a temperature measurement is also required, and in practice the temperature T_s is measured at the stagnation point. From T_s, p_s, and p_o, the velocity V_o may be handily calculated by using the second form of Eq. 13.11 between points O and S; this equation becomes (RT_s having been substituted for p_s/ρ_s)

$$\frac{V_o^2}{2} = c_p T_s\left[1 - \left(\frac{p_o}{p_s}\right)^{(k-1)/k}\right] \tag{13.15}$$ [1]

from which V_o may be calculated directly.

ILLUSTRATIVE PROBLEM 13.5

An airplane flies at 400 mph (586 ft/s) through still air at 13.0 psia and 0°F. Calculate pressure, temperature, and air density at the stagnation points (on nose of fuselage and leading edges of wings).

SOLUTION

Given the following data from Appendix 2:

$$c_p = 6\,000 \text{ ft}\cdot\text{lb/slug}\cdot{}^\circ\text{R}$$

$$R = 1\,715 \text{ ft}\cdot\text{lb/slug}\cdot{}^\circ\text{R} \qquad k = 1.40$$

[1]The use of Eqs. 13.13, 13.14 and 13.15 is restricted to $\mathbf{M}_o < 1$. For $\mathbf{M}_o > 1$ a shock wave exists across the streamline between points O and S and the flow is no longer isentropic. See Sections 13.12 and 14.7.

[2]The precision is excellent for small $\mathbf{M}_o$ but decreases with increasing $\mathbf{M}_o$ as the neglected terms of the binomial expansion become significant.

this solution involves a straightforward application of the previous equations as follows:

$$\rho = p/RT \quad \text{so} \quad \rho_o = \frac{13.0 \times 144}{1\,715 \times 460} = 0.002\,37 \text{ slug/ft}^3 \tag{1.3}$$

$$a = \sqrt{kRT} \quad \text{so} \quad a_o = \sqrt{1.4 \times 1\,715 \times 460} = 1\,052 \text{ ft/s} \tag{1.11}$$

$$\mathbf{M}_o = \frac{586}{1\,052} = 0.557$$

$$\frac{p_s}{p_o} = \left[1 + \mathbf{M}_o^2 \frac{k-1}{2}\right]^{k/(k-1)} \quad \text{so}$$

$$\text{(exact) } p_s = 13.0 \left[1 + (0.557)^2 \frac{1.4 - 1}{2}\right]^{3.5} = 16.18 \text{ psia} \bullet \tag{13.13}$$

$$p_s = p_o + \tfrac{1}{2}\rho_o V_o^2 [1 + \tfrac{1}{4}\mathbf{M}_o^2 + \ldots] \quad \text{so}$$

$$\text{(approx.) } p_s = 13.0 + \frac{1}{2}\frac{0.002\,37 \times (586)^2}{144}[1 + \tfrac{1}{4}(0.557)^2] = 16.05 \text{ psia} \bullet \tag{13.14}$$

$$c_p T_o + \frac{V_o^2}{2} = c_p T_s \quad \text{so}$$

$$\frac{(586)^2}{2} = 6\,000(T_s - 460) \qquad T_s = 488.5°\text{R} \qquad T_s = 28.5°\text{F} \bullet \tag{13.10}$$

$$\rho_s = \frac{16.18 \times 144}{488.5 \times 1\,715} = 0.002\,78 \text{ slug/ft}^3 \bullet \tag{1.3}$$

13.5 THE ONE-DIMENSIONAL ASSUMPTION

The foregoing equations of this chapter may be applied successfully to passages of finite cross section when the streamlines are essentially straight and parallel. Consistent with neglecting the difference in the z terms in the Euler and energy equations, the pressure is taken to be constant over the flow cross section (For the incompressible fluid $(p/\gamma + z)$ is constant over the flow cross section; see Section 5.3); the absence of friction permits no variation of velocity. With pressure and velocity constant throughout the flow cross section, constancy of temperature and density will follow from Eqs. 13.10 and 13.11.

When streamlines are not essentially straight and parallel, variations of pressure, velocity, temperature, and density are to be expected. As with incompressible flow (See Section 5.8), increase of pressure and decrease of velocity will be found with increasing distance from center of curvature; in compressible flow, such variations of pressure and velocity will produce variations of temperature and density as well. Although mean values of the variables may be visualized and computed, their use in the equations of one-dimensional flow is not recommended except for approximate calculations.

13.6 SUBSONIC AND SUPERSONIC VELOCITIES

Combination of the continuity and Euler equations yields information on the superficial shapes of passages required to produce changes of flow velocity when such velocities are subsonic or supersonic. These equations are

$$\frac{dA}{A} + \frac{d\rho}{\rho} + \frac{dV}{V} = 0 \tag{4.2}$$

$$\frac{dp}{\rho} + V\,dV = 0 \tag{13.8}$$

Multiplying the first term of Eq. 13.8 by $d\rho/d\rho$, a^2 is recognized (Eq. 1.9) as $dp/d\rho$, and $d\rho/\rho$ is obtained as $-V\,dV/a^2$ or $-V^2\,dV/a^2V$. Substituting this in Eq. 4.2, identifying V/a as the Mach number $\mathbf{M}$, and rearranging, we obtain

$$\frac{dA}{A} = \frac{dV}{V}(\mathbf{M}^2 - 1) \tag{13.16}$$

From this equation we can deduce some far-reaching and somewhat surprising conclusions. Analysis of the equation for $dV/V > 0$ shows that: for $\mathbf{M} < 1$, $dA/A < 0$; for $\mathbf{M} = 1$, $dA/A = 0$; for $\mathbf{M} > 1$, $dA/A > 0$. This means that, for subsonic flow ($\mathbf{M} < 1$), a reduction of cross-sectional area (Fig. 13.4) is required for an increase of velocity; however, for supersonic flow ($\mathbf{M} > 1$), *an increase of area is required to produce an increase of velocity.* For flow at sonic speed ($\mathbf{M} = 1$) the rate of change of area must be zero; that is, this might occur at a maximum or minimum cross section of the streamtube; use of the preceding conclusions will show this to be restricted to a minimum section (*a throat*) only. However, Eq. 13.16 does *not* allow the conclusion that a throat will always produce a flow of sonic velocity because $\mathbf{M}$ is *not necessarily unity* when $dA/A = 0$. If, at a throat ($dA/A = 0$) $\mathbf{M} \gtrless 1$, it follows that $dV/V = 0$, implying a maximum or minimum velocity there. Upstream from the throat $dA/A < 0$, so here $dV/V < 0$ for $\mathbf{M} > 1$ and $dV/V > 0$ for $\mathbf{M} < 1$. Accordingly it can be concluded that if the throat velocity is not sonic it will be a maximum in subsonic flow and a minimum in supersonic flow.

13.7 THE CONVERGENT NOZZLE

Consider now the frictionless flow of a gas from a large tank ($A_1 \sim \infty$, $V_1 \sim 0$) through a convergent nozzle (Fig. 13.5) into a region of pressure, p_2'. Using the second form of

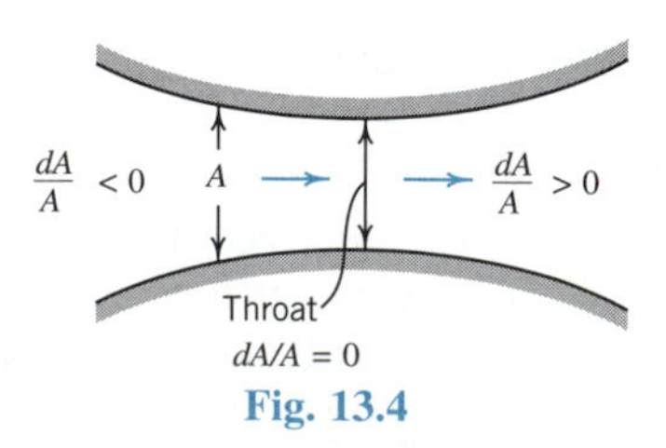

Fig. 13.4

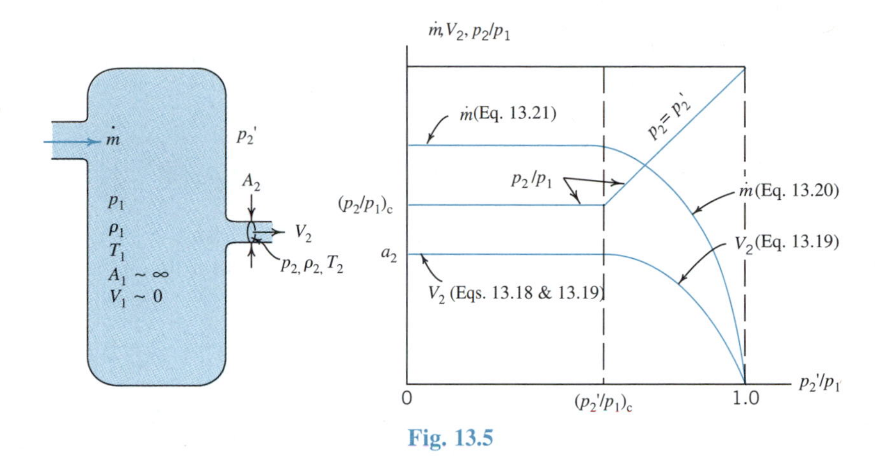

Fig. 13.5

Eq. 13.11 and substituting a_2^2 for kp_2/ρ_2 and $\mathbf{M}_2$ for V_2/a_2, we obtain

$$\mathbf{M}_2^2 = \frac{2}{k-1}\left[\left(\frac{p_1}{p_2}\right)^{(k-1)/k} - 1\right] \tag{13.17}$$

In view of the preceding development, supersonic velocities are not expected in this problem, because the fluid starts from rest and there are no divergent passages; accordingly $\mathbf{M}_2 \leqq 1$. If the pressure difference $(p_1 - p_2')$ is large enough to produce sonic velocity, this velocity must exist at the throat of the nozzle and $\mathbf{M}_2 = 1$; placing this value in Eq. 13.17 and solving for p_2/p_1 gives the critical pressure ratio $(p_2/p_1)_c$:

$$\left(\frac{p_2}{p_1}\right)_c = \left(\frac{2}{k+1}\right)^{k/(k-1)} \tag{13.18}$$

Thus, if the sonic velocity is attained by the fluid, the absolute pressure, p_2, in the minimum section is in fixed ratio to the absolute pressure, p_1, in the tank and *therefore independent of the pressure* p_2'. It follows that in a free jet moving at sonic velocity *the pressure within the jet at the nozzle exit is never less, and is usually more, than the pressure which surrounds it*; the outside pressure, p_2', tends to penetrate the jet at the sonic speed but cannot distribute itself over the whole jet cross section at the nozzle exit because the fluid there is moving at this same high velocity.

Of course, the sonic velocity is not attained unless the pressure drop between the inside and outside of the tank is large enough. For small pressure drops (i.e., for pressure ratios, p_2/p_1, above the critical) the pressures p_2' and p_2 are the same, and the velocity at the nozzle exit may be computed (from Eq. 13.11) by

$$\frac{V_2^2}{2} = \frac{p_1}{\rho_1}\frac{k}{k-1}\left[1 - \left(\frac{p_2}{p_1}\right)^{(k-1)/k}\right] \tag{13.19}$$

A graphic summary of these facts is given in Fig. 13.5.

In many problems (especially those of fluid metering) the flowrate is the most important quantity to be computed in flow through nozzles. The computation can be easily made by combining some of the foregoing equations.

If the pressure ratio p_2'/p_1 is more than the critical, the mass flowrate can be calculated from $\dot{m} = A_2\rho_2V_2$, using V_2 from Eq. 13.10 and the isentropic relation $p_2/p_1 = (\rho_2/\rho_1)^k$. The result is

$$\dot{m} = A_2\sqrt{\frac{2k}{k-1}p_1\rho_1\left[\left(\frac{p_2}{p_1}\right)^{2/k} - \left(\frac{p_2}{p_1}\right)^{(k+1)/k}\right]} \tag{13.20}$$

If the pressure ratio p_2'/p_1 is less than the critical, $p_2/p_1 = [2/(k+1)]^{k/(k-1)}$, and substituting this in Eq. 13.20, along with p_1/RT_1 for ρ_1, yields

$$\dot{m} = \frac{A_2p_1}{\sqrt{T_1}}\sqrt{\frac{k}{R}\left(\frac{2}{k+1}\right)^{(k+1)/(k-1)}} \tag{13.21}$$

in which the large square root is obviously a characteristic constant of the gas, dependent on R and k; this allows a simple calculation for flowrate and is good reason for selecting metering nozzles of such proportions that sonic velocities are produced.

ILLUSTRATIVE PROBLEM 13.6

Air discharges from a large tank, in which the pressure is 700 kPa and temperature 40°C, through a convergent nozzle of 25 mm tip diameter. Calculate the flowrates when the pressure outside the jet is (*a*) 200 kPa, and (*b*) 550 kPa, and the barometric pressure is 101.3 kPa. Also calculate the pressure, temperature, velocity, and sonic velocity at the nozzle tip for these flowrates.

SOLUTION

First, we convert the pressure and temperature data to absolute values:

$$p_1 = 700 \text{ kPa (gage)} = 801.3 \text{ kPa (abs)} \qquad T_1 = 40°\text{C} = 313 \text{ K}$$

Next, we calculate the density of the air in the tank and the critical pressure ratio, using Eqs. 1.3 and 13.18

$$\rho = p/RT \rightarrow \rho_1 = \frac{801.3 \times 10^3}{286.8 \times 313} = 8.92 \text{ kg/m}^3 \tag{1.3}$$

$$\left(\frac{p_2}{p_1}\right)_c = \left(\frac{2}{k+1}\right)^{k/(k-1)} \rightarrow \left(\frac{p_2}{p_1}\right)_c = \left(\frac{2}{2.4}\right)^{3.5} = 0.528 \tag{13.18}$$

(*a*) $p_2' = 200$ kPa (gage): $\quad \dfrac{p_2'}{p_1} = \dfrac{301.3}{801.3} = 0.376 < 0.528$

so

$$p_2 \text{ (abs.)} = 0.528 \times 801.3 = 423.1 \text{ kPa}$$

and substituting $k = 1.40$, $R = 286.8$, $A_2 = 4.91 \times 10^{-4}$, $p_1 = 801.3 \times 10^3$, and $T_1 = 313$ into Eq. 13.21

$$\dot{m} = \frac{A_2p_1}{\sqrt{T_1}}\sqrt{\frac{k}{R}\left(\frac{2}{k+1}\right)^{(k+1)/(k-1)}} \rightarrow \dot{m} = 0.9 \text{ kg/s.} \bullet$$

Using the first form of Eq. 13.11

$$\frac{V_2^2 - V_1^2}{2} = \frac{p_1}{\rho_1}\frac{k}{k-1}\left[1 - \left(\frac{p_2}{p_1}\right)^{(k-1)/k}\right]$$

with $V_1 = 0$ and $p_2/p_1 = 0.528$: $V_2 = 323.9$ m/s •, which is also the sonic velocity. Using Eq. 13.10

$$c_pT_1 + \frac{V_1^2}{2} = c_pT_2 + \frac{V_2^2}{2}$$

with $V_1 = 0$, $V_2 = 323.9$ m/s, and $T_1 = 313$ K: $T_2 = 261$ K. •

(b) $p_2' = 550$ kPa (gage): $\quad \dfrac{p_2'}{p_1} = \dfrac{651.3}{801.3} = 0.813 > 0.528$

so

$$p_2(\text{abs.}) = 651.3 \text{ kPa}$$

and, substituting $A_2 = 4.91 \times 10^{-4}$, $k = 1.40$, $p_1 = 801.3 \times 10^3$, $\rho_1 = 8.92$, and $p_2/p_1 = 0.813$ into Eq. 13.20

$$\dot{m} = A_2\sqrt{\frac{2k}{k-1}p_1\rho_1\left[\left(\frac{p_2}{p_1}\right)^{2/k} - \left(\frac{p_2}{p_1}\right)^{(k+1)/k}\right]} \rightarrow \dot{m} = 0.72 \text{ kg/s.} \; \bullet$$

Using the first form of Eq. 13.11 with $V_1 = 0$ and $p_2/p_1 = 0.813$ gives $V_2 = 190$ m/s. •
Using Eq. 13.10 with $V_1 = 0$, $V_2 = 190$ m/s, and $T_1 = 313$ K: $T_2 = 295$ K. •
Calculating the sonic velocity from Eq. 1.11 ($a = \sqrt{kRT}$) yields $a_2 = 344.2$ m/s, giving $\mathbf{M}_2 = 0.552$. •

13.8 CONSTRICTION IN STREAMTUBE

When a compressible fluid flows through a constriction in a streamtube or pipeline (Fig. 13.6), the equations of Section 13.7 are not applicable unless the constriction is very small (compared to the pipe). When the constriction is larger, the velocity V_1 is no longer negligible compared to V_2 (see Eq. 13.11), and adjustments must be made in the foregoing equations to account for this. However, when sonic velocities are attained in the constriction, it has usually been chosen small enough that V_1 is negligible and thus adjustment of Eq. 13.21 is not usually necessary and will not be discussed here.

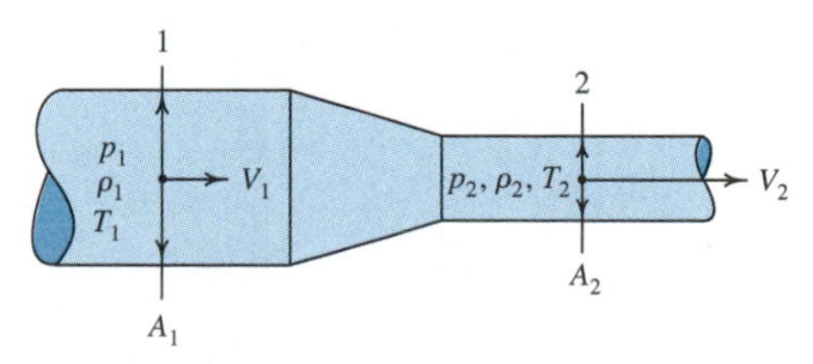

Fig. 13.6

For flow of gases well above the critical pressure ratio, however, V_1 usually cannot be neglected. For this case an equation for flowrate may be derived by simultaneous solution of

$$\frac{V_2^2 - V_1^2}{2} = \frac{p_1}{\rho_1}\frac{k}{k-1}\left[1 - \left(\frac{p_2}{p_1}\right)^{(k-1)/k}\right] \tag{13.11}$$

$$\dot{m} = A_1\rho_1 V_1 = A_2\rho_2 V_2 \tag{4.3}$$

$$p_2/p_1 = (\rho_2/\rho_1)^k \tag{1.7}$$

This yields

$$\dot{m} = \frac{A_2}{\sqrt{1 - \left(\frac{p_2}{p_1}\right)^{2/k}\left(\frac{A_2}{A_1}\right)^2}}\sqrt{\frac{2k}{k-1}p_1\rho_1\left[\left(\frac{p_2}{p_1}\right)^{2/k} - \left(\frac{p_2}{p_1}\right)^{(k+1)/k}\right]} \tag{13.22}$$

Comparison of this equation with Eq. 13.20 indicates that the effect of including V_1 is concentrated in the first square root, which can be seen to approach 1.00 rapidly as the area ratio, A_2/A_1, decreases.

Equation 13.22, because of its unwieldy form, is sometimes solved by the use of an *expansion factor*, Y, applied to the simpler and analogous solution for incompressible flow,[3] which is

$$Q = \frac{A_2}{\sqrt{1 - \left(\frac{A_2}{A_1}\right)^2}}\sqrt{2g_n\left(\frac{p_1 - p_2}{\gamma}\right)} \tag{13.23}$$

Y is defined as the factor that, when multiplied into the product of ρ_1 and Eq. 13.23, will yield 13.22. Thus

$$\dot{m} = \frac{YA_2\rho_1}{\sqrt{1 - \left(\frac{A_2}{A_1}\right)^2}}\sqrt{2g_n\left(\frac{p_1 - p_2}{\gamma_1}\right)} \tag{13.24}$$

An expression for Y can be derived by equating Eqs. 13.22 and 13.24; it is found to be a function of the three variables p_2/p_1, A_2/A_1, and k, and thus can be computed once for all and presented in tables or plots. The equation for Y and its tabulated values are found in Appendix 5.

ILLUSTRATIVE PROBLEM 13.7

Air flows through a 25 mm constriction in a 37.5 mm pipeline. In the pipe the pressure and temperature of the air are 689.5 kPa and 38°C, respectively. Calculate the flowrate if the pressure in the constriction is 551.6 kPa. Barometric pressure is 101.3 kPa.

[3]See Section 5.4

SOLUTION

To begin we calculate the absolute pressures.
$p_1 = 698.5 + 101.3 \text{ kPa} = 790.8 \text{ kPa}$ and $p_2 = 551.6 + 101.3 \text{ kPa} = 656.9 \text{ kPa}$. It follows that $p_2/p_1 = 0.825$ and from Eq. 1.3 [$\rho = p/RT$] that $\rho_1 = 8.86 \text{ kg/m}^3$. Substituting these values in Eq. 13.22, which is

$$\dot{m} = \frac{A_2}{\sqrt{1 - \left(\frac{p_2}{p_1}\right)^{2/k}\left(\frac{A_2}{A_1}\right)^2}} \sqrt{\frac{2k}{k-1} p_1 \rho_1 \left[\left(\frac{p_2}{p_1}\right)^{2/k} - \left(\frac{p_2}{p_1}\right)^{(k+1)/k}\right]} \tag{13.22}$$

yields a flowrate of 0.776 kg/s. ●

The flowrate may also be calculated by interpolating a value of Y (0.874) from Appendix 5 and using it in Eq. 13.24, which is

$$\dot{m} = \frac{YA_2\rho_1}{\sqrt{1 - \left(\frac{A_2}{A_1}\right)^2}} \sqrt{2g_n\left(\frac{p_1 - p_2}{\gamma_1}\right)} \tag{13.24}$$

Then,

$$\dot{m} = \frac{0.874 \times \frac{\pi}{4}(0.025)^2 \times 8.86}{\sqrt{1 - (2/3)^4}} \sqrt{2g_n \frac{137.9 \times 1\,000}{86.9}} = 0.776 \text{ kg/s} \bullet \tag{13.24}$$

13.9 THE CONVERGENT-DIVERGENT NOZZLE

From the study of the flow of compressible fluid through a convergent-divergent passage (De Laval nozzle), much may be learned of basic phenomena and engineering application. For simplicity consider the discharge from a large reservoir through such a passage with pressure p_1 and temperature T_1 in the reservoir (Fig. 13.7). If sonic velocity exists at the throat, the flowrate $\dot{m}$ is determined by Eq. 13.21, the throat area A_2 being known. This is also the condition under which the flowrate is a maximum and the nozzle is said to be *choked.* Assumption of the pressure p_3 in the jet at the nozzle exit allows computation of the exit area A_3 from Eq. 13.20, using A_3 for A_2 and p_3 for p_2; however, it will be found that *two very different p_3's will yield the same area A_3 and flowrate $\dot{m}$.* The higher of these pressures (p_3'') will cause subsonic velocity in the diverging part of the passage, the lower one (p_3''') will produce supersonic velocity there. The variations of pressure along the passage for these two conditions are shown in Fig. 13.7.

It is of fundamental importance to examine such a nozzle at the same reservoir conditions (p_1 and T_1) but with "back pressure" p_3' other than those used for the determination of A_3. Reduction of p_3' below p_3''' cannot affect conditions at the throat, so no change of flowrate is produced and no change of nozzle performance is to be expected (except that the compressed gas passing the nozzle exit will expand rapidly on emergence into a region of lower pressure). Raising the back pressure above p_3'', as shown in Fig. 13.8, causes a

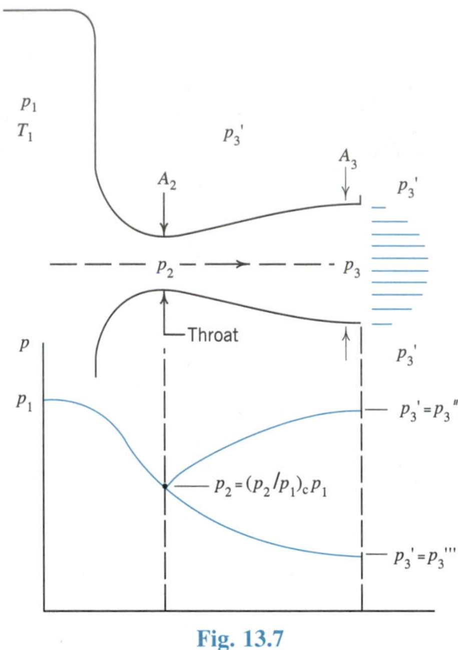

Fig. 13.7

reduction of flowrate, velocities throughout the nozzle become subsonic, and the upper pressure distribution exists in the nozzle.

Between p_3'' and p_3''' there are an infinite number of ''back pressures,'' none of which can satisfy the equations of isentropic flow; this is due to the formation of a *normal shock wave*[4] (with considerable internal friction and increase of entropy) in the divergent passage.

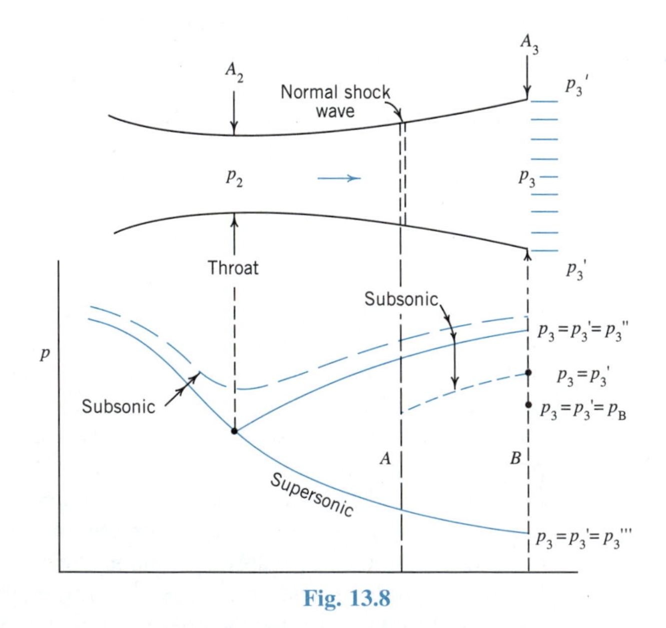

Fig. 13.8

[4]See Section 13.12.

Through such a wave the velocity drops abruptly from supersonic to subsonic and the pressure jumps abruptly about as shown on curve *A*. For the curve *A* and for others similar to it, the flow will emerge from the nozzle at subsonic velocity and the pressure p_3 will equal p_3'. At some lower value of p_3' (curve *B*) the shock wave will occur at the nozzle exit, which is the limiting case to be treated by one-dimensional methods; for pressures p_3' between p_B and p_3''' the jet will emerge from the nozzle with $p_3 = p_3'''$ and the shock wave will be in the flowfield downstream from the nozzle exit. Since such flowfields are either two- or three-dimensional, they cannot be described by the foregoing one-dimensional equations.

ILLUSTRATIVE PROBLEM 13.8

Air discharges through a convergent-divergent passage (of 25 mm throat diameter) into the atmosphere. The pressure and temperature in the reservoir are 700 kPa and 40°C, respectively; the barometric pressure is 101.3 kPa. Calculate the nozzle tip diameter required for $p_3 = 101.3$ kPa. Calculate the flow velocity, sonic velocity, and Mach number at the nozzle exit. Determine the pressure p_3'', which will yield the same flowrate, and the pressure p_B which will produce a normal shock wave at the nozzle exit.

SOLUTION

Since supersonic velocities will occur in the divergent portion of the nozzle, sonic velocity is expected at the throat. Because these are the conditions of Illustrative Problem 13.6 in Section 13.7, the flowrate $\dot{m}$ is 0.9 kg/s as before. In order to establish the pressure at the exit needed to produce a normal shock wave, we will need to extract the equation for the pressure jump across a shock wave from Section 13.12. With the given data, i.e.,

$$p_1 = 700 \text{ kPa (gage)} \qquad p_{\text{bar}} = 101.3 \text{ kPa} \qquad d_2 = 25 \text{ mm}$$

$$T_1 = 40°\text{C} \qquad p_3 = 101.3 \text{ kPa (abs.)}$$

we can calculate the velocity at cross-section 3 according to Eq. 13.19, written from point 1 to point 3 for the assumed supersonic flow:

$$\frac{V_3^2}{2} = \frac{p_1}{\rho_1}\frac{k}{k-1}\left[1 - \left(\frac{p_3'''}{p_1}\right)^{(k-1)/k}\right] \tag{13.19}$$

$$\frac{V_3^2}{2} = \frac{1.4}{0.4} \times \frac{801.3 \times 10^3}{8.92}\left[1 - \left(\frac{101.3}{801.3}\right)^{0.280}\right] \qquad V_3 = 525.8 \text{ m/s} \bullet \tag{13.19}$$

Then the temperature can be found by use of Eq. 13.10:

$$c_pT_1 + \frac{V_1^2}{2} = c_pT_2 + \frac{V_2^2}{2} \tag{13.10}$$

$$(525.8)^2/2 = 1\,003(T_1 - T_3) \qquad T_1 - T_3 = 137.8 \text{ K} \qquad T_3 = 175.2 \text{ K} \tag{13.10}$$

From Eq. 1.11, we have the speed of sound and the Mach number:

$$a_3 = \sqrt{1.4 \times 286.8 \times 175.2} = 265 \text{ m/s} \bullet \qquad \mathbf{M}_3 = \frac{525.8}{265} = 1.98 \bullet$$

The equation of state 1.3 yields

$$\rho = p/RT \rightarrow \rho_3 = \frac{101.3 \times 10^3}{286.8 \times 175.2} = 2.02 \text{ kg/m}^3 \tag{1.3}$$

From the mass flowrate equation 4.3 we obtain the cross-sectional area needed at section 3 and the diameter:

$$\dot{m} = \rho A V \rightarrow A_3 = \frac{0.9}{2.02 \times 525.8} = 8.5 \times 10^{-4}\text{m}^2 \qquad d_3 = 33 \text{ mm} \bullet \tag{4.3}$$

Finally, we obtain an equation to find the pressure p_3'' by using Eq. 13.20:

$$\dot{m} = A_3 \sqrt{\frac{2k}{k-1} p_1 \rho_1 \left[\left(\frac{p_3}{p_1}\right)^{2/k} - \left(\frac{p_3}{p_1}\right)^{(k+1)/k}\right]} \tag{13.20}$$

$$0.9 = 8.5 \times 10^{-4} \times \sqrt{\frac{2 \times 1.4}{0.4} 801.3 \times 10^3 \times 8.92 \left[\left(\frac{p_3''}{p_1}\right)^{1.43} - \left(\frac{p_3''}{p_1}\right)^{1.715}\right]} \tag{13.20}$$

Solving by trial, $\dfrac{p_3''}{p_1} = 0.91 \qquad p_3''(\text{abs.}) = 0.91 \times 801.3 \times 10^3 = 729.2 \text{ kPa} \bullet$

The pressure at the nozzle exit that just produces a shock wave is given by using the above results in Eq. 13.31, which is

$$\frac{p_B - p_3'''}{p_3'''} = \frac{2k}{k+1}(\mathbf{M}_1^2 - 1) \tag{13.31}$$

Thus,

$$p_B(\text{abs.}) = 101.3 \times 10^3 \left[1 + \frac{2 \times 1.4}{1.4 + 1}(1.98^2 - 1)\right] = 446.4 \text{ kPa} \bullet \tag{13.31}$$

Two-Dimensional Ideal Flow

The study of two-dimensional fields of compressible flow presents the same difficulties as those of incompressible flow (Sections 5.6 to 5.8) and a further complication: variation of density over the flowfield—which means another variable in the equations and increased difficulty of solution. Although formal mathematical solution of such problems is not possible, special techniques (particularly numerical solution of problems on digital computers) have been invented for the solution of certain problems of engineering interest; however, review of these methods is outside the scope of an elementary text. The intent of the following treatment is merely to develop the basic equations, to describe certain flowfields, and to discuss the applicability and limitations of the equations.

13.10 EULER'S EQUATIONS AND THEIR INTEGRATION

Euler's equations for the two-dimensional flow of an ideal compressible fluid are the same as those for the incompressible fluid except for the neglect of g_n, which was justified in Section 13.2. This neglection allows the flowfield to be considered to be in a horizontal $(x - y)$ plane and the Euler equations to be written as

$$-\frac{1}{\rho}\frac{\partial p}{\partial x} = u\frac{\partial u}{\partial x} + v\frac{\partial u}{\partial y} \tag{5.9a}$$

$$-\frac{1}{\rho}\frac{\partial p}{\partial y} = u\frac{\partial v}{\partial x} + v\frac{\partial v}{\partial y} \tag{5.9b}$$

They may be rearranged and combined in the same pattern as that of Section 5.7 to give

$$-\frac{dp}{\rho} = d\left(\frac{u^2 + v^2}{2}\right) + (u\,dy - v\,dx)\left(\frac{\partial v}{\partial x} - \frac{\partial u}{\partial y}\right)$$

in which $(\partial v/\partial x - \partial u/\partial y)$ is recognized as the vorticity ξ, and integrated to yield

$$\int\frac{dp}{\rho} + \frac{V^2}{2} = C + \int \xi(u\,dy - v\,dx)$$

which reduces, *for irrotational flow* ($\xi = 0$), to

$$\int\frac{dp}{\rho} + \frac{V^2}{2} = C$$

in which C is a constant for all points in the flow field (Fig. 13.9). For an isentropic flowfield this equation may be integrated by using the isentropic relation between p and ρ of Eq. 1.7; the result is

$$\frac{V_2^2 - V_1^2}{2} = \frac{k}{k-1}\frac{p_1}{\rho_1}\left[1 - \left(\frac{p_2}{p_1}\right)^{(k-1)/k}\right] \tag{13.25}$$

Although Eqs. 13.11 and 13.25 are identical, application of the former is restricted to points on the same streamline, whereas in Eq. 13.25 points 1 and 2 may be any points in

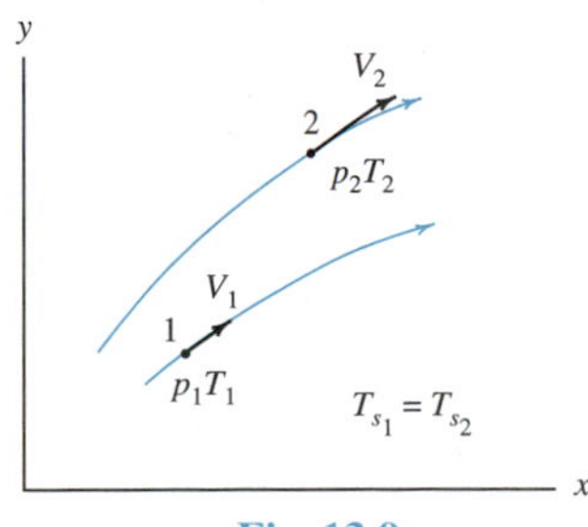

Fig. 13.9

the flowfield. Use of the isentropic relation between pressure and temperature in Eq. 13.25, along with $p_1/\rho_1 = RT_1$, yields

$$c_pT_1 + \frac{V_1^2}{2} = c_pT_2 + \frac{V_2^2}{2} = c_pT_s \tag{13.26}$$

which is identical to Eq. 13.10 but shows that stagnation temperature is constant not only along single streamlines but also at all points in an isentropic flowfield. The limitations of Eqs. 13.25 and 13.26 deserve emphasis: they have been derived for frictionless and adiabatic (i.e., isentropic) irrotational flow and hence cannot be expected to apply to real flowfields with heat transfer, boundary friction, and shock waves. They do, however, provide a useful method of approach to many compressible flow problems.

Consider now (for comparison with the foregoing) a particular compressible flowfield which is *nonisentropic* but throughout which the stagnation temperature is constant. With no change of stagnation temperature along the streamlines, Eq. 13.10 may be applied to conclude that there is no exchange of heat energy between adjacent streamtubes; however, frictional processes may vary from streamtube to streamtube with accompanying variability of entropy between them. This situation is closely approximated downstream from a curved shock wave. *Crocco's theorem,*[5] a classic synthesis of thermodynamic and fluid mechanics principles, shows that such a nonisentropic flowfield cannot be irrotational.

13.11 APPLICATION OF THE EQUATIONS

Application of Eqs. 13.25 and 13.26 to a flowfield is straightforward enough if all the velocities are known, along with the temperature and pressure at one point in the flowfield; from these, all pressures and temperatures throughout the field may be computed, and from pressure and temperature the fluid density at any point may be predicted.

ILLUSTRATIVE PROBLEM 13.9

Air approaches this streamlined object at the speed, pressure, and temperature shown. Calculate the pressures, temperatures, and Mach numbers at points A and B, where the velocities are 800 and 900 ft/s, respectively.

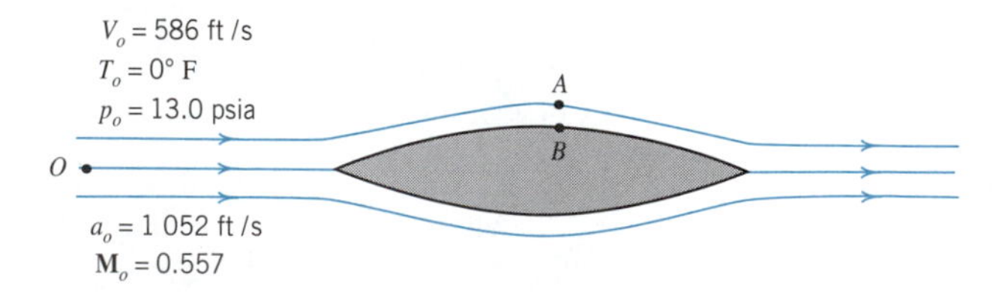

[5]Development and discussion of Crocco's theorem will be found in the references at the end of this chapter.

SOLUTION

Refer to Illustrative Problem 13.5 in Section 13.4 for some preliminary calculations and for the pressure and temperature at the stagnation point. Note that

$$R = 1\,715 \text{ ft·lb/slug·°R} \qquad c_p = 6\,000 \text{ ft·lb/slug·°R}$$

$$k = 1.4 \qquad V_A = 800 \text{ ft/s} \qquad V_B = 900 \text{ ft/s}$$

Then, using Eq. 13.26

$$c_p T_1 + \frac{V_1^2}{2} = c_p T_2 + \frac{V_2^2}{2} = c_p T_s \tag{13.26}$$

we obtain

$$\frac{(800)^2}{2} = 6\,000(488.5 - T_A) \qquad T_A = 435.3\text{°R} \bullet \tag{13.26}$$

$$\frac{(900)^2}{2} = 6\,000(488.5 - T_B) \qquad T_B = 421.0\text{°R} \bullet \tag{13.26}$$

From the isentropic relationship between pressure and temperature, i.e.,

$$p = \rho R T \tag{1.3}$$

$$\frac{p}{\rho^k} = \text{Constant} \tag{1.7}$$

we obtain

$$\frac{p_A}{16.18} = \left(\frac{435.3}{488.5}\right)^{3.5} \qquad \frac{p_B}{16.18} = \left(\frac{421.0}{488.5}\right)^{3.5}$$

$$p_A = 10.85 \text{ psia} \qquad p_B = 9.65 \text{ psia} \bullet$$

Then, since

$$a = \sqrt{kRT} \rightarrow a_A = \sqrt{1.4 \times 1\,715 \times 435.3} = 1\,021 \text{ ft/s}$$
$$a_B = \sqrt{1.4 \times 1\,715 \times 421.0} = 983 \text{ ft/s} \tag{1.11}$$

$$\mathbf{M}_A = \frac{800}{1\,021} = 0.783 \qquad \mathbf{M}_B = \frac{900}{983} = 0.915 \bullet$$

Formal mathematical solution of the inverse of the foregoing problem, when boundary conditions are specified and the flowfield through a passage or about an object is to be determined (i.e., velocities, pressures, etc., predicted), is impossible; answers to such problems are obtained by approximations, linearizations, numerical integration on a computer of the equations of the motion, and so on, most of which are beyond the scope of this elementary book. However, it is useful to consider such a problem and some of the methods and limitations of its solution.

The flowfield around the simple streamlined object of Fig. 13.10 is to be predicted. The shape and orientation of the object are known, and the *boundary condition* of velocity,

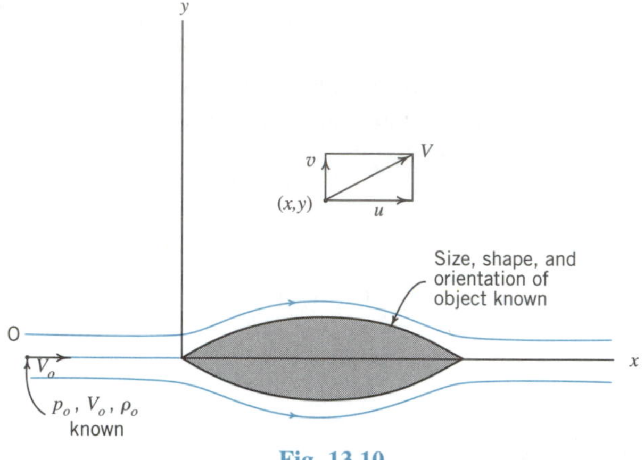

Fig. 13.10

pressure, and density of the fluid approaching the object is also known. The available independent equations are

$$\frac{\partial}{\partial x}(\rho u) + \frac{\partial}{\partial y}(\rho v) = 0 \tag{4.10}$$

$$-\frac{1}{\rho}\frac{\partial p}{\partial x} = u\frac{\partial u}{\partial x} + v\frac{\partial u}{\partial y} \tag{5.9a}$$

$$-\frac{1}{\rho}\frac{\partial p}{\partial y} = u\frac{\partial v}{\partial x} + v\frac{\partial v}{\partial y} \tag{5.9b}$$

$$\frac{p}{\rho^k} = \text{Constant} \tag{1.7}$$

The objective in solving these equations is to obtain u and v as functions of x and y. If such a solution can be obtained, pressures, temperatures, and fluid densities anywhere in the flowfield may also be predicted. The form and orientation of the object, of course, affect the flow picture; this enters the solution of the problem as a *boundary condition* expressing the fact that the velocity V along the surface of the body is everywhere tangent to the body. Mathematically, this means

$$\left[\frac{dy}{dx} = \frac{v}{u}\right]_{\text{at surface of body}}$$

The unknowns in the problem are seen to be u, v, x, y, ρ, and p, and with four equations and two boundary conditions the unknowns are seen to be determinable; however, analytic integration of such equations is not possible, so a formal mathematical solution of the problem cannot be obtained. However, it is possible to attack such problems by successive approximations in which the flow solution is improved systematically in a sequence of trials or by direct computer numerical solution of the so-called finite difference representations of the governing differential equations. The numerical methods are very powerful and have yielded a significant number of useful results and insights to the details of the motion.

The isentropic treatment of such compressible flowfields around solid objects is limited to the case where velocities are everywhere subsonic. If the approaching and leaving velocities (Fig. 13.10) are both subsonic and at a point on the body (near its midsection) the velocity becomes supersonic, a shock wave is to be expected downstream from this point where the velocity becomes subsonic again. Analysis of such problems is exceedingly complex because of the unknown position and extent of the shock wave, and its nonisentropic and discontinuous nature. Upstream from and out-board of the shock wave the flow of the ideal compressible fluid is (whether subsonic or supersonic) irrotational and isentropic, but downstream from such waves (in general) the flow is rotational and nonisentropic; however, if such shock waves are straight (or essentially so), the flowfield downstream from them may be shown to be irrotational and isentropic; this simplifies the problem somewhat.

Flowfields within passages present the same difficulties as the external flowfields described above, so further examples need not be cited here. In general, such internal flowfields are less amenable to the use of the ideal fluid because of the pervasiveness of wall frictional effects which render the isentropic (irrotational) assumption invalid. However, such methods may be effective for short passages, duct inlets, and in other cases where wall friction is of small importance.

Shock Waves

In Section 13.9 it was found that, for given entry conditions and configuration, there existed a range of exit pressures for which isentropic flow through a converging-diverging nozzle is not possible. Experiment shows that under these conditions the flow in the nozzle undergoes, at some point in the diverging section, an abrupt change from supersonic to subsonic velocity. This change is accompanied by large and abrupt rises in pressure, density, and temperature. The zone in which these changes take place is so thin that, for computations outside the zone, it may be considered to be a single line, that is, a discontinuity in the flow. This discontinuity is called a *normal shock wave* (that is, a shock wave perpendicular to the flow direction). The actual thickness of a normal shock wave is estimated to be of the order of the mean free path of the fluid molecules, that is, a micrometre (or between 10^{-4} and 10^{-5} inches). In this wave the gradients of velocity and density are so steep that viscous action, heat conduction, and mass diffusion are all appreciable with the result that the flow also undergoes a large entropy increase as it passes through the wave.

The normal shock wave is actually only a special case of the broader class of flow discontinuities called *oblique shock waves* that are found in most supersonic flows, both internal (in ducts, pipes, jet engine intakes, and compressors) and external (over the surfaces of wings, spacecraft, and so forth). These compression shock waves are analogous to the hydraulic jump and oblique standing waves, that were discussed in Section 6.3, and, comparable to the nozzle flow, occur because, for a given flowrate and downstream and upstream water depths, there exist no ideal flow (that is, isentropic) solutions to the governing equations. On the basis of this similarity, many meaningful model studies of supersonic gas flows have been made in water channels by use of the *so-called* hydraulic analogy.

13.12 THE NORMAL SHOCK WAVE

By applying the continuity principle to the normal shock wave of Fig. 13.11, we find that

$$\dot{m} = A_1\rho_1 V_1 = A_2\rho_2 V_2 \tag{13.27}$$

The impulse-momentum principle, applied as in the hydraulic jump, gives

$$\Sigma F_x = p_1 A_1 - p_2 A_2 = (V_2 - V_1)\dot{m}$$

After eliminating $\dot{m}$ and noting that $A_1 = A_2$, combination of these equations yields

$$p_2 - p_1 = (\rho_1 V_1^2 - \rho_2 V_2^2) \tag{13.28}$$

which allows the pressure jump $(p_2 - p_1)$ across the wave to be computed. However, this requires the use of another equation to obtain V_2 from p_1, γ_1, and V_1. The adiabatic energy equation

$$\frac{V_2^2}{2} - \frac{V_1^2}{2} = h_1 - h_2 \tag{13.9}$$

may be used, because, although internal friction (with increase of entropy) is to be expected in the shock wave, there is no flow of heat to or from a fluid control volume enclosing the shock. For perfect gases, (see Eq. 13.10) $h_1 - h_2 = c_p(T_1 - T_2)$. With $c_p = Rk/(k - 1)$ and $RT = p/\rho$, Eq. 13.9 can be written

$$\frac{V_2^2}{2} - \frac{V_1^2}{2} = \frac{k}{k-1}\left(\frac{p_1}{\rho_1} - \frac{p_2}{\rho_2}\right) \tag{13.29}$$

and solved simultaneously with Eq. 13.28 to yield, after some algebraic manipulation, a relationship between the Mach numbers $\mathbf{M}_1$ and $\mathbf{M}_2$. This is

$$\mathbf{M}_2^2 = \frac{1 + \dfrac{k-1}{2}\mathbf{M}_1^2}{k\mathbf{M}_1^2 - \dfrac{k-1}{2}} \tag{13.30}$$

from which $\mathbf{M}_2$ is found, given $\mathbf{M}_1$. As this equation is satisfied when $\mathbf{M}_1 = \mathbf{M}_2 = 1$, all other solutions must have the property: for $\mathbf{M}_1 > 1$, $\mathbf{M}_2 < 1$ and for $\mathbf{M}_1 < 1$, $\mathbf{M}_2 > 1$;

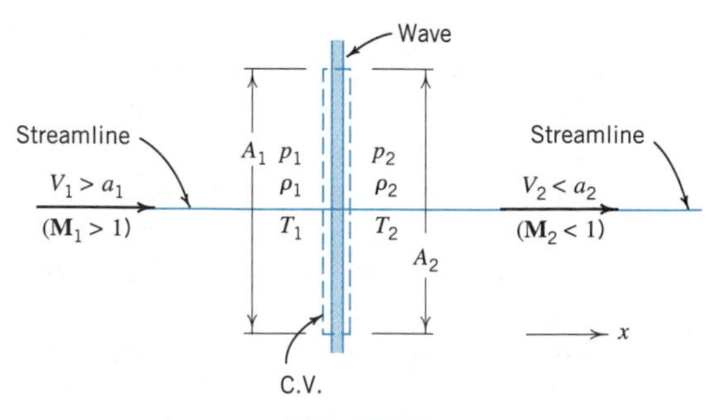

Fig. 13.11

however, it will be shown that only the first of these is physically possible—the second solution implies a *loss* of entropy through the wave and thus violates the Second Law of Thermodynamics (see Eq. 13.6). Accordingly it may be concluded that through a normal shock wave the velocity must fall from supersonic to subsonic.

The pressure jump through the shock wave may also be computed in terms of the *shock strength* $(p_2 - p_1)/p_1$ to yield

$$\frac{p_2 - p_1}{p_1} = \frac{2k}{k + 1}(\mathbf{M}_1^2 - 1) \tag{13.31}$$

and the velocity ratio, V_2/V_1,

$$\frac{V_2}{V_1} = \frac{(k - 1)\mathbf{M}_1^2 + 2}{(k + 1)\mathbf{M}_1^2} \tag{13.32}$$

and, from p_2, V_2, and $\mathbf{M}_2$, temperature, density, and sonic velocity downstream from the wave may be readily obtained.

A firm understanding of why a normal shock must always lead to $\mathbf{M}_2 < 1$ from $\mathbf{M}_1 > 1$ is simply obtained by constructing an enthalpy-entropy plot of the process. This exercise also leads to concepts that enjoy wide use for other compressible flow problems.[6] Such plots can be constructed for any gas; however, only perfect gases are considered here.

First, construct a line on an *h-s* plot (Fig. 13.12) for the adiabatic shock process of the locus of all states that satisfy only the continuity equation 13.27 and the energy equation 13.28. This is called a *Fanno* line. For given initial conditions p_1, V_1, ρ_1, T_1, and h_1, Eq. 13.9 gives

$$h_0 = h_1 + \frac{V_1^2}{2} = h + \frac{V^2}{2} \tag{13.9}$$

where h_0 is the stagnation (maximum) enthalpy. For a perfect gas $T = h/c_p$ and $p = \rho RT$. As the continuity equation 13.27 gives $\rho_1 V_1 = \rho V$ because the areas are the same on both sides of the shock, the entropy relation 13.7 for a perfect gas is

$$\begin{aligned} s - s_1 &= c_v \ln\left[\left(\frac{T}{T_1}\right)^k\left(\frac{p}{p_1}\right)^{1-k}\right] \\ &= c_v \ln\left[\left(\frac{h}{h_1}\right)\left(\frac{V_1}{V}\right)^{1-k}\right] \end{aligned} \tag{13.33}$$

Now, Eq. 13.9 gives V as a function of h and the initial (or equivalently, the stagnation) conditions; therefore, there is an explicit relation between s and h that can be plotted by varying h. A typical resulting Fanno line is sketched on Fig. 13.12.

Next, construct a line on Fig. 13.12 of the locus of all states that satisfy only the continuity equation 13.27 and the impulse-momentum equation 13.28. This is called a *Rayleigh* line. From Eq. 13.33

$$s - s_1 = c_v \ln\left[\left(\frac{h}{h_1}\right)^k\left(\frac{p}{p_1}\right)^{1-k}\right] \tag{13.34}$$

[6]See A. H. Shapiro, *The dynamics and thermodynamics of compressible fluid flow*, vol. I, The Ronald Press, 1953.

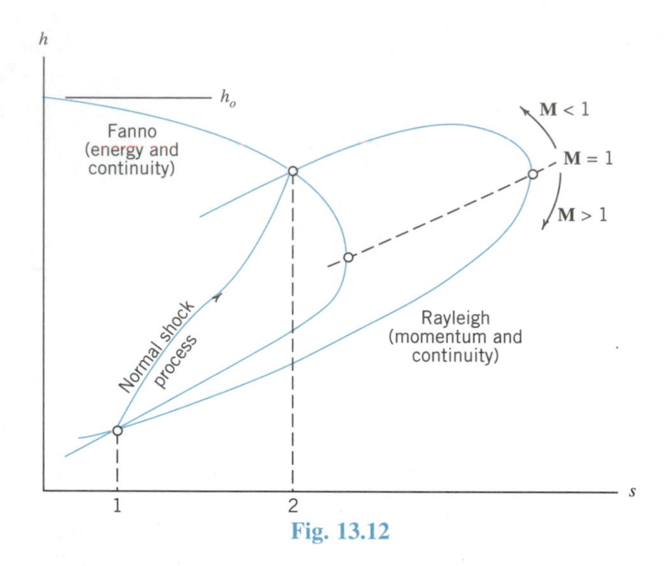

Fig. 13.12

As $\rho_1 V_1 = \rho V$ from continuity, Eq. 13.28 yields

$$\frac{p}{p_1} = 1 + \frac{\rho_1 V_1}{p_1}(V_1 - V)$$

Given an initial state (ρ_1, p_1, V_1, T_1 and $h_1 = c_p T_1$), then $\rho = \rho(V)$, $p/p_1 = f(V)$, and from the perfect gas relation $p/\rho = RT$ so $T = T(V)$ and $h = h(V)$. Accordingly, there is an explicit relationship between s and h that can be obtained by varying the parameter V. A typical resulting Rayleigh line is sketched on Fig. 13.12.

The continuity, impulse-momentum, and energy equations are all satisfied across a shock wave. Therefore, the normal shock process can occur only between the two states where the Rayleigh and Fanno lines intersect. Furthermore study of the above equations reveals that on each line the point where s is a maximum (where also $ds/dh = 0$) corresponds to $V = a$; that is, $\mathbf{M} = 1$. As the maximum value of h is h_0 when $V = 0$, the upper branches of the lines must correspond to $\mathbf{M} < 1$ and the lower branches to $\mathbf{M} > 1$. The Second Law of Thermodynamics (Section 13.1) requires that the entropy cannot decrease in an adiabatic process (Eq. 13.6); therefore, the normal shock process must go from $\mathbf{M} > 1$ (supersonic) to $\mathbf{M} < 1$ (subsonic) in every case.

ILLUSTRATIVE PROBLEM 13.10

Upstream from a normal shock wave in an airflow the pressure, velocity, and sonic velocity are 14.7 psia, 1 732 ft/s, and 862 ft/s, respectively. Calculate these quantities just downstream from the wave, and the rise in temperature through the wave.

SOLUTION

From the given data

$$p_1 = 14.7 \text{ psia} \qquad V_1 = 1\,732 \text{ ft/s} \qquad a_1 = 862 \text{ ft/s}$$

and the appropriate constants for air (see Appendix 2)

$$k = 1.4 \qquad R = 1\,715 \text{ ft·lb/slug·°R}$$

we can use Eq. 1.11

$$a = \sqrt{kRT} \tag{1.11}$$

and the definition of the Mach number to calculate

$$\mathbf{M}_1 = \frac{1\,732}{862} = 2.01$$

Then, Eq. 13.30

$$\mathbf{M}_2^2 = \frac{1 + \dfrac{k-1}{2}\mathbf{M}_1^2}{k\mathbf{M}_1^2 - \dfrac{k-1}{2}} \tag{13.30}$$

can be used to find

$$\mathbf{M}_2^2 = \frac{1 + 0.4(2.01)^2/2}{1.4(2.01)^2 - 0.4/2} = 0.331 \qquad \mathbf{M}_2 = 0.58 \tag{13.30}$$

Next, using Eqs. 13.31 and 13.32 yields

$$\frac{p_2 - p_1}{p_1} = \frac{2k}{k+1}(\mathbf{M}_1^2 - 1) \tag{13.31}$$

$$\rightarrow p_2 = 14.7\left[1 + \frac{2(1.4)}{2.4}(2.01^2 - 1)\right] = 66.8 \text{ psia } \bullet$$

$$\frac{V_2}{V_1} = \frac{(k-1)\mathbf{M}_1^2 + 2}{(k+1)\mathbf{M}_1^2} \rightarrow V_2 = 1\,732\left[\frac{0.4(2.01)^2 + 2}{2.4(2.01)^2}\right] = 646 \text{ ft/s } \bullet \tag{13.32}$$

Now,

$$a_2 = \frac{646}{0.58} = 1\,144 \text{ ft/s } \bullet$$

$$1\,144 = \sqrt{1.4 \times 1\,715 \times T_2} \qquad T_2 = 517\text{°R} \tag{1.11}$$

$$862 = \sqrt{1.4 \times 1\,715 \times T_1} \qquad T_1 = 310\text{°R} \tag{1.11}$$

The rise of temperature through the wave is $T_2 - T_1 = 207$°F. • The conditions which will produce this particular shock wave in a nozzle are shown in the Illustrative Problem 13.8 of Section 13.9.

13.13 THE OBLIQUE SHOCK WAVE

In an age of high-speed flight the reader has had casual acquaintance with the shock waves produced by objects traveling at supersonic speeds.[7] The geometry of such two-dimen-

[7]See Section 13.17 for photograph and analytical justification for such waves.

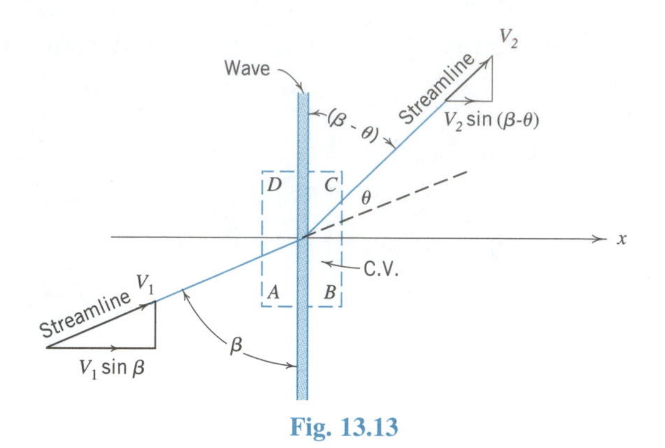

Fig. 13.13

sional waves may be studied by application of the continuity, impulse-momentum, and energy principles in the following manner.

Construct a control surface *ABCD* as indicated in Fig. 13.13, taking $BC = AD = 1$ unit, and also consider the control volume to be 1 unit deep perpendicular to the plane of the paper. From symmetry it is seen that the flowrates across *AB* and *CD* are equal; from this, it follows that the mass flowrates, $\dot{m}$, across *BC* and *AD* are also equal. These are

$$\dot{m} = \rho_1 V_1 \sin\beta = \rho_2 V_2 \sin(\beta - \theta) \tag{13.35}$$

Because the forces caused by the pressures on surfaces *AB* and *CD* are equal and opposite, the impulse-momentum equation applied in the *y*-direction discloses that the components of velocity parallel to the wave front are equal, giving

$$V_1 \cos\beta = V_2 \cos(\beta - \theta) \tag{13.36}$$

The impulse-momentum principle applied in the *x*-direction gives

$$p_1 - p_2 = [V_2 \sin(\beta - \theta) - V_1 \sin\beta]\dot{m} \tag{13.37}$$

The energy equation (without addition or extraction of heat) applied across the wave is (as in Section 13.12)

$$\frac{V_2^2}{2} - \frac{V_1^2}{2} = \frac{k}{k-1}\left(\frac{p_1}{\rho_1} - \frac{p_2}{\rho_2}\right) \tag{13.29}$$

These equations may be combined and manipulated to yield a variety of useful relationships, one of which is

$$\mathbf{M}_2^2 \sin^2(\beta - \theta) = \frac{1 + \dfrac{k-1}{2}\mathbf{M}_1^2 \sin^2\beta}{k\mathbf{M}_1^2 \sin^2\beta - \dfrac{k-1}{2}} \tag{13.38}$$

This equation reveals that (for $\mathbf{M}_1 > 1$): $\mathbf{M}_2 < 1$ for large β and $\mathbf{M}_2 > 1$ for small β. The so-called *strong shock* (featured by large β) is similar to the normal shock of Section 13.12 in that the flow changes from supersonic to subsonic in passing through the wave, whereas for small β (*weak shock*) the flow is supersonic on both sides of the wave. Through all such waves, however, there is internal friction with increase of entropy; the flow is rotational (see Section 13.10), and $\mathbf{M}_2 < \mathbf{M}_1$.

ILLUSTRATIVE PROBLEM 13.11

Upstream from an oblique shock wave in an airflow, the pressure, velocity, and sonic velocity are 14.7 psia, 1 732 ft/s, and 862 ft/s, respectively. The wave angle (β) is 40°. Calculate angle (θ) required to produce this wave, the pressure, velocity, and sonic velocity just downstream from the wave, and the rise of temperature through the wave.

SOLUTION

Given

$$p_1 = 14.7 \text{ psia} \qquad V_1 = 1\,732 \text{ ft/s} \qquad a_1 = 862 \text{ ft/s} \qquad \beta = 40°$$

$$R = 1\,715 \text{ ft·lb/slug·°R} \qquad c_p = 6\,000 \text{ ft·lb/slug·°R} \qquad k = 1.4$$

we obtain

$$\mathbf{M}_1 = \frac{1\,732}{862} = 2.01$$

and using Eq. 1.11

$$a = \sqrt{kRT} \rightarrow 862 = \sqrt{1.4 \times 1\,715 \times T_1} \qquad T_1 = 310°\text{R} \tag{1.11}$$

Applying Eq. 13.31, approximately adjusted with $\sin \beta = 0.642$, i.e.,

$$\frac{p_2 - p_1}{p_1} = \frac{2k}{k+1}(\mathbf{M}_1^2 \sin^2\beta - 1) \tag{13.31}$$

produces

$$p_2 = 14.7\left[1 + \frac{2(1.4)}{2.4}(2.01^2 \times 0.642^2 - 1)\right] = 26.0 \text{ psia} \bullet \tag{13.31}$$

Applying Eq. 13.32, appropriately adjusted with $\sin \beta$ and $\sin (\beta - \theta)$,

$$\frac{V_2 \sin (\beta - \theta)}{V_1 \sin \beta} = \frac{(k-1)\mathbf{M}_1^2 \sin^2\beta + 2}{(k+1)\mathbf{M}_1^2 \sin^2\beta} \tag{13.32}$$

produces

$$\frac{V_2 \sin (\beta - \theta)}{V_1 \sin \beta} = \frac{(k-1)\mathbf{M}_1^2 \sin^2\beta + 2}{(k+1)\mathbf{M}_1^2 \sin^2\beta} \tag{13.32}$$

and, using $V_2/V_1 = \cos \beta/\cos (\beta - \theta)$ from Eq. 13.36,

$$\frac{\tan (\beta - \theta)}{\tan 40°} = \frac{0.4(2.01 \times 0.642)^2 + 2}{2.4(2.01 \times 0.642)^2} \qquad \beta - \theta = 29.3° \qquad \theta = 10.7° \bullet$$

Then, Eq. 13.38

$$\mathbf{M}_2^2 \sin^2(\beta - \theta) = \frac{1 + \frac{k-1}{2}\mathbf{M}_1^2 \sin^2\beta}{k\mathbf{M}_1^2 \sin^2\beta - \frac{k-1}{2}} \tag{13.38}$$

gives

$$\mathbf{M}_2^2 = \frac{1}{(0.489)^2} \times \frac{1 + (0.4/2)(2.01 \times 0.642)^2}{1.4(2.01 \times 0.642)^2 - 0.4/2} = 2.635 \qquad (13.38)$$

$$\mathbf{M}_2 = 1.62$$

$$V_2 = \frac{1\,732 \cos 40^\circ}{\cos 29.3^\circ} = 1\,523 \text{ ft/s} \bullet \qquad (13.36)$$

$$a_2 = \frac{1\,523}{1.62} = 940 \text{ ft/s} \bullet$$

$$940 = \sqrt{1.4 \times 1\,715 \times T_2} \qquad T_2 = 367.5^\circ\text{R} \bullet \qquad (1.11)$$

or using Eq. 13.10

$$c_pT_1 + \frac{V_1^2}{2} = c_pT_2 + \frac{V_2^2}{2} \rightarrow \frac{(1\,732)^2 - (1\,523)^2}{2} = 6\,000(T_2 - T_1) \qquad (13.10)$$

$$T_2 - T_1 = 57.5^\circ \qquad T_2 = 367.5^\circ\text{R} \bullet$$

Real Fluid Flow in Pipes

13.14 PIPE FRICTION

The techniques in Chapter 9 are generally valid for flowing liquids. However, the analysis of the flow of gases in pipes often requires an account of compressibility effects. The calculation of pressure, velocity, temperature, and density changes caused by friction and heat transfer during the flow of gases in pipelines is, in general, a rather complex thermodynamic process and such problems cannot be treated exhaustively here. On the other hand, isothermal (constant temperature) and adiabatic flows of gas in pipes have practical applications and serve as a basis for both qualitative and quantitative examples of pipe flow situations with a fluid of varying density. The analysis is limited to conditions under which gases act as perfect gases and to one-dimensional flows.

To begin it is necessary to recall the four basic equations of compressible flow, that is, the fluid equation of state and the continuity, energy, and momentum equations which will be written in differential form now and then integrated. In addition the Second Law of Thermodynamics will be useful.

The equation of state for a perfect gas is

$$\frac{p}{\rho} = RT \qquad (1.3)$$

Furthermore, the specific heats c_p and c_v are constant and $k = c_p/c_v$ ($k = 1.4$ for air). The continuity equation 4.2 is

$$A\rho V = \text{constant} \qquad (4.2)$$

It follows that the weight flowrate $G = A\gamma V$ is constant as well as the mass flowrate $\dot{m} = A\rho V$. In such a flow the Reynolds number

$$\mathbf{R} = \frac{Vd\rho}{\mu} = \frac{G}{A\gamma}\frac{d\rho}{\mu} = \frac{Gd}{\mu g_n A} = \frac{\dot{m}d}{\mu A} \tag{13.39}$$

for a specific flow depends only on temperature because μ is a function of T while $\dot{m}$, d, and A are constants of the flow.

The energy equation 13.1 for steady compressible flow can be written as

$$\frac{d}{dl}(q_H) = \frac{d}{dl}h + \frac{d}{dl}\left(\frac{V^2}{2}\right) \tag{13.40}$$

for changes over a differential distance dl under the assumptions that there is no shear or shaft work and that the elevation terms can be neglected. Recall that q_H is the heat transfer per unit mass of fluid flowing and $h = p/\rho + ie = p/\rho + c_vT = c_pT$. Thus, Eq. 13.40 can be written as

$$c_p\frac{dT}{dl} + V\frac{dV}{dl} = \frac{d}{dl}(q_H) \tag{13.41}$$

for a perfect gas.

According to Eq. 7.33 the momentum equation for a differential length of pipe can be written in the form (because $\gamma = \rho g_n$, $R_h = d/4$, and elevation effects are negligible)

$$dp + \rho V\,dV + 4\tau_o\,dl/d = 0$$

Substituting $f\rho V^2/8$ for τ_o produces

$$dp + \rho V\,dV + f\rho V^2\,dl/2d = 0 \tag{13.42}$$

The Mach number $\mathbf{M} = V/a$ where (see Eq. 1.10) $a = \sqrt{kp/\rho} = \sqrt{kRT}$. Rewriting Eq. 13.42 in terms of $\mathbf{M}$ gives

$$\frac{dp}{p} + k\mathbf{M}^2\left(\frac{d\mathbf{M}}{\mathbf{M}} + f\frac{dl}{2d}\right) = 0 \tag{13.43}$$

when a and T are constant (isothermal flow). For fully developed pipe flow Shapiro[8] points out that experiments have shown that f values (cf., Fig. 9.10) for incompressible flow are applicable to compressible flows when $\mathbf{M} < 1$. However, for $\mathbf{M} > 1$ the compressible friction factors are about one-half of the corresponding incompressible value.

According to Eq. 4.2, $\rho V =$ constant in a pipe. Thus, as $\rho = p/RT$, $pV = RT$ for a perfect gas. Again, in the case $T =$ constant, $dp/p = -dV/V = -d\mathbf{M}/\mathbf{M}$ and Eq. 13.43 becomes

$$\frac{dp}{p} = -\frac{dV}{V} = -\frac{d\mathbf{M}}{\mathbf{M}} = -\frac{k\mathbf{M}^2}{1 - k\mathbf{M}^2}f\frac{dl}{2d} \tag{13.44}$$

for isothermal flow.

[8]A. H. Shapiro, *The dynamics and thermodynamics of compressible fluid flow*, vol. I, Ronald Press Co., 1953, pp. 184ff.

13.15 ISOTHERMAL PIPE FLOW

To secure isothermal flow in a pipe the heat transferred out of the fluid through the pipe walls and the energy converted into heat by the friction process must be adjusted so that the fluid temperature remains constant. Such an adjustment is approximated naturally in uninsulated pipes where velocities are low (well below sonic) and where temperatures inside and outside the pipe are of the same order; frequently the flow of gases in long pipelines may be treated isothermally.

In an isothermal flow the Reynolds number **R** (Eq. 13.39) is constant regardless of changes in V or ρ because the temperature and hence μ are constant. The friction factor is then constant also. Accordingly, dividing Eq. 13.42 by ρV^2, replacing ρ by p/RT and V^2 by $\dot{m}^2/A^2\rho^2$ gives

$$\frac{A^2}{\dot{m}^2RT}\,p\,dp + \frac{dV}{V} + \frac{f}{2d}\,dl = 0$$

Integrating this equation between the two points 1 and 2 of a control volume in the pipe of Fig. 13.14 produces

$$\frac{A^2}{\dot{m}^2RT}\int_{p_1}^{p_2} p\,dp + \int_{V_1}^{V_2}\frac{dV}{V} + \frac{f}{2d}\int_0^l dl = 0$$

or

$$\frac{A^2}{\dot{m}^2RT}\,\frac{p_2^2 - p_1^2}{2} + \ln\frac{V_2}{V_1} + f\,\frac{l}{2d} = 0$$

Thus,

$$p_1^2 - p_2^2 = \frac{\dot{m}^2RT}{A^2}\left(2\ln\frac{V_2}{V_1} + f\,\frac{l}{d}\right) \tag{13.45}$$

This equation can be rearranged by use of the continuity equation $A_1\rho_1V_1 = A_2\rho_2V_2$ in which $A_1 = A_2$, $\rho_1 = p_1/RT$, and $\rho_2 = p_2/RT$. Thus, by substitution, $p_1V_1 = p_2V_2$ and $V_2/V_1 = p_1/p_2$. Hence Eq. 13.45 may be written

$$p_1^2 - p_2^2 = \frac{\dot{m}^2RT}{A^2}\left(2\ln\frac{p_1}{p_2} + f\,\frac{l}{d}\right) \tag{13.46}$$

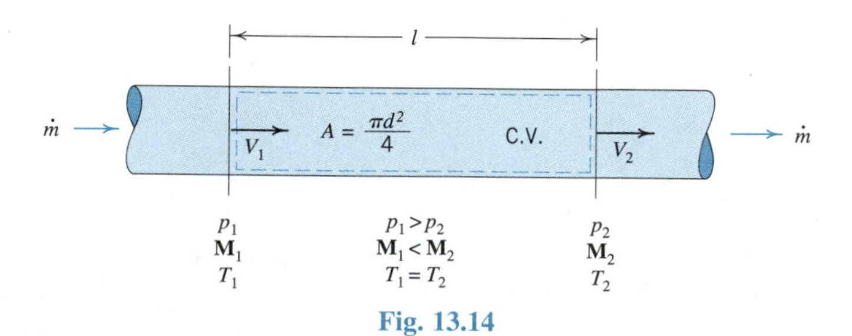

Fig. 13.14

Generally the solution of this equation for p_2 must be accomplished by trial, but frequently $2 \ln p_1/p_2$ is so small (in comparison with fl/d) that it may be neglected, thus allowing a direct solution.

While the actual pressure and velocity at any point in the pipe can be obtained by use of Eq. 13.46 and a set of prescribed values of p_1, $\mathbf{M}_1$, T_1, A, $\dot{m}$, f, l, and d, the qualitative aspects of the flow are obtained easily by examination of Eq. 13.44. Thus, if $\mathbf{M} < k^{-1/2}$ the pressure decreases, the velocity increases, the Mach number increases and (as $\rho = p/RT$) the density decreases with increases in l, that is, as point 2 moves downstream (positive l) from point 1. However, if $\mathbf{M}$ becomes greater than $k^{-1/2}$ these trends reverse. In principle then as the value of $\mathbf{M}$ passes $k^{-1/2}$, a reduction of l becomes necessary to achieve a further reduction of p in a subsonic flow. The pressure cannot drop below this point and thus Eq. 13.46 is applicable only between the pressure p_1 and the limiting value of p_2. Figure 13.15 gives an example of the pressure ratio p_2/p_1 as a function of fl/d for various Mach numbers $\mathbf{M}_1$ for isothermal flow of air ($k = 1.4$). The dotted lines indicate the limits of incompressible flow and $\mathbf{M} < 0.845$. In this case of isothermal pipe flow then, the limiting Mach number is not unity, rather the maximum flow occurs for $\mathbf{M} = k^{-1/2}$: this is the *choked* flow.

Equation 13.46 may also be written (in conformance with modern practice in gas dynamics) in terms of pressure ratio and Mach number, $\mathbf{M}$. Divide the equation by p_1^2 and multiply the right-hand side by k/k and substitute $A\rho_1 V_1$ for $\dot{m}$; whereupon $\dot{m}^2 RTk/kp_1^2 A^2$ will be found to be equal to $k\mathbf{M}_1^2$. Since the sonic velocity, a, is constant for isothermal flow, $V_2/V_1 = \mathbf{M}_2/\mathbf{M}_1$ and, since $V_2/V_1 = p_1/p_2$ (see above), $p_2/p_1 = \mathbf{M}_1/\mathbf{M}_2$ and Eq. 13.46 becomes

$$\frac{\mathbf{M}_1^2}{\mathbf{M}_2^2} = 1 - k\mathbf{M}_1^2\left(2 \ln \frac{\mathbf{M}_2}{\mathbf{M}_1} + f\frac{l}{d}\right) \tag{13.47}$$

Considering the flow process in terms of Mach number, $\mathbf{M}_1 \ll 1$ and $\mathbf{M}_2 > \mathbf{M}_1$. From the above-mentioned limitation of the equation the limiting value of $\mathbf{M}_2$ may be found by differentiating Eq. 13.47 with respect to l and setting $dp_2/dl = \infty$; the result is $\mathbf{M}_2 = \sqrt{1/k}$. Thus, as noted above, the equation is applicable only to that portion of the subsonic flow regime where $\mathbf{M}_1 < \mathbf{M}_2 \leq \sqrt{1/k}$. Such limitations are of far-reaching importance in engineering problems of gas flow.

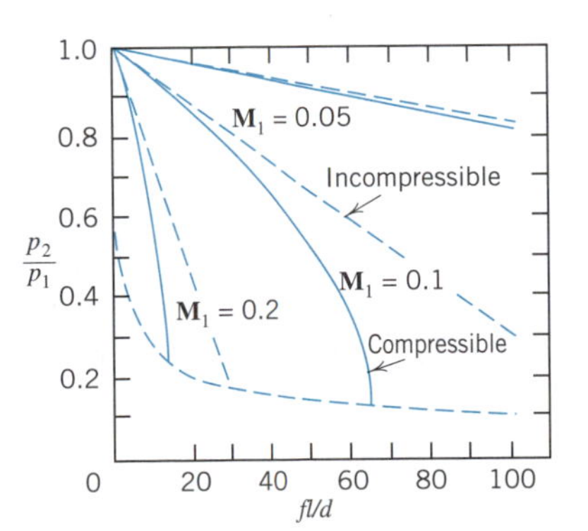

Fig. 13.15 Isothermal pipe flow pressure ratio for perfect gas ($k = 1.4$) as a function of initial conditions and fluid and pipe character.

ILLUSTRATIVE PROBLEM 13.12

If 18 kg/min of air flow isothermally through a smooth 75 mm pipeline at a temperature of 40°C, and the absolute pressure at a point in this line is 350 kPa, calculate the pressure in the line 600 m downstream from this point.

SOLUTION

The given data and fluid properties are

$$\dot{m} = 18 \text{ kg/min} \qquad d = 75 \text{ mm} \qquad T = 313 \text{ K}$$

$$p_1 = 350 \text{ kPa (abs)} \qquad l = 600 \text{ m} \qquad R = 286.8 \text{ J/kg·K}$$

$$k = 1.4 \qquad \mu = 1.92 \times 10^{-5} \text{ Pa·s}$$

Equation 13.39 yields the Reynolds number

$$\mathbf{R} = \frac{Vd\rho}{\mu} = \frac{\dot{m}d}{\mu A} = \frac{(18/60)(0.075)}{1.92 \times 10^{-5} \times 0.004\,4} = 266\,000 \tag{13.39}$$

The Moody diagram of Fig. 9.10 produces $f \cong 0.015$. Using Eqs. 1.3, 4.4 and 1.11, respectively, gives us:

$$\rho = p/RT \rightarrow \rho_1 = \frac{350 \times 10^3}{286.8 \times 313} = 3.90 \text{ kg/m}^3 \tag{1.3}$$

$$Q = AV \rightarrow V_1 = \frac{(18/60)}{0.004\,4 \times 3.9} = 17.5 \text{ m/s} \tag{4.4}$$

$$a = \sqrt{kRT} \rightarrow a = \sqrt{1.4 \times 286.8 \times 313} = 354 \text{ m/s} \tag{1.11}$$

$$\mathbf{M}_1 = \frac{17.5}{354} = 0.049$$

The limiting value of $\mathbf{M}_2$ is $\sqrt{1/1.4} = 0.845$. Calculating the limiting value of l for applicability of Eq. 13.46 from Eq. 13.47,

$$\left(\frac{0.049}{0.845}\right)^2 = 1 - 1.4(0.049)^2\left[2 \ln \frac{0.845}{0.049} + 0.015\frac{l}{0.075}\right]$$

$$l = 1\,454 \text{ m}$$

Since $600 < 1\,454$, Eq. 13.46 is applicable. Substituting values therein,

$$[(350 \times 10^3)^2 - p_2^2] = \frac{(18/60)^2 \times 286.8 \times 313}{(0.004\,4)^2} \times \left[2 \ln \frac{350 \times 10^3}{p_2} + 0.015\frac{600}{0.075}\right]$$

Solving for p_2 by trial, $p_2(\text{abs}) = 269 \times 10^3 \text{ Pa} = 269 \text{ kPa}$. ●

13.16 ADIABATIC PIPE FLOW WITH FRICTION

In short lengths of pipe or in fully insulated pipes the heat transfer is close to zero, and compressible flow in such pipes is taken to be adiabatic. Because there is no heat transfer and frictional (irreversible) processes are involved, it is clear that the entropy of the fluid must rise as the flow moves downstream. In Section 13.12 it was pointed out that the locus of all states on an enthalpy-entropy diagram for adiabatic flows (they satisfy the continuity and energy equations) is a Fanno line (see Fig. 13.12). From Fig. 13.12, it follows that in adiabatic flows, the Mach number increases downstream in subsonic flow, but decreases downstream in a supersonic adiabatic flow. Using the equations presented in Section 13.14 it is possible to prove these and other conclusions directly and to make useful flow computations.

In the case of adiabatic flow, it is convenient to assemble all the relevant differential equations. Writing Eq. 13.42 in terms of $\mathbf{M}$ gives when T is not constant

$$\frac{dp}{p} + k\mathbf{M}^2\left(\frac{dV}{V} + f\frac{dl}{2d}\right) = 0 \tag{13.48}$$

To obtain an expression involving only p, $\mathbf{M}$, and the parameter $f\,dl/d$, it is necessary to utilize relationships derived before. According to Eq. 4.2 $\rho V =$ constant so

$$\frac{d\rho}{\rho} = -\frac{dV}{V} \tag{13.49}$$

while from Eq. 1.3

$$\frac{dp}{p} = \frac{d\rho}{\rho} + \frac{dT}{T} \tag{13.50}$$

For $q_H = 0$ Eq. 13.41 yields

$$c_p\,dT + V\,dV = 0$$

Dividing by c_pT and using the expressions $c_p = Rk/(k-1)$ and $\mathbf{M}^2 = V^2/kRT$ leads to

$$\frac{dT}{T} = -(k-1)\mathbf{M}^2\frac{dV}{V} \tag{13.51}$$

Introducing Eqs. 13.49 and 13.51 into 13.50 now gives

$$\frac{dp}{p} = -[1 + (k-1)\mathbf{M}^2]\frac{dV}{V} \tag{13.52}$$

This result is used in Eq. 13.48 to obtain

$$\frac{dp}{p} = -\frac{k\mathbf{M}^2[1 + (k-1)\mathbf{M}^2]}{2(1-\mathbf{M}^2)} \cdot f\frac{dl}{d} \tag{13.53}$$

Then this result is used in Eq. 13.52 to obtain

$$\frac{dV}{V} = \frac{k\mathbf{M}^2}{2(1-\mathbf{M}^2)} \cdot f\frac{dl}{d} \tag{13.54}$$

As $\mathbf{M}^2 = V^2/kRT$,

$$\frac{d\mathbf{M}^2}{\mathbf{M}^2} = \frac{dV^2}{V^2} - \frac{dT}{T} = \frac{2\,dV}{V} - \frac{dT}{T}$$

Introducing Eq. 13.51 produces

$$\frac{d\mathbf{M}^2}{\mathbf{M}^2} = \frac{k\mathbf{M}^2\left[1 + \frac{(k-1)}{2}\mathbf{M}^2\right]}{1 - \mathbf{M}^2} \cdot f\frac{dl}{d} \tag{13.55}$$

when Eq. 13.54 is also used.

Noting that $(f/d)\,dl$ is positive in Eqs. 13.53, 13.54, and 13.55 leads to the following conclusions. If $\mathbf{M} < 1$, the pressure decreases, the velocity increases, and Mach number increases in the downstream direction in adiabatic flow. The flow accelerates in the subsonic case as a result of the friction effects. When $\mathbf{M} > 1$, the trends reverse. Thus the flow decelerates, but the pressure rises in the downstream direction. In adiabatic flow $\mathbf{M}$ tends to unity in either the subsonic or supersonic cases. Clearly $\mathbf{M} = 1$ is the limiting case. Given the conditions at a point in a pipe, the distance l downstream of that point can be increased until $\mathbf{M} = 1$ at the discharge point. No further increase in length can be made without either altering the given conditions or introducing a shock wave because of the reversal of flow trends at $\mathbf{M} = 1$.

To make a flow computation it is necessary to integrate Eq. 13.55

$$\frac{1}{d}\int_0^l f\,dl = \int_{\mathbf{M}_1^2}^{\mathbf{M}_2^2} \frac{1 - \mathbf{M}^2}{k\mathbf{M}^4\left[1 + \frac{(k-1)}{2}\mathbf{M}^2\right]}\,d\mathbf{M}^2 \tag{13.56}$$

On the left side of this equation one can either assume that f is a constant (often an adequate approximation, at high $\mathbf{R}$ in rough pipes for example where $f \approx$ constant) or define an average friction factor

$$\bar{f} = \frac{1}{l}\int_0^l f\,dl$$

In the subsonic flows considered here, $\bar{f}$ is assumed to be equal to f and to the usual incompressible value given by Fig. 9.10. Under these conditions Eq. 13.56 can be integrated to obtain a working relationship for $\mathbf{M}$ as a function of f, l, and d.

To find the maximum length of pipe l_{max} possible in a continuous flow Eq. 13.56 is integrated with $\mathbf{M}_2 = 1$ and $l = l_{max}$. The result is

$$\bar{f}\frac{l_{max}}{d} = \frac{1 - \mathbf{M}_1^2}{k\mathbf{M}_1^2} + \frac{k+1}{2k}\ln\left\{\frac{(k+1)\mathbf{M}_1^2}{2\left[1 + \frac{(k-1)}{2}\mathbf{M}_1^2\right]}\right\} \tag{13.57}$$

Attempts to use a larger value of l than l_{max} will reduce the flowrate, producing in effect a smaller $\mathbf{M}_1$. Here again the flow is said to be *choked*, but here the limiting Mach number is unity [cf., the converging–diverging nozzle (Section 13.9) and isothermal pipe flow (Section 13.15)].

ILLUSTRATIVE PROBLEM 13.13

Given $\bar{f} = f = 0.02$ find l_{max}/d as a function of $\mathbf{M}_1^2$ for the adiabatic flow of air.

SOLUTION

For air $k = 1.4$ and the friction factor is 0.02. For the adiabatic flow, Eq. 13.57

$$\bar{f}\frac{l_{max}}{d} = \frac{1 - \mathbf{M}_1^2}{k\mathbf{M}_1^2} + \frac{k+1}{2k}\ln\left\{\frac{(k+1)\mathbf{M}_1^2}{2\left[1 + \frac{(k-1)}{2}\mathbf{M}_1^2\right]}\right\} \tag{13.57}$$

applies and the working equation is therefore

$$\frac{l_{max}}{d} = 50\left\{\frac{1 - \mathbf{M}_1^2}{1.4\mathbf{M}_1^2} + 0.86\ln\left[\frac{1.2\mathbf{M}_1^2}{1 + 0.2\mathbf{M}_1^2}\right]\right\}$$

The results can be plotted as shown below.

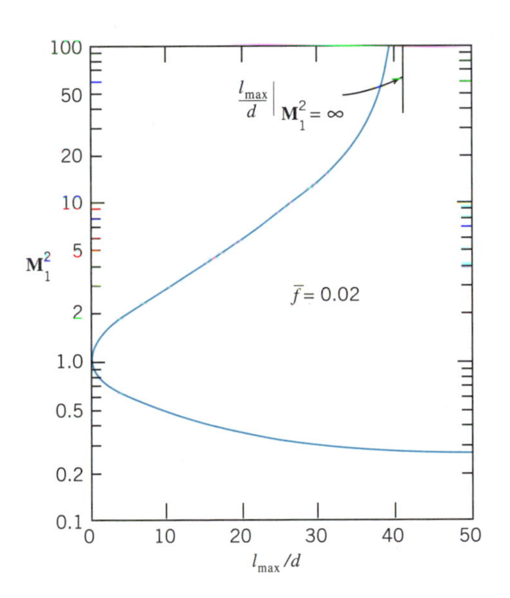

Drag and Lift

The problems of flow of compressible fluids about solid objects cover a vast field and are highly complex in their physical and mathematical aspects. Analytical solutions are available for only a small number of the less difficult problems, but steady progress is being made by theoretical research and through physical and numerical experiments. In view of the complexity and advanced nature of such problems, only the most rudimentary ideas and results can be presented in the following elementary treatment.

13.17 THE MACH WAVE

Consider the motion of a tiny source of disturbance such as a needle point or razor edge moving through a fluid (Fig. 13.16). If the fluid were truly incompressible its modulus of elasticity would be infinite, the velocity of propagation (Eqs. 1.9, 1.10, or 1.11) of the disturbance through the fluid would also be infinite, and the Mach number always zero. The effects of the disturbance would be felt instantaneously through the flowfield, and the fluid in front of the object would "know" of the presence of the object and adjust itself accordingly as the object approached. This situation is closely approximated in real fluid when the flow velocities are small compared with the sonic velocity (which is the velocity of propagation of the disturbance) and produces the conventional flowfields with which the reader is now familiar.

For the opposite case in which the velocity of motion of the disturbances is appreciably greater than the sonic velocity, a very different situation develops in which fluid upstream from the object is "unaware" of its existence; as the object encounters this fluid, the fluid must suddenly change its direction, producing the sharp discontinuities known as *shock waves* (Section 13.13). To investigate this, assume the point of Fig. 13.16 to move at a (supersonic) velocity V_o, and let it occupy positions 1, 2, 3, at times t_1, t_2, and t_3. At time t_1 the disturbance sends out an elastic wave from point 1 with a celerity a_o, and after the time $(t_3 - t_1)$ has elapsed the distance covered by the wave is $a_o\,(t_3 - t_1)$. In this elapsed time, however, the source of disturbance has moved to a point 3, and $l_1 = V_o(t_3 - t_1)$; eliminating $(t_3 - t_1)$, the distance covered by the waves is $a_o l_1/V_o$. Similarly the wave which started at point 2 has (when the disturbance reaches point 3) covered a distance of $a_o l_2/V_o$, and many other waves have done likewise from numerous intermediate positions. The result is a wave front (or *oblique shock wave*) which represents the line of advance of the elastic waves; the fluid to the left of this is at rest and "unaware" of the presence of the disturbance. This wave front, known as a *Mach wave*, will be conical (three-dimensional) if produced by the needle point, and wedge-shaped (two-dimensional) if produced by the razor edge. The *Mach angle*, β_o, may be seen from the figure to be defined by $\sin\beta_o = a_o/V_o$. This is useful in estimating from photographs of such wave fronts the velocity, V_o, by measuring β_o on the photograph and computing a_o from the fluid properties.

The same wave picture is, of course, produced when the fluid moves with velocity V_o

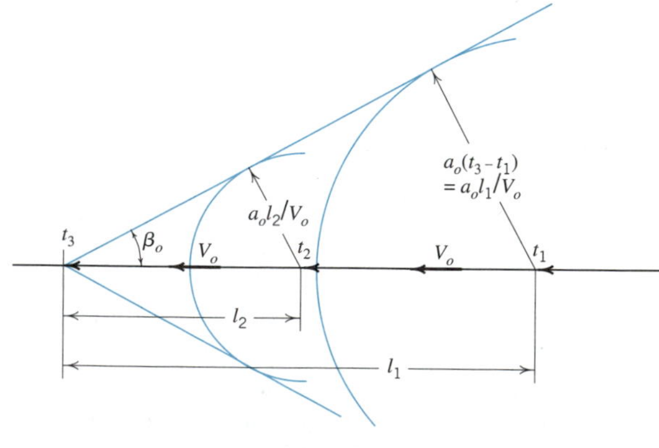

Fig. 13.16

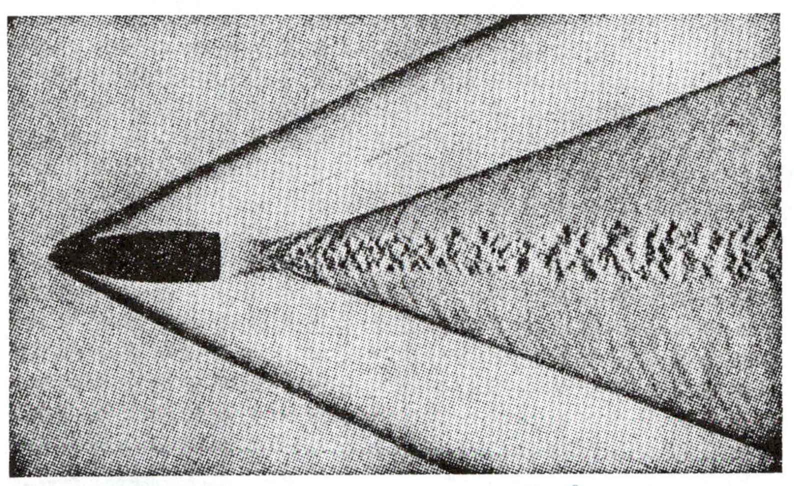

Fig. 13.17 Small-bore bullet in flight.[9]

past a source of disturbance at rest. On such a picture the streamlines may also be conveniently shown, and for a vanishingly small disturbance these will be straight parallel lines. For a disturbance of finite magnitude (Fig. 13.17) the velocity of propagation will be greater than a_o and the angle of the wave, therefore, greater than the Mach angle. Through a finite wave the streamlines will be deflected away from the object, velocities will diminish, and pressures increase (see Section 13.13), these changes being comparable in sense (not in magnitude)[10] to those occurring through the normal shock wave (Section 13.12). All these changes are very rapid as the shock wave is exceedingly thin; for most practical purposes it may be considered an abrupt discontinuity.

13.18 PHENOMENA AND DEFINITIONS

To visualize the flow phenomena in compressible flow about an object, consider the chain of events which occurs as the free stream velocity is increased from low subsonic to high subsonic to supersonic. At low subsonic speeds the fluid may be considered incompressible, resulting in a conventional streamline picture featuring a stagnation point, S, on the nose of the object and a point, m, of maximum velocity and minimum pressure on the upper side (Fig. 13.18*a*). Increasing the velocity of the free stream will, of course, raise the stagnation pressure, increase the maximum velocity, and lower the minimum pressure, even if compressibility of the fluid is neglected; inclusion of compressibility exaggerates these changes, however, and they continue until the velocity past m becomes equal to the *local sonic velocity.*[11] This means that the *local Mach number* at point m is now unity,

[9]From C. Cranz, *Lehrbuch der Ballistik*, vol. I, B. G. Teubner, Leipzig, 1917.

[10]The changes are largest in the case of the normal shock wave through which the velocity decreases from supersonic to subsonic; the velocities upstream and downstream from the oblique shock wave are both supersonic except in the case of relatively large β, where the oblique shock approaches the normal shock.

[11]It is very important to note that the pressure, density, and temperature at m are considerably less than those properties in the free stream and that the local sonic velocity (Eq. 1.11) is thus smaller than free stream sonic velocity.

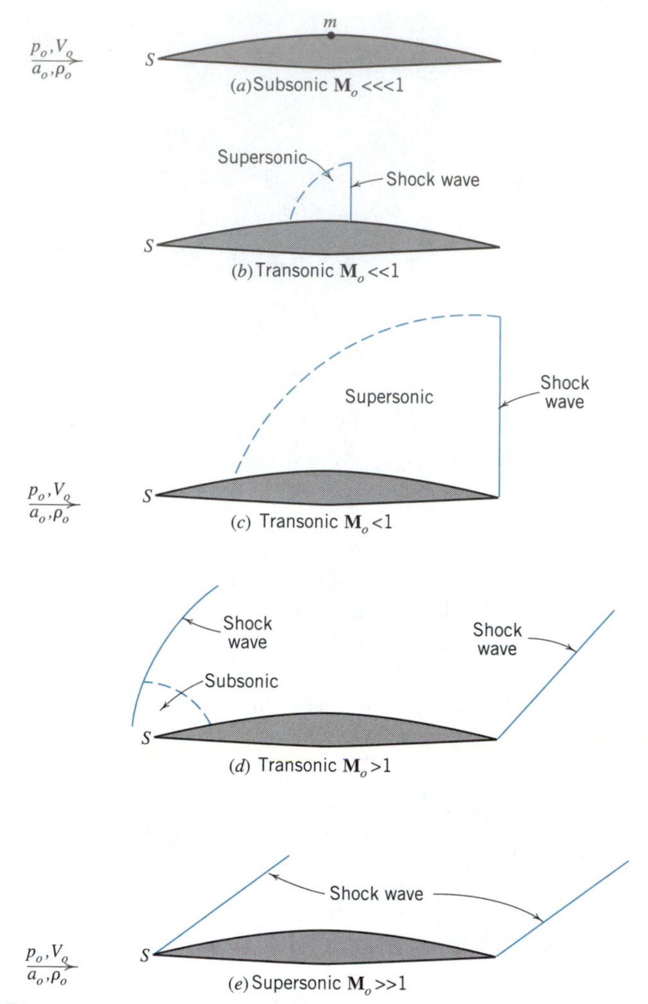

Fig. 13.18 Shock-wave phenomena on the upper side of an airfoil.

although the free stream Mach number is considerably less than this. In other words, the body is moving at subsonic speed, but sonic phenomena are beginning to appear at a point on its surface. This situation marks the end of the *subsonic* flow region (where all velocities in the flowfield are subsonic) and the beginning of the *transonic* regime (in which some velocities are subsonic and others supersonic); the transonic regime ends and the supersonic begins when all velocities in the field are supersonic.

A simplified[12] picture of a transonic flow is shown in Fig. 13.18*b*. Of particular interest is the large region of supersonic flow and the presence of the shock wave through which the velocity decreases from supersonic to subsonic. Through the shock wave the sudden

[12]The complex problem of shock wave-boundary layer interaction is omitted from these simplified sketches. In the boundary layer the velocities near the surface are subsonic while the flow outside the boundary layer and upstream from the shock wave is supersonic; the subsonic region allows transmission upstream of the adverse pressure gradient produced by the shock and thus spreads out the region of shock phenomena at the boundary surface.

jump in pressure produces an adverse pressure gradient in the boundary layer, which promotes separation and the usual effects on lift and drag. As the free stream velocity is further increased (but V_o maintained less than a_o) the shock wave is forced rearward (Fig. 13.18*c*) and the major portion of the flow immediately above the surface is supersonic although the free stream Mach number $\mathbf{M}_o$ is still less than 1.

Increasing the free stream velocity V_o to slightly more than a_o will produce the wave arrangement of Fig. 13.18*d*, featuring a new shock wave upstream from the body and the flowfield completely supersonic except in a small region[13] between this new shock wave and the nose of the object. The supersonic regime (Fig. 13.18*e*) will exist at higher free stream Mach numbers when the subsonic zone becomes of negligible size or vanishes entirely, as it will for a sharp-nosed object, to which the upstream shock wave will attach itself.

13.19 DRAG

In compressible fluid motion, drag results from energy dissipated in shock waves as well as from the skin friction and separation effects discussed in Sections 7.4 through 7.6 and 11.3. Skin friction drag may be computed approximately by the methods of Sections 7.4 through 7.6 up to free stream Mach numbers around 2, but boundary-layer thicknesses are much greater in compressible fluid motion and stability is greatly affected by the transfer of heat between solid surface and boundary layer. At higher Mach numbers, frictional heating in the boundary layer and adjacent surface becomes a serious problem. The approach to compressible fluid motion which assumes Mach number effects predominant and Reynolds number effects negligible implies that frictional effects and boundary layers are of little importance compared to shock phenomena. This should be recognized as an adequate working assumption but not an absolute fact, and many exceptions to this convenient rule are to be expected. For example, if methods are found for minimizing shock wave effects and preventing separation, a very large portion of total drag force will be composed of frictional drag, and Reynolds number effects will then by no means be negligible.

The variation of drag coefficients with Mach number (Fig. 13.19) is central to drag problems in compressible flows and may be examined fruitfully by considering the extreme cases of a streamlined airfoil or body of revolution and a blunt body such as a flat-nosed projectile. For the latter, separation of fluid from body is fixed in position by the geometry of the body, skin friction is small, and a steady increase in drag coefficient with Mach number results from the predominant compressibility effects on or near the flat nose of the projectile. Streamlining the tail of such an object will increase rather than reduce the drag coefficient, but pointing the nose, and thus reducing the frontal area near the stagnation point, will materially reduce the drag coefficient (see Fig. 13.20).

For a streamlined object the variation of drag coefficient is more interesting because of the changing position of the separation point which accompanies the formation of shock wave phenomena. Through the subsonic range (Fig. 13.18*a*) the slight decrease in drag coefficient with Reynolds number will appear on the plot (Fig. 13.19) of drag coefficient against Mach number since both of these numbers are directly proportional to V_o. As the

[13]Evidently the extent of this region depends on the nose geometry of the body since considerations of Section 13.17 have shown it to be nonexistent if the nose is sharply edged or pointed.

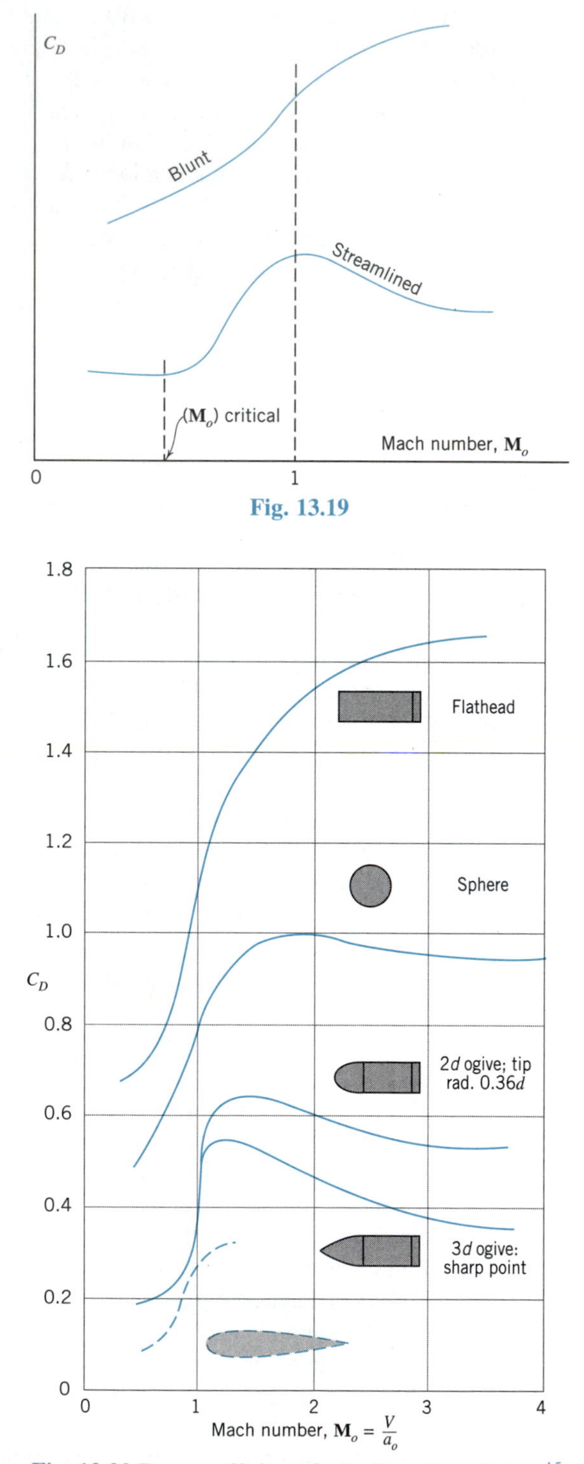

Fig. 13.19

Fig. 13.20 Drag coefficients for bodies of revolution.[15]

[15]Data from: F. R. W. Hunt, *The Mechanical Properties of Fluids*, p. 341, Blackie and Son, 1925; and A. C. Charters and R. N. Thomas, "The Aerodynamic Performance of Small Spheres from Subsonic to High Supersonic Velocities," *Jour. Aero. Sci.*, vol. 12, no. 4, October, 1945.

transonic range is entered (Fig. 13.18*b* and accompanying discussion), shock phenomena and separation appear in the vicinity of point *m* and drag coefficient begins to increase[14] rapidly. Toward the middle of the transonic range (Fig. 13.18*c*) the separation point has moved well to the rear, reducing the size of the wake, but shock phenomena have intensified and drag coefficient continues to rise but more slowly, eventually reaching a maximum value as these opposite effects cancel each other. In the supersonic range (Fig. 13.18*e*), the drag coefficient depends primarily on the energy dissipation through the inclined shock waves and decreases steadily with further increase in Mach number.

Of great practical interest is the conversion of airfoil data obtained for incompressible flow to flow at higher velocities where the effects of compressibility are significant. The appearance of shock phenomena prevents such conversion above the critical Mach number, but below this there is a considerable range of engineering interest, covering free stream velocities between 20 and (roughly) 80% of the velocity of sound. The higher figure is the critical Mach number and has been found to depend on the thickness of the airfoil, being larger for thinner airfoils. The method of conversion depends only on the Mach number and is known as the *Prandtl-Glauert rule*; the critical Mach number (the limit of application of the rule) may be established fairly reliably by application of elementary principles of compressible flow.

Pressure distribution data obtained for an airfoil at a certain angle of attack in incompressible flow are usually presented in dimensionless fashion as a plot of *pressure coefficient*, C_p, against the chord of the airfoil. The pressure coefficient is defined by

$$C_p = \frac{p - p_o}{\rho_o V_o^2/2} \tag{13.58}$$

in which p_o, V_o, and ρ_o are at free stream conditions and p is the pressure at any point on the airfoil. A typical plot of pressure coefficient is shown on Fig. 13.21, on which it is customary to plot C_p positive downward. The Prandtl-Glauert rule states[16] that such data may be corrected for use at higher Mach numbers by simply dividing C_p by $\sqrt{1 - \mathbf{M}_o^2}$, providing the critical Mach number for the wing is not exceeded. Space does not permit proof of the Prandtl–Glauert rule here, but its limit of applicability, the critical Mach number, may be easily found. Referring to the discussion of Fig. 13.18*a*, ignoring the

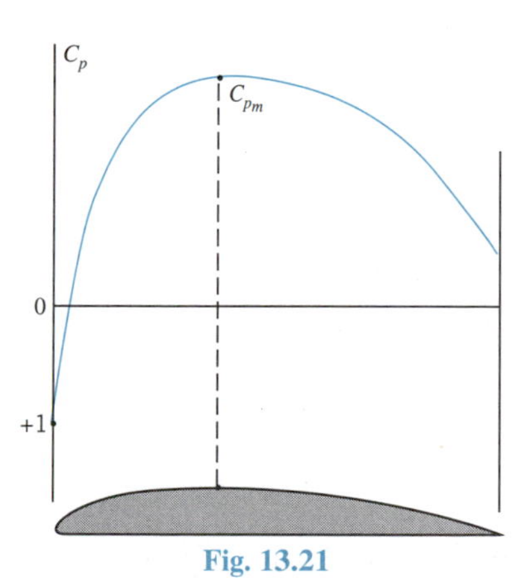

Fig. 13.21

[14]Calculations show the major part of this increase to be due more to the separation than to the energy dissipated in the shock waves.

[16]A more refined (and more complicated) rule is that of von Kármán and Tsien.

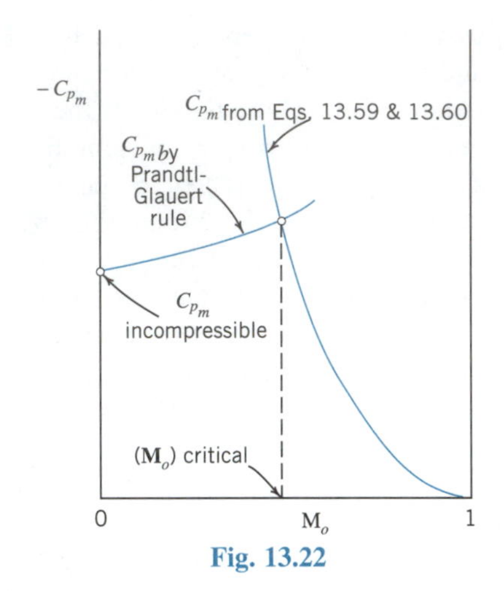

Fig. 13.22

boundary layer, and using Eq. 13.11 for an ideal compressible fluid between the free stream and point m, which is the point of minimum pressure, p_m, and minimum pressure coefficient, C_{p_m},

$$\frac{V_m^2 - V_o^2}{2} = \frac{p_o}{\rho_o}\frac{k}{k-1}\left[1 - \left(\frac{p_m}{p_o}\right)^{(k-1)/k}\right] \tag{13.11}$$

Shock phenomena begin to appear at m when V_m attains the *local sonic velocity* $\sqrt{kp_m/\rho_m}$. Substituting this for V_m, and a_o for $\sqrt{kp_o/\rho_o}$ this equation may be reduced to

$$\frac{p_m}{\rho_m}\frac{\rho_o}{p_o} - \mathbf{M}_o^2 = \frac{2}{k-1}\left[1 - \left(\frac{p_m}{p_o}\right)^{(k-1)/k}\right]$$

Substituting $(p_m/p_o)^{1/k}$ for ρ_o/ρ_m, and solving for p_m/p_o yields

$$\frac{p_m}{p_o} = \left[\frac{2 + (k-1)\mathbf{M}_o^2}{k+1}\right]^{k/(k-1)} \tag{13.59}$$

By dividing the numerator and denominator on the right-hand side of Eq. 13.58 by p_o and k, C_{p_m} becomes

$$C_{p_m} = \frac{p_m - p_o}{\rho_o V_o^2/2} = \frac{2(p_m/p_o - 1)}{k\rho_o V_o^2/p_o k} = \frac{2(p_m/p_o - 1)}{k\mathbf{M}_o^2} \tag{13.60}$$

into which p_m/p_o from Eq. 13.59 may be inserted to yield the limiting C_p below which values obtained by the Prandtl–Glauert rule are invalid. The solution for C_{p_m} and the resulting (critical) $\mathbf{M}_o$ may be done by trial or by plotting, as indicated on Fig. 13.22.

13.20 LIFT

Because of the steady decrease in pressure over the major portion of the upper side of an airfoil with increase of Mach number in the subsonic range (Fig. 13.18*a*), the lift coefficient at a given angle of attack will increase through this range. The Prandtl–Glauert rule shows that this may be computed approximately by dividing the lift coefficient of incompressible

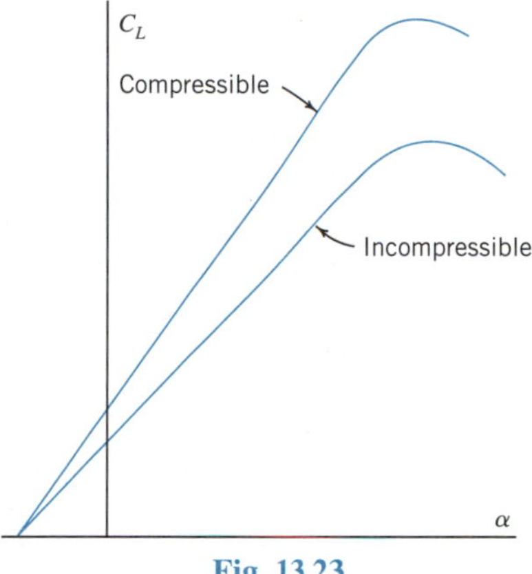

Fig. 13.23

flow by $\sqrt{1 - \mathbf{M}_o^2}$. The result may be seen on Fig. 13.23 and applies only up to the point of stall.

With increase of Mach number into the transonic range the lift coefficient continues to increase (for the same angle of attack) until the shock phenomena produce separation (Fig. 13.18*b*). The presence of the shock wave (through which there is a sudden rise in pressure) and the resulting separation produce a region of increased pressure on the upper side of the airfoil which will cause the lift coefficient to drop sharply. This so-called *shock-stall* usually occurs at Mach numbers slightly above the critical (Fig. 13.24).

The variation of the lift coefficient with Mach number through the transonic and into the supersonic range is pictured in Fig. 13.25 and related to the simplified flow pictures of Fig. 13.18. The first drop in lift coefficient brought about by the shock-stall is arrested by the formation of a shock wave and increase of pressure on the lower side of the airfoil and by the change in position of the shock wave on the upper side, which reduces the region of separation and high pressure there. The lift coefficient then increases with Mach number to another maximum until intense shock wave phenomena become predominant, after which the trend is steadily downward with increasing Mach number.

Although analytical methods are unavailable for treating problems in the transonic range, simple expressions for C_L have been worked out for the subsonic and supersonic flows about thin symmetrical airfoils of infinite span and small angle of attack, which have

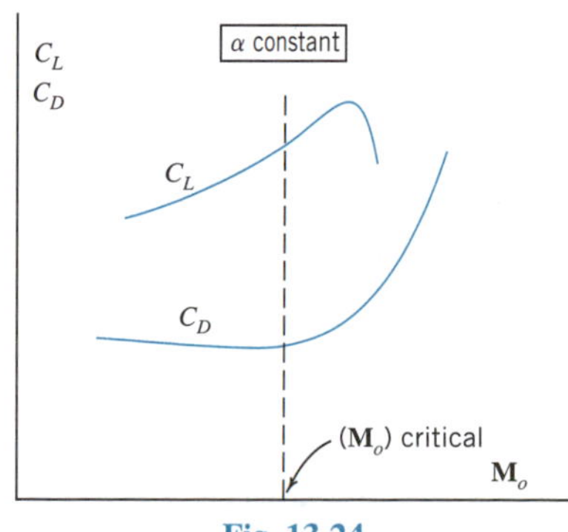

Fig. 13.24

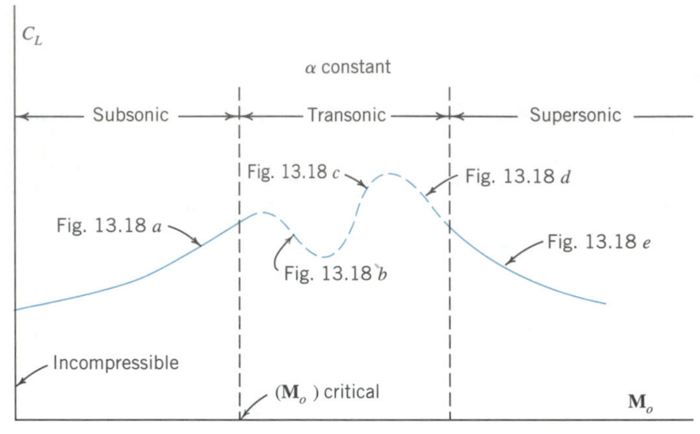

Fig. 13.25

been confirmed by experiment and may be used by engineers as rough guides. These expressions are

$$C_L = \frac{2\pi\alpha}{\sqrt{1 - \mathbf{M}_o^2}} \qquad \text{and} \qquad C_L = \frac{4\alpha}{\sqrt{\mathbf{M}_o^2 - 1}}$$

for the subsonic and supersonic cases, respectively. In spite of the tremendous advances in theory, however, reliable results for airfoils of finite length and arbitrary profile are still difficult to achieve. Sophisticated experiments and numerical simulations on powerful computers are required for an accurate design.

REFERENCES

Beckwith, I. E., and Miller, C. G., III. 1990. Aerodynamics and transition in high-speed wind tunnels at NASA Langley. *Annual Review of Fluid Mechanics* 22: 419–439.

Benedict, R. P. 1980. *Fundamentals of pipe flow.* New York: Wiley.

Burghardt, M. D., and Harbach, J. A. 1993. *Engineering thermodynamics.* 4th ed. New York: HarperCollins College.

Busemann, A. 1971. Compressible flow in the thirties. *Annual Review of Fluid Mechanics* 3: 1–12.

Chapman, A. J., and Walker, W. F. 1971. *Introductory gas dynamics.* New York: Holt, Rinehart and Winston.

Cheng, H. K. 1993. Perspectives on hypersonic viscous flow research. *Annual Review of Fluid Mechanics* 25: 455–484.

Durand, W. F., ed. 1963. *Aerodynamic theory.* New York: Dover Publications. 6 Vols.

Glauert, H. 1932. *Aerofoil and airscrew theory.* Cambridge: Cambridge Univ. Press.

Hatsopoulos, G. N., and Keenan, J. H. 1965. *Principles of general thermodynamics.* New York: Wiley.

Hayes, W. D., and Probstein, R. F. 1959. *Hypersonic flow theory.* New York: Academic Press.

Hoerner, S. F. 1965. *Fluid-dynamic drag.* 2nd ed. New Jersey: S. F. Hoerner.

Hornung, H. 1986. Regular and Mach reflection of shock waves. *Annual Review of Fluid Mechanics* 18: 33–58.

Jones, J. B., and Hawkins, G. A. 1986. *Engineering thermodynamics: an introductory textbook.* 2d ed. New York: Wiley.

Kuethe, A. M., and Chow, C.-Y. 1986. *Foundations of aerodynamics—bases of aerodynamic design.* 4th ed. New York: Wiley.

Liepmann, H. W., and Roshko, A. 1957. *Elements of gasdynamics.* New York: Wiley.

Mikhailov, V. V., Neiland, V. Ya., and Sychev, V. V. 1971. The theory of viscous hypersonic flow. *Annual Review of Fluid Mechanics* 3: 371–396.

Mises, R. von. 1958. *Mathematical theory of compressible fluid flow.* New York: Academic Press.

Moran, M. J., and Shapiro, H. N. 1992. *Fundamentals of engineering thermodynamics.* 2d ed. New York: Wiley.

Moretti, G. 1987. Computation of flows with shocks. *Annual Review of Fluid Mechanics* 19: 313–337.

Nieuwland, G. Y., and Spee, B. M. 1973. Transonic airfoils: recent developments in theory, experiment, and design. *Annual Review of Fluid Mechanics* 5: 119–150.

Oswatitsch, K. 1956. *Gas dynamics.* New York: Academic Press.

Prandtl, L., and Tietjens, O. G. 1934. *Fundamentals of hydro- and aeromechanics; Applied hydro- and aeromechanics.* New York: McGraw-Hill.

Reynolds, W. C., and Perkins, H. C. 1977. *Engineering thermodynamics.* 2nd ed. New York: McGraw-Hill.

Saad, M. A. 1993. *Compressible fluid flow.* 2d ed. Englewood Cliffs, N.J.: Prentice Hall.

Shapiro, A. H. 1953. *The dynamics and thermodynamics of compressible fluid flow.* Vols. I & II. New York: Ronald Press.

Sobieczky, H., and Seebass, A. R. 1984. Supercritical airfoil and wing design. *Annual Review of Fluid Mechanics* 16: 337–363.

Spalding, D. B., and Cole, E. H. 1973. *Engineering thermodynamics.* 3d ed. London: Edward Arnold.

Sutton, O. G. 1949. *The science of flight.* New York: Penguin Books.

Thompson, P. A. 1972. *Compressible-fluid dynamics.* New York: McGraw-Hill.

FILMS

Coles, D. Channel flow of a compressible fluid. NCFMF/EDC Film No. 21616, Encyclopaedia Britannica Educ. Corp.

Rouse, H. Mechanics of fluids: effects of fluid compressibility. Film No. 36960, Media Library, Audiovisual Center, Univ. of Iowa.

Shell Oil Co. Approaching the speed of sound; Beyond the speed of sound; Transonic flight. (3 films) Shell Film Library, 1433 Sadlier Circle, West Drive, Indianapolis, Ind. 46239.

PROBLEMS

13.1. Describe the processes that occur in the internal combustion engine. Are any close to being reversible? Can you find a cycle? Is the thermodynamic cycle the same as the cycle in a "four-stroke cycle" engine?

13.2. Calculate the energy delivered to the turbine per unit mass of airflow when the transfer in the heat exchanger is zero. Then, how does the energy delivered depend on q_H through the exchanger if all other conditions remain the same? Assume air is a perfect gas.

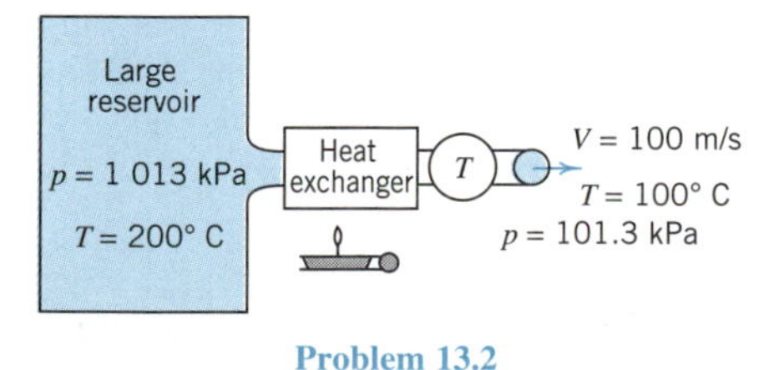

Problem 13.2

Problem 13.3

13.3. If hydrogen flows as a perfect gas without friction between stations 1 and 2 while $q_H = 7.5 \times 10^5$ J/kg, find V_2.

13.4. Air flows in a 2 in. pipe without friction. If the velocity and temperature at a section are 100 ft/s and 200°F, construct a relationship which shows the various amounts of heat or pump energy that must be added to produce a velocity and temperature of 90 ft/s and 205°F farther downstream.

13.5. The velocity and temperature are measured at two points on the same streamline in a flow of carbon dioxide and found to be 60 m/s, 40°C and 120 m/s, 35°C. This flow is *not* adiabatic. How much heat has been added to or extracted from the fluid between the two points?

13.6. Considering the following gases as perfect gases, calculate the entropy change in each process:

(*a*) Nitrogen	$T_1 = 250$ K	$p_1 = 150$ bar
	$T_2 = 330$ K	$p_2 = 10$ bar
(*b*) Air	$T_1 = 200$ K	$p_1 = 1$ bar
	$T_2 = 233$ K	$p_2 = 200$ bar

13.7. Derive the perfect gas equation for enthalpy change in an isentropic process.

13.8. Calculate the entropy change in problem 13.2.

13.9. Carbon dioxide flows at a speed of 10 m/s (30 ft/s) in a pipe and then through a nozzle where the velocity is 50 m/s (150 ft/s). What is the change in gas temperature between pipe and nozzle? Assume this is an adiabatic flow of a perfect gas.

13.10. Methane gas is flowing in an insulated pipe where the temperature and velocity are 30°C (86°F) and 22.5 m/s (74 ft/s), respectively. Using the perfect gas laws and assuming ideal flow, construct a graph of velocity versus temperature for other points in the line. Are there any limits on this graph?

13.11. The velocity and temperature at a point in an isentropic flow of helium are 90 m/s and 90°C, respectively. Predict the temperature on the same streamline where the velocity is 180 m/s. What is the ratio between the pressure at the two points?

13.12. At a point in an adiabatic flow of nitrogen the velocity is 200 m/s (650 ft/s). Between this point and another one on the same streamline the rise of temperature is 10°C (18°F). Calculate the velocity at the second point.

13.13. Derive Eq. 13.14 from Eq. 13.13.

13.14. Revise Eq. 13.14 by including the fourth term in the binomial expansion.

13.15. Carbon dioxide flows in a duct at a velocity of 90 m/s, absolute pressure 140 kPa, and temperature 90°C. Calculate pressure and temperature on the nose of a small object placed in this flow.

13.16. If nitrogen at 15°C is flowing and the stagnation temperature on the nose of a small object in the flow is measured as 38°C, what is the velocity in the pipe?

13.17. Oxygen flows in a passage at a pressure of 25 psia. The pressure and temperature on the nose of a small object in the flow are 28 psia and 150°F, respectively. What is the velocity in the passage?

13.18. Calculate the stagnation pressure in an airstream of absolute pressure, temperature, and velocity, 101.3 kPa, 15°C, and 320 m/s, respectively, using (*a*) Eq. 13.14 and (*b*) Eq. 13.13. Compare results.

13.19. What is the pressure on the nose of a bullet moving through standard sea level air at 300 m/s (985 ft/s) according to (*a*) Eq. 13.14 and (*b*) Eq. 13.13? Compare results.

13.20. With Eq. 13.13 invalid for $\mathbf{M}_o > 1$, derive an expression for the minimum allowable value of p_o/p_s which may be used for velocity computation in Eq. 13.15.

13.21. Assume air flows in a nozzle at 500 m/s, $\rho = 2$ kg/m^3 and $T = 300$°C. Is an increase or decrease in area required to further increase the flow velocity?

13.22. In a given duct flow $\mathbf{M} = 2.0$; the velocity undergoes a 20% decrease. What percent change in area was needed to accomplish this? What would be the answer if $\mathbf{M} = 0.5$?

13.23. Derive equations (*a*) 13.20 and (*b*) 13.21.

13.24. Nitrogen flows from a large tank, through a convergent nozzle of 2 in. tip diameter, into the atmosphere. The temperature in the tank is 200°F. Calculate pressure, velocity, temperature, and sonic velocity in the jet; and calculate the flowrate when the tank pressure is (*a*) 30 psia and (*b*) 25 psia. Barometric pressure is 15.0 psia. What is the lowest tank pressure that will produce sonic velocity in the jet? What is this velocity and what is the flowrate?

13.25. Air flows from the atmosphere into an evacuated tank through a convergent nozzle of 38 mm tip diameter. If atmospheric pressure and temperature are 101.3 kPa and 15°C, respectively, what vacuum must be maintained in the tank to produce sonic velocity in the jet? What is the flowrate? What is the flowrate when the vacuum is 254 mm of mercury?

13.26. Oxygen discharges from a tank through a convergent nozzle. The temperature and velocity in the jet are −20°C (0°F) and 270 m/s (900 ft/s), respectively. What is the temperature in the tank? What is the temperature on the nose of a small object in the jet?

13.27. Carbon dioxide discharges from a tank through a convergent nozzle into the atmosphere. If the tank temperature and pressure are 38°C and 140 kPa, respectively, what jet temperature, pressure, and velocity can be expected? Barometric pressure is 101.3 kPa.

13.28. In Illustrative Problem 13.6 of Section 13.7, calculate the pressure, temperature, velocity, and sonic velocity at a point in the nozzle where the diameter is 50 mm.

13.29. Air (at 100°F and 100 psia) in a large tank flows into a 6 in. pipe, whence it discharges to the atmosphere (15.0 psia) through a convergent nozzle of 4 in. tip diameter. Calculate pressure, temperature, and velocity in the pipe.

13.30. Carbon dioxide flows through a convergent nozzle in a wall between two large tanks in which the absolute pressures are 450 kPa and 210 kPa. Calculate the jet velocity if the jet temperature is 10°C. What is the temperature in the upstream tank?

13.31. Calculate the required diameter of a convergent nozzle to discharge 5.0 lb/s of air from a large tank (in which the temperatures is 100°F) to the atmosphere (14.7 psia) if the pressure in the tank is: (*a*) 25.0 psia, and (*b*) 30.0 psia.

13.32. Air flows from a 150 mm pipe, through a 25 mm convergent nozzle, into the atmosphere. Calculate the absolute pressure in the pipe when the absolute pressure in the jet is 138 kPa.

13.33. Derive Eq. 13.22.

13.34. Carbon dioxide flows through a 4 in. constriction in a 6 in. pipe. The pressures in pipe and constriction are 40 and 35 psia, respectively. The temperature in the pipe is 100°F. Calculate (*a*) flowrate, (*b*) temperature in the constriction, (*c*) sonic velocity in the constriction, and (*d*) velocities in pipe and constriction.

13.35. Nitrogen discharges from a 100 mm pipe through a 50 mm nozzle into the atmosphere. Barometric pressure is 101.3 kPa, and the pressure in the pipe is 0.3 m of water. The temperature in the pipe is 15.5°C. Calculate the flowrate, (*a*) neglecting expansion of the gas and (*b*) including expansion. What percent error is induced by neglecting expansion?

13.36. Five pounds of air per second discharge from a tank through a convergent-divergent nozzle into another tank where a vacuum of 10 in. of mercury is maintained. If the pressure and temperature in the upstream tank are 100 in. of mercury absolute and 100°F, respectively, what nozzle-exit diameter must be provided for full expansion? What throat diameter is required? Calculate pressure, temperature, velocity, and sonic velocity in throat and nozzle exits. Barometric pressure is 30 in. of mercury.

13.37. Carbon dioxide flows from a tank through a convergent-divergent nozzle of 25 mm throat and 50 mm exit diameter. The absolute pressure and temperature in the tank are 241.5 kPa and 37.8°C, respectively. Calculate the mass flowrate when the absolute pressure p_3' is (*a*) 172.5 kPa and (*b*) 221 kPa.

13.38. If the nozzle of Illustrative Problem 13.8 of Section 13.9 is a convergent one of 33 mm tip diameter, calculate velocity and sonic velocity in the jet, and the mass flowrate. Compare results, and note the effect of changing the shape of a nozzle without changing its exit diameter.

13.39. If the nozzle in Illustrative Problem 13.8 of Section 13.9 has a divergent passage 100 mm long, calculate the required passage diameter, at three equally spaced points between throat and nozzle tip, for a linear pressure drop through this part of the passage.

13.40. A convergent-divergent nozzle of 50 mm tip diameter discharges to the atmosphere (103.2 kPa) from a tank in which air is maintained at an absolute pressure and temperature of 690 kPa and 37.8°C, respectively. What is the maximum mass flowrate that can occur through this nozzle? What throat diameter must be provided to produce this mass flowrate?

13.41. Atmospheric air (at 98.5 kPa and 20°C) is drawn into a vacuum tank through a convergent-divergent nozzle of 50 mm throat diameter and 75 mm exit diameter. Calculate the largest mass flowrate that can be drawn through this nozzle under these conditions.

13.42. The exit section of a convergent-divergent nozzle is to be used for the test section of a supersonic wind tunnel. If the absolute pressure in the test section is to be 140 kPa (20 psia), what pressure is required in the reservoir to produce a Mach number of 5 in the test section? For the air temperature to be −20°C (0°F) in the test section, what temperature is required in the reservoir? What ratio of throat area to test section area is required to meet these conditions?

13.43. In the nozzle of problem 13.36, what range of pressures in the tank will cause a normal shock wave in the nozzle?

13.44. A perfect gas ($k = 2.0$, and $R = 3\,758$ ft·lb/slug °R) discharges from a tank through a convergent-divergent nozzle into the atmosphere (barometric pressure 28 in. of mercury). The gage pressure in the tank is 50 psi, and the temperature is 100°F. The throat and exit diameters of the nozzle are 1.00 in. and 1.05 in., respectively. Show calculations to prove whether a normal shock wave is to be expected in this nozzle.

13.45. Air discharges through a convergent-divergent nozzle which is attached to a large reservoir. At a point in the nozzle a normal shock wave is detected across which the absolute pressure jumps from 69 to 207 kPa (10 to 30 psia). Calculate the pressures in the throat of the nozzle and in the reservoir.

13.46. Derive Eqs. 13.30, 13.31, and 13.32.

13.47. A normal shock wave exists in an airflow. The absolute pressure, velocity, and temperature just upstream from the wave are 207 kPa, 610 m/s, and −17.8°C, respectively. Calculate the pressure, velocity, temperature, and sonic velocity just downstream from the shock wave.

13.48. If, through a normal shock wave (in air), the absolute pressure rises from 275 to 410 kPa (40 to 60 psia) and the velocity diminishes from 460 to 346 m/s (1 500 to 1 125 ft/s), what temperatures are to be expected upstream and downstream from the wave?

13.49. The stagnation temperature in an airflow is 149°C upstream and downstream from a normal shock wave. The absolute stagnation pressure downstream from the shock wave is 229.5 kPa. Through the wave the absolute pressure rises from 103.4 to 138 kPa. Determine the velocities upstream and downstream from the wave.

13.50. Assuming that air is a perfect gas, calculate the entropy change across the normal shock wave in Illustrative Problem 13.10 in Section 13.12.

13.51. Derive Eq. 13.38. Hint: Compare Eqs. 13.35 and 13.37 with their normal shock counterparts.

13.52. A wedge-shaped object in a supersonic wind tunnel is observed to produce an oblique shock wave of angle (β) 60°.

The absolute pressure through the wave rises from 210 to 280 kPa (30 to 40 psia) and the temperature of the air upstream from the object is 40°C (100°F). Calculate the velocity of the air at this point.

13.53. The velocities and absolute pressures upstream and downstream from an oblique shock wave in an airflow are found to be 610 m/s, 138 kPa, and 457 m/s, 207 kPa, respectively. Calculate the Mach numbers upstream and downstream from the wave if the wave angle (β) is 30°.

13.54. An airplane flies at twice the speed of sound in the U.S. Standard Atmosphere (Appendix 2) at an altitude of 10 km (30 000 ft). If the leading edge of the wing may be approximated by a wedge of angle (2θ) 20°, what pressure rise through the oblique shock wave is to be expected?

13.55. Air flows isothermally in a 75 mm pipeline at a mean velocity of 3 m/s, absolute pressure of 276 kPa, and a temperature of 10°C. If pressure is lost by friction, calculate the mean velocity where the absolute pressure is 207 kPa.

13.56. Through a horizontal section of 6 in. cast iron pipe 1 000 ft long in which the temperature is 60°F, 200 lb/min of air flow isothermally. If the pressure at the upstream end of this length is 30 psia, calculate the pressure at the downstream end and the mean velocities at these two points.

13.57. Carbon dioxide flows isothermally in a 50 mm horizontal wrought iron pipe, and at a certain point the mean velocity, absolute pressure, and temperature are 18 m/s, 400 kPa, and 25°C, respectively. Calculate the pressure and mean velocity 150 m downstream from this point.

13.58. The Mach number, V/a, at a point in an isothermal (38°C) flow of air in a 100 mm pipe is 0.20, and the absolute pressure at this point is 345 kPa. If the friction factor of the pipe is 0.020, what pressure and Mach number may be expected 60 m downstream from this point?

13.59. What are the limiting Mach numbers for isothermal flow of helium, carbon dioxide, and methane?

13.60. Construct a figure comparable to Fig. 13.15 for helium or carbon dioxide.

13.61. Nitrogen is being pumped adiabatically in an insulated 100 mm (4 in.) pipeline in which $f = 0.010$. If the Mach number at a point in the line is 0.3, how far downstream will the Mach number be 0.8?

13.62. What is the maximum possible distance downstream that the flow in problem 13.61 can continue before discharging from the pipe at sonic speed?

13.63. Construct a figure comparable to that accompanying Illustrative Problem 13.13 of Section 13.16 for nitrogen, carbon dioxide, or helium. Plot $\bar{f}\,(l_{max}/d)$ versus $\mathbf{M}_1^2$.

13.64. Carbon dioxide enters a pipe that is fully insulated at $\mathbf{M}_1 = 3.0$. Fifteen pipe diameters downstream $\mathbf{M}_2 = 1.5$. What is the effective $\bar{f}$ for the pipe?

13.65. Resolve problem 13.64 if the gas is helium.

13.66. A supersonic flow of helium gas in an insulated pipe is needed for a test. The pipe is 300 mm (12 in.) diameter. Assuming the flow is rough and that for a supersonic flow $\bar{f} = \frac{1}{2}f$ for any material, plot l_{max} versus f for $\mathbf{M}_1 = 2$. Indicate appropriate pipe materials corresponding to various l_{max} values.

13.67. Use Eqs. 13.53 and 13.55 to obtain

$$\frac{dp/p}{d\mathbf{M}^2/\mathbf{M}^2} = -\frac{1 + (k-1)\mathbf{M}^2}{2\left(1 + \dfrac{k-1}{2}\mathbf{M}^2\right)}$$

Integrate this result between sections where $p = p_1$, $\mathbf{M} = \mathbf{M}_1$, and where $\mathbf{M} = 1$ and $p = p_s$ to prove that

$$\frac{p_1}{p_s} = \frac{1}{\mathbf{M}_1}\left[\frac{k+1}{2\left(1 + \dfrac{k-1}{2}\mathbf{M}_1^2\right)}\right]^{1/2}$$

13.68. Use the results of problem 13.67 to compute the pressure ratio p_2/p_1 for problem 13.64 or 13.65.

13.69. A jet passes overhead at an altitude of 1 000 ft (300 m) but the sound is not heard until 0.5 s later. What is the jet's Mach number?

13.70. A supersonic airliner (1 800 mph or 2 900 km/h) passes a jumbo jet going in the opposite direction at 400 mph or 645 km/h at a distance of 1 mile or 1.61 km. When can the jumbo jet be expected to feel the shock wave?

13.71. If the pointed artillery projectile of Fig. 13.20 is of 12 in. (0.3 m) diameter and is moving at 2 000 ft/s (600 m/s) through standard sea level air (Appendix 2), what drag force is exerted on it?

13.72. What is the drag of the blunt-nosed projectile of Fig. 13.20 (if its diameter is 3 in.) when it travels at (*a*) 700 mph, and (*b*) 800 mph, through standard sea level air (Appendix 2)?

13.73. An airfoil travels through standard sea level air (Appendix 2) at 500 mph (805 km/h). Calculate the local pressure and velocity on the airfoil at which shock wave phenomena will be expected.

13.74. The minimum pressure coefficient for an airfoil at a certain angle of attack in incompressible flow is -0.7. Predict the critical Mach number ($\mathbf{M}_c$) for this airfoil at this angle of attack.

13.75. Plot lift coefficient against angle of attack for the Clark-Y airfoil of Fig. 11.19 for free stream Mach numbers of 0 and 4.

13.76. A thin airfoil at a small angle of attack produces a certain lift force at a free stream Mach number of 0.90. At what Mach number(s) will the same airfoil at the same angle of attack in the same air stream produce the same lift?

CHAPTER

14 FLUID MEASUREMENTS

In engineering and industrial practice one of the fluid mechanics problems most frequently encountered by engineers is the *measurement* of many of the variables and properties discussed in the foregoing chapters. Efficient and accurate measurements are also absolutely essential for correct conclusions in the various fields of fluid mechanics research. Whether the necessity for precise measurements is economic or scientific, engineers must be well-equipped with a knowledge of the fundamentals and existing methods of measuring various fluid properties and phenomena. The purpose of this chapter is to describe the principles and phenomena of fluid measurements; readers will find the details of installation and operation of the various measuring devices available in the abundant engineering literature. The literature is cited here in footnotes, when there is a specific and directed reference made, and otherwise in the references at the end of the chapter.

The sections of this chapter are essentially independent, relying only on principles derived and applied in the appropriate earlier sections of the book. Accordingly, readers can and should turn directly to the section of this chapter that is applicable to their need, e.g., to Section 14.12 when they are about to undertake a laboratory experiment to measure flowrate with a venturi meter. It is not necessary to work sequentially from the front of the chapter.

Before turning to measurement techniques, we introduce the concept of the uncertainty in reported results of measurements. In reading the literature or in conducting and interpreting their own measurements, readers will rapidly become familiar with the scatter in the results, compared to the theory or what was expected, or the differences between the results from different data sets from the same apparatus or different sets of measurements from ostensibly similar apparatus. Moffat[1] views uncertainty analysis as a powerful tool for the planning and development phases of an experiment, as well as an essential tool for the rational evaluation of the resulting data and comparison of one data set to another. He describes a technique, extending the seminal work of Kline and McClinstock,[2] that produces an estimate or error bound for the uncertainty caused by instrument calibration, unsteadiness [random fluctuation] and interpolation. According to Moffat, the Kline-McClintock root-sum-square equation has been shown to assess the uncertainty with good accuracy for most engineering cases.

Suppose that a set of data is obtained and a result R is calculated from that data. The result R is a function of all the bits of measured data x_i, and we can write that

$$R = \phi(x_1, x_2, x_3, \ldots, x_N)$$

Here, the N bits of measured data are designated by x_i, the best estimate of a bit of data is $\bar{x}_i$, the estimated uncertainty interval or bound is σ_{x_i} and we write $x_i = \bar{x}_i + \sigma_{xi}$. In addition, the relative uncertainty in the ith bit of data is $\sigma_{x_i}/\bar{x}_i$. The effect on R of the uncertainty in the ith bit of data is $(\partial R/\partial x_i)\sigma_{x_i}$; therefore, if the uncertainty in R is σ_R, it is given by

$$\sigma_R = \pm\left\{\left(\frac{\partial R}{\partial x_1}\sigma_{x_1}\right)^2 + \left(\frac{\partial R}{\partial x_2}\sigma_{x_2}\right)^2 + \cdots \left(\frac{\partial R}{\partial x_N}\sigma_{x_N}\right)^2\right\}^{1/2} \tag{14.1}$$

The relative uncertainty is then $U_R = \sigma_R/\bar{R}$.

In this concept the error is defined in terms of a confidence interval or odds, i.e., we are confident that 95% of the time the actual value will lie within the calculated uncertainty interval or, alternatively, we assert that the odds are 20/1 that the actual value will lie within the interval. The above equations are also based on the assumptions that the data bits are independent variables, that they come from a Gaussian distribution and that the odds are always the same.

ILLUSTRATIVE PROBLEM 14.1

Suppose that a set of measurements is made to determine the head loss in laminar flow in a pipe (for example, in Illustrative Problem 9.3). It is estimated that the velocity can be measured to within 5%, the kinematic viscosity is known to within 1%, the pipe length is measurable to within 1%, and its diameter to within 5%. What is the estimated percentage error bound on the head loss?

SOLUTION

The head loss for laminar flow is given by Eq. 9.9, i.e.,

$$h_L = \frac{32\mu lV}{\gamma d^2} = \frac{32\nu lV}{g_n d^2} \tag{9.9}$$

[1] R. J. Moffat, "Contributions to the theory of uncertainty analysis for single-sample experiments," *Complex Turbulent Flows*, vol. 1, Kline, Cantwell & Lilley, eds., Thermosciences Div., Mech. Engr. Dept., Stanford U., Stanford, CA, 1981, pp. 40–55.

[2] S. J. Kline and F. A. McClintock, "Describing Uncertainties in Single-Sample Experiments," *Mechanical Engineering*, ASME, January 1953.

By comparison to Eq. 14.1, it follows that by equating R and the head loss and then taking the appropriate partial derivatives (e.g., $\partial h_L/\partial \nu = 32V/g_n d^2 = 1 \times h_L/\nu$ since if $y = t^a$, then, $dy/dt = at^{a-1}$)

$$\frac{\sigma_{h_L}}{\bar{h}_L} = \pm\left\{(1 \times 0.01)^2 + (1 \times 0.01)^2 + (1 \times 0.05)^2 + (-2 \times 0.05)^2\right\}^{1/2}$$

Therefore, $\dfrac{\sigma_{h_L}}{\bar{h}_L} = \pm\left\{(0.01)^2 + (0.01)^2 + (0.05)^2 + (2 \times 0.05)^2\right\}^{1/2} = \pm 0.11$. The relative uncertainty is about 11%. ●

ILLUSTRATIVE PROBLEM 14.2

For the U-tube water and mercury (Hg) manometer setup shown, find the absolute and relative uncertainties of the pressure measured at A, if the following uncertainties are known:

$$\gamma_w = 9.798 \pm 0.002 \text{ kN/m}^3$$

$$\gamma_{Hg} = \rho g_n = 13\,555 \times 9.81 = 132.97 \pm 0.0245 \text{ kN/m}^3$$

$$MR = 0.300 \pm 0.005 \text{ m}$$

$$h = 1.200 \pm 0.002\,5 \text{ m}$$

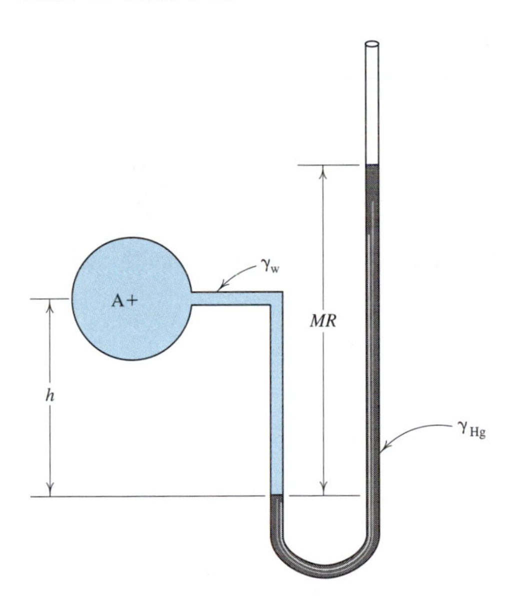

The fluid property uncertainties are based on the temperature being 15 ± 1°C.

SOLUTION

Based on the approach to manometry shown in Chapter 2, we have

$p_A + \gamma_w h - \gamma_{Hg} MR = 0$ (Gage); $p_A = \gamma_{Hg} MR - \gamma_w h$; so $p_A = f(\gamma_w, \gamma_{Hg}, MR, h)$

Thus,

$$\frac{\partial p_A}{\partial \gamma_w} = -h; \quad \frac{\partial p_A}{\partial \gamma_{Hg}} = MR; \quad \frac{\partial p_A}{\partial MR} = \gamma_{Hg}; \quad \frac{\partial p_A}{\partial h} = -\gamma_w$$

Now, using Eq. 14.1 we have

$$\sigma_{p_A} = \pm\left\{\left(\frac{\partial p_a}{\partial \gamma_w}\sigma_{\gamma_w}\right)^2 + \left(\frac{\partial p_a}{\partial \gamma_{Hg}}\sigma_{\gamma_{Hg}}\right)^2 + \left(\frac{\partial p_a}{\partial MR}\sigma_{MR}\right)^2 + \left(\frac{\partial p_a}{\partial h}\sigma_h\right)^2\right\}^{1/2} \quad (14.1)$$

$$= \pm\{(-h\sigma_{\gamma_w})^2 + (MR\sigma_{\gamma_{Hg}})^2 + (\gamma_{Hg}\sigma_{MR})^2 + (-\gamma_w\sigma_h)^2\}^{1/2}$$

Introducing the numerical values produces

$$\sigma_{p_A} = \pm\{(-1.2 \times 0.002)^2 + (0.300 \times 0.0245)^2 + (132.97 \times 0.005)^2 + (-9.798 \times 0.002\,5)^2\}^{1/2}$$

$$= \pm\ 0.665 \text{ kPa} \bullet$$

$$p_A = 132.97 \times 0.3 - 9.798 \times 1.2 = 28.13 \pm 0.665 \text{ kPa}$$

$$U_{p_A} = \pm\frac{\sigma_{p_A}}{p_A} = \pm 0.024 \approx 2.4\% \bullet$$

Fluid Properties Measurement

Of the fluid properties density, viscosity, elasticity, surface tension, and vapor pressure, the engineer is usually called on to measure only the first two. Since measurements of elasticity, surface tension, and vapor pressure are normally made by physicists and chemists, the various experimental techniques for measuring these properties are not reviewed here.

14.1 DENSITY

Density measurements of liquids may be made by the following methods, listed in approximate order of their accuracy: (1) weighing a known volume of liquid, (2) hydrostatic weighing, (3) Westphal balance, and (4) hydrometer.

To weigh accurately a known volume of liquid a device called a *pycnometer* is used. This is usually a glass vessel whose weight, volume, and variation of volume with temperature have been accurately determined. If the weight of the empty pycnometer is W_1, and the weight of the pycnometer, when containing a volume $\forall$ of liquid at temperature t, is W_2, the specific weight of the liquid, γ_t, at this temperature may be calculated directly from

$$\gamma_t \forall = W_2 - W_1$$

Density determination by hydrostatic weighing consists essentially in weighing a plummet of known volume (1) in air, and (2) in the liquid whose density is to be determined (Fig. 14.1*a*). If the weight of the plummet in air is W_a, its volume, $\forall$, and its weight when suspended in the liquid, W_l, the equilibrium of vertical forces on the plummet gives

$$W_l + \gamma_t \forall - W_a = 0$$

from which the specific weight, γ_t, at the temperature *t*, may be calculated directly.

Like the method of hydrostatic weighing, the Westphal balance (Fig. 14.1*b*) utilizes the buoyant force on a plummet as a measure of specific gravity. Balancing the scale beam with special riders placed at special points allows direct and precise reading of specific gravity.

Probably the most common means of obtaining liquid densities is with the hydrometer (Fig. 14.1*c*), whose operation is governed by the fact that a weighted tube will float with different immersions in liquids of different densities. To create a great variation of immersion for small density variation, and, thus, to provide a sensitive instrument, changes

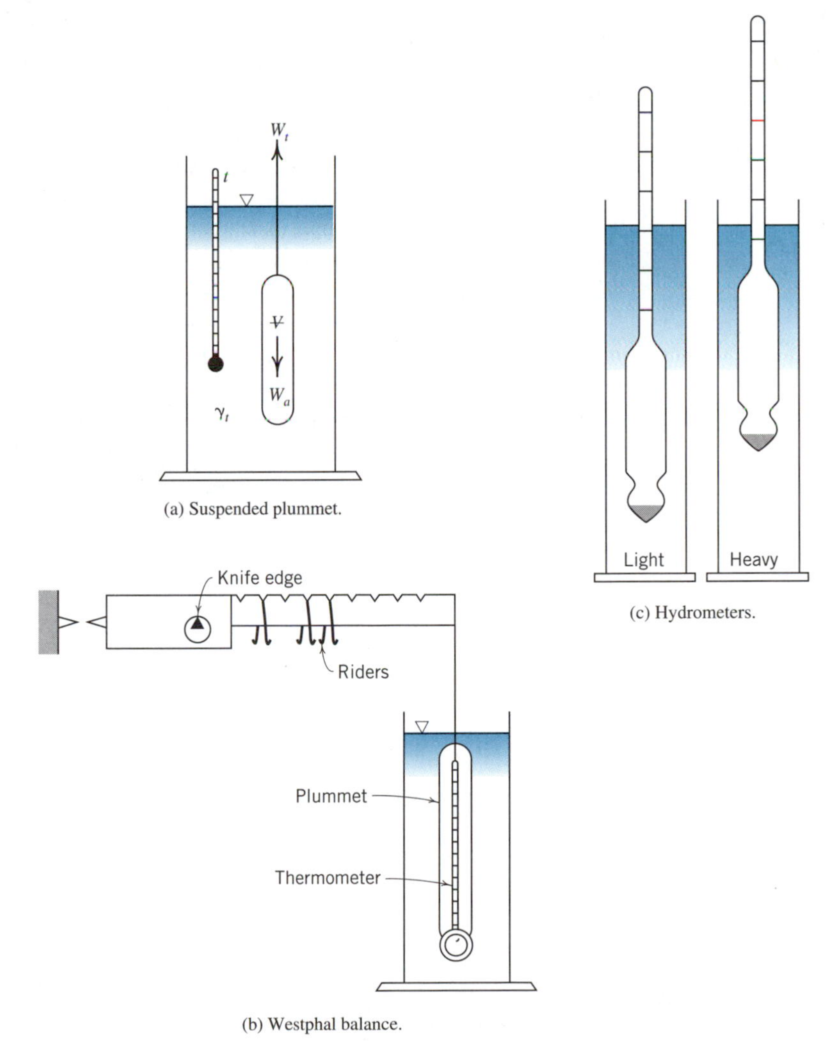

(a) Suspended plummet.

(b) Westphal balance.

(c) Hydrometers.

Fig. 14.1 Devices for density measurements.

in the immersion of the hydrometer occur along a slender tube, which is graduated to read the specific gravity of the liquid at the point where the liquid surface intersects the tube.

14.2 VISCOSITY

Viscosity measurements are made with devices known as *viscosimeters* or *viscometers*, the common varieties of which may be classified as *rotational* or *tube* devices. The operation of these viscometers *depends on the existence of laminar flow* (which has been seen in foregoing chapters to be dominated by viscous action) under certain controlled and reproducible conditions. In general, however, these conditions involve too many complexities to allow the constants of the viscometer to be calculated analytically, and they are, therefore, usually obtained by calibration with a liquid of known viscosity. Because of the variation of viscosity with temperature all viscometers *must be immersed in constant-temperature baths* and *provided with thermometers* for taking the temperatures at which the viscosity measurements are made.

Two instruments of the rotational type are the MacMichael and Stormer viscometers, whose essentials are shown diagrammatically in Fig. 14.2. Both consist of two concentric cylinders, with the space between containing the liquid whose viscosity is to be determined. In the MacMichael type, the outer cylinder is rotated at constant speed, and the rotational deflection of the inner cylinder (accomplished against a spring) becomes a measure of the liquid viscosity. In the Stormer instrument, the inner cylinder is rotated by a falling-weight mechanism, and the time necessary for a fixed number of revolutions becomes a measure of the liquid viscosity.

The measurement of viscosity by the above-mentioned variables may be justified by a simplified mechanical analysis, using the dimensions of Fig. 14.2. Assuming ΔR, Δh, and $\Delta R/R$ small, and the peripheral velocity of the moving cylinder to be V, the torque, T, may be calculated from the principles and methods of Section 1.6. The result is

$$T = \frac{2\pi R^2 h\mu V}{\Delta R} + \frac{\pi R^3 \mu V}{2\Delta h}$$

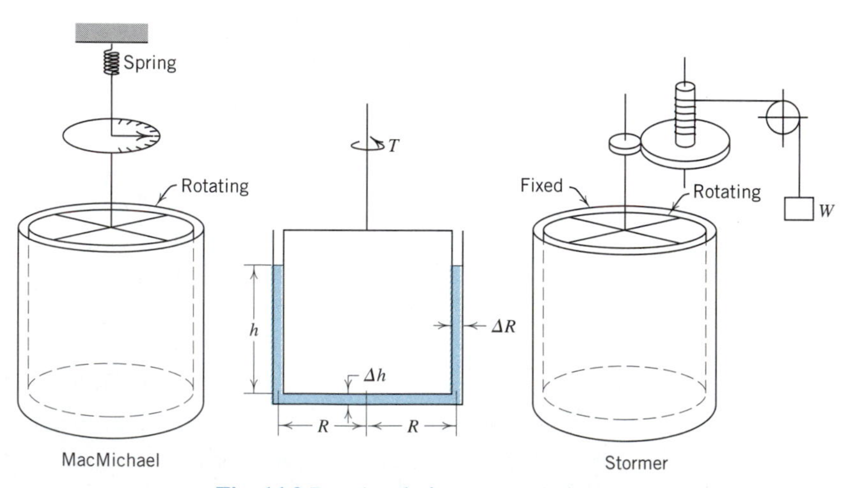

Fig. 14.2 Rotational viscometers (schematic).

in which the first term represents the torque due to viscous shear in the space between the cylinder walls, and the second term that between the ends of the cylinders. With R, h, ΔR, and Δh constants of the instrument, and the rotational speed (N) proportional to V, the equation may be written

$$T = K\mu N \qquad \text{or} \qquad \mu = \frac{T}{KN}$$

in which the constant K depends on the foregoing factors. For the MacMichael instrument the torque is proportional to the torsional deflection θ ($T = K_1\theta$) with the result

$$\mu = \frac{K_1\theta}{KN}$$

which shows that the viscosity may be obtained from deflection and speed measurements. In the Stormer viscometer the torque is constant since it is proportional to the weight W, which is a constant of the instrument; also, the time t required for a fixed number of revolutions is inversely proportional to N ($t = K_2/N$) with the result

$$\mu = \left(\frac{T}{KK_2}\right) t$$

and the time required for a fixed number of revolutions produced by the same falling weight thus becomes a direct measure of viscosity.

Typical tube-type viscometers are the Ostwald and Saybolt instruments of Fig. 14.3. The former is used typically for low-viscosity fluids, for example, water, while the latter is used for medium- to high-viscosity fluids. Similar to the Ostwald is the Bingham type, and similar to the Saybolt are the Redwood and Engler viscometers. All these instruments involve the *unsteady laminar flow* of a fixed volume of liquid through a small tube under standard head conditions. The time for the quantity of liquid to pass through the tube becomes a measure of the kinematic viscosity of the liquid.

The Ostwald viscometer is filled to level A, and the meniscus of the liquid in the right-hand tube is drawn up to a point above B and then released. The time for the meniscus to

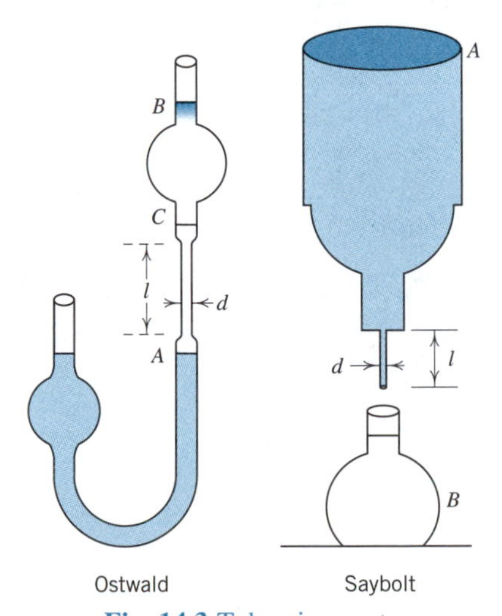

Fig. 14.3 Tube viscometers.

fall from B to C becomes a measure of the kinematic viscosity. In the Saybolt viscometer the outlet is plugged, and the reservoir filled to level A; the plug is then removed, and the time required to collect a fixed quantity of liquid in the vessel B is measured. This time then becomes a direct measure of the kinematic viscosity of the liquid.

The relation between time and kinematic viscosity for the tube-type viscometer may be indicated approximately by applying the Hagen-Poiseuille law for laminar flow in a circular tube (Section 9.2). The approximation involves the application of a law of steady established laminar motion to a condition of unsteady flow in a tube which may be too short for the establishment of laminar flow and, therefore, cannot be expected to give a complete or perfect relationship between efflux time and kinematic viscosity; it serves, however, to indicate elementary principles. From Eq. 9.8

$$Q = \frac{\pi d^4 \gamma h_L}{128 \mu l} \tag{14.2}$$

for steady laminar flow in a circular tube. But $Q = \forall / t$, in which $\forall$ is the volume of liquid collected in time t. Substituting this in Eq. 14.2 and solving for μ,

$$\mu = \left(\frac{\pi d^4 h_L}{128 \forall l} \right) \gamma t$$

The head loss, h_L, however, is nearly constant since it is approximately equal to the imposed head which varies between fixed limits. Since d, l, $\forall$, and h_L are constants of the instrument, the equation reduces to $\mu \cong K\gamma t = K\rho g_n t$, from which $\mu/\rho = \nu \cong K g_n t$, and kinematic viscosity is seen to depend almost linearly on measured time. A more exact (but empirical) equation relating ν and t for the Saybolt Universal viscometer is (for $t > 32$ s)

$$\nu \text{ (m}^2\text{/s)} = 10^{-4} \left(0.002\,197 t - \frac{1.798}{t} \right)$$

in which t is the time in seconds (called *Saybolt seconds*). This equation approaches the linear one predicted by the approximate analysis for large values of the time, t. The familiar S.A.E. numbers used for motor oils are indices of kinematic viscosity, as shown in Table 10.

Of the tube viscometers, the Saybolt, Engler, and Redwood are built of metal to rigid specifications and hence may be used without calibration. Since the dimensions of the glass viscometers such as the Bingham and Ostwald cannot be so perfectly controlled, these instruments must be calibrated before viscosity measurements are made.

TABLE 10[a] S.A.E. Viscosities

S.A.E. Viscosity No.	Saybolt seconds at 99°C	ν(mm^2/s) at 99°C
20	45 to 58	5.85 to 9.66
30	58 to 70	9.66 to 12.7
40	70 to 85	12.7 to 16.5

[a]Adapted from *S.A.E. Handbook*, Soc. Auto. Engrs., 1959.

Pressure Measurement

14.3 STATIC PRESSURE

The accurate measurement of pressure in a fluid at rest may be accomplished with comparative ease since it depends only on the accuracy of the gage or manometer used to record this pressure and is independent of the details of the connection between fluid and recording device. To measure the static pressure within a moving fluid with high accuracy is quite another matter, however, and depends on painstaking attention to the details of the connection between flowing fluid and measuring device.

For perfect measurement of static pressure in a flowing fluid a device is required which fits the streamline picture and causes no flow disturbance; it should contain a small smooth hole whose axis is normal to the direction of motion at the point where the static pressure is to be measured; to this opening is connected a manometer or pressure transducer. Meeting all these requirements is a virtual impossibility, but it is evident that the device must be as small as possible and constructed with great care. However, the most troublesome point in measuring the static pressure in a flowfield is the proper orientation or alignment of the device with the flow direction, which usually is not known in advance. Two basic designs (there are many adaptations) have solved this problem successfully. One of these is the thin disk of Fig. 14.4*a* containing separate piezometer openings in each side which lead to a differential manometer or transducer; this device is inserted in the flowfield and turned (about its stem) until the connected differential manometer reads zero, which shows the pressure on both sides of the disk to be the same and the disk thus aligned with the flow. After alignment is secured either pressure may be separately measured and taken to be the static pressure at the piezometer opening. A second device (for two-

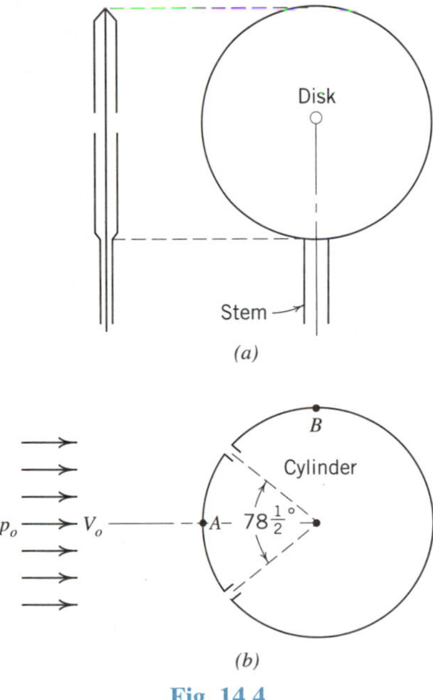

Fig. 14.4

dimensional flows) is the cylinder of Fig. 14.4*b* containing two separate piezometer openings connected to a differential manometer or transducer; the cylinder is turned about its own axis until the manometer balances, showing the direction of flow to be along the bisector of the angle between the openings. At the stagnation point A, $p_A > p_o$ and, at B, $p_B < p_o$ with continuous fall of pressure from p_A to p_B; accordingly, at some point between A and B on the surface of the cylinder, a local pressure equal to the static pressure will be found and may be measured on a separate manometer. Fechheimer[3] has shown that the appropriate angle between the piezometer openings should be 78.5° for incompressible flow, and Thrasher and Binder[4] have confirmed this for compressible flows of velocity up to one-fourth of the sonic (at higher velocities a larger angle is required). Once the flow direction has been found and the static pressure recorded, the cylinder may be rotated through the appropriate angle so that one of the piezometer openings is at the stagnation point, at which time the stagnation pressure may be measured and the velocity V_o obtained. Thus this simple and compact device serves a threefold purpose in the measurement of static pressure, flow direction, and velocity. A spherical counterpart of this device has been developed for use in three-dimensional flowfields.

In fields where the flow direction is known to fair accuracy, the *static tube* (Fig. 14.5) is the usual means of measuring static pressure. Such a tube is merely a small smooth cylinder with a rounded or pointed upstream end; in the side of the cylinder are piezometer holes or a circumferential slot through which pressure is transmitted to transducer or manometer. Assuming perfect alignment with the flow, the flow past the tube will be symmetrical and of mean velocity slightly larger than V_o; hence the pressure at the piezometer openings may be expected to be slightly less than p_o. This error is minimized by making the tube as small as possible and is usually negligible in engineering work. In experimental work or in the use of the static tube on aircraft, some misalignment of the tube with flow direction is to be expected; when this occurs the pressure on one side of the tube becomes larger than p_o, on the other side less than p_o, and some flow through the tube (from opening to opening) results. With such complexities, the pressure carried to gage or manometer cannot be exactly predicted but will be close to p_o for small angles of misalignment. A static tube which is insensitive to misalignment is desired by the experimentalist since, with larger (and incurable) errors of alignment, accurate values of static pressure may still be obtained.

The static pressures in the fluid passing over a solid surface (such as a pipe wall or the surface of an object; Fig. 14.6) may be measured successfully by small smooth piezometer holes drilled normal to the surface; such surfaces fit the flow perfectly since (assuming no separation) they are streamlines of the flow. These piezometer openings can measure only the local static pressures at their locations on the solid surface and cannot, in general, measure the pressures at a distance from this surface since such pressures differ

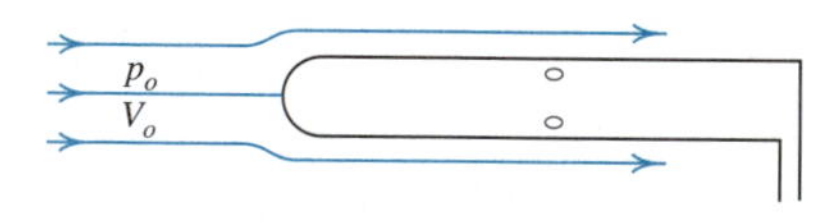

Fig. 14.5 Static tube.

[3]C. J. Fechheimer, "The Measurement of Static Pressure," *Trans. A.S.M.E.*, vol. 48, p. 965, 1926.

[4]L. W. Thrasher and R. C. Binder, "Influence of Compressibility on Cylindrical Pitot-Tube Measurements," *Trans. A.S.M.E.*, vol. 72, p. 647, 1950.

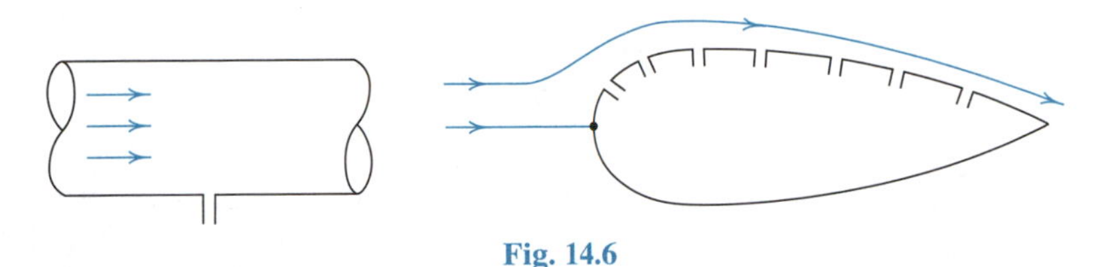

Fig. 14.6

from those at the surface owing to flow curvatures and accelerations. Where no flow curvatures exist, as in a straight passage, a wall piezometer opening will allow pressures throughout the cross section of the passage to be predicted.

In addition to the manometer (see Section 2.3), pressure may be sensed by a range of transducers that convert pressure or pressure differential to an electric output. A common form of transducer is a diaphragm gage (Fig. 14.7), which is essentially a differential pressure device (when port 1 is open to the atmosphere, gage pressure is measured). The diaphragm is an elastic (metal) element. If $p_1 = p_2$, the diaphragm is centered in the cavity; it is deflected to one side or the other depending on how p_1 and p_2 vary. Three types of diaphragm transducer are the (i) strain-gage, (ii) capacitance, and (iii) magnetic types.

If electrical-resistance strain gages are attached to the diaphragm, its deflection can be measured as an electrical output signal that, through a calibration process, can be correlated to the differential pressure. On the other hand, if the metal diaphragm is properly insulated so as to form one plate of a capacitor in an electric circuit, the changing capacitance caused by motion of the diaphragm can be correlated with pressure. In the magnetic or variable reluctance type of transducer, the magnetic diaphragm lies between magnetic output coils and its displacement by unequal pressures causes the reluctance ratio of the output coils to change and that is converted into an electrical output. These and other types of transducer are described by Benedict and Holman (see the end-of-chapter references).

14.4 SURFACE ELEVATION

The elevation of the surface of a liquid at rest may be determined by manometer, piezometer column, or pressure-gage readings (Sections 2.1–2.3).

The same methods may be applied to flowing liquids if the foregoing precautions for static pressure measurement (Section 14.3) are followed and if the piezometer method is used only where the streamlines of flow are *essentially* straight and parallel. Correct and incorrect measurements of a liquid surface by piezometer openings are illustrated in Fig. 14.8.

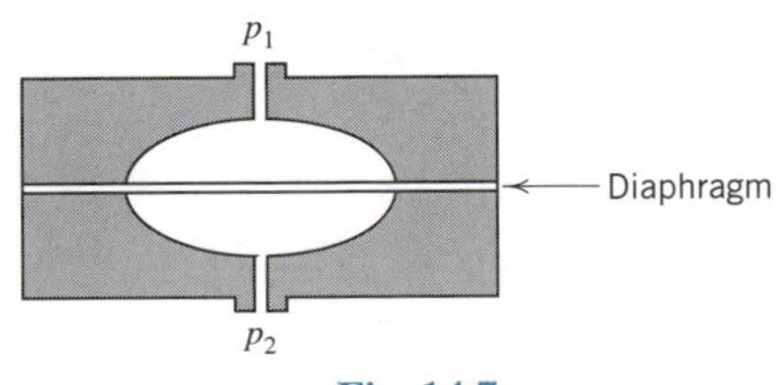

Fig. 14.7

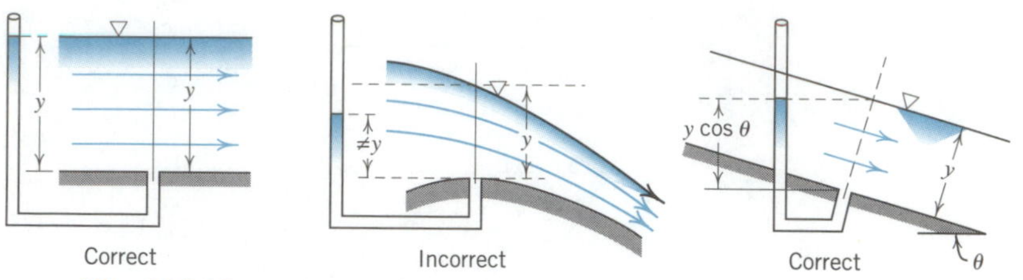

Fig. 14.8 Measurement of surface elevation with piezometer columns.

Floats are still used in connection with chronographic water-level recorders for measuring liquid-surface elevations. The arrangement of such floats is indicated schematically in Fig. 14.9. As the liquid level varies, the motion of the cable is measured on a scale, plotted automatically on a chronographic record sheet, or recorded electrically for later analysis.

There are several other methods of sensing the location of a liquid surface. Sonic devices can be used; they depend on the precise measurement of the time necessary for a sound pulse to travel to the surface and return to its source (cf., Section 14.9). Two very popular and cheap sensors are based on the introduction of either a single insulated wire or two bare conductors into the water as a component of an electrical system. In the first case a cylindrical capacitor is formed with the wire conductor being one plate, the water being the other plate, and the insulation acting as the dielectric between the plates. With the water as the ground terminal, a voltage applied to the capacitor encounters a capacitance proportional to the length of wire immersed in the water. The resistive sensor is created by applying a voltage across two bar wires which are partially immersed in the water. The resistance between the terminals is that which is offered by the water so that changes in the depth of immersion are indicated by the change in the resistance.

When these sensors are included in one arm of an electrical bridge of the appropriate type—capacitance or resistance—the changes due to the motion of the water surface are easily measured. For both types of sensors the output bridge voltage will be proportional to the change in depth if the change in the electrical quantity measured is small compared

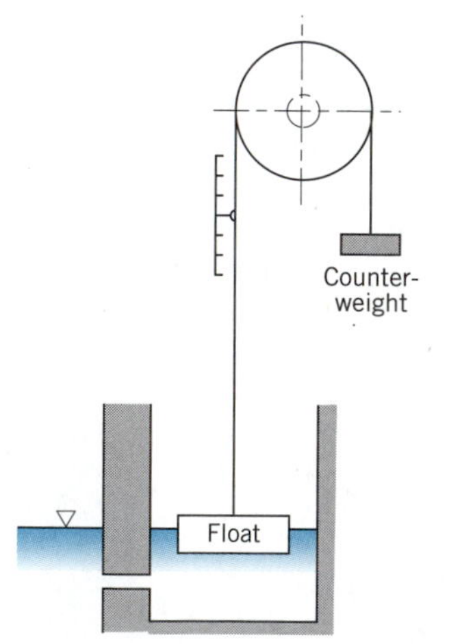

Fig. 14.9 Float for measurement of surface elevation (schematic).

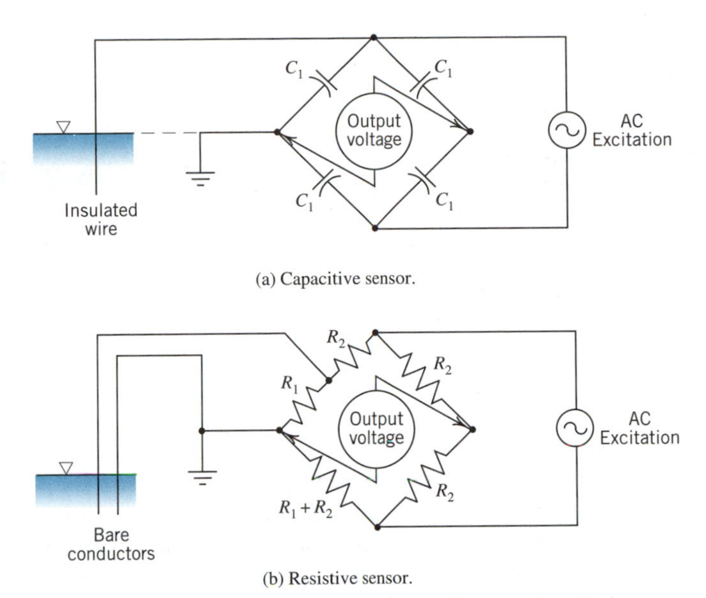

(a) Capacitive sensor.

(b) Resistive sensor.

Fig. 14.10 Schematic diagrams for surface sensing devices.

to that in the bridge arm to which the sensor is connected. Figure 14.10 illustrates the basic operational concepts of these sensors.

Staff gages give comparatively crude but direct measurements of liquid-surface elevation. From casual observation, the reader is familiar with their use as tide gages, in the measurement of reservoir levels, and in registering the draft of ships.

14.5 STAGNATION PRESSURE

From Sections 5.4 and 14.4, it is clear that stagnation pressure (also called *total pressure*) may be measured accurately by placing in the flow a small solid object having a small piezometer hole at the stagnation point. The piezometer opening may be easily located at the stagnation point if the hole is drilled along the axis of a symmetrical object such as a cylinder, cone, or hemisphere; with the axis of the objective properly aligned with the direction of flow, the piezometer opening is automatically located at the stagnation point, and the pressure there may be transferred through the opening to a recording device. Theoretically, the upstream end of solid objects for this purpose may be of any shape, since the shape of the object does not affect the magnitude of the stagnation pressure; in laboratory practice, however, the upstream end is usually made convergent (conical or hemispherical) in order to fix more precisely the location of the stagnation point.

When used on airplanes or in experimental work, misalignments of tube with flow are inevitable, but reliable measurements of stagnation pressure are still required; research[5] led to the development of a tube shielded by a special jacket which gives reliable measurements at angle of misalignment up to 45°.

[5]W. R. Russell, W. Gracey, W. Letko, and P. G. Fournier, ''Wind Tunnel Investigation of Six Shielded Total-Pressure Tubes at High Angles of Attack,'' *N.A.C.A. Tech. Note* 2530, 1951.

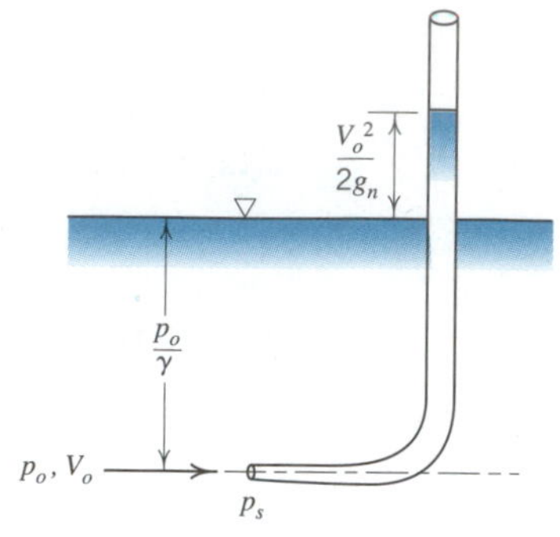

Fig. 14.11 Pitot tube.

The early experimental work of Henri Pitot (1732) provided the basis for the measurement of stagnation pressure by showing that a small tube with open end facing upstream (Fig. 14.11) provided a means for measuring velocity. He found that when such tubes (later called *pitot tubes*) were placed in an open flow where the velocity was V_o the liquid in the tube rose above the free surface a distance $V_o^2/2g_n$. From Fig. 14.11 the expression for stagnation pressure,

$$\frac{p_s}{\gamma} = \frac{p_o}{\gamma} + \frac{V_o^2}{2g_n} \quad \text{or} \quad p_s = p_o + \tfrac{1}{2}\rho V_o^2 \tag{5.3}$$

may be written directly, thus confirming the equations of Section 5.4.

In the use of pitot tubes for the measurement of velocity profiles in shear flows, there is an error due to the asymmetry of the flow near the tip of the tube. This error is small where the velocity gradient is low but increases near solid boundaries where the velocity gradient is high. The sense of the error is to move the effective center of the tube toward the region of lower velocity gradient; this means that the velocity being measured by stagnation pressure will be found at a point farther from the wall than the axis of the pitot tube; the magnitude of this error also depends on the detailed geometry of the tube.

Velocity Measurement

14.6 PITOT-STATIC TUBE—INCOMPRESSIBLE FLOW

From the stagnation pressure equation for an incompressible fluid (Eq. 5.3),

$$V_o = \sqrt{\frac{2(p_s - p_o)}{\rho}} = \sqrt{\frac{2g_n(p_s - p_o)}{\gamma}} \tag{14.3}$$

from which it is apparent that velocities may be calculated from measurements of stagnation and static pressures. It has been shown that stagnation pressures may be measured easily and accurately by a pitot tube, and static pressures by various methods such as tubes, flat plates, and wall piezometer openings. Therefore, a pitot tube can be used together with any static pressure device to obtain the necessary pressure difference $p_s - p_o$ from which

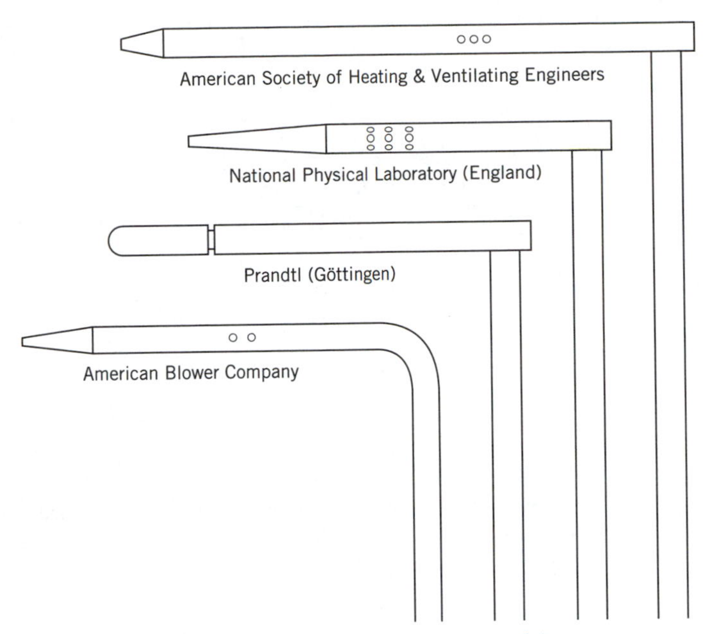

Fig. 14.12 Pitot-Static tubes (to scale).

V_o is deduced. An integral combination of stagnation- and static-pressure-measuring devices into a single device is known usually as a *pitot-static tube.*

Modern practice favors the combined type of pitot-static tube, several of which are illustrated in Fig. 14.12. Here the static tube jackets the stagnation pressure tube, resulting in a compact, efficient, velocity-measuring device. When connected to a differential pressure-measuring instrument, the pressure difference ($p_s - p_o$), which is seen from Eq. 14.3 to be a direct measure of the velocity V_o, may be read.

A static tube has been shown to record a pressure slightly less than the true static pressure, owing to the increase in velocity past the tube (Section 14.3). This means that Eq. 14.3 must be modified by an experimentally determined instrument coefficient, C_I, to

$$V_o = C_I\sqrt{\frac{2(p_s - p_o')}{\rho}} = C_I\sqrt{\frac{2g_n(p_s - p_o')}{\gamma}} \tag{14.4}$$

in which p_o' is the actual pressure measured by the static tube. Since p_o' is less than p_o, it is evident that C_I will always be less than unity. However, for most engineering problems the value of C_I may be taken as 1.00 for the conventional types of pitot-static tubes (Fig. 14.12), since the differences between p_o and p_o' are very small. Prandtl has designed a pitot-static tube in which the difference between p_o and p_o' is completely eliminated by ingenious location of the static-pressure opening. The opening is so located (Fig. 14.13) that the underpressure caused by the tube is exactly compensated by the overpressure due to the stagnation point on the leading edge of the stem, thus giving the true static pressure at the peizometer opening.

A practical aspect of velocity-measuring devices is their sensitivity to yaw or misalignment with the flow direction. Since perfect alignment is virtually impossible, it is advantageous for a pitot-static tube to produce minimum error when perfect alignment does not exist. Prandtl's pitot-static tube, designed to be insensitive to small angles of

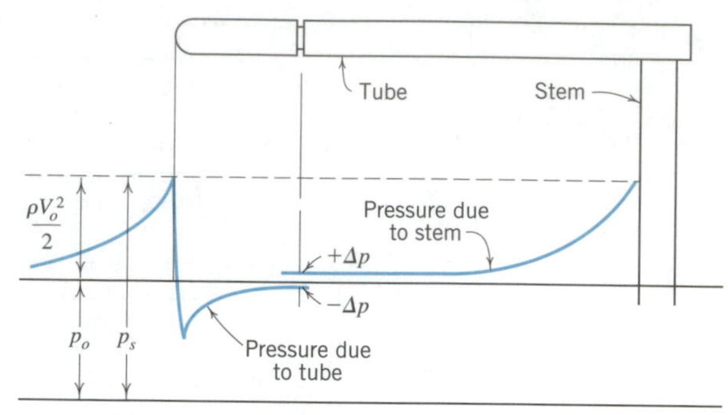

Fig. 14.13 Prandtl's pitot-static tube.

yaw, gives a variation of only 1% in its coefficient at an angle of yaw of 19°. For the same percentage variation in coefficient the American Society of Heating and Ventilating Engineers' pitot-static tube may have an angle of yaw of 12°, and that of the National Physical Laboratory only 7°.[6]

ILLUSTRATIVE PROBLEM 14.2

A pitot-static tube having a coefficient of 0.98 is placed at the center of a pipeline in which jet fuel (JP-4) is flowing. A manometer attached to the pitot-static tube contains mercury and jet fuel and shows a reading of 75 mm. Calculate the velocity at the centerline of the pipe.

SOLUTION

From Eq. 14.4,

$$V_o = C_I \sqrt{\frac{2(p_s - p_o')}{\rho}} = C_I \sqrt{\frac{2g_n(p_s - p_o')}{\gamma}} \tag{14.4}$$

From Appendix 2,

$$\text{s.g.}_{\text{Hg}} = 13.57 \qquad \text{s.g.}_{\text{fuel}} = 0.77$$

Given the manometer reading MR $= 0.075$ m and the tube coefficient $C_I = 0.98$, the use of manometer principles yields

$$\frac{p_s - p_o'}{\gamma} = 0.075 \times \frac{13.57 - 0.77}{0.77} = 1.25 \text{ m of fuel}$$

$$v_c = 0.98\sqrt{2g_n(1.25)} = 4.85 \text{ m/s} \bullet \tag{14.4}$$

[6]Data from K. G. Merriam and E. R. Spaulding, "Comparative Tests of Pitot-Static Tubes," *N.A.C.A. Tech. Note* 546, 1935.

14.7 PITOT-STATIC TUBE—COMPRESSIBLE FLOW

For velocity measurements in compressible flow, separate measurements of static pressure, stagnation pressure, and stagnation temperature are required. For subsonic flow ($\mathbf{M}_o < 1$), Eq. 13.15 may be used directly:

$$\frac{V_o^2}{2} = c_p T_s \left[1 - \left(\frac{p_o}{p_s} \right)^{(k-1)/k} \right] \tag{13.15}$$

with p_s obtained by pitot tube, p_o by static tube, and T_s by temperature probe. The temperature probe consists of a small thermocouple surrounded by a jacket with open upstream end and small holes at the rear; a stagnation point exists on the upstream end of the probe, and a temperature close to the stagnation temperature is measured by the thermocouple. Unfortunately, the measured temperature is not exactly the stagnation temperature, and calibration of the instrument is required.

For supersonic flow ($\mathbf{M}_o > 1$) a short section of normal shock wave (Section 13.12) will be found upstream from the stagnation point (Fig. 14.14) and velocity calculations are considerably more complicated. Applying Eq. 13.31 through the shock wave,

$$\frac{p_1}{p_o} = 1 + \frac{2k}{k+1} (\mathbf{M}_o^2 - 1) \tag{14.5}$$

Applying Eq. 13.13 between the downstream side of the shock wave and the stagnation point,

$$\frac{p_s}{p_1} = \left(1 + \frac{k-1}{2} \mathbf{M}_1^2 \right)^{k/(k-1)} \tag{14.6}$$

Multiplying these two equations together substituting the relation between $\mathbf{M}_o$ and $\mathbf{M}_1$ of Eq. 13.30 yields

$$\frac{p_s}{p_o} = \frac{k+1}{2} \mathbf{M}_o^2 \left[\frac{(k+1)^2 \mathbf{M}_o^2}{4k\mathbf{M}_o^2 - 2k + 2} \right]^{1/(k-1)} \tag{14.7}$$

which shows that measurements of p_s and p_o will allow prediction of the Mach number $\mathbf{M}_o$ of the undisturbed stream. Using the fact that the stagnation temperature is the same

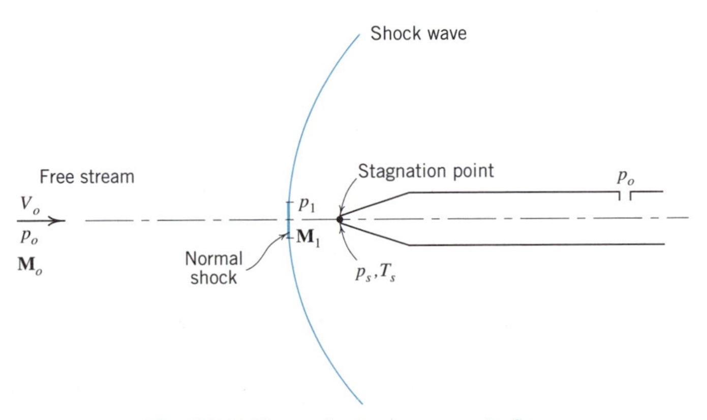

Fig. 14.14 Pitot tube in a supersonic flow.

on both sides of the shock wave,

$$V_o^2/2 = c_p(T_s - T_o) \tag{14.8}$$

However, from $a_o^2 = kRT_o$ and $\mathbf{M}_o = V_o/a_o$, $T_o = V_o^2/kR\mathbf{M}_o^2$, which when substituted in the foregoing equation yields

$$\frac{V_o^2}{2}\left(1 + \frac{2c_p}{kR\mathbf{M}_o^2}\right) = c_pT_s \tag{14.9}$$

Use of Eqs. 14.7 and 14.8 allows V_o to be calculated from measurements of T_s, p_s, and p_o.

From measurements of p_s and p_o the experimentalist may easily identify subsonic and supersonic flow and thus select the proper equation for velocity calculation. By inserting $\mathbf{M}_o = 1$ in Eq. 13.13 or 14.7, we obtain

$$\frac{p_s}{p_o} = \left(\frac{k+1}{2}\right)^{k/(k-1)} \tag{14.10}$$

Thus, for $\mathbf{M}_o < 1$, $p_s/p_o < [(k+1)/2]^{k/(k-1)}$ and, for $\mathbf{M}_o > 1$, $p_s/p_o > [(k+1)/2]^{k/(k-1)}$.

ILLUSTATIVE PROBLEM 14.3

A pitot-static tube in an air duct indicates an absolute stagnation pressure of 635 mm of mercury; the absolute stagnation pressure in the duct is 457 mm of mercury; a temperature probe shows a stagnation temperature of 66°C. What is the local velocity just upstream from the pitot-static tube?

SOLUTION

The first step is to determine if the flow is subsonic or supersonic; applying Eq. 14.10 with the given pressures $p_s = 635$ mm Hg and $p_o = 457$ mm Hg, and $k = 1.4$ for air (Appendix 2) yields with

$$\frac{p_s}{p_o} = \left(\frac{k+1}{k}\right)^{k/(k-1)} \quad \text{for sonic flow,} \quad \frac{635}{457} = 1.39 < \left(\frac{1.4+1}{1.4}\right)^{3.5} = 1.893 \tag{14.10}$$

The flow is subsonic and so Eq. 13.15 is appropriate. It is

$$\frac{V_o^2}{2} = c_pT_s\left[1 - \left(\frac{p_o}{p_s}\right)^{(k-1)/k}\right] \tag{13.15}$$

With the given pressures, $T_s = 66°\text{C}$, and $c_p = 1\,003$ J/kg·K,

$$V_o^2/2 = 1\,003(273 + 66)\left[1 - \left(\frac{457}{635}\right)^{0.286}\right] = 30\,529 \text{ m}^2/\text{s}^2 \text{ of air} \tag{13.15}$$

$$V_o = 247 \text{ m/s} \bullet$$

ILLUSTRATIVE PROBLEM 14.4

The instruments of a high-speed airplane flying at high altitude show a stagnation pressure of 20 in. of mercury absolute, a static pressure of 5 in. of mercury absolute, and a stagnation temperature of 150°F. Calculate the speed of this airplane.

SOLUTION

See Eq. 14.10 in the problem above; we use it to test whether the flow is subsonic or supersonic. Given that, from Appendix 2 and the problem statement, we have

$$p_s = 20 \text{ in. Hg} \qquad p_o = 5 \text{ in. Hg} \qquad T_s = 150°\text{F}$$

$$R = 1\,715 \text{ ft}\cdot\text{lb/slug}\cdot°\text{R} \qquad c_p = 6\,000 \text{ ft}\cdot\text{lb/slug}\cdot°\text{R} \qquad k = 1.4$$

then

$$\frac{20}{5} = 4 > \left(\frac{1.4 + 1}{2}\right)^{3.5} = 1.893 \tag{14.10}$$

Therefore Eqs. 14.7 et seq. should be used for velocity calculation:

First,

$$\frac{p_s}{p_o} = \frac{k+1}{2}\mathbf{M}_o^2\left[\frac{(k+1)^2\mathbf{M}_o^2}{4k\mathbf{M}_o^2 - 2k + 2}\right]^{1/(k-1)} \tag{14.7}$$

$$4 = \frac{1.4 + 1}{2}\mathbf{M}_o^2\left[\frac{(2.4)^2\mathbf{M}_o^2}{5.6\mathbf{M}_o^2 - 2.8 + 2}\right]^{2.5}$$

Solving (by trial),

$$\mathbf{M}_o^2 = 2.71 \qquad \mathbf{M}_o = 1.645$$

Then,

$$\frac{V_o^2}{2}\left(1 + \frac{2c_p}{kR\mathbf{M}_o^2}\right) = c_pT_s \tag{14.9}$$

$$\frac{V_o^2}{2}\left(1 + \frac{2 \times 6\,000}{1\,715 \times 1.4 \times 2.71}\right) = 6\,000(610) \qquad V_o = 1\,600 \text{ ft/s} \bullet$$

$$= 1\,091 \text{ mph} \bullet$$

14.8 MECHANICAL ANEMOMETERS AND CURRENT METERS

Mechanical devices of similar characteristics are utilized in the measurement of velocity in air and water. Surprisingly, these devices have not changed in their general form for many years, although digital and analog electronic recording of their output has become commonplace. Those for air are called *anemometers*; those for water, *current meters*. These devices consist essentially of a rotating element whose speed of rotation varies with the local velocity of flow, the relation between these variables being found by calibration. Anemometers and current meters fall into two main classes, depending on the design of the rotating elements; these are the cup type and the propeller type. Figure 14.15 shows samples of these basic types. Anemometers and current meters differ slightly in shape, ruggedness, and appurtenances because of the different conditions under which they are used. The anemometers are usually mounted on a rigid shaft. Current meters are also attached to towers or shafts, but are often suspended in a river, canal, bay or ocean flow by cables and hence must have empennages and weights to hold them in fixed positions in the flow. Wyngaard[7] describes in some detail the basic equations and the use of ane-

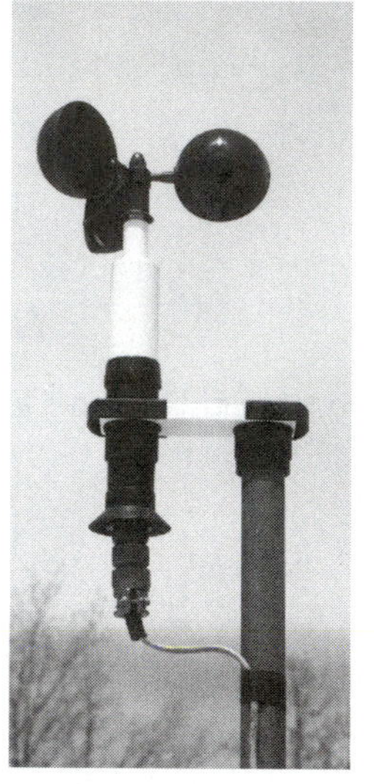

Cup type

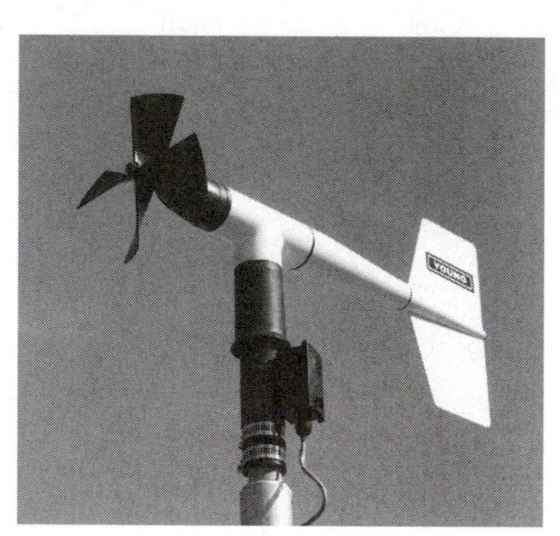

Propeller type

Anemometers
(Photos courtesy of R. M. Young Co., Traverse City, Michigan)

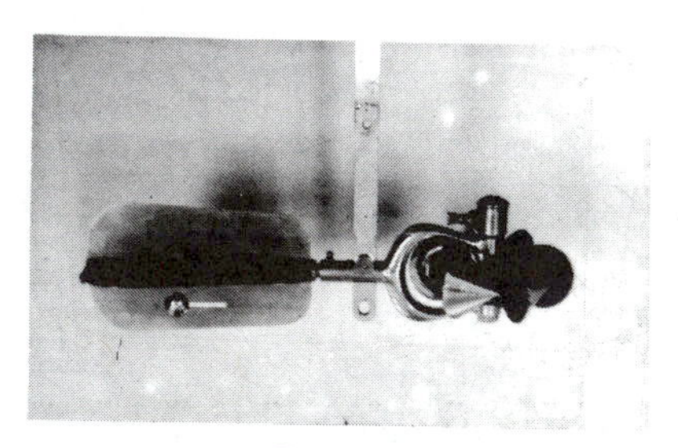

Cup type
(Picture courtesy of N.Y.U.)

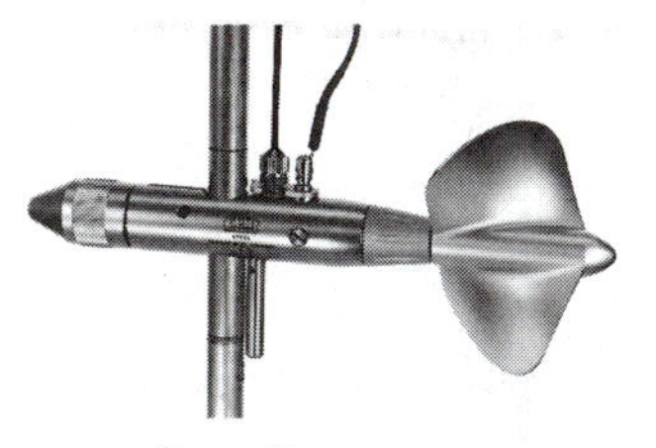

Propeller type
(Photo courtesy of Epic, Inc., New York and A. Ott, Kempten, Germany)

Current meters

Fig. 14.15 Anemometers and current meters.

[7] J. C. Wyngaard, "Cup, propeller, vane, and sonic anemometers in turbulence research," *Annual Review of Fluid Mechanics*, 13: 399–423, 1981.

mometers in meteorology. Fulford, et al.[8] present precision and accuracy assessments of 12 current meters in use throughout the world.

14.9 HOT-WIRE, HOT-FILM, SONIC, AND LASER-DOPPLER ANEMOMETERS

In this section, we mention briefly four types of non-mechanical anemometers, namely, hot-wire, hot-film, sonic (or acoustic), and laser-Doppler anemometers. All of these devices have benefited greatly from modern developments in data acquisition and analysis and are more useful and reliable as a result. In addition, all are used to make measurements in gas environments and all can be used to make measurements in liquids, but only the last two are likely to be used for liquids in field situations (outside the laboratory). The laser-Doppler and sonic anemometers also share two features, namely, (1) they can be installed in laboratory flows so that they do not disturb the flow and (2) no calibration is needed because the velocity is determined from the geometry and the characteristics of the acoustic or laser signal.

The *hot-wire* or *hot-film anemometers* may be treated together and are used for measuring both mean and instantaneous velocities. However, for accessible flows (those where a light or sound beam has a clear path in the flow or may enter or leave the flow through a free surface or [light or sound] transparent walls or ports), the sonic and laser-Doppler anemometers are the instruments of choice. Figure 14.16 shows the typical sensing elements and their supports for hot-wire and hot-film operation.

As seen in Fig. 14.16 the sensing element of the anemometer is either a thin wire or a metal film laid over a glass support and coated to protect the film. Because of the supporting rod and coating the hot-film sensor is mechanically superior and usable in contaminated environments. The sensor element is heated electrically by an electronic circuit which allows measurement of the power supplied to the element. The power supplied is related to the fluid velocity *normal to the sensor* through the laws of convective heat transfer between the sensor and the fluid. With the proper electronic control and compensation a thin hot-wire can respond to velocity fluctuations at frequencies up to 500 000 Hz.

The anemometer sensors are usually an electronic part of either a *constant temperature* or *constant current* circuit, the former being the system most often used. In a constant temperature anemometer (Fig. 14.17*a*) the sensor is one leg of a bridge circuit. The system of resistances is balanced at a no-flow condition by use of the variable resistor so that there is no unbalance-voltage output. Then flow past the sensor will cool it and decrease its resistance. The detector senses the unbalanced condition and changes its output voltage to increase the current flowing in the sensor, thus bringing its temperature back to the original value. The bridge voltage, V (the output voltage of the amplifier), is proportional to the current, I, flowing in the sensor because all resistances are kept constant. But, $P = I^2R$, so the square of the output voltage is proportional to the instantaneous heat transfer from the sensor.

In a constant current anemometer (Fig 14.17*b*), the current in the sensor is maintained constant by putting the sensor in series with a very large (relative to the sensor resistance) resistor. When a flow cools the sensor, its resistance decreases, thereby unbalancing the

[8] J. M. Fulford, K. G. Thibobeaux, and W. R. Kaehrle, ''Repeatability and oblique flow response characteristics of current meters,'' *Hydraulic Engineering 93*, New York: A.S.C.E., 1993, pp. 1452–1457.

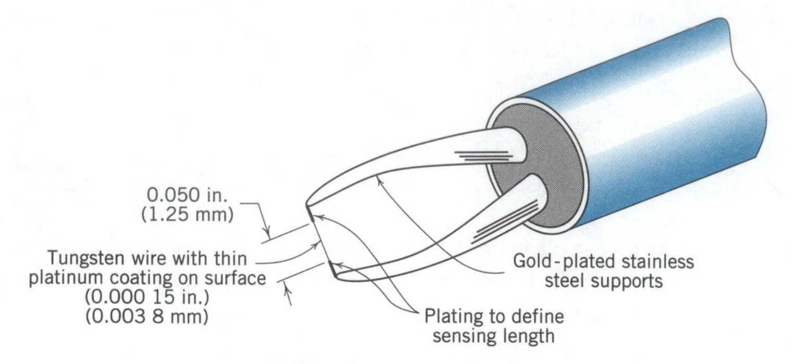

(a) Hot-wire sensor and support needles.

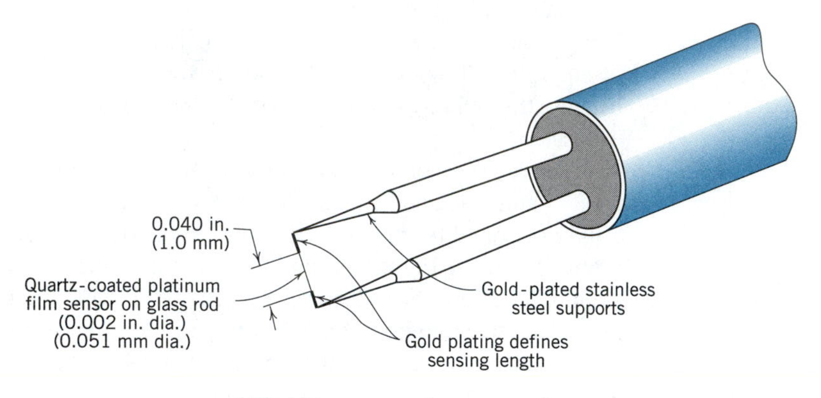

(b) Hot-film sensor and support needles.

Fig. 14.16 Anemometer sensors. (Reproduced from TB5, Thermo-Systems, Inc., 2500 Cleveland Avenue North, St. Paul, Minnesota, 55113.)

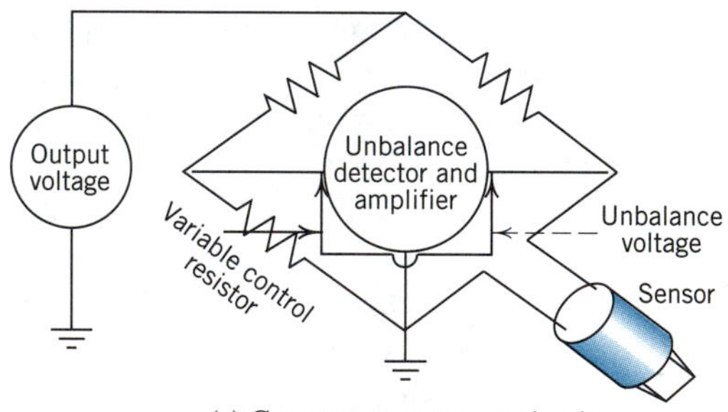

(a) Constant temperature circuit.

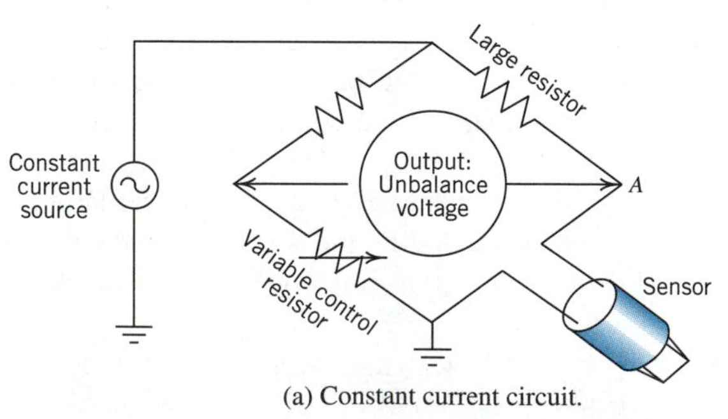

(a) Constant current circuit.

Fig. 14.17 Schematic of anemometers.

bridge by changing the voltage at point A. This unbalance voltage is the output which must be greatly amplified. Generally, a constant current system is more complex to operate and less accurate than a constant temperature system.

King[9] showed that the power transferred from a wire (or film) sensor can be expressed as

$$P = I^2 R_{\text{sensor}} = (a + b\sqrt{\rho v'})(T_{\text{sensor}} - T_{\text{fluid}})$$

where I is the current flowing through the sensor, v' is the instantaneous fluid velocity normal to the sensor, and a and b are empirical constants obtained by calibration of the wire (or film). For the constant temperature system, R_{sensor} and T_{sensor} are constant, so $V^2 \propto I^2$. Thus, provided T_{fluid} is constant,

$$V^2 = (A + B\sqrt{\rho v'}) \tag{14.11}$$

where V is the output voltage of the bridge. Figure 14.18 shows a typical calibration curve for a hot-film sensor.

Figure 7.4 shows a schematic of the variation of one component of the instantaneous velocity in a turbulent flow. Figure 14.19 gives a plot of experimental data, obtained at Stanford, for air flow over a smooth flat plate in a wind tunnel. Here a pair of hot-wires were mounted in an X-array. (Actually, the two wires lie next to each other in vertical parallel planes that are in line with the flow. One wire is inclined at $+45°$ to the horizontal and one at $-45°$.) In this configuration, the sum and difference of the velocities normal to each wire are used to obtain the horizontal and vertical components of the flow, respectively. With mean values removed, the remainder is used to calculate the intensities shown in Fig. 14.19.

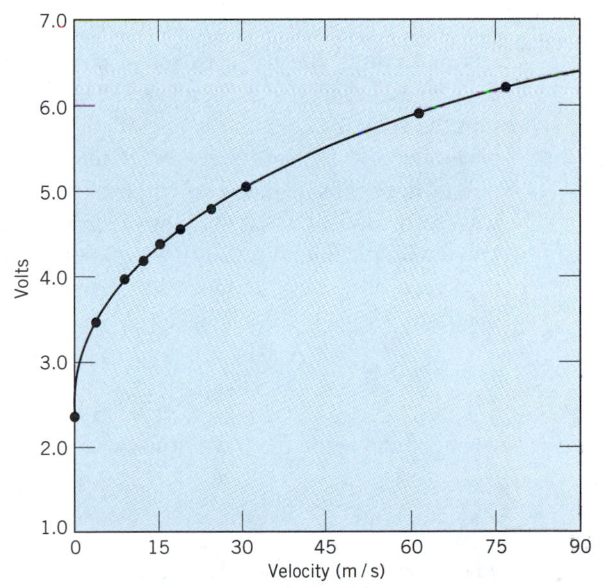

Fig. 14.18 Calibration for a 0.051 mm diameter hot-film sensor in atmospheric air: 0–90 m/s. (Reproduced from TB5, Thermo-Systems, Inc., 2500 Cleveland Avenue North, St. Paul, Minnesota 55113.)

[9]L. V. King, "On the Convection of Heat from Small Cylinders in a Stream of Fluid, with Applications to Hot-Wire Anemometry," *Phil. Trans. Roy. Soc. London*, vol. 214, no. 14, 1914, pp. 373–432.

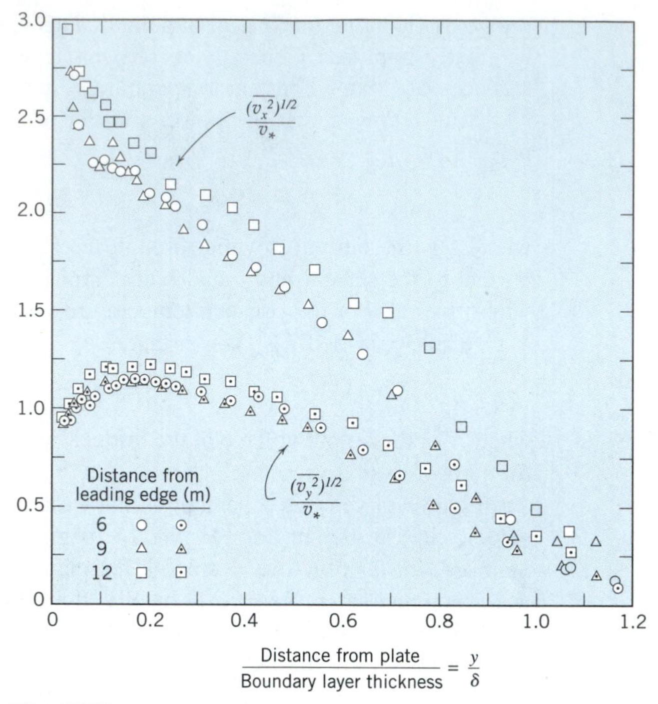

Fig. 14.19 Turbulent velocity intensities in air flow over a smooth flat path.

The sonic or acoustic anemometer has become a major instrument in turbulence research in meteorology. Wyngaard[10] describes the principles of operation of this anemometer and points out that it is free of the nonlinearities, time lag and other deficiencies of the cup and propeller anemometers. The acoustic anemometer in its simplest form measures the velocity along the acoustic path between two points. At each point there is a sound generator and a sound receiver. If the two opposing generators produce simultaneous pulses, the opposing receivers will receive these pulses at different times because the sound waves are slowed when they move against the fluid flow and are speeded up when they move with the fluid flow. Following Wyngaard, if the distance between the points is d and the speed of sound c is much larger than the velocity along the path V_p, then from the diagram of Fig. 14.20, we have

$$t_1 = \frac{d}{c + V_p} \qquad \text{and} \qquad t_2 = \frac{d}{c - V_p},$$

where t_1 and t_2 are the travel times of pulses from generator to receiver, and so

$$V_p = \frac{c^2 \, \Delta t}{2d}, \tag{14.12}$$

where the travel-time difference for the two pulses is Δt.

Measurements of several components of the velocity field are made by placing sets of single-axis sonic anemometers in arrays such that, for example, the horizontal axes of

[10]J. C. Wyngaard, "Cup, propeller, vane, and sonic anemometers in turbulence research," *Annual Review of Fluid Mechanics*, 13: 399–423, 1981.

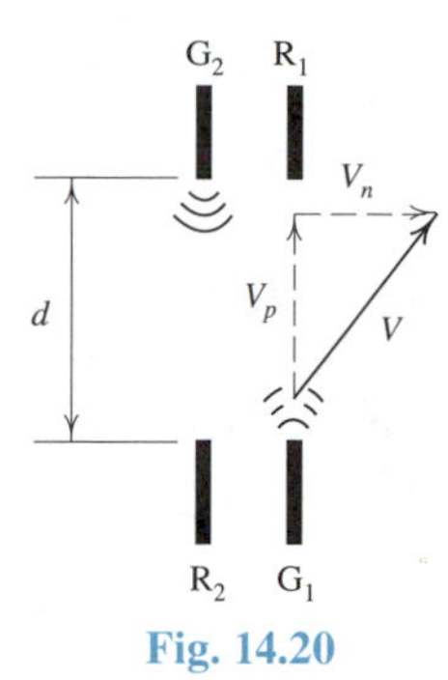

Fig. 14.20

two anemometers are separated by 120° while one anemometer lies on a vertical axis; the reader can confirm that this allows all three velocity components to be extracted from the three velocities measured along the individual anemometer axes. A typical value for the separation distance d is about 20 cm for a field installation.

The laser-Doppler anemometer is an optical device that allows the accurate measurement of fluctuating velocity fields without disturbing the flow. Both laboratory and field installations and use are common. This device makes use of the well-known fact that the radiation received at a fixed point from a moving body is frequency shifted (the Doppler effect). The actual construction and operation of a laser-Doppler anemometer (or LDA) are complex and exacting and require much experience; there is a broad literature on the topic and many papers have been written on various applications and techniques. On the other hand, the basic concept is quite simple and we describe it briefly here. Ignoring the niceties of optical theory, we can think of the LDA as follows:

(i) A powerful light source of a single frequency—a laser—provides a light beam that is passed through

(ii) Optics which might, for example, split the beam into two beams which are separated by a small distance and then focus these two beams on a very small measuring volume in the flow (in the experiments of Cheung and Koseff[11] for example, the measuring volume was 0.1 mm in diameter and 0.8 mm in length and the angle between the beams was 6.3°).

(iii) Particles passing though the measuring volume scatter light whose frequency is shifted due to the Doppler effect by which the velocity of the particles causes the scattered light to have a shift in frequency compared to the incident laser light. The size of the shift is proportional to the particle velocity component perpendicular to a line which bisects the angle between and lies in the plane of the converging beams. If the particles are very small, then they move essentially with the fluid and the fluid and particle velocities are equal.

(iv) Receiving optics and a photo detector collect the frequency-shifted scattered light. The time-varying frequency shift is recorded and analyzed automatically to obtain the instantaneous fluid velocity at the measuring volume.

If an anemometer is built with two intersecting beams as described here, then the fluid velocity perpendicular to the line bisecting the angle between the beams and lying in the

[11]T. K. Cheung and J. R. Koseff, ''Simultaneous backward-scatter forward-scatter laser Doppler anemometer measurements in an open channel flow,'' *DISA Information*, 28, pp. 3–9, Denmark, February 1983.

plane of these beams is given by

$$u_B = \frac{f_D}{2\sin(\theta_B/2)}\left(\frac{c}{f_B}\right) = \frac{f_D\lambda_B}{2\sin(\theta_B/2)} \tag{14.13}$$

where the fluid velocity component is u_B, the angle between the beams is θ_B, f_B and λ_B are the frequency (Hz) and wavelength (m) of the laser light, c is the speed of light, and f_D is the Doppler frequency shift of the scattered light. Notice that no calibration is needed; only the measurement of the beam angle, knowledge of the characteristics of the laser light, and recording of the Doppler frequency shift are required.

In a similar manner to that employed for the sonic anemometer, measurement of more than one velocity component is made possible with the LDA by using additional sets of crossing beams whose planes are at angles to the original set. For two components, the planes are usually perpendicular and the beams all cross in the same measuring volume.

Shear Measurement

14.10 SHEAR DETERMINED BY INFERENCE

No device has yet been invented which is capable of measuring the stress between moving layers of fluid. Shear measurements consist entirely of measurements of wall shear (τ_o) from which the shear between moving layers may be deduced from certain equations of fluid mechanics; such deductions may be of high or low accuracy, depending on the equations used and the approximations necessary for solving them.

The wall shear for a cylindrical pipe of uniform roughness and with established flow may be obtained easily and accurately from pressure measurements through the use of Eq. 7.34, which may be expressed as

$$\tau_o = \frac{\gamma d}{4l}\left[\frac{p_1}{\gamma} + z_1 - \frac{p_2}{\gamma} - z_2\right] \tag{14.14}$$

in which the bracketed term is recognized as the head loss between points 1 and 2. With all details of the flow axisymmetric and no variation of wall roughness, it may be safely assumed that τ_o is the same at all points on the pipe wall and its value deducible from Eq. 14.14. The same procedure may be applied to any prismatic conduit and the same equation may be used with d replaced by $4R_h$; here, however, the flow is not axisymmetric and the shear stress must be interpreted as the *mean value*. Although on any longitudinal element of such a conduit the wall shear may be presumed constant, the equation provides no information on whether the wall shear at any point is larger or smaller than the mean value.

14.11 WALL PROBES

Because of the foregoing limitations and for applications to problems of more complex boundary geometry, a more basic type of shear meter has been developed. This consists (Fig. 14.21) in replacing a small section of the wall by a movable plate mounted on elastic columns fastened to a rigid support. The columns are deflected slightly by the shearing

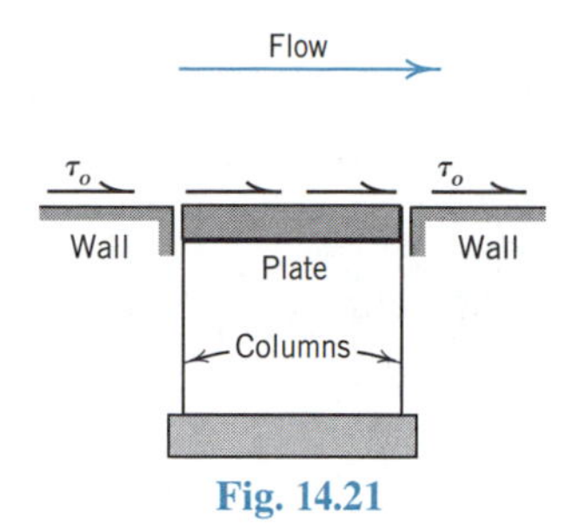

Fig. 14.21

force of fluid on plate, the small deflection measured by strain gages, and the shear stress deduced from this deflection.[12] Although the device is basically simple, it is costly, unwieldy, and by no means easy to operate and interpret because of the relatively small shear force to be measured and the relatively large extraneous forces that also contribute to the deflection of the columns.

Wall pitot tubes have been used successfully for the measurement of wall shear. Stanton[13] first used the design shown in Fig. 14.22, the wall forming one side of the pitot tube; calibration in laminar pipe flow and in the viscous sublayer of turbulent pipe flow showed τ_o to be a function of $(p_s - p_o)$. Its use for turbulent flows is thus restricted to measurements in the viscous sublayer covering the wall, where the velocity profile is essentially the same as that in the pipe for the same fluid viscosity and wall shear. Taylor[14] presented a dimensionless calibration for the Stanton tube which may be expressed by $\tau_o h^2/4\rho\nu^2$ as a function of $(p_s - p_o)h^2/4\rho\nu^2$; this relationship is shown in Fig. 14.22, which allows τ_o to be predicted from measurements of $(p_s - p_o)$, ρ, ν, and h, providing that h is smaller than the thickness of the viscous region.

More recently, Preston[15] has applied the foregoing idea to turbulent flow over smooth surfaces with a tube of simpler design. The Preston tube (Fig. 14.23) is not submerged in

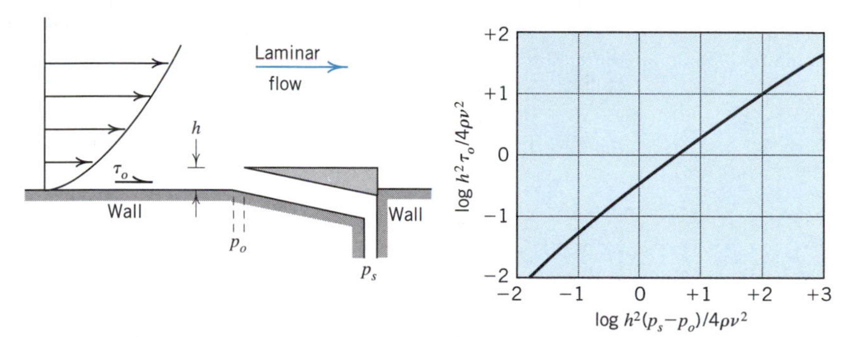

Fig. 14.22 Stanton tube.

[12]See Section 6.5.2.1 in W. H. Graf, *Hydraulics of Sediment Transport*, McGraw-Hill, 1971.

[13]T. E. Stanton, D. Marshal, and (Mrs.) C. N. Bryant, "On the Conditions at the Boundary of a Fluid in Turbulent Motion," *Proc. Roy. Soc.*, (A), vol. 97, 1920.

[14]G. I. Taylor, "Measurements with Half-Pitot Tubes," *Proc. Roy. Soc.*, (A), vol. 166, 1938.

[15]J. H. Preston, "The Determination of Turbulent Skin Friction by Means of Pitot Tubes," *Journ. Roy. Aero. Soc.*, vol. 58, 1954. See also E. Y. Hsu, "The Measurement of Local Turbulent Skin Friction by Means of Pitot Tubes," *David Taylor Model Basin* Rept. 957, August 1955, and V. C. Patel, "Calibration of the Preston Tube and Limitations on its Use in Pressure Gradients," *J. Fluid Mech.*, vol. 23, no. 1, 1965, pp. 185–208.

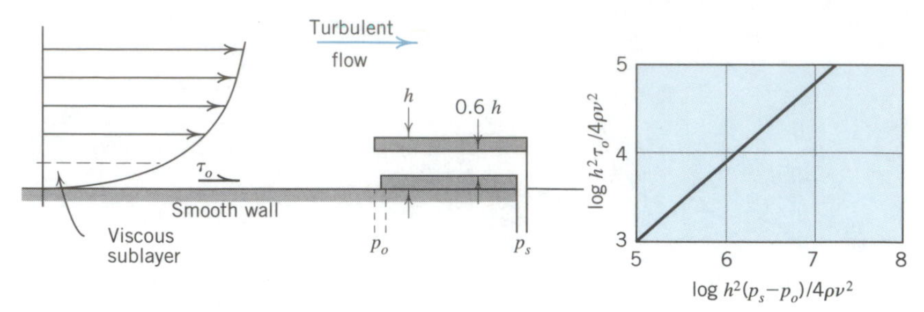

Fig. 14.23 Preston tube.

the viscous sublayer, and its performance depends on the similarity of the velocity profiles through the buffer[16] zone between the viscous sublayer and the turbulent region. The single calibration curve of Fig. 14.23 validates this similarity over the range indicated.

Flowrate Measurement

In some steady flows, the flowrate (Q or G) can be obtained by simple measurement of the total quantity of fluid collected in a measured time. Such collections can be made by weight or by volume and are a primary means of measuring fluid flow. However, they are usually practical only for small flows under laboratory conditions. For gas flows, volumetric measurements must be made under conditions of constant pressure and temperature.

For routine practical measurements there exist a plentiful supply of *flowmeters* and flow-measuring techniques. Among the flowmeters are positive displacement types (e.g., reciprocating piston meters, nutating disk meters, rotary piston and vane meters, etc.) and differential measurement systems (e.g., venturi meters, orifices, nozzles, and elbow meters). For open-channel flows, weirs are useful, while multiple current meter (Section 14.8) measurements across a channel section can be integrated to give flowrate. For many flow applications, magnetic and ultrasonic (acoustic) flow measurement devices[17] offer nonintrusive measurements.

No attempt is made here to describe all possible systems. A representative sample is discussed; the appropriate references at the end of the chapter contain detailed information.

14.12 VENTURI METERS

A constriction in a streamtube has been seen (Sections 5.4 and 13.8) to produce an accelerated flow and fall of hydraulic grade line or pressure which is directly related to flowrate, and thus is an excellent meter in which rate of flow may be calculated from pressure measurements. Such constrictions used as fluid meters are employed by Venturi meters, nozzles, and orifices.

[16]See Fig. 9.4 of Section 9.3.

[17]N. P. Cheremisinoff, *Applied Fluid Flow Measurement.* New York: M. Dekker, Inc., 1979. N. P. Cheremisinoff and P. N. Chreremisinoff, *Flow Measurement for Engineers and Scientists.* New York: M. Dekker, 1988.

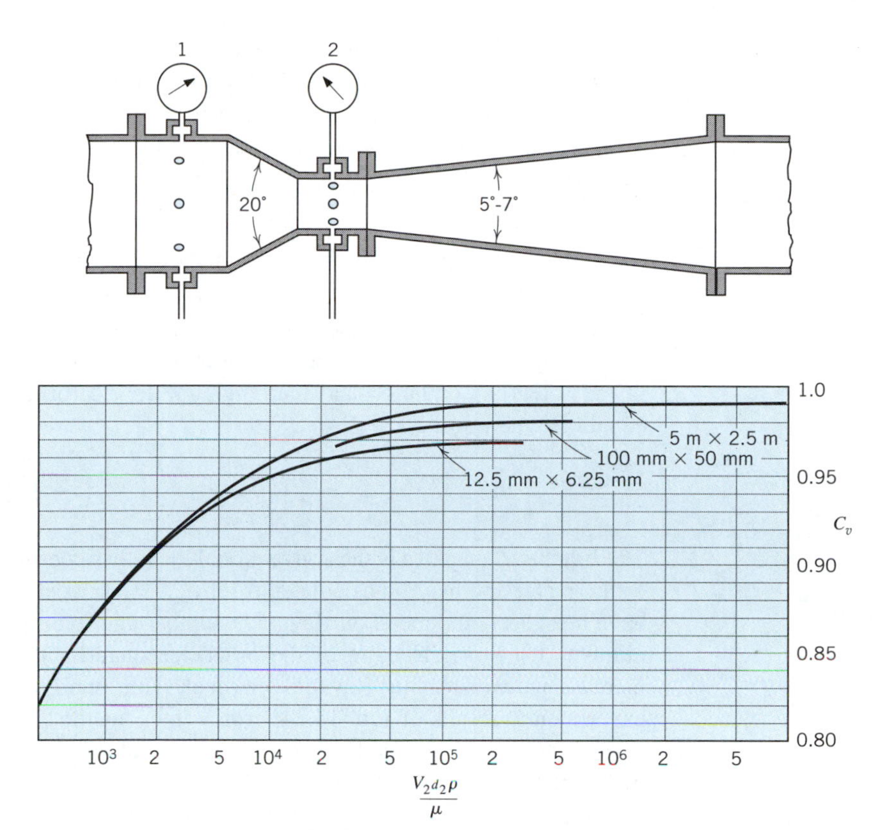

Fig. 14.24 Venturi meter and coefficients.

A Venturi meter is shown in Fig. 14.24. It consists of a smooth entrance cone of angle about 20°, a short cylindrical section, and a diffuser of 5° to 7° cone angle in order to minimize head loss (Section 9.9). For satisfactory operation of the meter the flow should be an established one as it passes section 1. To ensure this the meter should be installed downstream from a section of straight and uniform pipe, free from fittings, misalignments, and other sources of large-scale turbulence, and having a length of at least 30 (preferably 50) pipe diameters. Straightening vanes may also be placed upstream from the meter for reduction of rotational motion in the flow.

The pressures at the *base* of the meter (section 1) and at the *throat* (section 2) are obtained by piezometer rings, and the pressure difference is usually measured by differential manometer or pressure transducer.[18] For the metering of gases, separate measurements of pressure and temperature are required at the base of the meter, but for liquids the differential pressure reading alone will allow computation of the flowrate.

For ideal incompressible flow the flowrate may be obtained by solving simultaneously Eqs. 4.5 and 5.1 (see Section 5.4) to yield

$$Q = \frac{A_2}{\sqrt{1 - (A_2/A_1)^2}} \sqrt{2g_n \left(\frac{p_1}{\gamma} + z_1 - \frac{p_2}{\gamma} - z_2 \right)} \tag{14.15}$$

[18]In industrial practice, this pressure difference is frequently carried to an electromechanical or electronic device which calculates, records and/or plots a chronographic record of the flowrate.

However, for real-fluid flow and the same $(p_1/\gamma + z_1 - p_2/\gamma - z_2)$ the flowrate will be expected to be less than that given by Eq. 14.15 because of frictional effects and consequent head loss between sections 1 and 2; in metering practice it is customary to account for this by insertion of an experimentally determined coefficient C_v in Eq. 14.15,[19] which then becomes

$$Q = \frac{C_v A_2}{\sqrt{1 - (A_2/A_1)^2}} \sqrt{2g_n \left(\frac{p_1}{\gamma} + z_1 - \frac{p_2}{\gamma} - z_2\right)} \tag{14.16}$$

Since C_v is merely a convenient means of expressing the head loss $h_{L_{1-2}}$, an exact relation between these variables is to be expected; this may be found by substituting A_2V_2 for Q in the foregoing equation and rearranging it to the form of the conventional Bernoulli equation. The result is

$$h_{L_{1-2}} = \left[\left(\frac{1}{C_v^2} - 1\right)\left(1 - \left(\frac{A_2}{A_1}\right)^2\right)\right]\frac{V_2^2}{2g_n} \tag{14.17}$$

The bracketed quantity is the conventional local loss coefficient, K_L, for the entrance cone of the meter and may be calculated from C_v.[20] Loss coefficients and pipe friction factors have been seen to depend on Reynolds number and to diminish with increasing Reynolds number; from the structure of the bracketed quantity it can be predicted that C_v will increase with increasing Reynolds number. This prediction is borne out by the typical experimental results of Fig. 14.24[21] for Venturi meters of different size but of the same diameter ratio.[22] It should be observed in view of the principles of similitude (Chapter 8) that geometrically similar meters could be expected to give results falling on a single line on the plot, when installed in pipelines with established flow.

ILLUSTRATIVE PROBLEM 14.5

Air flows through a 150 mm by 75 mm Venturi meter. The gage pressure is 140 kPa and the fluid temperature 15°C at the base of the meter, and the differential manometer shows a reading of 150 mm of mercury. The barometric pressure is 101.3 kPa. Calculate the flowrate.

SOLUTION

This is a compressible flow so we employ Eq. 1.3 as an equation of state

$$\rho = p/RT \tag{1.3}$$

[19]For compressible flow C_v is inserted in Eqs. 13.20, 13.21, 13.22, and 13.24 in the same way.

[20]Nominal values of K_L and C_v in turbulent flow are 0.04 and 0.98, respectively. See Section 9.9.

[21]Test results for Simplex Valve and Meter Co. and Builders Iron Foundry Venturi meters. *Fluid Meters: Their Theory and Application*, 4th edition, A.S.M.E., 1937.

[22]Experiments at other diameter ratios (d_2/d_1) show a decrease of coefficient with increase of diameter ratio.

and calculate the mass flowrate according to Eq. 13.24

$$\dot{m} = \frac{YA_2\rho_1}{\sqrt{1 - \left(\frac{A_2}{A_1}\right)^2}} \sqrt{2g_n\left(\frac{p_1 - p_2}{\gamma_1}\right)} \qquad (13.24)$$

With the given data

$$p_1 = 140 \text{ kPa} \qquad p_{\text{atm}} = 101.3 \text{ kPa} \qquad \text{MR} = 150 \text{ mm Hg}$$

$$T_1 = 15°\text{C} = 288 \text{ K} \qquad d_1 = 0.150 \text{ m} \qquad d_2 = 0.075 \text{ m} \qquad R = 286.8 \text{ J/kg·K}$$

we calculate

$$\frac{p_2}{p_1} = \frac{140 + 101.3 - (150/760) \times 101.3}{140 + 101.3} = 0.917$$

$$\rho_1 = \frac{241.3 \times 10^3}{286.8 \times 288} = 2.92 \text{ kg/m}^3 \qquad (1.3)$$

$$\frac{A_2}{A_1} = 0.25 \qquad A_2 = 0.004\,42 \text{ m}^2$$

Then, from Appendix 5, $Y = 0.949$ and, assuming that $C_v = 0.98$, we find

$$\dot{m} = \frac{0.949 \times 0.98 \times 0.004\,42 \times 2.92}{\sqrt{1 - (0.25)^2}} \sqrt{2g_n(150 \times (101.3 \times 10^3/760))/2.92\, g_n}.$$

$$\dot{m} = 1.45 \text{ kg/s}$$

Checking $\mathbf{R}_2$ gives a value of 1 250 000, which (from Fig. 14.24) gives a better value of C_v of 0.981. The true flowrate is, therefore, $\dot{m} = 1.45$ kg/s (0.981/0.980) $= 1.45$ kg/s. •

14.13 NOZZLES

Nozzles are used in engineering practice for the creation of jets and streams for all purposes as well as for fluid metering; when placed in or at the end of a pipeline as metering devices they are generally termed *flow nozzles*. Since a thorough study of flow nozzles will develop certain general principles which may be applied to other special problems, only the flow nozzle will be treated here.

A typical flow nozzle is illustrated in Fig. 14.25. Such nozzles are designed to be clamped between the flanges of a pipe, generally possess rather abrupt curvatures of the converging surfaces, terminate in short cylindrical tips, and are essentially Venturi meters with the diffuser cones omitted. Since the diffuser cone exists primarily to minimize the head losses caused by the meter, it is obvious that larger head losses will result from flow nozzles than occur in Venturi meters and that herein lies a disadvantage of the flow nozzle; this disadvantage is somewhat offset, however, by the lower initial cost of the flow nozzle.

Extensive research on flow nozzles, sponsored by the American Society of Mechanical Engineers and the International Standards Association, has resulted in the accumulation of a large amount of reliable data on nozzle installation, specifications, and experimental

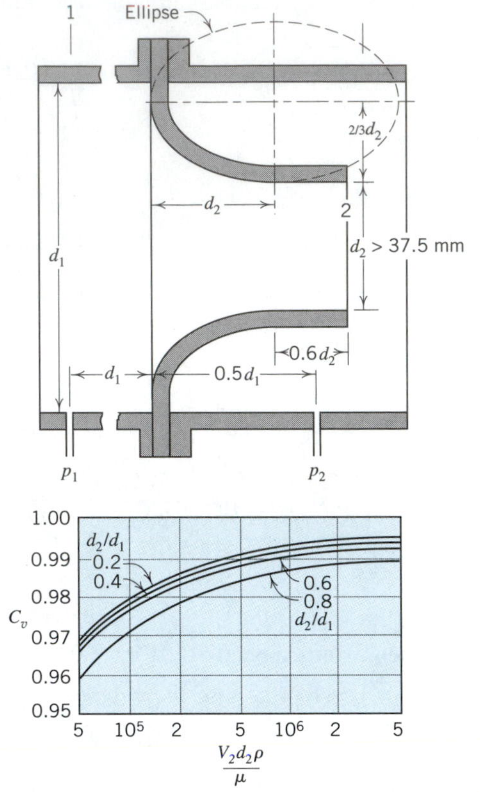

Fig. 14.25 A.S.M.E. flow nozzle and coefficients.[23]

coefficients. Only the barest outline of these results can be presented here; the reader is referred to the original papers of these societies for more detailed information.

The A.S.M.E. flow nozzle installation of Fig. 14.25 is typical of those employed in American practice, section 1 being taken one pipe diameter upstream and section 2 at the nozzle tip. It has been found that a pressure representative of that at the latter point may be obtained by a wall piezometer connection which leads, fortunately, to the simplification of the nozzle installation, since a wall piezometer is easier to construct than a direct connection to the tip of the nozzle. Pressures obtained in this manner are not, of course, the exact pressures existing in the live stream of flowing fluid passing section 2, but the slight deviations incurred are of no consequence since they are absorbed in the experimental coefficient, C_v. The variation of C_v with area ratio and Reynolds number is typical of the geometrically similar conditions specified; the constancy of C_v at high Reynolds number and decrease of C_v with decreasing Reynolds number is observed for the flow nozzle as for the Venturi meter. Coefficients for standardized flow nozzle installations at Reynolds numbers below 50 000 are available, too.[24]

The flow nozzle being essentially equivalent to the entrance cone of the Venturi meter, flowrates may be computed by Eqs. 13.24 and 14.16.

[23]Data from *Fluid Meters: Their Theory and Application,* 4th edition, A.S.M.E., 1937; and H. S. Bean, S. R. Beitler, and R. E. Sprenkle, "Discharge Coefficients of Long Radius Flow Nozzles When Used with Pipe Wall Pressure Taps," *Trans. A.S.M.E.,* p. 439, 1941.

[24]H. S. Bean, ed. *Fluid Meters—Their Theory and Application*, 6th ed, 2d printing, A.S.M.E., 1983.

ILLUSTRATIVE PROBLEM 14.6

An A.S.M.E. flow nozzle of 75 mm diameter is installed in a 150 mm waterline. The attached differential manometer contains mercury and water and shows a reading of 150 mm. Calculate the flowrate through the nozzle and the head loss caused by its installation. The water temperature is 15.6°C.

SOLUTION

Assembling the given data

$$d_1 = 0.150 \text{ m} \qquad d_2 = 0.075 \text{ m} \qquad T = 15.6°\text{C}$$

$$\text{s.g.}_{\text{Hg}} = 13.57 \qquad \text{s.g.}_{\text{water}} = 1.00 \qquad \text{MR} = 0.150 \text{ m}$$

selecting tentatively $C_v = 0.99$ from Fig. 14.25 and calculating A_2 as 0.004 56 m^2, we use Eq. 14.16

$$Q = \frac{C_v A_2}{\sqrt{1 - (A_2/A_1)^2}} \sqrt{2g_n \left(\frac{p_1}{\gamma} + z_1 - \frac{p_2}{\gamma} - z_2\right)} \tag{14.16}$$

to obtain

$$Q = \frac{0.99 \times 0.004\,56}{\sqrt{1 - (0.25)^2}} \sqrt{2g_n(0.15)(13.57 - 1)} = 0.028\,6 \text{ m}^3/\text{s} \tag{14.16}$$

Calculating $\mathbf{R}_2$ gives 423 000, which yields a better value of $C_v = 0.988$. Using this value in place of 0.99, $Q = (0.988/0.99)0.028\,6 = 0.028\,5$ m^3/s. •

Precise calculation of head loss caused by the nozzle installation is not possible or necessary, but adequate values may be obtained by computing $h_{L_{1-2}}$ from Eq. 14.17

$$h_{L_{1-2}} = \left[\left(\frac{1}{C_v^2} - 1\right)\left(1 - \left(\frac{A_2}{A_1}\right)^2\right)\right]\frac{V_2^2}{2g_n} = 0.03 \text{ m} \tag{14.17}$$

and adding it to the head loss caused by the flow deceleration downstream from the nozzle. Treating this as an abrupt enlargement, the head loss may be computed from Eq. 9.47 as $h_L = K_L(V_2 - V_1)^2/2g_n = 1.13$ m since $K_L \cong 1$. Thus the total head loss caused by the nozzle installation is about 1.16 m of water. •

14.14 ORIFICES

Like nozzles, orifices serve many purposes in engineering practice other than the metering of fluid flow, but the study of the orifice as a metering device will allow the application of principles to other problems, some of which will be treated subsequently.

The orifice for use as a metering device in a pipeline consists of a concentric square-edged circular hole in a thin plate which is clamped between the flanges of the pipe (Fig. 14.26). The flow characteristics of the orifice differ from those of the nozzle in that the minimum section of the streamtube occurs not within the orifice but downstream from it, owing to the formation of a *vena contracta* at section 2. The cross-sectional area at the

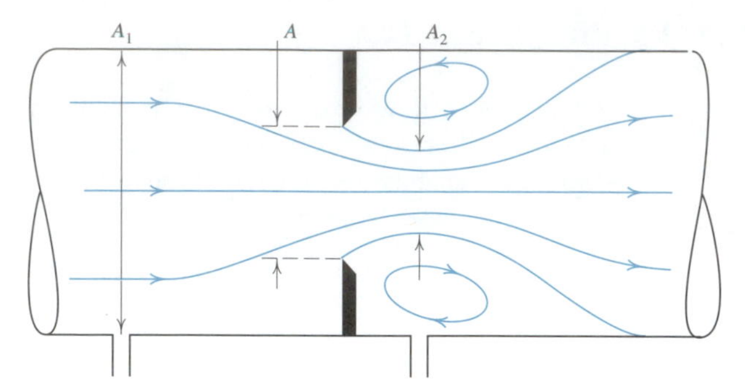

Fig. 14.26 Definition sketch for orifice meter.

vena contracta, A_2, is characterized by a coefficient of contraction,[25] C_c, and given by C_cA. Substituting this into Eq. 14.16,

$$Q = \frac{C_v C_c A}{\sqrt{1 - C_c^2(A/A_1)^2}} \sqrt{2g_n\left(\frac{p_1}{\gamma} + z_1 - \frac{p_2}{\gamma} - z_2\right)} \tag{14.18}$$

which is customarily written

$$Q = CA\sqrt{2g_n\left(\frac{p_1}{\gamma} + z_1 - \frac{p_2}{\gamma} - z_2\right)} \tag{14.19}$$

thus defining the orifice coefficient C as

$$C = \frac{C_v C_c}{\sqrt{1 - C_c^2(A/A_1)^2}} \tag{14.20}$$

which is thus dependent not only on C_v and C_c but on the shape of the installation (defined by A/A_1) as well.

In practice it is not feasible to locate the downstream pressure connection at the vena contracta because the location of the vena contracta depends on both Reynolds number and A/A_1. Accordingly, it is frequently located (as for the flow nozzle) at a fixed proportion of the pipe diameter downstream from the orifice plate, and the other connection one pipe diameter upstream. Any coefficient, C, will thus be dependent on, and associated with, the particular location of the pressure connections. Values of C over a wide range of Reynolds numbers may be obtained from Fig. 14.27; it is convenient and standard practice to define the Reynolds number on the basis of flowrate and orifice diameter as

$$\mathbf{R} = \frac{Qd}{(\pi d^2/4)\nu} = \frac{4Q}{\pi d\nu}$$

which is used as the abscissa for the plot. The trend of the coefficient with Reynolds number is of interest when compared with that of the Venturi meter and flow nozzle. The constancy of C at high Reynolds number is again noted, reflecting the substantial constancy of both C_v and C_c in this range. At lower Reynolds numbers an increase of C is noted, in spite of the expectation of a decrease of C_v in this region; evidently, increased viscous action not only lowers C_v but also raises C_c (by increasing the size of the vena contracta),

[25]The Weisbach values of Section 9.9 (Table 2) may be used as nominal at high Reynolds numbers.

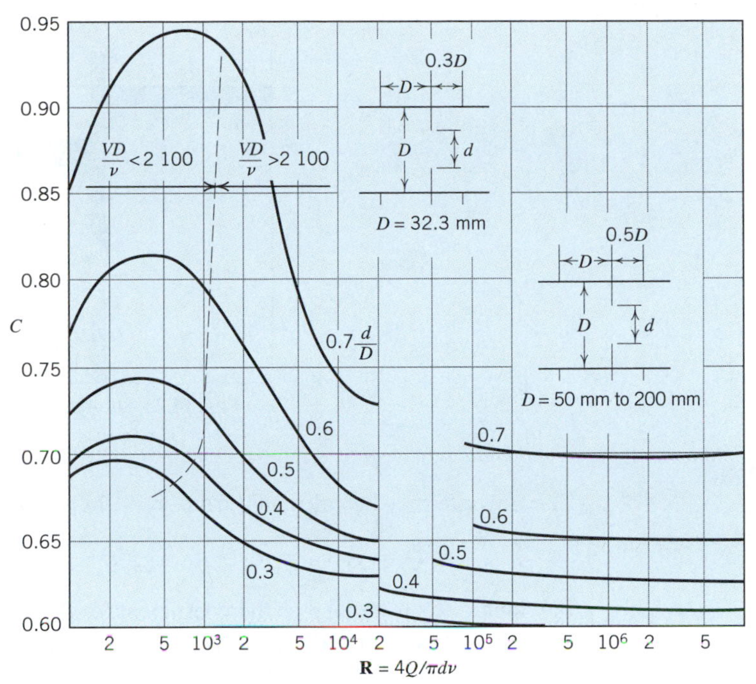

Fig. 14.27 Orifice meter coefficients.[26]

and the latter effect is predominant. In the range of very low Reynolds numbers the effect of viscous action of the vena contracta remains at a maximum (with C_c around 1), and the decrease of C with further decrease of Reynolds number reflects the steady decrease of C_v produced by viscous resistance.

In metering the flow of compressible fluids at high pressure ratios, Eq. 13.20 may be used but it strictly applies only when the downstream pressure connection is at the vena contracta. Values of Y from Appendix 5 may generally be used only as a first approximation; accurate values of Y for various locations of the pressure taps will be found in the A.S.M.E. Fluid Meters report, cited in the References at the end of the chapter.

An extension of the pipeline orifice of Fig. 14.26 is the *submerged orifice* of Fig. 14.28 featured by orifice discharge from one large reservoir into another. Here with A/A_1 virtually zero, C (of Eq. 14.20) becomes C_vC_c. Assuming a perfect fluid and applying the Bernoulli equation between the upstream reservoir and section 2,

$$h_1 = h_2 + \frac{V_2^2}{2g_n} \qquad \text{or} \qquad V = \sqrt{2g_n(h_1 - h_2)}$$

providing that the pressure distribution may be considered hydrostatic in the downstream reservoir, which is a valid assumption if h_2 is large compared to orifice size. For the real fluid, frictional effects will prevent the attainment of this velocity and C_v is introduced as before, so that

$$V_2 = C_v\sqrt{2g_n(h_1 - h_2)}$$

[26]Data from G. L. Tuve and R. E. Sprenkle, ''Orifice Discharge Coefficients for Viscous Liquids,'' *Instruments*, vol. 6, p. 201, 1933; vol. 8, pp. 202, 225, 232, 1935; and *Fluid Meters: Their Theory and Application*, 4th edition, A.S.M.E., 1937. See *Fluid Makers*, 6th ed., 2d printing, 1983 also.

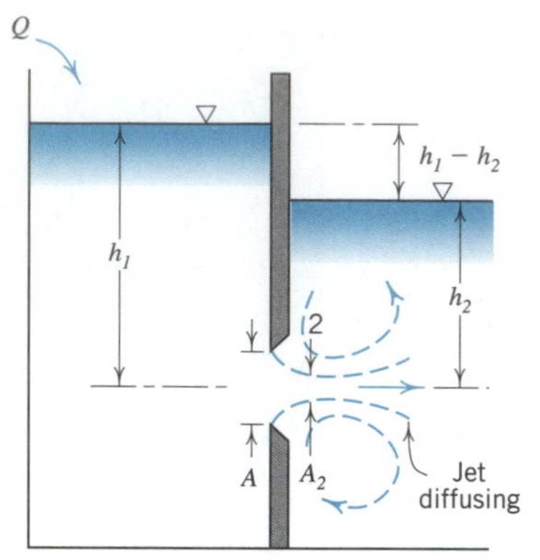

Fig. 14.28 Submerged orifice.

The flowrate may be calculated from A_2V_2, in which A_2 is replaced by C_cA:

$$Q = A_2V_2 = C_cC_vA\sqrt{2g_n(h_1 - h_2)} = CA\sqrt{2g_n(h_1 - h_2)} \qquad (14.21)$$

in which C_vC_c is defined as the coefficient (of discharge) of the orifice. When the orifice discharges freely into the atmosphere (Fig. 14.29), h_2 becomes zero and the equation reduces to

$$Q = C_cC_vA\sqrt{2g_nh} = CA\sqrt{2g_nh}$$

The dependence of the various orifice coefficients on shape of orifice is illustrated by Fig. 14.30. The coefficients given are nominal values for large orifices ($d > 1$ in. or 25 mm) operating under comparatively large heads of water ($h > 4$ ft or 1.2 m). Above these limits of head and size, various experiments have shown that the coefficients are practically constant. Coefficients for sharp-edged orifices over a wide range of Reynolds numbers are given in Fig. 14.31, which shows the same trend of values (for the same reasons) as that of Fig. 14.27. The plot of Fig. 14.31, although convenient and applicable to the flow of all fluids, has a certain limitation in orifice size caused by the action of surface tension. Surface-tension effects (although impossible to predict except in idealized situations) will increase with decreasing orifice size; the plotted values are valid only where such effects are negligible and, thus, cannot be applied to very small orifices.

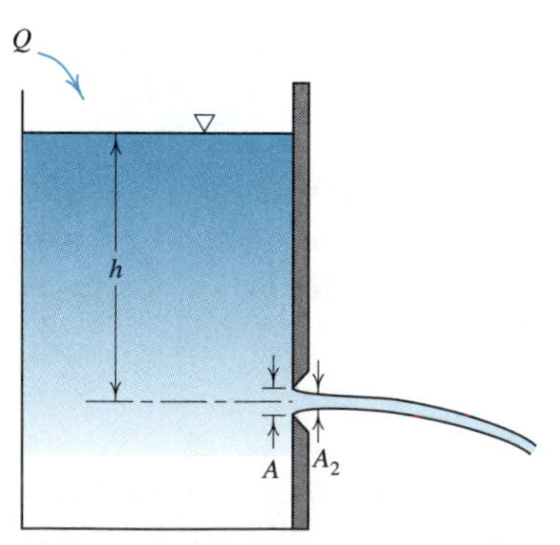

Fig. 14.29 Orifice discharging freely.

Orifices and their Nominal Coefficients				
	Sharp edged	Rounded	Short tube	Borda
C	0.61	0.98	0.80	0.51
C_c	0.62	1.00	1.00	0.52
C_v	0.98	0.98	0.80	0.98

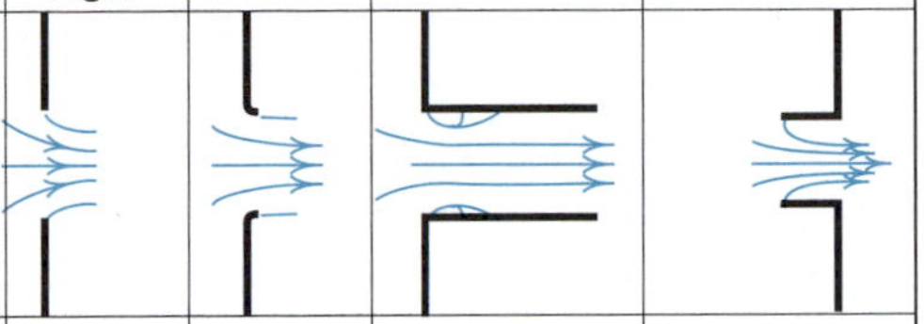

Fig. 14.30

The head lost between the reservoir and section 2 in an orifice operating under static head may be calculated from the coefficient of velocity and flowrate by Eq. 14.17. Since $A_2/A_1 \cong 0$ this equation reduces to

$$h_L = \left(\frac{1}{C_v^2} - 1\right)\frac{V_2^2}{2g_n}$$

One special problem of orifice flow is that of the two-dimensional sluice gate of Fig. 14.32 in which jet contraction occurs only on the top of the jet and pressure distribution

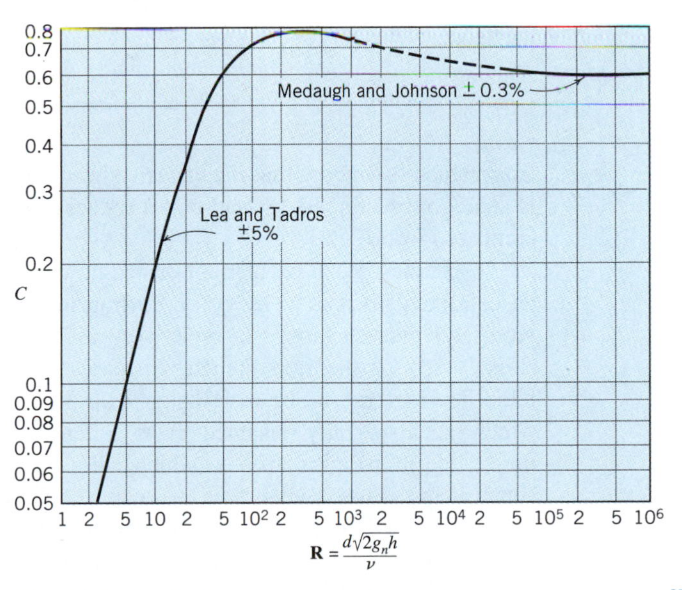

Fig. 14.31 Coefficients for sharp-edged orifices under static head[27] ($h/d > 5$).

[27]Data from F. C. Lea, *Hydraulics*, 6th edition, p. 87, Edward Arnold and Co., 1938; and F. W. Medaugh and G. D. Johnson, *Civil Eng.*, vol. 10, no. 7, p. 422, July, 1940.

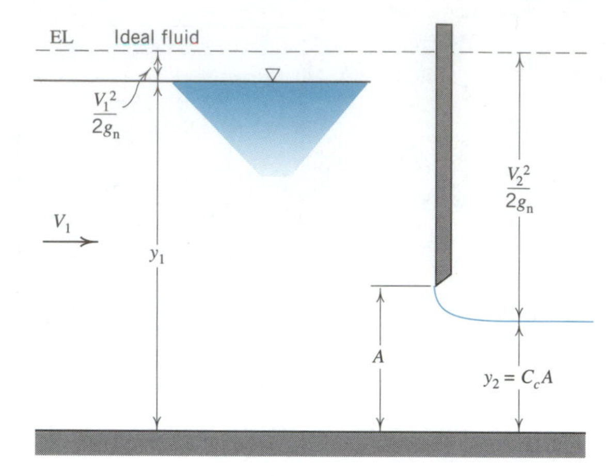

Fig. 14.32 Sluice gate.

in the vena contracta is hydrostatic. Assuming an ideal fluid,

$$y_1 + \frac{V_1^2}{2g_n} = y_2 + \frac{V_2^2}{2g_n}$$

and substituting $V_2 y_2 / y_1$ for V_1, and solving for V_2,

$$V_2 = \frac{1}{\sqrt{1 - (y_2/y_1)^2}} \sqrt{2g_n(y_1 - y_2)}$$

The actual velocity (allowing for head loss) is obtained by multiplying the above by C_v, and the flowrate by multiplying the result by C_cA. The flowrate through the sluice is, therefore,

$$q = \frac{C_v C_c A}{\sqrt{1 - (y_2/y_1)^2}} \sqrt{2g_n(y_1 - y_2)}$$

from which it is noted that the effective head on the sluice is $(y_1 - y_2)$, that the equation is analogous to Eq. 14.18, and that it reduces to Eq. 14.21 as the depth y_1 becomes large compared to y_2.

A second special problem of orifice flow is represented by the *rotameter*[28] which can be calibrated to read velocity or flowrate directly. Basically a rotameter consists of a precisely manufactured, tapered vertical tube through which fluid flows upward (Fig. 14.33). As the tube diameter increases upward, the fluid velocity in the tube, at any fixed flowrate, decreases with distance up the tube. Within the tube is placed a ''float,'' which has a specially designed shape, a density slightly greater than that of the flowing fluid, and (often) spiral grooves which cause the float to spin (and hence to remain roughly centered in the tube) when fluid is flowing. Accordingly, the flow around the float is quite like that through a needle valve or annular orifice.

When there is no flow, the float sits at the bottom of the tapered tube. When fluid is flowing, the float rises until the upward drag and buoyancy forces on it are balanced by its weight. Since the tube is tapered, the velocity past the float and, thus, the drag on it

[28]N. P. Cheremisinoff and P. N. Cheremisinoff, *Flow measurement for engineers and scientists*. New York: M. Dekker, 1988 or A. T. J. Hayward, *Flowmeters*. New York: Halsted Press [J. Wiley & Sons], 1979.

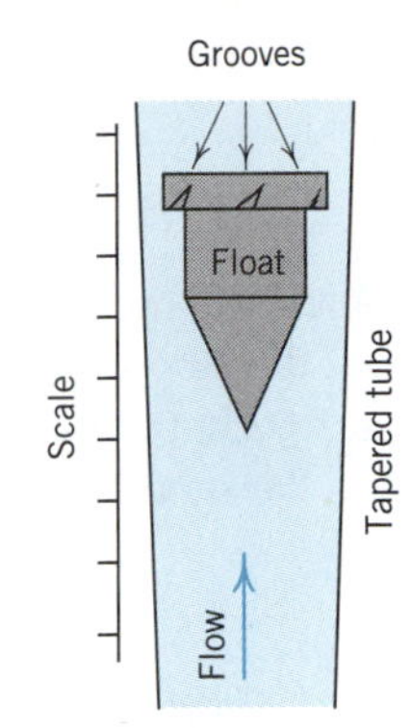

Fig. 14.33 Rotameter schematic (after Hayward[28]).

decrease as the float moves up in a constant rate flow. The points of equilibrium can be noted as a function of flowrate and, with a marked glass tube, the level of the float becomes a direct measure of flowrate. Rotameters are widely used in industrial applications where the visual output can be used by operators controlling flowrates in various processes.

14.15 ELBOW METERS

The orifice, nozzle, and Venturi meters as applied in the measurement of pipeline flow have been seen to be fundamentally methods of producing a regular and reproducible fall of the hydraulic grade line which is related to flowrate. Another meter of this type is the elbow meter of Fig. 14.34 which utilizes the difference in elevation of hydraulic grade lines between the outside and inside of a regular pipe bend (see Fig. 5.12 of Section 5.8). Analytical solutions of such problems are not feasible, and such devices are calibrated by determining experimentally the relation between difference in hydraulic grade lines and flowrate. Lansford[29] has done this for a variety of 90° flanged elbows, and it allows their

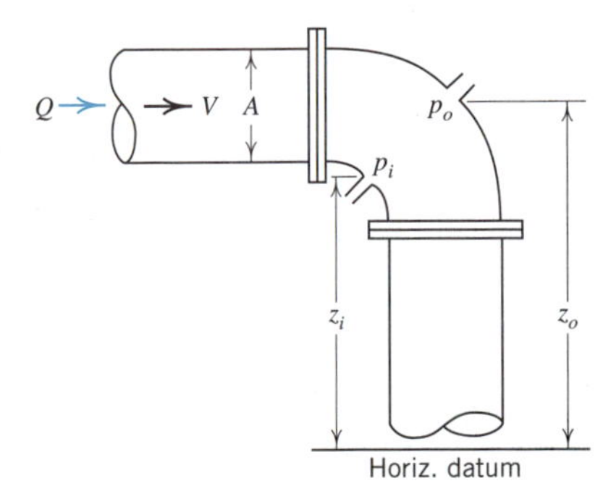

Fig. 14.34 Elbow meter.

[29]W. M. Lansford, "The Use of an Elbow in a Pipe Line for Determining the Flow in the Pipe," *Eng. Exp. Sta. Univ. Ill., Bull.*, 289, 1936.

use as accurate and economical flow meters; for a basic equation he proposes

$$\left(\frac{p}{\gamma} + z\right)_o - \left(\frac{p}{\gamma} + z\right)_i = C_k \frac{V^2}{2g_n}$$

with coefficient C_k ranging between 1.3 and 3.2, the magnitudes depending on the size and shape of the elbow. This equation may be solved for V and multiplied by A to obtain the flowrate, Q. If the coefficient of the meter is then defined as $\sqrt{1/C_k}$ the resulting equation has the same form as that for nozzles and orifices,

$$Q = CA\sqrt{2g_n\left(\frac{p_o}{\gamma} + z_o - \frac{p_i}{\gamma} - z_i\right)}$$

in which C will have values between 0.56 and 0.88.

14.16 WEIRS

For measuring large and small open flows in field or laboratory, the weir finds wide application. A weir may be defined in a general way as ''any regular obstruction *over* which flow occurs.'' Thus, for example, the overflow section (spillway) of a dam is a special type of weir and may be utilized for flow measurement. However, weirs for measuring purposes are usually of more simple and reproducible form, consisting of smooth, vertical, flat plates with upper edges sharpened. Such weirs, called *sharp-crested weirs*, appear in a variety of forms, the most popular of which is the rectangular weir; this type has a straight, horizontal crest and extends over the full width of the channel in which it is placed. The flow picture produced by such a weir is essentially two-dimensional and for this reason it will be used as a basis for the following discussion.

The flow of liquid over a sharp-crested weir is at best an exceedingly complex problem; however, the two-dimensional problem for the ideal fluid has been solved by application of potential theory and complex functions, while more complete solutions have been obtained by numerical simulation of the flow by use of the methods of finite differences, finite elements, or boundary integrals. An appreciation for the complexities, however, is necessary to an understanding of experimental results and of the deficiencies of simplified weir formulas. These complexities may be discovered by considering the flow over the sharp-crested weir of Fig. 14.35. Although it is obvious at once that the head, H,

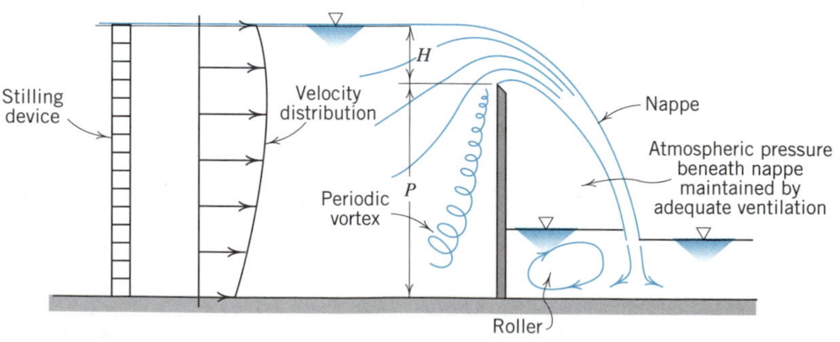

Fig. 14.35 Weir flow (actual).

on the weir is the primary factor causing the flow (Q) to occur, no simple relationship between these two variables can be derived, for two fundamental reasons: (1) the shape of the flow picture, and (2) the effect of turbulence and frictional processes cannot be calculated. The most important factors which affect the shape of the flow picture are the head on the weir, H, the weir height, P, and the extent of ventilation beneath the nappe. Although the effect of these factors may be found experimentally, there is no simple method of predicting the flow picture from the values of H, P, and pressure beneath the nappe. The effects of turbulence and friction cannot be predicted, nor can they be isolated for experimental measurement. It may be noted, however, that frictional resistance at the sidewalls will affect the flowrate to an increasing extent as the channel becomes narrower and the weir length, b (normal to the paper), smaller. Fluid turbulence and frictional processes at the sides and bottom of the approach channel also contribute to the velocity distribution in an unpredictable way. The effects of velocity distribution on weir flow are appreciable,[30] and an effort should be made in all weir installations to provide a good length of approach channel, with stilling devices such as racks and screens for the even distribution of turbulence and the prevention of abnormal velocity distributions. Another influence of frictional processes is the creation of a periodic helical secondary flow in the corners just upstream from the weir plate, resulting in a vortex (Fig. 14.35), which influences the flow in an unpredictable manner. The free liquid surfaces of weir flow also bring surface-tension effects into the problem, and these forces, although small, affect the flow picture appreciably at low heads and small flows.

In the light of the foregoing complexities, the derivation of any simple weir formula obviously requires drastic simplification of the problem which leads to an approximate result; however, by such methods the *form* of the relationship between flowrate and head can be found and an experimental coefficient defined. To derive a simple weir equation, let it be assumed that (1) velocity distribution upstream from the weir is uniform, (2) all fluid particles move horizontally as they pass the weir crest, (3) the pressure in the nappe is zero, and (4) the influence of viscosity, turbulence, secondary flows, and surface tension may be neglected. These assumptions produce the flow picture of Fig. 14.36. Taking section 1 in the approach channel well upstream from the weir and section 2 slightly downstream from the weir crest, Bernoulli's equation may be applied to a typical stream-

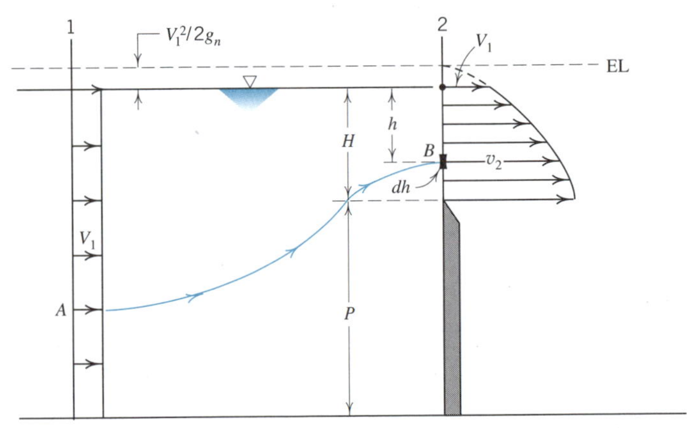

Fig. 14.36 Weir flow (simplified).

[30]P. Ackers, W. R. White, J. A. Perkins, and A. J. M. Harrison, 1978. *Weirs and flumes for flow measurement*. New York: Wiley.

line to find the velocity v_2. Using streamline AB as typical and taking the weir crest as datum,

$$H + \frac{V_1^2}{2g_n} = (H - h) + \frac{v_2^2}{2g_n}$$

from which we obtain

$$v_2 = \sqrt{2g_n\left(h + \frac{V_1^2}{2g_n}\right)}$$

which shows v_2 to be dependent on h. Taking dq to be the two-dimensional flowrate through a strip of height dh, $dq = v_2\, dh$ allowing integration to obtain the flowrate, q,

$$q = \int_0^H v_2\, dh = \sqrt{2g_n}\int_0^H \left(h + \frac{V_1^2}{2g_n}\right) dh$$

The result is

$$q = \tfrac{2}{3}\sqrt{2g_n}\left[\left(H + \frac{V_1^2}{2g_n}\right)^{3/2} - \left(\frac{V_1^2}{2g_n}\right)^{3/2}\right]$$

This equation is cumbersome since for given H and P, a trial-and-error solution is needed to obtain q. Fortunately, in many weir problems $P \gg H$ and V_1 is small, so $V_1^2/2g_n$ is customarily neglected and the equation further simplified to

$$q = \tfrac{2}{3}\sqrt{2g_n}H^{3/2} \tag{14.22}$$

which is the basic equation for rectangular weirs. Into this equation must be inserted an experimentally determined coefficient, C_w, which includes the effects of the many phenomena disregarded in the foregoing development and simplifications. For real weir flow the relation between flowrate and head then becomes

$$q = C_w \tfrac{2}{3}\sqrt{2g_n}H^{3/2} \tag{14.23}$$

showing that (to the extent C_w is constant), for rectangular weirs, $q \propto H^{3/2}$. The coefficient C_w is essentially a factor which transforms the simplified weir flow of Fig. 14.36 into the real weir flow of Fig. 14.35, and its magnitude is thus fixed by the most important difference between these flows—the shape of the flowfield. Thus, the weir coefficient is primarily a coefficient of contraction which expresses the extent of contraction of the true nappe below that assumed in the simplified analysis. Since the size of the weir coefficient depends primarily on the shape of the flowfield, the effect of other fluid properties and phenomena may usually be discovered by examining their influence on this shape.

Although a dimensional analysis of the weir problem must necessarily be incomplete because of the impossibility of including all the pertinent factors, it provides a rational basis for an understanding of some of the factors affecting weir coefficients. The expressible independent variables entering the two-dimensional weir problem are q, H, P, μ, σ, ρ, and g_n. Application of the Buckingham Π-theorem analysis of Section 8.2 to these variables shows that there are $7 - 3 = 4$ distinct dimensionless Π-groups. Using q, H, and ρ as the repeating variables produces

$$\Pi_1 = P/H \qquad \Pi_2 = \rho q/\mu = \mathbf{R} \qquad \Pi_3 = \rho q^2/\sigma H = \mathbf{W}$$

$$\Pi_4 = q/\sqrt{g_n}H^{3/2} = \mathbf{F}$$

Noting that, from Π_4, $q \propto \sqrt{g_n H}H$ allows Π_2 and Π_3 to be rewritten as

$$\Pi_2 = \mathbf{R} = \rho\sqrt{g_n}H^{3/2}/\mu \qquad \text{and} \qquad \Pi_3 = \mathbf{W} = \rho g_n H^2/\sigma$$

The relationship among the four dimensionless groups is typically written as

$$\frac{q}{\sqrt{g_n}H^{3/2}} = f\left(\mathbf{R}, \mathbf{W}, \frac{P}{H}\right)$$

in which the Froude number on the left-hand side is a direct measure of C_w; dimensional analysis thus shows the dependence of the weir coefficient on **R**, **W**, and P/H. Of these numbers P/H has been found to be the most important in determining the magnitude of C_w—as would be expected since this ratio, more than any of the others, has the greatest influence on the shape of the flowfield. The effect of **W** is negligible except at low heads where surface-tension effects may be large; the effect of **R** is small (except at low heads) since water is usually involved in weir flows, **R** is high, and viscous action small.

The experimental work of Rehbock[31] led to an empirical formula for the coefficient of well-ventilated, sharp-crested rectangular weirs for water measurement, which has an accuracy of better than 1 percent if care is taken with the details of installation. Rehbock's formula for C_w is

$$C_w = 0.602 + 0.08\frac{H}{P} + \frac{1}{900H} \tag{14.24}$$

which shows the strong influence of P/H, except where H (in m) is small and $1/900H$ of some significance. In this term, H is (for a liquid of constant physical properties) a direct measure of **R** and **W** (see the second forms of Π_2 and Π_3 above); this implies (as expected) that the influences of viscosity and surface tension on weir flow are strong only when H is small.

Weirs are reliable measuring devices only at heads above the range of strong action by viscosity and surface tension; in this range the formula shows that the coefficient can be expected to increase with increasing head and decreasing weir height. Although it is desirable to calibrate a new weir installation *in place*, this is frequently not possible and a formula must be selected for the weir coefficient. The Rehbock formula can be expected to give good results if such important details as adequate ventilation, stilling devices, crest sharpness, and smoothness of upstream face are not overlooked.

Broad-crested weirs (Fig. 14.37) have been shown (Section 10.8) to be critical-depth meters; here, for ideal flow,

$$q = \sqrt{g_n\left(\frac{2E}{3}\right)^3} = \left(\frac{2}{3}\right)^{3/2}\sqrt{g_n}E^{3/2} \tag{14.25}$$

The weir coefficient for the ideal broad-crested weir may be calculated by equating

[31]Th. Rehbock, ''Wassermessung mit scharfkantigen Uberfallwehren,'' *Zeitschrift des V.d.I.*, vol. 73, no. 24, June 15, 1929, and C. E. Kindsvater and R. W. Carter, ''Discharge Characteristics of Rectangular Thin Plate Weirs,'' *Trans. A.S.C.E.*, vol. 124, 1959. See P. Ackers, W. R., White, J. A. Perkins, and A. J. M. Harrison, *Weirs and flumes for flow measurement*, John Wiley & Sons, 1978, for a complete discussion of the many formulas proposed.

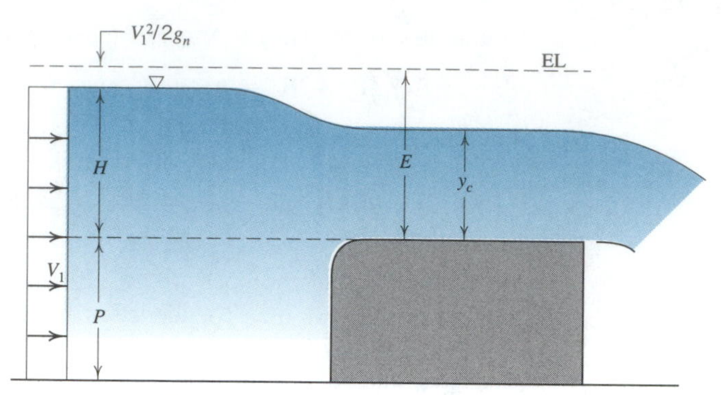

Fig. 14.37 Broad-crested weir.

Eqs. 14.23 and 14.25 to yield

$$C_w = \frac{1}{\sqrt{3}}\left(\frac{E}{H}\right)^{3/2}$$

For a very high weir $P/H \to \infty$, $E \to H$, $E/H \to 1$, and $C_w \to 1/\sqrt{3} = 0.577$ as shown in Section 10.8; for a lower weir $P/H < \infty$, $E > H$, $E/H > 1$, and $C_w > 0.577$. Hence the weir coefficient increases with decreasing P/H and thus exhibits the same trend as the Rehbock formula. Experimental measurements also substantiate this variation of C_w with P/H, but the values of C_w obtained from experiment are a few percent lower than those of ideal flow because of head loss accompanied by a falling energy line.

For small flowrates, *notch weirs* are widely used as measuring devices; of these the most popular is the triangular weir or V-notch (Fig. 14.38). A simplified analysis similar to that used on the rectangular weir (but neglecting velocity of approach) yields (after inserting the experimental coefficient) the fundamental formula

$$Q = C_w \tfrac{8}{15} \tan \alpha \sqrt{2g_n} H^{5/2} \tag{14.26}$$

Triangular weirs of 90° notch angle (2α) have coefficients (for water) near 0.59, but these are affected by viscosity, surface tension, and weir plate roughness; increases of any one of these tend to increase the coefficient. A comprehensive study of triangular weir flow has been made by Lenz,[32] who used many liquids in order to discover the effects of viscosity and surface tension on weir coefficients, thus extending the utility of the triangular

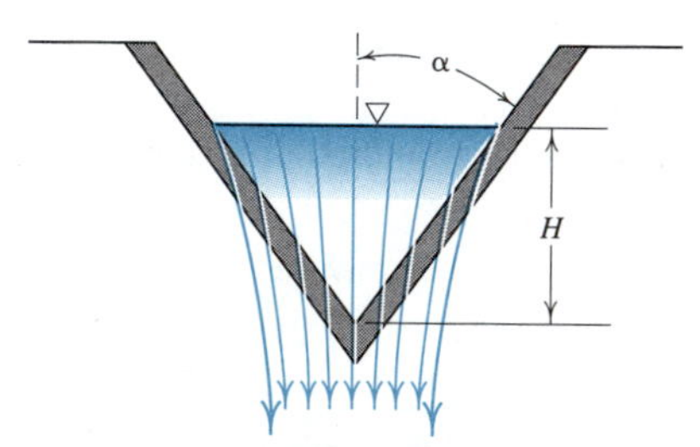

Fig. 14.38 Triangular weir.

[32] A. T. Lenz, "Viscosity and Surface-Tension Effects on V-Notch Weir Coefficients," *Trans. A.S.C.E.*, vol. 108, 1943.

weir as a reliable measuring device. For notch angles of 90° Lenz proposed that, if $\mathbf{R} = H\sqrt{g_n H}/\nu$, $\mathbf{W} = \rho g_n H^2/\sigma$, then

$$C_w = 0.56 + \frac{0.70}{\mathbf{R}^{0.165}\mathbf{W}^{0.170}}$$

applies to all liquids providing that the falling sheet of liquid does not cling to the weir plate and that $H > 0.06$ m, $\mathbf{R} > 300$, and $\mathbf{W} > 300$. The work of Lenz has not only broadened the field of application of the weir as a measuring device but has also documented the increase of coefficient with decreasing **R** and **W**, a characteristic of the coefficients for all sharp-crested weirs.

The crest of a *spillway structure* is shown in Fig. 14.39. Major considerations in the design of such a spillway are structural stability against hydrostatic pressure and other loads, and prevention of separation (Section 7.7) and reduced pressures on the surface of the structure. The rectangular weir equation may be applied to the spillway, the coefficient C_w ranging from 0.60 to 0.75. The relatively high value of C_w may be explained by a comparison of a sharp-crested weir (Fig. 14.35) and a spillway crest designed exactly to fit the curvature of the lower side of the nappe of this weir for a certain design head, H_D. In spite of the greater friction of the spillway, with a fixed reservoir surface the flow over the two structures must be approximately the same, but the heads for each structure will be measured from their respective crests and will, therefore, be quite different, the head on the weir being greater than the head on the spillway. Since for (about) the same flowrate the smaller head must be associated with a larger coefficient, the spillway coefficient is seen to be larger than that of the sharp-crested weir.

A spillway profile which will fit the flow of a sharp-crested weir and thus prevent harmful discontinuities of pressure is shown with its coefficients in Fig. 14.40. The results are both useful and instructive, since in defining the coefficients the head is taken as the vertical distance between spillway crest and energy line upstream from the structure and thus contains the velocity head of the approaching flow. After the design head and height of structure have been determined, the profile of the structure may be laid out and discharge coefficients accurately predicted for heads between 40% and 130% of the design head. For any weir height, P, a steady increase of C_w with H is as noted for the sharp-crested weir (Eq. 14.24). For the sharp-crested weir this trend was due to change in the overall shape of the flow picture; for the spillway crest it is due mostly to the steady decline in pressure over the crest with increasing head, which increases the effective head and is reflected in an increase of coefficient. This trend is beneficial up to the point where cavi-

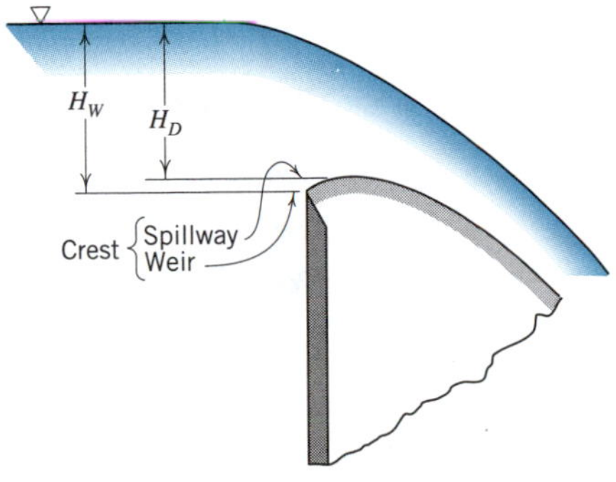

Fig. 14.39

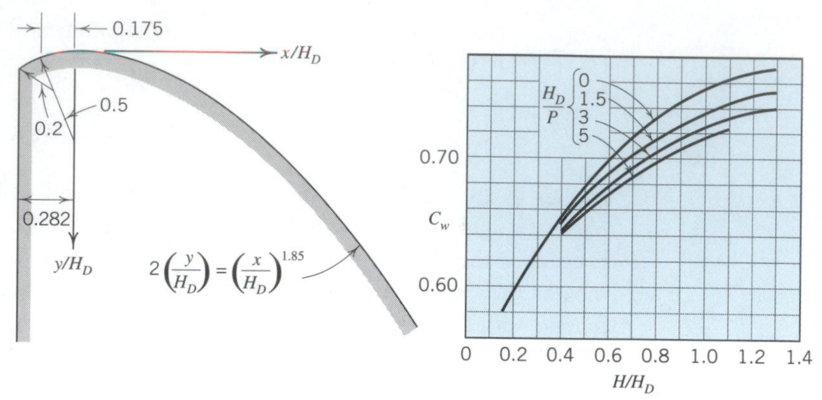

Fig. 14.40 Spillway profiles and coefficients.[33]

tation or separation occurs, inasmuch as an increase in C_w may be interpreted as an increase in the efficiency of the spillway.

The effects of submergence (Fig. 14.41) on weir flow are of some theoretical and practical interest. Sharp-crested weirs cannot be considered to be precise measuring devices when operating submerged, but the effect of submergence on broad-crested weirs is surprisingly small, making this type of weir a reliable measuring device even for high submergence. This reliability is due primarily to the straighter streamlines and essentially hydrostatic pressure distribution on the crest of the broad-crested weir which change little as the downstream water level rises above the crest of the weir. For the sharp-crested weir discharging freely the pressure distribution through the nappe is featured by zero pressure on top and bottom, and thus is far from hydrostatic; raising the downstream water level above the crest of the weir will drastically change this pressure distribution and immediately affect the whole flow picture. In all submerged weir problems the two flow situations (Fig. 14.41) of *plunging nappe* at low submergence and *surface nappe* at higher submergence are observed. Approximate results of investigations on submerged rectangular weirs may be seen on the sketch of Fig. 14.42 in which submergence ratio H_2/H_1 is plotted against the ratio of measured flowrate (Q_S) to that which would have existed with free flow (Q_F) for a head H_1. For accurate and detailed information see J. G. Woodburn, "Tests

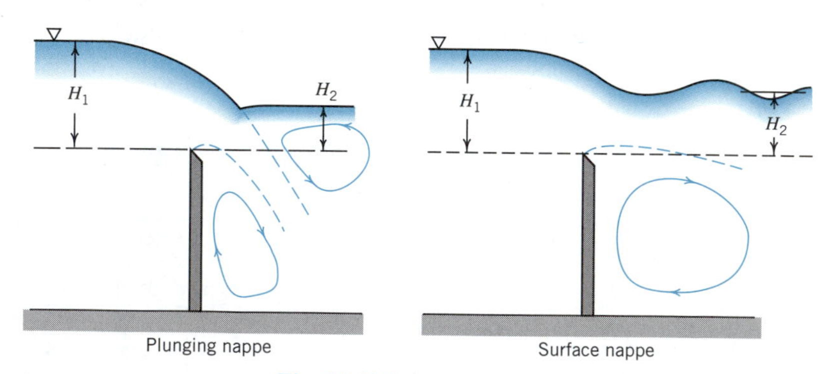

Fig. 14.41 Submerged weirs.

[33]F. R. Brown, "Hydraulic Models as an Aid to the Development of Design Criteria," *Waterways Expt. Sta., Corps of Engrs., Bull.*, 37, Vicksburg, Miss., June, 1951.

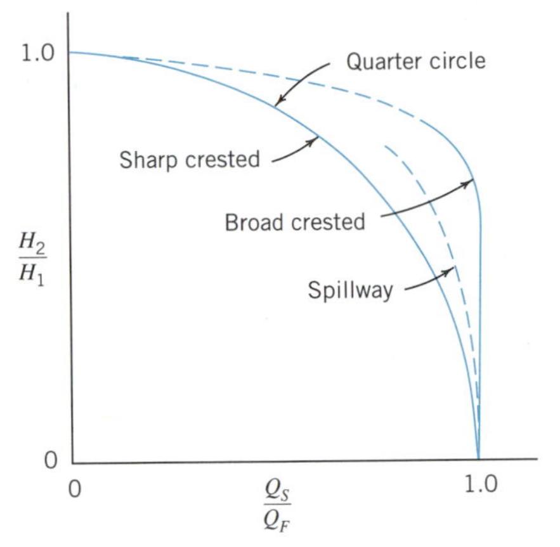

Fig. 14.42 Approximate effects of submergence.

of Broad Crested Weirs,'' *Trans. A.S.C.E.*, vol. 96, 1932; J. K. Vennard and R. F. Weston, ''Submergence Effect on Sharp-Crested Weirs,'' *Eng. News-Record*, June 3, 1943; J. R. Villemonte, ''Submerged Weir Discharge Studies,'' *Eng. News-Record*, December 25, 1947. Engineers require accurate information on weir submergence when flowrate measurements are needed in times of flood, in spillway design, and in situations where adequate vertical drop is unavailable for use of a (more precise) freely flowing weir.

14.17 CURRENT-METER MEASUREMENTS

The construction of a weir for measuring the flowrate in large canals, streams, or rivers is impractical for many obvious reasons, but existing spillways whose coefficients are known may frequently serve as measuring devices. The standard method of river flow measurement is to measure the velocity by means of a current meter (Section 14.8) and integrate the results to obtain the flowrate.

Fundamental to the use of a current meter is some knowledge of the general properties of velocity distribution in open flow. As in pipes, the velocities are reduced at the banks and bed of the channel, but it must be realized that in open flow the roughnesses and turbulences are of such great and irregular magnitudes that the velocity distribution problem cannot be placed on the precise basis which it enjoys in pipe flow. However, from long experience and thousands of measurements, the United States Geological Survey has established certain average characteristics of velocity distribution in streams and rivers which serve as a basis for current-meter measurements. These characteristics of velocity distribution in a vertical are shown in Fig. 14.43 and may be amplified by the following statements: (1) the curve may be assumed parabolic; (2) the location of the maximum velocity is from $0.05y$ to $0.25y$ below the water surface; (3) the mean velocity occurs at approximately $0.6y$ below the water surface; (4) the mean velocity is approximately 85 percent of the surface velocity; (5) a more accurate and reliable method of obtaining the mean velocity is to take a numerical average of the velocities at $0.2y$ and $0.8y$ below the water surface. These average values will, obviously, not apply perfectly to a particular stream or river, but numerous measurements with the current meter will tend toward ac-

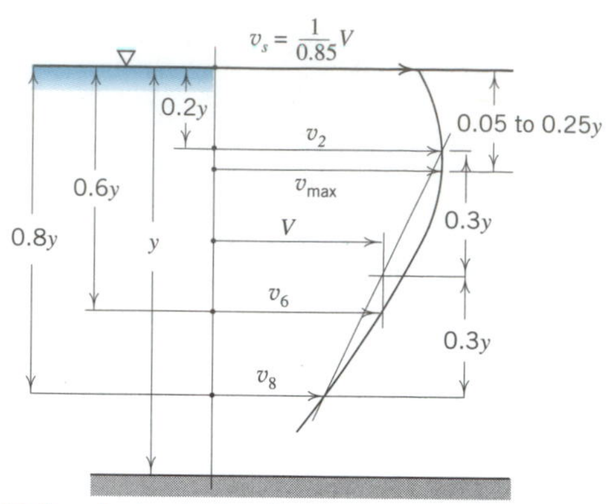

Fig. 14.43 Standard velocity distribution in a vertical in open flow.

curate results since deviations from the average values will tend to compensate, thus giving a greater accuracy than can be obtained from individual measurements.

Current-meter measurements for calculation of flowrate may be taken in the following manner. A reach of river having a fairly regular cross section is selected. This cross section is measured accurately by soundings. It is then divided into vertical strips (Fig. 14.44), the current meter is suspended, and velocities are measured at the two-tenths and eight-tenths points in each vertical (1, 2, 3, etc., Fig. 14.44). From these measurements the mean velocities (V_1, V_2, V_3, etc.) in each vertical may be calculated. The mean velocity through each vertical strip is taken as the average of the mean velocities in the two verticals which bound the strip, and thus the rates of flow (Q_{12}, Q_{23}, etc.) through the strips may be calculated from

$$Q_{12} = b_{12}\left(\frac{y_1 + y_2}{2}\right)\left(\frac{V_1 + V_2}{2}\right)$$

$$Q_{23} = b_{23}\left(\frac{y_2 + y_3}{2}\right)\left(\frac{V_2 + V_3}{2}\right)$$

and the flowrate in the stream may be calculated by totaling the flowrates through the various strips.

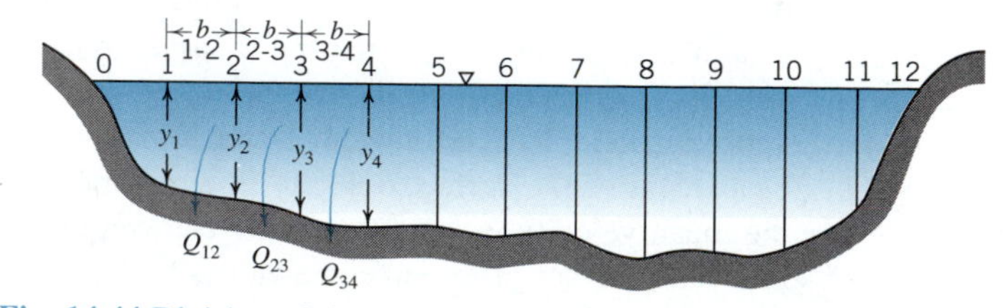

Fig. 14.44 Division of river cross section for current-meter measurements.

REFERENCES

Ackers, P., White, W. R., Perkins, J. A., and Harrison, A. J. M. 1978. *Weirs and flumes for flow measurement.* New York: Wiley.

Adrian, R. J. 1991. Particle-imaging techniques for experimental fluid mechanics. *Annual Review of Fluid Mechanics* 23: 261–304.

A.S.M.E. 1961. *Flowmeter computation handbook.*

Bean, H. S. Ed. 1983. *Fluid meters: their theory and application.* 6th ed. (2d printing) New York: A.S.M.E.

Benedict, R. P. 1984. *Fundamentals of temperature, pressure, and flow measurements.* 3rd ed. New York: Wiley.

Benedict, R. P. 1980. *Fundamentals of pipe flow.* New York: Wiley.

Buchhave, P., et al. 1979. The measurement of turbulence with the laser-Doppler anemometer. *Annual Review of Fluid Mechanics* 11: 443–503.

Cheremisinoff, N. P. 1979. *Applied fluid flow measurement.* New York: M. Dekker.

Cheremisinoff, N. P., and Cheremisinoff, P. N. 1988. *Flow measurement for engineers and scientists.* New York: M. Dekker.

Comte-Bellot, G. 1976. Hot-wire anemometry. *Annual Review of Fluid Mechanics* 8: 209–232.

Corbett, D. M., et al. 1943. Stream gaging procedure. *U.S. Geological Survey Water Supply Paper* 888.

Durst, F., Melling, A., and Whitelaw, J. H. 1981. *Principles and practice of laser-Doppler anemometry.* 2nd ed. New York: Academic Press.

Fried, E., and Idelchik, I. E. 1989. *Flow resistance: a guide for engineers.* New York: Hemisphere Pub. Corp.

Gad-el-Hak, M. Ed. 1989. *Advances in fluid mechanics measurements.* Lecture Notes in Engineering 45. New York: Springer-Verlag.

Hayward, A. T. J. 1979. *Flowmeters.* New York: Wiley.

Holman, J. P. 1978. *Experimental methods for engineers.* 3d ed. New York: McGraw-Hill.

Hydraulic Institute 1990. *Engineering data book.* 2d ed. Cleveland, OH: Hydraulic Institute.

Idelchik, I. E. 1986. *Handbook of hydraulic resistance.* 2d ed. New York: Hemisphere Pub. Corp.

Illustrated experiments in fluid mechanics. 1972. MIT Press.

Liepmann, H. W., and Roshko, A. 1957. *Elements of gasdynamics.* New York: Wiley, Chapter 6.

Linford, A. 1961. *Flow measurement & meters.* 2d ed. London: Spon.

Lowell, F. C., Jr., and Hirschfeld, F. 1979. Acoustic flowmeters for pipelines. *Mechanical Engineering* 101: 29–35.

Merzkirch, W. 1987. *Flow visualization.* 2d ed. New York: Academic Press.

Miller, R. W. 1989. *Flow measurement engineering handbook.* 2d ed. New York: McGraw-Hill.

Ower, E., and Pankhurst, R. C. 1977. *The measurement of air flow.* 5th ed. New York: Pergamon Press.

Sandborn, V. A. 1972. *Resistance temperature transducers.* Fort Collins: Metrology Press.

Schraub, F. A., et al. 1965. Use of hydrogen bubbles for quantitative determination of time-dependent velocity fields in low-speed water flows. *Journ. Basic Engineering, Trans. A.S.M.E.* 87, Ser. D, 2: 429–44.

Spink, L. K. 1967. *Principles and practice of flow meter engineering.* 9th ed. Foxboro: Foxboro.

U.S. Bureau of Reclamation. 1984. *Water measurement manual.* 2d ed., revised reprint.

Wilmarth, W. W. 1971. Unsteady force and pressure measurements. *Annual Review of Fluid Mechanics* 3: 147–70.

Wyngaard, J. C. 1981. Cup, propeller, vane, and sonic anemometers in turbulence research. *Annual Review of Fluid Mechanics* 13: 399–423.

FILMS

Abernathy, F. H. Fundamentals of boundary layers. NCFMF/EDC Film No. 21623, Encyclopaedia Britannica Educ. Corp.

Coles, D. Channel flow of a compressible fluid. NCFMF/EDC Film No. 21616, Encyclopaedia Britannica Educ. Corp.

Kline, S. J. Flow visualization. NCFMF/EDC Film No. 21607, Encyclopaedia Britannica Educ. Corp.

Shell Oil Co. Schlieren. Shell Film Library, 1433 Sadlier Circle, West Drive, Indianapolis, Ind. 46239.

PROBLEMS

14.1. The viscosity of a fluid may be determined by measuring the head loss over a section of a tube in which a laminar flow is occurring. If the head loss h_L and the flowrate Q are known to within $\pm 5\%$ at a 95% confidence level and the specific weight γ, diameter d, and section length l are known within $\pm 1\%$, what is the error bound on the viscosity μ?

14.2. A pycnometer weighs 0.220 7 lb when empty and 0.925 6 lb when filled with liquid. If its volume is 0.007 639 ft^3, calculate the specific gravity of the liquid. If there is an uncertainty of ± 5 in the last digit of each piece of data, estimate the uncertainty in the specific gravity.

14.3. A plummet weighs 4.00 N in air and 2.97 N in a liquid. If the volume of liquid displaced by the plummet is 1.29×10^{-4} m^3, what is the specific gravity of the liquid?

14.4. A cylindrical plummet weighing 0.44 N, of 25 mm diameter, and having a specific gravity of 7.70, is suspended in a liquid from the end of a balance arm, 150 mm from the knife edge. The arm is balanced by a weight of 0.53 N, 100 mm from the knife edge. What is the specific gravity of the liquid? Neglect the weight of the balance arm.

14.5. A crude hydrometer consists of a cylinder of 0.50 in. diameter and 2 in. length, surmounted by a cylindrical tube of 0.125 in. diameter and 8 in. long. Lead shot in the cylinder brings the hydrometer's total weight to 0.30 oz. What range of specific gravities can be measured by this hydrometer?

14.6. To what depth will the bottom of the hydrometer of the preceding problem sink in a liquid of specific gravity 1.10?

14.7. A hydrometer weighing 1.1 N and having a volume of 99 100 mm^3 is placed in a liquid of specific gravity 1.60. What percent of the volume remains above the liquid surface?

14.8. If the torque required to rotate the inner cylinder of problem 1.69 at a constant speed of 4 r/min is 2.7 N·m, calculate the approximate viscosity of the oil.

14.9. A Stormer-type viscometer consists of two cylinders, one of 75 mm outside diameter, the other of 77.5 mm inside diameter; both are 250 mm high. A 4.45 N weight falls 1.5 m in 10 s, its supporting wire unwinding from a spool of 50 mm diameter on the main shaft of the viscometer. If the space between the cylinders is filled with oil to a depth of 200 mm and the space between the ends of the cylinders is 1.25 mm, calculate the viscosity of the oil.

14.10. A Saybolt Universal viscometer has tube diameter and length of 1.76 mm and 12.25 mm, respectively. The internal diameter of the cylindrical reservoir is 30 mm, and the height from the tube outlet to rim of reservoir is 125 mm. Assuming that the loss of head may be taken as the average of the total heads on the tube outlet at the beginning and end of the run, derive an approximate relationship between ν (m^2/s) and t (Saybolt seconds), and compare with the exact equation relating these quantities. Volume collected is 60 000 mm^3. What relative uncertainty in ν would be caused by an error of the order of $\pm 2\%$ in the volume measurement?

14.11. If in 150 s a Saybolt viscometer discharges a standard volume of one oil, how long will it take to discharge the same volume of a second oil which is 10% denser and has a 10% smaller absolute viscosity than the first?

14.12. This viscometer is to be used for liquids of kinematic viscosity between 0.000 02 and 0.000 5 ft^2/s. Calculate the minimum tube length which will hold all Reynolds numbers below 1 500. Consider tube friction only.

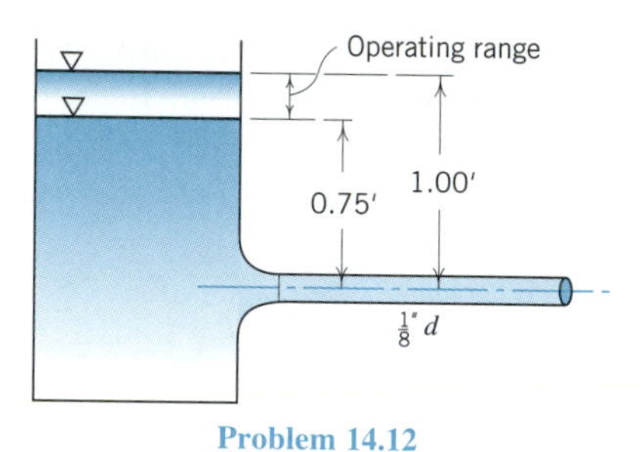

Problem 14.12

14.13. In this viscosity test the time from start (S) to finish (F) is 200 s. The oil has s.g. 0.87, the pipe is 2.1 m long and of 12.5 mm diameter, K_L for the entrance is 1.2, and for the exit 1.0. Calculate the approximate viscosity of the oil.

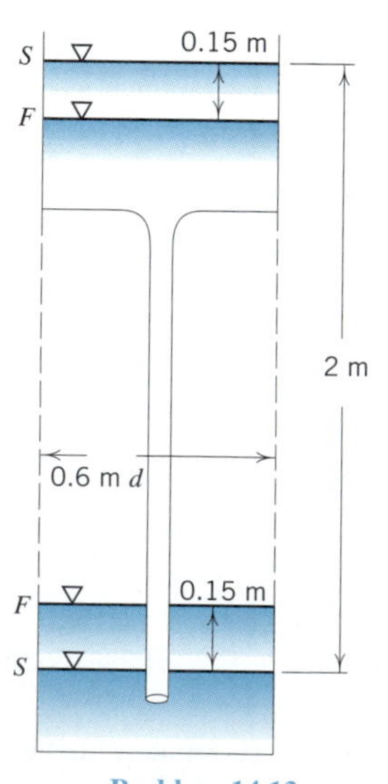

Problem 14.13

14.14. The disk of Fig. 14.4*a* and a pitot tube are placed in an airstream aligned properly with the flow, and connected to a U-tube containing water. If the difference of water elevation in the legs of the manometer is 100 mm, calculate the air velocity, assuming a specific weight of 12.0 N/m^3.

14.15. A pitot-static tube installed in an airduct shows a differential manometer reading of 2 in. of water. If the pressure and temperature in the duct are 15.0 psia and 80°F, respectively, what velocity is indicated? Neglect compressibility effects.

14.16. A differential manometer containing water is attached to the pitot-static tube on an airplane flying close to the ground through the U.S. Standard Atmosphere (Appendix 2) at 150 mph (67 m/s). What manometer reading can be expected? What velocity would be indicated if this manometer reading occurred at an altitude of 20 000 ft (6.1 km)? How much greater than the static pressure is the stagnation pressure?

14.17. A pitot-static tube is placed at the center of a 6 in. smooth pipe in which there is an established flow of jet fuel at 68°F. The attached differential manometer containing mercury and jet fuel shows a difference of 3 in. What flowrate exists in the line? If the uncertainty in the manometer is $\pm 1/4$ in., what is the uncertainty in the flowrate?

14.18. In a laminar flow in a 0.3 m pipe two pitot tubes are installed, one on the centerline, the other 75 mm from the centerline. If the specific weight of the liquid flowing is 7.86 kN/m^3 and the flowrate is 0.28 m^3/s, calculate the reading of a differential manometer connected to the two tubes when there is mercury in the bottom of the manometer.

14.19. The friction factor of this pipe is 0.025. Calculate the manometer reading when the mean velocity is 3 m/s.

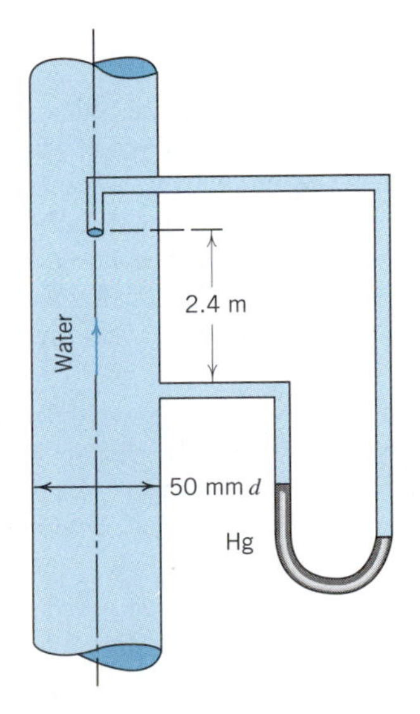

Problem 14.19

14.20. If the mean velocity is 8.02 ft/s and the fluid flowing has specific gravity and kinematic viscosity of 0.80 and 0.004 01 ft^2/s, respectively, calculate the manometer reading.

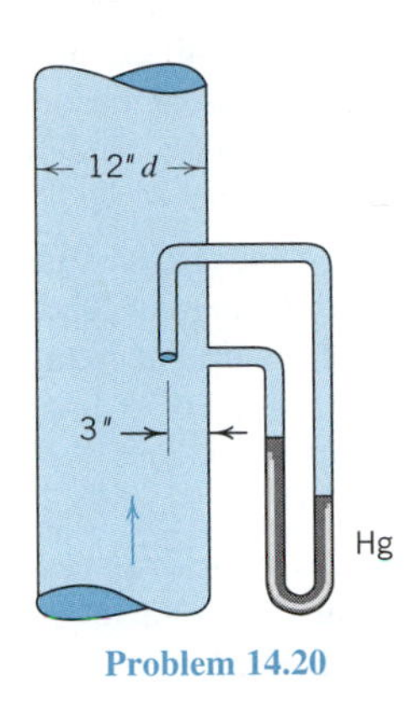

Problem 14.20

14.21. A liquid flows from left to right in this 150 mm diameter clean cast iron pipe at $\mathbf{R} > 10^6$. Calculate the head lost per 30 m of pipe.

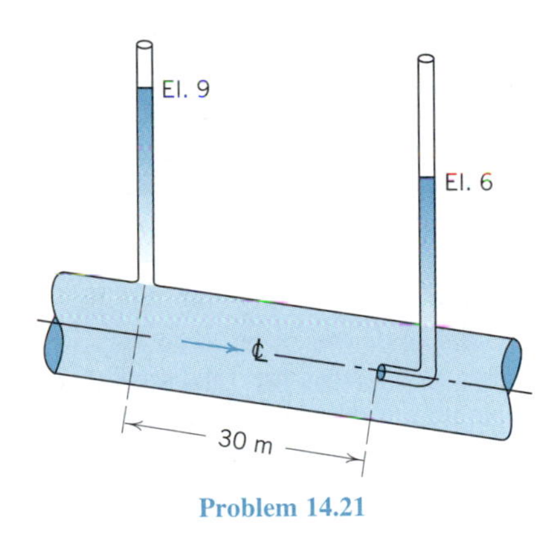

Problem 14.21

14.22. If the sensitivity of a diaphragm pressure transducer is 400 V/psi and the output voltmeter can be read to ± 0.001 volts, how accurately can the pressure be measured?

14.23. A diaphragm pressure transducer has a linear relation between output voltage and pressure. If a pair of readings are (1.0 V, 1.38 Pa) and (2.0 V, 2.76 Pa), what is the slope of the calibration curve?

14.24. Suppose a set of measurements is made to determine the pressure ratio of Eq. 14.7 for an air flow. What is the bound on the percentage error in this ratio if the Mach number measurement can be in error by $\pm 10\%$?

14.25. Calculate the velocity at a point in an airduct if the stagnation temperature is 150°F and the separate stagnation and static pressure manometers show readings of 8 in. and 15 in. of mercury vacuum, respectively. The barometric pressure is 28 in. of mercury.

14.26. Carbon dioxide flows in a pipe. At a point in the flow the stagnation temperature is 40°C, and the absolute static pressure 96.5 kPa. The differential manometer attached to the pitot-static tube shows a reading of 762 mm of mercury. Calculate the velocity at this point.

14.27. A pitot-static tube and temperature probe are installed in a duct where nitrogen is flowing. The stagnation pressure and temperature are 381 mm of mercury gage and 93.3°C, respectively. The static pressure is 75 mm of mercury vacuum. If the barometric pressure is 762 mm of mercury, what velocity is indicated?

14.28. Calculate the pressure and temperature on the nose of a projectile moving near the ground at 2 000 ft/s (610 m/s) through the U.S. Standard Atmosphere. (Appendix 2.)

14.29. The hot-film sensor of Fig. 14.18 produces an output voltage of 4.2 V; what is the indicated velocity? What increase in voltage will occur if the velocity is doubled?

14.30. For the sensor of Fig. 14.18 find the velocity-heat-transfer Eq. 14.11, that is, determine A and B. Assume room temperature is 20°C.

14.31. Prove that an X-array hot-wire can be used to obtain the turbulent components v_x and v_y in a two-dimensional flow. Assume the wires are inclined at $\pm 45°$ to the mean flow direction.

14.32. What is the error bound for the velocity measured by a sonic anemometer, according to Eq. 14.12, if the distance d is known to within $\pm 1\%$, the speed of sound to within $\pm 2\%$ and the travel-time difference to within $\pm 0.5\%$?

14.33. A sonic anemometer is set at an angle of 45° to the direction of flow of sea level air. If the separation distance of the pulse generators is 15 cm, derive an expression for the flow velocity as a function of the travel-time difference Δt.

14.34. If a sonic anemometer is required to measure the velocity along its path to within $\pm 1\%$ and the distance d can be measured to within ± 1 mm, what is the minimum acceptable value of d?

14.35. A Stanton tube of height h of 0.015 in. is installed in the wall of a 12 in. smooth pipe in which a fluid is flowing with a mean velocity of 10 ft/s. The fluid has viscosity and density of 0.002 lb·s/ft^2 and 2 slugs/ft^3, respectively. What pressure difference is to be expected? If this pressure difference is interpreted (as for a pitot-static tube) as $\rho v^2/2$, how far from the wall will this velocity be located?

14.36. A laminar flow occurs in a 300 mm smooth pipe at a Reynolds number of 1 000. In the pipe wall a Stanton tube having $h = 1$ mm is installed. Calculate the ratio between the expected pressure difference and the wall shear. If this pressure difference is presumed to equal $\rho v^2/2$, how far from the wall will this velocity be located?

14.37. A Preston tube of 0.5 in. (12.7 mm) outside diameter is attached to the hull of a ship to measure the local shear. When the ship moves through freshwater (68°F or 20°C) the pressure difference is found to be 75 psf (3.6 kPa). Calculate the local shear.

14.38. Water flows in a horizontal 0.3 m diameter pipe. A force of 0.2 N is exerted on a 50 mm × 50 mm square shear plate by the flow. What is the pressure gradient in the pipe?

14.39. Find the relative uncertainty in the flowrate determined in a venturi, according to Eq. 14.16, as a result of uncertainties of $\pm 5\%$ in the head measurements $\left(\dfrac{p}{\gamma} + z\right)$.

14.40. A 12 in. by 6 in. Venturi meter is installed in a horizontal waterline. The pressure gages read 30 and 20 psi. Calculate the flowrate if the water temperature is 68°F. Calculate the head lost between the base and throat of the meter. Calculate the total head lost by the meter if the diffuser tube has cone angle 7°. Calculate the flowrate if the pipe is vertical and the throat of the meter 2 ft below the base.

14.41. Crude oil flows through a horizontal 150 mm by 75 mm Venturi meter. What is the difference in pressure head between the base and throat of the meter when 7.6 litres/s flow at (*a*) 27°C and (*b*) 49°C? What is the head loss for each?

14.42. The maximum flowrate in a 250 mm waterline is expected to be 142 litres/s. To the Venturi meter is attached a mercury-under-water manometer 0.91 m long. Calculate the minimum throat diameter which should be specified.

14.43. If the head lost between base and throat of a 3 in by 6 in. Venturi meter is neglected, what coefficient of velocity would be required to allow for the change in velocity distribution between these points if α_1 and α_2 are 1.06 and 1.00, respectively (see Fig. 7.26)?

14.44. A pitot tube is installed at the center of the base of a 100 mm by 50 mm Venturi meter through which 28.1 litres/s of water is flowing. The pitot tube is connected to one side of a differential manometer (containing mercury and water); the other side of this manometer is connected to the throat piezometer ring. Calculate the manometer reading if V/v_c at sections 1 and 2 are 0.82 and 1.00, respectively. Let $C_v = 0.98$.

14.45. In Denver, Colorado, air ($\gamma = 0.066$ lb/ft^3) flows through a 4 in. by 2 in. frictionless Venturi meter. The pressures in the 4 in. and 2 in. sections are 0.25 psi and 0.300 in. of mercury vacuum, respectively. Calculate the flowrate, neglecting compressibility of the air.

14.46. Carbon dioxide flows through a 150 mm by 75 mm Venturi meter. Gages at the base and throat read 138 kPa and 96.5 kPa and temperature in the fluid at the base of the meter is 26.7°C. Calculate the weight flowrate, assuming standard barometer and $C_v = 0.99$.

14.47. An A.S.M.E. flow nozzle of 75 mm diameter is installed in a 150 mm water (20°C) line. The attached manometer contains mercury and water and registers a difference of 381 mm. Calculate the flowrate through the nozzle. Calculate the head lost by the nozzle installation.

14.48. A 3 in. nozzle is installed at the end of a 6 in. airduct in which the specific weight is 0.076 3 lb/ft^3. A differential manometer connected to a piezometer opening 6 in. upstream from the base of the nozzle and to a pitot tube in the jet shows a reading of 0.25 in. of water. Calculate the flowrate, assuming uniform velocity distribution in the jet and C_v of 0.97. Assume the air incompressible.

14.49. If air flows through the pipe and nozzle of problem 14.47 and if open mercury manometers at points 1 and 2 show positive gage pressures of 762 mm and 508 mm, and if the temperature of the air at point 1 is 15.6°C, calculate the weight flowrate, assuming standard barometric pressure.

14.50. A 1 in. nozzle has C_v of 0.98 and is attached to a 3 in. hose. What flowrate (water) will occur through the nozzle when the pressure in the hose is 60 psi? What is the velocity of the jet at the nozzle tip? How much head is lost through the nozzle? To what maximum height will this jet rise (neglect air friction)?

14.51. A sharp-edged orifice with conventional pressure connections is to be installed in a 300 mm waterline. For a flowrate of 0.28 m^3/s the maximum allowable head loss is 7.6 m. What is the smallest orifice that may be used? Since calculations are approximate, assume $C_v = 1$.

14.52. A 4 in. orifice at the end of a 6 in. line discharges 5.30 cfs of water. A pressure gage upstream from the orifice reads 58.0 psi and a gage connected to a pitot tube in the vena contracta reads 60.0 psi. Calculate C_c and C_v for this orifice assuming $\alpha_2 = 1.00$.

14.53. A 150 mm flow nozzle is installed in a 300 mm waterline. An orifice of what diameter will produce the same head loss as the nozzle? Assume C_v the same for nozzle and orifice.

14.54. If the coefficient of an orifice of 3 in. diameter installed in a 6 in. line is approximately 0.65 for the conventional piezometer connections of Fig. 14.26, what approximate coefficient can be expected if the downstream connection is made at a point where the expanding jet has a 4 in. diameter?

14.55. Predict the location of the water surface in the middle piezometer tube relative to one of the other water surfaces; C_v is 0.97.

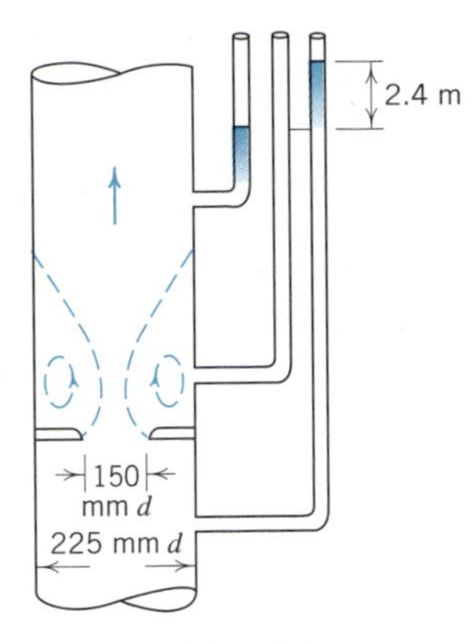

Problem 14.55

14.56. Find the ratio of the manometer readings for upward and downward flow of the same flowrate. The manometer liquids are the same and each downstream pressure connection is opposite the vena contracta.

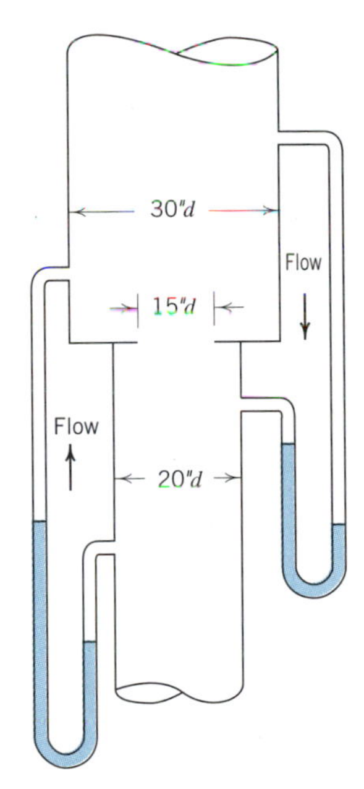

Problem 14.56

14.57. A 100 mm sharp-edged orifice at the end of a 150 mm waterline has C_v of 0.97. Calculate the flowrate when the pressure in the line is 275 kPa.

14.58. A conical nozzle of 50 mm tip diameter and having C_c of 0.85 and C_v of 0.97 is attached to the end of a 100 mm waterline. A manometer (containing carbon tetrachloride and water) is connected to a pitot tube in the vena contracta and to a piezometer ring at the base of the nozzle. Taking $\alpha_2 = 1.0$, calculate the flowrate and the pressure at the base of the nozzle. The manometer reads 610 mm.

14.59. A 2 in. conical nozzle having C_v of 0.98 and C_c of 0.80 is attached to a 4 in. pipeline and delivers water to an impulse turbine. The pipeline is 1 000 ft long and leaves a reservoir of surface elevation 450 at elevation 420. The nozzle is at elevation 25. Assuming a square-edged pipe entrance and a friction factor of 0.02, calculate (*a*) the flowrate through the pipe and nozzle, (*b*) the horsepower of the nozzle stream, and (*c*) the horsepower lost in line and nozzle.

14.60. This *inlet orifice* is used to meter the flow of air into the pipe. Assuming the downstream pressure connection to be in the plane of the vena contracta, predict the flowrate. Consider the air incompressible.

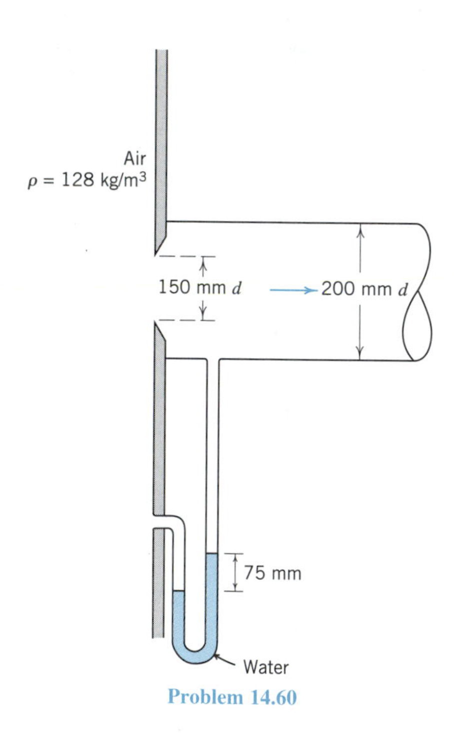

Problem 14.60

14.61. Calculate the flowrate if C_v for this entrance nozzle is 0.96.

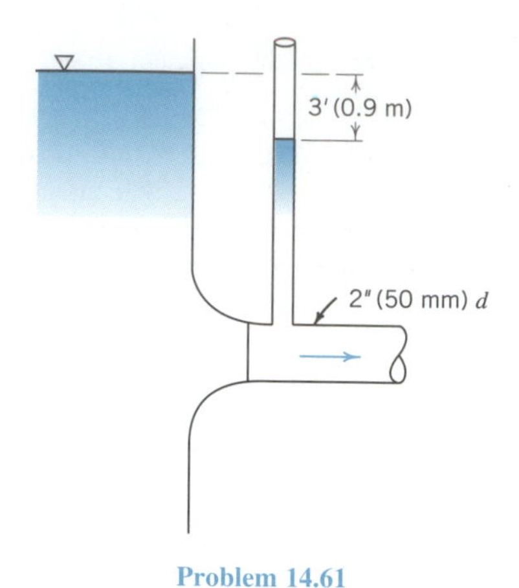

Problem 14.61

14.62. Water (20°C or 68°F) discharges into the atmosphere from a 37.5 mm or 1.5 in. sharp-edged orifice under a 1.5 m or 5 ft head. Calculate the flowrate. Repeat the calculation for crude oil at 20°C or 68°F.

14.63. Water flows from one tank to an adjacent one through a 75 mm sharp-edged orifice. The head of water on one side of the orifice is 1.8 m and on the other 0.6 m. Taking C_c as 0.62 and C_v as 0.95, calculate the flowrate.

14.64. A jet discharges vertically upward from a 2 in. sharp-edged orifice located in a horizontal plane. If the head (on the vena contracta) is 20 ft and the jet rises to a height of 19 ft above the vena contracta, what is the flowrate if the diameter of the vena contracta is 1.6 in.? Air friction on the jet may be neglected.

14.65. A conventional sharp-edged orifice of 50 mm diameter discharges into the atmosphere from a large tank. At a point in the jet the height of the energy line is measured by pitot tube and found to be 0.09 m below the free surface level in the tank. Calculate the flowrate and the head on the orifice. Air friction on the jet may be neglected.

14.66. A 3 in. (75 mm) sharp-edged orifice discharges vertically upward. At a point 10 ft (3 m) about the vena contracta, the diameter of the jet is 3 in. (75 mm). Under what head is the orifice discharging?

14.67. The flowrate is 5.4 l/s, and the head lost in the diffuser is 0.15 m. Predict the flowrate when the diffuser is removed.

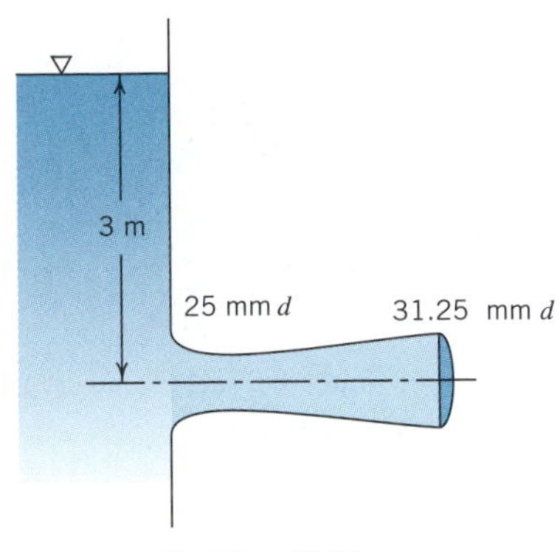

Problem 14.67

14.68. Under a 4.42 ft head, 0.056 cfs of water discharges from a 1 in. sharp-edged orifice in a vertical plane; 3.30 ft outward horizontally from the vena contracta the jet has dropped 0.65 ft below the centerline of the orifice. Calculate C, C_v, and C_c.

14.69. A 2 in. (50 mm) sharp-edged orifice discharges with a 20 ft (6 m) head on its vena contracta. To what height will the jet rise (above the vena contracta) if the jet discharges (*a*) vertically upward, and (*b*) upward at an angle of 45°? Neglect air friction on the jet.

14.70. Water discharges through a 1 in. diameter sharp-edged orifice under a 3 ft head. At what head will the same flowrate occur through a horizontal pipe 1 in. in diameter, 12 in. long (friction factor 0.020), and having a square-edged entrance?

14.71. A short tube of 25 mm diameter and 37.5 mm length may flow full or not full at its exit. Calculate the approximate ratio between the respective flowrates for the same head.

14.72. Predict the discharge coefficients of the standard short tube and re-entrant short tube from the loss coefficients of Fig. 9.17. Assume that the tubes are 4 diameters in length with friction factors of 0.020.

14.73. Predict the coefficient of velocity for this short tube with restricted entrance if the friction factor for the tube is 0.020.

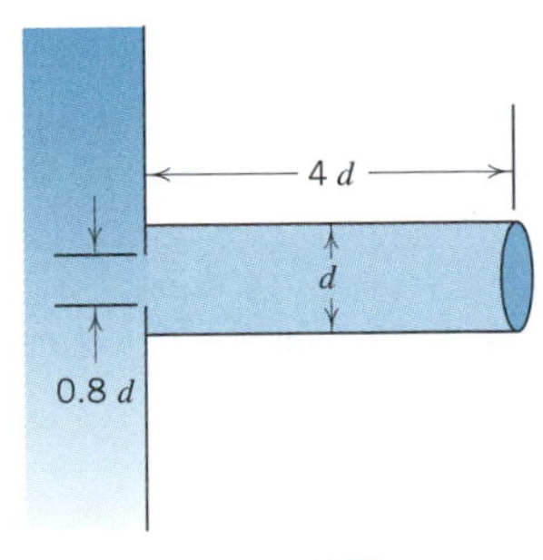

Problem 14.73

14.74. A sluice gate 4 ft (1.2 m) wide is open 3 ft (0.9 m) and discharges onto a horizontal surface. If the coefficient of contraction is 0.80 and the coefficient of velocity 0.90, calculate the flowrate if the upstream water surface is 4 ft (1.2 m) above the top of the gate opening.

14.75. This sluice gate extends the full width of a rectangular channel 5 ft wide. Assuming C_v is 0.96 and C_c is 0.75, estimate the flowrate, neglecting the dynamics of the roller.

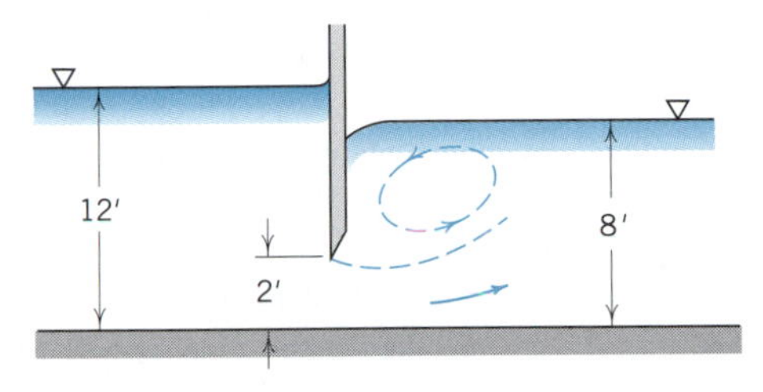

Problem 14.75

14.76. When this sluice gate is open 0.6 m, its vena contracta is in the plane of the brink of the outfall. If its coefficient of contraction is 0.75, channel and gate widths 2.4 m, and the flowrate 4.95 m^3/s, what is its coefficient of velocity?

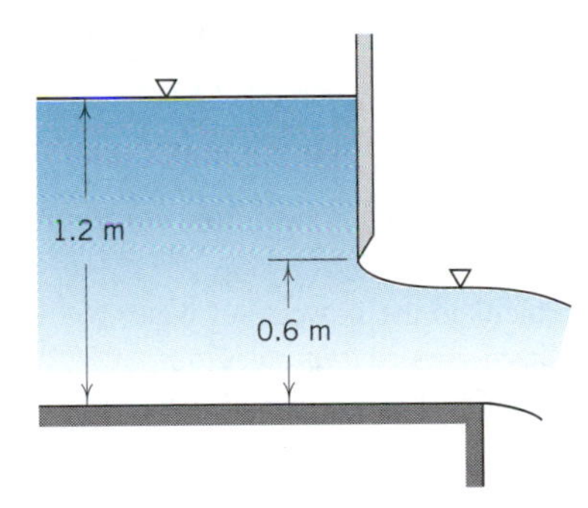

Problem 14.76

14.77. An elbow meter of 100 mm diameter has a coefficient of 0.815. What flowrate of water occurs through this meter when the attached manometer (containing mercury and water) shows a difference of 250 mm?

14.78. Carry out the dimensional analysis to show that the weir equation can be written as $q/(\sqrt{g_n}H^{3/2}) = f(\mathbf{R}, \mathbf{W}, P/H)$.

14.79. The head on a sharp-crested rectangular weir 1.2 m long and 0.9 m high is 100 mm. Calculate flowrate and velocity of approach. Repeat the calculation for a weir of 0.3 m height.

14.80. A certain flowrate passes over a sharp-crested rectangular weir 0.6 m (2 ft) high under a head of 0.3 m (1 ft). Calculate the head on a similar weir 0.3 m (1 ft) high for the same flowrate.

14.81. What depth of water must exist behind a rectangular sharp-crested weir 1.5 m long and 1.2 m high, when a flow of 0.28 m^3/s passes over it? What is the velocity of approach?

14.82. A rectangular channel 5.4 m (18 ft) wide carries a flowrate of 1.4 m^3/s (50 cfs). A rectangular sharp-crested weir is to be installed near the end of the channel to create a depth of 0.9 m (3 ft) upstream from the weir. Calculate the necessary weir height.

14.83. Across one end of a rectangular tank 0.9 m wide is a sharp-crested weir 1.2 m high. In the bottom of the tank is a sharp-edged orifice of 75 mm diameter. If 57 l/s flow into the tank, what depth of water will be attained?

14.84. Treating the upper edge of the pipe as a sharp weir crest, estimate the flowrate when the water depth in the basin is 0.75 m.

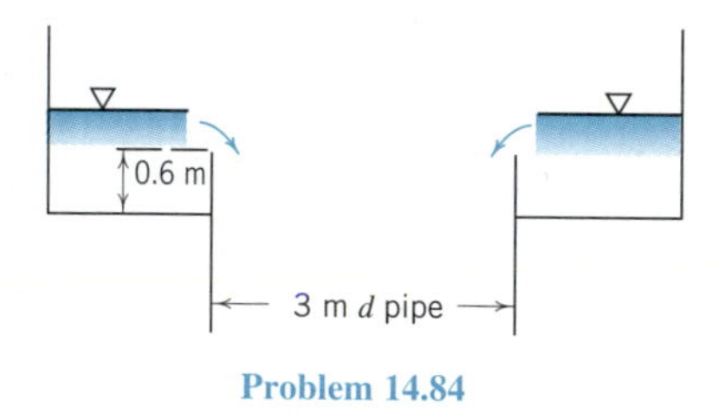

Problem 14.84

14.85. Derive the theoretical flow equation for the triangular weir.

14.86. Calculate the flowrate of water (68°F or 20°C) over a smooth sharp-crested triangular weir of 90° notch angle when operating under a head of 6 in. or 150 mm. Repeat the calculation for crude oil at the same temperature.

14.87. A triangular weir of 90° notch angle is to be used for measuring water flowrates up to 1.5 cfs or 42.5 l/s. What is the minimum depth of notch which will pass this flowrate?

14.88. A 90° triangular weir discharges water at a head of 0.5 ft (0.15 m) into a tank with a 2.5 in. (62.5 mm) sharp-edged orifice in the bottom. Predict the depth of water in the tank.

14.89. Calculate the approximate flowrate to be expected through this sharp-edged opening. Assume a weir coefficient of 0.62.

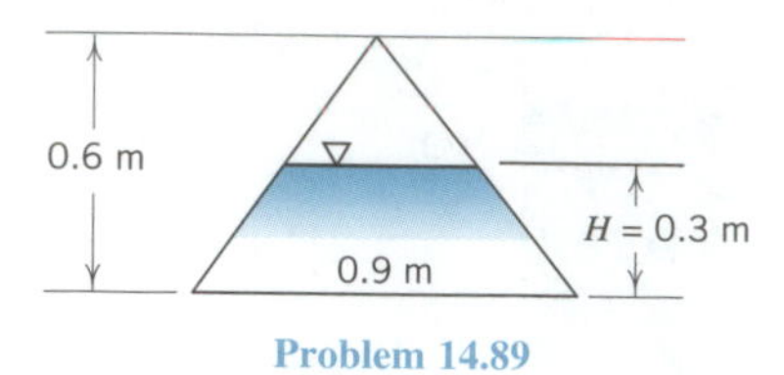

Problem 14.89

14.90. The depth of water behind the weir plate is 4.8 ft. Predict the flowrate over the weir.

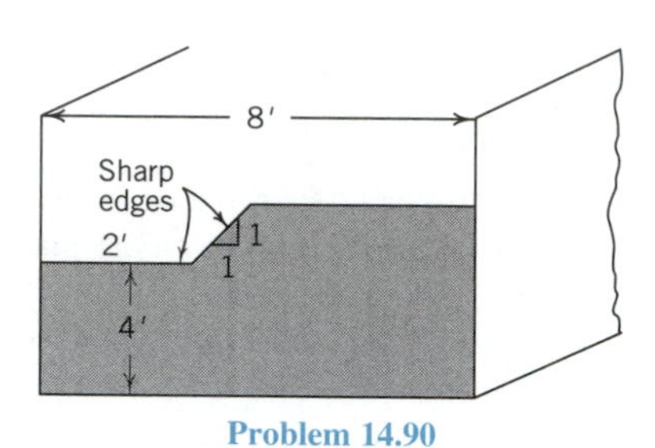

Problem 14.90

14.91. Estimate the flowrate over this sharp-crested weir, assuming a coefficient of 0.62.

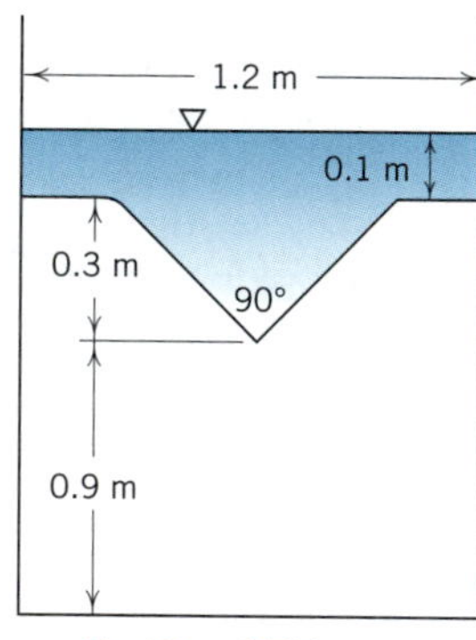

Problem 14.91

14.92. A rectangular channel 6 m wide carries 2.8 m^3/s at a depth of 0.9 m. What height of broad-crested rectangular weir must be installed to double the depth? Assume a weir coefficient of 0.56.

14.93. A broad-crested weir 0.9 m high has a flat crest and a coefficient of 0.55. If this weir is 6 m long and the head on it is 0.46 m, what flowrate will occur over it? What flowrate could be expected if the flow were frictionless?

14.94. A frictionless broad-crested weir 1.2 m high is built across a channel 2.4 m wide. If the energy line is 0.9 m above the weir crest, calculate the head, flowrate, and weir coefficient.

14.95. Using the specific energy and critical depth principles of Section 10.6, derive an expression for C_w as a function of P/H for the frictionless broad-crested weir of Fig. 14.37.

14.96. In order to justify the form of the relation between C_w and P/H of the Rehbock formula, calculate C_w for frictionless broad-crested weirs having values of P/H of 1, 5, and 9. Compare these with values obtained from the Rehbock formula, disregarding $1/900\ H$.

14.97. In a semi-empirical analysis of this square-edged broad-crested weir it is usually *assumed* that $y_2 = H/2$ and the pressure distribution on the weir face is hydrostatic as shown. Using these assumptions derive an expression for the weir coefficient as a function of P/H.

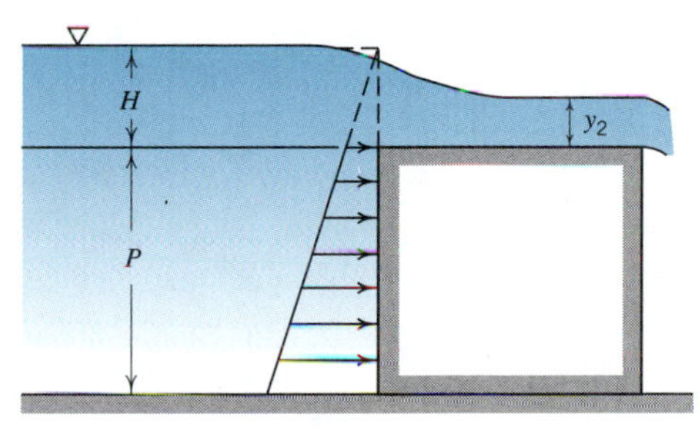

Problem 14.97

14.98. What flowrate will occur over a spillway of 500 ft (150 m) length when the head there is 4 ft (1.2 m), if the coefficient (referenced to the static head) of the spillway is 0.72?

14.99. A spillway 300 m long is found by model experiments to have a coefficient (referenced to the static head) of 0.68. It has a crest elevation of 30.0. When a flood flow of 1 400 m^3/s passes over the spillway, what is the elevation of the water surface in the reservoir just upstream from the spillway?

14.100. A spillway structure 20 ft high is designed by the methods of Fig. 14.40 for a head of 10 ft. Calculate the coefficients and flowrates for heads of 4, 10, and 12 ft. Calculate the corresponding coefficient for the design head when this is referenced to water surface rather than energy line.

14.101. Upstream and downstream from a sharp-crested rectangular weir 0.9 m high the water depths are 1.2 m and 1.05 m, respectively. Calculate the approximate flowrate.

14.102. The drop in the water surface in passing a submerged sharp-crested rectangular weir 1.2 m high is 76 mm. Calculate the approximate flowrate if the depth upstream from the weir is 1.5 m.

14.103. The following data are collected in a current-meter measurement at the river cross section of Fig. 14.44, which is 60 ft (18 m) wide at the water surface. Assume $V = 2.22 \times$ (r/s) [ft/s] or $0.677 \times$ (r/s) [m/s], and calculate the flowrate in the river.

Station Depth	0	1	2	3	4	5	6	7	8	9	10	11	12
ft	0.0	3.0	3.2	3.5	3.6	3.7	3.9	4.0	4.4	4.4	4.2	3.5	0.0
m	0.0	0.9	0.96	1.05	1.08	1.11	1.17	1.20	1.32	1.32	1.26	1.05	0.0
						rpm (r/min) of Rotating Element							
$0.2y$	—	40.0	53.5	58.6	63.0	66.7	61.5	56.3	54.0	52.6	50.0	45.0	—
$0.8y$	—	30.7	42.8	50.0	54.2	58.8	53.3	49.4	46.5	43.2	40.1	32.5	—

APPENDIX 1

SYMBOLS, UNITS AND DIMENSIONS

Units and Dimensions

Symbol	Quantity	SI Units	U.S. Customary Units	Mass-Length-Time Dimensions[a]
A	Area	m^2	ft^2	L^2
a	Wave speed or velocity	m/s	ft/s	L/t
a	Linear acceleration	m/s^2	ft/s^2	L/t^2
B	Bottom width (open channel)	m	ft	L
b	Surface width (open channel)	m	ft	L
b	Span of airfoil	m	ft	L
C	Chezy coefficient	$m^{1/2}/s$	$ft^{1/2}/s$	$L^{1/2}/t$
c_p	Specific heat at constant pressure	J/kg·K	ft·lb/slug·°R	L^2/Tt^2
c_v	Specific heat at constant volume	J/kg·K	ft·lb/slug·°R	L^2/Tt^2
D	Drag force	N	lb	ML/t^2
d	Diameter	m	ft	L
E	Modulus of elasticity of fluid	$Pa(N/m^2)$	lb/ft^2	M/Lt^2
E_p	Modulus of elasticity of pipe	$Pa(N/m^2)$	lb/ft^2	M/Lt^2
E	Unit energy	J/N	ft·lb/lb	L
e	Size of roughness	m	ft	L
e_p	Pipe wall thickness	m	ft	L
F	Force	$N(kg \cdot m/s^2)$	$lb(slug \cdot ft/s^2)$	ML/t^2
G	Weight flowrate	N/s	lb/s	ML/t^3
g_n	Gravitational acceleration	m/s^2	ft/s^2	L/t^2
H	Energy per unit mass	m^2/s^2	ft^2/s^2	L^2/t^2
H	Head on weir	m	ft	L
H	Total head	m	ft	L
h	Enthalpy	J/kg	ft·lb/slug	L^2/t^2
h	Head or height	m	ft	L
I	Second moment of an area	m^4	ft^4	L^4

[a]Mass (M), length (L), time (t), and thermodynamic temperature (T).

Units and Dimensions

Symbol	Quantity	SI Units	U.S. Customary Units	Mass-Length-Time Dimensions[a]
ie	Internal energy	J/kg	ft·lb/slug	L^2/t^2
L	Lift force	N	lb	ML/t^2
l	Length	m	ft	L
M	Mass	kg	slug	M
M	Moment	N·m	ft·lb	ML^2/t^2
$\dot{m}$	Mass flowrate	kg/s	slug/s	M/t
P	Power	W(J/s)	ft·lb/s	ML^2/t^3
P	Perimeter	m	ft	L
P	Weir height	m	ft	L
p	Pressure	Pa(N/m^2)	lb/ft^2(psf)	M/Lt^2
Q	Flowrate	m^3/s	ft^3/s(cfs)	L^3/t
q	Two-dimensional flowrate	m^2/s	ft^2/s	L^2/t
R	Radius	m	ft	L
R	Engineering gas constant	J/kg·K	ft·lb/slug·°R	L^2/Tt^2
R_h	Hydraulic radius	m	ft	L
r	Radial distance	m	ft	L
T	Torque	N·m	ft·lb	ML^2/t^2
T	Absolute temperature	K	°R	T
T	Temperature	°C	°F	T
t	Time (seconds)	s	s	t
u	Velocity (horizontal)	m/s	ft/s	L/t
V	Velocity	m/s	ft/s	L/t
$\forall$	Volume	m^3	ft^3	L^3
v	Velocity (horizontal)	m/s	ft/s	L/t
W	Weight	N	lb	ML/t^2
w	Velocity (vertical)	m/s	ft/s	L/t
x	Horizontal distance	m	ft	L
y	Depth (open channel)	m	ft	L
y	Distance from solid boundary	m	ft	L
z	Height above datum	m	ft	L
α	Specific volume	m^3/kg	ft^3/slug	L^3/M
Γ	Circulation	m^2/s	ft^2/s	L^2/t
γ	Specific weight	N/m^3	lb/ft^3	M/L^2t^2
δ	Boundary layer thickness	m	ft	L
δ_1	Displacement thickness	m	ft	L
δ_2	Momentum thickness	m	ft	L
δ_v	Viscous sublayer thickness	m	ft	L
ε	Eddy viscosity	Pa·s(N·s/m^2)	lb·s/ft^2	M/Lt
μ	Viscosity	Pa·s(N·s/m^2)	lb·s/ft^2	M/Lt

[a]Mass (M), length (L), time (t), and thermodynamic temperature (T).

Units and Dimensions

Symbol	Quantity	SI Units	U.S. Customary Units	Mass-Length-Time Dimensions[a]
ν	Kinematic viscosity	m^2/s	ft^2/s	L^2/t
ξ	Vorticity	s^{-1}	s^{-1}	$1/t$
ρ	Density	$kg/^3$	$slug/m^3$	M/L^3
σ	Surface tension	N/m	lb/ft	M/t^2
τ	Shear stress	Pa	lb/ft^2	M/Lt^2
ϕ	Velocity potential	m^2/s	ft^2/s	L^2/t
ψ	Stream function	m^2/s	ft^2/s	L^2/t
ω	Angular velocity	s^{-1}	s^{-1}	$1/t$

[a]Mass (M), length (L), time (t), and thermodynamic temperature (T).

Symbols for Dimensionless Quantities

Symbol	Quantity	Symbol	Quantity
$\mathbf{C}$	Cauchy number	n	Polytropic exponent; Manning's coefficient
C_c	Coefficient of contraction	$\mathbf{R}$	Reynolds number
C_D	Drag coefficient	S	Slope of energy line
C_f	Frictional drag coefficient	S_o	Bottom slope (open channel)
C_L	Lift coefficient	S_c	Critical slope
C_p	Pressure coefficient	$\mathbf{W}$	Weber number
C_v	Coefficient of velocity	Y	Expansion factor
C_w	Weir coefficient	α	Kinetic energy correction factor
$\mathbf{E}$	Euler number	β	Momentum correction factor
$\mathbf{F}$	Froude number	η	Efficiency
f	Friction factor	κ	von Kármán's turbulence "constant"
K_L	Loss coefficient	μ_p	Poisson's ratio for pipe material
k	Adiabatic exponent	Π	Dimensionless group
$\mathbf{M}$	Mach number	$\boldsymbol{\sigma}$	Cavitation number

Conversion Factor Table

Abbreviations used:

BTU = British Thermal Unit
cfs = cubic feet per second
ft/s = feet per second
ft = foot
gpm = gallons per minute
hp = horsepower
h = hour
Hz = hertz
in. = inch
J = joule = N·m
kg = kilogram = 10^3 gram
lb = pound force
m = metre (SI) = mile (U.S. Customary)
mb = millibar = 10^{-3} bar
mm = millimetre = 10^{-3} metre
mm^2 = square millimetre
mph = miles per hour
m/s = metres per second
N = newton
Pa = pascal = N/m^2
psi = pound per square inch
s = second
W = watt = J/s

Absolute viscosity: 1 Pa·s = 10 poises = 0.020 89 lb·s/ft^2
Acceleration due to gravity: 9.806 65 m/s^2 = 32.174 ft/s^2
Area: 1 m^2 = 10.76 ft^2; 1 mm^2 = 0.001 55 $in.^2$
Density: 1 kg/m^3 = 0.001 94 $slug/ft^3$
Energy: 1 N·m = 1 J = 0.737 5 ft·lb = 0.000 948 BTU
Flowrate: 1 m^3/s = 35.31 ft^3/s; 1 litre/s = 10^{-3} m^3/s = 0.022 83 mgd
1 ft^3/s = 449 gpm = 0.647 mgd
Force: 1 N = 0.224 8 lb
Kinematic viscosity: 1 m^2/s = 10^4 Stokes = 10.76 ft^2/s
Length: 1 mm = 0.039 4 in.; 1 m = 3.281 ft; 1 km = 0.622 miles
Mass: 1 kg = 0.068 5 slug
Power: 1 W = 1 J/s = 0.737 5 ft·lb/s; 1 kW = 1.341 hp = 737.5 ft·lb/s

Pressure: 1 kN/m^2 = 1 kPa = 0.145 psi; 1 mm Hg = 0.039 4 in. Hg = 133.3 Pa; 1 mm H_2O = 9.807 Pa; 101.325 kPa = 760 mm Hg = 29.92 in. Hg = 14.70 psi; 1 bar = 100 kPa = 14.504 psi
Specific Heat; Engineering Gas Constant: 1 J/kg·K = 5.98 ft·lb/slug·°R
Specific Volume: 1 m^3/kg = 515.5 $ft^3/slug$
Specific Weight: 1 N/m^3 = 0.006 365 lb/ft^3
Temperature: 1°C = 1 K = 1.8°F = 1.8°R (see Section 1.4)
Velocity: 1 m/s = 3.281 ft/s = 3.60 km/h = 2.28 mph; 1 knot = 0.515 5 m/s
Volume: 1 m^3 = 10^3 litres = 35.31 ft^3; 1 U.S. gallon = 3.785 litres; 1 U.K. gallon = 4.546 litres

APPENDIX

2 PHYSICAL PROPERTIES OF FLUIDS

TABLE A2.1 Approximate Properties of Some Common Liquids at Standard Atmospheric Pressure

	U.S. Customary Units						
	Temperature T, °F	Density, ρ, slug/ft^3	Specific Gravity, s.g., —	Modulus of Elasticity, E, psi	Viscosity $\mu \times 10^5$ lb·s/ft^2	Surface Tension[a] σ, lb/ft	Vapor Pressure p_v, psia
Benzene	68	1.70	0.88	150 000	1.37	0.002 0	1.45
Carbon tetrachloride	68	3.08	1.59	160 000	2.035	0.001 8	1.90
Crude oil	68	1.66	0.86	—	15.0	0.002	—
Ethyl alcohol	68	1.53	0.79	175 000	2.51	0.001 5	0.85
Freon-12	60	2.61	1.35	—	3.10	—	—
	−30	2.91	—	—	3.82	—	—
Gasoline	68	1.32	0.68	—	0.61	—	8.0
Glycerin	68	2.44	1.26	630 000	3 120	0.004 3	0.000 002
Hydrogen	−431	0.143	—	—	0.043 5	0.000 2	3.1
Jet fuel (JP-4)	60	1.50	0.77	—	1.82	0.002	1.3
Mercury	60	26.3	13.57	3 800 000	3.26	0.035	0.000 025
	600	24.9	12.8	—	1.88	—	6.85
Oxygen	−320	2.34	—	—	0.58	0.001	3.1
Sodium	600	1.70	—	—	0.690	—	—
	1000	1.60	—	—	0.472	—	—
Water[b]	68	1.936	1.00	318 000	2.10	0.005 0	0.34
Sea water[b]	68	1.99	1.03	336 000	2.25	0.005 0	0.34
	SI Units						
	T, °C	ρ, kg/m^3	s.g., —	E, kPa	$\mu \times 10^4$ Pa · s	σ, N/m	p_v, kPa
Benzene	20	876.2	0.88	1 034 250	6.56	0.029	10.0
Carbon tetrachloride	20	1 587.4	1.59	1 103 200	9.74	0.026	13.1
Crude oil	20	855.6	0.86	—	71.8	0.03	—

[a]In contact with air.

[b]The specific heat of liquid water is approximately 25 000 ft·lb/slug·°R or 4 180 J/kg·K.

TABLE A2.1 Approximate Properties of Some Common Liquids at Standard Atmospheric Pressure (cont.)

	SI Units						
	T, °C	ρ, kg/m^3	s.g., —	E, kPa	$\mu \times 10^4$ Pa · s	σ, N/m	p_v, kPa
Ethyl alcohol	20	788.6	0.79	1 206 625	12.0	0.022	5.86
Freon-12	15.6	1 345.2	1.35	—	14.8	—	—
	−34.4	1 499.8	—	—	18.3	—	—
Gasoline	20	680.3	0.68	—	2.9	—	55.2
Glycerin	20	1 257.6	1.26	4 343 850	14 939	0.063	0.000 014
Hydrogen	−257.2	73.7	—	—	0.21	0.002 9	21.4
Jet fuel (JP-4)	15.6	773.1	0.77	—	8.7	0.029	8.96
Mercury	15.6	13 555	13.57	26 201 000	15.6	0.51	0.000 17
	315.6	12 833	12.8	—	9.0	—	47.2
Oxygen	−195.6	1 206.0	—	—	2.78	0.015	21.4
Sodium	315.6	876.2	—	—	3.30	—	—
	537.8	824.6	—	—	2.26	—	—
Water[b]	20	998.2	1.00	2 170 500	10.0	0.073	2.34
Sea water[b]	20	1024.0	1.03	2 300 000	10.7	0.073	2.34

[b]The specific heat of liquid water is approximately 25 000 ft·lb/slug·°R or 4 180 J/kg·K.

TABLE A2.2 Approximate Properties of Some Common Gases

	U.S. Customary Units				
	Engineering Gas Constant, R, ft·lb/slug·°R	Universal Gas Constant, $\mathcal{R} = mR$ ft·lb/slug·°R	Adiabatic Exponent, k —	Specific Heat at Constant Pressure, c_p, ft·lb/slug·°R	Viscosity at 68°F (20°C), $\mu \times 10^5$ lb·s/ft^2
Carbon dioxide	1 123	49 419	1.28	5 132	0.030 7
Oxygen	1 554	49 741	1.40	5 437	0.041 9
Air	1 715	49 709	1.40	6 000	0.037 7
Nitrogen	1 773	49 644	1.40	6 210	0.036 8
Methane	3 098	49 644	1.31	13 095	0.028
Helium	12 419	49 677	1.66	31 235	0.041 1
Hydrogen	24 677	49 741	1.40	86 387	0.018 9

	SI Units				
	R, J/kg·K	$\mathcal{R} = mR$, J/kg·K	k —	c_p, J/kg·K	$\mu \times 10^5$, Pa·s
Carbon dioxide	187.8	8 264	1.28	858.2	1.47
Oxygen	259.9	8 318	1.40	909.2	2.01
Air	286.8	8 313	1.40	1 003	1.81
Nitrogen	296.5	8 302	1.40	1 038	1.76
Methane	518.1	8 302	1.31	2 190	1.34
Helium	2 076.8	8 307	1.66	5 223	1.97
Hydrogen	4 126.6	8 318	1.40	14 446	0.90

TABLE A2.3*a* Viscosities of Some Common Fluids (U.S. Customary Units)

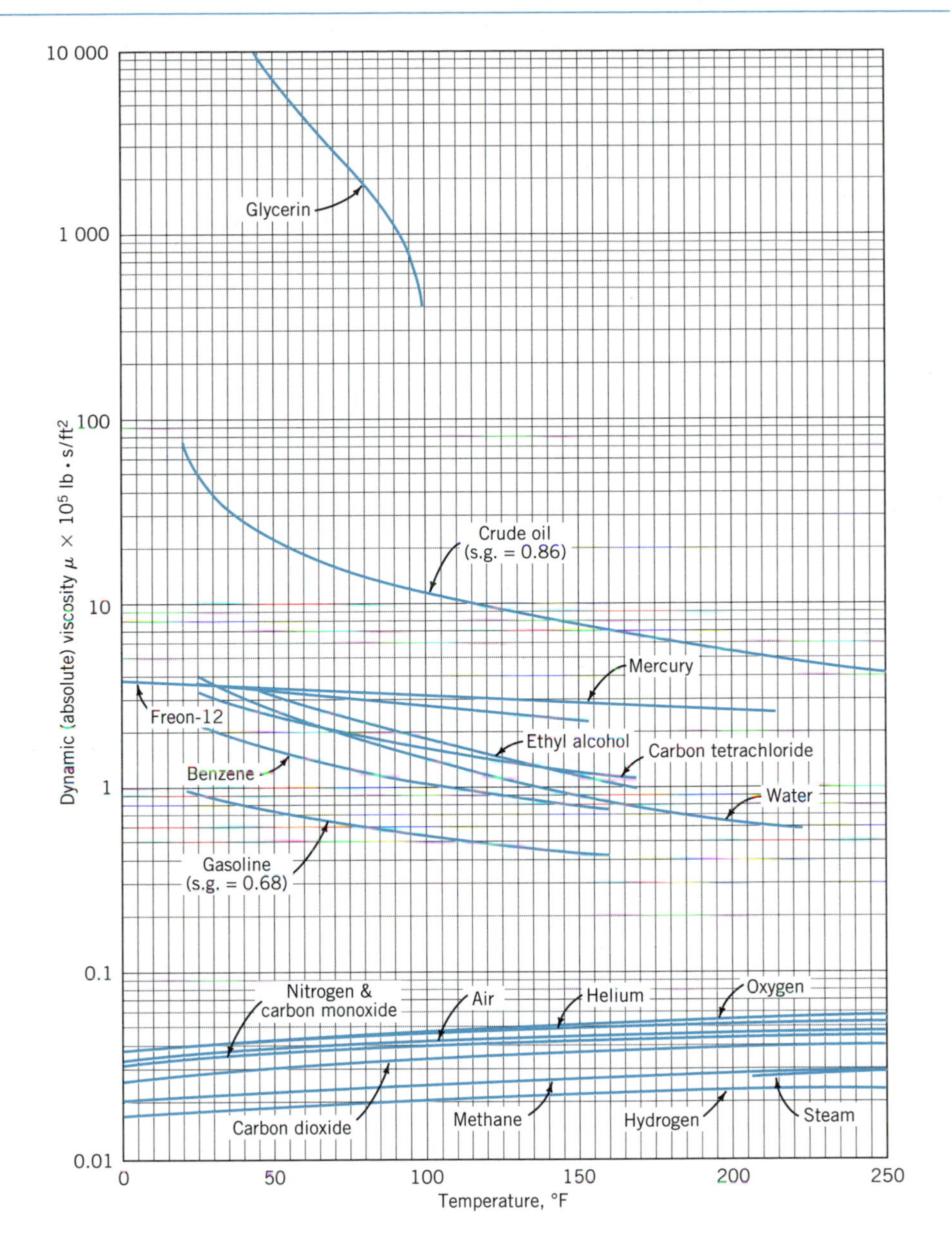

TABLE A2.3b Viscosities of Some Common Fluids (SI Units)

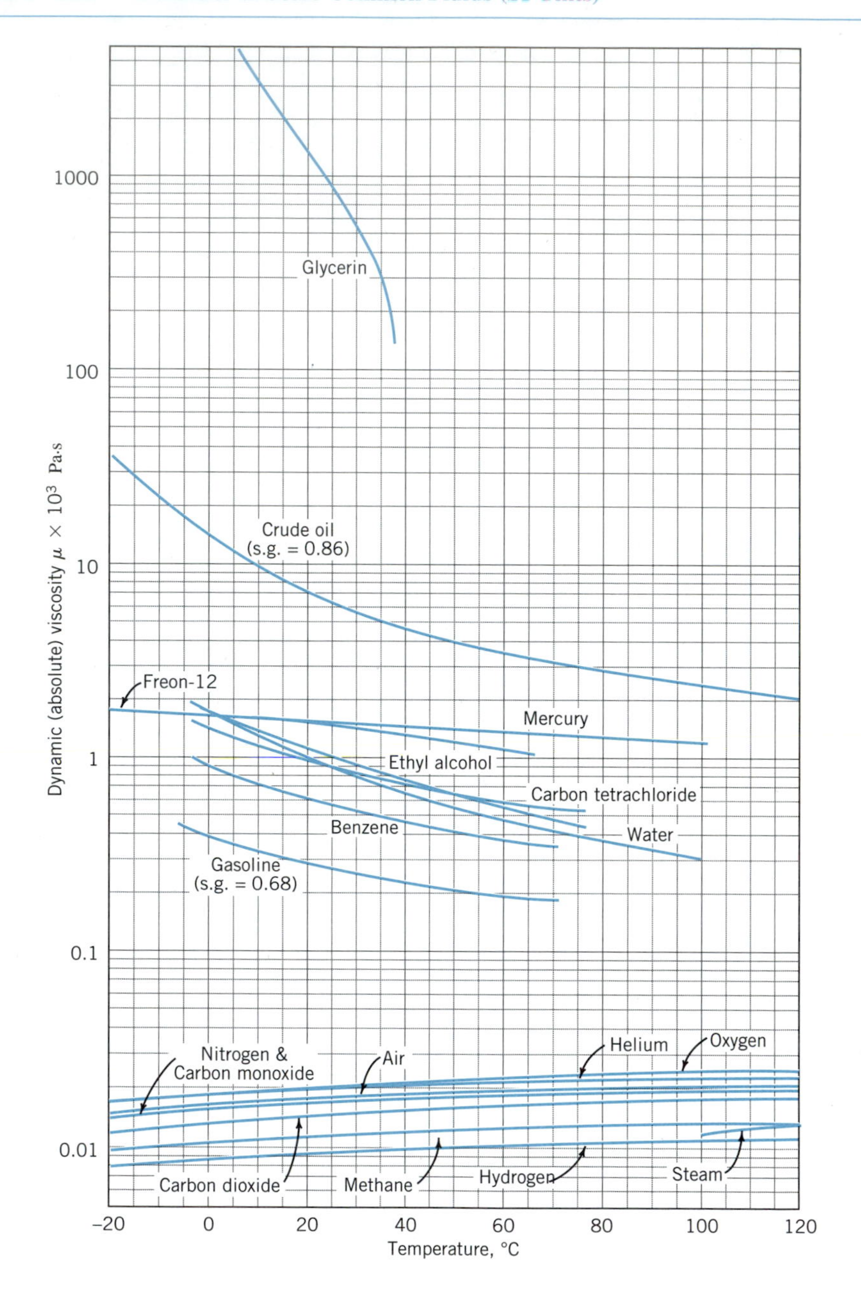

TABLE A2.4a Physical Properties of Water (U.S. Customary Units)[f]

Temperature, °F	Specific Weight,[a] γ, lb/ft^3	Density,[a] ρ, slug/ft^3	Modulus of Elasticity,[b,c] $E \times 10^{-3}$, psi	Viscosity,[a] $\mu \times 10^5$, lb·s/ft^2	Kinematic Viscosity,[a] $\nu \times 10^5$, ft^2/s	Surface Tension,[a,d] σ, lb/ft	Vapor Pressure,[e] p_v, psia
32	62.42	1.940	287	3.746	1.931	0.005 18	0.09
40	62.43	1.940	296	3.229	1.664	0.006 14	0.12
50	62.41	1.940	305	2.735	1.410	0.005 09	0.18
60	62.37	1.938	313	2.359	1.217	0.005 04	0.26
70	62.30	1.936	319	2.050	1.059	0.004 98	0.36
80	62.22	1.934	324	1.799	0.930	0.004 92	0.51
90	62.11	1.931	328	1.595	0.826	0.004 86	0.70
100	62.00	1.927	331	1.424	0.739	0.004 80	0.95
110	61.86	1.923	332	1.284	0.667	0.004 73	1.27
120	61.71	1.918	332	1.168	0.609	0.004 67	1.69
130	61.55	1.913	331	1.069	0.558	0.004 60	2.22
140	61.38	1.908	330	0.981	0.514	0.004 54	2.89
150	61.20	1.902	328	0.905	0.476	0.004 47	3.72
160	61.00	1.896	326	0.838	0.442	0.004 41	4.74
170	60.80	1.890	322	0.780	0.413	0.004 34	5.99
180	60.58	1.883	318	0.726	0.385	0.004 27	7.51
190	60.36	1.876	313	0.678	0.362	0.004 20	9.34
200	60.12	1.868	308	0.637	0.341	0.004 13	11.52
212	59.83	1.860	300	0.593	0.319	0.004 04	14.70

[a]From "Hydraulic Models," *A.S.C.E. Manual of Engineering Practice*, No. 25, A.S.C.E., 1942. See footnote 1.

[b]Approximate values averaged from many sources.

[c]At atmospheric pressure. See footnote 1.

[d]In contact with air.

[e]From J. H. Keenan and F. G. Keyes, *Thermodynamic Properties of Steam*, John Wiley & Sons, 1936.

[f]Compiled from many sources including those indicated, *Handbook of Chemistry and Physics*, 54th Ed., The CRC Press, 1973, and *Handbook of Tables for Applied Engineering Science*, The Chemical Rubber Co., 1970.

[1]Here, if $E \times 10^{-3} = 287$, then $E = 287 \times 10^3$ psi; while if $\mu \times 10^5 = 3.746$, then $\mu = 3.746 \times 10^{-5}$ lb·s/ft^2, and so on.

TABLE A2.4*b* Physical Properties of Water (SI Units)[f]

Temperature, °C	Specific Weight,[a] γ, kN/m^3	Density,[a] ρ, kg/m^3	Modulus of Elasticity,[b,c] $E \times 10^{-6}$, kPa	Viscosity,[a] $\mu \times 10^3$, Pa·s	Kinematic Viscosity,[a] $\nu \times 10^6$, m^2/s	Surface Tension,[a,d] σ, N/m	Vapor Pressure,[e] p_v, kPa
0	9.805	999.8	1.98	1.781	1.785	0.075 6	0.61
5	9.807	1 000.0	2.05	1.518	1.518	0.074 9	0.87
10	9.804	999.7	2.10	1.307	1.306	0.074 2	1.23
15	9.798	999.1	2.15	1.139	1.139	0.073 5	1.70
20	9.789	998.2	2.17	1.002	1.003	0.072 8	2.34
25	9.777	997.0	2.22	0.890	0.893	0.072 0	3.17
30	9.764	995.7	2.25	0.798	0.800	0.071 2	4.24
40	9.730	992.2	2.28	0.653	0.658	0.069 6	7.38
50	9.689	988.0	2.29	0.547	0.553	0.067 9	12.33
60	9.642	983.2	2.28	0.466	0.474	0.066 2	19.92
70	9.589	977.8	2.25	0.404	0.413	0.064 4	31.16
80	9.530	971.8	2.20	0.354	0.364	0.062 6	47.34
90	9.466	965.3	2.14	0.315	0.326	0.060 8	70.10
100	9.399	958.4	2.07	0.282	0.294	0.058 9	101.33

[a]From "Hydraulic Models," *A.S.C.E. Manual of Engineering Practice*, No. 25, A.S.C.E., 1942. See footnote 2.
[b]Approximate values averaged from many sources.
[c]At atmospheric pressure. See footnote 2.
[d]In contact with air.
[e]From J. H. Keenan and F. G. Keyes, *Thermodynamic Properties of Steam*, John Wiley & Sons, 1936.
[f]Compiled from many sources including those indicated, *Handbook of Chemistry and Physics*, 54th Ed., The CRC Press, 1973, and *Handbook of Tables for Applied Engineering Science*, The Chemical Rubber Co., 1970.

[2]Here, if $E \times 10^{-6} = 1.98$, then $E = 1.98 \times 10^6$ kPa, while if $\mu \times 10^3 = 1.781$, then $\mu = 1.781 \times 10^{-3}$ Pa·s, and so on.

TABLE A2.5a The U.S. Standard Atmosphere (U.S. Customary Units)[a]

Altitude, ft	Temperature, °F	Absolute Pressure, psia	Specific Weight, lb/ft^3	Density, slug/ft^3	Viscosity $\times 10^7$, lb·s/ft^2
0	59.00	14.696	0.076 47	0.002 377	3.737
5 000	41.17	12.243	0.065 87	0.002 048	3.637
10 000	23.36	10.108	0.056 43	0.001 756	3.534
15 000	5.55	8.297	0.048 07	0.001 496	3.430
20 000	−12.26	6.759	0.040 69	0.001 267	3.325
25 000	−30.05	5.461	0.034 18	0.001 066	3.217
30 000	−47.83	4.373	0.028 57	0.000 891	3.107
35 000	−65.61	3.468	0.023 67	0.000 738	2.995
40 000	−69.70	2.730	0.018 82	0.000 587	2.969
45 000	−69.70	2.149	0.014 81	0.000 462	2.969
50 000	−69.70	1.690	0.011 65	0.000 364	2.969
55 000	−69.70	1.331	0.009 17	0.000 287	2.969
60 000	−69.70	1.049	0.007 22	0.000 226	2.969
65 000	−69.70	0.826	0.005 68	0.000 178	2.969
70 000	−67.42	0.651	0.004 45	0.000 139	2.984
75 000	−64.70	0.514	0.003 49	0.000 109	3.001
80 000	−61.98	0.404	0.002 63	0.000 086	3.018
85 000	−59.26	0.322	0.002 15	0.000 067	3.035
90 000	−56.54	0.255	0.001 70	0.000 053	3.052
95 000	−53.82	0.203	0.001 34	0.000 042	3.070
100 000	−51.10	0.162	0.001 06	0.000 033	3.087

[a]Data from *U.S. Standard Atmosphere, 1962*, U.S. Government Printing Office, 1962. Data agree with ICAO standard atmosphere to 20 km and with ICAO proposed extension to 30 km. For atmospheric tables depicting conditions other than mid-latitude mean represented by standard atmosphere, see *U.S. Standard Atmosphere Supplements, 1966*, U.S. Government Printing Office, 1966.

TABLE A2.5*b* The U.S. Standard Atmosphere (SI Units)[a]

Altitude, km	Temperature, °C	Absolute Pressure, kPa	Specific Weight, N/m^3	Density, kg/m^3	Viscosity $\times 10^5$, Pa·s
0	15.00	101.33	12.01	1.225	1.789
2	2.00	79.50	9.86	1.007	1.726
4	−11.00	61.64	8.02	0.819	1.661
6	−23.96	47.22	6.46	0.660	1.595
8	−36.94	35.65	5.14	0.526	1.527
10	−49.90	26.50	4.04	0.414	1.458
12	−56.50	19.40	3.05	0.312	1.422
14	−56.50	14.17	2.22	0.228	1.422
16	−56.50	10.35	1.62	0.166	1.422
18	−56.50	7.57	1.19	0.122	1.422
20	−56.50	5.53	0.87	0.089	1.422
22	−54.58	4.05	0.63	0.065	1.432
24	−52.59	2.97	0.46	0.047	1.443
26	−50.61	2.19	0.33	0.034	1.454
28	−48.62	1.62	0.24	0.025	1.465
30	−46.64	1.20	0.18	0.018	1.475

[a]Data from *U.S. Standard Atmosphere, 1962*, U.S. Government Printing Office, 1962. Data agree with ICAO standard atmosphere to 20 km and with ICAO proposed extension to 30 km. For atmospheric tables depicting conditions other than mid-latitude mean represented by standard atmosphere, see *U.S. Standard Atmosphere Supplements, 1966*, U.S. Government Printing Office, 1966.

APPENDIX

3 PROPERTIES OF AREAS AND VOLUMES

	Sketch	Area or Volume	Location of Centroid	I or I_c
Rectangle		bh	$y_c = \frac{h}{2}$	$I_c = \frac{bh^3}{12}$
Triangle		$\frac{bh}{2}$	$y_c = \frac{h}{3}$	$I_c = \frac{bh^3}{36}$
Circle		$\frac{\pi d^2}{4}$	$y_c = \frac{d}{2}$	$I_c = \frac{\pi d^4}{64}$
Semicircle [1]		$\frac{\pi d^2}{8}$	$y_c = \frac{4r}{3\pi}$	$I_c = \frac{\pi d^4}{128}$
Ellipse		$\frac{\pi bh}{4}$	$y_c = \frac{h}{2}$	$I_c = \frac{\pi bh^3}{64}$
Semiellipse		$\frac{\pi bh}{4}$	$y_c = \frac{4h}{3\pi}$	$I = \frac{\pi bh^3}{16}$
Parabola		$\frac{2}{3}bh$	$y_c = \frac{3h}{5}$ $x_c = \frac{3b}{8}$	$I = \frac{2bh^3}{7}$
Segment of Square			$x_c = \frac{2}{3}\frac{r}{4-\pi}$	

[1]For the quarter-circle, the respective values are $\pi d^2/16$, $4r/3\pi$, and $\pi d^4/256$.

	Sketch	Area or Volume	Location of Centroid	I or I_c
Cylinder	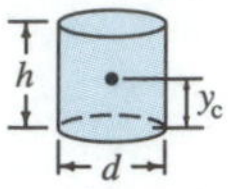	$\frac{\pi d^2 h}{4}$	$y_c = \frac{h}{2}$	
Cone	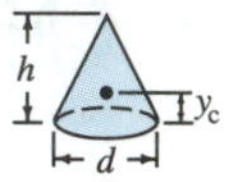	$\frac{1}{3}\left(\frac{\pi d^2 h}{4}\right)$	$y_c = \frac{h}{4}$	
Paraboloid of revolution	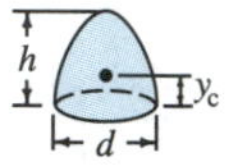	$\frac{1}{2}\left(\frac{\pi d^2 h}{4}\right)$	$y_c = \frac{h}{3}$	
Sphere	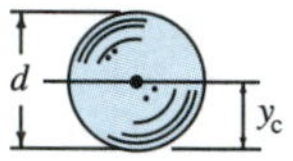	$\frac{\pi d^3}{6}$	$y_c = \frac{d}{2}$	
Hemisphere	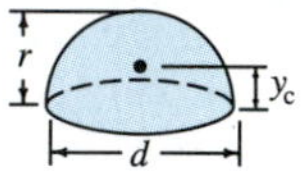	$\frac{\pi d^3}{12}$	$y_c = \frac{3r}{8}$	

APPENDIX

4 CAVITATION

The phenomenon of cavitation is of great importance in the design of high-speed hydraulic machinery such as turbines, pumps, and marine propellers, in the overflow and underflow structures of high dams, and in the high-speed motion of underwater bodies (such as submarines and hydrofoils). Cavitation also may be of critical significance in pipeline design and in certain problems of fluid metering. Typically considered a ''water-problem,'' cavitation has, in recent years, become a concern in a range of liquid flows, including liquid metal and cryogenic fluid pumps, and in a variety of areas, including medicine and industrial cleaning.

Cavitation may be expected in a flowing liquid[1] wherever the local pressure falls to the vapor pressure of the liquid. Local vaporization of the liquid will then result, causing development of a vapor-filled bubble or cavity in the flow. Cavitation is often accompanied by erosion (pitting) of solid boundary surfaces in machines or on hydrofoils, losses of efficiency, and serious vibration problems.

The nature of cavitation may be most easily observed by study of the ideal flow of a liquid through a constriction in a passage (Fig. A.1). With the valve partially open, the variation of pressure head through passage and constriction is given by hydraulic grade line A, the point of lowest pressure occurring at the minimum area, where the velocity is highest. Increase of valve opening (causing larger flowrate) produces hydraulic grade line B, for which the absolute pressure in the throat of the constriction falls to the vapor pressure of the liquid, causing the *inception* of cavitation. Further opening of the valve *does not*

[1]Cavitation is not possible in a gas because of its capacity for expansion.

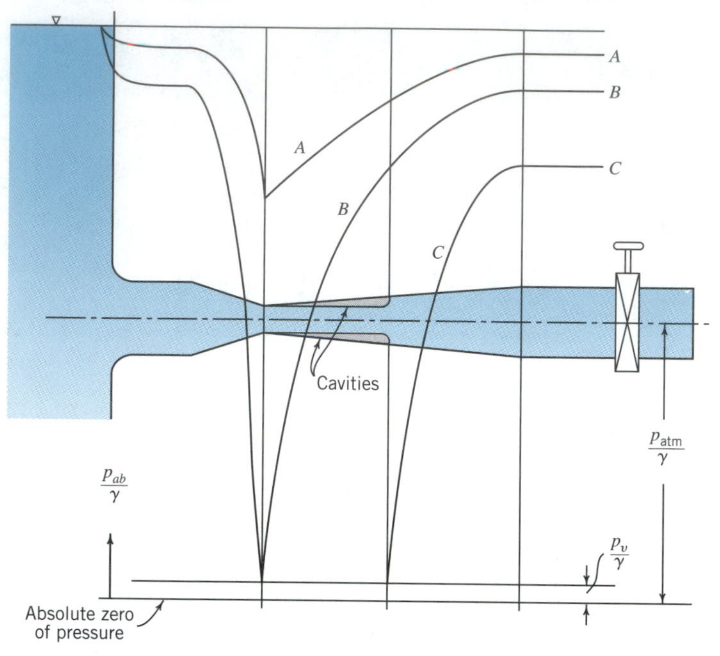

Fig. A.1

increase the flowrate but serves to extend the zone of vapor pressure downstream from the throat of the constriction; here the live stream of liquid separates from the boundary walls, producing a cavity in which the mean pressure is the vapor pressure of the liquid. The cavity contains a swirling mass of droplets and vapor and, although appearing steady to the naked eye, actually forms and reforms many times a second. The formation and disappearance of a single cavity are shown schematically in Fig. A.2, and the disappearance of the cavity is the clue to the destructive action caused by cavitation. The low-pressure cavity is swept swiftly downstream into a region of high pressure where it collapses suddenly, the surrounding liquid rushing in to fill the void. At the point of disappearance of the cavity the inrushing liquid comes together, momentarily raising the local pressure within the liquid to a very high value. If the point of collapse of the cavity is in contact with the boundary wall, the wall receives a blow as from a tiny hammer, and its surface may be stressed locally beyond its elastic limit, resulting in fatigue, probably enhanced chemical corrosion, and eventual destruction of the wall material (Fig. A.3).

Another form of cavitation is the *steady-state cavity* frequently observed in the tip vortices (Section 11.5) of marine propellers or surrounding high-speed hydrofoils. In the model test in a water tunnel shown in Fig. A.4, the cavitation number (cf., Eq. 8.16)

$$\boldsymbol{\sigma} = \frac{p_o - p_v}{\rho V_o^2}$$

is determined by the pressure p_o in the flow upstream from the propeller (to the right in the figure), the vapor pressure p_v existing in the cavities on the blades, the liquid density ρ, and the speed of advance V_o of the propeller. Because the propeller is fixed in the tunnel, V_o is the speed of the water in the tunnel upstream from the propeller. In the test shown, the propeller's effective-axial-speed V_o and r/min were held constant while p_o was reduced to increase cavitation. In Fig. A.4*a*, $\boldsymbol{\sigma}$ is relatively large and only a small amount of

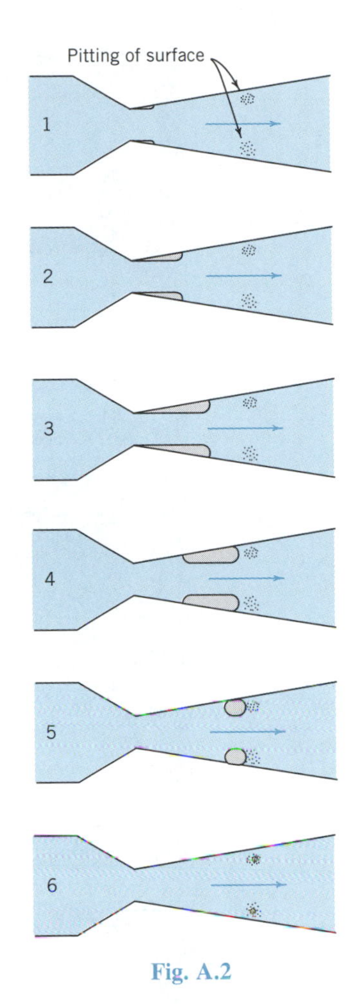

Fig. A.2

Fig. A.3 Pitting of brass plate after 5 hours' exposure to cavitation (magnification 10×).

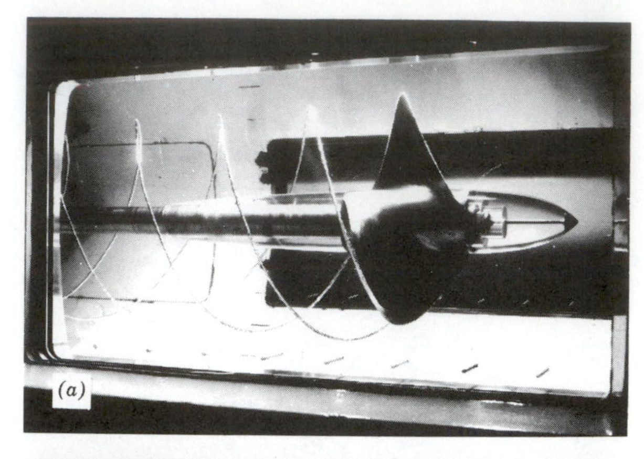

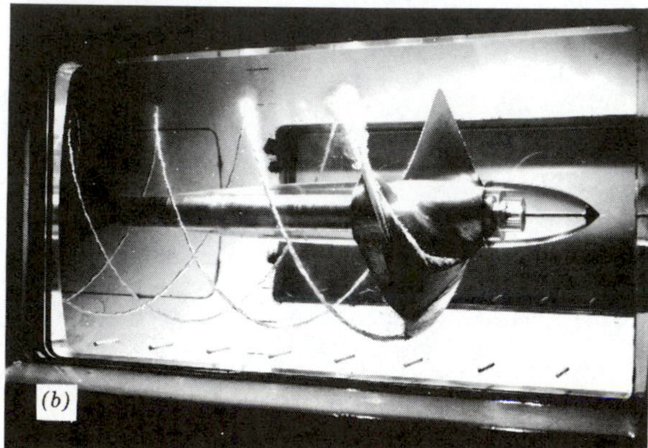

Fig. A.4 Cavitating propeller in a water tunnel. (Courtesy of National Physical Laboratory, Teddington, Middlesex, England; Crown Copyright reserved.)

cavitation occurs on the low pressure side of the blades and in the tip vortices. In Figs. A.4*a* and A.4*b*, the blade cavities collapse directly on the blade.

In Fig. A.4*c*, the propeller is operating close to its design condition as a fully cavitating or supercavitating propeller, and σ is very small. Although the downstream ends of such cavities exhibit certain unsteady phenomena, the large portion of the cavity is steady with its outer boundary acting as a streamline of the flowfield. Such a streamline is a *free streamline* (Section 5.8), because the pressure along it will be that in the cavity and equal to the vapor pressure of the liquid. Here the engineering problem is the prediction of cavity form and location of the separation points; the latter may be critically dependent on boundary-layer growth (Section 11.3) and on the fine details of vaporization in the flow as the pressure falls to the vapor pressure of the liquid.

One of the usual objectives in the design of hydraulic machinery and structures is the prevention of cavitation, which the designer accomplishes by improved forms of boundary surfaces and by setting limits beyond which the machine or structure should not be operated. In the case of high-speed ships and modern, high-performance pumps, there is little hope of preventing large-scale cavitation on propellers and pump impellers. Then, it is the objective of the design to predict the exact nature of the cavitation and to control its extent so, for example, large steady-state cavities exist that do not damage the machinery. As a result fully cavitating propellers and pumps can be made to have predictable and reliable performance.

REFERENCES

A.S.M.E. 1965. Symposium, *Cavitation in fluid machinery.*

Blevins, R. D. 1984. *Applied flid dynamics handbook.* New York: Van Nostrand Reinhold Co. Especially Chapters 6 and 10.

Eisenberg, P., and Tulin, M. P. 1961. Cavitation. Sec. 12, *Handbook of fluid dynamics.* New York: McGraw-Hill.

Falvey, H. T. 1990. Cavitation in chutes and spillways. *Engineering Monographs, 42.* Denver, CO: U. S. Department of Interior, Bureau of Reclamation.

IAHR, Institute of High Speed Mechanics (Sendai, Japan). 1962. Symposium, *Cavitation and hydraulic machinery.*

Knapp, R. T., Daily, J. W., and Hammitt, F. G. 1970. *Cavitation.* New York: McGraw-Hill.

Thomas, H. A., and Schuleen, E. P. 1942. Cavitation in outlet conduits of high dams. *Trans. A.S.C.E.* 107.

FILM

Eisenberg, P. Cavitation. NCFMF/EDC Film No. 21620, Encyclopaedia Britannica Educ. Corp.

APPENDIX 5

THE EXPANSION FACTOR,[1] Y

$\frac{A_2}{A_1}$	k	p_2/p_1				
		0.95	0.90	0.85	0.80	0.75
0	1.40	0.973	0.945	0.916	0.886	0.856
	1.30	0.971	0.941	0.910	0.878	0.846
	1.20	0.968	0.936	0.903	0.869	0.834
0.2	1.40	0.971	0.942	0.912	0.881	0.850
	1.30	0.969	0.938	0.906	0.873	0.839
	1.20	0.967	0.933	0.899	0.863	0.827
0.3	1.40	0.969	0.938	0.907	0.875	0.842
	1.30	0.967	0.934	0.900	0.866	0.831
	1.20	0.965	0.929	0.862	0.856	0.819
0.4	1.40	0.966	0.932	0.899	0.864	0.830
	1.30	0.964	0.928	0.891	0.855	0.818
	1.20	0.961	0.922	0.886	0.844	0.805
0.5	1.40	0.962	0.923	0.886	0.848	0.811
	1.30	0.959	0.918	0.878	0.839	0.799
	1.20	0.955	0.912	0.869	0.827	0.786
0.6	1.40	0.954	0.910	0.867	0.825	0.785
	1.30	0.951	0.904	0.858	0.814	0.772
	1.20	0.947	0.896	0.848	0.801	0.757

$$Y = \sqrt{\frac{1 - \left(\frac{A_2}{A_1}\right)^2}{1 - \left(\frac{A_2}{A_1}\right)^2 \left(\frac{p_2}{p_1}\right)^{2/k}} \; \frac{\frac{k}{k-1}\left(\frac{p_2}{p_1}\right)^2 \left[1 - \left(\frac{p_2}{p_1}\right)^{(k-1)/k}\right]}{1 - \frac{p_2}{p_1}}}$$

[1]The tabulated values were computed by J. P. Robb and R. E. Royer at the University of Wyoming.

APPENDIX

6 BASIC MATHEMATICAL OPERATIONS

If F is some function of x, y, and z, written $F(x, y, z)$, the partial derivative $\partial F/\partial x$ is obtained by differentiating F with respect to x, while y and z are held constant. To obtain $\partial F/\partial y$, the differentiation is performed with respect to y with x and z held constant, and so forth. Thus, partial derivatives give the rates of change of F in each of the coordinate directions. The *gradient* of F gives the maximum rate of change of F and has a direction perpendicular to lines of equal value of F, that is, perpendicular to contours of F drawn in space.

EXAMPLE

$$\text{Let } F = 2x^2 + 3xy + 4y^2 + 5z^2 + C$$

$$\frac{\partial F}{\partial x} = 4x + 3y$$

$$\frac{\partial F}{\partial y} = 3x + 8y$$

$$\frac{\partial F}{\partial z} = 10z$$

$$\frac{\partial^2 F}{\partial x\,\partial y} = 3 \qquad \frac{\partial^2 F}{\partial z^2} = 10$$

and so on.

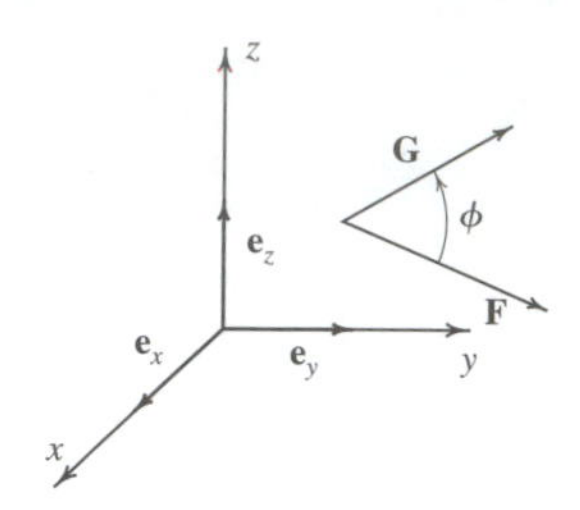

Fig. A.5

A *scalar quantity* or *scalar* is a quantity that can be described by a single number. A scalar has only a *magnitude* (size). A *vector* is a directed quantity that has both a *magnitude* and a *direction.* Pressure, density, and temperature are scalars. Velocity and force are vectors. A vector **F** can be expressed in terms of its components along the axes of the Cartesian system (x, y, z), that is,

$$\mathbf{F} = F_x\mathbf{e}_x + F_y\mathbf{e}_y + F_z\mathbf{e}_z$$

where F_x, F_y, F_z are the projections of the magnitude of **F** on the x, y, z axes, respectively. The unit vectors (their magnitude is unity) $\mathbf{e}_x$, $\mathbf{e}_y$, $\mathbf{e}_z$ are directed along the mutually perpendicular coordinate axes (see Fig. A5). The magnitude F of **F** is

$$F = |\mathbf{F}| = (F_x^2 + F_y^2 + F_z^2)^{1/2}$$

There are three important vector operations that may be new to the reader. The first is the *scalar or dot product*

$$S = \mathbf{F} \cdot \mathbf{G} = |\mathbf{F}|\,|\mathbf{G}| \cos \phi$$

where ϕ is the angle between the vectors **F** and **G** and S is a scalar. If **n** is a unit vector normal (perpendicular) to some surface or contour, the component of **F** along **n** is

$$F_n = \mathbf{F} \cdot \mathbf{n} = |\mathbf{F}| \cos \phi$$

that is, the scalar or dot product gives the projection of **F** on **n** (recall $|\mathbf{n}| = 1$). (See Fig. A6.)

The second operation is the *vector product*

$$\mathbf{V} = \mathbf{F} \times \mathbf{G}$$

where the magnitude of **V** is

$$|\mathbf{V}| = |\mathbf{F}|\,|\mathbf{G}| \sin \phi$$

and the direction of **V** is perpendicular to the plane of **F** and **G** and in accordance with the usual right-hand rule. (See Fig. A7.)

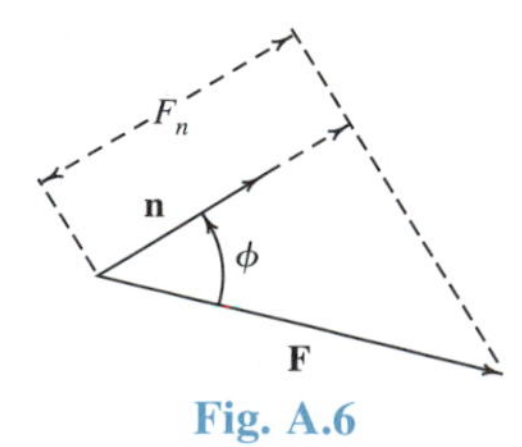

Fig. A.6

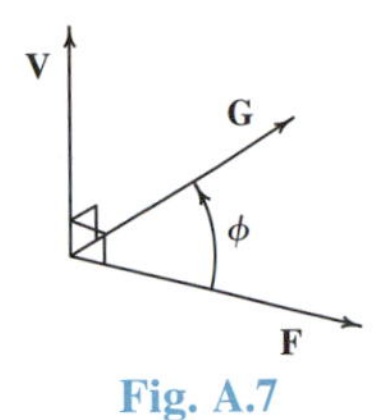

Fig. A.7

EXAMPLE

Calculate the moment **T** arising about 0 from the action of a force $\mathbf{F} = F\mathbf{e}_\theta$ at a point P shown in Fig. A8. In polar cylindrical coordinates the unit vectors are $\mathbf{e}_r$, $\mathbf{e}_\theta$, $\mathbf{e}_z$. Thus,

$$\mathbf{T} = \mathbf{r} \times \mathbf{F} = r\mathbf{e}_r \times F\mathbf{e}_\theta = rF\mathbf{e}_r \times \mathbf{e}_\theta$$

but $\mathbf{e}_r \times \mathbf{e}_\theta = \mathbf{e}_z$ according to the definition of the vector product. Therefore,

$$\mathbf{T} = T\mathbf{e}_z = rF\mathbf{e}_z$$

where $T = rF$ is the magnitude of **T**.

The third operation is the derivative of vectors. For example,

$$\frac{\partial \mathbf{F}}{\partial s} = \frac{\partial F_x}{\partial s}\mathbf{e}_x + \frac{\partial F_y}{\partial s}\mathbf{e}_y + \frac{\partial F_z}{\partial s}\mathbf{e}_z$$

(The Cartesian system's unit vectors have fixed magnitude and direction.) The gradient of F is

$$\text{grad } F = \frac{\partial F}{\partial x}\mathbf{e}_x + \frac{\partial F}{\partial y}\mathbf{e}_y + \frac{\partial F}{\partial z}\mathbf{e}_z = \nabla F$$

for a scalar $F(x, y, z)$. The divergence of **F** is defined as

$$\text{div } \mathbf{F} = \nabla \cdot \mathbf{F} = \left(\frac{\partial}{\partial x}\mathbf{e}_x + \frac{\partial}{\partial y}\mathbf{e}_y + \frac{\partial}{\partial z}\mathbf{e}_z\right) \cdot \mathbf{F}$$

$$= \frac{\partial F_x}{\partial x}\mathbf{e}_x + \frac{\partial F_y}{\partial y}\mathbf{e}_y + \frac{\partial F_z}{\partial z}\mathbf{e}_z$$

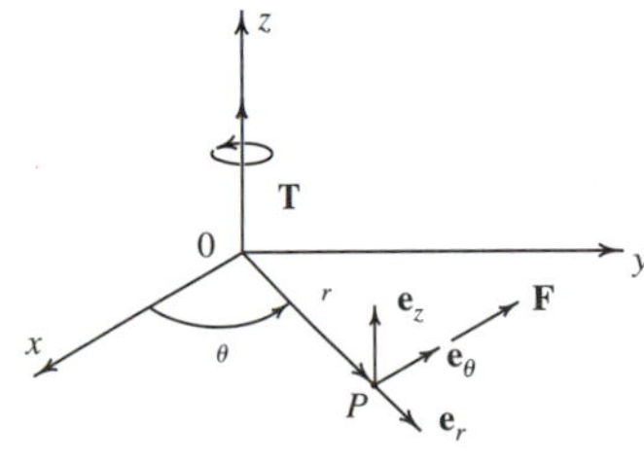

Fig. A.8

It is easy to show (why not try it?) that

$$\text{div (grad } F) = \nabla \cdot \nabla F = \nabla^2 F$$
$$= \frac{\partial^2 F}{\partial x^2} + \frac{\partial^2 F}{\partial y^2} + \frac{\partial^2 F}{\partial z^2}$$

for the scalar $F(x, y, z)$. The Laplace equation $\nabla^2 F = 0$ is accordingly a partial differential equation that is important in analyses of ideal fluid flows.

EXAMPLE

With

$$F = 2x^2 + 3xy + 4y^2 + 5z^2 + C$$

as before,

$$\text{div (grad } F) = 4 + 8 + 10 = 22 = \nabla^2 F$$

The total differential of $F(x, y, z)$ is

$$dF = \frac{\partial F}{\partial x} dx + \frac{\partial F}{\partial y} dy + \frac{\partial F}{\partial z} dz$$

EXAMPLE

Suppose the fluid density is given by $\rho(x, y, t)$, that is, is a function of space and time in a compressible flow. Then,

$$d\rho = \frac{\partial \rho}{\partial x} dx + \frac{\partial \rho}{\partial y} dy + \frac{\partial \rho}{\partial t} dt$$

and

$$\frac{d\rho}{dt} = \frac{\partial \rho}{\partial x}\frac{dx}{dt} + \frac{\partial \rho}{\partial y}\frac{dy}{dt} + \frac{\partial \rho}{\partial t}\frac{dt}{dt}$$

Because $u = dx/dt$ and $v = dy/dt$,

$$\frac{d\rho}{dt} = u\frac{\partial \rho}{\partial x} + v\frac{\partial \rho}{\partial y} + \frac{\partial \rho}{\partial t}$$

Often derivatives are known and the function is sought. A typical example (in polar coordinates) is to find $F(r, \theta)$ when

$$\frac{\partial F}{\partial r} = -\frac{\sin\theta}{r^2} \quad \text{and} \quad \frac{\partial F}{r\partial\theta} = \frac{\cos\theta}{r^2}$$

Here

$$dF = \left(\frac{\partial F}{r\partial\theta}\right) r\, d\theta + \left(\frac{\partial F}{\partial r}\right) dr$$

and the integral is

$$F = \int \frac{\cos\theta}{r^2} r\, d\theta - \int \frac{\sin\theta}{r^2} dr + C$$

However, $\cos\theta\, d\theta/r - \sin\theta\, dr/r^2$ may be recognized as $d[(\sin\theta)/r]$, so the solution is

$$F = \int d\left(\frac{\sin\theta}{r}\right) + C = \frac{\sin\theta}{r} + C$$

REFERENCES

Kreyszig, E. 1988. *Advanced engineering mathematics.* 6th ed. New York: Wiley.

McQuistan, R. B. 1965. *Scalar and vector fields.* New York: Wiley.

APPENDIX 7

COMPUTER PROGRAMS

The computer programs listed in this appendix are provided for the convenience of students and faculty to reduce the workload associated with solving some of the more numerically-challenging problems in the text. The source programs are written in Microsoft FORTRAN 5.1 and can be compiled and executed on any desktop computer with DOS as the operating system or on a Macintosh with Soft PC. Disks, which have the source listings and the executable elements, as well, are available with the solutions manual. To initiate execution of any one of the programs, just type the name of the program shown on line 1 of the source listing and key RETURN or ENTER. The program will query you for the data required for solution. In two cases (HARDY and HAMMER), an input data file created with the text editor will be required. Just respond to the prompt with the correct file name or data and press RETURN or ENTER.

The programs are not proprietary and may be copied without restriction. The authors assume no responsibility for the accuracy of the programs. Users of the programs do so at their own risk. However, the authors would appreciate hearing of any errors in the programs and any recommendations for improvement or additional programs which would be useful.

The programs available are as follows:

1. Program DARCY (Ref: Section 9.6 ff) This program uses Colebrook's equation to compute the Darcy-Weisbach friction factor for a given Reynolds number and relative roughness. The program requires the kinematic viscosity rather than the absolute viscosity and density.

2. Program HARDY (Ref: Section 9.11) This program is an elementary application of the Hardy Cross technique for solving simple network problems. It requires that the user create an input data file ahead of time before calling for program execution. An example of an input data file is listed after the program source listing.
3. Program HAMMER (Ref: Section 9.14) This program provides a water hammer analysis for the problem of a single constant diameter pipe conducting flow between two reservoirs with a valve at the downstream reservoir to control flow. The valve causes the water hammer by reducing the flow velocity uniformly from the steady state velocity to zero in a prescribed time. An example of an input data file is listed after the program source listing.
4. Program TRAP1 (Ref: Sections 10.6 and 10.7) This program helps the user to draw specific energy and force + momentum curves for rectangular and trapezoidal channels by generating the required data versus depth for the channel. The required input data are the shape, roughness and slope of the channel.
5. Program TRAP2 (Ref: Sections 10.6 and 10.7) This program provides the same results as TRAP1. The input data needed are the channel shape and flowrate.
6. Program QCURV1 (Ref: Section 10.6) This program helps the user to draw q-curves for rectangular channels by generating values of q versus depth for a given *flowrate* in a rectangular channel. The input data needed are the channel width, normal depth, and flowrate.
7. Program QCURV2 (Ref: Section 10.6) This program generates values of q versus depth for a given *specific energy* in a rectangular channel. The input data needed is the channel specific energy.
8. Program DIRSTEP (Ref: Section 10.10) This program along with the subroutines PROFIL and DEPTH calculates the water surface profile in a rectangular, trapezoidal or triangular channel using the direct step method. The input data required are the channel shape, slope, flowrate, roughness, starting depth, and the size and number of depth increments. The maximum distance for which calculations are to be carried is also required. This program is generally used when a profile over a specified *depth* range is required.
9. Program PRASAD (Ref: Section 10.10) This program calculates a water surface profile by integrating the varied flow equation for rectangular, trapezoidal, or triangular prismatic channels. It requires basically the same input data as DIRSTEP except a length increment is used instead of a depth increment. This program is generally used when a profile over a specified *length* range is required.

```
 1        PROGRAM DARCY
 2  ************************************************************************
 3  *   THIS PROGRAM USES COLEBROOK'S EQUATION TO COMPUTE THE DARCY-WEISBACH
 4  *   FRICTION FACTOR FOR A GIVEN REYNOLDS NUMBER AND RELATIVE ROUGHNESS
 5  ************************************************************************
 6  *
 7  *                       ** INPUT DATA DEFINITIONS **
 8  *                 PROGRAM ACCEPTS FREE-STYLE FORMAT INPUT DATA
 9  * V = PIPE VELOCITY - (FPS) OR (MPS)
10  * D = PIPE DIAMETER - (IN) OR (MM)
11  * NU = KINEMATIC VISCOSITY - (FT2/S) OR (M2/S)
12  * E = COMMERCIAL PIPE ROUGHNESS - (IN) OR (MM)
13  * G = ACCELERATION OF GRAVITY - (FT/S2) OR (M/S2)
14  *
15        REAL NU
16  *-----------------------------------------------------------------------
17        WRITE(*,100)
18        WRITE(*,101)
19        READ(*,*) V
20        WRITE(*,102)
21        READ(*,*) D
22        WRITE(*,103)
23        READ(*,*) NU
24        WRITE(*,104)
25        READ(*,*) E
26        WRITE(*,105)
27        READ(*,*) G
28  *-----------------------------------------------------------------------
29        ED=E/D
30        IF(G.GT.15.) R=V*D/(12.*NU)
31        IF(G.LT.15.) R=V*D/(1000.*NU)
32        F2=64./R
33        IF(R.LT.2100.) GO TO 20
34        F2=(1.0/(1.14-2.*LOG10(ED)))**2
35        IF((ED*R*SQRT(F2)).GT.200.) GO TO 20
36        F1=F2
37        DO 10 I=1,20
38        SQF=SQRT(F1)
39        BRACK=ED+9.28/(R*SQF)
40        FUNCT=1./SQF-1.14+2.*LOG10(BRACK)
41        FPRIME=-0.5/F1**1.5+(2.*9.28/(2.*R*F1**1.5))/BRACK
42        F2=F1-FUNCT/FPRIME
43        IF(ABS((F2-F1)/F2).LT.0.005) GO TO 20
44        F1=F2
45     10 CONTINUE
46        WRITE(*,200)
47        STOP
48     20 WRITE(*,201) R,ED,F2
49  *
```

```
50    100 FORMAT(//' ENTER THE APPROPRIATE NUMBER AFTER THE PROMPT AND PRESS
51       $ RETURN'/)
52    101 FORMAT(/' PIPE VELOCITY (FPS OR MPS) = 'ç)
53    102 FORMAT(/' PIPE DIAMETER (IN OR MM) = 'ç)
54    103 FORMAT(/' KINEMATIC VISCOSITY (FT/S2 OR M/S2) = 'ç)
55    104 FORMAT(/' PIPE ROUGHNESS (E-VALUE) (IN OR MM) = 'ç)
56    105 FORMAT(/' ACCELERATION OF GRAVITY (FT/S2 OR M/S2) = 'ç)
57    200 FORMAT(//' ITERATION HAS CYCLED 20 TIMES WITHOUT SUCCESS.'/
58       $' EXECUTION IS TERMINATED.')
59    201 FORMAT(//' REYNOLDS NUMBER =',E10.2/' RELATIVE ROUGHNESS =',
60       $F9.7/'     FRICTION FACTOR =',F6.4)
61        END
```

```
1         PROGRAM HARDY
2   ***********************************************************************
3   *           HARDY CROSS NETWORK ANALYSIS - SIMPLE VERSION
4   *           HEAD LOSS CALCULATED BY DARCY-WEISBACH EQUATION
5   *           USING EITHER SI OR US CUSTOMARY UNITS
6   ***********************************************************************
7   *
8   *               ** INPUT DATA DEFINITIONS **
9   *    PROGRAM ACCEPTS FREE-STYLE FORMAT INPUT DATA
10  *    PROGRAM IS PRESENTLY DIMENSIONED FOR 20 LOOPS AND 100 PIPES
11  *
12  * TITLE=JOB DESCRIPTION - ANY INFORMATION UP TO 80 COLUMNS MAXIMUM.
13  * NPIPES=NUMBER OF PIPES    NLOOPS=NUMBER OF LOOPS
14  * G=ACCELERATION OF GRAVITY - (FT/S2) OR (M/S2)
15  * ACC=ACCURACY OF ITERATION - (CFS) OR (CMS)
16  * NO()=PIPE NUMBER    D()=PIPE DIAMETER - (IN) OR (MM)
17  * L()=PIPE LENGTH - (FT) OR (M)  F()=PIPE DARCY-WEISBACH FRICTION FACTOR
18  * Q()=ESTIMATED PIPE FLOW RATE (WITH ALGEBRAIC SIGN) - (CFS) OR (CMS)
19  *    = (+) IF, WHEN TRAVELING CLOCKWISE AROUND THE PRIMARY LOOP, YOU ARE
20  *           GOING IN THE ASSUMED DIRECTION OF FLOW.
21  *    = (-) IF, WHEN TRAVELING CLOCKWISE AROUND THE PRIMARY LOOP, YOU ARE
22  *           OPPOSING THE ASSUMED DIRECTION OF FLOW.
23  * NP()=PRIMARY LOOP NUMBER    NS()=SECONDARY LOOP NUMBER IF THE PIPE
24  *                                   SHARES TWO LOOPS
25  *
26        DIMENSION Q(100),AK(100),F(100),L(100),D(100),HL(100)
27       $NS(100),NP(100),NO(100)
28        DIMENSION SN(20),SD(20),DEL(20)
29        REAL L
30        CHARACTER TITLE*80,USI*2,USL*2,USQ*3,SII*2,SIL*2,SIQ*3,DIMI*2,
31       $DIML*2,DIMQ*3,CH12
32        USI='IN'
33        USL='FT'
34        USQ='CFS'
35        SII='MM'
36        SIL='M '
37        SIQ='CMS'
38        DSCALE=12.
39  *
40  *----------------------------------------------------------------------
41        WRITE(*,799)
42        OPEN(5,FILE='        ')
43        READ(5,100) TITLE
44        READ(5,*) NPIPES,NLOOPS,G,ACC
45        READ(5,*) (NO(I),D(I),L(I),F(I),Q(I),NP(I),NS(I),I=1,NPIPES)
46        OPEN(6,FILE='        ',STATUS='NEW')
47  *----------------------------------------------------------------------
48  *
49        CH12=CHAR(12)
```

```
50          DIMI=USI
51          DIML=USL
52          DIMQ=USQ
53          IF(G.GT.15.) GO TO 1
54          DIMI=SII
55          DIML=SIL
56          DIMQ=SIQ
57          DSCALE=1000.
58        1 CONTINUE
59          WRITE(6,100) CH12
60          WRITE(6,101) TITLE
61          WRITE(6,102) DIMI,DIML,DIMQ
62          WRITE(6,103) (NO(I),D(I),L(I),F(I),Q(I),NP(I),NS(I),I=1,NPIPES)
63          IT=0
64          DO 3 I=1,NPIPES
65          AD=D(I)/DSCALE
66        3 AK(I)=F(I)*L(I)/(2.*G*(AD**5.)*.7854*.7854)
67        4 DO 2 I=1,NLOOPS
68          SN(I)=0.0
69        2 SD(I)=0.0
70          DO 8 I=1,NPIPES
71          QA=ABS(Q(I))
72          ANUM=AK(I)*Q(I)*QA
73          DENOM=2.AK(I)*QA
74          J=NP(I)
75          K=NS(I)
76          SN(J)=SN(J)+ANUM
77          SD(J)=SD(J)+DENOM
78          IF (K.LE.0) GO TO 8
79        6 SN(K)=SN(K)-ANUM
80          SD(K)=SD(K)+DENOM
81        8 CONTINUE
82          DO 20 I=1,NLOOPS
83          IF (SD(I).EQ.0.) GO TO 20
84       18 DEL(I)=SN(I)/SD(I)
85       20 CONTINUE
86          DO 25 I=1,NPIPES
87          J=NP(I)
88          K=NS(I)
89          IF (K.LE.0) GO TO 24
90       23 Q(I)=Q(I)+DEL(J)-DEL(K)
91          GO TO 25
92       24 Q(I)=Q(I)+DEL(J)
93       25 CONTINUE
94          IT=IT+1
95          IF (IT.GT.20) GO TO 50
96       28 DO 30 I=1,NLOOPS
97          IF (ABS(DEL(I)).GT.ACC) GO TO 4
98       30 CONTINUE
```

```
 99          DO 34 I=1,NPIPES
100       34 HL(I)=AK(I)*Q(I)*Q(I)
101          WRITE(6,100) CH12
102          WRITE(6,104) IT,ACC,DIMQ
103          WRITE(6,105) DIMQ,DIML
104          WRITE(6,106) (NO(I),AK(I),Q(I),HL(I),I=1,NPIPES)
105          GO TO 999
106       50 WRITE(6,109)
107   *
108      100 FORMAT(3A)
109      101 FORMAT(//5X,A)
110      102 FORMAT(  //' THE PIPE NETWORK IS COMPRISED OF THE FOLLOWING:' //
110         $' PIPE NO.',2X,' DIAM.—',A,2X,' LENGTH—',A,2X,' DARCY F',2X,
112         $' Q—',A,2X,' PRIMARY LOOP',2X,' SECOND. LOOP'/1X,8('—'),3X,
113         $8('—'),3X,9('—'),3X,7('—'),3X,5('—'),3X,12('—'),3X,12('—'))
114      103 FORMAT(I6,F13.1,F12.1,F10.4,F9.3,I10,I15)
115      104 FORMAT(///' THE RESULTS OF THE NETWORK ANALYSIS ARE:' ,//
116         $' NO. OF ITERATIONS IS',14,8X,'FLOW RATE ACCURATE TO',F7.5,1X,A)
117      105 FORMAT(///' PIPE NO.',7X,'    K', 8X,' Q—',A,6X,' HEAD LOSS—',A
118         $/1X,8('—'),7X,8('—'),7X,5('—'),7X,12('—'))
119      106 FORMAT(I5,E19.3,F12.3,F15.2/)
120      109 FORMAT(////' AFTER 20 ITERATIONS, ACCURACY IS STILL BELOW STANDARD
121         $')
122      799 FORMAT(//' YOU WILL BE REQUESTED TO SUPPLY THE NAMES OF YOUR'/
123         $' INPUT DATA FILE (UNIT 5) AND YOUR OUTPUT FILE (UNIT 6).'/
124         $' RESPOND TO THE PROMPT (?) WITH THE DISK DRIVE AND FILE NAME'/
125         $' - FOR EXAMPLE "DATA6.DAT" AND "JOB6.OUT"'//)
126      999 CONTINUE
127          END
```

Input Data File

```
DEMONSTRATION OF HARDY CROSS - ILLUSTRATIVE PROBLEM 9.21 - HCROSS.DAT


5   2      9.81     0.001
1   500     1000     .012   —0.5   1   0
2   400     1000     .012    0.5   1   0
3   400      100     .012    0.4   1   2
4   500     1000     .012   —0.9   2   0
5   300     1000     .013    0.1   2   0
```

```
 1          PROGRAM HAMMER
 2    ************************************************************************
 3    *    APPROXIMATE WATER HAMMER PROGRAM FOR A SINGLE PIPE WITH A RESERVOIR
 4    *    AT THE UPSTREAM END AND A VALVE AT THE DOWNSTREAM END. VELOCITY AT
 5    *    THE VALVE DECREASES LINEARLY WITH TIME FROM STEADY STATE TO ZERO.
 6    *    EITHER US CUSTOMARY OR SI UNITS ARE ACCEPTABLE ONLY WHEN USING THE
 7    *    DARCY-WEISBACH FORMULA.
 8    ************************************************************************
 9    *
10    *                    ** INPUT DATA DEFINITIONS **
11    *              PROGRAM ACCEPTS FREE-STYLE FORMAT INPUT DATA
12    *   TITLE = JOB DESCRIPTION. ANY INFORMATION UP TO 80 COLUMNS MAXIMUM
13    *   IOUT = PRINT OUTPUT INDEX. GIVES PRINTED OUTPUT EVERY IOUT-TH TIME
14    *          STEP. FOR EXAMPLE, IF IOUT=3 THEN OUTPUT IS PRINTED EVERY
15    *          THIRD TIME STEP.
16    *   NPARTS = NUMBER OF PIPE SEGMENTS INTO WHICH PIPE IS DIVIDED
17    *   D = PIPE DIAM - (IN) OR (MM)  L = PIPE LENGTH - FT OR (M)
18    *   F = DARCY-WEISBACH F-VALUE OR HAZEN-WILLIAMS C-VALUE (US CUSTOMARY
19    *       ONLY)
20    *   A = WAVE SPEED - (FPS) OR (MPS)
21    *   VZERO = INITIAL STEADY STATE VELOCITY - (FPS) OR (MPS)
22    *   HZERO = UPSTREAM RESERVOIR ELEVATION - (FT) OR (M)
23    *   ELEVUP = ELEVATION OF UPSTREAM END OF PIPE - (FT) OR (M)
24    *   ELEVDN = ELEVATION OF DOWNSTREAM END OF PIPE - (FT) OR (M)
25    *   TMAX = MAXIMUM REAL TIME OF SIMULATION - (SEC)
26    *   TCLOSE = TIME REQUIRED FOR VALVE CLOSURE - (SEC)
27    *   G = ACCELERATION OF GRAVITY - (FT/S2) OR (M/S2)
28    *
29          DIMENSION X°ALLOCATABLE§(:),V°ALLOCATABLE§(:),H°ALLOCATABLE§(:),
30         $HLOW°ALLOCATABLE§(:),HHIGH°ALLOCATABLE§(:),HEAD°ALLOCATABLE§(:)
31         $VNEW°ALLOCATABLE§(:),HNEW°ALLOCATABLE§(:),PIPEZ°ALLOCATABLE§(:),
32          REAL L,NEXP
33          CHARACTER TITLE*80,ROUGH*3,CH,CH12,CH78,CH79,CHN
34          CHARACTER USL*2,USI*2,USV*3,SIL*2,SII*2,SIV*3,DIML*2,DIMI*2,DIMV*3
35          NAMELIST /SPECS/ IOUT,NPARTS,D,L,F,A,VZERO,HZERO,ELEVUP,
36         $ELEVDN,TMAX,TCLOSE,G
37          USL='FT'
38          USI='IN'
39          USV='FPS'
40          SIL='M'
41          SII='MM'
42          SIV='MPS'
43    *
44    *-----------------------------------------------------------------------
45          WRITE(*,799)
46          OPEN(5,FILE='      ')
47          READ(5,100) TITLE
48          READ(5,SPECS)
49          OPEN(6,FILE='      ',STATUS='NEW')
50    *-----------------------------------------------------------------------
```

```
 51  *
 52        DIML=USL
 53        DIMI=USI
 54        DIMV=USV
 55        DIMD=12.
 56        IF(G.GT.15.) GO TO 10
 57        DIML=SIL
 58        DIMI=SII
 59        DIMV=SIV
 60        DIMD=1000.
        10 CONTINUE
 62        CH=CHAR(27)
 63        CH12=CHAR(12)
 67        CH78=CHAR(78)
 65        CH79=CHAR(79)
 66        CHN=CHAR(3)
 67        NP=NPARTS+1
 68        ALLOCATE (X(NP),V(NP),H(NP),HLOW(NP),HHIGH(NP),HEAD(NP),
 69       $VNEW(NP),HNEW(NP),PIPEZ(NP))
 70        WTT=L/A
 71        DELL=L/NPARTS
 72        T=0.
 73        NEXP=1.0
 74        ROUGH='F ='
 75        IF(F.GT.10.) NEXP=0.85
 76        IF(F.GT.10.) ROUGH='C ='
 77        DELT=DELL/A
 78        C=G/A
 79        INDEX=TMAX/DELT + 1
 80        DELEL=(ELEVDN-ELEVUP)/NPARTS
 81        NODES=NPARTS+1
 82        WRITE(6,101) CH,CH78,CHN
 83        WRITE(6,200)
 84        WRITE(6,203) TITLE
 85        WRITE(6,201) IOUT,NPARTS,L,DIML,A,DIMV,D,DIMI,ROUGH,F,VZERO,DIMV,
 86       $HZERO,DIML,ELEVUP,DIML,ELEVDN,DIML,WTT,TCLOSE,TMAX,DELT
 87        AK=DIMD*F*DELT/(2.0*D)
 88        IF(F.GT.10.) AK=DIMD*DELT*195./(2/0*D*(F**1.85)*(D/DIMD)**.17)
 89        IF(F.GT.10.) F=195./((F**1.85)*(VZERO**.15)*((D/DIMD)**.17))
 90        DELHF=DIMD*F*DELL*VZERO**2/(2.*G*D)
 91        DO 300 I=1,NODES
 92        V(I)=VZERO
 93        H(I)=HZERO-(I-1)*DELHF
 94        HLOW(I)=H(I)
 95        HHIGH(I)=H(I)
 96        X(I)=(I-1)*DELL/L
 97        PIPEZ(I)=ELEVUP+(I-1)*DELEL
 98        HEAD(I)=H(I)-PIPEZ(I)
 99    300 CONTINUE
100        WRITE(6,101) CH12
```

```
101         WRITE(6,202)
102         WRITE(6,204) DIML,DIML,DIMV,DIML,DIML,DIMV,T,
103        *(X(I),HEAD(I),H(I),V(I),I=1,NODES)
104         DO 99 II=1,INDEX
105         T=T+DELT
106   *  ** COMPUTE H AND V AT INTERIOR NODES **
107         DO 20 I=2,NPARTS
108         VNEW(I)=0.5*(V(I-1)+V(I+1)+C*(H(I-1)-H(I+1))-AK*(V(I-1)*ABS(V(I-1)
109        $)**NEXP+V(I+1)*ABS(V(I+1))**NEXP))
110      20 HNEW(I)=0.5*(H(I-1)+H(I+1)+(V(I-1)-V(I+1))/C-AK*(V(I-1)*ABS(V(I-1)
111        $)**NEXP-V(I+1)*ABS(V(I+1))**NEXP)/C)
112   **** COMPUTE H AND V AT UPSTREAM END **
113   *    -- THIS BOUNDARY CONDITION IS FOR A CONSTANT-HEAD RESERVOIR --
114         HNEW(1)=HZERO
115         VNEW(1)=V(2)+C*(HNEW(1)-H(2))-AK*V(2)*ABS(V(2))**NEXP
116   **** COMPUTE H AND V AT DOWNSTREAM END **
117          -- THIS BOUNDARY CONDITION IS FOR LINEARLY DECREASING VELOCITY --
118         IF(T.GT.TCLOSE) GO TO 30
119         VNEW(NODES)=VZERO*(1.-T/TCLOSE)
120         GO TO 31
121      30 VNEW(NODES)=0.0
122      31 HNEW(NODES)=H(NPARTS)+(V(NPARTS)-VNEW(NODES)-AK*V(NPARTS)*
123        $ABS(V(NPARTS))**NEXP)/C
124         DO 50 I=1,NODES
125         IF(HNEW(I).LT.HLOW(I)) HLOW(I)=HNEW(I)
126         IF(HNEW(I).GT.HHIGH(I)) HHIGH(I)=HNEW(I)
127      50 HEAD(I)=HNEW(I)-PIPEZ(I)
128         IF(MOD(II,IOUT).EQ.0) WRITE(6,204) DIML,DIML,DIMV,DIML,DIML,DIMV,
129        $T,(X(I),HEAD(I),HNEW(I),VNEW(I),I=1,NODES)
130         IF(T.GT.TMAX) GO TO 400
131         DO 40 I=1,NODES
132         V(I)=VNEW(I)
133      40 H(I)=HNEW(I)
134      99 CONTINUE
135     400 WRITE(6,101) CH12
136         WRITE(6,205)
137         DO 401 I=1,NODES
138         HEADMX=HHIGH(I)-PIPEZ(I)
139         HEADMN=HLOW(I)-PIPEZ(I)
140     401 WRITE(6,206) X(I),HEADMX,HEADMN,HHIGH(I),HLOW(I)
141         WRITE(6,101) CH,CH79
142   *
143     100 FORMAT(A)
144     101 FORMAT(3A)
145     200 FORMAT(///20X,33('*')/20X,'* WATER HAMMER IN A SIMPLE PIPE *'/
146        $20X,33('*'))
147     201 FORMAT(///29X,'INPUT DATA'/29X,10('-')//28X,'IOUT =',I3/26X,'NPART
148        $S =',I3//31X,'L =',F7.1,1X,A/31X,'A =',F7.1,1X,A/
149        $31X,'D =',F7.2,1X,A/31X,A,F7.4//
150        $27X,'VZERO =',F7.2,1X,A/27X,'HZERO =',F7.1,1X,A/
151        $26X,'ELEVUP =',F7.1,1X,A/26X,'ELEVDN =',F7.1,1X,A//
```

```
152          $29X,'L/A =',F7.3,' SEC'//26X,'TCLOSE =',F7.2,' SEC'/
153          $28X,'TMAX =',F7.2,' SEC'/28X,'DELT =',F7.3,' SEC')
154       202 FORMAT(///' PRESSURE HEADS, H—VALUES AND VELOCITIES AS FUNCTIONS
155          $OF TIME'/2X,60('—'))
156       203 FORMAT(//3X,A)
157       204 FORMAT(//18X,2(4X,'   X    HEAD—',A,2X,'H—',A,3X,'V—',A)/' TIME =',
158          $F7.3,' SEC',2(4X,'----- ------- ----- ------')/(18X,2(4X,F5.3,
159          $2F7.0,F8.2)))
160       205 FORMAT(////18X,27('*')/18X,'* TABLE OF EXTREME VALUES *'/18X,27('*
161          $')//13X,'X    MAX HEAD  MIN HEAD   MAX H    MIN H'/11X,5('—'),2X,8(
162          $'—'),2X,8('—'),2X,6('—'),2X,6('—'))
163       206 FORMAT(11X,F5.3,2X,F7.0,3X,F7.0,3X,F6.0,2X,F6.0)
164       799 FORMAT(//' YOU WILL BE REQUESTED TO SUPPLY THE NAMES OF YOUR'/
165          $' INPUT DATA FILE (UNIT 5) AND YOUR OUTPUT FILE (UNIT 6).'/
166          $' RESPOND TO THE PROMPT (?) WITH THE DISK DRIVE AND FILE NAME'/
167          $' -- FOR EXAMPLE "DATA6.DAT" AND "JOB6.OUT"'//)
168           END
```

Water Hammer Program Demonstration

A water line 18 500 ft long supplies water from a reservoir with a surface elevation of 4 218 ft to a lower reservoir with a surface elevation of 4 137 ft. The line enters and exits each reservoir 75 ft below the surface. The pipeline is constructed of T-30 transite (asbestos cement) pipe and has an inside diameter of 18.00 inches, an outside diameter of 19.70 inches, and a friction factor of 0.014.

A control valve at the entrance to the downstream reservoir can shut the flow off by reducing the flow linearly from its steady state value to zero in 20 s.

Assuming the pipe can be considered thin-walled, find the maximum and minimum pressure heads which will occur in the pipe and where they will occur. Use *NPARTS* = 5 in this instance and set *TMAX* to a value at least $6L/a$ sec beyond last valve movement.

Wave Speed Calculation

The wave speed is calculated from Eq. 9.74 using pipe properties from Table 4.

$$a = \frac{\sqrt{E/\rho}}{\sqrt{1 + \dfrac{E}{E_p}\dfrac{d}{e_p}(1 - \mu_p^2)}} = \frac{4\,860}{\sqrt{1 + \dfrac{318\,000}{3\,400\,000}\dfrac{18.00}{0.85}(1 - 0.3^2)}} = 2\,903 \text{ ft/s}$$

Steady State Hydraulics

The steady state velocity can be found by applying the work-energy equation between the two reservoirs. Neglecting local losses,

$$z_1 + \frac{p_1}{\gamma} + \frac{V_1^2}{2g_n} = z_2 + \frac{p_2}{\gamma} + \frac{V_2^2}{2g_n} + f\frac{l}{d}\frac{V^2}{2g_n}$$

$$4\,218 + 0 = 4\,137 + 0 + 0.014\frac{18\,500}{1.50}\frac{V^2}{2 \times 32.2}$$

$$V = 5.50 \text{ ft/s}$$

Input Data File

This data is incorporated into the input data file to give

```
DEMONSTRATION OF HAMMER - WHAMMER.DAT
&SPECS IOUT=1,NPARTS=5,D=18.0,L=18500.,F=0.014,A=2903.,VZERO=5.50,
       HZERO=4218.,ELEVUP=4143.,ELEVDN=4062.,TMAX=60.,TCLOSE=20.,G=32.2/
```

The results of the analysis give the maximum and minimum heads along the pipe as shown below.

```
                  ****************************
                  * TABLE OF EXTREME VALUES *
                  ****************************

      X     MAX HEAD   MIN HEAD    MAX H    MIN H
    -----   --------   --------   ------   ------
     .000      75.        75.      4218.    4218.
     .200     156.        29.      4282.    4156.
     .400     232.       -12.      4342.    4099.
     .600     302.       -14.      4396.    4080.
     .800     366.       -12.      4444.    4066.
    1.000     423.        -6.      4485.    4056.
```

The Table of Extreme Values shows that the maximum head occurs at the valve (X = 1.00) and the minimum head occurs 11 000 ft downstream of the upper reservoir (X = 0.600).

```
1        PROGRAM TRAP1
2  ***********************************************************************
3  *    THIS PROGRAM GENERATES VALUES OF SPECIFIC ENERGY AND FORCE+MOMENTUM
4  *    VS DEPTH FOR A GIVEN DEPTH, SHAPE, ROUGHNESS AND SLOPE OF A
5  *    RECTANGULAR OR TRAPEZOIDAL CHANNEL USING EITHER SI OR US CUSTOMARY
6  *    UNITS.
7  ***********************************************************************
8  *
9  *                 ** INPUT DATA DEFINITIONS **
10 * PROGRAM ACCEPTS FREE-STYLE FORMAT INPUT DATA
11 * YZERO=NORMAL DEPTH - (FT) OR (M)   B=CHANNEL WIDTH - (FT) OR (M)
12 * N=MANNING N-VALUE   Z=SIDE SLOPE   SZERO=CHANNEL BOTTOM SLOPE
13 * YB=STARTING DEPTH - (FT) OR (M)   DELY=DEPTH INCREMENT - (FT) OR (M)
14 * YE=ENDING DEPTH - (FT) OR (M)
15 * G=ACCELERATION OF GRAVITY - (FT/S2) OR (M/S2)
16 *
17       REAL M,N
18       CHARACTER USL*2,USQ*3,SIL*2,SIQ*3,DIML*2,DIMQ*3,CH12
19       USL='FT'
20       USQ='CFS'
21       SIL='M '
22       SIQ='CMS'
23       CH12=CHAR(12)
24 *-----------------------------------------------------------------------
25       WRITE(*,105)
26       WRITE(*,107
27       READ(*,*) YZERO
28       WRITE(*,108)
29       READ(*,*) B
30       WRITE(*,109)
31       READ(*,*) N
32       WRITE(*,110)
33       READ(*,*) Z
34       WRITE(*,111)
35       READ(*,*) SZERO
36       WRITE(*,112)
37       READ(*,*) YB
38       WRITE(*,113)
39       READ(*,*) DELY
40       WRITE(*,114)
41       READ(*,*) YE
42       WRITE(*,115)
43       READ(*,*) G
44       WRITE(*,106)
45       OPEN(6,FILE='      ',STATUS='NEW')
46 *-----------------------------------------------------------------------
47       DIML=USL
48       DIMQ=USQ
49       U=1.49
```

```
50          IF(G.LT.15.) DIML=SIL
51          IF(G.LT.15.) DIMQ=SIQ
52          IF(G.LT.15.) U=1.00
53          C=U*SQRT(SZERO)/N
54          Q=C*(B*YZERO+Z*YZERO*YZERO)**1.6667/(B+2.*YZERO*SQRT(1.+Z*Z))**0.6
55         $667
56          WRITE(6,100) CH12
57          WRITE(6,101) YZERO,DIML,Q,DIMQ,B,DIML,Z,N,SZERO
58          NN=(YE-YB)/DELY+1
59          IF(Z.GT.0.) GO TO 5
60          YC=(Q*Q/(B*B*G))**0.3333
61          WRITE(6,104) YC,DIML
62        5 CONTINUE
63          WRITE(6,103) DIML,DIML
64          Y=YB
65          DO 10 I=1,NN
66          A=B*Y+Z*Y*Y
67          P=B+2.*Y*SQRT(1.+Z*Z)
68          E=Y+Q*Q/(2.*G*A*A)
69          YBARA=Y*Y*(0.5*B+Z*Y/3.)
70          Q2A=Q*Q/(G*A)
71          M=YBARA+Q2A
72          IF(Z.LE.0) M=M/B
73          WRITE(6,102) Y,E,M
74       10 Y=Y+DELY
75    *
76      100 FORMAT(A)
77      101 FORMAT(' SPECIFIC ENERGY AND FORCE+MOMENTUM FOR A TRAPEZOIDAL CHAN
78         $NEL'//6X,'NORMAL DEPTH =',F6.2,1X,A/10X,'FLOWRATE =',F10.3,1X,A
79         $/6X,'BOTTOM WIDTH =',F7.2,1X,A/8X,'SIDE SLOPE =',F5.2/2X,'MANNING
80         $ N-VALUE =',F5.3/5X,'CHANNEL SLOPE =',F8.6/)
81      102 FORMAT(4X,F8.2,7X,F9.2,7X,F9.2
82      103 FORMAT(5X,'DEPTH-',A,2X,'SPECIFIC ENERGY-',A,2X,'FORCE+MOMENTUM'
83         $/5X,8('-'),2X,18('-'),2X,14('-'))
84      104 FORMAT(4X,'CRITICAL DEPTH =',F7.3,1X,A//)
85      105 FORMAT(//' ENTER THE APPROPRIATE NUMBER AFTER THE PROMPT AND PRESS
86         $ RETURN'/)
87      106 FORMAT(//' YOU ARE REQUIRED TO NAME AN OUTPUT FILE TO WHICH THE RE
88         $SULTS OF'/' THE COMPUTATION ARE SENT FOR YOUR REVIEW AND/OR PRINTI
89         $NG'//)
90      107 FORMAT(/' NORMAL DEPTH (FT OR M) = 'ç)
91      108 FORMAT(/' BOTTOM WIDTH (FT OR M) = 'ç)
92      109 FORMAT(/' MANNING N-VALUE = 'ç)
93      110 FORMAT(/' SIDE SLOPE Z = 'ç)
94      111 FORMAT(/' CHANNEL SLOPE = 'ç)
95      112 FORMAT(/' STARTING DEPTH (FT OR M) = 'ç)
96      113 FORMAT(/' DEPTH INCREMENT (FT OR M) = 'ç)
97      114 FORMAT(/' ENDING DEPTH (FT OR M) = 'ç)
98      115 FORMAT(/' ACCELERATION OF GRAVITY (FT/S2 OR M/S2) = 'ç)
99          END
```

```
1          PROGRAM TRAP2
2   ************************************************************************
3   *    THIS PROGRAM GENERATES VALUES OF SPECIFIC ENERGY AND FORCE+MOMENTUM
4   *    VS DEPTH FOR A GIVEN FLOWRATE AND SHAPE OF A RECTANGULAR OR
5   *    TRAPEZOIDAL CHANNEL USING EITHER SI OR US CUSTOMARY UNITS.
6   ************************************************************************
7   *
8   *                  ** INPUT DATA DEFINITIONS **
9   * PROGRAM ACCEPTS FREE-STYLE FORMAT INPUT DATA
10  * Q=FLOWRATE - (CFS) OR (CMS)  B=CHANNEL WIDTH - (FT) OR (M)
11  * YB=STARTING DEPTH - (FT) OR (M)   DELY=DEPTH INCREMENT - (FT) OR (M)
12  * YE=ENDING DEPTH - (FT) OR (M)   Z=SIDE SLOPE
13  * G=ACCELERATION OF GRAVITY - (FT/S2) OF (M/S2)
14  *
15         REAL M
16         CHARACTER USL*2,USQ*3,SIL*2,SIQ*3,DIML*2,DIMQ*3,CH12
17         USL='FT'
18         USQ='CFS'
19         SIL='M '
20         SIQ='CMS'
21         CH12=CHAR(12)
22  *-----------------------------------------------------------------------
23         WRITE(*,105)
24         WRITE(*,107)
25         READ(*,*) Q
26         WRITE(*,108)
27         READ(*,*) B
28         WRITE(*,109)
29         READ(*,*) Z
30         WRITE(*,110)
31         READ(*,*) YB
32         WRITE(*,111)
33         READ(*,*) DELY
34         WRITE(*,112)
35         READ(*,*) YE
36         WRITE(*,113)
37         READ(*,*) G
38         WRITE(*,106)
39         OPEN (6,FILE='        ',STATUS='NEW')
40  *-----------------------------------------------------------------------
41         DIML=USL
42         DIMQ=USQ
43         IF(G.LT.15.) DIML=SIL
44         IF(G.LT.15.) DIMQ=SIQ
45         WRITE(6,100) CH12
46         WRITE(6,101) Q,DIMQ,B,DIML,Z
47         NN=(YE-YB)/DELY+1
48         IF(Z.GT.0.) GO TO 5
49         YC=(Q*Q/(B*B*G))**0.3333
```

```
50        WRITE(6,104) YC,DIML
51      5 CONTINUE
52        WRITE(6,103) DIML,DIML
53        Y=YB
54        DO 10 I=1,NN
55        A=B*Y+Z*Y*Y
56        P=B+2.*Y*SQRT(1.+Z*Z)
57        E=Y+Q*Q/(2.*G*A*A)
58        YBARA=Y*Y*(0.5*B+Z*Y/3.)
59        Q2A=Q*Q/(G*A)
60        M=YBARA+Q2A
61        IF(Z.LE.0.) M=M/B
62        WRITE(6,102) Y,E,M
63     10 Y=Y+DELY
64  *
65    100 FORMAT(A)
66    101 FORMAT(' SPECIFIC ENERGY AND FORCE+MOMENTUM FOR THE CHANNEL'//
67       $10X,'FLOWRATE =',F10.3,1X,A/6X,'BOTTOM WIDTH =',F7.2,1X,A/8X,
68       $'SIDE SLOPE =',F5.2/)
69    102 FORMAT(4X,F8.2,7X,F9.2,7X,F9.2)
70    103 FORMAT(5X,'DEPTH—',A,2X,'SPECIFIC ENERGY—',A,2X,'FORCE+MOMENTUM'
71       $/5X,8('—'),2X,18('—'),2X,14('—'))
72    104 FORMAT(4X,'CRITICAL DEPTH =',F10.3,1X,A//)
73    105 FORMAT(//' ENTER THE APPROPRIATE NUMBER AFTER THE PROMPT AND PRESS
74       $ RETURN'/)
75    106 FORMAT(//' YOU ARE REQUIRED TO NAME AN OUTPUT FILE TO WHICH THE RE
76       $SULTS OF'/' THE COMPUTATION ARE SENT FOR YOUR REVIEW AND/OR PRINTI
77       $NG'//)
78    107 FORMAT(/' FLOWRATE (CFS OR CMS) = 'ç)
79    108 FORMAT(/' BOTTOM WIDTH (FT OR M) = 'ç)
80    109 FORMAT(/' SIDE SLOPE Z = 'ç)
81    110 FORMAT(/' STARTING DEPTH (FT OR M) = 'ç)
82    111 FORMAT(/' DEPTH INCREMENT (FT OR M) = 'ç)
83    112 FORMAT(/' ENDING DEPTH (FT OR M) = 'ç)
84    113 FORMAT(/' ACCELERATION OF GRAVITY (FT/S2 OR M/S2) = 'ç)
85        END
```

```
 1        PROGRAM QCURV1
 2  ************************************************************************
 3  *    THIS PROGRAM GENERATES VALUES OF q VS DEPTH FOR A GIVEN FLOWRATE IN
 4  *    A RECTANGULAR CHANNEL USING EITHER SI OR US CUSTOMARY UNITS.
 5  ************************************************************************
 6  *
 7  *                 ** INPUT DATA DEFINITIONS **
 8  * PROGRAM ACCEPTS FREE-STYLE FORMAT INPUT DATA
 9  * Q=FLOWRATE - (CFS) OR (CMS)  B=CHANNEL WIDTH - (FT) OR (M)
10  * YB=STARTING DEPTH - (FT) OR (M)   DELY=DEPTH INCREMENT - (FT) OR (M)
11  * YE=ENDING DEPTH - (FT) OR (M)  YZERO=NORMAL DEPTH - (FT) OR (M)
12  * G=ACCELERATION OF GRAVITY - (FT/S2) OF (M/S2)
13  *
14        CHARACTER USL*2,USQ*3,SIL*2,SIQ*3,DIML*2,DIMQ*3,CH12
15        USL='FT'
16        USQ='CFS'
17        SIL='M '
18        SIQ='CMS'
19        CH12=CHAR(12)
20  *-----------------------------------------------------------------------
21        WRITE(*,105)
22        WRITE(*,107)
23        READ(*,*) Q
24        WRITE(*,108)
25        READ(*,*) B
26        WRITE(*,109)
27        READ(*,*) YB
28        WRITE(*,110)
29        READ(*,*) DELY
30        WRITE(*,111)
31        READ(*,*) YE
32        WRITE(*,112)
33        READ(*,*) YZERO
34        WRITE(*,113)
35        READ(*,*) G
36        WRITE(*,106)
37        OPEN (6,FILE='       ',STATUS='NEW')
38  *-----------------------------------------------------------------------
39        DIML=USL
40        DIMQ=USQ
41        IF(G.LT.15.) DIML=SIL
42        IF(G.LT.15.) DIMQ=SIQ
43        A=B*YZERO
44        EZERO=YZERO+Q*Q/(2.*G*A*A)
45        WRITE(6,100) CH12
46        WRITE(6,101) Q,DIMQ,B,DIML,YZERO,DIML,EZERO,DIML
47        N=(YE-YB)/DELY+1
48        YC=(2./3.)*EZERO
49        QMAX=SQRT(G*YC**3)
```

```
50          WRITE(6,104) YC,DIML,QMAX,DIMQ,DIML
51          WRITE(6,103) DIML,DIMQ,DIML
52          Y=YB
53          DO 10 I=1,N
54          CHEK=EZERO-Y
55          IF(CHEK.LE.0.) GO TO 11
56          QQ=SQRT(2.*G*Y*Y*CHEK)
57          WRITE(6,102) Y,QQ
58       10 Y=Y+DELY
59       11 CONTINUE
60    *
61      100 FORMAT(A)
62      101 FORMAT(' q VS DEPTH FOR A RECTANGULAR CHANNEL'
63         $//10X,'FLOWRATE =',F10.3,1X,A/6X,'BOTTOM WIDTH =',F7.2,1X,A/6X,
64         $'NORMAL DEPTH =',F6.2,1X,A/12X,'E-ZERO =',F6.2,1X,A)
65      102 FORMAT(4X,F8.2,4X,F8.2)
66      103 FORMAT(5X,'DEPTH-',A,2X,'q-VALUE-',A,'/',A/5X,8('-'),2X,14('-')
67         $)
68      104 FORMAT(4X,'CRITICAL DEPTH =',F7.3,1X,A/9X,'MAXIMUM q =',F7.3,1X,
69         $A,'/',A//)
70      105 FORMAT(//' ENTER THE APPROPRIATE NUMBER AFTER THE PROMPT AND PRESS
71         $ RETURN'/)
72      106 FORMAT(//' YOU ARE REQUIRED TO NAME AN OUTPUT FILE TO WHICH THE RE
73         $SULTS OF'/' THE COMPUTATION ARE SENT FOR YOUR REVIEW AND/OR PRINTI
74         $NG'//)
75      107 FORMAT(/' FLOWRATE (CFS OR CMS) = 'ç)
76      108 FORMAT(/' BOTTOM WIDTH (FT OR M) = 'ç)
77      109 FORMAT(/' STARTING DEPTH (FT OR M) = 'ç)
78      110 FORMAT(/' DEPTH INCREMENT (FT OR M) = 'ç)
79      111 FORMAT(/' ENDING DEPTH (FT OR M) = 'ç)
80      112 FORMAT(/' NORMAL DEPTH (FT OR M) = 'ç)
81      113 FORMAT(/' ACCELERATION OF GRAVITY (FT/S2 OR M/S2) = 'ç)
82          END
```

```
1        PROGRAM QCURV2
2  ************************************************************************
3  *    THIS PROGRAM GENERATES VALUES OF q VS DEPTH FOR A GIVEN SPECIFIC
4  *    ENERGY IN A RECTANGULAR CHANNEL USING EITHER SI OR US CUSTOMARY UNITS.
5  ************************************************************************
6  *
7  *                ** INPUT DATA DEFINITIONS **
8  * PROGRAM ACCEPTS FREE-STYLE FORMAT INPUT DATA
9  * EZERO=SPECIFIC ENERGY - (FT) OR (M)
10 * YB=STARTING DEPTH - (FT) OR (M)   DELY=DEPTH INCREMENT - (FT) OR (M)
11 * YE=ENDING DEPTH - (FT) OR (M)
12 * G=ACCELERATION OF GRAVITY - (FT/S2) OF (M/S2)
13 *
14       CHARACTER USL*2,USQ*3,SIL*2,SIQ*3,DIML*2,DIMQ*3,CH12
15       USL='FT'
16       USQ='CFS'
17       SIL='M '
18       SIQ='CMS'
19       CH12=CHAR(12)
20 *-----------------------------------------------------------------------
21       WRITE(*,105)
22       WRITE(*,107)
23       READ(*,*) EZERO
24       WRITE(*,108)
25       READ(*,*) YB
26       WRITE(*,109)
27       READ(*,*) DELY
28       WRITE(*,110)
29       READ(*,*) YE
30       WRITE(*,111)
31       READ(*,*) G
32       WRITE(*,106)
33       OPEN (6,FILE='       ',STATUS='NEW')
34 *-----------------------------------------------------------------------
35       DIML=USL
36       DIMQ=USQ
37       IF(G.LT.15.) DIML=SIL
38       IF(G.LT.15.) DIMQ=SIQ
39       WRITE(6,100) CH12
40       WRITE(6,101) EZERO,DIML
41       N=(YE-YB)/DELY+1
42       YC=(2./3.)*EZERO
43       QMAX=SQRT(G*YC**3)
44       WRITE(6,104) YC,DIML,QMAX,DIMQ,DIML
45       WRITE(6,103) DIML,DIMQ,DIML
46       Y=YB
47       DO 10 I=1,N
48       CHEK=EZERO-Y
49       IF(CHEK.LT.0.) GO TO 11
```

```
50        QQ=SQRT(2.*G*Y*Y*CHEK)
51        WRITE(6,102) Y,QQ
52     10 Y=Y+DELY
53     11 CONTINUE
54  *
55    100 FORMAT(A)
56    101 FORMAT(' q VS DEPTH FOR A RECTANGULAR CHANNEL'
57       $//12X,'E-ZERO =',F7.3,1X,A//)
58    102 FORMAT(4X,F8.2,4X,F8.2)
59    103 FORMAT(5X,'DEPTH-',A,2X,'q-VALUE-',A,'/',A/5X,8('-'),2X,14('-'))
60    104 FORMAT(4X,'CRITICAL DEPTH =',F7.3,1X,A/9X,'MAXIMUM q =',F7.3,1X,
61       $A,'/',A//)
62    105 FORMAT(//' ENTER THE APPROPRIATE NUMBER AFTER THE PROMPT AND PRESS
63       $ RETURN'/)
64    106 FORMAT(//' YOU ARE REQUIRED TO NAME AN OUTPUT FILE TO WHICH THE RE
65       $SULTS OF'/' THE COMPUTATION ARE SENT FOR YOUR REVIEW AND/OR PRINTI
66       $NG'//)
67    107 FORMAT(/' E-ZERO VALUE (FT OR M) = 'ç)
68    108 FORMAT(/' STARTING DEPTH (FT OR M) = 'ç)
69    109 FORMAT(/' DEPTH INCREMENT (FT OR M) = 'ç)
70    110 FORMAT(/' ENDING DEPTH (FT OR M) = 'ç)
71    111 FORMAT(/' ACCELERATION OF GRAVITY (FT/S2 OR M/S2) = 'ç)
72        END
```

```
1          PROGRAM DIRSTEP
2   ***********************************************************************
3   *      THIS PROGRAM, ALONG WITH THE SUBROUTINES 'PROFIL' AND 'DEPTH',
4   *      CALCULATES THE WATER SURFACE PROFILE IN A TRAPEZOIDAL CHANNEL
5   *      USING THE DIRECT STEP METHOD FOR EITHER US CUSTOMARY OR SI UNITS.
6   ***********************************************************************
7   *
8   *                  ** INPUT DATA DEFINITIONS **
9   *      PROGRAM ACCEPTS FREE-STYLE FORMAT INPUT DATA
10  * Q=CHANNEL DISCHARGE - (CFS) OR (CMS)   N=MANNING ROUGHNESS COEFFICIENT
11  * B=BOTTOM WIDTH - (FT) OR (M)   Z=SIDE SLOPE   SZERO=BOTTOM SLOPE
12  * U=UNITS INDICATOR - U=1.49 FOR US CUSTOMARY UNITS OR 1.00 FOR SI UNITS
13  * LMAX=MAXIMUM DISTANCE TO CONTINUE CALCULATIONS - (FT) OR (M)
14  * YSTART=INITIAL DEPTH AT WHICH CALCULATIONS ARE BEGUN - (FT) OR (M)
15  * DELY=DEPTH INCREMENT FOR COMPUTATIONS - (+) FOR INCREASING DEPTH AND
16  *       (-) FOR DECREASING DEPTH - (FT) OR (M)
17  * DIR=DIRECTION OF COMPUTATION INDEX - DIR=1 TO COMPUTE IN THE UPSTREAM
18  *      DIRECTION AND DIR=0 TO COMPUTE IN THE DOWNSTREAM DIRECTION
19  * STEPS=NUMBER OF DEPTH INCREMENTS TO CARRY OUT
20  *
21         INTEGER STEPS,DIR
22         REAL N,LMAX
23  *-----------------------------------------------------------------------
24         WRITE(*,100)
25         WRITE(*,102)
26         READ(*,*) Q
27         WRITE(*,103)
28         READ(*,*) N
29         WRITE(*,104)
30         READ(*,*) B
31         WRITE(*,105)
32         READ(*,*) Z
33         WRITE(*,106)
34         READ(*,*) SZERO
35         WRITE(*,107)
36         READ(*,*) U
37         WRITE(*,108)
38         READ(*,*) LMAX
39         WRITE(*,109)
40         READ(*,*) YSTART
41         WRITE(*,110)
42         READ(*,*) DELY
43         WRITE(*,111)
44         READ(*,*) DIR
45         WRITE(*,112)
46         READ(*,*) STEPS
47         WRITE(*,101)
48         OPEN (6,FILE='        ',STATUS='NEW')
49  *-----------------------------------------------------------------------
```

```
50          AA=Q*N/(U*SZERO**0.50)
51          AA=(AA/(2.*SQRT(1.+Z*Z)-Z))*2.**0.6667
52          Y1=AA**(3./8.)
53          CALL DEPTH(Q,N,Y1,B,Z,SZERO,U,YZERO,YC)
54          CALL PROFIL(Q,B,Z,N,SZERO,U,YSTART,DELY,DIR,LMAX,YZERO,YC,STEPS
55    *
56      100 FORMAT(//' ENTER THE APPROPRIATE NUMBER AFTER THE PROMPT AND PRESS
57         $ RETURN'/)
58      101 FORMAT(//' YOU ARE REQUIRED TO NAME AN OUTPUT FILE TO WHICH THE RE
59         $SULTS OF'/' THE COMPUTATION ARE SENT FOR YOUR REVIEW AND/OR PRINTI
60         $NG'//)
61      102 FORMAT(/' FLOWRATE (CFS OR CMS) = 'ç)
62      103 FORMAT(/' MANNING N-VALUE = 'ç)
63      104 FORMAT(/' BOTTOM WIDTH (FT OR M) = 'ç)
64      105 FORMAT(/' SIDE SLOPE Z = 'ç)
65      106 FORMAT(/' CHANNEL SLOPE = 'ç)
66      107 FORMAT(/' U-FACTOR (1.49 FOR US CUSTOMARY OR 1.00 FOR SI) = 'ç)
67      108 FORMAT(/' MAX. CHANNEL LENGTH OVER WHICH TO MAK CALCULATIONS (FT
68         $OR M) = 'ç)
69      109 FORMAT(/' STARTING DEPTH (FT OR M) = 'ç)
70      110 FORMAT(/' DEPTH INCREMENT (FT OR M) = +/- 'ç)
71      111 FORMAT(/' DIRECTION OF COMPUTATION INDEX (1=UPSTREAM 0=DOWNSTREAM)
72         $ = 'ç)
73      112 FORMAT(/' NUMBER OF DEPTH INCREMENTS = 'ç)
74          END

 1          SUBROUTINE DEPTH(Q,N,Y1,B,Z,SZERO,U,YZERO,YC)
 2          REAL N
 3          G=32.2
 4          IF(U.LE.1.0) G=9.81
 5          AA=Q*N/(U*SZERO**.5)
 6          DO 10 I=1,20
 7          A=B*Y1+Z*Y1*Y1
 8          P=B+2.*Y1*SQRT(1.+Z*Z)
 9          R=A/P
10          RR=R**0.66667
11          F=A*RR-AA
12          APRIME=B+2.*Z*Y1
13          PPRIME=2.*SQRT(1.+Z*Z)
14          RPRIME=(P*APRIME-A*PPRIME)/(P*P)
15          FPRIME=A*0.6667*RPRIME/(R**0.3333)+RR*APRIME
16          Y2=Y1-F/FPRIME
17          IF(ABS((Y2-Y1)/Y2).LT.0.001) GO TO 11
18          Y1=Y2
19       10 CONTINUE
20          WRITE(*,201)
21          STOP
22       11 YZERO=Y2
23          Y1=0.10*YZERO
```

```
24          AA=Q*Q/G
25          DO 20 I=1,20
26          BB=B+2.*Z*Y1
27          A=B*Y1+Z*Y1*Y1
28          F=AA*(BB/A**3)-1.
29          FPRIME=AA*((-3.*BB**2+2.*Z*A)/A**4)
30          Y2=Y1-F/FPRIME
31          IF(ABS((Y2-Y1)/Y2).LT.0.001) GO TO 21
32          Y1=Y2
33       20 CONTINUE
34          WRITE(*,202)
35          STOP
36       21 YC=Y2
37    *
38      201 FORMAT(///' AFTER 20 ITERATIONS, THE TRIAL SOLUTION FOR NORMAL DEP
39         $TH IS STILL NOT'/' WITHIN 0.005 FT OR M.'/' EXECUTION IS TERMINATE
40         $D.')
41      202 FORMAT(' AFTER 20 ITERATIONS, THE TRIAL SOLUTION FOR CRITICAL DEPT
42         $H IS STILL NOT'/' WITHIN 0.005 FT OR M.'/' EXECUTION IS TERMINATED
43         $.')
44          RETURN
45          END

 1          SUBROUTINE PROFIL(Q,B,Z,N,SZERO,U,YSTART,DELY,DIRECT,LMAX,YZERO,
 2         $YC,STEPS)
 3          REAL N,L,LMAX
 4          INTEGER DIRECT, STEPS
 5          CHARACTER USL*2,USQ*3,USV*3,SIL*2,SIQ*3,SIV*3,DIML*2,DIMQ*3,DIMV*3
 6         $,CH12
 7          DIMENSION Y(50),A(50),P(50),R(50),V(50),E(50),VH(50)
 8          USL='FT'
 9          USQ='CFS'
10          USV='FPS'
11          SIL='M '
12          SIQ='CMS'
13          SIV='MPS'
14          DIML=USL
15          DIMQ=USQ
16          DIMV=USV
17          G=32.2
18          CH12=CHAR(12)
19          IF(U.NE.1.0) GO TO 1
20          DIML=SIL
21          DIMQ=SIQ
22          DIMV=SIV
23          G=9.81
24        1 WRITE(6,200) CH12
25          WRITE(6,201)
26          WRITE(6,202) Q,DIMQ,B,DIML,Z,N,SZERO,YZERO,DIML,YC,DIML
```

```
27          WRITE(6,203) DIML,DIML,DIML,DIML,DIML,DIML
28          L=0.
29          CHEK=YSTART-YZERO
30          Y(1)=YSTART
31          A(1)=B*Y(1)+Z*Y(1)*Y(1)
32          P(1)=B+2.*Y(1)*SQRT(1.+Z*Z)
33          R(1)=A(1)/P(1)
34          V(1)=Q/A(1)
35          VH(1)=V(1)*V(1)/(2.*G)
36          E(1)=Y(1)+VH(1)
37          FROUDE=SQRT(Q*Q*(B+2.*Z*YSTART)/(G*A(1)**3))
38          WRITE(6,204) Y(1),VH(1),E(1),L
39          DO 10 I=1,STEPS
40          Y(I+1)=Y(I)+DELY
41          IF((Y(I+1)-YZERO)*CHEK.LT.0.) GO TO 15
42          A(I+1)=B*Y(I+1)+Z*Y(I+1)**2
43          P(I+1)=B+2.*Y(I+1)*SQRT(1.+Z*Z)
44          F=SQRT(Q*Q*(B+2.*Z*Y(I+1))/(G*A(I+1)**3))
45          IF(F.GT.1.0.AND.FROUDE.LT.1.0) GO TO 14
46          IF(F.LT.1.0.AND.FROUDE.GT.1.0) GO TO 14
47          R(I+1)=A(I+1)/P(I+1)
48          V(I+1)=Q/A(I+1)
49          VH(I+1)=V(I+1)*V(I+1)/(2.*G)
50          E(I+1)=Y(I+1)+VH(I+1)
51          IF(DIRECT.EQ.1) GO TO 6
52          DELE=E(I+1)-E(I)
53          GO TO 7
54        6 DELE=E(I)-E(I+1)
55        7 VBAR=0.5*(V(I)+V(I+1))
56          RBAR=0.5*(R(I)+R(I+1))
57          REXP=RBAR**0.66667
58          SBAR=(VBAR*N/(U*REXP))**2
59          DELS=SZERO-SBAR
60          DELX=DELE/DELS
61          L=L+DELX
62          WRITE(6,205) DELE,SBAR,DELS, DELX
63          WRITE(6,204) Y(I+1),VH(I+1),E(I+1),L
64          Y1=Y(I)
65          Y2=Y(I+1)
66          IF(ABS(L).GT.LMAX) GO TO 13
67       10 CONTINUE
68          WRITE(6,206)
69          RETURN
70       15 WRITE(6,207)
71          RETURN
72       14 WRITE(6,208)
73          RETURN
74       13 WRITE(6,209)
75          RETURN
```

```
76   *
77      200 FORMAT(A)
78      201 FORMAT(/10X,'*** TABLE OF NONUNIFORM FLOW CALCULATIONS ***'//)
79      202 FORMAT(5X,'Q=',F9.2,1X,A,3X,'B =',F6.1,1X,A,3X,'Z =',F5.2,3X,
80         $'N =',F6.3/1X,'SZERO =',F8.6,4X.'YZERO =',F6.2,1X,A,2X,'YC =',
81         $F6.2,1X,A//)
82      203 FORMAT(' DEPTH-',A,' V-HEAD-',A,3X,'E-,A,3X,'DELE-',A,3X,
83         $'SBAR',5X,'DELS',4X,'DELX-',A,2X,'DISTANCE-',A/1X,8('-'),1X
84         $9('-'),2X,6('-'),2X,7('-'),1X,8('-'),1X,8('-'),2X,7('-'),2X,
85         $11('-'))
86      204 FORMAT(1X,F7.2,3X,F6.2,3X,F6.2,39X,F9.0)
87      205 FORMAT(29X,F7.3,1X,F8.6,1X,F8.6,1X,F8.1)
88      206 FORMAT(///' INCREMENTS IN DEPTH HAVE BEEN COMPLETED WITHOUT EXCEED
89         $ING THE "LMAX" VALUE')
90      207 FORMAT(///' THE CALCULATION HAS CROSSED NORMAL DEPTH')
91      208 FORMAT(///' THE CALCULATION HAS CROSSED CRITICAL DEPTH')
92      209 FORMAT(///' THE "LMAX" DISTANCE HAS BEEN EXCEEDED WITHOUT COMPLETI
93         $NG THE'/' NUMBER OF STEPS SPECIFIED')
94          END
```

```
 1         PROGRAM PRASAD
 2   ************************************************************************
 3   * THIS PROGRAM INTEGRATES THE VARIED FLOW EQUATION FOR PRISMATIC
 4   * CHANNELS OF RECTANGULAR, TRAPEZOIDAL, AND TRIANGLUAR CROSS-SECTION
 5   * USING EITHER SI OR US CUSTOMARY UNITS.
 6   ************************************************************************
 7   *
 8   *                 ** INPUT DATA DEFINITIONS **
 9   * PROGRAM ACCEPTS FREE-STYLE FORMAT INPUT DATA
10   * U = UNIT INDICATOR    U = 1.0 FOR SI    U = 1.49 FOR US
11   * L = TOTAL LENGTH OF CHANNEL - (FT) OR (M)
12   * SZERO = CHANNEL BOTTOM SLOPE
13   * N = MANNING N-VALUE
14   * B = CHANNEL BOTTOM WIDTH - (FT) OR (M)
15   * Z = SIDE SLOPE OF CHANNEL CROSS-SECTION
16   * DX = LENGTH INCREMENT FOR EACH ITERATION - (FT) OR (M)
17   * Q = CHANNEL DISCHARGE - (CFS) OR ( CMS)
18   * YI = INITIAL DEPTH OF WATER - (FT) OR (M)
19   * DIR = INDICATOR AS TO WHICH DIRECTION IN THE CHANNEL CALCULATIONS
20   *       ARE TO PROCEED. DIR=-1 CAUSES CALCULATIONS TO PROCEED UPSTREAM.
21   *       DIR=+1 CAUSES CALCULATIONS TO PROCEED DOWNSTREAM.
22   *
23         DIMENSION Y(200)
24         REAL L,N
25         INTEGER DIR
26         CHARACTER USL*2,USQ*3,SIL*2,SIQ*3,DIML*2,DIMQ*3,CH12
27         YPRIME(B,S,N,Z,U,G,Q,YY)=(S-((N*Q)**2*(B+2.*YY*SQRT(1.+Z*Z))**
28        $(4./3.))/((U*U)*((B+Z*YY)*YY)**(10./3.)))/(1.-((Q*Q*(B+2.*Z*
29        $YY))/(G*((B+Z*YY)*YY)**3)))
30         USL='FT'
31         USQ='CFS'
32         SIL='M '
33         SIQ='CMS'
34         CH12=CHAR(12)
35   *
36   *-----------------------------------------------------------------------
37         WRITE(*,215)
38         WRITE(*,205)
39         READ(*,*) Q
40         WRITE(*,206)
41         READ(*,*) N
42         WRITE(*,207)
43         READ(*,*) B
44         WRITE(*,208)
45         READ(*,*) Z
46         WRITE(*,209)
47         READ(*,*) SZERO
48         WRITE(*,210)
49         READ(*,*) U
```

```
50         WRITE(*,211)
51         READ(*,*) L
52         WRITE(*,212)
53         READ(*,*) DX
54         WRITE(*,213)
55         READ(*,*) YI
56         WRITE(*,214)
57         READ(*,*) DIR
58         WRITE(*,204)
59         OPEN (6,FILE='        ',STATUS='NEW')
49   *------------------------------------------------------------------------------
61   *
62         G=9.81
63         IF(U.GT.1.1) G=32.2
64         DIML=USL
65         DIMQ=USQ
66         IF(G.LT.15.) DIML=SIL
67         IF(G.LT.15.) DIMQ=SIQ
68         WRITE(6,199) CH12
69         WRITE(6,200) L,DIML,SZERO,N,B,DIML,Z,DX,DIML,Q,DIMQ,YI,DIML
70         NN=L/DX+0.1
71         N1=NN+1
72         Y(1)=Y1
73   *
74   *          ** ITERATION LOOP **
75         DO 150 I=1,NN
76         YPI=YPRIME(B,SZERO,N,Z,U,G,Q,Y(I))*DIR
77         YPJ=YPI
78         DO 160 J=1,15
79         TEMP=YPJ
80         Y(I+1)=Y(I)+(YPI+YPJ)*DX*0.5
81         IF(Y(I+1).LE.0.) GO TO 170
82         YPJ=YPRIME(B,SZERO,N,Z,U,G,Q,Y(I+1))*DIR
83         IF(ABS(TEMP-YPJ).LT.0.1E-06) GO TO 150
84     160 CONTINUE
85         GO TO 170
86     150 CONTINUE
87         GO TO 190
88     170 CONTINUE
89         XSTEP=-DX
90         DO 100 K=1,I
91         XSTEP=XSTEP+DX
92         STEP=DIR*XSTEP
93     100 WRITE(6,202) STEP, Y(K)
94         WRITE(6,201)
95         STOP
96     190 WRITE(6,203) DIML,DIML
97         XSTEP=-DX
98         DO 110 K=1,N1
```

```
 99          XSTEP=XSTEP+DX
100          STEP=DIR*XSTEP
101      110 WRITE(6,202) STEP,Y(K)
102   *
103      199 FORMAT(A)
104      200 FORMAT(////14X,'** INPUT DATA **'//11X,'TOTAL LENGTH =',F7.0,1X,A
105         $/18X,'SLOPE =',F8.6/16X,'N-VALUE =',F6.4/11X,'BOTTOM WIDTH =',F5.1
106         $,1X,A/13X,'SIDE SLOPE =',F5.2/11X,'REACH LENGTH =',F7.0,1X,A/14X,
107         $'FLOW RATE =',F7.1,1X,A/10X,'INITIAL DEPTH =',F6.2,1X,A//)
108      201 FORMAT(' ** DEPTH IS SMALL OR YPRIME IS LARGE (NEAR CRITICAL DEPTH
109         $) **')
110      202 FORMAT(10X,F10.1,F12.3)
111      203 FORMAT(/13X,'** ANALYSIS RESULTS **'//14X,'X (',A,')'4X,'Y(X) (',
112         $A,')'/13X,7('-'),4X,9('-'))
113      204 FORMAT(//' YOU ARE REQUIRED TO NAME AN OUTPUT FILE TO WHICH THE RE
114         $SULTS OF'/' THE COMPUTATION ARE SENT FOR YOUR REVIEW AND/OR PRINTI
115         $NG'//)
116      205 FORMAT(/' FLOWRATE (CFS OR CMS) = 'ç)
117      206 FORMAT(/' MANNING N-VALUE = 'ç)
118      207 FORMAT(/' BOTTOM WIDTH (FT OR M) = 'ç)
119      208 FORMAT(/' SIDE SLOPE Z = 'ç)
120      209 FORMAT(/' CHANNEL SLOPE = 'ç)
121      210 FORMAT(/' U-FACTOR (1.49 FOR US CUSTOMARY OR 1.00 FOR SI) = 'ç)
122      211 FORMAT(/' TOTAL LENGTH OF CHANNEL FOR COMPUTATIONS (FT OR M) = 'ç)
123      212 FORMAT(/' LENGTH INCREMENT (FT OR M) = 'ç)
124      213 FORMAT(/' STARTING DEPTH (FT OR M) = 'ç)
125      214 FORMAT(/' DIRECTION OF COMPUTATION INDEX (-1=UPSTREAM OR +1=DOWNST
126         $REAM) = 'ç)
127      215 FORMAT(//' ENTER THE APPROPRIATE NUMBER AFTER THE PROMPT AND PRESS
128         $ RETURN'/)
129          END
```

APPENDIX

8 ANSWERS TO SELECTED EVEN-NUMBERED PROBLEMS[1]

CHAPTER 1

1.4. 3 770 000 psi.

1.6. We're impressed; it was over 95°F.

1.8. 1 250 kg/m^3; 2.43 slug/ft^3; 8×10^{-4} m^3/kg; 0.412 ft^3/slug.

1.10. 62.1 slug or 907.2 kg on both earth and moon; 333.3 lbs; 1 483.3N.

1.12. 847 lb/ft^3; 802 lb/ft^3.

1.14. 0.826; 1.21×10^{-3} m^3/kg.

1.16. 0.818 m^3/kg.

1.18. 10.3 kg/m^3; 100.7 N/m^3; 0.097 m^3/kg.

1.20. 367.6 K.

1.22. 415 J/kg·K; 20.

1.26. L/t.

1.28. 0.997 m^3.

1.30. 3 180 psi or 21 700 kPa.

1.32. 175 000 psi.

1.34. 3 236 kPa; 596 K; 476 kPa; 4 530 kPa.

1.36. 1 087 ft/s; 331 m/s.

1.38. 0.945; 1.065.

1.40. 3.04×10^{-6} m^2/s.

1.42. 1.39×10^{-5} ft^2/s; 1.29×10^{-6} m^2/s.

1.44. 1.12 Pa·s.

1.46. ∞ s^{-1}; 14.9 s^{-1}; 5.90 s^{-1}; 0 s^{-1}.

1.48. ∞ s^{-1}; 4.25 s^{-1}; 3.35 s^{-1}.

1.52. 6.27×10^{-5} lb·s/ft^2.

1.54. Greater than 7°C.

1.56. 0.967 Pa·s; 1.93 Pa·s.

1.58. Greater than 66.9 N.

1.60. 0.002 N·m.

1.62. 6.65 N·m.

1.64. $w_1 - w_2 = 32\, Th/\pi\mu d^4$.

1.66. 0.024 ft/s.

1.68. 0.017 lb·s/ft^2.

1.70. 50 kW

1.74. $h = 2\,\sigma \cos(\beta)/\gamma d$.

1.76. 0.25 N/m.

1.78. $\Delta h = \sigma \cos(\theta)/\gamma_2 r$.

1.80. 1.9 psi; 13.1 Pa

1.82. 175°F; 78°C.

[1]Please note individuals will make different assumptions about data, interpolations, number of retained digits, etc. Thus, your answer(s) to a given problem may differ slightly from those given here.

CHAPTER 2

2.2. 288 lb/ft^3; 4.62; 46.67 kN/m^3; 4.77.

2.4. 6.28 psi; 45.8 kPa

2.6. 2 561 mm $CC1_4$; 5.15 m Alcohol.

2.8. 941 ft; 281.8 m.

2.10. 5.1% for density; 2.6% for pressure.

2.12. $p = \gamma_o h + (1/2)\ Kh^2$.

2.14. 5.5 psia; 0.001 072 $slug/ft^3$; 1.67 psia; 0.000 359 $slug/ft^3$; 35.6 kPa; 0.526 kg/m^3; 10.1 kPa; 0.163 kg/m^3.

2.20. $z = T_o/\alpha$

2.22. 238 kPa; 34.5 psia.

2.24. 121 mm Hg vac.; 4.8 in. Hg vac.

2.26. 52.3 psi; 360.4 kPa.

2.28. 12m; 39.4 ft; no.

2.30. 9.5 in. (241 mm).

2.32. 16.5 psi.

2.34. 61.9 kPa.

2.36. 462 mm.

2.38. 79 mm.

2.40. 6.45 m.

2.42. 21.6 in; 0.55 m.

2.44. 8 in. Hg abs.; 11.6 in.; 203 mm Hg abs.; 0.294 m.

2.46. 38 mm.

2.48. 7.2×10^6 lb; 3.19×10^4 kN.

2.50. $l_p - l_c = 0.19$ m; 173 kN.

2.52. 30 kN; $l_p - l_c = 0.043$ m.

2.54. -1.30 psi; 936 lb.

2.56. 337 lb/ft^2 (2.34 psi); 21 902 lb.

2.58. 23 996 lb at 45°; $l_p = 10.37$ ft.

2.60. (a) 132.2 kN; $l_p - l_c = 0.5$ m; 22 700 lb; $l_p - l_c = 1.5$ ft. (b) 1 455.3 kN; $l_p - l_c = 0.045$ m; 275 000 lb; $l_p - l_c = 0.124$ ft.

2.62. c.p. 1.054 ft below & 0.352 ft right of apex; 566.5 ft·lb.

2.64. 410 kN; $l_p - l_c = 0.047$ m.

2.68. 84.7 kN; $y_p - y_c = 0.34$ m; $x_p = 0.8$ m.

2.70. 24 200 lb.

2.72. 1.88 ft.

2.74. 397 kN.

2.76. 0.9 m.

2.78. 2.68 m.

2.82. 823.2 kN; $l_p - l_c = 0.137$ m; $x_p - x_{point\ on\ left} = 2.74$ m.

2.84. 0.60 m above gate center.

2.86. 13.97 kN.

2.88. 4.3 m^3.

2.90. 3 000 lb.

2.92. 43 250 lb at 40.7°.

2.94. 1/3.

2.96. 23.3 kN; 23.0 kN.

2.98. 68.3 kN·m.

2.100. 93 600 lb horizontal; 147 000 vertical.

2.102. 7.76 kN vertical; 8.56 kN horizontal.

2.104. 1.04 kN.

2.106. 6 800 lb; 30° downward.

2.108. 22.43 kN; 37.15 kN downward; 14.72 kN upward.

2.110. 34.9 kN horizontal; 67.3 kN vertical.

2.112. 0.32 ft^3; 0.009 1 m^3; 3.0.

2.114. 102 mm.

2.116. 96 lb.

2.118. 7.11 kN·m.

2.120. 9.3 m.

2.122. (a) no weight necessary; (b) 1 568 lb.

2.124. 4 504 lb.

2.126. 265 lb.

2.128. 31.6 kPa; 15.8 kN.

2.130. 22 kPa; 3.41 psi.

2.132. 40.2 lb.

2.134. 30.7°; 19.4°.

2.136. 0.55 m^3 spilled; 0.98 kPa; 7.06 kPa; 617 kN.

2.138. 79.6 rpm.

2.140. 4.18 MPa; 633 psi.

CHAPTER 3

3.2. $v = 4\pi(1 - e^{-t})$; $a_s = 4\pi e^{-t}$; $a_r = -8\pi^2(1 - e^{-t})^2$.

3.4. $u = 10t$; $v = -10/t^3$; $a_x = 10$; $a_y = 30/t^4$.

3.6. (b) $a_x = 0$; $a_y = 12$.
(d) $a_x = 12$; $a_y = 0$.
(f) $a_x = 16x$; $a_y = 16y$.
(l) $a_r = -c^2/r^3$; $a_t = 0$.

3.8. $-31.4\ ft^2/s$; $-2.82\ m^2/s$.

3.10. 2 s^{-1}

3.12. (a) $2v_c y/b^2$; (c) $v_c y/b^2(1 - y^2/b^2)^{1/2}$.
3.14. 0; the flow is irrotational.
3.16. $a_r = -\omega^2 r$; $a_t = 0$; $\xi = 2\omega$.

CHAPTER 4

4.2. 0.137 ft/s.
4.4. 8 m/s.
4.6. 10 m/s; 4:1.
4.8. $p_{20}V_{20} = 1.5 \times 10^6$ Pa·m/s; check Chap. 13 for compressible effects.
4.10. 0.202 m/s; 0.051 m/s; 0.013 m/s.
4.12. (a) 6.67 ft/s; 2 m/s; (c) 7.85 ft/s; 2.36 m/s.
4.14. (a) $V/v_c = n/(n + 1)$; (b) $V/v_c = 2n(1/(n + 1) - 1/(2n + 1))$.
4.16. 54.4 cfs/ft; 4.90 m^3/s/m.
4.18. 30.6 ft/s; 20.4 ft/s; 10.2 ft/s.
4.20. 1.78 m/s.
4.22. (a) $q = c \ln(r_2/r_1)$; (b) $q = \omega(r_2^2 - r_1^2)/2$.
4.26. Only (j) and (k) are not physically possible.
4.28. Yes.

CHAPTER 5

5.4. 9.8 kPa; 1.43 psi.
5.6. 389 kPa.
5.8. 1.5 m above B.
5.10. −11.6 ft H_2O, −5.02 psi.
5.12. 11.76 m; 78.8 kPa.
5.14. 3.07 psi.
5.16. 235 mm.
5.18. 0.248 ft^2.
5.20. 0.008 4 m^3/s; 0.32 ft^3/s.
5.22. 59.6 m/s.
5.24. 0.008 2 m^3/s; 398 mm.
5.26. 0.007 15 m^3/s.
5.28. 10 ft; 0.069 5 ft^2.
5.30. (a) 17.5 kPa; (b) −29.4 kPa = 221 mm Hg vac.
5.32. 0.033 m^3/s; −13.13 kPa = 99 mm Hg vac.
5.34. (a) 0.75 m; 5.2 m; 53.7 mm.
5.36. 0.223 m^3/s.
5.42. 0.99 m^3/s; 35.7 ft^3/s.
5.46. 0.06 m^3/s.
5.48. 13.1 kPa.
5.50. 4.29 ft^3/s.
5.52. 18 ft; 3.06 in.
5.54. 2 ft.
5.56. 246.4 mm.
5.58. 1.107.
5.60. 6.28 m; 20.5 ft.
5.62. 178 Pa; 0.024 psi.
5.64. 234.2 kPa; 34.5 psi.
5.66. (a) 7.88 m; (b) 1.55 m.
5.68. 0.024 m^3/s.
5.70. 16.0 ft.
5.72. 154.4 kPa.
5.74. 234 mm; up on left.
5.76 0.034 m^3/s & 39.1°; 1.24 ft^3/s & 39.2°.
5.78. 0.03 m^3/s.
5.80. 28.5 in.
5.82. 0.016 m^3/s.
5.84. 0.73 ft^3/s.
5.86. 668 ft^3/s.
5.88. 0.55 ft.
5.90. 1.97 m^3/s.
5.92. 2.39 m^3/s·m; 1.945 m^3/s·m; 0.443 m^3/s.
5.94. 6.6 kW; 9.22 hp.
5.96. 383 kW.
5.98. 8.38 kW.
5.100. 48 l/s.
5.102. 6.34 kW.
5.104. 33.0 kW.
5.106. 3.46 ft^3/s.
5.108. 1.77 kW.
5.110. 6.45 hp; 1.5 ft^3/s.
5.112. 3.94 ft^3/s; 23.6 hp added.
5.114. 53.1 ft/s; 78 hp.
5.116. 1 470 kW; 1 972 hp.
5.118. 6.67 ft.
5.120. 46.4 ft^3/s.
5.122. 9.9 kN.
5.124. 2 946 kW.
5.126. 10.3 kW.
5.128. −3.5 m & −5.5 m; 8.01 ft & 6.01 ft.
5.130. $z = h - c^2/2g_n r^2$.
5.132. 13 m/s; 5.7 m^3/s·m; 263 mm Hg vac.; 40

ft/s; 55.5 $ft^3/s{\cdot}ft$; 7.7 in. Hg vac.

5.134. 3.55 $m^3/s{\cdot}m$; 1.786 m; 17.5 kPa.

5.136. 9.46 $m^3/s{\cdot}m$; 29.5 kPa.

5.138. 24.1 $ft^3/s{\cdot}ft$.

5.140. 3.93 m/s; 13.0 ft/s.

5.142. 2.75 $m^3/s{\cdot}m$; 1.67 m.

5.148. 11.9 m/s; 39.0 ft/s.

5.150. $\psi = u_c[y - (4y^3/3b^2)]$.

CHAPTER 6

6.2. 1 341 N.

6.4. 17.4 N horizontally; 17.8 N vertically.

6.6. 0.30 m^3/s.

6.8. 6 582 N.

6.10. 29 280 N.

6.12. 1 266 N.

6.14. 37 040 N at 41° through 0.

6.16. 1 500 N at 82°.

6.18. 0.009 6 m^3.

6.20. $\cos\theta = \dfrac{d_1^2 - d_2^2}{d_1^2 + d_2^2}$

6.22. 0.707.

6.24. 25.5 kN on unit.

6.26. 490 N.

6.32. 2 670 lb

6.34. 6 313 N.

6.36. 22.1 kN/m.

6.38. 8 714 N/m horizontally, 14 928 N/m vertically.

6.40. 4 733 N.

6.42. 15 949 N.

6.44. −14.25 kPa; 11 970 N.

6.48. 9.13 m^3/s or 336 ft^3/s.

6.50. 1.54 m^3/s or 56.8 ft^3/s.

6.52. 2.44 m^3/s or 90 ft^3/s; 10.9°.

6.54. 14.7 kN·m or 11 300 ft·lb.

6.56. 261 N horizontally; 451 N vertically.

6.58. 1 579 N.

6.60. 696 N horizontally; 92 N vertically; 702 N.

6.62. 433 N.

6.64. 75% up the plate; 25% down.

6.66. 540.9 lb.

6.68. 9 224 lb.

6.70. 5.8 m^3.

6.72. 441 lb.

6.74. 69.3 N →; 113.3 N →.

6.76. 68.1 m/s; 82%; 9 435 N; 642.5 kW; 527 kW; 2.1 kPa.

6.78. 106.3 kW (140 hp).

6.80. (a) 85%; (b) 15%.

6.82. $P = Qp$.

6.84. 68.2 kN; 59.7 Pa; −36.5 Pa.

6.86. 220 kN.

6.88. 393 kg (25.1 slugs).

CHAPTER 7

7.2. Laminar flow.

7.4. 0.03 kg/s (0.22×10^{-3} slug/s).

7.6. $\mathbf{R} = 4Q/\pi d\nu$

7.8. 5.05×10^{-4} $m^3/s{\cdot}m$ (0.005 425 $ft^3/s{\cdot}ft$).

7.10. 60 Pa; 30 Pa.

7.12. $\varepsilon = \rho v_* Ky$; $\tau = \tau_o$.

7.18. 60 m.

7.20. 0.006 4 m (0.020 ft) in air; 0.000 44 m (0.0014 ft) in water.

7.22. Laminar.

7.30. (e) 0.157 N; (f) 71%.

7.32. 1 190 lb; 0.45 ft; 0.003 3 ft; 1.27 ft.

7.34. 1 157 lb; 925 hp.

7.36. 124.7 kN.

7.40. 4.93 m^3.

7.42. 71.1 lb/ft (1 021 N/m).

7.46. 4.3 lb/ft

7.48. 0.24 m/s (0.69 ft/s); 1.05 m (3.5 ft).

7.50. $Q = \pi\gamma R^4 h_L/8l\mu$

7.52. 329 N (76.6 lb).

7.54. $l \approx (g_n h_L R/2LA^2)^{1/2} y$.

7.56. 288.6 K; 277.5 K.

7.58. (a) 309.2 N; (b) 670 N.

7.60. (a) $\alpha = 2$; $\beta = 1.33$; (c) $\alpha = 1.345$; $\beta = 1.12$.

7.62. 0.166 m^3/s; $\alpha = 1.4$; $\beta = 1.12$; 1 746 N.

7.64. 34.6 $ft^3/s{\cdot}ft$; $\alpha = 1.08$; $\beta = 1.03$.

7.66. 10.95 lb/ft.

7.68. 48.3 kN/m.

7.70. 82.1 kPa.

7.72. 0.004 N; 0.000 97 lb.

7.74. 19.1 m; 63.9 ft.

7.76. $6V\mu/h^2$.

7.78. $\tau_o/(dp/dz)_{\text{at } r=d/2} = d/4$

CHAPTER 8

8.2. 28.1 m/s (93.8 ft/s); 0.20.

8.4. 3.64 m $\times$ 0.73 m; 3.65 N.

8.6. 10.8 m^3/s; 0.24 kPa.

8.8. 1.90 psi.

8.10. 1 750 kPa; 66°C.

8.12. (a) 33.9 km/h; 1 125 kN; (c) 0.096 km/h; 9N.

8.14. 962.2 kN.

8.16. 0.89 m/s; 1.05 $\times$ 10^{-6} Pa·s; 9.79 MN; 2.94 m/s; 212 $\times$ 10^{-10} lb·s/ft^2; 2.18 $\times$ 10^6 lb.

8.18. 0.000 266 ft^3/s; 7.08 $\times$ 10^{-7} lb·s/ft^2.

8.20. 0.079 m^3/s·m; expect cavitation in prototype.

8.22. 2.05 slug/ft^3; 0.0132 $\times$ 10^{-5} lb·s/ft^2; 0.586 $\times$ 10^{-5} lb/ft.

8.24. 0.096 m^3/s·m (1.08 ft^3/s·ft); 2.10; 0.53.

8.26. 141.4 r/min; $p_p/p_m = 1/6.77$; Froude's Law.

8.28. 1.17.

8.30. 0.078 m^3/s; 17.4 kPa absolute.

8.44. $V_w = K(q_n l)^{1/2}$.

8.58. $\Delta p/\rho V^2 = f(l/d, Vd\rho/\mu)$.

CHAPTER 9

9.2. 0.048 m^3/s.

9.4. 21.1 m.

9.6. 313 kW.

9.8. 84.8 kW.

9.10. $\tau_o = 8\mu V/d$; $(dv/dy)_o = 8V/d$; $v_* = (8\nu V/d)^{1/2}$.

9.12. 2.4 1/s (0.087 ft^3/s); 12.3 m (41.0 ft).

9.14. 9.6 Pa; 2.25 m/s.

9.16. 83.3 ft/s·ft.

9.18. For $y/R \leq 0.17$.

9.22. (a) 0.67 m^3/s; (b) 6 450 m/s·m; (c) 9.03 Pa; (d) 1 806 Pa.

9.24. 0.000 47 Pa·s.

9.26. 1.36 ft/s·ft.

9.28. 2.9 m/s.

9.30. 0.33 Pa.

9.32. 0.817.

9.34. $v_*\delta_v/\nu = 12.2$.

9.36. 0.003 68 m^3/s; 0.41 m.

9.38. 1.92 ft/s·ft.

9.40. $e/d = 0.017$.

9.42. Turbulent; rough.

9.48. 80 kPa (1 600 lb/ft^2); 33.3 Pa (0.67 lb/ft^2).

9.50. 23.0 m; 0.029; 0.237 m/s; 56.2 Pa.

9.52. 0.023.

9.54. $\Delta p \propto \rho V^2$

9.56. 62.8 m (234 ft.)

9.58. 15.3 m.

9.60. 0.032; 83 000.

9.62. 0.003 2 m^3/s (0.118 ft^3/s).

9.64. (25°C) 36 W; (40°C) 35 W.

9.66. 0.037.

9.68. 16.4 m; for concrete the flow is no longer smooth and headloss is at least twice as large as for smooth case, e.g., greater than 40 m.

9.70. 64/27.

9.72. (a) 17.9 m; (b) 27.5 m.

9.74. 1.15 $\times$ 10^{-3} Pa·s (2.95 $\times$ 10^{-5} lb·s/ft^2).

9.76. 0.014 kg/s.

9.78. 2.43 m.

9.80. 0.007 5 m/m (0.007 6 ft/ft).

9.82. 4.3 ft; 0.027; 0.003 5.

9.84. 28 in.

9.88. 150 kPa.

9.90. 8.9 in., up on right.

9.92. 65 mm, up on right.

9.94. 0.4 m; 0.1.

9.96. 10.6 kPa; 42 mm Hg vac.

9.98. 1.18 m; 0.95.

9.100. 0.17 m; 0.41.

9.102. 118 kPa; 58.7 kPa.

9.104. 0.056.

9.106. 2.5 m (8.33 ft); 1.25 m (4.17 ft).

9.108. 832 Pa.

9.110. (a) 0.004 ft^3/s; (b) 0.011 ft^3/s.

9.112. (a) 1.82 1/s; (b) 1.97 1/s.

9.114. (a) 0.24 m; (b) 0.22 m.

9.116. 5.97 ft^3/s; some cavitation is possible.

9.118. 0.05 m^3/s.

9.120. (a) 1.52 ft^3/s; (b) 20.2 psi.

9.122. 0.

9.124. 0.046 m^3/s; 331 mm Hg vac.

9.126. 62.6% reduction.

9.128. 2.40 cfs.

9.130. Elevation is 951 ft.

9.132. (a) 0.306 m^3/s; 2.88 kW; (b) 0.103 m^3/s; 28.2 kW.

9.134. (a) 0.28 m^3/s; (b) 47.6 kW; (c) 0.3 m yields 55.6 kW.

9.136. 666 kW; 747 W saved.

9.138. 113.3 kW; (a) 0.108 m^3/s; (b) 0.052 m^3/s.

9.140. Elevation minimum is 28.2 m; 1.56 l/s; 1.48 l/s.

9.142. 39.3 kW.

9.144. 148.5 kW.

9.146. 32.9 kW.

9.148. 240.7 hp; 13.8 ft/s; 0.318 lb/ft^2.

9.150. 4 hp.

9.154. 4 900 gpm.

9.156. 45 mm.

9.158. $V = [(V_2^2 - V_1^2)\ d/fl]^{1/2}$.

9.160. 26.5% increase.

9.162. 0.09 m^3/s (3.2 ft^3/s).

9.164. Flowrate decreases by about 2%.

9.166. 0.453 m^3/s; 0.468 m^3/s; 0.475 m^3/s.

9.168. 0.042 8 m^3/s; 0.039 6 m^3/s; 0.082 3 m^3/s; 0.008 6 m^3/s; 0.039 1 m^3/s; 0.048 m^3/s.

9.170. 405 kW; 261 mm.

9.172. 1.03 ft^3/s; 2.97 ft^3/s; 56 hp.

9.174. $Q_{AB} = 81$ l/s; $Q_{BC} = 26$ l/s; $Q_{AC} = 29$ l/s.

9.176. $\Delta_1 = 0.685$ cfs; $\Delta_2 = -0.500$ cfs; $Q_{AB} = -1.315$ cfs; $Q_{AC} = 1.685$ cfs; $Q_{BC} = -0.185$ cfs; $Q_{BD} = -2.50$ cfs; $Q_{CD} = 3.50$ cfs.

9.178. 588 kPa at junction below reservoir; 0.003 m^3/s to right in 100 mm *d* pipe at bottom of diagram.

9.180. 0.712 ft^3/s downwards in vertical pipe on left side; 2.21 ft^2/s toward right in 8″ *d* pipe on bottom.

9.182. 10.4 s.

9.184. (b) 90 ft head at valve at 100 s.

9.186. 38.6 s.

9.188. Maximum is 200 ft at the valve at 0 s; minimum is 112.5 ft at the valve at 10 s; $K_L = 96.6$.

9.190. Linear scheme cannot work because it requires the pressure head at the upstream reservoir to be greater than 200 ft.

9.192. (a) 3 502 ft/s; (b) 4 050 ft/s; (c) 3 370 ft/s; (d) 3 642 ft/s; (e) 1 068 ft/s.

9.194. Pipe volume change = 0.083%; Water density change = 0.168%

9.196. 400 ft^3.

9.198. (a) 329.6 psi; 330%; (b) 381.1 psi; 381%; (c) 317.2 psi; 317%; (d) 342.7 psi; 343%; (e) 100.5 psi; 101%.

9.200. 1.29 s.

9.202. 978 ft/s.

9.204. 53 psi.

CHAPTER 10

10.2. 6.54 Pa (0.139 lb/ft^2).

10.4. 23.4 Pa (0.496 lb/ft^2).

10.6. 0.018; 0.014 7; 0.007 8.

10.8. 6.23 m^3/s (230 ft^3/s).

10.10. 1.555 ft^3/s; 0.000 9; 0.020 8.

10.12. 13.52 m^3/s.

10.14. 1.25 m.

10.16. 7.05 ft.

10.18. 0.000 39.

10.20. 3.2 m (9.8 ft).

10.22. 1.97 cfs.

10.24. 0.417.

10.26. 0.000 79 $ft^3/s{\cdot}ft$; 79.

10.34. 0.92 m (2.88 ft).

10.36. (a) 6 m × 3 m; (b) 1.27 m × 2.7 m.

10.38. 0.54 m × 1.69 m (1.77 ft × 5.52 ft).

10.40. 0.000 194

10.42. 0.000 185.

10.46. (a) 4.99 ft; (b) 3.87 ft; (c) 6.22 ft.

10.48. 0.73 m or 2.85 m (2.44 ft or 9.50 ft).

10.52. Subcritical flow.

10.54. 4.96 ft; 832 ft^3/s.

10.56. 0.005 65 (0.005 35).

10.58. 3.64 $m^3/s \cdot m$; 2.60 m; 2.27 m; 1.92 m.

10.60. 2.7 m depth; 3.22 m width.

10.62. 4.61 ft.

10.64. 0.945 m.

10.66. 0.72 m; 0.73; 0.004 9.

10.68. Subcritical flow.

10.70. 0.004 4.

10.76. 262 ft^3/s.

10.78. 2.07 m (6.92 ft).

10.80. 2.07 m.

10.82. 7.82 ft^3/s

10.84. 1.65 ft.

10.88. 0.765 ft.

10.90. 0.023.

10.92. 465 ft^3/s.

10.94. 1.46 ft.

10.98. 0.87 m; 1.46 m.

10.100. 5.75 ft; 4.65 ft; 3.00 ft; 10.37 ft.

10.102. 2.55 ft; 2.12 ft.

10.104. 1.34 $m^3/s \cdot m$; 1.30 m or 0.30 m.

10.106. 2.2 m.

10.108. 1.32 m; 1.83 m.

10.110. 5.97 ft wide; 4.97 ft; 7.0 ft.

10.112. 31.2 m^3/s.

10.114. 0.90 m.

10.116. 35 ft.

10.122. (a) A3 & A2; (b) H2 & none; (c) None & S3; (d) M2; (e) M1; (f) none and S1.

10.124. 9.76 m is depth; drop is 0.24 m.

10.134. M1; Δx = 23 150 ft

10.136. (a) 7.4 ft; (b) 5.9 ft; (c) 12.5 ft; (d) 7.9 ft; (e) 4.6 ft; (f) 420 hp.

10.140. 4.35 ft.

10.142. 3.86 m^3/s.

10.144. 2 000 hp.

10.146. 145 ft^3/s.

10.148. 40.6 kN.

10.150. Jump occurs in Zone CD.

10.152. No; Δx = 57.3 ft.

10.154. Δx = 850 ft.

10.156. S1 profile upstream of jump; 5.27 ft; 1.7 ft·lb/lb loss in jump.

10.158. 2.74 to 4.25 m.

10.160. 4.0 ft; 2.67 ft; 3.16 ft.

10.162. 6.39 m.

CHAPTER 11

11.4. (a) 2.02 lb (8.66 N); 0.16 hp (115W); (b) 1 650 lb (7 068 N); 132 hp (94.2 kW).

11.6. 0.649; 0.043.

11.8. 0.58 lb.

11.12. 152.1 lb.

11.14. 0.27 ft/s.

11.16. 0.004 $lb \cdot s/ft^2$.

11.18. 1.82 N; 1.61 N; 8.94 m/s.

11.22. 2.4 m/s.

11.24. Larger (~6.5 m/s).

11.26. 0.75 m.

11.30. 55 N; 25.2 Hz; 1.626×10^9 N; no.

11.34. Air: 7.7 N/m; 15.4 N/m; 0;
Water: 6.27 kN/m; 12.5 kN/m; 0.

11.36. $C_L = 2\Gamma/cV$.

11.38. 692 N/m.

11.40. 1.31; 0.153; 11.3°.

11.44. Fowler flap (0.4c & 40°): 81% increase; 58% decrease.

11.46. 278 ft^2; 0.34 hp.

11.48. (a) Ball; (b) Ball; (c) Strike.

11.50. 3.31×10^6 lb (14.5 MN); Pyramid drag is about 11% less under similar conditions.

11.58. 8.5 & 24.4 kW; 7.6 & 20.3 kW; 7.10 & 19.9 kW.

11.62. Approximately 113 mph (182 km/h).

11.64. 302.6 kW (428 hp); about 21% faster.

11.66. $P_T = A\rho V_o^3 C_D/2f_D$.

CHAPTER 12

12.2. (a) $PD^4/\rho Q^3$; Q/ND^3; HD^4/Q^2; $\rho Q/\mu D$; $\rho Q^2/ED^4$; (b) $P/\rho QH$; $QN^2/H^{3/2}$; HD^4/Q^2; $\rho Q^{1/2} H^{1/4}/\mu$; $E/\rho H$.

12.4. $P/\rho\omega^3 D^5 = f(e/Dg, Q/\omega D^3, \omega D^2\rho/\mu)$.

12.6. 2 480 r/min; 0.012 4 m^3/s; 4.5 m.

12.8. $N_{new} = 0.707\ N_{old}$; $Q_{new} = 5.66\ Q_{old}$

12.10. 0.72 m (2.39 ft); 316 r/min; 48 m (159 ft).

12.12. 14.2 MW.

12.14. 0.93; centrifugal pump.

12.16. 2.8.

12.18. 99.1 m (325.2 ft); 94%.

12.20. 0.77 as a Francis turbine; 0.68 as a centrifugal pump.

12.22. (a) 10 m; (b) 3.1 m; (c) 55.6 m.

12.24. 10.1 m.

12.26. 4 600 N (1 096 lb) at 30°; 59.7 kW (86.4 hp).

12.30. 3 180 W (4.7 hp).

12.32. 2 693 N.

12.34. 2.40 lb; 0.56 hp; 128 ft/s.

12.36. 60°; 9 930N; 298 kW.

12.38. 184.5 ft/s (55.4 m/s); 76.5 ft/s (21.4 m/s); 250 hp (182.9 kW).

12.40. 3 374 W; 36.4°; 144°.

12.42. 12.3×10^6 ft·lb (16.7×10^6 N·m); 351 000 hp (262 MW); 387 ft (118.7 m).

12.44. 1 403 W.

12.46. 0.023 m^3/s (0.857 ft^3/s); (32.2 kPa); 4.8 psi; 24.15 N·m (19.37 ft·lb); 1.26 kW (1.84 hp).

12.48. 33.1 kN·m.

12.50. 2 960 gpm; 159.8 ft; 142 horsepower.

CHAPTER 13

13.2. 95.3 kJ/kg; 95.3 kJ/kg + q_H.

13.4. $q_H + g_n E_P = 29\,050$ ft·lb/slug.

13.6. (a) 1 091 J/K·kg; (b) −1 365 J/K·kg.

13.8. 422 J/K·kg.

13.10. Graph points: (100 K; 943 m/s); (250 K; 482 m/s) or (200°R; 3 029 ft/s); (500°R; 1 146 ft/s).

13.12. 138.7 m/s (446.0 ft/s).

13.16. 218.5 m/s.

13.18. (a) 178 kPa: (b) 179 kPa.

13.20. $(p_o/p_s)\text{min} = (2/(k+1))^{k/(k-1)}$

13.22. 60% decrease; 15% increase.

13.24. (a) 15.85 psia; 1 168 ft/s; 90°F; 1 168 ft/s; 0.060 slug/s; (b) 15.0 psia; 1 058 ft/s; 110°F; 1 189 ft/s; 0.049 slug/s; (c) 28.41 psia; 1 168 ft/s; 0.056 slug/s.

13.26. Both temperatures are 20.1°C (74.5°F).

13.28. 790 kPa; 38.7°C; 51.9 m/s; 353.8 m/s.

13.30. 260 m/s; 49.7°C.

13.32. 261.4 kPa.

13.34. (a) 0.32 slug/s; (b) 83°F; (c) 883 ft/s; (d) 177.8 ft/s and 443.5 ft/s.

13.36. 2.78 in.; 2.4 in.; 52.8 in. Hg abs; 6.5°F; 1 059 ft/s; 1 058 ft/s and 9.83 psia; −106.6°F; 1 574 ft/s; 921 ft/s.

13.38. 323.9 m/s; 323.9 m/s; 1.57 kg/s.

13.40. 2.0 kg/s; 40 mm.

13.42. 73 640 kPa abs. (10 520 psia); 1 245°C (2 298°F); 0.040.

13.44. Normal shock *not* expected.

13.48. 99°C (195°F); 139.6°C (275°F).

13.50. 473.6 ft·lb/°R·slug.

13.52. 464.4 m/s (1 520 ft/s).

13.54. 18.7 kPa (3.06 psi).

13.56. 19.7 psia; 108.5 ft/s; 165 ft/s.

13.58. 172.5 kPa; 0.40.

13.62. 53 m (177 ft).

13.64. 0.033.

13.66. For example, $l_{max}/d = 32.8$ and $f = 0.013$ for commercial steel.

13.68. 2.62; 3.02.

13.70. 4.5 s.

13.72. (a) 18 lb; (b) 44 lb.

13.74. 0.66.

13.76. 1.04.

CHAPTER 14

14.2. 1.48; uncertainty is about 0.12%

14.4. 1.575.

14.6. 8.5 in.

14.8. 4.09 Pa·s.

14.10. $\nu = 2.59 \times 10^{-7} t$; 2%.

14.12. 1.9 ft.

14.14. 40.0 m/s.

14.16. 11.1 in. (280 mm); 302 ft/s (91.3 m/s); 0.4 psi (2.76 kPa).

14.18. 87.7 mm.

14.20. 1.7 in.

14.22. $\pm 0.25 \times 10^{-5}$ psi.

14.24. Approximate bound for large Mach number is ±20%.

14.26. 279.5 m/s.

14.28. 68.0 psia (470 kPa); 390°F (201°C).

14.30. $A = 3.92$; $B = 3.42$.

14.32. 4.2%.

14.34. 0.1 m.

14.36. 6 mm.

14.38. 1.067 kPa/m.

14.40. 7.70 ft^3/s.

14.42. 110 mm.

14.44. 890 mm.

14.46. 23.33 N/s.

14.48. 27.5 ft^3/s.

14.50. 0.509 ft^3/s; 93.2 ft/s; 5.35 ft; 135 ft.

14.52. 0.642; 0.975.

14.54. 2.0.

14.56. 0.72.

14.58. 0.037 m^3/s; 244.6 kPa.

14.60. 0.370 m^3/s.

14.62. Water: 3.54 l/s (0.130 ft^3/s); Oil: 3.71 l/s (0.137 ft^3/s).

14.64. 0.49 ft^3/s

14.66. 17.0 ft (5.07 m).

14.68. 0.607; 0.973; 0.623.

14.70. 1.95 ft.

14.72. Short tube: 0.80; Re-entrant tube: 0.73.

14.74. 158 ft^3/s (4.29 m^3/s).

14.76. 0.97.

14.80. 0.288 m (0.96 ft).

14.82. 0.63 m (2.135 ft).

14.84. 1.02 m^3/s

14.86. Water: 0.440 ft^3/s (0.008 53 m^3/s); Oil: 0.444 ft^3/s (0.008 58 m^3/s).

14.88. 7.0 ft (2.10 m).

14.90. 5.45 ft^3/s.

14.92. 1.37 m.

14.94. 0.88 m; 3.49 m^3/s; 0.597.

14.98. 15 400 ft^3/s (419 m^3/s).

14.100. (0.650, 28 $ft^3/s\cdot ft$); (0.747, 126 $ft^3/s\cdot ft$); (0.762, 169 $ft^3/s\cdot ft$); 0.778.

14.102. 0.20 $m^3/s\cdot m$.

INDEX

(continued)

 (continued)

(continued)